LIST OF THE ELEMENTS WITH THEIR SYMBOLS AND ATOMIC MASSES*

Element	Symbol	Atomic Number	Atomic Mass†
Actinium	Ac	89	(227)
Aluminum	Al	13	26.98
Americium	Am	95	(243)
Antimony	Sb	51	121.8
Argon	Ar	18	39.95
Arsenic	As	33	74.92
Astatine	At	85	(210)
Barium	Ba	56	137.3
Berkelium	Bk	97	(247)
Beryllium	Be	4	9.012
Bismuth	Bi	83	209.0
Boron	B	5	10.81
Bromine	Br	35	79.90
Cadmium	Cd	48	112.4
Calcium	Ca	20	40.08
Californium	Cf	98	(249)
Carbon	C	6	12.01
Cerium	Ce	58	140.1
Cesium	Cs	55	132.9
Chlorine	Cl	17	35.45
Chromium	Cr	24	52.00
Cobalt	Co	27	58.93
Copper	Cu	29	63.55
Curium	Cm	96	(247)
Dysprosium	Dy	66	162.5
Einsteinium	Es	99	(254)
Erbium	Er	68	167.3
Europium	Eu	63	152.0
Fermium	Fm	100	(253)
Fluorine	F	9	19.00
Francium	Fr	87	(223)
Gadolinium	Gd	64	157.3
Gallium	Ga	31	69.72
Germanium	Ge	32	72.59
Gold	Au	79	197.0
Hafnium	Hf	72	178.5
Hahnium	Ha	105	(260)
Hassium	Hs	108	(265)
Helium	He	2	4.003
Holmium	Ho	67	164.9
Hydrogen	H	1	1.008
Indium	In	49	114.8
Iodine	I	53	126.9
Iridium	Ir	77	192.2
Iron	Fe	26	55.85
Krypton	Kr	36	83.80
Lanthanum	La	57	138.9
Lawrencium	Lr	103	(257)
Lead	Pb	82	207.2
Lithium	Li	3	6.941
Lutetium	Lu	71	175.0
Magnesium	Mg	12	24.31
Manganese	Mn	25	54.94
Meitnerium	Mt	109	(266)
Mendelevium	Md	101	(256)
Mercury	Hg	80	200.6
Molybdenum	Mo	42	95.94
Neodymium	Nd	60	144.2
Neon	Ne	10	20.18
Neptumium	Np	93	(237)
Nickel	Ni	28	58.69
Nielsbohrium	Ns	107	(262)
Niobium	Nb	41	92.91
Nitrogen	N	7	14.01
Nobelium	No	102	(253)
Osmium	Os	76	190.2
Oxygen	O	8	16.00
Palladium	Pd	46	106.4
Phosphorus	P	15	30.97
Platinum	Pt	78	195.1
Plutonium	Pu	94	(242)
Polonium	Po	84	(210)
Potassium	K	19	39.10
Praseodymium	Pr	59	140.9
Promethium	Pm	61	(147)
Protactinium	Pa	91	(231)
Radium	Ra	88	(226)
Radon	Rn	86	(222)
Rhenium	Re	75	186.2
Rhodium	Rh	45	102.9
Rubidium	Rb	37	85.47
Ruthenium	Ru	44	101.1
Rutherfordium	Rf	104	(257)
Samarium	Sm	62	150.4
Scandium	Sc	21	44.96
Seaborgium	Sg	106	(263)
Selenium	Se	34	78.96
Silicon	Si	14	28.09
Silver	Ag	47	107.9
Sodium	Na	11	22.99
Strontium	Sr	38	87.62
Sulfur	S	16	32.07
Tantalum	Ta	73	180.9
Technetium	Tc	43	(99)
Tellurium	Te	52	127.6
Terbium	Tb	65	158.9
Thallium	Tl	81	204.4
Thorium	Th	90	232.0
Thulium	Tm	69	168.9
Tin	Sn	50	118.7
Titanium	Ti	22	47.88
Tungsten	W	74	183.9
Uranium	U	92	238.0
Vanadium	V	23	50.94
Xenon	Xe	54	131.3
Ytterbium	Yb	70	173.0
Yttrium	Y	39	88.91
Zinc	Zn	30	65.39
Zirconium	Zr	40	91.22

*All atomic masses have four significant figures. These values are recommended by the Committee on Teaching of Chemistry, International Union of Pure and Applied Chemistry.

†Approximate values of atomic masses for radioactive elements are given in parentheses.

IMPORTANT:

HERE IS YOUR REGISTRATION CODE TO ACCESS YOUR PREMIUM McGRAW-HILL ONLINE RESOURCES.

For key premium online resources you need THIS CODE to gain access. Once the code is entered, you will be able to use the Web resources for the length of your course.

If your course is using **WebCT** or **Blackboard**, you'll be able to use this code to access the McGraw-Hill content within your instructor's online course.

Access is provided if you have purchased a new book. If the registration code is missing from this book, the registration screen on our Website, and within your WebCT or Blackboard course, will tell you how to obtain your new code.

Registering for McGraw-Hill Online Resources

To gain access to your McGraw-Hill web resources simply follow the steps below:

1. USE YOUR WEB BROWSER TO GO TO: **http:/www.mhhe.com/denniston**
2. CLICK ON **FIRST TIME USER**.
3. ENTER THE REGISTRATION CODE* PRINTED ON THE TEAR-OFF BOOKMARK ON THE RIGHT.
4. AFTER YOU HAVE ENTERED YOUR REGISTRATION CODE, CLICK **REGISTER**.
5. FOLLOW THE INSTRUCTIONS TO SET-UP YOUR PERSONAL UserID AND PASSWORD.
6. WRITE YOUR UserID AND PASSWORD DOWN FOR FUTURE REFERENCE.
   KEEP IT IN A SAFE PLACE.

TO GAIN ACCESS to the McGraw-Hill content in your instructor's **WebCT** or **Blackboard** course simply log in to the course with the UserID and Password provided by your instructor. Enter the registration code exactly as it appears in the box to the right when prompted by the system. You will only need to use the code the first time you click on McGraw-Hill content.

Thank you, and welcome to your McGraw-Hill online resources!

* YOUR REGISTRATION CODE CAN BE USED ONLY ONCE TO ESTABLISH ACCESS. IT IS NOT TRANSFERABLE.

0-07-246905-6 DENNISTON/TOPPING: GENERAL, ORGANIC, AND BIOCHEMISTRY, 4E

MCGRAW-HILL

ONLINE RESOURCES

REGISTRATION CODE

ectodermal-67139892

Higher Education

Mc Graw Hill

General, Organic, and Biochemistry

General, Organic, and Biochemistry

Fourth Edition

Katherine J. Denniston
Towson University

Joseph J. Topping
Towson University

Robert L. Caret
San José State University

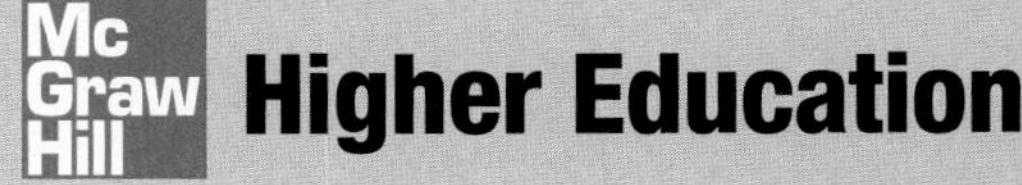

Boston Burr Ridge, IL Dubuque, IA Madison, WI New York San Francisco St. Louis
Bangkok Bogotá Caracas Kuala Lumpur Lisbon London Madrid Mexico City
Milan Montreal New Delhi Santiago Seoul Singapore Sydney Taipei Toronto

The McGraw·Hill Companies

GENERAL, ORGANIC, AND BIOCHEMISTRY, FOURTH EDITION

Published by McGraw-Hill, a business unit of The McGraw-Hill Companies, Inc., 1221 Avenue of the Americas, New York, NY 10020.

This book is printed on acid-free paper.

International 1 2 3 4 5 6 7 8 9 0 DOW/DOW 0 9 8 7 6 5 4 3
Domestic 1 2 3 4 5 6 7 8 9 0 DOW/DOW 0 9 8 7 6 5 4 3

ISBN 0–07–246905–6
ISBN 0–07–121451–8 (ISE)

Publisher: *Kent A. Peterson*
Sponsoring editor: *Thomas D. Timp*
Developmental editor: *Spencer J. Cotkin, Ph.D.*
Senior project manager: *Gloria G. Schiesl*
Lead production supervisor: *Sandy Ludovissy*
Lead media project manager: *Judi David*
Senior media technology producer: *Jeffry Schmitt*
Coordinator of freelance design: *Michelle D. Whitaker*
Cover designer: *Elise Lansdon*
Cover images: *Taxol illustration: Elise Lansdon; main image: © Mary Liz Austin/Getty Images, Inc.*
Senior photo research coordinator: *Lori Hancock*
Photo research: *David Tietz*
Supplement producer: *Brenda A. Ernzen*
Compositor: *GAC—Indianapolis*
Typeface: *10/12 Palatino*
Printer: *R. R. Donnelley Willard, OH*

The credits section for this book begins on page 853 and is considered an extension of the copyright page.

Library of Congress Cataloging-in-Publication Data

Denniston, K. J. (Katherine J.)
General, organic, and biochemistry. — 4th ed. / Katherine J. Denniston, Joseph J. Topping.
p. cm.
Includes index.
ISBN 0–07–246905–6 (acid-free paper)
1. Chemistry, Organic. 2. Biochemistry. I. Topping, Joseph J. II. Title.
QD253.2.D46 2004
547—dc21 2002044887
CIP

INTERNATIONAL EDITION ISBN 0–07–121451–8

www.mhhe.com

Brief Contents

Contents

3 Elements, Atoms, Ions, and the Periodic Table 53

4 Structure and Properties of Ionic and Covalent Compounds 77

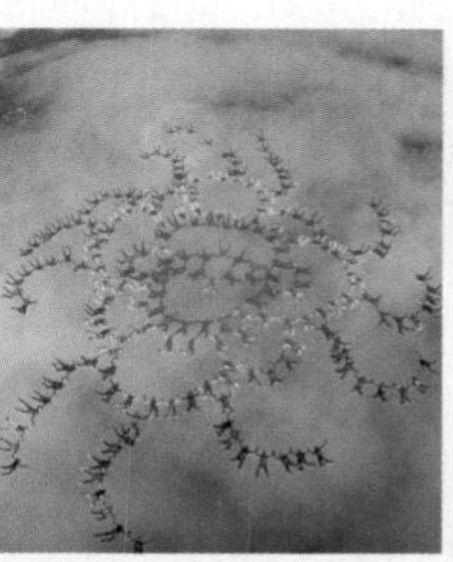

Chemistry Connections and Perspectives

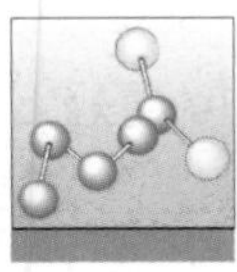

Chemistry Connection

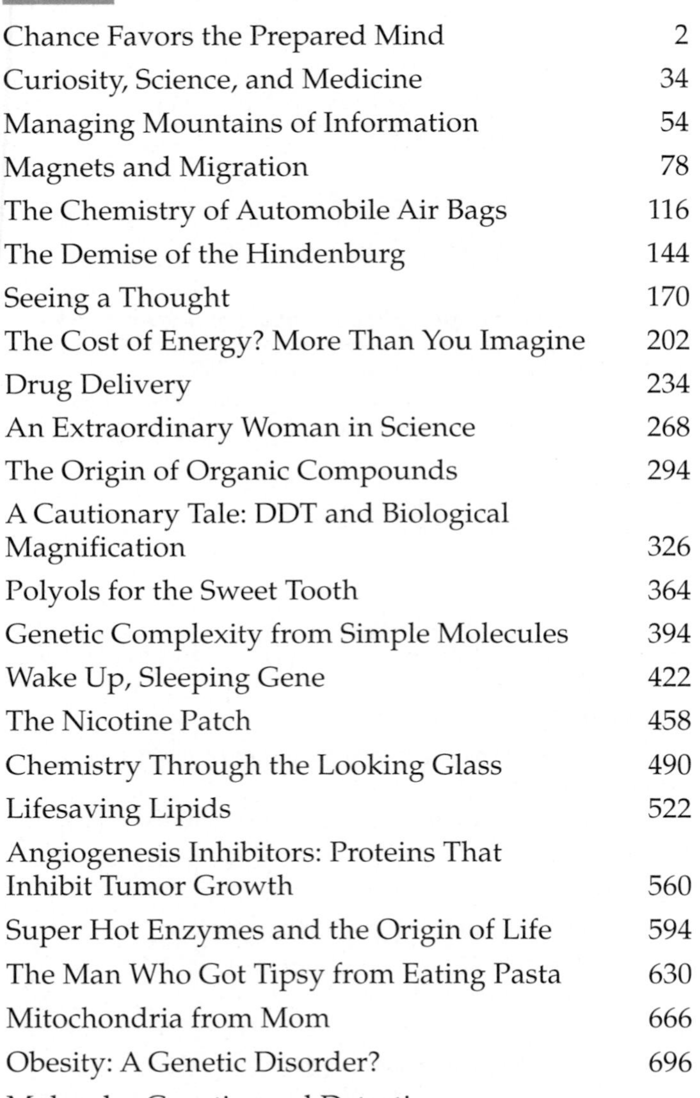

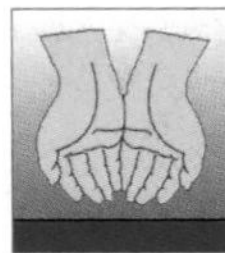

A Human Perspective

A Clinical Perspective

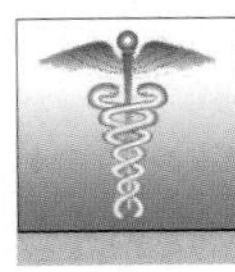

A Medical Perspective

An Environmental Perspective

Preface

The fourth edition of *General, Organic, and Biochemistry*, like our earlier editions, has been designed to help undergraduate majors in health-related fields understand key concepts and appreciate the significant connections between chemistry, health, and the treatment of disease. We have tried to strike a balance between theoretical and practical chemistry, while emphasizing material that is unique to health-related studies. We have written at a level intended for students whose professional goals do not include a mastery of chemistry, but for whom an understanding of the principles and practice of chemistry is a necessity.

While we have stressed the importance of chemistry to the health-related professions, this book was written for all students that need a one or two semester introduction to chemistry. Our focus on the relationship between chemistry, the environment, medicine, and the function of the human body is an approach that can engage students in a variety of majors.

In this text we treat the individual disciplines of inorganic, organic, and biological chemistry. Moreover, we have tried to integrate these areas to show the interrelatedness of these topics. This approach provides a sound foundation in chemistry and teaches students that life is not a magical property, but rather is the result of a set of chemical reactions that obey the scientific laws.

Key Features of the Fourth Edition

In the preparation of the fourth edition, we have been guided by the collective wisdom of over fifty reviewers who are experts in one of the three subdisciplines covered in the book and who represent a diversity of experience, including community colleges, and four-year colleges and universities. We have retained the core approach of our successful earlier editions, modernized material where necessary, and expanded or removed material consistent with retention of the original focus and mission of the book. Throughout the project, we have been careful to ensure that the final product is as student-oriented and readable as its predecessors.

Specifically, new features of the fourth edition include:

- Chapters 1 and 2 have been rearranged to more smoothly facilitate students' transition from descriptive to quantitative chemistry.
- Twenty new boxed elements, particularly in the Organic and Biochemistry parts, give students insight into the modern-day application of various topics.
- Approximately 200 new end-of-chapter questions will allow instructors greater flexibility in assigning problems and will give students more opportunity to test themselves.
- The website and other media supplements, as described later in this Preface, have been enhanced. Specifically, the Digital Content Manager, a CD-ROM, contains electronic files of text figures and tables as well as PowerPoint lecture slides.

We designed the fourth edition to promote student learning and facilitate teaching. It is important to engage students, to appeal to visual learners, and to provide a variety of pedagogical tools to help them organize and summarize information. We have utilized a variety of strategies to accomplish our goals.

Engaging Students

Students learn better when they can see a clear relationship between the subject material they are studying and real life. We wrote the text to help students make connections between the principles of chemistry and their previous life experiences and/or their future professional experiences. Our strategy to accomplish this integration includes the following:

- **Boxed Readings—"Chemistry Connection":** We have crafted introductory vignettes to allow the student to see the significance of chemistry in their daily lives and in their future professions.
- **Boxed Perspectives:** These short stories present real-world situations that involve one or more topics that students will encounter in the chapter. The "Medical Perspectives" and "Clinical Perspectives" relate the chemistry to a health concern or a diagnostic application. The "Environmental Perspectives" deal with issues, including the impact of chemistry on the ecosystem and the way in which these environmental changes affect human health. "Human Perspectives" delve into chemistry and society and include such topics as gender issues in science and historical viewpoints.

In the fourth edition, we have added 20 new boxed topics and have updated many of the earlier ones. We have tried to include topics, such as self-tanning lotions and sugar substitutes, which are of interest to students today. We have included the most recent strategies for treatment of AIDS and new information on the use of genetic engineering to treat a variety of genetic diseases.

Learning Tools

In designing the original learning system we asked ourselves the question, "If we were students, what would help us organize and understand the material covered in this chapter?" With valuable suggestions from our reviewers, we have made some modifications to improve the learning system. However, with the blessings of those reviewers, we have retained all of the elements of the previous edition, which have been shown to support student learning:

- **Learning Goals:** A set of chapter objectives at the beginning of each chapter previews concepts that will be covered in the chapter. Icons 1 locate text material that supports the learning goals.
- **Detailed Chapter Outline:** A detailed listing of topic headings is provided for each chapter. Topics are divided and subdivided in outline form to help students organize the material in their own minds.
- **Chapter Cross-References:** To help students locate the pertinent background material, references to previous chapters, sections, and perspectives are noted in the margins of the text. These marginal cross-references also alert students to upcoming topics that require an understanding of the information currently being studied.
- **Chapter Summary:** Each major topic of the chapter is briefly reviewed in paragraph form in the end-of-chapter summary. These summaries serve as a mini-study guide, covering the major concepts in the chapter.
- **Key Terms:** Key terms are printed in boldface in the text, defined immediately, and listed at the end of the chapter. Each end-of-chapter key term is accompanied by a section number for rapid reference.
- **Summary of Key Reactions:** In the organic chemistry chapters, each major reaction type is highlighted on a green background. These major reactions are summarized at the end of the chapter, facilitating review.
- **Glossary of Key Terms:** In addition to being listed at the end of the chapter, each key term from the text is also defined in the alphabetical glossary at the end of the book.
- **Appendix Material:** Each Appendix accomplishes one of two goals: remediation or expansion of information introduced in the chapter.

The Art Program

Today's students are much more visually oriented than any previous generation. Television and the computer represent alternate modes of learning. We have built upon this observation through expanded use of color, figures, and three-dimensional computer-generated models. This art program enhances the readability of the text and provides alternative pathways to learning.

- **Dynamic Illustrations:** Each chapter is amply illustrated using figures, tables, and chemical formulas. All of these illustrations are carefully annotated for clarity.
- **Color-Coding Scheme:** We have color-coded the reactions so that chemical groups being added or removed in a reaction can be quickly recognized. Each major organic reaction type is highlighted on a green background. The color-coding scheme is illustrated in the "Guided Tour" section of this book.
- **Computer-Generated Models:** The students' ability to understand the geometry and three-dimensional structure of molecules is essential to the understanding of organic and biochemical reactions. Computer-generated models are used throughout the text because they are both accurate and easily visualized.

Problem Solving and Critical Thinking

Perhaps the best preparation for a successful and productive career is the development of problem-solving and critical thinking skills. To this end, we created a variety of problems that require recall, fundamental calculations, and complex reasoning. In this edition, we have used suggestions from our reviewers, as well as from our own experience, to enhance the problem sets to include more practice problems for difficult concepts and further integration of the subject areas.

- **In-Chapter Examples, Solutions, and Problems:** Each chapter includes a number of examples that show the student, step-by-step, how to properly reach the correct solution to model problems. Whenever possible, they are followed by in-text problems that allow the students to test their mastery of information and to build self-confidence.
- **In-Chapter and End-of-Chapter Problems:** We have created a wide variety of paired concept problems. The answers to the odd-numbered questions are found in the back of the book as reinforcement for the students as they develop problem-solving skills. However, the students must then be able to apply the same principles to the related even-numbered problems.
- **Critical Thinking Problems:** Each chapter includes a set of critical thinking problems. These problems are intended to challenge the students to integrate concepts to solve more complex problems. They make a perfect complement to the classroom lecture, because they provide an opportunity for in-class discussion of complex problems dealing with daily life and the health care sciences.

Over the course of the last three editions, hundreds of reviewers have shared their knowledge and wisdom with us, as well as the reaction of their students to elements of this book. Their contributions, as well as our own continuing experience in the area of teaching and learning science, have resulted in a text that we are confident will provide a strong foundation in chemistry, while enhancing the learning experience of the students.

Supplementary Materials

This text has a complete support package for instructors and students. Several print and media supplements have been prepared to accompany the text and make learning as meaningful and up-to-date as possible.

For the Instructor

- **Digital Content Manager:** This is the primary instructor supplement and offers over 300 text images and nearly 300 PowerPoint lecture slides prepared by Kim Woodrum at the University of Kentucky. The text images are full color and can be readily incorporated into lecture presentations, exams, or classroom materials. The sets of PowerPoint slides cover all 24 chapters. These slides can be modified according to instructor preference.
- **Instructor's Manual:** Written by the authors and updated by Patricia DePra, this ancillary contains suggestions for organizing lectures, additional "Perspectives," and a list of each chapter's key problems and concepts. The Instructor's Manual also contains the test item file and solutions to the even-numbered problems.
- **Transparencies:** A set of 100 transparencies is available to help the instructor coordinate the lecture with key illustrations from the text.
- **Brownstone Diploma computerized classroom management system:** This service includes a database of test questions, reproducible student self-quizzes, and a grade-recording program.
- **Laboratory Resource Guide:** Written by Charles H. Henrickson, Larry C. Byrd, and Norman W. Hunter of Western Kentucky University, this helpful prep guide contains the hints that the authors have learned over the years to ensure students' success in the laboratory.
- **Online Learning Center:** A book-specific website is available to students and instructors using this text. The website will offer quizzes, key definitions, and interesting links for the students. The instructor will find a downloadable version of the Instructor's Manual. Also available for the instructor is PageOut, which allows the instructor to create his or her own personal course website. The address for the book-specific website is www.mhhe.com/denniston.

For the Students

- **Student Study Guide/Solutions Manual:** A separate Student Study Guide/Solutions Manual, prepared by the authors and updated by Patricia DePra, is available. It contains the answers and complete solutions for the odd-numbered problems. It also offers students a variety of exercises and keys for testing their comprehension of basic, as well as difficult, concepts.
- **Laboratory Manual:** Written by Charles H. Henrickson, Larry C. Byrd, and Norman W. Hunter, all of Western Kentucky University, *Experiments in General, Organic, and Biochemistry* carefully and safely guides students through the process of scientific inquiry. The manual features self-contained experiments that can easily be reorganized to suit individual course needs.
- **Schaum's Outline of General, Organic, and Biological Chemistry:** Written by George Odian and Ira Blei, this supplement provides students with over 1400 solved problems with complete solutions. It also teaches effective problem-solving techniques.
- **How to Study Science:** Written by Fred Drewes of Suffolk County Community College, this excellent workbook offers students helpful suggestions for meeting the considerable challenges of a science course. It offers tips on how to take notes and how to overcome science anxiety. The book's unique design helps to stir critical thinking skills, while facilitating careful note taking on the part of the student.
- **Online Learning Center:** A book-specific website is available to students and instructors using this text. The website offers quizzes, key definitions, and interesting links for the students. The address for the book-specific website is www.mhhe.com/denniston.

Acknowledgments

We are grateful to our families, whose patience and support made it possible for us to undertake this project. We are also grateful to our many colleagues at McGraw-Hill for their support, guidance, and assistance.

A revision cannot move forward without the feedback of professors teaching the course. The reviewers have our gratitude and assurance that their comments received serious consideration.

The following professors provided reviews, participated in a focus group, or gave valuable advice for the preparation of the fourth edition:

James Armstrong, *City College of San Francisco*
Satinder Bains, *Arkansas State University–Beebe*
Bal Barat, *Lake Michigan College*
Thomas Berke, *Brookdale Community College*
Lori Bolyard, *University of Evansville*
Sybil Burgess, *University of North Carolina–Wilmington*
Lynn Carlson, *University of Wisconsin–Kenosha*
Ralph E. Christensen, *University of Wisconsin–Stevens Point*
Don Colborn, *Missouri Baptist College*
Maria Derreck, *Berry College*
Son Do, *University of Southwestern Louisiana*
Kimberley Farah, *Lasell College*
J Farrar, *University of St. Francis–Fairfield*
Ted Fickel, *University of Judaism*
Edwin Geels, *Dordt College*
James Goll, *Glenville State College*
Meledath Govindan, *Fitchburg State College*
Larry Groth, *Lake Region State College*
Gamini Gunawardena, *Utah Valley State College–Orem*
Faith T. Halaweish, *South Dakota State University*
J. E. Hardcastle, *Texas Woman's University–Denton*
Klaus Himmeldirk, *Ohio University–Athens*
Carl Hoeger, *San Diego City College*
Ralph Jacobson, *Cal Poly–San Luis Obispo*
Beata Knoedler, *Springfield College*
Terri Lampe, *Georgia Perimeter College–Clarkston*
Mahela Leonida, *Fairleigh Dickinson University–Teaneck*
Scott Luaders, *Quincy University*
Tim Lubben, *Northwestern College*
Gina Mancini-Samuelson, *College of St. Catherine*
Jerome May, *Southeastern Louisiana University*
David Maynard, *California State University–San Bernardino*
Marcy McDonald, *University of Alabama–Tuscaloosa*
Harold McKone, *St. Joseph College*
Mel Mosher, *Missouri Southern College*
Lynda Nelson, *Westark Community College*
George S. Paul, *University of Central Arkansas*
Youyu Phillips, *Wilkes University*
Harold Pinnick, *West Virginia State College*
Wand Reiter, *Iowa Wesleyan College*
Gordon Sproul, *University of South Carolina–Beaufort*
Carrie Wolfe, *Union College*
Anne T. Wood, *University of Puget Sound*
Susan Yochum, *Seton Hall*
Qianhui Zhand, *Miami Dade Community College–North*

The following professors provided reviews and other valuable advice for the previous editions:

Hugh Akers, *Lamar University*
Catherine A. Anderson, *San Antonio College*
A. G. Andrewes, *Saginaw Valley State University*
Raymond D. Baechler, *Russell Sage College*
Satinder Bains, *Arkansas State University–Beebe*
Sister Marjorie Baird, O.P., *West Virginia Northern Community College*
Mark A. Benvenuto, *University of Detroit–Mercy*
Warren L. Bosch, *Elgin Community College*
Ronald Bost, *North Central Texas College*
James R. Braun, *Clayton College and State University*
Fred Brohn, *Oakland Community College*
Philip A. Brown, *Barton College*
Teresa L. Brown, *Rochester Community College*
Sister Helen Burke, *Chestnut Hill College*
Sharmaine S. Cady, *East Stroudsburg University*
Scott Carr, *Trinity Christian College*
Bernadette Corbett, *Metropolitan Community College*
Robert C. Costello, *University of South Carolina–Sumter*
Wayne B. Counts, *Georgia Southwestern State University*
Marianne Crocker, *Ozarks Technical Community College*
Peter DiMaria, *Delaware State University*
Robert P. Dixon, *Southern Illinois University*
Ronald Dunsdon, *Iowa Central Community College*
Donald R. Evers, *Iowa Central Community College*
Patrick Flash, *Kent State University–Ashtabula*
Wes Fritz, *College of DuPage*
Shelley Gaudia, *Lane Community College*

Edwin J. Geels, *Dordt College*
Deepa Godambe, *William Rainey Harper College*
W. M. Hemmerlin, *Pacific Union College*
Hildegard Hof, *College of Misericordia*
Rosalind Humerick, *St. Johns River Community College*
Devin Iimoto, *Whittier College*
Judith M. Iriarte-Gross, *Middle Tennessee State University*
T. G. Jackson, *University of South Alabama*
Michael A. Janusa, *Nicholls State University*
Paul G. Johnson, *Duquesne University*
Warren Johnson, *University of Wisconsin–Green Bay*
Donald R. Jones, *Ozarks Technical Community College*
Lidija Kampa, *Kean College of New Jersey*
Judith Kasperek, *Pitt Community College*
Kennan Kellaris, *Georgetown University*
James F. Kirby, *Quinnipiac College*
Roscoe E. Lancaster, *Golden West College*
Richard H. Langley, *Stephen F. Austin State University*
Julie E. Larson, *Bemidji State University*
Barid W. Lloyd, *Miami University*
K. W. Loach, *Plattsburgh State University*
Ralph Martinez, *Humboldt State University*
John Mazzella, *William Paterson University*
Lawrence McGahey, *College of St. Scholastica*
Cleon McKnight, *Hinds Community College*
Melvin Merken, *Worcester State College*
Robert Midden, *Bowling Green State University*
David Millsap, *South Plains College*
Ellen M. Mitchell, *Bridgewater College*
William Moeglein, *Northland Community College*
Jay Mueller, *Green River Community College*
Lynda P. Nelson, *Westark College*
Donal P. O'Mathuna, *Mount Carmel College of Nursing*
John A. Paparelli, *San Antonio College*
Richard E. Parent, *Housatonic Community Technical College*
Chetna Patel, *Aurora University*
Jeffrey A. Rahn, *Eastern Washington University*
B. R. Ramachandran, *Indiana State University*
Mona Y. Rampy, *Columbia Basin College*
John W. Reasoner, *Western Kentucky University*
Rill Ann Reuter, *Winona State University*
Terry Salerno, *Minnesota State University–Mankato*
Karen Sanchez, *Florida Community College at Jacksonville*
George Schwarzmann, Jr., *Southwest Technical College*
Sarah Selfe, *University of Washington*
Mary Selman, *Charminade University*
Kevin R. Siebenlist, *Marquette University*
Courtney J. Smith, *Tuskegee University*
Steven M. Socal, *Southern Utah University*
Gordon Sproul, *University of South Carolina–Beaufort*
Ronald H. Swisher, *Oregon Institute of Technology*
Pratibha Varma-Nelson, *Saint Xavier University*
C. G. Vlassis, *Keystone College*
Janet R. Waldeck, *The Ellis School*
Robert T. Wang, *Salem State University*
Steven Weitstock, *Indiana University*
Larry Williams, *Golden West College*
Catherine Woytowicz, *Loyola University*
Les Wynston, *California State University–Long Beach*
Gordon T. Yee, *University of Colorado–Boulder*
Jesse Yeh, *South Plains College*
Carolyn S. Yoder, *Millersville University*
Edward P. Zovinka, *St. Francis College*

The *General, Organic, and Biochemistry* Media System

The *General Organic, and Biochemistry* Media System provides both instructors and students with a variety of tools to augment teaching and learning.

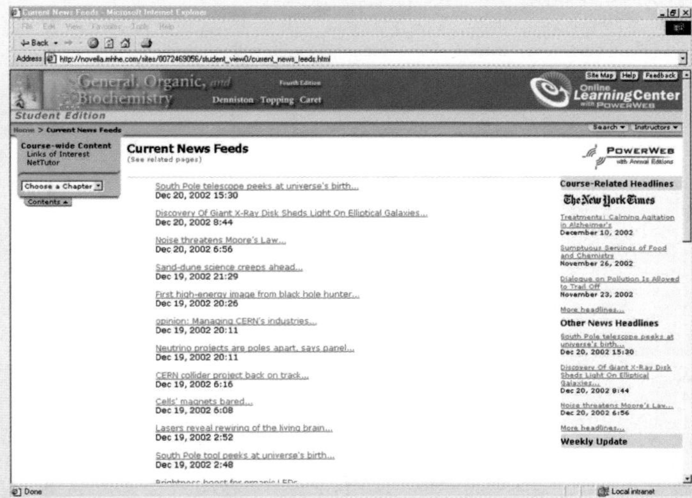

Online Learning Center

A book-specific website is available to students and instructors. Students will be able to take quizzes and access interesting links. Instructors will be able to download the Instructors Manual. The OLC is compatible with course management software such as WebCT and Blackboard.

Digital Content Manager

The DCM is the primary supplement for instructors. Contained on a CD-ROM, the DCM offers instructors full-color images of figures and tables from the text as well as PowerPoint lecture slides.

WebCT
Bb
Blackboard
www.blackboard.com
PageOut®
Create a custom course website with PageOut, free to instructors using a McGraw-Hill textbook.
To learn more, contact your McGraw-Hill publisher's representative or visit www.mhhe.com/solutions.

Course Management Systems: PageOut, WebCT, and Blackboard

The course cartridge that accompanies *General, Organic, and Biochemistry,* Fourth Edition, includes:

- All Online Learning Center content
- The entire test bank that accompanies this new edition

Brownstone Diploma

This computerized classroom management system includes test questions and a grade-recording program.

The *General, Organic, and Biochemistry* Learning System

The *General, Organic, and Biochemistry* Learning System is easy to follow, and will allow the student to excel in this course. The materials are presented in such a way that the student will effectively learn and retain the important information.

Clear Approach to Solving Problems

Because problem solving is most efficiently learned by a combination of studying examples and practicing, problems with step-by-step solutions are provided wherever appropriate. Examples are followed by a question requiring the student to integrate the newly learned material.

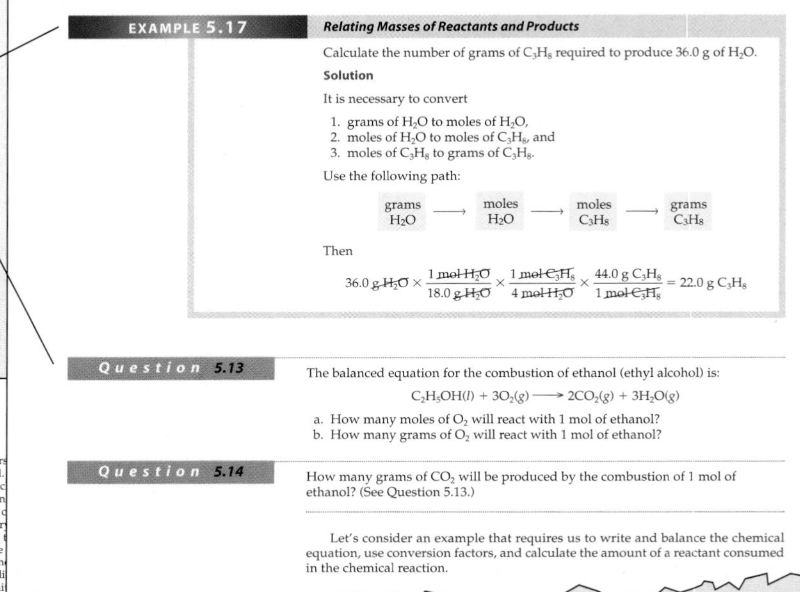

EXAMPLE 5.17 ***Relating Masses of Reactants and Products***

Calculate the number of grams of C_3H_8 required to produce 36.0 g of H_2O.

Solution

It is necessary to convert

1. grams of H_2O to moles of H_2O,
2. moles of H_2O to moles of C_3H_8, and
3. moles of C_3H_8 to grams of C_3H_8.

Use the following path:

$$\text{grams } H_2O \longrightarrow \text{moles } H_2O \longrightarrow \text{moles } C_3H_8 \longrightarrow \text{grams } C_3H_8$$

Then

$$36.0\ \text{g } H_2O \times \frac{1\ \text{mol } H_2O}{18.0\ \text{g } H_2O} \times \frac{1\ \text{mol } C_3H_8}{4\ \text{mol } H_2O} \times \frac{44.0\ \text{g } C_3H_8}{1\ \text{mol } C_3H_8} = 22.0\ \text{g } C_3H_8$$

Question 5.13 The balanced equation for the combustion of ethanol (ethyl alcohol) is:

$$C_2H_5OH(l) + 3O_2(g) \longrightarrow 2CO_2(g) + 3H_2O(g)$$

a. How many moles of O_2 will react with 1 mol of ethanol?
b. How many grams of O_2 will react with 1 mol of ethanol?

Question 5.14 How many grams of CO_2 will be produced by the combustion of 1 mol of ethanol? (See Question 5.13.)

Let's consider an example that requires us to write and balance the chemical equation, use conversion factors, and calculate the amount of a reactant consumed in the chemical reaction.

454 Chapter 15 Carboxylic Acids and Carboxylic Acid Derivatives

a. $CH_3CH_2CH_2CH_2{-}OH \xrightarrow{?} CH_3CH_2CH_2{-}C({=}O){-}Cl$

b. $CH_3CH_2{-}OH \xrightarrow{?} CH_3{-}C({=}O){-}O{-}C({=}O){-}CH_2CH_3$

c. Ethanol $\xrightarrow{?}$ ethanoic anhydride

15.53 Complete each of the following reactions by supplying the missing product:

a. $C_6H_5{-}C({=}O){-}Cl + H_2O \longrightarrow ?$

b. $CH_3{-}C({=}O){-}O{-}C({=}O){-}CH_3 + H_2O \xrightarrow{Heat} ?$

15.54 Use the I.U.P.A.C. Nomenclature System to name the products and reactants in Problem 15.53.

15.55 Write the condensed formula for each of the following compounds:
a. Decanoic anhydride
b. Acetic anhydride
c. Valeric anhydride
d. Benzoyl chloride

15.56 Write a condensed formula for each of the following compounds:
a. Propanoyl chloride
b. Heptanoyl chloride
c. Pentanoyl chloride

15.57 Describe the physical properties of acid chlorides.

15.58 Describe the physical properties of acid anhydrides.

15.59 Write an equation for the reaction of each of the following acid anhydrides with ethanol.
a. Propanoic anhydride
b. Ethanoic anhydride
c. Methanoic anhydride

15.60 Write an equation for the reaction of each of the following acid anhydrides with propanol. Name each of the products using the I.U.P.A.C. Nomenclature System.
a. Butanoic anhydride
b. Pentanoic anhydride
c. Methanoic anhydride

Phosphoesters and Thioesters

15.61 By reacting phosphoric acid with an excess of ethanol, it is possible to obtain the mono-, di-, and triesters of phosphoric acid. Draw all three of these products.

15.62 What is meant by a phosphoric anhydride bond?

15.63 We have described the molecule ATP as the body's energy storehouse. What do we mean by this designation? How does ATP actually store energy and provide it to the body as needed?

15.64 Write an equation for each of the following reactions:
a. Ribose + phosphoric acid
b. Methanol + phosphoric acid
c. Adenosine diphosphate + phosphoric acid

15.65 Draw the thioester bond between the acetyl group and coenzyme A.

15.66 Explain the significance of thioester formation in the metabolic pathways involved in fatty acid and carbohydrate breakdown.

15.67 It is also possible to form esters ... as sulfuric acid and nitric acid. ... product is nitroglycerine, whic... (explosive) and widely used in ... condition known as angina, a c... usually resulting from coronar... case its function is to alleviate ... angina. Nitroglycerine may be ... (usually placed just beneath th... salve or paste that can be appli... skin. Nitroglycerine is the trini... structure of nitroglycerine, usi...

$$H{-}C(H)(OH){-}C(H)(OH){-}C(H)(OH){-}H$$

Glycerol

15.68 Show the structure of the thioester that would be formed between coenzyme A and stearic acid.

Critical Thinking Problems

1. Radioactive isotopes of an element behave chemically in exactly the same manner as the nonradioactive isotopes. As a result, they can be used as tracers to investigate the details of chemical reactions. A scientist is curious about the origin of the bridging oxygen atom in an ester molecule. She has chosen to use the radioactive isotope oxygen-18 to study the following reaction:

$$CH_3CH_2OH + CH_3C({=}O){-}OH \xrightarrow{H^+,\ heat} CH_3C({=}O){-}O{-}CH_2CH_3 + H_2O$$

Design experiments using oxygen-18 that will demonstrate whether the oxygen in the water molecule came from the —OH of the alcohol or the —OH of the carboxylic acid.

2. Triglycerides are the major lipid storage form in the human body. They are formed in an esterification reaction between glycerol (1,2,3-propanetriol) and three fatty acids (long chain carboxylic acids). Write a balanced equation for the formation of a triglyceride formed in a reaction between glycerol and three molecules of decanoic acid.

3. Chloramphenicol is a very potent, broad-spectrum antibiotic. It is reserved for life-threatening bacterial infections because it is quite toxic. It is also a very bitter tasting chemical. As a result, children had great difficulty taking the antibiotic. A clever chemist found that the taste could be improved considerably by producing the palmitate ester. Intestinal enzymes hydrolyze the ester, producing chloramphenicol, which can then be absorbed. The following structure is the palmitate ester of chloramphenicol. Draw the structure of chloramphenicol.

$$O_2N{-}C_6H_4{-}CH(OH){-}CH(NHCCHCl_2)CH_2{-}O{-}C({=}O){-}(CH_2)_{14}CH_3$$

Chloramphenicol palmitate

15-34

A variety of questions and problems that range in level of difficulty help students measure their mastery of the chapter material. The odd-numbered questions are answered in the back of the text.

At the end of the chapter, the student will find several problems that require thought-provoking answers dealing with daily life and the health care sciences.

Dynamic Visuals

Many of the equations and reactions are color coded to help the student understand the chemical changes that occur in complex reactions. The student can easily recognize the chemical groups being added or removed in a reaction by the color coding. Green background illustrates an important equation or key reaction; yellow background illustrates energy in the general and biochemistry sections and reveals the parent chain of a compound in the organic section; red and blue lettering distinguish two or more compounds that appear similar.

The art program has been significantly updated with the use of molecular art and drawings. The students will gain a better perspective and understanding of a molecule with a Spartan computer-generated model.

7.6 Concentration-Dependent Solution Properties 187

We have already worked with one mole-based unit, *molarity*, and this concentration unit can be used to calculate either the freezing point depression or the boiling point elevation.

A second mole-based concentration unit, molality, is more commonly used in these types of situations. **Molality** (symbolized *m*) is defined as the number of moles of solute per kilogram of solvent in a solution:

$$m = \frac{\text{moles solute}}{\text{kg solvent}}$$

Molality does not vary with temperature, whereas molarity is temperature dependent. For this reason, molality is the preferred concentration unit for studies such as freezing point depression and boiling point elevation, in which measurement of *change* in temperature is critical.

Practical applications that take advantage of freezing point depression of solutions by solutes include the following:

Molarity is temperature dependent simply because it is expressed as mole/volume. Volume is temperature dependent—most liquids expand measurably when heated and contract when cooled. Molality is moles/mass; both moles and mass are temperature independent.

measurement of heat change in chemical reactions. The concentric Styrofoam cups insulate the system from its surroundings. Heat released by the chemical reaction enters the water, raising its temperature, which is measured by using a thermometer.

and the change in temperature (ΔT_s) of the solution as the reaction proceeds from the initial to final state.

The heat is calculated by using the following equation:

$$Q = m_s \times \Delta T_s \times SH_s$$

with units

$$\text{calories} = \text{gram} \times {}^\circ\text{C} \times \frac{\text{calories}}{\text{gram-}{}^\circ\text{C}}$$

Now add the substituents. In this example a bromine atom is bonded to carbon-1, and a methyl group is bonded to carbon-4:

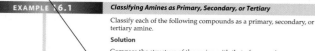

Finally, add the correct number of hydrogen atoms so that each carbon has four covalent bonds:

As a final check of your accuracy, use the I.U.P.A.C. system to name the compound that you have just drawn, and compare the name with that in

EXAMPLE 16.1 ***Classifying Amines as Primary, Secondary, or Tertiary***

Classify each of the following compounds as a primary, secondary, or tertiary amine.

Solution

Compare the structure of the amine with that of ammonia.

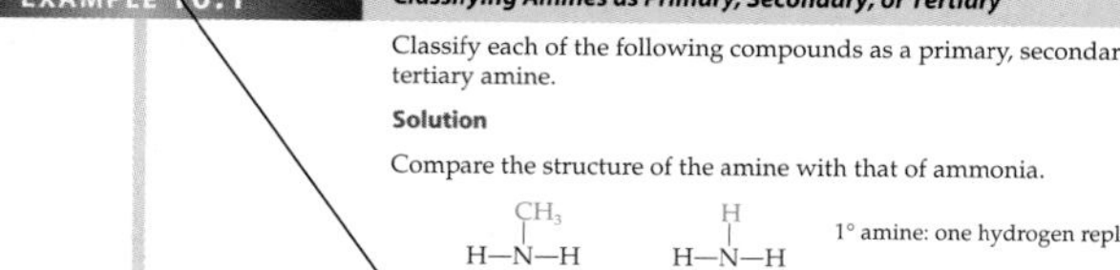

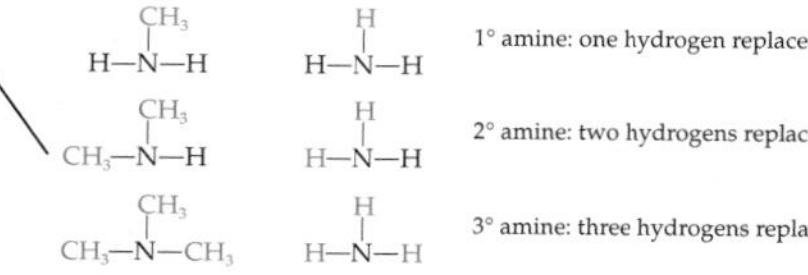

1° amine: one hydrogen replaced

2° amine: two hydrogens replaced

3° amine: three hydrogens replaced

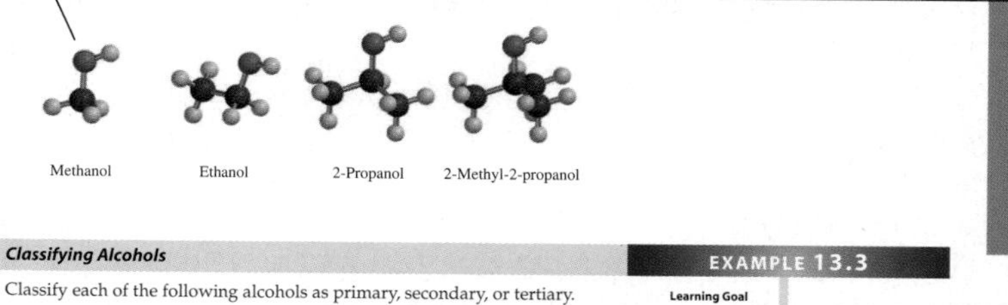

Classifying Alcohols **EXAMPLE 13.3**

Classify each of the following alcohols as primary, secondary, or tertiary.

Learning Goal

Solution

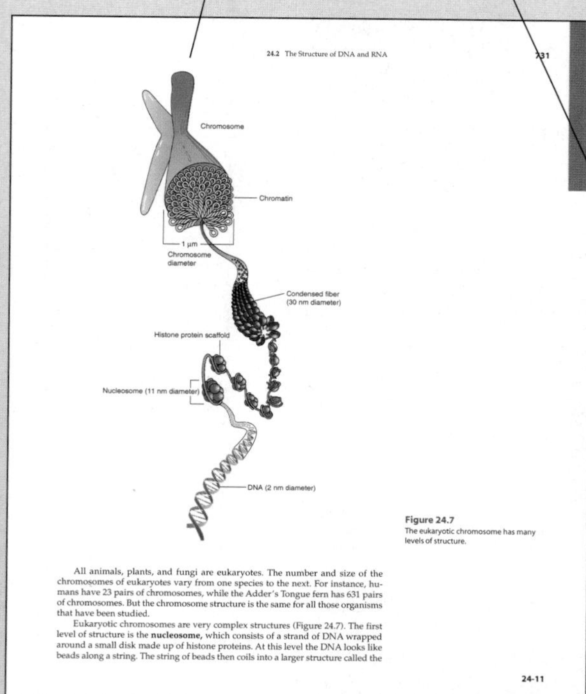
24.2 The Structure of DNA and RNA 731

Figure 24.7 The eukaryotic chromosome has many levels of structure.

All animals, plants, and fungi are eukaryotes. The number and size of the chromosomes of eukaryotes vary from one species to the next. For instance, humans have 23 pairs of chromosomes, while the Adder's Tongue fern has 631 pairs of chromosomes. But the chromosome structure is the same for all those organisms that have been studied.

Eukaryotic chromosomes are very complex structures (Figure 24.7). The first level of structure is the **nucleosome**, which consists of a strand of DNA wrapped around a small disk made up of histone proteins. At this level the DNA looks like beads along a string. The string of beads then coils into a larger structure called the

24-11

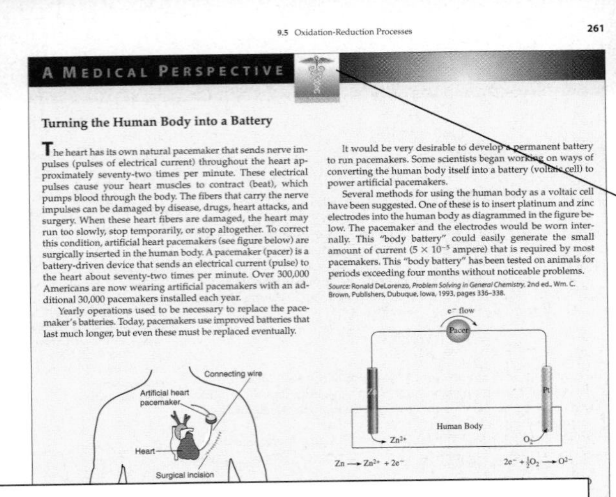

9.5 Oxidation-Reduction Processes 261

A MEDICAL PERSPECTIVE

Turning the Human Body into a Battery

The heart has its own natural pacemaker that sends nerve impulses (pulses of electrical current) throughout the heart approximately seventy-two times per minute. These electrical pulses cause your heart muscles to contract (beat), which pumps blood through the body. The fibers that carry the nerve impulses can be damaged by disease, drugs, heart attacks, and surgery. When these heart fibers are damaged, the heart may run too slowly, stop temporarily, or stop altogether. To correct this condition, artificial heart pacemakers (see figure below) are surgically inserted in the human body. A pacemaker (pacer) is a battery-driven device that sends an electrical current (pulse) to the heart about seventy-two times per minute. Over 300,000 Americans are now wearing artificial pacemakers with an additional 30,000 pacemakers installed each year.

Yearly operations used to be necessary to replace the pacemaker's batteries. Today, pacemakers use improved batteries that last much longer, but even these must be replaced eventually.

It would be very desirable to develop a permanent battery to run pacemakers. Some scientists began working on ways of converting the human body itself into a battery (voltaic cell) to power artificial pacemakers.

Several methods for using the human body as a voltaic cell have been suggested. One of these is to insert platinum and zinc electrodes into the human body as diagrammed in the figure below. The pacemaker and the electrodes would be worn internally. This "body battery" could easily generate the small amount of current (5×10^{-5} ampere) that is required by most pacemakers. This "body battery" has been tested on animals for periods exceeding four months without noticeable problems.

Source: Ronald DeLorenzo, *Problem Solving in General Chemistry*, 2nd ed., Wm. C. Brown, Publishers, Dubuque, Iowa, 1993, pages 336–338.

Connecting wire
Artificial heart pacemaker
Heart
Surgical incision

e^- flow
Pacer
Pt
Human Body
Zn^{2+}
O_2
$Zn \longrightarrow Zn^{2+} + 2e^-$
$2e^- + \frac{1}{2}O_2 \longrightarrow O^{2-}$

Health/Life Related Applications

There are four different Perspective boxes in the text. Chemistry Connections provide an introductory scenario for the chapter; Medical Perspectives and Clinical Perspectives demonstrate use of the chapter material in an allied health field; Environmental Perspectives demonstrate chapter concepts in ecological problems; and Human Perspectives demonstrate how important chemistry is in our day-to-day lives.

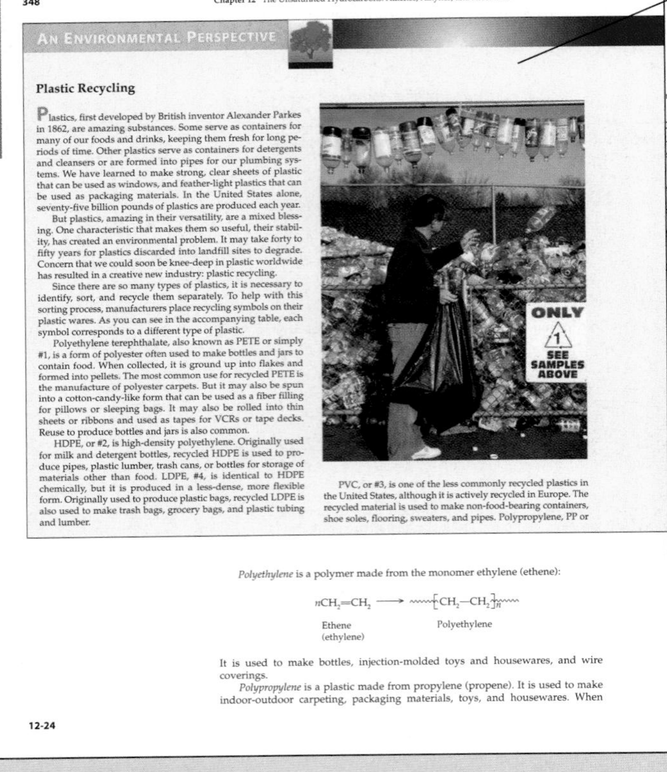

348 Chapter 12 The Unsaturated Hydrocarbons: Alkenes, Alkynes, and Aromatics

AN ENVIRONMENTAL PERSPECTIVE

Plastic Recycling

Plastics, first developed by British inventor Alexander Parkes in 1862, are amazing substances. Some serve as containers for many of our foods and drinks, keeping them fresh for long periods of time. Other plastics serve as containers for detergents and cleansers or are formed into pipes for our plumbing systems. We have learned to make strong, clear sheets of plastic that can be used as windows, and feather-light plastics that can be used as packaging materials. In the United States alone, seventy-five billion pounds of plastics are produced each year.

But plastics, amazing in their versatility, are a mixed blessing. One characteristic that makes them so useful, their stability, has created an environmental problem. It may take forty to fifty years for plastics discarded into landfill sites to degrade. Concern that we could soon be knee-deep in plastic worldwide has resulted in a creative new industry: plastic recycling.

Since there are so many types of plastics, it is necessary to identify, sort, and recycle them separately. To help with this sorting process, manufacturers place recycling symbols on their plastic wares. As you can see in the accompanying table, each symbol corresponds to a different type of plastic.

Polyethylene terephthalate, also known as PETE or simply #1, is a form of polyester often used to make bottles and jars to contain food. When collected, it is ground up into flakes and formed into pellets. The most common use for recycled PETE is the manufacture of polyester carpets. But it may also be spun into a cotton-candy-like form that can be used as a fiber filling for pillows or sleeping bags. It may also be rolled into thin sheets or ribbons and used as tapes for VCRs or tape decks. Reuse to produce bottles and jars is also common.

HDPE, or #2, is high-density polyethylene. Originally used for milk and detergent bottles, recycled HDPE is used to produce pipes, plastic lumber, trash cans, or bottles for storage of materials other than food. LDPE, #4, is identical to HDPE chemically, but it is produced in a less-dense, more flexible form. Originally used to produce plastic bags, recycled LDPE is also used to make trash bags, grocery bags, and plastic tubing and lumber.

ONLY
1
SEE SAMPLES ABOVE

PVC, or #3, is one of the less commonly recycled plastics in the United States, although it is actively recycled in Europe. The recycled material is used to make non-food-bearing containers, shoe soles, flooring, sweaters, and pipes. Polypropylene, PP or

Polyethylene is a polymer made from the monomer ethylene (ethene):

$$nCH_2{=}CH_2 \longrightarrow \sim\!\!\sim[CH_2{-}CH_2]_n\sim\!\!\sim$$

Ethene (ethylene) Polyethylene

It is used to make bottles, injection-molded toys and housewares, and wire coverings.

Polypropylene is a plastic made from propylene (propene). It is used to make indoor-outdoor carpeting, packaging materials, toys, and housewares. When

12-24

Clear Presentation

Each chapter begins with an outline that introduces students to the topics to be presented. This outline also provides the instructor with a quick topic summary to organize lecture material.

A list of learning goals, based on the major concepts covered in the chapter, enables students to preview the material and become aware of the topics they are expected to master.

This icon is found within the chapters wherever the associated learning goal is first presented, allowing the student to focus attention on the major concepts.

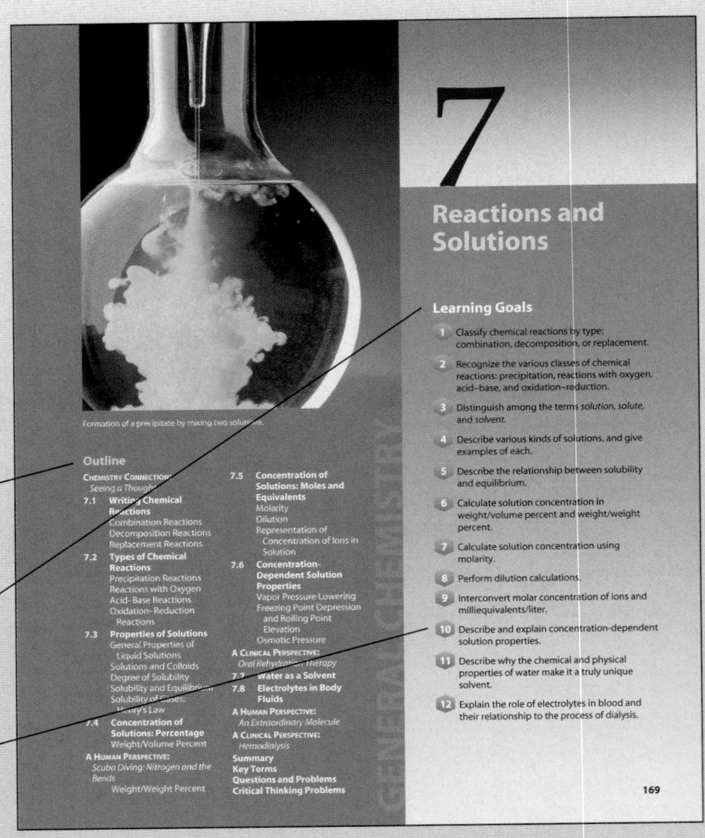

7

Reactions and Solutions

Formation of a precipitate by mixing two solutions.

Outline

CHEMISTRY CONNECTION: *Seeing a Thought*
7.1 **Writing Chemical Reactions**
Combination Reactions
Decomposition Reactions
Replacement Reactions
7.2 **Types of Chemical Reactions**
Precipitation Reactions
Reactions with Oxygen
Acid–Base Reactions
Oxidation–Reduction Reactions
7.3 **Properties of Solutions**
General Properties of Liquid Solutions
Solutions and Colloids
Degree of Solubility
Solubility and Equilibrium
Solubility of Gases: Henry's Law
7.4 **Concentration of Solutions: Percentage**
Weight/Volume Percent
A HUMAN PERSPECTIVE: *Scuba Diving: Nitrogen and the Bends*
Weight/Weight Percent
7.5 **Concentration of Solutions: Moles and Equivalents**
Molarity
Dilution
Representation of Concentration of Ions in Solution
7.6 **Concentration-Dependent Solution Properties**
Vapor Pressure Lowering
Freezing Point Depression and Boiling Point Elevation
Osmotic Pressure
A CLINICAL PERSPECTIVE: *Oral Rehydration Therapy*
7.7 **Water as a Solvent**
7.8 **Electrolytes in Body Fluids**
A HUMAN PERSPECTIVE: *An Extraordinary Molecule*
A CLINICAL PERSPECTIVE: *Hemodialysis*
Summary
Key Terms
Questions and Problems
Critical Thinking Problems

GENERAL CHEMISTRY

Learning Goals

1. Classify chemical reactions by type: combination, decomposition, or replacement.
2. Recognize the various classes of chemical reactions: precipitation, reactions with oxygen, acid–base, and oxidation–reduction.
3. Distinguish among the terms *solution, solute,* and *solvent.*
4. Describe various kinds of solutions, and give examples of each.
5. Describe the relationship between solubility and equilibrium.
6. Calculate solution concentration in weight/volume percent and weight/weight percent.
7. Calculate solution concentration using molarity.
8. Perform dilution calculations.
9. Interconvert molar concentration of ions and milliequivalents/liter.
10. Describe and explain concentration-dependent solution properties.
11. Describe why the chemical and physical properties of water make it a truly unique solvent.
12. Explain the role of electrolytes in blood and their relationship to the process of dialysis.

169

Additional Pedagogical Elements

Margin notes direct the student to a reference in the book for further material or review.

A dilute solution of phenol must be used because concentrated phenol causes severe burns and because phenol is not highly soluble in water.

Phenols are also widely used in health care as germicides. In fact, carbolic acid, a dilute solution of phenol, was used as an antiseptic and disinfectant by Joseph Lister in his early work to decrease postsurgical infections. He used carbolic acid to bathe surgical wounds and to "sterilize" his instruments. Other derivatives of phenol that are used as antiseptics and disinfectants include hexachlorophene, hexylresorcinol, and *o*-phenylphenol. The structures of these compounds are shown below:

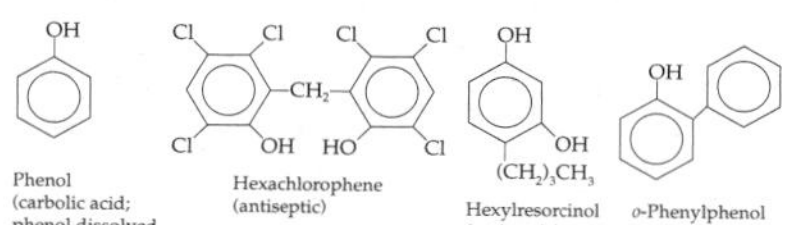

13.8 Ethers

Learning Goal 11

Ethers have the general formula R—O—R, and thus they are structurally related to alcohols (R—O—H). The C—O bonds of ethers are polar, so ether molecules are polar (Figure 13.5). However, ethers do not form hydrogen bonds to one another because there is no —OH group. Therefore they have much lower boiling points than alcohols of similar molecular weight but higher boiling points than alkanes of similar molecular weight. Compare the following examples:

$CH_3CH_2CH_2CH_3$	$CH_3—O—CH_2CH_3$	$CH_3CH_2CH_2OH$
Butane (butane) M.W. = 58 b.p. = −0.5°C	Methoxyethane (ethyl methyl ether) M.W. = 60 b.p. = 7.9°C	1-Propanol (propyl alcohol) M.W. = 60 b.p. = 97.2°C

Figure 13.5
Ball-and-stick model of the ether, methoxymethane (dimethyl ether).

In the I.U.P.A.C. system of naming ethers the —OR substituent is named as an alkoxy group. This is analogous to the name *hydroxy* for the —OH group. Thus $CH_3—O—$ is methoxy, $CH_3CH_2—O—$ is ethoxy, and so on.

An alkoxy group is an alkyl group bonded to an oxygen atom (—OR).

EXAMPLE 13.11 ***Using I.U.P.A.C. Nomenclature to Name an Ether***

Name the following ether using I.U.P.A.C. nomenclature.

Solution

$$\overset{O—CH_3}{CH_3CH_2CHCH_2CH_2CH_2CH_2CH_2CH_3}$$
1 2 3 4 5 6 7 8 9

Parent compound: nonane
Position of alkoxy group: carbon-3 (*not* carbon-7)
Substituents: 3-methoxy
Name: 3-Methoxynonane

All bold-faced terms in the chapter are listed at the end of each chapter and defined in the Glossary at the end of the text. The student can easily find important terms when reading and studying.

At the end of each chapter is a summary designed to help students more easily identify important concepts and help them review for quizzes and tests.

Summary

16.1 Amines

Amines are a family of organic compounds that contain an amino group or substituted amino group. A *primary amine* has the general formula RNH_2; a *secondary amine* has the general formula R_2NH; and a *tertiary amine* has the general formula R_3N. In the *Chemical Abstracts* nomenclature system, amines are named as *alkanamines*. In the I.U.P.A.C. system they are named as *aminoalkanes*. In the common system they are named as *alkylamines*. Amines behave as weak bases, forming *alkylammonium ions* in water and alkylammonium salts when they react with acids. *Quaternary ammonium salts* are ammonium salts that have four organic groups bonded to the nitrogen atom.

16.2 Heterocyclic Amines

Heterocyclic amines are cyclic compounds having at least one nitrogen atom in the ring structure. *Alkaloids* are natural plant products that contain at least one heterocyclic ring. Many alkaloids have powerful biological effects.

16.3 Amides

Amides are formed in a reaction between a carboxylic acid derivative and an amine (or ammonia). The *amide bond* is the bond between the carbonyl carbon of the *acyl group* and the nitrogen of the amine. In the I.U.P.A.C. Nomenclature System they are named by replacing the *-oic acid* ending of the carboxylic acid with the *-amide* ending. In the common system of nomenclature the *-ic acid* ending of the carboxylic acid is replaced by the *-amide* ending. Hydrolysis of an amide produces a carboxylic acid and an amine (or ammonia).

16.4 A Preview of Amino Acids, Proteins, and Protein Synthesis

Proteins are polymers of amino acids joined to one another by *amide bonds* called *peptide bonds*. During protein synthesis the *aminoacyl group* of one amino acid is transferred from a carrier molecule called a *transfer RNA* to the amino group nitrogen of another amino acid.

16.5 Neurotransmitters

Neurotransmitters are chemicals that carry messages, or signals, from a nerve cell to a target cell, which may be another nerve cell or a muscle cell. They may be inhibitory or excitatory and all are nitrogen-containing compounds. The catecholamines include dopamine, norepinephrine, and epinephrine. Too little dopamine results in Parkinson's disease. Too much is associated with schizophrenia. Dopamine is also associated with addictive behavior. A deficiency of serotonin is associated with depression and eating disorders. Serotonin is involved in pain perception, regulation of body temperature, and sleep. Histamine contributes to allergy symptoms. Antihistamines block histamines and provide relief from allergies. γ-Aminobutyric acid (GABA) and glycine are inhibitory neurotransmitters. It is believed that GABA is involved in control of aggressive behavior. Acetylcholine is a neurotransmitter that functions at the neuromuscular junction, carrying signals from the nerve to the muscle. Nitric oxide and glutamate function in a positive feedback loop that is thought to be involved in learning and the formation of memories.

Key Terms

acyl group (16.3)
alkaloid (16.2)
alkylammonium ion (16.1)
amide (16.3)
amide bond (16.3)
amine (16.1)
aminoacyl group (16.4)
analgesic (16.2)
anesthetic (16.2)
heterocyclic amine (16.2)
neurotransmitter (16.5)
peptide bond (16.4)
primary (1°) amine (16.1)
quaternary ammonium salt (16.1)
secondary (2°) amine (16.1)
tertiary (3°) amine (16.1)
transfer RNA (tRNA) (16.4)

Questions and Problems

Amines

16.15 For each pair of compounds predict which would have greater solubility in water. Explain your reasoning.
- a. Pentane or 1-butanamine
- b. Cyclohexane or 2-pentanamine

16.16 For each pair of compounds predict which would have the higher boiling point. Explain your reasoning.
- a. Ethanamine or ethanol
- b. Butane or 1-propanamine
- c. Methanamine or water
- d. Ethylmethylamine or butane

16.17 Explain why a tertiary amine such as triethylamine has a significantly lower boiling point than its primary amine isomer, 1-hexanamine.

16.18 Draw a diagram to illustrate your answer to Problem 16.17.

16.19 Use the *Chemical Abstracts* system of nomenclature to name each of the following amines:
- a. $CH_3CH_2\underset{CH_3}{CH}—NH_2$
- b. $CH_3CH_2CH_2\underset{NH_2}{CH}CH_2CH_3$
- c. (cyclopentane ring)—NH_2
- d. $(CH_3)_3C—NH_2$

16.20 Use the CA and common nomenclature systems to name each of the following amines:
- a. $CH_3CH_2CH_2CH_2CH_2CH_2CH_2CH_2—NH_2$
- b. Cl—(benzene ring)—NH_2

Name the forms of measurement that apply to this activity.

1 Chemistry: *Methods and Measurement*

GENERAL CHEMISTRY

Outline

Learning Goals

1. Describe the interrelationship of chemistry with other fields of science and medicine.
2. Discuss the approach to science, the scientific method.
3. Distinguish among the terms *hypothesis, theory,* and *scientific law*.
4. Describe the properties of the solid, liquid, and gaseous states.
5. Classify properties as chemical or physical.
6. Classify observed changes in matter as chemical or physical.
7. Provide specific examples of physical and chemical properties.
8. Distinguish between intensive and extensive properties.
9. Classify matter as element, compound, or mixture.
10. Distinguish between data and results.
11. Learn the major units of measure in the English and metric systems, and be able to convert from one system to another.
12. Report data and results using scientific notation and the proper number of significant figures.
13. Use appropriate experimental quantities in problem solving.
14. Calculate the density of an object from mass and volume data and calculate the specific gravity of an object from its density.

CHEMISTRY CONNECTION

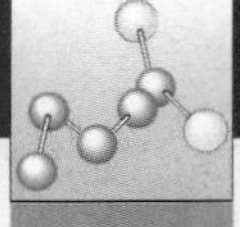

Chance Favors the Prepared Mind

Most of you have chosen a career in medicine because you want to help others. In medicine, helping others means easing pain and suffering by treating or curing diseases. One important part of the practice of medicine involves observation. The physician must carefully observe the patient and listen to his or her description of symptoms to arrive at a preliminary diagnosis. Then appropriate tests must be done to determine whether the diagnosis is correct. During recovery the patient must be carefully observed for changes in behavior or symptoms. These changes are clues that the treatment or medication needs to be modified.

These practices are also important in science. The scientist makes an observation and develops a preliminary hypothesis or explanation for the observed phenomenon. Experiments are then carried out to determine whether the hypothesis is correct. When performing the experiment and analyzing the data, the scientist must look for any unexpected results that indicate that the original hypothesis must be modified.

Several important discoveries in medicine and the sciences have arisen from accidental observations. A health care worker or scientist may see something quite unexpected. Whether this results in an important discovery or is ignored depends on the training and preparedness of the observer.

It was Louis Pasteur, a chemist and microbiologist, who said, "Chance favors the prepared mind." In the history of science and medicine there are many examples of individuals who have made important discoveries because they recognized the value of an unexpected observation.

One such example is the use of ultraviolet (UV) light to treat infant jaundice. Infant jaundice is a condition in which the skin and the whites of the eyes appear yellow because of high levels of the bile pigment bilirubin in the blood. Bilirubin is a breakdown product of the oxygen-carrying blood protein hemoglobin. If bilirubin accumulates in the body, it can cause brain damage and death. The immature liver of the baby cannot remove the bilirubin.

An observant nurse in England noticed that when jaundiced babies were exposed to sunlight, the jaundice faded. Research based on her observation showed that the UV light changes the bilirubin into another substance that can be excreted. To this day, jaundiced newborns are treated with UV light.

The Pap smear test for the early detection of cervical and uterine cancer was also developed because of an accidental observation. Dr. George Papanicolaou, affectionately called Dr. Pap, was studying changes in the cells of the vagina during the stages of the menstrual cycle. In one sample he recognized cells that looked like cancer cells. Within five years, Dr. Pap had perfected a technique for staining cells from vaginal fluid and observing them microscopically for the presence of any abnormal cells. The lives of countless women have been saved because a routine Pap smear showed early stages of cancer.

In this first chapter of your study of chemistry you will learn more about the importance of observation and accurate, precise measurement in medical practice and scientific study. You will also study the scientific method, the process of developing hypotheses to explain observations, and the design of experiments to test those hypotheses.

Introduction

When you awoke this morning, a flood of chemicals called neurotransmitters was sent from cell to cell in your nervous system. As these chemical signals accumulated, you gradually became aware of your surroundings. Chemical signals from your nerves to your muscles propelled you out of your warm bed to prepare for your day.

For breakfast you had a glass of milk, two eggs, and buttered toast, thus providing your body with needed molecules in the form of carbohydrates, proteins, lipids, vitamins, and minerals. As you ran out the door, enzymes of your digestive tract were dismantling the macromolecules of your breakfast. Other enzymes in your cells were busy converting the chemical energy of food molecules into adenosine triphosphate (ATP), the universal energy currency of all cells.

As you continue through your day, thousands of biochemical reactions will keep your cells functioning optimally. Hormones and other chemical signals will regulate the conditions within your body. They will let you know if you are hungry or thirsty. If you injure yourself or come into contact with a disease-causing microorganism, chemicals in your body will signal cells to begin the necessary repair or defense processes.

Life is an organized array of large, carbon-based molecules maintained by biochemical reactions. To understand and appreciate the nature of a living being, we must understand the principles of science and chemistry as they apply to biological molecules.

1.1 The Discovery Process

Chemistry

Chemistry is the study of matter, its chemical and physical properties, the chemical and physical changes it undergoes, and the energy changes that accompany those processes. **Matter** is anything that has mass and occupies space. The changes that matter undergoes always involve either gain or loss of energy. **Energy** is the ability to do work to accomplish some change. The study of chemistry involves matter, energy, and their interrelationship. Matter and energy are at the heart of chemistry.

Major Areas of Chemistry

Learning Goal

Chemistry is a broad area of study covering everything from the basic parts of an atom to interactions between huge biological molecules. Because of this, chemistry encompasses the following specialties.

Biochemistry is the study of life at the molecular level and the processes associated with life, such as reproduction, growth, and respiration. *Organic chemistry* is the study of matter that is composed principally of carbon and hydrogen. Organic chemists study methods of preparing such diverse substances as plastics, drugs, solvents, and a host of industrial chemicals. *Inorganic chemistry* is the study of matter that consists of all of the elements other than carbon and hydrogen and their combinations. Inorganic chemists have been responsible for the development of unique substances such as semiconductors and high-temperature ceramics for industrial use. *Analytical chemistry* involves the analysis of matter to determine its composition and the quantity of each kind of matter that is present. Analytical chemists detect traces of toxic chemicals in water and air. They also develop methods to analyze human body fluids for drugs, poisons, and levels of medication. *Physical chemistry* is a discipline that attempts to explain the way in which matter behaves. Physical chemists develop theoretical concepts and try to prove them experimentally. This helps us understand how chemical systems behave.

Over the last thirty years, the boundaries between the traditional sciences of chemistry and biology, mathematics, physics, and computer science have gradually faded. Medical practitioners, physicians, nurses, and medical technologists use therapies that contain elements of all these disciplines. The rapid expansion of the pharmaceutical industry is based on a recognition of the relationship between the function of an organism and its basic chemical makeup. Function is a consequence of changes that chemical substances undergo.

For these reasons, an understanding of basic chemical principles is essential for anyone considering a medically related career; indeed, a worker in any science-related field will benefit from an understanding of the principles and applications of chemistry.

The Scientific Method

Learning Goal

The **scientific method** is a systematic approach to the discovery of new information. How do we learn about the properties of matter, the way it behaves in nature, and how it can be modified to make useful products? Chemists do this by using the scientific method to study the way in which matter changes under carefully controlled conditions.

A HUMAN PERSPECTIVE

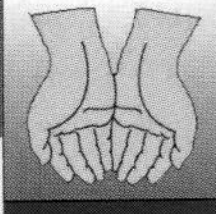

The Scientific Method

The discovery of penicillin by Alexander Fleming is an example of the scientific method at work. Fleming was studying the growth of bacteria. One day, his experiment was ruined because colonies of mold were growing on his plates. From this failed experiment, Fleming made an observation that would change the practice of medicine: Bacterial colonies could not grow in the area around the mold colonies. Fleming hypothesized that the mold was making a chemical compound that inhibited the growth of the bacteria. He performed a series of experiments designed to test this hypothesis.

The key to the scientific method is the design of carefully controlled experiments that will either support or disprove the hypothesis. This is exactly what Fleming did.

In one experiment he used two sets of tubes containing sterile nutrient broth. To one set he added mold cells. The second set (the control tubes) remained sterile. The mold was allowed to grow for several days. Then the broth from each of the tubes (experimental and control) was passed through a filter to remove any mold cells. Next, bacteria were placed in each tube. If Fleming's hypothesis was correct, the tubes in which the mold had grown would contain the chemical that inhibits growth, and the bacteria would not grow. On the other hand, the control tubes (which were never used to grow mold) would allow bacterial growth. This is exactly what Fleming observed.

Within a few years this *antibiotic*, penicillin, was being used to treat bacterial infections in patients.

The scientific method is not a "cookbook recipe" that, if followed faithfully, will yield new discoveries; rather, it is an organized approach to solving scientific problems. Every scientist brings his or her own curiosity, creativity, and imagination to scientific study. But scientific inquiry still involves some of the "cookbook approach."

Characteristics of the scientific process include the following:

1. *Observation.* The description of, for example, the color, taste, or odor of a substance is a result of observation. The measurement of the temperature of a liquid or the size or mass of a solid results from observation.
2. *Formulation of a question.* Humankind's fundamental curiosity motivates questions of why and how things work.
3. *Pattern recognition.* If a scientist finds a cause-and-effect relationship, it may be the basis of a generalized explanation of substances and their behavior.

Learning Goal

4. *Developing theories.* When scientists observe a phenomenon, they want to explain it. The process of explaining observed behavior begins with a hypothesis. A **hypothesis** is simply an attempt to explain an observation, or series of observations, in a commonsense way. If many experiments support a hypothesis, it may attain the status of a theory. A **theory** is a hypothesis supported by extensive testing (experimentation) that explains scientific facts and can predict new facts.
5. *Experimentation.* Demonstrating the correctness of hypotheses and theories is at the heart of the scientific method. This is done by carrying out carefully designed experiments that will either support or disprove the theory or hypothesis.
6. *Summarizing information.* A scientific **law** is nothing more than the summary of a large quantity of information. For example, the law of conservation of matter states that matter cannot be created or destroyed, only converted from one form to another. This statement represents a massive body of chemical information gathered from experiments.

The scientific method involves the interactive use of hypotheses, development of theories, and thorough testing of theories using well-designed experiments and is summarized in Figure 1.1.

Models in Chemistry

Hypotheses, theories, and laws are frequently expressed using mathematical equations. These equations may confuse all but the best of mathematicians. For this reason a *model* of a chemical unit or system is often used to make ideas more clear. A good model based on everyday experience, although imperfect, gives a great deal of information in a simple fashion. Consider the fundamental unit of methane, the major component of natural gas, which is composed of one carbon atom (symbolized by C) and four hydrogen atoms (symbolized by H).

A geometrically correct model of methane can be constructed from balls and sticks. The balls represent the individual units (atoms) of hydrogen and carbon, and the sticks correspond to the attractive forces that hold the hydrogen and carbon together. The model consists of four balls representing hydrogen symmetrically arranged around a center ball representing carbon. The "carbon" ball is attached to each "hydrogen" ball by sticks, as shown:

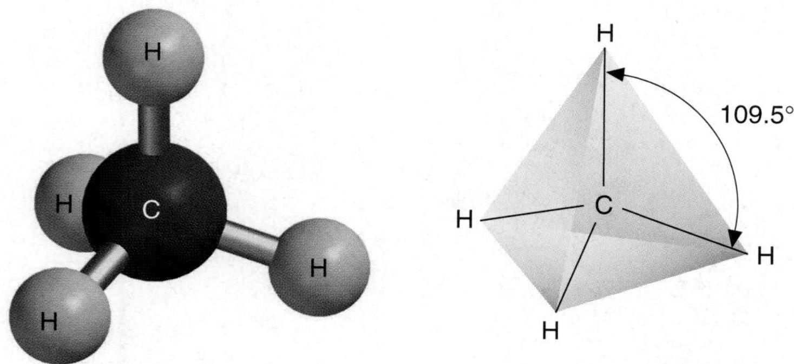

Color-coding the balls distinguishes one type of matter from another; the geometrical form of the model, all of the angles and dimensions of a tetrahedron, are the same for each methane unit found in nature. Methane is certainly not a collection of balls and sticks; but such models are valuable because they help us understand the chemical behavior of methane and other, more complex substances.

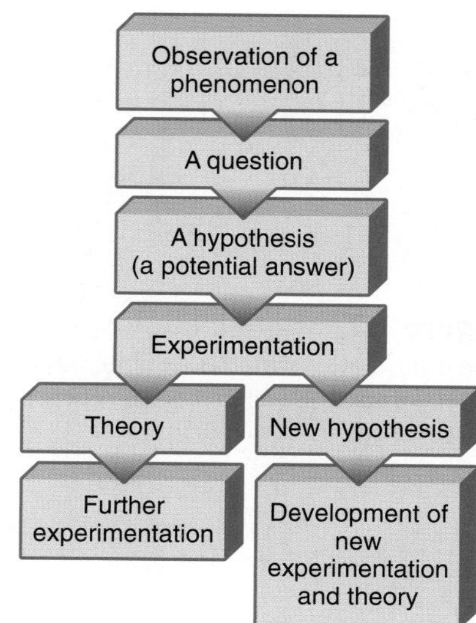

Figure 1.1
The scientific method, an organized way of doing science. A degree of trial and error is apparent here. If experimentation does not support the hypothesis, one must begin the cycle again.

1.2 Matter and Properties

Properties are characteristics of matter and are classified as either physical or chemical. In this section we will learn the meaning of physical and chemical properties and how they are used to characterize matter.

Matter and Physical Properties

There are three **states of matter:** the **gaseous state,** the **liquid state,** and the **solid state.** A gas is made up of particles that are widely separated. In fact, a gas will expand to fill any container; it has no definite shape or volume. In contrast, particles of a liquid are closer together; a liquid has a definite volume but no definite shape; it takes on the shape of its container. A solid consists of particles that are close together and that often have a regular and predictable pattern of particle arrangement (crystalline). A solid has both fixed volume and fixed shape. Attractive forces, which exist between all particles, are very pronounced in solids and much less so in gases.

Water is the most common example of a substance that can exist in all three states over a reasonable temperature range (Figure 1.2). Conversion of water from one state to another constitutes a *physical change.* A **physical change** produces a recognizable difference in the appearance of a substance without causing any change in its composition or identity. For example, we can warm an ice cube and it will melt, forming liquid water. Clearly its appearance has changed; it has been transformed from the solid to the liquid state. It is, however, still water; its

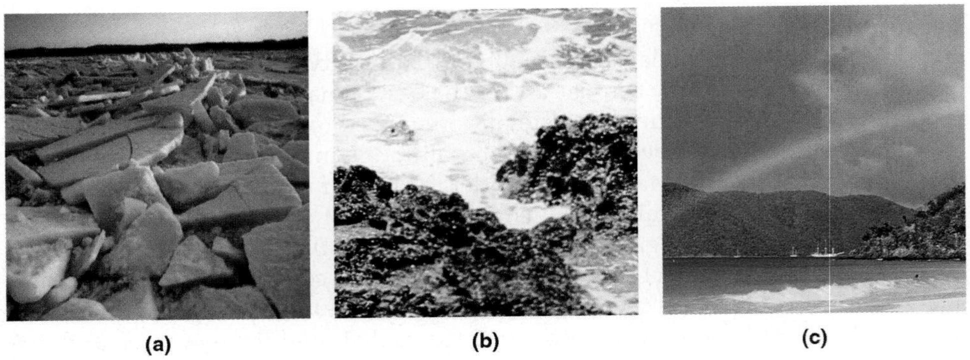

(a) (b) (c)

Figure 1.2
The three states of matter exhibited by water: (a) solid, as ice; (b) liquid, as ocean water; (c) gas, as humidity in the air.

Figure 1.3
An example of separation based on differences in physical properties. Magnetic iron is separated from other nonmagnetic substances. A large-scale version of this process is important in the recycling industry.

composition and identity remain unchanged. A physical change has occurred. We could in fact demonstrate the constancy of composition and identity by refreezing the liquid water, re-forming the ice cube. This melting and freezing cycle could be repeated over and over. This very process is a hallmark of our global weather changes. The continual interconversion of the three states of water in the environment (snow, rain, and humidity) clearly demonstrates the retention of the identity of water particles or *molecules.*

A **physical property** can be observed or measured without changing the composition or identity of a substance. As we have seen, melting ice is a physical change. We can measure the temperature when melting occurs; this is the *melting point* of water. We can also measure the *boiling point* of water, when liquid water becomes a gas. Both the melting and boiling points of water, and of any other substance, are physical properties.

A practical application of separation of materials based upon their differences in physical properties is shown in Figure 1.3.

Matter and Chemical Properties

Learning Goal 5 Learning Goal 6 Learning Goal 7

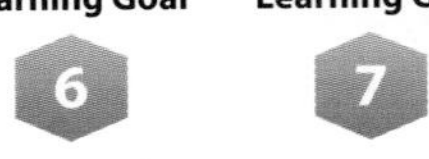

We have noted that physical properties can be exhibited, measured, or observed without any change in identity or composition. In contrast, **chemical properties** do result in a change in composition and can be observed only through chemical reactions. A **chemical reaction** is a process of rearranging, replacing, or adding atoms to produce new substances. For example, the process of photosynthesis can be shown as

$$\text{carbon dioxide} + \text{water} \xrightarrow[\text{Chlorophyll}]{\text{Light}} \text{sugar} + \text{oxygen}$$

Light is the energy needed to make the reaction happen. Chlorophyll is the energy absorber, converting light energy to chemical energy.

This chemical reaction involves the conversion of carbon dioxide and water (the **reactants**) to a sugar and oxygen (the **products**). The products and reactants are clearly different. We know that carbon dioxide and oxygen are gases at room temperature and water is a liquid at this temperature; the sugar is a solid white powder. A chemical property of carbon dioxide is its ability to form sugar under certain conditions. The process of formation of this sugar is the *chemical change*.

Chapter 8 discusses the role of energy in chemical reactions.

EXAMPLE 1.1

Identifying Properties

Can the process that takes place when an egg is fried be described as a physical or chemical change?

Solution

Examine the characteristics of the egg before and after frying. Clearly, some significant change has occurred. Furthermore the change appears irreversible. More than a simple physical change has taken place. A chemical reaction (actually, several) must be responsible; hence chemical change.

Question 1.1

Classify each of the following as either a chemical property or a physical property:

a. color
b. flammability
c. hardness
d. odor
e. taste

Question 1.2

Classify each of the following as either a chemical change or a physical change:

a. water boiling to become steam
b. butter becoming rancid
c. combustion of wood
d. melting of ice in spring
e. decay of leaves in winter

Intensive and Extensive Properties

It is important to recognize that properties can also be classified according to whether they depend on the size of the sample. Consequently, there is a fundamental difference between properties such as density and specific gravity and properties such as mass and volume.

See page 26 for a discussion of density and specific gravity.

An **intensive property** is a property of matter that is *independent* of the *quantity* of the substance. Density and specific gravity are intensive properties. For example, the density of one single drop of water is exactly the same as the density of a liter of water.

Learning Goal

An **extensive property** *depends* on the *quantity* of a substance. Mass and volume are extensive properties. There is an obvious difference between 1 g of silver and 1 kg of silver; the quantities and, incidentally, the value, differ substantially.

EXAMPLE 1.2

Differentiating between Intensive and Extensive Properties

Is temperature an extensive or intensive property?

Continued—

EXAMPLE 1.2 *—Continued*

Solution

Imagine two glasses each containing 100 g of water, and each at 25°C. Now pour the contents of the two glasses into a larger glass. You would predict that the mass of the water in the larger glass would be 200 g (100 g + 100 g) because mass is an extensive property, dependent on quantity. However, we would expect the temperature of the water to remain the same (not 25°C + 25°C); hence temperature is *an intensive property* . . . independent of quantity.

Classification of Matter

Learning Goal 9

Chemists look for similarities in properties among various types of materials. Recognizing these likenesses simplifies learning the subject and allows us to predict the behavior of new substances on the basis of their relationship to substances already known and characterized.

Many classification systems exist. The most useful system, based on composition, is described in the following paragraphs (see also Figure 1.4).

All matter is either a *pure substance* or a *mixture.* A **pure substance** is a substance that has only one component. Pure water is a pure substance. It is made up only of particles containing two hydrogen atoms and one oxygen atom, that is, water molecules (H_2O).

At present, more than one hundred elements have been characterized. A complete listing of the elements and their symbols is found on the inside front cover of this textbook.

There are different types of pure substances. Elements and compounds are both pure substances. An **element** is a pure substance that cannot be changed into a simpler form of matter by any chemical reaction. Hydrogen and oxygen, for example, are elements. Alternatively, a **compound** is a substance resulting from the combination of two or more elements in a definite, reproducible way. The elements hydrogen and oxygen, as noted earlier, may combine to form the compound water, H_2O.

A **mixture** is a combination of two or more pure substances in which each substance retains its own identity. Alcohol and water can be combined in a mixture. They coexist as pure substances because they do not undergo a chemical reaction; they exist as thoroughly mixed discrete molecules. This collection of dissimilar particles is the mixture. A mixture has variable composition; there are an infinite number of combinations of quantities of alcohol and water that can be mixed. For example, the mixture may contain a small amount of alcohol and a large amount of water or vice versa. Each is, however, an alcohol–water mixture.

A detailed discussion of solutions (homogeneous mixtures) and their properties is presented in Chapter 7.

A mixture may be either *homogeneous* or *heterogeneous* (Figure 1.5). A **homogeneous mixture** has uniform composition. Its particles are well mixed, or thoroughly intermingled. A homogeneous mixture, such as alcohol and water, is

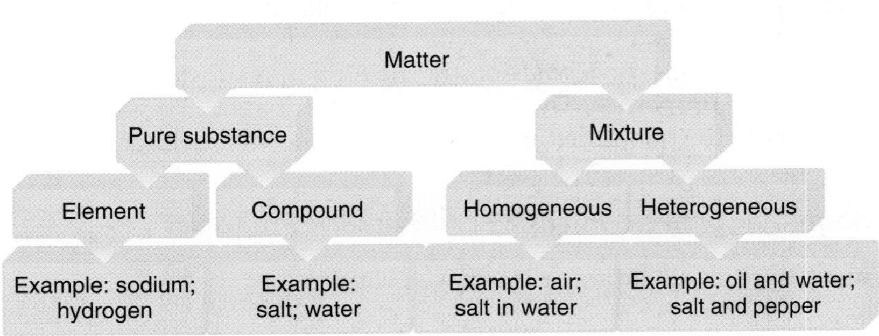

Figure 1.4
Classification of matter. All matter is either a pure substance or a mixture of pure substances. Pure substances are either elements or compounds, and mixtures may be either homogeneous (uniform composition) or heterogeneous (nonuniform composition).

described as a *solution*. Air, a mixture of gases, is an example of a gaseous solution. A **heterogeneous mixture** has a nonuniform composition. A mixture of salt and pepper is a good example of a heterogeneous mixture. Concrete is also composed of a heterogeneous mixture of materials (various types and sizes of stone and sand present with cement in a nonuniform mixture).

Categorizing Matter

EXAMPLE 1.3

Is seawater a pure substance, a homogeneous mixture, or a heterogeneous mixture?

Solution

Imagine yourself at the beach, filling a container with a sample of water from the ocean. Examine it. You would see a variety of solid particles suspended in the water: sand, green vegetation, perhaps even a small fish! Clearly, it is a mixture, and one in which the particles are not uniformly distributed throughout the water; hence a heterogeneous mixture.

Question 1.3

Is each of the following materials a pure substance, a homogeneous mixture, or a heterogeneous mixture?

a. ethyl alcohol
b. blood
c. Alka-Seltzer dissolved in water
d. oxygen in a hospital oxygen tank

Question 1.4

Is each of the following materials a pure substance, a homogeneous mixture, or a heterogeneous mixture?

a. air
b. paint
c. perfume
d. carbon monoxide

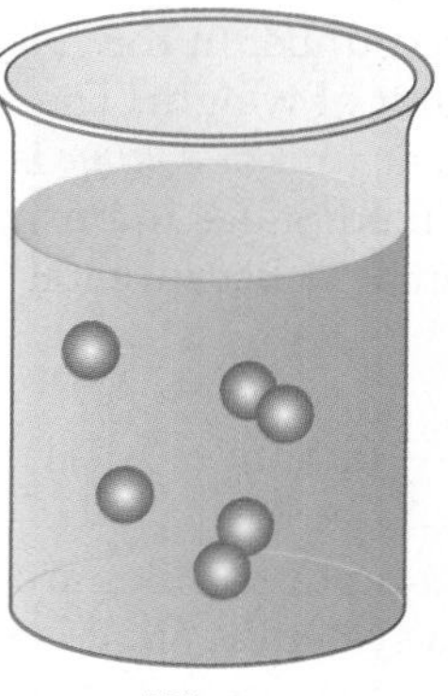

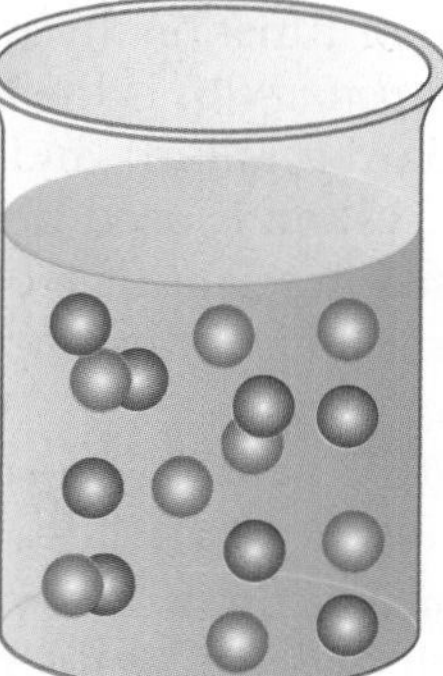

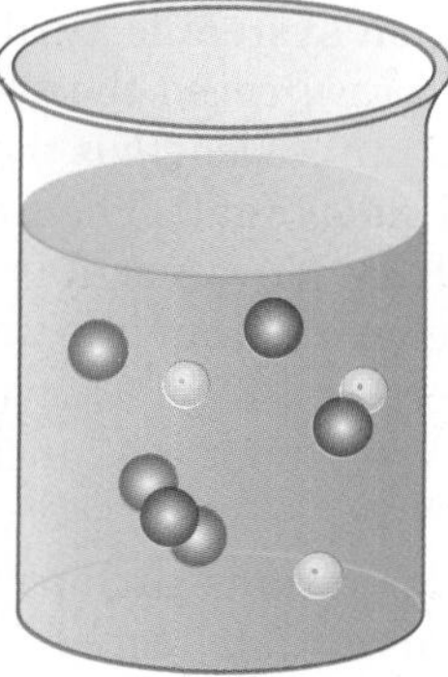

Figure 1.5
Schematic representation of some classes of matter. A pure substance (a) consists of a single component. A homogeneous mixture (b) has a uniform distribution of components. A heterogeneous mixture (c) has a nonuniform distribution of components.

1.3 Measurement in Chemistry

Data, Results, and Units

Learning Goal

A scientific experiment produces **data.** Each piece of data is the individual result of a single measurement or observation. Examples include the *mass* of a sample and the *time* required for a chemical reaction to occur. Mass, length, volume, time, temperature, and energy are common types of data obtained from chemical experiments.

Results are the outcome of an experiment. Data and results may be identical, but more often several related pieces of data are combined, and logic is used to produce a result.

EXAMPLE 1.4 ***Distinguishing between Data and Results***

In many cases, a drug is less stable if moisture is present, and excess moisture can hasten the breakdown of the active ingredient, leading to loss of potency. Therefore we may wish to know how much water a certain quantity of a drug gains when exposed to air. To do this experiment, we must first weigh the drug sample, then expose it to the air for a period and reweigh it. The change in weight,

$$[\text{weight}_{\text{final}} - \text{weight}_{\text{initial}}] = \text{weight difference}$$

indicates the weight of water taken up by the drug formulation. The initial and final weights are individual bits of *data;* by themselves they do not answer the question, but they do provide the information necessary to calculate the answer: the results. The difference in weight and the conclusions based on the observed change in weight are the *results* of the experiment.

The experiment described in Example 1.4 was really not a very good experiment because many other environmental conditions were not measured. Measurement of the temperature and humidity of the atmosphere and the length of time that the drug was exposed to the air (the creation of a more complete set of data) would make the results less ambiguous.

Proper use of units is central to all aspects of science. The following sections are designed to develop a fundamental understanding of this vital topic.

Any measurement made in the experiment must also specify the units of that measurement. An initial weight of three *ounces* is clearly quite different than three *pounds.* A **unit** defines the basic quantity of mass, volume, time, or whatever quantity is being measured. A number that is not followed by the correct unit usually conveys no useful information.

English and Metric Units

The **English system** is a collection of functionally unrelated units. In the *English system of measurement* the standard *pound* (lb) is the basic unit of weight. The fundamental unit of *length* is the standard *yard* (yd), and the basic unit of *volume* is the standard *gallon* (gal). The English system is used in the United States in business and industry. However, it is not used in scientific work, primarily because it is difficult to convert from one unit to another. For example,

$$1\text{ foot} = 12\text{ inches} = 0.33\text{ yard} = \frac{1}{5280}\text{ mile} = \frac{1}{6}\text{ fathom}$$

Clearly, operations such as the conversion of 1.62 yards to units of miles are not straightforward. In fact, the English "system" is not really a system at all. It is simply a collection of measures accumulated throughout English history. Because they have no functional relationship, it is not surprising that conversion from one unit to another is not straightforward.

Learning Goal

The United States, the last major industrial country to retain the English system, has begun efforts to convert to the metric system. The **metric system** is truly "systematic." It is composed of a set of units that are related to each other decimally, in other words, as powers of ten. Because the *metric system* is a decimal-based system, it is inherently simpler to use and less ambiguous. For example, the length of an object may be represented as

$$1 \text{ meter} = 10 \text{ decimeters} = 100 \text{ centimeters} = 1000 \text{ millimeters}$$

Only the decimal point moves in the conversion from one unit to another, simplifying many calculations.

The metric system was originally developed in France just before the French Revolution in 1789. The modern version of this system is the *Système International*, or *S.I. system*. Although the S.I. system has been in existence for over forty years, it has yet to gain widespread acceptance. To make the S.I. system truly systematic, it utilizes certain units, especially those for pressure, that many people find difficult to use.

In this text we will use the metric system, not the S.I. system, and we will use the English system only to the extent of converting *from* it to the more scientifically useful metric system.

In the metric system there are three basic units. Mass is represented as the *gram*, length as the *meter*, and volume as the *liter*. Any subunit or multiple unit contains one of these units preceded by a prefix indicating the power of ten by which the base unit is to be multiplied to form the subunit or multiple unit. The most common metric prefixes are shown in Table 1.1.

Other metric units, for time, temperature, and energy, will be treated in Section 1.5.

The same prefix may be used for volume, mass, length, time, and so forth. Consider the following examples:

$$1 \text{ milliliter (mL)} = \frac{1}{1000} \text{ liter} = 0.001 \text{ liter} = 10^{-3} \text{ liter}$$

See Appendix A for a review of the mathematics involved.

A volume unit is indicated by the base unit, liter, and the prefix *milli-*, which indicates that the unit is one thousandth of the base unit. In the same way,

$$1 \text{ milligram (mg)} = \frac{1}{1000} \text{ gram} = 0.001 \text{ gram} = 10^{-3} \text{ gram}$$

and

$$1 \text{ millimeter (mm)} = \frac{1}{1000} \text{ meter} = 0.001 \text{ meter} = 10^{-3} \text{ meter}$$

The representation of numbers as powers of ten may be unfamiliar to you. This useful notation is discussed in Section 1.4.

Table 1.1 Some Common Prefixes Used in the Metric System

Prefix	Multiple	Decimal Equivalent
mega (M)	10^6	1,000,000.
kilo (k)	10^3	1,000.
deka (da)	10^1	10.
deci (d)	10^{-1}	0.1
centi (c)	10^{-2}	0.01
milli (m)	10^{-3}	0.001
micro (μ)	10^{-6}	0.000001
nano (n)	10^{-9}	0.000000001

Unit Conversion: English and Metric Systems

To convert from one unit to another, we must have a **conversion factor** or series of conversion factors that relate two units. The proper use of these conversion factors is called the *factor-label method*. This method is also termed *dimensional analysis*.

This method is used for two kinds of conversions: to convert from one unit to another within the *same system* or to convert units from *one system to another*.

Conversion of Units within the Same System

Learning Goal 11

We know, for example, that in the English system,

$$1 \text{ gallon} = 4 \text{ quarts}$$

Because dividing both sides of the equation by the same term does not change its identity,

$$\frac{1 \text{ gallon}}{1 \text{ gallon}} = \frac{4 \text{ quarts}}{1 \text{ gallon}}$$

The expression on the left is equal to unity (1); therefore

$$1 = \frac{4 \text{ quarts}}{1 \text{ gallon}} \quad \text{or} \quad 1 = \frac{1 \text{ gallon}}{4 \text{ quarts}}$$

Now, multiplying any other expression by the ratio 4 quarts/1 gallon will not change the value of the term, because multiplication of any number by 1 produces the original value. However, there is one important difference: The units will have changed.

EXAMPLE 1.5 Using Conversion Factors

Convert 12 gallons to units of quarts.

Solution

$$12 \cancel{\text{gal}} \times \frac{4 \text{ qt}}{1 \cancel{\text{gal}}} = 48 \text{ qt}$$

The conversion factor, 4 qt/1 gal, serves as a bridge, or linkage, between the unit that was given (gallons) and the unit that was sought (quarts).

The conversion factor in Example 1.5 may be written as 4 qt/1 gal or 1 gal/4 qt, because both are equal to 1. However, only the first factor, 4 qt/1 gal, will give us the units we need to solve the problem. If we had set up the problem incorrectly, we would obtain

$$12 \text{ gal} \times \frac{1 \text{ gal}}{4 \text{ qt}} = 3 \frac{\text{gal}^2}{\text{qt}}$$

Incorrect units

Clearly, units of gal^2/qt are not those asked for in the problem, nor are they reasonable units. The factor-label method is therefore a self-indicating system; the correct units (those required by the problem) will result only if the factor is set up properly.

Table 1.2 lists a variety of commonly used English system relationships that may serve as the basis for useful conversion factors.

Conversion of units within the metric system may be accomplished by using the factor-label method as well. Unit prefixes that dictate the conversion factor facilitate unit conversion (refer to Table 1.1).

Table 1.2 Some Common Relationships Used in the English System

A. Weight	1 pound = 16 ounces
	1 ton = 2000 pounds
B. Length	1 foot = 12 inches
	1 yard = 3 feet
	1 mile = 5280 feet
C. Volume	1 gallon = 4 quarts
	1 quart = 2 pints
	1 quart = 32 fluid ounces

EXAMPLE 1.6 *Using Conversion Factors*

Convert 10.0 centimeters to meters.

Solution

First, recognize that the prefix *centi-* means 1/100 of the base unit, the meter. Thus our conversion factor is either

$$\frac{1 \text{ meter}}{100 \text{ cm}} \quad \text{or} \quad \frac{100 \text{ cm}}{1 \text{ meter}}$$

each being equal to 1. Only one, however, will result in proper cancellation of units, producing the correct answer to the problem. If we proceed as follows:

$$10.0 \cancel{\text{cm}} \times \frac{1 \text{ meter}}{100 \cancel{\text{cm}}} = 0.100 \text{ meter}$$

Data given — Conversion factor — Desired result

we obtain the desired units, meters (m). If we had used the conversion factor 100 cm/1 m, the resulting units would be meaningless and the answer would have been incorrect:

$$10.0 \text{ cm} \times \frac{100 \text{ cm}}{1 \text{ m}} = 1000 \frac{\text{cm}^2}{\text{m}}$$

Incorrect units

Question 1.5

Convert 1.0 liter to each of the following units, using the factor-label method:

a. milliliters
b. microliters
c. kiloliters
d. centiliters
e. dekaliters

Question 1.6

Convert 1.0 gram to each of the following units:

a. micrograms
b. milligrams
c. kilograms
d. centigrams
e. decigrams

Learning Goal

Table 1.3 Commonly Used "Bridging" Units for Intersystem Conversions

Quantity	English		Metric
Mass	1 pound	=	454 grams
	2.2 pounds	=	1 kilogram
Length	1 inch	=	2.54 centimeters
	1 yard	=	0.91 meter
Volume	1 quart	=	0.946 liter
	1 gallon	=	3.78 liters

Conversion of Units from One System to Another

English and metric conversions are shown in Tables 1.1 and 1.2.

The conversion of a quantity expressed in units of one system to an equivalent quantity in the other system (English to metric or metric to English) requires a *bridging* conversion unit. Examples are shown in Table 1.3.

The conversion may be represented as a three-step process:

1. Conversion from the units given in the problem to a bridging unit.
2. Conversion to the other system using the bridge.
3. Conversion within the desired system to units required by the problem.

EXAMPLE 1.7 ***Using Conversion Factors between Systems***

Convert 4.00 ounces to kilograms.

Solution

Step 1. A convenient bridging unit for mass is 1 lb = 454 grams. To use this conversion factor, we relate ounces (given in the problem) to pounds:

$$4.00 \cancel{\text{ounces}} \times \frac{1 \text{ pound}}{16 \cancel{\text{ounces}}} = 0.250 \text{ pound}$$

Step 2. Using the bridging unit conversion, we get

$$0.250 \cancel{\text{pound}} \times \frac{454 \text{ grams}}{1 \cancel{\text{pound}}} = 114 \text{ grams}$$

Step 3. Grams may then be directly converted to kilograms, the desired unit:

$$114 \cancel{\text{grams}} \times \frac{1 \text{ kilogram}}{1000 \cancel{\text{grams}}} = 0.114 \text{ kilogram}$$

The calculation may also be done in a single step by arranging the factors in a chain:

$$4.00 \cancel{\text{oz}} \times \frac{1 \cancel{\text{lb}}}{16 \cancel{\text{oz}}} \times \frac{454 \cancel{\text{g}}}{1 \cancel{\text{lb}}} \times \frac{1 \text{ kg}}{1000 \cancel{\text{g}}} = 0.114 \text{ kg}$$

Helpful Hint: Refer to the discussion of rounding off numbers on page 20.

Using Conversion Factors

EXAMPLE 1.8

Convert 1.5 meters² to centimeters².

Solution

The problem is similar to the conversion performed in Example 1.6. However, we must remember to include the exponent in the units. Thus

$$1.5 \text{ m}^2 \times \left(\frac{10^2 \text{ cm}}{1 \text{ m}}\right)^2 = 1.5 \cancel{\text{m}^2} \times \frac{10^4 \text{ cm}^2}{1 \cancel{\text{m}^2}} = 1.5 \times 10^4 \text{ cm}^2$$

Note: The exponent affects both the number *and* unit within the parentheses.

Question 1.7

a. Convert 0.50 inch to meters.
b. Convert 0.75 quart to liters.
c. Convert 56.8 grams to ounces.
d. Convert 1.5 cm² to m².

Question 1.8

a. Convert 0.50 inch to centimeters.
b. Convert 0.75 quart to milliliters.
c. Convert 56.8 milligrams to ounces.
d. Convert 3.6 m² to cm².

1.4 Significant Figures and Scientific Notation

Information-bearing figures in a number are termed *significant figures.* Data and results arising from a scientific experiment convey information about the way in which the experiment was conducted. The degree of uncertainty or doubt associated with a measurement or series of measurements is indicated by the number of figures used to represent the information.

Significant Figures

Learning Goal 12

Consider the following situation: A student was asked to obtain the length of a section of wire. In the chemistry laboratory, several different types of measuring devices are usually available. Not knowing which was most appropriate, the student decided to measure the object using each device that was available in the laboratory. The following data were obtained:

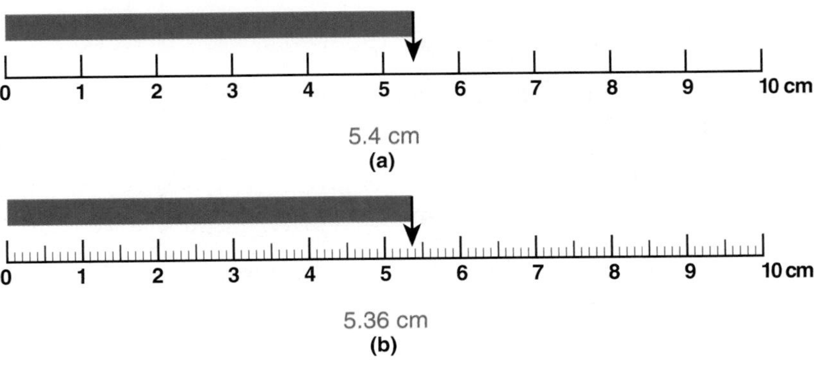

Two questions should immediately come to mind:

Are the two answers equivalent?

If not, which answer is correct?

In fact, the two answers are *not* equivalent, but *both* are correct. How do we explain this apparent contradiction?

The uncertain digit represents the degree of doubt in a single measurement.

The uncertain digit results from an estimation.

The data are not equivalent because each is known to a different degree of certainty. The answer 5.36 cm, containing three significant figures, specifies the length of the object more exactly than 5.4 cm, which contains only two significant figures. The term **significant figures** is defined to be all digits in a number representing data or results that are known with certainty *plus one uncertain digit.*

In case (a) we are certain that the object is at least 5 cm long and equally certain that it is *not* 6 cm long because the end of the object falls between the calibration lines 5 and 6. We can only estimate between 5 and 6, because there are no calibration indicators between 5 and 6. The end of the wire appears to be approximately four-tenths of the way between 5 and 6, hence 5.4 cm. The 5 is known with certainty, and 4 is estimated; there are two significant figures.

In case (b) the ruler is calibrated in tenths of centimeters. The end of the wire is at least 5.3 cm and not 5.4 cm. Estimation of the second decimal place between the two closest calibration marks leads to 5.36 cm. In this case, 5.3 is certain, and the 6 is estimated (or uncertain), leading to three significant digits.

Both answers are correct because each is consistent with the measuring device used to generate the data. An answer of 5.36 cm obtained from a measurement using ruler (a) would be *incorrect* because the measuring device is not capable of that exact specification. On the other hand, a value of 5.4 cm obtained from ruler (b) would be erroneous as well; in that case the measuring device is capable of generating a higher level of certainty (more significant digits) than is actually reported.

In summary, the number of significant figures associated with a measurement is determined by the measuring device. Conversely, the number of significant figures reported is an indication of the sophistication of the measurement itself.

Recognition of Significant Figures

Only *significant* digits should be reported as data or results. However, are all digits, as written, significant digits? Let's look at a few examples illustrating the rules that are used to represent data and results with the proper number of significant digits.

EXAMPLE 1.9

RULE: All nonzero digits are significant. ■

7.314 has *four* significant digits.

EXAMPLE 1.10

RULE: The number of significant digits is independent of the position of the decimal point. ■

73.14 has *four* significant digits.

EXAMPLE 1.11

RULE: Zeros located between nonzero digits are significant. ■

60.052 has *five* significant figures.

EXAMPLE 1.12

RULE: Zeros at the end of a number (often referred to as trailing zeros) are significant if the number contains a decimal point. ■

4.70 has *three* significant figures.

Helpful Hint: Trailing zeros are ambiguous; the next section offers a solution for this ambiguity.

EXAMPLE 1.13

RULE: Trailing zeros are insignificant if the number does not contain a decimal point and are significant if a decimal point is indicated. ■

100 has *one* significant figure; 100. has three significant figures.

EXAMPLE 1.14

RULE: Zeros to the left of the first nonzero integer are not significant; they serve only to locate the position of the decimal point. ■

0.0032 has *two* significant figures.

Question 1.9

How many significant figures are contained in each of the following numbers?

a. 7.26
b. 726
c. 700.2
d. 7.0
e. 0.0720

Question 1.10

How many significant figures are contained in each of the following numbers?

a. 0.042
b. 4.20
c. 24.0
d. 240
e. 204

Scientific Notation

Learning Goal

It is often difficult to express very large numbers to the proper number of significant figures using conventional notation. The solution to this problem lies in the use of **scientific notation,** also referred to as *exponential notation,* which involves the representation of a number as a power of ten.

See Appendix A for a review of the mathematics involved.

RULE: To convert a number greater than 1 to scientific notation, the original decimal point is moved x places to the left, and the resulting number is multiplied by 10^x. The exponent (x) is a *positive* number equal to the number of places the original decimal point was moved. ■

Scientific notation is also useful in representing numbers less than 1. For example, the mass of a single helium atom is

0.000000000000000000000006692 gram

a rather cumbersome number as written. Scientific notation would represent the mass of a single helium atom as 6.692×10^{-24} gram. The conversion is illustrated by using a simpler number:

$$0.0062 = 6.2 \times \frac{1}{1000} = 6.2 \times \frac{1}{10^3} = 6.2 \times 10^{-3}$$

or

$$0.0534 = 5.34 \times \frac{1}{100} = 5.34 \times \frac{1}{10^2} = 5.34 \times 10^{-2}$$

RULE: To convert a number less than 1 to scientific notation, the original decimal point is moved x places to the right, and the resulting number is multiplied by 10^{-x}. The exponent $(-x)$ is a *negative* number equal to the number of places the original decimal point was moved. ■

Question 1.11

Represent each of the following numbers in scientific notation, showing only significant digits:

a. 0.0024 b. 0.0180 c. 224

Question 1.12

Represent each of the following numbers in scientific notation, showing only significant digits:

a. 48.20 b. 480.0 c. 0.126

Significant Figures in Calculation of Results

Addition and Subtraction

If we combine the following numbers:

$$\begin{array}{rl} 37.68 & \text{liters} \\ 108.428 & \text{liters} \\ 6.71862 & \text{liters} \end{array}$$

our calculator will show a final result of

$$152.82662 \quad \text{liters}$$

Remember the distinction between the words *zero* and *nothing*. *Zero* is one of the ten digits and conveys as much information as 1, 2, and so forth. *Nothing* implies no information; the digits in the positions indicated by x's could be 0, 1, 2, or any other.

Clearly, the answer, with eight digits, defines the volume of total material much more accurately than *any* of the individual quantities being combined. This cannot be correct; *the answer cannot have greater significance than any of the quantities that produced the answer.* We rewrite the problem:

$$\begin{array}{rl} 37.68xxx & \text{liters} \\ 108.428xx & \text{liters} \\ +\ \ 6.71862 & \text{liters} \\ \hline 152.82662 & \text{(should be 152.83) liters} \end{array}$$

See rules for rounding off discussed on page 20.

where x = no information; x may be any integer from 0 to 9. Adding 2 to two unknown numbers (in the right column) produces no information. Similar logic prevails for the next two columns. Thus five digits remain, all of which are significant. Conventional rules for rounding off would dictate a final answer of 152.83.

Question 1.13

Report the result of each of the following to the proper number of significant figures:

a. 4.26 = 3.831 =
b. 8.321 − 2.4 =
c. 16.262 + 4.33 − 0.40 =

Question 1.14

Report the result of each of the following to the proper number of significant figures:

a. $7.939 + 6.26 =$
b. $2.4 - 8.321 =$
c. $2.333 + 1.56 - 0.29 =$

Multiplication and Division

In the preceding discussion of addition and subtraction the position of the decimal point in the quantities being combined has a bearing on the number of significant figures in the answer. In multiplication and division this is not the case. The decimal point position is irrelevant when determining the number of significant figures in the answer. It is the number of significant figures in the data that is important. Consider

$$\frac{4.237 \times 1.21 \times 10^{-3} \times 0.00273}{11.125} = 1.26 \times 10^{-6}$$

The answer is limited to three significant figures; the answer can have *only* three significant figures because two numbers in the calculation, 1.21×10^{-3} and 0.00273, have three significant figures and "limit" the answer. Remember, *the answer can be no more precise than the least precise number from which the answer is derived.* The *least precise number* is the number with the fewest significant figures.

Question 1.15

Report the results of each of the following operations using the proper number of significant figures:

a. $63.8 \times 0.80 =$
b. $\dfrac{63.8}{0.80} =$
c. $\dfrac{53.8 \times 0.90}{0.3025} =$

Question 1.16

Report the results of each of the following operations using the proper number of significant figures:

a. $\dfrac{27.2 \times 15.63}{1.84} =$
b. $\dfrac{13.6}{18.02 \times 1.6} =$
c. $\dfrac{12.24 \times 6.2}{18.02 \times 1.6} =$

Exponents

See Appendix A for a review of the mathematics involved.

Now consider the determination of the proper number of significant digits in the results when a value is multiplied by any power of ten. In each case the number of significant figures in the answer is identical to the number contained in the original term. Therefore

$$(8.314 \times 10^2)^3 = 574.7 \times 10^6 = 5.747 \times 10^8$$

and

$$(8.314 \times 10^2)^{1/2} = 2.883 \times 10^1$$

Each answer contains four significant figures.

It is important to note, in operating with significant figures, that defined or counted numbers do *not* determine the number of significant figures.

For example,

How many grams are contained in 0.240 kg?

$$0.245\,\cancel{kg} \times \frac{1000\text{ g}}{1\,\cancel{kg}} = 245\text{ g}$$

The "1" in the conversion factor is defined, or exact, and does not limit the number of significant digits.

Exact numbers are counting numbers or defined numbers. They have infinitely many significant figures. Consequently they do not limit the number of significant figures in the result of the calculation. You should *recognize* and *ignore* exact numbers when assigning significant figures.

A good rule of thumb to follow is: In the metric system the quantity being converted, not the conversion factor, generally determines the number of significant figures.

Rounding Off Numbers

The use of an electronic calculator generally produces more digits for a result than are justified by the rules of significant figures on the basis of the data input. For example, on your calculator,

$$3.84 \times 6.72 = 25.8048$$

The most correct answer would be 25.8, dropping 048.

RULE: When the number to be dropped is less than 5 the preceding number is not changed. When the number to be dropped is 5 or larger, the preceding number is increased by one unit. ■

EXAMPLE 1.15 ***Rounding Numbers***

Round off each of the following to three significant figures.

Solution

a. 63.6$\underline{6}$9 becomes 63.7. *Rationale:* $6 > 5$.
b. 8.77$\underline{1}$5 becomes 8.77. *Rationale:* $1 < 5$.
c. 2.22$\underline{4}$5 becomes 2.22. *Rationale:* $4 < 5$.
d. 0.000410$\underline{9}$ becomes 0.000411. *Rationale:* $9 > 5$.

Helpful Hint: Symbol $x > y$ implies "x greater than y." Symbol $x < y$ implies "x less than y."

Question 1.17

Round off each of the following numbers to three significant figures.

a. 61.40 b. 6.171 c. 0.066494

Question 1.18

Round off each of the following numbers to three significant figures.

a. 6.2262 b. 3895 c. 6.885

1.5 Experimental Quantities

Learning Goal

Thus far we have discussed the scientific method and its role in acquiring data and converting the data to obtain the results of the experiment. We have seen that such data must be reported in the proper units with the appropriate number of significant figures. The quantities that are most often determined include mass, length, volume, time, temperature, and energy. Now let's look at each of these quantities in more detail.

Mass

Mass describes the quantity of matter in an object. The terms *weight* and *mass,* in common usage, are often considered synonymous. They are not, in fact. **Weight** is the force of gravity on an object:

$$\text{Weight} = \text{mass} \times \text{acceleration due to gravity}$$

When gravity is constant, mass and weight are directly proportional. But gravity is not constant; it varies as a function of the distance from the center of the earth. Therefore weight cannot be used for scientific measurement because the weight of an object may vary from one place on the earth to the next.

Mass, on the other hand, is independent of gravity; it is a result of a comparison of an unknown mass with a known mass called a *standard mass.* Balances are instruments used to measure the mass of materials.

Examples of common balances used for the determination of mass are shown in Figure 1.6.

The common conversion units for mass are as follows:

$$1 \text{ gram (g)} = 10^{-3} \text{ kilogram (kg)} = \frac{1}{454} \text{ pound (lb)}$$

In chemistry, when we talk about incredibly small bits of matter such as individual atoms or molecules, units such as grams and even micrograms are much too large. We don't say that a 100-pound individual weighs 0.0500 ton; the unit

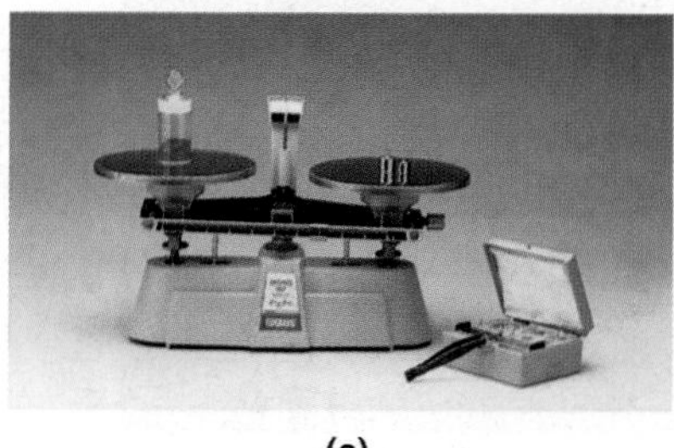

(a)

(b)

(c)

Figure 1.6
Illustration of three common balances that are useful for the measurement of mass. (a) A two-pan comparison balance for approximate mass measurement suitable for routine work requiring accuracy to 0.1 g (or perhaps 0.01 g). (b) A top-loading single-pan electronic balance that is similar in accuracy to (a) but has the advantages of speed and ease of operation. The revolution in electronics over the past twenty years has resulted in electronic balances largely supplanting the two-pan comparison balance in routine laboratory usage. (c) An analytical balance that is capable of precise mass measurement (three to five significant figures beyond the decimal point). A balance of this type is used when the highest level of precision and accuracy is required.

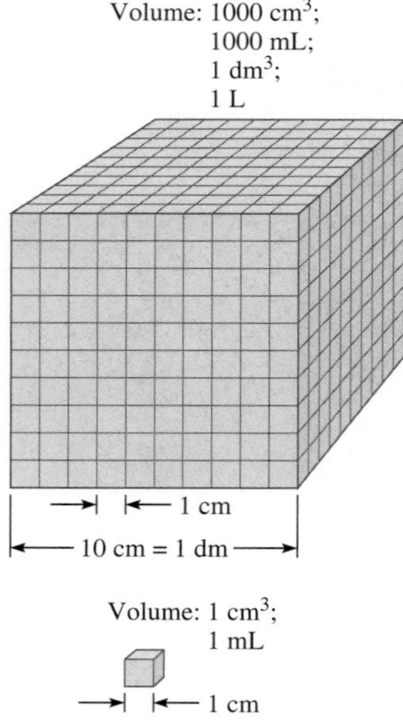

Figure 1.7
The relationship among various volume units.

The *milliliter* and the *cubic centimeter* are equivalent.

does not fit the quantity being described. Similarly, an atom of a substance such as hydrogen is very tiny. Its mass is only 1.661×10^{-24} gram.

One *atomic mass unit* (amu) is a more convenient way to represent the mass of one hydrogen atom, rather than 1.661×10^{-24} gram:

$$1 \text{ amu} = 1.661 \times 10^{-24} \text{ g}$$

Units should be chosen to suit the quantity being described. This can easily be done by choosing a unit that gives an exponential term closest to 10^0.

Length

The standard metric unit of *length,* the distance between two points, is the meter. Large distances are measured in kilometers; smaller distances are measured in millimeters or centimeters. Very small distances such as the distances between atoms on a surface are measured in *nanometers* (nm):

$$1 \text{ nm} = 10^{-7} \text{ cm} = 10^{-9} \text{ m}$$

Common conversions for length are as follows:

$$1 \text{ meter (m)} = 10^2 \text{ centimeters (cm)} = 3.94 \times 10^1 \text{ inch (in)}$$

Volume

The standard metric unit of *volume,* the space occupied by an object, is the liter. A liter is the volume occupied by 1000 grams of water at 4 degrees Celsius (°C). The volume, 1 liter, also corresponds to:

$$1 \text{ liter (L)} = 10^3 \text{ milliliters (mL)} = 1.06 \text{ quarts (qt)}$$

The relationship between the liter and the milliliter is shown in Figure 1.7.

Typical laboratory glassware used for volume measurement is shown in Figure 1.8. The volumetric flask is designed to *contain* a specified volume, and the graduated cylinder, pipet, and buret *dispense* a desired volume of liquid.

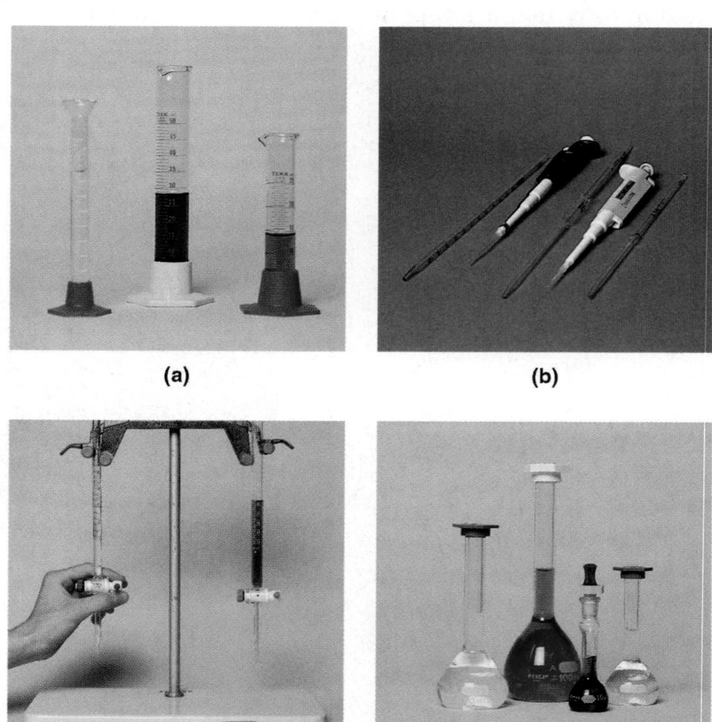

Figure 1.8
Common laboratory equipment used for the measurement of volume. Graduated cylinders (a), pipets (b), and burets (c) are used for the delivery of liquids; volumetric flasks (d) are used to contain a specific volume. A graduated cylinder is usually used for measurement of approximate volumes; it is less accurate and precise than either pipets or burets.

Time

The standard metric unit of time is the second. The need for accurate measurement of time by chemists may not be as apparent as that associated with mass, length, and volume. It is necessary, however, in many applications. In fact, matter may be characterized by measuring the time required for a certain process to occur. The rate of a chemical reaction is a measure of change as a function of time.

See Section 8.3, which discusses rates of reactions.

Temperature

Temperature is the degree of "hotness" of an object. This may not sound like a very "scientific" definition, and, in a sense, it is not. We know intuitively the difference between a "hot" and a "cold" object, but developing a precise definition to explain this is not easy. We may think of the temperature of an object as a measure of the amount of heat in the object. However, this is not strictly true. An object increases in temperature because its heat content has increased and vice versa; however, the relationship between heat content and temperature depends on the composition of the material.

Section 8.1 describes the distinction between heat and temperature.

Many substances, such as mercury, expand as their temperature increases, and this expansion provides us with a way to measure temperature and temperature changes. If the mercury is contained within a sealed tube, as it is in a thermometer, the height of the mercury is proportional to the temperature. A mercury thermometer may be calibrated, or scaled, in different units, just as a ruler can be. Three common temperature scales are *Fahrenheit (°F), Celsius (°C),* and *Kelvin (K).* Two convenient reference temperatures that are used to calibrate a thermometer are the freezing and boiling temperatures of water. Figure 1.9 shows the relationship between the scales and these reference temperatures.

The Kelvin scale is of particular importance because it is directly related to molecular motion. As molecular speed increases, the Kelvin temperature proportionately increases.

Although Fahrenheit temperature is most familiar to us, Celsius and Kelvin temperatures are used exclusively in scientific measurements. It is often necessary to convert a temperature reading from one scale to another. To convert from Fahrenheit to Celsius, we use the following formula:

$$°\text{C} = \frac{°\text{F} - 32}{1.8}$$

See Appendix A for a review of the mathematics involved.

To convert from Celsius to Fahrenheit, we use the formula

$$°\text{F} = 1.8°\text{C} + 32$$

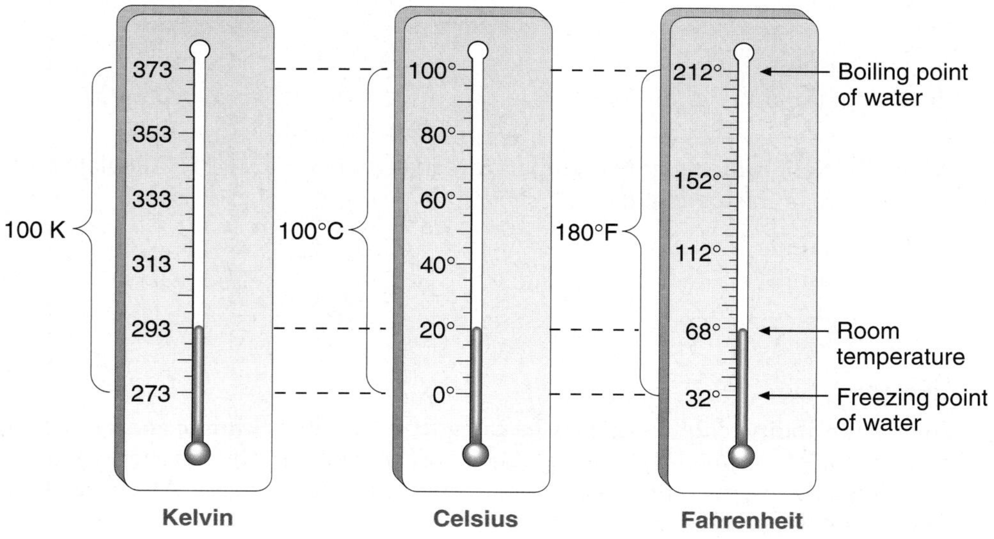

Figure 1.9
The freezing point and boiling point of water expressed in the three common units of temperature.

The Kelvin symbol does not have a degree sign. The degree sign implies a value that is *relative* to some standard. Kelvin is an *absolute* scale.

To convert from Celsius to Kelvin, we use the formula

$$K = °C + 273.15$$

EXAMPLE 1.16 ***Converting from Fahrenheit to Celsius and Kelvin***

Normal body temperature is 98.6°F. Calculate the corresponding temperature in degrees Celsius:

Solution

Using the expression relating °C and °F,

$$°C = \frac{°F - 32}{1.8}$$

Substituting the information provided,

$$= \frac{98.6 - 32}{1.8} = \frac{66.6}{1.8}$$

results in:

$$= 37.0°C$$

Calculate the corresponding temperature in Kelvin units:

Solution

Using the expression relating K and °C,

$$K = °C + 273.15$$

substituting the value obtained in the first part,

$$= 37.0 + 273.15$$

results in:

$$= 310.2\ K$$

Question 1.19

The freezing temperature of water is 32°F. Calculate the freezing temperature of water in:

a. Celsius units
b. Kelvin units

Question 1.20

When a patient is ill, his or her temperature may increase to 104°F. Calculate the temperature of this patient in:

a. Celsius units
b. Kelvin units

Energy

Energy, the ability to do work, may be categorized as either **kinetic energy,** the energy of motion, or **potential energy,** the energy of position. Kinetic energy may be considered as energy in process; potential energy is stored energy. All energy is either kinetic or potential.

A HUMAN PERSPECTIVE

Food Calories

The body gets its energy through the processes known collectively as metabolism, which will be discussed in detail in subsequent chapters on biochemistry and nutrition. The primary energy sources for the body are carbohydrates, fats, and proteins, which we obtain from the foods we eat. The amount of energy available from a given foodstuff is related to the Calories (C) available in the food. Calories are a measure of the energy and heat content that can be derived from the food. One (food) Calorie (symbolized by C) equals 1000 (metric) calories (symbolized by c):

$$1 \text{ Calorie} = 1000 \text{ calories} = 1 \text{ kilocalorie}$$

The energy available in food can be measured by totally burning the food; in other words, we are using the food as a fuel. The energy given off in the form of heat is directly related to the amount of chemical energy, energy stored in chemical bonds, that is available in the food and that the food could provide to the body through the various metabolic pathways.

The classes of food molecules are not equally energy rich. For instance, when oxidized via metabolic pathways, carbohydrates and proteins provide the cell with four Calories per gram, whereas fats generate approximately nine Calories per gram.

In addition, as with all processes, not all the available energy can be efficiently extracted from the food; a certain percentage is always lost. The average person requires between 2000 and 3000 Calories per day to maintain normal body functions such as the regulation of body temperature, muscle movement, and so on. If a person takes in more Calories than the body uses, the Calorie-containing substances will be stored as fat, and the person will gain weight. Conversely, if a person uses more Calories than are ingested, the individual will lose weight.

Excess Calories are stored in the form of fat, the form that provides the greatest amount of energy per gram. Too many Calories lead to too much fat. Similarly, a lack of Calories (in the form of food) forces the body to raid its storehouse, the fat. Weight is lost in this process as the fat is consumed. Unfortunately, it always seems easier to add fat to the storehouse than to remove it.

The "rule of thumb" is that 3500 Calories are equivalent to approximately 1 pound of body weight. You have to take in 3500 Calories more than you use to gain a pound, and you have to expend 3500 Calories more than you normally use to lose a pound. If you eat as little as 100 Calories a day above your body's needs, you could gain about 10–11 pounds per year:

$$\frac{100\ \cancel{C}}{\cancel{day}} \times \frac{365\ \cancel{day}}{1\ \text{year}} \times \frac{1\ \text{lb}}{3500\ \cancel{C}} = \frac{10.4\ \text{lb}}{\text{year}}$$

A frequently recommended procedure for increasing the rate of weight loss involves a combination of dieting (taking in fewer Calories) and exercise. The numbers of Calories used in several activities are:

Activity	Energy Output (C/min)
Running	19.4
Swimming	11.0
Jogging	10.0
Bicycling	8.0
Tennis	7.1
Walking	5.2
Golfing	5.0
Driving a car	2.8
Standing or sitting	1.9
Sleeping	1.0

Another useful way of classifying energy is by form. The principal forms of energy include light, heat, electrical, mechanical, and chemical energy. All of these forms of energy share the following set of characteristics:

- In chemical reactions, energy cannot be created or destroyed.
- Energy may be converted from one form to another.
- Energy conversion always occurs with less than 100% efficiency.
- All chemical reactions involve either a "gain" or a "loss" of energy.

The *kilocalorie* (kcal) is the familiar nutritional calorie. It is also known as the large Calorie; note that in this term the *C* is uppercase to distinguish it from the normal calorie. The large calorie is 1000 small calories. Refer to Section 8.2 and A Human Perspective: Food Calories for more information.

Energy absorbed or liberated in chemical reactions is usually in the form of heat energy. Heat energy may be represented in units of *calories* or *joules*, their relationship being

$$1 \text{ calorie (cal)} = 4.18 \text{ joules (J)}$$

One calorie is defined as the amount of heat energy required to increase the temperature of 1 gram of water 1°C.

Water in the environment (lakes, oceans, and streams) has a powerful effect on the climate because of its ability to store large quantities of energy. In summer, water stores heat energy, moderating temperatures of the surrounding area. In winter, some of this stored energy is released to the air as the water temperature falls; this prevents the surroundings from experiencing extreme changes in temperature.

Heat energy measurement is a quantitative measure of heat content. It is an extensive property, dependent upon the quantity of material. Temperature, as we have mentioned, is an intensive property, independent of quantity.

Not all substances have the same capacity for holding heat; 1 gram of iron and 1 gram of water, even if they are at the same temperature, do *not* contain the same amount of heat energy. One gram of iron will absorb and store 0.108 calorie of heat energy when the temperature is raised 1°C. In contrast, 1 gram of water will absorb almost ten times as much energy, 1.00 calorie, when the temperature is increased an equivalent amount.

Units for other forms of energy will be introduced in later chapters.

Concentration

Concentration is a measure of the number of particles of a substance, or the mass of those particles, that are contained in a specified volume. Concentration is a widely used way of representing mixtures of different substances. Examples include:

- The concentration of oxygen in the air
- Pollen counts, given during the hay fever seasons, which are simply the number of grains of pollen contained in a measured volume of air
- The amount of an illegal drug in a certain volume of blood, indicating the extent of drug abuse
- The proper dose of an antibiotic, based on a patient's weight.

We will describe many situations in which concentration is used to predict useful information about chemical reactions (Sections 7.6 and 9.2, for example). In Chapter 7 we calculate a numerical value for concentration from experimental data.

Figure 1.10
Density (mass/volume) is a unique property of a material. A mixture of wood, water, brass, and mercury is shown, with the cork—the least dense—floating on water. Additionally, brass, with a density greater than water but less than liquid mercury, floats on the interface between these two liquids.

Density and Specific Gravity

Both mass and volume are a function of the *amount* of material present (extensive property). **Density,** the ratio of mass to volume,

$$d = \frac{\text{mass}}{\text{volume}} = \frac{m}{V}$$

is *independent* of the amount of material (intensive property). Density is a useful way to characterize a substance because each substance has a unique density (Figure 1.10).

Learning Goal

Intensive and extensive properties are described on page 7.

One milliliter of air and one milliliter of iron do not weigh the same amount. There is much more mass in 1 milliliter of iron; its density is greater.

Density measurements were used to discriminate between real gold and "fool's gold" during the gold rush era. Today the measurement of the density of a substance is still a valuable analytical technique. The densities of a number of common substances are shown in Table 1.4.

In density calculations the mass is usually represented in grams, and volume is given in either milliliters (mL) or cubic centimeters (cm^3 or cc):

$$1 \text{ mL} = 1 \text{ cm}^3 = 1 \text{ cc}$$

The unit of density would therefore be g/mL, g/cm^3, or g/cc.

EXAMPLE 1.17 ***Calculating the Density of a Solid***

2.00 cm^3 of aluminum are found to weigh 5.40 g. Calculate the density of aluminum in units of g/cm^3.

Continued—

EXAMPLE 1.17 —*Continued*

Solution

The density expression is:

$$d = \frac{m}{V} = \frac{g}{cm^3}$$

Substituting the information given in the problem,

$$= \frac{5.40\ g}{2.00\ cm^3}$$

results in:

$$= 2.70\ g/cm^3$$

EXAMPLE 1.18

Calculating the Mass of a Gas from Its Density

Air has a density of 0.0013 g/mL. What is the mass of a 6.0-L sample of air?

Solution

$$0.0013\ g/mL = 1.3 \times 10^{-3}\ g/mL$$

(The decimal point is moved three positions to the right.) This problem can be solved by using conversion factors:

$$6.0\ \cancel{L\ air} \times \frac{10^3\ \cancel{mL\ air}}{1\ \cancel{L\ air}} \times \frac{1.3 \times 10^{-3}\ g\ air}{\cancel{mL\ air}} = 7.8\ g\ air$$

Table 1.4 Densities of Some Common Materials

Substance	Density (g/mL)	Substance	Density (g/mL)
Air	0.00129 (at 0°C)	Methyl alcohol	0.792
Ammonia	0.00771 (at 0°C)	Milk	1.028–1.035
Benzene	0.879	Oxygen	0.00143 (at 0°C)
Bone	1.7–2.0	Rubber	0.9–1.1
Carbon dioxide	0.01963 (at 0°C)	Turpentine	0.87
Ethyl alcohol	0.789	Urine	1.010–1.030
Gasoline	0.66–0.69	Water	1.000 (at 4°C)
Gold	19.3	Water	0.998 (at 20°C)
Hydrogen	0.00090 (at 0°C)	Wood	0.3–0.98
Kerosene	0.82	(balsa, least dense; ebony	
Lead	11.3	and teak, most dense)	
Mercury	13.6		

EXAMPLE 1.19 ***Using the Density to Calculate the Mass of a Liquid***

Calculate the mass, in grams, of 10.0 mL of mercury (symbolized Hg) if the density of mercury is 13.6 g/mL.

Solution

Using the density as a conversion factor from volume to mass, we have

$$m = (10.0\ \cancel{\text{mL Hg}})\left(13.6\ \frac{\text{g Hg}}{\cancel{\text{mL Hg}}}\right)$$

Cancellation of units results in:

$$= 136\ \text{g Hg}$$

EXAMPLE 1.20 ***Using the Density to Calculate the Volume of a Liquid***

Calculate the volume, in milliliters, of a liquid that has a density of 1.20 g/mL and a mass of 5.00 grams.

Solution

Using the density as a conversion factor from mass to volume, we have

$$V = (5.00\ \cancel{\text{g liquid}})\left(\frac{1\ \text{mL liquid}}{1.20\ \cancel{\text{g liquid}}}\right)$$

Cancellation of units results in:

$$= 4.17\ \text{mL liquid}$$

Question 1.21

The density of ethyl alcohol (200 proof, or pure alcohol) is 0.789 g/mL at 20°C. Calculate the mass of a 30.0-mL sample.

Question 1.22

Calculate the volume, in milliliters, of 10.0 g of a saline solution that has a density of 1.05 g/mL.

Specific gravity is frequently referenced to water at 4°C, its temperature of maximum density (1.000 g/mL). Other reference temperatures may be used. However, the temperature must be specified.

For convenience, values of density are often related to a standard, well-known reference, the density of pure water at 4°C. This "referenced" density is called the **specific gravity,** the ratio of the density of the object in question to the density of pure water at 4°C.

$$\text{specific gravity} = \frac{\text{density of object (g/mL)}}{\text{density of water (g/mL)}}$$

Specific gravity is a *unitless* term. Because the density of water at 4.0°C is 1.00 g/mL, the numerical values for the density and specific gravity of a substance are equal. That is, an object with a density of 2.00 g/mL has a specific gravity of 2.00 at 4°C.

Routine hospital tests involving the measurement of the specific gravity of urine and blood samples are frequently used as diagnostic tools. For example, diseases such as kidney disorders and diabetes change the composition of urine. This compositional change results in a corresponding change in the specific gravity. This change is easily measured and provides the basis for a quick preliminary diagnosis. This topic is discussed in greater detail in A Clinical Perspective: Diagnosis Based on Waste.

A Clinical Perspective

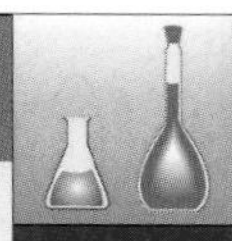

Diagnosis Based on Waste

Any archaeologist would say that you can learn a great deal about the activities and attitudes of a society by finding the remains of their dump sites and studying their waste.

Similarly, urine, a waste product consisting of a wide variety of metabolites, may be analyzed to indicate abnormalities in various metabolic processes or even unacceptable behavior (recall the steroid tests in Olympic competition).

Many of these tests must be performed by using sophisticated and sensitive instrumentation. However, a very simple test, the measurement of the specific gravity of urine, can be an indicator of diabetes mellitus or Bright's disease. The normal range for human urine specific gravity is 1.010–1.030.

A hydrometer, a weighted glass bulb inserted in a liquid, may be used to determine specific gravity. The higher it floats in the liquid, the more dense the liquid. A hydrometer that is calibrated to indicate the specific gravity of urine is called a urinometer.

Although hydrometers have been replaced by more modern measuring devices that use smaller samples, these newer instruments operate on the same principles as the hydrometer.

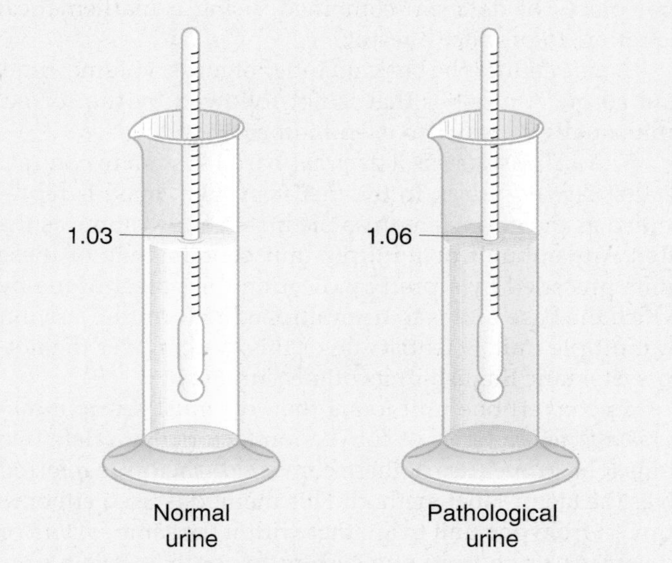

A hydrometer, used in the measurement of the specific gravity of urine.

Summary

1.1 The Discovery Process

Chemistry is the study of matter and the changes that matter undergoes. *Matter* is anything that has mass and occupies space. The changes that matter undergoes always involve either gain or loss of energy. *Energy* is the ability to do work (to accomplish some change). Thus a study of chemistry involves matter, energy, and their interrelationship.

The major areas of chemistry include *biochemistry, organic chemistry, inorganic chemistry, analytical chemistry,* and *physical chemistry.*

The *scientific method* consists of six distinct processes: observation, questioning, pattern recognition, development of theories from hypotheses, experimentation, and summarizing information. A *law* summarizes a large quantity of information.

The development of the scientific method has played a major role in civilization's rapid growth during the past two centuries.

1.2 Matter and Properties

Properties (characteristics) of matter may be classified as either physical or chemical. Physical properties can be observed without changing the chemical composition of the sample. *Chemical properties* result in a change in composition and can be observed only through *chemical reactions. Intensive properties* are independent of the quantity of the substance. *Extensive properties* depend on the quantity of a substance.

Three states of matter exist (solid, liquid, and gas); these states of matter are distinguishable by differences in physical properties.

All matter is classified as either a *pure substance* or a *mixture.* A pure substance is a substance that has only one component. A mixture is a combination of two or more pure substances in which the combined substances retain their identity.

A *homogeneous mixture* has uniform composition. Its particles are well mixed. A *heterogeneous mixture* has a nonuniform composition.

An *element* is a pure substance that cannot be converted into a simpler form of matter by any chemical reaction. A *compound* is a substance produced from the combination of two or more elements in a definite, reproducible fashion.

1.3 Measurement in Chemistry

Science is the study of humans and their environment. Its tool is experimentation. A scientific experiment produces *data*. Each piece of data is the individual result of a single measurement. Mass, length, volume, time, temperature, and energy are the most common types of data obtained from chemical experiments.

Results are the outcome of an experiment. Usually, several pieces of data are combined, using a mathematical equation, to produce a result.

A *unit* defines the basic quantity of mass, volume, time, and so on. A number that is not followed by the correct unit usually conveys no useful information.

The *metric system* is a decimal-based system in contrast to the *English system*. In the metric system, mass is represented as the gram, length as the meter, and volume as the liter. Any subunit or multiple unit contains one of these units preceded by a prefix indicating the power of ten by which the base unit is to be multiplied to form the subunit or multiple unit. Scientists favor this system over the not-so-systematic English units of measurement.

To convert one unit to another, we must set up a *conversion factor* or series of conversion factors that relate two units. The proper use of these conversion factors is referred to as the factor-label method. This method is used either to convert from one unit to another within the same system or to convert units from one system to another. It is a very useful problem-solving tool.

1.4 Significant Figures and Scientific Notation

Significant figures are all digits in a number representing data or results that are known with certainty plus the first uncertain digit. The number of significant figures associated with a measurement is determined by the measuring device. Results should be rounded off to the proper number of significant figures.

Very large and very small numbers may be represented with the proper number of significant figures by using *scientific notation*.

1.5 Experimental Quantities

Mass describes the quantity of matter in an object. The terms weight and mass are often used interchangeably, but they are not equivalent. *Weight* is the force of gravity on an object. The fundamental unit of mass in the metric system is the gram. One atomic mass unit (amu) is equal to 1.661×10^{-24} g.

The standard metric unit of length is the meter. Large distances are measured in kilometers; smaller distances are measured in millimeters or centimeters. Very small distances (on the atomic scale) are measured in nanometers (nm). The standard metric unit of volume is the liter. A liter is the volume occupied by 1000 grams of water at 4 degrees Celsius. The standard metric unit of time is the second, a unit that is used in the English system as well.

Temperature is the degree of "hotness" of an object. Many substances, such as liquid mercury, expand as their temperature increases, and this expansion provides us with a way to measure temperature and temperature changes. Three common temperature scales are Fahrenheit (°F), Celsius (°C), and Kelvin (K).

Energy, the ability to do work, may be categorized as either *kinetic energy,* the energy of motion, or *potential energy,* the energy of position. The principal forms of energy are light, heat, mechanical, electrical, nuclear, and chemical energy.

Energy absorbed or liberated in chemical reactions is most often in the form of heat energy. Heat energy may be represented in units of calories or joules: 1 calorie (cal) = 4.18 joules (J). One calorie is defined as the amount of heat energy required to change the temperature of 1 gram of water 1°C.

Concentration is a measure of the number of particles of a substance, or the mass of those particles, that are contained in a specified volume. Concentration is a widely used way of representing relative quantities of different substances in a mixture of those substances.

Density is the ratio of mass to volume and is a useful way of characterizing a substance. Values of density are often related to a standard reference, the density of pure water at 4°C. This "referenced" density is the *specific gravity,* the ratio of the density of the object in question to the density of pure water at 4°C.

Key Terms

chemical property (1.2)
chemical reaction (1.2)
chemistry (1.1)
compound (1.2)
concentration (1.5)
conversion factor (1.3)
data (1.3)
density (1.5)
element (1.2)
energy (1.1)
English system (1.3)
extensive property (1.2)
gaseous state (1.2)
heterogeneous mixture (1.2)
homogeneous mixture (1.2)
hypothesis (1.1)
intensive property (1.2)
kinetic energy (1.5)
law (1.1)
liquid state (1.2)
mass (1.5)
matter (1.1)
metric system (1.3)
mixture (1.2)
physical change (1.2)
physical property (1.2)
potential energy (1.5)
product (1.2)
properties (1.2)
pure substance (1.2)
reactant (1.2)
results (1.3)
scientific method (1.1)
scientific notation (1.4)
significant figures (1.4)
solid state (1.2)
specific gravity (1.5)
states of matter (1.2)
temperature (1.5)
theory (1.1)
unit (1.3)
weight (1.5)

Questions and Problems

The Discovery Process

1.23 Define each of the following terms:
a. chemistry
b. matter
c. energy
1.24 Define each of the following terms:
a. hypothesis
b. theory
c. law
1.25 Define each of the following terms:
a. potential energy
b. kinetic energy
c. data
1.26 Define each of the following terms:
a. results
b. mass
c. weight
1.27 Give the base unit for each of the following in the metric system:
a. mass
b. volume
c. length
1.28 Give the base unit for each of the following in the metric system:
a. time
b. temperature
c. energy
1.29 Discuss the difference between the terms *mass* and *weight*.
1.30 Discuss the difference between the terms *data* and *results*.
1.31 Distinguish between specific gravity and density.
1.32 Distinguish between kinetic energy and potential energy.
1.33 Discuss the meaning of the term *scientific method*.
1.34 Describe an application of reasoning involving the scientific method that has occurred in your day-to-day life.

Matter and Properties

1.35 Describe what is meant by a physical property.
1.36 Describe what is meant by a physical change.
1.37 Label each of the following as either a physical change or a chemical reaction:
a. An iron nail rusts.
b. An ice cube melts.
c. A limb falls from a tree.
1.38 Label each of the following as either a physical change or a chemical reaction:
a. A puddle of water evaporates.
b. Food is digested.
c. Wood is burned.
1.39 Label each of the following properties of sodium as either a physical property or a chemical property:
a. Sodium is a soft metal (can be cut with a knife).
b. Sodium reacts violently with water to produce hydrogen gas and sodium hydroxide.
1.40 Label each of the following properties of sodium as either a physical property or a chemical property:
a. When exposed to air, sodium forms a white oxide.
b. Sodium melts at 98°C.
c. The density of sodium metal at 25°C is 0.97 g/cm^3.
1.41 Describe several chemical properties of matter.
1.42 Describe what is meant by a chemical reaction.
1.43 Distinguish between a pure substance and a mixture.
1.44 Label each of the following as either a pure substance or a mixture:
a. water
b. table salt (sodium chloride)
c. blood
d. sucrose (table sugar)
e. orange juice
1.45 Distinguish between a homogeneous mixture and a heterogeneous mixture.
1.46 Label each of the following as either a homogeneous mixture or a heterogeneous mixture:
a. a soft drink
b. a saline solution
c. gelatin
d. gasoline
e. vegetable soup
1.47 Describe the general properties of the gaseous state.
1.48 Contrast the physical properties of the gaseous and solid states.
1.49 Distinguish between an intensive property and an extensive property.
1.50 Label each of the following as either an intensive property or an extensive property.
a. mass
b. volume
c. density
d. specific gravity
1.51 Describe the difference between the terms *atom* and *element*.
1.52 Give at least one example of each of the following:
a. an element
b. a pure substance
c. a homogeneous mixture
d. a heterogeneous mixture

Measurement in Chemistry

1.53 Convert 2.0 pounds to:
a. ounces
b. tons
c. grams
d. milligrams
e. dekagrams
1.54 Convert 5.0 quarts to:
a. gallons
b. pints
c. liters
d. milliliters
e. microliters
1.55 Convert 3.0 grams to:
a. pounds
b. ounces
c. kilograms
d. centigrams
e. milligrams
1.56 Convert 3.0 meters to:
a. yards
b. inches
c. feet
d. centimeters
e. millimeters
1.57 Convert 50.0°F to:
a. °C
b. K
1.58 Convert −10.0°F to:
a. °C
b. K
1.59 Convert 20.0°C to:
a. K
b. °F
1.60 Convert 300.0 K to:
a. °C
b. °F
1.61 A 150-lb adult has approximately 9 pints of blood. How many liters of blood does the individual have?

1.62 If a drop of blood has a volume of 0.05 mL, how many drops of blood are in the adult described in Problem 1.61?

1.63 A patient's temperature is found to be 38.5°C. To what Fahrenheit temperature does this correspond?

1.64 A newborn is 21 inches in length and weighs 6 lb 9 oz. Describe the baby in metric units.

Significant Figures and Scientific Notation

1.65 How many significant figures are contained in each of the following numbers?
a. 10.0
b. 0.214
c. 0.120
d. 2.062
e. 10.50
f. 1050

1.66 How many significant figures are contained in each of the following numbers?
a. 3.8×10^{-3}
b. 5.20×10^{2}
c. 0.00261
d. 24
e. 240
f. 2.40

1.67 Round the following numbers to three significant figures:
a. 3.873×10^{-3}
b. 5.202×10^{-2}
c. 0.002616
d. 24.3387
e. 240.1
f. 2.407

1.68 Round the following numbers to three significant figures:
a. 123700
b. 0.00285792
c. 1.421×10^{-3}
d. 53.2995
e. 16.96
f. 507.5

1.69 Perform each of the following arithmetic operations, reporting the answer with the proper number of significant figures:
a. (23)(657)
b. 0.00521 + 0.236
c. $\frac{18.3}{3.0576}$
d. 1157.23 − 17.812
e. $\frac{(1.987)(298)}{0.0821}$

1.70 Perform each of the following arithmetic operations, reporting the answer with the proper number of significant figures:
a. $\frac{(16.0)(0.1879)}{45.3}$
b. $\frac{(76.32)(1.53)}{0.052}$
c. (0.0063)(57.8)
d. 18 + 52.1
e. 58.17 − 57.79

1.71 Express the following numbers in scientific notation (use the proper number of significant figures):
a. 12.3
b. 0.0569
c. −1527
d. 0.000000789
e. 92,000,000
f. 0.005280
g. 1.279
h. −531.77

1.72 Using scientific notation, express the number two thousand in terms of:
a. one significant figure
b. two significant figures
c. three significant figures
d. four significant figures
e. five significant figures

1.73 Express each of the following numbers in decimal notation:
a. 3.24×10^{3}
b. 1.50×10^{-4}
c. 4.579×10^{-1}
d. -6.83×10^{5}
e. -8.21×10^{-2}
f. 2.9979×10^{8}
g. 1.50×10^{0}
h. 6.02×10^{23}

1.74 Which of the following numbers have two significant figures? Three significant figures? Four significant figures?
a. 327
b. 1.049×10^{4}
c. 1.70
d. 0.000570
e. 7.8×10^{3}
f. 1507
g. 4.8×10^{2}
h. 7.389×10^{15}

Experimental Quantities

1.75 Calculate the density of a 3.00×10^{2}-g object that has a volume of 50.0 mL.

1.76 What volume, in liters, will 8.00×10^{2} g of air occupy if the density of air is 1.29 g/L?

1.77 What is the mass, in grams, of a piece of iron that has a volume of 1.50×10^{2} mL and a density of 7.20 g/mL?

1.78 What is the mass of a femur (leg bone) having a volume of 118 cm^3? The density of bone is 1.8 g/cm^3.

1.79 You are given a piece of wood that is maple, teak, or oak. The piece of wood has a volume of 1.00×10^{2} cm^3 and a mass of 98 g. The densities of maple, teak, and oak are as follows:

Wood	Density (g/cm^3)
Maple	0.70
Teak	0.98
Oak	0.85

What is the identity of the piece of wood?

1.80 The specific gravity of a patient's urine sample was measured to be 1.008. Given that the density of water is 1.000 g/mL at 4°C, what is the density of the urine sample?

1.81 The density of grain alcohol is 0.789 g/mL. Given that the density of water at 4°C is 1.00 g/mL, what is the specific gravity of grain alcohol?

1.82 The density of mercury is 13.6 g/mL. If a sample of mercury weighs 272 g, what is the volume of the sample in milliliters?

1.83 You are given three bars of metal. Each is labeled with its identity (lead, uranium, platinum). The lead bar has a mass of 5.0×10^{1} g and a volume of 6.36 cm^3. The uranium bar has a mass of 75 g and a volume of 3.97 cm^3. The platinum bar has a mass of 2140 g and a volume of 1.00×10^{2} cm^3. Which of these metals has the lowest density? Which has the greatest density?

1.84 Refer to Problem 1.83. Suppose that each of the bars had the same mass. How could you determine which bar had the lowest density or highest density?

Critical Thinking Problems

1. An instrument used to detect metals in drinking water can detect as little as one microgram of mercury in one liter of water. Mercury is a toxic metal; it accumulates in the body and is responsible for the deterioration of brain cells. Calculate the number of mercury atoms you would consume if you drank one liter of water that contained only one microgram of mercury. (The mass of one mercury atom is 3.3×10^{-22} grams.)
2. Yesterday's temperature was 40°F. Today it is 80°F. Bill tells Sue that it is twice as hot today. Sue disagrees. Do you think Sue is correct or incorrect? Why or why not?
3. Aspirin has been recommended to minimize the chance of heart attacks in persons who have already had one or more occurrences. If a patient takes one aspirin tablet per day for ten years, how many pounds of aspirin will the patient consume? (Assume that each tablet is approximately 325 mg.)
4. Design an experiment that will allow you to measure the density of your favorite piece of jewelry.
5. The diameter of an aluminum atom is 250 picometers (1 picometer = 10^{-12} meters). How many aluminum atoms must be placed end to end to make a "chain" of aluminum atoms one foot long?

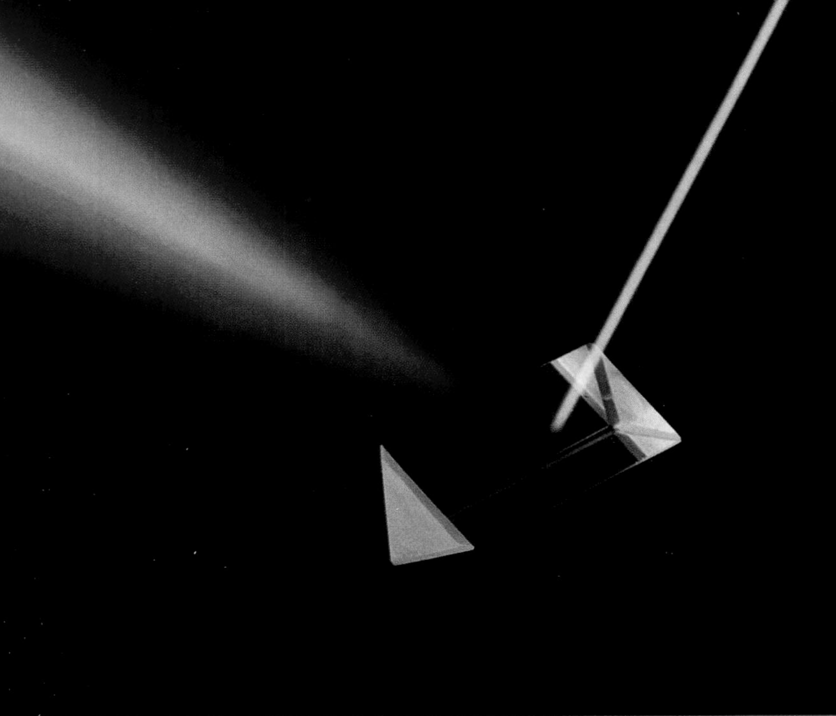

The spectrum of visible light displayed by a prism.

2

The Composition and Structure of the Atom

Outline

Learning Goals

1. Recognize the interrelationship of the structure of matter and its physical and chemical properties.
2. Describe the important properties of protons, neutrons, and electrons.
3. Calculate the number of protons, neutrons, and electrons in any atom.
4. Distinguish among atoms, ions, and isotopes.
5. Trace the history of the development of atomic theory, beginning with Dalton.
6. Summarize the experimental basis for the discovery of charged particles and the nucleus.
7. Explain the critical role of spectroscopy in the development of atomic theory *and* in our everyday lives.
8. State the basic postulates of Bohr's theory.
9. Compare and contrast Bohr's theory and the more sophisticated "wave-mechanical" approach.

CHEMISTRY CONNECTION

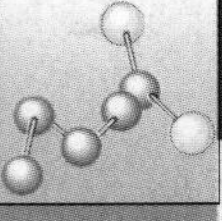

Curiosity, Science, and Medicine

Curiosity is one of the most important human traits. Small children constantly ask "why?". As we get older, our questions become more complex, but the curiosity remains.

Curiosity is also the basis of the scientific method. A scientist observes an event, wonders why it happens, and sets out to answer the question. Dr. Michael Zasloff's curiosity may lead to the development of an entirely new class of antibiotics. When he was a geneticist at the National Institutes of Health, his experiments involved the surgical removal of the ovaries of African clawed frogs. After surgery he sutured (sewed up) the incision and put the frogs back in their tanks. These water-filled tanks were teeming with bacteria, but the frogs healed quickly, and the incisions did not become infected!

Of all the scientists to observe this remarkable healing, only Zasloff was curious enough to ask whether there were chemicals in the frogs' skin that defended the frogs against bacterial infections—a new type of antibiotic. All currently used antibiotics are produced by fungi or are synthesized in the laboratory. One big problem in medicine today is more and more pathogenic (disease-causing) bacteria are becoming resistant to these antibiotics. Zasloff hoped to find an antibiotic that worked in an entirely new way so the current problems with antibiotic resistance might be overcome.

Dr. Zasloff found two molecules in frog skin that can kill bacteria. Both are small proteins. Zasloff named them *magainins*, from the Hebrew word for shield. Most of the antibiotics that we now use enter bacteria and kill them by stopping some biochemical process inside the cell. Magainins are more direct; they simply punch holes in the bacterial membrane, and the bacteria explode.

One of the magainins, now chemically synthesized in the laboratory so that no frogs are harmed, may be available to the public in the near future. This magainin can kill a wide variety of bacteria (broad-spectrum antibiotic), and it has passed the Phase I human trials. If this compound passes all the remaining tests, it will be used in treating deep infected wounds and ulcers, providing an alternative to traditional therapy.

The curiosity that enabled Zasloff to advance the field of medicine also catalyzed the development of atomic structure. We will see the product of this fundamental human characteristic in the work of Crookes, Bohr, and others throughout this chapter.

Introduction

In Chapter 1 we described chemistry as the study of matter and the changes that matter undergoes. In this chapter we will expand and enhance our understanding of matter. We can deal with visible quantities of matter such as an ounce of silver or a pint of blood. However, we can also describe matter at the level of individual particles that make it up. For instance, one atom of silver is the smallest amount of silver that retains the properties of the bulk material. This is important because the description of matter at the atomic level can be used to explain the behavior of the larger, visible quantities of the same material.

Why does ice float on water? Why don't oil and water mix? Why does blood transport oxygen to our cells, whereas carbon monoxide inhibits this process? Questions such as these are best explained by understanding the behavior of substances at the atomic level.

2.1 Matter and Structure

An Overview

Chemists and physicists have used the observed properties of matter to develop models of the individual units of matter. These models collectively make up what we now know as the atomic theory of matter.

As we saw in Section 1.1, models allow us to make ideas more clear and enable us to predict behavior; this is their main value.

These models have developed from experimental observations over the past two hundred years. Thus theory and experiment reinforce each other. We must gain some insight into atomic structure to appreciate the behavior of the atoms themselves as well as larger aggregates of atoms: compounds.

Figure 2.1
Examples of technology originating from scientific inquiry: (a) synthesis of a new drug, (b) solar energy cells, (c) preparation of solid-state electronics, (d) use of a gypsy moth sex attractant for insect control.

Learning Goal

The structure–properties concept has advanced so far that compounds are designed and synthesized in the laboratory with the hope that they will perform very specific functions, such as curing diseases that have been resistant to other forms of treatment. Figure 2.1 shows some of the variety of modern technology that has its roots in the understanding of the atom.

Atomic Structure

Section 1.1 discusses the scientific method.

The theory of atomic structure has progressed rapidly, from a very primitive level to its present point of sophistication, in a relatively short time. We briefly summarize our present knowledge of the composition of the atom in Section 2.2. Then in Section 2.3 we present an outline of the significant scientific discoveries that spurred development of atomic theory. Before we proceed, let us insert a note of caution. We must not think of the present picture of the atom as final. Scientific inquiry continues, and we should view the present theory as a step in an evolutionary process. *Theories are subject to constant refinement,* as was noted in our discussion of the scientific method.

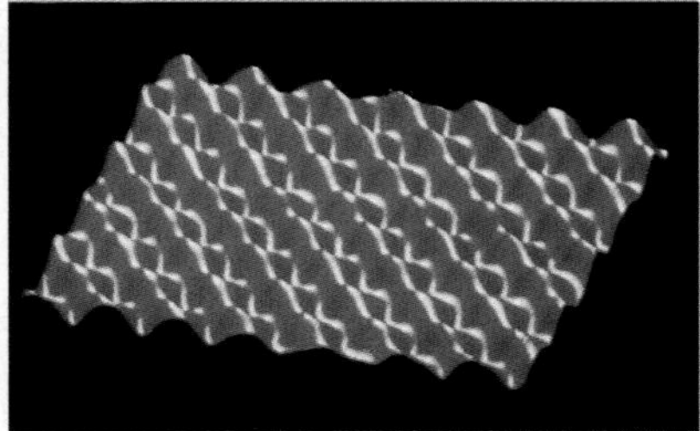

Figure 2.2
Sophisticated techniques, such as scanning tunneling electron microscopy, provide visual evidence for the structure of atoms and molecules. The planar nature of graphite, a commonly used lubricant, is shown here; the peaks are images of carbon atoms.

2.2 Composition of the Atom

The basic structural unit of an element is the **atom,** which is the smallest unit of an element that retains the chemical properties of that element. A tiny sample of the element copper, too small to be seen by the naked eye, is composed of billions of copper atoms arranged in some orderly fashion. Each atom is incredibly small. Only recently have we been able to "see" atoms using modern instruments such as the scanning tunneling microscope (Figure 2.2).

Table 2.1 Selected Properties of the Three Basic Subatomic Particles

Name	Charge	Mass (amu)	Mass (grams)
Electron (e)	−1	5.4×10^{-4}	9.1095×10^{-28}
Proton (p)	+1	1.00	1.6725×10^{-24}
Neutron (n)	0	1.00	1.6750×10^{-24}

Electrons, Protons, and Neutrons

Radioactivity and radioactive decay are discussed in Chapter 10.

We know from experience that certain kinds of atoms can "split" into smaller particles and release large amounts of energy; this process is *radioactive decay.* We also know that the atom is composed of three primary particles: the *electron,* the *proton,* and the *neutron.* Although other subatomic fragments with unusual names (neutrinos, gluons, quarks, and so forth) have also been discovered, we shall concern ourselves only with the primary particles: the protons, neutrons, and electrons.

We can consider the atom to be composed of two distinct regions:

Learning Goal

1. The **nucleus** is a small, dense, positively charged region in the center of the atom. The nucleus is composed of positively charged **protons** and uncharged **neutrons.**
2. Surrounding the nucleus is a diffuse region of negative charge populated by **electrons,** the source of the negative charge. Electrons are very low in mass in contrast to the protons and neutrons.

The properties of these particles are summarized in Table 2.1.

Atoms of various types differ in their number of protons, neutrons, and electrons. The number of protons determines the identity of the atom. As such, the number of protons is *characteristic* of the element. When the number of protons is equal to the number of electrons, the atom is neutral because the charges are balanced and effectively cancel one another.

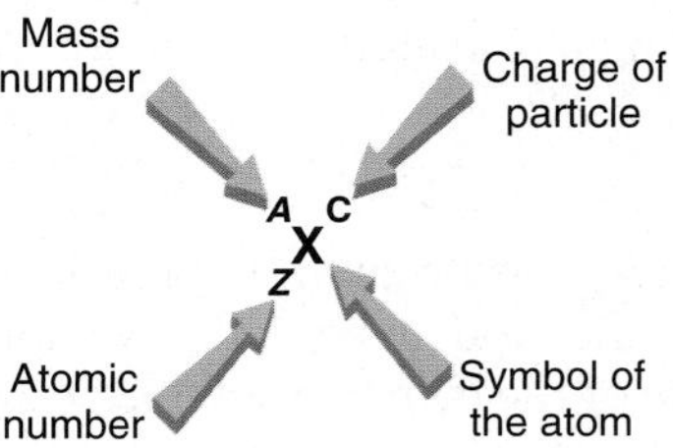

We may represent an element symbolically as follows:

Learning Goal

The **atomic number** (Z) is equal to the number of protons in the atom, and the **mass number** (A) is equal to the *sum* of the number of protons and neutrons (the mass of the electrons is so small as to be insignificant in comparison to that of the nucleus).

If

$$\text{number of protons} + \text{number of neutrons} = \text{mass number}$$

then, if the number of protons is subtracted from each side,

$$\text{number of neutrons} = \text{mass number} - \text{number of protons}$$

or, because the number of protons equals the atomic number,

$$\text{number of neutrons} = \text{mass number} - \text{atomic number}$$

For an atom, in which positive and negative charges cancel, the number of protons and electrons must be equal and identical to the atomic number.

EXAMPLE 2.1

Determining the Composition of an Atom

Calculate the numbers of protons, neutrons, and electrons in an atom of fluorine. The atomic symbol for the fluorine atom is ${}^{19}_{9}\mathrm{F}$.

Solution

The mass number 19 tells us that the total number of protons + neutrons is 19. The atomic number, 9, represents the number of protons. The difference, 19 − 9, or 10, is the number of neutrons. The number of electrons must be the same as the number of protons, hence 9, for a neutral fluorine atom.

Question 2.1

Calculate the number of protons, neutrons, and electrons in each of the following atoms:

a. ${}^{32}_{16}\mathrm{S}$
b. ${}^{23}_{11}\mathrm{Na}$

Question 2.2

Calculate the number of protons, neutrons, and electrons in each of the following atoms:

a. ${}^{1}_{1}\mathrm{H}$
b. ${}^{244}_{94}\mathrm{Pu}$

Isotopes

Learning Goal

Isotopes are atoms of the same element having different masses *because they contain different numbers of neutrons.* In other words, isotopes have different mass numbers. For example, all of the following are isotopes of hydrogen:

${}^{1}_{1}\mathrm{H}$	${}^{2}_{1}\mathrm{H}$	${}^{3}_{1}\mathrm{H}$
Hydrogen	Deuterium	Tritium
(Hydrogen-1)	(Hydrogen-2)	(Hydrogen-3)

Isotopes are often written with the name of the element followed by the mass number. For example, the isotopes ${}^{12}_{6}\mathrm{C}$ and ${}^{14}_{6}\mathrm{C}$ may be written as carbon-12 (or C-12) and carbon-14 (or C-14), respectively.

Certain isotopes (radioactive isotopes) of elements emit particles and energy that can be used to trace the behavior of biochemical systems. These isotopes otherwise behave identically to any other isotope of the same element. Their chemical behavior is identical; it is their nuclear behavior that is unique. As a result, a radioactive isotope can be substituted for the "nonradioactive" isotope, and its biochemical activity can be followed by monitoring the particles or energy emitted by the isotope as it passes through the body.

A detailed discussion of the use of radioactive isotopes in the diagnosis and treatment of disease is found in Chapter 10.

The existence of isotopes explains why the average masses, measured in atomic mass units (amu), of the various elements are not whole numbers. This is contrary to what we would expect from proton and neutron masses, which are whole numbers to three significant figures.

See Section 1.5 for the definition of the atomic mass unit.

Consider, for example, the mass of one chlorine atom, containing 17 protons (atomic number) and 18 neutrons:

$$17 \text{ ~~protons~~} \times \frac{1.00 \text{ amu}}{\text{~~proton~~}} = 17.00 \text{ amu}$$

$$18 \text{ ~~neutrons~~} \times \frac{1.00 \text{ amu}}{\text{~~neutron~~}} = 18.00 \text{ amu}$$

$$17.00 \text{ amu} + 18.00 \text{ amu} = 35.00 \text{ amu (mass of chlorine atom)}$$

Inspection of the periodic table reveals that the mass number of chlorine is actually 35.45 amu, *not* 35.00 amu. The existence of isotopes accounts for this difference. A natural sample of chlorine is composed principally of two isotopes, chlorine-35 and chlorine-37, in approximately a 3:1 ratio, and the tabulated mass is the *weighted average* of the two isotopes. In our calculation the chlorine atom referred to was the isotope that has a mass number of 35 amu.

The weighted average is not a true average but is corrected by the relative amounts (the weighting factor) of each isotope present in nature.

The weighted average of the masses of all of the isotopes of an element is the **atomic mass** and should be distinguished from the mass number, which is the sum of the number of protons and neutrons in a single isotope of the element.

Example 2.2 demonstrates the calculation of the atomic mass of chlorine.

EXAMPLE 2.2 ***Determining Atomic Mass***

Calculate the atomic mass of naturally occurring chlorine if 75.77% of chlorine atoms are $^{35}_{17}Cl$ (chlorine-35) and 24.23% of chlorine atoms are $^{37}_{17}Cl$ (chlorine-37).

Solution

Step 1. Convert each percentage to a decimal fraction.

$$75.77\% \text{ chlorine-35} \times \frac{1}{100\%} = 0.7577 \text{ chlorine-35}$$

$$24.23\% \text{ chlorine-37} \times \frac{1}{100\%} = 0.2423 \text{ chlorine-37}$$

Step 2. Multiply the decimal fraction of each isotope by the mass of that isotope to determine the isotopic contribution to the average atomic mass.

contribution to atomic mass by chlorine-35	=	fraction of all Cl atoms that are chlorine-35	×	mass of a chlorine-35 atom
	=	0.7577	×	35.00 amu
	=	26.52 amu		
contribution to atomic mass by chlorine-37	=	fraction of all Cl atoms that are chlorine-37	×	mass of a chlorine-37 atom
	=	0.2423	×	37.00 amu
	=	8.965 amu		

Continued—

EXAMPLE 2.2 *—Continued*

Step 3. The weighted average is:

atomic mass of naturally occurring Cl	=	contribution of chlorine-35	+	contribution of chlorine-37
	=	26.52 amu	+	8.965 amu
	=	35.49 amu		

which is very close to the tabulated value of 35.45 amu. An even more exact value would be obtained by using a more exact value of the mass of the proton and neutron (experimentally known to a greater number of significant figures).

Whenever you do calculations such as those in Example 2.2, before even beginning the calculation you should look for an approximation of the value sought. Then do the calculation and see whether you obtain a reasonable number (similar to your anticipated value). In the preceding problem, if the two isotopes have masses of 35 and 37, the atomic mass must lie somewhere between the two extremes. Furthermore, because the majority of a naturally occurring sample is chlorine-35 (about 75%), the value should be closer to 35 than to 37. An analysis of the results often avoids problems stemming from untimely events such as pushing the wrong button on a calculator.

A hint for numerical problem solving: Estimate (at least to an order of magnitude) your answer before beginning the calculation using your calculator.

EXAMPLE 2.3

Determining Atomic Mass

Calculate the atomic mass of naturally occurring carbon if 98.90% of carbon atoms are $^{12}_{6}C$ (carbon-12) with a mass of 12.00 amu and 1.11% are $^{13}_{6}C$ (carbon-13) with a mass of 13.00 amu. (Note that a small amount of $^{14}_{6}C$ is also present but is small enough to ignore in a calculation involving three or four significant figures.)

Solution

Step 1. Convert each percentage to a decimal fraction.

$$98.90\%\ \text{carbon-12} \times \frac{1}{100\%} = 0.9890\ \text{carbon-12}$$

$$1.11\%\ \text{carbon-13} \times \frac{1}{100\%} = 0.0111\ \text{carbon-13}$$

Step 2.

contribution to atomic mass by carbon-12	=	fraction of all C atoms that are carbon-12	×	(mass of a carbon-12 atom)
	=	0.9890	×	12.00 amu
	=	11.87 amu		

Continued—

EXAMPLE 2.3 —*Continued*

contribution to atomic mass by carbon-13	=	(fraction of all C atoms that are carbon-13)	×	(mass of a carbon-13 atom)
	=	0.0111	×	13.00 amu
	=	0.144 amu		

Step 3. The weighted average is:

atomic mass of naturally occurring carbon	=	(contribution of carbon-12)	+	(contribution of carbon-13)
	=	11.87 amu	+	0.144 amu
	=	12.01 amu		

Helpful Hint: Because most of the carbon is carbon-12, with very little carbon-13 present, the atomic mass should be very close to that of carbon-12. Approximations, before performing the calculation, provide another check on the accuracy of the final result.

Question 2.3

The element neon has three naturally occurring isotopes. One of these has a mass of 19.99 amu and a natural abundance of 90.48%. A second isotope has a mass of 20.99 amu and a natural abundance of 0.27%. A third has a mass of 21.99 amu and a natural abundance of 9.25%. Calculate the atomic mass of neon.

Question 2.4

The element nitrogen has two naturally occurring isotopes. One of these has a mass of 14.003 amu and a natural abundance of 99.63%; the other isotope has a mass of 15.000 amu and a natural abundance of 0.37%. Calculate the atomic mass of nitrogen.

Ions

Learning Goal

Ions are electrically charged particles that result from a gain of one or more electrons by the parent atom (forming negative ions, or **anions**) or a loss of one or more electrons from the parent atom (forming positive ions, or **cations**).

Formation of an anion may occur as follows:

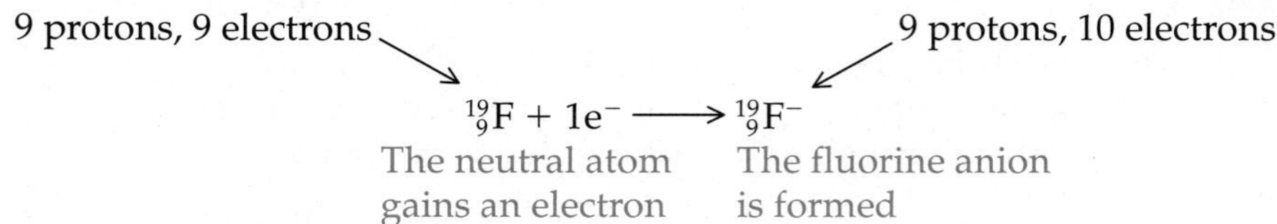

Alternatively, formation of a cation of sodium may proceed as follows:

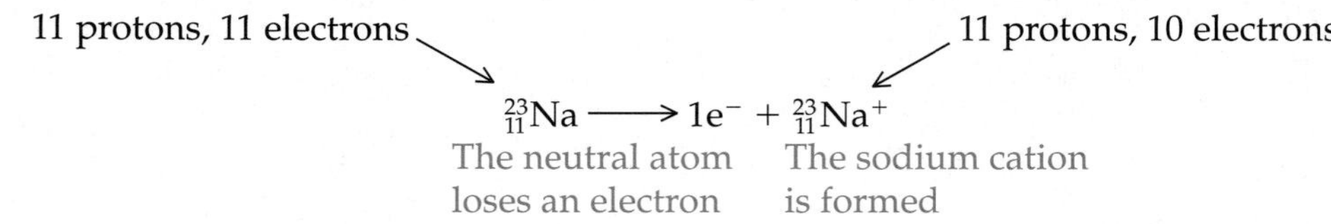

Note that the electrons gained are written to the left of the reaction arrow (they are reactants), whereas the electrons lost are written as products to the right of the reaction arrow. For simplification the atomic and mass numbers are often omitted, because they do not change during ion formation. For example, the sodium cation would be written as Na^+ and the anion of fluorine as F^-.

2.3 Development of Atomic Theory

With this overview of our current understanding of the structure of the atom, we now look at a few of the most important scientific discoveries that led to modern atomic theory.

Learning Goal

Dalton's Theory

The first experimentally based theory of atomic structure was proposed by John Dalton, an English schoolteacher, in the early 1800s. Dalton proposed the following description of atoms:

1. All matter consists of tiny particles called atoms.
2. An atom cannot be created, divided, destroyed, or converted to any other type of atom.
3. Atoms of a particular element have identical properties.
4. Atoms of different elements have different properties.
5. Atoms of different elements combine in simple whole-number ratios to produce compounds (stable aggregates of atoms).
6. Chemical change involves joining, separating, or rearranging atoms.

Although Dalton's theory was founded on meager and primitive experimental information, we regard much of it as correct today. Postulates 1, 4, 5, and 6 are currently regarded as true. The discovery of the processes of nuclear fusion, fission ("splitting" of atoms), and radioactivity has disproved the postulate that atoms cannot be created or destroyed. Postulate 3, that all the atoms of a particular element are identical, was disproved by the discovery of isotopes.

Fusion, fission, radioactivity, and isotopes are discussed in some detail in Chapter 10. Figure 2.3 uses a simple model to illustrate Dalton's theory.

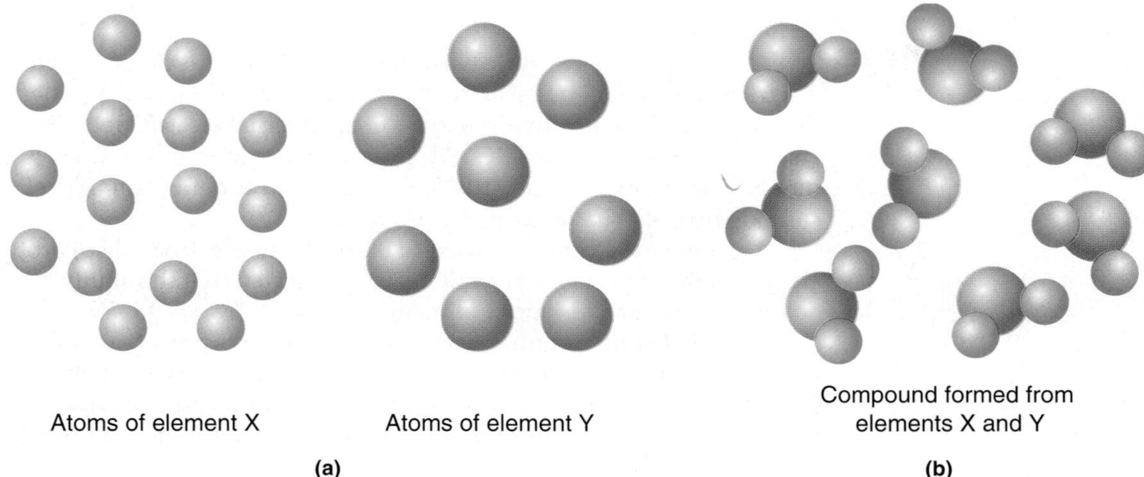

Figure 2.3
An illustration of John Dalton's atomic theory. (a) Atoms of the same element are identical, but different from atoms of any other element. (b) Atoms combine in whole-number ratios to form compounds.

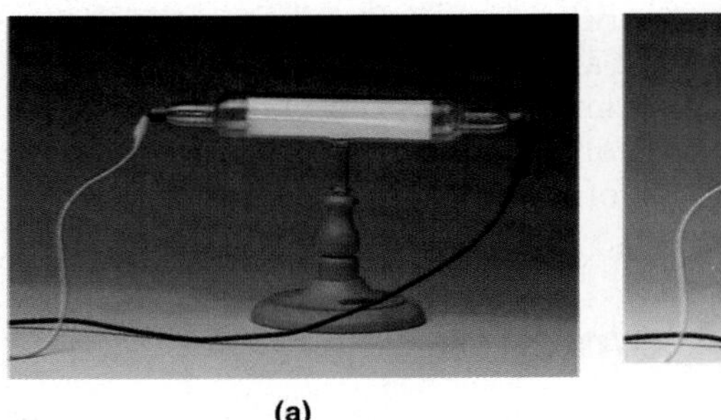

(a)

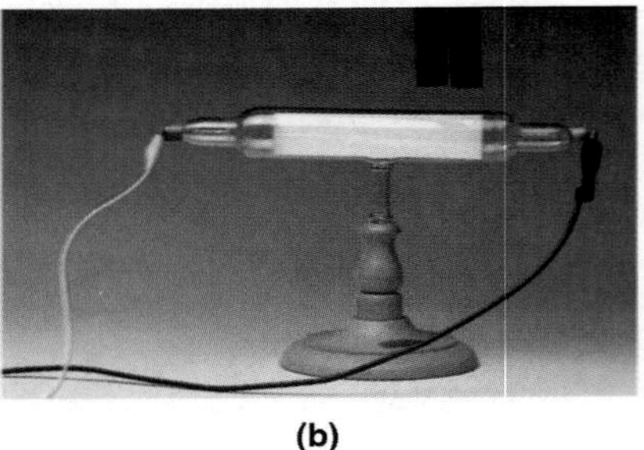

(b)

Figure 2.4
Illustration of Crookes's experiment. (a) When a high voltage was applied between two electrodes in a sealed, evacuated tube, a cathode ray (an electron track) was observed between the electrodes, originating at the negative electrode, the cathode. (b) When a magnetic field is applied, the path of the cathode ray shifts, indicating that the cathode ray has magnetic properties.

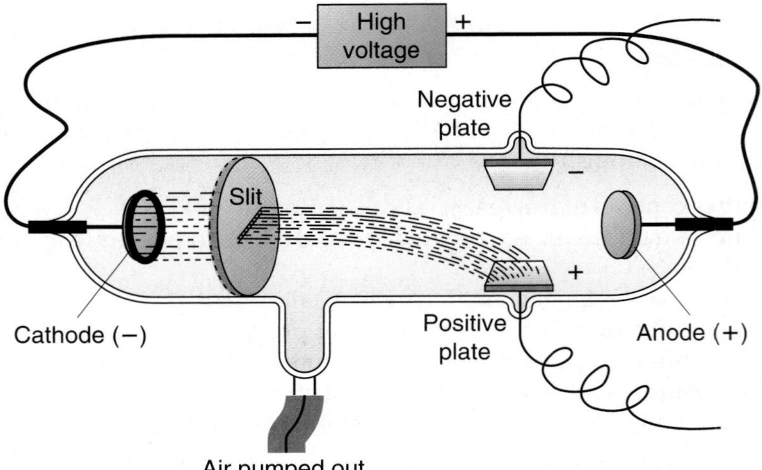

Figure 2.5
Illustration of an experiment demonstrating the charge of cathode rays. The application of an external electric field causes the electron beam to deflect toward a positive charge, implying that the cathode ray is negative.

Evidence for Subatomic Particles: Electrons, Protons, and Neutrons

Learning Goal 6

Crookes's cathode ray tube was the forerunner of the computer screen (often called CRT) and the television.

The next major discoveries occurred almost a century later (1879–1897). Although Dalton pictured atoms as indivisible, various experiments, particularly those of William Crookes and Eugene Goldstein, indicated that the atom is composed of charged (+ and −) particles.

Crookes connected two metal electrodes (metal discs connected to a source of electricity) at opposite ends of a sealed glass vacuum tube. When the electricity was turned on, rays of light were observed to travel between the two electrodes. They were called **cathode rays** because they traveled from the **cathode** (the negative electrode) to the **anode** (the positive electrode). A diagram of the apparatus is shown in Figure 2.4.

Later experiments by J. J. Thomson, an English scientist, demonstrated the electrical and magnetic properties of cathode rays (Figure 2.5). The rays were deflected toward the positive electrode of an external electric field. Because opposite charges attract, this indicates the negative character of the rays. Similar experiments with an external magnetic field showed a deflection as well; hence these cathode rays also have magnetic properties.

A change in the material used to fabricate the electrode discs brought about no change in the experimental results. This suggested that the ability to produce cathode rays is a characteristic of all materials.

In 1897, Thomson announced that cathode rays are streams of negative particles of energy. These particles are *electrons*. Similar experiments, conducted by Goldstein, led to the discovery of particles that are equal in charge to the electron but opposite in sign. These particles, much heavier than electrons (actually 1837 times as heavy), are called *protons*.

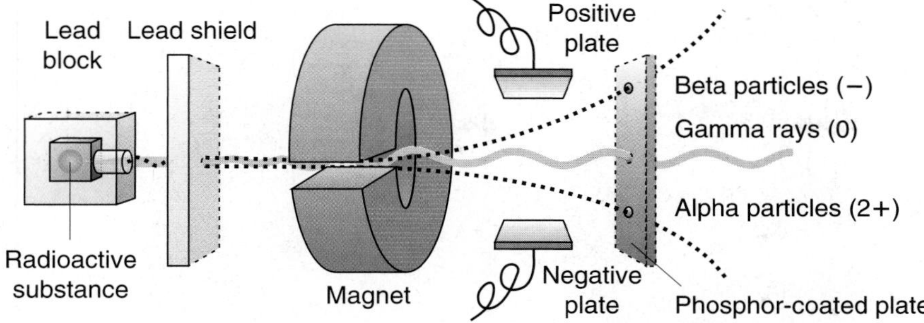

Figure 2.6
Types and characteristics of radioactive emissions. The direction taken by the radioactive emissions indicates the presence of three types of emissions: positive, negative, and neutral components.

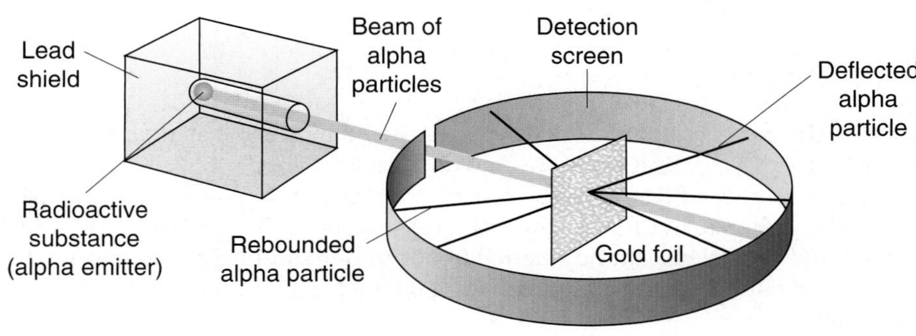

Figure 2.7
The alpha particle scattering experiment.

As we have seen, the third fundamental atomic particle is the *neutron.* It has a mass virtually identical (it is less than 1% heavier) to that of the proton and has zero charge. The neutron was first postulated in the early 1920s, but it was not until 1932 that James Chadwick demonstrated its existence with a series of experiments involving the use of small particle bombardment of nuclei.

Evidence for the Nucleus

Learning Goal

In the early 1900s it was believed that protons and electrons were uniformly distributed throughout the atom. However, an experiment by Hans Geiger led Ernest Rutherford (in 1911) to propose that the majority of the mass and positive charge of the atom was actually located in a small, dense region, the *nucleus,* with small, negatively charged electrons occupying a much larger volume outside of the nucleus.

To understand how Rutherford's theory resulted from the experimental observations of Geiger, let us examine this experiment in greater detail. Rutherford and others had earlier demonstrated that some atoms spontaneously "decay" to produce three types of radiation: alpha (α), beta (β), and gamma (γ) radiation. This process is known as **natural radioactivity** (Figure 2.6). Geiger used radioactive materials, such as *radium,* as projectile sources, "firing" alpha particles at a thin metal foil target (gold leaf). He then observed the interaction of the metal and alpha particles with a detection screen (Figure 2.7) and found that:

a. Most alpha particles pass through the foil without being deflected.
b. A small fraction of the particles were deflected, some even *directly back to the source.*

Rutherford interpreted this to mean that most of the atom is empty space, because most alpha particles were not deflected. Further, most of the mass and positive charge must be located in a small, dense region; collision of the heavy and positively charged alpha particle with this small dense and positive region (the nucleus) caused the great deflections. Rutherford summarized his astonishment at

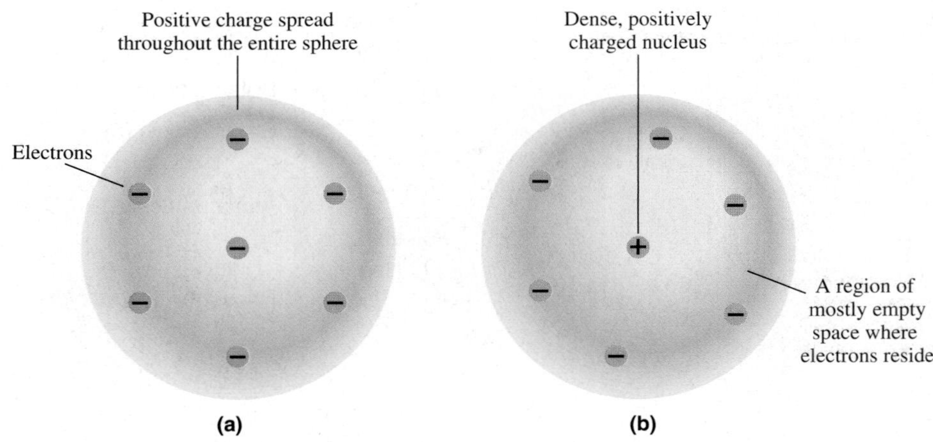

Figure 2.8
(a) A model of the atom (credited to Thomson) prior to the work of Geiger and Rutherford. This was termed the "plum pudding" model. (b) A model of the atom supported by the alpha-particle scattering experiments of Geiger and Rutherford.

observing the deflected particles: "It was almost as incredible as if you fired a 15-inch shell at a piece of tissue and it came back and hit you."

The significance of Rutherford's contribution cannot be overstated. It caused a revolutionary change in the way that scientists pictured the atom (Figure 2.8). His discovery of the nucleus is fundamental to our understanding of chemistry. Chapter 10 will provide much more information on a special branch of chemistry: nuclear chemistry.

2.4 The Relationship between Light and Atomic Structure

The Rutherford atom leaves us with a picture of a tiny, dense, positively charged nucleus containing protons and surrounded by electrons. The electron arrangement, or configuration, is not clearly detailed. More information is needed regarding the relationship of the electrons to each other and to the nucleus. In dealing with dimensions on the order of 10^{-8} cm (the atomic level), conventional methods for measurement of location and distance of separation become impossible. An alternative approach involves the measurement of *energy* rather than the *position* of the atomic particles to determine structure. For example, information obtained from the absorption or emission of *light* by atoms (energy changes) can yield valuable insight into structure. Such studies are referred to as **spectroscopy.**

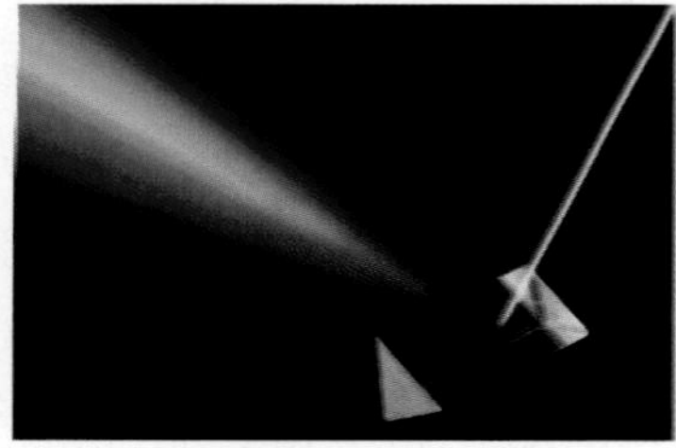

Figure 2.9
The visible spectrum of light. Light passes through a prism, producing a continuous spectrum.

In a general sense we refer to light as **electromagnetic radiation.** Electromagnetic radiation travels in *waves* from a source. The most recognizable source of this radiation is the sun. We are aware of a rainbow, in which visible white light from the sun is broken up into several characteristic bands of different colors. Similarly, visible white light, when passed through a glass prism, is separated into its various component colors (Figure 2.9). These various colors are simply light (electromagnetic radiation) of differing *wavelengths.*

All electromagnetic radiation travels at a speed of 3.0×10^8 m/s, the **speed of light.** However, each wavelength of light, although traveling with identical velocity, has its own characteristic energy. A collection of all electromagnetic radiation, including each of these wavelengths, is referred to as the **electromagnetic spectrum.** For convenience in discussing this type of radiation we subdivide electromagnetic radiation into various spectral regions, which are characterized by physical properties of the radiation, such as its *wavelength* or its *energy* (Figure 2.10). Some of these regions are quite familiar to us from our everyday experiences; the visible and microwave regions are two common examples.

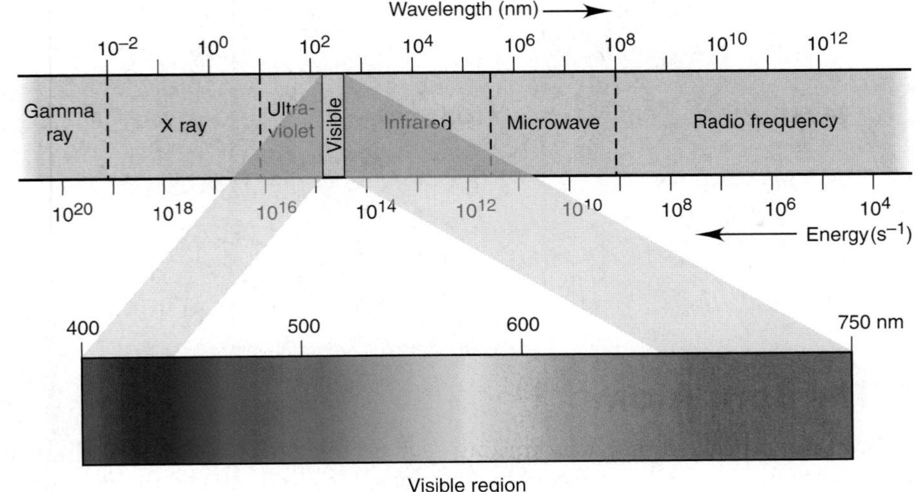

Figure 2.10
The electromagnetic spectrum. Note that the visible spectrum is only a small part of the total electromagnetic spectrum.

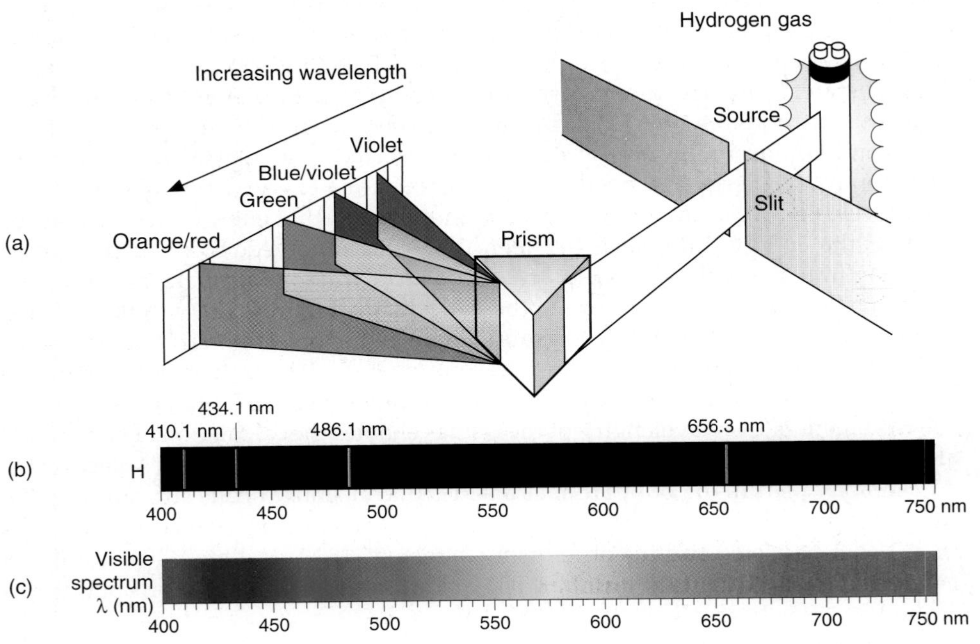

Figure 2.11
(a) The emission spectrum of hydrogen. Certain wavelengths of light, characteristic of the atom, are emitted upon electrical excitation. (b) The line spectrum of hydrogen is compared with the spectrum of visible light (c).

Light of shorter wavelength has higher energy; this means that the magnitude of the energy and wavelength is inversely proportional. The wavelength of a particular type of light can be measured, and from this the energy may be calculated.

If we take a sample of some element, such as hydrogen, in the gas phase, place it in an evacuated glass tube containing a pair of electrodes, and pass an electrical charge (cathode ray) through the hydrogen gas, light is emitted. Not all wavelengths (or energies) of light are emitted—only certain wavelengths that are characteristic of the gas under study. This is referred to as an *emission spectrum* (Figure 2.11). If a different gas, such as helium, is used, a different spectrum (different wavelengths of light) is observed. The reason for this phenomenon was explained by Niels Bohr.

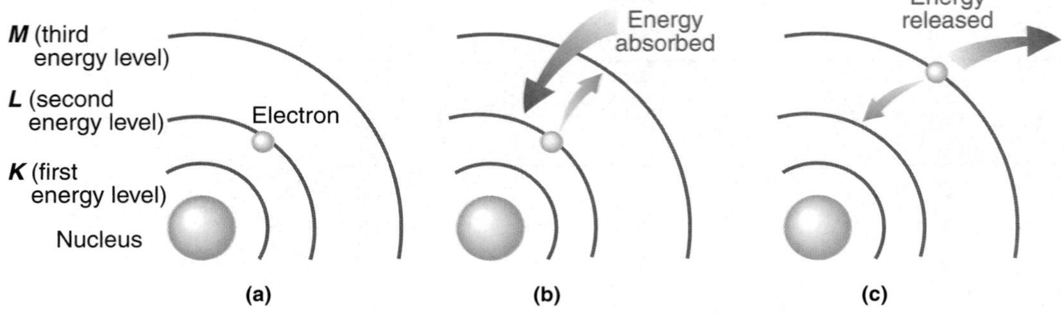

Figure 2.12
(a) The Bohr atom. (b) Excitation involves promotion of an electron to a higher energy level by absorption of energy. (c) Relaxation is the reverse process, whereby an atom returns to the ground state as the electron moves to a lower energy level and energy is released. (Orbits are not drawn to scale.)

2.5 The Bohr Atom

Learning Goal 8

Niels Bohr hypothesized that surrounding each atomic nucleus were certain fixed **energy levels** that could be occupied by electrons. He also believed that each level was defined by a spherical **orbit** around the nucleus, located at a specific distance from the nucleus. The concept of certain fixed energy levels is referred to as the **quantization** of energy. The implication is that only these orbits, or **quantum levels,** as described by Max Planck, are allowed locations for electrons. If an atom *absorbs* energy, an electron undergoes **promotion** from an orbit closer to the nucleus (lower energy) to one farther from the nucleus (higher energy), creating an **excited state.** Similarly, the release of energy by an atom, or **relaxation,** results from an electron falling into an orbit closer to the nucleus (lower energy level).

Promotion and relaxation processes are referred to as **electronic transitions.** The amount of energy absorbed in jumping from one energy level to a higher energy level is a precise quantity (hence, quantum), and that energy corresponds exactly to the energy differences between the orbits involved. Electron promotion resulting from absorption of energy results in an *excited state atom;* the process of relaxation allows the atom to return to the *ground state* (Figure 2.12) with the simultaneous release of light energy. The **ground state** is the lowest possible energy state. This emission process, such as the release of energy after excitation of hydrogen atoms by an electric arc, produces the series of emission lines (emission spectrum). Measurement of the wavelengths of these lines enables the calculation of energy levels in the atom. These energy levels represent the location of the atom's electrons.

We may picture the Bohr atom as a series of concentric orbits surrounding the nucleus. The orbits are identified by two different systems, one using numbers ($n = 1, 2, 3, \ldots$, etc.) and the other using letters (K, L, M, . . . , etc.). The number n is referred to as a **quantum number.** The quantum number $n = 1$ corresponds to a K shell, $n = 2$ is L, and so forth.

The hydrogen spectrum consists of four lines in the visible region of the spectrum. Electronic transitions, calculated from the Bohr theory, account for each of these lines. Table 2.2 gives a summary of the hydrogen spectrum.

Table 2.2 Electronic Transitions Responsible for the Hydrogen Spectrum

Line Color	Wavelength Emitted (nm)	Electronic Transition $n =$	to	$n =$
Red	656.4	3		2
Green	486.3	4		2
Blue	434.2	5		2
Violet	410.3	6		2

A HUMAN PERSPECTIVE

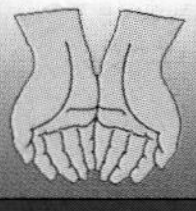

Atomic Spectra and the Fourth of July

At one time or another we have all marveled at the bright, multicolored display of light and sound that is a fireworks display. These sights and sounds are produced by a chemical reaction that generates the energy necessary to excite a variety of elements to their higher-energy electronic states. Light emission results from relaxation of the excited atoms to the ground state. Each atom releases light of specific wavelengths. The visible wavelengths are seen as colored light.

Fireworks need a chemical reaction to produce energy. We know from common experience that oxygen and a fuel will release energy. The fuel in most fireworks preparations is sulfur or aluminum. Each reacts slowly with oxygen; a more potent solid-state source of oxygen is potassium perchlorate ($KClO_4$). The potassium perchlorate reacts with the fuel (an oxidation–reduction reaction), producing a bright white flash of light. The heat produced excites the various elements packaged with the fuel and oxidant.

Sodium salts, such as sodium chloride, furnish sodium ions, which, when excited, produce yellow light (a wavelength of 589 nm). Red colors arise from salts of strontium, which emit several shades of red corresponding to wavelengths in the 600- to 700-nm region of the visible spectrum. Copper salts produce blue radiation, because copper emits in the 400- to 500-nm spectral region.

A fireworks display is a dramatic illustration of light emission by excited atoms.

The beauty of fireworks is a direct result of the skill of the manufacturer. Selection of the proper oxidant, fuel, and color-producing elements is critical to the production of a spectacular display. Packaging these chemicals in proper quantities so that they can be stored and used safely is an equally important consideration.

A summary of the major features of the Bohr theory is as follows:

- Atoms can absorb and emit energy via *promotion* of electrons to higher energy levels and *relaxation* to lower levels.
- Energy that is emitted upon relaxation is observed as a single wavelength of light.
- These *spectral lines* are a result of electron transitions between *allowed levels* in the atom.
- The allowed levels are quantized energy levels, or orbits.
- Electrons are found only in these energy levels.
- The highest-energy orbits are located farthest from the nucleus.
- Atoms absorb energy by excitation of electrons to higher energy levels.
- Atoms release energy by relaxation of electrons to lower energy levels.
- Energy differences may be calculated from the wavelengths of light emitted.

2.6 Modern Atomic Theory

Learning Goal

The Bohr model was an immensely important contribution to the understanding of atomic structure. The idea that electrons exist in specific energy states and that transitions between states involve quanta of energy provided the linkage between atomic structure and atomic spectra. However, some limitations of this model quickly became apparent. Although it explained the hydrogen spectrum, it provided only a crude approximation of the spectra for more complex atoms. Subsequent development of more sophisticated experimental techniques demonstrated that there are problems with the Bohr theory even in the case of hydrogen.

AN ENVIRONMENTAL PERSPECTIVE

Electromagnetic Radiation and Its Effects on Our Everyday Lives

From the preceding discussion of the interaction of electromagnetic radiation with matter—spectroscopy—you might be left with the impression that the utility of such radiation is limited to theoretical studies of atomic structure. Although this is a useful application that has enabled us to learn a great deal about the structure and properties of matter, it is by no means the only application. Useful, everyday applications of the theories of light energy and transmission are all around us. Let's look at just a few examples.

Transmission of sound and pictures is conducted at radio frequencies or radio wavelengths. We are immersed in radio waves from the day we are born. A radio or television is our "detector" of these waves. Radio waves are believed to cause no physical harm because of their very low energy, although some concern for people who live very close to transmission towers has resulted from recent research.

X rays are electromagnetic radiation, and they travel at the speed of light just like radio waves. However, because of their higher energy, they can pass through the human body and leave an image of the body's interior on a photographic film. X-ray photographs are invaluable for medical diagnosis. However, caution is advised in exposing oneself to X rays, because the high energy can remove electrons from biological molecules, causing subtle and potentially harmful changes in their chemistry.

The sunlight that passes through our atmosphere provides the basis for a potentially useful technology for providing heat and electricity: *solar energy.* Light is captured by absorbers, referred to as *solar collectors,* which convert the light energy into heat energy. This heat can be transferred to water circulating beneath the collectors to provide heat and hot water for homes or industry. Wafers of a silicon-based material can convert light energy to electrical energy; many believe that if the efficiency of these processes can be improved, such approaches may provide at least a partial solution to the problems of rising energy costs and pollution associated with our fossil fuel–based energy economy.

The intensity of infrared radiation from a solid or liquid is an indicator of relative temperature. This has been used to advantage in the design of infrared cameras, which can obtain images without the benefit of the visible light that is necessary for conventional cameras. The infrared photograph shows the coastline surrounding the city of San Francisco.

Max Planck noted that in certain situations, energy possessed particlelike properties. A French physicist, Louis deBroglie, hypothesized that the reverse could be true as well: Electrons could, at times, behave as waves rather than particles. This is known today as deBroglie's *wave–particle duality.*

Werner Heisenberg, a German physicist, building on deBroglie's hypothesis, argued that it would be impossible to exactly specify the location of a particle (such as the electron) because of its wavelike character (a wave travels indefinitely in space in contrast to a particle that has fixed dimensions). This hypothesis in turn led to the *Heisenberg uncertainty principle* (1927), which states that it is impossible to specify both the location and the momentum (momentum is the product of mass and velocity) of an electron in an atom at the same time.

The work of deBroglie and Heisenberg represents a departure from the Bohr theory and paved the way for the development of modern atomic theory. Although Bohr's concept of principal energy levels is still valid, restriction of electrons to fixed orbits is too rigorous in light of Heisenberg's principle. All current evidence shows that electrons do *not,* in fact, orbit the nucleus. We now speak of

Microwave radiation for cooking, *infrared* lamps for heating and remote sensing, *ultraviolet* lamps used to kill microorganisms on environmental surfaces, *gamma radiation* from nuclear waste, the *visible* light from the lamp you are using to read this chapter—all are forms of the same type of energy that, for better or worse, plays such a large part in our twenty-first century technological society.

Electromagnetic radiation and spectroscopy also play a vital role in the field of diagnostic medicine. They are routinely used as diagnostic and therapeutic tools in the detection and treatment of disease.

The radiation therapy used in the treatment of many types of cancer has been responsible for saving many lives and extending the span of many others. When radiation is used as a treatment, it destroys cancer cells. This topic will be discussed in detail in Chapter 10.

As a diagnostic tool, spectroscopy has the benefit of providing data quickly and reliably; it can also provide information that might not be available through any other means. Additionally, spectroscopic procedures are often nonsurgical, outpatient procedures. Such procedures are safer, can be more routinely performed, and are more acceptable to the general public than surgical procedures. The potential cost savings because of the elimination of many unnecessary surgical procedures is an added benefit.

The most commonly practiced technique uses the CT scanner, an acronym for *computer-accentuated tomography.* In this technique, X rays are directed at the tissue of interest. As the X rays pass through the tissue, detectors surrounding the tissue gather the signal, compare it to the original X-ray beam, and, using the computer, produce a three-dimensional image of the tissue.

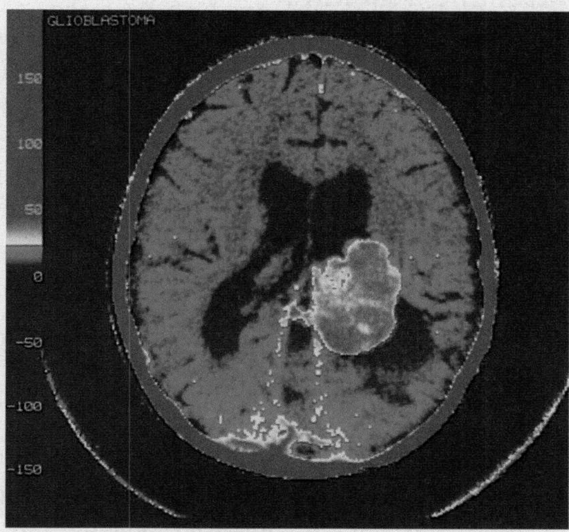

An image of a tumor detected by a CT scan.

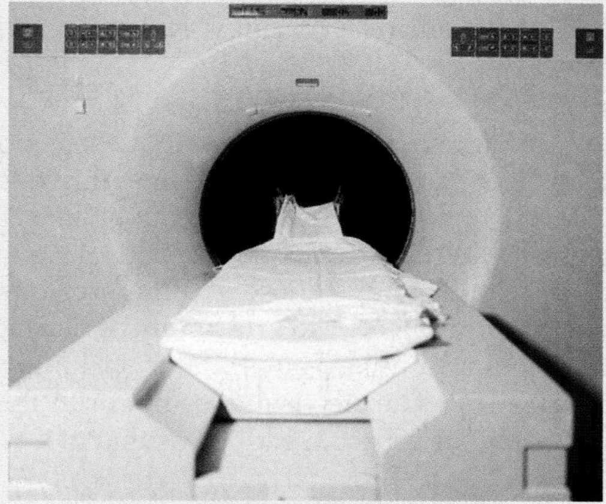
The CT scanner is a device used for diagnostic purposes.

the *probability* of finding an electron in a *region* of space within the principal energy level, referred to as an **atomic orbital.** The rapid movement of the electron spreads the charge into a *cloud* of charge. This cloud is more dense in certain regions, the **electron density** being proportional to the probability of finding the electron at any point in time. Insofar as these atomic orbitals are part of the principal energy levels, they are referred to as sublevels. In Chapter 3 we will see that the orbital model of the atom can be used to predict how atoms can bond together to form compounds. Furthermore, electron arrangement in orbitals enables us to predict various chemical and physical properties of these compounds.

Question 2.5

What was deBroglie's new way of considering matter?

Question 2.6

Why is it not possible to know both the exact energy and the location of an electron?

Summary

2.1 Matter and Structure

Observation of properties has enabled scientists to design models of matter that explain behavior. The sophistication and predictive properties of such models has increased markedly over the last two hundred years. This has enabled a technological revolution. We are reminded that this evolutionary process continues today. Theories are continually undergoing modification and refinement.

2.2 Composition of the Atom

The basic structural unit of an element is the *atom,* which is the smallest unit of an element that retains the chemical properties of that element. The atom is composed of three primary particles: the electron, the proton, and the neutron.

The atom has two distinct regions. The *nucleus* is a small, dense, positively charged region in the center of the atom composed of positively charged *protons* and uncharged *neutrons.* Surrounding the nucleus is a diffuse region of negative charge occupied by *electrons,* the source of the negative charge. Electrons are very low in mass in comparison to protons and neutrons.

The *atomic number* (*Z*) is equal to the number of protons in the atom. The *mass number* (*A*) is equal to the sum of the protons and neutrons (the mass of the electrons is insignificant).

Isotopes are atoms of the same element that have different masses because they have different numbers of neutrons (different mass numbers). Isotopes have chemical behavior identical to that of any other isotope of the same element.

Ions are electrically charged particles that result from a gain or loss of one or more electrons by the parent atom. *Anions,* negative ions, are formed by a gain of one or more electrons by the parent atom. *Cations,* positive ions, are formed by a loss of one or more electrons from the parent atom.

2.3 Development of Atomic Theory

The first experimentally based theory of atomic structure was proposed by John Dalton. Although Dalton pictured atoms as indivisible, the experiments of William Crookes, Eugene Goldstein, and J. J. Thomson indicated that the atom is composed of charged particles: protons and electrons. The third fundamental atomic particle is the neutron. An experiment conducted by Hans Geiger led Ernest Rutherford to propose that the majority of the mass and positive charge of the atom is located in a small, dense region, the *nucleus,* with small, negatively charged electrons occupying a much larger, diffuse space outside of the nucleus.

2.4 The Relationship between Light and Atomic Structure

The study of the interaction of *light* and *matter* is termed *spectroscopy.* Light, *electromagnetic radiation,* travels at a speed of 3.0×10^8 m/s, the *speed of light.* Light is made up of many wavelengths. Collectively, they comprise the *electromagnetic spectrum.* Samples of elements emit certain wavelengths of light when an electrical current is passed through the sample. Different elements emit a different pattern (different wavelengths) of light.

2.5 The Bohr Atom

Niels Bohr proposed an atomic model that described the atom as a nucleus surrounded by fixed *energy levels* (or *quantum levels*) that can be occupied by electrons. He believed that each level was defined by a spherical *orbit* located at a specific distance from the nucleus.

Promotion and relaxation processes are referred to as *electronic transitions.* Electron promotion resulting from absorption of energy results in an *excited state* atom; the process of relaxation allows the atom to return to the *ground state* by emitting a certain wavelength of light.

2.6 Modern Atomic Theory

The modern view of the atom describes the probability of finding an electron in a region of space within the principal energy level, referred to as an *atomic orbital.* The rapid movement of the electrons spreads them into a cloud of charge. This cloud is more dense in certain regions, the density being proportional to the probability of finding the electron at any point in time. The orbital is strikingly different from Bohr's orbit. The electron does not orbit the nucleus; rather, its behavior is best described as that of a wave.

Key Terms

anion (2.2)
anode (2.3)
atom (2.2)
atomic mass (2.2)
atomic number (2.2)
atomic orbital (2.6)
cathode (2.3)
cathode rays (2.3)
cation (2.2)
electromagnetic radiation (2.4)
electromagnetic spectrum (2.4)
electron (2.2)
electron density (2.6)
electronic transitions (2.5)
energy level (2.5)
excited state (2.5)
ground state (2.5)
ion (2.2)
isotope (2.2)
mass number (2.2)
natural radioactivity (2.3)
neutron (2.2)
nucleus (2.2)
orbit (2.5)

promotion (2.5)
proton (2.2)
quantization (2.5)
quantum levels (2.5)
quantum number (2.5)
relaxation (2.5)
spectroscopy (2.4)
speed of light (2.4)

Questions and Problems

Matter and Structure

2.7 Make a list of everyday items that were developed (or developed more rapidly) because of our understanding of atomic structure and composition.

2.8 Make a list of technological developments in your major area of study that were developed as a consequence of our understanding of atomic structure and composition.

Composition of the Atom

2.9 In what way(s) are protons and neutrons similar?

2.10 In what way(s) are protons and neutrons different?

2.11 Calculate the number of protons, neutrons, and electrons in:
a. ${}^{16}_{8}\text{O}$
b. ${}^{31}_{15}\text{P}$

2.12 Calculate the number of protons, neutrons, and electrons in:
a. ${}^{136}_{56}\text{Ba}$
b. ${}^{209}_{84}\text{Po}$

2.13 Why are isotopes useful in medicine?

2.14 Describe the similarities and differences among carbon-12, carbon-13, and carbon-14.

2.15 State the mass and charge of the:
a. electron
b. proton
c. neutron

2.16 Calculate the number of protons, neutrons, and electrons in:
a. ${}^{37}_{17}\text{Cl}$
b. ${}^{23}_{11}\text{Na}$
c. ${}^{84}_{36}\text{Kr}$

2.17 a. What is an ion?
b. What process results in the formation of a cation?
c. What process results in the formation of an anion?

2.18 a. What are isotopes?
b. What is the major difference among isotopes of an element?
c. What is the major similarity among isotopes of an element?

2.19 How many protons are in the nucleus of the isotope Rn-220?

2.20 How many neutrons are in the nucleus of the isotope Rn-220?

2.21 Selenium-80 is a naturally occurring isotope. It is found in over-the-counter supplements.
a. How many protons are found in one atom of selenium-80?
b. How many neutrons are found in one atom of selenium-80?

2.22 Iodine-131 is an isotope used in thyroid therapy.
a. How many protons are found in one atom of iodine-131?
b. How many neutrons are found in one atom of iodine-131?

2.23 Write symbols for each isotope:
a. Each atom contains 1 proton and 0 neutrons.
b. Each atom contains 6 protons and 8 neutrons.

2.24 Write symbols for each isotope:
a. Each atom contains 1 proton and 2 neutrons.
b. Each atom contains 92 protons and 146 neutrons.

2.25 In what way do isotopes of the same element differ?

2.26 In what way do atoms of different elements differ?

2.27 Fill in the blanks:

Symbol	No. of Protons	No. of Neutrons	No. of Electrons	Charge
Example:				
${}^{40}_{20}\text{Ca}$	20	20	20	0
${}^{23}_{11}\text{Na}$	11	_____	11	0
${}^{32}_{16}\text{S}^{2-}$	16	16	_____	2−
_____	8	8	8	0
${}^{24}_{12}\text{Mg}^{2+}$	_____	12	_____	2+
_____	19	20	18	_____

2.28 Fill in the blanks:

Atomic Symbol	No. of Protons	No. of Neutrons	No. of Electrons	Charge
Example:				
${}^{27}_{13}\text{Al}$	13	14	13	0
${}^{39}_{19}\text{K}$	19	_____	19	0
${}^{31}_{15}\text{P}^{3-}$	15	16	_____	_____
_____	29	34	27	2+
${}^{55}_{26}\text{Fe}^{2+}$	_____	29	_____	2+
_____	8	8	10	_____

2.29 Fill in the blanks:
a. An isotope of an element differs in mass because the atom has a different number of _____.
b. The atomic number gives the number of _____ in the nucleus.
c. The mass number of an atom is due to the number of _____ and _____ in the nucleus.
d. A charged atom is called a(n) _____.
e. Electrons surround the _____ and have a _____ charge.

2.30 Label each of the following statements as true or false:
a. An atom with an atomic number of 7 and a mass of 14 is identical to an atom with an atomic number of 6 and a mass of 14.
b. Neutral atoms have the same number of electrons as protons.
c. The mass of an atom is due to the sum of the number of protons, neutrons, and electrons.

Development of Atomic Theory

2.31 What are the major postulates of Dalton's atomic theory?

2.32 What points of Dalton's theory are no longer current?

2.33 Note the major accomplishment of each of the following:
a. Chadwick
b. deBroglie

2.34 Note the major accomplishment of each of the following:
a. Geiger
b. Bohr

2.35 Note the major accomplishment of each of the following:
a. Dalton
b. Crookes

2.36 Note the major accomplishment of each of the following:
a. Thomson
b. Rutherford

2.37 Describe the experiment that provided the basis for our understanding of the nucleus.

2.38 Describe the process that occurs when electrical energy is applied to a sample of hydrogen gas.
2.39 Describe the series of experiments that characterized the electron.
2.40 List at least three properties of the electron.
2.41 What is a cathode ray? Which subatomic particle is detected?
2.42 Pictured is a cathode ray tube. Show the path that an electron would follow in the tube.

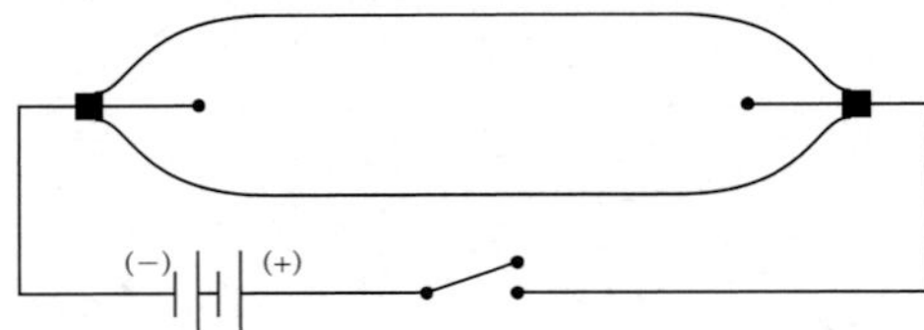

The Relationship between Light and Atomic Structure

2.43 Rank the various regions of the electromagnetic spectrum in order of increasing wavelength.
2.44 Rank the various regions of the electromagnetic spectrum in order of increasing energy.
2.45 Which form of radiation has greater energy, microwave or infrared?
2.46 Which form of radiation has the longer wavelength, ultraviolet or infrared?
2.47 What is meant by the term *spectroscopy?*
2.48 What is meant by the term *electromagnetic spectrum?*

The Bohr Atom

2.49 Critique this statement: Electrons can exist in any position outside of the nucleus.
2.50 Critique this statement: Promotion of electrons is accompanied by a release of energy.
2.51 What are the most important points of the Bohr theory?
2.52 Give two reasons why the Bohr theory did not stand the test of time.
2.53 What was the major contribution of Bohr's atomic model?
2.54 What was the major deficiency of Bohr's atomic model?

Modern Atomic Theory

2.55 Describe the meaning of the deBroglie hypothesis.
2.56 Describe the meaning of Heisenberg's Uncertainty Principle.
2.57 What is meant by the term *electron density?*
2.58 How do *orbits* and *orbitals* differ?

Critical Thinking Problems

1. A natural sample of chromium, taken from the ground, will contain four isotopes: Cr-50, Cr-52, Cr-53, and Cr-54. Predict which isotope is in greatest abundance. Explain your reasoning.
2. Copper, silver, and gold are termed coinage metals; they are often used to mint coins (see Figure 5.1). List all the chemical and physical properties that you believe would make these elements suitable for coins. Now, choose three metals that would not be suitable for coinage, and explain why.
3. Rutherford's theory of the nucleus was based on the measurement of the results of a series of interactions. Explain how the process of reading this page involves similar principles.
4. Crookes's cathode ray tube experiment inadvertently supplied the basic science for a number of modern high-tech devices. List a few of these devices and describe how they involve one or more aspects of this historic experiment.

Organization and understanding go hand-in-hand.

3

Elements, Atoms, Ions, and the Periodic Table

GENERAL CHEMISTRY

Outline

Learning Goals

1. Recognize the important subdivisions of the periodic table: periods, groups (families), metals, and nonmetals.
2. Use the periodic table to obtain information about an element.
3. Describe the relationship between the electronic structure of an element and its position in the periodic table.
4. Write electron configurations for atoms of the most commonly occurring elements.
5. Know the meaning of the octet rule and its predictive usefulness.
6. Use the octet rule to predict the charge of common cations and anions.
7. Utilize the periodic table and its predictive power to estimate the relative sizes of atoms and ions, as well as relative magnitudes of ionization energy and electron affinity.
8. Use values of ionization energies and electron affinities to predict ion formation.

CHEMISTRY CONNECTION

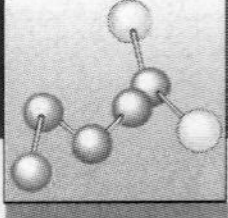

Managing Mountains of Information

Recall for a moment the first time that you sat down in front of a computer. Perhaps it was connected to the Internet; somewhere in its memory was a word processor program, a spreadsheet, a few games, and many other features with strange-sounding names. Your challenge, very simply, was to use this device to access and organize information. Several manuals, all containing hundreds of pages of bewilderment, were your only help. How did you overcome this seemingly impossible task?

We are quite sure that you did not succeed without doing some reading and talking to people who had experience with computers. Also, you did not attempt to memorize every single word in each manual.

Success with a computer or any other storehouse of information results from developing an overall understanding of the way in which the system is organized. Certain facts must be memorized, but seeing patterns and using these relationships allows us to accomplish a wide variety of tasks that involve similar logic.

The study of chemistry is much like "real life." Just as it is impossible to memorize every single fact that will allow you to run a computer or drive an automobile in traffic, it is equally impossible to learn every fact in chemistry. Knowing the organization and logic of a process, along with a few key facts, makes a task manageable.

One powerful organizational device in chemistry is the periodic table. Its use in organizing and predicting the behavior of all of the known elements (and many of the compounds formed from these elements) is the subject of this chapter.

Introduction

We discussed some of the early experiments that established the existence of fundamental atomic particles (protons, neutrons, and electrons) and their relationship within the atom in Chapter 2. Let us now consider the relationships among the elements themselves. The unifying concept is called the *periodic law,* and it gives rise to an organized "map" of the elements that relates their structure to their chemical and physical properties. This "map" is the *periodic table.*

As we study the periodic law and periodic table, we shall see that the chemical and physical properties of elements follow directly from the electronic structure of the atoms that make up these elements. A thorough familiarity with the arrangement of the periodic table is vital to the study of chemistry. It not only allows us to predict the structure and properties of the various elements, but it also serves as the basis for developing an understanding of chemical bonding, or the process of forming molecules. Additionally, the properties and behavior of these larger units on a macroscopic scale (bulk properties) are fundamentally related to the properties of the atoms that comprise them.

3.1 The Periodic Law and the Periodic Table

Learning Goal

In 1869, Dmitri Mendeleev, a Russian, and Lothar Meyer, a German, working independently, found ways of arranging elements in order of increasing atomic mass such that elements with similar properties were grouped together in a *table of elements.* The **periodic law** is embodied by Mendeleev's statement, "the elements if arranged according to their atomic weights (masses), show a distinct *periodicity* (regular variation) of their properties." The *periodic table* (Figure 3.1) is a visual representation of the periodic law.

Chemical and physical properties of elements correlate with the electronic structure of the atoms that make up these elements. In turn, the electronic structure correlates with position on the periodic table.

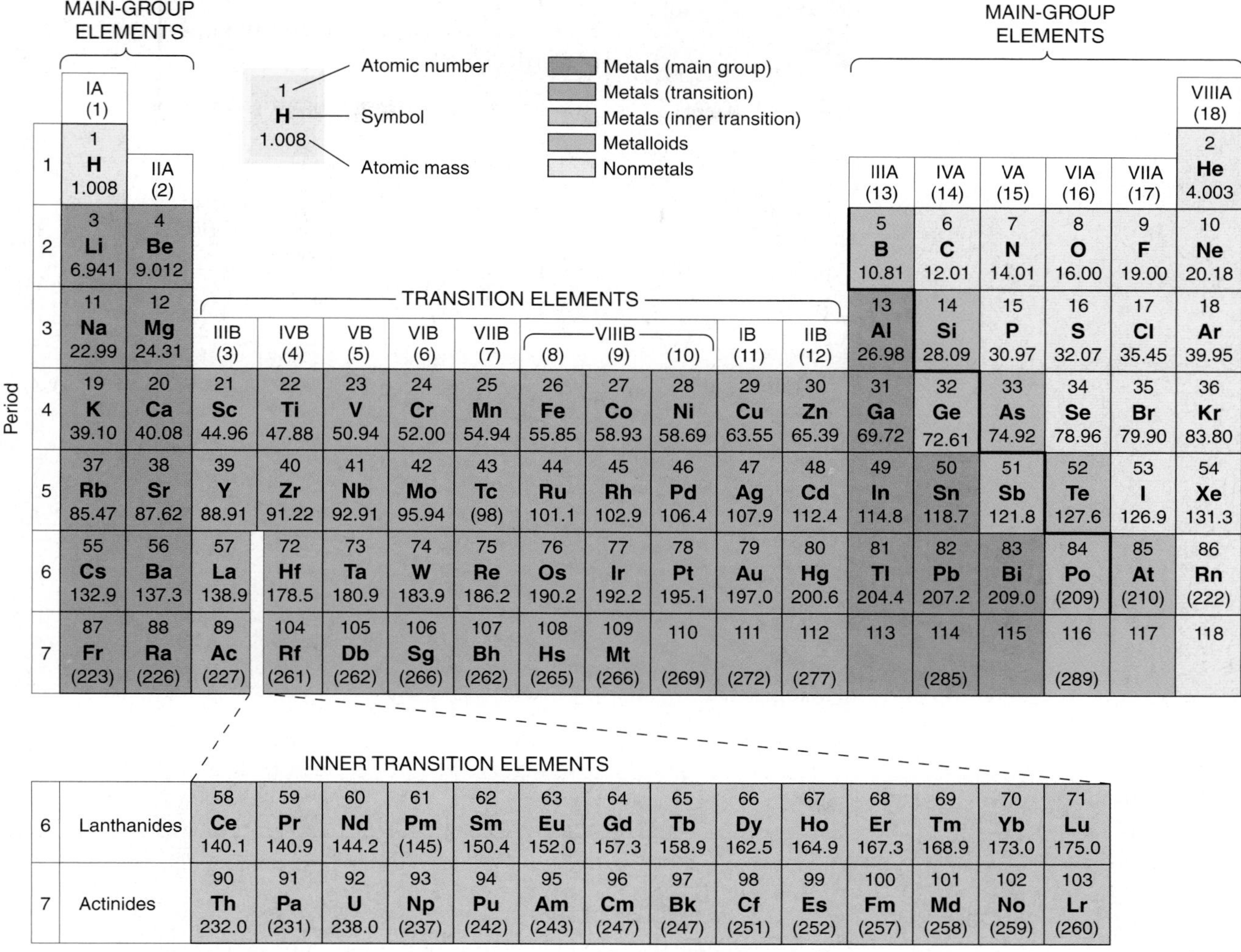

Figure 3.1
Classification of the elements: the periodic table.

A thorough familiarity with the arrangement of the periodic table allows us to predict electronic structure and physical and chemical properties of the various elements. It also serves as the basis for understanding chemical bonding.

The concept of "periodicity" may be illustrated by examining a portion of the modern periodic table (see Figure 3.1). The elements in the second row (beginning with lithium, Li, and proceeding to the right) show a marked difference in properties. However, sodium (Na) has properties similar to those of lithium, and sodium is therefore placed below lithium; once sodium is fixed in this position, the elements Mg through Ar have properties remarkably similar (though not identical) to those of the elements just above them. The same is true throughout the complete periodic table.

Mendeleev arranged the elements in his original periodic table in order of increasing atomic mass. However, as our knowledge of atomic structure increased, atomic numbers became the basis for the organization of the table.

The modern periodic law states that *the physical and chemical properties of the elements are periodic functions of their atomic numbers.* If we arrange the elements in order of increasing number of protons, the properties of the elements repeat at regular intervals.

Not all of the elements are of equal importance to an introductory study of chemistry. Table 3.1 lists twenty of the elements that are most important to biological systems, along with their symbols and a brief description of their functions.

Table 3.1 Summary of the Most Important Elements in Biological Systems

Element	Symbol	Significance
Hydrogen	H	Components of major biological molecules
Carbon	C	
Oxygen	O	
Nitrogen	N	
Phosphorus	P	
Sulfur	S	
Potassium	K	Produce electrolytes responsible for fluid balance and nerve transmission
Sodium	Na	
Chlorine	Cl	
Calcium	Ca	Bones, nerve function
Magnesium	Mg	
Zinc	Zn	Essential trace metals in human metabolism
Strontium	Sr	
Iron	Fe	
Copper	Cu	
Cobalt	Co	
Manganese	Mn	
Cadmium	Cd	"Heavy metals" toxic to living systems
Mercury	Hg	
Lead	Pb	

We will use the periodic table as our "map," just as a traveler would use a road map. A short time spent learning how to read the map (and remembering to carry it along on your trip!) is much easier than memorizing every highway and intersection. The information learned about one element relates to an entire family of elements grouped as a recognizable unit within the table.

Numbering Groups in the Periodic Table

The periodic table created by Mendeleev has undergone numerous changes over the years. These modifications occurred as more was learned about the chemical and physical properties of the elements. The labeling of groups with Roman numerals followed by the letter *A* (representative elements) or *B* (transition elements) was standard, until 1983, in North America and Russia. However, in other parts of the world, the letters *A* and *B* were used in a different way. Consequently, two different periodic tables were in widespread use. This certainly created some confusion.

The International Union of Pure and Applied Chemistry (IUPAC), in 1983, recommended that a third system, using numbers 1–18 to label the groups, replace both of the older systems. Unfortunately, multiple systems now exist and this can cause confusion for both students and experienced chemists.

The periodic tables in this textbook are "double labeled." Both the old (Roman numeral) and new (1–18) systems are used to label the groups. The label that you use is simply a guide to reading the table; the real source of information is in the structure of the table itself. The following sections will show you how to extract useful information from this structure.

Periods and Groups

Learning Goal

A **period** is a horizontal row of elements in the periodic table. The periodic table consists of seven periods containing 2, 8, 8, 18, 18, and 32 elements. The seventh period is still incomplete but potentially holds 32 elements. Note that the **lanthanide series,** a collection of 14 elements that are chemically and physically similar to the element lanthanum, is a part of period six. It is written separately for convenience of presentation and is inserted between lanthanum (La), atomic number 57, and hafnium (Hf), atomic number 72. Similarly, the **actinide series,** consisting of 14 elements similar to the element actinium, is inserted between actinium, atomic number 89, and rutherfordium, atomic number 104.

Groups or *families* are columns of elements in the periodic table. The elements of a particular group or family share many similarities, as in a human family. The similarities extend to physical and chemical properties that are related to similarities in electronic structure (that is, the way in which electrons are arranged in an atom).

Group A elements are called **representative elements,** and Group B elements are **transition elements.** Certain families also have common names. For example, Group IA (or 1) elements are also known as the **alkali metals;** Group IIA (or 2), the **alkaline earth metals;** Group VIIA (or 17), the **halogens;** and Group VIIIA (or 18), the **noble gases.**

Many metals, as positive ions, are essential nutrients in biological systems. A Medical Perspective: Copper Deficiency and Wilson's Disease gives but one example.

A **metal** is a substance whose atoms tend to lose electrons during chemical change, forming positive ions. A **nonmetal,** on the other hand, is a substance whose atoms may gain electrons, forming negative ions.

Metals and Nonmetals

Learning Goal

A closer inspection of the periodic table reveals a bold zigzag line running from top to bottom, beginning to the left of boron (B) and ending between polonium (Po) and astatine (At). This line acts as the boundary between *metals,* to the left, and *nonmetals,* to the right. Elements straddling the boundary have properties intermediate between those of metals and nonmetals. These elements are referred to as **metalloids.** The metalloids include boron (B), silicon (Si), germanium (Ge), arsenic (As), antimony (Sb), tellurium (Te), polonium (Po), and astatine (At).

Note that aluminum (Al) is classified as a metal, not a metalloid.

Atomic Number and Atomic Mass

Learning Goal

The atomic number is the number of protons in the nucleus of an atom of an element. It also corresponds to the nuclear charge, the positive charge from the nucleus. Both the atomic number and atomic mass of each element are readily available from the periodic table. For example,

20 ⟵ atomic number
Ca ⟵ symbol
calcium ⟵ name
40.08 ⟵ atomic mass

More detailed periodic tables may also include such information as the electron arrangement, relative sizes of atoms and ions, and most probable ion charges.

Question 3.1

Refer to the periodic table (Figure 3.1) and find the following information:

a. the symbol of the element with an atomic number of 40
b. the mass of the element sodium (Na)
c. the element whose atoms contain 24 protons
d. the known element that should most resemble the as-yet undiscovered element with an atomic number of 117

A MEDICAL PERSPECTIVE

Copper Deficiency and Wilson's Disease

An old adage tells us that we should consume all things in moderation. This is very true of many of the trace minerals, such as copper. Too much copper in the diet causes toxicity and too little copper results in a serious deficiency disease.

Copper is extremely important for the proper functioning of the body. It aids in the absorption of iron from the intestine and facilitates iron metabolism. It is critical for the formation of hemoglobin and red blood cells in the bone marrow. Copper is also necessary for the synthesis of collagen, a protein that is a major component of the connective tissue. It is essential to the central nervous system in two important ways. First, copper is needed for the synthesis of norepinephrine and dopamine, two chemicals that are necessary for the transmission of nerve signals. Second, it is required for the deposition of the myelin sheath (a layer of insulation) around nerve cells. Release of cholesterol from the liver depends on copper, as does bone development and proper function of the immune and blood clotting systems.

The estimated safe and adequate daily dietary intake (ESADDI) for adults is 1.5–3.0 mg. Meats, cocoa, nuts, legumes, and whole grains provide significant amounts of copper. The accompanying table shows the amount of copper in some common foods.

Copper in One-Cup Portions of Food

Food	Mass of copper (mg)
Sesame seeds	5.88
Cashews	3.04
Oysters	2.88
Sunflower seeds	2.52
Peanuts, roasted	1.85
Crabmeat	1.71
Walnuts	1.28
Almonds	1.22
Cereal, All Bran	0.98
Tuna fish	0.93
Wheat germ	0.70
Prunes	0.69
Kidney beans	0.56
Dried apricots	0.56
Lentils, cooked	0.54
Sweet potato, cooked	0.53
Dates	0.51
Whole milk	0.50
Raisins	0.45
Cereal, C. W. Post, Raisins	0.40
Grape Nuts	0.38
Whole-wheat bread	0.34
Cooked cereal, Roman Meal	0.32

Source: From David C. Nieman, Diane E. Butterworth, and Catherine N. Nieman, *Nutrition*, Revised First Edition. Copyright 1992 Wm. C. Brown Communications, Inc., Dubuque, Iowa. All Rights Reserved. Reprinted by permission.

Although getting enough copper in the diet would appear to be relatively simple, it is estimated that Americans often ingest only marginal levels of copper, and we absorb only 25–40% of that dietary copper. Despite these facts, it appears that copper deficiency is not a serious problem in the United States.

Individuals who are at risk for copper deficiency include people who are recovering from abdominal surgery, which causes decreased absorption of copper from the intestine. Others at risk are premature babies and people who are sustained solely by intravenous feedings that are deficient in copper. In addition, people who ingest high doses of antacids or take excessive supplements of zinc, iron, or vitamin C can develop copper deficiency because of reduced copper absorption. Because copper is involved in so many processes in the body, it is not surprising that the symptoms of copper deficiency are many and diverse. They include anemia; decreased red and white blood cell counts; heart disease; increased levels of serum cholesterol; loss of bone; defects in the nervous system, immune system, and connective tissue; and abnormal hair.

Some of these symptoms are seen among people who suffer from the rare genetic disease known as Menkes' kinky hair syndrome. The symptoms of this disease, which is caused by a defect in the ability to absorb copper from the intestine, include very low copper levels in the serum, kinky white hair, slowed growth, and degeneration of the brain.

Just as too little copper causes serious problems, so does an excess of copper. At doses greater than about 15 mg, copper causes toxicity that results in vomiting. The effects of extended exposure to excess copper are apparent when we look at Wilson's disease. This is a genetic disorder in which excess copper cannot be removed from the body and accumulates in the cornea of the eye, liver, kidneys, and brain. The symptoms include a greenish ring around the cornea, cirrhosis of the liver, copper in the urine, dementia and paranoia, drooling, and progressive tremors. As a result of the condition, the victim generally dies in early adolescence. Wilson's disease can be treated with moderate success if it is recognized early, before permanent damage has occurred to any tissues. The diet is modified to reduce the intake of copper; for instance, such foods as chocolate are avoided. In addition, the drug penicillamine is administered. This compound is related to the antibiotic penicillin but has no antibacterial properties; rather it has the ability to bind to copper in the blood and enhance its excretion by the kidneys into the urine. In this way the brain degeneration and tissue damage that are normally seen with the disease can be lessened.

Question 3.2

Refer to the periodic table (Figure 3.1) and find the following information:

a. the symbol of the noble gas in period 3
b. the lightest element in Group IVA
c. the only metalloid in Group IIIA
d. the element whose atoms contain 18 protons

Question 3.3

For each of the following element symbols, give the name of the element, its atomic number, and its atomic mass:

a. He
b. F
c. Mn

Question 3.4

For each of the following element symbols, give the name of the element, its atomic number, and its atomic mass:

a. Mg
b. Ne
c. Se

3.2 Electron Arrangement and the Periodic Table

Learning Goal

A primary objective of studying chemistry is to understand the way in which atoms join together to form chemical compounds. The most important factor in this *bonding process* is the arrangement of the electrons in the atoms that are combining. The **electronic configuration** describes the arrangement of electrons in atoms. The periodic table is helpful because it provides us with a great deal of information about the electron arrangement or electronic configuration of atoms.

Valence Electrons

If we picture two spherical objects that we wish to join together, perhaps with glue, the glue can be applied to the surface, and the two objects can then be brought into contact. We can extend this analogy to two atoms that are modeled as spherical objects. Although this is not a perfect analogy, it is apparent that the surface interaction is of primary importance. Although the positively charged nucleus and "interior" electrons certainly play a role in bonding, we can most easily understand the process by considering only the outermost electrons. We refer to these as *valance electrons.* **Valence electrons** are the outermost electrons in an atom, which are involved, or have the potential to become involved in the bonding process.

Metals tend to have fewer valence electrons, and nonmetals tend to have more valence electrons.

For representative elements the number of valence electrons in an atom corresponds to the number of the *group* or *family* in which the atom is found. For example, elements such as hydrogen and sodium (in fact, all alkali metals, Group IA or 1) have a valence of 1 (or one valence electron). From left to right in period 2, beryllium, Be (Group IIA or 2), has two valence electrons; boron, B (Group IIIA or 3), has three; carbon, C (Group IVA or 4), has four; and so forth.

We have seen that an atom may have several energy levels, or regions where electrons are located. These energy levels are symbolized by n, the lowest energy level being assigned a value of $n = 1$. Each energy level may contain up to a fixed maximum number of electrons. For example, the $n = 1$ energy level may contain a maximum of two electrons. Thus hydrogen (atomic number = 1) has one electron and helium (atomic number = 2) has two electrons in the $n = 1$ level. Only these elements have electrons *exclusively* in the first energy level:

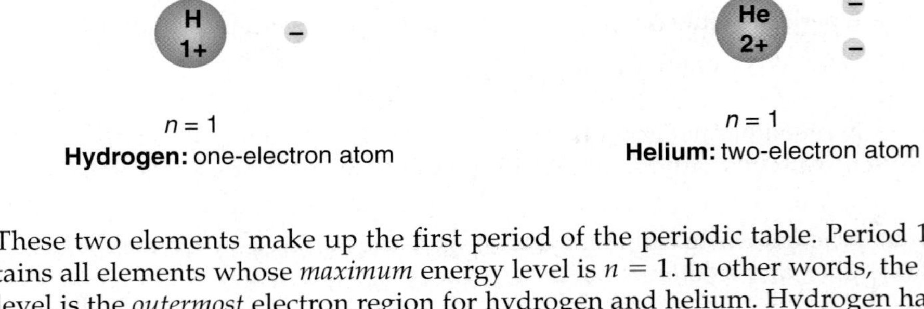

These two elements make up the first period of the periodic table. Period 1 contains all elements whose *maximum* energy level is $n = 1$. In other words, the $n = 1$ level is the *outermost* electron region for hydrogen and helium. Hydrogen has one electron and helium has two electrons in the $n = 1$ level.

The valence electrons of elements in the second period are in the $n = 2$ energy level. (Remember that you must fill the $n = 1$ level with two electrons before adding electrons to the next level.) The third electron of lithium (Li) and the remaining electrons of the second period elements must be in the $n = 2$ level and are considered the valence electrons for lithium and the remaining second period elements.

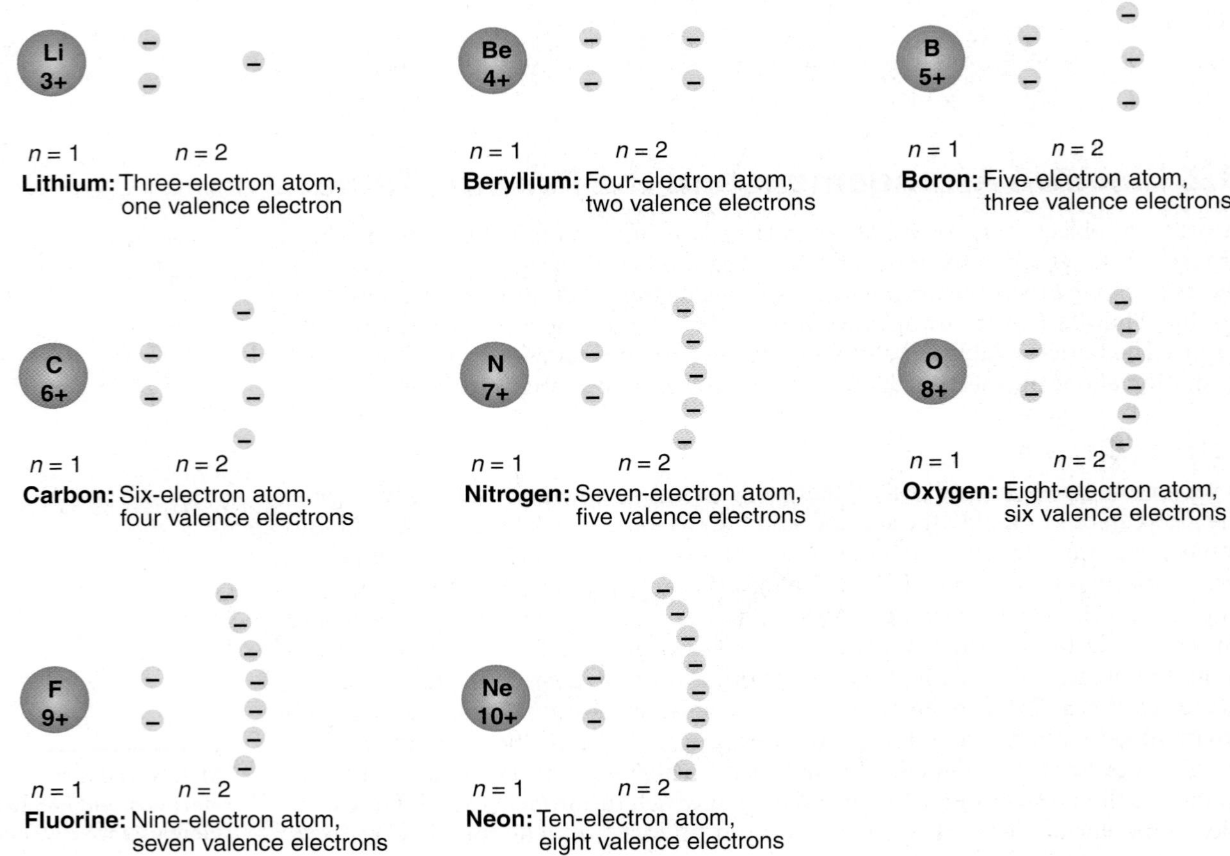

The electron distribution (arrangement) of the first twenty elements of the periodic table is given in Table 3.2.

Two general rules of electron distribution are based on the periodic law:

RULE 1: The number of valence electrons in an atom equals the *group* number for all representative (A group) elements. ■

RULE 2: The energy level ($n = 1, 2$, etc.) in which the valence electrons are located corresponds to the *period* in which the element may be found. ■

Table 3.2 **The Electron Distribution for the First Twenty Elements of the Periodic Table**

Element Symbol and Name	Total Number of Electrons (Valence Electrons)	Electrons in $n = 1$	Electrons in $n = 2$	Electrons in $n = 3$	Electrons in $n = 4$
H, hydrogen	1 (1)	1	0	0	0
He, helium	2 (2)	2	0	0	0
Li, lithium	3 (1)	2	1	0	0
Be, beryllium	4 (2)	2	2	0	0
B, boron	5 (3)	2	3	0	0
C, carbon	6 (4)	2	4	0	0
N, nitrogen	7 (5)	2	5	0	0
O, oxygen	8 (6)	2	6	0	0
F, fluorine	9 (7)	2	7	0	0
Ne, neon	10 (8)	2	8	0	0
Na, sodium	11 (1)	2	8	1	0
Mg, magnesium	12 (2)	2	8	2	0
Al, aluminum	13 (3)	2	8	3	0
Si, silicon	14 (4)	2	8	4	0
P, phosphorus	15 (5)	2	8	5	0
S, sulfur	16 (6)	2	8	6	0
Cl, chlorine	17 (7)	2	8	7	0
Ar, argon	18 (8)	2	8	8	0
K, potassium	19 (1)	2	8	8	1
Ca, calcium	20 (2)	2	8	8	2

For example,

Group IA	*Group IIA*	*Group IIIA*	*Group VIIA*
Li	Ca	Al	Br
one valence electron in $n = 2$ energy level; period 2	two valence electrons in $n = 4$ energy level; period 4	three valence electrons in $n = 3$ energy level; period 3	seven valence electrons in $n = 4$ energy level; period 4

Determining Electron Arrangement — EXAMPLE 3.1

Provide the total number of electrons, total number of valence electrons, and energy level in which the valence electrons are found for the silicon (Si) atom.

Solution

Step 1. Determine the position of silicon in the periodic table. Silicon is found in Group IVA and period 3 of the table. Silicon has an atomic number of 14.

Continued—

EXAMPLE 3.1 —*Continued*

***Step* 2.** The atomic number provides the number of electrons in an atom. Silicon therefore has 14 electrons.

***Step* 3.** Because silicon is in Group IV, only 4 of the 14 electrons are valence electrons.

***Step* 4.** Silicon has 2 electrons in $n = 1$, 8 electrons in $n = 2$, and 4 electrons in the $n = 3$ level.

Question 3.5

For each of the following elements, provide the *total* number of electrons and *valence* electrons in its atom:

a. Na
b. Mg
c. S
d. Cl
e. Ar

Question 3.6

For each of the following elements, provide the *total* number of electrons and *valence* electrons in its atom:

a. K
b. F
c. P
d. O
e. Ca

The Quantum Mechanical Atom

As we noted at the end of Chapter 2, the success of Bohr's theory was short-lived. Emission spectra of multi-electron atoms (recall that the hydrogen atom has only one electron) could not be explained by Bohr's theory. DeBroglie's statement that electrons have wave properties served to intensify the problem. Bohr stated that electrons in atoms had very specific locations. The very nature of waves, spread out in space, defies such an exact model of electrons in atoms. Furthermore, the exact model is contradictory to Heisenberg's Uncertainty Principle.

The basic concept of the Bohr theory, that the energy of an electron in an atom is quantized, was refined and expanded by an Austrian physicist, Erwin Schröedinger. He described electrons in atoms in probability terms, developing equations that emphasize the wavelike character of electrons. Although Schröedinger's approach was founded on complex mathematics, we can readily use models of electron probability regions that enable us to gain a reasonable insight into atomic structure without the need to understand the underlying mathematics.

Schröedinger's theory, often described as quantum mechanics, incorporates Bohr's principal energy levels ($n = 1, 2$, and so forth); however, it is proposed that each of these levels is made up of one or more sublevels. Each sublevel, in turn, contains one or more atomic orbitals. In the following section we shall look at each of these regions in more detail and learn how to predict the way that electrons are arranged in stable atoms.

Energy Levels and Sublevels

Principal Energy Levels

The principal energy levels are designated $n = 1, 2, 3$, and so forth. The number of possible sublevels in a principal energy level is also equal to n. When $n = 1$, there can be only one sublevel; $n = 2$ allows two sublevels, and so forth.

The total electron capacity of a principal level is $2(n)^2$. For example:

$n = 1$	$2(1)^2$	Capacity $= 2e^-$
$n = 2$	$2(2)^2$	Capacity $= 8e^-$
$n = 3$	$2(3)^2$	Capacity $= 18e^-$

Sublevels

The sublevels, or subshells, are symbolized as s, p, d, f, and so forth; they increase in energy in the following order:

$$s < p < d < f$$

We specify both the principal energy level and type of sublevel when describing the location of an electron—for example, 1*s*, 2*s*, 2*p*. Energy level designations for the first four principal energy levels follow:

- The first principal energy level ($n = 1$) has one possible sublevel: 1*s*.
- The second principal energy level ($n = 2$) has two possible sublevels: 2*s* and 2*p*.
- The third principal energy level ($n = 3$) has three possible sublevels: 3*s*, 3*p*, and 3*d*.
- The fourth principal energy level ($n = 4$) has four possible sublevels: 4*s*, 4*p*, 4*d*, and 4*f*.

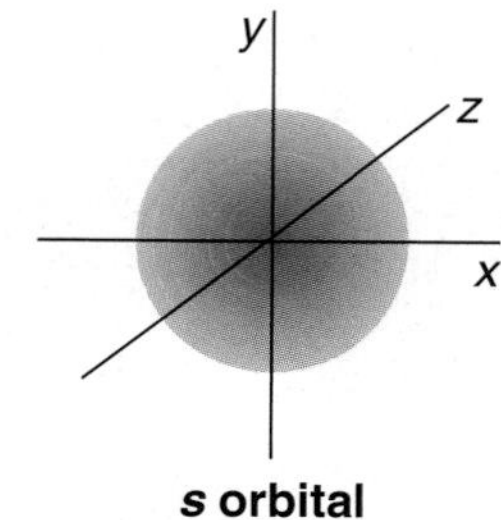

Figure 3.2
Representation of an *s* orbital.

Orbitals

An **orbital** is a specific region of a sublevel containing a maximum of two electrons.

Figure 3.2 depicts a model of an *s* orbital. It is spherically symmetrical, much like a Ping-Pong ball. Its volume represents a region where there is a high probability of finding electrons of similar energy. A close inspection of Figure 3.2 shows that this probability decreases as we approach the outer region of the atom (the decreasing color density in the model represents a decrease in the electron density). The nucleus is at the center of the *s* orbital. At that point the probability of finding the electron is zero; electrons cannot reside in the nucleus. Only one *s* orbital can be found in any *n* level. Atoms with many electrons, occupying a number of *n* levels, have an *s* orbital in each *n* level. Consequently 1*s*, 2*s*, 3*s*, and so forth are possible orbitals.

The shape and orientation of atomic orbitals strongly influence the structure and properties of compounds.

Figure 3.3 describes the shapes of the three possible *p* orbitals within a given level. Each has the same shape, and that shape appears much like a dumbbell; these three orbitals differ only in the direction they extend into space. Imaginary coordinates *x*, *y*, and *z* are superimposed on these models to emphasize this fact. These three orbitals, termed p_x, p_y, and p_z, may coexist in a single atom. Their arrangement is shown in Figure 3.4.

In a similar fashion, five possible *d* orbitals and seven possible *f* orbitals exist. The *d* orbitals exist only in $n = 3$ and higher principal energy levels; *f* orbitals exist only in $n = 4$ and higher principal energy levels. Because of their complexity, we will not consider the shapes of *d* and *f* orbitals.

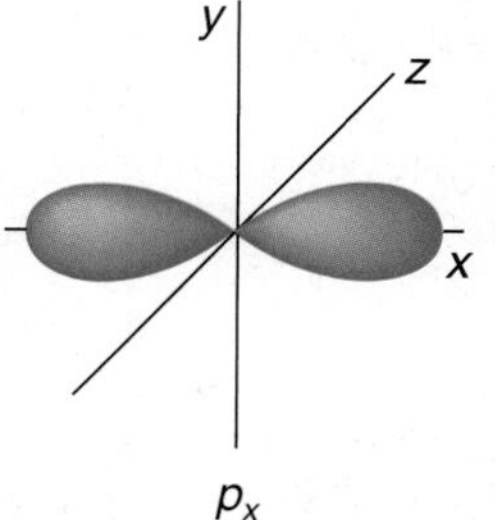

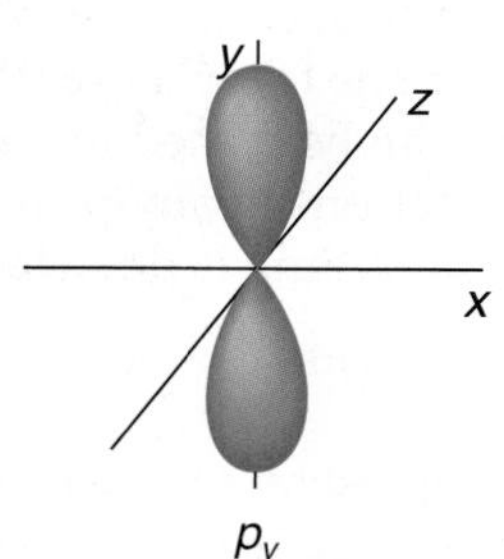

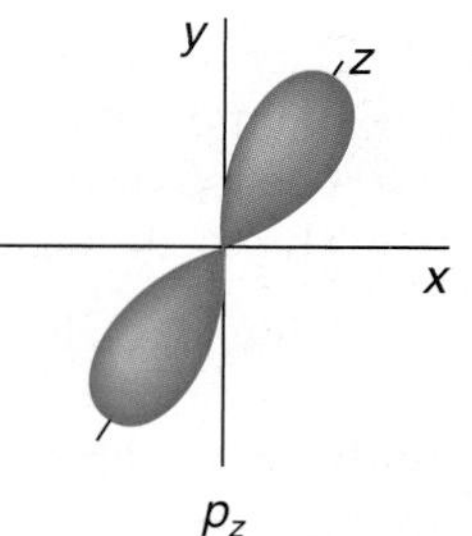

Figure 3.3
Representation of the three *p* orbitals, p_x, p_y, and p_z.

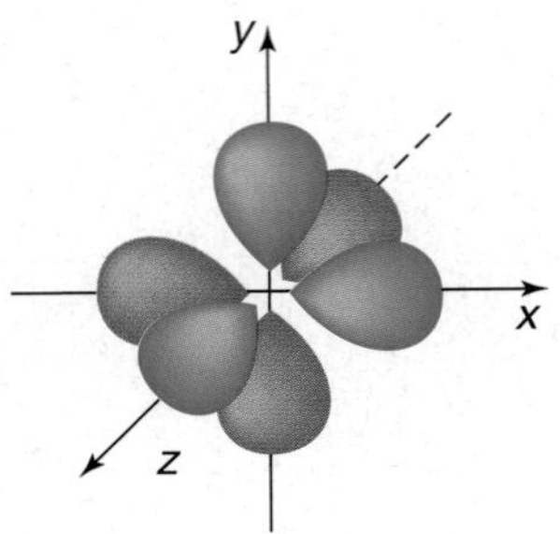

Figure 3.4
The orientation of three *p* orbitals in space.

Electrons in Sublevels

We can deduce the maximum electron capacity of each sublevel based on the information just given.

For the *s* sublevel:

$$1 \text{ }\cancel{\text{orbital}} \times \frac{2e^- \text{ capacity}}{\cancel{\text{orbital}}} = 2e^- \text{ capacity}$$

For the *p* sublevel:

$$3 \text{ }\cancel{\text{orbitals}} \times \frac{2e^- \text{ capacity}}{\cancel{\text{orbital}}} = 6e^- \text{ capacity}$$

For the *d* sublevel:

$$5 \text{ }\cancel{\text{orbitals}} \times \frac{2e^- \text{ capacity}}{\cancel{\text{orbital}}} = 10e^- \text{ capacity}$$

For the *f* sublevel:

$$7 \text{ }\cancel{\text{orbitals}} \times \frac{2e^- \text{ capacity}}{\cancel{\text{orbital}}} = 14e^- \text{ capacity}$$

Section 2.3 discusses the properties of electrons demonstrated by Thomson.

Electron Spin

As we have noted, each atomic orbital has a maximum capacity of two electrons. The electrons are perceived to *spin* on an imaginary axis, and the two electrons in the same orbital must have opposite spins: clockwise and counterclockwise. Their behavior is analogous to two ends of a magnet. Remember, electrons have magnetic properties. The electrons exhibit sufficient magnetic attraction to hold themselves together despite the natural repulsion that they "feel" for each other, owing to their similar charge (remember, like charges repel). Electrons must therefore have opposite spins to coexist in an orbital. A pair of electrons in one orbital that possess opposite spins are referred to as *paired* electrons.

Electron Configuration and the Aufbau Principle

Learning Goal 4

The arrangement of electrons in atomic orbitals is referred to as the atom's electron configuration. The *aufbau*, or building up, *principle* helps us to represent the electron configuration of atoms of various elements. According to this principle, electrons fill the lowest-energy orbital that is available first. We should also recall that the maximum capacity of an *s* level is two, that of a *p* level is six, that of a *d* level is ten, and that of an *f* level is fourteen electrons. Consider the following guidelines for writing electron configurations:

Rules for Writing Electron Configurations

- Obtain the total number of electrons in the atoms from the atomic number found on the periodic table.
- Electrons in atoms occupy the lowest energy orbitals that are available, beginning with 1*s*.
- Each principal energy level, *n*, can contain only *n* subshells.
- Each sublevel is composed of one (*s*) or more (three *p*, five *d*, seven *f*) orbitals.
- No more than two electrons can be placed in any orbital.
- The maximum number of electrons in any principal energy level is $2(n)^2$.
- The theoretical order of orbital filling is depicted in Figure 3.5.

Now let us look at several elements:

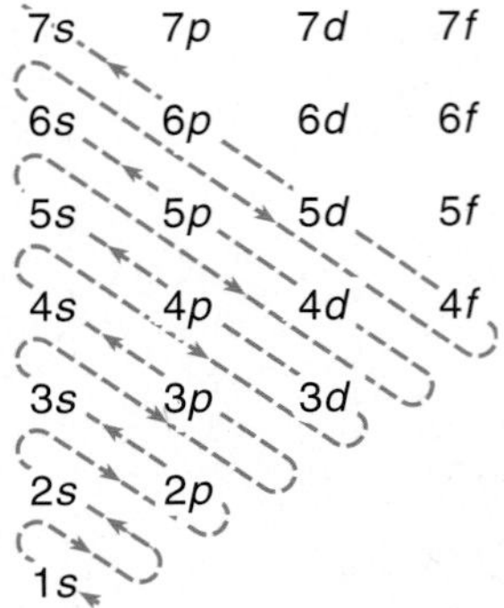

Figure 3.5
A useful way to remember the filling order for electrons in atoms.

Hydrogen: Hydrogen is the simplest atom; it has only one electron. That electron must be in the lowest principal energy level ($n = 1$) and the lowest orbital (*s*). We indicate the number of electrons in a region with a *superscript*, so we write $1s^1$.

Helium: Helium has two electrons, which will fill the lowest energy level. The ground state (lowest energy) electron configuration for helium is $1s^2$.

Lithium: Lithium has three electrons. The first two are configured as helium. The third must go into the orbital of the lowest energy in the second principal energy level; therefore the configuration is $1s^2\ 2s^1$.

Beryllium Through Neon: The second principal energy level can contain eight electrons $[2(2)^2]$, two in the *s* level and six in the *p* level. The "building up" process results in

Be	$1s^2\ 2s^2$
B	$1s^2\ 2s^2\ 2p^1$
C	$1s^2\ 2s^2\ 2p^2$
N	$1s^2\ 2s^2\ 2p^3$
O	$1s^2\ 2s^2\ 2p^4$
F	$1s^2\ 2s^2\ 2p^5$
Ne	$1s^2\ 2s^2\ 2p^6$

Sodium Through Argon: Electrons in these elements retain the basic $1s^2\ 2s^2\ 2p^6$ arrangement of the preceding element, neon; new electrons enter the third principal energy level:

Na	$1s^2\ 2s^2\ 2p^6\ 3s^1$
Mg	$1s^2\ 2s^2\ 2p^6\ 3s^2$
Al	$1s^2\ 2s^2\ 2p^6\ 3s^2\ 3p^1$
Si	$1s^2\ 2s^2\ 2p^6\ 3s^2\ 3p^2$
P	$1s^2\ 2s^2\ 2p^6\ 3s^2\ 3p^3$
S	$1s^2\ 2s^2\ 2p^6\ 3s^2\ 3p^4$
Cl	$1s^2\ 2s^2\ 2p^6\ 3s^2\ 3p^5$
Ar	$1s^2\ 2s^2\ 2p^6\ 3s^2\ 3p^6$

By knowing the order of filling of atomic orbitals, lowest to highest energy, you may write the electron configuration for any element. The order of orbital filling can be represented by the diagram in Figure 3.5. Such a diagram provides an easy way of predicting the electron configuration of the elements. Remember that the diagram is based on an energy scale, with the lowest energy orbital at the beginning of the "path" and the highest energy orbital at the end of the "path." An alternative way of representing orbital energies is through the use of an energy level diagram, such as the one in Figure 3.6.

Writing the Electron Configuration of Tin — EXAMPLE 3.2

Tin, Sn, has an atomic number of 50; thus we must place fifty electrons in atomic orbitals. We must also remember the total electron capacities of orbital types: *s*, 2; *p*, 6; *d*, 10; and *f*, 14. The electron configuration is as follows:

$$1s^2\ 2s^2\ 2p^6\ 3s^2\ 3p^6\ 4s^2\ 3d^{10}\ 4p^6\ 5s^2\ 4d^{10}\ 5p^2$$

As a check, count electrons in the electron configuration to see that we have accounted for all fifty electrons of the Sn atom.

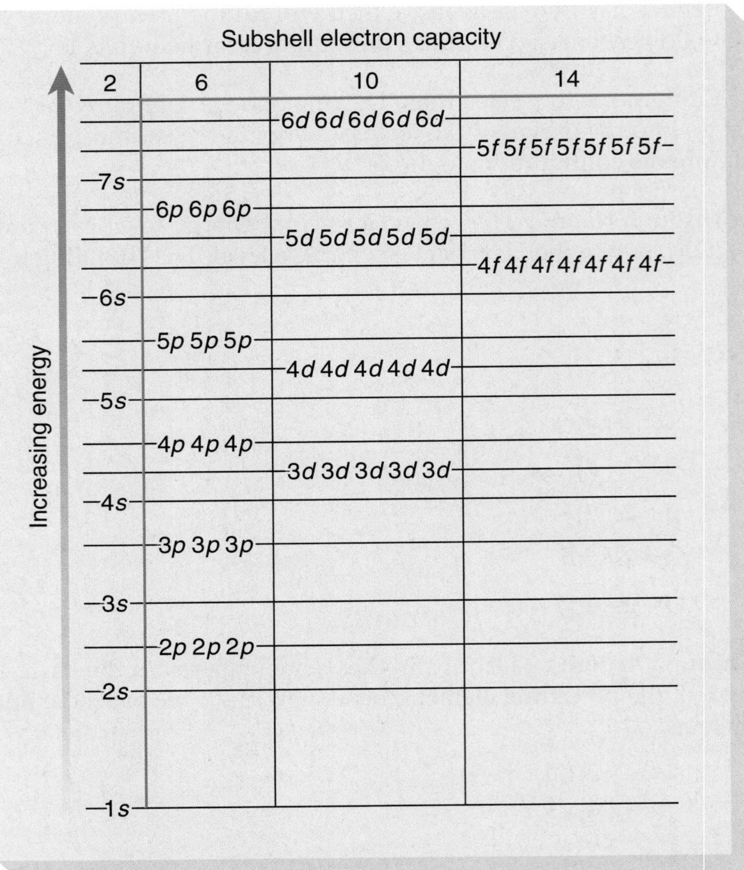

Figure 3.6
An orbital energy-level diagram. Electrons fill orbitals in the order of increasing energy.

Question 3.7

Give the electron configuration for an atom of:
a. sulfur
b. calcium

Question 3.8

Give the electron configuration for an atom of:
a. potassium
b. phosphorus

Abbreviated Electron Configurations

As we noted earlier the electron configuration for the sodium atom (Na, atomic number 11) is

$$1s^2\ 2s^2\ 2p^6\ 3s^1$$

The electron configuration for the preceding noble gas, neon (Ne, atomic number 10), is

$$1s^2\ 2s^2\ 2p^6$$

The electron configuration for sodium is really the electron configuration of Ne, with $3s^1$ added to represent one additional electron. So it is permissible to write

[Ne] $3s^1$ as equivalent to $1s^2\ 2s^2\ 2p^6\ 3s^1$

[Ne] $3s^1$ is the *abbreviated electron configuration* for sodium.

Similarly,

[Ne] $3s^2$ representing Mg

[Ne] $3s^2$ $3p^5$ representing Cl

[Ar] $4s^1$ representing K

are valid electron configurations.

The use of abbreviated electron configurations, in addition to being faster and easier to write, serves to highlight the valence electrons, those electrons involved in bonding. The symbol of the noble gas represents the *core*, nonvalence electrons and the valence electron configuration follows the noble gas symbol.

Question 3.9 Give the *abbreviated* electron configuration for each atom in Question 3.7.

Question 3.10 Give the *abbreviated* electron configuration for each atom in Question 3.8.

3.3 The Octet Rule

Learning Goal

Elements in the last family, the noble gases, have either two valence electrons (helium) or eight valence electrons (neon, argon, krypton, xenon, and radon). These elements are extremely stable and were often termed *inert gases* because they do not readily bond to other elements, although they can be made to do so under extreme experimental conditions. A full $n = 1$ energy level (as in helium) or an outer *octet* of electrons (eight valence electrons, as in all of the other noble gases) is responsible for this unique stability.

Atoms of elements in other groups are more reactive than the noble gases because in the process of chemical reaction they are trying to achieve a more stable "noble gas" configuration by gaining or losing electrons. This is the basis of the **octet rule.** Elements usually react in such a way as to attain the electron configuration of the noble gas closest to them in the periodic table (a stable octet of electrons). In chemical reactions they will gain, lose, or share the minimum number of electrons necessary to attain this more stable energy state. The octet rule, although simple in concept, is a remarkably reliable predictor of chemical change, especially for representative elements.

We may think of stability as a type of contentment; a noble gas atom does not need to rearrange its electrons or lose or gain any electrons to get to a more stable, lower energy, or more "contented" configuration.

Ion Formation and the Octet Rule

Learning Goal

Metals and nonmetals differ in the way in which they form ions. Metallic elements (located at the left of the periodic table) tend to form positively charged ions called **cations.** Positive ions are formed when an atom loses one or more electrons, for example,

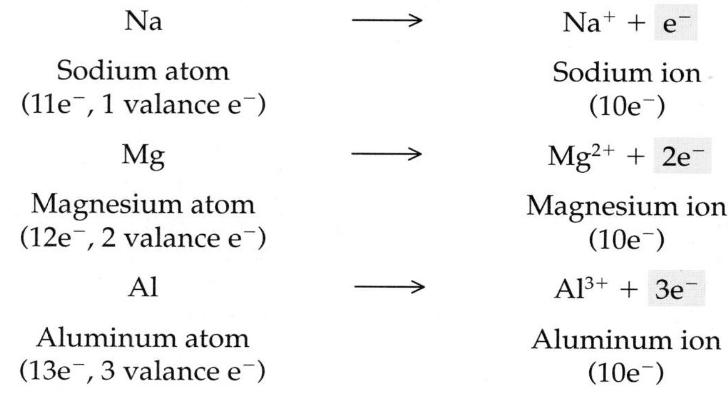

$$Na \longrightarrow Na^+ + e^-$$

Sodium atom (11e⁻, 1 valance e⁻) → Sodium ion (10e⁻)

$$Mg \longrightarrow Mg^{2+} + 2e^-$$

Magnesium atom (12e⁻, 2 valance e⁻) → Magnesium ion (10e⁻)

$$Al \longrightarrow Al^{3+} + 3e^-$$

Aluminum atom (13e⁻, 3 valance e⁻) → Aluminum ion (10e⁻)

In each of these cases the atom has lost *all* of its valence electrons. The resulting ion has the same number of electrons as the nearest noble gas atom:

Na^+ ($10e^-$) and Mg^{2+} ($10e^-$) and Al^{3+} ($10e^-$) are all isoelectronic with Ne ($10e^-$).

Recall that the prefix *iso* (Greek *isos*) means equal.

These ions are particularly stable. Each ion is **isoelectronic** (that is, it has the same number of electrons) with its nearest noble gas neighbor and has an octet of electrons in its outermost energy level.

Sodium is typical of each element in its group. Knowing that sodium forms a 1+ ion leads to the prediction that H, Li, K, Rb, Cs, and Fr also will form 1+ ions. Furthermore, magnesium, which forms a 2+ ion, is typical of each element in its group; Be^{2+}, Ca^{2+}, Sr^{2+}, and so forth are the resulting ions.

Nonmetallic elements, located at the right of the periodic table, tend to gain electrons to become isoelectronic with the nearest noble gas element, forming negative ions called **anions.**

Section 4.2 discusses the naming of ions.

Consider:

The ion of fluorine is the *fluoride ion;* the ion of oxygen is the *oxide ion;* and the ion of nitrogen is the *nitride ion.*

$F + 1e^-$	$\longrightarrow$	F^-	(isoelectronic with Ne, $10e^-$)
Fluorine atom ($9e^-$, 7 valence e^-)		Fluoride ion ($10e^-$)	
$O + 2e^-$	$\longrightarrow$	O^{2-}	(isoelectronic with Ne, $10e^-$)
Oxygen atom ($8e^-$, 6 valence e^-)		Oxide ion ($10e^-$)	
$N + 3e^-$	$\longrightarrow$	N^{3-}	(isoelectronic with Ne, $10e^-$)
Nitrogen atom ($7e^-$, 5 valence e^-)		Nitride ion ($10e^-$)	

As in the case of positive ion formation, each of these negative ions has an octet of electrons in its outermost energy level.

The element fluorine, forming F^-, indicates that the other halogens, Cl, Br, and I, behave as a true family and form Cl^-, Br^-, and I^- ions. Also, oxygen and the other nonmetals in its group form 2^- ions; nitrogen and phosphorus form 3^- ions.

Question 3.11

Give the charge of the most probable ion resulting from each of the following elements. With what element is the ion isoelectronic?

a. Ca
b. Sr
c. S
d. Mg
e. P

Question 3.12

Which of the following pairs of atoms and ions are isoelectronic?

a. Cl^-, Ar
b. Na^+, Ne
c. Mg^{2+}, Na^+
d. Li^+, Ne
e. O^{2-}, F^-

The transition metals tend to form positive ions by losing electrons, just like the representative metals. Metals, whether representative or transition, share this characteristic. However, the transition elements are characterized as "variable valence" elements; depending on the type of substance with which they react, they may form more than one stable ion. For example, iron has two stable ionic forms:

$$Fe^{2+} \text{ and } Fe^{3+}$$

A CLINICAL PERSPECTIVE

Dietary Calcium

"Drink your milk!" "Eat all of your vegetables!" These imperatives are almost universal memories from our childhood. Our parents knew that calcium, present in abundance in these foods, was an essential element for the development of strong bones and healthy teeth.

Recent studies, spanning the fields of biology, chemistry, and nutrition science indicate that the benefits of calcium go far beyond bones and teeth. This element has been found to play a role in the prevention of disease throughout our bodies.

Calcium is the most abundant mineral (metal) in the body. It is ingested as the calcium ion (Ca^{2+}) either in its "free" state or "combined," as a part of a larger compound; calcium dietary supplements often contain ions in the form of calcium carbonate. The acid naturally present in the stomach produces the calcium ion:

$CaCO_3$	+	$2H^+$	$\longrightarrow$	Ca^{2+}	+	H_2O	+	CO_2
calcium carbonate		stomach acid		calcium ion		water		carbon dioxide

Vitamin D serves as the body's regulator of calcium ion uptake, release, and transport in the body (see Appendix E.3).

Calcium is responsible for a variety of body functions including:

- transmission of nerve impulses
- release of "messenger compounds" that enable communication among nerves
- blood clotting
- hormone secretion
- growth of living cells throughout the body

The body's storehouse of calcium is bone tissue. When the supply of calcium from external sources, the diet, is insufficient, the body uses a mechanism to compensate for this shortage. With vitamin D in a critical role, this mechanism removes calcium from bone to enable other functions to continue to take place. It is evident then that prolonged dietary calcium deficiency can weaken the bone structure. Unfortunately, current studies show that as many as 75% of the American population may not be consuming sufficient amounts of calcium. Developing an understanding of the role of calcium in premenstrual syndrome, cancer, and blood pressure regulation is the goal of three current research areas.

Calcium and premenstrual syndrome (PMS). Dr. Susan Thys-Jacobs, a gynecologist at St. Luke's-Roosevelt Hospital Center in New York City, and colleagues at eleven other medical centers are conducting a study of calcium's ability to relieve the discomfort of PMS. They believe that women with chronic PMS have calcium blood levels that are normal only because calcium is continually being removed from the bone to maintain an adequate supply in the blood. To complicate the situation, vitamin D levels in many young women are very low (as much as 80% of a person's vitamin D is made in the skin, upon exposure to sunlight; many of us now minimize our exposure to the sun because of concerns about ultraviolet radiation and skin cancer). Because vitamin D plays an essential role in calcium metabolism, even if sufficient calcium is consumed, it may not be used efficiently in the body.

Colon cancer. The colon is lined with a type of cell (epithelial cell) that is similar to those that form the outer layers of skin. Various studies have indicated that by-products of a high-fat diet are irritants to these epithelial cells and produce abnormal cell growth in the colon. Dr. Martin Lipkin, Rockefeller University in New York, and his colleagues have shown that calcium ions may bind with these irritants, reducing their undesirable effects. It is believed that a calcium-rich diet, low in fat, and perhaps use of a calcium supplement can prevent or reverse this abnormal colon cell growth, delaying or preventing the onset of colon cancer.

Blood pressure regulation. Dr. David McCarron, a blood pressure specialist at the Oregon Health Sciences University, believes that dietary calcium levels may have a significant influence on hypertension (high blood pressure). Preliminary studies show that a diet rich in low-fat dairy products, fruits, and vegetables, all high in calcium, may produce a significant lowering of blood pressure in adults with mild hypertension.

The take-home lesson appears clear: a high calcium, low fat diet promotes good health in many ways. Once again, our parents were right!

Copper can exist as

$$Cu^+ \text{ and } Cu^{2+}$$

and elements such as vanadium, V, and manganese, Mn, each can form four different stable ions.

Predicting the charge of an ion or the various possible ions for a given transition metal is not an easy task. Energy differences between valence electrons of transition metals are small and not easily predicted from the position of the element in the periodic table. In fact, in contrast to representative metals, the transition metals show great similarities within a *period* as well as within a *group.*

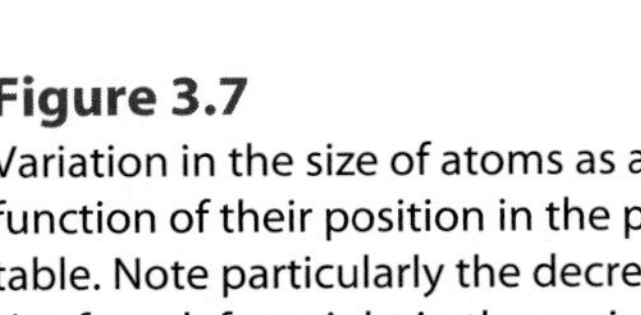

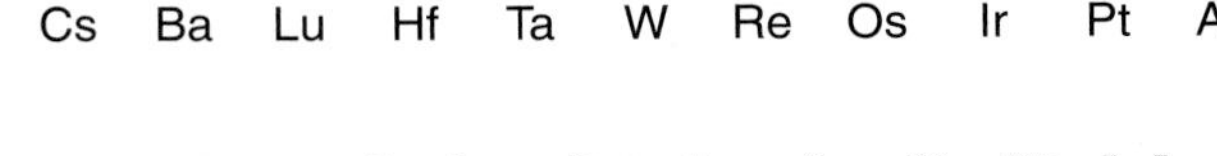

Figure 3.7
Variation in the size of atoms as a function of their position in the periodic table. Note particularly the decrease in size from left to right in the periodic table and the increase in size as we proceed down the table, although some exceptions do exist. (Lanthanide and actinide elements are not included here.)

3.4 Trends in the Periodic Table

The radius of an atom is traditionally defined as one-half of the distance between atoms in a covalent bond. The covalent bond is discussed in Section 4.1.

If our model of the atom is a tiny sphere whose radius is determined by the distance between the center of the nucleus and the boundary of the region where the valence electrons have a probability of being located, the size of the atom will be determined principally by two factors.

Atomic Size

Learning Goal

1. The energy level (n level) in which the outermost electron(s) is (are) found increases as we go *down* a group. (Recall that the outermost n level correlates with period number.) Thus the size of atoms should increase from top to bottom of the periodic table as we fill successive energy levels of the atoms with electrons (Figure 3.7).
2. As the magnitude of the positive charge of the nucleus increases, its "pull" on all of the electrons increases, and the electrons are drawn closer to the nucleus. This results in a contraction of the atomic radius and therefore a decrease in atomic size. This effect is apparent as we go *across* the periodic table within a period. Atomic size decreases from left to right in the periodic table (see Figure 3.7). See how many exceptions you can find in Figure 3.7.

Ion Size

Learning Goal 7

Positive ions (cations) are smaller than the parent atom. The cation has more protons than electrons (an increased nuclear charge). The excess nuclear charge pulls the remaining electrons closer to the nucleus. Also, cation formation often results in the loss of all outer-shell electrons, resulting in a significant decrease in radius.

Negative ions (anions) are larger than the parent atom. The anion has more electrons than protons. Owing to the excess negative charge, the nuclear "pull" on each individual electron is reduced. The electrons are held less tightly, resulting in a larger anion radius in contrast to the neutral atom.

Ions with multiple positive charge (such as Cu^{2+}) are even *smaller* than their corresponding monopositive ion (Cu^{+}); ions with multiple negative charge (such as O^{2-}) are *larger* than their corresponding less negative ion.

Figure 3.8 depicts the relative sizes of several atoms and their corresponding ions.

Ionization Energy

Learning Goal Learning Goal

The energy required to remove an electron from an isolated atom is the **ionization energy.** The process for sodium is represented as follows:

$$\text{ionization energy} + \text{Na} \longrightarrow \text{Na}^{+} + e^{-}$$

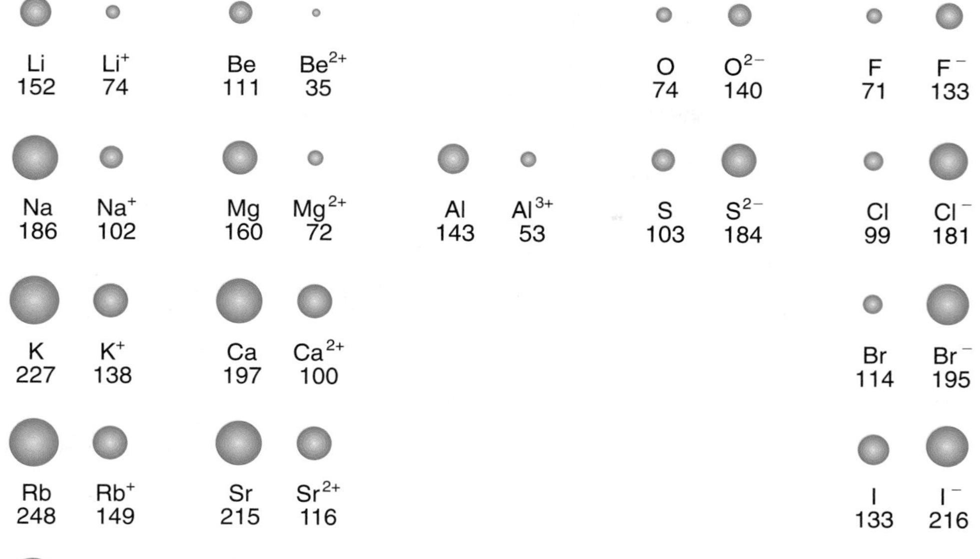

Figure 3.8
Relative size of ions and their parent atoms. Atomic radii are provided in units of picometers.

The magnitude of the ionization energy should correlate with the strength of the attractive force between the nucleus and the outermost electron.

- Reading *down* a group, note that the ionization energy decreases, because the atom's size is increasing. The outermost electron is progressively farther from the nuclear charge, hence easier to remove.
- Reading *across* a period, note that atomic size decreases, because the outermost electrons are closer to the nucleus, more tightly held, and more difficult to remove. Therefore the ionization energy generally increases.

Remember: ionization energy and electron affinity (below) are predictable from *trends* in the periodic table. As with most trends, exceptions occur.

A correlation does indeed exist between trends in atomic size and ionization energy. Atomic size generally *decreases* from the bottom to top of a group and from left to right in a period. Ionization energies generally *increase* in the same periodic way. Note also that ionization energies are highest for the noble gases (Figure 3.9). A high value for ionization energy means that it is difficult to remove electrons

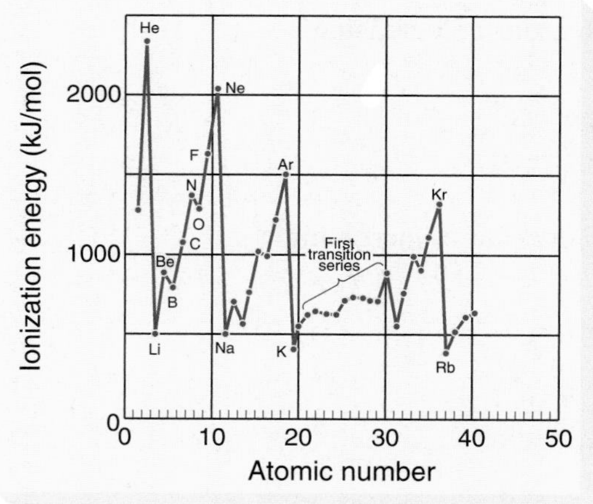

Figure 3.9
The ionization energies of the first forty elements versus their atomic numbers. Note the very high values for elements located on the right in the periodic table, and low values for those on the left. Some exceptions to the trends are evident.

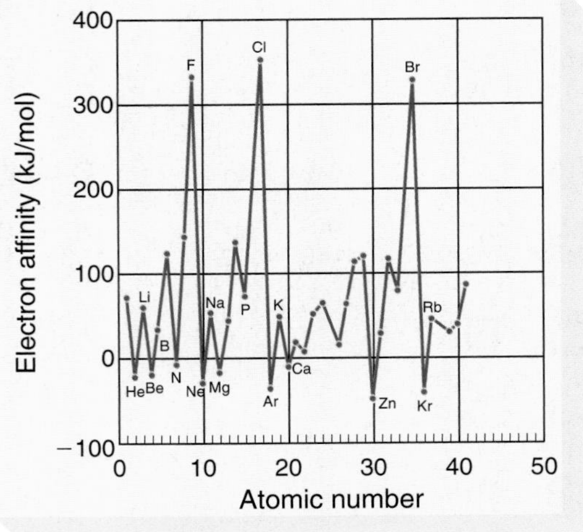

Figure 3.10
The periodic variation of electron affinity. Note the very low values for the noble gases and the elements on the far left of the periodic table. These elements do not form negative ions. In contrast, F, Cl, and Br readily form negative ions.

from the atom, and this, in part, accounts for the extreme stability and nonreactivity of the noble gases.

Electron Affinity

Learning Goal 7 Learning Goal 8

The energy released when a single electron is added to an isolated atom is the **electron affinity.** If we consider ionization energy in relation to positive ion formation (remember that the magnitude of the ionization energy tells us the ease of *removal* of an electron, hence the ease of forming positive ions), then electron affinity provides a measure of the ease of forming negative ions. A large electron affinity (energy released) indicates that the atom becomes more stable as it becomes a negative ion (through gaining an electron). Consider the gain of an electron by a bromine atom:

$$\text{Br} + e^- \longrightarrow \text{Br}^- + \text{energy}$$

Electron affinity

Periodic trends for electron affinity are as follows:

- Electron affinities generally decrease down a group.
- Electron affinities generally increase across a period.

Remember these trends are not absolute. Exceptions exist, as seen in the irregularities in Figure 3.10.

Question 3.13

Rank Be, N, and F in order of increasing

a. atomic size
b. ionization energy
c. electron affinity

Question 3.14

Rank Cl, Br, I, and F in order of increasing

a. atomic size
b. ionization energy
c. electron affinity

Summary

3.1 The Periodic Law and the Periodic Table

The *periodic law* is an organized "map" of the elements that relates their structure to their chemical and physical properties. It states that the elements, when arranged according to their atomic numbers, show a distinct periodicity (regular variation) of their properties. The periodic table is the result of the periodic law.

The modern periodic table exists in several forms. The most important variation is in group numbering. The tables in this text use the two most commonly accepted numbering systems.

A horizontal row of elements in the periodic table is referred to as a *period*. The periodic table consists of seven periods. The *lanthanide series* is a part of period 6; the *actinide series* is a part of period 7.

The columns of elements in the periodic table are called *groups* or *families*. The elements of a particular family share many similarities in physical and chemical properties because of the similarities in electronic structure. Some of the most important groups are named; for example, the *alkali metals* (IA or 1), *alkaline earth metals* (IIA or 2), the *halogens* (VIIA or 17), and the *noble gases* (VIII or 18).

Group A elements are called *representative elements;* Group B elements are *transition elements.* A bold zigzag line runs from top to bottom of the table, beginning to the left of boron (B) and ending between polonium (Po) and astatine (At). This line acts as the boundary between *metals* to the left and *nonmetals* to the right. Elements straddling the boundary, *metalloids,* have properties intermediate between those of metals and nonmetals.

3.2 Electron Arrangement and the Periodic Table

The outermost electrons in an atom are *valence electrons.* For representative elements the number of valence electrons in an atom corresponds to the group or family number (old numbering system using Roman numerals). Metals tend to have fewer valence electrons than nonmetals.

Electron configuration of the elements is predictable, using the aufbau principle. Knowing the electron configuration, we can identify valence electrons and begin to predict the kinds of reactions that the elements will undergo.

Elements in the last family, the noble gases, have either two valence electrons (helium) or eight valence electrons (neon, argon, krypton, xenon, and radon). Their most important properties are their extreme stability and lack of reactivity. A full valence level is responsible for this unique stability.

3.3 The Octet Rule

The *octet rule* tells us that in chemical reactions, elements will gain, lose, or share the minimum number of electrons necessary to achieve the electron configuration of the nearest noble gas.

Metallic elements tend to form cations. The ion is *isoelectronic* with its nearest noble gas neighbor and has a stable octet of electrons in its outermost energy level. Nonmetallic elements tend to gain electrons to become isoelectronic with the nearest noble gas element, forming anions.

3.4 Trends in the Periodic Table

Atomic size decreases from left to right and from bottom to top in the periodic table. *Cations* are smaller than the parent atom. *Anions* are larger than the parent atom. Ions with multiple positive charge are even smaller than their corresponding monopositive ion; ions with multiple negative charge are larger than their corresponding less negative ion.

The energy required to remove an electron from the atom is the *ionization energy.* Down a group, the ionization energy generally decreases. Across a period, the ionization energy generally increases.

The energy released when a single electron is added to a neutral atom in the gaseous state is known as the *electron affinity.* Electron affinities generally decrease proceeding down a group and increase proceeding across a period.

Key Terms

actinide series (3.1)
alkali metal (3.1)
alkaline earth metal (3.1)
anion (3.3)
cation (3.3)
electron affinity (3.4)
electronic configuration (3.2)
group (3.1)
halogen (3.1)
ionization energy (3.4)
isoelectronic (3.3)
lanthanide series (3.1)
metal (3.1)
metalloid (3.1)
noble gas (3.1)
nonmetal (3.1)
octet rule (3.3)
orbital (3.2)
period (3.1)
periodic law (3.1)
representative element (3.1)
transition element (3.1)
valence electron (3.2)

Questions and Problems

The Periodic Law and the Periodic Table

3.15 Define each of the following terms:
- **a.** periodic law
- **b.** period
- **c.** group
- **d.** ion

3.16 Define each of the following terms:
- **a.** electron configuration
- **b.** octet rule
- **c.** ionization energy
- **d.** isoelectronic

3.17 Label each of the following statements as true or false:
a. Elements of the same group have similar properties.
b. Atomic size decreases from left to right across a period.

3.18 Label each of the following statements as true or false:
a. Ionization energy increases from top to bottom within a group.
b. Representative metals are located on the left in the periodic table.

3.19 For each of the elements Na, Ni, Al, P, Cl, and Ar, provide the following information:
a. Which are metals?
b. Which are representative metals?
c. Which tend to form positive ions?
d. Which are inert or noble gases?

3.20 For each of the elements Ca, K, Cu, Zn, Br, and Kr provide the following information:
a. Which are metals?
b. Which are representative metals?
c. Which tend to form positive ions?
d. Which are inert or noble gases?

3.21 Provide the name of the element represented by each of the following symbols:
a. Na
b. K
c. Mg

3.22 Provide the name of the element represented by each of the following symbols:
a. Ca
b. Cu
c. Co

3.23 Which group of the periodic table is known as the alkali metals? List them.

3.24 Which group of the periodic table is known as the alkaline earth metals? List them.

3.25 Which group of the periodic table is known as the halogens? List them.

3.26 Which group of the periodic table is known as the noble gases? List them.

3.27 What are the major differences between the early and modern periodic tables?

3.28 Provide the name of the element represented by each of the following symbols:
a. B
b. Si
c. As

3.29 What is meant by the term *metalloid?*

3.30 Give three examples of elements that are:
a. metals
b. metalloids
c. nonmetals

Electron Arrangement and the Periodic Table

3.31 How many valence electrons are found in an atom of each of the following elements?
a. H **d.** F
b. Na **e.** Ne
c. B **f.** He

3.32 How many valence electrons are found in an atom of each of the following elements?
a. Mg **d.** Br
b. K **e.** Ar
c. C **f.** Xe

3.33 What is the common feature of the electron configurations of elements in Group IA (1)?

3.34 What is the common feature of the electron configurations of elements in Group VIIIA (18)?

3.35 How do we calculate the electron capacity of a principal energy level?

3.36 What sublevels would be found in each of the following principal energy levels?
a. $n = 1$ **c.** $n = 3$
b. $n = 2$ **d.** $n = 4$

3.37 Distinguish between a principal energy level and a sublevel.

3.38 Distinguish between a sublevel and an orbital.

3.39 Sketch a diagram and describe our current model of an *s* orbital.

3.40 How is a 2*s* orbital different from a 1*s* orbital?

3.41 How many *p* orbitals can exist in a given principal energy level?

3.42 Sketch diagrams of a set of *p* orbitals. How does a p_x orbital differ from a p_y orbital? From a p_z orbital?

3.43 How does a 3*p* orbital differ from a 2*p* orbital?

3.44 What is the maximum number of electrons that an orbital can hold?

3.45 What is the maximum number of electrons in each of the following energy levels?
a. $n = 1$
b. $n = 2$
c. $n = 3$

3.46 **a.** What is the maximum number of *s* electrons that can exist in any one principal energy level?
b. How many *p* electrons?
c. How many *d* electrons?
d. How many *f* electrons?

3.47 In which orbital is the highest-energy electron located in each of the following elements?
a. Al **d.** Ca
b. Na **e.** Fe
c. Sc **f.** Cl

3.48 Using only the periodic table or list of elements, write the electron configuration of each of the following atoms:
a. B **d.** V
b. S **e.** Cd
c. Ar **f.** Te

3.49 Which of the following electron configurations are not possible? Why?
a. $1s^2\ 1p^2$
b. $1s^2\ 2s^2\ 2p^2$
c. $1s^2, 2s^2, 2p^6, 2d^1$
d. $1s^2, 2s^3$

3.50 For each incorrect electron configuration in Question 3.49, assume that the number of electrons is correct, identify the element, and write the correct electron configuration.

The Octet Rule

3.51 Give the most probable ion formed from each of the following elements:
a. Li **d.** Br
b. O **e.** S
c. Ca **f.** Al

3.52 Using only the periodic table or list of elements, write the electron configuration of each of the following ions:
a. I^-
b. Ba^{2+}
c. Se^{2-}
d. Al^{3+}

3.53 Which of the following pairs of atoms and/or ions are isoelectronic with one another?
a. O^{2-}, Ne
b. S^{2-}, Cl^-

3.54 Which of the following pairs of atoms and/or ions are isoelectronic with one another?

a. F^-, Cl^-
b. K^+, Ar

3.55 Why do Group IA (1) metals form only one ion (1+)? Does the same hold true for Group IIA (2): Can they form only a 2+ ion?

3.56 Why are noble gases so nonreactive?

3.57 Which species in each of the following groups would you expect to find in nature?
a. Na, Na^+, Na^-
b. S^{2-}, S^-, S^+
c. Cl, Cl^-, Cl^+

3.58 Which atom or ion in each of the following groups would you expect to find in nature?
a. K, K^+, K^-
b. O^{2-}, O, O^{2+}
c. Br, Br^-, Br^+

3.59 Write the electron configuration of each of the following biologically important ions:
a. Ca^{2+}
b. Mg^{2+}

3.60 Write the electron configuration of each of the following biologically important ions:
a. K^+
b. Cl^-

Trends in the Periodic Table

3.61 Arrange each of the following lists of elements in order of increasing atomic size:
a. N, O, F
b. Li, K, Cs
c. Cl, Br, I

3.62 Arrange each of the following lists of elements in order of increasing atomic size:
a. Al, Si, P, Cl, S
b. In, Ga, Al, B, Tl
c. Sr, Ca, Ba, Mg, Be
d. P, N, Sb, Bi, As

3.63 Which of the elements has the highest electron affinity?

3.64 Which of the elements has the highest ionization energy?

3.65 Arrange each of the following lists of elements in order of increasing ionization energy:
a. N, O, F
b. Li, K, Cs
c. Cl, Br, I

3.66 Arrange each of the following lists of elements in order of decreasing electron affinity:
a. Na, Li, K
b. Br, F, Cl
c. S, O, Se

3.67 Explain why a positive ion is always smaller than its parent atom.

3.68 Explain why a negative ion is always larger than its parent atom.

3.69 Explain why a fluoride ion is commonly found in nature but a fluorine atom is not.

3.70 Explain why a sodium ion is commonly found in nature but a sodium atom is not.

Critical Thinking Problems

1. Name five elements that you came in contact with today. Were they in combined form or did they exist in the form of atoms? Were they present in pure form or in mixtures? If mixtures, were they heterogeneous or homogeneous? Locate each in the periodic table by providing the group and period designation, for example: Group IIA (2), period 3.
2. The periodic table is incomplete. It is possible that new elements will be discovered from experiments using high-energy particle accelerators. Predict as many properties as you can that might characterize the element that would have an atomic number of 118. Can you suggest an appropriate name for this element?
3. The element titanium is now being used as a structural material for bone and socket replacement (shoulders, knees). Predict properties that you would expect for such applications; go to the library or internet and look up the properties of titanium and evaluate your answer.
4. Imagine that you have undertaken a voyage to an alternate universe. Using your chemical skills, you find a collection of elements quite different than those found here on earth. After measuring their properties and assigning symbols for each, you wish to organize them as Mendeleev did for our elements. Design a periodic table using the information you have gathered:

Symbol	Mass (amu)	Reactivity	Electrical Conductivity
A	2.0	High	High
B	4.0	High	High
C	6.0	Moderate	Trace
D	8.0	Low	0
E	10.0	Low	0
F	12.0	High	High
G	14.0	High	High
H	16.0	Moderate	Trace
I	18.0	Low	0
J	20.0	None	0
K	22.0	High	High
L	24.0	High	High

Predict the reactivity and conductivity of an element with a mass of 30.0 amu. What element in our universe does this element most closely resemble?

5. Why does the octet rule not work well for compounds of lanthanide and actinide elements? Suggest a number other than eight that may be more suitable.

The pattern formed depends on the courage and skill of the individuals.

4 Structure and Properties of Ionic and Covalent Compounds

Outline

Learning Goals

1. Classify compounds as having ionic, covalent, or polar covalent bonds.
2. Write the formulas of compounds when provided with the name of the compound.
3. Name common inorganic compounds using standard conventions and recognize the common names of frequently used substances.
4. Predict the differences in physical state, melting and boiling points, solid-state structure, and solution chemistry that result from differences in bonding.
5. Draw Lewis structures for covalent compounds and polyatomic ions.
6. Describe the relationship between stability and bond energy.
7. Predict the geometry of molecules and ions using the octet rule and Lewis structure.
8. Understand the role that molecular geometry plays in determining the solubility and melting and boiling points of compounds.
9. Use the principles of VSEPR theory and molecular geometry to predict relative melting points, boiling points, and solubilities of compounds.

CHEMISTRY CONNECTION

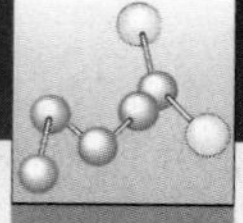

Magnets and Migration

All of us, at one time or another, have wondered at the magnificent sight of thousands of migrating birds, flying in formation, heading south for the winter and returning each spring.

Less visible, but no less impressive, are the schools of fish that travel thousands of miles, returning to the same location year after year. Almost instantly, when faced with some external stimulus such as a predator, they snap into a formation that rivals an army drill team for precision.

The questions of how these life-forms know when and where they are going and how they establish their formations have perplexed scientists for many years. The explanations so far are really just hypotheses.

Some clues to the mystery may be hidden in very tiny particles of magnetite, Fe_3O_4. Magnetite contains iron that is naturally magnetic, and collections of these particles behave like a compass needle; they line up in formation aligned with the earth's magnetic field.

Magnetotactic bacteria contain magnetite in the form of magnetosomes, small particles of Fe_3O_4. Fe_3O_4 is a compound whose atoms are joined by chemical bonds. The electrons in the iron atoms are present in an electron configuration that results in single electrons (not pairs of electrons) occupying orbitals. These single electrons impart magnetic properties to the compound.

The normal habitat of magnetotactic bacteria is either fresh water or the ocean; the bacteria orient themselves to the earth's magnetic field and swim to the nearest pole (north or south). This causes them to swim into regions of nutrient-rich sediment.

Could the directional device, the simple F_3O_4 unit, also be responsible for direction finding in higher organisms in much the same way that an explorer uses a compass? Perhaps so! Recent studies have shown evidence of magnetosomes in the brains of birds, tuna, green turtles, and dolphins.

Most remarkably, at least one study has shown evidence that magnetite is present in the human brain.

These preliminary studies offer hope of unraveling some of the myth and mystery of guidance and communication in living systems. The answers may involve a very basic compound that is like those we will study in this chapter.

Introduction

A chemical compound is formed when two or more atoms of different elements are joined by attractive forces called chemical bonds. These bonds result from either a transfer of electrons from one atom to another (the ionic bond) or a sharing of electrons between two atoms (the covalent bond). The elements, once converted to a compound, cannot be recovered by any physical process. A chemical reaction must take place to regenerate the individual elements. The chemical and physical *properties* of a compound are related to the *structure* of the compound, and this structure is, in turn, determined by the arrangement of electrons in the atoms that produced the compounds. Properties such as solubility, boiling point, and melting point correlate well with the shape and charge distribution in the individual units of the compound.

We need to learn how to properly name and write formulas for ionic and covalent compounds. We should become familiar with some of their properties and be able to relate these properties to the structure and bonding of the compounds.

4.1 Chemical Bonding

When two or more atoms form a chemical compound, the atoms are held together in a characteristic arrangement by attractive forces. The **chemical bond** is the force of attraction between any two atoms in a compound. The attraction is the force that overcomes the repulsion of the positively charged nuclei of the two atoms.

Interactions involving valence electrons are responsible for the chemical bond. We shall focus our attention on these electrons and the electron arrangement of atoms both before and after bond formation.

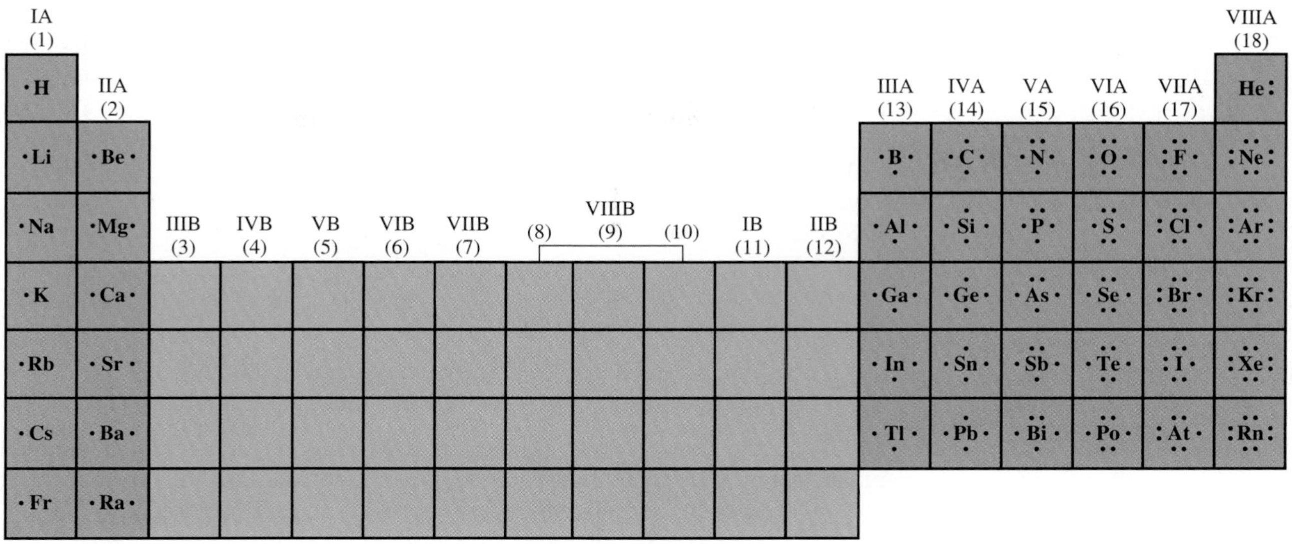

IA (1)	IIA (2)	IIIB (3)	IVB (4)	VB (5)	VIB (6)	VIIB (7)	VIIIB (8)	VIIIB (9)	VIIIB (10)	IB (11)	IIB (12)	IIIA (13)	IVA (14)	VA (15)	VIA (16)	VIIA (17)	VIIIA (18)
·H																	He:
·Li	·Be·											·B·	·C·	·N·	·O·	:F·	:Ne:
·Na	·Mg·											·Al·	·Si·	·P·	·S·	:Cl·	:Ar:
·K	·Ca·											·Ga·	·Ge·	·As·	·Se·	:Br·	:Kr:
·Rb	·Sr·											·In·	·Sn·	·Sb·	·Te·	:I·	:Xe:
·Cs	·Ba·											·Tl·	·Pb·	·Bi·	·Po·	:At·	:Rn:
·Fr	·Ra·																

Figure 4.1
Lewis dot symbols for representative elements. Each unpaired electron is a potential bond.

Lewis Symbols

The **Lewis symbol,** or Lewis structure, developed by G. N. Lewis early in this century, is a convenient way of representing atoms singly or in combination. Its principal advantage is that *only* valence electrons (those that may participate in bonding) are shown.

To draw Lewis structures, we first write the chemical symbol of the atom; this symbol represents the nucleus and all of the lower energy nonvalence electrons. The valence electrons are indicated by dots arranged around the atomic symbol. For example:

Recall that the number of valence electrons can be determined from the position of the element in the periodic table (see Figure 3.1).

H ·	He :
Hydrogen	Helium
Li ·	· Be ·
Lithium	Beryllium
· B ·	· C ·
Boron	Carbon
· N ·	· O ·
Nitrogen	Oxygen
: F ·	: Ne :
Fluorine	Neon

Note particularly that the number of dots corresponds to the number of valence electrons in the outermost shell of the atoms of the element. Each unpaired dot (representing an unpaired electron) is available to form a chemical bond with another element, producing a compound. Figure 4.1 depicts the Lewis dot structures for the representative elements.

Principal Types of Chemical Bonds: Ionic and Covalent

Two principal classes of chemical bonds exist: ionic and covalent. Both involve valence electrons.

Ionic bonding involves a transfer of one or more electrons from one atom to another, leading to the formation of an ionic bond. **Covalent bonding** involves a sharing of electrons resulting in the covalent bond.

Before discussing each type, we should recognize that the distinction between ionic and covalent bonding is not always clear-cut. Some compounds are clearly ionic, and some are clearly covalent, but many others possess both ionic and covalent characteristics.

Ionic Bonding

Learning Goal

Consider the reaction of a sodium atom and a chlorine atom to produce sodium chloride:

$$\text{Na} + \text{Cl} \longrightarrow \text{NaCl}$$

Recall that the sodium atom has

Refer to Section 3.4 for a discussion of ionization energy and electron affinity.

- a low ionization energy (it readily loses an electron) and
- a low electron affinity (it does not want more electrons).

If sodium loses its valence electron, it will become isoelectronic (same number of electrons) with neon, a very stable noble gas atom. This tells us that the sodium atom would be a good electron donor, forming the sodium ion:

$$\text{Na}\cdot \longrightarrow \text{Na}^+ + e^-$$

Recall that the chlorine atom has

- a high ionization energy (it will not easily give up an electron) and
- a high electron affinity (it readily accepts another electron).

Chlorine will gain one more electron. By doing so, it will complete an octet (eight outermost electrons) and be isoelectronic with argon, a stable noble gas. Therefore chlorine behaves as a willing electron acceptor, forming a chloride ion:

$$:\ddot{\underset{\cdot\cdot}{\text{Cl}}}\cdot + e^- \longrightarrow [:\ddot{\underset{\cdot\cdot}{\text{Cl}}}:]^-$$

The electron released by sodium (*electron donor*) is the electron received by chlorine (*electron acceptor*):

$$\text{Na}\cdot \longrightarrow \text{Na}^+ + e^-$$

$$e^- + \cdot\ddot{\underset{\cdot\cdot}{\text{Cl}}}: \longrightarrow [:\ddot{\underset{\cdot\cdot}{\text{Cl}}}:]^-$$

The resulting ions of opposite charge, Na^+ and Cl^-, are attracted to each other (opposite charges attract) and held together by this *electrostatic force* as an **ion pair:** Na^+Cl^-.

This electrostatic force, the attraction of opposite charges, is quite strong and holds the ions together. It is the ionic bond.

The essential features of ionic bonding are the following:

- Atoms of elements with low ionization energy and low electron affinity tend to form positive ions.
- Atoms of elements with high ionization energy and high electron affinity tend to form negative ions.
- Ion formation takes place by an electron transfer process.
- The positive and negative ions are held together by the electrostatic force between ions of opposite charge in an ionic bond.
- Reactions between representative metals and nonmetals (elements far to the left and right, respectively, in the periodic table) tend to result in ionic bonds.

Covalent Bonding

Learning Goal

Consider the bond formed between two hydrogen atoms, producing the diatomic form of hydrogen: H_2. Individual hydrogen atoms are not stable, and two hydrogen atoms readily combine to produce diatomic hydrogen:

$$\text{H} + \text{H} \longrightarrow \text{H}_2$$

If a hydrogen atom were to gain a second electron, it would be isoelectronic with the stable electron configuration of helium. However, because two identical hydrogen atoms have an equal tendency to gain or lose electrons, an electron transfer from one atom to the other is unlikely to occur under normal conditions. Each atom may attain a noble gas structure only by *sharing* its electron with the other, as shown with Lewis symbols:

$$H\cdot + \cdot H \longrightarrow H:H$$

A diatomic compound is one that is composed of two atoms joined by a covalent bond.

When electrons are shared rather than transferred, the *shared electron pair* is referred to as a *covalent bond*. Compounds characterized by covalent bonding are called *covalent compounds*. Covalent bonds tend to form between atoms with similar tendencies to gain or lose electrons. The most obvious examples are the diatomic molecules H_2, N_2, O_2, F_2, Cl_2, Br_2, and I_2. Bonding in these molecules is *totally covalent* because there can be no net tendency for electron transfer between identical atoms. The formation of F_2, for example, may be represented as

$$:\ddot{\underset{\cdot\cdot}{F}}\cdot + \cdot\ddot{\underset{\cdot\cdot}{F}}: \longrightarrow :\ddot{\underset{\cdot\cdot}{F}}:\ddot{\underset{\cdot\cdot}{F}}:$$

Fourteen valence electrons are arranged in such a way that each fluorine atom is surrounded by eight electrons. The octet rule is satisfied for each fluorine atom.

As in H_2, a single covalent bond is formed. The bonding electron pair is said to be *localized*, or largely confined to the region between the two fluorine nuclei.

Two atoms do not have to be identical to form a covalent bond. Consider compounds such as the following:

$H:\ddot{\underset{\cdot\cdot}{F}}:$	$H:\ddot{\underset{\cdot\cdot}{O}}:H$	$H:\overset{H}{\underset{H}{\ddot{\underset{\cdot\cdot}{C}}}}:H$	$H:\overset{H}{\ddot{\underset{\cdot\cdot}{N}}}:H$
Hydrogen fluoride	**Water**	**Methane**	**Ammonia**
$7e^-$ from F	$6e^-$ from O	$4e^-$ from C	$5e^-$ from N
$1e^-$ from H	$2e^-$ from 2H	$4e^-$ from 4H	$3e^-$ from 3H
$8e^-$ for F	$8e^-$ for O	$8e^-$ for C	$8e^-$ for N
$2e^-$ for H	$2e^-$ for H	$2e^-$ for H	$2e^-$ for H

In each of these cases, bond formation satisfies the octet rule. A total of eight electrons surround each atom other than hydrogen. Hydrogen has only two electrons (corresponding to the electronic structure of helium).

Polar Covalent Bonding and Electronegativity

The Polar Covalent Bond

Learning Goal

Covalent bonding is the sharing of an electron pair by two atoms. However, just as we may observe in our day-to-day activities, sharing is not always equal. In a molecule like H_2 (or N_2, or any other diatomic molecule composed of only one element), the electrons, on average, spend the same amount of time in the vicinity of each atom; the electrons have no preference because both atoms are identical.

Now consider a diatomic molecule composed of two different elements; HF is a common example. It has been experimentally shown that the electrons in the H—F bond are not equally shared; the electrons spend more time in the vicinity of the fluorine atom. This unequal sharing can be described in various ways:

Partial electron transfer: This describes the bond as having both covalent and ionic properties.

Unequal electron density: The density of electrons around F is greater than the density of electrons around H.

Below 1.0 | 1.0–3.0 | Above 3.0

IA (1)	IIA (2)	IIIB (3)	IVB (4)	VB (5)	VIB (6)	VIIB (7)	VIIIB (8)	VIIIB (9)	VIIIB (10)	IB (11)	IIA (12)	IIIA (13)	IVA (14)	VA (15)	VIA (16)	VIIA (17)
H 2.1																
Li 1.0	Be 1.5											B 2.0	C 2.5	N 3.0	O 3.5	F 4.0
Na 0.9	Mg 1.2											Al 1.5	Si 1.8	P 2.1	S 2.5	Cl 3.0
K 0.8	Ca 1.0	Sc 1.3	Ti 1.5	V 1.6	Cr 1.6	Mn 1.5	Fe 1.8	Co 1.9	Ni 1.9	Cu 1.9	Zn 1.6	Ga 1.6	Ge 1.8	As 2.0	Se 2.4	Br 2.8
Rb 0.8	Sr 1.0	Y 1.2	Zr 1.4	Nb 1.6	Mo 1.8	Tc 1.9	Ru 2.2	Rh 2.2	Pd 2.2	Ag 1.9	Cd 1.7	In 1.7	Sn 1.8	Sb 1.9	Te 2.1	I 2.5
Cs 0.7	Ba 0.9	La* 1.1	Hf 1.3	Ta 1.5	W 1.7	Re 1.9	Os 2.2	Ir 2.2	Pt 2.2	Au 2.4	Hg 1.9	Tl 1.8	Pb 1.9	Bi 1.9	Po 2.0	At 2.2
Fr 0.7	Ra 0.9	Ac† 1.1														

Alkali metals

***Lathanides: 1.1 – 1.3**
†Actinides: 1.1 – 1.5

Figure 4.2
Electronegativities of the elements.

Polar covalent bond is the preferred term for a bond made up of unequally shared electron pairs. One end of the bond (in this case, the F atom) is more electron rich (higher electron density), hence, more negative. The other end of the bond (in this case, the H atom) is less electron rich (lower electron density), hence, more positive. These two ends, one somewhat positive and the other somewhat negative may be described as electronic poles, hence the term polar covalent bonds.

Once again, we can use the predictive power of the periodic table to help us determine whether a particular bond is polar or nonpolar covalent. We already know that elements that tend to form negative ions (by gaining electrons) are found to the right of the table whereas positive ion formers (that may lose electrons) are located on the left side of the table. Elements whose atoms strongly attract electrons are described as electronegative elements. Linus Pauling, a chemist noted for his theories on chemical bonding, developed a scale of relative electronegativities that correlates reasonably well with the positions of the elements in the periodic table.

Linus Pauling is the only person to receive two Nobel Prizes in very unrelated fields; the chemistry award in 1954 and eight years later, the Nobel Peace Prize. His career is a model of interdisciplinary science, with important contributions ranging from chemical physics to molecular biology.

Electronegativity

Electronegativity (E_n) is a measure of the ability of an atom to attract electrons in a chemical bond. Elements with high electronegativity have a greater ability to attract electrons than do elements with low electronegativity. Pauling developed a method to assign values of electronegativity to many of the elements in the periodic table. These values range from a low of 0.7 to a high of 4.0, 4.0 being the most electronegative element.

Figure 4.2 shows that the most electronegative elements (excluding the nonreactive noble gas elements) are located in the upper right corner of the periodic table, whereas the least electronegative elements are found in the lower left corner of the table. In general, electronegativity values increase as we proceed left to right and bottom to top of the table. Like other periodic trends, numerous exceptions occur.

If we picture the covalent bond as a competition for electrons between two positive centers, it is the difference in electronegativity, ΔE_n, that determines the extent of polarity. Consider: H_2 or H—H

$$\Delta E_n = \begin{bmatrix}\text{Electronegativity}\\ \text{of hydrogen}\end{bmatrix} - \begin{bmatrix}\text{Electronegativity}\\ \text{of hydrogen}\end{bmatrix}$$

$$\Delta E_n = 2.1 - 2.1 = 0$$

The bond in H_2 is nonpolar covalent. Bonds between *identical* atoms are *always* nonpolar covalent. Also, Cl_2 or Cl—Cl

$$\Delta E_n = \begin{bmatrix}\text{Electronegativity}\\ \text{of chlorine}\end{bmatrix} - \begin{bmatrix}\text{Electronegativity}\\ \text{of chlorine}\end{bmatrix}$$

$$\Delta E_n = 3.0 - 3.0 = 0$$

The bond in Cl_2 is nonpolar covalent. Now consider HCl or H—Cl

$$\Delta E_n = \begin{bmatrix}\text{Electronegativity}\\ \text{of chlorine}\end{bmatrix} - \begin{bmatrix}\text{Electronegativity}\\ \text{of hydrogen}\end{bmatrix}$$

$$\Delta E_n = 3.0 - 2.1 = 0.9$$

The bond in HCl is polar covalent.

By convention, the electronegativity difference is calculated by subtracting the less electronegative element's value from the value for the more electronegative element. In this way, negative numbers are avoided.

A ΔE_n value of 1.9 is generally accepted as the boundary between a polar covalent and an ionic compound.

4.2 Naming Compounds and Writing Formulas of Compounds

Nomenclature is the assignment of a correct and unambiguous name to each and every chemical compound. Assignment of a name to a structure or deducing the structure from a name is a necessary first step in any discussion of these compounds.

Ionic Compounds

The "shorthand" symbol for a compound is its formula—for example,

NaCl and $MgBr_2$

The formula identifies the number and type of the various atoms that make up the compound unit. The number of like atoms in the unit is shown by the use of a subscript. The presence of only one atom is understood when no subscript is present.

The formula NaCl indicates that each ion pair consists of one sodium cation (Na^+) and one chloride anion (Cl^-). Similarly, the formula $MgBr_2$ indicates that one magnesium ion and two bromide ions combine to form the compound.

In Chapter 3 we learned that positive ions are formed from elements that

- are located at the left of the periodic table,
- are referred to as *metals,* and
- have low ionization energies, low electron affinities, and hence easily *lose* electrons.

Elements that form negative ions, on the other hand,

- are located at the right of the periodic table (but exclude the noble gases),
- are referred to as *nonmetals,* and
- have high ionization energies, high electron affinities, and hence easily *gain* electrons.

In short, metals and nonmetals usually react to produce ionic compounds resulting from the transfer of one or more electrons from the metal to the nonmetal.

Although ionic compounds are sometimes referred to as ion pairs, in the solid state these ion pairs do not actually exist as individual units. The positive ions exert attractive forces on several negative ions, and the negative ions are attracted to

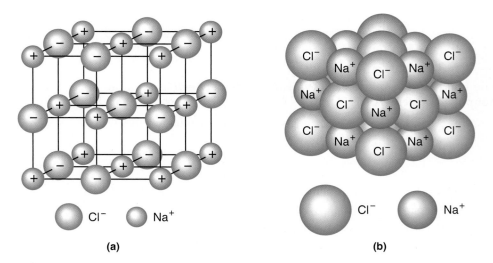

Figure 4.3
The arrangement of ions in a crystal of NaCl (sodium chloride, table salt). (a) Microscopic arrangement of ions as point charges. (b) Microscopic arrangement of the spherical ions in crystal lattice.

several positive centers. Positive and negative ions arrange themselves in a regular three-dimensional repeating array to produce a stable arrangement known as a **crystal lattice.** The lattice structure for sodium chloride is shown from two different perspectives in Figure 4.3. The **formula** of an ionic compound is the smallest whole-number ratio of ions in the crystal.

Writing Formulas of Ionic Compounds from the Identities of the Component Ions

Learning Goal

It is important to be able to write the formula of an ionic compound when provided with the identities of the ions that make up the compound. The charge of each ion can usually be determined from the group (family) of the periodic table in which the parent element is found. The cations and anions must combine in such a way that the resulting formula unit has a net charge of zero.

Consider the following examples.

EXAMPLE 4.1 *Predicting the Formula of an Ionic Compound*

Predict the formula of the ionic compound formed from the reaction of sodium and oxygen atoms.

Solution

Sodium is in group IA (or 1); it has *one* valence electron. Loss of this electron produces Na^+. Oxygen is in group VIA (or 16); it has *six* valence electrons. A gain of two electrons (to create a stable octet) produces O^{2-}. Two positive charges are necessary to counterbalance two negative charges on the oxygen anion. Because each sodium ion carries a 1+ charge, two sodium ions are needed for each O^{2-}. The subscript 2 is used to indicate that the formula unit contains two sodium ions. Thus the formula of the compound is Na_2O.

EXAMPLE 4.2 *Predicting the Formula of an Ionic Compound*

Predict the formula of the compound formed by the reaction of aluminum and oxygen atoms.

Continued—

EXAMPLE 4.2 —*Continued*

Solution

Aluminum is in group IIIA (or 13) of the periodic table; we predict that it has three valence electrons. Loss of these electrons produces Al^{3+}. Oxygen is in group VIA (or 16) of the periodic table and has six valence electrons. A gain of two electrons (to create a stable octet) produces O^{2-}. How can we combine Al^{3+} and O^{2-} to yield a unit of zero charge? It is necessary that *both* the cation and anion be multiplied by factors that will result in a zero net charge:

$$2 \times (+3) = +6 \quad \text{and} \quad 3 \times (-2) = -6$$

$$2 \times Al^{3+} = +6 \quad \text{and} \quad 3 \times O^{2-} = -6$$

Hence the formula is Al_2O_3.

Question 4.1

Predict the formulas of the compounds formed from the combination of ions of the following elements:

a. lithium and bromine
b. calcium and bromine
c. calcium and nitrogen

Question 4.2

Predict the formulas of the compounds formed from the combination of ions of the following elements:

a. potassium and chlorine
b. magnesium and bromine
c. magnesium and nitrogen

Writing Names of Ionic Compounds from the Formula of the Compound

Learning Goal

Nomenclature, the way in which compounds are named, is based on their formulas. The name of the cation appears first, followed by the name of the anion. The positive ion has the name of the element; the negative ion is named by using the *stem* of the name of the element joined to the suffix *-ide*. Some examples follow.

Formula	*cation*	*and*	*anion stem*	+	*ide*	=	*Compound name*
NaCl	sodium		chlor	+	ide		sodium chloride
Na_2O	sodium		ox	+	ide		sodium oxide
Li_2S	lithium		sulf	+	ide		lithium sulfide
$AlBr_3$	aluminum		brom	+	ide		aluminum bromide
CaO	calcium		ox	+	ide		calcium oxide

If the cation and anion exist in only one common charged form, there is no ambiguity between formula and name. Sodium chloride *must be* NaCl, and lithium sulfide *must be* Li_2S, so that the sum of positive and negative charges is zero. With many elements, such as the transition metals, several ions of different charge may exist. Fe^{2+}, Fe^{3+} and Cu^{+}, Cu^{2+} are two common examples. Clearly, an ambiguity exists if we use the name iron for both Fe^{2+} and Fe^{3+} or copper for both Cu^{+} and Cu^{2+}. Two systems have been developed to avoid this problem: the *Stock system* and the *common nomenclature system.*

Table 4.1 Systematic (Stock) and Common Names for Iron and Copper Ions

For systematic name:			
Formula	***+ Ion Charge***	***Cation Name***	***Compound Name***
$FeCl_2$	2+	Iron(II)	Iron(II) chloride
$FeCl_3$	3+	Iron(III)	Iron(III) chloride
Cu_2O	1+	Copper(I)	Copper(I) oxide
CuO	2+	Copper(II)	Copper(II) oxide
For common nomenclature:			
Formula	***+ Ion Charge***	***Cation Name***	***Common -ous/ic Name***
$FeCl_2$	2+	Ferr*ous*	Ferrous chloride
$FeCl_3$	3+	Ferr*ic*	Ferric chloride
Cu_2O	1+	Cupr*ous*	Cuprous oxide
CuO	2+	Cupr*ic*	Cupric oxide

Table 4.2 Common Monatomic Cations and Anions

Cation	Name	Anion	Name
H^+	Hydrogen ion	H^-	Hydride ion
Li^+	Lithium ion	F^-	Fluoride ion
Na^+	Sodium ion	Cl^-	Chloride ion
K^+	Potassium ion	Br^-	Bromide ion
Cs^+	Cesium ion	I^-	Iodide ion
Be^{2+}	Beryllium ion	O^{2-}	Oxide ion
Mg^{2+}	Magnesium ion	S^{2-}	Sulfide ion
Ca^{2+}	Calcium ion	N^{3-}	Nitride ion
Ba^{2+}	Barium ion	P^{3-}	Phosphide ion
Al^{3+}	Aluminum ion		
Ag^+	Silver ion		

Note: The ions of principal importance are highlighted in blue.

In the Stock system (systematic name), a Roman numeral indicates the magnitude of the cation's charge. In the older common nomenclature system, the suffix *-ous* indicates the lower ionic charge, and the suffix *-ic* indicates the higher ionic charge. Consider the examples in Table 4.1.

Systematic names are easier and less ambiguous than common names. Whenever possible, we will use this system of nomenclature. The older, common names (-ous, -ic) are less specific; furthermore, they often use the Latin names of the elements (for example, iron compounds use *ferr-*, from *ferrum,* the Latin word for iron).

Monatomic ions are ions consisting of a single atom. Common monatomic ions are listed in Table 4.2. The ions that are particularly important in biological systems are highlighted in blue.

Polyatomic ions, such as the hydroxide ion, OH^-, are composed of two or more atoms bonded together. These ions, although bonded to other ions with ionic bonds, are themselves held together by covalent bonds.

The polyatomic ion has an *overall* positive or negative charge. Some common polyatomic ions are listed in Table 4.3. The formulas, charges, and names of these polyatomic ions, especially those highlighted in blue, should be memorized.

Table 4.3 Common Polyatomic Cations and Anions

Ion	Name
NH_4^+	Ammonium
NO_2^-	Nitrite
NO_3^-	Nitrate
SO_3^{2-}	Sulfite
SO_4^{2-}	Sulfate
HSO_4^-	Hydrogen sulfate
OH^-	Hydroxide
CN^-	Cyanide
PO_4^{3-}	Phosphate
HPO_4^{2-}	Hydrogen phosphate
$H_2PO_4^-$	Dihydrogen phosphate
CO_3^{2-}	Carbonate
HCO_3^-	Bicarbonate
ClO^-	Hypochlorite
ClO_2^-	Chlorite
ClO_3^-	Chlorate
ClO_4^-	Perchlorate
CH_3COO^- (or $C_2H_3O_2^-$)	Acetate
MnO_4^-	Permanganate
$Cr_2O_7^{2-}$	Dichromate
CrO_4^{2-}	Chromate
O_2^{2-}	Peroxide

Note: The most commonly encountered ions are highlighted in blue.

The following examples are formulas of several compounds containing polyatomic ions.

Formula	Cation	Anion	Name
NH_4Cl	NH_4^+	Cl^-	ammonium chloride
$Ca(OH)_2$	Ca^{2+}	OH^-	calcium hydroxide
Na_2SO_4	Na^+	SO_4^{2-}	sodium sulfate
$NaHCO_3$	Na^+	HCO_3^-	sodium bicarbonate

Sodium bicarbonate may also be named sodium hydrogen carbonate, a preferred and less ambiguous name. Likewise, Na_2HPO_4 is named sodium hydrogen phosphate, and other ionic compounds are named similarly.

Question 4.3

Name each of the following compounds:

a. KCN
b. MgS
c. $Mg(CH_3COO)_2$

Question 4.4

Name each of the following compounds:

a. Li_2CO_3
b. $FeBr_2$
c. $CuSO_4$

Writing Formulas of Ionic Compounds from the Name of the Compound

Learning Goal

It is also important to be able to write the correct formula when given the compound name. To do this, we must be able to predict the charge of monatomic ions and remember the charge and formula of polyatomic ions. Equally important, the relative number of positive and negative ions in the unit must result in a net (compound) charge of zero. The compounds are electrically neutral. Two examples follow.

EXAMPLE 4.3 ***Writing a Formula When Given the Name of the Compound***

Write the formula of sodium sulfate.

Solution

Step 1. The sodium ion is Na^+, a group I (or 1) element. The sulfate ion is SO_4^{2-} (from Table 4.3).

Step 2. Two positive charges, two sodium ions, are needed to cancel the charge on one sulfate ion (two negative charges).

Hence the formula is Na_2SO_4.

EXAMPLE 4.4 ***Writing a Formula When Given the Name of the Compound***

Write the formula of ammonium sulfide.

Solution

Step 1. The ammonium ion is NH_4^+ (from Table 4.3). The sulfide ion is S^{2-} (from its position on the periodic table).

Step 2. Two positive charges are necessary to cancel the charge on one sulfide ion (two negative charges).

Hence the formula is $(NH_4)_2S$.

Note that parentheses must be used whenever a subscript accompanies a polyatomic ion.

Question 4.5

Write the formula for each of the following compounds:

a. calcium carbonate
b. sodium bicarbonate
c. copper(I) sulfate

Question 4.6

Write the formula for each of the following compounds:

a. sodium phosphate
b. potassium bromide
c. iron(II) nitrate

Covalent Compounds

Naming Covalent Compounds

Learning Goal

Most covalent compounds are formed by the reaction of nonmetals. **Molecules** are compounds characterized by covalent bonding. We saw earlier that ionic

Table 4.4 Prefixes Used to Denote Numbers of Atoms in a Compound

Prefix	Number of Atoms
Mono-	1
Di-	2
Tri-	3
Tetra-	4
Penta-	5
Hexa-	6
Hepta-	7
Octa-	8
Nona-	9
Deca-	10

compounds are not composed of single units but are a part of a massive three-dimensional crystal structure in the solid state. Covalent compounds exist as discrete molecules in the solid, liquid, and gas states. This is a distinctive feature of covalently bonded substances.

The conventions for naming covalent compounds follow:

1. The names of the elements are written in the order in which they appear in the formula.
2. A prefix (Table 4.4) indicating the number of each kind of atom found in the unit is placed before the name of the element.
3. If only one atom of a particular kind is present in the molecule, the prefix mono- is usually omitted from the first element.
4. The stem of the name of the last element is used with the suffix -ide.
5. The final vowel in a prefix is often dropped before a vowel in the stem name.

By convention the prefix *mono-* is often omitted from the second element as well (dinitrogen oxide, not dinitrogen monoxide). In other cases, common usage retains the prefix (carbon monoxide, not carbon oxide).

Naming a Covalent Compound **EXAMPLE 4.5**

Name the covalent compound N_2O_4.

Solution

Step 1. two nitrogen atoms   four oxygen atoms

Step 2. di-   tetra-

Step 3. dinitrogen   tetr(a)oxide

The name is dinitrogen tetroxide.

The following are examples of other covalent compounds.

Formula	Name
N_2O	dinitrogen monoxide
NO_2	nitrogen dioxide
SiO_2	silicon dioxide
CO_2	carbon dioxide
CO	carbon monoxide

Question 4.7

Name each of the following compounds:

a. B_2O_3
b. NO
c. ICl
d. PCl_3

Question 4.8

Name each of the following compounds:

a. H_2S
b. CS_2
c. PCl_5
d. P_2O_5

Writing Formulas of Covalent Compounds

Learning Goal 2

Many compounds are so familiar to us that their *common names* are generally used. For example, H_2O is water, NH_3 is ammonia, C_2H_5OH (ethanol) is alcohol, and $C_6H_{12}O_6$ is glucose. It is useful to be able to correlate both systematic and common names with the corresponding molecular formula and vice versa.

When common names are used, formulas of covalent compounds can be written *only* from memory. You *must* remember that water is H_2O, ammonia is NH_3, and so forth. This is the major disadvantage of common names. Because of their widespread use, however, they cannot be avoided and must be memorized.

Compounds named by using Greek prefixes are easily converted to formulas. Consider the following examples.

EXAMPLE 4.6 ***Writing the Formula of a Covalent Compound***

Write the formula of nitrogen monoxide.

Solution

Nitrogen has no prefix; one is understood. Oxide has the prefix *mono*—one oxygen. Hence the formula is NO.

EXAMPLE 4.7 ***Writing the Formula of a Covalent Compound***

Write the formula of dinitrogen tetroxide.

Solution

Nitrogen has the prefix *di*—two nitrogen atoms. Oxygen has the prefix *tetr(a)*—four oxygen atoms. Hence the formula is N_2O_4.

Question 4.9

Write the formula of each of the following compounds:

a. diphosphorus pentoxide
b. silicon dioxide

Question 4.10

Write the formula of each of the following compounds:

a. nitrogen trifluoride
b. carbon monoxide

A HUMAN PERSPECTIVE

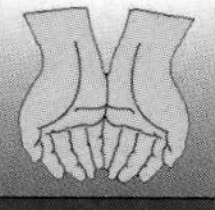

Origin of the Elements

The current, most widely held theory of the origin of the universe is the "big bang" theory. An explosion of very dense matter was followed by expansion into space of the fragments resulting from this explosion. This is one of the scenarios that have been created by scientists fascinated by the origins of matter, the stars and planets, and life as we know it today.

The first fragments, or particles, were protons and neutrons moving with tremendous velocity and possessing large amounts of energy. Collisions involving these high-energy protons and neutrons formed deuterium atoms (^2H), which are isotopes of hydrogen. As the universe expanded and cooled, tritium (^3H), another hydrogen isotope, formed as a result of collisions of neutrons with deuterium atoms. Subsequent capture of a proton produced helium (He). Scientists theorize that a universe that was principally composed of hydrogen and helium persisted for perhaps 100,000 years until the temperature decreased sufficiently to allow the formation of a simple molecule, hydrogen, two atoms of hydrogen bonded together (H_2).

Many millions of years later, the effect of gravity caused these small units to coalesce, first into clouds and eventually into stars, with temperatures of millions of degrees. In this setting, these small collections of protons and neutrons combined to form larger atoms such as carbon (C) and oxygen (O), then sodium (Na), neon (Ne), magnesium (Mg), silicon (Si), and so forth. Subsequent explosions of stars provided the conditions that formed many larger atoms. These fragments, gathered together by the force of gravity, are the most probable origin of the planets in our own solar system.

The reactions that formed the elements as we know them today were a result of a series of *fusion reactions,* the joining of nuclei to produce larger atoms at very high temperatures (millions of degrees Celsius). These fusion reactions are similar to processes that are currently being studied as a possible alternative source of nuclear power. We shall study such nuclear processes in more detail in Chapter 10.

Nuclear reactions of this type do not naturally occur on the earth today. The temperature is simply too low. As a result we have, for the most part, a collection of stable elements existing as chemical compounds, atoms joined together by chemical bonds while retaining their identity even in the combined state. Silicon exists all around us as sand and soil in a combined form, silicon dioxide; most metals exist as a part of a chemical compound, such as iron ore. We are learning more about the structure and properties of these compounds in this chapter.

4.3 Properties of Ionic and Covalent Compounds

Learning Goal

The differences in ionic and covalent bonding result in markedly different properties for ionic and covalent compounds. Because covalent molecules are distinct units, they have less tendency to form an extended structure in the solid state. Ionic compounds, with ions joined by electrostatic attraction, do not have definable units but form a crystal lattice composed of enormous numbers of positive and negative ions in an extended three-dimensional network. The effects of this basic structural difference are summarized in this section.

Physical State

All ionic compounds (for example, NaCl, KCl, and $NaNO_3$) are solids at room temperature; covalent compounds may be solids (sugar), liquids (H_2O, ethanol), or gases (carbon monoxide, carbon dioxide). The three-dimensional crystal structure that is characteristic of ionic compounds holds them in a rigid, solid arrangement, whereas molecules of covalent compounds may be fixed, as in a solid, or more mobile, a characteristic of liquids and gases.

Melting and Boiling Points

The **melting point** is the temperature at which a solid is converted to a liquid and the **boiling point** is the temperature at which a liquid is converted to a gas at a specified pressure. Considerable energy is required to break apart an ionic crystal lattice with uncountable numbers of ionic interactions and convert the ionic substance to a liquid or a gas. As a result, the melting and boiling temperatures for ionic compounds are generally higher than those of covalent compounds, whose molecules interact less strongly in the solid state. A typical ionic compound, sodium chloride, has a melting point of 801°C; methane, a covalent compound, melts at −182°C. Exceptions to this general rule do exist; diamond, a covalent solid with an extremely high melting point, is a well-known example.

Structure of Compounds in the Solid State

Ionic solids are *crystalline*, characterized by a regular structure, whereas covalent solids may either be crystalline or have no regular structure. In the latter case they are said to be *amorphous*.

Solutions of Ionic and Covalent Compounds

In Chapter 2 we saw that mixtures are either heterogeneous or homogeneous. A homogeneous mixture is a solution. Many ionic solids dissolve in solvents, such as water. An ionic solid, if soluble, will form positive and negative ions in solution by **dissociation.**

The role of the solvent in the dissolution of solids is discussed in Section 4.5.

Because ions in water are capable of carrying (conducting) a current of electricity, we refer to these compounds as **electrolytes,** and the solution is termed an **electrolytic solution.** Covalent solids dissolved in solution usually retain their neutral (molecular) character and are **nonelectrolytes.** The solution is not an electrical conductor.

4.4 Drawing Lewis Structures of Molecules and Polyatomic Ions

Lewis Structures of Molecules

Learning Goal

In Section 4.1, we used Lewis structures of individual atoms to help us understand the bonding process. To begin to explain the relationship between molecular structure and molecular properties, we will first need a set of guidelines to help us write Lewis structures for more complex molecules.

The skeletal structure indicates only the relative positions of atoms in the molecule or ion. Bonding information results from the Lewis structure.

The central atom is often the element farthest to the left and/or lowest in the periodic table.
The central atom is often the element in the compound for which there is only one atom.
Hydrogen is *never* the central atom.

1. *Use chemical symbols for the various elements to write the skeletal structure of the compound.* To accomplish this, place the bonded atoms next to one another. This is relatively easy for simple compounds; however, as the number of atoms in the compound increases, the possible number of arrangements increases dramatically. We may be told the pattern of arrangement of the atoms in advance; if not, we can make an intelligent guess and see if a reasonable Lewis structure can be constructed. Three considerations are very important here:
   - the least electronegative atom will be placed in the central position (the central atom),
   - hydrogen and fluorine (and the other halogens) often occupy terminal positions,
   - carbon often forms chains of carbon-carbon covalent bonds.
2. *Determine the number of valence electrons associated with each atom; combine them to determine the total number of valence electrons in the compound.* However, if we are representing polyatomic cations or anions, we must account for the charge on the ion. Specifically:
   - for polyatomic cations, subtract one electron for each unit of positive charge. This accounts for the fact that the positive charge arises from electron loss.
   - for polyatomic anions, add one electron for each unit of negative charge. This accounts for excess negative charge resulting from electron gain.
3. *Connect the central atom to each of the surrounding atoms using electron pairs.* Then complete the octets of all of the atoms bonded to the central atom. Recall that hydrogen needs only two electrons to complete its valence shell.

A CLINICAL PERSPECTIVE

Blood Pressure and the Sodium Ion/Potassium Ion Ratio

When you have a physical exam, the physician measures your blood pressure. This indicates the pressure of blood against the walls of the blood vessels each time the heart pumps. A blood pressure reading is always characterized by two numbers. With every heartbeat there is an increase in pressure; this is the systolic blood pressure. When the heart relaxes between contractions, the pressure drops; this is the diastolic pressure. Thus the blood pressure is expressed as two values—for instance, 117/72—measured in millimeters of mercury. Hypertension is simply defined as high blood pressure. To the body it means that the heart must work too hard to pump blood, and this can lead to heart failure or heart disease.

Heart disease accounts for 50% of all deaths in the United States. Epidemiological studies correlate the following major risk factors with heart disease: heredity, sex, race, age, diabetes, cigarette smoking, high blood cholesterol, and hypertension. Obviously, we can do little about our age, sex, and genetic heritage, but we can stop smoking, limit dietary cholesterol, and maintain a normal blood pressure.

The number of Americans with hypertension is alarmingly high: 60 million adults and children. More than 10 million of these individuals take medication to control blood pressure, at a cost of nearly $2.5 billion each year. In many cases, blood pressure can be controlled without medication by increasing physical activity, losing weight, decreasing consumption of alcohol, and limiting intake of sodium.

It has been estimated that the average American ingests 7.5–10 g of salt (NaCl) each day. Because NaCl is about 40% (by mass) sodium ions, this amounts to 3–4 g of sodium daily. Until 1989 the Food and Nutrition Board of the National Academy of Sciences National Research Council's defined *e*stimated *s*afe and *a*dequate *d*aily *d*ietary *i*ntake (ESADDI) of sodium ion was 1.1–3.3 g. Clearly, Americans exceed this recommendation.

Recently, studies have shown that excess sodium is not the sole consideration in the control of blood pressure. More important is the sodium ion/potassium ion (Na^+/K^+) ratio. That ratio should be about 0.6; in other words, our diet should contain about 67% more potassium than sodium. Does the typical American diet fall within this limit? Definitely not! Young American males (25–30 years old) consume a diet with a $Na^+/K^+ = 1.07$, and the diet of females of the same age range has a $Na^+/K^+ = 1.04$. It is little wonder that so many Americans suffer from hypertension.

How can we restrict sodium in the diet, while increasing the potassium? The following table lists a variety of foods that are low in sodium and high in potassium. These include fresh fruits and vegetables and fruit juices, a variety of cereals, unsalted nuts, and cooked dried beans (legumes). The table also notes some high-sodium, low-potassium foods. Notice that most of these are processed or prepared foods. This points out how difficult it can be to control sodium in the diet. The majority of the sodium that we ingest comes from commercially prepared foods. The consumer must read the nutritional information printed on cans and packages to determine whether the sodium levels are within acceptable limits.

Low Sodium Ion, High Potassium Ion Foods

Food Category	Examples
Fruit and fruit juices	Pineapple, grapefruit, pears, strawberries, watermelon, raisins, bananas, apricots, oranges
Low-sodium cereals	Oatmeal (unsalted), Roman Meal Hot Cereal, shredded wheat
Nuts (unsalted)	Hazelnuts, macadamia nuts, almonds, peanuts, cashews, coconut
Vegetables	Summer squash, zucchini, eggplant, cucumber, onions, lettuce, green beans, broccoli
Beans (dry, cooked)	Great Northern beans, lentils, lima beans, red kidney beans

High Sodium Ion, Low Potassium Ion Foods

Food Category	Examples
Fats	Butter, margarine, salad dressings
Soups	Onion, mushroom, chicken noodle, tomato, split pea
Breakfast cereals	Many varieties; consult the label for specific nutritional information.
Breads	Most varieties
Processed meats	Most varieties
Cheese	Most varieties

Electrons not involved in bonding must be represented as lone pairs and the total number of electrons in the structure must equal the number of valence electrons computed in our second step.

4. *If the octet rule is not satisfied for the central atom, move one or more electron pairs* from the surrounding atoms to create double or triple bonds until all atoms have an octet.

5. *After you are satisfied with the Lewis structure that you have constructed, perform a final electron count* verifying that the total number of electrons **and** the number around each atom are correct.

Now, let us see how these guidelines are applied in the examples that follow.

EXAMPLE 4.8 ***Drawing Lewis Structures of Covalent Compounds***

Draw the Lewis structure of carbon dioxide, CO_2.

Solution

Draw a skeletal structure of the molecule, arranging the atoms in their most probable order.

For CO_2, two possibilities exist:

$$\text{C—O—O} \quad \text{and} \quad \text{O—C—O}$$

Referring to Figure 4.2, we find that the electronegativity of oxygen is 3.5 whereas that of carbon is 2.5. Our strategy dictates that the least electronegative atom, in this case carbon, is the central atom. Hence the skeletal structure O—C—O may be presumed correct.

Next, we want to determine the number of valence electrons on each atom and add them to arrive at the total for the compound.

For CO_2,

$$1 \text{ C atom} \times 4 \text{ valence electrons} = 4\ e^-$$
$$2 \text{ O atoms} \times 6 \text{ valence electrons} = 12\ e^-$$
$$16\ e^- \text{ total}$$

Now, use electron pairs to connect the central atom, C, to each oxygen with a single bond.

$$\text{O:C:O}$$

Distribute the electrons around the atoms (in pairs if possible) in an attempt to satisfy the octet rule, eight electrons around each element.

$$:\ddot{\underset{..}{\text{O}}}:\text{C}:\ddot{\underset{..}{\text{O}}}:$$

This structure satisfies the octet rule for each oxygen atom, but not the carbon atom (only four electrons surround the carbon).

However, when this structure is modified by moving two electrons from each oxygen atom to a position between C and O, each oxygen and carbon atom is surrounded by eight electrons. The octet rule is satisfied, and the structure below is the most probable Lewis structure for CO_2.

$$\ddot{\underset{..}{\text{O}}}::\text{C}::\ddot{\underset{..}{\text{O}}}$$

In this structure, four electrons (two electron pairs) are located between C and each O, and these electrons are shared in covalent bonds. Because a **single bond** is composed of two electrons (one electron pair) and because four electrons "bond" the carbon atom to each oxygen atom in this structure, there must be two bonds between each oxygen atom and the carbon atom, a **double bond:**

The notation for a single bond : is equivalent to — (one pair of electrons).

The notation for a double bond : : is equivalent to = (two pairs of electrons).

Continued—

EXAMPLE 4.8 —*Continued*

We may write CO_2 as shown above or, replacing dots with dashes to indicate bonding electron pairs,

$$\ddot{\underset{..}{O}}=C=\ddot{\underset{..}{O}}$$

As a final step, let us do some "electron accounting." There are eight electron pairs, and they correspond to sixteen valence electrons (8 pair $\times$ $2e^-$/pair). Furthermore, there are eight electrons around each atom and the octet rule is satisfied. Therefore

$$\ddot{\underset{..}{O}}=C=\ddot{\underset{..}{O}}$$

is a satisfactory way to depict the structure of CO_2.

Drawing Lewis Structures of Covalent Compounds

EXAMPLE 4.9

Draw the Lewis structure of ammonia, NH_3.

Solution

When trying to implement the first step in our strategy we may be tempted to make H our central atom because it is less electronegative than N. But, remember the margin note in this section:

"Hydrogen is *never* the central atom"

Hence:

```
   H
   |
H—N—H
```

is our skeletal structure.

Applying our strategy to determine the total valence electrons for the molecule, we find that there are five valence electrons in nitrogen and one in each of the three hydrogens, for a total of eight valence electrons.

Applying our strategy for distribution of valence electrons results in the following Lewis diagram:

```
    H
    ..
H : N : H
    ..
```

This satisfies the octet rule for nitrogen (eight electrons around N) and hydrogen (two electrons around each H) and is an acceptable structure for ammonia. Ammonia may also be written:

```
   H
   |
H—N—H
  ..
```

Note the pair of nonbonding electrons on the nitrogen atom. These are often called a **lone pair,** or *unshared* pair, of electrons. As we will see later in this section, lone pair electrons have a profound effect on molecular geometry. The geometry, in turn, affects the reactivity of the molecule.

Question 4.11

Draw a Lewis structure for each of the following covalent compounds:

a. H_2O (water)
b. CH_4 (methane)

Question 4.12

Draw a Lewis structure for each of the following covalent compounds:

a. C_2H_6 (ethane)
b. N_2 (nitrogen gas)

Lewis Structures of Polyatomic Ions

Learning Goal

The strategies for writing the Lewis structures of polyatomic ions are similar to those for neutral compounds. There is, however, one major difference: The charge on the ion must be accounted for when computing the total number of valence electrons.

EXAMPLE 4.10 ***Drawing Lewis Structures of Polyatomic Cations***

Draw the Lewis structure of the ammonium ion, NH_4^+.

Solution

The ammonium ion has the following skeletal structure and charge:

$$\left[\begin{array}{ccc} & H & \\ & | & \\ H— & N & —H \\ & | & \\ & H & \end{array}\right]^+$$

The total number of valence electrons is determined by subtracting one electron for each unit of positive charge.

$$\begin{array}{lcr} 1 \text{ N atom} \times 5 \text{ valence electrons} & = & 5\ e^- \\ 4 \text{ H atoms} \times 1 \text{ valence electron} & = & 4\ e^- \\ -\ 1 \text{ electron for } +1 \text{ charge} & = & -1\ e^- \\ \hline & & 8\ e^- \text{ total} \end{array}$$

Distribute these eight electrons around our skeletal structure:

$$\left[\begin{array}{c} H \\ H : \ddot{\underset{\cdot\cdot}{N}} : H \\ H \end{array}\right]^+ \quad \text{or} \quad \left[\begin{array}{ccc} & H & \\ & | & \\ H— & N & —H \\ & | & \\ & H & \end{array}\right]^+$$

A final check shows eight total electrons, eight around the central atom, nitrogen, and two electrons associated with each hydrogen. Hence the structure is satisfactory.

EXAMPLE 4.11 ***Drawing Lewis Structures of Polyatomic Anions***

Draw the Lewis structure of the sulfate ion, SO_4^{2-}.

Solution

Sulfur is less electronegative than oxygen. Therefore sulfur is the central atom. The sulfate ion has the following skeletal structure and charge:

Continued—

EXAMPLE 4.11 —*Continued*

$$\left[\begin{array}{ccc} & O & \\ & | & \\ O— & S & —O \\ & | & \\ & O & \end{array}\right]^{2-}$$

The total number of valence electrons is determined by adding one electron for each unit of negative charge:

$$\begin{array}{lcl} 1\ \text{S atom} \times 6\ \text{valence electrons} & = & 6\ e^- \\ 4\ \text{O atoms} \times 6\ \text{valence electrons} & = & 24\ e^- \\ +\ 2\ \text{negative charges} & = & 2\ e^- \\ \hline & & 32\ e^-\ \text{total} \end{array}$$

Distributing the electron dots around the central sulfur atom (forming four bonds) and around the surrounding oxygen atoms in an attempt to satisfy the octet rule results in the final structure:

$$\left[\begin{array}{ccc} & :\ddot{O}: & \\ :\ddot{O}: & S & :\ddot{O}: \\ & :\underset{..}{O}: & \end{array}\right]^{2-} \quad \text{or} \quad \left[\begin{array}{ccc} & :\ddot{O}: & \\ & | & \\ :\ddot{O}— & S & —\ddot{O}: \\ & | & \\ & :\underset{..}{O}: & \end{array}\right]^{2-}$$

A final check shows thirty-two electrons and eight electrons around each atom. Hence the structure is satisfactory.

Drawing Lewis Structures of Polyatomic Anions **EXAMPLE 4.12**

Draw the Lewis structure of the acetate ion, CH_3COO^-.

Solution

A commonly encountered anion, the acetate ion has a skeletal structure that is more complex than any of the examples that we have studied thus far. Which element should we choose as the central atom? We have three choices H, O, and C. H is eliminated because hydrogen can never be the central atom. Oxygen is more electronegative than carbon, so carbon must be the central atom. There are two carbon atoms; often they are joined. Further clues are obtained from the formula itself; CH_3COO^- implies three hydrogen atoms attached to the first carbon atom and two oxygen atoms joined to the second carbon. A plausible skeletal structure is:

$$\left[\begin{array}{cccc} & H & O & \\ & | & | & \\ H— & C— & C— & O \\ & | & & \\ & H & & \end{array}\right]^{-}$$

The pool of valence electrons for anions is determined by adding one electron for each unit of negative charge:

Continued—

EXAMPLE 4.12 —*Continued*

$$
\begin{array}{lcl}
2\text{ C atoms} \times 4\text{ valence electrons} & = & 8\ e^- \\
3\text{ H atoms} \times 1\text{ valence electron} & = & 3\ e^- \\
2\text{ O atoms} \times 6\text{ valence electrons} & = & 12\ e^- \\
+\ 1\text{ negative charge} & = & 1\ e^- \\
\hline
 & & 24\ e^-\text{ total}
\end{array}
$$

Distributing these twenty-four electrons around our skeletal structure gives

$$\left[\begin{array}{c} \text{H:O:} \\ \text{H:}\ddot{\text{C}}\text{:}\ddot{\text{C}}\text{:}\ddot{\text{O}}\text{:} \\ \text{H} \end{array}\right]^- \quad \text{or} \quad \left[\begin{array}{c} \text{H}\quad |\text{O}| \\ |\quad \| \\ \text{H—C—C—}\overline{\text{O}}| \\ | \\ \text{H} \end{array}\right]^-$$

This Lewis structure satisfies the octet rule for carbon and oxygen and surrounds each hydrogen with two electrons. All twenty-four electrons are used in this process.

Question 4.13

Draw the Lewis structure for each of the following ions:

a. H_3O^+ (the hydronium ion)
b. OH^- (the hydroxide ion)

Question 4.14

Draw the Lewis structure for each of the following ions:

a. CN^- (the cyanide ion)
b. CO_3^{2-} (the carbonate ion)

Question 4.15

Write a Lewis structure describing the bonding in each of the following polyatomic ions:

a. the bicarbonate ion, HCO_3^-
b. the phosphate ion, PO_4^{3-}

Question 4.16

Write a Lewis structure describing the bonding in each of the following polyatomic ions:

a. the hydrogen sulfide ion, HS^-
b. the peroxide ion, O_2^{2-}

Lewis Structure, Stability, Multiple Bonds, and Bond Energies

Learning Goal

Hydrogen, oxygen, and nitrogen are present in the atmosphere as diatomic gases, H_2, O_2, and N_2. All are covalent molecules. Their stability and reactivity, however, are quite different. Hydrogen is an explosive material, sometimes used as a fuel. Oxygen, although more stable than hydrogen, reacts with fuels in combustion. The explosion of the space shuttle *Challenger* resulted from the reaction of massive amounts of hydrogen and oxygen. Nitrogen, on the other hand, is extremely nonreactive. Because nitrogen makes up about 80% of the atmosphere, it dilutes the oxygen, which accounts for only about 20% of the atmosphere.

Breathing pure oxygen for long periods, although necessary in some medical situations, causes the breakdown of nasal and lung tissue over time. Oxygen diluted with nonreactive nitrogen is an ideal mixture for humans and animals to breathe.

Why is there such a great difference in reactivity among these three gases? We can explain this, in part, in terms of their bonding characteristics. The Lewis structure for H_2 (two valence electrons) is

$$H_2 \quad \text{or} \quad H:H \quad \text{or} \quad H\text{—}H$$

For oxygen (twelve valence electrons, six on each atom), the only Lewis structure that satisfies the octet rule is

$$O_2 \quad \text{or} \quad \ddot{\underset{\cdot\cdot}{O}}::\ddot{\underset{\cdot\cdot}{O}} \quad \text{or} \quad \ddot{\underset{\cdot\cdot}{O}}=\ddot{\underset{\cdot\cdot}{O}}$$

The Lewis structure of N_2 (ten total valence electrons) must be

$$N_2 \quad \text{or} \quad :N:::N: \quad \text{or} \quad :N\equiv N:$$

Therefore

N_2 has a *triple bond* (six bonding electrons).

O_2 has a *double bond* (four bonding electrons).

H_2 has a *single bond* (two bonding electrons).

The term *bond order* is sometimes used to distinguish among single, double, and triple bonds. A bond order of 1 corresponds to a single bond, 2 corresponds to a double bond, and 3 corresponds to a triple bond.

A **triple bond,** in which three pairs of electrons are shared by two atoms, is very stable. More energy is required to break a triple bond than a double bond, and a double bond is stronger than a single bond. Stability is related to the bond energy. The **bond energy** is the amount of energy, in units of kilocalories or kilojoules, required to break a bond holding two atoms together. Bond energy is therefore a *measure* of stability. The values of bond energies decrease in the order *triple bond > double bond > single bond.*

The bond length is related to the presence or absence of multiple bonding. The distance of separation of two nuclei is greatest for a single bond, less for a double bond, and still less for a triple bond. The *bond length* decreases in the order *single bond > double bond > triple bond.*

Question 4.17

Contrast a single and double bond with regard to:

a. distance of separation of the bonded nuclei
b. strength of the bond

How are these two properties related?

Question 4.18

Two nitrogen atoms in a nitrogen molecule are held together more strongly than the two chlorine atoms in a chlorine molecule. Explain this fact by comparing their respective Lewis structures.

Lewis Structures and Resonance

In some cases we find that it is possible to write more than one Lewis structure that satisfies the octet rule for a particular compound. Consider sulfur dioxide, SO_2. Its skeletal structure is

$$O\text{—}S\text{—}O$$

Total valence electrons may be calculated as follows:

$$
\begin{array}{rr}
1 \text{ sulfur atom} \times 6 \text{ valence e}^-/\text{atom} = & 6 \text{ e}^- \\
+\ 2 \text{ oxygen atoms} \times 6 \text{ valence e}^-/\text{atom} = & 12 \text{ e}^- \\
\hline
 & 18 \text{ e}^- \text{ total}
\end{array}
$$

The resulting Lewis structures are

$$\underset{\cdot\cdot}{\ddot{\text{O}}}::\ddot{\text{S}}:\underset{\cdot\cdot}{\ddot{\text{O}}}: \qquad \text{and} \qquad :\underset{\cdot\cdot}{\ddot{\text{O}}}:\ddot{\text{S}}::\underset{\cdot\cdot}{\ddot{\text{O}}}$$

Both satisfy the octet rule. However, experimental evidence shows no double bond in SO_2. The two sulfur-oxygen bonds are equivalent. Apparently, neither structure accurately represents the structure of SO_2, and neither actually exists. The actual structure is said to be an average or *hybrid* of these two Lewis structures. When a compound has two or more Lewis structures that contribute to the real structure, we say that the compound displays the property of **resonance.** The contributing Lewis structures are **resonance forms.** The true structure, a hybrid of the resonance forms, is known as a **resonance hybrid** and may be represented as:

$$\underset{\cdot\cdot}{\ddot{\text{O}}}::\ddot{\text{S}}:\underset{\cdot\cdot}{\ddot{\text{O}}}: \qquad \leftrightarrow \qquad :\underset{\cdot\cdot}{\ddot{\text{O}}}:\ddot{\text{S}}::\underset{\cdot\cdot}{\ddot{\text{O}}}$$

A common analogy might help to clarify this concept. A horse and a donkey may be crossbred to produce a hybrid, the mule. The mule doesn't look or behave exactly like either parent, yet it has attributes of both. The resonance hybrid of a molecule has properties of each resonance form but is not identical to any one form. Unlike the mule, resonance hybrids *do not actually exist.* Rather, they comprise a model that results from the failure of any one Lewis structure to agree with experimentally obtained structural information.

The presence of resonance enhances molecular stability. The more resonance forms that exist, the greater is the stability of the molecule they represent. This concept is important in understanding the chemical reactions of many complex organic molecules and is used extensively in organic chemistry.

EXAMPLE 4.13 ***Drawing Resonance Hybrids of Covalently Bonded Compounds***

Draw the possible resonance structures of the nitrate ion, NO_3^-, and represent them as a resonance hybrid.

Solution

Nitrogen is less electronegative than oxygen; therefore, nitrogen is the central atom and the skeletal structure is:

$$\left[\begin{array}{c} \text{O} \\ | \\ \text{O—N—O} \end{array}\right]^-$$

The pool of valence electrons for anions is determined by adding one electron for each unit of negative charge:

$$
\begin{array}{lr}
1 \text{ N atom} \times 5 \text{ valence electrons} & = \ \ 5 \text{ e}^- \\
3 \text{ O atoms} \times 6 \text{ valence electrons} & = 18 \text{ e}^- \\
+\ 1 \text{ negative charge} & = \ \ 1 \text{ e}^- \\
\hline
 & 24 \text{ e}^- \text{ total}
\end{array}
$$

Distributing the electrons throughout the structure results in the legitimate Lewis structures:

Continued—

EXAMPLE 4.13 —*Continued*

$\left[\begin{array}{c} :\ddot{O}: \\ \ddot{O}::\ddot{N}:\ddot{O}: \end{array}\right]^-$ and $\left[\begin{array}{c} :O: \\ :\ddot{O}:\ddot{N}:\ddot{O}: \end{array}\right]^-$ and $\left[\begin{array}{c} :\ddot{O}: \\ :\ddot{O}:\ddot{N}::\ddot{O} \end{array}\right]^-$

All contribute to the true structure of the nitrate ion, represented as a resonance hybrid.

$\left[\begin{array}{c} :\ddot{O}: \\ \ddot{O}::\ddot{N}:\ddot{O}: \end{array}\right]^- \leftrightarrow \left[\begin{array}{c} :O: \\ :\ddot{O}:\ddot{N}:\ddot{O}: \end{array}\right]^- \leftrightarrow \left[\begin{array}{c} :\ddot{O}: \\ :\ddot{O}:\ddot{N}::\ddot{O} \end{array}\right]^-$

Question 4.19

SeO_2, like SO_2, has two resonance forms. Draw their Lewis structures.

Question 4.20

Explain any similarities between the structures for SeO_2 and SO_2 (in Question 4.19) in light of periodic relationships.

Lewis Structures and Exceptions to the Octet Rule

The octet rule is remarkable in its ability to realistically model bonding and structure in covalent compounds. But, like any model, it does not adequately describe all systems. Beryllium, boron, and aluminum, in particular, tend to form compounds in which they are surrounded by fewer than eight electrons. This situation is termed an *incomplete octet.* Other molecules, such as nitric oxide:

$$\dot{\ddot{N}}=\ddot{\underset{..}{O}}$$

are termed *odd electron* molecules. Note that it is impossible to pair all electrons to achieve an octet simply because the compound contains an odd number of valence electrons. Elements in the third period and beyond may involve *d* orbitals and form an *expanded octet,* with ten or even twelve electrons surrounding the central atom. Examples 4.14 and 4.15 illustrate common exceptions to the octet rule.

EXAMPLE 4.14

Drawing Lewis Structures of Covalently Bonded Compounds That Are Exceptions to the Octet Rule

Draw the Lewis structure of beryllium hydride, BeH_2.

Solution

A reasonable skeletal structure of BeH_2 is:

H—Be—H

The total number of valence electrons in BeH_2 is:

$$\begin{array}{l} 1 \text{ beryllium atom} \times 2 \text{ valence } e^-/\text{atom} = 2\ e^- \\ \underline{2 \text{ hydrogen atoms} \times 1 \text{ valence } e^-/\text{atom} = 2\ e^-} \\ \qquad\qquad\qquad\qquad\qquad\qquad 4\ e^- \text{ total} \end{array}$$

Continued—

EXAMPLE 4.14 *—Continued*

The resulting Lewis structure must be:

$$\mathrm{H:Be:H} \qquad \text{or} \qquad \mathrm{H{-}Be{-}H}$$

It is apparent that there is no way to satisfy the octet rule for Be in this compound. Consequently, BeH_2 is an exception to the octet rule. It contains an incomplete octet.

EXAMPLE 4.15 ***Drawing Lewis Structures of Covalently Bonded Compounds That Are Exceptions to the Octet Rule***

Draw the Lewis structure of phosphorus pentafluoride.

Solution

A reasonable skeletal structure of PF_5 is:

```
    F
 F\ | /F
    P
  /   \
 F     F
```

Phosphorus is a third-period element; it may have an expanded octet.

The total number of valence electrons is:

$$\begin{array}{r} 1 \text{ phosphorus atom} \times 5 \text{ valence e}^-/\text{atom} = 5 \text{ e}^- \\ \underline{5 \text{ fluorine atoms} \times 7 \text{ valence e}^-/\text{atom} = 35 \text{ e}^-} \\ 40 \text{ e}^- \text{ total} \end{array}$$

Distributing the electrons around each F in the skeletal structure results in the Lewis structure:

```
        ..
       :F: ..
        | /F:
  ..    |/ ..
 :F — P
  ..    |\ ..
        | \F:
       :F: ..
        ..
```

PF_5 is an example of a compound with an expanded octet.

Lewis Structures and Molecular Geometry; VSEPR Theory

Learning Goal 7

The shape of a molecule plays a large part in determining its properties and reactivity. We may predict the shapes of various molecules by inspecting their Lewis structures for the orientation of their electron pairs. The covalent bond, for instance, in which bonding electrons are localized between the nuclear centers of the atoms, is *directional;* the bond has a specific orientation in space between the bonded atoms. Electrostatic forces in ionic bonds, in contrast, are *nondirectional;* they have no specific orientation in space. The specific orientation of electron pairs in covalent molecules imparts a characteristic shape to the molecules. Consider the following series of molecules whose Lewis structures are shown.

BeH_2	H : Be : H
BF_3	:F̤: (above) :F̤ : B : F̤:
CH_4	H (above) H : C : H, H (below)
NH_3	H : N̤ : H, H (below)
H_2O	H : O̤ : H

The electron pairs around the central atom of the molecule arrange themselves to minimize electronic repulsion. This means that the electron pairs arrange themselves so that they can be as far as possible from each other. We may use this fact to predict molecular shape. This approach is termed the ***v*alence *s*hell *e*lectron *p*air *r*epulsion (VSEPR) theory.**

Let's see how the VSEPR theory can be used to describe the bonding and structure of each of the preceding molecules.

BeH_2

As we saw in Example 4.14, beryllium hydride has two shared electron pairs around the beryllium atom. These electron pairs have minimum repulsion if they are located as far apart as possible while still bonding the hydrogen to the central atom. This condition is met if the electron pairs are located on opposite sides of the molecule, resulting in a **linear structure,** 180° apart:

Only four electrons surround the beryllium atom in BeH_2. Consequently, BeH_2 is a stable exception to the octet rule.

H : Be : H

or

H—Be—H (180°)

The *bond angle,* the angle between H—Be and Be—H bonds, formed by the two bonding pairs is 180° (Figure 4.4).

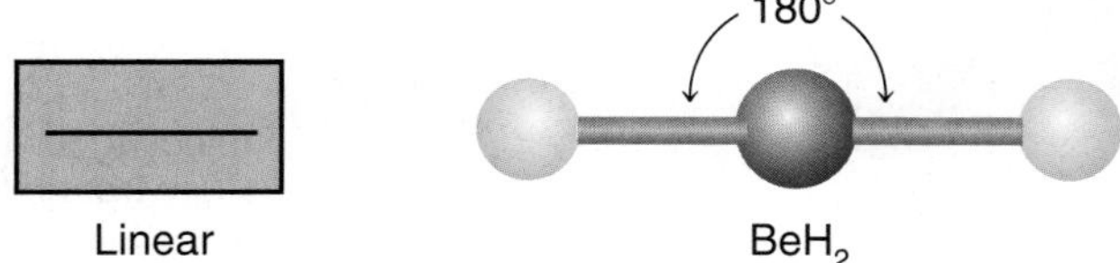

Figure 4.4
Bonding and geometry in beryllium hydride, BeH_2.

BF_3

Boron trifluoride has three shared electron pairs around the central atom. Placing the electron pairs in a plane, forming a triangle, minimizes the electron pair repulsion in this molecule, as depicted in Figure 4.5 and the following sketches:

BF_3 has only six electrons around the central atom, B. It is one of a number of stable compounds that are exceptions to the octet rule.

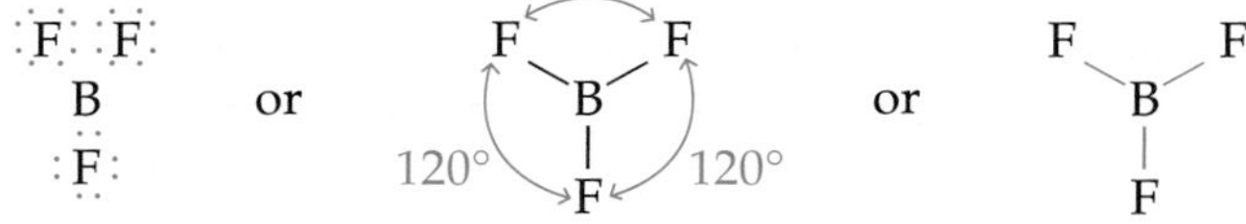

Such a structure is *trigonal planar,* and each F—B—F bond angle is 120°. We also find that compounds with central atoms in the same group of the periodic table have similar geometry. Aluminum, in the same group as boron, produces compounds such as AlH_3, which is also trigonal planar.

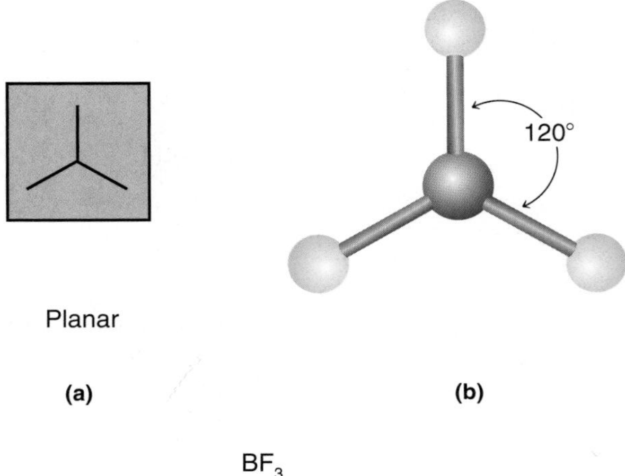

Figure 4.5
Representation of the two-dimensional structure of boron trifluoride, BF_3. (a) Trigonal planar structure. (b) Ball and stick model of trigonal planar BF_3.

CH_4, NH_3, and H_2O all have eight electrons around their central atoms; all obey the octet rule.

CH_4

Methane has four shared pairs of electrons. Here, minimum electron repulsion is achieved by arranging the electrons at the corners of a tetrahedron (Figure 4.6). Each H—C—H bond angle is 109.5°. Methane has a three-dimensional **tetrahedral structure.** Silicon, in the same group as carbon, forms compounds such as $SiCl_4$ and SiH_4 that also have tetrahedral structures.

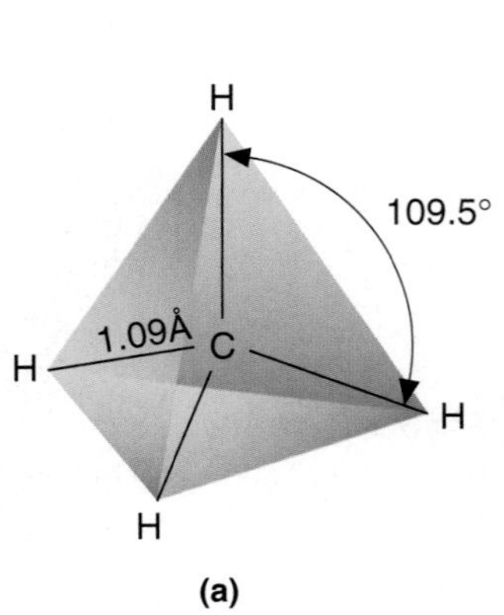

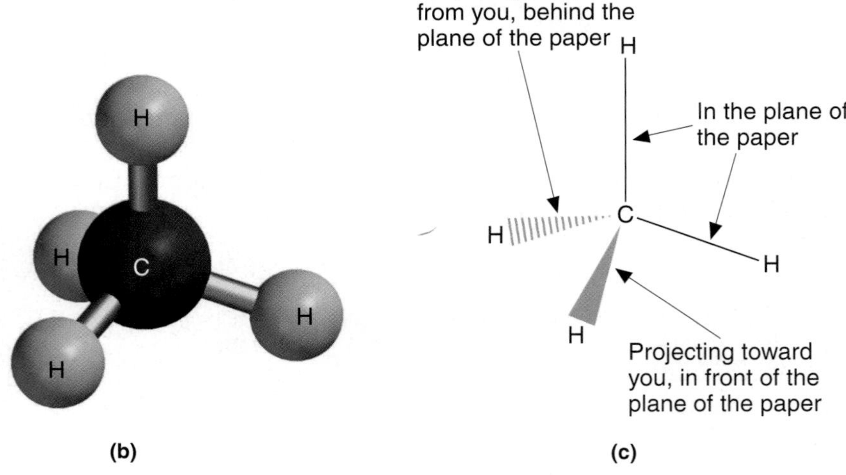

Figure 4.6
Representations of the three-dimensional structure of methane, CH_4. (a) Tetrahedral methane structure. (b) Ball and stick model of tetrahedral methane. (c) Three-dimensional representation of structure (b).

NH_3

Ammonia also has four electron pairs about the central atom. In contrast to methane, in which all four pairs are bonding, ammonia has three pairs of bonding electrons and one nonbonding lone pair of electrons. We might expect CH_4 and NH_3 to have electron pair arrangements that are similar but not identical. The lone pair in ammonia is more negative than the bonding pairs; some of the negative charge on the bonding pairs is offset by the presence of the hydrogen atoms with their positive nuclei. Thus the arrangement of electron pairs in ammonia is distorted. The resulting distribution appears as

The hydrogen atoms in ammonia are pushed closer together than in methane (Figure 4.7). The bond angle, *b*, is 107° because lone pair–bond pair repulsions are greater than bond pair–bond pair repulsions. The structure or shape is termed *trigonal pyramidal* and the molecule is termed a **trigonal pyramidal molecule.**

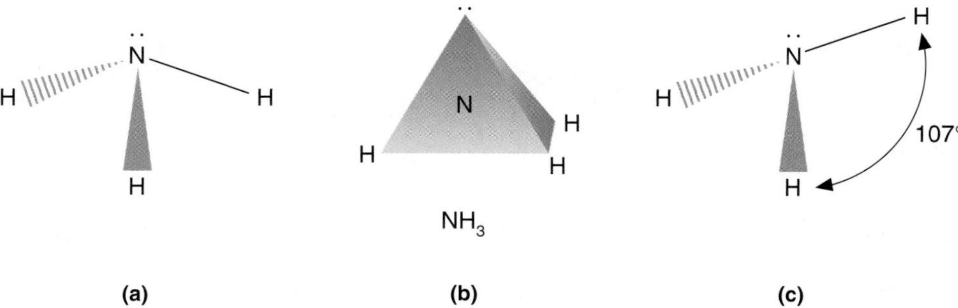

Figure 4.7
The structure of the ammonia molecule. (a) A three-dimensional sketch. (b) Pyramidal ammonia structure. (c) The H—N—H bond angle in ammonia.

H_2O

Water also has four electron pairs around the central atom; two pairs are bonding, and two pairs are nonbonding. These four electron pairs are approximately tetrahedral to each other; however, because of the difference between bonding and nonbonding electrons, noted earlier, the tetrahedral relationship is only approximate.

The **angular** (or *bent*) **structure** has a bond angle of 104.5°, which is 5° smaller than the tetrahedral angle, because of the repulsive effects of the lone pairs of electrons (as shown in Figure 4.8).

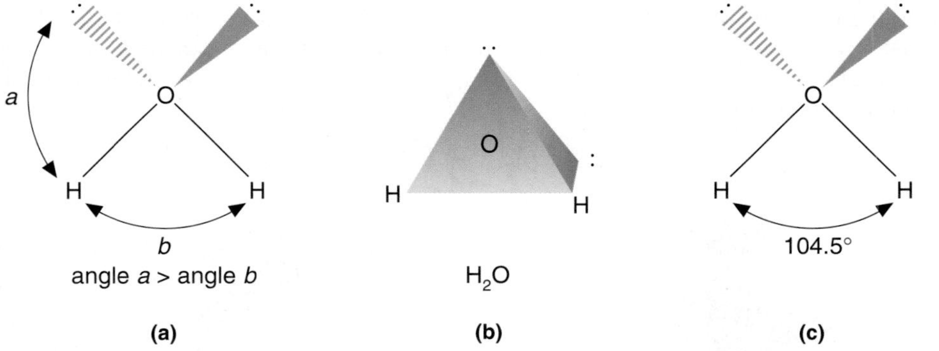

Figure 4.8
The structure of the water molecule. (a) A three-dimensional sketch. (b) Angular water structure. (c) The H—O—H bond angle in water.

The characteristics of linear, trigonal planar, angular, and tetrahedral structures are summarized in Table 4.5.

Periodic Structural Relationships

The molecules considered above contain the central atoms Be (Group IIA), B (Group IIIA), C (Group IVA), N (Group VA), and O (Group VIA). We may expect that a number of other compounds, containing the same central atom, will have structures with similar geometries. This is an approximation, not always true, but still useful in expanding our ability to write reasonable, geometrically accurate structures for a large number of compounds.

Molecules with five and six electron pairs also exist. They may have structures that are *trigonal bipyramidal* (forming a six-sided figure) or *octahedral* (forming an eight-sided figure).

The periodic similarity of group members is also useful in predictions involving bonding. Consider Group VI, oxygen, sulfur, and selenium (Se). Each has six valence electrons. Each needs two more electrons to complete its octet. Each should react with hydrogen, forming H_2O, H_2S, and H_2Se.

Table 4.5 Molecular Structure: The Geometry of a Molecule Is Affected by the Number of Nonbonded Electron Pairs Around the Central Atom and the Number of Bonded Atoms

Bonded Atoms	Nonbonding Electron Pairs	Bond Angle	Molecular Structure	Example
2	0	180°	Linear	CO_2
3	0	120°	Trigonal planar	SO_3
2	1	120°	Angular	SO_2
4	0	~109°	Tetrahedral	CH_4
3	1	<107°	Trigonal pyramidal	NH_3
2	2	<104.5°	Angular	H_2O

If we recall that H_2O is an angular molecule with the following Lewis structure,

H : Ö :
H

it follows that H_2S and H_2Se would also be angular molecules with similar Lewis structures, or

H : S̤̈ : and H : S̤̈e :
H H

This logic applies equally well to the other representative elements.

More Complex Molecules

A molecule such as dimethyl ether, CH_3—O—CH_3, has two different central atoms: oxygen and carbon. We could picture the parts of the molecule containing the CH_3 group (commonly referred to as the *methyl group*) as exhibiting tetrahedral geometry (analogous to methane):

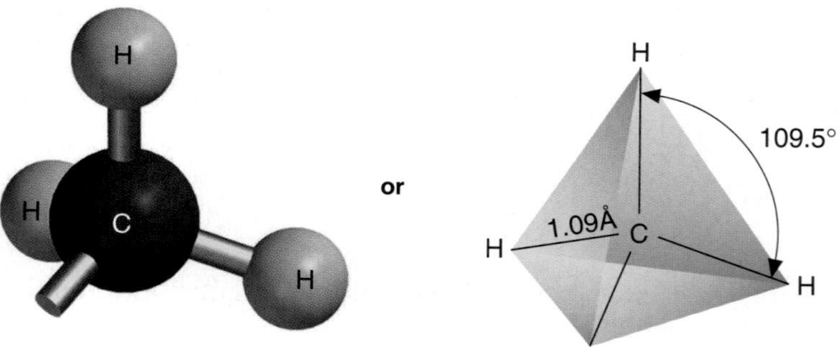

The part of the molecule connecting these two methyl groups (the oxygen) would have a bond angle similar to that of water (in which oxygen is also the central atom), approximately 104°, as seen in Figure 4.9. This is a reasonable way to represent the molecule dimethyl ether.

Trimethylamine, $(CH_3)_3N$, is a member of the amine family. As in the case of ether, two different central atoms are present. Carbon and nitrogen determine the geometry of amines. In this case the methyl group should assume the tetrahedral geometry of methane, and the nitrogen atom should have the methyl groups in a pyramidal arrangement, similar to the hydrogen atoms in ammonia, as seen in

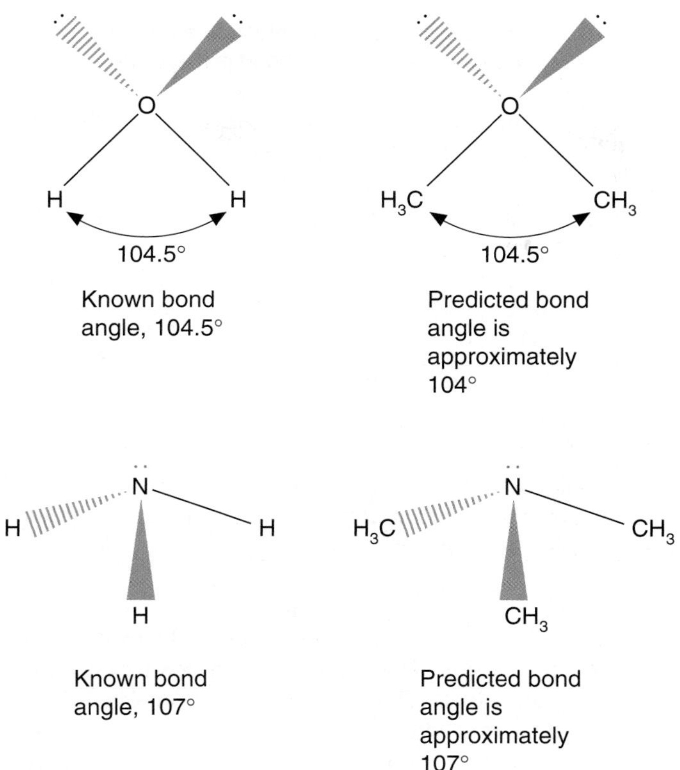

Figure 4.9
A comparison of the bonding in water and dimethyl ether.

Figure 4.10
A comparison of the bonding in ammonia and trimethylamine.

Figure 4.10. This creates a pyramidal geometry around nitrogen. H—N—H bond angles in ammonia are 107°; experimental information shows a very similar C—N—C bond angle in trimethylamine.

Question 4.21

Sketch the geometry of each of the following molecules (basing your structure on the Lewis electron dot representation of the molecule):

a. PH_3
b. SiH_4

Question 4.22

Sketch the geometry of each of the following molecules (basing your structure on the Lewis electron dot representation of the molecule):

a. C_2H_4
b. C_2H_2

It is essential to represent the molecule in its correct geometric form, using the Lewis and VSEPR theories, to understand its physical and chemical behavior. In the rest of this section we use these models to predict molecular behavior.

Lewis Structures and Polarity

A molecule is *polar* if its centers of positive and negative charges do not coincide. Molecules whose positive and negative charges are separated when the molecules are placed in an electric field align themselves with the field. The molecule behaves as a *dipole* (having two "poles" or ends, one pole is more negative and the other pole is more positive) and is said to be polar.

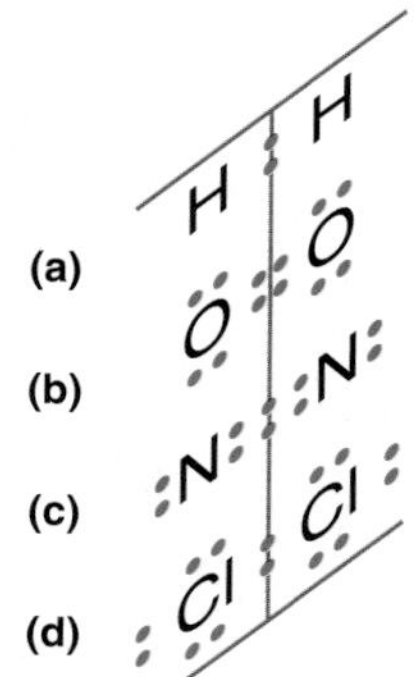

The "pulling power" or electron attracting power of the two atoms composing the bond is the same.

The electrons making up the bond are symmetrically distributed about the molecule.

One side of the molecule is a "mirror image" of the other.

Figure 4.11
Nonpolar bonds. Molecules are depicted to emphasize their symmetry: (a) the hydrogen molecule, (b) the oxygen molecule, (c) the nitrogen molecule, (d) the chlorine molecule.

Nonpolar molecules will not align with the electric field because their positive and negative centers are the same; no dipole exists. These molecules are nonpolar.

The hydrogen molecule is the simplest nonpolar molecule:

$$H:H \qquad \text{or} \qquad H\text{—}H$$

Both electrons, on average, are located at the center of the molecule and positively charged nuclei are on either side. The center of both positive and negative charge is at the center of the molecule; therefore the bond is nonpolar.

Remember: Electronegativity deals with atoms in molecules, whereas electron affinity and ionization energy deal with isolated atoms.

We may arrive at the same conclusion by considering the equality of electron sharing between the atoms being bonded. Electron sharing is related to the concept of electronegativity introduced in Section 4.1.

The atoms that comprise H_2 are identical; their electronegativity (electron attracting power) is the same. Thus the electrons remain at the center of the molecule, and the molecule is nonpolar.

Similarly, O_2, N_2, Cl_2, and F_2 are nonpolar molecules with nonpolar bonds. Arguments analogous to those made for hydrogen explain these observations as well (Figure 4.11).

Let's next consider hydrogen fluoride, HF. Fluorine is more electronegative than hydrogen. This indicates that the electrons are more strongly attracted to a fluorine atom than they are to a hydrogen atom. This results in a bond and molecule that are polar. The symbol

$$\text{Less electronegative part of bond} \nearrow \; +\!\!\longrightarrow \; \nwarrow \text{More electronegative part of bond}$$

placed below a bond indicates the direction of polarity. The more negative end of the bond is near the head of the arrow, and the less negative end of the bond is next to the tail of the arrow. Symbols using the Greek letter *delta* may also be used to designate polarity. In this system the more negative end of the bond is designated δ^- (partial negative), and the less negative end is designated δ^+ (partial positive). The symbols are applied to the hydrogen fluoride molecule as follows:

$$^{\delta+}\,H\text{—}F\,^{\delta-}$$

$$\text{Less electronegative end of bond} \nearrow \; +\!\!\longrightarrow \; \nwarrow \text{More electronegative end of bond}$$

The word *partial* implies less than a unit charge. Thus δ^+ and δ^- do not imply overall charge on a unit (such as the + or − sign on an ion); they are meant to show only the relative distribution of charge within a unit. The HF molecule shown is *neutral* but has an unequal charge distribution *within* the molecule.

HF is a **polar covalent molecule** characterized by **polar covalent bonding.** This implies that the electrons are shared unequally.

A molecule containing all nonpolar bonds must also be nonpolar. In contrast, a molecule containing polar bonds may be either polar or nonpolar depending on the relative arrangement of the bonds and any lone pairs of electrons.

Let's now examine the bonding in methane. All four bonds of CH_4 are polar because of the electronegativity difference between C and H. However, because of the symmetrical arrangement of the four C—H bonds, their polarities cancel, and the molecule is nonpolar covalent:

Now look at H_2O. Because of its angular (bent) structure, the polar bonds do not cancel, and the molecule is polar covalent:

The electron density is shifted away from the hydrogens toward oxygen in the water molecule. In methane, equal electron "pull" in all directions results in a nonpolar covalent molecule.

Question 4.23

Predict which of the following bonds are polar, and, if polar, in which direction the electrons are pulled:

a. O—S
b. C≡N
c. Cl—Cl
d. I—Cl

Question 4.24

Predict which of the following bonds are polar, and, if polar, in which direction the electrons are pulled:

a. Si—Cl
b. S—Cl
c. H—C
d. C—C

Question 4.25

Predict whether each of the following molecules is polar:

a. BCl_3
b. NH_3
c. HCl
d. $SiCl_4$

Question 4.26

Predict whether each of the following molecules is polar:

a. CO_2
b. SCl_2
c. BrCl
d. CS_2

4.5 Properties Based on Electronic Structure and Molecular Geometry

Learning Goal

Learning Goal

Intramolecular forces are attractive forces *within* molecules. They are the chemical bonds that determine the shape and polarity of individual molecules. **Intermolecular forces,** on the other hand, are forces *between* molecules.

It is important to distinguish between these two kinds of forces. It is the *intermolecular* forces that determine such properties as the solubility of one substance in another and the freezing and boiling points of liquids. But, at the same time we must realize that these forces are a direct consequence of the *intramolecular* forces in the individual units, the molecules.

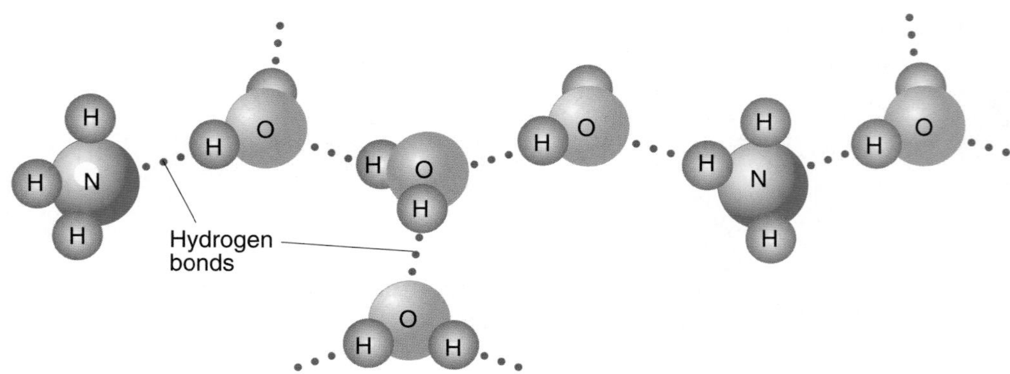

Figure 4.12
The interaction of polar covalent water molecules (the solvent) with polar covalent solute molecules such as ammonia, resulting in the formation of a solution.

In the following section we will see some of the consequences of bonding that are directly attributable to differences in intermolecular forces (solubility, boiling and melting points). In Section 6.2 we will investigate, in some detail, the nature of the intermolecular forces themselves.

Solubility

The solute is the substance that is present in lesser quantity, and the solvent is the substance that is present in the greater amount (see Section 7.3).

Solubility is defined as the maximum amount of solute that dissolves in a given amount of solvent at a specified temperature. Polar molecules are most soluble in polar solvents, whereas nonpolar molecules are most soluble in nonpolar solvents. This is the rule of *"like dissolves like."* Substances of similar polarity are mutually soluble, and large differences in polarity lead to insolubility.

Case I: Ammonia and Water

The interaction of water and ammonia is an example of a particularly strong intermolecular force, the hydrogen bond; this phenomenon is discussed in Chapter 6.

Ammonia is soluble in water because both ammonia and water are polar molecules:

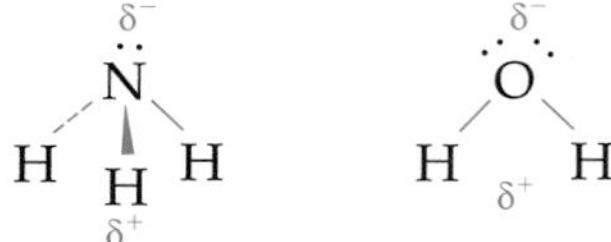

Dissolution of ammonia in water is a consequence of the intermolecular forces present among the ammonia and water molecules. The δ^- end (a nitrogen) of the ammonia molecule is attracted to the δ^+ end (a hydrogen) of the water molecule; at the same time the δ^+ end (a hydrogen) of the ammonia molecule is attracted to the δ^- end (an oxygen) of the water molecule. These attractive forces thus "pull" ammonia into water (and water into ammonia), and the ammonia molecules are randomly distributed throughout the solvent, forming a homogeneous solution (Figure 4.12).

Figure 4.13
The interaction of polar water molecules and nonpolar oil molecules. The familiar salad dressing, oil and vinegar, forms two layers. The oil does not dissolve in vinegar, an aqueous solution of acetic acid.

Case II: Oil and Water

Oil and water do not mix; oil is a nonpolar substance composed primarily of molecules containing carbon and hydrogen. Water molecules, on the other hand, are quite polar. The potential solvent, water molecules, have partially charged ends, whereas the molecules of oil do not. As a result, water molecules exert their attractive forces on other water molecules, not on the molecules of oil; the oil remains insoluble, and because it is less dense than water, the oil simply floats on the surface of the water. This is illustrated in Figure 4.13.

Boiling Points of Liquids and Melting Points of Solids

Boiling a liquid requires energy. The energy is used to overcome the intermolecular attractive forces in the liquid, driving the molecules into the less associated gas phase. The amount of energy required to accomplish this is related to the boiling temperature. This, in turn, depends on the strength of the intermolecular attractive

Table 4.6 Melting and Boiling Points of Selected Compounds in Relation to Their Bonding Type

Formula (Name)	Bonding Type	M.P. (°C)	B.P. (°C)
N_2 (nitrogen)	Nonpolar covalent	−210	−196
O_2 (oxygen)	Nonpolar covalent	−219	−183
NH_3 (ammonia)	Polar covalent	−78	−33
H_2O (water)	Polar covalent	0	100
NaCl (sodium chloride)	Ionic	801	1413
KBr (potassium bromide)	Ionic	730	1435

forces in the liquid, which parallel the polarity. This is not the only determinant of boiling point. Molecular mass is also an important consideration. The larger the mass of the molecule, the more difficult it becomes to convert the collection of molecules to the gas phase.

A similar argument can be made for the melting points of solids. The ease of conversion of a solid to a liquid also depends on the magnitude of the attractive forces in the solid. The situation actually becomes very complex for ionic solids because of the complexity of the crystal lattice.

As a general rule, polar compounds have strong attractive (intermolecular) forces, and their boiling and melting points tend to be higher than those of nonpolar substances of similar molecular mass.

Melting and boiling points of a variety of substances are included in Table 4.6.

See Section 6.3, the solid state.

Question 4.27

Predict which compound in each of the following groups should have the higher melting and boiling points:

a. H_2O and C_2H_4
b. CO and CH_4
c. NH_3 and N_2
d. Cl_2 and ICl

Question 4.28

Predict which compound in each of the following groups should have the higher melting and boiling points:

a. C_2H_6 and CH_4
b. CO and NO
c. F_2 and Br_2
d. $CHCl_3$ and CCl_4

Summary

4.1 Chemical Bonding

When two atoms are joined to make a chemical compound, the force of attraction between the two atoms is the *chemical bond*. *Ionic bonding* is characterized by an electron transfer process occurring before bond formation, forming an *ion pair*. In *covalent bonding*, electrons are shared between atoms in the bonding process. *Polar covalent bonding*, like covalent bonding, is based on the concept of electron sharing; however, the sharing is unequal and based on the *electronegativity* difference between joined atoms. The *Lewis symbol*, showing only valence electrons, is a convenient way of representing atoms singly or in combination.

4.2 Naming Compounds and Writing Formulas of Compounds

The "shorthand" symbol for a compound is its *formula*. The formula identifies the number and type of atoms in the compound.

An ion that consists of only a single atom is said to be *monatomic*. *Polyatomic ions*, such as the hydroxide ion, OH^-, are composed of two or more atoms bonded together.

Names of ionic compounds are derived from the names of their ions. The name of the cation appears first, followed by the name of the anion. In the Stock system for naming an ion (the systematic name), a Roman numeral indicates the charge of the cation. In the older common nomenclature system, the suffix *-ous* indicates the lower of the ionic charges, and the suffix *-ic* indicates the higher ionic charge.

Most covalent compounds are formed by the reaction of nonmetals. Covalent compounds exist as *molecules*.

The convention used for naming covalent compounds is as follows:

- The names of the elements are written in the order in which they appear in the formula.
- A prefix indicating the number of each kind of atom found in the unit is placed before the name of the element.
- The stem of the name of the last element is used with the suffix *-ide*.

Many compounds are so familiar to us that their common names are used. It is useful to be able to correlate both systematic and common names with the corresponding molecular formula.

4.3 Properties of Ionic and Covalent Compounds

Covalently bonded molecules are discrete units, and they have less tendency to form an extended structure in the solid state. Ionic compounds, with ions joined by electrostatic attraction, do not have definable units but form a *crystal lattice* composed of positive and negative ions in an extended three-dimensional network.

The *melting point* is the temperature at which a solid is converted to a liquid; the *boiling point* is the temperature at which a liquid is converted to a gas. Melting and boiling temperatures for ionic compounds are generally higher than those of covalent compounds.

Ionic solids are crystalline, whereas covalent solids may be either crystalline or amorphous.

Many ionic solids dissolve in water, *dissociating* into positive and negative ions (an *electrolytic solution*). Because these ions can carry (conduct) a current of electricity, they are called *electrolytes*. Covalent solids in solution usually retain their neutral character and are *nonelectrolytes*.

4.4 Drawing Lewis Structures of Molecules and Polyatomic Ions

The procedure for drawing Lewis structures of molecules involves writing a skeletal structure of the molecule, arranging the atoms in their most probable order, determining the number of valence electrons on each atom, and combining them to get the total for the compound. The electrons are then distributed around the atoms (in pairs if possible) to satisfy the octet rule. At this point electron pairs may be moved, creating double or triple bonds, in an effort to satisfy the octet rule for all atoms and produce the final structure.

Stability of a covalent compound is related to the *bond energy*. The magnitude of the bond energy decreases in the order *triple bond* > *double bond* > *single bond*. The bond length decreases in the order *single bond* > *double bond* > *triple bond*.

The valence shell electron pair repulsion theory states that electron pairs around the central atom of the molecule arrange themselves to minimize electronic repulsion; the electrons orient themselves as far as possible from each other. Two electron pairs around the central atom lead to a *linear* arrangement of the attached atoms; three indicate a *trigonal planar* arrangement, and four result in a *tetrahedral geometry*. Both lone pair and bonding pair electrons must be taken into account when predicting structure. Molecules with fewer than four and as many as five or six electron pairs around the central atom also exist. They are exceptions to the octet rule.

A molecule is polar if its centers of positive and negative charges do not coincide. A *polar covalent* molecule has at least one *polar covalent bond*. An understanding of the concept of *electronegativity*, the relative electron-attracting power of atoms in molecules, helps us to assess the polarity of a bond.

A molecule containing all nonpolar bonds must be nonpolar. A molecule containing polar bonds may be either polar or nonpolar, depending on the relative position of the bonds.

4.5 Properties Based on Electronic Structure and Molecular Geometry

Attractions between molecules are called *intermolecular forces*. *Intramolecular forces*, on the other hand, are the attractive forces within molecules. It is the intermolecular forces that determine such properties as the solubility of one substance in another and the freezing and boiling points of liquids.

Solubility is the maximum amount of solute that dissolves in a given amount of solvent at a specified temperature. Polar molecules are most soluble in polar solvents; nonpolar molecules are most soluble in nonpolar solvents. This is the rule of "like dissolves like."

As a general rule, polar compounds have strong intermolecular forces, and their boiling and melting points tend to be higher than nonpolar compounds of similar molecular mass.

Key Terms

angular structure (4.4)
boiling point (4.3)
bond energy (4.4)
chemical bond (4.1)
covalent bonding (4.1)
crystal lattice (4.2)
dissociation (4.3)
double bond (4.4)

electrolyte (4.3)
electrolytic solution (4.3)
electronegativity (4.1)
formula (4.2)
intermolecular force (4.5)
intramolecular force (4.5)
ionic bonding (4.1)
ion pair (4.1)
Lewis symbol (4.1)
linear structure (4.4)
lone pair (4.4)
melting point (4.3)
molecule (4.2)
monatomic ion (4.2)
nomenclature (4.2)
nonelectrolyte (4.3)
polar covalent bonding (4.4)
polar covalent molecule (4.4)
polyatomic ion (4.2)
resonance (4.4)
resonance form (4.4)
resonance hybrid (4.4)
single bond (4.4)
solubility (4.5)
tetrahedral structure (4.4)
trigonal pyramidal molecule (4.4)
triple bond (4.4)
valence shell electron pair repulsion (VSEPR) theory (4.4)

Questions and Problems

Chemical Bonding

4.29 Classify each of the following compounds as ionic or covalent:
a. $MgCl_2$ c. H_2S
b. CO_2 d. NO_2

4.30 Classify each of the following compounds as ionic or covalent:
a. NaCl c. ICl
b. CO d. H_2

4.31 Using Lewis symbols, write an equation predicting the product of the reaction of:
a. Li + Br
b. Mg + Cl

4.32 Using Lewis symbols, write an equation predicting the product of the reaction of:
a. Na + O
b. Na + S

4.33 Give the Lewis structure for each of the following compounds:
a. NCl_3 b. CH_3OH c. CS_2

4.34 Give the Lewis structure for each of the following compounds:
a. HNO_3 b. CCl_4 c. PBr_3

Naming Compounds and Writing Formulas of Compounds

4.35 Name each of the following ions:
a. Na^+
b. Cu^+
c. Mg^{2+}
d. Fe^{2+}
e. Fe^{3+}

4.36 Name each of the following ions:
a. S^{2-}
b. Cl^-
c. CO_3^{2-}
d. NH_4^+
e. CH_3COO^-

4.37 Write the formula for each of the following monatomic ions:
a. the potassium ion
b. the bromide ion
c. the calcium ion
d. the chromium(VI) ion

4.38 Write the formula for each of the following complex ions:
a. the sulfate ion
b. the nitrate ion
c. the phosphate ion
d. the bicarbonate ion

4.39 Write the correct formula for each of the following:
a. sodium chloride
b. magnesium bromide
c. copper(II) oxide
d. iron(III) oxide
e. aluminum chloride

4.40 Write the correct formula for each of the following:
a. silver cyanide
b. ammonium chloride
c. silver oxide
d. magnesium carbonate
e. magnesium bicarbonate

4.41 Name each of the following compounds:
a. $MgCl_2$
b. $AlCl_3$
c. CaS
d. Na_2O
e. $Fe(OH)_3$

4.42 Name each of the following covalent compounds:
a. NO_2
b. SO_3
c. PCl_3
d. N_2O_4
e. CCl_4

4.43 Predict the formula of a compound formed from:
a. aluminum and oxygen
b. lithium and sulfur
c. boron and hydrogen
d. magnesium and phosphorus

4.44 Predict the formula of a compound formed from:
a. carbon and oxygen
b. sulfur and hydrogen
c. calcium and oxygen
d. silicon and hydrogen

4.45 Write a suitable formula for each of the following:
a. sodium nitrate
b. magnesium nitrate
c. aluminum nitrate
d. ammonium nitrate

4.46 Write a suitable formula for each of the following:
a. ammonium iodide
b. ammonium sulfate
c. ammonium acetate
d. ammonium cyanide

4.47 Name each of the following:
a. CuS
b. $CuSO_4$
c. $Cu(OH)_2$
d. CuO

4.48 Name each of the following:
a. NaClO
b. $NaClO_2$
c. $NaClO_3$
d. $NaClO_4$

Properties of Ionic and Covalent Compounds

4.49 Contrast ionic and covalent compounds with respect to their relative boiling points.

4.50 Contrast ionic and covalent compounds with respect to their relative melting points.

4.51 Would KCl or CCl_4 be expected to be a solid at room temperature? Why?

4.52 Would H_2O or CCl_4 be expected to have a higher boiling point? Why?

Drawing Lewis Structures of Molecules and Polyatomic Ions

4.53 Draw the appropriate Lewis structure for each of the following atoms:
a. H **b.** He **c.** C **d.** N

4.54 Draw the appropriate Lewis structure for each of the following atoms:
a. Be **b.** B **c.** F **d.** S

4.55 Draw the appropriate Lewis structure for each of the following ions:
a. Li^+ **b.** Mg^{2+} **c.** Cl^- **d.** P^{3-}

4.56 Draw the appropriate Lewis structure for each of the following ions:
a. Be^{2+} **b.** Al^{3+} **c.** O^{2-} **d.** S^{2-}

4.57 Discuss the concept of resonance, being certain to define the terms *resonance, resonance form,* and *resonance hybrid.*

4.58 Why is resonance an important concept in bonding?

4.59 Ethanol (ethyl alcohol or grain alcohol) has a molecular formula of C_2H_5OH. Represent the structure of ethanol using the Lewis electron dot approach.

4.60 Formaldehyde, H_2CO, in water solution has been used as a preservative for biological specimens. Represent the Lewis structure of formaldehyde.

4.61 Acetone, C_3H_6O, is a common solvent. It is found in such diverse materials as nail polish remover and industrial solvents. Draw its Lewis structure if its skeletal structure is

```
   O
   |
C—C—C
```

4.62 Ethylamine is an example of an important class of organic compounds. The molecular formula of ethylamine is $CH_3CH_2NH_2$. Draw its Lewis structure.

4.63 Predict whether the bond formed between each of the following pairs of atoms would be ionic, nonpolar, or polar covalent:
a. S and O
b. Si and P
c. Na and Cl
d. Na and O
e. Ca and Br

4.64 Predict whether the bond formed between each of the following pairs of atoms would be ionic, nonpolar, or polar covalent:
a. Cl and Cl
b. H and H
c. C and H
d. Li and F
e. O and O

4.65 Draw an appropriate covalent Lewis structure formed by the simplest combination of atoms in Problem 4.63 for each solution that involves a nonpolar or polar covalent bond.

4.66 Draw an appropriate covalent Lewis structure formed by the simplest combination of atoms in Problem 4.64 for each solution that involves a nonpolar or polar covalent bond.

Properties Based on Electronic Structure and Molecular Geometry

4.67 What is the relationship between the polarity of a bond and the polarity of the molecule?

4.68 What effect does polarity have on the solubility of a compound in water?

4.69 What effect does polarity have on the melting point of a pure compound?

4.70 What effect does polarity have on the boiling point of a pure compound?

Critical Thinking Problems

1. Predict differences in our global environment that may have arisen if the freezing point and boiling point of water were 20°C higher than they are.
2. Would you expect the compound $C_2S_2H_4$ to exist? Why or why not?
3. Draw the resonance forms of the carbonate ion. What conclusions, based on this exercise, can you draw about the stability of the carbonate ion?
4. Which of the following compounds would be predicted to have the higher boiling point? Explain your reasoning.

```
  H H
  | |
H—C—C—O—H
  | |
  H H
```
Ethanol

```
  H H
  | |
H—C—C—H
  | |
  H H
```
Ethane

Careful measurements validate chemical equations.

5

Calculations and the Chemical Equation

GENERAL CHEMISTRY

Outline

Learning Goals

1. Know the relationship between the mole and Avogadro's number, and the usefulness of these quantities.
2. Perform calculations using Avogadro's number and the mole.
3. Write chemical formulas for common inorganic substances.
4. Calculate the formula weight and molar mass of a compound.
5. Know the major function served by the chemical equation, the basis for chemical calculations.
6. Balance chemical equations given the identity of products and reactants.
7. Calculate the number of moles of product resulting from a given number of moles of reactants or the number of moles of reactant needed to produce a certain number of moles of product.
8. Calculate theoretical and percent yield.

CHEMISTRY CONNECTION

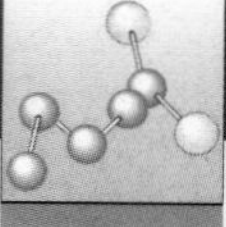

The Chemistry of Automobile Air Bags

Each year, thousands of individuals are killed or seriously injured in automobile accidents. Perhaps most serious is the front-end collision. The car decelerates or stops virtually on impact; the momentum of the passengers, however, does not stop, and the driver and passengers are thrown forward toward the dashboard and the front window. Suddenly, passive parts of the automobile, such as control knobs, the rearview mirror, the steering wheel, the dashboard, and the windshield, become lethal weapons.

Automobile engineers have been aware of these problems for a long time. They have made a series of design improvements to lessen the potential problems associated with front-end impact. Smooth switches rather than knobs, recessed hardware, and padded dashboards are examples. These changes, coupled with the use of lap and shoulder belts, which help to immobilize occupants of the car, have decreased the frequency and severity of the impact and lowered the death rate for this type of accident.

An almost ideal protection would be a soft, fluffy pillow, providing a cushion against impact. Such a device, an air bag inflated only on impact, is now standard equipment for the protection of the driver and front-seat passenger.

How does it work? Ideally, it inflates only when severe front-end impact occurs; it inflates very rapidly (in approximately 40 milliseconds), then deflates to provide a steady deceleration, cushioning the occupants from impact. A remarkably simple chemical reaction makes this a reality.

When solid sodium azide (NaN_3) is detonated by mechanical energy produced by an electric current, it decomposes to form solid sodium and nitrogen gas:

$$2NaN_3(s) \longrightarrow 2Na(s) + 3N_2(g)$$

The nitrogen gas inflates the air bag, cushioning the driver and front-seat passenger.

The solid sodium azide has a high density (characteristic of solids) and thus occupies a small volume. It can easily be stored in the center of a steering wheel or in the dashboard. The rate of the detonation is very rapid. In milliseconds it produces three moles of N_2 gas for every two moles of NaN_3. The N_2 gas occupies a relatively large volume because its density is low. This is a general property of gases.

Figuring out how much sodium azide is needed to produce enough nitrogen to properly inflate the bag is an example of a practical application of the chemical arithmetic that we are learning in this chapter.

Introduction

The calculation of chemical quantities based on chemical equations, termed *stoichiometry,* is the application of logic and arithmetic to chemical systems to answer questions such as the following:

A pharmaceutical company wishes to manufacture 1000 kg of a product next year. How much of each of the starting materials must be ordered? If the starting materials cost $20/g, how much money must be budgeted for the project?

We often need to predict the quantity of a product produced from the reaction of a given amount of material. This calculation is possible. It is equally possible to calculate how much of a material would be necessary to produce a desired amount of product. One of many examples was shown in the preceding Chemistry Connection: the need to solve a very practical problem.

What is required is a recipe: a procedure to follow. The basis for our recipe is the *chemical equation.* A properly written chemical equation provides all of the necessary information for the chemical calculation. That critical information is the *combining ratio* of elements or compounds that must occur to produce a certain amount of product or products.

In this chapter we define the mole, the fundamental unit of measure of chemical arithmetic, learn to write and balance chemical equations, and use these tools to perform calculations of chemical quantities.

5.1 The Mole Concept and Atoms

Atoms are exceedingly small, yet their masses have been experimentally determined for each of the elements. The unit of measurement for these determinations is the **atomic mass unit,** abbreviated amu:

$$1 \text{ amu} = 1.661 \times 10^{-24} \text{ g}$$

The Mole and Avogadro's Number

Learning Goal

The exact value of the atomic mass unit is defined in relation to a standard, just as the units of the metric system represent defined quantities. The carbon-12 isotope has been chosen and is assigned a mass of exactly 12 atomic mass units. Hence this standard reference point defines an atomic mass unit as exactly one-twelfth the mass of a carbon-12 atom.

The term atomic weight is not correct but is a fixture in common usage. Just remember that atomic weight is really "average atomic mass."

The periodic table provides atomic weights in atomic mass units. These atomic weights are average values, based on the contribution of all the isotopes of the particular element. For example, the average mass of a carbon atom is 12.01 amu and

$$\frac{12.01 \cancel{\text{amu C}}}{\text{C atom}} \times \frac{1.661 \times 10^{-24} \text{ g C}}{1 \cancel{\text{amu C}}} = 1.995 \times \frac{10^{-23} \text{ g C}}{\text{C atom}}$$

The average mass of a helium atom is 4.003 amu and

$$\frac{4.003 \cancel{\text{amu He}}}{\text{He atom}} \times \frac{1.661 \times 10^{-24} \text{ g He}}{1 \cancel{\text{amu He}}} = 6.649 \times \frac{10^{-24} \text{ g He}}{\text{He atom}}$$

In everyday work, chemists use much larger quantities of matter (typically, grams or kilograms). A more practical unit for defining a "collection" of atoms is the **mole:**

$$1 \text{ mol of atoms} = 6.022 \times 10^{23} \text{ atoms of an element}$$

This number is **Avogadro's number.** Amedeo Avogadro, a nineteenth-century scientist, conducted a series of experiments that provided the basis for the mole concept.

Learning Goal

The practice of defining a unit for a quantity of small objects is common; a *dozen* eggs, a *ream* of paper, and a *gross* of pencils are well-known examples. Similarly, a mole is 6.022×10^{23} individual units of anything. We could, if we desired, speak of a mole of eggs or a mole of pencils. However, in chemistry we use the mole to represent a specific quantity of atoms, ions, or molecules.

The mole (mol) and the atomic mass unit (amu) are related. The atomic mass of an element corresponds to the average mass of a single atom in amu *and* the mass of a mole of atoms in grams.

The mass of 1 mol of atoms, in grams, is defined as the **molar mass.** Consider this relationship for sodium in Example 5.1.

Relating Avogadro's Number to Molar Mass **EXAMPLE 5.1**

Calculate the mass, in grams, of Avogadro's number of sodium atoms.

Solution

The average mass of one sodium atom is 22.99 amu. This may be formatted as the conversion factor:

$$\frac{22.99 \text{ amu Na}}{1 \text{ atom Na}}$$

Continued—

EXAMPLE 5.1 —*Continued*

As previously noted, 1 amu is 1.661×10^{-24} g, and 6.022×10^{23} atoms of sodium is Avogadro's number. Similarly, these relationships may be formatted as:

$$1.661 \times 10^{-24} \frac{\text{g Na}}{\text{amu}} \text{ and } 6.022 \times 10^{23} \frac{\text{atoms Na}}{\text{mol Na}}$$

Formatting this information as a series of conversion factors, using the factor-label method, we have

$$22.99 \frac{\cancel{\text{amu Na}}}{\cancel{\text{atom Na}}} \times 1.661 \times 10^{-24} \frac{\text{g Na}}{\cancel{\text{amu Na}}} \times 6.022 \times 10^{23} \frac{\cancel{\text{atoms Na}}}{\text{mol Na}} = 22.99 \frac{\text{g Na}}{\text{mol Na}}$$

The average mass of one *atom* of sodium, in units of amu, is *numerically identical* to the mass of *Avogadro's number of atoms*, expressed in units of grams. Hence the molar mass of sodium is 22.99 g Na/mol.

Helpful Hint: Section 1.3 discusses the use of conversion factors.

The sodium example is not unique. The relationship holds for every element in the periodic table.

Because Avogadro's number of particles (atoms) is 1 mol, it follows that

the average mass of one atom of hydrogen is 1.008 amu

and

the mass of 1 mol of hydrogen atoms is 1.008 g

or

the average mass of one atom of carbon is 12.01 amu

and

the mass of 1 mol of carbon atoms is 12.01 g

and so forth. One mole of atoms of *any element* contains the same number, Avogadro's number, of atoms, 6.022×10^{23} atoms.

The difference in mass of a mole of two different elements can be quite striking (Figure 5.1). For example, a mole of hydrogen atoms is 1.008 g, and a mole of lead atoms is 207.19 g.

Figure 5.1
The comparison of approximately one mole each of silver (as Morgan and Peace dollars), gold (as Canadian Maple Leaf coins), and copper (as pennies) shows the considerable difference in mass (as well as economic value) of equivalent moles of different substances.

Question 5.1

Calculate the mass, in grams, of Avogadro's number of aluminum atoms.

Question 5.2

Calculate the mass, in grams, of Avogadro's number of mercury atoms.

Calculating Atoms, Moles, and Mass

Learning Goal

Performing calculations based on a chemical equation requires a facility for relating the number of atoms of an element to a corresponding number of moles of that element and ultimately to their mass in grams. Such calculations involve the use of conversion factors. This type of calculation was first described in Chapter 1. Some examples follow.

Converting Moles to Atoms

EXAMPLE 5.2

How many iron atoms are present in 3.0 mol of iron metal?

Solution

The calculation is based on choosing the appropriate conversion factor. The relationship

$$\frac{6.022 \times 10^{23} \text{ atoms Fe}}{1 \text{ mol Fe}}$$

follows directly from

$$1 \text{ mol Fe} = 6.022 \times 10^{23} \text{ atoms Fe}$$

Using this conversion factor, we have

$$\text{number of atoms of Fe} = 3.0 \cancel{\text{mol Fe}} \times \frac{6.022 \times 10^{23} \text{ atoms Fe}}{1 \cancel{\text{mol Fe}}}$$

$$= 18 \times 10^{23} \text{ atoms of Fe, or}$$

$$= 1.8 \times 10^{24} \text{ atoms of Fe}$$

Converting Atoms to Moles

EXAMPLE 5.3

Calculate the number of moles of sulfur represented by 1.81×10^{24} atoms of sulfur.

Solution

$$1.81 \times 10^{24} \cancel{\text{atoms S}} \times \frac{1 \text{ mol S}}{6.022 \times 10^{23} \cancel{\text{atoms S}}} = 3.01 \text{ mol S}$$

Note that this conversion factor is the inverse of that used in Example 5.2. Remember, the conversion factor must cancel units that should not appear in the final answer.

Converting Moles of a Substance to Mass in Grams

EXAMPLE 5.4

What is the mass, in grams, of 3.01 mol of sulfur?

Solution

We know from the periodic table that 1 mol of sulfur has a mass of 32.06 g. Setting up a suitable conversion factor between grams and moles results in

$$3.01 \cancel{\text{mol S}} \times \frac{32.06 \text{ g S}}{1 \cancel{\text{mol S}}} = 96.5 \text{ g S}$$

Converting Kilograms to Moles

EXAMPLE 5.5

Calculate the number of moles of sulfur in 1.00 kg of sulfur.

Continued—

EXAMPLE 5.5 —*Continued*

Solution

$$1.00 \text{ kg S} \times \frac{10^3 \text{ g S}}{1 \text{ kg S}} \times \frac{1 \text{ mol S}}{32.06 \text{ g S}} = 31.2 \text{ mol S}$$

EXAMPLE 5.6 ***Converting Grams to Number of Atoms***

Calculate the number of atoms of sulfur in 1.00 g of sulfur.

Solution

$$1.00 \text{ g S} \times \frac{1 \text{ mol S}}{32.06 \text{ g S}} \times \frac{6.022 \times 10^{23} \text{ atoms S}}{1 \text{ mol S}} = 1.88 \times 10^{22} \text{ atoms S}$$

The preceding examples demonstrate the use of a sequence of conversion factors to proceed from the information *provided* in the problem to the information *requested* by the problem.

It is generally useful to map out a pattern for the required conversion. In Example 5.6 we are given the number of grams and need the number of atoms that correspond to that mass.

Begin by "tracing a path" to the answer:

grams sulfur $\xrightarrow{\text{Step 1}}$ moles sulfur $\xrightarrow{\text{Step 2}}$ atoms sulfur

Two transformations, or conversions, are required:

Step 1. Convert grams to moles.

Step 2. Convert moles to atoms.

For the first conversion we could consider either

$$\frac{1 \text{ mol S}}{32.06 \text{ g S}}$$

or the inverse

$$\frac{32.06 \text{ g S}}{1 \text{ mol S}}$$

If we want grams to cancel, $\frac{1 \text{ mol S}}{32.06 \text{ g S}}$ is the correct choice, resulting in

$$\text{g S} \times \frac{1 \text{ mol S}}{32.06 \text{ g S}} = \text{value in mol S}$$

For the second conversion, moles to atoms, the moles of S must cancel; therefore

$$\text{mol S} \times \frac{6.022 \times 10^{23} \text{ atoms S}}{1 \text{ mol S}} = \text{number of atoms S}$$

which are the units desired in the solution.

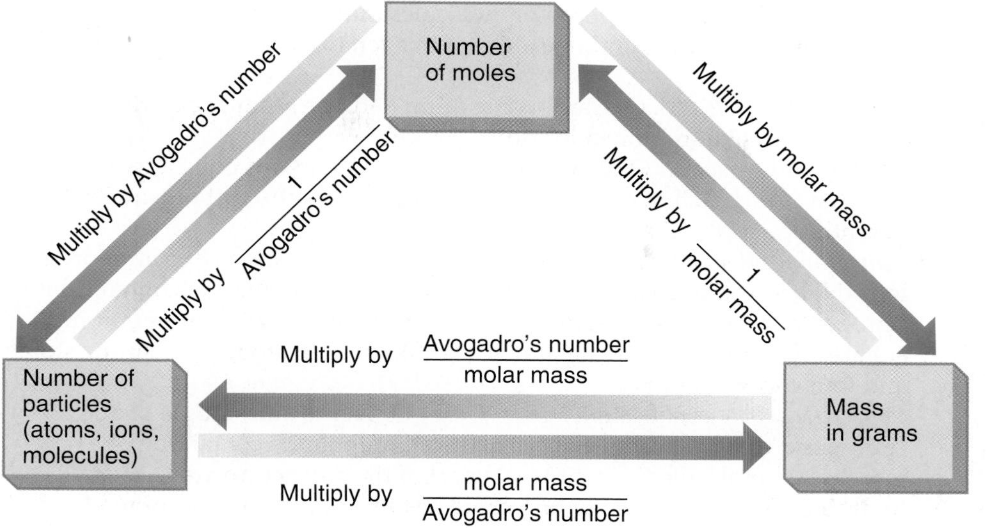

Figure 5.2
Interconversion between numbers of moles, particles, and grams. The mole concept is central to chemical calculations involving measured quantities of matter.

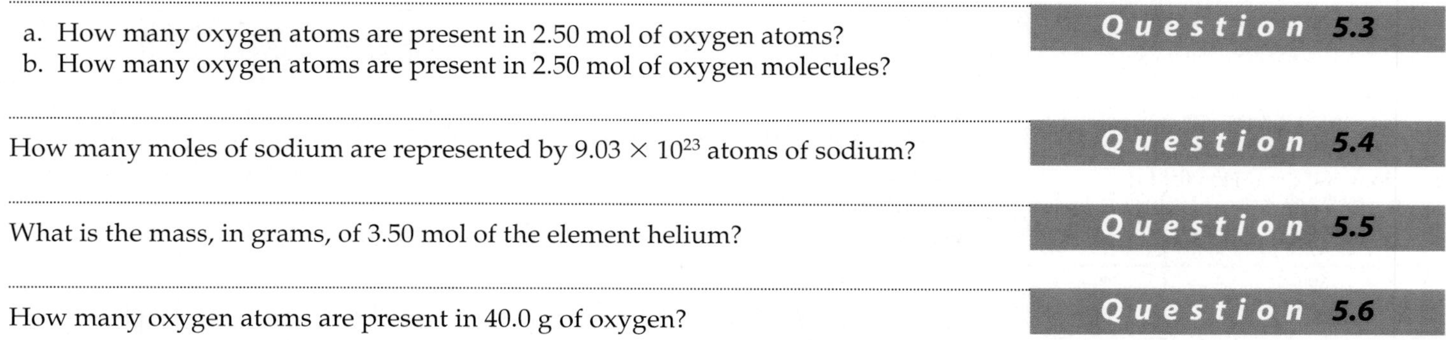

Question 5.3

a. How many oxygen atoms are present in 2.50 mol of oxygen atoms?
b. How many oxygen atoms are present in 2.50 mol of oxygen molecules?

Question 5.4

How many moles of sodium are represented by 9.03×10^{23} atoms of sodium?

Question 5.5

What is the mass, in grams, of 3.50 mol of the element helium?

Question 5.6

How many oxygen atoms are present in 40.0 g of oxygen?

The conversion between the three principal measures of quantity of matter—the number of grams (mass), the number of moles, and the number of individual particles (atoms, ions, or molecules)—is essential to the art of problem solving in chemistry. Their interrelationship is depicted in Figure 5.2.

5.2 Compounds

The Chemical Formula

Learning Goal  3

Compounds are pure substances. They are composed of two or more elements that are chemically combined. A **chemical formula** is a combination of symbols of the various elements that make up the compound. It serves as a convenient way to represent a compound. The chemical formula is based on the formula unit. The **formula unit** is the smallest collection of atoms that provides two important pieces of information:

- the identity of the atoms present in the compound and
- the relative numbers of each type of atom.

Let's look at the following formulas:

- *Hydrogen gas*, H_2. This indicates that two atoms of hydrogen are chemically bonded forming a molecule, hence the subscript 2.

Figure 5.3
The marked difference in color of (a) hydrated and (b) anhydrous copper sulfate is clear evidence that they are, in fact, different compounds.

- *Water,* H_2O. Water is composed of molecules that contain two atoms of hydrogen (subscript 2) and one atom of oxygen (lack of a subscript means *one* atom).
- *Sodium chloride,* NaCl. One atom of sodium and one atom of chlorine combine to make sodium chloride.
- *Calcium hydroxide,* $Ca(OH)_2$. Calcium hydroxide contains one atom of calcium and two atoms each of oxygen and hydrogen. The subscript outside the parentheses applies to *all* atoms inside the parentheses.
- *Ammonium sulfate,* $(NH_4)_2SO_4$. Ammonium sulfate contains two ammonium ions (NH_4^+) and one sulfate ion (SO_4^{2-}). Each ammonium ion contains one nitrogen and four hydrogen atoms. The formula shows that ammonium sulfate contains two nitrogen atoms, eight hydrogen atoms, one sulfur atom, and four oxygen atoms.
- *Copper(II) sulfate pentahydrate,* $CuSO_4 \cdot 5H_2O$. This is an example of a compound that has water in its structure. Compounds containing one or more water molecules as an integral part of their structure are termed **hydrates.** Copper sulfate pentahydrate has five units of water (or ten H atoms and five O atoms) in addition to one copper atom, one sulfur atom, and four oxygen atoms for a total atomic composition of:

1 copper atom
1 sulfur atom
9 oxygen atoms
10 hydrogen atoms

It is possible to determine the correct molecular formula of a compound from experimental data. This useful application of molar quantities is discussed in Appendix C.

Note that the symbol for water is preceded by a dot, indicating that, although the water is a formula unit capable of standing alone, in this case it is a part of a larger structure. Copper sulfate also exists as a structure free of water, $CuSO_4$. This form is described as anhydrous (no water) copper sulfate. The physical and chemical properties of a hydrate often differ markedly from the anhydrous form (Figure 5.3).

5.3 The Mole Concept Applied to Compounds

Learning Goal

Just as the atomic weight of an element is the average atomic mass for one atom of the naturally occurring element, expressed in atomic mass units, the **formula weight** of a compound is the sum of the atomic weights of all atoms in the compound, as represented by its formula. To calculate the formula weight of a compound we *must* know the correct formula. The formula weight is expressed in atomic mass units.

When working in the laboratory, we do not deal with individual molecules; instead, we use units of moles or grams. Eighteen grams of water (less than one ounce) contain approximately Avogadro's number of molecules (6.022×10^{23} molecules). Defining our working units as moles and grams makes good chemical sense.

We earlier concluded that the atomic mass of an element in amu from the periodic table corresponds to the mass of a mole of atoms of that element in units of grams. It follows that *molar mass,* the mass of a mole of compound, is numerically equal to the formula weight in atomic mass units.

EXAMPLE 5.7 ***Calculating Formula Weight and Molar Mass***

Calculate the formula weight and molar mass of water, H_2O.

Solution

Each water molecule contains two hydrogen atoms and one oxygen atom. The formula weight is

Continued—

EXAMPLE 5.7 *—Continued*

$$\begin{array}{lrl} \text{2 atoms of hydrogen} & \times\ 1.008\ \text{amu/atom} = & 2.016\ \text{amu} \\ \underline{\text{1 atom of oxygen}} & \underline{\times\ 16.00\ \text{amu/atom} =} & \underline{16.00\ \text{amu}} \\ & & 18.02\ \text{amu} \end{array}$$

The average mass of a single molecule of H_2O is 18.02 amu and is the formula weight. Therefore the mass of a mole of H_2O is 18.02 g or 18.02 g/mol.

Helpful Hint: Adding 2.016 and 16.00 shows a result of 18.016 on your calculator. Proper use of significant figures (Chapter 1) dictates rounding that result to 18.02.

Calculating Formula Weight and Molar Mass

EXAMPLE 5.8

Calculate the formula weight and molar mass of sodium sulfate.

Solution

The sodium ion is Na^+, and the sulfate ion is SO_4^{2-}. Two sodium ions must be present to neutralize the negative charges on the sulfate ion. The formula is Na_2SO_4. Sodium sulfate contains two sodium atoms, one sulfur atom, and four oxygen atoms. The formula weight is

$$\begin{array}{lrl} \text{2 atoms of sodium} & \times\ 22.99\ \text{amu/atom} = & 45.98\ \text{amu} \\ \text{1 atom of sulfur} & \times\ 32.06\ \text{amu/atom} = & 32.06\ \text{amu} \\ \underline{\text{4 atoms of oxygen}} & \underline{\times\ 16.00\ \text{amu/atom} =} & \underline{64.00\ \text{amu}} \\ & & 142.04\ \text{amu} \end{array}$$

The average mass of a single unit of Na_2SO_4 is 142.04 amu and is the formula weight. Therefore the mass of a mole of Na_2SO_4 is 142.04 g, or 142.04 g/mol.

In Example 5.8, Na_2SO_4 is an ionic compound. As we have seen, it is not technically correct to describe ionic compounds as molecules; similarly, the term *molecular weight* is not appropriate for Na_2SO_4. The term *formula weight* may be used to describe ions, ion pairs, or molecules. We shall use the term *formula weight* in a general way to represent each of these species.

Figure 5.4 illustrates the difference between molecules and ion pairs.

Calculating Formula Weight and Molar Mass

EXAMPLE 5.9

Calculate the formula weight and molar mass of calcium phosphate.

Solution

The calcium ion is Ca^{2+}, and the phosphate ion is PO_4^{3-}. To form a neutral unit, $3Ca^{2+}$ must combine with $2PO_4^{3-}$; $[3 \times (+2)]$ calcium ion charges are balanced by $[2 \times (-3)]$, the phosphate ion charge. Thus for calcium phosphate, $Ca_3(PO_4)_2$, the subscript 2 for phosphate dictates that there are two phosphorus atoms and eight oxygen atoms (2×4) in the formula unit. Therefore

Continued—

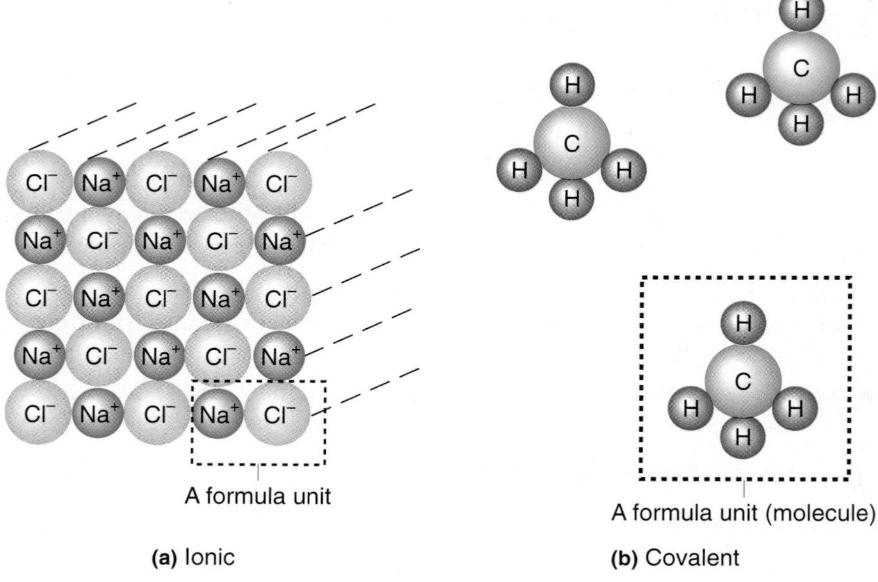

Figure 5.4
Formula units of (a) sodium chloride, an ionic compound, and (b) methane, a covalent compound.

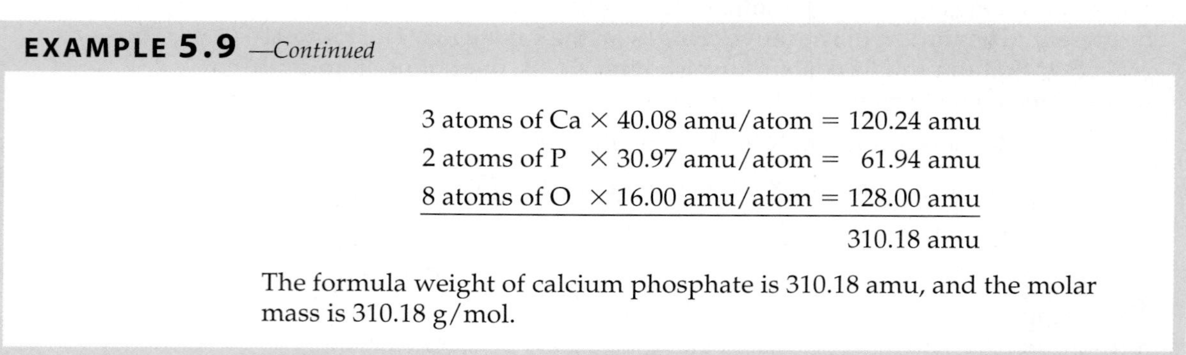

EXAMPLE 5.9 —*Continued*

$$\begin{array}{lr} \text{3 atoms of Ca} \times 40.08 \text{ amu/atom} = & 120.24 \text{ amu} \\ \text{2 atoms of P} \times 30.97 \text{ amu/atom} = & 61.94 \text{ amu} \\ \text{8 atoms of O} \times 16.00 \text{ amu/atom} = & 128.00 \text{ amu} \\ \hline & 310.18 \text{ amu} \end{array}$$

The formula weight of calcium phosphate is 310.18 amu, and the molar mass is 310.18 g/mol.

Question 5.7

Calculate the formula weight and molar mass of each of the following compounds:

a. NH_3 (ammonia)
b. $C_6H_{12}O_6$ (a sugar, glucose)
c. $CoCl_2 \cdot 6H_2O$ (cobalt chloride hexahydrate)

Question 5.8

Calculate the formula weight and molar mass of each of the following compounds:

a. $C_2F_2Cl_4$ (a Freon gas)
b. C_3H_7OH (isopropyl alcohol, rubbing alcohol)
c. CH_3Br (bromomethane, a pesticide)

5.4 The Chemical Equation and the Information It Conveys

A Recipe for Chemical Change

Learning Goal

5

The **chemical equation** is the shorthand notation for a chemical reaction. It describes all of the substances that react and all the products that form. **Reactants,** or starting materials, are all substances that undergo change in a chemical reaction; **products** are substances produced by a chemical reaction.

The chemical equation also describes the physical state of the reactants and products as solid, liquid, or gas. It tells us whether the reaction occurs and identifies the solvent and experimental conditions employed, such as heat, light, or electrical energy added to the system.

Most important, the relative number of moles of reactants and products appears in the equation. According to the **law of conservation of mass,** matter cannot be either gained or lost in the process of a chemical reaction. The total mass of the products must be equal to the total mass of the reactants. In other words, the law of conservation of mass tells us that we must have a balanced chemical equation.

Features of a Chemical Equation

Consider the decomposition of calcium carbonate:

$$\underset{\text{Calcium carbonate}}{CaCO_3(s)} \xrightarrow{\Delta} \underset{\text{Calcium oxide}}{CaO(s)} + \underset{\text{Carbon dioxide}}{CO_2(g)}$$

This equation reads: One mole of solid calcium carbonate decomposes upon heating to produce one mole of solid calcium oxide and one mole of gaseous carbon dioxide.

The factors involved in writing equations of this type are described as follows:

1. ***The identity of products and reactants must be specified using chemical symbols.*** In some cases it is possible to predict the products of a reaction. More often, the reactants and products must be verified by chemical analysis. (Generally, you will be given information regarding the identity of the reactants and products.)
2. ***Reactants are written to the left of the reaction arrow (→), and products are written to the right.*** The direction in which the arrow points indicates the direction in which the reaction proceeds. In the decomposition of calcium carbonate, the reactant on the left ($CaCO_3$) is converted to products on the right ($CaO + CO_2$) during the course of the reaction.
3. ***The physical states of reactants and products may be shown in parentheses.*** For example:
   - $Cl_2(g)$ means that chlorine is in the gaseous state.
   - $Mg(s)$ indicates that magnesium is a solid.
   - $Br_2(l)$ indicates that bromine is present as a liquid.
   - $NH_3(aq)$ tells us that ammonia is present as an aqueous solution (dissolved in water).
4. ***The symbol Δ over the reaction arrow means that heat energy is necessary for the reaction to occur.*** Often, other special conditions are noted above or below the reaction arrow. For example, "light" means that a light source provides energy necessary for the reaction. Such reactions are termed photochemical reactions.
5. ***The equation must be balanced.*** All of the atoms of every reactant must also appear in the products, although in different compounds. We will treat this topic in detail later in this chapter.

According to the factors outlined, the equation for the decomposition of calcium carbonate may now be written as

$$CaCO_3(s) \xrightarrow{\Delta} CaO(s) + CO_2(g)$$

The Experimental Basis of a Chemical Equation

The chemical equation must represent a chemical change: One or more substances are changed into new substances, with different chemical and physical properties. Evidence for the reaction may be based on observations such as

- the release of carbon dioxide gas when an acid is added to a carbonate,
- the formation of a solid (or precipitate) when solutions of iron ions and hydroxide ions are mixed,

See A Clinical Perspective: Hot and Cold Packs in Chapter 8.

- the production of heat when using hot packs for treatment of injury, and
- the change in color of a solution upon addition of a second substance.

Many reactions are not so obvious. Sophisticated instruments are now available to the chemist. These instruments allow the detection of subtle changes in chemical systems that would otherwise go unnoticed. Such instruments may measure

- heat or light absorbed or emitted as the result of a reaction,
- changes in the way the sample behaves in an electric or magnetic field before and after a reaction, and
- changes in electrical properties before and after a reaction.

Whether we use our senses or a million dollar computerized instrument, the "bottom line" is the same: We are measuring a change in one or more chemical or physical properties in an effort to understand the changes taking place in a chemical system.

Disease can be described as a chemical system (actually a biochemical system) gone awry. Here, too, the underlying changes may not be obvious. Just as technology has helped chemists see subtle chemical changes in the laboratory, medical diagnosis has been revolutionized in our lifetimes using very similar technology. Some of these techniques are described in the Clinical Perspective, Magnetic Resonance Imaging, in Chapter 10.

5.5 Balancing Chemical Equations

Learning Goal

The chemical equation shows the *molar quantity* of reactants needed to produce a certain *molar quantity* of products.

The relative number of moles of each product and reactant is indicated by placing a whole-number *coefficient* before the formula of each substance in the chemical equation. A coefficient of 2 (for example, 2NaCl) indicates that 2 mol of sodium chloride are involved in the reaction. Also, $3NH_3$ signifies 3 mol of ammonia; it means that 3 mol of nitrogen atoms and 3×3, or 9, mol of hydrogen atoms are involved in the reaction. The coefficient 1 is understood, not written. Therefore H_2SO_4 would be interpreted as 1 mol of sulfuric acid, or 2 mol of hydrogen atoms, 1 mol of sulfur atoms, and 4 mol of oxygen atoms.

The equation

$$CaCO_3(s) \xrightarrow{\Delta} CaO(s) + CO_2(g)$$

is balanced as written. On the reactant side we have

1 mol Ca

1 mol C

3 mol O

On the product side there are

1 mol Ca

1 mol C

3 mol O

The coefficients indicate *relative* numbers of moles: 10 mol of $CaCO_3$ produce 10 mol of CaO; 0.5 mol of $CaCO_3$ produce 0.5 mol of CaO; and so forth.

Therefore the law of conservation of mass is obeyed, and the equation is balanced as written.

Now consider the reaction of aqueous hydrogen chloride with solid calcium metal in aqueous solution:

$$HCl(aq) + Ca(s) \longrightarrow CaCl_2(aq) + H_2(g)$$

The equation, as written, is not balanced.

Reactants	Products
1 mol H atoms	2 mol H atoms
1 mol Cl atoms	2 mol Cl atoms
1 mol Ca atoms	1 mol Ca atoms

We need 2 mol of both H and Cl on the left, or reactant, side. An *incorrect* way of balancing the equation is as follows:

$$H_2Cl_2(aq) + Ca(s) \longrightarrow CaCl_2(aq) + H_2(g)$$

NOT a correct equation

The equation satisfies the law of conservation of mass; however, we have altered one of the reacting species. Hydrogen chloride is HCl, not H_2Cl_2. We must remember that *we cannot alter any chemical substance in the process of balancing the equation.* We can *only* introduce coefficients into the equation. Changing subscripts changes the identity of the chemicals involved, and that is not permitted. The equation must represent the reaction accurately. The correct equation is

$$2HCl(aq) + Ca(s) \longrightarrow CaCl_2(aq) + H_2(g)$$

Correct equation

This process is illustrated in Figure 5.5.

Many equations are balanced by trial and error. After the identity of the products and reactants, the physical state, and the reaction conditions are known, the following steps provide a method for correctly balancing a chemical equation:

Step 1. Count the number of moles of atoms of each element on both product and reactant side.

Step 2. Determine which elements are not balanced.

Step 3. Balance one element at a time using coefficients.

Step 4. After you believe that you have successfully balanced the equation, check, as in Step 1, to be certain that mass conservation has been achieved.

Let us apply these steps to the reaction of calcium with aqueous hydrogen chloride:

$$HCl(aq) + Ca(s) \longrightarrow CaCl_2(aq) + H_2(g)$$

Coefficients placed in front of the formula indicate the relative number of moles of compound (represented by the formula) that are involved in the reaction. Subscripts placed to the lower right of the atomic symbol indicate the relative number of atoms in the compound.

Water (H_2O) and hydrogen peroxide (H_2O_2) illustrate the effect a subscript can have. The two compounds show marked differences in physical and chemical properties.

H Cl + Ca → Ca Cl + H

(a) Incorrect equation

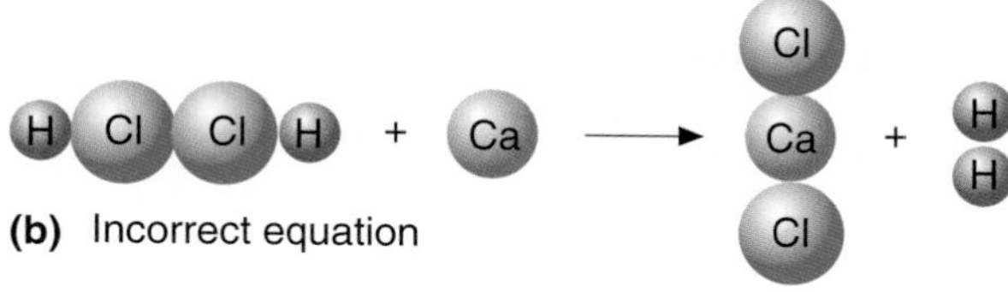

(b) Incorrect equation

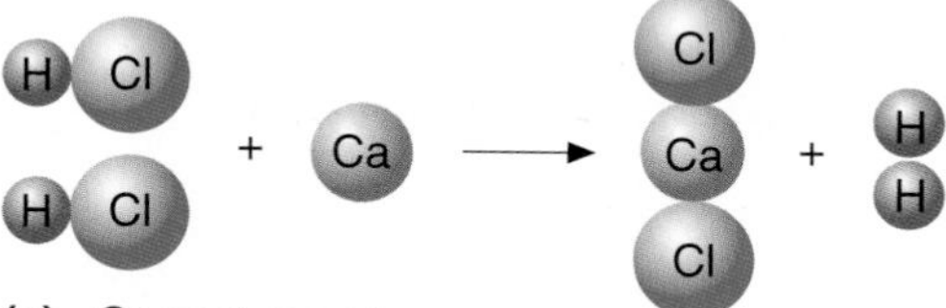

(c) Correct equation

Figure 5.5

Balancing the equation HCl + Ca → $CaCl_2$ + H_2. (a) Neither product is the correct chemical species. (b) The reactant, HCl, is incorrectly represented as H_2Cl_2. (c) This equation is correct; all species are correct, and the law of conservation of mass is obeyed.

Step 1.

Reactants	Products
1 mol H atoms	2 mol H atoms
1 mol Cl atoms	2 mol Cl atoms
1 mol Ca atoms	1 mol Ca atoms

Step 2. The numbers of moles of H and Cl are not balanced.

Step 3. Insertion of a 2 before HCl on the reactant side should balance the equation:

$$2HCl(aq) + Ca(s) \longrightarrow CaCl_2(aq) + H_2(g)$$

Step 4. Check for mass balance:

Reactants	Products
2 mol H atoms	2 mol H atoms
2 mol Cl atoms	2 mol Cl atoms
1 mol Ca atoms	1 mol Ca atoms

Hence the equation is balanced.

EXAMPLE 5.10 Balancing Equations

Balance the following equation: Hydrogen gas and oxygen gas react explosively to produce gaseous water.

Solution

Recall that hydrogen and oxygen are diatomic gases; therefore

$$H_2(g) + O_2(g) \longrightarrow H_2O(g)$$

Note that the moles of hydrogen atoms are balanced but that the moles of oxygen atoms are not; therefore we must first balance the moles of oxygen atoms:

$$H_2(g) + O_2(g) \longrightarrow 2H_2O(g)$$

Balancing moles of oxygen atoms creates an imbalance in the number of moles of hydrogen atoms, so

$$2H_2(g) + O_2(g) \longrightarrow 2H_2O(g)$$

The equation is balanced, with 4 mol of hydrogen atoms and 2 mol of oxygen atoms on each side of the reaction arrow.

EXAMPLE 5.11 Balancing Equations

Balance the following equation: Propane gas, C_3H_8, a fuel, reacts with oxygen gas to produce carbon dioxide and water vapor. The reaction is

$$C_3H_8(g) + O_2(g) \longrightarrow CO_2(g) + H_2O(g)$$

Solution

First, balance the carbon atoms; there are 3 mol of carbon atoms on the left and only 1 mol of carbon atoms on the right. We need $3CO_2$ on the right side of the equation:

$$C_3H_8(g) + O_2(g) \longrightarrow 3CO_2(g) + H_2O(g)$$

Continued—

EXAMPLE 5.11 *—Continued*

Next, balance the hydrogen atoms; there are 2 mol of hydrogen atoms on the right and 8 mol of hydrogen atoms on the left. We need $4H_2O$ on the right:

$$C_3H_8(g) + O_2(g) \longrightarrow 3CO_2(g) + 4H_2O(g)$$

There are now 10 mol of oxygen atoms on the right and 2 mol of oxygen atoms on the left. To balance, we must have $5O_2$ on the left side of the equation:

$$C_3H_8(g) + 5O_2(g) \longrightarrow 3CO_2(g) + 4H_2O(g)$$

Remember: In every case, be sure to check the final equation for mass balance.

EXAMPLE 5.12

Balancing Equations

Balance the following equation: Butane gas, C_4H_{10}, a fuel used in pocket lighters, reacts with oxygen gas to produce carbon dioxide and water vapor. The reaction is

$$C_4H_{10}(g) + O_2(g) \longrightarrow CO_2(g) + H_2O(g)$$

Solution

First, balance the carbon atoms; there are 4 mol of carbon atoms on the left and only 1 mol of carbon atoms on the right:

$$C_4H_{10}(g) + O_2(g) \longrightarrow 4CO_2(g) + H_2O(g)$$

Next, balance hydrogen atoms; there are 10 mol of hydrogen atoms on the left and only 2 mol of hydrogen atoms on the right:

$$C_4H_{10}(g) + O_2(g) \longrightarrow 4CO_2(g) + 5H_2O(g)$$

There are now 13 mol of oxygen atoms on the right and only 2 mol of oxygen atoms on the left. Therefore a coefficient of 6.5 is necessary for O_2.

$$C_4H_{10}(g) + 6.5O_2(g) \longrightarrow 4CO_2(g) + 5H_2O(g)$$

Fractional or decimal coefficients are often needed and used. However, the preferred form requires all integer coefficients. Multiplying each term in the equation by a suitable integer (2, in this case) satisfies this requirement. Hence

$$2C_4H_{10}(g) + 13O_2(g) \longrightarrow 8CO_2(g) + 10H_2O(g)$$

The equation is balanced, with 8 mol of carbon atoms, 20 mol of hydrogen atoms, and 26 mol of oxygen atoms on each side of the reaction arrow.

Helpful Hint: When balancing equations, we find that it is often most efficient to begin by balancing the atoms in the most complicated formulas.

EXAMPLE 5.13

Balancing Equations

Balance the following equation: Aqueous ammonium sulfate reacts with aqueous lead nitrate to produce aqueous ammonium nitrate and solid lead sulfate. The reaction is

Continued—

A CLINICAL PERSPECTIVE

Carbon Monoxide Poisoning: A Case of Combining Ratios

A fuel, such as methane, CH_4, burned in an excess of oxygen produces carbon dioxide and water:

$$CH_4(g) + 2O_2(g) \longrightarrow CO_2(g) + 2H_2O(g)$$

The same combustion in the presence of insufficient oxygen produces carbon monoxide and water:

$$2CH_4(g) + 3O_2(g) \longrightarrow 2CO(g) + 4H_2O(g)$$

The combustion of methane, repeated over and over in millions of gas furnaces, is responsible for heating many of our homes in the winter. The furnace is designed to operate under conditions that favor the first reaction and minimize the second; excess oxygen is available from the surrounding atmosphere. Furthermore, the vast majority of exhaust gases (containing principally CO, CO_2, H_2O, and unburned fuel) are removed from the home through the chimney. However, if the chimney becomes obstructed, or the burner malfunctions, carbon monoxide levels within the home can rapidly reach hazardous levels.

Why is exposure to carbon monoxide hazardous? Hemoglobin, an iron-containing compound, binds with O_2 and transports it throughout the body. Carbon monoxide also combines with hemoglobin, thereby blocking oxygen transport. The binding affinity of hemoglobin for carbon monoxide is about two hundred times as great as for O_2. Therefore, to maintain O_2 binding and transport capability, our exposure to carbon monoxide must be minimal. Proper ventilation and suitable oxygen-to-fuel ratio are essential for any combustion process in the home, automobile, or workplace. In recent years carbon monoxide sensors have been developed. These sensors sound an alarm when toxic levels of CO are reached. These warning devices have helped to create a safer indoor environment.

The example we have chosen is an illustration of what is termed the *law of multiple proportions.* This law states that identical reactants may produce different products, depending on their combining ratio. The experimental conditions (in this case, the quantity of available oxygen) determine the preferred path of the chemical reaction. In Section 5.6 we will learn how to use a properly balanced equation, representing the chemical change occurring, to calculate quantities of reactants consumed or products produced.

EXAMPLE 5.13 —*Continued*

$$(NH_4)_2SO_4(aq) + Pb(NO_3)_2(aq) \longrightarrow NH_4NO_3(aq) + PbSO_4(s)$$

Solution

In this case the polyatomic ions remain as intact units. Therefore we can balance them as we would balance molecules rather than as atoms.

There are two ammonium ions on the left and only one ammonium ion on the right. Hence

$$(NH_4)_2SO_4(aq) + Pb(NO_3)_2(aq) \longrightarrow 2NH_4NO_3(aq) + PbSO_4(s)$$

No further steps are necessary. The equation is now balanced. There are two ammonium ions, two nitrate ions, one lead ion, and one sulfate ion on each side of the reaction arrow.

Question 5.9

Balance each of the following chemical equations:

a. $Fe(s) + O_2(g) \longrightarrow Fe_2O_3(s)$
b. $C_6H_6(l) + O_2(g) \longrightarrow CO_2(g) + H_2O(g)$

Question 5.10

Balance each of the following chemical equations:

a. $S_2Cl_2(s) + NH_3(g) \longrightarrow N_4S_4(s) + NH_4Cl(s) + S_8(s)$
b. $C_2H_5OH(l) + O_2(g) \longrightarrow CO_2(g) + H_2O(g)$

5.6 Calculations Using the Chemical Equation

General Principles

Learning Goal 7

The calculation of quantities of products and reactants based on a balanced chemical equation is important in many fields. The synthesis of drugs and other complex molecules on a large scale is conducted on the basis of a balanced equation. This minimizes the waste of expensive chemical compounds used in these reactions. Similarly, the ratio of fuel and air in a home furnace or automobile must be adjusted carefully, according to their combining ratio, to maximize energy conversion, minimize fuel consumption, and minimize pollution.

In carrying out chemical calculations we apply the following guidelines.

1. The chemical formulas of all reactants and products must be known.
2. The basis for the calculations is a balanced equation because the conservation of mass must be obeyed. If the equation is not properly balanced, the calculation is meaningless.
3. The calculations are performed in terms of moles. The coefficients in the balanced equation represent the relative number of moles of products and reactants.

We have seen that the number of moles of products and reactants often differs in a balanced equation. For example,

$$C(s) + O_2(g) \longrightarrow CO_2(g)$$

is a balanced equation. Two moles of reactants combine to produce one mole of product:

$$1 \text{ mol C} + 1 \text{ mol } O_2 \longrightarrow 1 \text{ mol } CO_2$$

However, 1 mol of C *atoms* and 2 mol of O *atoms* produce 1 mol of C *atoms* and 2 mol of O *atoms*. In other words, the number of moles of reactants and products may differ, but the number of moles of atoms cannot. The formation of CO_2 from C and O_2 may be described as follows:

$$C(s) + O_2(g) \longrightarrow CO_2(g)$$

$$1 \text{ mol C} + 1 \text{ mol } O_2 \longrightarrow 1 \text{ mol } CO_2$$

$$12.0 \text{ g C} + 32.0 \text{ g } O_2 \longrightarrow 44.0 \text{ g } CO_2$$

The mole is the basis of our calculations. However, moles are generally measured in grams (or kilograms). A facility for interconversion of moles and grams is fundamental to chemical arithmetic (see Figure 5.2). These calculations are reviewed in Example 5.14.

Use of Conversion Factors

Conversion between Moles and Grams

Conversion from moles to grams, and vice versa, requires only the formula weight of the compound of interest. Consider the following examples.

Converting between Moles and Grams — EXAMPLE 5.14

a. Convert 1.00 mol of oxygen gas, O_2, to grams.

Solution

Use the following path:

Continued—

EXAMPLE 5.14 —*Continued*

moles of oxygen ⟶ grams of oxygen

The molar mass of oxygen (O_2) is

$$\frac{32.0 \text{ g } O_2}{1 \text{ mol } O_2}$$

Therefore

$$1.00 \text{ mol } O_2 \times \frac{32.0 \text{ g } O_2}{1 \text{ mol } O_2} = 32.0 \text{ g } O_2$$

b. How many grams of carbon dioxide are contained in 10.0 mol of carbon dioxide?

Solution

Use the following path:

moles of carbon dioxide ⟶ grams of carbon dioxide

The formula weight of CO_2 is

$$\frac{44.0 \text{ g } CO_2}{1 \text{ mol } CO_2}$$

and

$$10.0 \text{ mol } CO_2 \times \frac{44.0 \text{ g } CO_2}{1 \text{ mol } CO_2} = 4.40 \times 10^2 \text{ g } CO_2$$

c. How many moles of sodium are contained in 1 lb (454 g) of sodium metal?

Solution

Use the following path:

grams of sodium ⟶ moles of sodium

The number of moles of sodium atoms is

$$454 \text{ g Na} \times \frac{1 \text{ mol Na}}{22.99 \text{ g Na}} = 19.7 \text{ mol Na}$$

Helpful Hint: Note that each factor can be inverted producing a second possible factor. Only one will allow the appropriate unit cancellation.

Question 5.11

Perform each of the following conversions:

a. 5.00 mol of water to grams of water
b. 25.0 g of LiCl to moles of LiCl

Question 5.12

Perform each of the following conversions:

a. 1.00×10^{-5} mol of $C_6H_{12}O_6$ to micrograms of $C_6H_{12}O_6$
b. 35.0 g of $MgCl_2$ to moles of $MgCl_2$

Conversion of Moles of Reactants to Moles of Products

In Example 5.11 we balanced the equation for the reaction of propane and oxygen as follows:

$$C_3H_8(g) + 5O_2(g) \rightarrow 3CO_2(g) + 4H_2O(g)$$

In this reaction, 1 mol of C_3H_8 corresponds to, or results in,

5 mol of O_2 being consumed and

3 mol of CO_2 being formed and

4 mol of H_2O being formed.

This information may be written in the form of a conversion factor or ratio:

$$1 \text{ mol } C_3H_8/5 \text{ mol } O_2$$

Translated: One mole of C_3H_8 reacts with five moles of O_2.

$$1 \text{ mol } C_3H_8/3 \text{ mol } CO_2$$

Translated: One mole of C_3H_8 produces three moles of CO_2.

$$1 \text{ mol } C_3H_8/4 \text{ mol } H_2O$$

Translated: One mole of C_3H_8 produces four moles of H_2O.

Conversion factors, based on the chemical equation, permit us to perform a variety of calculations.

Let us look at a few examples, based on the combustion of propane and the equation that we balanced in Example 5.11.

Calculating Reacting Quantities EXAMPLE 5.15

Calculate the number of grams of O_2 that will react with 1.00 mol of C_3H_8.

Solution

Two conversion factors are necessary to solve this problem:

1. conversion from moles of C_3H_8 to moles of O_2 and
2. conversion of moles of O_2 to grams of O_2.

Therefore our path is

moles C_3H_8 ⟶ moles O_2 ⟶ grams O_2

and

$$1.00 \text{ mol } C_3H_8 \times \frac{5 \text{ mol } O_2}{1 \text{ mol } C_3H_8} \times \frac{32.0 \text{ g } O_2}{1 \text{ mol } O_2} = 1.60 \times 10^2 \text{ g } O_2$$

EXAMPLE 5.16 ***Calculating Grams of Product from Moles of Reactant***

Calculate the number of grams of CO_2 produced from the combustion of 1.00 mol of C_3H_8.

Solution

Employ logic similar to that used in Example 5.15 and use the following path:

$$\text{moles } C_3H_8 \longrightarrow \text{moles } CO_2 \longrightarrow \text{grams } CO_2$$

Then

$$1.00 \text{ mol } C_3H_8 \times \frac{3 \text{ mol } CO_2}{1 \text{ mol } C_3H_8} \times \frac{44.0 \text{ g } CO_2}{1 \text{ mol } CO_2} = 132 \text{ g } CO_2$$

EXAMPLE 5.17 ***Relating Masses of Reactants and Products***

Calculate the number of grams of C_3H_8 required to produce 36.0 g of H_2O.

Solution

It is necessary to convert

1. grams of H_2O to moles of H_2O,
2. moles of H_2O to moles of C_3H_8, and
3. moles of C_3H_8 to grams of C_3H_8.

Use the following path:

$$\text{grams } H_2O \longrightarrow \text{moles } H_2O \longrightarrow \text{moles } C_3H_8 \longrightarrow \text{grams } C_3H_8$$

Then

$$36.0 \text{ g } H_2O \times \frac{1 \text{ mol } H_2O}{18.0 \text{ g } H_2O} \times \frac{1 \text{ mol } C_3H_8}{4 \text{ mol } H_2O} \times \frac{44.0 \text{ g } C_3H_8}{1 \text{ mol } C_3H_8} = 22.0 \text{ g } C_3H_8$$

Question 5.13

The balanced equation for the combustion of ethanol (ethyl alcohol) is:

$$C_2H_5OH(l) + 3O_2(g) \longrightarrow 2CO_2(g) + 3H_2O(g)$$

a. How many moles of O_2 will react with 1 mol of ethanol?
b. How many grams of O_2 will react with 1 mol of ethanol?

Question 5.14

How many grams of CO_2 will be produced by the combustion of 1 mol of ethanol? (See Question 5.13.)

Let's consider an example that requires us to write and balance the chemical equation, use conversion factors, and calculate the amount of a reactant consumed in the chemical reaction.

Calculating a Quantity of Reactant

EXAMPLE 5.18

Calcium hydroxide may be used to neutralize (completely react with) aqueous hydrochloric acid. Calculate the number of grams of hydrochloric acid that would be neutralized by 0.500 mol of solid calcium hydroxide.

Solution

The formula for calcium hydroxide is $Ca(OH)_2$ and that for hydrochloric acid is HCl. The unbalanced equation produces calcium chloride and water as products:

$$Ca(OH)_2(s) + HCl(aq) \longrightarrow CaCl_2(aq) + H_2O(l)$$

First, balance the equation:

$$Ca(OH)_2(s) + 2HCl(aq) \longrightarrow CaCl_2(aq) + 2H_2O(l)$$

Next, determine the necessary conversion:

1. moles of $Ca(OH)_2$ to moles of HCl and
2. moles of HCl to grams of HCl.

Use the following path:

moles $Ca(OH)_2$ ⟶ moles HCl ⟶ grams HCl

$$0.500 \text{ mol } Ca(OH)_2 \times \frac{2 \text{ mol HCl}}{1 \text{ mol } Ca(OH)_2} \times \frac{36.5 \text{ g HCl}}{1 \text{ mol HCl}} = 36.5 \text{ g HCl}$$

This reaction is illustrated in Figure 5.6.

Helpful Hints:

1. The reaction between an acid and a base produces a salt and water (Chapter 9).
2. Remember to balance the chemical equation; the proper coefficients are essential parts of the subsequent calculations.

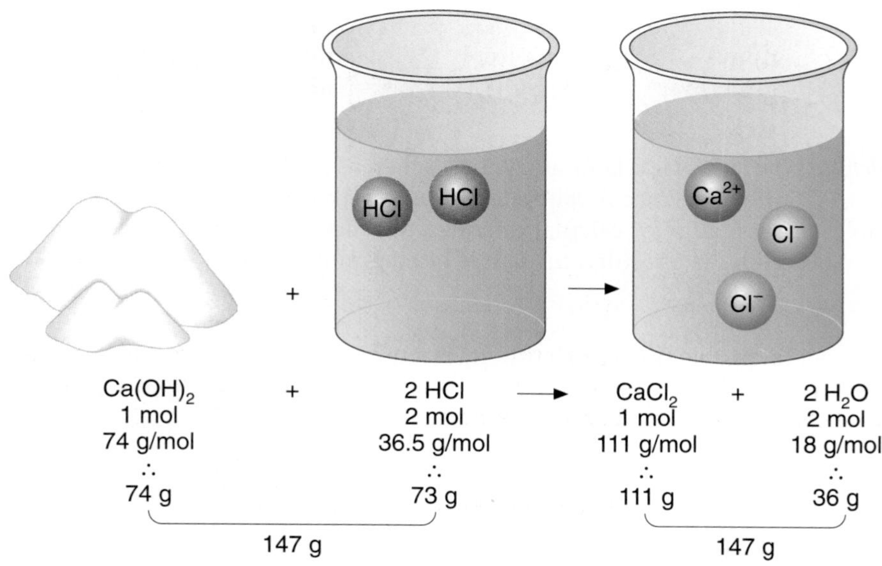

Figure 5.6
An illustration of the law of conservation of mass. In this example, 1 mol of calcium hydroxide and 2 mol of hydrogen chloride react to produce 3 mol of product (2 mol of water and 1 mol of calcium chloride). The total mass, in grams, of reactant(s) consumed is equal to the total mass, in grams, of product(s) formed. *Note:* In reality, HCl does not exist as discrete molecules in water. The HCl separates to form H^+ and Cl^-. Ionization in water will be discussed with the chemistry of acids and bases in Chapter 9.

EXAMPLE 5.19 Calculating Reactant Quantities

What mass of sodium hydroxide, NaOH, would be required to produce 8.00 g of the antacid milk of magnesia, $Mg(OH)_2$, by the reaction of $MgCl_2$ with NaOH?

Solution

$$MgCl_2(aq) + 2NaOH(aq) \longrightarrow Mg(OH)_2(s) + 2NaCl(aq)$$

The equation tells us that 2 mol of NaOH form 1 mol of $Mg(OH)_2$. If we calculate the number of moles of $Mg(OH)_2$ in 8.00 g of $Mg(OH)_2$, we can determine the number of moles of NaOH necessary and then the mass of NaOH required:

$$\text{mass } Mg(OH)_2 \longrightarrow \text{moles } Mg(OH)_2 \longrightarrow \text{moles NaOH} \longrightarrow \text{mass NaOH}$$

$$58.3 \text{ g } Mg(OH)_2 = 1 \text{ mol } Mg(OH)_2$$

Therefore

$$8.00 \text{ g } Mg(OH)_2 \times \frac{1 \text{ mol } Mg(OH)_2}{58.3 \text{ g } Mg(OH)_2} = 0.137 \text{ mol } Mg(OH)_2$$

Two moles of NaOH react to give one mole of $Mg(OH)_2$. Therefore

$$0.137 \text{ mol } Mg(OH)_2 \times \frac{2 \text{ mol NaOH}}{1 \text{ mol } Mg(OH)_2} = 0.274 \text{ mol NaOH}$$

40.0 g of NaOH = 1 mol of NaOH. Therefore

$$0.274 \text{ mol NaOH} \times \frac{40.0 \text{ g NaOH}}{1 \text{ mol NaOH}} = 11.0 \text{ g NaOH}$$

The calculation may be done in a single step:

$$8.00 \text{ g } Mg(OH)_2 \times \frac{1 \text{ mol } Mg(OH)_2}{58.3 \text{ g } Mg(OH)_2} \times \frac{2 \text{ mol NaOH}}{1 \text{ mol } Mg(OH)_2} \times \frac{40.0 \text{ g NaOH}}{1 \text{ mol NaOH}} = 11.0 \text{ g NaOH}$$

Note once again that we have followed a logical and predictable path to the solution:

$$\text{grams } Mg(OH)_2 \longrightarrow \text{moles } Mg(OH)_2 \longrightarrow \text{moles NaOH} \longrightarrow \text{grams NaOH}$$

Helpful Hint: Mass is a laboratory unit, whereas moles is a calculation unit. The laboratory balance is calibrated in units of mass (grams). Although moles are essential for calculation, often the starting point and objective are in mass units. As a result, our path is often grams → moles → grams.

A general problem-solving strategy is summarized in Figure 5.7. By systematically applying this strategy, you will be able to solve virtually any problem requiring calculations based on the chemical equation.

Question 5.15

Metallic iron reacts with O_2 gas to produce iron(III) oxide.

a. Write and balance the equation.
b. Calculate the number of grams of iron needed to produce 5.00 g of product.

For the reaction:

$$A + B \longrightarrow C$$

(a) Given a specified number of grams of *A*, calculate moles of *C*.

$$\text{g } A \longrightarrow \text{mol } A \longrightarrow \text{mol } C$$

$$\times \left(\frac{1 \text{ mol } A}{\text{g } A}\right) \quad \times \left(\frac{\text{mol } C}{\text{mol } A}\right)$$

(b) Given a specified number of grams of *A*, calculate grams of *C*.

$$\text{g } A \longrightarrow \text{mol } A \longrightarrow \text{mol } C \longrightarrow \text{g } C$$

$$\times \left(\frac{1 \text{ mol } A}{\text{g } A}\right) \quad \times \left(\frac{\text{mol } C}{\text{mol } A}\right) \quad \times \left(\frac{\text{g } C}{\text{mol } C}\right)$$

(c) Given a volume of *A* in milliliters, calculate grams of *C*.

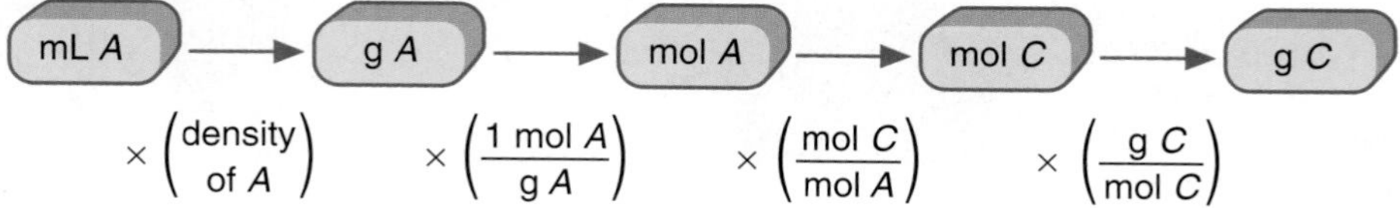

Figure 5.7
A general problem-solving strategy, using molar quantities.

Question 5.16

Barium carbonate decomposes upon heating to barium oxide and carbon dioxide.

a. Write and balance the equation.
b. Calculate the number of grams of carbon dioxide produced by heating 50.0 g of barium carbonate.

Theoretical and Percent Yield

Learning Goal

The **theoretical yield** is the *maximum* amount of product that can be produced (in an ideal world). In the "real" world it is difficult to produce the amount calculated as the theoretical yield. This is true for a variety of reasons. Some experimental error is unavoidable. Moreover, many reactions simply are not complete; some amount of reactant remains at the end of the reaction. We will study these processes, termed *equilibrium reactions* in Chapter 8.

A **percent yield,** the ratio of the actual and theoretical yields multiplied by 100%, is often used to show the relationship between predicted and experimental quantities. Thus

$$\% \text{ yield} = \frac{\text{actual yield}}{\text{theoretical yield}} \times 100\%$$

In Example 5.16, the theoretical yield of CO_2 is 132 g. For this reaction let's assume that a chemist actually obtained 125 g CO_2. This is the actual yield and would normally be provided as a part of the data in the problem.

Calculate the percent yield as follows:

$$\% \text{ yield} = \frac{\text{actual yield}}{\text{theoretical yield}} \times 100\%$$

A Medical Perspective

Pharmaceutical Chemistry: The Practical Significance of Percent Yield

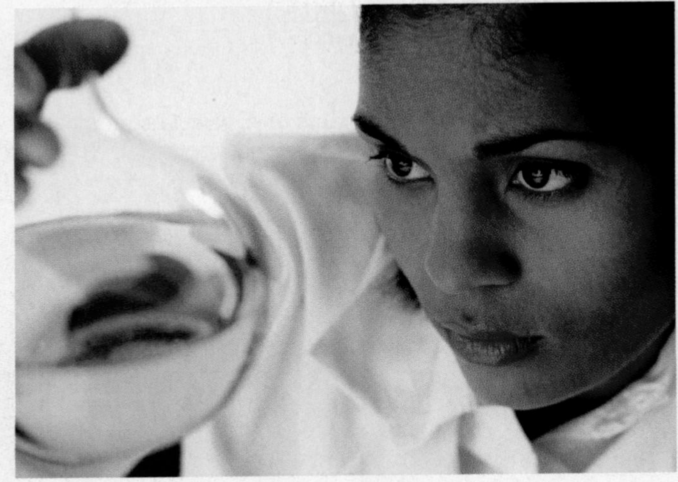

In recent years the major pharmaceutical industries have introduced a wide variety of new drugs targeted to cure or alleviate the symptoms of a host of diseases that afflict humanity.

The vast majority of these drugs are synthetic; they are made in a laboratory or by an industrial process. These substances are complex molecules that are patiently designed and constructed from relatively simple molecules in a series of chemical reactions. A series of ten to twenty "steps," or sequential reactions, is not unusual to put together a final product that has the proper structure, geometry, and reactivity for efficacy against a particular disease.

Although a great deal of research occurs to ensure that each of these steps in the overall process is efficient (having a large percent yield), the overall process is still very inefficient (low percent yield). This inefficiency, and the research needed to minimize it, at least in part determines the cost and availability of both prescription and over-the-counter preparations.

Consider a hypothetical five-step sequential synthesis. If each step has a percent yield of 80% our initial impression might be that this synthesis is quite efficient. However, on closer inspection we find quite the contrary to be true.

The overall yield of the five-step reaction is the product of the decimal fraction of the percent yield of each of the sequential reactions. So, if the decimal fraction corresponding to 80% is 0.80:

$$0.80 \times 0.80 \times 0.80 \times 0.80 \times 0.80 = 0.33$$

Converting the decimal fraction to percentage:

$$0.33 \times 100\% = 33\% \text{ yield}$$

Many reactions are considerably less than 80% efficient, especially those that are used to prepare large molecules with complex arrangements of atoms. Imagine a more realistic scenario in which one step is only 20% efficient (a 20% yield) and the other four steps are 50%, 60%, 70%, and 80% efficient. Repeating the calculation with these numbers (after conversion to a decimal fraction):

$$0.20 \times 0.50 \times 0.60 \times 0.70 \times 0.80 = 0.0336$$

Converting the decimal fraction to a percentage:

$$0.0336 \times 100\% = 3.36\% \text{ yield}$$

a very inefficient process.

If we apply this logic to a fifteen- or twenty-step synthesis we gain some appreciation of the difficulty of producing modern pharmaceutical products. Add to this the challenge of predicting the most appropriate molecular structure that will have the desired biological effect and be relatively free of side effects. All these considerations give new meaning to the term *wonder drug* that has been attached to some of the more successful synthetic products.

We will study some of the elementary steps essential to the synthesis of a wide range of pharmaceutical compounds in later chapters, beginning with Chapter 11.

$$= \frac{125 \text{ g } CO_2 \text{ actual}}{132 \text{ g } CO_2 \text{ theoretical}} \times 100\% = 94.7\%$$

EXAMPLE 5.20 ***Calculation of Percent Yield***

Assume that the theoretical yield of iron in the process

$$2Al(s) + Fe_2O_3(s) \longrightarrow Al_2O_3(l) + 2Fe(l)$$

was 30.0 g.

Continued—

EXAMPLE 5.20 —*Continued*

If the actual yield of iron were 25.0 g in the process, calculate the percent yield.

Solution

$$\% \text{ yield} = \frac{\text{actual yield}}{\text{theoretical yield}} \times 100\%$$

$$= \frac{25.0 \text{ g}}{30.0 \text{ g}} \times 100\%$$

$$= 83.3\%$$

Question 5.17

Given the reaction represented by the balanced equation

$$Sn(s) + 2HF(aq) \longrightarrow SnF_2(s) + H_2(g)$$

a. Calculate the number of grams of SnF_2 produced by mixing 100.0 g Sn with excess HF.
b. If only 5.00 g SnF_2 were produced, calculate the % yield.

Question 5.18

Given the reaction represented by the balanced equation

$$CH_4(g) + 3Cl_2(g) \longrightarrow 3HCl(g) + CHCl_3(g)$$

a. Calculate the number of grams of $CHCl_3$ produced by mixing 105 g Cl_2 with excess CH_4.
b. If 10.0 g $CHCl_3$ were produced, calculate the % yield.

Summary

5.1 The Mole Concept and Atoms

Atoms are exceedingly small, yet their masses have been experimentally determined for each of the elements. The unit of measurement for these determinations is the *atomic mass unit*, abbreviated amu:

$$1 \text{ amu} = 1.661 \times 10^{-24} \text{ g}$$

The periodic table provides atomic masses in atomic mass units.

A more practical unit for defining a "collection" of atoms is the *mole:*

$$1 \text{ mol of atoms} = 6.022 \times 10^{23} \text{ atoms of an element}$$

This number is referred to as *Avogadro's number.*

The mole and the atomic mass unit are related. The atomic mass of a given element corresponds to the average mass of a single atom in atomic mass units and the mass of a mole of atoms in grams. One mole of atoms of any element contains the same number, Avogadro's number, of atoms.

5.2 Compounds

Compounds are pure substances that are composed of two or more elements that are chemically combined. They are represented by their *chemical formula,* a combination of symbols of the various elements that make up the compounds. The chemical formula is based on the *formula unit.* This is the smallest collection of atoms that provides the identity of the atoms present in the compound and the relative numbers of each type of atom.

5.3 The Mole Concept Applied to Compounds

Just as a mole of atoms is based on the atomic mass, a mole of a compound is based on the formula mass or *formula weight*. The formula weight is calculated by addition of the masses of all the atoms or ions of which the unit is composed. To calculate the formula weight, the formula unit must be known.

5.4 The Chemical Equation and the Information It Conveys

The *chemical equation* is the shorthand notation for a chemical reaction. It describes all of the substances that react to produce the product(s). *Reactants,* or starting materials, are all substances that undergo change in a chemical reaction; *products* are substances produced by a chemical reaction.

According to the *law of conservation of mass,* matter can neither be gained nor lost in the process of a chemical reaction. The law of conservation of mass states that we must have a balanced chemical equation.

Features of a suitable equation include the following:

- The identity of products and reactants must be specified.
- Reactants are written to the left of the reaction arrow (→) and products to the right.
- The physical states of reactants and products are shown in parentheses.
- The symbol Δ over the reaction arrow means that heat energy is necessary for the reaction to occur.
- The equation must be balanced.

5.5 Balancing Chemical Equations

The chemical equation enables us to determine the quantity of reactants needed to produce a certain molar quantity of products. The chemical equation expresses these quantities in terms of moles.

The number of moles of each product and reactant is indicated by placing a whole-number coefficient before the formula of each substance in the chemical equation.

Many equations are balanced by trial and error. If the identity of the products and reactants, the physical state, and the reaction conditions are known, the following steps provide a method for correctly balancing a chemical equation:

- Count the number of atoms of each element on both product and reactant sides.
- Determine which atoms are not balanced.
- Balance one element at a time using coefficients.
- After you believe that you have successfully balanced the equation, check to be certain that mass conservation has been achieved.

5.6 Calculations Using the Chemical Equation

Calculations involving chemical quantities are based on the following requirements:

- The basis for the calculations is a balanced equation.
- The calculations are performed in terms of moles.
- The conservation of mass must be obeyed.

The mole is the basis for calculations. However, masses are generally measured in grams (or kilograms). Therefore you must be able to interconvert moles and grams to perform chemical arithmetic.

Key Terms

atomic mass unit (5.1)
Avogadro's number (5.1)
chemical equation (5.4)
chemical formula (5.2)
formula unit (5.2)
formula weight (5.3)
hydrate (5.2)
law of conservation of mass (5.4)
molar mass (5.1)
mole (5.1)
percent yield (5.6)
product (5.4)
reactant (5.4)
theoretical yield (5.6)

Questions and Problems

The Mole Concept and Atoms

5.19 What is the mass in grams of 1.00 mol of helium atoms?

5.20 What is the mass in grams of 1.00 mol of nitrogen atoms?

5.21 Calculate the number of moles corresponding to:
- **a.** 20.0 g He
- **b.** 0.040 kg Na
- **c.** 3.0 g Cl_2

5.22 Calculate the number of moles corresponding to:
- **a.** 0.10 g Ca
- **b.** 4.00 g Fe
- **c.** 2.00 kg N_2

5.23 What is the mass, in grams, of 15.0 mol of silver?

5.24 What is the mass, in grams, of 15.0 mol of carbon?

Compounds

5.25 Distinguish between the terms *molecule* and *ion pair.*

5.26 Distinguish between the terms *formula weight* and *molecular weight.*

5.27 Calculate the formula weight, in grams per mole, of each of the following formula units:
- **a.** NaCl
- **b.** Na_2SO_4
- **c.** $Fe_3(PO_4)_2$

5.28 Calculate the formula weight, in grams per mole, of each of the following formula units:
- **a.** S_8
- **b.** $(NH_4)_2SO_4$
- **c.** CO_2

5.29 Calculate the formula weight, in grams per mole, of oxygen gas, O_2.

5.30 Calculate the formula weight, in grams per mole, of ozone, O_3.

The Mole Concept Applied to Compounds

5.31 Calculate the number of moles corresponding to:
- **a.** 15.0 g NaCl
- **b.** 15.0 g Na_2SO_4

5.32 Calculate the number of moles corresponding to:
- **a.** 15.0 g NH_3
- **b.** 16.0 g O_2

5.33 Calculate the mass in grams corresponding to:
- **a.** 1.000 mol H_2O
- **b.** 2.000 mol NaCl

5.34 Calculate the mass in grams corresponding to:
- **a.** 0.400 mol NH_3
- **b.** 0.800 mol $BaCO_3$

5.35 Calculate the mass in grams corresponding to:
- **a.** 10.0 mol He
- **b.** 1.00×10^2 mol H_2

5.36 Calculate the mass in grams corresponding to:
a. 2.00 mol CH_4
b. 0.400 mol $Ca(NO_3)_2$

5.37 How many grams are required to have 0.100 mol of each of the following compounds?
a. Mg
b. $CaCO_3$
c. $C_6H_{12}O_6$ (glucose)
d. NaCl

5.38 How many grams are required to have 0.100 mol of each of the following compounds?
a. NaOH
b. H_2SO_4
c. C_2H_5OH (ethanol)
d. $Ca_3(PO_4)_2$

5.39 How many moles are in 50.0 g of each of the following substances?
a. KBr
b. $MgSO_4$
c. Br_2
d. NH_4Cl

5.40 How many moles are in 50.0 g of each of the following substances?
a. CS_2
b. $Al_2(CO_3)_3$
c. $Sr(OH)_2$
d. $LiNO_3$

The Chemical Equation and the Information It Conveys

5.41 What law is the ultimate basis for a correct chemical equation?

5.42 List the general types of information that a chemical equation provides.

5.43 What is the meaning of the subscript in a chemical formula?

5.44 What is the meaning of the coefficient in a chemical equation?

Balancing Chemical Equations

5.45 Balance each of the following equations.
a. $C_6H_{10}(g) + O_2(g) \longrightarrow H_2O(g) + CO_2(g)$
b. $Au_2S_3(s) + H_2(g) \longrightarrow Au(s) + H_2S(g)$
c. $Al(OH)_3(s) + HCl(aq) \longrightarrow AlCl_3(aq) + H_2O(l)$
d. $(NH_4)_2Cr_2O_7(s) \longrightarrow Cr_2O_3(s) + N_2(g) + H_2O(g)$
e. $C_2H_5OH(l) + O_2(g) \longrightarrow CO_2(g) + H_2O(g)$

5.46 Balance each of the following equations:
a. $Fe_2O_3(s) + CO(g) \longrightarrow Fe_3O_4(s) + CO_2(g)$
b. $C_6H_6(l) + O_2(g) \longrightarrow CO_2(g) + H_2O(g)$
c. $I_4O_9(s) + I_2O_6(s) \longrightarrow I_2(s) + O_2(g)$
d. $KClO_3(s) \longrightarrow KCl(s) + O_2(g)$
e. $C_6H_{12}O_6(s) \longrightarrow C_2H_6O(l) + CO_2(g)$

5.47 Write a balanced equation for each of the following reactions:
a. Ammonia is formed by the reaction of nitrogen and hydrogen.
b. Hydrochloric acid reacts with sodium hydroxide to produce water and sodium chloride.

5.48 Write a balanced equation for each of the following reactions:
a. Nitric acid reacts with calcium hydroxide to produce water and calcium nitrate.
b. Butane (C_4H_{10}) reacts with oxygen to produce water and carbon dioxide.

5.49 Write a balanced equation for each of the following reactions:
a. Glucose, a sugar, $C_6H_{12}O_6$, is oxidized in the body to produce water and carbon dioxide.
b. Sodium carbonate, upon heating, produces sodium oxide and carbon dioxide.

5.50 Write a balanced equation for each of the following reactions:
a. Sulfur, present as an impurity in coal, is burned in oxygen to produce sulfur dioxide.
b. Hydrofluoric acid (HF) reacts with glass (SiO_2) in the process of etching to produce silicon tetrafluoride and water.

Calculations Using the Chemical Equation

5.51 How many grams of boron oxide, B_2O_3, can be produced from 20.0 g diborane (B_2H_6)?

$$B_2H_6(l) + 3O_2(g) \longrightarrow B_2O_3(s) + 3H_2O(l)$$

5.52 How many grams of Al_2O_3 can be produced from 15.0 g Al?

$$4Al(s) + 3O_2(g) \longrightarrow 2Al_2O_3(s)$$

5.53 Calculate the amount of $CrCl_3$ that could be produced from 50.0 g Cr_2O_3 according to the equation

$$Cr_2O_3(s) + 3CCl_4(l) \longrightarrow 2CrCl_3(s) + 3COCl_2(aq)$$

5.54 A 3.5-g sample of water reacts with PCl_3 according to the following equation:

$$3H_2O(l) + PCl_3(g) \longrightarrow H_3PO_3(aq) + 3HCl(aq)$$

How many grams of H_3PO_3 are produced?

5.55 For the reaction

$$N_2(g) + H_2(g) \longrightarrow NH_3(g)$$

a. Balance the equation.
b. How many moles of H_2 would react with 1 mol of N_2?
c. How many moles of product would form from 1 mol of N_2?
d. If 14.0 g of N_2 were initially present, calculate the number of moles of H_2 required to react with all of the N_2.
e. For conditions outlined in part (d), how many grams of product would form?

5.56 Aspirin (acetylsalicylic acid) may be formed from salicylic acid and acetic acid as follows:

$$\underset{\text{Salicylic acid}}{C_7H_6O_3(aq)} + \underset{\text{Acetic acid}}{CH_3COOH(aq)} \longrightarrow \underset{\text{Aspirin}}{C_9H_8O_4(s)} + H_2O(l)$$

a. Is this equation balanced? If not, complete the balancing.
b. How many moles of aspirin may be produced from 1.00×10^2 mol salicylic acid?
c. How many grams of aspirin may be produced from 1.00×10^2 mol salicylic acid?
d. How many grams of acetic acid would be required to react completely with the 1.00×10^2 mol salicylic acid?

5.57 The proteins in our bodies are composed of molecules called amino acids. One amino acid is methionine; its molecular formula is $C_5H_{11}NO_2S$. Calculate:
a. the formula weight of methionine
b. the number of oxygen atoms in a mole of this compound
c. the mass of oxygen in a mole of the compound
d. the mass of oxygen in 50.0 g of the compound

5.58 Triglycerides (Chapters 18 and 23) are used in biochemical systems to store energy; they can be formed from glycerol and fatty acids. The molecular formula of glycerol is $C_3H_8O_3$. Calculate:
a. the formula weight of glycerol
b. the number of oxygen atoms in a mole of this compound
c. the mass of oxygen in a mole of the compound
d. the mass of oxygen in 50.0 g of the compound

5.59 Joseph Priestley discovered oxygen in the eighteenth century by using heat to decompose mercury(II) oxide:

$$2HgO(s) \xrightarrow{\Delta} 2Hg(l) + O_2(g)$$

How much oxygen is produced from 1.00×10^2 g HgO?

5.60 Dinitrogen monoxide (also known as nitrous oxide and used as an anesthetic) can be made by heating ammonium nitrate:

$$NH_4NO_3(s) \longrightarrow N_2O(g) + 2H_2O(g)$$

How much dinitrogen monoxide can be made from 1.00×10^2 g of ammonium nitrate?

5.61 The burning of acetylene (C_2H_2) in oxygen is the reaction in the oxyacetylene torch. How much oxygen is needed to burn 20.0 kg of acetylene? The unbalanced equation is

$$C_2H_2(g) + O_2(g) \longrightarrow CO_2(g) + H_2O(g)$$

5.62 The reaction of calcium hydride with water can be used to prepare hydrogen gas:

$$CaH_2(s) + 2H_2O(l) \longrightarrow Ca(OH)_2(aq) + 2H_2(g)$$

How many moles of hydrogen gas are produced in the reaction of 1.00×10^2 g calcium hydride with water?

5.63 Various members of a class of compounds, alkenes (Chapter 12), react with hydrogen to produce a corresponding alkane (Chapter 11). Termed hydrogenation, this type of reaction is used to produce products such as margarine. A typical hydrogenation reaction is

$$\underset{\text{Decene}}{C_{10}H_{20}(l)} + H_2(g) \longrightarrow \underset{\text{Decane}}{C_{10}H_{22}(s)}$$

How much decane can be produced in a reaction of excess decene with 1.00 g hydrogen?

5.64 The Human Perspective: Alcohol Consumption and the Breathalyzer Test (Chapter 13), describes the reaction between the dichromate ion and ethanol to produce acetic acid. How much acetic acid can be produced from a mixture containing excess of dichromate ion and 1.00×10^{-1} g of ethanol?

5.65 A rocket can be powered by the reaction between dinitrogen tetroxide and hydrazine:

$$N_2O_4(l) + 2N_2H_4(l) \longrightarrow 3N_2(g) + 4H_2O(g)$$

An engineer designed the rocket to hold 1.00 kg N_2O_4 and excess N_2H_4. How much N_2 would be produced according to the engineer's design?

5.66 A 4.00-g sample of Fe_3O_4 reacts with O_2 to produce Fe_2O_3:

$$4Fe_3O_4(s) + O_2(g) \longrightarrow 6Fe_2O_3(s)$$

Determine the number of grams of Fe_2O_3 produced.

5.67 If the actual yield of decane in Problem 5.63 was 65.4 g, what is the % yield?

5.68 If the actual yield of acetic acid in problem 5.64 was 0.110 g, what is the % yield?

5.69 If the % yield of nitrogen gas in problem 5.65 was 75.0%, what was the actual yield of nitrogen?

5.70 If the % yield of Fe_2O_3 in problem 5.66 was 90.0%, what was the actual yield of Fe_2O_3?

Critical Thinking Problems

1. Which of the following has fewer moles of carbon: 100 g of $CaCO_3$ or 0.5 mol of CCl_4?
2. Which of the following has fewer moles of carbon: 6.02×10^{22} molecules of C_2H_6 or 88 g of CO_2?
3. How many molecules are found in each of the following?
   a. 1.0 lb of sucrose, $C_{12}H_{22}O_{11}$ (table sugar)
   b. 1.57 kg of N_2O (anesthetic)
4. How many molecules are found in each of the following?
   a. 4×10^5 tons of SO_2 (produced by the 1980 eruption of the Mount St. Helens volcano)
   b. 25.0 lb of SiO_2 (major constituent of sand)

An exciting application of the gas laws.

6

States of Matter: Gases, Liquids, and Solids

Outline

GENERAL CHEMISTRY

Learning Goals

1. Describe the behavior of gases expressed by the gas laws: Boyle's law, Charles's law, combined gas law, Avogadro's law, the ideal gas law, and Dalton's law.
2. Use gas law equations to calculate conditions and changes in conditions of gases.
3. Describe the major points of the kinetic molecular theory of gases.
4. Explain the relationship between the kinetic molecular theory and the physical properties of macroscopic quantities of gases.
5. Describe properties of the liquid state in terms of the properties of the individual molecules that comprise the liquid.
6. Describe the processes of melting, boiling, evaporation, and condensation.
7. Describe the dipolar attractions known collectively as van der Waals forces.
8. Describe hydrogen bonding and its relationship to boiling and melting temperatures.
9. Relate the properties of the various classes of solids (ionic, covalent, molecular, and metallic) to the structure of these solids.

Chemistry Connection

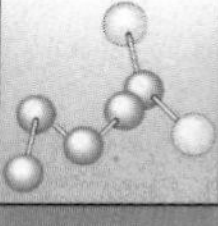

The Demise of the Hindenburg

One of the largest and most luxurious airships of the 1930s, the Hindenburg, completed thirty-six transatlantic flights within a year after its construction. It was the flagship of a new era of air travel. But, on May 6, 1937, while making a landing approach near Lakehurst, New Jersey, the hydrogen-filled airship exploded and burst into flames. In this tragedy, 37 of the 96 passengers were killed and many others were injured.

We may never know the exact cause. Many believe that the massive ship (it was more than 800 feet long) struck an overhead power line. Others speculate that lightning ignited the hydrogen and some believe that sabotage may have been involved.

In retrospect, such an accident was inevitable. Hydrogen gas is very reactive, it combines with oxygen readily and rapidly, and this reaction liberates a large amount of energy. An explosion is the result of rapid, energy-releasing reactions.

Why was hydrogen chosen? Hydrogen is the lightest element. One mole of hydrogen has a mass of two grams. Hydrogen can be easily prepared in pure form, an essential requirement; more than seven million cubic feet of hydrogen were needed for each airship. Hydrogen has a low density; hence it provides great lift. The lifting power of a gas is based on the difference in density of the gas and the surrounding air (air is composed of gases with much greater molar masses; N_2 is 28 g and O_2 is 32 g). Engineers believed that the hydrogen would be safe when enclosed by the hull of the airship.

Today, airships are filled with helium (its molar mass is 4 g) and are used principally for advertising and television. A Goodyear blimp can be seen hovering over almost every significant outdoor sporting event.

In this chapter we will study the relationships that predict the behavior of gases in a wide variety of applications from airships to pressurized oxygen for respiration therapy.

Introduction

We have learned that the major differences between solids, liquids, and gases are due to the relationships among particles. These relationships include:

- the average distance of separation of particles in each state,
- the kinds of interactions between the particles, and
- the degree of organization of particles.

Section 2.1 introduces the properties of the three states of matter.

We have already discovered that the solid state is the most organized, with particles close together, allowing significant interactions among the particles. This results in high melting and boiling points for solid substances. Large amounts of energy are needed to overcome the attractive forces and disrupt the orderly structure.

Substances that are gases at room temperature and atmospheric pressure, on the other hand, are disordered, with particles widely separated and weak interactions between particles. Their melting and boiling points are relatively low. Gases at room temperature must be cooled a great deal for them to liquefy or solidify. For example, the melting and boiling points of N_2 are $-210°C$ and $-196°C$, respectively.

Liquids are intermediate in character. The molecules of a liquid are close together, like those of solids. However, the molecules of a liquid are disordered, like those of a gas.

Table 6.1 A Comparison of Physical Properties of Gases, Liquids, and Solids

	Gas	Liquid	Solid
Volume and Shape	Expands to fill the volume of its container; consequently takes the shape of the container	Has a fixed volume at a given mass and temperature; volume principally dependent on its mass and secondarily on temperature; it assumes the shape of its container	Has a fixed volume; volume principally dependent on its mass and secondarily on temperature; it has a definite shape
Density	Low	High	High
Compressibility	High	Very low	Virtually incompressible
Particle Motion	Virtually free	Molecules or atoms "slide" past each other	Vibrate about a fixed position
Intermolecular Distance	Very large	Molecules or atoms are close to each other	Molecules, ions, or atoms are close to each other

Changes in state are described as physical changes. When a substance undergoes a change in state, many of its physical properties change. For example, when ice forms from liquid water, changes occur in density and hardness, but it is still water. Table 6.1 summarizes the important differences in physical properties among gases, liquids, and solids.

6.1 The Gaseous State

Ideal Gas Concept

An **ideal gas** is simply a model of the way that particles (molecules or atoms) behave at the microscopic level. The behavior of the individual particles can be inferred from the macroscopic behavior of samples of real gases. We can easily measure temperature, volume, pressure, and quantity (mass) of real gases. Similarly, when we systematically change one of these properties, we can determine the effect on each of the others. For example, putting more molecules in a balloon (the act of blowing up a balloon) causes its volume to increase in a predictable way. In fact, careful measurements show a direct proportionality between the quantity of molecules and the volume of the balloon, an observation made by Amadeo Avogadro more than 200 years ago.

We owe a great deal of credit to the efforts of scientists Boyle, Charles, Avogadro, Dalton, and Gay-Lussac, whose careful work elucidated the relationships among the gas properties. Their efforts are summarized in the ideal gas law and are the subject of the first section of this chapter.

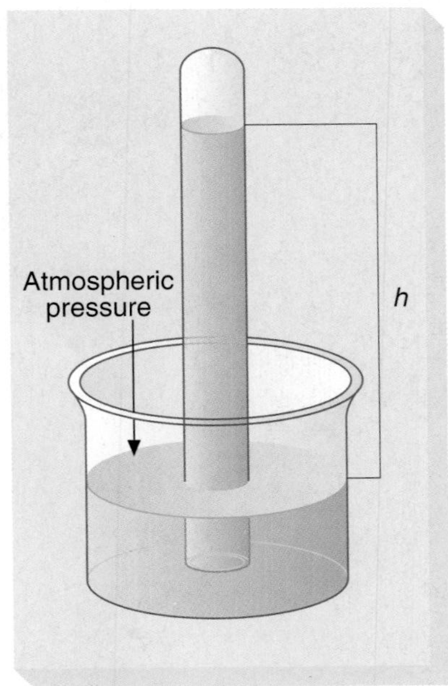

Figure 6.1
A mercury barometer of the type invented by Torricelli. The height of the column of mercury (*h*) is a function of the magnitude of the surrounding atmospheric pressure. The mercury in the tube is supported by atmospheric pressure.

Measurement of Gases

The most important gas laws (Boyle's law, Charles's law, Dalton's law, and the ideal gas law) involve the relationships between pressure (P), volume (V), temperature (T), and number of moles (n) of gas. We are already familiar with the measurement of temperature and quantity from our laboratory experience. Measurement of pressure is perhaps not as obvious.

Gas pressure is a result of the force exerted by the collision of particles with the walls of the container. **Pressure** is force per unit area. The pressure of a gas may be measured with a **barometer,** invented by Evangelista Torricelli in the mid-1600s. The most common type of barometer is the mercury barometer depicted in Figure 6.1. A tube, sealed at one end, is filled with mercury and inverted in a dish of mercury. The pressure of the atmosphere pushing down on the mercury surface in the dish supports the column of mercury. The height of the column is proportional to the atmospheric pressure. The tube can be calibrated to give a numerical reading in millimeters, centimeters, or inches of mercury. A commonly used unit of measurement is the atmosphere (atm). One standard atmosphere (1 atm) of pressure is equivalent to a height of mercury that is equal to

760 mm Hg (millimeters of mercury)

76.0 cm Hg (centimeters of mercury)

1 mm of Hg is also = 1 torr, in honor of Torricelli.

The English system equivalent is a pressure of 14.7 lb/in^2 (pounds per square inch) or 29.9 in Hg (inches of mercury). A recommended, yet less frequently used, systematic unit is the pascal (or kilopascal), named in honor of Blaise Pascal, a seventeenth-century French mathematician and scientist:

$$1 \text{ atm} = 1.01 \times 10^5 \text{ Pa (pascal)} = 101 \text{ kPa (kilopascal)}$$

Atmospheric pressure is due to the cumulative force of the air molecules (N_2 and O_2, for the most part) that are attracted to the earth's surface by gravity.

Question 6.1

Express each of the following in units of atmospheres:

a. 725 mm Hg
b. 29.0 cm Hg
c. 555 torr

Question 6.2

Express each of the following in units of atmospheres:

a. 10.0 torr
b. 61.0 cm Hg
c. 275 mm Hg

Boyle's Law

Learning Goal

The Irish scientist Robert Boyle found that the volume of a gas varies *inversely* with the pressure exerted by the gas if the number of moles and temperature of gas are held constant. This relationship is known as **Boyle's law.**

Mathematically, the *product* of pressure (P) and volume (V) is a constant:

$$PV = k_1$$

This relationship is illustrated in Figure 6.2.

Boyle's law is often used to calculate the volume resulting from a pressure change or vice versa. We consider

$$P_iV_i = k_1$$

the *initial* condition and

$$P_fV_f = k_1$$

the final condition. Because PV, initial or final, is constant and is equal to k_1,

$$P_iV_i = P_fV_f$$

Consider a gas occupying a volume of 10.0 L at 1.00 atm of pressure. The product, PV = (10.0 L)(1.00 atm), is a constant, k_1. Doubling the pressure, to 2.0 atm, decreases the volume to 5.0 L:

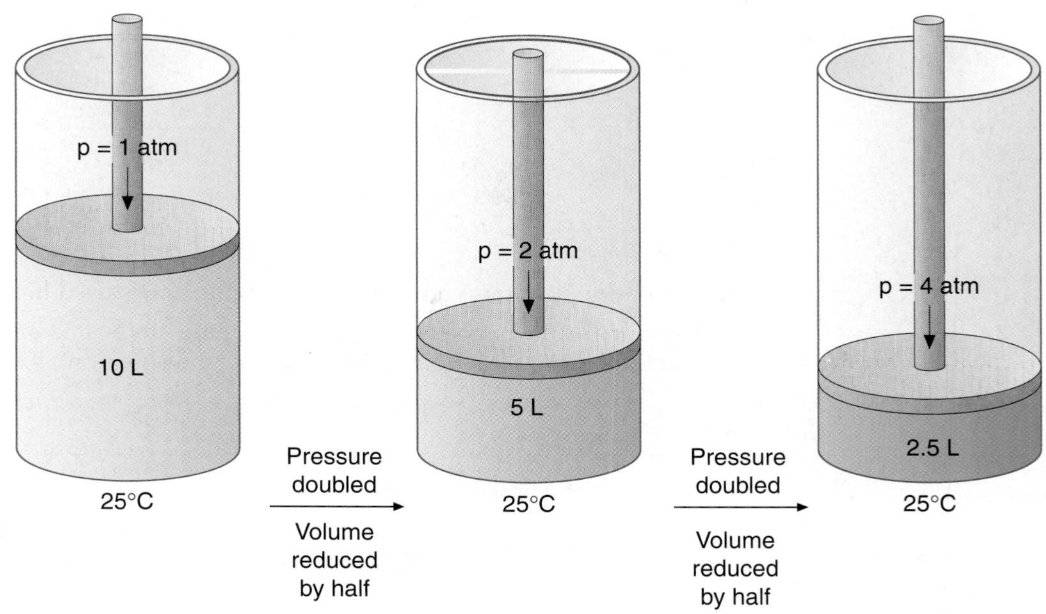

Figure 6.2
An illustration of Boyle's law. Note the inverse relationship of pressure and volume.

$$(2.0 \text{ atm})(V_x) = (10.0 \text{ L})(1.00 \text{ atm})$$

$$V_x = 5.0 \text{ L}$$

Tripling the pressure decreases the volume by a factor of 3:

$$(3.0 \text{ atm})(V_x) = (10.0 \text{ L})(1.00 \text{ atm})$$

$$V_x = 3.3 \text{ L}$$

Appendix A contains a review of the mathematics used here.

Calculating a Final Pressure

EXAMPLE 6.1

Learning Goal

A sample of oxygen, at 25°C, occupies a volume of 5.00×10^2 mL at 1.50 atm pressure. What pressure must be applied to compress the gas to a volume of 1.50×10^2 mL, with no temperature change?

Solution

Boyle's law applies directly, because there is no change in temperature or number of moles (no gas enters or leaves). Begin by identifying each term in the Boyle's law expression:

$$P_i = 1.50 \text{ atm}$$

$$V_i = 5.00 \times 10^2 \text{ mL}$$

$$V_f = 1.50 \times 10^2 \text{ mL}$$

$$P_i V_i = P_f V_f$$

and solve

$$P_f = \frac{P_i V_i}{V_f}$$

Continued—

EXAMPLE 6.1 —*Continued*

$$= \frac{(1.50 \text{ atm})(5.00 \times 10^2 \text{ mL})}{1.50 \times 10^2 \text{ mL}}$$

$$= 5.00 \text{ atm}$$

Helpful Hints:

1. Appendix A contains a review of the mathematics used here.
2. The calculation can be done with any volume units. It is important only that the units be the *same* on both sides of the equation.

Question 6.3

Complete the following table:

	Initial Pressure (atm)	Final Pressure (atm)	Initial Volume (L)	Final Volume (L)
a.	*X*	5.0	1.0	7.5
b.	5.0	*X*	1.0	0.20

Question 6.4

Complete the following table:

	Initial Pressure (atm)	Final Pressure (atm)	Initial Volume (L)	Final Volume (L)
a.	1.0	0.50	*X*	0.30
b.	1.0	2.0	0.75	*X*

Charles's Law

Learning Goal

Jacques Charles, a French scientist, studied the relationship between gas volume and temperature. This relationship, **Charles's law,** states that the volume of a gas varies *directly* with the absolute temperature (K) if pressure and number of moles of gas are constant.

Temperature is a measure of the energy of molecular motion. The Kelvin scale is *absolute,* that is, directly proportional to molecular motion. Celsius and Fahrenheit are simply numerical scales based on the melting and boiling points of water. It is for this reason that Kelvin is used for energy-dependent relationships such as the gas laws.

Mathematically, the *ratio* of volume (V) and temperature (T) is a constant:

$$\frac{V}{T} = k_2$$

In a way analogous to Boyle's law, we may establish a set of initial conditions,

$$\frac{V_i}{T_i} = k_2$$

and final conditions,

$$\frac{V_f}{T_f} = k_2$$

Because k_2 is a constant, we may equate them, resulting in

$$\frac{V_i}{T_i} = \frac{V_f}{T_f}$$

and use this expression to solve some practical problems.

Consider a gas occupying a volume of 10.0 L at 273 K. The ratio V/T is a constant, k_2. Doubling the temperature, to 546 K, increases the volume to 20.0 L as shown here:

Appendix A contains a review of the mathematics used here.

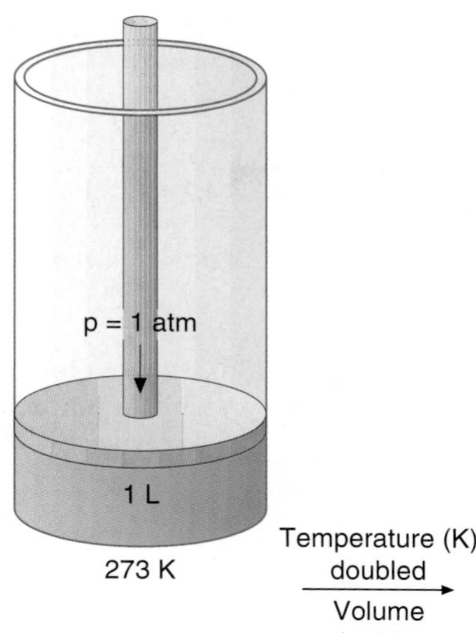

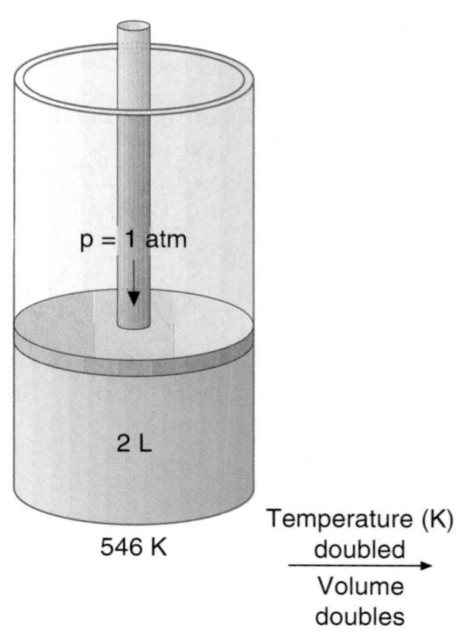

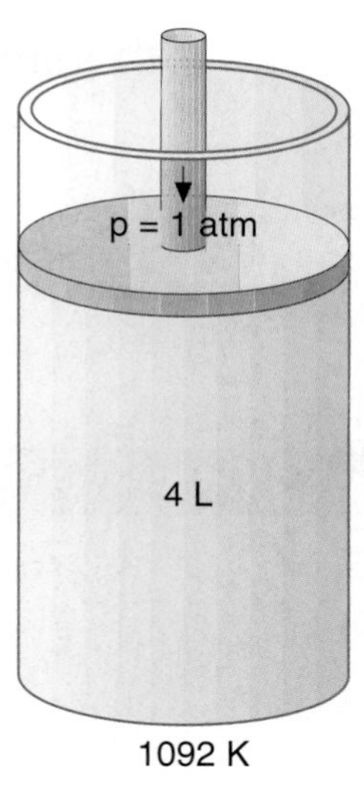

Figure 6.3
An illustration of Charles's law. Note the direct relationship between volume and temperature.

$$\frac{10.0 \text{ L}}{273 \text{ K}} = \frac{V_f}{546 \text{ K}}$$

$$V_f = 20.0 \text{ L}$$

Tripling the temperature, to 819 K, increases the volume by a factor of 3:

$$\frac{10.0 \text{ L}}{273 \text{ K}} = \frac{V_f}{819 \text{ K}}$$

$$V_f = 30.0 \text{ L}$$

These relationships are illustrated in Figure 6.3.

Calculating a Final Volume

EXAMPLE 6.2

Learning Goal

A balloon filled with helium has a volume of 4.0×10^3 L at 25°C. What volume will the balloon occupy at 50°C if the pressure surrounding the balloon remains constant?

Solution

Remember, the temperature must be converted to Kelvin before Charles's law is applied:

$$T_i = 25°\text{C} + 273 = 298 \text{ K}$$

$$T_f = 50°\text{C} + 273 = 323 \text{ K}$$

$$V_i = 4.0 \times 10^3 \text{ L}$$

$$V_f = ?$$

Using

$$\frac{V_i}{T_i} = \frac{V_f}{T_f}$$

Continued—

EXAMPLE 6.2 —*Continued*

and substituting our data, we get

$$V_f = \frac{(V_i)(T_f)}{T_i} = \frac{(4.0 \times 10^3 \text{ L})(323 \text{ K})}{298 \text{ K}} = 4.3 \times 10^3 \text{ L}$$

Question 6.5

A sample of nitrogen gas has a volume of 3.00 L at 25°C. What volume will it occupy at each of the following temperatures if the pressure and number of moles are constant?

a. 100°C
b. 150°F
c. 273 K

Question 6.6

A sample of nitrogen gas has a volume of 3.00 L at 25°C. What volume will it occupy at each of the following temperatures if the pressure and number of moles are constant?

a. 546 K
b. 0.00°C
c. 373 K

Figure 6.4
Charles's law predicts that the volume of air in the balloon will increase when heated. We assume that the volume of the balloon is fixed; consequently, some air will be pushed out. The air remaining in the balloon is less dense (same volume, less mass) and the balloon will rise. When the heater is turned off the air cools, the density increases, and the balloon returns to earth.

The behavior of a hot-air balloon is a commonplace consequence of Charles's law. The balloon rises because air expands when heated (Figure 6.4). The volume of the balloon is fixed because the balloon is made of an inelastic material; as a result, when the air expands some of the air must be forced out. Hence the density of the remaining air is less (less mass contained in the same volume), and the balloon rises. Turning down the heat reverses the process, and the balloon descends.

Boyle's law describes the inverse proportional relationship between volume and pressure; Charles's law shows the direct proportional relationship between volume and temperature.

Combined Gas Law

Often, a sample of gas (a fixed number of moles of gas) undergoes change involving volume, pressure, and temperature simultaneously. It would be useful to have one equation that describes such processes.

Learning Goal

The **combined gas law** is such an equation. It can be derived from Boyle's law and Charles's law and takes the form:

$$\frac{P_iV_i}{T_i} = \frac{P_fV_f}{T_f}$$

Let's look at two examples that use this expression.

EXAMPLE 6.3 ***Using the Combined Gas Law***

Learning Goal

Calculate the volume of N_2 that results when 0.100 L of the gas is heated from 300. K to 350. K at 1.00 atm.

Continued—

A CLINICAL PERSPECTIVE

Autoclaves and the Gas Laws

Jacques Charles and Joseph Gay-Lussac were eighteenth-century chemists and physicists. They were also balloon enthusiasts. It is clear that their hobby and their scientific pursuits were intertwined. Charles's law is actually attributed to the work of both men. The observation that the pressure and temperature of a gas are directly proportional

$$P \propto T$$

follows directly from Charles's law and Boyle's law. The equality

$$P = kT$$

is often termed *Gay-Lussac's law.* You can readily verify this proportionality by observing the fate of a balloon when it is heated or cooled (try putting an inflated balloon in the refrigerator, remove it, and allow it to return to room temperature).

A very practical application of Gay-Lussac's law is the autoclave, a piece of equipment commonly found in hospital, clinical, and biological laboratories. It is designed and used to sterilize laboratory materials such as glassware, surgical instruments, and so forth. It uses steam at high temperatures and pressures and takes advantage of the exceptionally high heat capacity (energy storage capability) of steam.

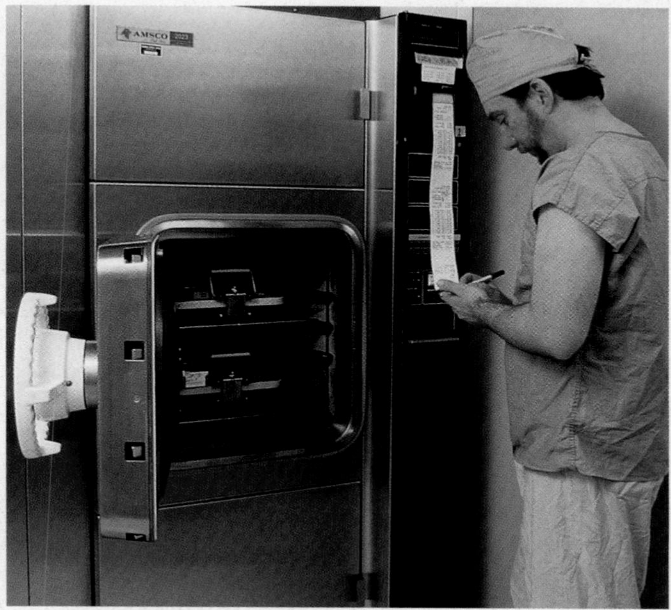

The autoclave kills microorganisms by using the heat energy in steam. However, steam has a temperature of 100°C at normal atmospheric pressure; this temperature is too low (insufficient energy) to kill all harmful bacteria. Gay-Lussac's law predicts that, at a constant volume (the volume of the autoclave), if the steam is heated further, both the pressure and temperature of the steam will increase. At the maximum safe operating pressure of the autoclave, temperatures may reach as much as 150°C; when maintained for a sufficient interval, this temperature is sufficiently high to kill most microorganisms.

EXAMPLE 6.3 —*Continued*

Solution

Summarize the data:

$P_i = 1.00$ atm $\quad P_f = 1.00$ atm

$V_i = 0.100$ L $\quad V_f = ?$ L

$T_i = 300.$ K $\quad T_f = 350.$ K

$$\frac{P_iV_i}{T_i} = \frac{P_fV_f}{T_f}$$

that can be rearranged as

$$P_fV_fT_i = P_iV_iT_f$$

and

$$V_f = \frac{P_iV_iT_f}{P_fT_i}$$

Continued—

EXAMPLE 6.3 —*Continued*

Because $P_i = P_f$

$$V_f = \frac{V_i T_f}{T_i}$$

Substituting gives

$$V_f = \frac{(0.100 \text{ L})(350. \text{ K})}{300. \text{ K}}$$

$$= 0.117 \text{ L}$$

Helpful Hints:

1. Appendix A contains a review of the mathematics used here.
2. In this case, because the pressure is constant, the combined gas law reduces to Charles's law.

EXAMPLE 6.4 ***Using the Combined Gas Law***

Learning Goal

A sample of helium gas has a volume of 1.27 L at 149 K and 5.00 atm. When the gas is compressed to 0.320 L at 50.0 atm, the temperature increases markedly. What is the final temperature?

Solution

Summarize the data:

$P_i = 5.00$ atm $P_f = 50.0$ atm

$V_i = 1.27$ L $V_f = 0.320$ L

$T_i = 149$ K $T_f = ?$ K

The combined gas law expression is

$$\frac{P_i V_i}{T_i} = \frac{P_f V_f}{T_f}$$

which we rearrange as

$$P_f V_f T_i = P_i V_i T_f$$

and

$$T_f = \frac{P_f V_f T_i}{P_i V_i}$$

Substituting yields

$$T_f = \frac{(50.0 \text{ atm})(0.320 \text{ L})(149 \text{ K})}{(5.00 \text{ atm})(1.27 \text{ L})}$$

$$= 375 \text{ K}$$

Helpful Hint: Appendix A contains a review of the mathematics used here.

Question 6.7

Hydrogen sulfide, H_2S, has the characteristic odor of rotten eggs. If a sample of H_2S gas at 760. torr and 25.0°C in a 2.00-L container is allowed to expand into a 10.0-L container at 25.0°C, what is the pressure in the 10.0-L container?

Question 6.8

Cyclopropane, C_3H_6, is used as a general anesthetic. If a sample of cyclopropane stored in a 2.00-L container at 10.0 atm and 25.0°C is transferred to a 5.00-L container at 5.00 atm, what is the resulting temperature?

Avogadro's Law

Learning Goal

The relationship between the volume and number of moles of a gas at constant temperature and pressure is known as **Avogadro's law.** It states that equal volumes of any ideal gas contain the same number of moles if measured under the same conditions of temperature and pressure.

Mathematically, the *ratio* of volume (V) to number of moles (n) is a constant:

$$\frac{V}{n} = k_3$$

Consider 1 mol of gas occupying a volume of 10.0 L; using logic similar to the application of Boyle's and Charles's laws, 2 mol of the gas would occupy 20.0 L, 3 mol would occupy 30.0 L, and so forth. As we have done with the previous laws, we can formulate a useful expression relating initial and final conditions:

$$\frac{V_i}{n_i} = \frac{V_f}{n_f}$$

Appendix A contains a review of the mathematics used here.

EXAMPLE 6.5

Using Avogadro's Law

Learning Goal

If 5.50 mol of CO occupy 20.6 L, how many liters will 16.5 mol of CO occupy at the same temperature and pressure?

Solution

The quantities moles and volume are related through Avogadro's law. Summarizing the data:

$V_i = 20.6$ L $\qquad V_f = ?$ L

$n_i = 5.50$ mol $\qquad n_f = 16.5$ mol

Using the mathematical expression for Avogadro's law:

$$\frac{V_i}{n_i} = \frac{V_f}{n_f}$$

and rearranging as

$$V_f = \frac{V_i n_f}{n_i}$$

then substituting yields

$$V_f = \frac{(20.6 \text{ L})(16.5 \text{ mol})}{(5.50 \text{ mol})}$$

$$= 61.8 \text{ L of CO}$$

Question 6.9

1.00 mole of hydrogen gas occupies 22.4 L. How many moles of hydrogen are needed to fill a 100.0 L container at the same pressure and temperature?

Question 6.10

How many moles of hydrogen are needed to triple the volume occupied by 0.25 mol of hydrogen, assuming no changes in pressure or temperature?

Molar Volume of a Gas

The volume occupied by *1 mol* of any gas is referred to as its **molar volume.** At **standard temperature and pressure (STP)** the molar volume of any gas is 22.4 L. STP conditions are defined as follows:

$$T = 273 \text{ K (or } 0^{\circ}\text{C)}$$

$$P = 1 \text{ atm}$$

Thus 1 mol of N_2, O_2, H_2, or He all occupy the *same volume, 22.4 L,* at STP.

Gas Densities

It is also possible to compute the density of various gases at STP. If we recall that density is the mass/unit volume,

$$d = \frac{m}{V}$$

and that 1 mol of helium weighs 4.00 g,

$$d_{\text{He}} = \frac{4.00 \text{ g}}{22.4 \text{ L}} = 0.178 \text{ g/L at STP}$$

or, because 1 mol of nitrogen weighs 28.0 g, then

$$d_{\text{N}_2} = \frac{28.0 \text{ g}}{22.4 \text{ L}} = 1.25 \text{ g/L at STP}$$

Heating a gas, such as air, will decrease its density and have a lifting effect as well.

The large difference in gas densities of helium and nitrogen (which makes up about 80% of the air) accounts for the lifting power of helium. A balloon filled with helium will rise through a predominantly nitrogen atmosphere because its gas density is less than 15% of the density of the surrounding atmosphere:

$$\frac{0.178 \text{ g/L}}{1.25 \text{ g/L}} \times 100\% = 14.2\%$$

The Ideal Gas Law

Learning Goal

Boyle's law (relating volume and pressure), Charles's law (relating volume and temperature), and Avogadro's law (relating volume to the number of moles) may be combined into a single expression relating all four terms. This expression is the **ideal gas law:**

$$PV = nRT$$

Remember that 0.0821 L-atm/K mol is identical to 0.0821 L-atm K^{-1} mol^{-1}.

in which R, based on k_1, k_2, and k_3 (Boyle's, Charles's, and Avogadro's law constants), is a constant and is referred to as the *ideal gas constant:*

$$R = 0.0821 \text{ L-atm K}^{-1} \text{ mol}^{-1}$$

if the units

P in atmospheres

V in liters

n in number of moles

T in Kelvin

are used.

Consider some examples of the application of the ideal gas equation.

Calculating a Molar Volume

EXAMPLE 6.6

Learning Goal

Demonstrate that the molar volume of oxygen gas at STP is 22.4 L.

Solution

$$PV = nRT$$

$$V = \frac{nRT}{P}$$

At standard temperature and pressure,

$$T = 273 \text{ K}$$

$$P = 1.00 \text{ atm}$$

The other constants are

$$n = 1.00 \text{ mol}$$

$$R = 0.0821 \text{ L-atm K}^{-1} \text{ mol}^{-1}$$

Then

$$V = \frac{(1.00 \text{ mol})(0.0821 \text{ L-atm K}^{-1} \text{ mol}^{-1})(273 \text{ K})}{(1.00 \text{ atm})}$$

$$= 22.4 \text{ L}$$

Calculating the Number of Moles of a Gas

EXAMPLE 6.7

Learning Goal 2

Calculate the number of moles of helium in a 1.00-L balloon at 27°C and 1.00 atm of pressure.

Solution

$$PV = nRT$$

$$n = \frac{PV}{RT}$$

If

$P = 1.00$ atm

$V = 1.00$ L

$T = 27°\text{C} + 273 = 300.$ K

$R = 0.0821 \text{ L-atm K}^{-1} \text{ mol}^{-1}$

Continued—

EXAMPLE 6.7 —*Continued*

then

$$n = \frac{(1.00 \text{ atm})(1.00 \text{ L})}{(0.0821 \text{ L-atm K}^{-1} \text{ mol}^{-1})(300. \text{ K})}$$

$$n = 0.0406 \text{ or } 4.06 \times 10^{-2} \text{ mol}$$

EXAMPLE 6.8 ***Converting Mass to Volume***

Learning Goal

Oxygen used in hospitals and laboratories is often obtained from cylinders containing liquefied oxygen. If a cylinder contains 1.00×10^2 kg of liquid oxygen, how many liters of oxygen can be produced at 1.00 atm of pressure at room temperature (20.0°C)?

Solution

$$PV = nRT$$

$$V = \frac{nRT}{P}$$

Using conversion factors, we obtain

$$n_{O_2} = 1.00 \times 10^2 \cancel{\text{kg O}_2} \times \frac{10^3 \cancel{\text{g O}_2}}{1 \cancel{\text{kg O}_2}} \times \frac{1 \text{ mol O}_2}{32.0 \cancel{\text{g O}_2}}$$

Then

$$n = 3.13 \times 10^3 \text{ mol O}_2$$

and

$$T = 20.0°\text{C} + 273 = 293 \text{ K}$$

$$P = 1.00 \text{ atm}$$

then

$$V = \frac{(3.13 \times 10^3 \text{ mol})(0.0821 \text{ L-atm K}^{-1} \text{ mol}^{-1})(293 \text{ K})}{1.00 \text{ atm}}$$

$$= 7.53 \times 10^4 \text{ L}$$

Question 6.11 What volume is occupied by 10.0 g N_2 at 30.0°C and a pressure of 750 torr?

Question 6.12 A 20.0-L gas cylinder contains 4.80 g H_2 at 25°C. What is the pressure of this gas?

Question 6.13 How many moles of N_2 gas will occupy a 5.00-L container at standard temperature and pressure?

Question 6.14 At what temperature will 2.00 mol He fill a 2.00-L container at standard pressure?

AN ENVIRONMENTAL PERSPECTIVE

The Greenhouse Effect and Global Warming

A greenhouse is a bright, warm, and humid environment for growing plants, vegetables, and flowers even during the cold winter months. It functions as a closed system in which the concentration of water vapor is elevated and visible light streams through the windows; this creates an ideal climate for plant growth.

Some of the visible light is absorbed by plants and soil in the greenhouse and radiated as infrared radiation. This radiated energy is blocked by the glass or absorbed by water vapor and carbon dioxide (CO_2). This trapped energy warms the greenhouse and is a form of solar heating: light energy is converted to heat energy.

On a global scale, the same process takes place. Although more than half of the sunlight that strikes the earth's surface is reflected back into space, the fraction of light that is absorbed produces sufficient heat to sustain life. How does this happen? Greenhouse gases, such as CO_2, trap energy radiated from the earth's surface and store it in the atmosphere. This moderates our climate. The earth's surface would be much colder and more inhospitable if the atmosphere was not able to capture some reasonable amount of solar energy.

Can we have too much of a good thing? It appears so. Since 1900 the atmospheric concentration of CO_2 has increased from 296 parts per million (ppm) to over 350 ppm (approximately 17% increase). The energy demands of technological and population growth have caused massive increases in the combustion of organic matter and carbon-based fuels (coal, oil, and natural gas), adding over 50 billion tons of CO_2 to that already present in the atmosphere. Photosynthesis naturally removes CO_2 from the atmosphere. However, the removal of forestland to create living space and cropland decreases the amount of vegetation available to consume atmospheric CO_2 through photosynthesis. The rapid destruction of the Amazon rain forest is just the latest of many examples.

If our greenhouse model is correct, an increase in CO_2 levels should produce global warming, perhaps changing our climate in unforeseen and undesirable ways.

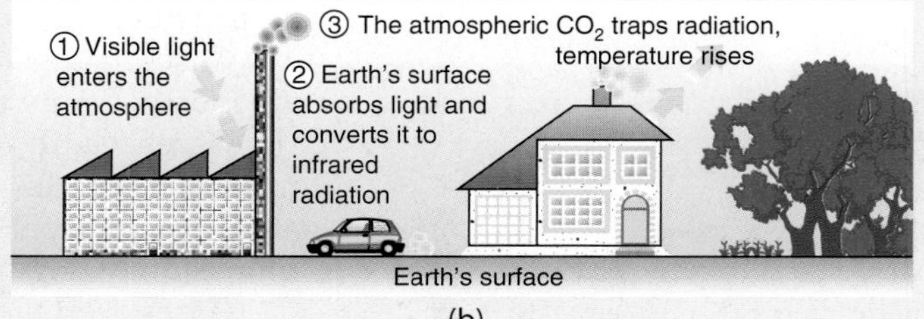

(a) A greenhouse traps solar radiation as heat. (b) Our atmosphere also acts as a solar collector. Carbon dioxide, like the windows of a greenhouse, allows the visible light to enter and traps the heat.

Dalton's Law of Partial Pressures

Learning Goal

Our discussion of gases so far has presumed that we are working with a single pure gas. A *mixture* of gases exerts a pressure that is the *sum* of the pressures that each gas would exert if it were present alone under the same conditions. This is known as **Dalton's law** of partial pressures.

Stated another way, the total pressure of a mixture of gases is the sum of the **partial pressures.** That is,

$$P_i = p_1 + p_2 + p_3 + \ldots$$

in which P_t = total pressure and $p_1, p_2, p_3, \ldots$, are the partial pressures of the component gases. For example, the total pressure of our atmosphere is equal to the sum of the pressures of N_2 and O_2 (the principal components of air):

$$P_{air} = p_{N_2} + p_{O_2}$$

Other gases, such as argon (Ar), carbon dioxide (CO_2), carbon monoxide (CO), and methane (CH_4) are present in the atmosphere at very low partial pressures. However, their presence may result in dramatic consequences; one such gas is carbon dioxide. Classified as a "greenhouse gas," it exerts a significant effect on our climate. Its role is described in An Environmental Perspective: The Greenhouse Effect and Global Warming.

Kinetic Molecular Theory of Gases

Learning Goal

The kinetic molecular theory of gases provides a reasonable explanation of the behavior of gases that we have studied in this chapter. The macroscopic properties result from the action of the individual molecules comprising the gas.

The **kinetic molecular theory** can be summarized as follows:

1. Gases are made up of small atoms or molecules that are in constant, random motion.
2. The distance of separation among these atoms or molecules is very large in comparison to the size of the individual atoms or molecules. In other words, a gas is mostly empty space.
3. All of the atoms and molecules behave independently. No attractive or repulsive forces exist between atoms or molecules in a gas.
4. Atoms and molecules collide with each other and with the walls of the container without *losing* energy. The energy is *transferred* from one atom or molecule to another.
5. The average kinetic energy of the atoms or molecules increases or decreases in proportion to absolute temperature.

Learning Goal

We know that gases are easily *compressible.* The reason is that a gas is mostly empty space, providing space for the particles to be pushed closer together.

Gases will *expand* to fill any available volume because they move freely with sufficient energy to overcome their attractive forces.

Gases have a *low density.* Density is defined as mass per volume. Because gases are mostly empty space, they have a low mass per volume.

Gases readily *diffuse* through each other simply because they are in continuous motion and paths are readily available because of the large space between adjacent atoms or molecules. Light molecules diffuse rapidly; heavier molecules diffuse more slowly (Figure 6.5).

Kinetic energy (K.E.) is equal to $\frac{1}{2}mv^2$, in which m = mass and v = velocity. Thus increased velocity at higher temperature correlates with an increase in kinetic energy.

Gases exert *pressure* on their containers. Pressure is a force per unit area resulting from collisions of gas particles with the walls of their container.

Gases behave most *ideally at low pressures and high temperatures.* At low pressures, the average distance of separation among atoms or molecules is greatest, minimizing interactive forces. At high temperatures, the atoms and molecules are in rapid motion and are able to overcome interactive forces more easily.

Ideal Gases Versus Real Gases

To this point we have assumed, in both theory and calculations, that all gases behave as ideal gases. However, in reality there is no such thing as an ideal gas. As we noted at the beginning of this section, the ideal gas is a model (a very useful one) that describes the behavior of individual atoms and molecules; this behavior translates to the collective properties of measurable quantities of these atoms and molecules. Limitations of the model arise from the fact that interactive forces, even between the widely spaced particles of gas, are not totally absent in any sample of gas.

Attractive forces are present in gases composed of polar molecules. Nonuniform charge distribution on polar molecules creates positive and negative regions, resulting in electrostatic attraction and deviation from ideality.

See Sections 4.5 and 6.2 for a discussion of interactions of polar molecules.

Calculations involving polar gases such as HF, NO, and SO_2 based on ideal gas equations (which presume no such interactions) are approximations. However, at low pressures, such approximations certainly provide useful information. Nonpolar molecules, on the other hand, are only weakly attracted to each other and behave much more ideally in the gas phase.

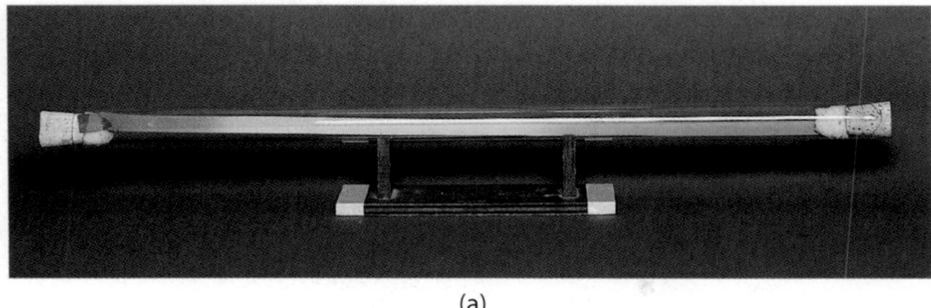

(a)

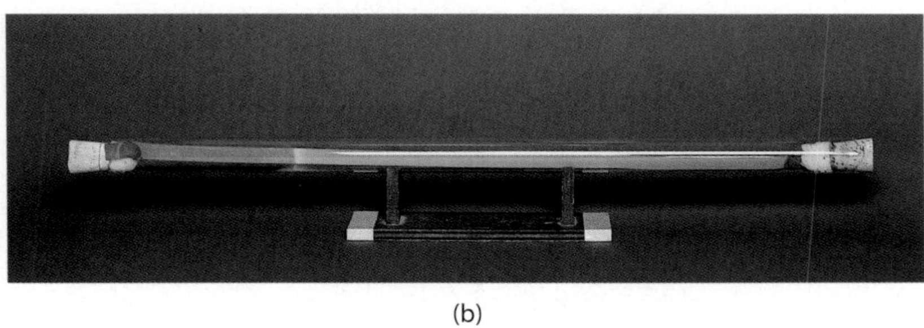

(b)

Figure 6.5
Gaseous diffusion. (a) Ammonia (17.0 g/mol) and hydrogen chloride (36.5 g/mol) are introduced into the ends of a glass tube containing indicating paper. Red indicates the presence of hydrogen chloride and blue indicates ammonia. (b) Note that ammonia has diffused much farther than hydrogen chloride in the same amount of time. This is a verification of the kinetic molecular theory. Light molecules move faster than heavier molecules at a specified temperature.

6.2 The Liquid State

Learning Goal

Molecules in the liquid state are close to one another. Attractive forces are large enough to keep the molecules together in contrast to gases, whose cohesive forces are so low that a gas expands to fill any volume. However, these attractive forces in a liquid are not large enough to restrict movement, as in solids. Because each liquid has a different molecular structure, we would expect their properties to differ as well. Let's look at the various properties of liquids in more detail.

Compressibility

Liquids are practically incompressible. In fact, the molecules are so close to one another that even the application of many atmospheres of pressure does not significantly decrease the volume. This makes liquids ideal for the transmission of force, as in the brake lines of an automobile. The force applied by the driver's foot on the brake pedal does not compress the brake fluid in the lines; rather, it transmits the force directly to the brake pads, and the friction between the brake pads and rotors (that are attached to the wheel) stops the car.

Viscosity

The **viscosity** of a liquid is a measure of its resistance to flow. Viscosity is a function of both the attractive forces between molecules and molecular geometry.

Molecules with complex structures, which do not "slide" smoothly past each other, and polar molecules, tend to have higher viscosity than less structurally complex, less polar liquids. Glycerol, which is used in a variety of skin treatments, has the structural formula:

```
    H
    |
H — C — O — H
    |
H — C — O — H
    |
H — C — O — H
    |
    H
```

A CLINICAL PERSPECTIVE

Blood Gases and Respiration

Respiration must deliver oxygen to cells and the waste product, carbon dioxide, to the lungs to be exhaled. Dalton's law of partial pressures helps to explain the way in which this process occurs.

Gases (such as O_2 and CO_2) move from a region of higher partial pressure to one of lower partial pressure in an effort to establish an equilibrium. At the interface of the lung, the membrane barrier between the blood and the surrounding atmosphere, the following situation exists: Atmospheric O_2 partial pressure is high, and atmospheric CO_2 partial pressure is low. The reverse is true on the other side of the membrane (blood). Thus CO_2 is efficiently removed from the blood, and O_2 is efficiently moved into the bloodstream.

At the other end of the line, capillaries are distributed in close proximity to the cells that need to expel CO_2 and gain O_2. The partial pressure of CO_2 is high in these cells, and the partial pressure of O_2 is low, having been used up by the energy-harvesting reaction, the oxidation of glucose:

$$C_6H_{12}O_6 + 6O_2 \longrightarrow 6CO_2 + 6H_2O + \text{energy}$$

The O_2 diffuses into the cells (from a region of high to low partial pressure), and the CO_2 diffuses from the cells to the blood (again from a region of high to low partial pressure).

The net result is a continuous process proceeding according to Dalton's law. With each breath we take, oxygen is distributed to the cells and used to generate energy, and the waste product, CO_2, is expelled by the lungs.

It is quite viscous, owing to its polar nature and its ability to hydrogen bond to other glycerol molecules. This is certainly desirable in a skin treatment because its viscosity keeps it on the area being treated. Gasoline, on the other hand, is much less viscous and readily flows through the gas lines of your auto; it is composed of nonpolar molecules.

Viscosity generally decreases with increasing temperature. The increased kinetic energy at higher temperatures overcomes some of the intermolecular attractive forces. The temperature effect is an important consideration in the design of products that must remain fluid at low temperatures, such as motor oils and transmission fluids found in automobiles.

Surface Tension

The **surface tension** of a liquid is a measure of the attractive forces exerted among molecules at the surface of a liquid. It is only the surface molecules that are not totally surrounded by other liquid molecules (the top of the molecule faces the atmosphere). These surface molecules are surrounded and attracted by fewer liquid molecules than those below and to each side. Hence the net attractive forces on surface molecules pull them downward, into the body of the liquid. As a result, the surface molecules behave as a "skin" that covers the interior.

This increased surface force is responsible for the spherical shape of drops of liquid. Drops of water "beading" on a polished surface, such as a waxed automobile, illustrate this effect.

Because surface tension is related to the attractive forces exerted among molecules, surface tension generally decreases with an increase in temperature or a decrease in the polarity of molecules that make up the liquid.

Substances known as **surfactants** can be added to a liquid to decrease surface tension. Common surfactants include soaps and detergents that reduce water's surface tension; this promotes the interaction of water with grease and dirt, making it easier to remove.

Vapor Pressure of a Liquid

Learning Goal

Evaporation, condensation, and the meaning of the term *boiling point* are all related to the concept of liquid vapor pressure. Consider the following example. A liquid, such as water, is placed in a sealed container. After a time the contents of the container are analyzed. Both liquid water and water vapor are found at room temperature, when we might expect water to be found only as a liquid. In this closed system, some of the liquid water was converted to a gas:

$$\text{energy} + H_2O(l) \longrightarrow H_2O(g)$$

How did this happen? The temperature is too low for conversion of a liquid to a gas by boiling. According to the kinetic theory, liquid molecules are in continuous motion, with their *average* kinetic energy directly proportional to the Kelvin temperature. The word *average* is the key. Although the average kinetic energy is too low to allow "average" molecules to escape from the liquid phase to the gas phase, there exists a range of molecules with different energies, some low and some high, that make up the "average" (Figure 6.6). Thus some of these high-energy molecules possess sufficient energy to escape from the bulk liquid.

At the same time a fraction of these gaseous molecules lose energy (perhaps by collision with the walls of the container) and return to the liquid state:

$$H_2O(g) \longrightarrow H_2O(l) + \text{energy}$$

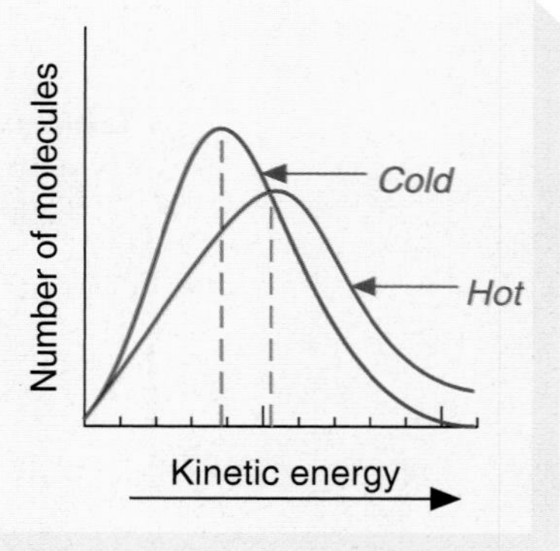

Figure 6.6
The temperature dependence of liquid vapor pressure is illustrated. The average molecular kinetic energy increases with temperature. Note that the average values are indicated by dashed lines. The small number of high-energy molecules may evaporate.

The process of conversion of liquid to gas, at a temperature too low to boil, is **evaporation.** The reverse process, conversion of the gas to the liquid state, is **condensation.** After some time the rates of evaporation and condensation become *equal,* and this sets up a dynamic equilibrium between liquid and vapor states. The **vapor pressure of a liquid** is defined as the pressure exerted by the vapor *at equilibrium.*

$$H_2O(g) \rightleftharpoons H_2O(l)$$

The equilibrium process of evaporation and condensation of water is depicted in Figure 6.7.

The boiling point of a liquid is defined as the temperature at which the vapor pressure of the liquid becomes equal to the atmospheric pressure. The "normal" atmospheric pressure is 760 torr, or 1 atm, and the **normal boiling point** is the temperature at which the vapor pressure of the liquid is equal to 1 atm.

The process of evaporation of perspiration from the skin produces a cooling effect, because heat is stored in the evaporating molecules.

It follows from the definition that the boiling point of a liquid is not constant. It depends on the atmospheric pressure. At high altitudes, where the atmospheric pressure is low, the boiling point of a liquid, such as water, is lower than the normal boiling point (for water, 100°C). High atmospheric pressure increases the boiling point.

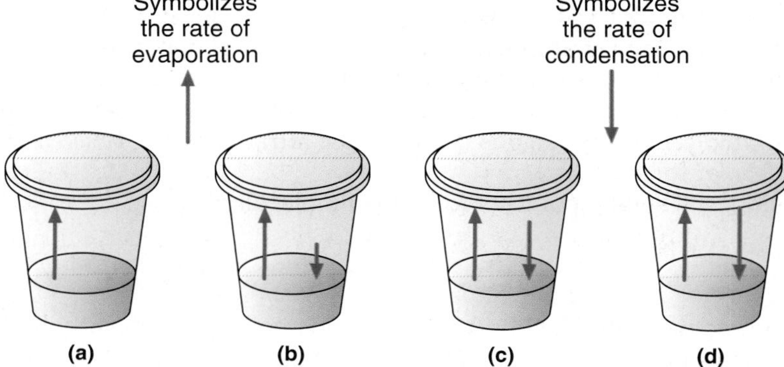

Figure 6.7
Liquid water in equilibrium with water vapor. (a) Initiation: process of evaporation exclusively. (b, c) After a time, both evaporation and condensation occur, but evaporation predominates. (d) Dynamic equilibrium established. Rates of evaporation and condensation are equal.

Apart from its dependence on the surrounding atmospheric pressure, the boiling point depends on the nature of the attractive forces between the liquid molecules. Polar liquids, such as water, with large intermolecular attractive forces have *higher* boiling points than nonpolar liquids, such as gasoline, which exhibit weak attractive forces.

Van der Waals Forces

Learning Goal

Physical properties of liquids, such as those discussed in the previous section, can be explained in terms of their intermolecular forces. We have seen (see Section 4.5) that attractive forces between polar molecules, **dipole-dipole interactions,** significantly decrease vapor pressure and increase the boiling point. However, nonpolar substances can exist as liquids as well; many are liquids and even solids at room temperature. What is the nature of the attractive forces in these nonpolar compounds?

In 1930 Fritz London demonstrated that he could account for a weak attractive force between any two molecules, whether polar or nonpolar. He postulated that the electron distribution in molecules is not fixed; electrons are in continuous motion, relative to the nucleus. So, for a short time a nonpolar molecule could experience an *instantaneous dipole,* a short-lived polarity caused by a temporary dislocation of the electron cloud. These temporary dipoles could interact with other temporary dipoles, just as permanent dipoles interact in polar molecules. We now call these intermolecular forces **London forces.**

London forces and dipole-dipole interactions are collectively known as **van der Waals forces.** London forces exist among polar and nonpolar molecules because electrons are in constant motion in all molecules. Dipole-dipole attractions occur only among polar molecules. In the next section we will see a special type of dipole-dipole force, the *hydrogen bond.*

Hydrogen Bonding

Learning Goal

Typical forces in polar liquids, discussed above, are only about 1–2% as strong as ionic and covalent bonds. However, certain liquids have boiling points that are much higher than we would predict from these dipolar interactions alone. This indicates the presence of some strong intermolecular force. This attractive force is due to **hydrogen bonding.** Molecules in which a hydrogen atom is bonded to a small, highly electronegative atom such as nitrogen, oxygen, or fluorine exhibit this effect. The presence of a highly electronegative atom bonded to a hydrogen atom creates a large dipole:

δ^- N bonded to three H (δ^+, δ^+, δ^+) $\qquad$ δ^- O bonded to two H (δ^+, δ^+) $\qquad$ δ^- F bonded to H (δ^+)

Recall that the most electronegative elements are in the upper right corner of the periodic table, and these elements exert strong electron attraction in molecules as described in Chapter 4.

This arrangement of atoms produces a very polar bond, often resulting in a polar molecule with strong intermolecular attractive forces. Although the hydrogen bond is weaker than bonds formed *within* molecules (covalent and polar covalent *intra*molecular forces), it is the strongest attractive force *between* molecules (intermolecular force).

Consider the boiling points of four small molecules:

CH_4	NH_3	H_2O	HF
−161°C	−33°C	+100°C	+19.5°C

Clearly, ammonia, water, and hydrogen fluoride boil at significantly higher temperatures than methane. The N—H, O—H, and F—H bonds are far more polar than the C—H bond, owing to the high electronegativity of N, O, and F.

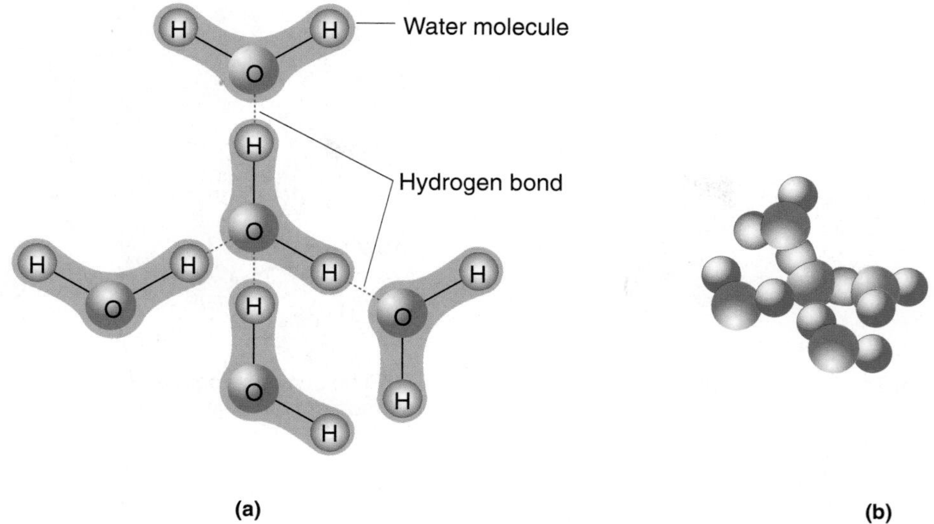

Figure 6.8
(a) An illustration of hydrogen bonding in water. The red dotted lines represent the hydrogen bonds between the hydrogen (δ^+) and oxygen (δ^-) ends of the water molecule. (b) Water has a unique structure and properties owing to the extensive hydrogen bonding. In this model, the blue units represent H atoms, and the violet units represent O atoms.

It is interesting to note that the boiling points increase as the electronegativity of the element bonded to hydrogen increases, with one exception: Fluorine, with the highest electronegativity should cause HF to have the highest boiling point. This is not the case. The order of boiling points is

water > hydrogen fluoride > ammonia > methane

not

hydrogen fluoride > water > ammonia > methane

Why? To answer this question we must look at the *number of potential bonding sites* in each molecule. Water has two partial positive sites (located at each hydrogen atom) and two partial negative sites (two lone pairs of electrons on the oxygen atom); it can form hydrogen bonds at each site. This results in a complex network of attractive forces among water molecules in the liquid state and the strength of the forces holding this network together accounts for water's unusually high boiling point. This network is depicted in Figure 6.8.

Ammonia and hydrogen fluoride can form only one hydrogen bond per molecule. Ammonia has three partial positive sites (three hydrogen atoms bonded to nitrogen) but only one partial negative site (the lone pair); the single lone pair is the limiting factor. One positive site and one negative site are needed for each hydrogen bond. Hydrogen fluoride has only one partial positive site and one partial negative site. It too can form only one hydrogen bond per molecule. Consequently, the network of attractive forces in ammonia and hydrogen fluoride is much less extensive than that found in water, and their boiling points are considerably lower than that of water.

Hydrogen bonding has an extremely important influence on the behavior of many biological systems. Molecules such as proteins and DNA require extensive hydrogen bonding to maintain their structures and hence functions. DNA (deoxyribonucleic acid, Section 24.2) is a giant among molecules with intertwined chains of atoms held together by thousands of hydrogen bonds.

Intramolecular hydrogen bonding between polar regions helps keep proteins folded in their proper three-dimensional structure. See Chapter 19.

6.3 The Solid State

The close packing of the particles of a solid results from attractive forces that are strong enough to restrict motion. This occurs because the kinetic energy of the particles is low enough for the attractive forces to dominate. The particles are "locked" together in a defined and highly organized fashion. This results in fixed shape and volume, although, at the atomic level, vibrational motion is observed.

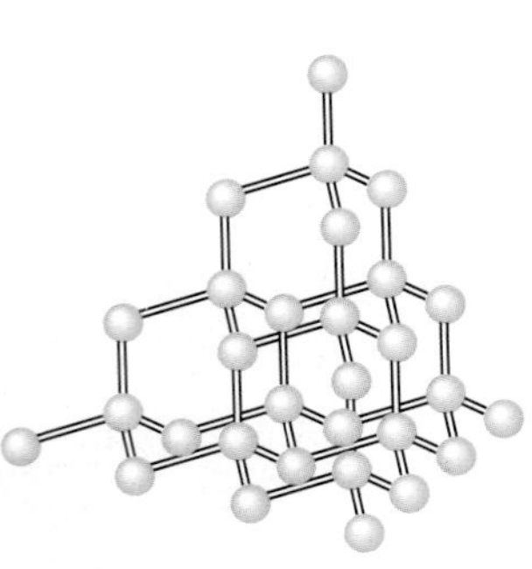

(a) The crystal structure of diamond.

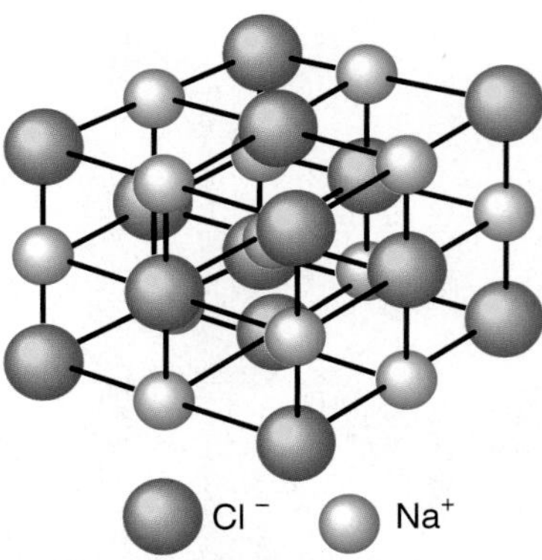

(b) The crystal structure of sodium chloride.

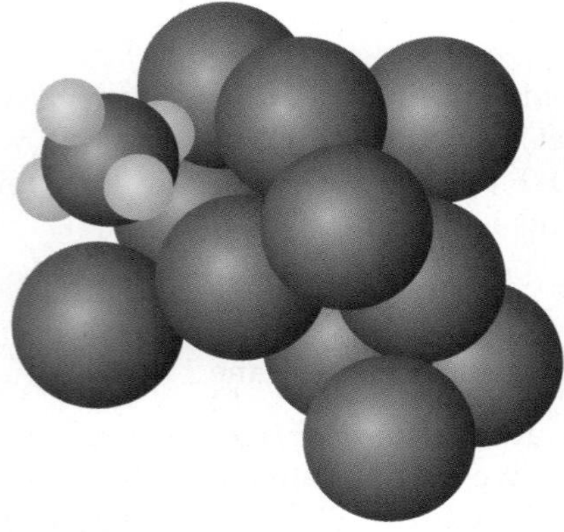

(c) A molecular solid, frozen methane. Only one methane molecule is shown in detail.

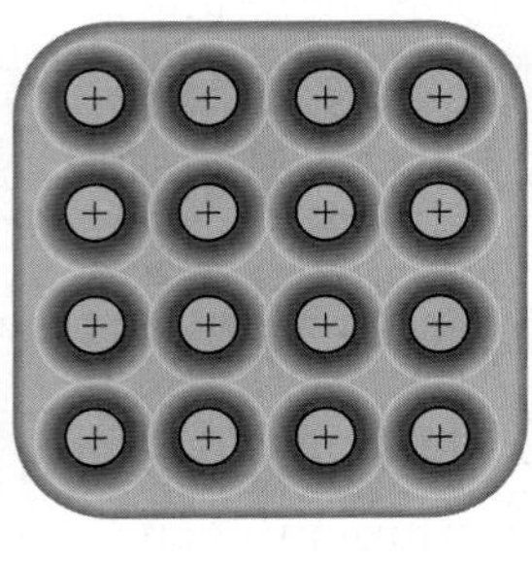

(d) A metallic solid. The gray area represents mobile electrons around fixed metal ions.

Figure 6.9
Crystalline solids.

Properties of Solids

Learning Goal

Solids are virtually incompressible, owing to the small distance between particles. Most will convert to liquids at a higher temperature, when the increased heat energy overcomes some of the attractive forces within the solid. The temperature at which a solid is converted to the liquid phase is its **melting point.** The melting point depends on the strength of the attractive forces in the solid, hence its structure. As we might expect, polar solids have higher melting points than nonpolar solids of the same molecular weight.

A solid may be a **crystalline solid,** having a regular repeating structure, or an **amorphous solid,** having no organized structure. Diamond and sodium chloride (Figure 6.9) are examples of crystalline substances; glass, plastic, and concrete are examples of amorphous solids.

Types of Crystalline Solids

Crystalline solids may exist in one of four general groups:

1. *Ionic solids.* The units that comprise an **ionic solid** are positive and negative ions. Electrostatic forces hold the crystal together. They generally have high melting points, and are hard and brittle. A common example of an ionic solid is sodium chloride.

2. *Covalent solids.* The units that comprise a **covalent solid** are atoms held together by covalent bonds. They have very high melting points (1200°C to 2000°C or more is not unusual) and are extremely hard. They are insoluble in most solvents. Diamond is a covalent solid composed of covalently bonded carbon atoms. Diamonds are used for industrial cutting because they are so hard and as gemstones because of their crystalline beauty.
3. *Molecular solids.* The units that comprise a **molecular solid,** molecules, are held together by intermolecular attractive forces (London forces, dipole-dipole interactions, and hydrogen bonding). Molecular solids are usually soft and have low melting points. They are frequently volatile and are poor electrical conductors. A common example is ice (solid water).
4. *Metallic solids.* The units that comprise a **metallic solid** are metal atoms held together by metallic bonds. **Metallic bonds** are formed by the overlap of orbitals of metal atoms, resulting in regions of high electron density surrounding the positive metal nuclei. Electrons in these regions are extremely mobile, resulting in the high *conductivity* (ability to carry electrical current) exhibited by many metallic solids. Silver and copper are common examples of metallic solids. Metals are easily shaped and are used for a variety of purposes. Most of these are practical applications such as hardware, cookware, and surgical and dental tools. Others are purely for enjoyment and decoration, such as silver and gold jewelry.

Intermolecular forces are also discussed in Sections 4.5 and 6.2.

Summary

6.1 The Gaseous State

The *kinetic molecular theory* describes an *ideal gas* in which gas particles exhibit no interactive or repulsive forces and the volumes of the individual gas particles are assumed to be negligible.

Boyle's law states that the volume of a gas varies inversely with the pressure exerted by the gas if the number of moles and temperature of gas are held constant ($PV = k_1$).

Charles's law states that the volume of a gas varies directly with the absolute temperature (K) if pressure and number of moles of gas are constant ($V/T = k_2$).

Avogadro's law states that equal volumes of any gas contain the same number of moles if measured at constant temperature and pressure ($V/n = k_3$).

The volume occupied by 1 mol of any gas is its *molar volume.* At *standard temperature and pressure* (STP) the molar volume of any ideal gas is 22.4 L. STP conditions are defined as 273 K (or 0°C) and 1 atm pressure.

Boyle's law, Charles's law, and Avogadro's law may be combined into a single expression relating all four terms, the *ideal gas law:* $PV = nRT$. R is the ideal gas constant (0.0821 L-atm K^{-1} mol^{-1}) if the units P (atmospheres), V (liters), n (number of moles), and T (Kelvin) are used.

The *combined gas law* provides a convenient expression for performing gas law calculations involving the most common variables: pressure, volume, and temperature.

Dalton's law of *partial pressures* states that a mixture of gases exerts a pressure that is the sum of the pressures that each gas would exert if it were present alone under similar conditions ($P_t = p_1 + p_2 + p_3 + \ldots$).

6.2 The Liquid State

Liquids are practically incompressible because of the closeness of the molecules. The *viscosity* of a liquid is a measure of its resistance to flow. Viscosity generally decreases with increasing temperature. The *surface tension* of a liquid is a measure of the attractive forces at the surface of a liquid. *Surfactants* decrease surface tension.

The conversion of liquid to vapor at a temperature below the boiling point of the liquid is *evaporation.* Conversion of the gas to the liquid state is *condensation.* The *vapor pressure of the liquid* is defined as the pressure exerted by the vapor at equilibrium at a specified temperature. The *normal boiling point* of a liquid is the temperature at which the vapor pressure of the liquid is equal to 1 atm.

Molecules in which a hydrogen atom is bonded to a small, highly electronegative atom such as nitrogen, oxygen, or fluorine exhibit *hydrogen bonding.* Hydrogen bonding in liquids is responsible for lower than expected vapor pressures and higher than expected boiling points. The presence of *van der Waals forces* and hydrogen bonds significantly affects the boiling points of liquids as well as the melting points of solids.

6.3 The Solid State

Solids have fixed shapes and volumes. They are *incompressible,* owing to the closeness of the particles. Solids may be *crystalline,* having a regular, repeating structure, or *amorphous,* having no organized structure.

Crystalline solids may exist as *ionic solids, covalent solids, molecular solids,* or *metallic solids.* Electrons in metallic solids are extremely mobile, resulting in the high *conductivity* (ability to carry electrical current) exhibited by many metallic solids.

Key Terms

amorphous solid (6.3)
Avogadro's law (6.1)
barometer (6.1)
Boyle's law (6.1)
Charles's law (6.1)
combined gas law (6.1)
condensation (6.2)
covalent solid (6.3)
crystalline solid (6.3)
Dalton's law (6.1)
dipole-dipole interactions (6.2)
evaporation (6.2)
hydrogen bonding (6.2)
ideal gas (6.1)
ideal gas law (6.1)
ionic solid (6.3)
kinetic molecular theory (6.1)
London forces (6.2)
melting point (6.3)
metallic bond (6.3)
metallic solid (6.3)
molar volume (6.1)
molecular solid (6.3)
normal boiling point (6.2)
partial pressure (6.1)
pressure (6.1)
standard temperature and pressure (STP) (6.1)
surface tension (6.2)
surfactant (6.2)
van der Waals forces (6.2)
vapor pressure of a liquid (6.2)
viscosity (6.2)

Questions and Problems

Boyle's Law

A sample of helium gas was placed in a cylinder and the volume of the gas was measured as the pressure was slowly increased. The results of this experiment are shown graphically.

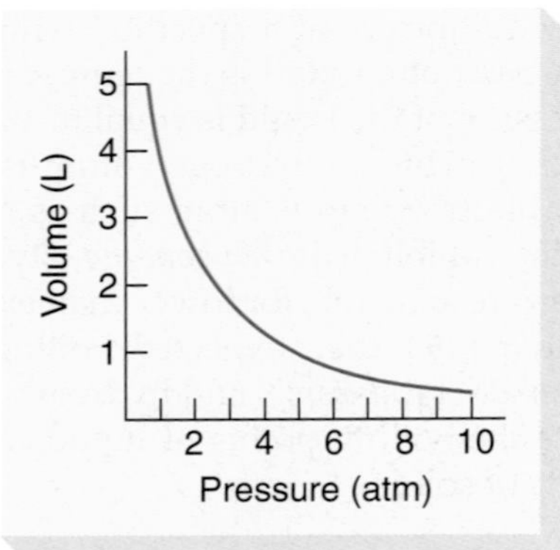

Questions 6.15–6.18 are based on this experiment.

6.15 At what pressure does the gas occupy a volume of 5 L?
6.16 What is the volume of the gas at a pressure of 5 atm?
6.17 Calculate the Boyle's law constant at a volume of 2 L.
6.18 Calculate the Boyle's law constant at a pressure of 2 atm.
6.19 Calculate the pressure, in atmospheres, required to compress a sample of helium gas from 20.9 L (at 1.00 atm) to 4.00 L.
6.20 A balloon filled with helium gas at 1.00 atm occupies 15.6 L. What volume would the balloon occupy in the upper atmosphere, at a pressure of 0.150 atm?

Charles's Law

6.21 State Charles's law in words.
6.22 State Charles's law in equation form.
6.23 Determine the change in volume that takes place when a 2.00-L sample of $N_2(g)$ is heated from 250°C to 500°C.
6.24 Determine the change in volume that takes place when a 2.00-L sample of $N_2(g)$ is heated from 250 K to 500 K.
6.25 A balloon containing a sample of helium gas is warmed in an oven. If the balloon measures 1.25 L at room temperature (20°C), what is its volume at 80°C?
6.26 The balloon described in Problem 6.25 was then placed in a refrigerator at 39°F. Calculate its new volume.

Combined Gas Law

6.27 Will the volume of gas increase, decrease, or remain the same if the temperature is increased and the pressure is decreased? Explain.
6.28 Will the volume of gas increase, decrease, or remain the same if the temperature is decreased and the pressure is increased? Explain.

Use the combined gas law,

$$\frac{P_iV_i}{T_i} = \frac{P_fV_f}{T_f}$$

to answer Questions 6.29 and 6.30.

6.29 Solve the combined gas law expression for the final volume.
6.30 Solve the combined gas law expression for the final temperature.
6.31 If 2.25 L of a gas at 16°C and 1.00 atm is compressed at a pressure of 125 atm at 20°C, calculate the new volume of the gas.
6.32 A balloon filled with helium gas occupies 2.50 L at 25°C and 1.00 atm. When released, it rises to an altitude where the temperature is 20°C and the pressure is only 0.800 atm. Calculate the new volume of the balloon.

Avogadro's Law

6.33 If 5.00 g helium gas is added to a 1.00 L balloon containing 1.00 g of helium gas, what is the new volume of the balloon? Assume no change in temperature or pressure.
6.34 How many grams of helium must be added to a balloon containing 8.00 g helium gas to double its volume? Assume no change in temperature or pressure.
6.35 State Avogadro's law in words.
6.36 State Avogadro's law in equation form.

Molar Volume and the Ideal Gas Law

6.37 A sample of argon (Ar) gas occupies 65.0 mL at 22°C and 750 torr. What is the volume of this Ar gas sample at STP?
6.38 A sample of O_2 gas occupies 257 mL at 20°C and 1.20 atm. What is the volume of this O_2 gas sample at STP?
6.39 Calculate the molar volume of Ar gas at STP.
6.40 Calculate the molar volume of O_2 gas at STP.
6.41 Calculate the volume of 4.00 mol Ar gas at 8.25 torr and 27°C.

6.42 Calculate the volume of 6.00 mol O_2 gas at 30 cm Hg and 72°F.
6.43 1.75 g of O_2 gas occupy 2.00 L at 1.00 atm. What is the temperature (°C) of the gas?
6.44 How many grams of O_2 gas occupy 10.0 L at STP?

Kinetic Molecular Theory, Ideal and Real Gases

6.45 Do gases exhibit more ideal behavior at low or high pressures? Why?
6.46 Do gases exhibit more ideal behavior at low or high temperatures? Why?
6.47 Use the kinetic molecular theory to explain why dissimilar gases mix more rapidly at high temperatures than at low temperatures.
6.48 Use the kinetic molecular theory to explain why aerosol cans carry instructions warning against heating or disposing of the container in a fire.

Dalton's Law

6.49 State Dalton's law in words.
6.50 State Dalton's law in equation form.
6.51 A gas mixture has three components: N_2, F_2, and He. Their partial pressures are 0.40 atm, 0.16 atm, and 0.18 atm, respectively. What is the pressure of the gas mixture?
6.52 A gas mixture has a total pressure of 0.56 atm and consists of He and Ne. If the partial pressure of the He in the mixture is 0.27 atm, what is the partial pressure of the Ne in the mixture?

The Liquid State

6.53 Compare the strength of intermolecular forces in liquids with those in gases.
6.54 Compare the strength of intermolecular forces in liquids with those in solids.
6.55 What is the relationship between the temperature of a liquid and the vapor pressure of that liquid?
6.56 What is the relationship between the strength of the attractive forces in a liquid and its vapor pressure?
6.57 Distinguish between the terms *evaporation* and *condensation*.
6.58 Distinguish between the terms *evaporation* and *boiling*.
6.59 Describe the process occurring at the molecular level that accounts for the property of viscosity.
6.60 Describe the process occurring at the molecular level that accounts for the property of surface tension.

The Solid State

6.61 Explain why solids are essentially incompressible.
6.62 Distinguish between amorphous and crystalline solids.
6.63 Describe one property that is characteristic of:
   a. ionic solids
   b. covalent solids
6.64 Describe one property that is characteristic of:
   a. molecular solids
   b. metallic solids
6.65 Predict whether beryllium or carbon would be a better conductor of electricity in the solid state. Why?
6.66 Why is diamond used as an industrial cutting tool?

Critical Thinking Problems

1. An elodea plant, commonly found in tropical fish aquaria, was found to produce 5.0×10^{22} molecules of oxygen per hour. What volume of oxygen (STP) would be produced in an eight-hour period?
2. A chemist measures the volume of 1.00 mol of helium gas at STP and obtains a value of 22.4 L. After changing the temperature to 137 K, the experimental value was found to be 11.05 L. Verify the chemist's results using the ideal gas law and explain any apparent discrepancies.
3. A chemist measures the volumes of 1.00 mol of H_2 and 1.00 mol of CO and finds that they differ by 0.10 L. Which gas produced the larger volume? Do the results contradict the ideal gas law? Why or why not?
4. A 100.0-g sample of water was decomposed using an electric current (electrolysis) producing hydrogen gas and oxygen gas. Write the balanced equation for the process and calculate the volume of each gas produced (STP). Explain any relationship you may observe between the volumes obtained and the balanced equation for the process.
5. An autoclave is used to sterilize surgical equipment. It is far more effective than steam produced from boiling water in the open atmosphere because it generates steam at a pressure of 2 atm. Explain why an autoclave is such an efficient sterilization device.

Formation of a precipitate by mixing two solutions.

7

Reactions and Solutions

GENERAL CHEMISTRY

Outline

Learning Goals

1. Classify chemical reactions by type: combination, decomposition, or replacement.
2. Recognize the various classes of chemical reactions: precipitation, reactions with oxygen, acid–base, and oxidation–reduction.
3. Distinguish among the terms *solution, solute,* and *solvent.*
4. Describe various kinds of solutions, and give examples of each.
5. Describe the relationship between solubility and equilibrium.
6. Calculate solution concentration in weight/volume percent and weight/weight percent.
7. Calculate solution concentration using molarity.
8. Perform dilution calculations.
9. Interconvert molar concentration of ions and milliequivalents/liter.
10. Describe and explain concentration-dependent solution properties.
11. Describe why the chemical and physical properties of water make it a truly unique solvent.
12. Explain the role of electrolytes in blood and their relationship to the process of dialysis.

CHEMISTRY CONNECTION

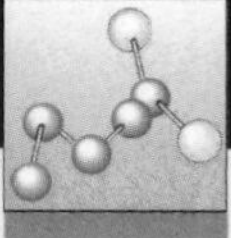

Seeing a Thought

At one time, not very long ago, mental illness was believed to be caused by some failing of the human spirit. Thoughts are nonmaterial (you can't hold a thought in your hand), and the body is quite material. No clear relationship, other than the fact that thoughts somehow come from the brain, could be shown to link the body and the spirit.

A major revolution in the diagnosis and treatment of mental illness has taken place in the last two decades. Several forms of depression, paranoia, and schizophrenia have been shown to have chemical and genetic bases. Remarkable improvement in behavior often results from altering the chemistry of the brain by using chemical therapy. Similar progress may result from the use of gene therapy (discussed in Chapter 24).

Although a treatment of mental illness, as well as of memory and logic failures, may occasionally arise by chance, a cause-and-effect relationship, based on the use of scientific methodology, certainly increases the chances of developing successful treatment. If we understand the chemical reactions involved in the thought process, we can perhaps learn to "repair" them when, for whatever reason, they go astray.

Recently, scientists at Massachusetts General Hospital in Boston have developed sophisticated versions of magnetic resonance imaging devices (MRI, discussed in Clinical Perspective in Chapter 10). MRI is normally used to locate brain tumors and cerebral damage in patients. The new generation of instruments is so sensitive that it is able to detect chemical change in the brain resulting from an external stimulus. A response to a question or the observation of a flash of light produces a measurable signal. This signal is enhanced with the aid of a powerful computer that enables the location of the signal to be determined with pinpoint accuracy. So there is evidence not only for the chemical basis of thought, but for its location in the brain as well.

In this chapter and throughout your study of chemistry you will be introduced to a wide variety of chemical reactions, some rather ordinary, some quite interesting. All are founded on the same principles that power our thoughts and actions.

Introduction

A tremendous variety of chemical reactions occurs in biological systems, in industry, and in the environment. Because of this it is useful to classify chemical reactions into a few general types, emphasizing similarities rather than differences. We recognize various patterns that reactions follow, and this helps us to write reasonable equations describing their behavior. These patterns are *combination* of reactants to produce product, *decomposition* of reactant into products, or *replacement* of one or more elements in a compound to yield a new product.

Reactions may share similar characteristics because they involve the same process, such as combustion (burning), formation of a solid (precipitate) from an aqueous solution, or the transfer of a proton (H^+) or an electron.

Proton transfer reactions, known as *acid-base reactions,* and *electron transfer reactions,* also called *oxidation-reduction reactions,* will be discussed in detail after we develop an understanding of the substances (acids, bases, oxidizing and reducing agents) that undergo these reactions.

Many chemical reactions, and virtually all important organic and biochemical reactions, take place as reactants dissolved in solution. For this reason the major emphasis of this chapter will be on *aqueous solution reactions.*

We will see that the properties of solutions depend not only on the types of substances that make up the solution but also on the amount of each substance that is contained in a certain volume of the solution. The latter is termed the *concentration* of the solution.

7.1 Writing Chemical Reactions

Learning Goal

Chemical reactions, whether they involve the formation of precipitate, reaction with oxygen, acids and bases, or oxidation–reduction, generally follow one of a few simple patterns: combination, decomposition, and single- or double- replacement. Recognizing the underlying pattern will improve your ability to write and understand chemical reactions.

Combination Reactions

Combination reactions involve the joining of two or more elements or compounds, producing a product of different composition. The general form of a combination reaction is

$$A + B \longrightarrow AB$$

in which *A* and *B* represent reactant elements or compounds and *AB* is the product.

Examples include

1. combination of a metal and a nonmetal to form a salt,

$$Ca(s) + Cl_2(g) \longrightarrow CaC1_2(s)$$

2. combination of hydrogen and chlorine molecules to produce hydrogen chloride,

$$H_2(g) + Cl_2(g) \longrightarrow 2HC1(g)$$

3. formation of water from hydrogen and oxygen molecules,

$$2H_2(g) + O_2(g) \longrightarrow 2H_2O(g)$$

4. reaction of magnesium oxide and carbon dioxide to produce magnesium carbonate,

$$MgO(s) + CO_2(g) \longrightarrow MgCO_3(s)$$

Decomposition Reactions

Decomposition reactions produce two or more products from a single reactant. The general form of these reactions is the reverse of a combination reaction:

$$AB \longrightarrow A + B$$

Some examples are

1. the heating of calcium carbonate to produce calcium oxide and carbon dioxide,

$$CaCO_3(s) \longrightarrow CaO(s) + CO_2(g)$$

2. the removal of water from a hydrated material (a *hydrate* is a substance that has water molecules incorporated in its structure),

$$CuSO_4 \cdot 5H_2O(s) \longrightarrow CuSO_4(s) + 5H_2O(g)$$

Section 5.2 describes hydrated compounds.

Replacement Reactions

Replacement reactions include both *single-replacement* and *double-replacement.* In a **single-replacement reaction,** one atom replaces another in the compound, producing a new compound

$$A + BC \longrightarrow AC + B$$

Examples include

1. the replacement of copper by zinc in copper sulfate,

$$Zn(s) + CuSO_4(aq) \longrightarrow ZnSO_4(aq) + Cu(s)$$

2. the replacement of aluminum by sodium in aluminum nitrate,

$$3Na(s) + Al(NO_3)_3(aq) \longrightarrow 3NaNO_3(aq) + Al(s)$$

A **double-replacement reaction,** on the other hand, involves *two compounds* undergoing a "change of partners." Two compounds react by exchanging atoms to produce two new compounds:

$$AB + CD \longrightarrow AD + CB$$

Examples include

1. the reaction of an acid (hydrochloric acid) and a base (sodium hydroxide) to produce water and salt, sodium chloride,

$$HCl(aq) + NaOH(aq) \longrightarrow H_2O(l) + NaCl(aq)$$

2. the formation of solid barium sulfate from barium chloride and potassium sulfate,

$$BaCl_2(aq) + K_2SO_4(aq) \longrightarrow BaSO_4(s) + 2KCl(aq)$$

Question 7.1

Classify each of the following reactions as decomposition (D), combination (C), single-replacement (SR), or double-replacement (DR):

a. $HNO_3(aq) + KOH(aq) \longrightarrow KNO_3(aq) + H_2O(aq)$
b. $Al(s) + 3NiNO_3(aq) \longrightarrow Al(NO_3)_3(aq) + 3Ni(s)$
c. $KCN(aq) + HCl(aq) \longrightarrow HCN(aq) + KCl(aq)$
d. $MgCO_3(s) \longrightarrow MgO(s) + CO_2(g)$

Question 7.2

Classify each of the following reactions as decomposition (D), combination (C), single-replacement (SR), or double-replacement (DR):

a. $2Al(OH)_3(s) \xrightarrow{\Delta} Al_2O_3(s) + 3H_2O(g)$
b. $Fe_2S_3(s) \xrightarrow{\Delta} 2Fe(s) + 3S(s)$
c. $Na_2CO_3(aq) + BaCl_2(aq) \longrightarrow BaCO_3(s) + 2NaCl(aq)$
d. $C(s) + O_2(g) \xrightarrow{\Delta} CO_2(g)$

7.2 Types of Chemical Reactions

Precipitation Reactions

Learning Goal

Precipitation reactions include any chemical change in solution that results in one or more insoluble product(s). For aqueous solution reactions the product is insoluble in water.

An understanding of precipitation reactions is useful in many ways. They may explain natural phenomena, such as the formation of stalagmites and stalactites in caves; they are simply precipitates in rocklike form. Kidney stones may result from the precipitation of calcium oxalate (CaC_2O_4). The routine act of preparing a solution requires that none of the solutes will react to form a precipitate.

How do you know whether a precipitate will form? Readily available solubility tables, such as Table 7.1, make prediction rather easy.

The following example illustrates the process.

Table 7.1 Solubilities of Some Common Ionic Compounds

Solubility Predictions
Sodium, potassium, and ammonium compounds are generally *soluble.*
Nitrates and acetates are generally *soluble.*
Chlorides, bromides, and iodides (halides) are generally *soluble.* However, halide compounds containing lead(II), silver(I), and mercury(I) are *insoluble.*
Carbonates and phosphates are generally *insoluble.* Sodium, potassium, and ammonium carbonates and phosphates are, however, *soluble.*
Hydroxides and sulfides are generally *insoluble.* Sodium, potassium, calcium, and ammonium compounds are, however, *soluble.*

EXAMPLE 7.1

Predicting Whether Precipitation Will Occur

Will a precipitate form if two solutions of the soluble salts NaCl and $AgNO_3$ are mixed?

Solution

Two soluble salts, if they react to form a precipitate, will probably "exchange partners":

$$NaCl(aq) + AgNO_3(aq) \longrightarrow AgCl(?) + NaNO_3(?)$$

Next, refer to Table 7.1 to determine the solubility of AgCl and $NaNO_3$. We predict that $NaNO_3$ is soluble and AgCl is not:

$$NaCl(aq) + AgNO_3(aq) \longrightarrow AgCl(s) + NaNO_3(aq)$$

The fact that the solid AgCl is predicted to form classifies this reaction as a precipitation reaction.

Helpful Hints:

1. See Section 7.1 for various strategies for writing chemical reactions.
2. (aq) indicates a soluble species. (s) indicates an insoluble species.

Question 7.3

Predict whether the following reactants, when mixed in aqueous solution, undergo a precipitation reaction. Write a balanced equation for each precipitation reaction.

a. potassium chloride and silver nitrate
b. potassium acetate and silver nitrate

Question 7.4

Predict whether the following reactants, when mixed in aqueous solution, undergo a precipitation reaction. Write a balanced equation for each precipitation reaction.

a. sodium hydroxide and ammonium chloride
b. sodium hydroxide and iron(II) chloride

Reactions with Oxygen

Energetics of reactions is discussed in Section 8.1.

Many substances react with oxygen. These reactions are generally energy releasing. The combustion of gasoline is used for transportation. Fossil fuel combustion is used to heat homes and provide energy for industry. Reactions involving oxygen provide energy for all sorts of biochemical processes.

When organic (carbon-containing) compounds react with the oxygen in air (burning), carbon dioxide is usually produced. If the compound contains hydrogen, water is the other product.

The reaction between oxygen and methane, CH_4, the major component of natural gas, is

$$CH_4(g) + 2O_2(g) \longrightarrow CO_2(g) + 2H_2O(g)$$

See An Environmental Perspective: The Greenhouse Effect and Global Warming, Chapter 6.

CO_2 and H_2O are waste products, and CO_2 may contribute to the greenhouse effect and global warming. The really important, unseen product is heat energy. That is why we use this reaction in our furnaces!

Inorganic substances also react with oxygen and produce heat, but these reactions usually proceed more slowly. *Corrosion* (rusting iron) is a familiar example:

$$4Fe(s) + 3O_2(g) \longrightarrow \underset{\text{Rust}}{2Fe_2O_3(s)}$$

Some reactions of metals with oxygen are very rapid. A dramatic example is the reaction of magnesium with oxygen (see Figure 8.12):

$$2Mg(s) + O_2(g) \longrightarrow 2MgO(s)$$

Acid-Base Reactions

Another approach to the classification of chemical reactions is based on a consideration of charge transfer. **Acid-base reactions** involve the transfer of a *hydrogen ion*, H^+, from one reactant to another.

A common example of an acid-base reaction involves hydrochloric acid and sodium hydroxide:

$$\underset{\text{Acid}}{HCl(aq)} + \underset{\text{Base}}{NaOH(aq)} \longrightarrow \underset{\text{Water}}{H_2O(l)} + \underset{\text{Salt}}{Na^+(aq) + Cl^-(aq)}$$

A hydrogen ion is transferred from the acid to the base, producing water and a salt in solution.

Oxidation-Reduction Reactions

Another important reaction type, **oxidation-reduction,** takes place because of the transfer of negative charge (one or more *electrons*) from one reactant to another.

The reaction of zinc metal with copper(II) ions is one example of oxidation-reduction:

$$\underset{\substack{\uparrow \\ \text{Substance} \\ \text{to be oxidized}}}{Zn(s)} + \underset{\substack{\uparrow \\ \text{Substance to be} \\ \text{reduced}}}{Cu^{2+}(aq)} \longrightarrow \underset{\substack{\uparrow \\ \text{Oxidized} \\ \text{product}}}{Zn^{2+}(aq)} + \underset{\substack{\uparrow \\ \text{Reduced} \\ \text{product}}}{Cu(s)}$$

Zinc metal atoms each donate two electrons to copper(II) ions; consequently zinc atoms become zinc(II) ions and copper(II) ions become copper atoms. Zinc is oxidized (increased positive charge) and copper is reduced (decreased positive charge) as a result of electron transfer.

The principles and applications of acid-base reactions will be discussed in Sections 9.1 through 9.4, and oxidation-reduction processes will be discussed in Section 9.5.

7.3 Properties of Solutions

Learning Goal

A **solution** is a homogeneous (or uniform) mixture of two or more substances. A solution is composed of one or more *solutes,* dissolved in a *solvent.* The **solute** is a compound of a solution that is present in lesser quantity than the solvent. The **solvent** is the solution component present in the largest quantity. For example, when sugar (the solute) is added to water (the solvent), the sugar dissolves in the water to produce a solution. In those instances in which the solvent is water, we refer to the homogeneous mixture as an **aqueous solution,** from the Latin *aqua,* meaning water.

The dissolution of a solid in a liquid is perhaps the most common example of solution formation. However, it is also possible to form solutions in gases and solids as well as in liquids. For example:

- Air is a gaseous mixture, but it is also a solution; oxygen and a number of trace gases are dissolved in the gaseous solvent, nitrogen.
- Alloys, such as brass and silver and the gold used to make jewelry, are also homogeneous mixtures of two or more kinds of metal atoms in the solid state.

Although solid and gaseous solutions are important in many applications, our emphasis will be on *liquid solutions* because so many important chemical reactions take place in liquid solutions.

General Properties of Liquid Solutions

Learning Goal

Section 4.3 discusses properties of compounds.

Liquid solutions are clear and transparent with no visible particles of solute. They may be colored or colorless, depending on the properties of the solute and solvent. Note that the terms *clear* and *colorless* do not mean the same thing; a clear solution has only one state of matter that can be detected; *colorless* simply means the absence of color.

Recall that solutions of **electrolytes** are formed from solutes that are soluble ionic compounds. These compounds dissociate in solution to produce ions that behave as charge carriers. Solutions of electrolytes are good conductors of electricity. For example, sodium chloride dissolving in water:

$$NaCl(s) \xrightarrow{H_2O} Na^+(aq) + Cl^-(aq)$$

Particles in electrolyte solutions are ions, making the solution an electrical conductor.

In contrast, solutions of **nonelectrolytes** are formed from nondissociating *molecular* solutes (nonelectrolytes), and these solutions are nonconducting. For example, dissolving sugar in water:

$$\underset{\text{Solid glucose}}{C_6H_{12}O_6(s)} \xrightarrow{H_2O} \underset{\text{Dissolved glucose}}{C_6H_{12}O_6(aq)}$$

Particles in solution are individual molecules. No ions are formed in the dissolution process.

A *true solution* is a homogeneous mixture with uniform properties throughout. In a true solution the solute cannot be isolated from the solution by filtration. The particle size of the solute is about the same as that of the solvent, and solvent and solute pass directly through the filter paper. Furthermore, solute particles will not "settle out" after a time. All of the molecules of solute and solvent are intimately mixed. The continuous particle motion in solution maintains the homogeneous, random distribution of solute and solvent particles.

Recall that matter in solution, as in gases, is in continuous, random motion (Section 6.1).

Section 4.5 relates properties and molecular geometry.

Volumes of solute and solvent are not additive; 1L of alcohol mixed with 1L of water does not result in exactly 2L of solution. The volume of pure liquid is determined by the way in which the individual molecules "fit together." When two or more kinds of molecules are mixed, the interactions become more complex. Solvent interacts with solvent, solute interacts with solvent, and solute may interact with other solute. This will be important to remember when we solve concentration problems later.

Solutions and Colloids

How can you recognize a solution? A beaker containing a clear liquid may be a pure substance, a true solution, or a colloid. Only chemical analysis, determining the identity of all substances in the liquid, can distinguish between a pure substance and a solution. A pure substance has *one* component, pure water being an example. A true solution will contain more than one substance, with the tiny particles homogeneously intermingled.

See Section 7.2 for more information on precipitates.

A **colloidal suspension** also consists of solute particles distributed throughout a solvent. However, the distribution is not completely homogeneous, owing to the size of the colloidal particles. Particles with diameters of 1×10^{-9} m (1 nm) to 2×10^{-7} m (200 nm) are colloids. Particles smaller than 1 nm are solution particles; those larger than 200 nm are precipitates (solid in contact with solvent).

To the naked eye, a colloidal suspension and a true solution appear identical; neither solute nor colloid can be seen by the naked eye. However, a simple experiment, using only a bright light source, can readily make the distinction based upon differences in their interaction with light. Colloid particles are large enough to scatter light; solute particles are not. When a beam of light passes through a colloidal suspension, the particles are large enough to scatter light, and the liquid appears hazy. We see this effect in sunlight passing through fog. The haze is light scattered by droplets of water. You may have noticed that your automobile headlights are not very helpful in foggy weather. Visibility becomes worse rather than better because light scattering increases.

The light-scattering ability of colloidal suspensions is termed the *Tyndall effect.* True solutions, with very tiny particles, do not scatter light—no haze is observed—and true solutions are easily distinguished from colloidal suspensions by observing their light-scattering properties (Figure 7.1).

A **suspension** is a heterogeneous mixture that contains particles much larger than a colloidal suspension; over time, these particles may settle, forming a second phase. A suspension is not a true solution, nor is it a precipitate.

Figure 7.1
The Tyndall effect. The beaker on the left contains a colloidal suspension, which scatters the light. This scattered light is visible as a haze. The beaker on the right contains a true solution; no scattered light is observed.

Degree of Solubility

In our discussion of the relationship of polarity and solubility, the rule *"like dissolves like"* was described as the fundamental condition for solubility. Polar solutes are soluble in polar solvents, and nonpolar solutes are soluble in nonpolar solvents. Thus, knowing a little bit about the structure of the molecule enables us to predict qualitatively the solubility of the compound.

Section 4.5 describes solute-solvent interactions in detail.

The degree of **solubility,** *how much* solute can dissolve in a given volume of solvent, is a quantitative measure of solubility. It is difficult to predict the solubility of each and every compound. However, general solubility trends are based on the following considerations:

The term *qualitative* implies identity, and the term *quantitative* relates to quantity.

- *The magnitude of difference between polarity of solute and solvent.* The greater the difference, the less soluble is the solute.
- *Temperature.* An increase in temperature usually, but not always, increases solubility. Often, the effect is dramatic. For example, an increase in temperature from 0°C to 100°C increases the water solubility of KCl from 28 g/100 mL to 58 g/100 mL.
- *Pressure.* Pressure has little effect on the solubility of solids and liquids in liquids. However, the solubility of a gas in liquid is directly proportional to the applied pressure. Carbonated beverages, for example, are made by dissolving carbon dioxide in the beverage under high pressure (hence the term *carbonated*).

When a solution contains all the solute that can be dissolved at a particular temperature, it is a **saturated solution.** When solubility values are given—for example, 13.3 g of potassium nitrate in 100 mL of water at 24°C—they refer to the concentration of a saturated solution.

As we have already noted, *increasing* the temperature generally increases the amount of solute a given solution may hold. Conversely, *cooling* a saturated solution often results in a decrease in the amount of solute in solution. The excess solute falls to the bottom of the container as a **precipitate** (a solid in contact with the solution). Occasionally, on cooling, the excess solute may remain in solution for a time. Such a solution is described as a **supersaturated solution.** This type of solution is inherently unstable. With time, excess solute will precipitate, and the solution will revert to a saturated solution, which is stable.

Solubility and Equilibrium

Learning Goal

When an excess of solute is added to a solvent, it begins to dissolve and continues until it establishes a *dynamic equilibrium* between dissolved and undissolved solute.

Initially, the rate of dissolution is large. After a time the rate of the reverse process, precipitation, increases. The rates of dissolution and precipitation eventually become equal, and there is no further change in the composition of the solution. There is, however, a continual exchange of solute particles between solid and liquid phases because particles are in constant motion. The solution is saturated. The most precise definition of a saturated solution is a solution that is in equilibrium with undissolved solute.

The concept of equilibrium was introduced in Section 6.2 and will be discussed in detail in Section 8.4.

Solubility of Gases: Henry's Law

When a liquid and a gas are allowed to come to equilibrium, the amount of gas dissolved in the liquid reaches some maximum level. This quantity can be predicted from a very simple relationship. **Henry's law** states that the number of moles of a gas dissolved in a liquid at a given temperature is proportional to the partial pressure of the gas. In other words, the gas solubility is directly proportional to the pressure of that gas in the atmosphere that is in contact with the liquid.

The concept of partial pressure is a consequence of Dalton's law, discussed in Section 6.1.

Carbonated beverages are bottled at high pressures of carbon dioxide. When the cap is removed, the fizzing results from the fact that the partial pressure of carbon dioxide in the atmosphere is much less than that used in the bottling process. As a result, the equilibrium quickly shifts to one of lower gas solubility.

Gases are most soluble at low temperatures, and the gas solubility decreases markedly at higher temperatures. This explains many common observations. For example, a chilled container of carbonated beverage that is opened quickly goes flat as it warms to room temperature. As the beverage warms up, the solubility of the carbon dioxide decreases.

The exchange of O_2 and CO_2 in the lungs and other tissues is a complex series of events described in greater detail in Section 19.9.

See A Clinical Perspective: Blood Gases and Respiration, Chapter 6.

Henry's law helps to explain the process of respiration. Respiration depends on a rapid and efficient exchange of oxygen and carbon dioxide between the atmosphere and the blood. This transfer occurs through the lungs. The process, oxygen entering the blood and carbon dioxide released to the atmosphere, is accomplished in air sacs called *alveoli,* which are surrounded by an extensive capillary system. Equilibrium is quickly established between alveolar air and the capillary blood. The temperature of the blood is effectively constant. Therefore the equilibrium concentration of both oxygen and carbon dioxide are determined by the partial pressures of the gases (Henry's law). The oxygen is transported to cells, a variety of reactions takes place, and the waste product of respiration, carbon dioxide, is brought back to the lungs to be expelled into the atmosphere.

7.4 Concentration of Solutions: Percentage

Learning Goal

Solution **concentration** is defined as the amount of solute dissolved in a given amount of solution. The concentration of a solution has a profound effect on the properties of a solution, both *physical* (melting and boiling points) and *chemical* (solution reactivity). Solution concentration may be expressed in many different units. Here we consider concentration units based on percentage.

Weight/Volume Percent

The concentration of a solution is defined as the amount of solute dissolved in a specified amount of solution,

$$\text{concentration} = \frac{\text{amount of solute}}{\text{amount of solution}}$$

If we define the amount of solute as the *mass* of solute (in grams) and the amount of solution in *volume* units (milliliters), concentration is expressed as the ratio

$$\text{concentration} = \frac{\text{grams of solute}}{\text{milliliters of solution}}$$

This concentration can then be expressed as a percentage by multiplying the ratio by the factor 100%. This results in

$$\%\ \text{concentration} = \frac{\text{grams of solute}}{\text{milliliters of solution}} \times 100\%$$

The percent concentration expressed in this way is called **weight/volume percent,** or **% (W/V).** Thus

$$\%\left(\frac{\text{W}}{\text{V}}\right) = \frac{\text{grams of solute}}{\text{milliliters of solution}} \times 100\%$$

Consider the following examples.

A HUMAN PERSPECTIVE

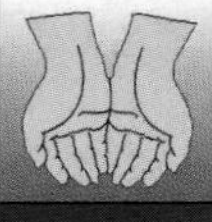

Scuba Diving: Nitrogen and the Bends

A deep-water diver's worst fear is the interruption of the oxygen supply through equipment malfunction, forcing his or her rapid rise to the surface in search of air. If a diver must ascend too rapidly, he or she may suffer a condition known as "the bends."

Key to understanding this problem is recognition of the tremendous increase in pressure that divers withstand as they descend, because of the weight of the water above them. At the surface the pressure is approximately 1 atm. At a depth of 200 feet the pressure is approximately six times as great.

At these pressures the solubility of nitrogen in the blood increases dramatically. Oxygen solubility increases as well, although its effect is less serious (O_2 is 20% of air, N_2 is 80%). As the diver quickly rises, the pressure decreases rapidly, and the nitrogen "boils" out of the blood, stopping blood flow and impairing nerve transmission. The joints of the body lock in a bent position, hence the name of the condition: the bends.

To minimize the problem, scuba tanks are often filled with mixtures of helium and oxygen rather than nitrogen and oxygen. Helium has a much lower solubility in blood and, like nitrogen, is inert.

EXAMPLE 7.2 Calculating Weight/Volume Percent

Calculate the percent composition, or % (W/V), of 3.00×10^2 mL of solution containing 15.0 g of glucose.

Solution

There are 15.0 g of glucose, the solute, and 3.00×10^2 mL of total solution. Therefore, substituting in our expression for weight/volume percent:

$$\% \left(\frac{W}{V}\right) = \frac{15.0 \text{ g glucose}}{3.00 \times 10^2 \text{ mL solution}} \times 100\%$$

$$= 5.00\% \left(\frac{W}{V}\right) \text{ glucose}$$

EXAMPLE 7.3 Calculating the Weight of Solute from a Weight/Volume Percent

Calculate the number of grams of NaCl in 5.00×10^2 mL of a 10.0% solution.

Solution

Begin by substituting the data from the problem:

$$10.0\% \left(\frac{W}{V}\right) = \frac{X \text{ g NaCl}}{5.00 \times 10^2 \text{ mL solution}} \times 100\%$$

Continued—

EXAMPLE 7.3 *—Continued*

Cross-multiplying to simplify:

$$X \text{ g NaCl} \times 100\% = \left(10.0\% \frac{W}{V}\right)(5.00 \times 10^2 \text{ mL solution})$$

Dividing both sides by 100% to isolate grams NaCl on the left side of the equation:

$$X = 50.0 \text{ g NaCl}$$

Section 1.3 discusses units and unit conversion.

If the units of mass are other than grams, or if the solution volume is in units other than milliliters, the proper conversion factor must be used to arrive at the units used in the equation.

Question 7.5

Calculate the % (W/V) of 0.0600 L of solution containing 10.0 g NaCl.

Question 7.6

Calculate the volume (in milliliters) of a 25.0% (W/V) solution containing 10.0 g NaCl.

Question 7.7

Calculate the % (W/V) of 0.200 L of solution containing 15.0 g KCl.

Question 7.8

Calculate the mass (in grams) of sodium hydroxide required to make 2.00 L of a 1.00% (W/V) solution.

Weight/Weight Percent

The **weight/weight percent,** or **% (W/W),** is most useful for mixtures of solids, whose weights (masses) are easily obtained. The expression used to calculate weight/weight percentage is analogous in form to % (W/V):

$$\%\left(\frac{W}{V}\right) = \frac{\text{grams solute}}{\text{grams solution}} \times 100\%$$

EXAMPLE 7.4 ***Calculating Weight/Weight Percent***

Calculate the % (W/W) of platinum in a gold ring that contains 14.00 g gold and 4.500 g platinum.

Solution

Using our definition of weight/weight percent

$$\%\left(\frac{W}{V}\right) = \frac{\text{grams solute}}{\text{grams solution}} \times 100\%$$

Substituting,

$$= \frac{4.500 \text{ g platinum}}{4.500 \text{ g platinum} + 14.00 \text{ g gold}} \times 100\%$$

Continued—

EXAMPLE 7.4 —*Continued*

$$= \frac{4.500 \text{ g}}{18.50 \text{ g}} \times 100\%$$

$$= 24.32\% \text{ platinum}$$

Question 7.9

20.0 g of oxygen gas are diluted with 80.0 g of nitrogen gas in a 78.0-L container at standard temperature and pressure. Calculate the % (W/V) of oxygen gas.

Question 7.10

50.0 g of argon gas are diluted with 80.0 g of helium gas in a 476-L container at standard temperature and pressure. Calculate the % (W/V) of argon gas.

Question 7.11

Calculate the % (W/W) of oxygen gas in Question 7.9.

Question 7.12

Calculate the % (W/W) of argon gas in Question 7.10.

7.5 Concentration of Solutions: Moles and Equivalents

In our discussion of the chemical arithmetic of reactions in Chapter 5, we saw that the chemical equation represents the relative number of *moles* of reactants producing products. When chemical reactions occur in solution, it is most useful to represent their concentrations on a *molar* basis.

Molarity

Learning Goal

7

The most common mole-based concentration unit is molarity. **Molarity,** symbolized M, is defined as the number of moles of solute per liter of solution, or

$$M = \frac{\text{moles solute}}{\text{L solution}}$$

EXAMPLE 7.5

Calculating Molarity from Moles

Calculate the molarity of 2.0 L of solution containing 5.0 mol NaOH.

Solution

Using our expression for molarity

$$M = \frac{\text{moles solute}}{\text{L solution}}$$

Continued—

EXAMPLE 7.5 —*Continued*

Substituting,

$$M_{NaOH} = \frac{5.0 \text{ mol solute}}{2.0 \text{ L solution}}$$

$$= 2.5\ M$$

Section 1.3 discussed units and unit conversion.

Remember the need for conversion factors to convert from mass to number of moles. Consider the following example.

EXAMPLE 7.6 Calculating Molarity from Mass

If 5.00 g glucose are dissolved in 1.00×10^2 mL of solution, calculate the molarity, *M*, of the glucose solution.

Solution

To use our expression for molarity it is necessary to convert from units of grams of glucose to moles of glucose. The molar mass of glucose is 1.80×10^2 g/mol. Therefore

$$5.00 \cancel{g} \times \frac{1 \text{ mol}}{1.80 \times 10^2 \cancel{g}} = 2.78 \times 10^{-2} \text{ mol glucose}$$

and we must convert mL to L:

$$1.00 \times 10^2 \cancel{mL} \times \frac{1 \text{ L}}{10^3 \cancel{mL}} = 1.00 \times 10^{-1} \text{ L}$$

Substituting these quantities:

$$M_{glucose} = \frac{2.78 \times 10^{-2} \text{ mol}}{1.00 \times 10^{-1} \text{L}}$$

$$= 2.78 \times 10^{-1}\ M$$

EXAMPLE 7.7 Calculating Volume from Molarity

Calculate the volume of a 0.750 *M* sulfuric acid (H_2SO_4) solution containing 0.120 mol of solute.

Solution

Substituting in our basic expression for molarity, we obtain

$$0.750\ M\ H_2SO_4 = \frac{0.120 \text{ mol } H_2SO_4}{X \text{ L}}$$

$$X \text{ L} = 0.160 \text{ L}$$

Question 7.13

Calculate the number of moles of solute in 5.00×10^2 mL of 0.250 *M* HCl.

Question 7.14

Calculate the number of grams of silver nitrate required to prepare 2.00 L of 0.500 *M* $AgNO_3$.

Dilution

Learning Goal

Laboratory reagents are often purchased as concentrated solutions (for example, 12 *M* HCl or 6 *M* NaOH) for reasons of safety, economy, and space limitations. We must often *dilute* such a solution to a larger volume to prepare a less concentrated solution for the experiment at hand. The approach to such a calculation is as follows.

We define

M_1 = molarity of solution *before* dilution

M_2 = molarity of solution *after* dilution

V_1 = volume of solution *before* dilution

V_2 = volume of solution *after* dilution

and

$$M = \frac{\text{moles solute}}{\text{L solution}} \quad \text{or moles solute} = (M)\text{ (L solution)}$$

The number of moles of solute *before* and *after* dilution is unchanged, because dilution involves only addition of extra solvent:

$$\text{moles}_1 \text{ solute} = \text{moles}_2 \text{ solute}$$

Initial condition — Final condition

or

$$(M_1)(\text{L}_1 \text{ solution}) = (M_2)(\text{L}_2 \text{ solution})$$

$$(M_1)(V_1) = (M_2)(V_2)$$

Knowing any three of these terms enables us to calculate the fourth.

EXAMPLE 7.8

Calculating Molarity after Dilution

Calculate the molarity of a solution made by diluting 0.050 L of 0.10 *M* HCl solution to a volume of 1.0 L.

Solution

Summarize the information provided in the problem:

$$M_1 = 0.10\,M$$

$$M_2 = X\,M$$

$$V_1 = 0.050\text{ L}$$

$$V_2 = 1.0\text{ L}$$

Then, using the dilution expression:

$$(M_1)\,(V_1) = (M_2)\,(V_2)$$

Continued—

EXAMPLE 7.8 —*Continued*

Solve for M_2, the final solution concentration:

$$M_2 = \frac{(M_1)\,(V_1)}{V_2}$$

Substituting,

$$X\,M = \frac{(0.10\ M)\,(0.050\ \text{L})}{(1.0\ \text{L})}$$

$$= 0.0050\ M \text{ or } 5.0 \times 10^{-3}\ M\ \text{HCl}$$

EXAMPLE 7.9 ***Calculating a Dilution Volume***

Calculate the volume, in liters, of water that must be added to dilute 20.0 mL of 12.0 *M* HCl to 0.100 *M* HCl.

Solution

Summarize the information provided in the problem:

$$M_1 = 12.0\ M$$

$$M_2 = 0.100\ M$$

$$V_1 = 20.0\ \text{mL}\ (0.0200\ \text{L})$$

$$V_2 = V_{\text{final}}$$

Then, using the dilution expression:

$$(M_1)\,(V_1) = (M_2)\,(V_2)$$

Solve for V_2, the final volume:

$$V_2 = \frac{(M_1)\,(V_1)}{(M_2)}$$

Substituting,

$$V_{\text{final}} = \frac{(12.0\ M)\,(0.0200\ \text{L})}{0.100\ M}$$

$$= 2.40\ \text{L solution}$$

Note that this is the *total final volume.* The amount of water added equals this volume *minus* the original solution volume, or

$$2.40\ \text{L} - 0.0200\ \text{L} = 2.38\ \text{L water}$$

Question 7.15

How would you prepare 1.0×10^2 mL of 2.0 *M* HCl, starting with concentrated (12.0 *M*) HCl?

Question 7.16

What volume of 0.200 *M* sugar solution can be prepared from 50.0 mL of 0.400 *M* solution?

The dilution equation is valid with any concentration units, such as % (W/V) *as well as* molarity, which was used in Examples 7.8 and 7.9. However, you must use the same units for both initial *and* final concentration values. Only in this way can you cancel units properly.

Representation of Concentration of Ions in Solution

Learning Goal

The concentration of ions in solution may be represented in a variety of ways. The most common include moles per liter (molarity) and equivalents per liter.

When discussing solutions of ionic compounds, molarity emphasizes the number of individual ions. A one molar solution of Na^+ contains Avogadro's number, 6.022×10^{23}, of Na^+ per liter. In contrast, equivalents per liter emphasize charge; one equivalent of Na^+ contains Avogadro's number of positive charge.

We defined 1 mol as the number of grams of an atom, molecule, or ion corresponding to Avogadro's number of particles. One **equivalent** of an ion is the number of grams of the ion corresponding to Avogadro's number of electrical charges. Some examples follow:

$$1 \text{ mol } Na^+ = 1 \text{ equivalent } Na^+ \qquad (\text{one } Na^+ = 1 \text{ unit of charge/ion})$$
$$1 \text{ mol } Cl^- = 1 \text{ equivalent } Cl^- \qquad (\text{one } Cl^- = 1 \text{ unit of charge/ion})$$
$$1 \text{ mol } Ca^{2+} = 2 \text{ equivalents } Ca^{2+} \qquad (\text{one } Ca^{2+} = 2 \text{ units of charge/ion})$$
$$1 \text{ mol } CO_3^{2-} = 2 \text{ equivalents } CO_3^{2-} \qquad (\text{one } CO_3^{2-} = 2 \text{ units of charge/ion})$$
$$1 \text{ mol } PO_4^{3-} = 3 \text{ equivalents } PO_4^{3-} \qquad (\text{one } PO_4^{3-} = 3 \text{ units of charge/ion})$$

Changing from moles per liter to equivalents per liter (or the reverse) can be accomplished by using conversion factors.

Milliequivalents (meq) or milliequivalents/liter (meq/L) are often used when describing small amounts or low concentration of ions. These units are routinely used when describing ions in blood, urine, and blood plasma.

Calculating Ion Concentration — EXAMPLE 7.10

Calculate the number of equivalents per liter (eq/L) of phosphate ion, PO_4^{3-}, in a solution that is 5.0×10^{-3} *M* phosphate.

Solution

It is necessary to use two conversion factors:

$$\text{mol } PO_4^{3-} \longrightarrow \text{mol charge}$$

and

$$\text{mol charge} \longrightarrow \text{eq } PO_4^{3-}$$

Arranging these factors in sequence yields:

$$\frac{5.0 \times 10^{-3} \cancel{\text{mol } PO_4^{3-}}}{1L} \times \frac{3 \cancel{\text{mol charge}}}{1 \cancel{\text{mol } PO_4^{3-}}} \times \frac{1 \text{ eq}}{1 \cancel{\text{mol charge}}} = \frac{1.5 \times 10^{-2} \text{ eq } PO_4^{3-}}{L}$$

7.6 Concentration-Dependent Solution Properties

Learning Goal

Colligative properties are solution properties that depend on the *concentration of the solute particles*, rather than the *identity of the solute*.

There are four colligative properties of solutions:

1. vapor pressure lowering,
2. boiling point elevation,

3. freezing point depression, and
4. osmotic pressure

Each of these properties has widespread practical application. We look at each in some detail in the following sections.

Vapor Pressure Lowering

Recall that the concept of liquid vapor pressure was discussed in Section 6.2.

Raoult's law states that, when a nonvolatile solute is added to a solvent, the vapor pressure of the solvent decreases in proportion to the concentration of the solute.

Perhaps the most important consequence of Raoult's law is the effect of the solute on the freezing and boiling points of a solution.

When a nonvolatile solute is added to a solvent, the freezing point of the resulting solution decreases (a lower temperature is required to convert the liquid to a solid). The boiling point of the solution is found to increase (it requires a higher temperature to form the gaseous state).

Raoult's law may be explained in molecular terms by using the following logic: Vapor pressure of a solution results from the escape of solvent molecules from the liquid to the gas phase, thus increasing the partial pressure of the gas phase solvent molecules until the equilibrium vapor pressure is reached. Presence of solute molecules hinders the escape of solvent molecules, thus lowering the equilibrium vapor pressure (Figure 7.2).

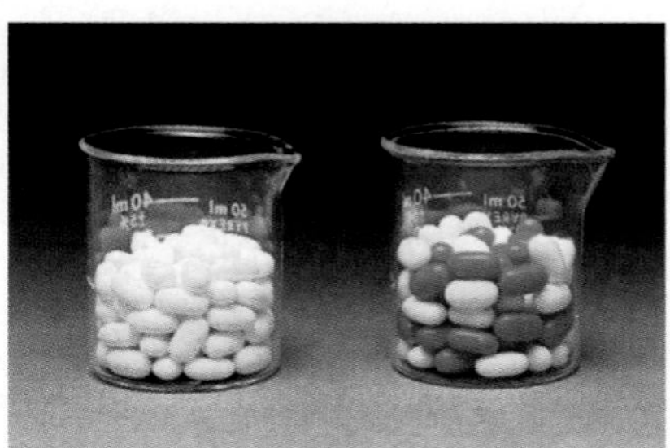

Figure 7.2
An illustration of Raoult's law: lowering of vapor pressure by addition of solute molecules. White units represent solvent molecules, and red units are solute molecules. Solute molecules present a barrier to escape of solvent molecules, thus decreasing the vapor pressure.

Freezing Point Depression and Boiling Point Elevation

The freezing point depression may be explained by equilibrium considerations. At the freezing point, ice is in equilibrium with liquid water:

$$H_2O\ (l) \underset{(r)}{\overset{(f)}{\rightleftharpoons}} H_2O\ (s)$$

The solute molecules interfere with the rate at which liquid water molecules associate to form the solid state. For a true equilibrium, the rate of the forward (f) and reverse (r) processes must be equal. Lowering the temperature eventually slows the rate of the reverse (r) process sufficiently. At the lower temperature, equilibrium is established, and the solution freezes.

Section 8.4 discusses equilibrium.

The boiling point elevation can be explained by considering the definition of the boiling point, that is, the temperature at which the vapor pressure of the liquid equals the atmospheric pressure. Raoult's law states that the vapor pressure of a solution is decreased by the presence of a solute. Therefore a higher temperature is necessary to raise the vapor pressure to the atmospheric pressure, hence the boiling point elevation.

The extent of the freezing point depression (ΔT_f) is proportional to the solute concentration over a limited range of concentration:

$$\Delta T_f = k_f \times \text{(solute concentration)}$$

The boiling point elevation (ΔT_b) is also proportional to the solute concentration:

$$\Delta T_b = k_b \times \text{(solute concentration)}$$

If the value of the proportionality factor (k_f or k_b) is known for the solvent of interest, the magnitude of the freezing point depression or boiling point elevation can be calculated for a solution of known concentration.

Solute concentration must be in *mole*-based units. The number of particles (molecules or ions) is critical here, not the mass of solute. One *heavy* molecule will have exactly the same effect on the freezing or boiling point as one *light* molecule. A mole-based unit, because it is related directly to Avogadro's number, will correctly represent the number of particles in solution.

We have already worked with one mole-based unit, *molarity*, and this concentration unit can be used to calculate either the freezing point depression or the boiling point elevation.

A second mole-based concentration unit, molality, is more commonly used in these types of situations. **Molality** (symbolized m) is defined as the number of moles of solute per kilogram of solvent in a solution:

$$m = \frac{\text{moles solute}}{\text{kg solvent}}$$

Molality does not vary with temperature, whereas molarity is temperature dependent. For this reason, molality is the preferred concentration unit for studies such as freezing point depression and boiling point elevation, in which measurement of *change* in temperature is critical.

Molarity is temperature dependent simply because it is expressed as mole/volume. Volume is temperature dependent—most liquids expand measurably when heated and contract when cooled. Molality is moles/mass; both moles and mass are temperature independent.

Practical applications that take advantage of freezing point depression of solutions by solutes include the following:

- Salt is spread on roads to melt ice in winter. The salt lowers the freezing point of the water, so it exists in the liquid phase below its normal freezing point, 0°C or 32°F.
- Solutes such as ethylene glycol, "antifreeze," are added to car radiators in the winter to prevent freezing by lowering the freezing point of the coolant.

We have referred to the concentration of *particles* in our discussion of colligative properties. Why did we stress this term? The reason is that there is a very important difference between electrolytes and nonelectrolytes. That difference is the way in which they behave when they dissolve. For example, if we dissolve 1 mol of glucose ($C_6H_{12}O_6$) in 1L of water,

$$1\ C_6H_{12}O_6(s) \xrightarrow{H_2O} 1\ C_6H_{12}O_6(aq)$$

1 mol (Avogadro's number, 6.022×10^{23} particles) of glucose is present in solution. *Glucose is a covalently bonded nonelectrolyte.* Dissolving 1 mol of sodium chloride in 1L of water,

$$1\ NaCl(s) \xrightarrow{H_2O} 1\ Na^+(aq) + 1\ Cl^-(aq)$$

produces 2 mol of particles (1 mol of sodium ions and 1 mol of chloride ions). *Sodium chloride is an ionic electrolyte.*

$$1 \text{ mol glucose} \longrightarrow 1 \text{ mol of particles in solution}$$

$$1 \text{ mol sodium chloride} \longrightarrow 2 \text{ mol of particles in solution}$$

It follows that 1 mol of sodium chloride will decrease the vapor pressure, increase the boiling point, or depress the freezing point of 1L of water *twice as much* as 1 mol of glucose in the same quantity of water.

Osmotic Pressure

Certain types of thin films, or *membranes*, although appearing impervious to matter, actually contain a network of small holes or pores. These pores may be large enough to allow small *solvent* molecules, such as water, to move from one side of the membrane to the other. On the other hand, *solute* molecules cannot cross the membrane because they are too large to pass through the pores. **Semipermeable membranes** are membranes that allow the solvent, but not solute, to diffuse from one side of the membrane to the other. Examples of semipermeable membranes range from synthetics, such as cellophane, to membranes of cells. When the pores are so small that only water molecules can pass through, they are called *osmotic membranes*.

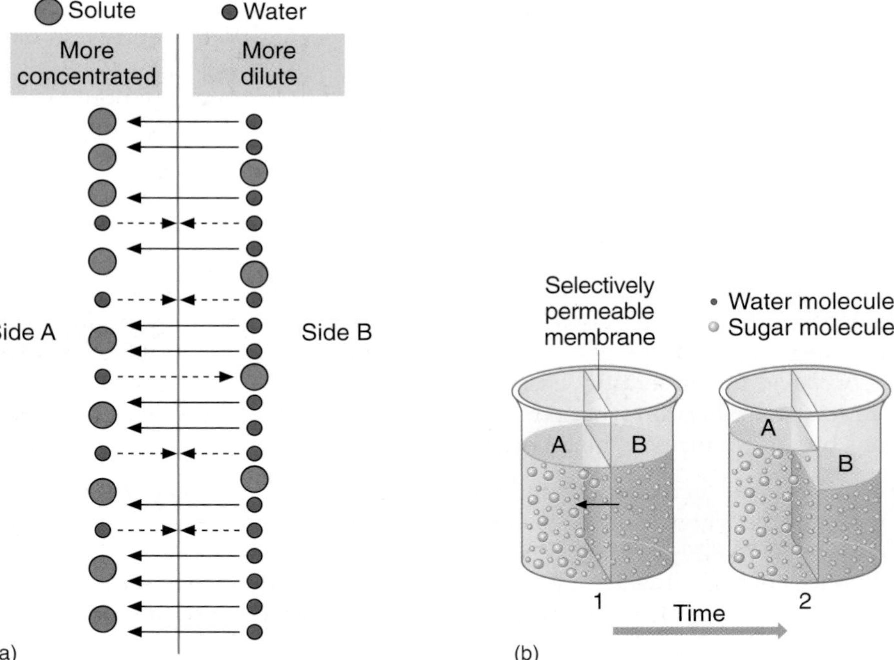

Figure 7.3
(a) Attainment of equilibrium by osmosis. Note that the solutions attain equilibrium when sufficient solvent has passed from the more dilute side (side B) to equalize the concentrations on both sides of the membrane. Side A becomes more dilute, and side B becomes more concentrated. (b) An illustration of osmosis. A semipermeable membrane separates a solution of sugar in water from the pure solvent, water. Over time, water diffuses from side B to side A in an attempt to equalize the concentration in the two compartments. The water level in side A will rise at the expense of side B because the net flow of water is from side B to side A.

Osmosis is the movement of solvent from a *dilute solution* to a more *concentrated solution* through a *semipermeable membrane.* Pressure must be applied to the more concentrated solution to stop this flow. **Osmotic pressure** is the amount of pressure required to stop the flow.

The process of osmosis occurring between pure water and glucose (sugar) solution is illustrated in Figure 7.3. Note that the "driving force" for the osmotic process is the need to establish an equilibrium between the solutions on either side of the membrane. Pure solvent enters the more concentrated solution in an effort to dilute it. If this process is successful, and concentrations on both sides of the membrane become equal, the "driving force," or concentration difference, disappears. A dynamic equilibrium is established, and the osmotic pressure difference between the two sides is zero.

The osmotic pressure, like the pressure exerted by a gas, may be treated quantitatively. Osmotic pressure, symbolized by π, follows the same form as the ideal gas equation:

The term *selectively permeable* or *differentially permeable* is used to describe biological membranes because they restrict passage of particles based both on size and charge. Even small ions, such as H^+, cannot pass freely across a cell membrane.

Ideal Gas	**Osmotic Pressure**
$PV = nRT$	$\pi V = nRT$
or	or
$P = \frac{n}{V}RT$	$\pi = \frac{n}{V}RT$
and since	and since
$M = \frac{n}{V}$	$M = \frac{n}{V}$
then	then
$P = MRT$	$\pi = MRT$

The osmotic pressure can be calculated from the solution concentration at any temperature. How do we determine "solution concentration"? Recall that osmosis is a colligative property, dependent on the concentration of solute particles. Again, it becomes necessary to distinguish between solutions of electrolytes and nonelectrolytes. For example, a 1 M glucose solution consists of 1 mol of particles per liter; glucose is a nonelectrolyte. A solution of 1 M NaCl produces 2 mol of particles per liter (1 mol of Na^+ and 1 mol of Cl^-). A 1 M $CaCl_2$ solution is 3 M in particles (1 mol of Ca^{2+} and 2 mol of Cl^- per liter).

Osmolarity, the molarity of particles in solution, and abbreviated osmol, is used for osmotic pressure calculations.

Calculating Osmolarity — EXAMPLE 7.11

Determine the osmolarity of 5.0×10^{-3} M Na_3PO_4.

Solution

Na_3PO_4 is an ionic compound and produces an electrolytic solution:

$$Na_3PO_4 \xrightarrow{H_2O} 3Na^+ + PO_4^{3-}$$

1 mol of Na_3PO_4 yields four product ions; consequently

$$5.0 \times 10^{-3} \frac{\cancel{\text{mol } Na_3PO_4}}{\text{L}} \times \frac{4 \text{ mol particles}}{1 \cancel{\text{mol } Na_3PO_4}} = 2.0 \times 10^{-2} \frac{\text{mol particles}}{\text{L}}$$

and, using our expression for osmolarity,

$$2.0 \times 10^{-2} \frac{\text{mol particles}}{\text{L}} = 2.0 \times 10^{-2} \text{ osmol}$$

Question 7.17

Determine the osmolarity of the following solution:

5.0×10^{-3} M NH_4NO_3 (electrolyte)

Question 7.18

Determine the osmolarity of the following solution:

5.0×10^{-3} M $C_6H_{12}O_6$ (nonelectrolyte)

Calculating Osmotic Pressure — EXAMPLE 7.12

Calculate the osmotic pressure of a 5.0×10^{-2} M solution of NaCl at 25°C (298 K).

Solution

Using our definition of osmotic pressure, π:

$$\pi = MRT$$

M should be represented as osmolarity as we have shown in Example 7.11

$$M = 5.0 \times 10^{-2} \frac{\cancel{\text{mol NaCl}}}{\text{L}} \times \frac{2 \text{ mol particles}}{1 \cancel{\text{mol NaCl}}} = 1.0 \times 10^{-1} \frac{\text{mol particles}}{\text{L}}$$

Continued—

EXAMPLE 7.12 —*Continued*

and substituting in our osmotic pressure expression:

$$\pi = 1.0 \times 10^{-1}\,\frac{\text{mol particles}}{\text{L}} \times 0.0821\,\frac{\text{L-atm}}{\text{K-mol}} \times 298\,\text{K}$$

$$= 2.4 \text{ atm}$$

Question 7.19

Calculate the osmotic pressure of the solution described in Question 7.17. (Assume a temperature of 25°C.)

Question 7.20

Calculate the osmotic pressure of the solution described in Question 7.18. (Assume a temperature of 25°C.)

Living cells contain aqueous solution (intracellular fluid) and the cells are surrounded by aqueous solution (intercellular fluid). Cell function (and survival!) depend on maintaining approximately the same osmotic pressure inside and outside the cell. If the solute concentration of the fluid surrounding red blood cells is higher than that inside the cell (a **hypertonic solution**), water flows from the cell, causing it to collapse. This process is **crenation.** On the other hand, if the solute concentration of this fluid is too low relative to the solution within the cell (a **hypotonic solution**), water will flow into the cells, causing the cell to rupture, **hemolysis.** To prevent either of these effects from taking place when fluids are administered to a patient intravenously, aqueous fluids [0.9% (W/W) NaCl, also referred to as *physiological saline,* or 5.0% (W/W) glucose] are prepared in such a way as to be *isotonic solutions* with intracellular fluids (Figure 7.4).

Two solutions are **isotonic solutions** if they have identical osmotic pressures. In that way the osmotic pressure differential across the cell membrane is zero, and no cell disruption occurs.

Practical examples of osmosis abound, including the following:

- A sailor, lost at sea in a lifeboat, dies of dehydration while surrounded by water. Seawater, because of its high salt concentration, dehydrates the cells of the body as a result of the large osmotic pressure difference between itself and intracellular fluids.

Figure 7.4
The effect of hypertonic and hypotonic solutions on the cell. (a) Crenation occurs when blood cells are surrounded by a hypertonic solution (water leaving > water entering). (b) Cell rupture occurs when cells are surrounded by a hypotonic solution (water entering > water leaving). (c) Cell size remains unchanged when surrounded by an isotonic solution (water entering = water leaving).

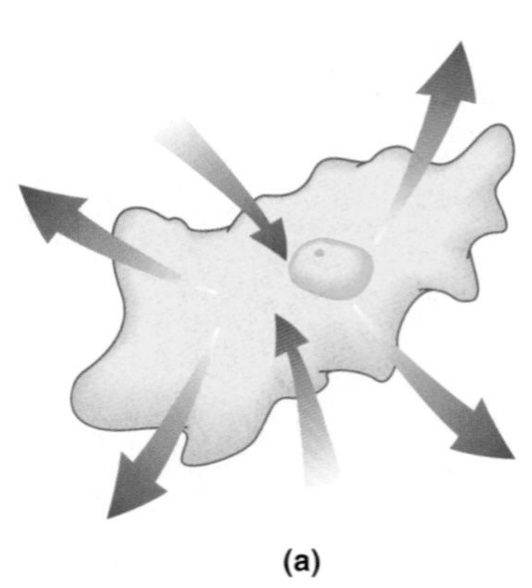

(a)

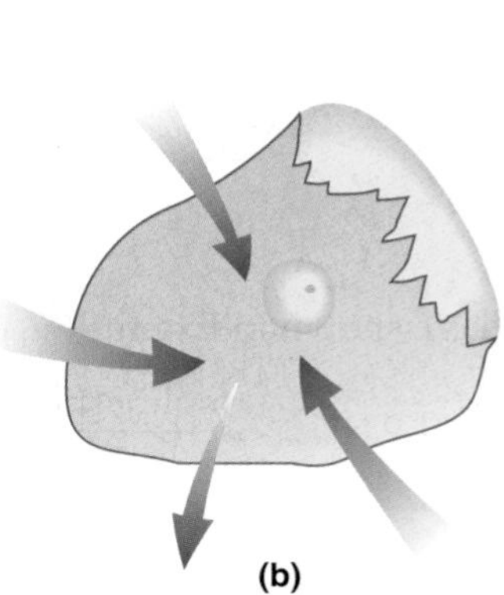

(b)

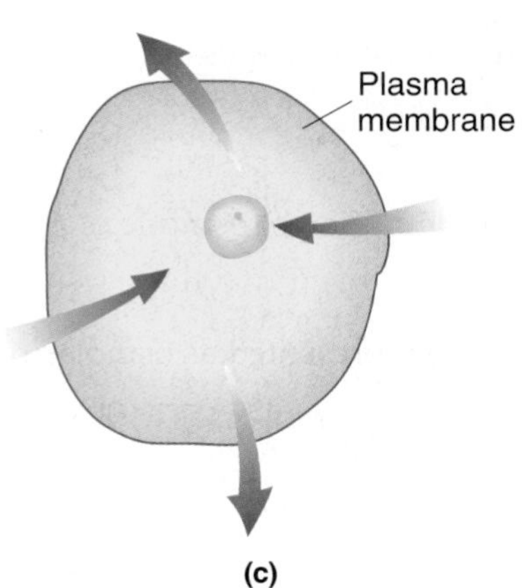

(c)

A CLINICAL PERSPECTIVE

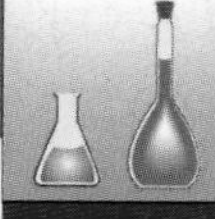

Oral Rehydration Therapy

Diarrhea kills millions of children before they reach the age of five years. This is particularly true in third world countries where sanitation, water supplies, and medical care are poor. In the case of diarrhea, death results from fluid loss, electrolyte imbalance, and hypovolemic shock (multiple organ failure due to insufficient perfusion). Cholera is one of the best-understood bacterial diarrheas. The organism *Vibrio cholera*, seen in the micrograph below, survives passage through the stomach and reproduces in the intestine, where it produces a toxin called choleragen. The toxin causes the excessive excretion of Na^+, Cl^-, and HCO_3^- from epithelial cells lining the intestine. The increased ion concentration (hypertonic solution) outside the cell results in movement of massive quantities of water into the intestinal lumen. This causes the severe, abundant, clear vomit and diarrhea that can result in the loss of 10–15 L of fluid per day. Over the four- to six-day progress of the disease, a patient may lose from one to two times his or her body mass!

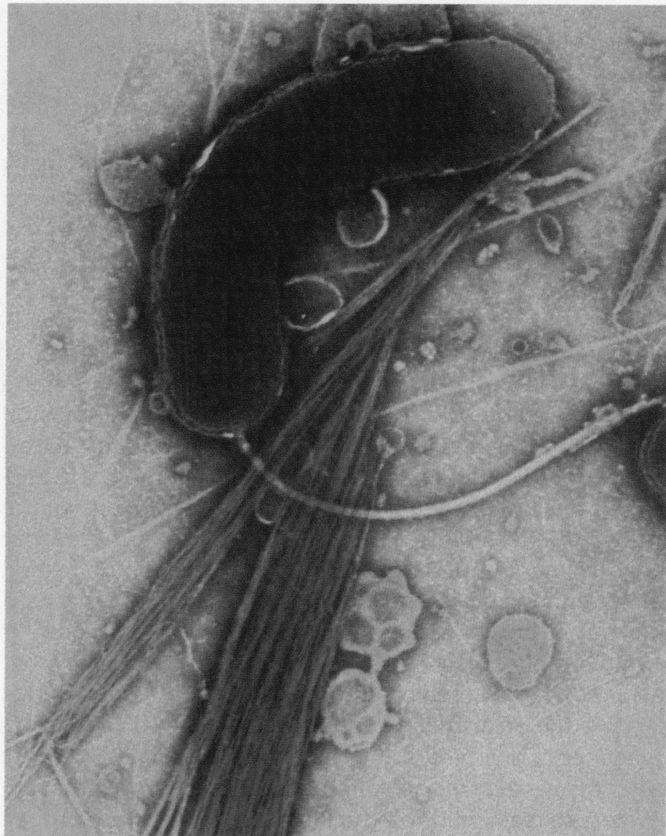

The need for fluid replacement is obvious. Oral rehydration is preferred over intravenous administration of fluids and electrolytes since it is noninvasive. In many third world countries, it is the only therapy available in remote areas. The rehydration formula includes 50–80 g/L rice (or other starch), 3.5 g/L sodium chloride, 2.5 g/L sodium bicarbonate, and 1.5 g/L potassium chloride. Oral rehydration takes advantage of the cotransport of Na^+ and glucose across the cells lining the intestine. Thus, the channel protein brings glucose into the cells, and Na^+ is carried along. Movement of these materials into the cells will help alleviate the osmotic imbalance, reduce the diarrhea, and correct the fluid and electrolyte imbalance.

The disease runs its course in less than a week. In fact, antibiotics are not used to combat cholera. The only effective therapy is oral rehydration, which reduces mortality to less than 1%. A much better option is prevention. In the photo below, a woman is shown filtering water through sari cloth. This simple practice has been shown to reduce the incidence of cholera significantly.

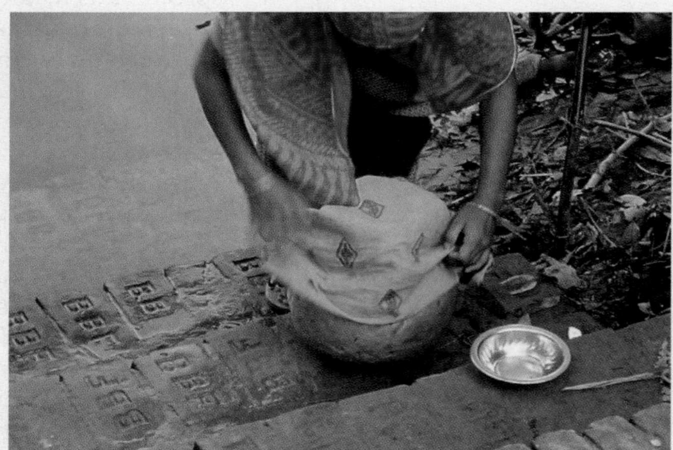

- A cucumber, soaked in brine, shrivels into a pickle. The water in the cucumber is drawn into the brine (salt) solution because of a difference in osmotic pressure (Figure 7.5).
- Clinical Perspective: Oral Rehydration Therapy, above, describes one of the most lethal and pervasive examples of cellular fluid imbalance.

(a) (b)

Figure 7.5
A cucumber (a) in an acidic salt solution undergoes considerable shrinkage on its way to becoming a pickle (b) because of osmosis.

7.7 Water as a Solvent

Learning Goal

11

See: A Human Perspective: An Extraordinary Molecule, in this chapter.

Refer to Sections 4.4 and 4.5 for a more complete description of the bonding, structure, and polarity of water.

Recall the discussion of intermolecular forces in Chapters 4 and 6.

Water is by far the most abundant substance on earth. It is an excellent solvent for most inorganic substances. In fact, it is often referred to as the "universal solvent" and is the principal biological solvent. Approximately 60% of the adult human body is water, and maintenance of this level is essential for survival. These characteristics are a direct consequence of the molecular structure of water.

As we saw in our previous discussion, water is a bent molecule with a 104.5° bond angle. This angular structure, resulting from the effect of the two lone pairs of electrons around the oxygen atom, is responsible for the polar nature of water. The polarity, in turn, gives water its unique properties.

Because water molecules are polar, water is an excellent solvent for other polar substances ("like dissolves like"). Because much of the matter on earth is polar, hence at least somewhat water soluble, water has been described as the universal solvent. It is readily accessible and easily purified. It is nontoxic and quite nonreactive. The high boiling point of water, 100°C, compared with molecules of similar size such as N_2 (b.p. = −196°C), is also explained by water's polar character. Strong dipole-dipole interactions between a δ^+ hydrogen of one molecule and δ^- oxygen of a second, referred to as hydrogen bonding, create an interactive molecular network in the liquid phase (see Figure 6.8a). The strength of these interactions requires more energy (higher temperature) to cause water to boil. The higher than expected boiling point enhances water's value as a solvent; often, reactions are carried out at higher temperatures to increase their rate. Other solvents, with lower boiling points, would simply boil away, and the reaction would stop.

This idea is easily extended to our own chemistry—because 60% of our bodies is water, we should appreciate the polarity of water on a hot day. As a biological solvent in the human body, water is involved in the transport of ions, nutrients, and waste into and out of cells. Water is also the solvent for biochemical reactions in cells and the digestive tract. Water is a reactant or product in some biochemical processes.

EXAMPLE 7.13 ***Predicting Structure from Observable Properties***

Sucrose is a common sugar and we know that it is used as a sweetener when dissolved in many beverages. What does this allow us to predict about the structure of sucrose?

Continued—

EXAMPLE 7.13 —*Continued*

Solution

Sucrose is used as a sweetener in teas, coffee, and a host of soft drinks. The solvent in all of these beverages is water, a polar molecule. The rule "like dissolves like" implies that sucrose must also be a polar molecule. Without even knowing the formula or structure of sucrose, we can infer this important information from a simple experiment—dissolving sugar in our morning cup of coffee.

Question 7.21

Predict whether carbon monoxide or carbon dioxide would be more soluble in water. Explain your answer. (Hint: Refer to Section 6.2, the discussion of interactions in the liquid state.)

Question 7.22

Predict whether ammonia or methane would be more soluble in water. Explain your answer. (Hint: Refer to Section 6.2, the discussion of interactions in the liquid state.)

7.8 Electrolytes in Body Fluids

Learning Goal

The concentrations of cations, anions, and other substances in biological fluids are critical to health. Consequently, the osmolarity of body fluids is carefully regulated by the kidney.

The two most important cations in body fluids are Na^+ and K^+. Sodium ion is the most abundant cation in the blood and intercellular fluids whereas potassium ion is the most abundant intracellular cation. In blood and intercellular fluid, the Na^+ concentration is 135 milliequivalents/L and the K^+ concentration is 3.5–5.0 meq/L. Inside the cell the situation is reversed. The K^+ concentration is 125 meq/L and the Na^+ concentration is 10 meq/L.

If osmosis and simple diffusion were the only mechanisms for transporting water and ions across cell membranes, these concentration differences would not occur. One positive ion would be just as good as any other. However, the situation is more complex than this. Large protein molecules embedded in cell membranes actively pump sodium ions to the outside of the cell and potassium ions into the cell. This is termed *active transport* because cellular energy must be expended to transport those ions. Proper cell function in the regulation of muscles and the nervous system depends on the sodium ion/potassium ion ratio inside and outside of the cell.

If the Na^+ concentration in the blood becomes too low, urine output decreases, the mouth feels dry, the skin becomes flushed, and a fever may develop. The blood level of Na^+ may be elevated when large amounts of water are lost. Diabetes, certain high-protein diets, and diarrhea may cause elevated blood Na^+ level. In extreme cases, elevated Na^+ levels may cause confusion, stupor, or coma.

Concentrations of K^+ in the blood may rise to dangerously high levels following any injury that causes large numbers of cells to rupture, releasing their intracellular K^+. This may lead to death by heart failure. Similarly, very low levels of K^+ in the blood may also cause death from heart failure. This may occur following

A HUMAN PERSPECTIVE

An Extraordinary Molecule

Think for a moment. What is the only common molecule that exists in all three physical states of matter (solid, liquid, and gas) under natural conditions on earth? This molecule is absolutely essential for life; in fact, life probably arose in this substance. It is the most abundant molecule in the cells of living organisms (70–95%) and covers 75% of the earth's surface. Without it, cells quickly die, and without it the earth would not be a fit environment in which to live. By now you have guessed that we are talking about the water molecule. It is so abundant on earth that we take this deceptively simple molecule for granted.

What are some of the properties of water that cause it to be essential to life as we know it? Water has the ability to stabilize temperatures on the earth and in the body. This ability is due in part to the energy changes that occur when water changes physical state; but ultimately, this ability is due to the polar nature of the water molecule.

Life can exist only within a fairly narrow range of temperatures. Above or below that range, the chemical reactions necessary for life, and thus life itself, will cease. Water can moderate temperature fluctuation and maintain the range necessary for life, and one property that allows it to do so is its unusually high specific heat, 1 cal/g °C. This means that water can absorb or lose more heat energy than many other substances without a significant temperature change. This is because in the liquid state, every water molecule is hydrogen bonded to other water molecules. Because a temperature increase is really just a measure of increased (more rapid) molecular movement, we must get the water molecules moving more rapidly, independent of one another, to register a temperature increase. Before we can achieve this independent, increased activity, the hydrogen bonds between molecules must be broken. Much of the heat energy that water absorbs is involved in breaking hydrogen bonds and is *not* used to increase molecular movement. Thus a great deal of heat is needed to raise the temperature of water even a little bit.

Water also has a very high heat of vaporization. It takes 540 calories to change 1 g of liquid water at 100°C to a gas and even more, 603 cal/g, when the water is at 37°C, human body temperature. That is about twice the heat of vaporization of alcohol. As water molecules evaporate, the surface of the liquid cools because only the highest-energy (or "hottest") molecules leave as a gas. Only the "hottest" molecules have enough energy to break the hydrogen bonds that bind them to other water molecules. Indeed, evaporation of water molecules from the surfaces of lakes and oceans helps to maintain stable temperatures in those bodies of water. Similarly, evaporation of perspiration from body surfaces helps to prevent overheating on a hot day or during strenuous exercise.

Even the process of freezing helps stabilize and moderate temperatures. This is especially true in the fall. Water releases heat when hydrogen bonds are formed. This is an example of an exothermic process. Thus, when water freezes, solidifying into ice, additional hydrogen bonds are formed, and heat is released into the environment. As a result, the temperature change between summer and winter is more gradual, allowing organisms to adjust to the change.

One last feature that we take for granted is the fact that when we put ice in our iced tea on a hot summer day, the ice floats. This means that the solid state of water is actually *less* dense than the liquid state! In fact, it is about 10% less dense, having an open lattice structure with each molecule hydrogen bonded to the maximum of four other water molecules. What would happen if ice did sink? All bodies of water, including the mighty oceans would eventually freeze solid, killing all aquatic and marine plant and animal life. Even in the heat of summer, only a few inches of ice at the surface would thaw. Instead, the ice forms at the surface and provides a layer of insulation that prevents the water below from freezing.

As we continue our study of chemistry, we will refer again and again to this amazing molecule. In other Human Perspective features we will examine other properties of water that make it essential to life.

prolonged exercise that results in excessive sweating. When this happens, both body fluids and electrolytes must be replaced. Salt tablets containing both NaCl and KCl taken with water and drinks such as Gatorade effectively provide water and electrolytes and prevent serious symptoms.

A CLINICAL PERSPECTIVE

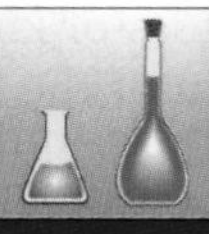

Hemodialysis

As we have seen in Section 7.8, blood is the medium for exchange of both nutrients and waste products. The membranes of the kidneys remove waste materials such as urea and uric acid (Chapter 22) and excess salts and large quantities of water. This process of waste removal is termed **dialysis,** a process similar in function to osmosis (Section 7.6). Semipermeable membranes in the kidneys, dialyzing membranes, allow small molecules (principally water and urea) and ions in solution to pass through and ultimately collect in the bladder. From there they can be eliminated from the body.

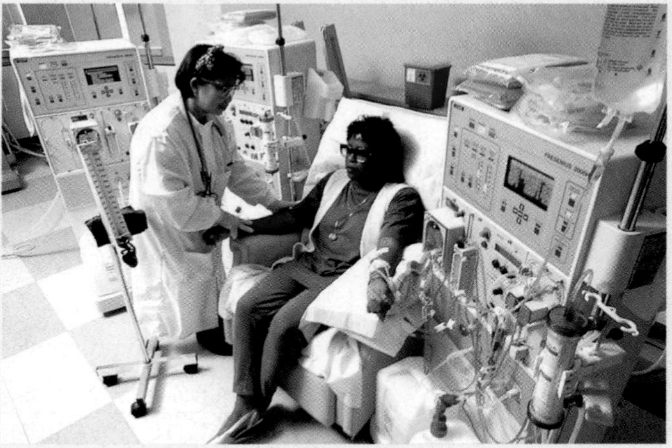

Unfortunately, a variety of diseases can cause partial or complete kidney failure. Should the kidneys fail to perform their primary function, dialysis of waste products, urea and other waste products rapidly increase in concentration in the blood. This can become a life-threatening situation in a very short time.

The most effective treatment of kidney failure is the use of a machine, an artificial kidney, that mimics the function of the kidney. The artificial kidney removes waste from the blood using the process of hemodialysis (blood dialysis). The blood is pumped through a long semipermeable membrane, the dialysis membrane. The dialysis process is similar to osmosis. However, in addition to water molecules, larger molecules (including the waste products in the blood) and ions can pass across the membrane from the blood into a dialyzing fluid. The dialyzing fluid is isotonic with normal blood; it also is similar in its concentration of all other essential blood components. The waste materials move across the dialysis membrane (from a higher to a lower concentration, as in diffusion). A successful dialysis procedure selectively removes the waste from the body without upsetting the critical electrolyte balance in the blood.

Hemodialysis, although lifesaving, is not by any means a pleasant experience. The patient's water intake must be severely limited to minimize the number of times each week that treatment must be used. Many dialysis patients require two or three treatments per week and each session may require one-half (or more) day of hospitalization, especially when the patient suffers from complicating conditions such as diabetes.

Improvements in technology, as well as the growth and sophistication of our health care delivery systems over the past several years, have made dialysis treatment much more patient friendly. Dialysis centers, specializing in the treatment of kidney patients, are now found in most major population centers. Smaller, more automated dialysis units are available for home use, under the supervision of a nursing practitioner. With the remarkable progress in kidney transplant success, dialysis is becoming, more and more, a temporary solution, sustaining life until a suitable kidney donor match can be found.

The cationic charge in blood is neutralized by two major anions, Cl^- and HCO_3^-. The chloride ion plays a role in acid-base balance, maintenance of osmotic pressure within an acceptable range, and oxygen transport by hemoglobin. The bicarbonate anion is the form in which most waste CO_2 is carried in the blood.

A variety of proteins is also found in the blood. Because of their larger size, they exist in colloidal suspension. These proteins include blood clotting factors, immunoglobulins (antibodies) that help us fight infection, and albumins that act as carriers of nonpolar, hydrophobic substances (fatty acids and steroid hormones) that cannot dissolve in water.

Additionally, blood is the medium for exchange of nutrients and waste products. Nutrients, such as the polar sugar glucose, enter the blood from the intestine or the liver. Because glucose molecules are polar, they dissolve in body fluids and are circulated to tissues throughout the body. As noted above, nonpolar nutrients are transported with the help of carrier proteins. Similarly, nitrogen-containing waste products, such as urea, are passed from cells to the blood. They are continuously and efficiently removed from the blood by the kidneys.

In cases of loss of kidney function, mechanical devices—dialysis machines—mimic the action of the kidney. The process of blood dialysis—hemodialysis—is discussed in A Clinical Perspective: Hemodialysis on page 195.

EXAMPLE 7.14 ***Calculating Electrolyte Concentrations***

A typical concentration of calcium ion in blood plasma is 4 meq/L. Represent this concentration in moles/L.

Solution

The calcium ion has a 2+ charge (recall that calcium is in Group IIA of the periodic table; hence, a 2+ charge on the calcium ion).

We will need three conversion factors:

$$\text{meq (milliequivalents)} \longrightarrow \text{eq (equivalents)}$$

$$\text{eq (equivalents)} \longrightarrow \text{moles of charge}$$

$$\text{moles of charge} \longrightarrow \text{moles of calcium ion}$$

Using dimensional analysis as in Example 7.10,

$$\frac{4\ \cancel{\text{meq Ca}^{2+}}}{1\text{L}} \times \frac{1\ \cancel{\text{eq Ca}^{2+}}}{10^3\ \cancel{\text{meq Ca}^{2+}}} \times \frac{1\ \cancel{\text{mol charge}}}{1\ \cancel{\text{eq Ca}^{2+}}} \times \frac{1\ \text{mol Ca}^{2+}}{2\ \cancel{\text{mol charge}}} = \frac{2 \times 10^{-3}\,\text{mol Ca}^{2+}}{\text{L}}$$

Question 7.23

Sodium chloride [0.9% (W/V)] is a solution administered intravenously to replace fluid loss. It is frequently used to avoid dehydration. The sodium ion concentration is 15.4 meq/L. Calculate the sodium ion concentration in moles/L.

Question 7.24

A potassium chloride solution that also contains 5% (W/V) dextrose is administered intravenously to treat some forms of malnutrition. The potassium ion concentration in this solution is 40 meq/L. Calculate the potassium ion concentration in moles/L.

Summary

7.1 Writing Chemical Reactions

Chemical reactions involve the *combination* of reactants to produce products, the *decomposition* of reactant(s) into products, or the *replacement* of one or more elements in a compound to yield products. Replacement reactions are subclassified as either *single-* or *double*-replacement.

7.2 Types of Chemical Reactions

Reactions that produce products with similar characteristics are often classified as a single group. The formation of an insoluble solid, a *precipitate,* is very common. Such reactions are precipitation reactions.

Chemical reactions that have a common reactant may be grouped together. Reactions involving oxygen, *combustion reactions,* are such a class.

Another approach to the classification of chemical reactions is based on charge transfer. *Acid-base reactions* involve the transfer of a hydrogen ion, H^+, from one reactant to another. Another important reaction type, *oxidation-reduction,* takes place because of the transfer of negative charge, one or more electrons, from one reactant to another.

7.3 Properties of Solutions

A majority of chemical reactions, and virtually all important organic and biochemical reactions, take place not as a combination of two or more pure substances, but rather as reactants dissolved in solution, *solution reactions.*

A *solution* is a homogeneous (or uniform) mixture of two or more substances. A solution is composed of one or more *solutes*, dissolved in a *solvent*. When the solvent is water, the solution is called an *aqueous solution*.

Liquid solutions are clear and transparent with no visible particles of solute. They may be colored or colorless, depending on the properties of the solute and solvent.

In solutions of *electrolytes* the solutes are ionic compounds that dissociate in solution to produce ions. They are good conductors of electricity. Solutions of *nonelectrolytes* are formed from nondissociating molecular solutes (nonelectrolytes), and their solutions are nonconducting.

The rule "like dissolves like" is the fundamental condition for solubility. Polar solutes are soluble in polar solvents, and nonpolar solutes are soluble in nonpolar solvents.

The degree of solubility depends on the difference between the polarity of solute and solvent, the temperature, and the pressure. Pressure considerations are significant only for solutions of gases.

When a solution contains all the solute that can be dissolved at a particular temperature, it is *saturated*. Excess solute falls to the bottom of the container as a *precipitate*. Occasionally, on cooling, the excess solute may remain in solution for a time before precipitation. Such a solution is a *supersaturated solution*. When excess solute, the precipitate, contacts solvent, the dissolution process reaches a state of dynamic equilibrium. *Colloidal suspensions* have particle sizes between those of true solutions and precipitates. A *suspension* is a heterogeneous mixture that contains particles much larger than a colloidal suspension. Over time, these particles may settle, forming a second phase.

Henry's law describes the solubility of gases in liquids. At a given temperature the solubility of a gas is proportional to the partial pressure of the gas.

7.4 Concentration of Solutions: Percentage

The amount of solute dissolved in a given amount of solution is the solution *concentration*. The more widely used percentage-based concentration units are *weight/volume percent* and *weight/weight percent*.

7.5 Concentration of Solutions: Moles and Equivalents

Molarity, symbolized *M*, is defined as the number of moles of solute per liter of solution.

Dilution is often used to prepare less concentrated solutions. The expression for this calculation is $(M_1)(V_1) = (M_2)(V_2)$. Knowing any three of these terms enables one to calculate the fourth. The concentration of ions in solution may be represented as moles per liter (molarity) or any other suitable concentration units. However, both concentrations must be in the same units when using the dilution equation.

When discussing solutions of ionic compounds, molarity emphasizes the number of individual ions. A 1 *M* solution of Na^+ contains Avogadro's number of sodium ions. In contrast, equivalents per liter emphasizes charge; a solution containing one equivalent of Na^+ per liter contains Avogadro's number of positive charge.

One *equivalent* of an ion is the number of grams of the ion corresponding to Avogadro's number of electrical charges. Changing from moles per liter to equivalents per liter (or the reverse) is done using conversion factors.

7.6 Concentration-Dependent Solution Properties

Solution properties that depend on the concentration of solute particles, rather than the identity of the solute, are *colligative properties*.

There are four colligative properties of solutions, all of which depend on the concentration of *particles* in solution.

1. *Vapor pressure lowering*. *Raoult's law* states that when a solute is added to a solvent, the vapor pressure of the solvent decreases in proportion to the concentration of the solute.
2. and 3. *Freezing point depression and boiling point elevation*. When a nonvolatile solid is added to a solvent, the freezing point of the resulting solution decreases, and the boiling point increases. The magnitudes of both the freezing point depression (ΔT_f) and the boiling point elevation (ΔT_b) are proportional to the solute concentration over a limited range of concentrations. The mole-based concentration unit, molality, is more commonly used in calculations involving colligative properties. This is due to the fact that molality is temperature independent. *Molality* (symbolized *m*) is defined as the number of moles of solute per kilogram of solvent in a solution.
4. *Osmosis and osmotic pressure*. *Osmosis* is the movement of solvent from a dilute solution to a more concentrated solution through a *semipermeable membrane*. The pressure that must be applied to the more concentrated solution to stop this flow is the *osmotic pressure*. The osmotic pressure, like the pressure exerted by a gas, may be treated quantitatively by using an equation similar in form to the ideal gas equation: $\pi = MRT$. By convention the molarity of particles that is used for osmotic pressure calculations is termed *osmolarity (osmol)*.

In biological systems, if the concentration of the fluid surrounding red blood cells is higher than that inside the cell (a *hypertonic* solution), water flows from the cell, causing it to collapse (*crenation*). Too low a concentration of this fluid relative to the solution within the cell (a *hypotonic* solution) will cause cell rupture (*hemolysis*).

Two solutions are *isotonic* if they have identical osmotic pressures. In that way the osmotic pressure differential across the cell is zero, and no cell disruption occurs.

7.7 Water as a Solvent

The role of water in the solution process deserves special attention. It is often referred to as the "universal solvent" because of the large number of ionic and polar covalent compounds that are at least partially soluble in water. It is the principal biological solvent. These characteristics are a direct consequence of the molecular geometry and structure of water and its ability to undergo hydrogen bonding.

7.8 Electrolytes in Body Fluids

The concentrations of cations, anions, and other substances in biological fluids are critical to health. As a result, the osmolarity of body fluids is carefully regulated by the kidney using the process of *dialysis*.

Key Terms

acid-base reaction (7.2)
aqueous solution (7.3)
colligative property (7.6)
colloidal suspension (7.3)
combination reaction (7.1)
concentration (7.4)
crenation (7.6)
decomposition reaction (7.1)
dialysis (7.8)
double-replacement reaction (7.1)
electrolyte (7.3)
equivalent (7.5)
hemolysis (7.6)
Henry's law (7.3)
hypertonic solution (7.6)
hypotonic solution (7.6)
isotonic solution (7.6)
molality (7.6)
molarity (7.5)
nonelectrolyte (7.3)
osmolarity (7.6)
osmosis (7.6)
osmotic pressure (7.6)
oxidation-reduction reaction (7.2)
precipitate (7.3)
Raoult's law (7.6)
saturated solution (7.3)
semipermeable membrane (7.6)
single-replacement reaction (7.1)
solubility (7.3)
solute (7.3)
solution (7.3)
solvent (7.3)
supersaturated solution (7.3)
suspension (7.3)
weight/volume percent (% [W/V]) (7.4)
weight/weight percent (% [W/W]) (7.4)

Questions and Problems

Chemical Reactions

7.25 Give an example of:
a. a decomposition reaction
b. a single-replacement reaction

7.26 Give an example of:
a. a combination reaction
b. a double-replacement reaction

7.27 Give an example of a precipitate-forming reaction.

7.28 Give an example of a reaction in which oxygen is a reactant.

7.29 Balance each of the following equations:
a. $C_2H_6(g) + O_2(g) \longrightarrow CO_2(g) + H_2O(g)$
b. $K_2O(s) + P_4O_{10}(s) \longrightarrow K_3PO_4(s)$
c. $MgBr_2(aq) + H_2SO_4(aq) \longrightarrow HBr(g) + MgSO_4(aq)$

7.30 Balance each of the following equations:
a. $C_6H_{12}O_6(s) + O_2(g) \longrightarrow CO_2(g) + H_2O(g)$
b. $H_2O(l) + P_4O_{10}(s) \longrightarrow H_3PO_4(aq)$
c. $PCl_5(g) + H_2O(l) \longrightarrow HCl(aq) + H_3PO_4(aq)$

7.31 Complete, then balance, each of the following equations:
a. $Ca(s) + F_2(g) \longrightarrow$
b. $Mg(s) + O_2(g) \longrightarrow$
c. $H_2(g) + N_2(g) \longrightarrow$

7.32 Complete, then balance, each of the following equations:
a. $Li(s) + O_2(g) \longrightarrow$
b. $Ca(s) + N_2(g) \longrightarrow$
c. $Al(s) + S(s) \longrightarrow$

Concentration of Solutions: Percentage

7.33 Calculate the composition of each of the following solutions in weight/volume %:
a. 20.0 g NaCl in 1.00 L solution
b. 33.0 g sugar, $C_6H_{12}O_6$, in 5.00×10^2 mL solution

7.34 Calculate the composition of each of the following solutions in weight/volume %:
a. 0.700 g KCl per 1.00 mL
b. 1.00 mol $MgCl_2$ in 2.50×10^2 mL solution

7.35 Calculate the composition of each of the following solutions in weight/volume %:
a. 50.0 g ethanol dissolved in 1.00 L solution
b. 50.0 g ethanol dissolved in 5.00×10^2 mL solution

7.36 Calculate the composition of each of the following solutions in weight/volume %:
a. 20.0 g acetic acid dissolved in 2.50 L solution
b. 20.0 g benzene dissolved in 1.00×10^2 mL solution

7.37 Calculate the composition of each of the following solutions in weight/weight %:
a. 21.0 g NaCl in 1.00×10^2 g solution
b. 21.0 g NaCl in 5.00×10^2 mL solution ($d = 1.12$ g/mL)

7.38 Calculate the composition of each of the following solutions in weight/weight %:
a. 1.00 g KCl in 1.00×10^2 g solution
b. 50.0 g KCl in 5.00×10^2 mL solution ($d = 1.14$ g/mL)

7.39 How many grams of solute are needed to prepare each of the following solutions?
a. 2.50×10^2 g of 0.900% (W/W) NaCl
b. 2.50×10^2 g of 1.25% (W/V) $NaC_2H_3O_2$ (sodium acetate)

7.40 How many grams of solute are needed to prepare each of the following solutions?
a. 2.50×10^2 g of 5.00% (W/W) NH_4Cl (ammonium chloride)
b. 2.50×10^2 g of 3.50% (W/V) Na_2CO_3

Concentration of Solutions: Moles and Equivalents

7.41 Calculate the molarity of each solution in Problem 7.33.

7.42 Calculate the molarity of each solution in Problem 7.34.

7.43 Calculate the number of grams of solute that would be needed to make each of the following solutions:
a. 2.50×10^2 mL of 0.100 *M* NaCl
b. 2.50×10^2 mL of 0.200 *M* $C_6H_{12}O_6$ (glucose)

7.44 Calculate the number of grams of solute that would be needed to make each of the following solutions:
a. 2.50×10^2 mL of 0.100 *M* NaBr
b. 2.50×10^2 mL of 0.200 *M* KOH

7.45 Calculate the molarity of a sucrose (table sugar, $C_{12}H_{22}O_{11}$) solution that contains 50.0 g of sucrose per liter.

7.46 A saturated silver chloride solution is 1.58×10^{-4} g of silver chloride per 1.00×10^2 mL of solution. What is the molarity of this solution?

7.47 It is desired to prepare 0.500 L of a 0.100 *M* solution of NaCl from a 1.00 *M* stock solution. How many milliliters of the stock solution must be taken for the dilution?

7.48 50.0 mL of a 0.250 *M* sucrose solution was diluted to 5.00×10^2 mL. What is the molar concentration of the resulting solution?

7.49 A 50.0-mL portion of a stock solution was diluted to 500.0 mL. If the resulting solution was 2.00 *M*, what was the molarity of the original stock solution?

7.50 A 6.00-mL portion of an 8.00 *M* stock solution is to be diluted to 0.400 *M*. What will be the final volume after dilution?

Concentration-Dependent Solution Properties

7.51 What is meant by the term *colligative property?*

7.52 Name and describe four colligative solution properties.

7.53 Explain, in terms of solution properties, why salt is used to melt ice in the winter.

7.54 Explain, in terms of solution properties, why a wilted plant regains its "health" when watered.

Answer questions 7.55 to 7.60 by comparing two solutions: 0. 50 *M* sodium chloride (an ionic compound) and 0.50 *M* sucrose (a covalent compound).

7.55 Which solution has the higher melting point?

7.56 Which solution has the higher boiling point?

7.57 Which solution has the higher vapor pressure?

7.58 Each solution is separated from water by a semipermeable membrane. Which solution has the higher osmotic pressure?

7.59 Calculate the osmotic pressure of 0.50 *M* sodium chloride.

7.60 Calculate the osmotic pressure of 0.50 *M* sucrose.

Water as a Solvent

7.61 What properties make water such a useful solvent?

7.62 Sketch the "interactive network" of water molecules in the liquid state.

7.63 Solutions of ammonia in water are sold as window cleaner. Why do these solutions have a long "shelf life"?

7.64 Why does water's abnormally high boiling point help to make it a desirable solvent?

7.65 Sketch the interaction of a water molecule with a sodium ion.

7.66 Sketch the interaction of a water molecule with a chloride ion.

Electrolytes in Body Fluids

7.67 Explain why a dialysis solution must have a low sodium ion concentration if it is designed to remove excess sodium ion from the blood.

7.68 Explain why a dialysis solution must have an elevated potassium ion concentration when loss of potassium ion from the blood is a concern.

7.69 Describe the clinical effects of elevated concentrations of sodium ion in the blood.

7.70 Describe the clinical effects of depressed concentrations of potassium ion in the blood.

7.71 Describe conditions that can lead to elevated concentrations of sodium in the blood.

7.72 Describe conditions that can lead to dangerously low concentrations of potassium in the blood.

Critical Thinking Problems

1. Which of the following compounds would cause the greater freezing point depression, per mole, in H_2O: $C_6H_{12}O_6$ (glucose) or NaCl?
2. Which of the following compounds would cause the greater boiling point elevation, per mole, in H_2O: $MgCl_2$ or $HOCH_2CH_2OH$ (ethylene glycol, antifreeze)? (Hint: $MgCl_2$ is ionic; $HOCH_2CH_2OH$ is covalent.)
3. Analytical chemists often take advantage of differences in solubility to separate ions. For example, adding Cl^- to a solution of Cu^{2+} and Ag^+ causes AgCl to precipitate; Cu^{2+} remains in solution. Filtering the solution results in a separation. Design a scheme to separate the cations Ca^{2+} and Pb^{2+}.
4. Using the strategy outlined in the above problem, design a scheme to separate the anions S^{2-} and CO_3^{2-}.
5. Design an experiment that would enable you to measure the degree of solubility of a salt such as KI in water.
6. How could you experimentally distinguish between a saturated solution and a supersaturated solution?
7. Blood is essentially an aqueous solution, but it must transport a variety of nonpolar substances (hormones, for example). Colloidal proteins, termed *albumins*, facilitate this transport. Must these albumins be polar or nonpolar? Why?

A rapid, exothermic chemical reaction.

8

Chemical and Physical Change: *Energy, Rate, and Equilibrium*

Outline

GENERAL CHEMISTRY

Learning Goals

1. Correlate the terms *endothermic* and *exothermic* with heat flow between a *system* and its *surroundings.*
2. State the meaning of the terms *enthalpy, entropy,* and *free energy* and know their implications.
3. Describe experiments that yield thermochemical information and calculate fuel values based on experimental data.
4. Describe the concept of reaction rate and the role of kinetics in chemical and physical change.
5. Describe the importance of *activation energy* and the *activated complex* in determining reaction rate.
6. Predict the way reactant structure, concentration, temperature, and catalysis affect the rate of a chemical reaction.
7. Write rate equations for elementary processes.
8. Recognize and describe equilibrium situations.
9. Write equilibrium-constant expressions and use these expressions to calculate equilibrium constants.
10. Use LeChatelier's principle to predict changes in equilibrium position.

CHEMISTRY CONNECTION

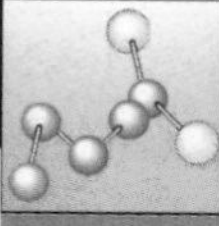

The Cost of Energy? More Than You Imagine

When we purchase gasoline for our automobiles or oil for the furnace, we are certainly buying matter. That matter is only a storage device; we are really purchasing the energy stored in the chemical bonds. Combustion, burning in oxygen, releases the stored potential energy in a form suited to its function: mechanical energy to power a vehicle or heat energy to warm a home.

Energy release is a consequence of change. In fuel combustion, this change results in the production of waste products that may be detrimental to our environment. This necessitates the expenditure of time, money, and *more* energy to clean up our surroundings.

If we are paying a considerable price for our energy supply, it would be nice to believe that we are at least getting full value for our expenditure. Even that is not the case. Removal of energy from molecules also extracts a price. For example, a properly tuned automobile engine is perhaps 30% efficient. That means that less than one-third of the available energy actually moves the car. The other two-thirds is released into the atmosphere as wasted energy, mostly heat energy. The law of conservation of energy tells us that the energy is not destroyed, but it is certainly not available to us in a useful form.

Can we build a 100% efficient energy transfer system? Is there such a thing as cost-free energy? No, on both counts. It is theoretically impossible, and the laws of thermodynamics, which we discuss in this chapter, tell us why this is so.

Introduction

In Chapter 5 we calculated quantities of matter involved in chemical change, assuming that all of the reacting material was consumed and that only products of the reaction remain at the end of the reaction. Often this is not true. Furthermore, not all chemical reactions take place at the same speed; some occur almost instantaneously (explosions), whereas others may proceed for many years (corrosion).

Two concepts play important roles in determining the extent and speed of a chemical reaction: *thermodynamics,* which deals with energy changes in chemical reactions, and *kinetics,* which describes the rate or speed of a chemical reaction.

Although both thermodynamics and kinetics involve energy, they are two separate considerations. A reaction may be thermodynamically favored but very slow; conversely, a reaction may be very fast because it is kinetically favorable yet produce very little (or no) product because it is thermodynamically unfavorable.

In this chapter we investigate the fundamentals of thermodynamics and kinetics, with an emphasis on the critical role that energy changes play in chemical reactions. We consider physical change and chemical change, including the conversions that take place among the states of matter (solid, liquid, and gas). We use these concepts to explain the behavior of reactions that do not go to completion, *equilibrium reactions.* We develop the *equilibrium-constant expression* and demonstrate how equilibrium composition can be altered using *LeChatelier's principle.*

8.1 Thermodynamics

Thermodynamics is the study of energy, work, and heat. It may be applied to chemical change, such as the calculation of the quantity of heat obtainable from the combustion of one gallon of fuel oil. Similarly, energy released or consumed in physical change, such as the boiling or freezing of water, may be determined.

There are three basic laws of thermodynamics, but only the first two will be of concern here. They help us to understand why some chemical reactions occur readily and others do not. For instance, a mixture of concentrated solutions of

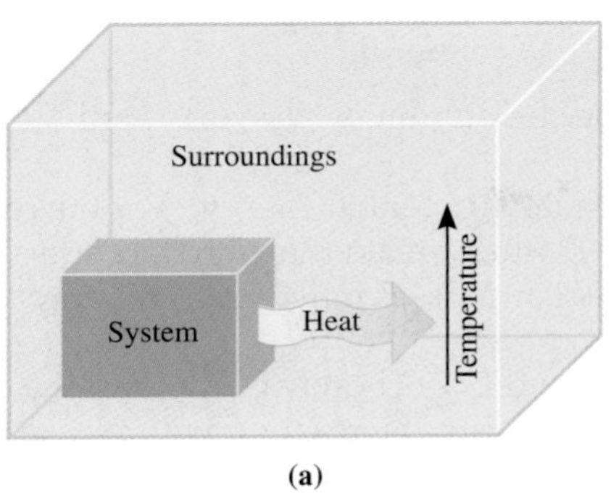

(a)

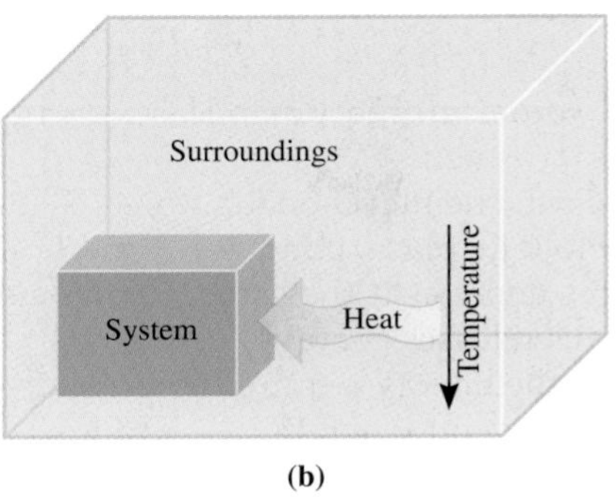

(b)

Figure 8.1
Illustration of heat flow in (a) exothermic and (b) endothermic reactions.

hydrochloric acid and sodium hydroxide reacts violently releasing a large quantity of heat. On the other hand, nitrogen and oxygen have coexisted in the atmosphere for thousands of years with no significant chemical reaction occurring.

The Chemical Reaction and Energy

John Dalton believed that chemical change involved joining, separating, or rearranging atoms. Two centuries later, this statement stands as an accurate description of chemical reactions. However, we now know much more about the nonmaterial energy changes that are an essential part of every reaction.

Throughout the discussion of thermodynamics and kinetics it will be useful to remember the basic ideas of the kinetic molecular theory (Section 6.1):

- molecules and atoms in a reaction mixture are in constant, random motion;
- these molecules and atoms frequently collide with each other;
- only some collisions, those with sufficient energy, will break bonds in molecules; and
- when reactant bonds are broken, new bonds may be formed and products result.

It is worth noting that we cannot measure an absolute value for energy stored in a chemical system. We can only measure the *change* in energy (energy absorbed or released) as a chemical reaction occurs. Also, it is often both convenient and necessary to establish a boundary between the *system* and its *surroundings*.

The **system** is the process under study. The **surroundings** encompass the rest of the universe. Energy is lost from the system to the surroundings or energy may be gained by the system at the expense of the surroundings. This energy change, in the form of heat, may be measured because the temperature of the system or surroundings may change and this property can be measured. This process is illustrated in Figure 8.1.

Consider the combustion of methane in a Bunsen burner, the system. The temperature of the air surrounding the burner increases, indicating that heat energy of the system (methane and oxygen) is being lost to the surroundings.

Now, an exact temperature measurement of the air before and after the reaction is difficult. However, if we could insulate a portion of the surroundings, to isolate and trap the heat, we could calculate a useful quantity, the heat of the reaction. Experimental strategies for measuring temperature change and calculating heats of reactions, termed *calorimetry,* are discussed in Section 8.2.

Exothermic and Endothermic Reactions

Learning Goal

The first law of thermodynamics states that the energy of the universe is constant. It is the law of conservation of energy. The study of energy changes that occur in chemical reactions is a very practical application of the first law. Consider, for example, the generalized reaction:

$$A—B + C—D \longrightarrow A—D + C—B$$

An **exothermic reaction** releases energy to the surroundings. The surroundings become warmer.

Each chemical bond is stored chemical energy (potential energy). For the reaction to take place, bond *A—B* and bond *C—D* must break; this process *always* requires energy. At the same time, bonds *A—D* and *C—B* must form; this process always releases energy.

If the energy required to break the *A—B* and *C—D* bonds is *less* than the energy given off when the *A—D* and *C—B* bonds form, the reaction will release the excess energy. The energy is a *product*, and the reaction is called an exothermic (*Gr. exo*, out, and *Gr. therm*, heat) reaction. This conversion of chemical energy to heat energy is represented in Figure 8.2a.

An example of an exothermic reaction is the combustion of methane:

$$CH_4(g) + 2O_2(g) \longrightarrow CO_2(g) + 2H_2O(g) + 211 \text{ kcal}$$

Exothermic reaction

In an exothermic reaction, heat is released *from the system* to the surroundings. In an endothermic reaction, heat is absorbed *by the system* from the surroundings.

An **endothermic reaction** absorbs energy from the surroundings. The surroundings become colder.

If the energy required to break the *A—B* and *C—D* bonds is *greater* than the energy released when the *A—D* and *C—B* bonds form, the reaction will need an external supply of energy (perhaps from a Bunsen burner). Insufficient energy is available in the system to initiate the bond-breaking process. Such a reaction is called an endothermic (*Gr. endo*, to take on, and *Gr. therm*, heat) reaction, and energy is a *reactant*. The conversion of heat energy into chemical energy is represented in Figure 8.2b.

The decomposition of ammonia into nitrogen and hydrogen is one example of an endothermic reaction:

$$22 \text{ kcal} + 2NH_3(g) \longrightarrow N_2(g) + 3H_2(g)$$

Endothermic reaction

The examples used here show the energy absorbed or released as heat energy. Depending on the reaction and the conditions under which the reaction is run, the energy may take the form of light energy or electrical energy. A firefly releases energy as a soft glow of light on a summer evening. An electrical current results from a chemical reaction in a battery, enabling your car to start.

Enthalpy

Learning Goal

Enthalpy is the term used to represent heat energy. The *change in enthalpy* is the energy difference between the products and reactants of a chemical reaction and is symbolized as $\Delta H°$. By convention, energy released is represented with a negative

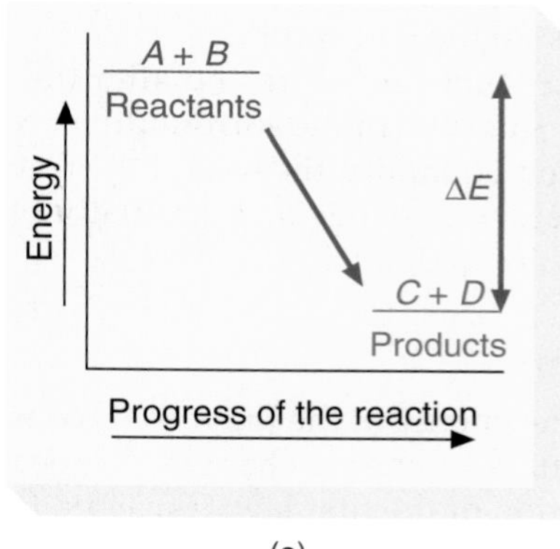

(a)

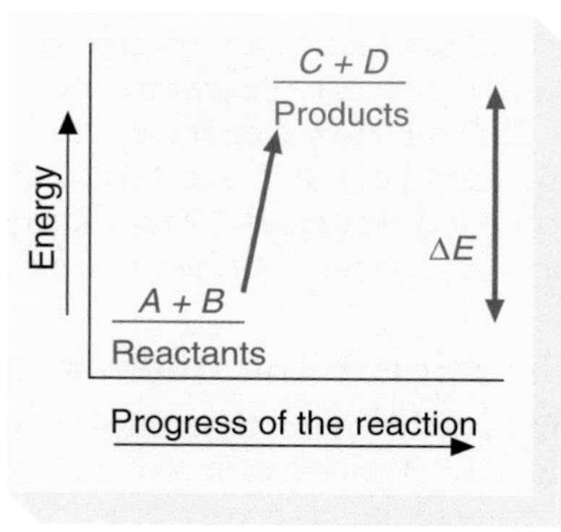

(b)

Figure 8.2
(a) An exothermic reaction. ΔE represents the energy released during the progress of the exothermic reaction: $A + B \longrightarrow C + D + \Delta E$. (b) An endothermic reaction. ΔE represents the energy absorbed during the progress of the endothermic reaction: $\Delta E + A + B \longrightarrow C + D$.

sign (indicating an exothermic reaction), and energy absorbed is shown with a positive sign (indicating an endothermic reaction).

For the combustion of methane, an exothermic process,

$$\Delta H^\circ = -211 \text{ kcal}$$

For the decomposition of ammonia, an endothermic process,

$$\Delta H^\circ = +22 \text{ kcal}$$

Spontaneous and Nonspontaneous Reactions

Spontaneous reactions are just that: they occur without any external energy input. Nonspontaneous reactions must be persuaded; they need an input of energy.

It seems that all exothermic reactions should be spontaneous. After all, an external supply of energy does not appear to be necessary; in fact, energy is a product of the reaction. It also seems that all endothermic reactions should be nonspontaneous: energy is a reactant that we must provide. However, these hypotheses are not supported by experimentation.

Experimental measurement has shown that most *but not all* exothermic reactions are spontaneous. Likewise, most *but not all*, endothermic reactions are not spontaneous. There must be some factor in addition to enthalpy that will help us to explain the less obvious cases of nonspontaneous exothermic reactions and spontaneous endothermic reactions. This other factor is entropy.

Entropy

Learning Goal

The first law of thermodynamics considers the enthalpy of chemical reactions. The second law states that the universe spontaneously tends toward increasing disorder or randomness.

A measure of the randomness of a chemical system is its **entropy.** The entropy of a substance is represented by the symbol S°. A random, or disordered, system is characterized by *high entropy;* a well-organized system has *low entropy.*

A *system* is a part of the universe upon which we wish to focus our attention. For example, it may be a beaker containing reactants and products.

What do we mean by disorder in chemical systems? Disorder is simply the absence of a regular repeating pattern. Disorder or randomness increases as we convert from the solid to the liquid to the gaseous state. As we have seen, solids often have an ordered crystalline structure, liquids have, at best, a loose arrangement, and gas particles are virtually random in their distribution. Therefore gases have high entropy, and crystalline solids have very low entropy. Figures 8.3 and 8.4 illustrate properties of entropy in systems.

Chapter 6 compares the physical properties of solids, liquids, and gases.

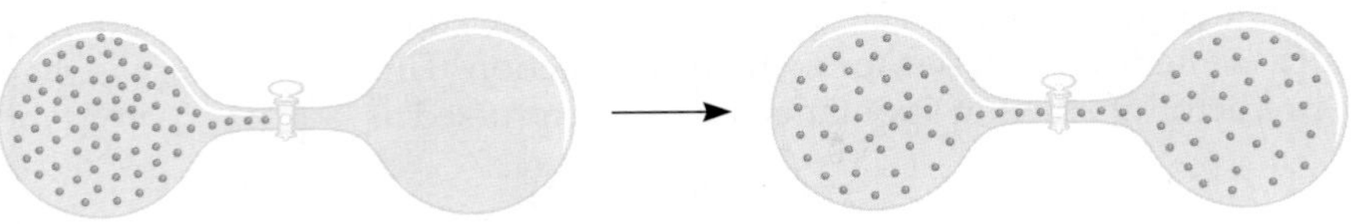

Spontaneous Process

(a)

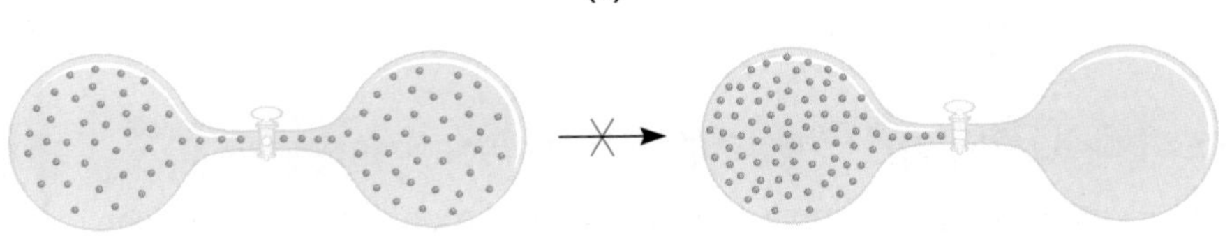

Nonspontaneous Process

(b)

Figure 8.3

(a) Gas particles, trapped in the left chamber, spontaneously diffuse into the right chamber, initially under vacuum, when the valve is opened. (b) It is unimaginable that the gas particles will rearrange themselves and reverse the process to create a vacuum. This can only be accomplished using a pump, that is, by doing work on the system.

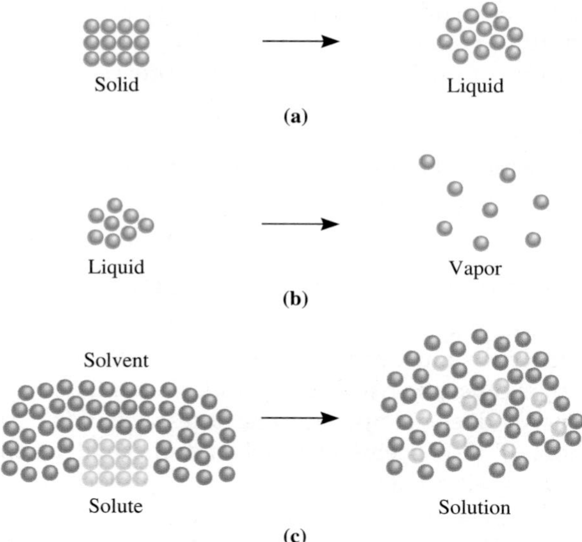

Figure 8.4
Processes such as (a) melting, (b) vaporization, and (c) dissolution increase entropy, or randomness, of the particles.

The second law describes the entire universe. On a more personal level, we all fall victim to the law of increasing disorder. Chaos in our room or workplace is certainly not our intent! It happens almost effortlessly. However, reversal of this process requires work and energy. The same is true at the molecular level. The gradual deterioration of our cities' infrastructure (roads, bridges, water mains, and so forth) is an all-too-familiar example. Millions of dollars (translated into energy and work) are needed annually just to try to maintain the status quo.

The entropy of a reaction is measured as a difference, $\Delta S°$, between the entropies, $S°$, of products and reactants.

The drive toward increased entropy, along with a tendency to achieve a lower potential energy, is responsible for spontaneous chemical reactions. Reactions that are exothermic and whose products are more disordered (higher in entropy) will occur spontaneously, whereas endothermic reactions producing products of lower entropy will not be spontaneous. If they are to take place at all, they will need some energy input.

Free Energy

Learning Goal

The two situations described above are clear-cut and unambiguous. In any other situation the reaction may or may not be spontaneous. It depends on the relative size of the enthalpy and entropy values.

Free energy, symbolized by $\Delta G°$, represents the combined contribution of the enthalpy *and* entropy values for a chemical reaction. Thus free energy is the ultimate predictor of reaction spontaneity and is expressed as

$$\Delta G° = \Delta H° - T\Delta S°$$

$\Delta H°$ represents the change in enthalpy between products and reactants, $\Delta S°$ represents the change in entropy between products and reactants, and T is the Kelvin temperature of the reaction. A reaction with a negative value $\Delta G°$ will *always* be spontaneous. Reactions with a positive $\Delta G°$ will *always* be nonspontaneous.

EXAMPLE 8.1 ***Determining Whether a Process is Exothermic or Endothermic***

An ice cube is dropped into a glass of water at room temperature. The ice cube melts. Is the melting of the ice exothermic or endothermic?

Continued—

A HUMAN PERSPECTIVE

Triboluminescence: Sparks in the Dark with Candy

Generations of children have inadvertently discovered the phenomenon of triboluminescence. Crushing a wintergreen candy (Lifesavers) with the teeth in a dark room (in front of a few friends or a mirror) or simply rubbing two pieces of candy together may produce the effect—transient sparks of light!

Triboluminescence is simply the production of light upon fracturing a solid. It is easily observed and straightforward to describe but difficult to explain. It is believed to result from charge separation produced by the disruption of a crystal lattice. The charge separation has a very short lifetime. When the charge distribution returns to equilibrium, energy is released, and that energy is the light that is observed.

Dr. Linda M. Sweeting and several other groups of scientists are trying to reproduce these events under controlled circumstances. Crystals similar to the sugars in wintergreen candy are prepared with a very high level of purity. Some theories attribute the light emission to impurities in a crystal rather than to the crystal itself. Devices have been constructed that will crush the crystal with a uniform and reproducible force. Light-measuring devices, spectrophotometers, accurately measure the various wavelengths of light and the intensity of the light at each wavelength.

Through the application of careful experimentation and measurement of light-emitting properties of a variety of related compounds, these scientists hope to develop a theory of light emission from fractured solids.

This is one more example of the scientific method improving our understanding of everyday occurrences.

Charles Schulz's "Peanuts" vision of triboluminescence.

EXAMPLE 8.1 —*Continued*

Solution

Consider the ice cube to be the system and the water, the surroundings. For the cube to melt, it must gain energy and its energy source must be the water. The heat flow is from surroundings to system. The system gains energy (+energy); hence, the melting process (physical change) is endothermic.

Question 8.1

Are the following processes exothermic or endothermic?

a. Fuel oil is burned in a furnace.
b. $C_6H_{12}O_6(s) \longrightarrow 2C_2H_5OH(l) + 2CO_2(g)$, $\Delta H° = -16$ kcal
c. $N_2O_5(g) + H_2O(l) \longrightarrow 2HNO_3(l) + 18.3$ kcal

Question 8.2

Are the following processes exothermic or endothermic?

a. When solid NaOH is dissolved in water, the solution gets hotter.
b. $S(s) + O_2(g) \longrightarrow SO_2(g)$, $\Delta H° = -71$ kcal
c. $N_2(g) + 2O_2(g) + 16.2\text{ kcal} \longrightarrow 2NO_2(g)$

8.2 Experimental Determination of Energy Change in Reactions

The measurement of heat energy changes in a chemical reaction is **calorimetry.** This technique involves the measurement of the change in the temperature of a quantity of water or solution that is in contact with the reaction of interest and isolated from the surroundings. A device used for these measurements is a *calorimeter,* which measures heat changes in calories.

A Styrofoam coffee cup is a simple design for a calorimeter, and it produces surprisingly accurate results. It is a good insulator, and, when filled with solution, it can be used to measure temperature changes taking place as the result of a chemical reaction occurring in that solution (Figure 8.5). The change in the temperature of the solution, caused by the reaction, can be used to calculate the gain or loss of heat energy for the reaction.

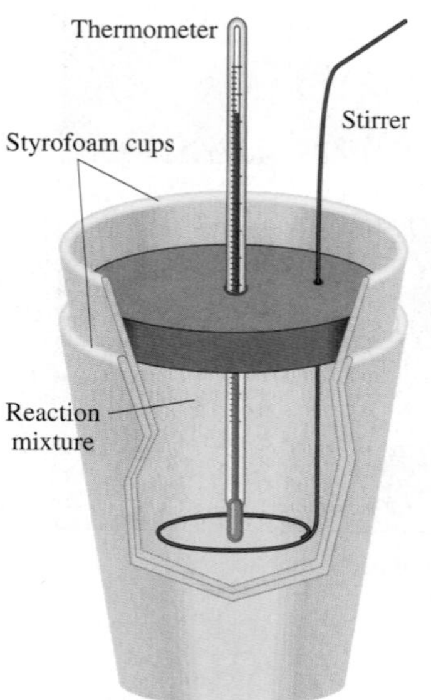

Figure 8.5
A "coffee cup" calorimeter used for the measurement of heat change in chemical reactions. The concentric Styrofoam cups insulate the system from its surroundings. Heat released by the chemical reaction enters the water, raising its temperature, which is measured by using a thermometer.

For an exothermic reaction, heat released by the reaction is absorbed by the surrounding solution. For an endothermic reaction, the reactants absorb heat from the solution.

The **specific heat** of a substance is defined as the number of calories of heat needed to raise the temperature of 1 g of the substance 1 degree Celsius. Knowing the specific heat of the water or the aqueous solution along with the total number of grams of solution and the temperature increase (measured as the difference between the final and initial temperatures of the solution), enables the experimenter to calculate the heat released during the reaction.

The solution behaves as a "trap" or "sink" for energy released in the exothermic process. The temperature increase indicates a gain in heat energy. Endothermic reactions, on the other hand, take heat energy away from the solution, lowering its temperature.

The quantity of heat absorbed or released by the reaction (Q) is the product of the mass of solution in the calorimeter (m_s), the specific heat of the solution (SH_s), and the change in temperature (ΔT_s) of the solution as the reaction proceeds from the initial to final state.

The heat is calculated by using the following equation:

$$Q = m_s \times \Delta T_s \times SH_s$$

with units

$$\text{calories} = \cancel{\text{gram}} \times \cancel{^\circ\text{C}} \times \frac{\text{calories}}{\cancel{\text{gram-}^\circ\text{C}}}$$

The details of the experimental approach are illustrated in Example 8.2.

EXAMPLE 8.2 ***Calculating Energy Involved in Calorimeter Reactions***

If 0.050 mol of hydrochloric acid (HCl) is mixed with 0.050 mol of sodium hydroxide (NaOH) in a "coffee cup" calorimeter, the temperature of 1.00×10^2 g of the resulting solution increases from 25.0°C to 31.5°C. If the specific heat of the solution is 1.00 cal/g H_2O °C, calculate the quantity of energy involved in the reaction. Also, is the reaction endothermic or exothermic?

Solution

The change in temperature is

$$\Delta T_s = T_{s\,\text{final}} - T_{s\,\text{initial}}$$
$$= 31.5^\circ\text{C} - 25.0^\circ\text{C} = 6.5^\circ\text{C}$$

Continued—

EXAMPLE 8.2 —*Continued*

$$Q = m_s \times \Delta T_s \times SH_s$$

$$Q = 1.00 \times 10^2 \cancel{\text{g solution}} \times 6.5\cancel{^\circ\text{C}} \times \frac{1.00 \text{ cal}}{\cancel{\text{g solution}}\ \cancel{^\circ\text{C}}}$$

$$= 6.5 \times 10^2 \text{ cal}$$

6.5×10^2 cal (or 0.65 kcal) of heat energy were released by this acid-base reaction to the surroundings, the solution; the reaction is exothermic.

EXAMPLE 8.3

Calculating Energy Involved in Calorimeter Reactions

If 0.10 mol of ammonium chloride (NH_4Cl) is dissolved in water producing 1.00×10^2 g solution, the water temperature decreases from 25.0°C to 18.0°C. If the specific heat of the resulting solution is 1.00 cal/g-°C, calculate the quantity of energy involved in the process. Also, is the dissolution of ammonium chloride endothermic or exothermic?

Solution

The change in temperature is

$$\Delta T = T_{s \text{ final}} - T_{s \text{ initial}}$$

$$= 18.0^\circ\text{C} - 25.0^\circ\text{C} = -7.0^\circ\text{C}$$

$$Q = m_s \times \Delta T_s \times SH_s$$

$$Q = 1.00 \times 10^2 \cancel{\text{g solution}} \times (-7.0\cancel{^\circ\text{C}}) \times \frac{1.00 \text{ cal}}{\cancel{\text{g solution}}\ \cancel{^\circ\text{C}}}$$

$$= -7.0 \times 10^2 \text{ cal}$$

7.0×10^2 cal (or 0.70 kcal) of heat energy were absorbed by the dissolution process because the solution lost (− sign) 7.0×10^2 cal of heat energy to the system. The reaction is endothermic.

Question 8.3

Refer to Example 8.2 and calculate the temperature change that would have been observed if 50.0 g solution were in the calorimeter instead of 1.00×10^2 g solution.

Question 8.4

Refer to Example 8.2 and calculate the temperature change that would have been observed if 1.00×10^2 g of another liquid, producing a solution with a specific heat of 0.800 cal/g-°C, was substituted for the water in the calorimeter.

Question 8.5

Convert the energy released in Example 8.2 to joules (recall the conversion factor for calories and joules, Chapter 1).

Question 8.6

Convert the energy absorbed in Example 8.3 to joules (recall the conversion factor for calories and joules, Chapter 1).

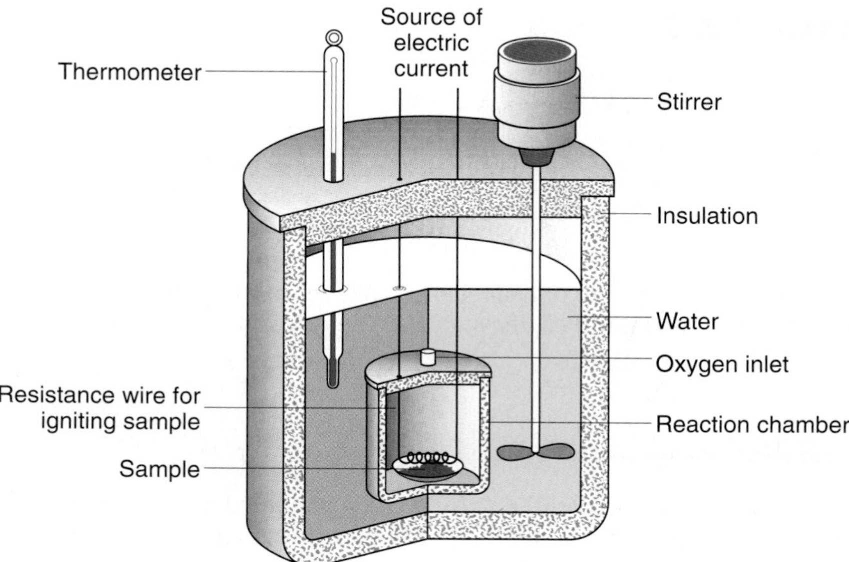

Figure 8.6
A bomb calorimeter that may be used to measure heat released upon combustion of a sample. This device is commonly used to determine the fuel value of foods. The bomb calorimeter is similar to the "coffee cup" calorimeter. However, note the electrical component necessary to initiate the combustion reaction.

Note: Refer to A Human Perspective: Food Calories, Section 1.6.

Many chemical reactions that produce heat are combustion reactions. In our bodies many food substances (carbohydrates, proteins, and fats, Chapters 22 and 23) are oxidized to release energy. **Fuel value** is the amount of energy per gram of food.

Learning Goal

The fuel value of food is an important concept in nutrition science. The fuel value is generally reported in units of *nutritional Calories.* One **nutritional Calorie** is equivalent to one kilocalorie (1000 calories). It is also known as the *large Calorie* (uppercase *C*).

Energy necessary for our daily activity and bodily function comes largely from the "combustion" of carbohydrates. Chemical energy from foods that is not used to maintain normal body temperature or in muscular activity is stored as fat. Thus "high-calorie" foods are implicated in obesity.

A special type of calorimeter, a *bomb calorimeter,* is useful for the measurement of the fuel value (Calories) of foods. Such a device is illustrated in Figure 8.6. Its design is similar, in principle, to that of the "coffee cup" calorimeter discussed earlier. It incorporates the insulation from the surroundings, solution pool, reaction chamber, and thermometer. Oxygen gas is added as one of the reactants, and an electrical igniter is inserted to initiate the reaction. However, it is not open to the atmosphere. In the sealed container the reaction continues until the sample is completely oxidized. All of the heat energy released during the reaction is captured in the water.

EXAMPLE 8.4 ***Calculating the Fuel Value of Foods***

One gram of glucose (a common sugar or carbohydrate) was burned in a bomb calorimeter. The temperature of 1.00×10^3 g H_2O was raised from 25.0°C to 28.8°C ($\Delta T_w = 3.8°C$). Calculate the fuel value of glucose.

Solution

Recall that the fuel value is the number of nutritional Calories liberated by the combustion of 1 g of material and 1 g of material was burned in the calorimeter. Then

Continued—

EXAMPLE 8.4 —*Continued*

$$\text{Fuel value} = Q = m_w \times \Delta T_w \times SH_w$$

Water is the surroundings in the calorimeter; it has a specific heat equal to 1.00 cal/g H_2O °C.

$$\text{Fuel value} = Q = \text{g } H_2O \times {}^\circ C \times \frac{1.00 \text{ cal}}{\text{g } H_2O \text{ }{}^\circ C}$$

$$= 1.00 \times 10^3 \cancel{\text{g } H_2O} \times 3.8 \cancel{{}^\circ C} \times \frac{1.00 \text{ cal}}{\cancel{\text{g } H_2O \text{ }{}^\circ C}}$$

$$= 3.8 \times 10^3 \text{ cal}$$

and

$$3.8 \times 10^3 \cancel{\text{cal}} \times \frac{1 \text{ nutritional Calorie}}{10^3 \cancel{\text{cal}}} = 3.8 \text{ C (nutritional Calories, or kcal)}$$

The fuel value of glucose is 3.8 kcal/g.

Question 8.7

A 1.0-g sample of a candy bar (which contains lots of sugar!) was burned in a bomb calorimeter. A 3.0°C temperature increase was observed for 1.00×10^3 g of water. The entire candy bar weighed 2.5 ounces. Calculate the fuel value (in nutritional Calories) of the sample and the total caloric content of the candy bar.

Question 8.8

If the fuel value of 1.00 g of a certain carbohydrate (sugar) is 3.00 nutritional Calories, how many grams of water must be present in the calorimeter to record a 5.00°C change in temperature?

8.3 Kinetics

Learning Goal

The first two laws of thermodynamics help us to decide whether a chemical reaction will take place. Knowing that a reaction can occur tells us nothing about the time that it may take.

Chemical **kinetics** is the study of the **rate** (or speed) **of chemical reactions.** Kinetics also gives an indication of the *mechanism* of a reaction, a step-by-step description of how reactants become products. Kinetic information may be represented as the *disappearance* of reactants or *appearance* of product over time. A typical graph of concentration versus time is shown in Figure 8.7.

Information about the rate at which various chemical processes occur is useful. For example, what is the "shelf life" of processed foods? When will slow changes in composition make food unappealing or even unsafe? Many drugs lose their potency with time because the active ingredient decomposes into other substances. The rate of hardening of dental filling material (via a chemical reaction) influences the dentist's technique. Our very lives depend on the efficient transport of oxygen to each of our cells and the rapid use of the oxygen for energy-harvesting reactions.

The diagram in Figure 8.8 is a useful way of depicting the kinetics of a reaction at the molecular level.

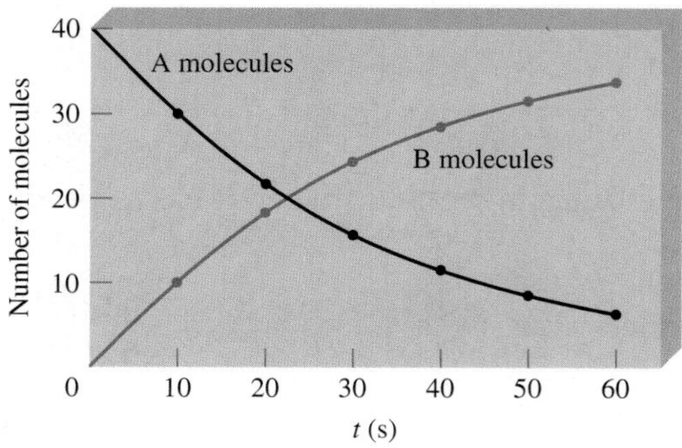

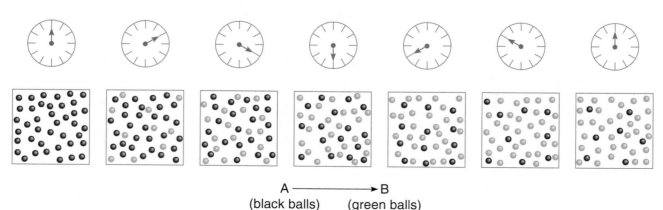

Figure 8.7
For a hypothetical reaction $A \longrightarrow B$ the concentration of *A* molecules (reactant molecules) decreases over time and *B* molecules (product molecules) increase in concentration over time.

Often a color change, over time, can be measured. Such changes are useful in assessing the rate of a chemical reaction (Figure 8.9).

Let's see what actually happens when two chemical compounds "react" and what experimental conditions affect the rate of a reaction.

The Chemical Reaction

Consider the exothermic reaction that we discussed in Section 8.1:

$$CH_4(g) + 2O_2(g) \longrightarrow CO_2(g) + 2H_2O(l) + \text{211 kcal}$$

For the reaction to proceed, C—H and O—O bonds must be broken, and C—O and H—O bonds must be formed. Sufficient energy must be available to cause the bonds to break if the reaction is to take place. This energy is provided by the collision of molecules. If sufficient energy is available at the temperature of the reaction, one or more bonds will break, and the atoms will recombine in a lower energy arrangement, in this case as carbon dioxide and water. A collision producing product molecules is termed an *effective collision*. Only effective collisions lead to chemical reaction.

Activation Energy and the Activated Complex

Learning Goal

The minimum amount of energy required to produce a chemical reaction is called the **activation energy** for the reaction.

We can picture the chemical reaction in terms of the changes in potential energy that occur during the reaction. Figure 8.10a graphically shows these changes for an exothermic reaction. Important characteristics of this graph include the following:

- The reaction proceeds from reactants to products through an extremely unstable state that we call the **activated complex.** The activated complex cannot be isolated from the reaction mixture but may be thought of as a

Figure 8.8
An alternate way of representing the information contained in Figure 8.7.

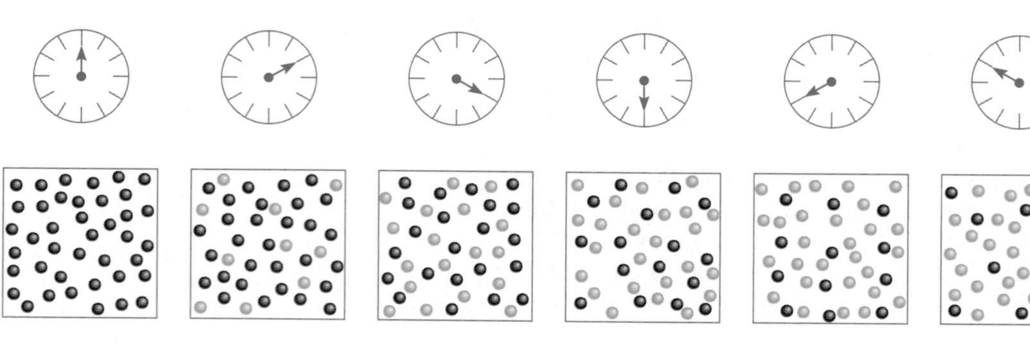

Figure 8.9
The conversion of reddish brown Br_2 in solution to colorless Br^- over time.

short-lived group of atoms structured in such a way that it quickly and easily breaks apart into the products of the reaction.

- Formation of the activated complex requires energy. The difference between the energy of reactants and that of the activated complex is the activation energy. This energy must be provided by the collision of the reacting molecules or atoms at the temperature of the reaction.
- Because this is an exothermic reaction, the overall energy change must be a *net* release of energy. The *net* release of energy is the difference in energy between products and reactants.

For an endothermic reaction, such as the decomposition of water,

$$\text{energy} + 2H_2O(l) \longrightarrow 2H_2(g) + O_2(g)$$

the change of potential energy with reaction time is shown in Figure 8.10b.

This reaction will take place when an electrical current is passed through water. The process is called *electrolysis*.

Liquid water is *stable* because the decomposition products are less stable (higher energy) than the reactant. Furthermore, the reaction takes place slowly because of the large activation energy required for the conversion of water into the elements hydrogen and oxygen.

Question 8.9

The act of striking a match illustrates the role of activation energy in a chemical reaction. Explain.

Question 8.10

Distinguish between the terms *net energy* and *activation energy*.

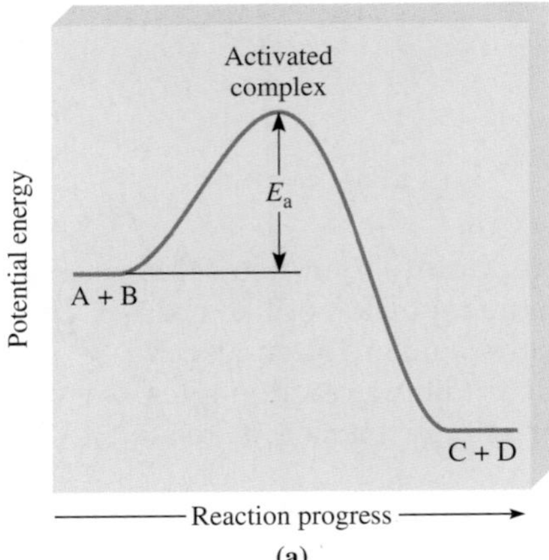

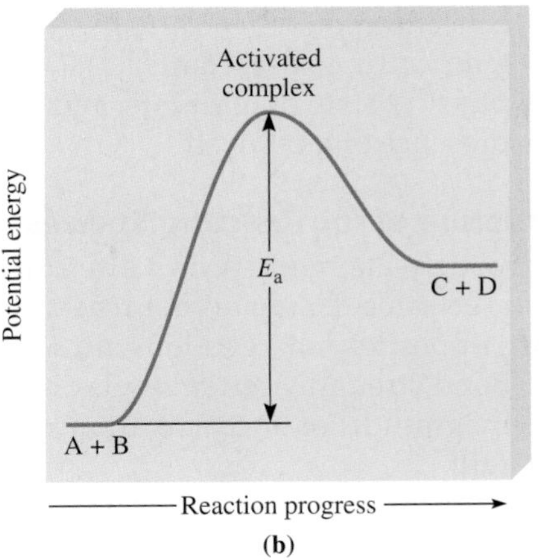

Figure 8.10
(a) The change in potential energy as a function of reaction time for an exothermic chemical reaction. Note particularly the energy barrier associated with the formation of the activated complex. This energy barrier (E_a) *is* the activation energy. (b) The change in potential energy as a function of reaction time for an endothermic chemical reaction. In contrast to the exothermic reaction in (a), the energy of the products is greater than the energy of the reactants.

A Clinical Perspective

Hot and Cold Packs

Hot packs provide "instant warmth" for hikers and skiers and are used in treatment of injuries such as pulled muscles. Cold packs are in common use today for the treatment of injuries and the reduction of swelling.

These useful items are an excellent example of basic science producing a technologically useful product. (Recall our discussion in Chapter 1 of the relationship of science and technology.)

Both hot and cold packs depend on large energy changes taking place during a chemical reaction. Cold packs rely on an endothermic reaction, and hot packs generate heat energy from an exothermic reaction.

A cold pack is fabricated as two separate compartments within a single package. One compartment contains NH_4NO_3, and the other contains water. When the package is squeezed, the inner boundary between the two compartments ruptures, allowing the components to mix, and the following reaction occurs:

$$6.7 \text{ kcal/mol} + NH_4NO_3(s) \longrightarrow NH_4^+(aq) + NO_3^-(aq)$$

This reaction is endothermic; heat taken from the surroundings produces the cooling effect.

The design of a hot pack is similar. Here, finely divided iron powder is mixed with oxygen. Production of iron oxide results in the evolution of heat:

$$4Fe + 3O_2 \longrightarrow 2Fe_2O_3 + 198 \text{ kcal/mol}$$

This reaction occurs via an oxidation-reduction mechanism (see Chapter 9). The iron atoms are oxidized, O_2 is reduced. Electrons are transferred from the iron atoms to O_2 and Fe_2O_3 forms exothermically. The rate of the reaction is slow; therefore the heat is liberated gradually over a period of several hours.

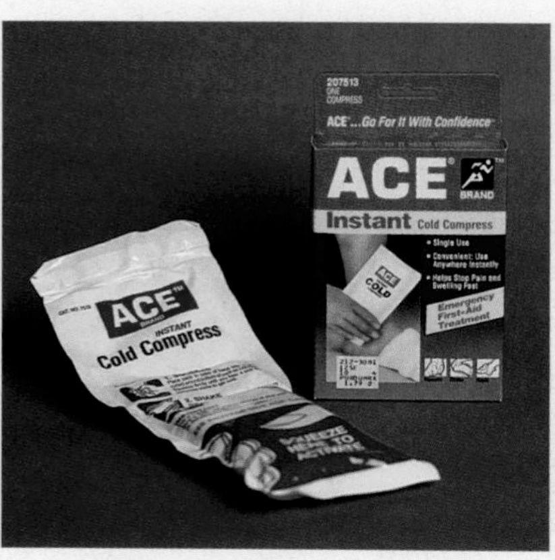

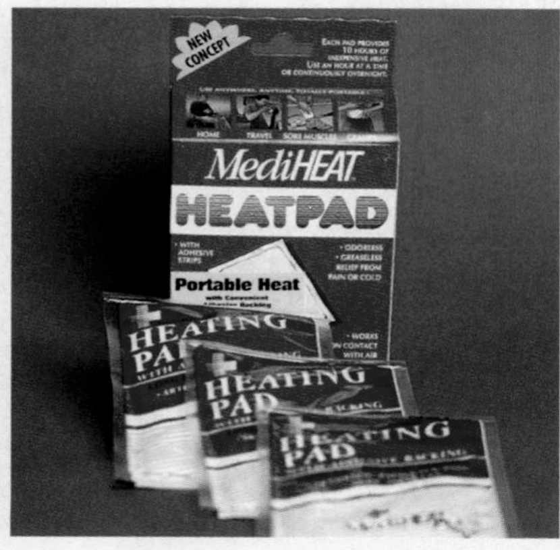

Factors That Affect Reaction Rate

Learning Goal

Five major factors influence reaction rate:

- structure of the reacting species,
- concentration of reactants,
- temperature of reactants,
- physical state of reactants, and
- presence of a catalyst.

Structure of the Reacting Species

Oppositely charged species often react more rapidly than neutral species. Ions with the same charge do not react, owing to the repulsion of like charges. In contrast, oppositely charged ions attract one another and are often reactive.

Bond strengths certainly play a role in determining reaction rates as well, for the magnitude of the activation energy, or energy barrier, is related to bond strength.

The size and shape of reactant molecules influence the rate of the reaction. Large molecules, containing bulky groups of atoms, may block the reactive part of the molecule from interacting with another reactive substance, causing the reaction to proceed slowly.

The Concentration of Reactants

The rate of a chemical reaction is often a complex function of the concentration of one or more of the reacting substances. The rate will generally *increase* as concentration *increases* simply because a higher concentration means more reactant molecules in a given volume and therefore a greater number of collisions per unit time. If we assume that other variables are held constant, a larger number of collisions leads to a larger number of effective collisions. The explosion (very fast exothermic reaction) of gunpowder is a dramatic example of a rapid rate at high reactant concentration.

Concentration is introduced in Section 1.6, and units and calculations are discussed in Sections 7.4 and 7.5.

The Temperature of Reactants

The rate of a reaction *increases* as the temperature increases, because the kinetic energy of the reacting particles is directly proportional to the Kelvin temperature. Increasing the speed of particles increases the likelihood of collision, and the higher kinetic energy means that a higher percentage of these collisions will result in product formation (effective collisions). A 10°C rise in temperature has often been found to double the reaction rate.

The Physical State of Reactants

The rate of a reaction depends on the physical state of the reactants: solid, liquid, or gas. For a reaction to occur the reactants must collide frequently and have sufficient energy to react. In the solid state, the atoms, ions, or compounds are restricted in their motion. In the gaseous state, the particles are free to move, but the spacing between particles is so great that collisions are relatively infrequent. In the liquid state the particles have both free motion and proximity to each other. Hence reactions tend to be fastest in the liquid state and slowest in the solid state.

These factors were considered in our discussion of the states of matter (Chapter 6).

The Presence of a Catalyst

A **catalyst** is a substance that *increases* the reaction rate. If added to a reaction mixture, the catalytic substance undergoes no net change, nor does it alter the outcome of the reaction. However, the catalyst interacts with the reactants to create an alternative pathway for production of products. This alternative path has a lower activation energy. This makes it easier for the reaction to take place and thus increases the rate. This effect is illustrated in Figure 8.11.

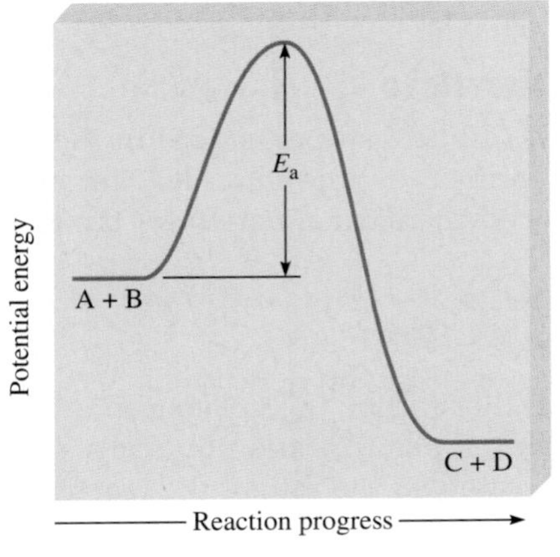

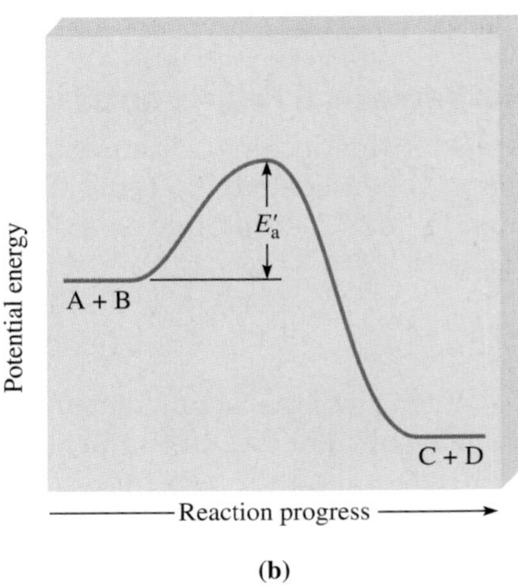

Figure 8.11
The effect of a catalyst on the magnitude of the activation energy of a chemical reaction. Note that the presence of a catalyst decreases the activation energy ($E'_a < E_a$), thus increasing the rate of the reaction.

Figure 8.12
The rapid reaction of magnesium metal and oxygen (in air) is a graphic example of a highly exothermic reaction.

Sections 12.4 and 18.2 describe the role of catalysis in organic reactions.

Catalysis is important industrially; it may often make the difference between profit and loss in the sale of a product. For example, catalysis is useful in converting double bonds to single bonds. An important application of this principle involves the process of hydrogenation. Hydrogenation converts one or more of the carbon-carbon double bonds of unsaturated fats (e.g., corn oil, olive oil) to single bonds characteristic of saturated fats (such as margarine). The use of a metal catalyst, such as nickel, in contact with the reaction mixture dramatically increases the rate of the reaction.

Sections 20.1 through 20.6 describe enzyme catalysis.

Thousands of essential biochemical reactions in our bodies are controlled and speeded up by biological catalysts called *enzymes*.

An example of a chemical reaction that is exothermic *and* very fast is shown in Figure 8.12. A molecular level view of the action of a solid catalyst widely used in industrial synthesis is presented in Figure 8.13.

Question 8.11

Would you imagine that a substance might act as a poison if it interfered with the function of an enzyme? Why?

Question 8.12

Bacterial growth decreases markedly in a refrigerator. Why?

Mathematical Representation of Reaction Rate

Learning Goal 7

Consider the decomposition reaction of N_2O_5 (dinitrogen pentoxide) in the gas phase. When heated, N_2O_5 decomposes and forms two products: NO_2 (nitrogen dioxide) and O_2 (diatomic oxygen). The balanced chemical equation for the reaction is

$$2N_2O_5(g) \xrightarrow{\Delta} 4NO_2(g) + O_2(g)$$

When all of the factors that affect the rate of the reaction (except concentration) are held constant (i.e., the nature of the reactant, temperature and physical state of the reactant, and the presence or absence of a catalyst) the rate of the reaction is proportional to the concentration of N_2O_5.

$$\text{rate} \propto \text{concentration } N_2O_5$$

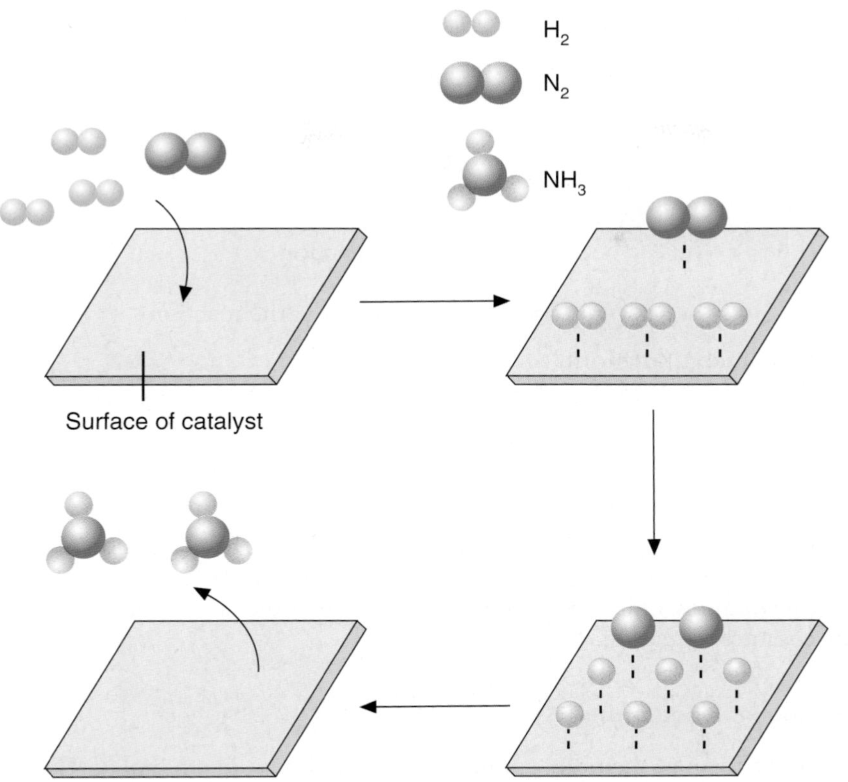

Figure 8.13
The synthesis of ammonia, an important industrial product, is facilitated by a solid phase catalyst (the Haber process). H_2 and N_2 bind to the surface, their bonds are weakened, dissociation and re-formation as ammonia occur, and the newly formed ammonia molecules leave the surface. This process is repeated over and over, with no change in the catalyst.

We will represent the concentration of N_2O_5 in units of molarity and represent molar concentration using brackets.

$$\text{concentration } N_2O_5 = [N_2O_5]$$

Then,

$$\text{rate} \propto [N_2O_5]$$

Laboratory measurement shows that the rate of the reaction depends on the molar concentration raised to an experimentally determined exponent that we will symbolize as n

$$\text{rate} \propto [N_2O_5]^n$$

In expressions such as the one shown, the proportionality symbol, $\propto$, may be replaced by an equality sign and a proportionality constant that we represent as k, the **rate constant.**

$$\text{rate} = k[N_2O_5]^n$$

The exponent, n, is the **order of the reaction.** For the reaction described here, which has been studied in great detail, n is numerically equal to 1, hence the reaction is first order in N_2O_5 and the rate equation for the reaction is:

$$\text{rate} = k[N_2O_5]$$

Equations that follow this format, the rate being equal to the rate constant multiplied by the reactant concentration raised to an exponent that is the order, are termed **rate equations.**

Note that the exponent, n, in the rate equation is not the same as the coefficient of N_2O_5 in the balanced equation. However, in many elementary reactions the coefficient in the balanced equation and the exponent n (the order of the reaction) are numerically the same.

In general, the rate of reaction for an equation of the general form:

$$A \longrightarrow product$$

is

$$rate = k[A]^n$$

in which

n = order of the reaction

k = the rate constant of the reaction

An equation of the form

$$A + B \longrightarrow products$$

has a rate expression

$$rate = k[A]^n[B]^{n'}$$

Both the value of the rate constant and the order of the reaction are deduced from a series of experiments. We cannot predict them by simply looking at the chemical equation. Only the *form* of the rate expression can be found by inspection of the chemical equation.

EXAMPLE 8.5 Writing Rate Equations

Write the form of the rate equation for the oxidation of ethanol (C_2H_5OH). The reaction has been experimentally determined to be first order in ethanol and third order in oxygen (O_2).

Solution

The rate expression involves only the reactants, C_2H_5OH and O_2. Depict their concentration as

$$[C_2H_5OH][O_2]$$

and raise each to an exponent corresponding to its experimentally determined order

$$[C_2H_5OH][O_2]^3$$

and this is proportional to the rate:

$$rate \propto [C_2H_5OH][O_2]^3$$

or

$$rate = k[C_2H_5OH][O_2]^3$$

is the rate expression. (Remember that 1 is understood as an exponent; $[C_2H_5OH]$ is correct and $[C_2H_5OH]^1$ is not.)

Question 8.13

Write the general form of the rate equation for each of the following processes.

a. $N_2(g) + O_2(g) \longrightarrow 2NO(g)$
b. $2C_4H_6(g) \longrightarrow C_8H_{12}(g)$

Question 8.14

Write the general form of the rate equation for each of the following processes.

a. $CH_4(g) + 2O_2(g) \longrightarrow CO_2(g) + 2H_2O(g)$
b. $2NO_2(g) \longrightarrow 2NO(g) + O_2(g)$

Knowledge of the form of the rate equation, coupled with the experimental determination of the value of the rate constant, k, and the order, n, are valuable in a number of ways. Industrial chemists use this information to establish optimum conditions for preparing a product in the shortest practical time. The design of an entire manufacturing facility may, in part, depend on the rates of the critical reactions.

In Section 8.4 we will see how the rate equation forms the basis for describing equilibrium reactions.

8.4 Equilibrium

Rate and Reversibility of Reactions

Learning Goal

We have assumed that most chemical and physical changes considered thus far proceed to completion. A complete reaction is one in which all reactants have been converted to products. However, many important chemical reactions do not go to completion. As a result, after no further obvious change is taking place, measurable quantities of reactants and products remain. Reactions of this type (incomplete reactions) are called **equilibrium reactions.**

Examples of physical and chemical equilibria abound in nature. Many environmental systems depend on fragile equilibria. The amount of oxygen dissolved in a certain volume of lake water (the oxygen concentration) is governed by the principles of equilibrium. The lives of plants and animals within this system are critically related to the levels of dissolved oxygen in the water.

The very form and function of the earth is a consequence of a variety of complex equilibria. Stalactite and stalagmite formations in caves are made up of solid calcium carbonate ($CaCO_3$). They owe their existence to an equilibrium process described by the following equation:

$$Ca^{2+}(aq) + 2HCO_3^-(aq) \rightleftharpoons CaCO_3(s) + CO_2(aq) + H_2O(l)$$

Physical Equilibrium

A physical equilibrium, such as sugar dissolving in water, is a reversible reaction. A **reversible reaction** is a process that can occur in both directions. It is indicated by using a double arrow ($\rightleftharpoons$) symbol.

Dissolution of sugar in water is a convenient illustration of a state of *dynamic equilibrium.*

A **dynamic equilibrium** is a situation in which the rate of the forward process in a reversible reaction is exactly balanced by the rate of the reverse process.

Let's now look at the sugar and water equilibrium in more detail.

Sugar in Water

Imagine that you mix a small amount of sugar (2 or 3g) in 100 mL of water. After you have stirred it for a short time, all of the sugar dissolves; there is no residual solid sugar because the sugar has dissolved *completely.* The reaction clearly has converted all solid sugar to its dissolved state, an aqueous solution of sugar, or

$$\text{sugar}(s) \longrightarrow \text{sugar}(aq)$$

Now, suppose that you add a very large amount of sugar (100 g), more than can possibly dissolve, to the same volume of water. As you stir the mixture you observe more and more sugar dissolving. After some time the amount of solid sugar remaining in contact with the solution appears constant. Over time, you observe no further change in the amount of dissolved sugar. At this point, although nothing further appears to be happening, in reality a great deal of activity is taking place!

An equilibrium situation has been established. Over time the amount of sugar dissolved in the measured volume of water (the concentration of sugar in water) does not change. Hence the amount of undissolved sugar remains the same. However, if you could look at the individual sugar molecules, you would see something quite amazing. Rather than sugar molecules in the solid simply staying in place, you would see them continuing to leave the solid state and go into solution. At the same time, a like number of dissolved sugar molecules would leave the water and form more solid. This active process is described as a *dynamic equilibrium.* The reaction is proceeding in a forward (left to right) and a reverse (right to left) direction at the same time and is a reversible reaction:

$$\text{sugar}(s) \rightleftharpoons \text{sugar}(aq)$$

The double arrow serves as

- an indicator of a reversible process,
- an indicator of an equilibrium process, and
- a reminder of the dynamic nature of the process.

How can we rationalize the apparent contradiction: continuous change is taking place yet no observable change in the amount of sugar in either the solid or dissolved form is observed.

The only possible explanation is that the rate of the forward process

$$\text{sugar}(s) \longrightarrow \text{sugar}(aq)$$

must be equal to the rate of the reverse process

$$\text{sugar}(s) \longleftarrow \text{sugar}(aq)$$

Under this condition, the number of sugar molecules leaving the solid in a given time interval is identical to the number of sugar molecules returning to the solid state.

If we use symbols:

$$\text{rate}_f = \text{forward rate}$$

$$\text{rate}_r = \text{reverse rate}$$

then, at equilibrium,

$$\text{rate}_f = \text{rate}_r$$

For our sugar example, the rate expression for the forward reaction is

$$\text{rate}_f = k_f[\text{sugar}(s)]$$

and the rate expression for the reverse reaction is

$$\text{rate}_r = k_r[\text{sugar}(aq)]$$

At equilibrium

$$\text{rate}_f = \text{rate}_r$$

$$k_f[\text{sugar}(s)] = k_r[\text{sugar}(aq)]$$

Appendix A reviews the mathematical steps used here.

Rearranging this equation by first dividing both sides of the equation by k_r,

$$\frac{k_f}{k_r}[\text{sugar}(s)] = [\text{sugar}(aq)]$$

and then dividing both sides by [sugar(*s*)] yields

$$\frac{k_f}{k_r} = \frac{[\text{sugar}(aq)]}{[\text{sugar}(s)]}$$

The ratio of the two rate constants is itself a constant, the **equilibrium constant,** represented by K_{eq}

$$\frac{k_f}{k_r} = K_{eq}$$

and

$$K_{eq} = \frac{[\text{sugar}(aq)]}{[\text{sugar}(s)]}$$

The concentration of a solid is constant, no matter how much solid is present. Consequently, in actual practice, the equation on the left is reduced to:

$$K_{eq} = [\text{sugar(aq)}]$$

See page 223

We can measure concentrations at equilibrium for any physical change, whether it is a simple case of sugar and water or a complex problem encountered in the purification of a drug product. These concentrations can be used to calculate a numerical value for the equilibrium constant. This number allows us to predict the behavior of systems and compare these systems to any other.

Question 8.15

Construct an example of a dynamic equilibrium using a subway car at rush hour.

Question 8.16

A certain change in reaction conditions for a process was found to increase the rate of the forward reaction much more than that of the reverse reaction. Did the equilibrium constant increase, decrease, or remain the same? Why?

Chemical Equilibrium

The Reaction of N_2 and H_2

When we mix nitrogen gas (N_2) and hydrogen gas (H_2) at an elevated temperature (perhaps 500°C), some of the molecules will collide with sufficient energy to break N—N and H—H bonds. Rearrangement of the atoms will produce the product (NH_3):

$$N_2(g) + 3H_2(g) \rightleftharpoons 2NH_3(g)$$

Beginning with a mixture of hydrogen and nitrogen, the rate of the reaction is initially rapid, because the reactant concentration is high; as the reaction proceeds, the concentration of reactants decreases. At the same time the concentration of the product, ammonia, is increasing. At equilibrium the *rate of depletion* of hydrogen and nitrogen *is equal to* the *rate of depletion* of ammonia. In other words, *the rates of the forward and reverse reactions are equal.*

Dynamic equilibrium can be particularly dangerous for living cells because it represents a situation in which nothing is getting done. There is no gain. Let's consider an exothermic reaction designed to produce a net gain of energy for the cell. In a dynamic equilibrium the rate of the forward (energy-releasing) reaction is equal to the rate of the backward (energy-requiring) reaction. Thus there is no net gain of energy to fuel cellular activity, and the cell will die.

The concentration of the various species is fixed at equilibrium because product is being *consumed and formed at the same rate.* In other words, the reaction continues indefinitely (dynamic), but the concentration of products and reactants is fixed (equilibrium). This is a *dynamic equilibrium.* The composition of this reaction mixture as a function of time is depicted in Figure 8.14.

The equilibrium constant expression can be obtained in the following way:

$$N_2(g) + 3H_2(g) \rightleftharpoons 2NH_3(g)$$

We may subdivide this expression into two reactions: a forward reaction,

$$N_2(g) + 3H_2(g) \longrightarrow 2NH_3(g)$$

and a reverse reaction,

$$2NH_3(g) \longrightarrow N_2(g) + 3H_2(g)$$

The rate expressions for these reactions are

$$\text{rate}_f = k_f[N_2]^n[H_2]^{n'}$$

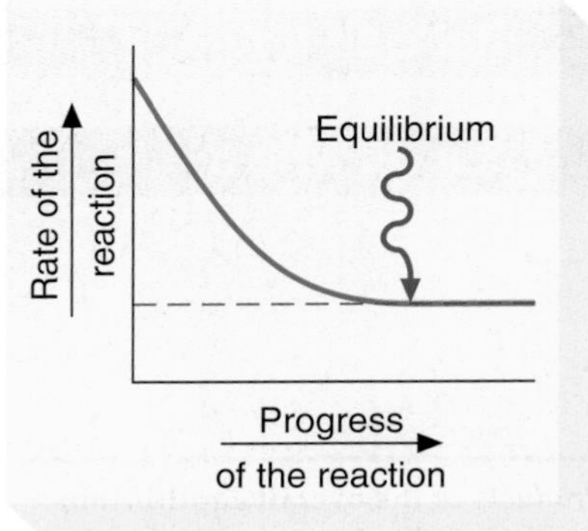

Figure 8.14
The change of the rate of reaction as a function of time. The rate of reaction, initially rapid, decreases as the concentration of reactant decreases and approaches a limiting value at equilibrium.

and

$$\text{rate}_r = k_r[\text{NH}_3]^{n''}$$

At equilibrium, the forward and reverse rates become equal:

$$rate_f = rate_r$$

Consequently,

$$k_f[\text{N}_2]^n[\text{H}_2]^{n'} = k_r[\text{NH}_3]^{n''}$$

Rearranging this equation as we have previously

$$\frac{k_f}{k_r} = \frac{[\text{NH}_3]^{n''}}{[\text{N}_2]^n[\text{H}_2]^{n'}}$$

and using the definition

$$\frac{k_f}{k_r} = K_{eq}$$

yields the equilibrium-constant expression

$$K_{eq} = \frac{[\text{NH}_3]^{n''}}{[\text{N}_2]^n[\text{H}_2]^{n'}}$$

As noted earlier, the exponents n, n', and n'' are experimentally determined. However, for elementary, single-step reactions, the exponents in the rate expression are numerically equal to the coefficients in the balanced chemical equation. Consequently, for this situation, and all other equilibrium situations that we shall encounter in this book, we shall assume that the exponents in the rate expressions are equal to the coefficients in the balanced chemical equation. Hence,

$$K_{eq} = \frac{[\text{NH}_3]^2}{[\text{N}_2][\text{H}_2]^3}$$

It does not matter what initial amounts (concentrations) of reactants or products we choose. When the system reaches equilibrium, the calculated value of K_{eq} will not change. The magnitude of K_{eq} can be altered only by changing the temperature. Thus K_{eq} is temperature dependent. The chemical industry uses this fact to advantage by choosing a reaction temperature that will maximize the yield of a desired product.

Question 8.17

How could one determine when a reaction has reached equilibrium?

Question 8.18

Does the attainment of equilibrium imply that no further change is taking place in the system?

The Generalized Equilibrium-Constant Expression for a Chemical Reaction

Products **of the overall equilibrium reaction are in the numerator, and** ***reactants*** **are in the denominator.**

We write the general form of an equilibrium chemical reaction as

$$aA + bB \rightleftharpoons cC + dD$$

in which A and B represent reactants, C and D represent products, and a, b, c, and d are the coefficients of the balanced equation. The equilibrium constant expression for this general case is

[ ] represents molar concentration, *M*.

$$K_{eq} = \frac{[C]^c[D]^d}{[A]^a[B]^b}$$

Writing Equilibrium-Constant Expressions

Learning Goal

An equilibrium-constant expression can be written only after a correct, balanced chemical equation that describes the equilibrium system has been developed. A balanced equation is essential because the *coefficients* in the equation become the *exponents* in the equilibrium-constant expression.

The exponents correspond to the *coefficients* of the balanced equation.

Each chemical reaction has a unique equilibrium constant value at a specified temperature. Equilibrium constants listed in the chemical literature are often reported at 25°C, to allow comparison of one system with any other. For any equilibrium reaction, the value of the equilibrium constant changes with temperature.

The brackets represent molar concentration or molarity; recall that molarity has units of mol/L. Although the equilibrium constant may have units (owing to the units on each concentration term), by convention units are usually not used. In our discussion of equilibrium, all equilibrium constants are shown as *unitless.*

A properly written equilibrium-constant expression may not include all of the terms in the chemical equation upon which it is based. Only the concentration of gases and substances in solution are shown, because their concentrations can change. Concentration terms for pure liquids and solids are *not* shown. The concentration of a pure liquid is constant. A solid also has a fixed concentration and, for solution reactions, is not really a part of the solution. When a solid is formed it exists as a solid phase in contact with a liquid phase (the solution).

Writing an Equilibrium-Constant Expression — EXAMPLE 8.6

Write an equilibrium-constant expression for the reversible reaction:

$$H_2(g) + F_2(g) \rightleftharpoons 2HF(g)$$

Solution

Inspection of the chemical equation reveals that no solids or pure liquids are present. Hence all reactants and products appear in the equilibrium-constant expression:

The numerator term is the product term $[HF]^2$.

The denominator terms are the reactants $[H_2]$ and $[F_2]$.

Note that each term contains an exponent identical to the corresponding coefficient in the balanced equation. Arranging the numerator and denominator terms as a fraction and setting the fraction equal to K_{eq} yields

$$K_{eq} = \frac{[HF]^2}{[H_2][F_2]}$$

Writing an Equilibrium-Constant Expression — EXAMPLE 8.7

Write an equilibrium-constant expression for the reversible reaction:

$$MnO_2(s) + 4HCl(aq) \rightleftharpoons MnCl_2(aq) + Cl_2(g) + 2H_2O(l)$$

Solution

MnO_2 is a solid and H_2O is a pure liquid. Thus they are not written in the equilibrium-constant expression.

$$\boxed{MnO_2(s)} + 4HCl(aq) \rightleftharpoons MnCl_2(aq) + Cl_2(g) + \boxed{2H_2O(l)}$$

(arrows from $MnO_2(s)$ and $2H_2O(l)$:) Not a part of the K_{eq} expression

Continued—

EXAMPLE 8.7 —*Continued*

The numerator term includes the remaining products:

$$[MnCl_2] \quad \text{and} \quad [Cl_2]$$

The denominator term includes the remaining reactant:

$$[HCl]^4$$

Note that the exponent is identical to the corresponding coefficient in the chemical equation.

Arranging the numerator and denominator terms as a fraction and setting the fraction equal to K_{eq} yields

$$K_{eq} = \frac{[MnCl_2][Cl_2]}{[HCl]^4}$$

Question 8.19

Write an equilibrium-constant expression for each of the following reversible reactions.

a. $2NO_2(g) \rightleftharpoons N_2(g) + 2O_2(g)$
b. $2H_2O(l) \rightleftharpoons 2H_2(g) + O_2(g)$

Question 8.20

Write an equilibrium-constant expression for each of the following reversible reactions.

a. $2HI(g) \rightleftharpoons H_2(g) + I_2(g)$
b. $PCl_5(s) \rightleftharpoons PCl_3(l) + Cl_2(g)$

Interpreting Equilibrium Constants

What utility does the equilibrium constant have? The reversible arrow in the chemical equation alerts us to the fact that an equilibrium exists. Some measurable quantity of the product and reactant remain. However, there is no indication whether products predominate, reactants predominate, or significant concentrations of both products and reactants are present at equilibrium.

The numerical value of the equilibrium constant provides additional information. It tells us the extent to which reactants have converted to products. This is important information for anyone who wants to manufacture and sell the product. It also is important to anyone who studies the effect of equilibrium reactions on environmental systems and living organisms.

Although an absolute interpretation of the numerical value of the equilibrium constant depends on the form of the equilibrium-constant expression, the following generalizations are useful:

- K_{eq} greater than 1×10^2. A large numerical value of K_{eq} indicates that the numerator (product term) is much larger than the denominator (reactant term) and that at equilibrium mostly product is present.
- K_{eq} less than 1×10^{-2}. A small numerical value of K_{eq} indicates that the numerator (product term) is much smaller than the denominator (reactant term) and that at equilibrium mostly reactant is present.
- K_{eq} between 1×10^{-2} and 1×10^2. In this case the equilibrium mixture contains significant concentrations of both reactants and products.

Question 8.21

At a given temperature, the equilibrium constant for a certain reaction is 1×10^{20}. Does this equilibrium favor products or reactants? Why?

Question 8.22

At a given temperature, the equilibrium constant for a certain reaction is 1×10^{-18}. Does this equilibrium favor products or reactants? Why?

Learning Goal

Section 8.3 discusses rates of reaction and Section 7.5 describes molar concentration.

Calculating Equilibrium Constants

The magnitude of the equilibrium constant for a chemical reaction is determined experimentally. The reaction under study is allowed to proceed until the composition of products and reactants no longer changes (Figure 8.15). This may be a matter of seconds, minutes, hours, or even months or years, depending on the rate of the reaction. The reaction mixture is then analyzed to determine the molar concentration of each of the products and reactants. These concentrations are substituted in the equilibrium-constant expression and the equilibrium constant is calculated. The following example illustrates this process.

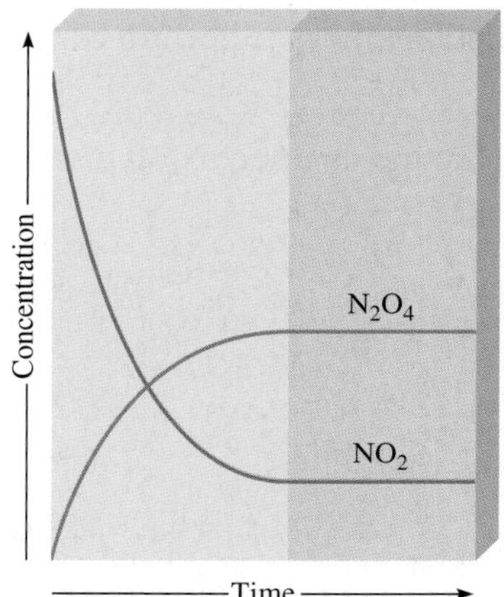

Figure 8.15
The combination reaction of NO_2 molecules produces N_2O_4. Initially, the concentration of reactant (NO_2) diminishes rapidly while the N_2O_4 concentration builds. Eventually, the concentrations of both reactant and product become constant over time (blue area). The equilibrium condition has been attained.

Calculating an Equilibrium Constant — EXAMPLE 8.8

Hydrogen iodide is placed in a sealed container and allowed to come to equilibrium. The equilibrium reaction is:

$$2HI(g) \rightleftharpoons H_2(g) + I_2(g)$$

and the equilibrium concentrations are:

$$[HI] = 0.54\,M$$

$$[H_2] = 1.72\,M$$

$$[I_2] = 1.72\,M$$

Calculate the equilibrium constant.

Solution

First, write the equilibrium-constant expression:

$$K_{eq} = \frac{[H_2][I_2]}{[HI]^2}$$

Then substitute the equilibrium concentrations of products and reactants to obtain

$$K_{eq} = \frac{[1.72][1.72]}{[0.54]^2} = \frac{2.96}{0.29}$$

$$= 10.1 \text{ or } 1.0 \times 10^1 \text{ (two significant figures)}$$

Question 8.23

A reaction chamber contains the following mixture at equilibrium:

$$[NH_3] = 0.25\ M$$

$$[N_2] = 0.11\ M$$

$$[H_2] = 1.91\ M$$

If the reaction is:

$$N_2(g) + 3H_2(g) \rightleftharpoons 2NH_3(g)$$

Calculate the equilibrium constant.

Question 8.24

A reaction chamber contains the following mixture at equilibrium:

$$[H_2] = 0.22\ M$$

$$[S_2] = 1.0 \times 10^{-6}\ M$$

$$[H_2S] = 0.80\ M$$

If the reaction is:

$$2H_2(g) + S_2(g) \rightleftharpoons 2H_2S(g)$$

Calculate the equilibrium constant.

LeChatelier's Principle

Learning Goal

In the nineteenth century the French chemist LeChatelier discovered that changes in equilibrium depend on the amount of "stress" applied to the system. The stress may take the form of an increase or decrease of the temperature of the system at equilibrium or perhaps a change in the amount of reactant or product present in a fixed volume (the concentration of reactant or product).

LeChatelier's principle states that if a stress is placed on a system at equilibrium, the system will respond by altering the equilibrium composition in such a way as to minimize the stress.

Consider the equilibrium situation discussed earlier:

$$N_2(g) + 3H_2(g) \rightleftharpoons 2NH_3(g)$$

If the reactants and products are present in a fixed volume (such as 1L) and more NH_3 (the *product*) is introduced into the container, the system will be stressed—the equilibrium will be disturbed. The system will try to alleviate the stress (as we all do) by *removing* as much of the added material as possible. How can it accomplish this? By converting some NH_3 to H_2 and N_2. The equilibrium shifts to the left, and the dynamic equilibrium is soon reestablished.

Adding extra H_2 or N_2 would apply the stress to the other side of the equilibrium. To minimize the stress, the system would "use up" some of the excess H_2 or N_2 to make product, NH_3. The equilibrium would shift to the right.

In summary,

Addition of products or reactants may have a profound effect on the composition of a reaction mixture but *does not* affect the value of the equilibrium constant.

$$N_2(g) + 3H_2(g) \rightleftharpoons 2NH_3(g)$$

Product introduced: Equilibrium shifted $\longleftarrow$

Reactant introduced: Equilibrium shifted $\longrightarrow$

Figure 8.16
The effect of concentration on equilibrium position of the reaction:

$$\underset{\text{(red)}}{FeSCN^{2+}(aq)} \rightleftharpoons \underset{\text{(yellow)}}{Fe^{3+}(aq)} + \underset{\text{(colorless)}}{SCN^{-}(aq)}$$

Solution (a) represents this reaction at equilibrium; addition of excess SCN^- shifts the equilibrium to the left (b) intensifying the red color. Removal of SCN^- shifts the equilibrium to the right (c) shown by the disappearance of the red color.

What would happen if some of the ammonia molecules were *removed* from the system? The loss of ammonia represents a stress on the system; to relieve that stress, the ammonia would be replenished by the reaction of hydrogen and nitrogen. The equilibrium would shift to the right.

Effect of Concentration

Addition of extra product or reactant to a fixed reaction volume is just another way of saying that we have increased the concentration of product or reactant. Removal of material from a fixed volume decreases the concentration. Therefore changing the concentration of one or more components of a reaction mixture is a way to alter the equilibrium composition of an equilibrium mixture (Figure 8.16). Let's look at some additional experimental variables that may change equilibrium composition.

Effect of Heat

The change in equilibrium composition caused by the addition or removal of heat from an equilibrium mixture can be explained by treating heat as a product or reactant. The reaction of nitrogen and hydrogen is an exothermic reaction:

$$N_2(g) + 3H_2(g) \rightleftharpoons 2NH_3(g) + 22 \text{ kcal}$$

Adding heat to the reaction is similar to increasing the amount of product. The equilibrium will shift to the left, increasing the amounts of N_2 and H_2 and decreasing the amount of NH_3. If the reaction takes place in a fixed volume, the concentrations of N_2 and H_2 increase and the NH_3 concentration decreases.

Removal of heat produces the reverse effect. More ammonia is produced from N_2 and H_2, and the concentrations of these reactants must decrease.

In the case of an endothermic reaction such as

$$39 \text{ kcal} + 2N_2(g) + O_2(g) \rightleftharpoons 2N_2O(g)$$

addition of heat is analogous to the addition of reactant, and the equilibrium shifts to the right. Removal of heat would shift the reaction to the left, favoring the formation of reactants.

The dramatic effect of heat on the position of equilibrium is shown in Figure 8.17.

Effect of Pressure

Only gases are affected significantly by changes in pressure because gases are free to expand and compress in accordance with Boyle's law. However, liquids and solids are not compressible, so their volumes are unaffected by pressure.

Expansion and compression of gases and Boyle's law are discussed in Section 6.1.

Therefore pressure changes will alter equilibrium composition only in reactions that involve a gas or variety of gases as products and/or reactants. Again, consider the ammonia example,

Figure 8.17
The effect of heat on equilibrium position. For the reaction:

$$CoCl_4^{2-}(aq) + 6H_2O(l) \rightleftharpoons Co(H_2O)_6^{2+}(aq) + 4Cl^-(aq)$$

(blue) (pink)

Heating the solution favors the blue $CoCl_4^{2-}$ species; cooling favors the pink $Co(H_2O)_6^{2+}$ species.

$$N_2(g) + 3H_2(g) \rightleftharpoons 2NH_3(g)$$

The industrial process for preparing ammonia, the Haber process, uses pressures of several hundred atmospheres to increase the yield.

One mole of N_2 and three moles of H_2 (total of four moles of reactants) convert to two moles of NH_3 (two moles of product). An increase in pressure favors a decrease in volume and formation of product. This decrease in volume is made possible by a shift to the right in equilibrium composition. Two moles of ammonia require less volume than four moles of reactant.

A decrease in pressure allows the volume to expand. The equilibrium composition shifts to the left and ammonia decomposes to form more nitrogen and hydrogen.

In contrast, the decomposition of hydrogen iodide,

$$2HI(g) \rightleftharpoons H_2(g) + I_2(g)$$

is unaffected by pressure. The number of moles of gaseous product and reactant are identical. No volume advantage is gained by a shift in equilibrium composition.

In summary:

- Pressure affects the equilibrium composition only of reactions that involve at least one gaseous substance.
- Additionally, the relative number of moles of gaseous products and reactants must differ.
- The equilibrium composition will shift to increase the number of moles of gas when the pressure decreases; it will shift to decrease the number of moles of gas when the pressure increases.

Effect of a Catalyst

A catalyst has no effect on the equilibrium composition. A catalyst increases the rates of both forward and reverse reactions to the same extent. The equilibrium composition *and* equilibrium concentration do not change when a catalyst is used, but the equilibrium composition is achieved in a shorter time. The role of a solid-phase catalyst in the synthesis of ammonia is shown in Figure 8.13.

EXAMPLE 8.9 *Predicting Changes in Equilibrium Composition*

Earlier in this section we considered the geologically important reaction that occurs in rock and soil.

$$Ca^{2+}(aq) + 2HCO_3^-(aq) \rightleftharpoons CaCO_3(s) + CO_2(aq) + H_2O(l)$$

Continued—

EXAMPLE 8.9 —*Continued*

Predict the effect on the equilibrium composition for each of the following changes.

a. The $[Ca^{2+}]$ is increased.
b. The amount of $CaCO_3$ is increased.
c. The amount of H_2O is increased.
d. The $[HCO_3^-]$ is decreased.
e. A catalyst is added.

Solution

a. The concentration of reactant increases; the equilibrium shifts to the right, and more products are formed.
b. $CaCO_3$ is a solid; solids are not written in the equilibrium-constant expression, so there is no effect on the equilibrium composition.
c. H_2O is a pure liquid; it is not written in the equilibrium expression, so the equilibrium composition is unaffected.
d. The concentration of reactant decreases; the equilibrium shifts to the left, and more reactants are formed.
e. A catalyst has no effect on the equilibrium composition.

Question 8.25

For the hypothetical equilibrium reaction

$$A(g) + B(g) \rightleftharpoons C(g) + D(g)$$

predict whether the amount of *A* in a 5.0-L container would increase, decrease, or remain the same for each of the following changes.

a. Addition of excess *B*
b. Addition of excess *C*
c. Removal of some *D*
d. Addition of a catalyst

Question 8.26

For the hypothetical equilibrium reaction

$$A(g) + B(g) \rightleftharpoons C(g) + D(g)$$

predict whether the amount of *A* in a 5.0-L container would increase, decrease, or remain the same for each of the following changes.

a. Removal of some *B*
b. Removal of some *C*
c. Addition of excess *D*
d. Removal of a catalyst

Summary

8.1 Thermodynamics

Thermodynamics is the study of energy, work, and heat. Thermodynamics can be applied to the study of chemical reactions because we can measure the heat flow (by measuring the temperature change) between the *system* and the *surroundings*. *Exothermic reactions* release energy and products that are lower in energy than the reactants. *Endothermic reactions* require energy input. Heat energy is represented as *enthalpy*, *H*°. The energy gain or loss is the change in enthalpy, $\Delta H°$, and is one factor that is useful in predicting whether a reaction is spontaneous or nonspontaneous.

Entropy, S°, is a measure of the randomness of a system. A random, or disordered system has high entropy; a well-ordered system has low entropy. The change in entropy in a chemical reaction, $\Delta S°$, is also a factor in predicting reaction spontaneity.

Free energy, $\Delta G°$, incorporates both factors, enthalpy and entropy; as such, it is an absolute predictor of the spontaneity of a chemical reaction.

8.2 Experimental Determination of Energy Change in Reactions

A *calorimeter* measures heat changes (in calories or joules) that occur in chemical reactions.

The *specific heat* of a substance is the number of calories of heat needed to raise the temperature of 1 g of the substance 1 degree Celsius.

The amount of energy per gram of food is referred to as its *fuel value*. Fuel values are commonly reported in units of *nutritional Calories* (1 nutritional Calorie = 1 kcal). A bomb calorimeter is useful for measurement of the fuel value of foods.

8.3 Kinetics

Chemical *kinetics* is the study of the *rate* or speed of a chemical reaction. Energy for reactions is provided by molecular collisions. If this energy is sufficient, bonds may break, and atoms may recombine in a different arrangement, producing product. A collision producing one or more product molecules is termed an effective collision.

The minimum amount of energy needed for a reaction is the *activation energy*. The reaction proceeds from reactants to products through an intermediate state, the *activated complex*.

Experimental conditions influencing the reaction rate include the structure of the reacting species, the concentration of reactants, the temperature of reactants, the physical state of reactants, and the presence or absence of a catalyst.

A *catalyst* increases the rate of a reaction. The catalytic substance undergoes no net change in the reaction, nor does it alter the outcome of the reaction.

8.4 Equilibrium

Many chemical reactions do not completely convert reactants to products. A mixture of products and reactants exists, and its composition will remain constant until the experimental conditions are changed. This mixture is in a state of *chemical equilibrium*. The reaction continues indefinitely (dynamic), but the concentrations of products and reactants are fixed (equilibrium) because the rates of the forward and reverse reactions are equal. This is a *dynamic equilibrium*.

LeChatelier's principle states that if a stress is placed on an equilibrium system, the system will respond by altering the equilibrium in such a way as to minimize the stress.

Key Terms

activated complex (8.3)
activation energy (8.3)
calorimetry (8.2)
catalyst (8.3)
dynamic equilibrium (8.4)
endothermic reaction (8.1)
enthalpy (8.1)
entropy (8.1)
equilibrium constant (8.4)
equilibrium reaction (8.4)
exothermic reaction (8.1)
free energy (8.1)
fuel value (8.2)
kinetics (8.3)
LeChatelier's principle (8.4)
nutritional Calorie (8.2)
order of the reaction (8.3)
rate constant (8.3)
rate equation (8.3)
rate of chemical reaction (8.3)
reversible reaction (8.4)
specific heat (8.2)
surroundings (8.1)
system (8.1)
thermodynamics (8.1)

Questions and Problems

Energy and Thermodynamics

8.27 Define or explain each of the following terms:
- **a.** exothermic reaction
- **b.** endothermic reaction
- **c.** calorimeter

8.28 Define or explain each of the following terms:
- **a.** free energy
- **b.** specific heat
- **c.** fuel value

8.29 Explain what is meant by the term *enthalpy*.

8.30 Explain what is meant by the term *entropy*.

8.31 5.00 g of octane are burned in a bomb calorimeter containing 2.00×10^2 g H_2O. How much energy, in calories, is released if the water temperature increases 6.00°C?

8.32 0.0500 mol of a nutrient substance is burned in a bomb calorimeter containing 2.00×10^2 g H_2O. If the formula weight of this nutrient substance is 114 g/mol, what is the fuel value (in nutritional Calories) if the temperature of the water increased 5.70°C.

8.33 Calculate the energy released, in joules, in Question 8.31 (recall conversion factors, Chapter 1).

8.34 Calculate the fuel value, in kilojoules, in Question 8.32 (recall conversion factors, Chapter 1).

8.35 Predict whether each of the following processes increases or decreases entropy, and explain your reasoning.
- **a.** melting of a solid metal
- **b.** boiling of water

8.36 Predict whether each of the following processes increases or decreases entropy, and explain your reasoning.
- **a.** burning a log in a fireplace
- **b.** condensation of water vapor on a cold surface

8.37 Explain why an exothermic reaction produces products that are more stable than the reactants.

8.38 Provide an example of entropy from your own experience.

8.39 Isopropyl alcohol, commonly known as rubbing alcohol, feels cool when applied to the skin. Explain why.

8.40 Energy is required to break chemical bonds during the course of a reaction. When is energy released?

Kinetics

8.41 Define the term *activated complex* and explain its significance in a chemical reaction.
8.42 Define and explain the term *activation energy* as it applies to chemical reactions.
8.43 Sketch a potential energy diagram for a reaction that shows the effect of a catalyst on an exothermic reaction.
8.44 Sketch a potential energy diagram for a reaction that shows the effect of a catalyst on an endothermic reaction.
8.45 Give at least two examples from life sciences in which the rate of a reaction is critically important.
8.46 Give at least two examples from everyday life in which the rate of a reaction is an important consideration.
8.47 Describe how an increase in the concentration of reactants increases the rate of a reaction.
8.48 Describe how an increase in the temperature of reactants increases the rate of a reaction.
8.49 Write the rate expression for the single-step reaction:

$$N_2O_4(g) \rightleftharpoons 2NO_2(g)$$

8.50 Write the rate expression for the single-step reaction:

$$H_2S(aq) + Cl_2(aq) \rightleftharpoons S(s) + 2HCl(aq)$$

8.51 Describe how a catalyst speeds up a chemical reaction.
8.52 Explain how a catalyst can be involved in a chemical reaction without being consumed in the process.

Equilibrium

8.53 Describe the meaning of the term *dynamic equilibrium*.
8.54 What is the relationship between the forward and reverse rates for a reaction at equilibrium?
8.55 Write a valid equilibrium constant for the reaction shown in Question 8.49.
8.56 Write a valid equilibrium constant for the reaction shown in Question 8.50.
8.57 Distinguish between a physical equilibrium and a chemical equilibrium.
8.58 Distinguish between the rate constant and the equilibrium constant for a reaction.
8.59 For the reaction

$$CH_4(g) + Cl_2(g) \rightleftharpoons CH_3Cl(g) + HCl(g) + 26.4 \text{ kcal}$$

predict the effect on the equilibrium (will it shift to the left or to the right, or will there be no change?) for each of the following changes.

a. The temperature is increased.
b. The pressure is increased by decreasing the volume of the container.
c. A catalyst is added.

8.60 For the reaction

$$47 \text{ kcal} + 2SO_3(g) \rightleftharpoons 2SO_2(g) + O_2(g)$$

predict the effect on the equilibrium (will it shift to the left or to the right, or will there be no change?) for each of the following changes.

a. The temperature is increased.
b. The pressure is increased by decreasing the volume of the container.
c. A catalyst is added.

8.61 Label each of the following statements as true or false and explain why.

a. A slow reaction is an incomplete reaction.
b. The rates of forward and reverse reactions are never the same.

8.62 Label each of the following statements as true or false and explain why.

a. A reaction is at equilibrium when no reactants remain.
b. A reaction at equilibrium is undergoing continual change.

8.63 Use LeChatelier's principle to predict whether the amount of PCl_3 in a 1.00-L container is increased, is decreased, or remains the same for the equilibrium

$$PCl_3(g) + Cl_2(g) \rightleftharpoons PCl_5(g) + \text{heat}$$

when each of the following changes is made.

a. PCl_5 is added.
b. Cl_2 is added.
c. PCl_5 is removed.
d. The temperature is decreased.
e. A catalyst is added.

8.64 Use LeChatelier's principle to predict the effects, if any, of each of the following changes on the equilibrium system, described below, in a closed container.

$$C(s) + 2H_2(g) \rightleftharpoons CH_4(g) + 18 \text{ kcal}$$

a. adding more C.
b. adding more H_2.
c. removing CH_4.
d. increasing the temperature.
e. adding a catalyst.

8.65 Will an increase in pressure increase, decrease, or have no effect on the concentration of $H_2(g)$ in the reaction:

$$C(s) + H_2O(g) \rightleftharpoons CO(g) + H_2(g)$$

8.66 Will an increase in pressure increase, decrease, or have no effect on the concentration of $NO(g)$ in the reaction:

$$N_2(g) + O_2(g) \rightleftharpoons 2NO(g)$$

8.67 Write the equilibrium-constant expression for the reaction described in Question 65.
8.68 Write the equilibrium-constant expression for the reaction described in Question 66.
8.69 True or false: The equilibrium will shift to the right when a catalyst is added to the mixture described in Question 65. Explain your reasoning.
8.70 True or false: The equilibrium for an endothermic reaction will shift to the right when the reaction mixture is heated. Explain your reasoning.

Critical Thinking Problems

1. For the reaction:

$$2NH_3(g) \rightleftharpoons N_2(g) + 3H_2(g)$$

What is the relationship among the equilibrium concentrations of NH_3, N_2, and H_2 for each of the following situations:

- We begin with 2 mol of NH_3 in a 1-L container.
- We begin with 1 mol of N_2 and 3 mol of H_2 in a 1-L container.

Explain your reasoning.

2. Can the following statement ever be true? "Heating a reaction mixture increases the rate of a certain reaction but decreases the yield of product from the reaction." Explain why or why not.
3. Molecules must collide for a reaction to take place. Sketch a model of the orientation and interaction of HI and Cl that is most favorable for the reaction:

$$HI(g) + Cl(g) \longrightarrow HCl(g) + I(g)$$

4. Silver ion reacts with chloride ion to form the precipitate, silver chloride:

$$Ag^+(aq) + Cl^-(aq) \rightleftharpoons AgCl(s)$$

After the reaction reached equilibrium, the chemist filtered 99% of the solid silver chloride from the solution, hoping to shift the equilibrium to the right, to form more product. Critique the chemist's experiment.

5. Human behavior often follows LeChatelier's principle. Provide one example and explain in terms of LeChatelier's principle.

6. A clever device found in some homes is a figurine that is blue on dry, sunny days and pink on damp, rainy days. These figurines are coated with substances containing chemical species that undergo the following equilibrium reaction:

$$Co(H_2O)_6{}^{2+}(aq) + 4Cl^-(aq) \rightleftharpoons CoCl_4{}^{2-}(aq) + 6H_2O(l)$$

a. Which substance is blue?
b. Which substance is pink?
c. How is LeChatelier's principle applied here?

7. You have spent the entire morning in a 20°C classroom. As you ride the elevator to the cafeteria, six persons enter the elevator after being outside on a subfreezing day. You suddenly feel chilled. Explain the heat flow situation in the elevator in thermodynamic terms.

Solution properties, including color, are often pH dependent.

9

Charge-Transfer Reactions:

Acids and Bases and Oxidation-Reduction

GENERAL CHEMISTRY

Outline

Learning Goals

1. Identify acids and bases and acid-base reactions.
2. Write equations describing acid-base dissociation and label the conjugate acid-base pairs.
3. Describe the role of the solvent in acid-base reactions, and explain the meaning of the term *pH*.
4. Calculate pH from concentration data.
5. Calculate hydronium and/or hydroxide ion concentration from pH data.
6. Provide examples of the importance of pH in chemical and biochemical systems.
7. Describe the meaning and utility of neutralization reactions.
8. State the meaning of the term *buffer* and describe the applications of buffers to chemical and biochemical systems, particularly blood chemistry.
9. Describe *oxidation* and *reduction,* and describe some practical examples of redox processes.
10. Diagram a voltaic cell and describe its function.
11. Compare and contrast voltaic and electrolytic cells.

CHEMISTRY CONNECTION

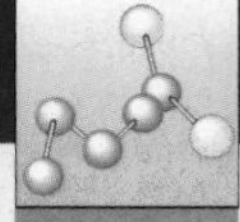

Drug Delivery

When a doctor prescribes medicine to treat a disease or relieve its symptoms, the medication may be administered in a variety of ways. Drugs may be taken orally, injected into a muscle or a vein, or absorbed through the skin. Specific instructions are often provided to regulate the particular combination of drugs that can or cannot be taken. The diet, both before and during the drug therapy, may be of special importance.

To appreciate why drugs are administered in a specific way, it is necessary to understand a few basic facts about medications and how they interact with the body.

Drugs function by undergoing one or more chemical reactions in the body. Few compounds react in only one way, to produce a limited set of products, even in the simple environment of a beaker or flask. Imagine the number of possible reactions that a drug can undergo in a complex chemical factory like the human body. In many cases a drug can react in a variety of ways other than its intended path. These alternative paths are side reactions, sometimes producing *side effects* such as nausea, vomiting, insomnia, or drowsiness. Side effects may be unpleasant and may actually interfere with the primary function of the drug.

The development of safe, effective medication, with minimal side effects, is a slow and painstaking process and determining the best drug delivery system is a critical step. For example, a drug that undergoes an unwanted side reaction in an acidic solution would not be very effective if administered orally. The acidic digestive fluids in the stomach could prevent the drug from even reaching the intended organ, let alone retaining its potency. The drug could be administered through a vein into the blood; blood is not acidic, in contrast to digestive fluids. In this way the drug may be delivered intact to the intended site in the body, where it is free to undergo its primary reaction.

Drug delivery has become a science in its own right. Pharmacology, the study of drugs and their uses in the treatment of disease, has a goal of creating drugs that are highly selective. In other words, they will undergo only one reaction, the intended reaction. Encapsulation of drugs, enclosing them within larger molecules or collections of molecules, may protect them from unwanted reactions as they are transported to their intended site.

In this chapter we will explore the fundamentals of solutions and solution reactions, including acid-base and oxidation-reduction reactions. Knowing a few basic concepts that govern reactions in beakers will help us to understand the conditions that affect the reactivity of a host of biochemically interesting molecules that we will encounter in later chapters.

Introduction

In this chapter we will learn about two general classes of chemical change: acid-base reactions and oxidation-reduction reactions. Although superficially quite different, their underlying similarity is that both are essentially charge-transfer processes. An *acid-base reaction* involves the transfer of one or more positively charged units, protons or hydrogen ions; an *oxidation-reduction reaction* involves the transfer of one or more negatively charged particles, electrons.

Acids and bases include some of the most important compounds in nature. Historically, it was recognized that certain compounds, acids, had a sour taste, were able to dissolve some metals, and caused vegetable dyes to change color. Bases have long been recognized by their bitter taste, slippery feel, and corrosive nature. Bases react strongly with acids and cause many metal ions in solution to form a solid precipitate.

The effects of pH on enzyme activity are discussed in Chapter 20.

Buffers are discussed in Section 9.4.

Digestion of proteins is aided by stomach acid (hydrochloric acid) and many biochemical processes such as enzyme catalysis depend on the proper level of acidity. Indeed, a wide variety of chemical reactions critically depend on the acid-base composition of the solution (Figure 9.1). This is especially true of the biochemical reactions occurring in the cells of our bodies. For this reason the level of acidity must be very carefully regulated. This is done with substances called *buffers*.

Oxidation-reduction processes are also common in living systems. Respiration is driven by oxidation-reduction reactions. Additionally, oxidation-reduction reactions generate heat that warms our homes and workplaces and fuels our industrial civilization. Moreover, oxidation-reduction is the basis for battery design. Batteries are found in automobiles and electronic devices such as cameras and radios, and are even implanted in the human body to regulate heart rhythm.

9.1 Acids and Bases

Learning Goal

The properties of acids and bases are related to their chemical structure. All acids have common characteristics that enable them to increase the hydrogen ion concentration in water. All bases lower the hydrogen ion concentration in water.

Two theories, one developed from the other, help us to understand the unique chemistry of acids and bases.

Arrhenius Theory of Acids and Bases

One of the earliest definitions of acids and bases is the **Arrhenius theory.** According to this theory, an acid, dissolved in water, dissociates to form *hydrogen ions or protons* (H^+), and a base, dissolved in water, dissociates to form *hydroxide ions* (OH^-). For example, hydrochloric acid dissociates in solution according to the reaction

$$HCl(aq) \longrightarrow H^+(aq) + Cl^-(aq)$$

Sodium hydroxide, a base, produces hydroxide ions in solution:

$$NaOH(aq) \longrightarrow Na^+(aq) + OH^-(aq)$$

The Arrhenius theory satisfactorily explains the behavior of many acids and bases. However, a substance such as ammonia, NH_3, has basic properties but cannot be an Arrhenius base, because it contains no OH^-. The **Brønsted-Lowry theory** explains this mystery and gives us a broader view of acid-base theory by considering the central role of the solvent in the dissociation process.

Brønsted-Lowry Theory of Acids and Bases

The Brønsted-Lowry theory defines an acid as a proton (H^+) donor and a base as a proton acceptor.

Hydrochloric acid in solution *donates* a proton to the solvent water thus behaving as a Brønsted-Lowry acid:

$$HCl(aq) + H_2O(l) \longrightarrow H_3O^+(aq) + Cl^-(aq)$$

H_3O^+ is referred to as the hydrated proton or **hydronium ion.**

The basic properties of ammonia are clearly accounted for by the Brønsted-Lowry theory. Ammonia *accepts* a proton from the solvent water, producing OH^-. An equilibrium mixture of NH_3, H_2O, NH_4^+, and OH^- results.

$$\begin{array}{c}H-\ddot{N}-H \\ | \\ H\end{array} + \begin{array}{c}H-\ddot{\underset{..}{O}}: \\ | \\ H\end{array} \rightleftharpoons \left[\begin{array}{c}H \\ | \\ H-N-H \\ | \\ H\end{array}\right]^+ + H-\ddot{\underset{..}{O}}:^-$$

$$\underset{\text{base}}{NH_3(aq)} + \underset{\text{acid}}{H{-}OH(l)} \rightleftharpoons \underset{\text{acid}}{NH_4^+(aq)} + \underset{\text{base}}{OH^-(aq)}$$

Figure 9.1
The yellow solution on the left, containing CrO_4^{2-} (chromate ion), was made acidic producing the reddish brown solution on the right. The principal component in solution is now $Cr_2O_7^{2-}$. Addition of base to this solution removes H^+ ions and regenerates the yellow CrO_4^{2-}. This is an example of an acid-base dependent chemical equilibrium.

For aqueous solutions, the Brønsted-Lowry theory adequately describes the behavior of acids and bases. We shall limit our discussion of acid-base chemistry to aqueous solutions and use the following definitions:

An **acid** is a proton donor.

A **base** is a proton acceptor.

Conjugate Acids and Bases

Learning Goal

The Brønsted-Lowry theory contributed several fundamental ideas that broadened our understanding of solution chemistry. First of all, an acid-base reaction is a charge-transfer process. Second, the transfer process usually involves the solvent. Water may, in fact, accept or donate a proton. Last, and perhaps most important, the acid-base reaction is seen as a reversible process. This leads to the possibility of a reversible, dynamic equilibrium (see Section 8.4).

Consequently, any acid-base reaction can be represented by the general equation

$$\underset{\text{(acid)}}{HA} + \underset{\text{(base)}}{B} \rightleftharpoons BH^+ + A^-$$

In the forward reaction, the acid (HA) donates a proton (H^+) to the base (B) leading to the formation of BH^+ and A^-. However, in the reverse reaction, it is the BH^+ that behaves as an acid; it donates its "extra" proton to A^-. A^- is therefore a base in its own right because it accepts the proton.

These *product* acids and bases are termed *conjugate acids and bases.*

A **conjugate acid** is the species formed when a base accepts a proton.

A **conjugate base** is the species formed when an acid donates a proton.

The acid and base on the opposite sides of the equation are collectively termed a **conjugate acid-base pair.** In the above equation:

BH^+ is the conjugate acid of the base B.

A^- is the conjugate base of the acid HA.

B and BH^+ constitute a conjugate acid-base pair.

HA and A^- constitute a conjugate acid-base pair.

Rewriting our model equation:

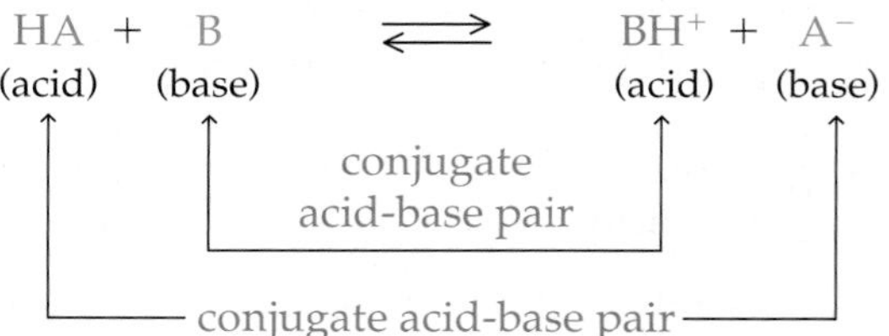

Although we show the forward and reverse arrows to indicate the reversibility of the reaction, seldom are the two processes "equal but opposite." One reaction, either forward or reverse, is usually favored. Consider the reaction of hydrochloric acid in water:

Forward reaction: significant

$$HCl(aq) + H_2O(l) \rightleftharpoons H_3O^+(aq) + Cl^-(aq)$$

Reverse reaction: not significant

HCl is a much better proton donor than H_3O^+. Consequently the forward reaction predominates, the reverse reaction is inconsequential, and hydrochloric acid is termed a *strong acid.* As we learned in Chapter 8, reactions in which the forward reaction is strongly favored have large equilibrium constants. The dissociation of hydrochloric acid is so favorable that we describe it as 100% dissociated and use only a single forward arrow to represent its behavior in water:

$$HCl(aq) + H_2O(l) \longrightarrow H_3O^+(aq) + Cl^-(aq)$$

The degree of dissociation, or strength, of acids and bases has a profound influence on their aqueous chemistry. For example, vinegar (a 5% [w/v] solution of acetic acid in water) is a consumable product; aqueous hydrochloric acid in water is not. Why? Acetic acid is a *weak* acid and, as a result, a dilute solution does no damage to the mouth and esophagus. The following section looks at the strength of acids and bases in solution in more detail.

Question 9.1

Write an equation for the reaction of each of the following with water:

a. HF (a weak acid)
b. NH_3 (a weak base)

Question 9.2

Write an equation for the reaction of each of the following with water:

a. HBr (a weak acid)
b. H_2S (a weak acid)

Question 9.3

Select the conjugate acid-base pairs for each reaction in Question 9.1.

Question 9.4

Select the conjugate acid-base pairs for each reaction in Question 9.2.

Acid-Base Properties of Water

Learning Goal

The role that the solvent, water, plays in acid-base reactions is noteworthy. In the example above, the water molecule accepts a proton from the HCl molecule. The water is behaving as a proton acceptor, a base.

However, when water is a solvent for ammonia (NH_3), a base, the water molecule donates a proton to the ammonia molecule. The water, in this situation, is acting as a proton donor, an acid.

Water, owing to the fact that it possesses *both* acid and base properties, is termed **amphiprotic.** The solvent properties of water are a consequence of this ability to either accept or donate protons. Water is the most commonly used solvent for acids and bases. These interactions promote solubility and dissociation of acids and bases.

Acid and Base Strength

The terms *acid or base strength* and *acid or base concentration* are easily confused. *Strength* is a measure of the *degree of dissociation* of an acid or base in solution, independent of its concentration. Concentration, as we have learned, refers to the amount of solute (in this case, the amount of acid or base) per quantity of solution.

Concentration of solutions is discussed in Sections 7.4 and 7.5.

The concentration of an acid or base does affect the degree of dissociation. However, the major factor in determining the degree of dissociation is the strength of the acid or base.

The strength of acids and bases in water depends on the extent to which they react with the solvent, water. Acids and bases are classified as *strong* when the reaction with water is virtually 100% complete and as *weak* when the reaction with water is much less than 100% complete (perhaps as little as 2–3%).

Important strong acids include:

Hydrochloric acid $HCl(aq) + H_2O(l) \longrightarrow H_3O^+(aq) + Cl^-(aq)$

Nitric acid $HNO_3(aq) + H_2O(l) \longrightarrow H_3O^+(aq) + NO_3^-(aq)$

Sulfuric acid $H_2SO_4(aq) + H_2O(l) \longrightarrow H_3O^+(aq) + HSO_4^-(aq)$

Note that the equation for the dissociation of each of these acids is written with a single arrow. This indicates that the reaction has little or no tendency to

Reversibility of reactions is discussed in Section 8.4.

proceed in the reverse direction to establish equilibrium. All of the acid molecules are dissociated to form ions.

All common strong bases are *metal hydroxides.* Strong bases completely dissociate in aqueous solution to produce hydroxide ions and metal cations. Of the common metal hydroxides, only NaOH and KOH are soluble in water and are readily usable strong bases:

Sodium hydroxide	$NaOH(aq) \longrightarrow Na^+(aq) + OH^-(aq)$
Potassium hydroxide	$KOH(aq) \longrightarrow K^+(aq) + OH^-(aq)$

Weak acids and weak bases dissolve in water principally in the molecular form. Only a small percentage of the molecules dissociate to form the hydronium or hydroxide ion.

The double arrow implies an equilibrium between dissociated and undissociated species.

Two important weak acids are:

Acetic acid	$CH_3COOH(aq) + H_2O(l) \rightleftharpoons H_3O^+(aq) + CH_3COO^-(aq)$
Carbonic acid	$H_2CO_3(aq) + H_2O(l) \rightleftharpoons H_3O^+(aq) + HCO_3^-(aq)$

Many organic compounds have acid or base properties. The chemistry of organic acids and bases will be discussed in Chapters 15 (Carboxylic Acids and Carboxylic Acid Derivatives) and 16 (Amines and Amides).

We have already mentioned the most common weak base, ammonia. Many organic compounds function as weak bases. Several examples of weak bases follow:

Pyridine	$C_5H_5N(aq) + H_2O(l) \rightleftharpoons C_5H_5NH^+(aq) + OH^-(aq)$
Aniline	$C_6H_5NH_2(aq) + H_2O(l) \rightleftharpoons C_6H_5NH_3^+(aq) + OH^-(aq)$
Methylamine	$CH_3NH_2(aq) + H_2O(l) \rightleftharpoons CH_3NH_3^+(aq) + OH^-(aq)$

The fundamental chemical difference between strong and weak acids or bases is their equilibrium ion concentration. A strong acid, such as HCl, does not, in aqueous solution, exist to any measurable degree in equilibrium with its ions, H_3O^+ and Cl^-. On the other hand, a weak acid, such as acetic acid, establishes a dynamic equilibrium with its ions, H_3O^+ and CH_3COO^-.

The relative strength of an acid or base is determined by the ease with which it donates or accepts a proton. Acids with the greatest proton-donating capability (strongest acids) have the weakest conjugate bases. Good proton acceptors (strong bases) have weak conjugate acids. This relationship is clearly indicated in Figure 9.2. This figure can be used to help us compare and predict relative acid-base strength.

EXAMPLE 9.1 Predicting Relative Acid-Base Strengths

a. Write the conjugate acid of HS^-.

Solution

The conjugate acid may be constructed by adding a proton (H^+) to the base structure, consequently, H_2S.

b. Using Figure 9.2 identify the stronger base, HS^- or F^-.

Solution

HS^- is the stronger base because it is located farther down the right-hand column.

c. Using Figure 9.2 identify the stronger acid, H_2S or HF.

Solution

HF is the stronger acid because its conjugate base is weaker *and* because it is located farther up the left-hand column.

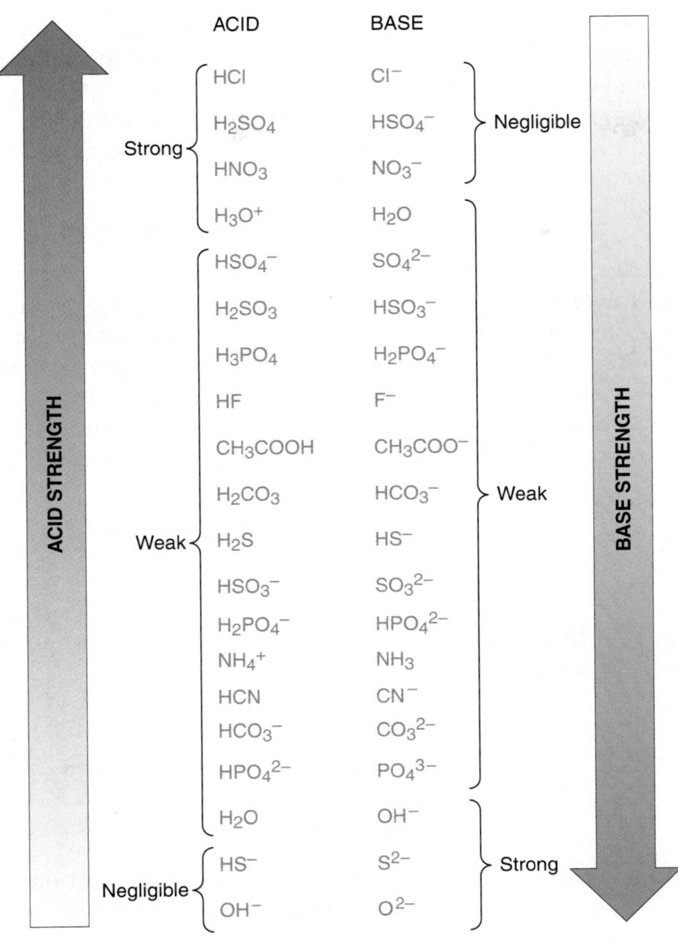

Figure 9.2
Conjugate acid-base pairs. Strong acids have weak conjugate bases; strong bases have weak conjugate acids.

Question 9.5

In each pair, identify the stronger acid.

a. H_2O or NH_4^+
b. H_2SO_4 or H_2SO_3

Question 9.6

In each pair, identify the stronger base.

a. CO_3^{2-} or PO_4^{3-}
b. HCO_3^- or HPO_4^{2-}

Solutions of acids and bases used in the laboratory must be handled with care. Acids burn because of their exothermic reaction with water present on and in the skin. Bases react with proteins, which are principal components of the skin and eyes.

Such solutions are more hazardous if they are strong or concentrated. A strong acid or base produces more H_3O^+ or OH^- than does the corresponding weak acid or base. More-concentrated acids or bases contain more H_3O^+ or OH^- than do less-concentrated solutions of the same strength.

The Dissociation of Water

Aqueous solutions of acids and bases are electrolytes. The dissociation of the acid or base produces ions that can conduct an electrical current. As a result of the

Solutions of electrolytes are discussed in Section 7.3.

differences in the degree of dissociation, *strong acids and bases are strong electrolytes; weak acids and bases are weak electrolytes.* The conductivity of these solutions is principally dependent on the solute and not the solvent (water).

Although pure water is virtually 100% molecular, a small number of water molecules do ionize. This process occurs by the transfer of a proton from one water molecule to another, producing a hydronium ion and a hydroxide ion:

$$H_2O(l) + H_2O(l) \rightleftharpoons H_3O^+(aq) + OH^-(aq)$$

This process is the **autoionization,** or self-ionization, of water. Water is therefore a *very* weak electrolyte and a very poor conductor of electricity. Water has *both* acid and base properties; dissociation produces both the hydronium and hydroxide ion.

Pure water at room temperature has a hydronium ion concentration of $1.0 \times 10^{-7}\,M$. One hydroxide ion is produced for each hydronium ion. Therefore, the hydroxide ion concentration is also $1.0 \times 10^{-7}\,M$. Molar equilibrium concentration is conveniently indicated by brackets around the species whose concentration is represented:

$$[H_3O^+] = 1.0 \times 10^{-7}\,M$$

$$[OH^-] = 1.0 \times 10^{-7}\,M$$

The product of hydronium and hydroxide ion concentration in pure water is referred to as the **ion product for water.**

$$\text{ion product} = [H_3O^+][OH^-]$$

$$= [1.0 \times 10^{-7}][1.0 \times 10^{-7}]$$

$$= 1.0 \times 10^{-14}$$

The ion product is constant because its value does not depend on the nature or concentration of the solute, as long as the temperature does not change. The ion product is a temperature-dependent quantity.

The nature and concentration of the solutes added to water do alter the relative concentrations of H_3O^+ and OH^- present, but the product, $[H_3O^+][OH^-]$, always equals 1.0×10^{-14} at 25°C. This relationship is the basis for a scale that is useful in the measurement of the level of acidity or basicity of solutions. This scale, the pH scale, is discussed next.

9.2 pH: A Measurement Scale for Acids and Bases

A Definition of pH

Learning Goal

The **pH scale** relates the hydronium ion concentration to a number, the pH, that serves as a useful indicator of the degree of acidity or basicity of a solution. The pH scale is somewhat analogous to the temperature scale used for assignment of relative levels of hot or cold. The temperature scale was developed to allow us to indicate how cold or how hot an object is. The pH scale specifies how acidic or how basic a solution is. The pH scale has values that range from 0 (very acidic) to 14 (very basic). A pH of 7, the middle of the scale, is neutral, neither acidic nor basic.

To help us to develop a concept of pH, let's consider the following:

- Addition of an acid (proton donor) to water *increases* the $[H_3O^+]$ and decreases the $[OH^-]$.
- Addition of a base (proton acceptor) to water *decreases* the $[H_3O^+]$ by increasing the $[OH^-]$.
- $[H_3O^+] = [OH^-]$ when *equal* amounts of acid and base are present.
- In all three cases, $[H_3O^+][OH^-] = 1.0 \times 10^{-14}$ = the ion product for water at 25°C.

(a)

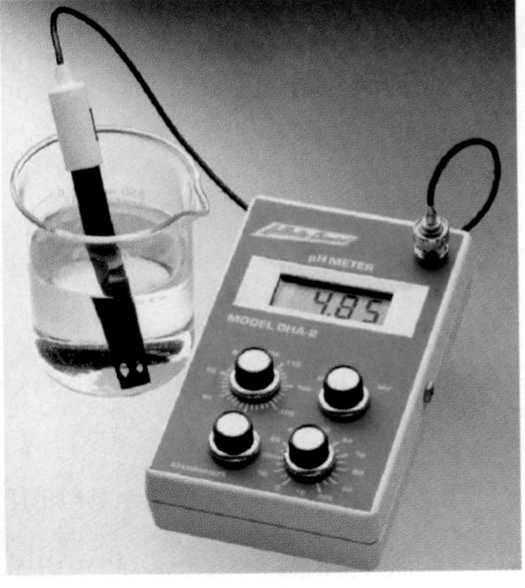

(b)

Figure 9.3
The measurement of pH. (a) A strip of test paper impregnated with indicator (a material that changes color as the acidity of the surroundings changes) is put in contact with the solution of interest. The resulting color is matched with a standard color chart (colors shown as a function of pH) to obtain the approximate pH. (b) A pH meter uses a sensor (a pH electrode) that develops an electrical potential that is proportional to the pH of the solution.

Measuring pH

The pH of a solution can be calculated if the concentration of either H_3O^+ or OH^- is known. Alternatively, measurement of pH allows the calculation of H_3O^+ or OH^- concentration. The pH of aqueous solutions may be approximated by using indicating paper (pH paper) that develops a color related to the solution pH. Alternatively, a pH meter can give us a much more exact pH measurement. A sensor measures an electrical property of a solution that is proportional to pH (Figure 9.3).

Calculating pH

One of our objectives in this chapter is to calculate the pH of a solution when the hydronium or hydroxide ion concentration is known, and to calculate $[H_3O^+]$ or $[OH^-]$ from the pH. We will develop this skill using two different approaches, one requiring a working knowledge of logarithms, the other using decimal logic.

Learning Goal  4

Learning Goal  5

Approach I: Logarithm-Based pH Calculations

The pH of a solution is defined as the negative logarithm of the molar concentration of the hydronium ion:

$$pH = -\log [H_3O^+]$$

EXAMPLE 9.2

Calculating pH from Acid Molarity

Calculate the pH of a 1.0×10^{-3} *M* solution of HCl.

Solution

HCl is a strong acid. If 1 mol HCl dissolves and dissociates in 1L of aqueous solution, it produces 1 mol H_3O^+ (a 1 *M* solution of H_3O^+). Therefore a 1.0×10^{-3} *M* HCl solution has $[H_3O^+] = 1.0 \times 10^{-3}$ *M*, and

$$pH = -\log [H_3O^+]$$
$$= -\log [1.0 \times 10^{-3}]$$

Consider the concentration term as composed of two parts, 1.0 and 10^{-3}. The logarithm of 1.0 = 0, and the logarithm of 10^{-3} is simply the exponent, −3. Therefore

$$pH = -[\log 1.0 + \log 10^{-3}]$$
$$= -[0 - 3.00]$$
$$= -[-3.00] = 3.00$$

EXAMPLE 9.3 ***Calculating $[H_3O^+]$ from pH***

Calculate the $[H_3O^+]$ of a solution of hydrochloric acid with pH = 4.00.

Solution

We use the pH expression:

$$\text{pH} = -\log [H_3O^+]$$
$$4.00 = -\log [H_3O^+]$$

Multiplying both sides of the equation by -1, we get

$$-4.00 = \log [H_3O^+]$$

Taking the antilogarithm of both sides (the reverse of a logarithm), we have

$$\text{antilog } -4.00 = [H_3O^+]$$

The antilog is the exponent of 10; therefore

$$1.0 \times 10^{-4}\,M = [H_3O^+]$$

EXAMPLE 9.4 ***Calculating the pH of a Base***

Calculate the pH of a $1.0 \times 10^{-5}\,M$ solution of NaOH.

Solution

NaOH is a strong base. If 1 mol NaOH dissolves and dissociates in 1L of aqueous solution, it produces 1 mol OH^- (a 1 M solution of OH^-). Therefore a $1.0 \times 10^{-5}\,M$ NaOH solution has $[OH^-] = 1.0 \times 10^{-5}\,M$. To calculate pH, we need $[H_3O^+]$. Recall that

$$[H_3O^+][OH^-] = 1.0 \times 10^{-14}$$

Solving this equation for $[H_3O^+]$,

$$[H_3O^+] = \frac{1.0 \times 10^{-14}}{[OH^-]}$$

substituting the information provided in the problem,

$$= \frac{1.0 \times 10^{-14}}{1.0 \times 10^{-5}}$$

$$= 1.0 \times 10^{-9}\,M$$

The solution is now similar to that in Example 9.2:

$$\text{pH} = -\log [H_3O^+]$$
$$= -\log [1.0 \times 10^{-9}]$$
$$= -(\log 1.0 + \log 10^{-9}) = -[0 + (-9.00)]$$
$$= 9.00$$

Calculating Both Hydronium and Hydroxide Ion Concentrations from pH

EXAMPLE 9.5

Calculate the $[H_3O^+]$ and $[OH^-]$ of a sodium hydroxide solution with a pH = 10.00.

Solution

First, calculate $[H_3O^+]$:

$$\begin{aligned} pH &= -\log [H_3O^+] \\ 10.00 &= -\log [H_3O^+] \\ -10.00 &= \log [H_3O^+] \\ \text{antilog } -10 &= [H_3O^+] \\ 1.0 \times 10^{-10}\, M &= [H_3O^+] \end{aligned}$$

To calculate the $[OH^-]$, we need to solve for $[OH^-]$ by using the following expression:

$$[H_3O^+][OH^-] = 1.0 \times 10^{-14}$$

$$[OH^-] = \frac{1.0 \times 10^{-14}}{[H_3O^+]}$$

Substituting the $[H_3O^+]$ from the first part, we have

$$[OH^-] = \frac{1.0 \times 10^{-14}}{[1.0 \times 10^{-10}]}$$

$$= 1.0 \times 10^{-4}\, M$$

Often, the pH or $[H_3O^+]$ will not be a whole number (pH = 1.5, pH = 5.3, $[H_3O^+] = 1.5 \times 10^{-3}$ and so forth). With the advent of inexpensive and versatile calculators, calculations with noninteger numbers pose no great problems. Consider Examples 9.6 and 9.7.

Calculating pH with Noninteger Numbers

EXAMPLE 9.6

Calculate the pH of a sample of lake water that has a $[H_3O^+] = 6.5 \times 10^{-5}\, M$.

Solution

$$\begin{aligned} pH &= -\log[H_3O^+] \\ &= -\log[6.5 \times 10^{-5}] \\ &= 4.19 \end{aligned}$$

The pH, 4.19, is low enough to suspect acid rain. (See An Environmental Perspective: Acid Rain in this chapter.)

EXAMPLE 9.7

Calculating $[H_3O^+]$ from pH

The measured pH of a sample of lake water is 6.40. Calculate $[H_3O^+]$.

Solution

An alternative mathematical form of

$$pH = -\log [H_3O^+]$$

is the expression

$$[H_3O^+] = 10^{-pH}$$

which we will use when we must solve for $[H_3O^+]$.

$$[H_3O^+] = 10^{-6.40}$$

Performing the calculation on your calculator results in 3.98×10^{-7} or $4.0 \times 10^{-7}\ M = [H_3O^+]$.

Examples 9.2–9.7 illustrate the most frequently used pH calculations. It is important to remember that in the case of a base you must convert the $[OH^-]$ to $[H_3O^+]$, using the expression for the ion product for the solvent, water.

Approach II: Decimal-Based pH Calculation

Learning Goal 4

Learning Goal 5

If you do not have a facility with logarithms or a calculator available, it is still possible to approximate the pH of a solution of an acid or base and determine acid or base concentration from the pH. To do this, remember the following facts:

1. The pH of a 1 *M* solution of any strong acid is 0.
2. The pH of a 1 *M* solution of any strong base is 14.
3. Each tenfold change in concentration changes the pH by one unit. A tenfold change in concentration is equivalent to moving the decimal point one place.
4. A *decrease* in acid concentration *increases* the pH.
5. A *decrease* in base concentration *decreases* the pH.

For a strong acid:

HCl molarity	*pH*
1.0×10^{0}	0.00
1.0×10^{-1}	1.00
1.0×10^{-2}	2.00
1.0×10^{-3}	3.00
1.0×10^{-4}	4.00
1.0×10^{-5}	5.00
1.0×10^{-6}	6.00
1.0×10^{-7}	7.00

More acidic

For a strong acid the exponent, with the sign changed, is the pH.

For a strong base:

NaOH molarity	*pH*
1.0×10^{0}	14.00
1.0×10^{-1}	13.00
1.0×10^{-2}	12.00
1.0×10^{-3}	11.00
1.0×10^{-4}	10.00
1.0×10^{-5}	9.00
1.0×10^{-6}	8.00
1.0×10^{-7}	7.00

More basic

For a strong base the exponent, algebraically added to 14, is the pH.

Question 9.7

Use the decimal-based system to solve the problems posed in Examples 9.2 and 9.3.

Question 9.8

Use the decimal-based system to solve the problems posed in Examples 9.4 and 9.5.

Question 9.9

Calculate the pH corresponding to a solution of sodium hydroxide with a $[OH^-]$ of 1.0×10^{-2} *M*.

Question 9.10

Calculate the pH corresponding to a solution of sodium hydroxide with a $[OH^-]$ of 1.0×10^{-6} *M*.

Question 9.11

Use the most suitable method to calculate the $[H_3O^+]$ corresponding to pH = 8.50.

Question 9.12

Use the most suitable method to calculate the $[H_3O^+]$ corresponding to pH = 4.50.

The Importance of pH and pH Control

Learning Goal

Solution pH and pH control play a major role in many facets of our lives. Consider a few examples:

- *Agriculture:* Crops grow best in a soil of proper pH. Proper fertilization involves the maintenance of a suitable pH.
- *Physiology:* If the pH of our blood were to shift by one unit, we would die. Many biochemical reactions in living organisms are extremely pH dependent.
- *Industry:* From manufacture of processed foods to the manufacture of automobiles, industrial processes often require rigorous pH control.
- *Municipal services:* Purification of drinking water and treatment of sewage must be carried out at their optimum pH.
- *Acid rain:* Nitric acid and sulfuric acid, resulting largely from the reaction of components of vehicle emissions (nitrogen and sulfur oxides) with water, are carried down by precipitation and enter aquatic systems (lakes and streams), lowering the pH of the water. A less than optimum pH poses serious problems for native fish populations.

See An Environmental Perspective: Acid Rain in this chapter.

The list could continue on for many pages. However, in summary, any change that takes place in aqueous solution generally has at least some pH dependence.

9.3 Reactions between Acids and Bases

Neutralization

Learning Goal

The reaction of an acid with a base to produce a salt and water is referred to as **neutralization.** In the strictest sense, neutralization requires equal numbers of moles of H_3O^+ and OH^- to produce a neutral solution (no excess acid or base).

Consider the reaction of a solution of hydrochloric acid and sodium hydroxide:

$$\underset{\text{Acid}}{HCl(aq)} + \underset{\text{Base}}{NaOH(aq)} \longrightarrow \underset{\text{Salt}}{NaCl(aq)} + \underset{\text{Water}}{H_2O(l)}$$

Our objective is to make the balanced equation represent the process actually occurring. We recognize that HCl, NaOH, and NaCl are dissociated in solution:

Equation balancing is discussed in Chapter 5.

$$H^+(aq) + Cl^-(aq) + Na^+(aq) + OH^-(aq) \longrightarrow Na^+(aq) + Cl^-(aq) + H_2O(l)$$

We further know that Na^+ and Cl^- are unchanged in the reaction. If we write only those components that actually change, we produce a *net, balanced ionic equation:*

$$H^+(aq) + OH^-(aq) \longrightarrow H_2O(l)$$

If we realize that the H^+ occurs in aqueous solution as the hydronium ion, H_3O^+, the most correct form of the net, balanced ionic equation is

$$H_3O^+(aq) + OH^-(aq) \longrightarrow 2H_2O(l)$$

The equation for any strong acid/strong base neutralization reaction is the same as this equation.

A neutralization reaction may be used to determine the concentration of an unknown acid or base solution. The technique of **titration** involves the addition of measured amounts of a **standard solution** (one whose concentration is known with certainty) to neutralize the second, unknown solution. From the volumes of the two solutions and the concentration of the standard solution the concentration of the unknown solution may be determined. Consider the following application.

EXAMPLE 9.8 Determining the Concentration of a Solution of Hydrochloric Acid

Step 1. A known volume (perhaps 25.00 mL) of the unknown acid is measured into a flask using a pipet.

Step 2. An **indicator,** a substance that changes color as the solution reaches a certain pH (Figures 9.4 and 9.5), is added to the unknown solution.

Step 3. A solution of sodium hydroxide (perhaps 0.1000 *M*) is carefully added to the unknown solution using a **buret** (Figure 9.6), which is a long glass tube calibrated in milliliters. A stopcock at the bottom of the buret regulates the amount of liquid dispensed. The standard solution is added until the indicator changes color.

Step 4. At this point, the **equivalence point,** the number of moles of hydroxide ion added is equal to the number of moles of hydronium ion present in the unknown acid.

Step 5. The volume dispensed by the buret (perhaps 35.00 mL) is measured and used in the calculation of the unknown acid concentration.

Step 6. The calculation is as follows:

Pertinent information for this titration includes:

Volume of the unknown acid solution, 25.00 mL

Volume of sodium hydroxide solution added, 35.00 mL

Concentration of the sodium hydroxide solution, 0.1000 *M*

Furthermore, from the balanced equation, we know that HCl and NaOH react in a 1:1 combining ratio.

Using a strategy involving conversion factors

$$35.00\ \text{m}\cancel{\text{L NaOH}} \times \frac{1\ \text{L}\ \cancel{\text{NaOH}}}{10^3\ \text{m}\cancel{\text{L NaOH}}} \times \frac{0.1000\ \text{mol NaOH}}{\text{L}\ \cancel{\text{NaOH}}} = 3.500 \times 10^{-3}\ \text{mol NaOH}$$

Continued—

EXAMPLE 9.8 —*Continued*

Knowing that HCl and NaOH undergo a 1:1 reaction,

$$3.500 \times 10^{-3}\ \cancel{\text{mol NaOH}} \times \frac{1\ \text{mol HCl}}{1\ \cancel{\text{mol NaOH}}} = 3.500 \times 10^{-3}\ \text{mol HCl}$$

3.500×10^{-3} mol HCl are contained in 25.00 mL of HCl solution.

Thus,

$$\frac{3.500 \times 10^{-3}\ \text{mol HCl}}{25.00\ \cancel{\text{mL HCl soln}}} \times \frac{10^3\ \cancel{\text{mL HCl soln}}}{1\ \text{L HCl soln}} = 1.400 \times 10^{-1}\ \text{mol HCl/L HCl soln}$$

$$= 0.1400\ M$$

The titration of an acid with a base is depicted in Figure 9.6.

Figure 9.4
The color of the petals of the hydrangea is formed by molecules that behave as acid-base indicators. The color is influenced by the pH of the soil in which the hydrangea is grown.

Figure 9.5
The relationship between pH and color of a variety of compounds commonly used as acid-base indicators. Many indicators are naturally occurring substances.

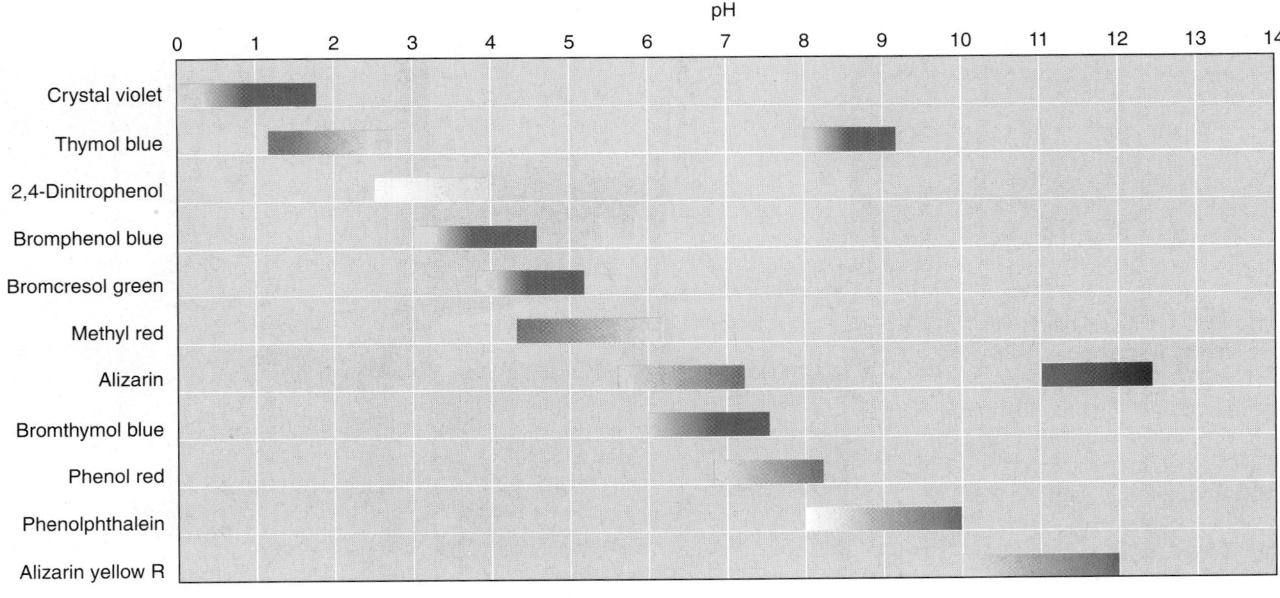

AN ENVIRONMENTAL PERSPECTIVE

Acid Rain

Acid rain is a global environmental problem that has raised public awareness of the chemicals polluting the air through the activities of our industrial society. Normal rain has a pH of about 5.6 as a result of the chemical reaction between carbon dioxide gas and water in the atmosphere. The following equation shows this reaction:

$$\underset{\text{Carbon dioxide}}{CO_2(g)} + \underset{\text{Water}}{H_2O(l)} \rightleftharpoons \underset{\text{Carbonic acid}}{H_2CO_3(aq)}$$

Acid rain refers to conditions that are much more acidic than this. In upstate New York the rain has as much as 25 times the acidity of normal rainfall. One rainstorm, recorded in West Virginia, produced rainfall that measured 1.5 on the pH scale. This is approximately the pH of stomach acid or about ten thousand times more acidic than "normal rain" (remember that the pH scale is logarithmic).

Acid rain is destroying life in streams and lakes. More than half the highland lakes in the western Adirondack Mountains have no native game fish. In addition to these 300 lakes, 140 lakes in Ontario have suffered a similar fate. It is estimated that 48,000 other lakes in Ontario and countless others in the northeastern and central United States are threatened. Our forests are endangered as well. The acid rain decreases soil pH, which in turn alters the solubility of minerals needed by plants. Studies have shown that about 40% of the red spruce and maple trees in New England have died. Increased acidity of rainfall appears to be the major culprit.

What is the cause of this acid rain? The combustion of fossil fuels (gas, oil, and coal) by power plants produces oxides of sulfur and nitrogen. Nitrogen oxides, in excess of normal levels, arise mainly from conversion of atmospheric nitrogen to nitrogen oxides in the engines of gasoline and diesel powered

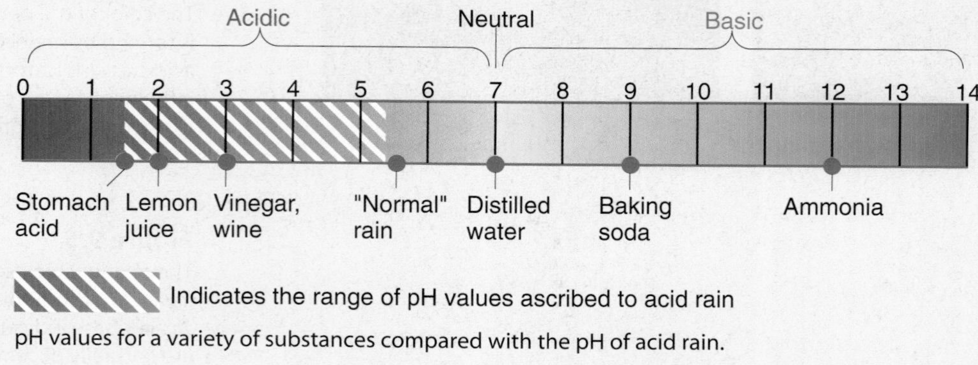

pH values for a variety of substances compared with the pH of acid rain.

(a)

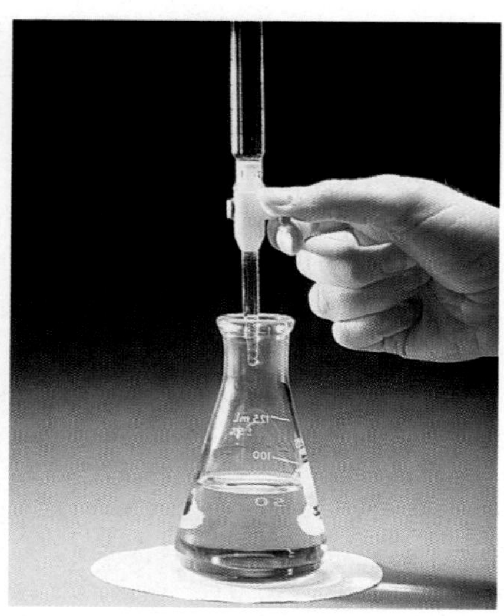

(b)

Figure 9.6
An acid-base titration. (a) An exact volume of a standard solution (in this example, a base) is added to a solution of unknown concentration (in this example, an acid). (b) From the volume (read from the buret) and concentration of the standard solution, coupled with the mass or volume of the unknown, the concentration of the unknown may be calculated.

vehicles. Sulfur oxides result from the oxidation of sulfur in fossil fuels. The sulfur atoms were originally a part of the amino acids and proteins of plants and animals that became, over the millenia, our fuel. These react with water, as does the CO_2 in normal rain, but the products are strong acids: sulfuric and nitric acids. Let's look at the equations for these processes.

In the atmosphere, nitric oxide (NO) can react with oxygen to produce nitrogen dioxide as shown:

$$\underset{\text{Nitric oxide}}{2NO(g)} + \underset{\text{Oxygen}}{O_2(g)} \longrightarrow \underset{\text{Nitrogen dioxide}}{2NO_2(g)}$$

Nitrogen dioxide (which causes the brown color of smog) then reacts with water to form nitric acid:

$$3NO_2(g) + H_2O(l) \longrightarrow 2HNO_3(aq) + NO(g)$$

A similar chemistry is seen with the sulfur oxides. Coal may contain as much as 3% sulfur. When the coal is burned, the sulfur also burns. This produces choking, acrid sulfur dioxide gas:

$$S(s) + O_2(g) \longrightarrow SO_2(g)$$

By itself, sulfur dioxide can cause serious respiratory problems for people with asthma or other lung diseases, but matters are worsened by the reaction of SO_2 with atmospheric oxygen:

$$2SO_2(g) + O_2(g) \longrightarrow 2SO_3(g)$$

Sulfur trioxide will react with water in the atmosphere:

$$SO_3(g) + H_2O(l) \longrightarrow H_2SO_4(aq)$$

The product, sulfuric acid, is even more irritating to the respiratory tract. When the acid rain created by the reactions shown above falls to earth, the impact is significant.

It is easy to balance these chemical equations, but decades could be required to balance the ecological systems that we have disrupted by our massive consumption of fossil fuels. A sudden decrease of even 25% in the use of fossil fuels would lead to worldwide financial chaos. Development of alternative fuel sources, such as solar energy and safe nuclear power, will help to reduce our dependence on fossil fuels and help us to balance the global equation.

Damage caused by acid rain.

Question 9.13

Calculate the molar concentration of a sodium hydroxide solution if 40.00 mL of this solution were required to neutralize 20.00 mL of a 0.2000 *M* solution of hydrochloric acid.

Question 9.14

Calculate the molar concentration of a sodium hydroxide solution if 36.00 mL of this solution were required to neutralize 25.00 mL of a 0.2000 *M* solution of hydrochloric acid.

Polyprotic Substances

Not all acid-base reactions occur in a 1:1 combining ratio (as hydrochloric acid and sodium hydroxide in the previous example). Acid-base reactions with other than 1:1 combining ratios occur between what are termed *polyprotic substances.* **Polyprotic substances** donate (as acids) or accept (as bases) more than one proton per formula unit.

Reactions of Polyprotic Substances

HCl dissociates to produce one H^+ ion for each HCl. For this reason, it is termed a *monoprotic acid.* Its reaction with sodium hydroxide is:

$$HCl(aq) + NaOH(aq) \longrightarrow H_2O(l) + Na^+(aq) + Cl^-(aq)$$

Sulfuric acid, in contrast, is a *diprotic acid.* Each unit of H_2SO_4 produces two H^+ ions (the prefix *di-* indicating two). Its reaction with sodium hydroxide is:

$$H_2SO_4(aq) + 2NaOH(aq) \longrightarrow 2H_2O(l) + 2Na^+(aq) + SO_4^{2-}(aq)$$

Phosphoric acid is a *triprotic acid.* Each unit of H_3PO_4 produces three H^+ ions. Its reaction with sodium hydroxide is:

$$H_3PO_4(aq) + 3NaOH(aq) \longrightarrow 3H_2O(l) + 3Na^+(aq) + PO_4^{3-}(aq)$$

Dissociation of Polyprotic Substances

Sulfuric acid, and other diprotic acids, dissociate in two steps:

Step 1. $H_2SO_4(aq) + H_2O(l) \longrightarrow H_3O^+(aq) + HSO_4^-(aq)$

Step 2. $HSO_4^-(aq) + H_2O(l) \rightleftharpoons H_3O^+(aq) + SO_4^{2-}(aq)$

Notice that H_2SO_4 behaves as a strong acid (Step 1) and HSO_4^- behaves as a weak acid, indicated by a double arrow (Step 2).

Phosphoric acid dissociates in three steps, all forms behaving as weak acids.

Step 1. $H_3PO_4(aq) + H_2O(l) \rightleftharpoons H_3O^+(aq) + H_2PO_4^-(aq)$

Step 2. $H_2PO_4^-(aq) + H_2O(l) \rightleftharpoons H_3O^+(aq) + HPO_4^{2-}(aq)$

Step 3. $HPO_4^{2-}(aq) + H_2O(l) \rightleftharpoons H_3O^+(aq) + PO_4^{3-}(aq)$

Bases exhibit this property as well.

NaOH produces one OH^- ion per formula unit:

$$NaOH(aq) \longrightarrow Na^+(aq) + OH^-(aq)$$

$Ba(OH)_2$, barium hydroxide, produces two OH^2 ions per formula unit:

$$Ba(OH)_2(aq) \longrightarrow Ba^{2+}(aq) + 2OH^-(aq)$$

9.4 Acid-Base Buffers

Learning Goal

A **buffer solution** contains components that enable the solution to resist large changes in pH when either acids or bases are added. Buffer solutions may be prepared in the laboratory to maintain optimum conditions for a chemical reaction. Buffers are routinely used in commercial products to maintain optimum conditions for product behavior (Figure 9.7).

Buffer solutions also occur naturally. Blood, for example, is a complex natural buffer solution maintaining a pH of approximately 7.4, optimum for oxygen transport. The major buffering agent in blood is the mixture of carbonic acid (H_2CO_3) and bicarbonate ions (HCO_3^-).

The Buffer Process

The basis of buffer action is the establishment of an equilibrium between either a weak acid and its conjugate base or a weak base and its conjugate acid. Let's consider the case of a weak acid and its salt.

A common buffer solution may be prepared from acetic acid (CH_3COOH) and sodium acetate (CH_3COONa). Sodium acetate is a salt that is the source of the conjugate base CH_3COO^-. An *equilibrium* is established in solution between the weak acid and the conjugate base.

We ignore Na^+ in the description of the buffer. Na^+ does not actively participate in the reaction.

$$\underset{\text{(weak acid)}}{\underset{\text{Acetic acid}}{CH_3COOH(aq)}} + \underset{\text{Water}}{H_2O(l)} \rightleftharpoons \underset{\text{Hydronium ion}}{H_3O^+(aq)} + \underset{\text{(conjugate base)}}{\underset{\text{Acetate ion}}{CH_3COO^-(aq)}}$$

The acetate ion is the conjugate base of acetic acid.

A buffer solution functions in accordance with LeChatelier's principle, which states that an equilibrium system, when stressed, will shift its equilibrium to relieve that stress. This principle is illustrated by the following examples.

Figure 9.7
Commercial products that claim improved function owing to their ability to control pH.

Addition of Base (OH^-) to a Buffer Solution

Addition of a basic substance to a buffer solution causes the following changes.

- OH^- from the base reacts with H_3O^+ producing water.
- Molecular acetic acid *dissociates* to replace the H_3O^+ consumed by the base, maintaining the pH close to the initial level.

This is an example of LeChatelier's principle, because the loss of H_3O^+ (the *stress*) is compensated by the dissociation of acetic acid to produce more H_3O^+.

Addition of Acid (H_3O^+) to a Buffer Solution

Addition of an acidic solution to a buffer results in the following changes.

- H_3O^+ from the acid increases the overall $[H_3O^+]$.
- The system reacts to this stress, in accordance with LeChatelier's principle, to form more molecular acetic acid; the acetate ion combines with H_3O^+. Thus, the H_3O^+ concentration and therefore, the pH, remain close to the initial level.

These effects may be summarized as follows:

$$CH_3COOH(aq) + H_2O(l) \rightleftharpoons H_3O^+(aq) + CH_3COO^-(aq)$$

$$\xrightarrow{\text{OH}^- \text{ added, equilibrium shifts to the right}}$$

$$\xleftarrow{\text{H}_3\text{O}^+ \text{ added, equilibrium shifts to the left}}$$

Buffer capacity is a measure of the ability of a solution to resist large changes in pH when a strong acid or strong base is added. More specifically, buffer capacity is described as the amount of strong acid or strong base that a buffer can neutralize without significantly changing its pH. Buffering capacity against base is a function of the concentration of the weak acid (in this case CH_3COOH). Buffering capacity against acid is dependent on the concentration of the anion of the salt, the conjugate base (CH_3COO^- in this example).

Question 9.15

Explain how the molar concentration of H_2CO_3 in the blood would change if the partial pressure of CO_2 in the lungs were to increase. (Refer to A Clinical Perspective: Control of Blood pH on page 255.)

Question 9.16

Explain how the molar concentration of H_2CO_3 in the blood would change if the partial pressure of CO_2 in the lungs were to decrease. (Refer to A Clinical Perspective: Control of Blood pH on page 255.)

Question 9.17

Explain how the molar concentration of hydronium ion in the blood would change under each of the conditions described in Questions 9.15 and 9.16.

Question 9.18

Explain how the pH of blood would change under each of the conditions described in Questions 9.15 and 9.16.

Preparation of a Buffer Solution

It is useful to understand how to prepare a buffer solution and how to determine the pH of the resulting solution. Many chemical reactions produce the largest amount of product only when they are run at a known, constant pH. The study of biologically important processes in the laboratory often requires conditions that approximate the composition of biological fluids. A constant pH would certainly be essential.

The buffer process is an equilibrium reaction and is described by an equilibrium constant expression. For acids, the equilibrium constant is represented as K_a, the subscript *a* implying an acid equilibrium. For example, the acetic acid/sodium acetate system is described by

$$CH_3COOH(aq) + H_2O(l) \rightleftharpoons H_3O^+(aq) + CH_3COO^-(aq)$$

and

$$K_a = \frac{[H_3O^+][CH_3COO^-]}{[CH_3COOH]}$$

Using a few mathematical maneuvers we can turn this equilibrium-constant expression into one that will allow us to calculate the pH of the buffer if we know how much acid (acetic acid) and salt (sodium acetate) are present in a known volume of the solution.

First, multiply both sides of the equation by the concentration of acetic acid, $[CH_3COOH]$. This will eliminate the denominator on the right side of the equation.

$$[CH_3COOH]K_a = \frac{[H_3O^+][CH_3COO^-][CH_3COOH]}{[CH_3COOH]}$$

or

$$[CH_3COOH]K_a = [H_3O^+][CH_3COO^-]$$

The calculation of pH from $[H_3O^+]$ is discussed in Section 9.2.

Now, dividing both sides of the equation by the acetate ion concentration $[CH_3COO^-]$ will give us an expression for the hydronium ion concentration $[H_3O^+]$

$$\frac{[CH_3COOH]K_a}{[CH_3COO^-]} = [H_3O^+]$$

Once we know the value for $[H_3O^+]$, we can easily find the pH.

To use this equation:

- assume that $[CH_3COOH]$ represents the concentration of the acid component of the buffer.
- assume that $[CH_3COO^-]$ represents the concentration of the conjugate base (principally from the dissociation of the salt, sodium acetate) component of the buffer.

$$\begin{matrix} \text{acid} \longrightarrow \\ \text{conjugate base} \longrightarrow \end{matrix} \frac{[CH_3COOH]K_a}{[CH_3COO^-]} = [H_3O^+]$$

$$\frac{[\text{acid}]K_a}{[\text{conjugate base}]} = [H_3O^+]$$

Let's look at an example of a practical application of this equation.

Calculating the pH of a Buffer Solution

EXAMPLE 9.9

Calculate the pH of a buffer solution in which both the acetic acid and sodium acetate concentrations are 1.00×10^{-1} *M*. The equilibrium constant, K_a, for acetic acid is 1.75×10^{-5}.

Solution

Acetic acid is the acid; [acid] = 1.00×10^{-1} *M*
Sodium acetate is the salt, furnishing the conjugate base; [conjugate base] = 1.00×10^{-1} *M*

The equilibrium is

$$\underset{\text{acid}}{CH_3COOH(aq)} + H_2O(l) \rightleftharpoons H_3O^+(aq) + \underset{\text{conjugate base}}{CH_3COO^-(aq)}$$

and the hydronium ion concentration,

$$[H_3O^+] = \frac{[\text{acid}]K_a}{[\text{conjugate base}]}$$

Substituting the values given in the problem

$$[H_3O^+] = \frac{[1.00 \times 10^{-1}]1.75 \times 10^{-5}}{[1.00 \times 10^{-1}]}$$

$$[H_3O^+] = 1.75 \times 10^{-5}$$

and because

$$pH = -\log [H_3O^+]$$

$$pH = -\log 1.75 \times 10^{-5}$$

$$= 4.76$$

The pH of the buffer solution is 4.76.

Calculating the pH of a Buffer Solution

EXAMPLE 9.10

Calculate the pH of a buffer solution similar to that described in Example 9.9 except that the acid concentration is doubled, while the salt concentration remains the same.

Solution

Acetic acid is the acid; [acid] = 2.00×10^{-1} *M* (remember, the acid concentration is twice that of Example 9.9; $2 \times [1.00 \times 10^{-1}] = 2.00 \times 10^{-1}$ *M*

Sodium acetate is the salt, furnishing the conjugate base; [conjugate base] = 1.00×10^{-1} *M*

The equilibrium is

$$\underset{\text{acid}}{CH_3COOH(aq)} + H_2O(l) \rightleftharpoons H_3O^+(aq) + \underset{\text{conjugate base}}{CH_3COO^-(aq)}$$

Continued—

EXAMPLE 9.10 —*Continued*

and the hydronium ion concentration,

$$[H_3O^+] = \frac{[\text{acid}]K_a}{[\text{conjugate base}]}$$

Substituting the values given in the problem

$$[H_3O^+] = \frac{[2.00 \times 10^{-1}]1.75 \times 10^{-5}}{[1.00 \times 10^{-1}]}$$

$$[H_3O^+] = 3.50 \times 10^{-5}$$

and because

$$\text{pH} = -\log [H_3O^+]$$
$$\text{pH} = -\log 3.50 \times 10^{-5}$$
$$= 4.46$$

The pH of the buffer solution is 4.46.

A comparison of the two solutions described in Examples 9.9 and 9.10 demonstrates a buffer solution's most significant attribute: the ability to stabilize pH. Although the acid concentration of these solutions differs by a factor of two, the difference in their pH is only 0.301 units.

Question 9.19

A buffer solution is prepared in such a way that the concentrations of propanoic acid and sodium propanoate are each 2.00×10^{-1} *M*. If the buffer equilibrium is described by

$$C_2H_5COOH(aq) + H_2O(l) \rightleftharpoons H_3O^+(aq) + C_2H_5COO^-(aq)$$

Propanoic acid — Propanoate anion

with $K_a = 1.34 \times 10^{-5}$, calculate the pH of the solution.

Question 9.20

Calculate the pH of the buffer solution in Question 9.19 if the concentration of the sodium propanoate were doubled while the acid concentration remained the same.

The Henderson-Hasselbalch Equation

The solution of the equilibrium-constant expression and the pH are sometimes combined into one operation. The combined expression is termed the *Henderson-Hasselbalch* equation.

pKa = −log Ka, analogous to pH = −log [H₃O⁺].

For the acetic acid/sodium acetate buffer system,

$$CH_3COOH(aq) + H_2O(l) \rightleftharpoons H_3O^+(aq) + CH_3COO^-(aq)$$

the Henderson-Hasselbalch expression is:

$$\text{pH} = \text{p}K_a - \log \frac{[CH_3COOH]}{[CH_3COO^-]}$$

or its equivalent form:

$$\text{pH} = \text{p}K_a + \log \frac{[CH_3COO^-]}{[CH_3COOH]}$$

A CLINICAL PERSPECTIVE

Control of Blood pH

A pH of 7.4 is maintained in blood partly by a carbonic acid–bicarbonate buffer system based on the following equilibrium:

$$H_2CO_3(aq) + H_2O(l) \rightleftharpoons H_3O^+(aq) + HCO_3^-(aq)$$

Carbonic acid (weak acid) — Bicarbonate ion (salt)

The regulation process based on LeChatelier's principle is similar to the acetic acid–sodium acetate buffer, which we have already discussed.

Red blood cells transport O_2, bound to hemoglobin, to the cells of body tissue. The metabolic waste product, CO_2, is picked up by the blood and delivered to the lungs.

The CO_2 in the blood also participates in the carbonic acid–bicarbonate buffer equilibrium. Carbon dioxide reacts with water in the blood to form carbonic acid:

$$CO_2(aq) + H_2O(l) \rightleftharpoons H_2CO_3(aq)$$

As a result the buffer equilibrium becomes more complex:

$$CO_2(aq) + 2H_2O(l) \rightleftharpoons H_2CO_3(aq) + H_2O(l) \rightleftharpoons H_3O^+(aq) + HCO_3^-(aq)$$

Through this sequence of relationships the concentration of CO_2 in the blood affects the blood pH.

Higher than normal CO_2 concentrations shift the above equilibrium to the right (LeChatelier's principle), increasing $[H_3O^+]$ and lowering the pH. The blood becomes too acidic, leading to numerous medical problems. A situation of high blood CO_2 levels and low pH is termed *acidosis.* Respiratory acidosis results from various diseases (emphysema, pneumonia) that restrict the breathing process, causing the buildup of waste CO_2 in the blood.

Lower than normal CO_2 levels, on the other hand, shift the equilibrium to the left, decreasing $[H_3O^+]$ and making the pH more basic. This condition is termed *alkalosis* (from "alkali," implying basic). Hyperventilation, or rapid breathing, is a common cause of respiratory alkalosis.

The form of this equation is especially amenable to buffer problem calculations. In this expression, $[CH_3COOH]$ represents the molar concentration of the weak acid and $[CH_3COO^-]$ is the molar concentration of the conjugate base of the weak acid. The generalized expression is:

$$pH = pK_a + \log \frac{[\text{conjugate base}]}{[\text{weak acid}]}$$

Substituting concentrations along with the value for the pK_a of the acid allows the calculation of the pH of the buffer solution in problems such as those shown in Examples 9.9 and 9.10 as well as Questions 9.19 and 9.20.

Question 9.21

Solve the problem in Example 9.9 using the Henderson-Hasselbalch equation.

Question 9.22

Solve the problem in Example 9.10 using the Henderson-Hasselbalch equation.

Question 9.23

Solve Question 9.19 using the Henderson-Hasselbalch equation.

Question 9.24

Solve Question 9.20 using the Henderson-Hasselbalch equation.

9.5 Oxidation-Reduction Processes

Learning Goal

Oxidation-reduction processes are responsible for many types of chemical change. Corrosion, the operation of a battery, and biochemical energy-harvesting reactions are a few examples. In this section we explore the basic concepts underlying this class of chemical reactions.

A CLINICAL PERSPECTIVE

Oxidizing Agents for Chemical Control of Microbes

Before the twentieth century, hospitals were not particularly sanitary establishments. Refuse, including human waste, was disposed of on hospital grounds. Because many hospitals had no running water, physicians often cleaned their hands and instruments by wiping them on their lab coats and then proceeded to treat the next patient! As you can imagine, many patients died of infections in hospitals.

By the late nineteenth century a few physicians and microbiologists had begun to realize that infectious diseases are transmitted by microbes, including bacteria and viruses. To decrease the number of hospital-acquired infections, physicians like Joseph Lister and Ignatz Semmelweis experimented with chemicals and procedures that were designed to eliminate pathogens from environmental surfaces and from wounds.

Many of the common disinfectants and antiseptics are oxidizing agents. A disinfectant is a chemical that is used to kill or inhibit the growth of pathogens, disease-causing microorganisms, on environmental surfaces. An antiseptic is a milder chemical that is used to destroy pathogens associated with living tissue.

Hydrogen peroxide is an effective antiseptic that is commonly used to cleanse cuts and abrasions. We are all familiar with the furious bubbling that occurs as the enzyme catalase from our body cells catalyzes the breakdown of H_2O_2:

$$2H_2O_2(aq) \longrightarrow 2H_2O(l) + O_2(g)$$

A highly reactive and deadly form of oxygen, the superoxide radical (O_2^-), is produced during this reaction. This molecule inactivates proteins, especially critical enzyme systems.

At higher concentrations (3–6%), H_2O_2 is used as a disinfectant. It is particularly useful for disinfection of soft contact lenses, utensils, and surgical implants because there is no residual toxicity. Concentrations of 6–25% are even used for complete sterilization of environmental surfaces.

Benzoyl peroxide is another powerful oxidizing agent. Ointments containing 5–10% benzoyl peroxide have been used as antibacterial agents to treat acne. The compound is currently found in over-the-counter facial scrubs because it is also an exfoliant, causing sloughing of old skin and replacement with smoother-looking skin. A word of caution is in order: in sensitive individuals, benzoyl peroxide can cause swelling and blistering of tender facial skin.

Chlorine is a very widely used disinfectant and antiseptic. Calcium hypochlorite [$Ca(OCl)_2$] was first used in hospital maternity wards in 1847 by the pioneering Hungarian physician Ignatz Semmelweis. Semmelweis insisted that hospital workers cleanse their hands in a $Ca(OCl)_2$ solution and dramatically reduced the incidence of infection. Today, calcium hypochlorite is more commonly used to disinfect bedding, clothing, restaurant eating utensils, slaughterhouses, barns, and dairies.

Sodium hypochlorite (NaOCl), sold as Clorox, is used as a household disinfectant and deodorant but is also used to disinfect swimming pools, dairies, food-processing equipment, and kidney dialysis units. It can be used to treat drinking water of questionable quality. Addition of 1/2 teaspoon of household bleach (5.25% NaOCl) to 2 gallons of clear water renders it drinkable after 1/2 hour. The Centers for Disease Control even recommend a 1:10 dilution of bleach as an effective disinfectant against human immunodeficiency virus, the virus that causes acquired immune deficiency syndrome (AIDS).

Chlorine gas (Cl_2) is used to disinfect swimming pool water, sewage, and municipal water supplies. This treatment has successfully eliminated epidemics of waterborne diseases. However, chlorine is inactivated in the presence of some organic materials and, in some cases, may form toxic chlorinated organic compounds. For these reasons, many cities are considering the use of ozone (O_3) rather than chlorine.

Ozone is produced from O_2 by high-voltage electrical discharges. (That fresh smell in the air after an electrical storm is ozone.) Several European cities use ozone to disinfect drinking water. It is a more effective killing agent than chlorine, especially with some viruses: less ozone is required for disinfection; there is no unpleasant residual odor or flavor; and there appear to be fewer toxic by-products. However, ozone is more expensive than chlorine, and maintaining the required concentration in the water is more difficult. Nonetheless, the benefits seem to outweigh the drawbacks, and many U.S. cities may soon follow the example of European cities and convert to the use of ozone for water treatment.

Oxidation and Reduction

Oxidation is defined as a loss of electrons, loss of hydrogen atoms, or gain of oxygen atoms. *Sodium metal,* is, for example, oxidized to a *sodium ion,* losing one electron when it reacts with a nonmetal such as chlorine:

$$Na \longrightarrow Na^+ + e^-$$

Reduction is defined as a gain of electrons, gain of hydrogen atoms, or loss of oxygen atoms. A *chlorine atom* is reduced to a *chloride ion* by gaining one electron when it reacts with a metal such as sodium:

$$Cl + e^- \longrightarrow Cl^-$$

Oxidation and reduction are complementary processes. The *oxidation half-reaction* produces an electron that is the reactant for the *reduction half-reaction.* The combination of two half-reactions, one oxidation and one reduction, produces the complete reaction:

Oxidation half-reaction:	$Na \longrightarrow Na^+ + e^-$
Reduction half-reaction:	$Cl + e^- \longrightarrow Cl^-$
Complete reaction:	$Na + Cl \longrightarrow Na^+ + Cl^-$

Half-reactions, one oxidation and one reduction, are exactly that: one-half of a complete reaction. The two half-reactions combine to produce the complete reaction. Note that the electrons cancel: in the electron transfer process, no free electrons remain.

In the preceding reaction, sodium metal is the **reducing agent.** It releases electrons for the reduction of chlorine. Chlorine is the **oxidizing agent.** It accepts electrons from the sodium, which is oxidized.

The reducing agent becomes oxidized and the oxidizing agent becomes reduced.

The characteristics of oxidizing and reducing agents may be summarized as follows:

Oxidizing Agent	**Reducing Agent**
• Is reduced	• Is oxidized
• Gains electrons	• Loses electrons
• Causes oxidation	• Causes reduction

Applications of Oxidation and Reduction

Learning Goal

Oxidation-reduction processes are important in many areas as diverse as industrial manufacturing and biochemical processes.

Corrosion

The deterioration of metals caused by an oxidation-reduction process is termed **corrosion.** Metal atoms are converted to metal ions; the structure, hence the properties, changes dramatically, and usually for the worse (Figure 9.8).

At the same time that iron is oxidized, O_2 is being reduced to O^{2-} and is incorporated into the structure of iron(III) oxide. Electrons lost by iron reduce oxygen. This again shows that oxidation and reduction processes go hand in hand.

Millions of dollars are spent annually in an attempt to correct the damage resulting from corrosion. A current area of chemical research is concerned with the development of corrosion-inhibiting processes. In one type of corrosion, elemental iron is oxidized to iron(III) oxide (rust):

$$4Fe(s) + 3O_2(g) \longrightarrow 2Fe_2O_3(s)$$

Figure 9.8
The rust (an oxide of iron) that diminishes structural strength and ruins the appearance of automobiles, bridges, and other iron-based objects is a common example of an oxidation-reduction reaction.

A CLINICAL PERSPECTIVE

Electrochemical Reactions in the Statue of Liberty and in Dental Fillings

Throughout history, we have suffered from our ignorance of basic electrochemical principles. For example, during the Middle Ages, our chemistry ancestors (alchemists) placed an iron rod into a blue solution of copper sulfate. They noticed that bright shiny copper plated out onto an iron rod and they thought that they had changed a base metal, iron, into copper. What actually happened was the redox reaction shown in Equation 1.

$$2Fe(s) + 3Cu^{2+}(aq) \longrightarrow 2Fe^{3+}(aq) + 3Cu(s) \quad (1)$$

This misunderstanding encouraged them to embark on a futile, one-thousand-year attempt to change base metals into gold.

Over one hundred years ago, France presented the United States with the Statue of Liberty. Unfortunately, the French did not anticipate the redox reaction shown in Equation 1 when they mounted the copper skin of the statue on iron support rods. Oxygen in the atmosphere oxidized the copper skin to produce copper ions. Then, because iron is more active than copper, the displacement shown in Equation 1 aided the corrosion of the support bars. As a result of this and other reactions, the statue needed refurbishing before we celebrated its one hundredth anniversary in 1986.

Sometimes dentists also overlook possible redox reactions when placing gold caps over teeth next to teeth with amalgam fillings. The amalgam in tooth fillings is an alloy of mercury, silver, tin, and copper. Atmospheric oxygen oxidizes some of the gold cap to gold ions. Because the metals in the amalgam are more active than gold, contact between the amalgam fillings and gold ions results in redox reactions such as the following.*

$$3Sn(s) + 2Au^{3+}(aq) \longrightarrow 3Sn^{2+}(aq) + 2Au(s) \quad (2)$$

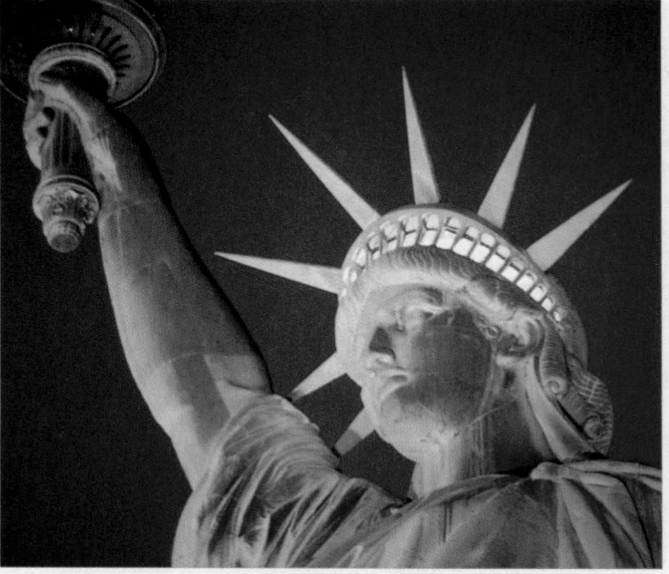

As a result, the dental fillings dissolve and the patients are left with a constant metallic taste in their mouths.

These examples show that like our ancestors, we continue to experience unfortunate results because of a lack of understanding of basic electrochemical principles.

Source: Ronald DeLorenzo, *Journal of Chemical Education*, May 1985, pages 424–425.

*Equation 2 is oversimplified to illustrate more clearly the basic displacement of gold ions by metallic tin atoms. Actually, only complex ions of gold and tin can exist in aqueous solutions, not the simple cations that are shown.

Combustion of Fossil Fuels

Burning fossil fuel is an extremely exothermic process. Energy is released to heat our homes, offices, and classrooms. The simplest fossil fuel is methane, CH_4, and its oxidation reaction is written:

$$CH_4(g) + 2O_2(g) \longrightarrow CO_2(g) + 2H_2O(g)$$

Methane is a hydrocarbon. The complete oxidation of any hydrocarbon (including those in gasoline, heating oil, liquid propane, and so forth) produces carbon dioxide and water. The energy released by these reactions is of paramount importance. The water and carbon dioxide are viewed as waste products, and the carbon dioxide contributes to the greenhouse effect (see An Environmental Perspective: The Greenhouse Effect and Global Warming on page 157).

Bleaching

Bleaching agents are most often oxidizing agents. Sodium hypochlorite (NaOCl) is the active ingredient in a variety of laundry products. It is an effective oxidizing agent. Products containing NaOCl are advertised for their stain-removing capabilities.

Stains are a result of colored compounds adhering to surfaces. Oxidation of these compounds produces products that are not colored or compounds that are subsequently easily removed from the surface, thus removing the stain.

Biological Processes

Respiration

There are many examples of biological oxidation-reduction reactions. For example, the electron-transport chain of aerobic respiration involves the reversible oxidation and reduction of iron atoms in cytochrome *c*,

$$\text{cytochrome } c\ (Fe^{3+}) + e^- \longrightarrow \text{cytochrome } c\ (Fe^{2+})$$

The reduced iron ion transfers an electron to an iron ion in another protein, called cytochrome *c* oxidase, according to the following reaction:

$$\text{cytochrome } c\ (Fe^{2+}) + \text{cytochrome } c \text{ oxidase } (Fe^{3+}) \longrightarrow \text{cytochrome } c\ (Fe^{3+}) + \text{cytochrome } c \text{ oxidase } (Fe^{2+})$$

Cytochrome *c* oxidase eventually passes four electrons to O_2, the final electron acceptor of the chain:

$$O_2 + 4e^- + 4H^+ \longrightarrow 2H_2O$$

See Chapters 22 and 23 for the details of these energy-harvesting cellular oxidation-reduction reactions.

Metabolism

When ethanol is metabolized in the liver, it is oxidized to acetaldehyde (the molecule partially responsible for hangovers). Continued oxidation of acetaldehyde produces acetic acid, which is eventually oxidized to CO_2 and H_2O. These reactions, summarized as follows, are catalyzed by liver enzymes.

$$\underset{\text{Ethanol}}{CH_3CH_2\text{—}OH} \longrightarrow \underset{\text{Acetaldehyde}}{CH_3\overset{\overset{\large O}{\|}}{C}\text{—}H} \longrightarrow \underset{\text{Acetic acid}}{CH_3\overset{\overset{\large O}{\|}}{C}\text{—}OH} \longrightarrow CO_2 + H_2O$$

It is more difficult to recognize these reactions as oxidations because neither the product nor the reactant carries a charge. In previous examples we looked for an increase in positive charge as an indication that an oxidation had occurred. A decrease in positive charge (or increased negative charge) would signify reduction.

Alternative descriptions of oxidation and reduction are useful in identifying these reactions.

Oxidation is the *gain* of oxygen or *loss* of hydrogen.

Reduction is the *loss* of oxygen or *gain* of hydrogen.

In the conversion of ethanol to acetaldehyde, ethanol has six hydrogen atoms per molecule; the product acetaldehyde has four hydrogen atoms per molecule. This represents a loss of two hydrogen atoms per molecule. Therefore, ethanol has been oxidized to acetaldehyde, based on the interpretation of the above-mentioned rules.

This strategy is most useful for recognizing oxidation and reduction of *organic compounds* and organic compounds of biological interest, *biochemical compounds*. Organic compounds and their structures and reactivity are the focus of Chapters 11 through 16 and biochemical compounds are described in Chapters 17 through 24.

Voltaic Cells

When zinc metal is dipped into a copper(II) sulfate solution, zinc atoms are oxidized to zinc ions and copper(II) ions are reduced to copper metal, which deposits on the surface of the zinc metal (Figure 9.9). This reaction is summarized as follows:

Learning Goal

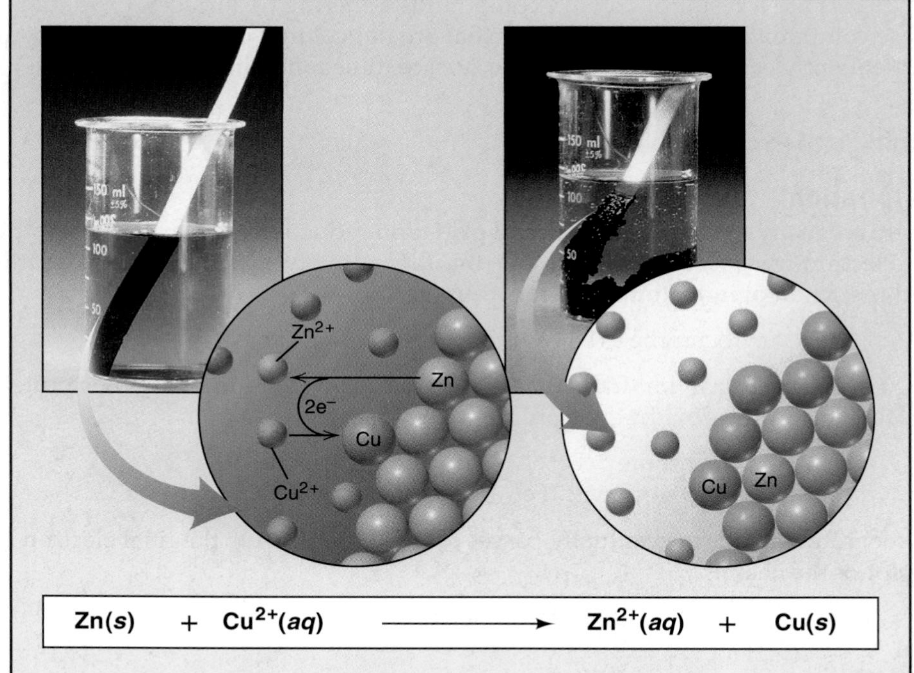

Figure 9.9
The spontaneous reaction of zinc metal and Cu^{2+} ions is the basis of the cell depicted in Figure 9.10.

$$\overbrace{}^{\text{Oxidation/e}^- \text{ loss}}\quad Zn(s) + Cu^{2+}(aq) \longrightarrow Zn^{2+}(aq) + Cu(s) \quad \underbrace{}_{\text{Reduction/e}^- \text{ gain}}$$

In the reduction of aqueous copper(II) ions by zinc metal, electrons flow from the zinc rod directly to copper(II) ions in the solution. If electron transfer from the zinc rod to the copper ions in solution could be directed through an external electrical circuit, this spontaneous oxidation-reduction reaction could be used to produce an electrical current that could perform some useful function.

Recall that solutions of ionic salts are good conductors of electricity (Chapter 7).

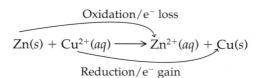

However, when zinc metal in one container is connected by a copper wire with a copper(II) sulfate solution in a separate container, no current flows through the wire. A complete, or continuous circuit is necessary for current to flow. To complete the circuit, we connect the two containers with a tube filled with a solution of an electrolyte such as potassium chloride. This tube is described as a *salt bridge.*

Current now flows through the external circuit (Figure 9.10). The device shown in Figure 9.10 is an example of a *voltaic cell.* A **voltaic cell** is an *electrochemical* cell that converts stored *chemical* energy into *electrical* energy.

This cell consists of two *half-cells.* The oxidation half-reaction occurs in one half-cell and the reduction half-reaction occurs in the other half-cell. The sum of the two half-cell reactions is the overall oxidation-reduction reaction that describes the cell. The electrode at which oxidation occurs is called the **anode,** and the electrode at which reduction occurs is the **cathode.** In the device shown in Figure 9.10, the zinc metal is the anode. At this electrode the zinc atoms are oxidized to zinc ions:

$$\text{Anode half-reaction: } Zn(s) \longrightarrow Zn^{2+}(aq) + 2e^-$$

Electrons released at the anode travel through the external circuit to the cathode (the copper rod) where they are transferred to copper(II) ions in the solution. Copper(II) ions are reduced to copper atoms that deposit on the copper metal surface, the cathode:

A Medical Perspective

Turning the Human Body into a Battery

The heart has its own natural pacemaker that sends nerve impulses (pulses of electrical current) throughout the heart approximately seventy-two times per minute. These electrical pulses cause your heart muscles to contract (beat), which pumps blood through the body. The fibers that carry the nerve impulses can be damaged by disease, drugs, heart attacks, and surgery. When these heart fibers are damaged, the heart may run too slowly, stop temporarily, or stop altogether. To correct this condition, artificial heart pacemakers (see figure below) are surgically inserted in the human body. A pacemaker (pacer) is a battery-driven device that sends an electrical current (pulse) to the heart about seventy-two times per minute. Over 300,000 Americans are now wearing artificial pacemakers with an additional 30,000 pacemakers installed each year.

Yearly operations used to be necessary to replace the pacemaker's batteries. Today, pacemakers use improved batteries that last much longer, but even these must be replaced eventually.

It would be very desirable to develop a permanent battery to run pacemakers. Some scientists began working on ways of converting the human body itself into a battery (voltaic cell) to power artificial pacemakers.

Several methods for using the human body as a voltaic cell have been suggested. One of these is to insert platinum and zinc electrodes into the human body as diagrammed in the figure below. The pacemaker and the electrodes would be worn internally. This "body battery" could easily generate the small amount of current (5×10^{-5} ampere) that is required by most pacemakers. This "body battery" has been tested on animals for periods exceeding four months without noticeable problems.

Source: Ronald DeLorenzo, *Problem Solving in General Chemistry*, 2nd ed., Wm. C. Brown, Publishers, Dubuque, Iowa, 1993, pages 336–338.

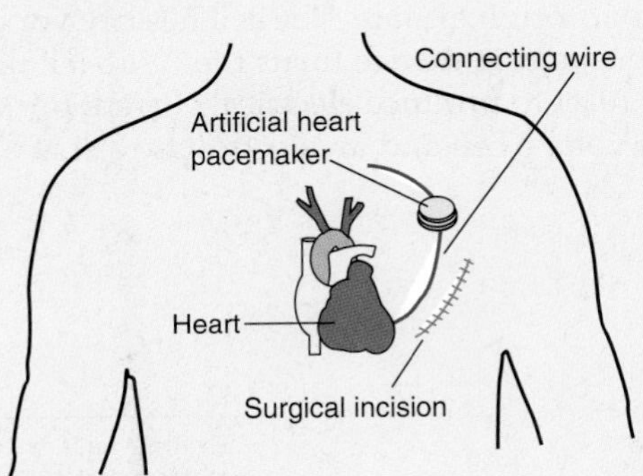

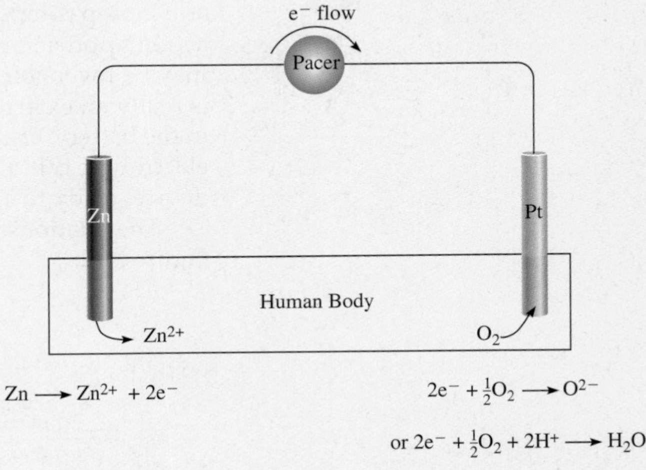

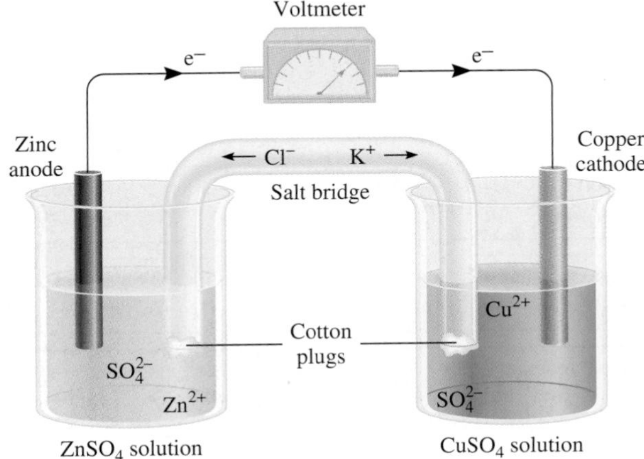

Figure 9.10
A voltaic cell generating electrical current by the reaction:

$$Zn(s) + Cu^{2+}(aq) \longrightarrow Zn^{2+}(aq) + Cu(s)$$

Each electrode consists of the pure metal, zinc or copper. Zinc is oxidized, releasing electrons that flow to the copper, reducing Cu^{2+} to Cu. The salt bridge completes the circuit and the voltmeter displays the voltage (or chemical potential) associated with the reaction.

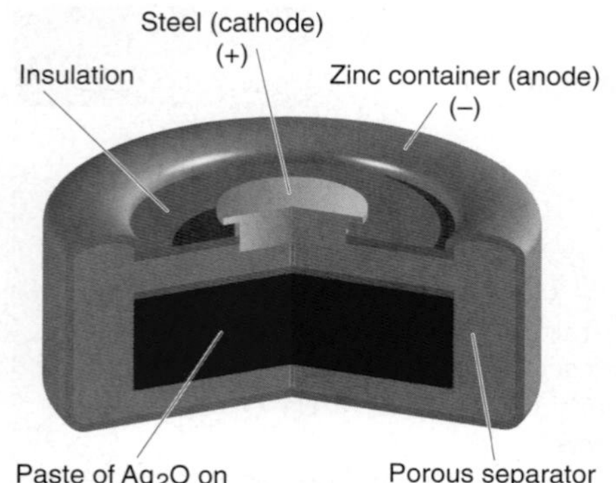

Figure 9.11
A silver battery used in cameras, heart pacemakers, and hearing aids. This battery is small, stable, and nontoxic (hence implantable in the human body).

$$\text{Cathode half-reaction: } Cu^{2+}(aq) + 2e^- \longrightarrow Cu(s)$$

The sum of these half-cell reactions is the cell reaction:

$$Zn(s) + Cu^{2+}(aq) \longrightarrow Zn^{2+}(aq) + Cu(s)$$

Voltaic cells are found in many aspects of our life, as convenient and reliable sources of electrical energy, the battery. Batteries convert stored chemical energy to an electrical current to power a wide array of different commercial appliances: radios, portable televisions and computers, flashlights, a host of other useful devices.

Technology has made modern batteries smaller, safer, and more dependable than our crudely constructed copper-zinc voltaic cell. In fact, the silver cell (Figure 9.11) is sufficiently safe and nontoxic that it can be implanted in the human body as a part of a pacemaker circuit that is used to improve heart rhythm. A rather futuristic potential application of voltaic cells is noted in A Medical Perspective: Turning the Human Body into a Battery on page 261.

Electrolysis

Learning Goal

Electrolysis reactions use electrical energy to cause nonspontaneous oxidation-reduction reactions to occur. They are the reverse of voltaic cells. One common application is the rechargeable battery. When it is being used to power a device, such as a laptop computer, it behaves as a voltaic cell. After some time, the chemical reaction approaches completion and the voltaic cell "runs down." The cell reaction may be reversible. If so, the battery is plugged into a battery charger. The charger is really an external source of electrical energy that reverses the chemical reaction in the battery, bringing it back to its original state. The cell has been operated as an electrolytic cell. Removal of the charging device turns the cell back into a voltaic device, ready to spontaneously react to produce electrical current once again.

The relationship between a voltaic cell and an electrolytic cell is illustrated in Figure 9.12.

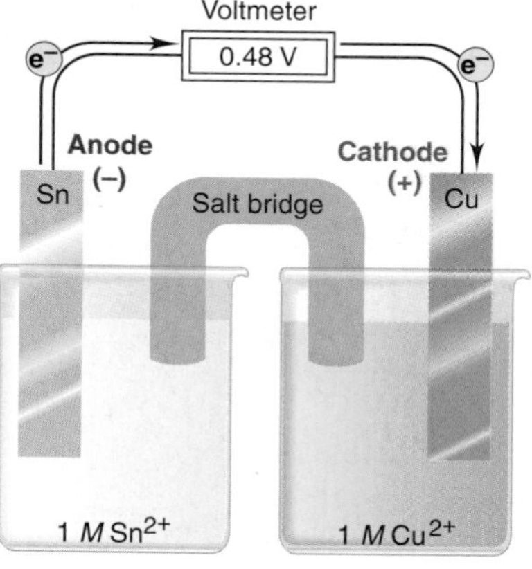

Oxidation half-reaction
$Sn(s) \longrightarrow Sn^{2+}(aq) + 2e^-$

Reduction half-reaction
$Cu^{2+}(aq) + 2e^- \longrightarrow Cu(s)$

Overall (cell) reaction
$Sn(s) + Cu^{2+}(aq) \longrightarrow Sn^{2+}(aq) + Cu(s)$

(a) Voltaic cell

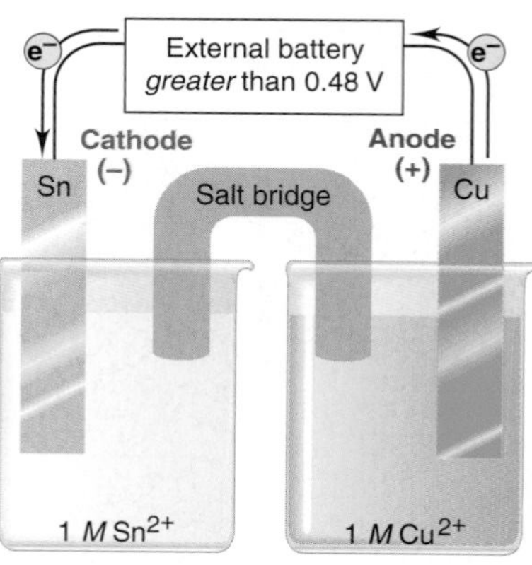

Oxidation half-reaction
$Cu(s) \longrightarrow Cu^{2+}(aq) + 2e^-$

Reduction half-reaction
$Sn^{2+}(aq) + 2e^- \longrightarrow Sn(s)$

Overall (cell) reaction
$Cu(s) + Sn^{2+}(aq) \longrightarrow Cu^{2+}(aq) + Sn(s)$

(b) Electrolytic cell

Figure 9.12
A voltaic cell (a) is converted to an electrolytic cell (b) by attaching a battery with a voltage sufficiently large to reverse the reaction. This process underlies commercially available rechargeable batteries.

Summary

9.1 Acids and Bases

One of the earliest definitions of acids and bases is the *Arrhenius theory.* According to this theory, an acid dissociates to form hydrogen ions, H^+, and a base dissociates to form hydroxide ions, OH^-. The *Brønsted-Lowry theory* defines an acid as a proton (H^+) donor and a base as a proton acceptor.

Water, the solvent in many acid-base reactions, is *amphiprotic.* It has both acid and base properties.

The strength of acids and bases in water depends on their degree of dissociation, the extent to which they react with the solvent, water. Acids and bases are strong when the reaction with water is virtually 100% complete and weak when the reaction with water is much less than 100% complete.

Weak acids and weak bases dissolve in water principally in the molecular form. Only a small percentage of the molecules dissociate to form the *hydronium* ion or *hydroxide* ion.

Aqueous solutions of acids and bases are electrolytes. The dissociation of the acid or base produces ions, which conduct an electrical current. Strong acids and bases are strong electrolytes. Weak acids and bases are weak electrolytes.

Although pure water is virtually 100% molecular, a small number of water molecules do ionize. This process occurs by the transfer of a proton from one water molecule to another, producing a hydronium ion and a hydroxide ion. This process is the *autoionization,* or self-ionization, of water.

Pure water at room temperature has a hydronium ion concentration of 1.0×10^{-7} *M*. One hydroxide ion is produced for each hydronium ion. Therefore, the hydroxide ion concentration is also 1.0×10^{-7} *M*. The product of hydronium and hydroxide ion concentration (1.0×10^{-14}) is the *ion product for water.*

9.2 pH: A Measurement Scale for Acids and Bases

The *pH scale* correlates the hydronium ion concentration with a number, the pH, that serves as a useful indicator of the degree of acidity or basicity of a solution. The pH of a solution is defined as the negative logarithm of the molar concentration of the hydronium ion ($pH = -\log [H_3O^+]$).

9.3 Reactions between Acids and Bases

The reaction of an acid with a base to produce a salt and water is referred to as *neutralization.* Neutralization requires equal numbers of moles of H_3O^+ and OH^- to produce a neutral solution (no excess acid or base). A neutralization reaction may be used to determine the concentration of an unknown acid or base solution. The technique of *titration* involves the addition of measured amounts of a *standard solution* (one whose concentration is known) from a *buret* to neutralize the second, unknown solution. The *equivalence point* is signaled by an *indicator.*

9.4 Acid-Base Buffers

A *buffer solution* contains components that enable the solution to resist large changes in pH when acids or bases are added. The basis of buffer action is an equilibrium between either a weak acid and its salt or a weak base and its salt.

A buffer solution follows LeChatelier's principle, which states that an equilibrium system, when stressed, will shift its equilibrium to alleviate that stress.

Buffering against base is a function of the concentration of the weak acid for an acidic buffer. Buffering against acid is dependent on the concentration of the anion of the salt.

A buffer solution can be described by an equilibrium-constant expression. The equilibrium-constant expression for an acidic system can be rearranged and solved for $[H_3O^+]$. In that way, the pH of a buffer solution can be obtained, if the composition of the solution is known. Alternatively, the *Henderson-Hasselbalch equation,* derived from the equilibrium constant expression, may be used to calculate the pH of a buffer solution.

9.5 Oxidation-Reduction Processes

Oxidation is defined as a loss of electrons, loss of hydrogen atoms, or gain of oxygen atoms. *Reduction* is defined as a gain of electrons, gain of hydrogen atoms, or loss of oxygen atoms.

Oxidation and reduction are complementary processes. The oxidation half-reaction produces an electron that is the reactant for the reduction half-reaction. The combination of two half-reactions, one oxidation and one reduction, produces the complete reaction.

The *reducing agent* releases electrons for the reduction of a second substance to occur. The *oxidizing agent* accepts electrons, causing the oxidation of a second substance to take place.

A *voltaic cell* is an electrochemical cell that converts chemical energy into electrical energy. *Electrolysis* is the opposite of a battery. It converts electrical energy into chemical potential energy.

Key Terms

acid (9.1)
amphiprotic (9.1)
anode (9.5)
Arrhenius theory (9.1)
autoionization (9.1)
base (9.1)
Brønsted-Lowry theory (9.1)
buffer capacity (9.4)
buffer solution (9.4)
buret (9.3)
cathode (9.5)
conjugate acid (9.1)
conjugate acid-base pair (9.1)
conjugate base (9.1)
corrosion (9.5)
electrolysis (9.5)
equivalence point (9.3)
Henderson-Hasselbalch equation (9.4)
hydronium ion (9.1)
indicator (9.3)

ion product for water (9.1)
neutralization (9.3)
oxidation (9.5)
oxidizing agent (9.5)
pH scale (9.2)
polyprotic substance (9.3)
reducing agent (9.5)
reduction (9.5)
standard solution (9.3)
titration (9.3)
voltaic cell (9.5)

Questions and Problems

Acids and Bases

9.25 a. Define an acid according to the Arrhenius theory.
b. Define an acid according to the Brønsted-Lowry theory.

9.26 a. Define a base according to the Arrhenius theory.
b. Define a base according to the Brønsted-Lowry theory.

9.27 What are the essential differences between the Arrhenius and Brønsted-Lowry theories?

9.28 Why is ammonia described as a Brønsted-Lowry base and not an Arrhenius base?

9.29 Write an equation for the reaction of each of the following with water:
a. HNO_2
b. HCN

9.30 Write an equation for the reaction of each of the following with water:
a. HNO_3
b. HCOOH

9.31 Select the conjugate acid-base pairs for each reaction in Question 9.29.

9.32 Select the conjugate acid-base pairs for each reaction in Question 9.30.

9.33 Label each of the following as a strong or weak acid (consult Figure 9.2, if necessary):
a. H_2SO_3
b. H_2CO_3
c. H_3PO_4

9.34 Label each of the following as a strong or weak base (consult Figure 9.2, if necessary):
a. KOH
b. CN^-
c. SO_4^{2-}

9.35 Identify the conjugate acid-base pairs in each of the following chemical equations:
a. $NH_4^+ (aq) + CN^- (aq) \rightleftharpoons NH_3 (aq) + HCN(aq)$
b. $CO_3^{2-} (aq) + HCl (aq) \rightleftharpoons HCO_3^- (aq) + Cl^- (aq)$

9.36 Identify the conjugate acid-base pairs in each of the following chemical equations:
a. $HCOOH (aq) + NH_3 (aq) \rightleftharpoons HCOO^- (aq) + NH_4^+ (aq)$
b. $HCl (aq) + OH^- (aq) \rightleftharpoons H_2O (l) + Cl^- (aq)$

9.37 Distinguish between the terms acid-base *strength* and acid-base *concentration*.

9.38 Of the diagrams shown here, which one represents:
a. a concentrated strong acid
b. a dilute strong acid
c. a concentrated weak acid
d. a dilute weak acid

H^+ H–X X^- H–X H–X H–X H^+ H–X H–X H–X X^- H X H–X

I

H^+ X^- H^+ X^- X^- H^+ X^- H^+ H^+ X^- H^+ X^- X^- H^+ X^- H^+ H^+ X^- H^+ X^- X^- H^+

II

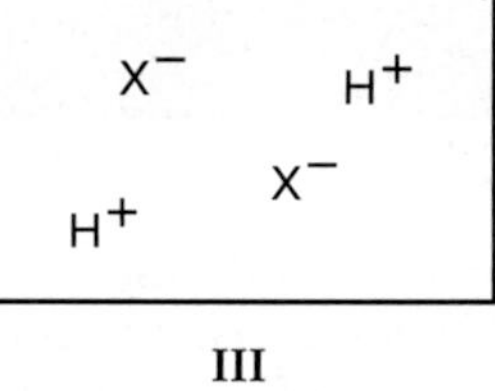

III

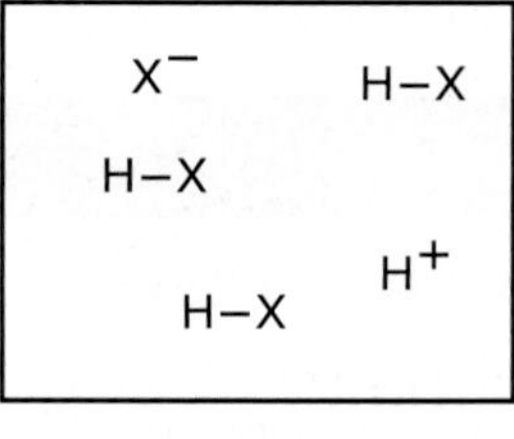

IV

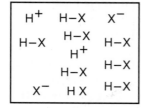

pH of Acid and Base Solutions

9.39 Calculate the $[H_3O^+]$ of an aqueous solution that is:
a. 1.0×10^{-7} *M* in OH^-
b. 1.0×10^{-3} *M* in OH^-

9.40 Calculate the $[H_3O^+]$ of an aqueous solution that is:
a. 1.0×10^{-9} *M* in OH^-
b. 1.0×10^{-5} *M* in OH^-

9.41 Label each solution in Problem 9.39 as acidic, basic, or neutral.

9.42 Label each solution in Problem 9.40 as acidic, basic, or neutral.

9.43 Calculate the pH of a solution that has
a. $[H_3O^+] = 1.0 \times 10^{-7}$
b. $[OH^-] = 1.0 \times 10^{-9}$

9.44 Calculate the pH of a solution that has:
a. $[H_3O^+] = 1.0 \times 10^{-10}$
b. $[OH^-] = 1.0 \times 10^{-5}$

9.45 Calculate *both* $[H_3O^+]$ and $[OH^-]$ for a solution that is:
a. pH = 1.00
b. pH = 9.00

9.46 Calculate *both* $[H_3O^+]$ and $[OH^-]$ for a solution that is:
a. pH = 5.00
b. pH = 7.20

9.47 Calculate *both* $[H_3O^+]$ and $[OH^-]$ for a solution that is:
a. pH = 1.30
b. pH = 9.70

9.48 Calculate *both* $[H_3O^+]$ and $[OH^-]$ for a solution that is:
a. pH = 5.50
b. pH = 7.00

9.49 What is a neutralization reaction?

9.50 Describe the purpose of a titration.

9.51 The pH of urine may vary between 4.5 and 8.2. Determine the H_3O^+ concentration and OH^- concentration if the measured pH is:
a. 6.00
b. 5.20
c. 7.80

9.52 The hydronium ion concentration in blood of three different patients was:

Patient	$[H_3O^+]$
A	5.0×10^{-8}
B	3.1×10^{-8}
C	3.2×10^{-8}

What is the pH of each patient's blood? If the normal range is 7.30–7.50, which, if any, of these patients have an abnormal blood pH?

9.53 Determine how many times more acidic a solution is at:
a. pH 2 relative to pH 4
b. pH 7 relative to pH 11
c. pH 2 relative to pH 12

9.54 Determine how many times more basic a solution is at:
a. pH 6 relative to pH 4
b. pH 10 relative to pH 9
c. pH 11 relative to pH 6

9.55 What is the H_3O^+ concentration of a solution with a pH of:
a. 5.0
b. 12.0
c. 5.50

9.56 What is the OH^- concentration of each solution in question 9.55?

9.57 Calculate the pH of a solution with a H_3O^+ concentration of:
a. 1.0×10^{-6} *M*
b. 1.0×10^{-8} *M*
c. 5.6×10^{-4} *M*

9.58 What is the OH^- concentration of each solution in question 9.57?

Buffer Solutions

9.59 Which of the following are capable of forming a buffer solution?
a. NH_3 and NH_4Cl
b. HNO_3 and KNO_3

9.60 Which of the following are capable of forming a buffer solution?
a. HBr and $MgCl_2$
b. H_2CO_3 and $NaHCO_3$

9.61 Define:
a. buffer solution
b. acidosis (refer to A Clinical Perspective: Control of Blood pH on page 255)

9.62 Define:
a. alkalosis (refer to A Clinical Perspective: Control of Blood pH on page 255)
b. standard solution

9.63 For the equilibrium situation involving acetic acid,

$$CH_3COOH(aq) + H_2O(l) \rightleftharpoons CH_3COO^-(aq) + H_3O^+(aq)$$

explain the equilibrium shift occurring for the following changes:
a. A strong acid is added to the solution.
b. The solution is diluted with water.

9.64 For the equilibrium situation involving acetic acid,

$$CH_3COOH(aq) + H_2O(l) \rightleftharpoons CH_3COO^-(aq) + H_3O^+(aq)$$

explain the equilibrium shift occurring for the following changes:
a. A strong base is added to the solution.
b. More acetic acid is added to the solution.

9.65 What is $[H_3O^+]$ for a buffer solution that is 0.200 *M* in acid and 0.500 *M* in the corresponding salt if the weak acid $K_a = 5.80 \times 10^{-7}$?

9.66 What is the pH of the solution described in Question 9.65?

Oxidation-Reduction Reactions

9.67 Define:
a. oxidation
b. oxidizing agent

9.68 Define:
a. reduction
b. reducing agent

9.69 During an oxidation process in an oxidation-reduction reaction, the species oxidized __________ electrons.

9.70 During an oxidation-reduction reaction, the species __________ is the oxidizing agent.

9.71 During an oxidation-reduction reaction, the species __________ is the reducing agent.

9.72 Metals tend to be good __________ agents.

9.73 In the following reaction, identify the oxidized species, reduced species, oxidizing agent, and reducing agent:

$$Cl_2(aq) + 2KI(aq) \longrightarrow 2KCl(aq) + I_2(aq)$$

9.74 In the following reaction, identify the oxidized species, reduced species, oxidizing agent, and reducing agent:

$$Zn(s) + Cu^{2+}(aq) \longrightarrow Zn^{2+}(aq) + Cu(s)$$

9.75 Explain the relationship between oxidation-reduction and voltaic cells.

9.76 Compare and contrast a battery and electrolysis.

9.77 Describe one application of voltaic cells.

9.78 Describe one application of electrolytic cells.

Critical Thinking Problems

1. Acid rain is a threat to our environment because it can increase the concentration of toxic metal ions, such as Cd^{2+} and Cr^{3+}, in rivers and streams. If cadmium and chromium are present in sediment as $Cd(OH)_2$ and $Cr(OH)_3$, write reactions that demonstrate the effect of acid rain. Use the library to find the properties of cadmium and chromium responsible for their environmental impact.
2. Aluminum carbonate is more soluble in acidic solution, forming aluminum cations. Write a reaction (or series of reactions) that explains this observation.
3. Carbon dioxide reacts with the hydroxide ion to produce the bicarbonate anion. Write the Lewis dot structures for each reactant and product. Label each as a Brønsted acid or base. Explain the reaction using the Brønsted theory. Why would the Arrhenius theory provide an inadequate description of this reaction?
4. Maalox is an antacid composed of $Mg(OH)_2$ and $Al(OH)_3$. Explain the origin of the trade name Maalox. Write chemical reactions that demonstrate the antacid activity of Maalox.
5. Acid rain has been described as a regional problem, whereas the greenhouse effect is a global problem. Do you agree with this statement? Why or why not?

Madame Marie Curie, a Nobel Prize winning contributor to our understanding of the nucleus and radioactivity.

10

The Nucleus, Radioactivity, and Nuclear Medicine

Outline

GENERAL CHEMISTRY

Learning Goals

1. Enumerate the characteristics of alpha, beta, and gamma radiation.
2. Write balanced equations for common nuclear processes.
3. Calculate the amount of radioactive substance remaining after a specified number of half-lives.
4. Describe the various ways in which nuclear energy may be used to generate electricity: fission, fusion, and the breeder reactor.
5. Explain the process of radiocarbon dating.
6. Cite several examples of the use of radioactive isotopes in medicine.
7. Describe the use of ionizing radiation in cancer therapy.
8. Discuss the preparation and use of radioisotopes in diagnostic imaging studies.
9. Explain the difference between natural and artificial radioactivity.
10. Describe the characteristics of radioactive materials that relate to radiation exposure and safety.
11. Be familiar with common techniques for the detection of radioactivity.
12. Know the common units in which radiation intensity is represented: the curie, roentgen, rad, and rem.

CHEMISTRY CONNECTION

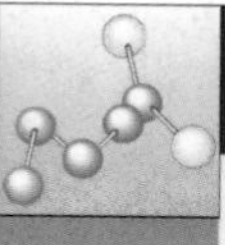

An Extraordinary Woman in Science

The path to a successful career in science, or any other field for that matter, is seldom smooth or straight. That was certainly true for Madame Marie Sklodowska Curie. Her lifelong ambition was to raise a family and do something interesting for a career. This was a lofty goal for a nineteenth-century woman.

The political climate in Poland, coupled with the prevailing attitudes toward women and careers, especially careers in science, certainly did not make it any easier for Mme. Curie. To support herself and her sister, she toiled at menial jobs until moving to Paris to resume her studies.

It was in Paris that she met her future husband and fellow researcher, Pierre Curie. Working with crude equipment in a laboratory that was primitive, even by the standards of the time, she and Pierre made a most revolutionary discovery only two years after Henri Becquerel discovered radioactivity. Radioactivity, the emission of energy from certain substances, was released from *inside* the atom and was independent of the molecular form of the substance. The absolute proof of this assertion came only after the Curies processed over one *ton* of a material (pitchblende) to isolate less than a gram of pure radium. The difficult conditions under which this feat was accomplished are perhaps best stated by Sharon Bertsch McGrayne in her book *Nobel Prize Women in Science* (Birch Lane Press, New York, p. 23):

> The only space large enough at the school was an abandoned dissection shed. The shack was stifling hot in summer and freezing cold in winter. It had no ventilation system for removing poisonous fumes, and its roof leaked. A chemist accustomed to Germany's modern laboratories called it "a cross between a stable and a potato cellar and, if I had not seen the work table with the chemical apparatus, I would have thought it a practical joke." This ramshackle shed became the symbol of the Marie Curie legend.

The pale green glow emanating from the radium was beautiful to behold. Mme. Curie would go to the shed in the middle of the night to bask in the light of her accomplishment. She did not realize that this wonderful accomplishment would, in time, be responsible for her death.

Mme. Curie received not one, but two Nobel Prizes, one in physics and one in chemistry. She was the first woman in France to earn the rank of professor.

As you study this chapter, the contributions of Mme. Curie, Pierre Curie, and the others of that time will become even more clear. Ironically, the field of medicine has been a major beneficiary of advances in nuclear and radiochemistry, despite the toxic properties of those same radioactive materials.

Introduction

Our discussion of the atom and atomic structure revealed a nucleus containing protons and neutrons surrounded by electrons. Until now, we have treated the nucleus as simply a region of positive charge in the center of the atom. The focus of our interest has been the electrons and their arrangement around the nucleus. Electron arrangement is an essential part of a discussion of bonding or chemical change.

Chapter 2 describes the electronic structure of atoms.

In this chapter we consider the nucleus and nuclear properties. The behavior of nuclei may have as great an effect on our everyday lives as any of the thousands of synthetic compounds developed over the past several decades. Examples of nuclear technology range from everyday items (smoke detectors) to sophisticated instruments for medical diagnosis and treatment and electrical power generation (nuclear power plants).

Beginning in 1896 with Becquerel's discovery of radiation emitted from uranium ore, the technology arising from this and related findings has produced both risks and benefits. Although early discoveries of radioactivity and its properties expanded our fundamental knowledge and brought fame to the investigators, it was not accomplished without a price. Several early investigators died prematurely of cancer and other diseases caused by the radiation they studied.

Even today, the existence of nuclear energy and its associated technology is a mixed blessing. On one side, the horrors of Nagasaki and Hiroshima, the fear of nuclear war, and potential contamination of populated areas resulting from the peaceful application of nuclear energy are critical problems facing society. Conversely, hundreds of thousands of lives have been saved because of the early detection of disease by X-ray diagnosis and the cure of cancer using cobalt-60 treatment. Furthermore, nuclear energy is an alternative energy source, providing an opportunity for us to compensate for the depletion of oil reserves.

10.1 Natural Radioactivity

Radioactivity is the process by which atoms emit energetic particles or rays. These particles or rays are termed *radiation*. Nuclear radiation occurs as a result of an alteration in nuclear composition or structure. This process occurs in a nucleus that is unstable and hence radioactive. Radioactivity is a nuclear event: *matter and energy released during this process come from the nucleus.*

We shall designate the nucleus using *nuclear symbols*, analogous to the *atomic symbols* that were introduced in Section 2.1. The nuclear symbols consist of the *elemental symbol*, the *atomic number* (the number of protons in the nucleus), and the *mass number*, which is defined as the sum of neutrons and protons in the nucleus.

Be careful not to confuse the mass number (a simple count of the neutrons and protons) with the atomic mass, which includes the contribution of electrons and is a true *mass* figure.

With the use of nuclear symbols, the fluorine nucleus is represented as

Mass number ⟶ $^{19}_{9}F$ ⟵ Atomic symbol
Atomic number ⟶
(or nuclear charge)

This symbol is equivalent to writing *fluorine-19*. This alternative representation is frequently used to denote specific isotopes of elements.

Isotopes are introduced in Section 2.2.

Not all nuclei are unstable. Only unstable nuclei undergo change and produce radioactivity, the process of radioactive decay. Recall that different atoms of the same element having different masses exist as *isotopes*. One isotope of an element may be radioactive, whereas others of the same element may be quite stable.

The terms *isotope* and *nuclide* may be used interchangeably.

Many elements in the periodic table occur in nature as mixtures of isotopes. Two common examples include carbon,

$^{12}_{6}C$	$^{13}_{6}C$	$^{14}_{6}C$
Carbon-12	Carbon-13	Carbon-14

and hydrogen,

$^{1}_{1}H$	$^{2}_{1}H$	$^{3}_{1}H$
Hydrogen-1	Hydrogen-2	Hydrogen-3
Protium	Deuterium (symbol D)	Tritium (symbol T)

Protium is a stable isotope and makes up more than 99.9% of naturally occurring hydrogen. Deuterium (D) can be isolated from hydrogen; it can form compounds such as "heavy water," D_2O. Heavy water is a potential source of deuterium for fusion processes. Tritium (T) is unstable, hence radioactive, and is a waste product of nuclear reactors.

In writing the symbols for a nuclear process, it is essential to indicate the particular isotope involved. This is why the mass number and atomic number are used. These values tell us the number of neutrons in the species, hence the isotope's identity.

Alpha, beta, and gamma radiation have widespread use in the field of medicine. Other radiation particles, such as neutrinos and deuterons, will not be discussed here.

Three types of natural radiation emitted by unstable nuclei are *alpha particles*, *beta particles*, and *gamma rays*.

Alpha Particles

Learning Goal

An **alpha particle** (α) contains two protons and two neutrons. An alpha particle is identical to the nucleus of the helium atom (He) or a *helium ion* (He^{2+}), which also contains two protons (atomic number = 2) and two neutrons (mass number − atomic number = 2). Having no electrons to counterbalance the nuclear charge, the alpha particle may be symbolized as

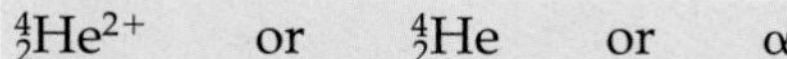

$${}^{4}_{2}He^{2+} \quad \text{or} \quad {}^{4}_{2}He \quad \text{or} \quad \alpha$$

Alpha particles have a relatively large mass compared to other nuclear particles. Consequently, alpha particles emitted by radioisotopes are relatively slow-moving particles (approximately 10% of the speed of light), and they are stopped by barriers as thin as a few pages of this book.

Beta Particles

The **beta particle** (β), in contrast, is a fast-moving electron traveling at approximately 90% of the speed of light as it leaves the nucleus. It is formed in the nucleus by the conversion of a neutron into a proton. The beta particle is represented as

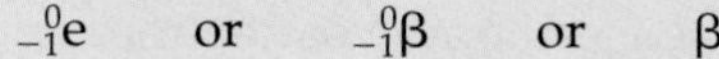

$${}^{0}_{-1}e \quad \text{or} \quad {}^{0}_{-1}\beta \quad \text{or} \quad \beta$$

The subscript −1 is written in the same position as the atomic number and, like the atomic number (number of protons), indicates the charge of the particle.

Beta particles are smaller and faster than alpha particles. They are more penetrating and are stopped only by more dense materials such as wood, metal, or several layers of clothing.

Gamma Rays

A **gamma ray** (γ) is pure energy (part of the electromagnetic spectrum, see Section 2.3), resulting from nuclear processes; alpha radiation and beta radiation are matter. Because pure energy has no protons, neutrons, or electrons, the symbol for a gamma ray is simply

$$\gamma$$

Gamma radiation is highly energetic and is the most penetrating form of nuclear radiation. Barriers of lead, concrete, or, more often, a combination of the two are required for protection from this type of radiation.

Properties of Alpha, Beta, and Gamma Radiation

Learning Goal

Important properties of alpha, beta, and gamma radiation are summarized in Table 10.1.

Alpha, beta, and gamma radiation are collectively termed *ionizing radiation.* **Ionizing radiation** produces a trail of ions throughout the material that it penetrates. The ionization process changes the chemical composition of the material. When the material is living tissue, radiation-induced illness may result (Section 10.7).

The penetrating power of alpha radiation is very low. Damage to internal organs from this form of radiation is negligible except when an alpha particle emitter is actually ingested. Beta particles are higher in energy; still, they have limited penetrating power. They cause skin and eye damage and, to a lesser extent, damage to internal organs. Shielding is required in working with beta emitters. Pregnant women must take special precautions. The great penetrating power and high energy of gamma radiation make it particularly difficult to shield. Hence, it can damage internal organs.

Anyone working with any type of radiation must take precautions. Radiation safety is required, monitored, and enforced in the United States under provisions of the Occupational Safety and Health Act (OSHA).

Table 10.1 A Summary of the Major Properties of Alpha, Beta, and Gamma Radiation

Name and Symbol	Identity	Charge	Mass (amu)	Velocity	Penetration
Alpha (α)	Helium nucleus	+2	4.0026	5–10% of the speed of light	Low
Beta (β)	Electron	−1	0.000549	Up to 90% of the speed of light	Medium
Gamma (γ)	Radiant energy	0	0	Speed of light	High

Question 10.1

Gamma radiation is a form of *electromagnetic radiation*. Provide examples of other forms of electromagnetic radiation.

Question 10.2

How does the energy of gamma radiation compare with that of other regions of the electromagnetic spectrum?

10.2 Writing a Balanced Nuclear Equation

Learning Goal

Nuclear equations represent nuclear change in much the same way as chemical equations represent chemical change.

A **nuclear equation** can be used to represent the process of radioactive decay. In radioactive decay an isotope breaks down, producing a *new isotope, smaller particles, and/or energy*. The concept of mass balance, required when writing chemical equations, is also essential for nuclear equations. When writing a balanced equation, remember that:

- the total mass on each side of the reaction arrow must be identical, and
- the sum of the atomic numbers on each side of the reaction arrow must be identical.

Alpha Decay

Consider the decay of one isotope of uranium, ${}^{238}_{92}U$, into thorium and an alpha particle. Because an alpha particle is lost in this process, this decay is called *alpha decay*.

Examine the balanced equation for this nuclear reaction:

$$ {}^{238}_{92}\text{U} \longrightarrow {}^{234}_{90}\text{Th} + {}^{4}_{2}\text{He} $$

Uranium-238 Thorium-234 Helium-4

The sum of the mass numbers on the right (234 + 4 = 238) are equal to the mass number on the left. The atomic numbers on the right (90 + 2 = 92) are equal to the atomic number on the left.

Beta Decay

Beta decay is illustrated by the decay of one of the less-abundant nitrogen isotopes, ${}^{16}_{7}N$. Upon decomposition, nitrogen-16 produces oxygen-16 and a beta particle. The reaction is represented as

$$^{16}_{7}N \longrightarrow ^{16}_{8}O + ^{0}_{-1}e$$

Recall that the mass of an electron is approximately 0.0005 of the mass of the proton or neutron. Hence, making the mass of a β particle equal to zero is a good approximation.

Note that the mass number of the beta particle is zero, because the electron includes no protons or neutrons. Sixteen nuclear particles are accounted for on both sides of the reaction arrow.

The atomic number on the left (+7) is counterbalanced by [8 + (−1)] or (+7) on the right. Therefore the equation is correctly balanced.

Gamma Production

If *gamma radiation* were the only product of nuclear decay, there would be no measurable change in the mass or identity of the radioactive nuclei. This is so because the gamma emitter has simply gone to a lower energy state. An example of an isotope that decays in this way is technetium-99m. It is described as a **metastable isotope,** meaning that it is unstable and increases its stability through gamma decay without change in the mass or charge of the isotope. The letter *m* is used to denote a metastable isotope. The decay equation for $^{99m}_{43}Tc$ is

$$^{99m}_{43}Tc \longrightarrow ^{99}_{43}Tc + \gamma$$

More often, gamma radiation is produced along with other products. For example, iodine-131 decays as follows:

$^{131}_{53}I$	$\longrightarrow$	$^{131}_{54}Xe$	+	$^{0}_{-1}\beta$	+	γ
Iodine-131		Xenon-131		Beta particle		Gamma ray

This reaction may also be represented as

$$^{131}_{53}I \longrightarrow ^{131}_{54}Xe + ^{0}_{-1}e + \gamma$$

An isotope of xenon, a beta particle, and gamma radiation are produced.

Predicting Products of Nuclear Decay

It is possible to use a nuclear equation to predict one of the products of a nuclear reaction if the others are known. Consider the following example, in which we represent the unknown product as ?:

$$^{40}_{19}K \longrightarrow ? + ^{0}_{-1}e$$

Step 1. The mass number of this isotope of potassium is 40. Therefore the sum of the mass number of the products must also be 40, and ? must have a mass number of 40.

Step 2. Likewise, the atomic number on the left is 19, and the sum of the unknown atomic number plus the charge of the beta particle (−1) must equal 19.

Step 3. The unknown atomic number must be 20, because [20 + (−1) = 19]. The unknown is

$$^{40}_{20}?$$

If we consult the periodic table, the element that has atomic number 20 is calcium; therefore ? = $^{40}_{20}Ca$.

EXAMPLE 10.1 ***Predicting the Products of Radioactive Decay***

Determine the identity of the unknown product of the alpha decay of curium-245:

Continued—

EXAMPLE 10.1 —*Continued*

$$^{245}_{96}\text{Cm} \longrightarrow \, ^{4}_{2}\text{He} + ?$$

Solution

Step 1. The mass number of the curium isotope is 245. Therefore the sum of the mass numbers of the products must also be 245, and ? must have a mass number of 241.

Step 2. Likewise, the atomic number on the left is 96, and the sum of the unknown atomic number plus the atomic number of the alpha particle (2) must equal 96.

Step 3. The unknown atomic number must be 94, because [94 + 2 = 96]. The unknown is

$$^{241}_{94}?$$

Referring to the periodic table, we find that the element that has atomic number 94 is plutonium; therefore ? = $^{241}_{94}\text{Pu}$.

Question 10.3

Complete each of the following nuclear equations:

a. $^{85}_{36}\text{Kr} \longrightarrow ? + \, ^{0}_{-1}\text{e}$
b. $? \longrightarrow \, ^{4}_{2}\text{He} + \, ^{222}_{86}\text{Rn}$

Question 10.4

Complete each of the following nuclear equations:

a. $^{239}_{92}\text{U} \longrightarrow ? + \, ^{0}_{-1}\text{e}$
b. $^{11}_{5}\text{B} \longrightarrow \, ^{7}_{3}\text{Li} + ?$

10.3 Properties of Radioisotopes

Why are some isotopes radioactive but others are not? Do all radioactive isotopes decay at the same rate? Are all radioactive materials equally hazardous? We address these and other questions in this section.

Nuclear Structure and Stability

The energy that holds the protons, neutrons, and other particles together in the nucleus is the **binding energy** of the nucleus. This binding energy must be very large, because identically charged protons in the nucleus exert extreme repulsive forces on one another. These forces must be overcome if the nucleus is to be stable. When an isotope decays, some of this binding energy is released. This released energy is the source of the high-energy radiation emitted and the basis for all nuclear technology.

Why are some isotopes more stable than others? The answer to this question is not completely clear. Evidence obtained so far points to several important factors that describe stable nuclei:

- Nuclear stability correlates with the ratio of neutrons to protons in the isotope. For example, for light atoms a neutron:proton ratio of 1 characterizes a stable atom.

- Nuclei with large numbers of protons (84 or more) tend to be unstable.
- Naturally occurring isotopes containing 2, 8, 20, 50, 82, or 126 protons or neutrons are stable. These *magic numbers* seem to indicate the presence of energy levels in the nucleus, analogous to electronic energy levels in the atom.
- Isotopes with even numbers of protons or neutrons are generally more stable than those with odd numbers of protons or neutrons.
- All isotopes (except hydrogen-1) with more protons than neutrons are unstable. However, the reverse is not true.

Half-Life

Learning Goal

The **half-life ($t_{1/2}$)** is the time required for one-half of a given quantity of a substance to undergo change. Not all radioactive isotopes decay at the same rate. The rate of nuclear decay is generally represented in terms of the half-life of the isotope. Each isotope has its own characteristic half-life that may be as short as a few millionths of a second or as long as a billion years. Half-lives of some naturally occurring and synthetic isotopes are given in Table 10.2.

Refer to the discussion of radiation exposure and safety in Sections 10.7, 10.8, and 10.9.

The stability of an isotope is indicated by the isotope's half-life. Isotopes with short half-lives decay rapidly; they are very unstable. This is not meant to imply that substances with long half-lives are less hazardous. Often, just the reverse is true.

Imagine that we begin with 100 mg of a radioactive isotope that has a half-life of 24 hours. After one half-life, or 24 hours, 1/2 of 100 mg will have decayed to other products, and 50 mg remain. After two half-lives (48 hours), 1/2 of the remaining material has decayed, leaving 25 mg, and so forth:

$$100\text{ mg} \xrightarrow[\substack{\text{Half-life} \\ (24\text{ h})}]{\text{One}} 50\text{ mg} \xrightarrow[\substack{\text{Half-life} \\ (48\text{ h total})}]{\text{A Second}} 25\text{ mg} \longrightarrow \text{etc.}$$

Decay of a radioisotope that has a reasonably short $t_{1/2}$ is experimentally determined by following its activity as a function of time. Graphing the results produces a radioactive decay curve as shown in Figure 10.1.

The mass of any radioactive substance remaining after a period may be calculated with a knowledge of the initial mass and the half-life of the isotope, following the scheme just outlined.

Table 10.2 Half-Lives of Selected Radioisotopes

Name	Symbol	Half-Life
Carbon-14	$^{14}_{6}C$	5730 years
Cobalt-60	$^{60}_{27}Co$	5.3 years
Hydrogen-3	$^{3}_{1}H$	12.3 years
Iodine-131	$^{131}_{53}I$	8.1 days
Iron-59	$^{59}_{26}Fe$	45 days
Molybdenum-99	$^{99}_{42}Mo$	67 hours
Sodium-24	$^{24}_{11}Na$	15 hours
Strontium-90	$^{90}_{38}Sr$	28 years
Technetium-99m	$^{99m}_{43}Tc$	6 hours
Uranium-235	$^{235}_{92}U$	710 million years

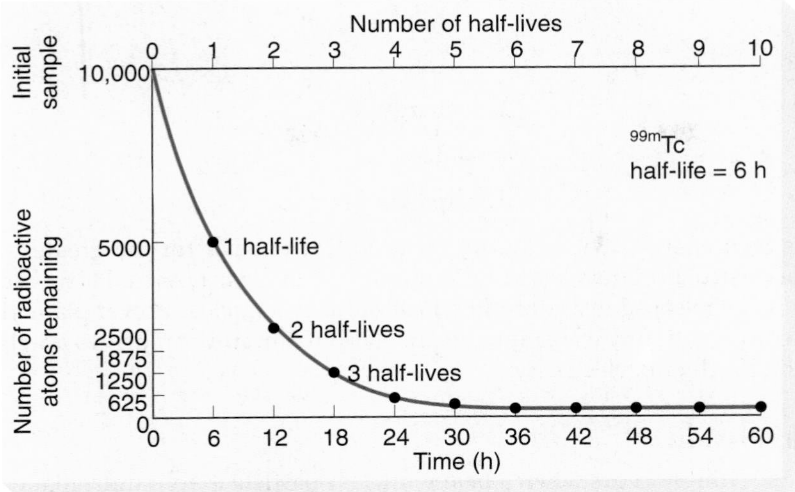

Figure 10.1
The decay curve for the medically useful radioisotope technetium-99m. Note that the number of radioactive atoms remaining—hence the radioactivity—approaches zero.

EXAMPLE 10.2

Predicting the Extent of Radioactive Decay

A 50.0-mg supply of iodine-131, used in hospitals in the treatment of hyperthyroidism, was stored for 32.4 days. If the half-life of iodine-131 is 8.1 days, how many milligrams remain?

Solution

First calculate n, the number of half-lives elapsed using the half-life as a conversion factor:

$$n = 32.4 \cancel{\text{days}} \times \frac{1 \text{ half-life}}{8.1 \cancel{\text{days}}} = 4.0 \text{ half-lives}$$

Then calculate the amount remaining:

$$50.0 \text{ mg} \xrightarrow[\text{half-life}]{\text{first}} 25.0 \text{ mg} \xrightarrow[\text{half-life}]{\text{second}} 12.5 \text{ mg} \xrightarrow[\text{half-life}]{\text{third}} 6.25 \text{ mg} \xrightarrow[\text{half-life}]{\text{fourth}} 3.13 \text{ mg}$$

Hence, 3.13 mg of iodine-131 remain after 32.4 days.

Question 10.5

A 100.0-ng sample of sodium-24 was stored in a lead-lined cabinet for 2.5 days. How much sodium-24 remained? See Table 10.2 for the half-life of sodium-24.

Question 10.6

If a patient is administered 10 ng of technetium-99m, how much will remain one day later, assuming that no technetium has been eliminated by any other process? See Table 10.2 for the half-life of technetium-99m.

10.4 Nuclear Power

Energy Production

Learning Goal

4

Einstein predicted that a small amount of nuclear mass corresponds to a very large amount of energy that is released when the nucleus breaks apart. Einstein's equation is

$$E = mc^2$$

in which

$$E = \text{energy}$$
$$m = \text{mass}$$
$$c = \text{speed of light}$$

This heat energy, when rapidly released, is the basis for the greatest instruments of destruction developed by humankind, nuclear bombs. However, when heat energy is released in a controlled fashion, as in a nuclear power plant, the heat energy converts liquid water into steam. The steam, in turn, drives an electrical generator, producing electricity.

Nuclear Fission

Fission (splitting) occurs when a heavy nuclear particle is split into smaller nuclei by a smaller nuclear particle (such as a neutron). This splitting process is accompanied by the release of large amounts of energy.

A nuclear power plant uses a fissionable material (capable of undergoing fission), such as uranium-235, as fuel. The energy released by the fission process in the nuclear core heats water in an adjoining chamber, producing steam. The high pressure of the steam drives a generator, or turbine, which converts this heat energy into electricity. The energy transformation may be summarized as follows:

nuclear energy	$\longrightarrow$	heat energy	$\longrightarrow$	mechanical energy	$\longrightarrow$	electrical energy
Nuclear reactor		Steam		Turbine		Electricity

The fission reaction, once initiated, is self-perpetuating. For example, neutrons are used to initiate the reaction:

$$\underset{\text{Fuel}}{{}^{1}_{0}\text{n} + {}^{235}_{92}\text{U}} \longrightarrow \underset{\text{Unstable}}{{}^{236}_{92}\text{U}} \longrightarrow \underset{\text{Products of reaction}}{{}^{92}_{36}\text{Kr} + {}^{141}_{56}\text{Ba} + 3{}^{1}_{0}\text{n} + \text{energy}}$$

Note that three neutrons are released as product for each single reacting neutron. Each of the three neutrons produced is available to initiate another fission process. Nine neutrons are released from this process. These, in turn, react with other nuclei. The fission process continues and intensifies, producing very large amounts of energy (Figure 10.2). This process of intensification is referred to as a **chain reaction.**

To maintain control over the process and to prevent dangerous overheating, rods fabricated from carbon or boron are inserted into the core. These rods, which are controlled by the reactor's main operating system, absorb free neutrons as needed, thereby moderating the reaction.

A nuclear fission reactor may be represented as a series of energy transfer zones, as depicted in Figure 10.3. A view of the core of a fission reactor is shown in Figure 10.4.

Nuclear Fusion

Fusion (meaning *to join together*) results from the combination of two small nuclei to form a larger nucleus with the concurrent release of large amounts of energy. The best example of a fusion reactor is the sun. Continuous fusion processes furnish our solar system with light and heat.

An example of a fusion reaction is the combination of two isotopes of hydrogen, deuterium (${}^{2}_{1}\text{H}$) and tritium (${}^{3}_{1}\text{H}$), to produce helium, a neutron, and energy:

$${}^{2}_{1}\text{H} + {}^{3}_{1}\text{H} \longrightarrow {}^{4}_{2}\text{He} + {}^{1}_{0}\text{n} + \text{energy}$$

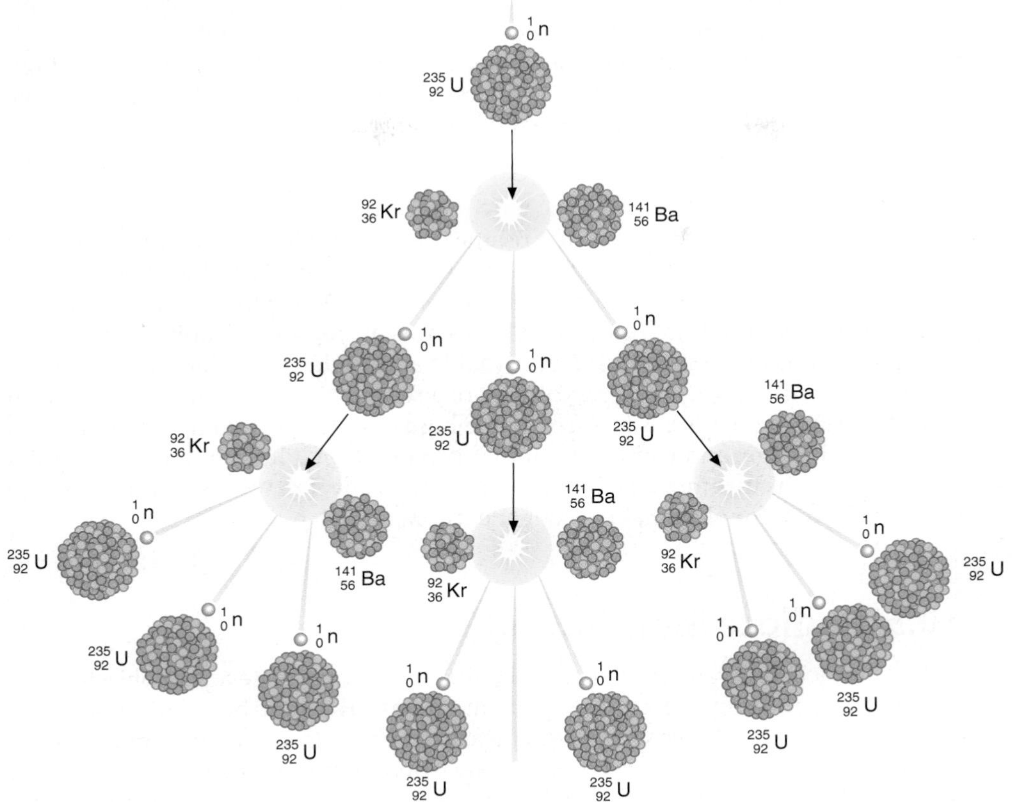

Figure 10.2
The fission of uranium-235 producing a chain reaction. Note that the number of available neutrons, which "trigger" the decomposition of the fissionable nuclei to release energy, increases at each step in the "chain." In this way the reaction builds in intensity. Control rods stabilize (or limit) the extent of the chain reaction to a safe level.

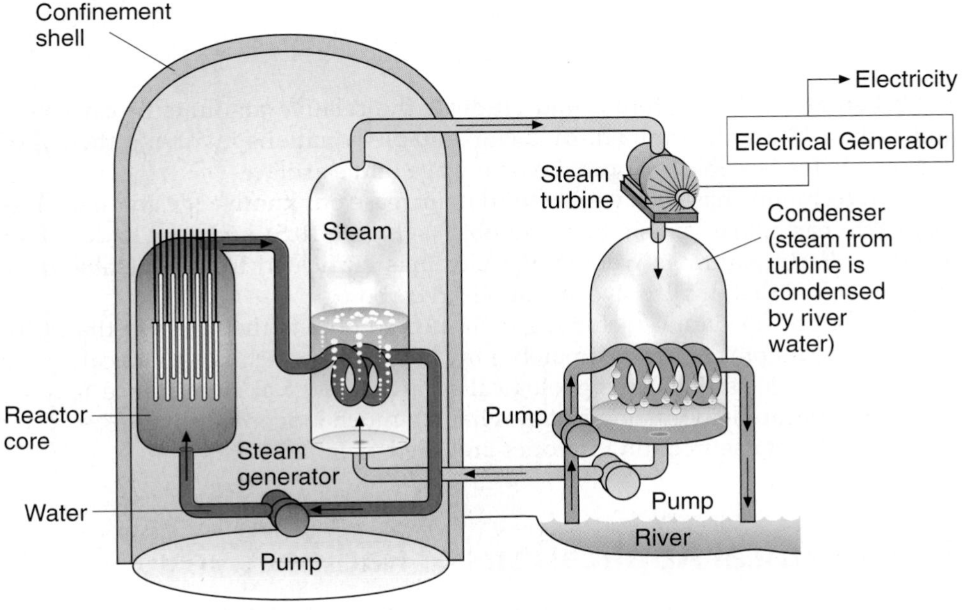

Figure 10.3
A representation of the "energy zones" of a nuclear reactor. Heat produced by the reactor core is carried by water in a second zone to a boiler. Water in the boiler (third zone) is converted to steam, which drives a turbine to convert heat energy to electrical energy. The isolation of these zones from each other allows heat energy transfer without actual physical mixing. This minimizes the transport of radioactive material into the environment.

Figure 10.4
The core of a nuclear reactor located at the Oak Ridge National Laboratories in Tennessee.

Although fusion is capable of producing tremendous amounts of energy, no commercially successful fusion plant exists in the United States. Safety concerns relating to problems of containment of the reaction, resulting directly from the high temperatures required to sustain a fusion process (millions of degrees), have slowed the technological development of fusion reactors.

Breeder Reactors

A **breeder reactor** is a variation of a fission reactor that literally manufactures its own fuel. A perceived shortage of fissionable isotopes makes the breeder an attractive alternative to conventional fission reactors. A breeder reactor uses ${}^{238}_{92}U$, which is abundant but nonfissionable. In a series of steps, the uranium-238 is converted to plutonium-239, which *is* fissionable and undergoes a fission chain reaction, producing energy. The attractiveness of a reactor that makes its own fuel from abundant starting materials is offset by the high cost of the system, potential environmental damage, and fear of plutonium proliferation. Plutonium can be readily used to manufacture nuclear bombs. Currently only France and Japan operate breeder reactors for electrical power generation.

10.5 Radiocarbon Dating

Learning Goal

Natural radioactivity is useful in establishing the approximate age of objects of archaeological, anthropological, or historical interest. **Radiocarbon dating** is the estimation of the age of objects through measurement of isotopic ratios of carbon.

Radiocarbon dating is based on the measurement of the relative amounts (or ratio) of ${}^{14}_{6}C$ and ${}^{12}_{6}C$ present in an object. The ${}^{14}_{6}C$ is formed in the upper atmosphere by the bombardment of ${}^{14}_{7}N$ by high-speed neutrons from the sun:

$${}^{14}_{7}N + {}^{1}_{0}n \longrightarrow {}^{14}_{6}C + {}^{1}_{1}H$$

The carbon-14, along with the more abundant carbon-12, is converted into living plant material by the process of photosynthesis. Carbon proceeds up the food chain as the plants are consumed by animals, including humans. When a plant or animal dies, the uptake of both carbon-14 and carbon-12 ceases. However, the amount of carbon-14 slowly decreases because carbon-14 is radioactive ($t_{1/2} = 5730$ years). Carbon-14 decay produces nitrogen:

$${}^{14}_{6}C \longrightarrow {}^{14}_{7}N + {}^{0}_{-1}e$$

When an artifact is found and studied, the relative amounts of carbon-14 and carbon-12 are determined. By using suitable equations involving the $t_{1/2}$ of carbon-14, it is possible to approximate the age of the artifact.

This technique has been widely used to increase our knowledge about the history of the earth, to establish the age of objects (Figure 10.5), and even to detect art forgeries. Early paintings were made with inks fabricated from vegetable dyes (plant material that, while alive, metabolized carbon).

Figure 10.5
Radiocarbon dating was used in the authentication study of the Shroud of Turin. It is an essentially nondestructive technique and is valuable in estimating the age of historical artifacts.

The carbon-14 dating technique is limited to objects that are less than fifty thousand years old, or approximately nine half-lives, which is a practical upper limit. Older objects that have geological or archaeological significance may be dated with naturally occurring isotopes having much longer half-lives.

Examples of useful dating isotopes are listed in Table 10.3.

10.6 Medical Applications of Radioactivity

Learning Goal

The use of radiation in the treatment of various forms of cancer, as well as the newer area of **nuclear medicine,** the use of radioisotopes in diagnosis, has become

An Environmental Perspective

Nuclear Waste Disposal

Nuclear waste arises from a variety of sources. A major source is the spent fuel from nuclear power plants. Medical laboratories generate significant amounts of low-level waste from tracers and therapy. Even household items with limited lifetimes, such as certain types of smoke detectors, use a tiny amount of radioactive material.

Virtually everyone is aware, through television and newspapers, of the problems of solid waste (nonnuclear) disposal that our society faces. For the most part, this material will degrade in some reasonable amount of time. Still, we are disposing of trash and garbage at a rate that far exceeds nature's ability to recycle it.

Now imagine the problem with nuclear waste. We cannot alter the rate at which it decays. This is defined by the half-life. We can't heat it, stir it, or add a catalyst to speed up the process as we can with chemical reactions. Furthermore, the half-lives of many nuclear waste products are very long: plutonium, for example, has a half-life in excess of 24,000 years. Ten half-lives are required for the radioactivity of a substance to reach background levels. So we are talking about a *very* long storage time.

Where on earth can something so very hazardous be contained and stored with reasonable assurance that it will lie undisturbed for a quarter of a million years? Perhaps this is a rhetorical question. Scientists, engineers, and politicians have debated this question for almost fifty years. As yet, no permanent disposal site has been agreed upon. Most agree that the best solution is burial in a stable rock formation, but there is no firm agreement on the location. Fear of earthquakes, which may release large quantities of radioactive materials into the underground water system, is the most serious consideration. Such a disaster could render large sections of the country unfit for habitation.

Many argue for the continuation of temporary storage sites with the hope that the progress of science and technology will, in the years ahead, provide a safer and more satisfactory long-term solution.

The nuclear waste problem, important for its own sake, also affects the development of future societal uses of nuclear chemistry. Before we can enjoy its benefits, we must learn to use and dispose of it safely.

A photograph of the earth, taken from the moon, clearly illustrates the limits of resources and the limits to waste disposal.

Table 10.3 Isotopes Useful in Radioactive Dating

Isotope	Half-Life (years)	Upper Limit (years)	Dating Applications
Carbon-14	5730	5×10^4	Charcoal, organic material, artwork
Tritium (^3_1H)	12.3	1×10^2	Aged wines, artwork
Potassium-40	1.3×10^9	Age of earth (4×10^9)	Rocks, planetary material
Rhenium-187	4.3×10^{10}	Age of earth (4×10^9)	Meteorites
Uranium-238	4.5×10^9	Age of earth (4×10^9)	Rocks, earth's crust

widespread in the past quarter century. Let's look at the properties of radiation that make it an indispensable tool in modern medical care.

Cancer Therapy Using Radiation

Learning Goal

When high-energy radiation, such as gamma radiation, passes through a cell, it may collide with one of the molecules in the cell and cause it to lose one or more electrons, causing a series of events that result in the production of ion pairs. For this reason, such radiation is termed *ionizing radiation* (Section 10.1).

Ions produced in this way may damage biological molecules and cause changes in cellular biochemical processes. This may result in diminished or altered cell function or, in extreme cases, the death of the cell.

An organ that is cancerous is composed of both healthy cells and malignant cells. Tumor cells are more susceptible to the effects of gamma radiation than normal cells because they are undergoing cell division more frequently. Therefore exposure of the tumor area to carefully targeted and controlled dosages of high-energy gamma radiation from cobalt-60 (a high-energy gamma ray source) will kill a higher percentage of abnormal cells than normal cells. If the dosage is administered correctly, a sufficient number of malignant cells will die, destroying the tumor, and enough normal cells will survive to maintain the function of the affected organ.

Gamma radiation can cure cancer. Paradoxically, the exposure of healthy cells to gamma radiation can actually cause cancer. For this reason, radiation therapy for cancer is a treatment that requires unusual care and sophistication.

Nuclear Medicine

Learning Goal

The diagnosis of a host of biochemical irregularities or diseases of the human body has been made routine through the use of radioactive tracers. Medical **tracers** are small amounts of radioactive substances used as probes to study internal organs. Medical techniques involving tracers are **nuclear imaging** procedures.

A small amount of the tracer, an isotope of an element that is known to be attracted to the organ of interest, is administered to the patient. For a variety of reasons, such as ease of administration of the isotope to the patient and targeting the organ of interest, the isotope is often a part of a larger molecule or ion. Because the isotope is radioactive, its path may be followed by using suitable detection devices. A "picture" of the organ is obtained, often far more detailed than is possible with conventional X rays. Such techniques are noninvasive; that is, surgery is not required to investigate the condition of the internal organ, eliminating the risk associated with an operation.

The radioactive isotope of an element chosen for tracer studies has exactly the same chemical behavior as any other isotope of the same element. For example, iodine-127, the most abundant nonradioactive isotope of iodine, tends to concentrate in the thyroid gland. Both radioactive iodine-131 and iodine-125 behave in the same way and are used to study the thyroid. The rate of uptake of the radioactive isotope gives valuable information regarding underactivity or overactivity (hypoactive or hyperactive thyroid).

Isotopes with short half-lives are preferred for tracer studies. These isotopes emit their radiation in a more concentrated burst (short half-life materials have greater activity), facilitating their detection. If the radioactive decay is easily detected, the method is more sensitive and thus capable of providing more information. Furthermore, an isotope with a short half-life decays to background more rapidly. This is a mechanism for removal of the radioactivity from the body. If the radioactive element is also rapidly metabolized and excreted, this is obviously beneficial as well.

The following examples illustrate the use of imaging procedures for diagnosis of disease.

- *Bone disease and injury.* The most widely used isotope for bone studies is technetium-99m, which is incorporated into a variety of ions and molecules that direct the isotope to the tissue being investigated. Technetium compounds containing phosphate are preferentially adsorbed on the surface of bone. New bone formation (common to virtually all bone injuries) increases the incorporation of the technetium compound. As a result, an enhanced image appears at the site of the injury. Bone tumors behave in a similar fashion.
- *Cardiovascular diseases.* Thallium-201 is used in the diagnosis of coronary artery disease. The isotope is administered intravenously and delivered to the heart muscle in proportion to the blood flow. Areas of restricted flow are observed as having lower levels of radioactivity, indicating some type of blockage.
- *Pulmonary disease.* Xenon is one of the noble gases. Radioactive xenon-133 may be inhaled by the patient. The radioactive isotope will be transported from the lungs and distributed through the circulatory system. Monitoring the distribution, as well as the reverse process, the removal of the isotope from the body (exhalation), can provide evidence of obstructive pulmonary disease, such as cancer or emphysema.

Examples of useful isotopes and the organ(s) in which they tend to concentrate are summarized in Table 10.4.

For many years, imaging with radioactive tracers was used exclusively for diagnosis. Recent applications have expanded to other areas of medicine. Imaging is now used extensively to guide surgery, assist in planning radiation therapy, and support the technique of angioplasty.

Table 10.4 Isotopes Commonly Used in Nuclear Medicine

Area of Body	Isotope	Use
Blood	Red blood cells tagged with chromium-51	Determine blood volume in body
Bone	*Technetium-99m, barium-131	Allow early detection of the extent of bone tumors and active sites of rheumatoid arthritis
Brain	*Technetium-99m	Detect and locate brain tumors and stroke
Coronary artery	Thallium-201	Determine the presence and location of obstructions in coronary arteries
Heart	*Technetium-99m	Determine cardiac output, size, and shape
Kidney	*Technetium-99m	Determine renal function and location of cysts; a common follow-up procedure for kidney transplant patients
Liver-spleen	*Technetium-99m	Determine size and shape of liver and spleen; location of tumors
Lung	Xenon-133	Determine whether lung fills properly; locate region of reduced ventilation and tumors
Thyroid	Iodine-131	Determine rate of iodine uptake by thyroid

*The destination of this isotope is determined by the identity of the compound in which it is incorporated.

Question 10.7

Technetium-99m is used in diagnostic imaging studies involving the brain. What fraction of the radioisotope remains after 12 hours have elapsed? See Table 10.2 for the half-life of technetium-99m.

Question 10.8

Barium-131 is a radioisotope used to study bone formation. A patient ingested barium-131. How much time will elapse until only one-fourth of the barium-131 remains, assuming that none of the isotope is eliminated from the body through normal processes? The half-life of barium-131 is 11.6 minutes.

Making Isotopes for Medical Applications

Learning Goal  Learning Goal 

In early experiments with radioactivity, the radioactive isotopes were naturally occurring. For this reason the radioactivity produced by these unstable isotopes is described as **natural radioactivity.** If, on the other hand, a normally stable, nonradioactive nucleus is made radioactive, the resulting radioactivity is termed **artificial radioactivity.** The stable nucleus is made unstable by the introduction of "extra" protons, neutrons, or both.

The process of forming radioactive substances is often accomplished in the core of a **nuclear reactor,** in which an abundance of small nuclear particles, particularly neutrons, is available. Alternatively, extremely high-velocity charged particles (such as alpha and beta particles) may be produced in **particle accelerators,** such as a cyclotron. Accelerators are extremely large and use magnetic and electric fields to "push and pull" charged particles toward their target at very high speeds. A portion of the accelerator at the Brookhaven National Laboratory is shown in Figure 10.6.

Many isotopes that are useful in medicine are produced by particle bombardment. A few examples include the following:

- Gold-198, used as a tracer in the liver, is prepared by neutron bombardment.

$$^{197}_{79}\text{Au} + ^{1}_{0}\text{n} \longrightarrow ^{198}_{79}\text{Au}$$

- Gallium-67, used in the diagnosis of Hodgkin's disease, is prepared by proton bombardment.

$$^{66}_{30}\text{Zn} + ^{1}_{1}\text{p} \longrightarrow ^{67}_{31}\text{Ga}$$

Some medically useful isotopes, with short half-lives, must be prepared near the site of the clinical test. Preparation and shipment from a reactor site would waste time and result in an isotopic solution that had already undergone significant decay, resulting in diminished activity.

A common example is technetium-99m. It has a half-life of only six hours. It is prepared in a small generator, often housed in a hospital's radiology laboratory (Figure 10.7). The generator contains radioactive molybdate ion ($MoO_4{}^{2-}$). Molybdenum-99 is more stable than technetium-99m; it has a half-life of 67 hours.

The molybdenum in molybdate ion decays according to the following nuclear equation:

$$^{99}_{42}\text{Mo} \longrightarrow ^{99\text{m}}_{43}\text{Tc} + ^{0}_{-1}\text{e}$$

Chemically, radioactive molybdate $MoO_4{}^{2-}$ converts to radioactive pertechnetate ion ($TcO_4{}^{-}$). The radioactive $TcO_4{}^{-}$ is removed from the generator when needed. It is administered to the patient as an aqueous salt solution that has an osmotic pressure identical to that of human blood.

Figure 10.6
A portion of a linear accelerator located at Brookhaven National Laboratory in New York. Particles can be accelerated at velocities close to the speed of light and accurately strike small "target" nuclei. At such facilities, rare isotopes can be synthesized and their properties studied.

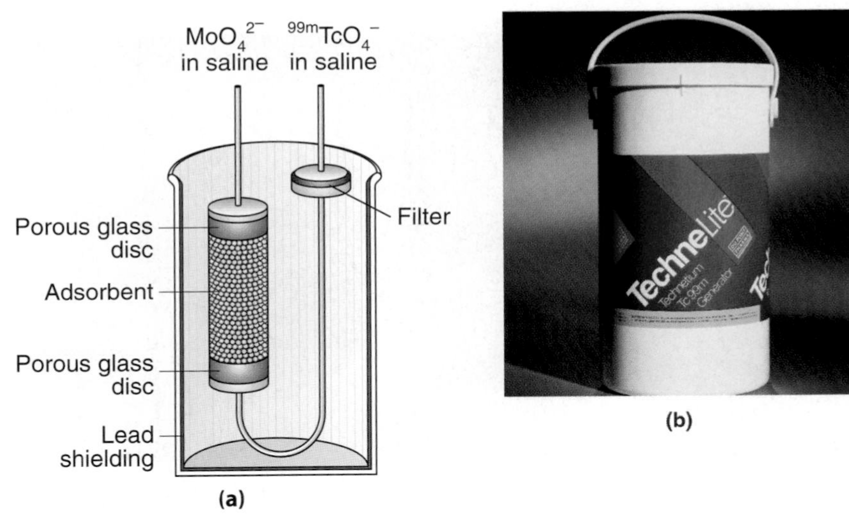

Figure 10.7
Preparation of technetium-99m. (a) A diagram depicting the conversion of $^{99}MoO_4^{2-}$ to $^{99m}TcO_4^-$ through radioactive decay. The radioactive pertechnetate ion is periodically removed from the generator in saline solution and used in tracer studies. (b) A photograph of a commercially available technetium-99m generator suitable for use in a hospital laboratory.

10.7 Biological Effects of Radiation

It is necessary to use suitable precautions in working with radioactive substances. The chosen protocol is based on an understanding of the effects of radiation, dosage levels and "tolerable levels," the way in which radiation is detected and measured, and the basic precepts of radiation safety.

Radiation Exposure and Safety

In working with radioactive materials, the following factors must be considered.

Learning Goal

The Magnitude of the Half-Life

In considering safety, isotopes with short half-lives have, at the same time, one major disadvantage and one major advantage.

A CLINICAL PERSPECTIVE

Magnetic Resonance Imaging

It has been known for some time that nuclei, like electrons, exist in different energy states or energy levels. Furthermore, under the influence of electromagnetic radiation, transitions involving absorption of radiation can occur between the various nuclear states. This is analogous to the behavior of the electron in an atom. However, electronic transitions occur under the influence of ultraviolet and visible radiation, whereas the nuclear transitions occur in the radio frequency region of the electromagnetic spectrum under the influence of a magnetic field. The nuclei of hydrogen atoms may be affected in different ways, depending on their position in a molecule. These differences give rise to unique patterns of energy absorption (called absorption spectra), and the technique of nuclear magnetic resonance (NMR) has become a useful tool for the study of molecules containing hydrogen.

Human organs and tissue are made up of compounds containing hydrogen atoms. In the 1970s and 1980s the NMR experimental technique was extended beyond tiny laboratory samples of pure compounds to the most complex sample possible—the human body. The result of these experiments is termed *magnetic resonance imaging (MRI)*.

MRI is noninvasive to the body, requires no use of radioactive substances, and is quick, safe, and painless. A person is placed in a cavity surrounded by a magnetic field, and an image (based on the extent of radio frequency energy absorption) is generated, stored, and sorted in a computer. Differences between normal and malignant tissue, atherosclerotic thickening of an aortal wall, and a host of other problems may be seen clearly in the final image.

Advances in MRI technology have provided medical practitioners with a powerful tool in diagnostic medicine. This is but one more example of basic science leading to technological advancement.

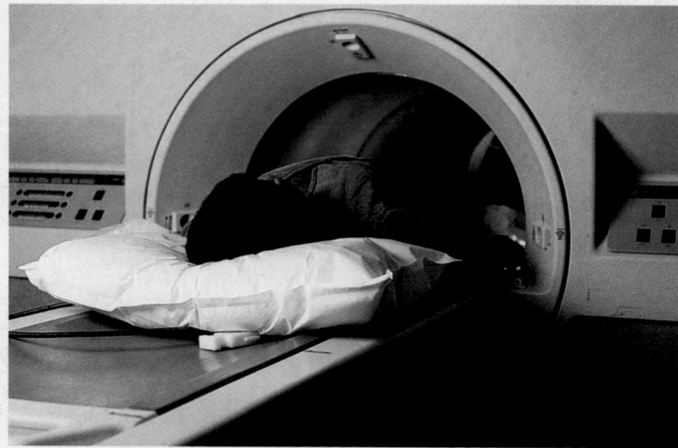

A patient entering an MRI scanner.

Dr. Paul Barnett of the Greater Baltimore Medical Center studies images obtained using MRI.

Higher levels of exposure in a short time produce clearer images.

On one hand, short-half-life radioisotopes produce a larger amount of radioactivity per unit time than a long-half-life substance. For example, consider equal amounts of hypothetical isotopes that produce alpha particles. One has a half-life of ten days; the other has a half-life of one hundred days. After one half-life, each substance will produce exactly the same number of alpha particles. However, the first substance generates the alpha particles in only one-tenth of the time, hence emits ten times as much radiation per unit time. Equal exposure times will result in a higher level of radiation exposure for substances with short half-lives, and lower levels for substances with long half-lives.

On the other hand, materials with short half-lives (weeks, days, or less) may be safer to work with, especially if an accident occurs. Over time (depending on the magnitude of the half-life) radioactive isotopes will decay to **background radiation** levels. This is the level of radiation attributable to our surroundings on a day-to-day basis.

All matter is composed of both radioactive and nonradioactive isotopes. Small amounts of radioactive material in the air, water, soil, and so forth make up a part of the background levels. Cosmic rays from outer space continually bombard us with radiation, contributing to the total background. Owing to the inevitability of background radiation, there can be no such thing as "zero" radiation!

An isotope with a short half-life, for example 5.0 min, may decay to background in as few as ten half-lives

$$10 \cancel{\text{half-lives}} \times \frac{5.0 \text{ min}}{1 \cancel{\text{half-life}}} = 50 \text{ min}$$

A spill of such material could be treated by waiting ten half-lives, perhaps by going to lunch. When you return to the laboratory, the material that was spilled will be no more radioactive than the floor itself. An accident with plutonium-239, which has a half-life of 24,000 years, would be quite a different matter! After fifty minutes, virtually all of the plutonium-239 would still remain. Long-half-life isotopes, by-products of nuclear technology, pose the greatest problems for safe disposal. Finding a site that will remain undisturbed "forever" is quite a formidable task.

See An Environmental Perspective: Nuclear Waste Disposal on page 279.

Question 10.9

Describe the advantage of using isotopes with short half-lives for tracer applications in a medical laboratory.

Question 10.10

Can you think of any disadvantage associated with the use of isotopes described in Question 10.9? Explain.

Shielding

Alpha and beta particles, being relatively low in energy, require low-level **shielding.** A lab coat and gloves are generally sufficient protection from this low-penetration radiation. On the other hand, shielding made of lead, concrete, or both is required for gamma rays (and X-rays, which are also high-energy radiation). Extensive manipulation of gamma emitters is often accomplished in laboratory and industrial settings by using robotic control: computer-controlled mechanical devices that can be programmed to perform virtually all manipulations normally carried out by humans.

Distance from the Radioactive Source

Radiation intensity varies *inversely* with the *square* of the distance from the source. Doubling the distance from the source *decreases* the intensity by a factor of four (2^2). Again, the use of robot manipulators is advantageous, allowing a greater distance between the operator and the radioactive source.

Time of Exposure

The effects of radiation are cumulative. Generally, potential damage is directly proportional to the time of exposure. Workers exposed to moderately high levels of radiation on the job may be limited in the time that they can perform that task. For example, workers involved in the cleanup of the Three Mile Island nuclear plant, incapacitated in 1979, observed strict limits on the amount of time that they could be involved in the cleanup activities.

Figure 10.8
Photograph of the construction of one-million-gallon capacity storage tanks for radioactive waste. Located in Hanford, Washington, they are now covered with 6–8 ft of earth.

Types of Radiation Emitted

Alpha and beta emitters are generally less hazardous than gamma emitters, owing to differences in energy and penetrating power that require less shielding. However, ingestion or inhalation of an alpha emitter or beta emitter can, over time, cause serious tissue damage; the radioactive substance is in direct contact with sensitive tissue. (On page 288, An Environmental Perspective: Radon and Indoor Air Pollution expands on this problem.)

Waste Disposal

Virtually all applications of nuclear chemistry create radioactive waste and, along with it, the problems of safe handling and disposal. Most disposal sites, at present, are considered temporary, until a long-term safe solution can be found. Figure 10.8 conveys a sense of the enormity of the problem. Also, An Environmental Perspective: Nuclear Waste Disposal on page 279, examines this problem in more detail.

10.8 Measurement of Radiation

The changes that take place when radiation interacts with matter (such as photographic film) provide the basis of operation for various radiation detection devices.

The principal detection methods involve the use of either photographic film to create an image of the location of the radioactive substance or a counter that allows the measurement of intensity of radiation emitted from some source by converting the radiation energy to an electrical signal.

Learning Goal

11

Nuclear Imaging

This approach is often used in nuclear medicine. An isotope is administered to a patient, perhaps iodine-131, which is used to study the thyroid gland, and the isotope begins to concentrate in the organ of interest. Nuclear images (photographs) of that region of the body are taken at periodic intervals using a special type of film. The emission of radiation from the radioactive substance creates the image, in much the same way as light causes the formation of images on conventional film in a camera. Upon development of the series of photographs, a record of the organ's uptake of the isotope over time enables the radiologist to assess the condition of the organ.

Computer Imaging

The coupling of rapid developments in the technology of television and computers, resulting in the marriage of these two devices, has brought about a versatile alternative to photographic imaging.

A specialized television camera, sensitive to emitted radiation from a radioactive substance administered to a patient, develops a continuous and instantaneous record of the voyage of the isotope throughout the body. The signal, transmitted to the computer, is stored, sorted, and portrayed on a monitor. Advantages include increased sensitivity, allowing a lower dose of the isotope, speed through elimination of the developing step, and versatility of application, limited perhaps only by the creativity of the medical practitioners.

CT represents *Computer-aided Tomography:* the computer reconstructs a series of measured images of tissue density (tomography). Small differences in tissue density may indicate the presence of a tumor.

A particular type of computer imaging, useful in diagnostic medicine, is the CT scanner. The CT scanner gathers huge amounts of data and processes the data to produce detailed information, all in a relatively short time. Such a device may be less hazardous than conventional X-ray techniques because it generates more useful information per unit of radiation. It often produces a superior image. A photograph of a CT scanner is shown in Figure 10.9, and an image of a damaged spinal bone, taken by a CT scanner, is shown in Figure 10.10.

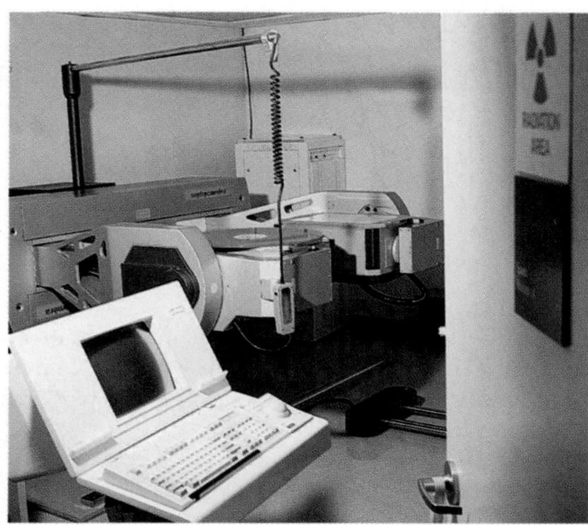

Figure 10.9
An imaging laboratory at the Greater Baltimore Medical Center.

The Geiger Counter

A Geiger counter is an instrument that detects ionizing radiation. Ions, produced by radiation passing through a tube filled with an ionizable gas, can conduct an electrical current between two electrodes. This current flow can be measured and is proportional to the level of radiation (Figure 10.11). Such devices, which were routinely used in laboratory and industrial monitoring, have been largely replaced by more sophisticated devices, often used in conjunction with a computer.

Film Badges

A common sight in any hospital or medical laboratory or any laboratory that routinely uses radioisotopes is the film badge worn by all staff members exposed in any way to low-level radioactivity.

Recall that the time of exposure to radiation is one critical hazard consideration (Section 10.7).

A film badge is merely a piece of photographic film that is sensitive to energies corresponding to radioactive emissions. It is shielded from light, which would interfere, and mounted in a clip-on plastic holder that can be worn throughout the workday. The badges are periodically collected and developed. The degree of darkening is proportional to the amount of radiation to which the worker has been exposed, just as a conventional camera produces images on film in proportion to the amount of light that it "sees."

Proper record keeping thus allows the laboratory using radioactive substances to maintain an ongoing history of each individual's exposure and, at the same time, promptly pinpoint any hazards that might otherwise go unnoticed.

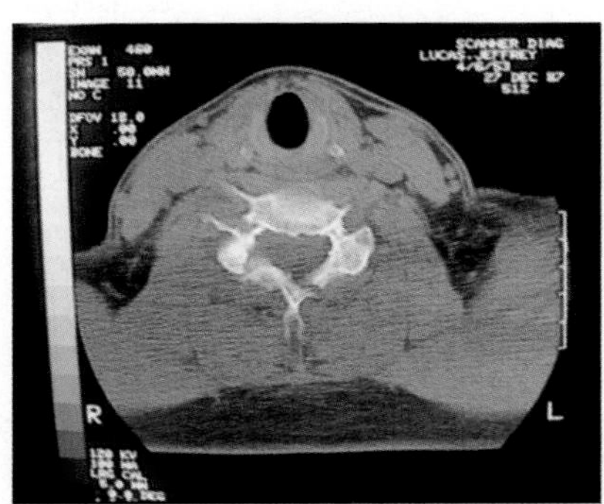

Figure 10.10
Damage observed in a spinal bone on a CT scan image.

Units of Radiation Measurement

The amount of radiation emitted by a source or received by an individual is reported in a variety of ways, using units that describe different aspects of radiation. The *curie* and the *roentgen* describe the intensity of the emitted radiation, whereas the *rad* and the *rem* describe the biological effects of radiation.

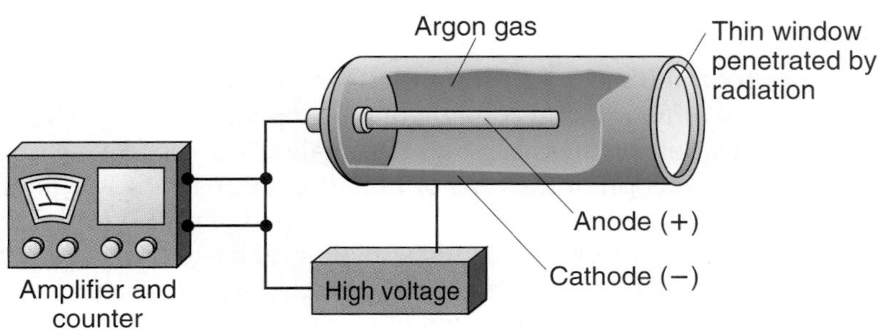

Figure 10.11
The design of a Geiger counter used for the measurement of radioactivity.

An Environmental Perspective

Radon and Indoor Air Pollution

Marie and Pierre Curie first discovered that air in contact with radium compounds became radioactive. Later experiments by Ernest Rutherford and others isolated the radioactive substance from the air. This substance was an isotope of the noble gas radon (Rn).

We now know that radium (Ra) produces radon by spontaneous decay:

$$^{226}_{88}\text{Ra} \longrightarrow {}^{4}_{2}\text{He} + {}^{222}_{86}\text{Rn}$$

Radium in trace quantities is found in the soil and rock and is unequally distributed in the soil. The decay product, radon, is emitted from the soil to the surrounding atmosphere. Radon is also found in higher concentrations when uranium is found in the soil. This is not surprising, because radium is formed as a part of the stepwise decay of uranium.

If someone constructs a building over soil or rock that has a high radium content (or uses stone with a high radium content to build the foundation!), the radon gas can percolate through the basement and accumulate in the house. Couple this with the need to build more energy-efficient, well-insulated dwellings, and the radon levels in buildings in some regions of the country can become quite high.

Radon itself is radioactive; however, its radiation is not the major problem. Because it is a gas and chemically inert, it is rapidly exhaled after breathing. However, radon decays to polonium:

$$^{222}_{86}\text{Rn} \longrightarrow {}^{4}_{2}\text{He} + {}^{218}_{84}\text{Po}$$

This polonium isotope is radioactive and is a nonvolatile heavy metal that can attach itself to bronchial or lung tissue, emitting hazardous radiation and producing other isotopes that are also radioactive.

In the United States, homes are now being tested and monitored for radon. Studies continue to attempt to find reasonable solutions to the problem. Current recommendations include sealing cracks and openings in basements, increasing ventilation, and evaluating sites before construction of buildings. Debate continues within the scientific community regarding a safe and attainable indoor air quality standard for radon.

Learning Goal 12

The Curie

The **curie** is a measure of the amount of radioactivity in a radioactive source. The curie is independent of the nature of the radiation (alpha, beta, or gamma) and its effect on biological tissue. A curie is defined as the amount of radioactive material that produces 3.7×10^{10} atomic disintegrations per second.

The Roentgen

The **roentgen** is a measure of very high energy ionizing radiation (X ray and gamma ray) only. The roentgen is defined as the amount of radiation needed to produce 2×10^{9} ion pairs when passing through one cm^3 of air at 0°C. The roentgen is a measure of radiation's interaction with air and gives no information about the effect on biological tissue.

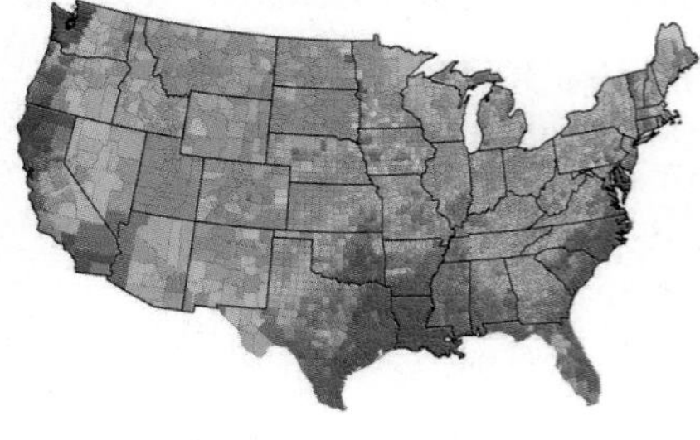

Figure 10.12
Relative yearly radiation dosages for individuals in the continental United States. Red, yellow, and green shading indicates higher levels of background radiation. Blue shading indicates regions of lower background exposure.

The Rad

The **rad,** or *radiation absorbed dosage,* provides more meaningful information than either of the previous units of measure. It takes into account the nature of the absorbing material. It is defined as the dosage of radiation able to transfer 2.4×10^{-3} cal of energy to one kg of matter.

The Rem

The **rem,** or *roentgen equivalent for man,* describes the biological damage caused by the absorption of different kinds of radiation by the human body. The *rem* is obtained by multiplication of the *rad* by a factor called the *relative biological effect (RBE).* The RBE is a function of the type of radiation (alpha, beta, or gamma). Although a beta particle is more penetrating than an alpha particle, an alpha particle is approximately ten times more damaging to biological tissue. As a result, the RBE is ten for alpha particles and one for beta particles. Relative yearly radiation dosages received by Americans are shown in Figure 10.12.

The **lethal dose (LD_{50})** of radiation is defined as the acute dosage of radiation that would be fatal for 50% of the exposed population within 30 days. An estimated lethal dose is 500 rems. Some biological effects, however, may be detectable at a level as low as 25 rem.

Question 10.11

From a clinical standpoint, what advantages does expressing radiation in rems have over the use of other radiation units?

Question 10.12

Is the roentgen unit used in the measurement of alpha particle radiation? Why or why not?

Summary

10.1 Natural Radioactivity

Radioactivity is the process by which atoms emit energetic, ionizing particles or rays. These particles or rays are termed radiation. Nuclear radiation occurs because the nucleus is unstable, hence radioactive. Nuclear symbols consist of the elemental symbol, the atomic number, and the mass number.

Not all nuclei are unstable. Only unstable nuclei undergo change and produce radioactivity in the process of radioactive decay. Three types of natural radiation emitted by unstable nuclei are *alpha particles, beta particles,* and *gamma rays.* This radiation is collectively termed *ionizing radiation.*

10.2 Writing a Balanced Nuclear Equation

A *nuclear equation* represents a nuclear process such as radioactive decay. The total of the mass numbers on each side of the reaction arrow must be identical, and the sum of the atomic numbers of the reactants must equal the sum of the atomic numbers of the products. Nuclear equations can be used to predict products of nuclear reactions.

10.3 Properties of Radioisotopes

The energy that holds nuclear particles together in the nucleus is the *binding energy* of the nucleus. When an isotope decays, some of this binding energy is released. Nuclear stability correlates with the ratio of neutrons to protons in the isotope. Nuclei with large numbers of protons tend to be unstable, and isotopes containing 2, 8, 20, 50, 82, or 126 protons or neutrons (magic numbers) are stable. Also, isotopes with even numbers of protons or neutrons are generally more stable than those with odd numbers of protons or neutrons.

The *half-life, $t_{1/2}$,* is the time required for one-half of a given quantity of a substance to undergo change. Each isotope has its own characteristic half-life. The degree of stability of an isotope is indicated by the isotope's half-life. Isotopes with short half-lives decay rapidly; they are very unstable.

10.4 Nuclear Power

Einstein predicted that a small amount of nuclear mass would convert to a very large amount of energy when the nucleus breaks apart. Fission reactors are used to generate electrical power. Technological problems with *fusion* and *breeder* reactors have prevented their commercialization in the United States.

10.5 Radiocarbon Dating

Radiocarbon dating is based on the measurement of the relative amounts of carbon-12 and carbon-14 present in an object. The ratio of the masses of these isotopes changes slowly over time, making it useful in determining the age of objects containing carbon.

10.6 Medical Applications of Radioactivity

The use of radiation in the treatment of various forms of cancer, and in the newer area of *nuclear medicine,* has become widespread in the past quarter century.

Ionizing radiation causes changes in cellular biochemical processes that may damage or kill the cell. A cancerous organ is composed of both healthy and malignant cells. Exposure of the tumor area to controlled dosages of high-energy gamma radiation from cobalt-60 will kill a higher percentage of abnormal cells than normal cells and is a valuable cancer therapy.

The diagnosis of a host of biochemical irregularities or diseases of the human body has been made routine through the use of radioactive tracers. *Tracers* are small amounts of radioactive substances used as probes to study internal organs. Because the isotope is radioactive, its path may be followed by using suitable detection devices. A "picture" of the organ is obtained, far more detailed than is possible with conventional X rays.

The radioactivity produced by unstable isotopes is described as *natural radioactivity.* A normally stable, nonradioactive nucleus can be made radioactive, and this is termed *artificial radioactivity* (the process produces synthetic isotopes). Synthetic isotopes are often used in clinical situations. Isotopic synthesis may be carried out in the core of a *nuclear reactor* or in a *particle accelerator.* Short-lived isotopes, such as technetium-99m, are often produced directly at the site of the clinical testing.

10.7 Biological Effects of Radiation

Safety considerations are based on the magnitude of the *half-life, shielding,* distance from the radioactive source, time of exposure, and type of radiation emitted. We are never entirely free of the effects of radioactivity. *Background radiation* is normal radiation attributable to our surroundings.

Virtually all applications of nuclear chemistry create radioactive waste and, along with it, the problems of safe handling and disposal. Most disposal sites are considered temporary, until a long-term safe solution can be found.

10.8 Measurement of Radiation

The changes that take place when radiation interacts with matter provide the basis for various radiation detection devices. Photographic imaging, computer imaging, the Geiger counter, and film badges represent the most frequently used devices for detecting and measuring radiation.

Commonly used radiation units include the *curie,* a measure of the amount of radioactivity in a radioactive source; the *roentgen,* a measure of high-energy radiation (X ray and gamma ray); the *rad* (radiation absorbed dosage), which takes into account the nature of the absorbing material; and the *rem* (roentgen equivalent for man), which describes the biological damage caused by the absorption of different kinds of radiation by the human body. The *lethal dose* of radiation, LD_{50}, is defined as the dose that would be fatal for 50% of the exposed population within thirty days.

Key Terms

alpha particle (10.1)
artificial radioactivity (10.6)
background radiation (10.7)
beta particle (10.1)
binding energy (10.3)
breeder reactor (10.4)
chain reaction (10.4)
curie (10.8)
fission (10.4)
fusion (10.4)
gamma ray (10.1)
half-life ($t_{1/2}$) (10.3)
ionizing radiation (10.1)
lethal dose (LD_{50}) (10.8)
metastable isotope (10.2)
natural radioactivity (10.6)
nuclear equation (10.2)
nuclear imaging (10.6)
nuclear medicine (10.6)
nuclear reactor (10.6)
particle accelerator (10.6)
rad (10.8)
radioactivity (10.1)
radiocarbon dating (10.5)
rem (10.8)
roentgen (10.8)
shielding (10.7)
tracer (10.6)

Questions and Problems

Natural Radioactivity

10.13 Define or describe each of the following terms:
a. natural radioactivity
b. background radiation

10.14 Define or describe each of the following terms:
a. alpha particle
b. alpha decay

10.15 Define or describe each of the following terms:
a. beta particle
b. gamma radiation

10.16 Define or describe each of the following terms:
a. beta decay
b. artificial radioactivity

10.17 Write the nuclear symbol for each of the following:
a. an alpha particle
b. a beta particle

10.18 Write the nuclear symbol for each of the following:
a. a proton
b. uranium-235

10.19 Write the nuclear symbol for each of the following:
a. deuterium (hydrogen-2)
b. tritium (hydrogen-3)

10.20 Write the nuclear symbol for each of the following:
a. nitrogen-15
b. carbon-14

10.21 Compare and contrast the three major types of radiation produced by nuclear decay.

10.22 Rank the three major types of radiation in order of size, speed, and penetrating power.

10.23 How does an α particle differ from a helium atom?

10.24 What is the major difference between β and δ radiation?

Writing a Balanced Nuclear Equation

10.25 Write a nuclear reaction to represent cobalt-60 decaying to nickel-60 plus a beta particle plus a gamma ray.

10.26 Write a nuclear reaction to represent radium-226 decaying to radon-222 plus an alpha particle.

10.27 Complete the following nuclear reaction:

$$^{23}_{11}Na + ^{2}_{1}H \longrightarrow ? + ^{1}_{1}H$$

10.28 Complete the following nuclear reaction:

$$^{238}_{92}U + ^{14}_{7}N \longrightarrow ? + 6^{1}_{0}n$$

10.29 Complete the following nuclear reaction:

$$^{24}_{10}Ne \longrightarrow \beta + ?$$

10.30 Complete the following nuclear reaction:

$$^{190}_{78}Pt \longrightarrow \alpha + ?$$

10.31 Complete the following nuclear reaction:

$$? \longrightarrow ^{140}_{56}Ba + ^{0}_{-1}e$$

10.32 Complete the following nuclear reaction:

$$? \longrightarrow ^{214}_{90}Th + ^{4}_{2}He$$

Properties of Radioisotopes

10.33 What is the difference between natural radioactivity and artificial radioactivity?

10.34 Is the fission of uranium-235 an example of natural or artificial radioactivity?

10.35 Summarize the major characteristics of nuclei for which we predict a high degree of stability.

10.36 Explain why the binding energy of a nucleus is expected to be large.

10.37 If 3.2 mg of the radioisotope iodine-131 is administered to a patient, how much will remain in the body after 24 days, assuming that no iodine has been eliminated from the body by any other process? (See Table 10.2 for the half-life of iodine-131.)

10.38 A patient receives 10.0 ng of a radioisotope with a half-life of 12 hours. How much will remain in the body after 2.0 days, assuming that radioactive decay is the only path for removal of the isotope from the body?

10.39 A sample containing 1.00×10^2 mg of iron-59 is stored for 135 days. What mass of iron-59 will remain at the end of the storage period? (See Table 10.2 for the half-life of iron-59.)

10.40 An instrument for cancer treatment containing a cobalt-60 source was manufactured in 1978. In 1995 it was removed from service and, in error, was buried in a landfill with the source still in place. What percentage of its initial radioactivity will remain in the year 2010? (See Table 10.2 for the half-life of cobalt-60.)

Nuclear Power

10.41 Which type of nuclear process splits nuclei to release energy?

10.42 Which type of nuclear process combines small nuclei to release energy?

10.43 a. Describe the process of fission.
b. How is this reaction useful as the basis for the production of electrical energy?

10.44 a. Describe the process of fusion.
b. How could this process be used for the production of electrical energy?

10.45 Write a balanced nuclear equation for a fusion reaction.

10.46 What are the major disadvantages of a fission reactor for electrical energy production?

10.47 What is meant by the term *breeder reactor?*

10.48 What are the potential advantages and disadvantages of breeder reactors?

10.49 Describe what is meant by the term *chain reaction.*

10.50 Why are carbon rods used in a fission reaction?

10.51 What is the greatest barrier to development of fusion reactors?

10.52 What type of nuclear reaction fuels our solar system?

Radiocarbon Dating

10.53 Describe the process used to determine the age of the wooden coffin of King Tut.

10.54 What property of carbon enables us to assess the age of a painting?

Medical Applications of Radioactivity

10.55 The isotope indium-111 is used in medical laboratories as a label for blood platelets. To prepare indium-111, silver-108 is bombarded with an alpha particle, forming an intermediate isotope of indium. Write a nuclear equation for the process, and identify the intermediate isotope of indium.

10.56 Radioactive molybdenum-99 is used to produce the tracer isotope, technetium-99m. Write a nuclear equation for the formation of molybdenum-99 from stable molybdenum-98 bombarded with neutrons.

10.57 Describe an application of each of the following isotopes:
a. technetium-99m
b. xenon-133

10.58 Describe an application of each of the following isotopes:
a. iodine-131
b. thallium-201

10.59 Why is radiation therapy an effective treatment for certain types of cancer?

10.60 Describe how medically useful isotopes may be prepared.

Biological Effects of Radiation

10.61 What is the effect on a person's level of radiation exposure resulting from:
a. increasing distance from the source?
b. wearing gloves?

10.62 What is the effect on a person's level of radiation exposure resulting from:
a. using concrete instead of wood paneling for the walls of a radiation laboratory?
b. wearing a lab apron lined with thin sheets of lead?

10.63 What is the source of background radiation?

10.64 Why do high-altitude jet flights increase a person's exposure to background radiation?

Measurement of Radiation

10.65 X-ray technicians often wear badges containing photographic film. How is this film used to indicate exposure to X rays?

10.66 Why would a Geiger counter be preferred to film for assessing the immediate danger resulting from a spill of some solution containing a radioisotope?

10.67 What is meant by the term *relative biological effect?*

10.68 What is meant by the term *lethal dose* of radiation?

10.69 Define each of the following units:
a. curie
b. roentgen

10.70 Define each of the following radiation units:
a. rad
b. rem

Critical Thinking Problems

1. Isotopes used as radioactive tracers have chemical properties that are similar to those of a nonradioactive isotope of the same element. Explain why this is a critical consideration in their use.
2. A chemist proposes a research project to discover a catalyst that will speed up the decay of radioactive isotopes that are waste products of a medical laboratory. Such a discovery would be a potential solution to the problem of nuclear waste disposal. Critique this proposal.
3. A controversial solution to the disposal of nuclear waste involves burial in sealed chambers far below the earth's surface. Describe potential pros and cons of this approach.
4. What type of radioactive decay is favored if the number of protons in the nucleus is much greater than the number of neutrons? Explain.
5. If the proton-to-neutron ratio in question 4 (above) were reversed, what radioactive decay process would be favored? Explain.
6. Radioactive isotopes are often used as "tracers" to follow an atom through a chemical reaction, and the following is an example. Acetic acid reacts with methyl alcohol by eliminating a molecule of water to form methyl acetate. Explain how you would use the radioactive isotope oxygen-18 to show whether the oxygen atom in the water product comes from the —OH of the acid or the —OH of the alcohol.

$$\underset{\text{Acetic acid}}{H_3C{-}\overset{\overset{\displaystyle O}{\|}}{C}{-}OH} + \underset{\text{Methyl alcohol}}{HOCH_3} \longrightarrow \underset{\text{Methyl acetate}}{H_3C{-}\overset{\overset{\displaystyle O}{\|}}{C}{-}O{-}CH_3} + H_2O$$

The origins of fossil fuels.

11

An Introduction to Organic Chemistry:

The Saturated Hydrocarbons

ORGANIC CHEMISTRY

Outline

Learning Goals

1. Compare and contrast organic and inorganic compounds.
2. Draw structures that represent each of the families of organic compounds.
3. Write the names and draw the structures of the common functional groups.
4. Write condensed and structural formulas for saturated hydrocarbons.
5. Describe the relationship between the structure and physical properties of saturated hydrocarbons.
6. Use the basic rules of the I.U.P.A.C. Nomenclature System to name alkanes and substituted alkanes.
7. Draw constitutional (structural) isomers of simple organic compounds.
8. Write the names and draw the structures of simple cycloalkanes.
9. Draw *cis* and *trans* isomers of cycloalkanes.
10. Describe conformations of alkanes.
11. Draw the chair and boat conformations of cyclohexane.
12. Write equations for combustion reactions of alkanes.
13. Write equations for halogenation reactions of alkanes.

CHEMISTRY CONNECTION

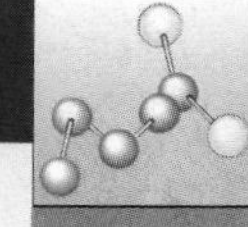

The Origin of Organic Compounds

About 425 million years ago, mountain ranges rose, and enormous inland seas emptied, producing new and fertile lands. In the next 70 million years the simple aquatic plants evolved into land plants, and huge forests of ferns, trees, and shrubs flourished. Reptiles roamed the forests. During the period between 360 and 280 million years ago the seas rose and fell at least fifty times. During periods of flood, the forests were buried under sediments. When the seas fell again, the forests were reestablished. The cycle was repeated over and over. Each flood period deposited a new layer of peat—partially decayed, sodden, compressed plant matter. These layers of peat were compacted by the pressure of the new sediments forming above them. Much of the sulfur and hydrogen was literally squeezed out of the peat, increasing the percentage of carbon. Slowly, the peat was compacted into seams of coal, which is 55–95% carbon. Oil, consisting of a variety of hydrocarbons, formed on the bottoms of ancient oceans from the remains of marine plants and animals.

Together, coal and oil are the "fossil fuels" that we use to generate energy for transportation, industry, and our homes. In the last two centuries we have extracted many of the known coal reserves from the earth and have become ever more dependent on the world's oil reserves. Coal and oil are products of the chemical reactions of photosynthesis that occurred over millions of years of the earth's history. Our society must recognize that they are nonrenewable resources. We must actively work to conserve the supply that remains and to develop alternative energy sources for the future.

In this chapter we take a closer look at the structure and properties of the hydrocarbons, such as those that make up oil. In this and later chapters we will study the amazing array of organic molecules (molecules made up of carbon, hydrogen, and a few other elements), many of which are essential to life. As we will see, all the structural and functional molecules of the cell, including the phospholipids that make up the cell membrane and the enzymes that speed up biological reactions, are organic molecules. Smaller organic molecules, such as the sugars glucose and fructose, are used as fuel by our cells, whereas others, such as penicillin and aspirin, are useful in the treatment of disease. All these organic compounds, and many more, are the subject of the remaining chapters of this text.

Introduction

Organic chemistry is the study of carbon-containing compounds. The term *organic* was coined in 1807 by the Swedish chemist Jöns Jakob Berzelius. At that time it was thought that all organic compounds, such as fats, sugars, coal, and petroleum, were formed by living or once living organisms. All early attempts to synthesize these compounds in the laboratory failed, and it was thought that a *vital force,* available only in living cells, was needed for their formation.

This idea began to change in 1828 when a twenty-seven-year-old German physician, whose first love was chemistry, synthesized the organic molecule urea from inorganic starting materials. This man was Friedrich Wöhler, the "father of organic chemistry."

As a child, Wöhler didn't do particularly well in school because he spent so much time doing chemistry experiments at home. Eventually, he did earn his medical degree, but he decided to study chemistry in the laboratory of Berzelius rather than practice medicine.

After a year he returned to Germany to teach and, as it turned out, to do the experiment that made him famous. The goal of the experiment was to prepare ammonium cyanate from a mixture of potassium cyanate and ammonium sulfate. He heated a solution of the two salts and crystallized the product. But the product didn't look like ammonium cyanate. It was a white crystalline material that looked exactly like urea! Urea is a waste product of protein breakdown in the body and is excreted in the urine. Wöhler recognized urea crystals because he had previously purified them from the urine of dogs and humans. Excited about his accidental discovery, he wrote to his teacher and friend Berzelius, "I can make urea without the necessity of a kidney, or even an animal, whether man or dog."

$NH_4^+ [N{=}C{=}O]^-$ — Ammonium cyanate (inorganic salt)

$H_2N{-}C({=}O){-}NH_2$ — Urea (organic compound)

Ironically, Wöhler, the first person to synthesize an organic compound from inorganic substances, devoted the rest of his career to inorganic chemistry. However, other chemists continued this work, and as a result, the "vital force theory" was laid to rest, and modern organic chemistry was born.

11.1 The Chemistry of Carbon

The number of possible carbon-containing compounds is almost limitless. The importance of these organic compounds is reflected in the fact that over half of this book is devoted to the study of molecules made with this single element.

Why are there so many organic compounds? There are several reasons. First, carbon can form *stable, covalent* bonds with other carbon atoms. Consider three of the *allotropic forms* of elemental carbon: graphite, diamond, and buckminsterfullerene. Models of these allotropes are shown in Figure 11.1.

Allotropes are forms of an element that have the same physical state but different properties.

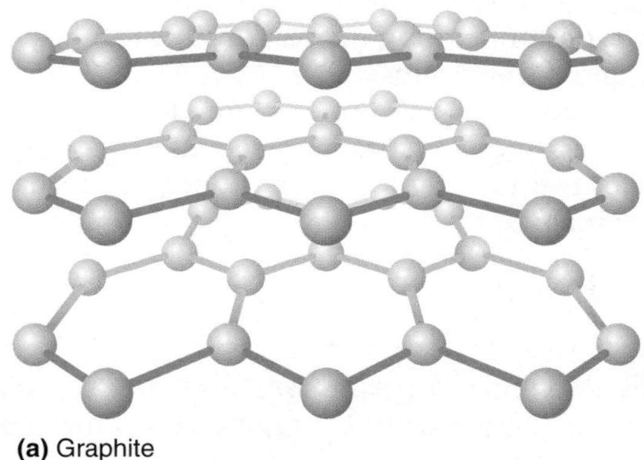

(a) Graphite

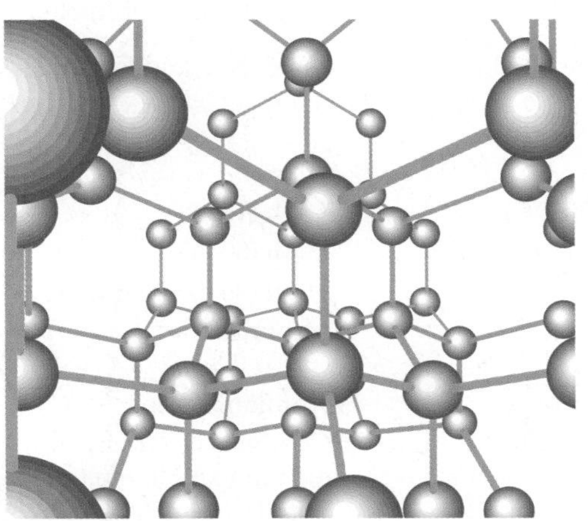

(b) Diamond

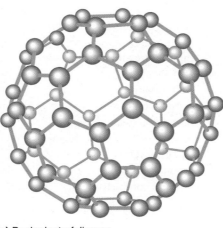

(c) Buckminsterfullerene

Figure 11.1
Three allotropic forms of elemental carbon.

Graphite consists of planar layers in which all carbon-to-carbon bonds extend in two dimensions. Because the planar units can slide over one another, graphite is an excellent lubricant. In contrast, diamond consists of a large, three-dimensional network of carbon-to-carbon bonds. As a result, it is an extremely hard substance used in jewelry and cutting tools.

The third allotropic form of carbon is buckminsterfullerene, affectionately called the *buckey ball*. The buckey ball consists of sixty carbon atoms in the shape of a soccer ball. Discovered in the 1980s, buckminsterfullerene was named for Buckminster Fuller, who used such shapes in the design of geodesic domes.

In future chapters we will discuss families of organic molecules containing oxygen atoms (alcohols, aldehydes, ketones, carboxylic acids, ethers, and esters), nitrogen atoms (amides and amines), sulfur atoms, and halogen atoms.

A second reason for the vast number of organic compounds is that carbon atoms can form stable bonds with other elements. Several families of organic compounds (alcohols, aldehydes, ketones, esters, and ethers) contain oxygen atoms bonded to carbon. Others contain nitrogen, sulfur, or halogens. The presence of these elements confers a wide variety of new chemical and physical properties on an organic compound.

Third, carbon can form double or triple bonds with other carbon atoms to produce a variety of organic molecules with very different properties. Finally, the number of ways in which carbon and other atoms can be arranged is nearly limitless. In addition to linear chains of carbon atoms, ring structures and branched chains are common. Two organic compounds may even have the same number and kinds of atoms but completely different structures and thus different properties. Such organic molecules are referred to as *isomers*.

Important Differences between Organic and Inorganic Compounds

Learning Goal

The bonds between carbon and another atom are almost always *covalent bonds*, whereas the bonds in many inorganic compounds are *ionic bonds*. Differences between these two types of bonding are responsible for most of the differences between inorganic and organic substances (Table 11.1). Ionic bonds result from the *transfer* of one or more electrons from one atom to another. Thus, ionic bonds are electrostatic, resulting from the attraction between the positive and negative ions formed by the electron transfer. Covalent bonds are formed by sharing one or more pairs of electrons.

Table 11.1 Comparison of the Major Properties of a Typical Organic and an Inorganic Compound: Butane Versus Sodium Chloride

Property	Organic Compounds (e.g., Butane)	Inorganic Compounds (e.g., Sodium Chloride)
Bonding	Covalent (C_4H_{10})	Ionic (Na^+ and Cl^- ions)
Physical state at room temperature and atmospheric pressure	Gas	Solid
Boiling point	Low (−0.4°C)	High (1433°C)
Melting point	Low (−139°C)	High (801°C)
Solubility in water	Insoluble	High (36 g/100 mL)
Solubility in organic solvents (e.g., hexane)	High	Insoluble
Flammability	Flammable	Nonflammable
Electrical conductivity	Nonconductor	Conducts electricity in solution and in molten liquid

An Environmental Perspective

Frozen Methane: Treasure or Threat?

Methane is the simplest hydrocarbon, but it has some unusual behaviors. One of these is the ability to form a clathrate, which is an unusual type of matter in which molecules of one substance form a cage around molecules of another substance. For instance, water molecules can form a latticework around methane molecules to form frozen methane hydrate, possibly one of the biggest reservoirs of fossil fuel on earth.

Typically we wouldn't expect a nonpolar molecule, such as methane, to interact with a polar molecule, such as water. So, then, how is this structure formed? As we have studied earlier, water molecules interact with one another by strong hydrogen bonding. In the frozen state, these hydrogen-bonded water molecules form an open latticework. The nonpolar methane molecule is simply trapped inside one of the spaces within the lattice.

Frozen methane is found on the ocean floor. Formed by animals and decaying plant life, there are large pockets of oil and natural gas all over the earth. Methane hydrate forms when methane from one of these pockets under the ocean seeps up through the sea sediments. When the gas reaches the ocean floor, it expands and freezes. Vast regions of the ocean floor are covered by such ice fields.

Fascinating communities of methanogens, organisms that can use methane for a food source, have developed on these ice fields. These creatures live under great pressures, at extremely low temperatures, and with no light. But these unusual communities are not the major focus of interest in the methane ice fields. Scientists would like to "mine" this ice to use the methane as a fuel. In fact, the U.S. Geological Survey estimates that the amount of methane hydrate in the United States is worth over two hundred times the conventional natural gas resources in this country!

But is it safe to harvest the methane from this ice? Caution will certainly be required. Methane is flammable, and, like carbon dioxide, it is a greenhouse gas. In fact, it is about twenty times more efficient at trapping heat than carbon dioxide. So the U.S. Department of Energy, which is working with industry to develop ways to harvest the methane, must figure out how to do that without releasing much into the atmosphere where it could intensify global warming.

It may be that a huge release of methane from these frozen reserves was responsible for a major global warming that occurred fifty-five million years ago and lasted for one hundred thousand years. NASA scientists using computer simulations hypothesize that a shift of the continental plates may have released vast amounts of methane gas from the ocean floor. This methane raised the temperature of earth by about 13°F. In fact the persistence of the methane in the atmosphere warmed earth enough to melt the ice in the oceans and at polar caps and completely change the global climate. This theory, if it turns out to be true, highlights the importance of controlling the amount of methane, as well as carbon dioxide, that we release into the air. Certainly, harvesting the frozen methane of the oceans, if we chose to do it, must be done with great care.

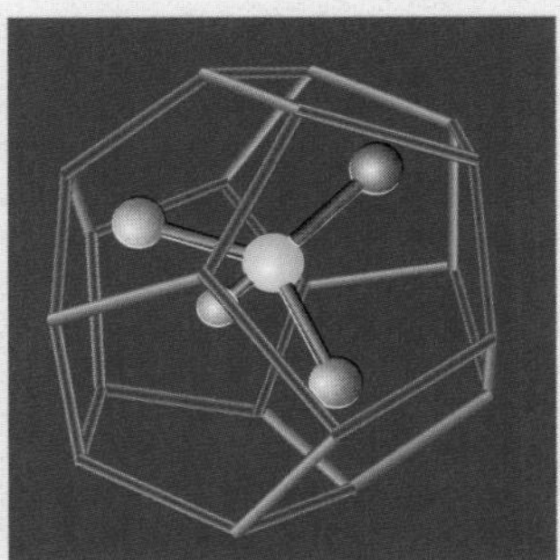

Three-dimensional structure of methane hydrate.

Ionic compounds often form three-dimensional crystals made up of many positive and negative ions. Covalent compounds exist as individual units called *molecules.* Water-soluble ionic compounds often dissociate in water to form ions and are called electrolytes. Most covalent compounds are nonelectrolytes, keeping their identity in solution.

As a result of these differences, ionic substances usually have much higher melting and boiling points than covalent compounds. They are more likely to dissolve in water than in a less-polar solvent, whereas organic compounds, which are typically nonpolar or only moderately polar, are less soluble, or insoluble in water.

Polar covalent compounds, such as HCl, dissociate in water and, thus, are electrolytes. Carboxylic acids, the family of organic compounds we will study in Chapter 15, are weak electrolytes when dissolved in water.

Question 11.1

A student is presented with a sample of an unknown substance and asked to classify the substance as organic or inorganic. What tests should the student carry out?

Question 11.2

What results would the student expect if the sample in Question 11.1 is an inorganic compound? What results would the student expect if it is an organic compound?

Families of Organic Compounds

Learning Goal

The most general classification of organic compounds divides them into hydrocarbons and substituted hydrocarbons. A **hydrocarbon** molecule contains only carbon and hydrogen. A **substituted hydrocarbon** is one in which one or more hydrogen atoms is replaced by another atom or group of atoms.

The hydrocarbons can be further subdivided into aliphatic and aromatic hydrocarbons (Figure 11.2). The three families of **aliphatic hydrocarbons** are the alkanes, alkenes, and alkynes.

Alkanes contain only carbon-to-carbon single bonds (C—C); alkenes have at least one carbon-to-carbon double bond (C═C); and alkynes have at least one carbon-to-carbon triple bond (C≡C).

Alkanes are **saturated hydrocarbons** because they contain only carbon and hydrogen and have only carbon-to-hydrogen and carbon-to-carbon single bonds. The alkenes and alkynes are **unsaturated hydrocarbons** because they contain at least one carbon-to-carbon double or triple bond, respectively.

Recall that each of the lines in these structures represents a shared pair of electrons. See Chapter 4.

```
    H   H   H               H  H   H
    |   |   |               |  |   |
H—C — C — C—H           H—C—C = C—H
    |   |   |               |
    H   H   H               H
```

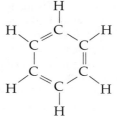

Saturated Hydrocarbon   Unsaturated Hydrocarbon

Some hydrocarbons are cyclic. Cycloalkanes consist of carbon atoms bonded to one another to produce a ring. **Aromatic hydrocarbons** contain a benzene ring or a derivative of the benzene ring.

We will learn a more accurate way to represent benzene in Section 12.6.

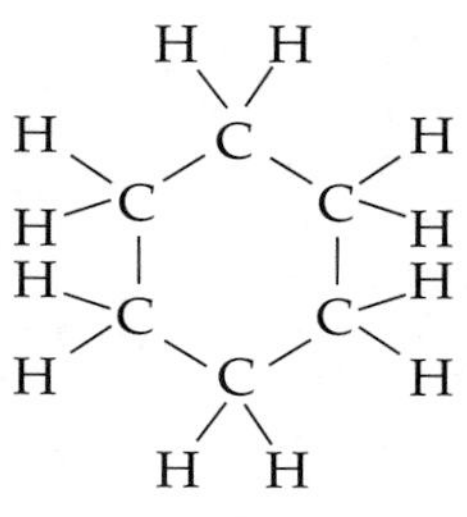

A Cycloalkane (Cyclohexane)

Benzene

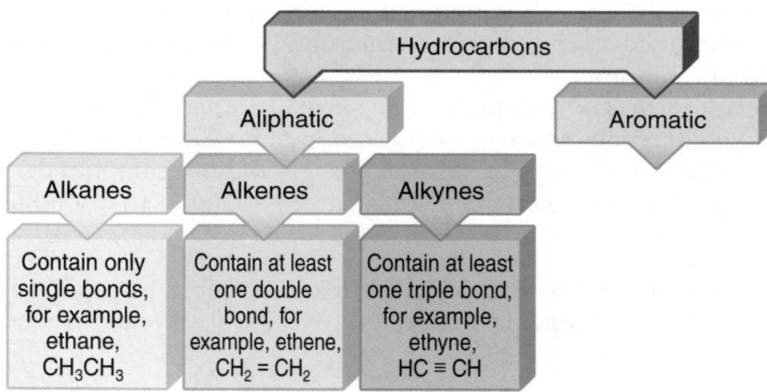

Figure 11.2
The family of hydrocarbons is divided into two major classes: aliphatic and aromatic. The aliphatic hydrocarbons are further subdivided into three major subclasses: alkanes, alkenes, and alkynes.

A substituted hydrocarbon is produced by replacing one or more hydrogen atoms with a functional group. A **functional group** is an atom or group of atoms arranged in a particular way that is primarily responsible for the chemical and physical properties of the molecule in which it is found. The importance of functional groups becomes more apparent when we consider that hydrocarbons have little biological activity. However, the addition of a functional group confers unique and interesting properties that give the molecule important biological or medical properties.

Learning Goal

All compounds that have a particular functional group are members of the same family. For instance, all compounds having the hydroxyl group (—OH) are classified as alcohols. The common functional groups are shown in Table 11.2.

The chemistry of organic and biological molecules is usually controlled by the functional group found in the molecule. Just as members of the same family of the periodic table exhibit similar chemistry, organic molecules with the same functional group exhibit similar chemistry. Although it would be impossible to learn the chemistry of each organic molecule, it is relatively easy to learn the chemistry

This is analogous to the classification of the elements within the periodic table. See Chapter 3.

Learning Goal

Table 11.2 Common Functional Groups

Functional Group	Name	Family of Organic Compounds
>C=C<	Carbon-carbon double bond	Alkene
—C≡C—	Carbon-carbon triple bond	Alkyne
C_6H_6 ring (or) benzene hexagon	Benzene ring	Aromatic
—C—X (X = F, Cl, Br, I)	Halogen atom	Alkyl halide
—C—OH	Hydroxyl group	Alcohol
—C—O—R*	Alkoxy group	Ether
>C=O	Carbonyl group	Aldehyde or ketone
—C(=O)—OH	Carboxyl group	Carboxylic acid
—C(=O)—G (G = Cl, OR*, and others)	Acyl group	Carboxylic acid derivatives
—C—N—	Amino group	Amine

*R is an abbreviation for any alkyl or aryl group; aryl is used for aromatic compounds in the same way that alkyl is used for aliphatic compounds (for example, methyl, ethyl, isopropyl). An aryl group is an aromatic compound with one hydrogen removed (for example, phenyl—the phenyl group is benzene with one hydrogen removed).

of each functional group. In this way you can learn the chemistry of all members of a family of organic compounds, or biological molecules, just by learning the chemistry of its characteristic functional group or groups.

11.2 Alkanes

Alkanes are saturated hydrocarbons; that is, alkanes contain only carbon and hydrogen bonded together through carbon-hydrogen and carbon-carbon single bonds. C_nH_{2n+2} is the general formula for alkanes. In this formula, n is the number of carbon atoms in the molecule.

Structure and Physical Properties

Learning Goal  Learning Goal 

Four types of formulas, each providing different information, are used in organic chemistry: the molecular formula, the structural formula, the condensed formula, and the line formula.

The **molecular formula** tells the kind and number of each type of atom in a molecule but does not show the bonding pattern. Consider the molecular formulas for simple alkanes:

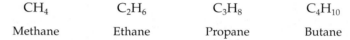

For the first three compounds, there is only one possible arrangement of the atoms. However, for C_4H_{10} there are two possible arrangements. How do we know which is correct? The problem is solved by using the **structural formula,** which shows each atom and bond in a molecule. The following are the structural formulas for methane, ethane, propane, and the two isomers of butane:

Recall that a covalent bond, representing a pair of shared electrons, can be drawn as a line between two atoms. For the structure to be correct, each carbon atom must show four pairs of shared electrons.

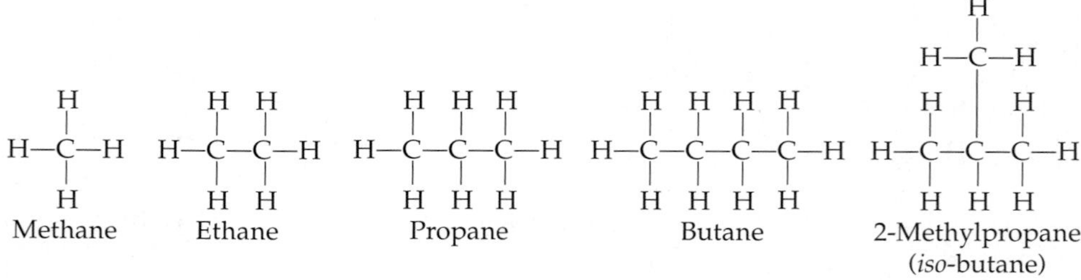

The advantage of a structural formula is that it shows the complete structure, but for large molecules it is time-consuming to draw and requires too much space. The compromise is the **condensed formula.** It shows all the atoms in a molecule and places them in a sequential order that indicates which atoms are bonded to which. The following are the condensed formulas for the preceding five compounds.

CH_4	CH_3CH_3	$CH_3CH_2CH_3$	$CH_3(CH_2)_2CH_3$	$(CH_3)_3CH$
Methane	Ethane	Propane	Butane	2-Methylpropane (*iso*-butane)

The names and formulas of the first ten straight-chain alkanes are shown in Table 11.3.

The simplest representation of a molecule is the **line formula.** In the line formula we assume that there is a carbon atom at any location where two or more lines intersect. We also assume that there is a carbon at the end of any line and that each carbon in the structure is bonded to the correct number of hydrogen atoms. Compare the structural and line formulas for butane and 2-methylpropane, shown here:

```
                                   H
                                   |
                                 H-C-H
   H  H  H  H                    H |  H
   |  |  |  |                    | |  |
H--C--C--C--C--H = /\/         H-C-C-C-H = >-
   |  |  |  |                    | |  |
   H  H  H  H                    H H  H
      Butane                2-Methylpropane
```

Each carbon atom forms four single covalent bonds, but each hydrogen atom has only a single covalent bond. Although a carbon atom may be involved in single, double, or triple bonds, it always shares four pairs of electrons. When carbon is involved in four single bonds, the *bond angle*, the angle between two atoms or substituents attached to carbon, is 109.5°, as predicted by the valence shell electron pair repulsion (VESPR) theory. Thus, alkanes contain carbon atoms that have tetrahedral geometry.

Molecular geometry is described in Section 4.4

A tetrahedron is a geometric solid having the structure shown in Figure 11.3a. There are many different ways to draw the tetrahedral carbon (Figures 11.3b–11.3d). In Figure 11.3b, solid lines, dashes, and wedges are used to represent the structure of methane. Dashes go back into the page away from you; wedges come out of the page toward you; and solid lines are in the plane of the page. The structure in Figure 11.3c is the same as that in Figure 11.3b; it just leaves a lot more to the imagination. Figure 11.3d is a ball-and-stick model of the methane molecule. Three-dimensional drawings of two other simple alkanes are shown in Figure 11.4.

All hydrocarbons are nonpolar molecules. As a result they are not water soluble but are soluble in nonpolar organic solvents. Furthermore, they have relatively low melting points and boiling points and are generally less dense than water. In general, the longer the hydrocarbon chain (greater the molecular weight), the higher the melting and boiling points and the greater the density (see Table 11.3).

See Section 6.2 for a discussion of the forces responsible for the physical properties of a substance.

Table 11.3 Names and Formulas of the First Ten Straight-Chain Alkanes

Name	Molecular Formula	Condensed Formula	Melting Point, °C	Boiling Point, °C
Alkanes	C_nH_{2n+2}			
Methane	CH_4	CH_4	−182.5	−160
Ethane	C_2H_6	CH_3CH_3	−183.6	−88.7
Propane	C_3H_8	$CH_3CH_2CH_3$	−187.6	−42.2
Butane	C_4H_{10}	$CH_3CH_2CH_2CH_3$ or $CH_3(CH_2)_2CH_3$	−139.0	−0.4
Pentane	C_5H_{12}	$CH_3CH_2CH_2CH_2CH_3$ or $CH_3(CH_2)_3CH_3$	−129.9	36.0
Hexane	C_6H_{14}	$CH_3CH_2CH_2CH_2CH_2CH_3$ or $CH_3(CH_2)_4CH_3$	−94.5	68.8
Heptane	C_7H_{16}	$CH_3CH_2CH_2CH_2CH_2CH_2CH_3$ or $CH_3(CH_2)_5CH_3$	−90.6	98.4
Octane	C_8H_{18}	$CH_3CH_2CH_2CH_2CH_2CH_2CH_2CH_3$ or $CH_3(CH_2)_6CH_3$	−56.9	125.6
Nonane	C_9H_{20}	$CH_3CH_2CH_2CH_2CH_2CH_2CH_2CH_2CH_3$ or $CH_3(CH_2)_7CH_3$	−53.6	150.7
Decane	$C_{10}H_{22}$	$CH_3CH_2CH_2CH_2CH_2CH_2CH_2CH_2CH_2CH_3$ or $CH_3(CH_2)_8CH_3$	−29.7	174.0

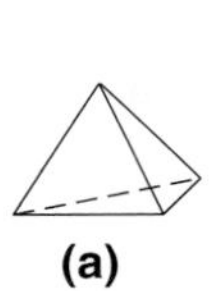
(a)

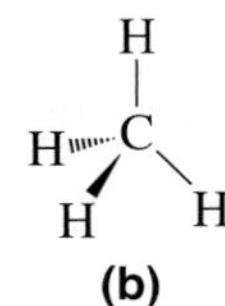

(b)

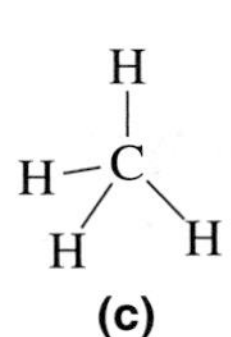

(c)

(d)

Figure 11.3
The tetrahedral carbon atom: (a) a tetrahedron; (b) the tetrahedral carbon drawn with dashes and wedges; (c) the stick drawing of the tetrahedral carbon atom; (d) ball-and-stick model of methane.

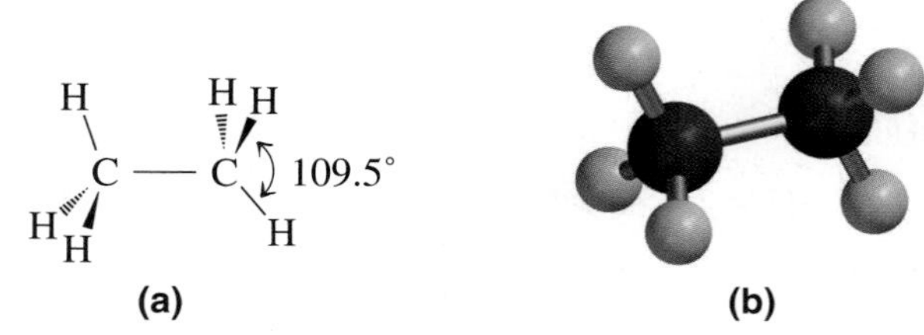

(c) (d)

Figure 11.4
(a) Drawing and (b) ball-and-stick model of ethane. All the carbon atoms have a tetrahedral arrangement, and all bond angles are approximately 109.5°.
(c) Drawing and (d) ball-and-stick model of a more complex alkane, butane.

EXAMPLE 11.1 *Using Different Types of Formulas to Represent Organic Compounds*

Draw the structural and condensed formulas of the following line structure:

Solution

Remember that each intersection of lines represents a carbon atom and that each line ends in a carbon atom. This gives us the following carbon skeleton:

```
            C       C
            |       |
C—C—C—C———C—C—C
```

By adding the correct number of hydrogen atoms to the carbon skeleton, we are able to complete the structural formula of this compound.

```
              H        H
              |        |
            H—C—H    H—C—H
    H  H  H   |     H  |  H
    |  |  |   |     |  |  |
  H—C—C—C—C———————C—C—C—H
    |  |  |  |      |  |  |
    H  H  H  H      H  H  H
```

From the structural formula we can write the condensed formula as follows:

$$CH_3CH_2CH_2CH(CH_3)CH_2CH(CH_3)CH_3$$

Table 11.4 **Names and Formulas of the First Five Continuous-Chain Alkyl Groups**

Alkyl Group Structure	Name
CH_3—	Methyl
CH_3CH_2—	Ethyl
$CH_3CH_2CH_2$—	Propyl
$CH_3CH_2CH_2CH_2$—	Butyl
$CH_3CH_2CH_2CH_2CH_2$—	Pentyl

Alkyl Groups

Alkyl groups result when a hydrogen atom is removed from an alkane. The name of the alkyl group is derived from the name of the alkane containing the same number of carbon atoms. The *-ane* ending of the alkane name is replaced by the *-yl* ending. Thus, CH_3— is a methyl group and CH_3CH_2— is an ethyl group. The dash at the end of these two structures represents the point at which the alkyl group can bond to another atom. The first five continuous-chain alkyl groups are presented in Table 11.4.

Carbon atoms are classified according to the number of other carbon atoms to which they are attached. A **primary carbon (1°)** is directly bonded to one other carbon. A **secondary carbon (2°)** is bonded to two other carbon atoms; a **tertiary carbon (3°)** is bonded to three other carbon atoms, and a **quaternary carbon (4°)** to four.

Alkyl groups are classified according to the number of carbons attached to the carbon atom that joins the alkyl group to a molecule.

```
   H              C              C
   |              |              |
C—C—           C—C—           C—C—
   |              |              |
   H              H              C
```

Primary alkyl group | Secondary alkyl group | Tertiary alkyl group

All of the continuous-chain alkyl groups are primary alkyl groups (see Table 11.4). Several branched-chain alkyl groups are shown in Table 11.5. Notice that the isopropyl and *sec*-butyl groups are secondary alkyl groups; the isobutyl group is a primary alkyl group; and the *t*-butyl (*tert*-butyl) is a tertiary alkyl group.

Nomenclature

Learning Goal

Historically, organic compounds were named by the chemist who discovered them. Often the names reflected the source of the compound. For instance, the antibiotic penicillin is named for the mold *Penicillium notatum,* which produces it. The pain reliever aspirin was made by adding an acetate group to a compound purified from the bark of a willow tree, hence the name aspirin: *a*- (acetate) and *spirin* (genus of willow, *Spirea*).

These names are easy for us to remember because we come into contact with these compounds often. However, as the number of compounds increased, organic chemists realized that historical names were not adequate because they revealed nothing about the structure of a compound. Thousands of such compounds and their common names had to be memorized! What was needed was a set of nomenclature (naming) rules that would produce a unique name for every organic compound. Furthermore, the name should be so descriptive that, by knowing the name, a student or scientist could write the structure.

Table 11.5 Structures and Names of Some Branched-Chain Alkyl Groups

Structure	Classification	Name
CH_3CH— (with CH_3 below CH)	2°	Isopropyl*
CH_3 (above CH) CH_3CHCH_2—	1°	Isobutyl*
CH_3 (above CH) CH_3CH_2CH—	2°	*sec*-Butyl†
CH_3 (above C) CH_3C— CH_3 (below C)	3°	*t*-Butyl or *tert*-Butyl‡

*The prefix iso- (isomeric) is used when there are two methyl groups at the end of the alkyl group.
†The prefix *sec*- (secondary) indicates that there are two carbons bonded to the carbon that attaches the alkyl group to the parent molecule.
‡The prefix *t*- or *tert*- (tertiary) means that three carbons are attached to the carbon that attaches the alkyl group to the parent molecule.

The International Union of Pure and Applied Chemistry (I.U.P.A.C.) is the organization responsible for establishing and maintaining a standard, universal system for naming organic compounds. The system of nomenclature developed by this group is called the **I.U.P.A.C. Nomenclature System.** The following rules are used for naming alkanes by the I.U.P.A.C. system.

It is important to learn the prefixes for the carbon chain lengths. We will use them in the nomenclature for all organic molecules.

1. Determine the name of the **parent compound,** the longest continuous carbon chain in the compound. Refer to Tables 11.3 and 11.6 to determine the parent name. Notice that these names are made up of a prefix related to the number of carbons in the chain and the suffix *-ane,* indicating that the molecule is an alkane (Table 11.6). Write down the name of the parent compound, leaving space before the name to identify the substituents. Parent chains are highlighted in yellow in the following examples:

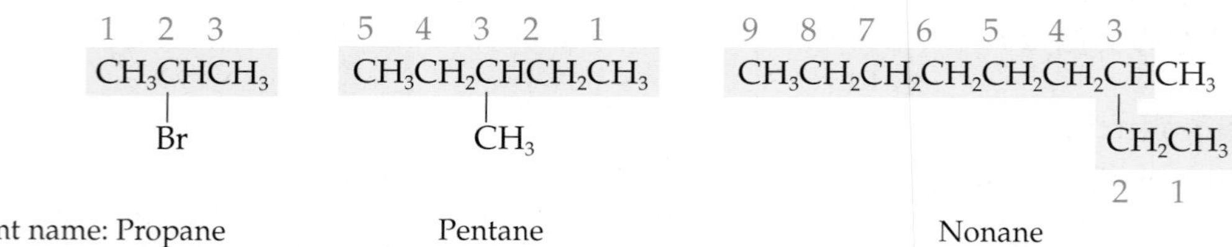

2. Number the parent chain to give the lowest number to the carbon bonded to the first group encountered on the parent chain, regardless of the numbers that result for the other substituents.
3. Name and number each atom or group attached to the parent compound. The number tells you the position of the group on the main chain, and the name tells you what type of substituent is present at that position. For example, it may be one of the halogens [F-(fluoro), Cl-(chloro), Br-(bromo), and I-(iodo)] or an alkyl group (Tables 11.4 and 11.5). In the following examples the parent chain is highlighted in yellow:

Table 11.6 Carbon Chain Length and Prefixes Used in the I.U.P.A.C. Nomenclature System

Carbon Chain Length	Prefix	Alkane Name
1	Meth-	Meth*ane*
2	Eth-	Eth*ane*
3	Prop-	Prop*ane*
4	But-	But*ane*
5	Pent-	Pent*ane*
6	Hex-	Hex*ane*
7	Hept-	Hept*ane*
8	Oct-	Oct*ane*
9	Non-	Non*ane*
10	Dec-	Dec*ane*

Structure	$\overset{1}{C}H_3\overset{2}{C}H(Br)\overset{3}{C}H_3$	$CH_3\overset{3}{C}H(\overset{4}{C}H_2\overset{5}{C}H_3)\overset{2}{C}H_2\overset{1}{C}H_3$	$\overset{1}{C}H_3\overset{2}{C}H_2\overset{3}{C}H_2\overset{4}{C}H(\overset{5}{C}H_2\overset{6}{C}H_2\overset{7}{C}H_2\overset{8}{C}H_3)CH_2CH_3$
Substituent:	2-Bromo	3-Methyl	4-Ethyl
I.U.P.A.C. name:	2-Bromopropane	3-Methylpentane	4-Ethyloctane

4. If the same substituent occurs more than once in the compound, a separate position number is given for each, and the prefixes *di-*, *tri-*, *tetra-*, *penta-*, and so forth are used, as shown in the following examples:

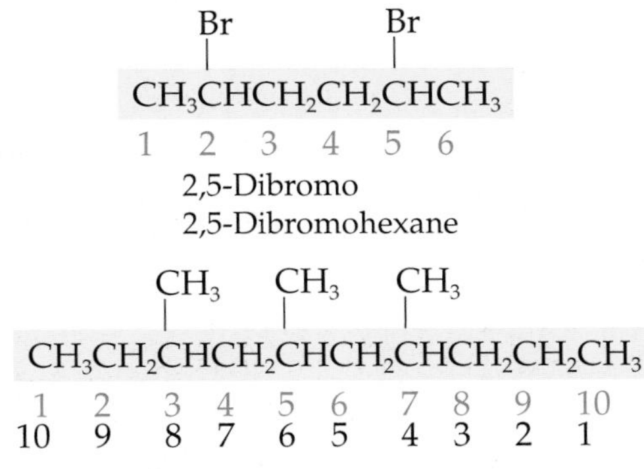

2,5-Dibromo
2,5-Dibromohexane

3,5,7-Trimethyldecane
NOT 4,6,8-Trimethyldecane

5. Place the names of the substituents in alphabetical order before the name of the parent compound, which you wrote down in Step 1. Numbers are separated by commas, and numbers are separated from names by hyphens. By convention, halogen substituents are placed before alkyl substituents.

$\overset{1}{C}H_3\overset{2}{C}H(Br)\overset{3}{C}(CH_3)_2\overset{4}{C}H_2\overset{5}{C}H_3$

2-Bromo-3,3-dimethylpentane

Throughout this book we will primarily use the I.U.P.A.C. Nomenclature System. When common names are used, they will be shown in parentheses beneath the I.U.P.A.C. name.

EXAMPLE 11.2 Naming Substituted Alkanes Using the I.U.P.A.C. System

Name the following alkanes using I.U.P.A.C. nomenclature.

Solution

$$\overset{6}{C}H_3\overset{5}{C}H_2\overset{4}{C}H_2\overset{3}{C}H_2\overset{2}{C}H(Br)\overset{1}{C}H_3$$

Parent chain: hexane
Substituent: 2-bromo (*not* 5-bromo)
Name: 2-Bromohexane

$$\overset{5}{C}H_3\overset{4}{C}H_2-\overset{3}{C}H(CH_3)-\overset{2}{C}(CH_3)_2-\overset{1}{C}H_3$$

Parent chain: pentane
Substituent: 2,2,3-trimethyl (*not* 3,4,4-trimethyl)
Name: 2,2,3-Trimethylpentane

$$\overset{1}{C}H_3\overset{2}{C}H(CH_3)\overset{3}{C}H_2\overset{4}{C}H_2\overset{5}{C}H_2\overset{6}{C}H(CH_3)\overset{7}{C}H_2\overset{8}{C}H_2\overset{9}{C}H_3$$

Parent chain: nonane
Substituent: 2,6-dimethyl
(*not* 4,8-dimethyl)
Name: 2,6-Dimethylnonane

Question 11.3

Name the following compounds, using the I.U.P.A.C. Nomenclature System:

a. $CH_3-CH(CH_3)-CH(CH_3)-CH_3$

b. $CH_3-C(CH_2CH_2CH_3)(CH_3)-CH_3$

c. $CH_3-C(CH_3)_2-CH_3$

d. $CH_2(Br)-CH(Br)-CH_2(Br)$

Question 11.4

Name the following compounds, using the I.U.P.A.C. Nomenclature System:

a. $CH_3CH_2CH_2CH_2CH(CH_2CH_3)CH_3$

b. $CH_3-C(CH_2Br)(CH_3)-CH_2-Br$

c. $CH_3CH(I)CH_2CH_2CH(CH_3)CH_2-Br$

d. $CH_3CH(CH_3)CH(CH_2CH_3)CH_2CH_2CH_2-Cl$

Having learned to name a compound using the I.U.P.A.C. system, we can easily write the structural formula of a compound, given its name. First, draw and number the parent carbon chain. Add the substituent groups to the correct carbon and finish the structure by adding the correct number of hydrogen atoms.

EXAMPLE 11.3 Drawing the Structure of a Compound using the I.U.P.A.C. Name

Draw the structural formula for 1-bromo-4-methylhexane.

Continued—

EXAMPLE 11.3 —*Continued*

Solution

Begin by drawing the six-carbon parent chain and indicating the four bonds for each carbon atom.

```
 |  |  |  |  |  |
—C—C—C—C—C—C—
 |  |  |  |  |  |
```

Next, number each carbon atom:

```
 |  |  |  |  |  |
—C—C—C—C—C—C—
 |  |  |  |  |  |
 1  2  3  4  5  6
```

Now add the substituents. In this example a bromine atom is bonded to carbon-1, and a methyl group is bonded to carbon-4:

```
 Br
 |  |  |  |  |  |
—C—C—C—C—C—C—
 |  |  |  |  |  |
          |
        H—C—H
          |
          H
 1  2  3  4  5  6
```

Finally, add the correct number of hydrogen atoms so that each carbon has four covalent bonds:

```
  Br H  H  H  H  H
  |  |  |  |  |  |
H—C—C—C—C—C—C—H
  |  |  |  |  |  |
  H  H  H  |  H  H
           |
         H—C—H
           |
           H
  1  2  3  4  5  6
```

As a final check of your accuracy, use the I.U.P.A.C. system to name the compound that you have just drawn, and compare the name with that in the original problem.

The molecular formula and condensed formula can be written from the structural formula shown. The molecular formula is $C_7H_{15}Br$, and the condensed formula is $BrCH_2CH_2CH_2CH(CH_3)CH_2CH_3$.

Constitutional or Structural Isomers

Learning Goal

As we saw earlier, there are two arrangements of the atoms represented by the molecular formula C_4H_{10}: butane and 2-methylpropane. Molecules having the same molecular formula but a different arrangement of atoms are called **constitutional,** or **structural, isomers.** These isomers are unique compounds because of their structural differences, and they have different physical and chemical properties. Butane and 2-methylpropane both have molecular weights of 58.1 g/mol, but they differ slightly in their melting points and boiling points:

$$CH_3CH_2CH_2CH_3 \qquad\qquad \begin{array}{c} CH_3 \\ | \\ CH_3CHCH_3 \end{array}$$

Butane	2-Methylpropane
b.p. = 0.4°C	b.p. = −12°C
m.p. = −139°C	m.p. = −145°C

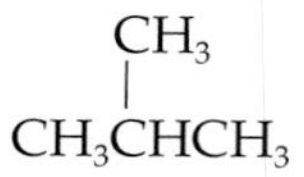

EXAMPLE 11.4

Drawing Constitutional or Structural Isomers of Alkanes

Write all the constitutional isomers having the molecular formula C_6H_{14}.

Solution

1. Begin with the continuous six-carbon chain structure:

$$\begin{array}{c} 1 \quad\ 2 \quad\ 3 \quad\ 4 \quad\ 5 \quad\ 6 \\ CH_3{-}CH_2{-}CH_2{-}CH_2{-}CH_2{-}CH_3 \end{array}$$

Isomer A

2. Now try five-carbon chain structures with a methyl group attached to one of the internal carbon atoms of the chain:

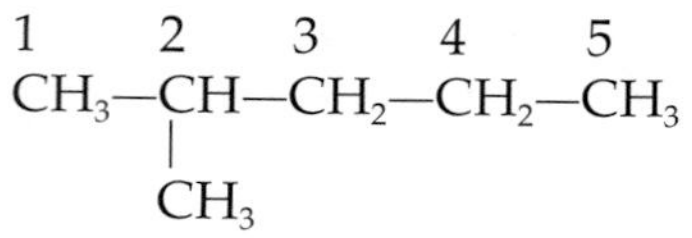

$$\begin{array}{l} 1 \quad\ 2 \quad\ 3 \quad\ 4 \quad\ 5 \\ CH_3{-}\underset{\displaystyle CH_3}{\underset{|}{CH}}{-}CH_2{-}CH_2{-}CH_3 \end{array} \quad \text{and} \quad \begin{array}{l} 1 \quad\ 2 \quad\ 3 \quad\ 4 \quad\ 5 \\ CH_3{-}CH_2{-}\underset{\displaystyle CH_3}{\underset{|}{CH}}{-}CH_2{-}CH_3 \end{array}$$

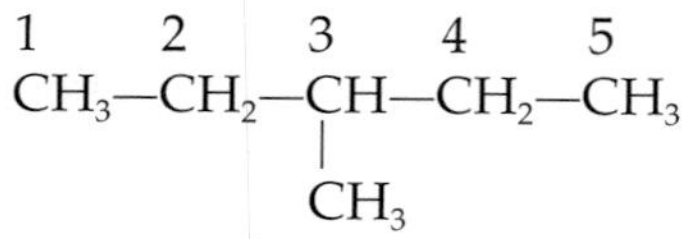

Isomer B | Isomer C

3. Next consider the possibilities for a four-carbon structure to which two methyl groups ($-CH_3$) may be attached:

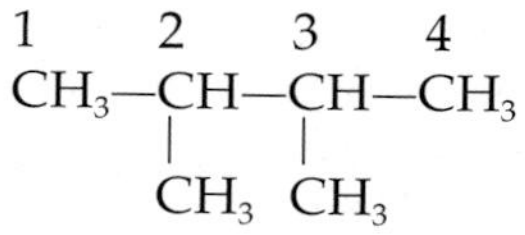

$$\begin{array}{l} 1 \quad\ 2 \quad\ 3 \quad\ 4 \\ CH_3{-}\underset{\displaystyle CH_3}{\underset{|}{CH}}{-}\underset{\displaystyle CH_3}{\underset{|}{CH}}{-}CH_3 \end{array} \quad \text{and} \quad \begin{array}{l} 1 \quad\ 2 \quad\ 3 \quad\ 4 \\ CH_3{-}\overset{\displaystyle CH_3}{\overset{|}{\underset{\displaystyle CH_3}{\underset{|}{C}}}}{-}CH_2{-}CH_3 \end{array}$$

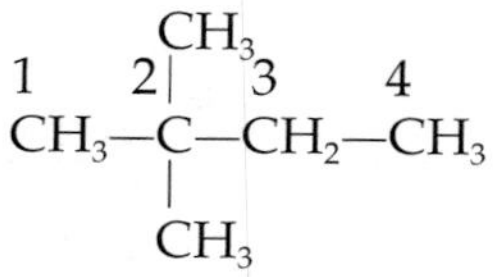

Isomer D | Isomer E

These are the five possible constitutional isomers of C_6H_{14}. At first it may seem that other isomers are also possible. But careful comparison will show that they are duplicates of those already constructed. For example, rather than add two methyl groups, a single ethyl group ($-CH_2CH_3$) could be added to the four-carbon chain:

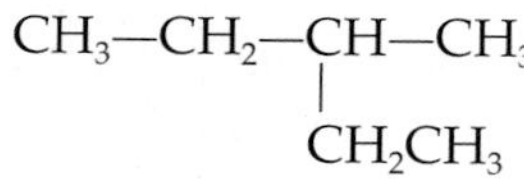

$$CH_3{-}CH_2{-}\underset{\displaystyle CH_2CH_3}{\underset{|}{CH}}{-}CH_3$$

But close examination will show that this is identical to isomer C. Perhaps we could add one ethyl group and one methyl group to a three-carbon parent chain, with the following result:

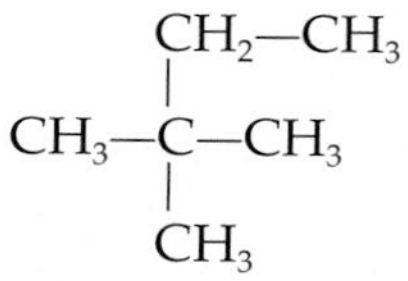

$$CH_3{-}\overset{\displaystyle CH_2{-}CH_3}{\overset{|}{\underset{\displaystyle CH_3}{\underset{|}{C}}}}{-}CH_3$$

Continued—

An Environmental Perspective

Oil-Eating Bacteria

Our highly industrialized society has come to rely more and more on petroleum as a source of energy and a raw material source for the manufacture of plastics, drugs, and a host of other consumables. Over 50% of the petroleum consumed in the United States is imported, and the major carrier is the supertanker.

Well-publicized oil spills, such as that from the *Exxon Valdez* (in Alaska in 1989), have fueled research to develop clean-up methods that will help to preserve the fragile aquatic environment.

It has been known for some time that there are strains of bacteria that will accelerate the oxidation of many of the compounds present in unrefined petroleum. These bacteria have been termed *oil-eating bacteria.*

Recently, oceanographers at the University of Texas have developed strains of bacteria that will actually "eat" a wide variety of crude oils. At the same time these bacteria have a very short lifetime. It appears that they die shortly after they have operated on an oil slick.

This latter characteristic, a short lifetime in water, is particularly appealing to scientists, who fear that the introduction of nonindigenous (nonnative) bacteria into natural water systems may disrupt the ecology of the water.

Some also fear that the products of these reactions, in which some of the oil is converted to fatty acids, may disperse in water and cause more problems than the original oil spill.

Obviously, a great deal of research involving biodegradation remains to be done. Such technologies offer hope for alleviating many land-based solid waste disposal problems, in addition to petroleum spills.

EXAMPLE 11.4 —*Continued*

Again we find that this structure is the same as one of the isomers we have already identified, isomer E.

To check whether you have accidentally made duplicate isomers, name them using the I.U.P.A.C. system. All isomers must have different I.U.P.A.C. names. So if two names are identical, the structures are also identical. Use the I.U.P.A.C. system to name the isomers in this example, and prove to yourself that the last two structures are simply duplicates of two of the original five isomers.

Question 11.5

Draw a complete structural formula for each of the straight-chain isomers of the following alkanes:

a. C_4H_9Br b. $C_4H_8Br_2$

Question 11.6

Name all of the isomers that you obtained in Question 11.5.

11.3 Cycloalkanes

The **cycloalkanes** are a family having C—C single bonds in a ring structure. They have the general molecular formula C_nH_{2n} and thus have two fewer hydrogen atoms than the corresponding alkane (C_nH_{2n+2}). The relationship that exists between an alkane and a cycloalkane is shown for hexane and cyclohexane.

$$\begin{array}{c} CH_2{-}CH_2 \\ / \quad\quad \backslash \\ CH_2 \quad\quad\quad CH_2 \\ \backslash \quad\quad / \\ CH_2 \quad CH_2 \\ | \quad\quad\quad | \\ H \quad\quad\quad H \end{array} \xrightarrow[\text{2 H's}]{\text{loss of}} \begin{array}{c} CH_2{-}CH_2 \\ / \quad\quad \backslash \\ CH_2 \quad\quad\quad CH_2 \\ \backslash \quad\quad / \\ CH_2{-}CH_2 \end{array}$$

Hexane	Cyclohexane
Alkane	*Cycloalkane*
C_nH_{2n+2}	C_nH_{2n}

The structures and names of some simple cycloalkanes are shown in Figure 11.5.

In the I.U.P.A.C. system the cycloalkanes are named by applying the following simple rules.

- Determine the name of the alkane with the same number of carbon atoms as there are within the ring and add the prefix *cyclo-*. For example, cyclopentane is the cycloalkane that has five carbon atoms.
- If the cycloalkane is substituted, place the names of the groups in alphabetical order before the name of the cycloalkane. No number is needed if there is only one substituent.
- If more than one group is present, use numbers that result in the *lowest possible position numbers*.

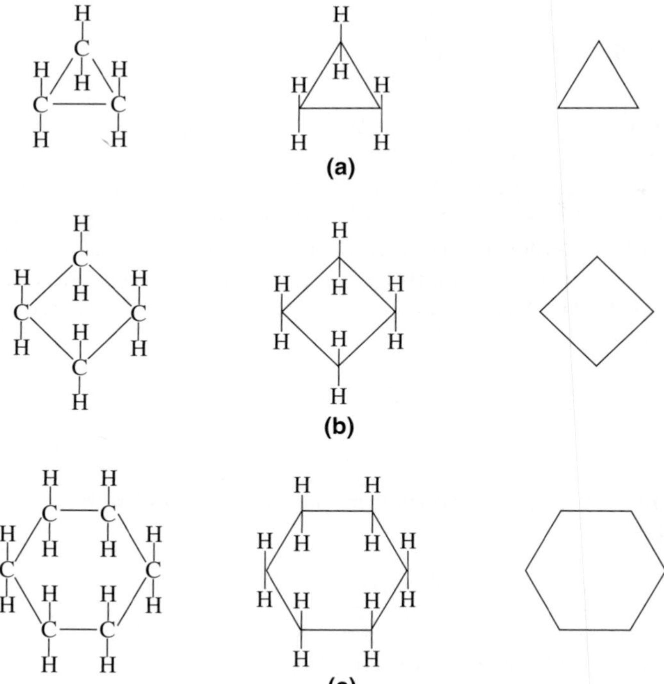

Figure 11.5
Cycloalkanes: (a) cyclopropane; (b) cyclobutane; (c) cyclohexane. All of the cycloalkanes are shown using structural formulas (left column), condensed structural formulas (center column), and line formulas (right column).

EXAMPLE 11.5 ***Naming a Substituted Cycloalkane Using the I.U.P.A.C. Nomenclature System***

Name the following cycloalkanes using I.U.P.A.C. nomenclature.

Continued—

EXAMPLE 11.5 —*Continued*

Solution

Parent chain: cyclohexane
Substituent: chloro (no number is required because there is only one substituent)
Name: Chlorocyclohexane

Parent chain: cyclopentane
Substituent: methyl (no number is required because there is only one substituent)
Name: Methylcyclopentane

These cycloalkanes could also be shown as line formulas, as shown below. Each line represents a carbon-carbon bond. A carbon atom and the correct number of hydrogen atoms are assumed to be at the point where the lines meet and at the end of a line.

Chlorocyclohexane

Methylcyclopentane

cis-trans Isomerism in Cycloalkanes

Learning Goal

Atoms of an alkane can rotate freely around the carbon-carbon single bond, resulting in an unlimited number of arrangements. However, rotation around the bonds in a cyclic structure is limited by the fact that the carbons of the ring are all bonded to another carbon within the ring. The formation of ***cis-trans*** **isomers,** or **geometric isomers,** is a consequence of the absence of free rotation. Geometric isomers are a type of *stereoisomer*. The *cis-trans* isomers of cycloalkanes are stereoisomers that differ from one another in the arrangement of substituents in space. Consider the following examples:

Stereoisomers are molecules that have the same structural formulas and bonding patterns but different arrangements of atoms in space. They are discussed in greater detail in Section 17.3 and Appendix D, Stereochemistry and Stereoisomers Revisited.

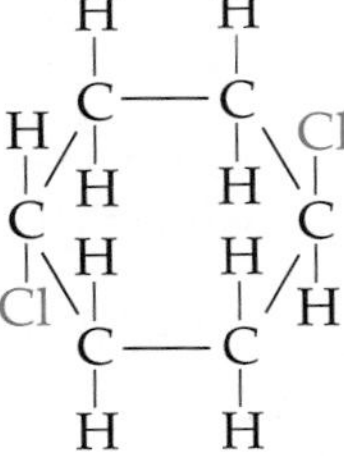

trans-1,4-Dichlorocyclohexane

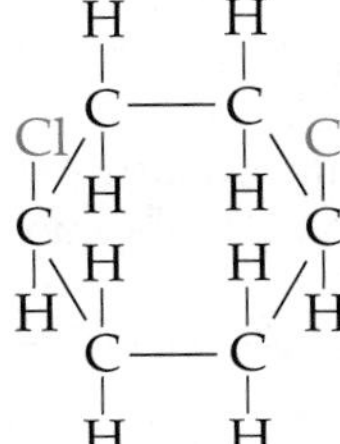

cis-1,4-Dichlorocyclohexane

Imagine that you are viewing the ring structures as if an edge were projecting toward you. In the structure on the right, both Cl atoms are above the ring. They are termed *cis* (L., "on the same side"). The complete name for this compound is *cis*-1,4-dichlorocyclohexane. In the structure on the left, one Cl is above the ring and the other is below it. They are said to be *trans* (*L.*, "across from") to one another and the complete name of this compound is *trans*-1,4-dichlorocyclohexane.

Geometric isomers do not readily interconvert. The cyclic structure prevents unrestricted free rotation and, thus, prevents interconversion. Only by breaking carbon-carbon bonds of the ring could interconversion occur. As a result, geometric isomers may be separated from one another in the laboratory.

EXAMPLE 11.6 Naming cis-trans Isomers of Substituted Cycloalkanes

Determine whether the following substituted cycloalkanes are *cis* or *trans* isomers and write the complete name for each.

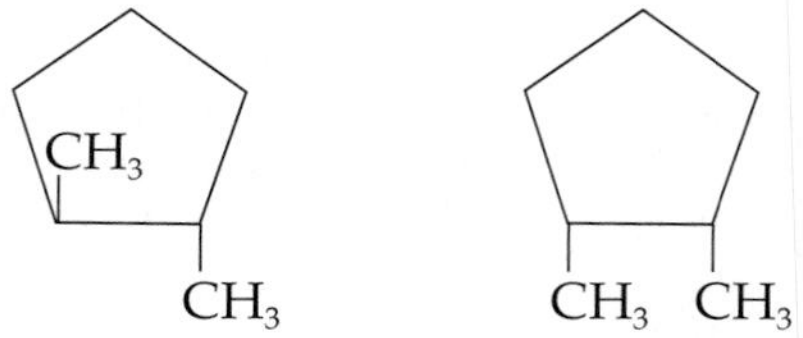

CH_3 CH_3 CH_3 CH_3

Solution

Both molecules are cyclopentanes having two methyl group substituents. Thus both would be named 1,2-dimethylcyclopentane. In the structure on the left, one methyl group is above the ring and the other is below the ring; they are in the *trans* configuration, and the structure is named *trans*-1,2-dimethylcyclopentane. In the structure on the right, both methyl groups are on the same side of the ring (below it, in this case); they are *cis* to one another, and the complete name of this compound is *cis*-1,2-dimethylcyclopentane.

EXAMPLE 11.7 Naming a Cycloalkane Having Two Substituents Using the I.U.P.A.C. Nomenclature System

Name the following cycloalkanes using I.U.P.A.C. nomenclature.

Solution

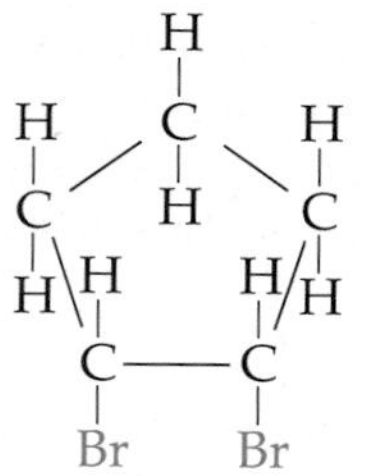

Parent chain: cyclopentane
Substituent: 1,2-dibromo
Isomer: *cis*
Name: *cis*-1,2-Dibromocyclopentane

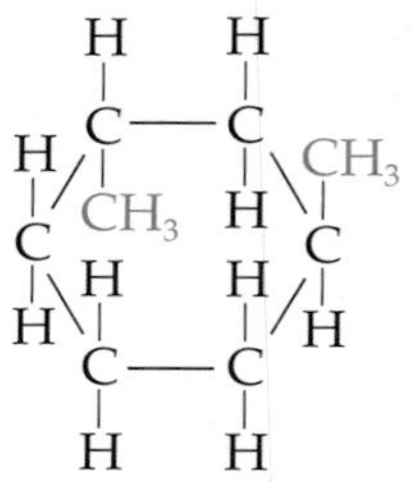

Parent chain: cyclohexane
Substituent: 1,3-dimethyl
Isomer: *trans*
Name: *trans*-1,3-Dimethylcyclohexane

Question 11.7

Name each of the following substituted cycloalkanes using the I.U.P.A.C. Nomenclature System:

a. (cyclobutane ring: C with H, H; C with H, H; C with H, Br; C with H, CH_2CH_3)

b. (cyclopropane ring: C with CH_3, H; C with H, H; C with H, CH_3)

c. (cyclohexane ring: one C with H, $CH_2CH_2CH_3$; remaining C atoms with H, H)

Question 11.8

Name each of the following substituted cycloalkanes using the I.U.P.A.C. Nomenclature System:

a. (cyclohexane ring bearing two Cl substituents, remaining positions H)

b. (cyclopentane ring bearing two CH_3 substituents, remaining positions H)

c. (cyclobutane ring bearing CH_3 and CH_2CH_3 substituents, remaining positions H)

Question 11.9

There are four isomers of dichlorocyclopropane. Use a set of molecular models to construct the isomers and to contrast their differences. Draw all these isomers.

Question 11.10

How many isomers of dibromocyclobutane can you construct? As in Question 11.9, use a set of molecular models to construct the isomers and then draw them.

Learning Goal 10

Learning Goal 11

11.4 Conformations of Alkanes and Cycloalkanes

Because there is *free rotation* around a carbon-carbon single bond, even a very simple alkane, like ethane, can exist in an unlimited number of forms. These different arrangements are called **conformations,** or **conformers.**

Figure 11.6 shows two conformations of a more complex alkane, butane. In addition to these two conformations, an infinite number of intermediate conformers exist. Keep in mind that all these conformations are simply different forms of the

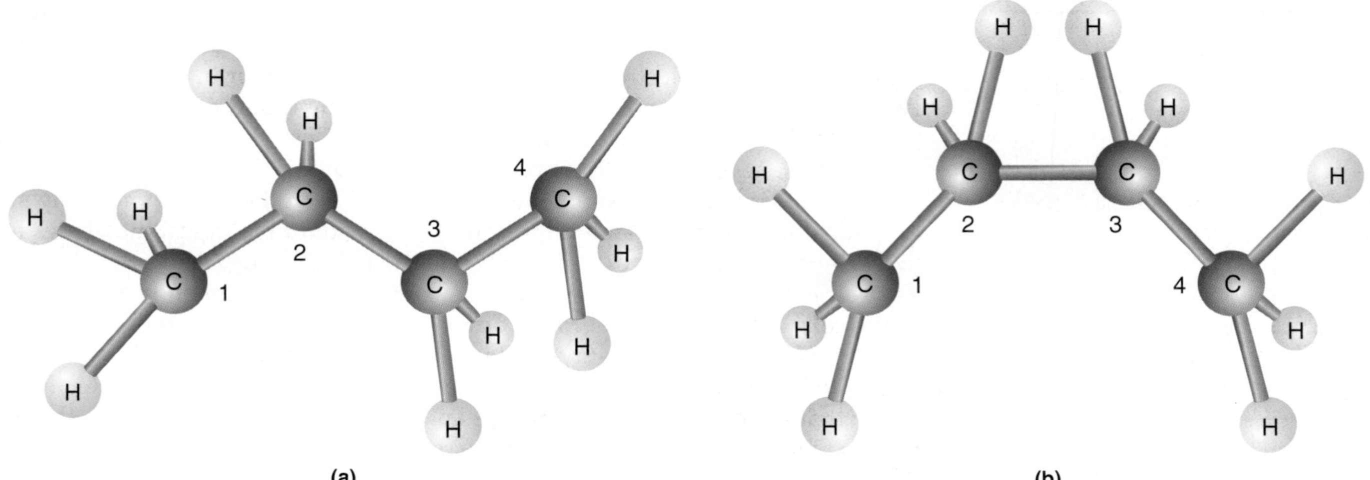

Figure 11.6
Conformational isomers of butane. The hydrogen atoms are much more crowded in the conformation depicted in (b) compared with the conformation shown in (a). The form shown in (a) is energetically favored.

AN ENVIRONMENTAL PERSPECTIVE

The Petroleum Industry and Gasoline Production

Petroleum consists primarily of alkanes and small amounts of alkenes and aromatic hydrocarbons. Substituted hydrocarbons, such as phenol, are also present in very small quantities. Although the composition of petroleum varies with the source of the petroleum (United States, Persian Gulf, etc.), the mixture of hydrocarbons can be separated into its component parts on the basis of differences in the boiling points of various hydrocarbons (distillation). Often several successive distillations of various fractions of the original mixture are required to completely purify the desired component. In the first distillation, the petroleum is separated into several fractions, each of which consists of a mix of hydrocarbons. Each fraction can be further purified by successive distillations. On an industrial scale, these distillations are carried out in columns that may be hundreds of feet in height.

The gasoline fraction of petroleum, called straight-run gasoline, consists primarily of alkanes and cycloalkanes with six to twelve carbon atoms in the skeleton. This fraction has very poor fuel performance. In fact, branched-chain alkanes are superior to straight-chain alkanes as fuels because they are more volatile, burn less rapidly in the cylinder, and thus reduce "knocking." Alkenes and aromatic hydrocarbons are also good fuels. Methods have been developed to convert hydrocarbons of higher and lower molecular weights than gasoline to the appropriate molecular weight range and to convert straight-chain hydrocarbons into branched ones. *Catalytic cracking* fragments a large reactant into smaller products. *Catalytic reforming* results in the rearrangement of a reactant into a more useful form.

The antiknock quality of a fuel is measured as its octane rating. Heptane is a very poor fuel and is given an octane rating of zero. 2,2,4-Trimethylpentane (commonly called isooctane) is an excellent fuel and is given an octane rating of one hundred. Gasoline octane ratings are experimentally determined by comparison with these two compounds in test engines.

Mining the sea for hydrocarbons.

Make a ball-and-stick model of butane and demonstrate these rotational changes for yourself.

same molecule produced by rotation around the carbon-carbon single bonds. Even at room temperature these conformers interconvert rapidly. As a result, they cannot be separated from one another.

Although all conformations can be found in a population of molecules, the staggered conformation (see Figure 11.6a) is the most common. One reason for this is that the bonding electrons are farthest from one another in this conformation. This minimizes the repulsion between these bonding electrons.

Cycloalkanes also exist in different conformations. The only exception to this is cyclopropane. Because it has only three carbon atoms, it is always planar.

The structure of glucose is found in Section 17.4. The physiological roles of glucose are discussed in Chapters 21 and 23.

The conformations of six-member rings have been the most thoroughly studied. One reason is that many important and abundant biological molecules have six-member ring structures. Among these is the simple sugar glucose, also called *blood sugar.* Glucose is the most important sugar in the human body. It is absorbed by the cells of the body and broken down to provide energy for the cells.

This conformation gets its name because it resembles a lawn chair.

The most energetically favorable conformation for a six-member ring is the **chair conformation.** In this conformation the hydrogen atoms are perfectly staggered; that is, they are as far from one another as possible. In addition, the bond angle between carbons is 109.5°, exactly the angle expected for tetrahedral carbon atoms.

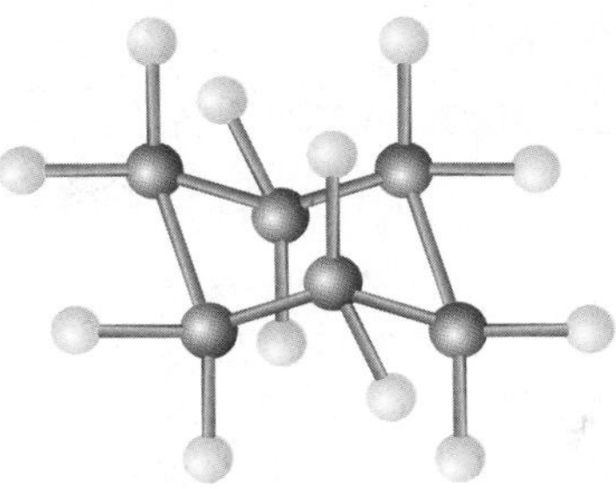

Six-member rings can also exist in a **boat conformation,** so-called because it resembles a rowboat. This form is much less stable than the chair conformation because the hydrogen atoms are not perfectly staggered.

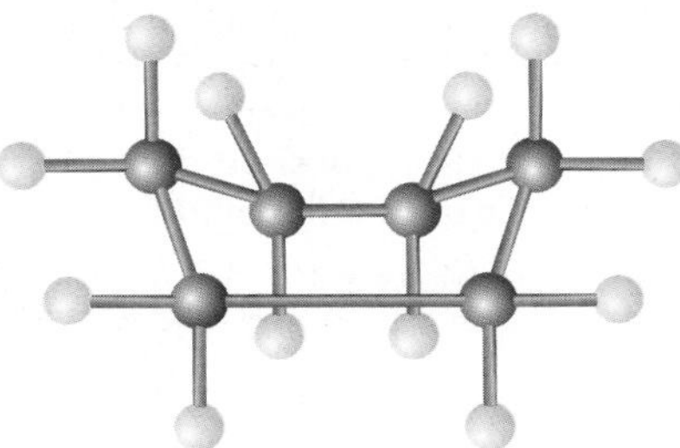

The hydrogen atoms of cyclohexane are described according to their position relative to the ring. Those that lie above or below the ring are said to be **axial atoms.** Those that lie roughly in the plane of the ring are called **equatorial atoms.**

Question 11.11

Describe the positions of the six axial hydrogens of the chair conformation of cyclohexane.

Question 11.12

Describe the positions of the six equatorial hydrogens of the chair conformation of cyclohexane.

11.5 Reactions of Alkanes and Cycloalkanes

Combustion

Learning Goal

Alkanes, cycloalkanes, and other hydrocarbons can be oxidized (by burning) in the presence of excess molecular oxygen. In this reaction, called **combustion,** they burn at high temperatures, producing carbon dioxide and water and releasing large amounts of energy as heat.

C_nH_{2n+2}	+	O_2	$\longrightarrow$	CO_2	+	H_2O	+	heat energy
Alkane		Oxygen	$\longrightarrow$	Carbon dioxide		Water		

A MEDICAL PERSPECTIVE

Polyhalogenated Hydrocarbons Used as Anesthetics

Polyhalogenated hydrocarbons are hydrocarbons containing two or more halogen atoms. Some polyhalogenated compounds are notorious for the problems they have caused humankind. For instance, some insecticides such as DDT, chlordane, kepone, and lindane do not break down rapidly in the environment. As a result, these toxic compounds accumulate in biological tissue of a variety of animals, including humans, and may cause neurological damage, birth defects, or even death.

Other halogenated hydrocarbons are very useful in medicine. They were among the first anesthetics (pain relievers) used routinely in medical practice. These chemicals played a central role as the studies of medicine and dentistry advanced into modern times.

$CH_3CH_2—Cl$
Chloroethane
(ethyl chloride)

$CH_3—Cl$
Chloromethane
(methyl chloride)

Chloroethane and chloromethane are local anesthetics. A local anesthetic deadens the feeling in a portion of the body. Applied topically (on the skin), chloroethane and chloromethane numb the area. Rapid evaporation of these anesthetics lowers the skin temperature, deadening the local nerve endings. They act rapidly, but the effect is brief, and feeling is restored quickly.

$CHCl_3$
Trichloromethane
(chloroform)

In the past, chloroform was used as both a general and a local anesthetic. When administered through inhalation, it rapidly causes loss of consciousness. However, the effects of this powerful anesthetic are of short duration. Chloroform is no longer used because it was shown to be carcinogenic.

$CH_3—CH(Cl)—Br$
1-Bromo-1-chloroethane
(Halothane)

Halothane is a general anesthetic that is administered by inhalation. It is considered to be a very safe anesthetic and is widely used.

Combustion reactions are discussed in Section 7.2.

The following examples show a combustion reaction for a simple alkane and a simple cycloalkane:

$$CH_4 + 2O_2 \longrightarrow CO_2 + 2H_2O + \text{heat energy}$$
Methane

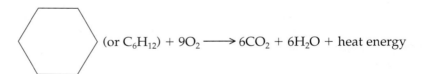

$$\text{(Cyclohexane hexagon)} \text{ (or } C_6H_{12}) + 9O_2 \longrightarrow 6CO_2 + 6H_2O + \text{heat energy}$$
Cyclohexane

See An Environmental Perspective: The Greenhouse Effect and Global Warming in Chapter 6.

The energy released, along with their availability and relatively low cost, makes hydrocarbons very useful as fuels. In fact, combustion is essential to our very existence. It is the process by which we heat our homes, run our cars, and generate electricity. Although combustion of fossil fuels is vital to industry and society, it also represents a threat to the environment. The buildup of CO_2 may contribute to global warming and change the face of the earth in future generations.

Other pollutants are formed as a result of *incomplete combustion.* If not enough oxygen is present, partial combustion produces compounds such as carbon monoxide, formaldehyde, and acetic acid. The following equations show some incomplete combustion reactions that contribute to air pollution:

$$2CH_4 + 3O_2 \longrightarrow 2CO + 4H_2O$$
Methane — Carbon monoxide

A MEDICAL PERSPECTIVE

Chloroform in Your Swimming Pool?

Has anyone ever asked you to take a shower before swimming in an indoor pool? Perhaps that sounds a bit silly, but there may be a very good reason for doing just that. A team of British researchers has identified the suspected carcinogen chloroform and some other potentially hazardous compounds called trihalomethanes (THMs) in indoor swimming pools in London.

Several questions arise. How did these trihalomethanes get into the pool and why are they a problem in indoor pools? THMs are products of chemical reactions between the chlorine used to disinfect the pool and organic substances from the swimmers themselves. Skin cells are shed into the pool, along with lotions and other body care products. Organic molecules from these substances are chlorinated to produce the trihalomethanes. THMs are volatile compounds. In an outdoor pool, they would evaporate and be blown away by the breeze. In an indoor pool, where there is less air circulation, THMs tend to build to higher concentrations in the air above the pool. These fumes are inhaled by the swimmers and the THMs diffuse into the blood.

It seems logical that some of the factors favoring the production of these compounds include warmer water temperatures and larger numbers of swimmers, which means more organic material in the water. In some of the public pools the level of THMs was as high as 132 $\mu g/L$. Compare this to the levels the researchers found in drinking water, only about 3.5 $\mu g/L$.

In the past, chloroform was used as an anesthetic. This is no longer the case, since many safer alternatives are available. One reason for discontinuing the use of chloroform as an anesthetic is the determination by the U.S. Department of Health and Human Services that chloroform is a potential carcinogen. Rats and mice exposed to chloroform in their food or water developed liver and kidney cancers.

How concerned should we be about these levels of THMs? Studies have shown that a one-hour swim can increase the blood concentrations of chloroform as much as tenfold. Animal studies have shown that miscarriages occurred in rats and mice that breathed air containing 30–300 parts per million of chloroform during pregnancy. Some studies have suggested that miscarriages, birth defects, and low birthrate might be associated with human exposure to chloroform, as well. But at the current time, researchers feel that these results are inconsistent and that further research needs to be done before any conclusions can be reached. They do recommend, however, that the amount of THMs be reduced as much as possible, while maintaining a high enough level of chlorine to control waterborne infectious diseases. Keeping the water at cooler temperatures and asking swimmers to shower before entering the pool are effective measures to help reduce the production of toxic THMs.

Swimmers enjoying an indoor pool.

$$CH_4 + O_2 \longrightarrow H{-}\overset{\overset{\displaystyle O}{\|}}{C}{-}H + H_2O$$

Methane → Methanal (formaldehyde)

$$2C_2H_6 + 3O_2 \longrightarrow 2CH_3COOH + 2H_2O$$

Ethane → Ethanoic acid (acetic acid)

Halogenation

Learning Goal  13

Alkanes and cycloalkanes can also react with a halogen (usually chlorine or bromine) in a reaction called **halogenation.** Halogenation is a **substitution reaction,** that is, a reaction that results in the replacement of one group for another. In this reaction a halogen atom is substituted for one of the hydrogen atoms in the alkane. The products of this reaction are an **alkyl halide** or *haloalkane* and a hydrogen halide. Alkanes are not very reactive molecules. However, alkyl halides are very useful reactants for the synthesis of other organic compounds. Thus, the halogenation reaction is of great value because it converts unreactive alkanes into versatile starting materials for the synthesis of desired compounds. This is important in the pharmaceutical industry for the synthesis of some drugs. In addition, alkyl halides having two or more halogen atoms are useful solvents, refrigerants, insecticides, and herbicides.

Halogenation can occur only in the presence of heat and/or light, as indicated by the reaction conditions noted over the reaction arrows. The general equation for the halogenation of an alkane follows. The R in the general structure for the alkane may be either a hydrogen atom or an alkyl group.

$$R{-}CH_2{-}H + X_2 \xrightarrow{\text{Light or heat}} R{-}CH_2{-}X + H{-}X$$

Alkane Halogen Alkyl halide Hydrogen halide

$$H{-}CH_2{-}H + Br_2 \xrightarrow{\text{Light or heat}} H{-}CH_2{-}Br + H{-}Br$$

Methane Bromine Bromomethane Hydrogen bromide

$$CH_3CH_3 + Cl_2 \xrightarrow{\text{Light}} CH_3CH_2{-}Cl + H{-}Cl$$

Ethane Chlorine Chloroethane Hydrogen chloride

$$C_6H_{12} + Cl_2 \xrightarrow{\text{Heat}} C_6H_{11}Cl + HCl$$

Cyclohexane Chlorine Chlorocyclohexane Hydrogen chloride

The alkyl halide may continue to react forming a mixture of products substituted at multiple sites or substituted multiple times at the same site.

If the halogenation reaction is allowed to continue, the alkyl halide formed may react with other halogen atoms. When this happens, a mixture of products may be formed. For instance, bromination of methane will produce bromomethane (CH_3Br), dibromomethane (CH_2Br_2), tribromomethane ($CHBr_3$), and tetrabromomethane (CBr_4).

In more complex alkanes, halogenation can occur to some extent at all positions to give a mixture of monosubstituted products. For example, bromination of propane produces a mixture of 1-bromopropane and 2-bromopropane.

Question 11.13

Write a balanced equation for each of the following reactions. Show all possible products.

a. the complete combustion of cyclobutane
b. the monobromination of propane
c. the complete combustion of ethane
d. the monochlorination of butane

Question 11.14

Write a balanced equation for each of the following reactions. Show all possible products.

a. the complete combustion of decane
b. the monochlorination of cyclobutane
c. the monobromination of pentane
d. the complete combustion of hexane

Question 11.15

Provide the I.U.P.A.C. names for the products of the reactions in Question 11.13b and 11.13d.

Question 11.16

Provide the I.U.P.A.C. names for the products of the reactions in Question 11.14b and 11.14c.

Summary of Reactions

Reactions of Alkanes

Combustion:

$$\underset{\text{Alkane}}{C_nH_{2n+2}} + \underset{\text{Oxygen}}{O_2} \longrightarrow \underset{\text{Carbon dioxide}}{CO_2} + \underset{\text{Water}}{H_2O} + \text{heat energy}$$

Halogenation:

$$\underset{\text{Alkane}}{R\text{—}\overset{\overset{\Large H}{|}}{\underset{\underset{\Large H}{|}}{C}}\text{—}H} + \underset{\text{Halogen}}{X_2} \xrightarrow{\text{Light or heat}} \underset{\text{Alkyl halide}}{R\text{—}\overset{\overset{\Large H}{|}}{\underset{\underset{\Large H}{|}}{C}}\text{—}X} + \underset{\text{Hydrogen halide}}{H\text{—}X}$$

Summary

11.1 The Chemistry of Carbon

The modern science of organic chemistry began with Wöhler's synthesis of urea in 1828. At that time, people believed that it was impossible to synthesize an organic molecule outside of a living system. We now define organic chemistry as the study of carbon-containing compounds. The differences between the ionic bond, which is characteristic of many inorganic substances, and the covalent bond in organic compounds are responsible for the great contrast in properties and reactivity between organic and inorganic compounds. All organic compounds are classified as either *hydrocarbons* or *substituted hydrocarbons*. In substituted hydrocarbons a hydrogen atom is replaced by a functional group. A *functional group* is an atom or group of atoms arranged in a particular way that imparts specific chemical or physical properties to a molecule. The major families of organic molecules are defined by the specific functional groups that they contain.

11.2 Alkanes

The *alkanes* are *saturated hydrocarbons*, that is, hydrocarbons that have only carbon and hydrogen atoms that are bonded together by carbon-carbon and carbon-hydrogen single bonds. They have the general molecular formula C_nH_{2n+2} and are nonpolar, water-insoluble compounds with low melting and boiling points. In the *I.U.P.A.C. Nomenclature System* the alkanes are named by determining the number of carbon atoms in the parent compound and numbering the carbon chain to provide the lowest possible number for all substituents. The substituent names and numbers are used as prefixes before the name of the parent compound.

Constitutional or *structural isomers* are molecules that have the same molecular formula but different structures. They have different physical and chemical properties because the atoms are bonded to one another in different patterns.

11.3 Cycloalkanes

Cycloalkanes are a family of organic molecules having C—C single bonds in a ring structure. They are named by adding the prefix *cyclo-* to the name of the alkane parent compound.

A *cis-trans isomer* is a type of stereoisomer. Stereoisomers are molecules that have the same structural formula and bonding pattern but different arrangements of atoms in space. A cycloalkane is in the *cis* configuration if two substituents are on the same side of the ring (either both above or both below). A cycloalkane is in the *trans* configuration when one substituent is above the ring and the other is below the ring. The *cis-trans* isomers are not interconvertible.

11.4 Conformations of Alkanes and Cycloalkanes

As a result of *free rotation* around carbon-carbon single bonds, infinitely many *conformations* or *conformers* exist for any alkane.

Limited rotation around the carbon-carbon single bonds of cycloalkanes also results in a variety of conformations of cycloalkanes. In cyclohexane the *chair conformation* is the most energetically favored. Another conformation is the *boat conformation.*

11.5 Reactions of Alkanes and Cycloalkanes

Alkanes can participate in *combustion* reactions. In complete combustion reactions they are oxidized to produce carbon dioxide, water, and heat energy. They can also undergo *halogenation* reactions to produce *alkyl halides*.

Key Terms

aliphatic hydrocarbon (11.1)
alkane (11.2)
alkyl group (11.2)
alkyl halide (11.5)
aromatic hydrocarbon (11.1)
axial atom (11.4)
boat conformation (11.4)
chair conformation (11.4)
***cis-trans* isomers (11.3)**
combustion (11.5)
condensed formula (11.2)
conformations (11.4)
conformers (11.4)
constitutional isomers (11.2)
cycloalkane (11.3)
equatorial atom (11.4)
functional group (11.1)
geometric isomers (11.3)
halogenation (11.5)
hydrocarbon (11.1)
I.U.P.A.C. Nomenclature System (11.2)
line formula (11.2)
molecular formula (11.2)
parent compound (11.2)
primary (1°) carbon (11.2)
quaternary (4°) carbon (11.2)
saturated hydrocarbon (11.1)
secondary (2°) carbon (11.2)
structural formula (11.2)
structural isomer (11.2)
substituted hydrocarbon (11.1)
substitution reaction (11.5)
tertiary (3°) carbon (11.2)
unsaturated hydrocarbon (11.1)

Questions and Problems

The Chemistry of Carbon

11.17 Consider the differences between organic and inorganic compounds as you answer each of the following questions.
a. Which compounds make good electrolytes?
b. Which compounds exhibit ionic bonding?
c. Which compounds have lower melting points?
d. Which compounds are more likely to be soluble in water?
e. Which compounds are flammable?

11.18 Describe the major differences between ionic and covalent bonds.

11.19 Give the structural formula for each of the following:

```
     CH3   CH3
     |     |
a. CH3CHCH2CHCH3

     Br Br
     |  |
b. CH3CHCHCH3

                      CH3
                      |
c. CH3CH2CH—CH2CHCH2CH3
          |
          CH3

                          Br
                          |
d. CH3CH2CH2CH2CH2CH
                          |
                          CH3
```

11.20 Condense each of the following structural formulas:

```
      H  H  H  H
      |  |  |  |
a. H—C—C—C—C—H
      |  |  |  |
      H  H  |  H
            |
         H—C—H
            |
            H

            H           H
            |           |
         H—C—H     H—C—H
      H  H  |  H  H  |  H
      |  |  |  |  |  |  |
b. H—C—C—C—C—C—C—C—H
      |  |  |  |  |  |  |
      H  H  |  H  H  H  H
            |
         H—C—H
            |
            H

         H
         |
      H—C—H
      H  |  H  H  H
      |  |  |  |  |
c. H—C—C—C—C—C—H
      |  |  |  |  |
      H  H  H  H  H

         H
         |
      H—C—H
      H  |  H
      |  |  |
d. H—C—C—C—H
      |  |  |
      H  |  H
      H—C—H
         |
         H
```

11.21 Which of the following structures are not possible? State your reasons.

```
     CH3
     |
a. CH3CHCH2CH3

     CH3
     |
b. CH3CHCH2CH3
     |
     CH3

     CH3   CH3
     |     |
c. CH3CHCH2CHCH3

      CH3
      |
d. CH3CH2CH2CH2CH3

e. CH2CH3CH2CH3
      |
      CH3

     CH2CH3
     |
f. CH3CH2CH2CH3
```

11.22 Using the octet rule, explain why carbon forms four bonds in a stable compound.

11.23 Using structural formulas, draw a typical alcohol, aldehyde, ketone, carboxylic acid, and amine. (*Hint:* Refer to Table 11.2).

11.24 Name the functional group in each of the following molecules:

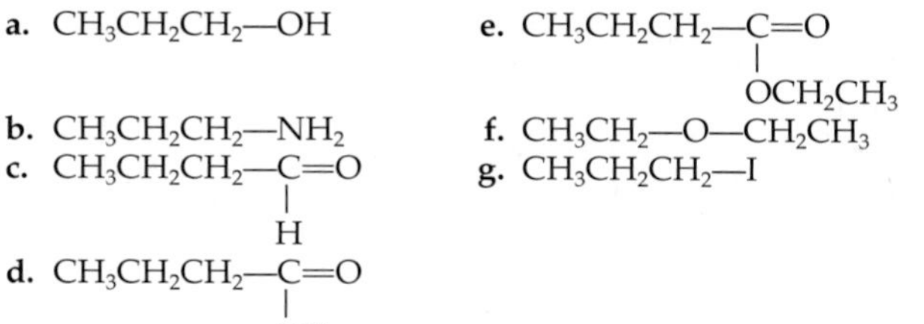

11.25 Give the general formula for each of the following:

a. An alkane
b. An alkyne
c. An alkene
d. A cycloalkane
e. A cycloalkene

11.26 Of the classes of compounds listed in Problem 11.25, which are saturated? Which are unsaturated?

11.27 What major structural feature distinguishes the alkanes, alkenes, and alkynes? Give examples.

11.28 What is the major structural feature that distinguishes between saturated and unsaturated hydrocarbons?

11.29 Give an example, using structural formulas, of each of the following families of organic compounds. Each of your examples should contain a minimum of three carbons. (*Hint:* Refer to Table 11.2.)

a. A carboxylic acid
b. An amine
c. An alcohol
d. An ether

11.30 Folic acid is a vitamin required by the body for nucleic acid synthesis. The structure of folic acid is given below. Circle and identify as many functional groups as possible.

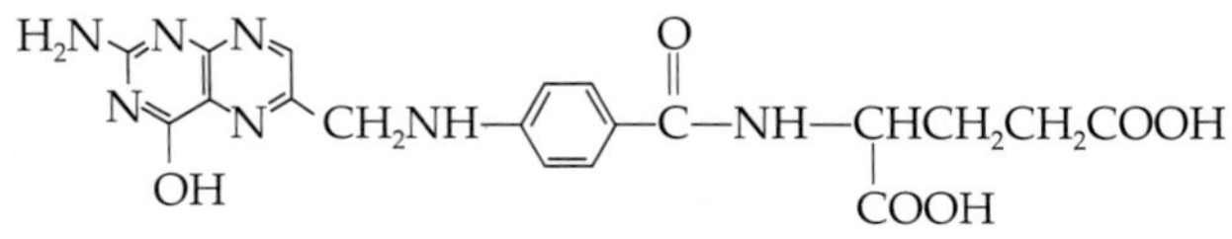

Folic acid

Alkanes

11.31 Draw each of the following:

a. 2-Bromobutane
b. 2-Chloro-2-methylpropane
c. 2,2-Dimethylhexane
d. Dichlorodiiodomethane
e. 1,4-Diethylcyclohexane
f. 2-Iodo-2,4,4-trimethylpentane

11.32 Draw each of the following compounds using structural formulas:

a. 2,2-Dibromobutane
b. 2-Iodododecane
c. 1,2-Dichloropentane
d. 1-Bromo-2-methylpentane
e. 1,1,1-Trichlorodecane
f. 1,2-Dibromo-1,1,2-trifluoroethane
g. 3,3,5-Trimethylheptane
h. 1,3,5-Trifluoropentane

11.33 Name each of the following using the I.U.P.A.C. Nomenclature System:

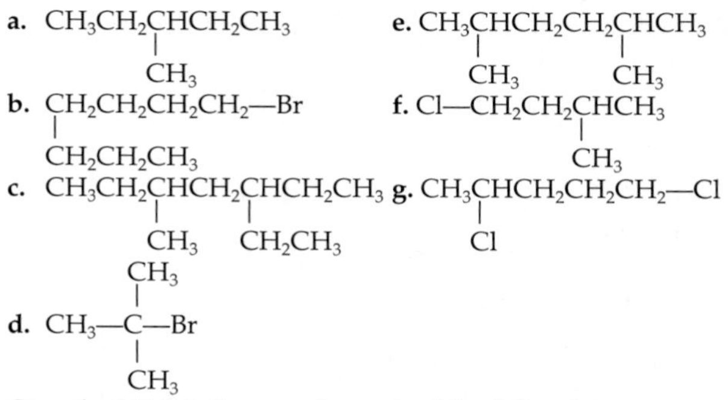

11.34 Give the I.U.P.A.C. name for each of the following:

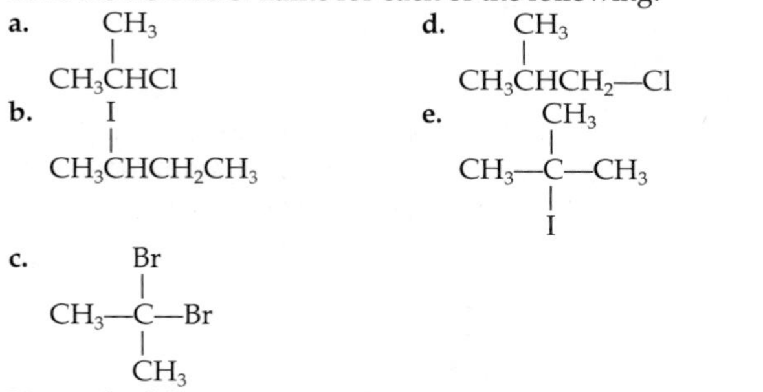

11.35 Name the following using the I.U.P.A.C. Nomenclature System:

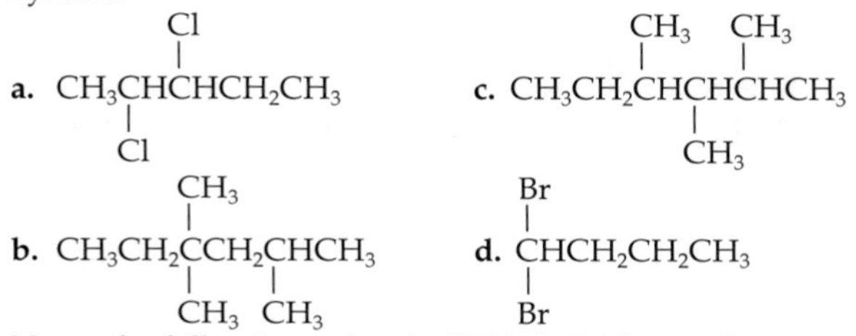

11.36 Name the following using the I.U.P.A.C. Nomenclature System:

a. $CH_3(CH_2)_3CH(Cl)CH_3$ **c.** $CH_3CH_2CH(Cl)CH_2CH_3$
b. $CH_2(Br)(CH_2)_2CH_2Br$ **d.** $CH_3CH(CH_3)(CH_2)_4CH_3$

11.37 Which of the following pairs of compounds are identical? Which are constitutional isomers? Which are completely unrelated?

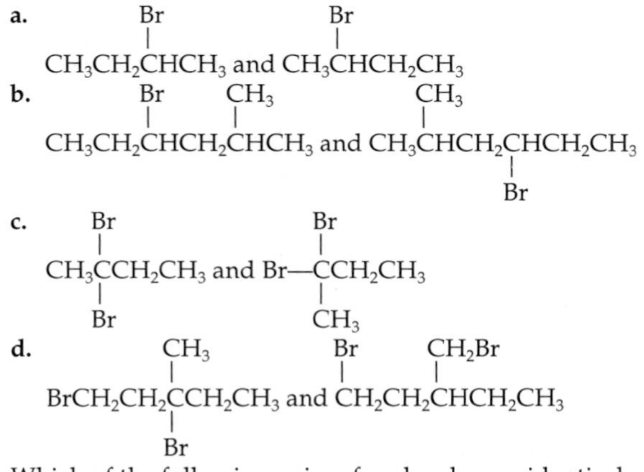

11.38 Which of the following pairs of molecules are identical compounds? Which are constitutional isomers?

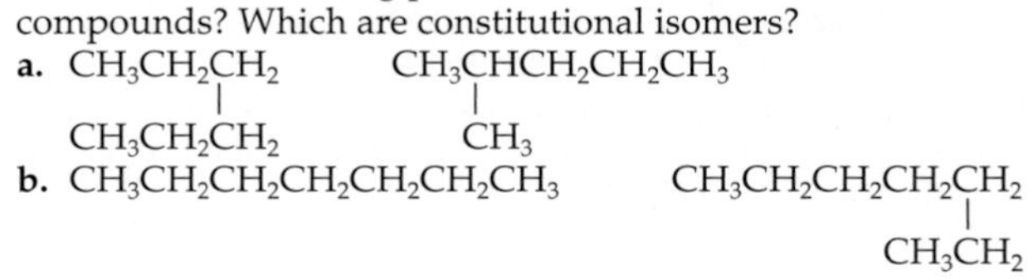

11.39 Which of the following structures are incorrect?

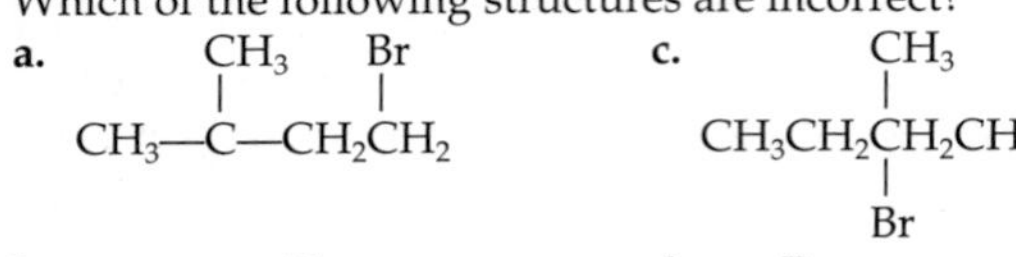

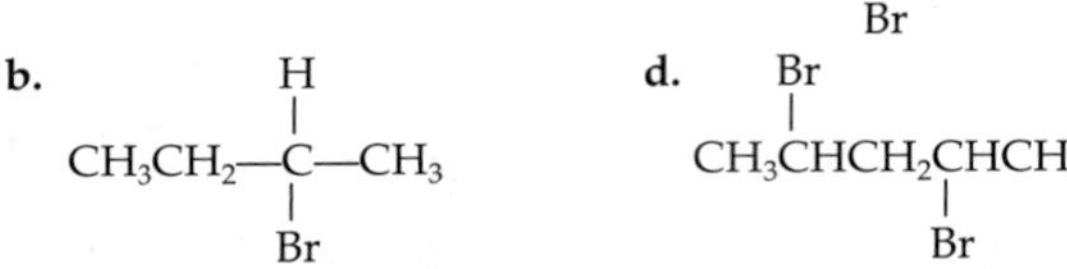

11.40 Are the following names correct or incorrect? If they are incorrect, give the correct name.
a. 1,3-Dimethylpentane
b. 2-Ethylpropane
c. 3-Butylbutane
d. 3-Ethyl-4-methyloctane

11.41 In your own words, describe the steps used to name a compound, using I.U.P.A.C. nomenclature.

11.42 Draw the structures of the following compounds. Are the names provided correct or incorrect? If they are incorrect, give the correct name.
a. 2,4-Dimethylpentane
b. 1,3-Dimethylpentane
c. 1,5-Diiodopentane
d. 1,4-Diethylheptane
e. 1,6-Dibromo-6-methyloctane

Cycloalkanes

11.43 Name each of the following cycloalkanes, using the I.U.P.A.C. system:

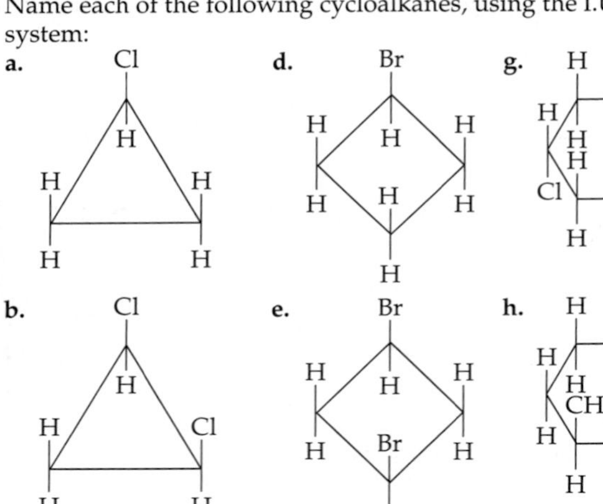

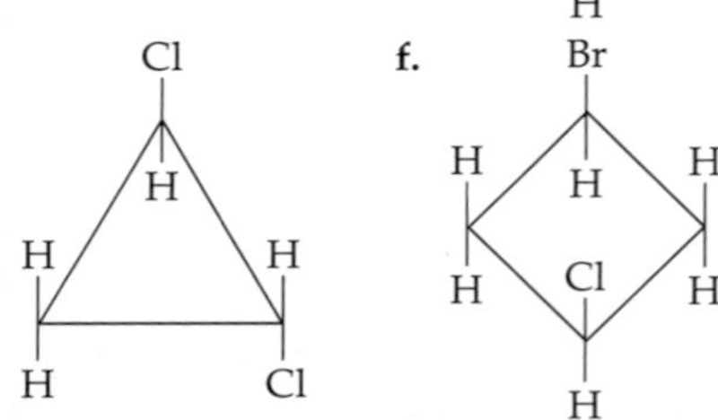

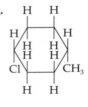

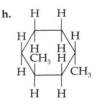

11.44 Draw the structure of each of the following cycloalkanes:
a. 1-Bromo-2-methylcyclobutane
b. Iodocyclopropane
c. 1-Bromo-3-chlorocyclopentane
d. 1,2-Dibromo-3-methycyclohexane

11.45 What is the general formula for a cycloalkane?

11.46 How does the general formula of a cycloalkane compare to that of an alkane?

11.47 Which of the following names are correct and which are incorrect? If incorrect, write the correct name.
a. 2,3-Dibromocyclobutane
b. 1,4-Diethylclobutane
c. 1,2-Dimethylcyclopropane
d. 4,5,6-Trichlorocyclohexane

11.48 Which of the following names are correct and which are incorrect? If incorrect, write the correct name.
a. 1,4,5-Tetrabromocyclohexane
b. 1,3-Dimethylcyclobutane
c. 1,2-Dichlorocyclopentane
d. 3-Bromocyclopentane

11.49 Draw the structures of each of the following compounds:
a. *cis*-1,3-Dibromocyclopentane
b. *trans*-1,2-Dimethylcyclobutane
c. *cis*-1,2-Dichlorocyclopropane
d. *trans*-1,4-Diethylcyclohexane

11.50 Draw the structures of each of the following compounds:
a. *trans*-1,4-Dimethylcyclooctane
b. *cis*-1,3-Dichlorocyclohexane
c. *cis*-1,3-Dibromocyclobutane

11.51 Name each of the following compounds:

11.52 Name each of the following compounds:

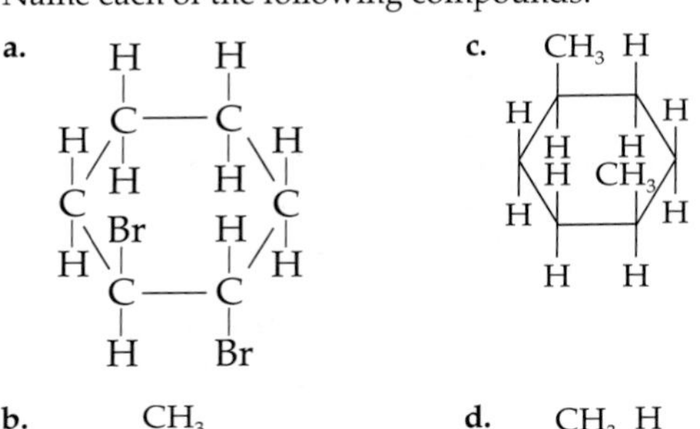

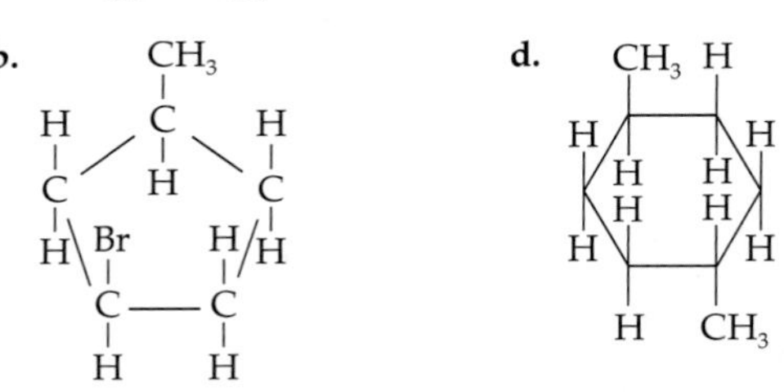

Conformations of Alkanes and Cycloalkanes

11.53 Make a model of cyclohexane and compare the boat and chair conformations. Use your model to explain why the chair conformation is more energetically favored.

11.54 Why would the ethyl group of ethylcyclohexane generally be found in the equatorial position?

11.55 Why can't conformations be separated from one another?

11.56 What is meant by free rotation around a carbon-carbon single bond?

11.57 Explain why one conformation is more stable than another. (*Hint:* refer to Figure 11.6).

11.58 Explain why a substituent on a cyclohexane ring would tend to be located in the equatorial position.

Reactions of Alkanes and Cycloalkanes

11.59 Complete each of the following reactions by supplying the missing reactant or product as indicated by a question mark:

a. $2CH_3CH_2CH_2CH_3 + 13O_2 \xrightarrow{\text{Heat}}$? (Complete combustion)

b. 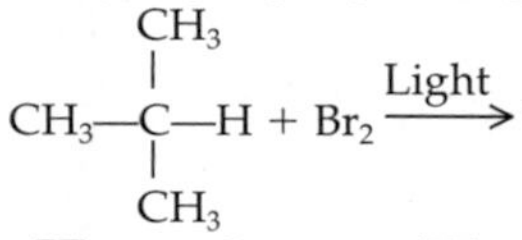 $CH_3—C(CH_3)_2—H + Br_2 \xrightarrow{\text{Light}}$? (Give all possible monobrominated products)

c. 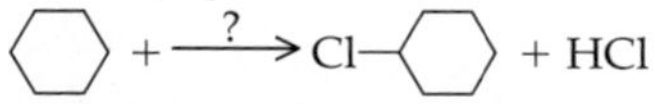 cyclohexane + ? $\xrightarrow{?}$ Cl–cyclohexane + HCl

11.60 Give all the possible monochlorinated products for the following reaction:

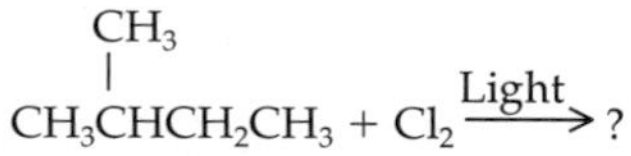

$$CH_3CH(CH_3)CH_2CH_3 + Cl_2 \xrightarrow{\text{Light}} ?$$

Name the products, using I.U.P.A.C. nomenclature.

11.61 Draw the constitutional isomers of molecular formula C_6H_{14} and name each using the I.U.P.A.C. system:

a. Which one gives two and only two monobromo derivatives when it reacts with Br_2 and light? Name the products, using the I.U.P.A.C. system.

b. Which give three and only three monobromo products? Name the products, using the I.U.P.A.C. system.

c. Which give four and only four monobromo products? Name the products, using the I.U.P.A.C. system.

11.62 a. Draw and name all of the isomeric products obtained from the monobromination of propane with Br_2/light. If halogenation were a completely random reaction and had an equal probability of occurring at any of the C—H bonds in a molecule, what percentage of each of these monobromo products would be expected?

b. Answer part (a) using 2-methylpropane as the starting material.

11.63 A mole of hydrocarbon formed eight moles of CO_2 and eight moles of H_2O upon combustion. Determine the molecular formula of the hydrocarbon and give the balanced combustion reaction.

11.64 Highly substituted alkyl fluorides, called perfluoroalkanes, are often used as artificial blood substitutes. These perfluoroalkanes have the ability to transport O_2 through the bloodstream as blood does. Some even have twice the O_2 transport capability and are used to treat gangrenous tissue. The structure of perfluorodecalin is shown below. How many moles of fluorine must be reacted with one mole of decalin to produce perfluorodecalin?

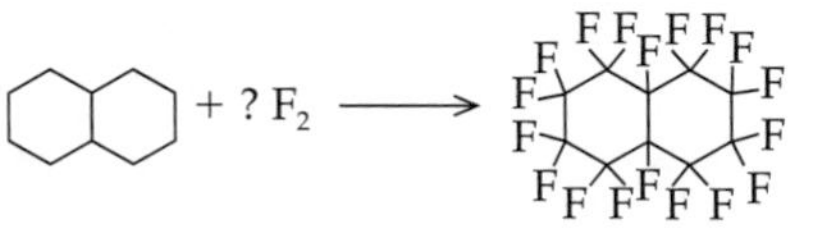

Decalin + ? F_2 ⟶ Perfluorodecalin

Critical Thinking Problems

1. You are given two unlabeled bottles, each of which contains a colorless liquid. One contains hexane and the other contains water. What physical properties could you use to identify the two liquids? What chemical property could you use to identify them?

2. You are given two beakers, each of which contains a white crystalline solid. Both are soluble in water. How would you determine which of the two solids is an ionic compound and which is a covalent compound?

3. Chlorofluorocarbons (CFCs) are man-made compounds made up of carbon and the halogens fluorine and chlorine. One of the most widely used is Freon-12 (CCl_2F_2). It was introduced as a refrigerant in the 1930s. This was an important advance because Freon-12 replaced ammonia and sulfur dioxide, two toxic chemicals that were previously used in refrigeration systems. Freon-12 was hailed as a perfect replacement because it has a boiling point of −30°C and is almost completely inert. To what family of organic molecules do CFCs belong? Design a strategy for the synthesis of Freon-12.

4. Over time, CFC production increased dramatically as their uses increased. They were used as propellants in spray cans, as gases to expand plastic foam, and in many other applications. By 1985 production of CFCs reached 850,000 tons. Much of this leaked into the atmosphere and in that year the concentration of CFCs reached 0.6 parts per billion. Another observation was made by groups of concerned scientists: as the level of CFCs rose, the ozone level in the upper atmosphere declined. Does this correlation between CFC levels and ozone levels prove a relationship between these two phenomena? Explain your reasoning.

5. Although manufacture of CFCs was banned on December 31, 1995, the C—F and C—Cl bonds of CFCs are so strong that the molecules may remain in the atmosphere for 120 years. Within 5 years they diffuse into the upper stratosphere where ultraviolet photons can break the C—Cl bonds. This process releases chlorine atoms, as shown here for Freon-12:

$$CCl_2F_2 + \text{photon} \longrightarrow CClF_2 + Cl$$

The chlorine atoms are extremely reactive because of their strong tendency to acquire a stable octet of electrons. The following reactions occur when a chlorine atom reacts with an ozone molecule (O_3). First, chlorine pulls an oxygen atom away from ozone:

$$Cl + O_3 \longrightarrow ClO + O_2$$

Then ClO, a highly reactive molecule, reacts with an oxygen atom:

$$ClO + O \longrightarrow Cl + O_2$$

Write an equation representing the overall reaction (sum of the two reactions). How would you describe the role of Cl in these reactions?

Each year millions of dollars are lost because of crop damage by pests.

12

The Unsaturated Hydrocarbons:

Alkenes, Alkynes, and Aromatics

Outline

Learning Goals

1. Describe the physical properties of alkenes and alkynes.
2. Draw the structures and write the I.U.P.A.C. names for simple alkenes and alkynes.
3. Write the names and draw the structures of simple geometric isomers of alkenes.
4. Write equations predicting the products of addition reactions of alkenes: hydrogenation, halogenation, hydration, and hydrohalogenation.
5. Apply Markovnikov's rule to predict the major and minor products of the hydration and hydrohalogenation reactions of unsymmetrical alkenes.
6. Write equations representing the formation of addition polymers of alkenes.
7. Draw the structures and write the names of common aromatic hydrocarbons.
8. Write equations for substitution reactions involving benzene.
9. Describe heterocyclic aromatic compounds and list several biological molecules in which they are found.

CHEMISTRY CONNECTION

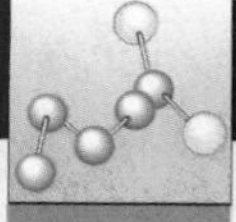

A Cautionary Tale: DDT and Biological Magnification

We have heard the warnings for years: Stop using nonbiodegradable insecticides because they are killing many animals other than their intended victims! Are these chemicals not specifically targeted to poison insects? How then can they be considered a threat to humans and other animals?

DDT, a polyhalogenated hydrocarbon, was discovered in the early 1940s by Paul Müller, a Swiss chemist. Müller showed that DDT is a nerve poison that causes convulsions, paralysis, and eventually death in insects. From the 1940s until 1972, when it was banned in the United States, DDT was sprayed on crops to kill insect pests, sprayed on people as a delousing agent, and sprayed in and on homes to destroy mosquitoes carrying malaria. At first, DDT appeared to be a miraculous chemical, saving literally millions of lives and millions of dollars in crops. However, as time went by, more and more evidence of a dark side of DDT use accumulated. Over time, the chemical had to be sprayed in greater and greater doses as the insect populations evolved to become more and more resistant to it. In 1962, Rachel Carson published her classic work, *Silent Spring,* which revealed that DDT was accumulating in the environment. In particular, high levels of DDT in birds interfered with their calcium metabolism. As a result, the egg shells produced by the birds were too thin to support development of the chick within. It was feared that in spring, when the air should have been filled with bird song, there would be silence. This is the "silent spring" referred to in the title of Carson's book.

DDT: Dichlorodiphenyltrichloroethane

DDT is not biodegradable; furthermore, it is not water-soluble, but it is soluble in nonpolar solvents. Thus if DDT is ingested by an animal, it will dissolve in fat tissue and accumulate there, rather than being excreted in the urine. When DDT is introduced into the food chain, which is inevitable when it is sprayed over vast areas of the country, the result is *biological magnification.* This stepwise process begins when DDT applied to crops is ingested by insects. The insects, in turn, are eaten by birds, and the birds are eaten by a hawk. We can imagine another food chain: Perhaps the insects are eaten by mice, which are in turn eaten by a fox, which is then eaten by an owl. Or to make it more personal, perhaps the grass is eaten by a steer, which then becomes your dinner. With each step up one of these food chains, the concentration of DDT in the tissues becomes higher and higher because it is not degraded, it is simply stored. Eventually, the concentration may reach toxic levels in some of the animals in the food chain.

Consider for a moment the series of events that occurred in Borneo in 1955. The World Health Organization elected to spray DDT in Borneo because 90% of the inhabitants were infected with malaria. As a result of massive spraying, the mosquitoes bearing the malaria parasite were killed. If this sounds like the proverbial happy ending, read on. This is just the beginning of the story. In addition to the mosquitoes, millions of other household insects were killed. In tropical areas it is common for small lizards to live in homes, eating insects found there. The lizards ate the dead and dying DDT-contaminated insects and were killed by the neurotoxic effects of DDT. The house cats ate the lizards, and they, too, died. The number of rats increased dramatically because there were no cats to control the population. The rats and their fleas carried sylvatic plague, a form of bubonic plague. With more rats in contact with humans came the threat of a bubonic plague epidemic. Happily, cats were parachuted into the affected areas of Borneo, and the epidemic was avoided.

The story has one further twist. Many of the islanders lived in homes with thatched roofs. The vegetation used to make these roofs was the preferred food source for a caterpillar that was not affected by DDT. Normally, the wasp population preyed on these caterpillars and kept the population under control. Unfortunately, the wasps were killed by the DDT. The caterpillars prospered, devouring the thatched roofs, which collapsed on the inhabitants.

Every good story has a moral, and this one is not difficult to decipher. The introduction of large amounts of any chemical into the environment, even to eradicate disease, has the potential for long-term and far-reaching effects that may be very difficult to predict. We must be cautious with our fragile environment. Our well-intentioned intervention all too often upsets the critical balance of nature, and in the end we inadvertently do more harm than good.

Introduction

Unsaturated hydrocarbons are those that contain at least one carbon-carbon double or triple bond. They include the alkenes, alkynes, and aromatic compounds. All alkenes have at least one carbon-carbon double bond; all alkynes have

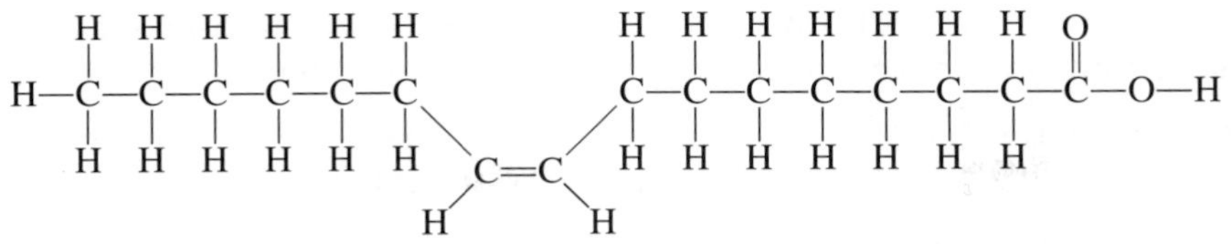

Palmitoleic acid

(a)

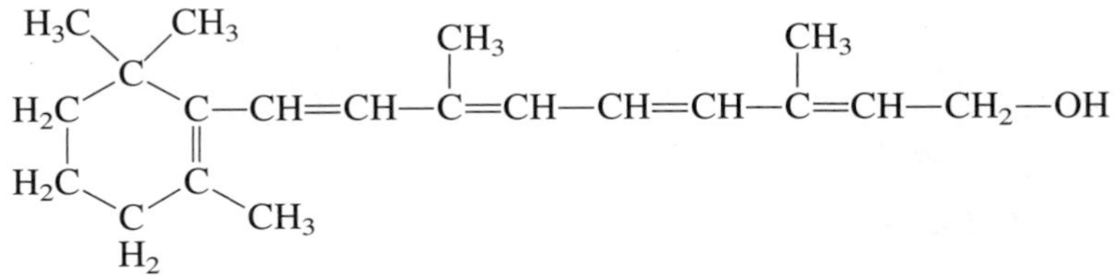

Vitamin A

(b)

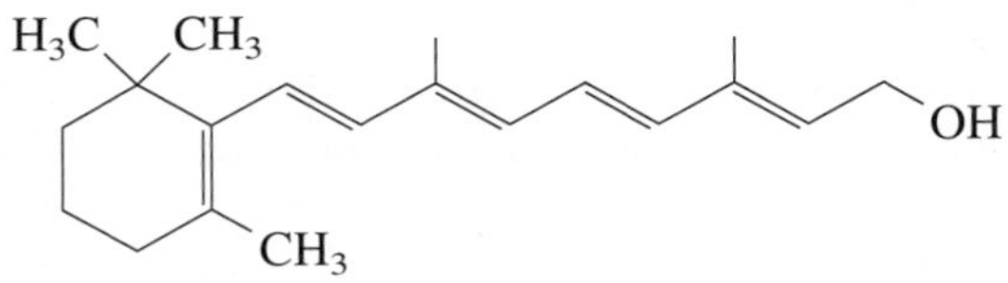

Vitamin A

(c)

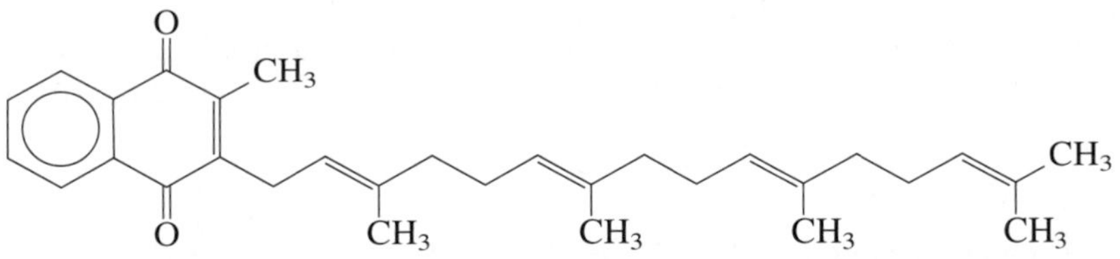

Vitamin K

(d)

Figure 12.1

(a) Structural formula of the sixteen-carbon monounsaturated fatty acid palmitoleic acid. (b) Condensed formula of vitamin A, which is required for vision. Notice that the carbon chain of vitamin A is a conjugated system of double bonds. (c) Line formula of vitamin A. In the line formula, each line represents a carbon-carbon bond, each double line represents a carbon-carbon double bond. A carbon atom and the appropriate number of hydrogen atoms are assumed to be at the point where two lines meet. The vertical lines are assumed to terminate in a methyl group. (d) Line formula of vitamin K, a lipid-soluble vitamin required for blood clotting. The six-member ring with the circle represents a benzene ring. See Figure 12.6 for other representations of the benzene ring.

at least one carbon-carbon triple bond. Aromatic compounds are particularly stable cyclic compounds and sometimes are depicted as having alternating single and double carbon-carbon bonds. This arrangement of alternating single and double bonds is called a conjugated system of double bonds.

Fatty acids are long hydrocarbon chains having a carboxyl group at the end. Thus by definition they are carboxylic acids. See Chapters 15 and 18.

Many important biological molecules are characterized by the presence of double bonds or a linear or cyclic conjugated system of double bonds (Figure 12.1). For instance, we classify fatty acids as either *monounsaturated* (having one double bond), *polyunsaturated* (having two or more double bonds), or *saturated* (having single bonds only). Vitamin A (retinol), a vitamin required for vision, contains a ten-carbon conjugated hydrocarbon chain. Vitamin K, a vitamin required for blood clotting, contains an aromatic ring.

These vitamins are discussed in detail in Appendix E, Lipid-Soluble Vitamins.

12.1 Alkenes and Alkynes: Structure and Physical Properties

Learning Goal

Alkenes and **alkynes** are unsaturated hydrocarbons. The characteristic functional group of an alkene is the carbon-carbon double bond. The functional group that characterizes the alkynes is the carbon-carbon triple bond. The following general formulas compare the structures of alkanes, alkenes, and alkynes.

	Alkane	Alkene	Alkyne
General formulas:	C_nH_{2n+2}	C_nH_{2n}	C_nH_{2n-2}
Structural formulas:	$H{-}\overset{H}{\underset{H}{C}}{-}\overset{H}{\underset{H}{C}}{-}H$	$H_2C{=}CH_2$ (planar, H's at 120°)	$H{-}C{\equiv}C{-}H$
	Ethane (ethane)	Ethene (ethylene)	Ethyne (acetylene)
Molecular formulas:	C_2H_6	C_2H_4	C_2H_2
Condensed formulas:	CH_3CH_3	$H_2C{=}CH_2$	$HC{\equiv}CH$

These compounds have the same number of carbon atoms but differ in the number of hydrogen atoms, a feature of all alkanes, alkenes, and alkynes that contain the same number of carbon atoms. Alkenes contain two fewer hydrogens than the corresponding alkanes, and alkynes contain two fewer hydrogens than the corresponding alkenes.

In alkanes the four bonds to the central carbon have tetrahedral geometry. When carbon is bonded by one double bond and two single bonds, as in ethene (an alkene), the molecule is *planar,* because all atoms lie in a single plane. Each bond angle is approximately 120°. When two carbon atoms are bonded by a triple bond, as in ethyne (an alkyne), each bond angle is 180°. Thus, the molecule is linear, and all atoms are positioned in a straight line (Figure 12.2).

The physical properties of alkenes, alkynes, and aromatic compounds are very similar to those of alkanes. They are nonpolar. As a result of the "like dissolves like" rule, they are not soluble in water but are very soluble in nonpolar solvents such as other hydrocarbons. They also have relatively low boiling points and melting points.

12.2 Alkenes and Alkynes: Nomenclature

Learning Goal

To determine the name of an alkene or alkyne using the I.U.P.A.C. Nomenclature System, use the following simple rules:

- Name the parent compound using the longest continuous carbon chain containing the double bond (alkenes) or triple bond (alkynes).
- Replace the *-ane* ending of the alkane with the *-ene* ending for an alkene or the *-yne* ending for an alkyne. For example:

$CH_3{-}CH_3$	$CH_2{=}CH_2$	$CH{\equiv}CH$
Ethane	Ethene	Ethyne
$CH_3{-}CH_2{-}CH_3$	$CH_2{=}CH{-}CH_3$	$CH{\equiv}C{-}CH_3$
Propane	Propene	Propyne

- Number the chain to give the lowest number for the first of the two carbons containing the double bond or triple bond. For example:

$$\overset{4}{C}H_3\overset{3}{C}H_2\overset{2}{C}H{=}\overset{1}{C}H_2$$

1-Butene
(*not* 3-butene)

$$\overset{1}{C}H{\equiv}\overset{2}{C}{-}\overset{3}{C}H_2\overset{4}{C}H_2\overset{5}{C}H_3$$

1-Pentyne
(*not* 4-pentyne)

Tetrahedral

All bond angles approximately 109.5°

Ethane

Planar

All bond angles approximately 120°

Ethene

Linear

H—C≡C—H

Bond angles 180°

Ethyne

(a)

A long-chain alkane (pentane)

A long-chain alkene (1-pentene)

A long-chain alkyne (1-pentyne)

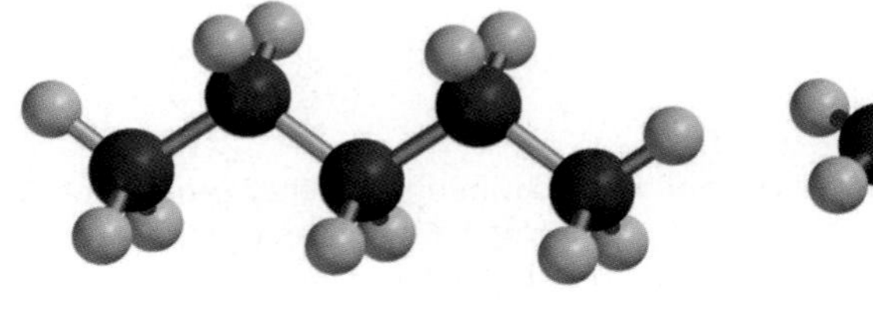

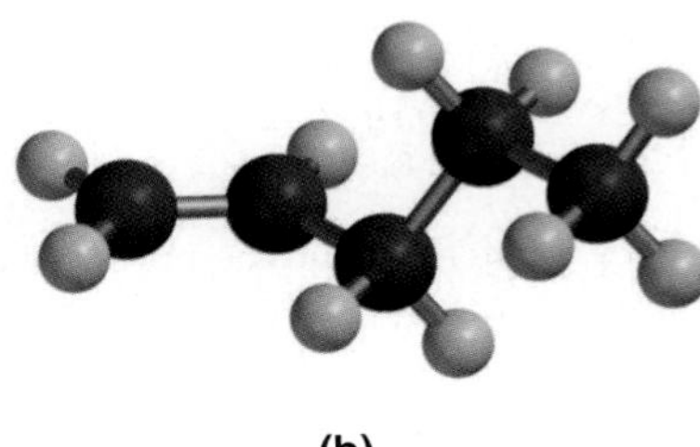

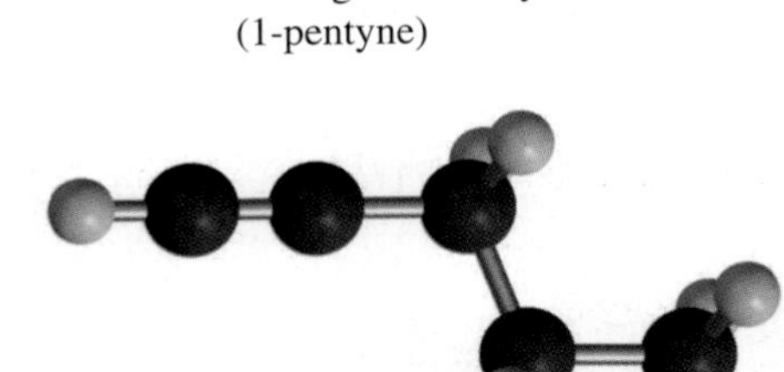

(b)

Figure 12.2
(a)Three-dimensional drawings and ball-and-stick models of ethane, ethene, and ethyne. (b) Examples of typical long-chain hydrocarbons.

Remember, it is the position of the double bond, not the substituent, that determines the numbering of the carbon chain.

- Name and number all groups bonded to the parent alkene or alkyne, and place the name and number in front of the name of the parent compound. Remember that with alkenes and alkynes the double or triple bond takes precedence over a halogen or alkyl group, as shown in the following examples:

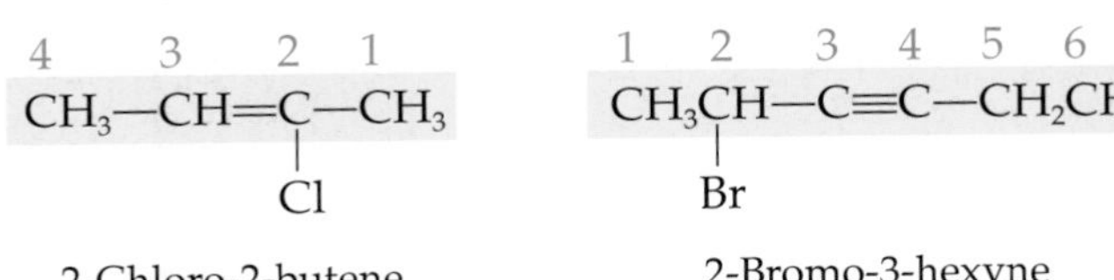

2-Chloro-2-butene

2-Bromo-3-hexyne

A MEDICAL PERSPECTIVE

Killer Alkynes in Nature

There are many examples of alkynes that are beneficial to humans. Among these are *parasalamide,* a pain reliever, *pargyline,* an antihypertensive, and *17-ethynylestradiol,* a synthetic estrogen that is used as an oral contraceptive.

Parasalamide

Paragyline

17-Ethynylestradiol

Alkynes used for medicinal purposes.

But in addition to these medically useful alkynes, there are in nature a number that are toxic. Some are extremely toxic to mammals, including humans; others are toxic to fungi, fish, or insects. All of these compounds are plant products that may help protect the plant from destruction by predators.

Capillin is produced by the oriental wormwood plant. Research has shown that a dilute solution of capillin inhibits the growth of certain fungi. Since fungal growth can damage or destroy a plant, the ability to make capillin may provide a survival advantage to the plants. Perhaps it may one day be developed to combat fungal infections in humans.

Ichthyotherol is a fast-acting poison commonly found in plants referred to as fish-poison plants. Ichthyotherol is a very toxic polyacetylenic alcohol that inhibits energy production in the mitochondria. Latin American native tribes use these plants to coat the tips of the arrows used to catch fish. Although ichthyotherol is poisonous to the fish, fish caught by this method are quite safe for human consumption.

An extract of the leaves of English ivy has been reported to have antibacterial, analgesic, and sedative effects. The compound thought to be responsible for these characteristics, as well as antifungal activity, is *falcarinol.* Falcarinol, isolated from a tree in Panama, also has been reported by the Molecular Targets Drug Discovery Program, to have antitumor activity. Perhaps one day this compound, or a derivative of it, will be useful in treating cancer in humans.

$C_6H_5-C(=O)-C{\equiv}C-C{\equiv}CCH_3$

Capillin

$CH_3-C{\equiv}C-C{\equiv}C-C{\equiv}C-CH{=}CH-C_6H_{10}(OH)$

Ichthyotherol

$CH_2{=}CH-CH(OH)-C{\equiv}C-C{\equiv}C-CH_2-CH{=}CH-(CH_2)_7CH_3$

Falcarinol

$HOCH_2-CH_2-CH_2-CH_2-C{\equiv}C-C{\equiv}C-CH{=}CH-CH{=}CH-CH{=}CH-CH(OH)-CH_2-CH_2-CH_3$

Cicutoxin

Alkynes that exhibit toxic activity.

Cicutoxin has been described as the most lethal toxin native to North America. It is a neurotoxin that is produced by the water hemlock (*Cicuta maculata*), which is in the same family of plants as parsley, celery, and carrots. Cicutoxin is present in all parts of the plants, but is most concentrated in the root. Eating a portion as small as 2–3 cm^2 can be fatal to adults. Cicutoxin acts directly on the nervous system. Signs and symptoms of cicutoxin poisoning include dilation of pupils, muscle twitching, rapid pulse and breathing, violent convulsions, coma, and death. Onset of symptoms is rapid and death may occur within two to three hours. No antidote exists for cicutoxin poisoning. The only treatment involves controlling convulsions and seizures in order to preserve normal heart and lung function. Fortunately, cicutoxin poisoning is a very rare occurrence. Occasionally animals may graze on the plants in the spring, resulting in death within fifteen minutes. Humans seldom come into contact with the water hemlock. The most recent cases have involved individuals foraging for wild ginseng, or other wild roots, and mistaking the water hemlock root for an edible plant.

Cicuta maculata, or water hemlock, produces the most deadly toxin indigenous to North America.

Alkenes with many double bonds are often referred to as polyenes (*poly*—many *enes*—double bonds).

- Alkenes having more than one double bond are called alkadienes (two double bonds) or alkatrienes (three double bonds), as seen in these examples:

$CH_3CH{=}CH{-}CH{=}CHCH_3$
2,4-Hexadiene

$CH_2{=}CHCH_2CH{=}CH_2$
1,4-Pentadiene

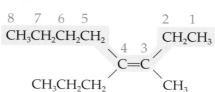

3-Methyl-1,
4-cyclohexadiene

EXAMPLE 12.1 Naming Alkenes and Alkynes Using the I.U.P.A.C. Nomenclature System

Learning Goal 2

Name the following alkene and alkyne using I.U.P.A.C. nomenclature.

Solution

$CH_3CH_2CH_2CH_2(C8{-}C5)(CH_3CH_2CH_2)C{=}C(CH_3)CH_2CH_3$ (carbons numbered 8 7 6 5 4 3 2 1)

Longest chain containing the double bond: octene

Position of double bond: 3-octene (*not* 5-octene)

Substituents: 3-methyl and 4-propyl

Name: 3-Methyl-4-propyl-3-octene

$CH_3CH_2{-}C{\equiv}C{-}C(CH_3)_2{-}CH_3$ (carbons numbered 6 5 4 3 2 1)

Longest chain containing the triple bond: hexyne

Position of triple bond: 3-hexyne (must be!)

Substituents: 2,2-dimethyl

Name: 2,2-Dimethyl-3-hexyne

EXAMPLE 12.2 Naming Cycloalkenes Using I.U.P.A.C. Nomenclature

Learning Goal 2

Name the following cycloalkenes using I.U.P.A.C. nomenclature.

Solution

Parent chain: cyclohexene

Position of double bond: carbon-1 (carbons of the double bond are numbered 1 and 2)

Substituents: 4-chloro

Name: 4-Chlorocyclohexene

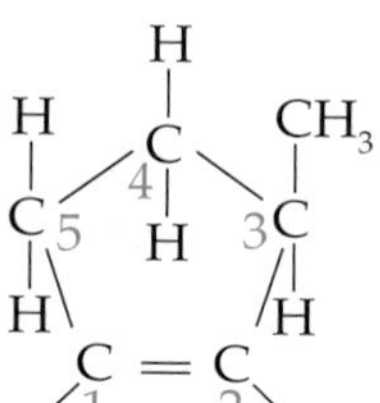

Parent chain: cyclopentene

Position of double bond: carbon-1

Substituent: 3-methyl

Name: 3-Methylcyclopentene

Question 12.1

Draw a complete structural formula for each of the following compounds:

a. 1-Bromo-3-hexyne
b. 2-Butyne
c. Dichloroethyne
d. 9-Iodo-1-nonyne

Question 12.2

Name the following compounds using the I.U.P.A.C. Nomenclature System:

a. $CH_3—C≡C—CH_2CH_3$

b.
```
CH3CH2CHCHCH2C≡CH
      |  |
      Br Br
```

c.
```
           Br CH3
           |  |
CH3CH—C=C—CHCH3
     |        |
     CH3      CH3
```

d.
```
     CH2CH3
     |
CH3CH—C≡C—CHCH3
            |
            Br
```

12.3 Geometric Isomers: A Consequence of Unsaturation

Learning Goal 

Restricted rotation around double bonds is partially responsible for the conformation and hence the activity of many biological molecules that we will study later.

The carbon-carbon double bond is rigid because of the shapes of the orbitals involved in its formation. The electrons of one of the two carbon-carbon bonds lie in a line between the two nuclei. This is called a sigma (σ) bond. The second bond is formed between two *p* orbital electrons and is called a pi (π) bond. The two electrons of the π bond lie in the region above and below the two carbon atoms as shown in the following diagram.

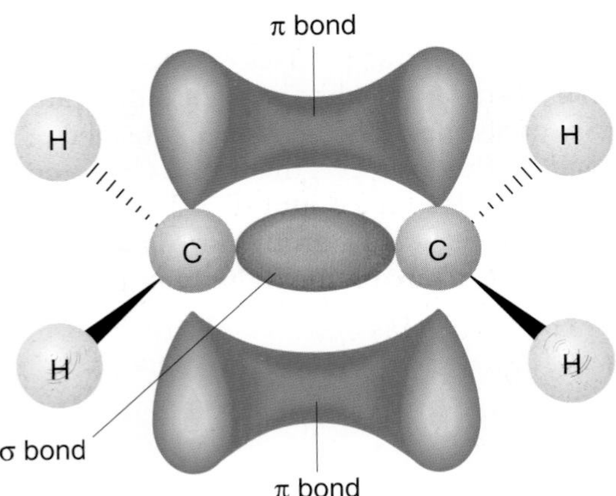

The electron charge cloud associated with the two electrons making up the σ bond (in red) is concentrated between the two nuclei. The electron charge cloud associated with the two electrons of the π bond (in blue) is concentrated in two regions above and below the σ bond framework of the molecule.

Rotation around the double bond is restricted because the π bond would have to be broken to allow rotation. Thus, the double bond is rigid.

In Section 11.3, we observed that the rotation around the carbon-carbon bonds of cycloalkanes was restricted. The consequence of the absence of free rotation was the formation of geometric or *cis-trans* isomers. The *cis* isomers of cycloalkanes had substituent groups on the same side of the ring (L., *cis*, "on the same side"). The *trans* isomers of cycloalkanes had substituent groups located on opposite sides of the ring (L., *trans*, "across from").

In the alkenes, **geometric isomers** differ from one another by the location of groups on the same or opposite sides of the double bond. Because the double bond of the alkenes is also rigid and there is no free rotation around it, geometric isomers are formed when there are two different groups on each of the carbon atoms attached by the double bond. If both groups are on the same side of the double bond, the molecule is a *cis* isomer. If the groups are on opposite sides of the double bond, the molecule is a *trans* isomer.

Consider the two isomers of 1,2-dichloroethene:

```
H       H          H       Cl
 \     /            \     /
  C=C                C=C
 /     \            /     \
Cl      Cl         Cl      H
```

cis-1, 2-Dichloroethene   *trans*-1, 2-Dichloroethene

If one of the two carbon atoms of the double bond has two identical substituents, there are no *cis-trans* isomers for that molecule. Consider the example of 1,1-dichloroethene:

```
H       Cl
 \     /
  C=C
 /     \
H       Cl
```

1, 1-Dichloroethene

EXAMPLE 12.3 *Identifying* cis *and* trans *Isomers of Alkenes*

Learning Goal

Two isomers of 2-butene are shown below. Which is the *cis* isomer and which is the *trans* isomer?

```
  H       H             H       CH3
   \     /               \     /
    C=C                   C=C
   /     \               /     \
H3C       CH3         H3C       H
```

Solution

As we saw with cycloalkanes, the prefixes *cis* and *trans* refer to the placement of the substituents attached to a bond that cannot undergo free rotation. In the case of alkenes, it is the groups attached to the carbon-carbon double bond (in this example, the H and CH_3 groups). When the groups are on the same side of the double bond, as in the structure on the left, the prefix *cis* is used. When the groups are on the opposite sides of the double bond, as in the structure on the right, *trans* is the appropriate prefix.

```
  H       H             H       CH3
   \     /               \     /
    C=C                   C=C
   /     \               /     \
H3C       CH3         H3C       H
```

cis-2-Butene   *trans*-2-Butene

EXAMPLE 12.4 *Naming* cis *and* trans *Compounds*

Learning Goal

Name the following geometric isomers.

Continued—

EXAMPLE 12.4 —*Continued*

Solution

The longest chain of carbon atoms in each of the following molecules is highlighted in yellow. *The chain must also contain the carbon-carbon double bond.* The location of functional groups relative to the double bond is used in determining the appropriate prefix, *cis* or *trans*, to be used in naming each of the molecules.

Parent chain: heptene

Position of double bond: 3-

Configuration: *trans*

Substituents: 3,4-dichloro

Name: *trans*-3,4-Dichloro-3-heptene

Cl, CH_2CH_3 / C=C / $CH_3CH_2CH_2$, Cl

Parent chain: octene

Position of double bond: 3-

Configuration: *cis*

Substituents: 3,4-dimethyl

Name: *cis*-3,4-Dimethyl-3-octene

CH_3CH_2, $CH_2CH_2CH_2CH_3$ / C=C / CH_3, CH_3

Question 12.3

In each of the following pairs of molecules, identify the *cis* isomer and the *trans* isomer.

a. H, H / C=C / CH_3CH_2, CH_2CH_3 H, CH_2CH_3 / C=C / CH_3CH_2, H

b. Br, CH_3 / C=C / H_3C, Br   Br, Br / C=C / H_3C, CH_3

Question 12.4

Provide the complete I.U.P.A.C. name for each of the compounds in Question 12.3.

EXAMPLE 12.5

Identifying Geometric Isomers

Determine whether each of the following molecules can exist as *cis-trans* isomers: (1) 1-pentene, (b) 3-ethyl-3-hexene, and (c) 3-methyl-2-pentene.

Solution

a. Examine the structure of 1-pentene,

H, $CH_2CH_2CH_3$ / C=C / H, H

Continued—

EXAMPLE 12.5 —*Continued*

We see that carbon-1 is bonded to two hydrogen atoms, rather than to two different substituents. In this case there can be no *cis-trans* isomers.

b. Examine the structure of 3-ethyl-3-hexene:

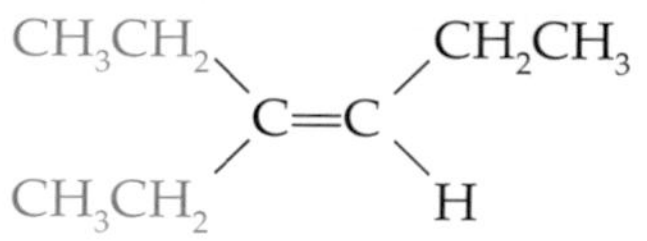

$$\begin{array}{ccc} CH_3CH_2 & & CH_2CH_3 \\ & C{=}C & \\ CH_3CH_2 & & H \end{array}$$

We see that one of the carbons of the carbon-carbon double bond is bonded to two ethyl groups. As in example (a), because this carbon is bonded to two identical groups, there can be no *cis* or *trans* isomers of this compound.

c. Finally, examination of the structure of 3-methyl-2-pentene reveals that both a *cis* and *trans* isomer can be drawn.

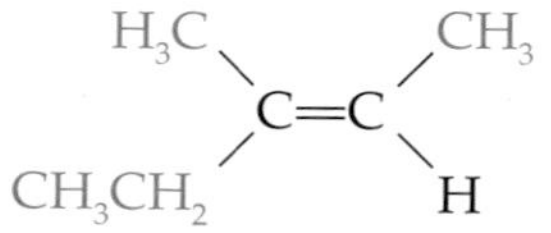

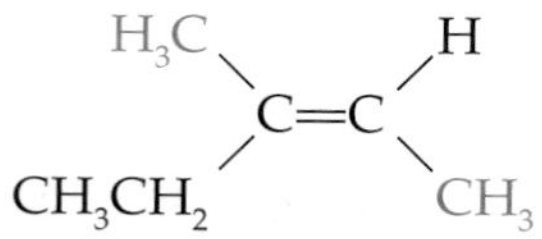

$$\begin{array}{ccc} H_3C & & CH_3 \\ & C{=}C & \\ CH_3CH_2 & & H \end{array} \qquad \begin{array}{ccc} H_3C & & H \\ & C{=}C & \\ CH_3CH_2 & & CH_3 \end{array}$$

Each of the carbon atoms involved in the double bond is attached to two different groups. As a result, we can determine which is the *cis* isomer and which is the *trans* isomer based on the positions of the methyl groups relative to the double bond.

Question 12.5

Draw condensed formulas for each of the following compounds:

a. *cis*-3-Octene
b. *trans*-5-Chloro-2-hexene
c. *trans*-2,3-Dichloro-2-butene

Question 12.6

Name each of the following compounds, using the I.U.P.A.C. system. Be sure to indicate *cis* or *trans* where applicable.

a. 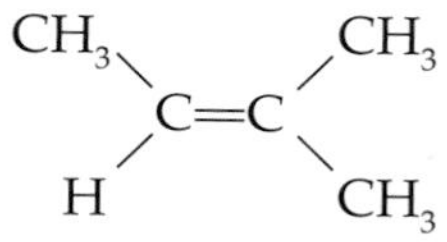 $$\begin{array}{ccc} CH_3 & & CH_3 \\ & C{=}C & \\ H & & CH_3 \end{array}$$

b. $$\begin{array}{ccc} CH_3CH_2 & & CH_2CH_3 \\ & C{=}C & \\ CH_3 & & H \end{array}$$

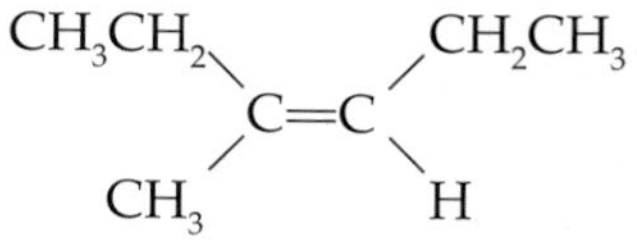

c. $$\begin{array}{ccc} CH_3 & & H \\ & C{=}C & \\ H & & CH_2C(CH_3)_3 \end{array}$$

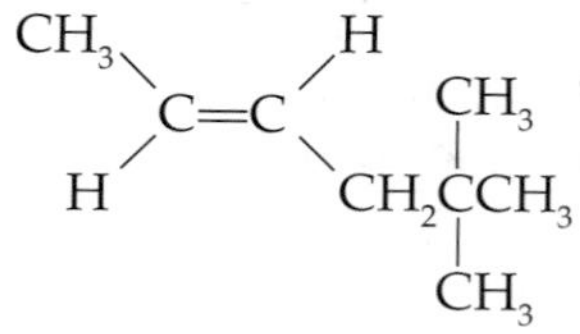

12.4 Alkenes in Nature

Folklore tells us that placing a ripe banana among green tomatoes will speed up the ripening process. In fact, this phenomenon has been demonstrated experimentally. The key to the reaction is *ethene,* the simplest alkene. Ethene, produced by ripening fruit, is a plant growth substance. It is produced in the greatest abundance in areas of the plant where cell division is occurring. It is produced during fruit ripening, during leaf fall and flower senescence, as well as under conditions of stress, including wounding, heat, cold, or water stress, and disease.

There are a surprising number of polyenes, alkenes with several double bonds, found in nature. These molecules, which have wildly different properties and functions, are built from one or more five-carbon units called *isoprene.*

$$\begin{array}{c} \quad CH_3 \\ \quad | \\ CH_2{=}C{-}CH{=}CH_2 \end{array}$$

Isoprene

The molecules that are produced are called *isoprenoids,* or *terpenes* (Figure 12.3). Terpenes include the steroids; chlorophyll and carotenoid pigments that function in photosynthesis, and the lipid soluble vitamins A, D, and K (Figure 12.1).

Many other terpenes are plant products familiar to us because of their distinctive aromas. *Geraniol,* the familiar scent of geraniums, is a molecule made up of two isoprene units. Purified from plant sources, geraniol is the active ingredient in several natural insect repellants. These can be applied directly to the skin to provide four hours of protection against a variety of insects, including mosquitoes, ticks, and fire ants.

D-Limonene is the most abundant component of the oil extracted from the rind of citrus fruits. Because of its pleasing orange aroma, D-limonene is used as a flavor and fragrance additive in foods. However, the most rapidly expanding use of the compound is as a solvent. In this role, D-limonene can be used in place of more toxic solvents, such as mineral spirits, methyl ethyl ketone, acetone, toluene, and fluorinated and chlorinated organic solvents. It can also be formulated as a water-based cleaning product, such as Orange Glo, that can be used in place of more caustic cleaning solutions. There is a form of limonene that is a molecular mirror image of D-limonene. It is called L-limonene and has a pine or turpentine aroma.

The terpene *myrcene* is found in bayberry. It is used in perfumes and scented candles because it adds a refreshing, spicy aroma to them. Trace amounts of myrcene may be used as a flavor component in root beer.

Farnesol is a terpene found in roses, orange blossom, wild cyclamen, and lily of the valley. Cosmetics companies began to use farnesol in skin care products in the early 1990s. It is claimed that farnesol smoothes wrinkles and increases skin elasticity. It is also thought to reduce skin aging by promoting regeneration of cells and activation of the synthesis of molecules, such as collagen, that are required for healthy skin.

Another terpene, *retinol,* is a form of vitamin A (Figure 12.1). It is able to penetrate the outer layers of skin and stimulate the formation of collagen and elastin. This reduces wrinkles by creating skin that is firmer and smoother.

12.5 Reactions Involving Alkenes

Learning Goal

Reactions of alkenes involve the carbon-carbon double bond. The key reaction of the double bond is the **addition reaction.** This involves the addition of two atoms or groups of atoms to a double bond. The major alkene addition reactions include addition of hydrogen (H_2), halogens (Cl_2 or Br_2), water (HOH), or hydrogen halides (HBr or HCl). A generalized addition reaction is:

$$\begin{array}{c} R \quad R \\ \diagdown \ \diagup \\ C \\ \| \\ C \\ \diagup \ \diagdown \\ R \quad R \end{array} + \begin{array}{c} A \\ | \\ B \end{array} \longrightarrow \begin{array}{c} R \\ | \\ R{-}C{-}A \\ | \\ R{-}C{-}B \\ | \\ R \end{array}$$

Geraniol
(Rose and geraniums)

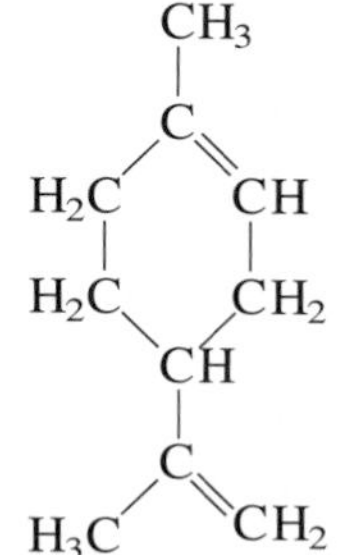

Limonene
(Oil of lemon and orange)

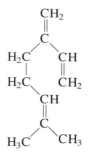

Myrcene
(Oil of bayberry)

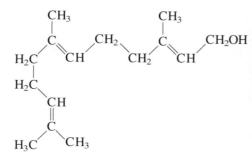

Farnesol
(Lily of the valley)

Figure 12.3
Many plant products, familiar to us because of their distinctive aromas, are isoprenoids, which are alkenes having several double bonds.

Note that the double bond is replaced by a single bond. The former double bond carbons receive a new single bond to a new atom, producing either an alkane or a substituted alkane. This involves breaking the π bond between the two carbons of the double bond and forming a new σ bond to each of these carbons.

Hydrogenation: Addition of H_2 to an Alkene

Hydrogenation is the addition of a molecule of hydrogen (H_2) to a carbon-carbon double bond to give an alkane. In this reaction the double bond is broken, and two new C—H single bonds result. Platinum, palladium, or nickel is needed as a catalyst to speed up the reaction. Heat and/or pressure may also be required.

Recall that a catalyst itself undergoes no net change in the course of a chemical reaction (see Section 8.3).

$$R_2C{=}CR_2 + H{-}H \xrightarrow[\text{Heat or pressure}]{\text{Pt, Pd, or Ni}} R_2CH{-}CHR_2$$

Alkene   Hydrogen   Alkane

Remember that the R in these general formulas represents an alkyl group.

Note that the alkene is gaining two hydrogens. Thus, hydrogenation is a reduction reaction (see Sections 9.5 and 13.6).

EXAMPLE 12.6

Writing Equations for the Hydrogenation of Alkenes

Learning Goal

Write a balanced equation showing the hydrogenation of (a) 1-pentene and (b) *trans*-2-pentene.

Solution

(a) Begin by drawing the structure of 1-pentene and of diatomic hydrogen (H_2) and indicating the catalyst.

$$CH_3CH_2CH_2(H)C{=}CH_2 + H{-}H \xrightarrow{\text{Ni}}$$

1-Pentene   Hydrogen

Knowing that one hydrogen atom will form a covalent bond with each of the carbon atoms of the carbon-carbon double bond, we can write the product and complete the equation.

$$CH_3CH_2CH_2(H)C{=}CH_2 + H{-}H \xrightarrow{\text{Ni}} CH_3CH_2CH_2{-}CH_2{-}CH_3$$

1-Pentene   Hydrogen   Pentane

(b) Begin by drawing the structure of *trans*-2-pentene and of diatomic hydrogen (H_2) and indicating the catalyst.

$$CH_3CH_2(H)C{=}C(H)CH_3 + H{-}H \xrightarrow{\text{Ni}}$$

trans-2-Pentene   Hydrogen

Continued—

EXAMPLE 12.6 —*Continued*

Knowing that one hydrogen atom will form a covalent bond with each of the carbon atoms of the carbon-carbon double bond, we can write the product and complete the equation.

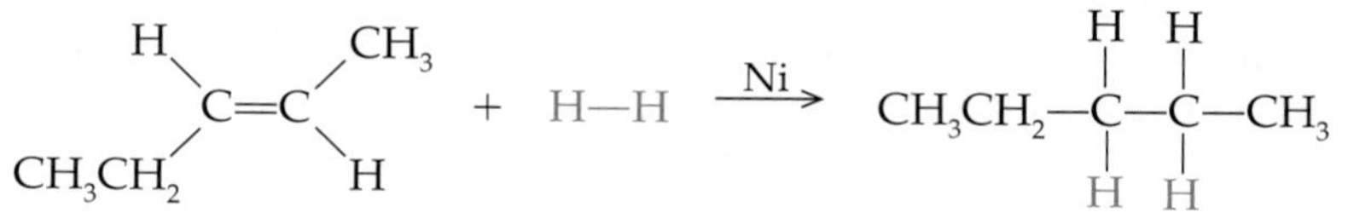

$$\underset{\textit{trans}\text{-2-Pentene}}{(H)(CH_3CH_2)C{=}C(CH_3)(H)} + \underset{\text{Hydrogen}}{H{-}H} \xrightarrow{Ni} \underset{\text{Pentane}}{CH_3CH_2{-}CH_2{-}CH_2{-}CH_3}$$

Question 12.7

The *trans* isomer of 2-pentene was used in Example 12.6. Would the result be any different if the *cis* isomer had been used?

Question 12.8

Write balanced equations for the hydrogenation of 1-butene and *cis*-2-butene.

Saturated and unsaturated dietary fats are discussed in Section 18.2.

Hydrogenation is used in the food industry to produce margarine, which is a mixture of hydrogenated vegetable oils (Figure 12.4). Vegetable oils are unsaturated, that is, they contain many double bonds and as a result have low melting points and are liquid at room temperature. The hydrogenation of these double bonds to single bonds increases the melting point of these oils and results in a fat, such as Crisco, that remains solid at room temperature. Through further processing they may be converted to margarine, such as corn oil or sunflower oil margarines.

Halogenation: Addition of X_2 to an Alkene

Chlorine (Cl_2) or bromine (Br_2) can be added to a double bond. This reaction, called **halogenation,** proceeds readily and does not require a catalyst:

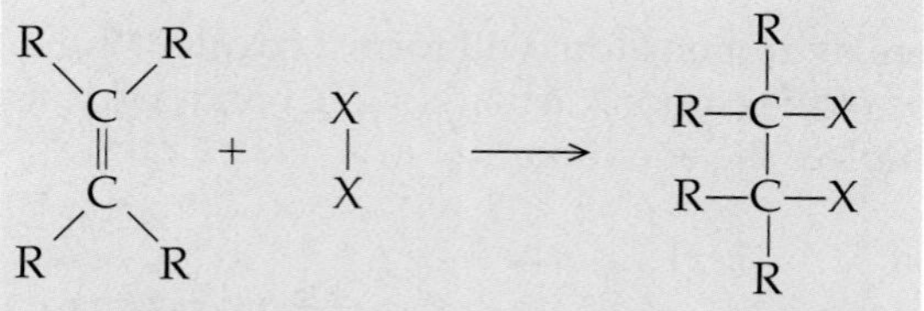

$$\underset{\text{Alkene}}{R_2C{=}CR_2} + \underset{\text{Halogen}}{X{-}X} \longrightarrow \underset{\text{Alkyl dihalide}}{R{-}CR(X){-}CR(X){-}R}$$

Figure 12.4
The conversion of a typical oil to a fat involves hydrogenation. In this example, triolein (an oil) is converted to tristearin (a fat).

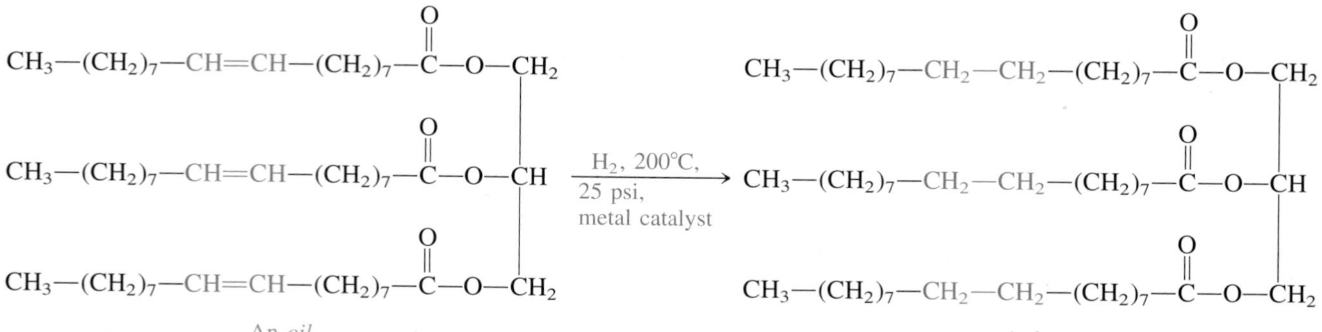

$$\underset{\text{An } \textit{oil}}{\begin{matrix} CH_3{-}(CH_2)_7{-}CH{=}CH{-}(CH_2)_7{-}\overset{O}{\overset{\|}{C}}{-}O{-}CH_2 \\ CH_3{-}(CH_2)_7{-}CH{=}CH{-}(CH_2)_7{-}\overset{O}{\overset{\|}{C}}{-}O{-}CH \\ CH_3{-}(CH_2)_7{-}CH{=}CH{-}(CH_2)_7{-}\overset{O}{\overset{\|}{C}}{-}O{-}CH_2 \end{matrix}} \xrightarrow[\text{25 psi, metal catalyst}]{H_2,\ 200^\circ C,} \underset{\text{A } \textit{fat}}{\begin{matrix} CH_3{-}(CH_2)_7{-}CH_2{-}CH_2{-}(CH_2)_7{-}\overset{O}{\overset{\|}{C}}{-}O{-}CH_2 \\ CH_3{-}(CH_2)_7{-}CH_2{-}CH_2{-}(CH_2)_7{-}\overset{O}{\overset{\|}{C}}{-}O{-}CH \\ CH_3{-}(CH_2)_7{-}CH_2{-}CH_2{-}(CH_2)_7{-}\overset{O}{\overset{\|}{C}}{-}O{-}CH_2 \end{matrix}}$$

Writing Equations for the Halogenation of Alkenes

EXAMPLE 12.7

Learning Goal

Write a balanced equation showing (a) the chlorination of 1-pentene and (b) the bromination of *trans*-2-butene.

Solution

(a) Begin by drawing the structure of 1-pentene and of diatomic chlorine (Cl_2).

$$CH_3CH_2CH_2(H)C{=}CH_2 + Cl{-}Cl \longrightarrow$$

1-Pentene Chlorine

Knowing that one chlorine atom will form a covalent bond with each of the carbon atoms of the carbon-carbon double bond, we can write the product and complete the equation.

$$CH_3CH_2CH_2(H)C{=}CH_2 + Cl{-}Cl \longrightarrow CH_3CH_2CH_2{-}\underset{Cl}{\overset{H}{C}}{-}\underset{Cl}{\overset{H}{C}}{-}H$$

1-Pentene Chlorine 1,2-Dichloropentane

(b) Begin by drawing the structure of *trans*-2-butene and of diatomic bromine (Br_2).

$$H_3C(H)C{=}C(H)CH_3 + Br{-}Br \longrightarrow$$

trans-2-Butene Bromine

Knowing that one bromine atom will form a covalent bond with each of the carbon atoms of the carbon-carbon double bond, we can write the product and complete the equation.

$$H_3C(H)C{=}C(H)CH_3 + Br{-}Br \longrightarrow CH_3{-}\underset{Br}{\overset{H}{C}}{-}\underset{Br}{\overset{H}{C}}{-}CH_3$$

trans-2-Butene Bromine 2,3-Dibromobutane

Below we see an equation representing the bromination of 1-pentene. Notice that the solution of reactants is red because of the presence of bromine. However, the product is colorless (Figure 12.5).

$$CH_3CH_2CH_2CH{=}CH_2 + Br_2 \longrightarrow CH_3CH_2CH_2\underset{Br}{CH}\underset{Br}{CH_2}$$

1-Pentene (colorless) Bromine (red) 1,2-Dibromopentane (colorless)

This bromination reaction can be used to show the presence of double bonds in an organic compound. The reaction mixture is red because of the presence of dissolved bromine. If the red color is lost, the bromine has been consumed. Thus bromination has occurred, and the compound must have had a carbon-carbon

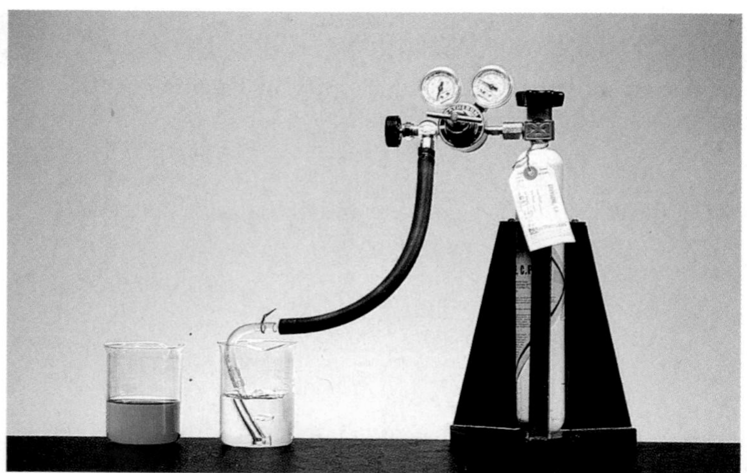

Figure 12.5
Bromination of an alkene. The solution of reactants is red because of the presence of bromine. When the bromine has been used in the reaction, the solution becomes colorless.

double bond. The greater the amount of bromine that must be added to the reaction, the more unsaturated the compound is, that is, the greater the number of carbon-carbon double bonds.

Hydration: Addition of H_2O to an Alkene

A water molecule can be added to an alkene. This reaction, termed **hydration,** requires a trace of acid (H^+) as a catalyst. The product is an alcohol, as shown in the following reaction:

$$\underset{\text{Alkene}}{R_2C{=}CR_2} + \underset{\text{Water}}{H{-}OH} \xrightarrow{H^+} \underset{\text{Alcohol}}{R_2C(H){-}C(OH)R_2}$$

The following equation shows the hydration of ethene to produce ethanol.

$$\underset{\text{Ethene}}{H_2C{=}CH_2} + \underset{\text{Water}}{H{-}OH} \xrightarrow{H^+} \underset{\text{Ethanol (ethyl alcohol)}}{H_3C{-}CH_2{-}OH}$$

Learning Goal 5

With alkenes in which the groups attached to the two carbons of the double bond are different (unsymmetrical alkenes), two products are possible. For example:

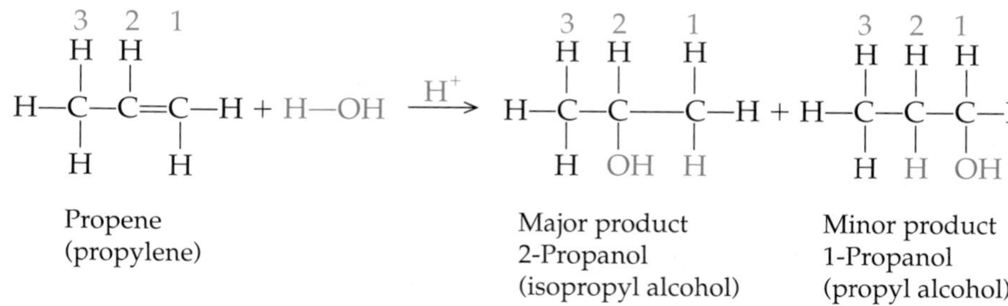

A HUMAN PERSPECTIVE

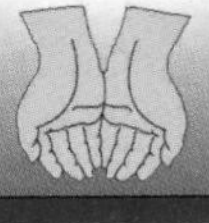

Folklore, Science, and Technology

For many years it was suspected that there existed a gas that stimulated fruit ripening and had other effects on plants. The ancient Chinese observed that their fruit ripened more quickly if incense was burned in the room. Early in this century, shippers realized that they could not store oranges and bananas on the same ships because some "emanation" given off by the oranges caused the bananas to ripen too early.

Puerto Rican pineapple growers and Philippine mango growers independently developed a traditional practice of building bonfires near their crops. They believed that the smoke caused the plants to bloom synchronously.

In the mid–nineteenth century, streetlights were fueled with natural gas. Occasionally the pipes leaked, releasing gas into the atmosphere. On some of these occasions, the leaves fell from all the shade trees in the region surrounding the gas leak.

What is the gas responsible for these diverse effects on plants? In 1934, R. Gane demonstrated that the simple alkene ethylene was the "emanation" responsible for fruit ripening. More recently, it has been shown that ethylene induces and synchronizes flowering in pineapples and mangos, induces senescence (aging) and loss of leaves in trees, and effects a wide variety of other responses in various plants.

We can be grateful to ethylene for the fresh, unbruised fruits that we can purchase at the grocery store. These fruits are picked when they are not yet ripe, while they are still firm. They then can be shipped great distances and gassed with ethylene when they reach their destination. Under the influence of the ethylene, the fruit ripens and is displayed in the store.

The history of the use of ethylene to bring fresh ripe fruits and vegetables to markets thousands of miles from the farms is an interesting example of the scientific process and its application for the benefit of society. Scientists began with the curious observations of Chinese, Puerto Rican, and Filipino farmers. Through experimentation they came to understand the phenomenon that caused the observations. Finally, through technology, scientists have made it possible to harness the power of ethylene so that grocers can "artificially" ripen the fruits and vegetables they sell to us.

When hydration of an unsymmetrical alkene, such as propene, is carried out in the laboratory, one product (2-propanol) is favored over the other. The Russian chemist Vladimir Markovnikov studied many such reactions and came up with a rule that can be used to predict the major product of such a reaction. **Markovnikov's rule** tells us that the carbon of the carbon-carbon double bond that originally has more hydrogen atoms receives the hydrogen atom being added to the double bond. The remaining carbon forms a bond with the —OH. Simply stated, "the rich get richer"—the carbon with the most hydrogens gets the new one as well. In the preceding example, carbon-1 has two C—H bonds originally, and carbon-2 has only one. The major product, 2-propanol, results from the new C—H bond forming on carbon-1 and the new C—OH bond on carbon-2.

Addition of water to a double bond is a reaction that we find in several biochemical pathways. For instance, the citric acid cycle is a key metabolic pathway for the complete oxidation of the sugar glucose and the release of the majority of the energy used by the body. It is also the source of starting materials for the synthesis of the biological molecules needed for life. The next-to-last reaction in the citric acid cycle is the hydration of a molecule of fumarate to produce a molecule called malate.

We have seen that hydration of a double bond requires a trace of acid as a catalyst. In the cell, this reaction is catalyzed by an enzyme, or biological catalyst, called fumarase.

$$\underset{\text{Fumarate}}{{}^{-}OOC{-}CH{=}CH{-}COO^{-}} + H_2O \xrightarrow{\text{Fumarase}} \underset{\text{Malate}}{{}^{-}OOC{-}CH(OH){-}CH_2{-}COO^{-}}$$

Hydration of a double bond also occurs in the β-oxidation pathway (see Section 23.2). This pathway carries out the oxidation of dietary fatty acids. Like the citric acid cycle, β-oxidation harvests the energy of the food molecules to use as fuel for body functions.

EXAMPLE 12.8 Writing Equations for the Hydration of Alkenes

Learning Goal

Learning Goal

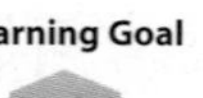

Write an equation showing all the products of the hydration of 1-pentene.

Solution

Begin by drawing the structure of 1-pentene and of water and indicating the catalyst.

$$\underset{\text{1-Pentene}}{CH_3CH_2CH_2CH{=}CH_2} + \underset{\text{Water}}{H{-}OH} \xrightarrow{H^+}$$

Markovnikov's rule tells us that the carbon atom that is already bonded to the greater number of hydrogen atoms is more likely to receive the hydrogen atom from the water molecule. The other carbon atom is more likely to become bonded to the hydroxyl group. Thus we can predict that the major product of this reaction will be 2-pentanol and that the minor product will be 1-pentanol. Now we write the equation showing the reactants and products:

$$\underset{\text{1-Pentene}}{CH_3CH_2CH_2CH{=}CH_2} + \underset{\text{Water}}{H{-}OH} \xrightarrow{H^+} \underset{\text{2-Pentanol (major product)}}{CH_3CH_2CH_2{-}CH(OH){-}CH_2{-}H} + \underset{\text{1-Pentanol (minor product)}}{CH_3CH_2CH_2{-}CH_2{-}CH(OH){-}H}$$

Question 12.9

Write a balanced equation for the hydration of each of the following alkenes. Predict the major product of each of the reactions.

a. $CH_3CH{=}CHCH_3$

b. $CH_2{=}CHCH_2CH_2CH(CH_3)CH_3$

c. $CH_3CH_2CH_2CH{=}CHCH_2CH_3$

d. $CH_3CHClCH{=}CHCHClCH_3$

Question 12.10

Write a balanced equation for the hydration of each of the following alkenes. Predict the major product of each of the reactions.

a. $CH_2{=}CHCH_2CH{=}CH_2$

b. $CH_3CH_2CH_2CH{=}CHCH_3$

c. $CH_3CHBrCH_2CH{=}CHCH_2Cl$

d. $CH_3CH_2CH_2CH_2CH_2CH{=}CHCH_3$

Hydrohalogenation: Addition of HX to an Alkene

A hydrogen halide (HBr, HCl, or HI) also can be added to an alkene. The product of this reaction, called **hydrohalogenation,** is an alkyl halide:

$$R_2C{=}CR_2 + H{-}X \longrightarrow R_2CH{-}CR_2X$$

Alkene Hydrogen halide Alkyl halide

$$H_2C{=}CH_2 + H{-}Br \longrightarrow H_3C{-}CH_2Br$$

Ethene Hydrogen bromide Bromoethane

This reaction also follows Markovnikov's rule. That is, if HX is added to an unsymmetrical alkene, the hydrogen atom will be added preferentially to the carbon atom that originally had the most hydrogen atoms. Consider the following example:

$$CH_3{-}CH{=}CH_2 + H{-}Br \longrightarrow CH_3{-}CHBr{-}CH_3 + CH_3{-}CH_2{-}CH_2Br$$

Propene; Major product 2-Bromopropane; Minor product 1-Bromopropane

Writing Equations for the Hydrohalogenation of Alkenes

EXAMPLE 12.9

Learning Goal

Learning Goal

Write an equation showing all the products of the hydrohalogenation of 1-pentene with HCl.

Solution

Begin by drawing the structure of 1-pentene and of hydrochloric acid.

$$CH_3CH_2CH_2(H)C{=}CH_2 + H{-}Cl \longrightarrow$$

1-Pentene Hydrochloric acid

Markovnikov's rule tells us that the carbon atom that is already bonded to the greater number of hydrogen atoms is more likely to receive the hydrogen atom of the hydrochloric acid molecule. The other carbon atom is more likely to become bonded to the chlorine atom. Thus we can predict

Continued—

EXAMPLE 12.9 —*Continued*

that the major product of this reaction will be 2-chloropentane and that the minor product will be 1-chloropentane. Now we write the equation showing the reactants and products:

$$\underset{\text{1-Pentene}}{CH_3CH_2CH_2(H)C{=}CH_2} + \underset{\text{Hydrochloric acid}}{H{-}Cl} \longrightarrow \underset{\substack{\text{2-Chloropentane}\\\text{(major product)}}}{CH_3CH_2CH_2{-}CH(Cl){-}CH_2{-}H} + \underset{\substack{\text{1-Chloropentane}\\\text{(minor product)}}}{CH_3CH_2CH_2{-}CH_2{-}CH_2(Cl)}$$

Question 12.11

Predict the major product in each of the following reactions. Name the alkene reactant and the product, using I.U.P.A.C. nomenclature.

a. $CH_3(H)C{=}C(H)CH_3$ (both CH_3 groups on the same side) $+ H_2 \xrightarrow{Pd} ?$

b. $CH_3CH_2CH{=}CH_2 + H_2O \xrightarrow{H+} ?$

c. $CH_3CH{=}CHCH_3 + Cl_2 \longrightarrow ?$

d. $CH_3CH_2CH_2CH{=}CH_2 + HBr \longrightarrow ?$

Question 12.12

Predict the major product in each of the following reactions. Name the alkene reactant and the product, using I.U.P.A.C. names.

a. $CH_3(H)C{=}C(H)CH_3$ (CH_3 groups on opposite sides) $+ H_2 \xrightarrow{Ni} ?$

b. $CH_3{-}C(CH_3){=}CHCH_2CH_2CH_3 + H_2O \xrightarrow{H^+} ?$

c. $CH_3C(CH_3){=}CHCH_3 + Br_2 \longrightarrow ?$

d. $CH_3{-}C(CH_3)_2{-}CH{=}CH_2 + HCl \longrightarrow ?$

Addition Polymers of Alkenes

Learning Goal

Polymers are macromolecules composed of repeating structural units called **monomers.** A polymer may be made up of several thousand monomers. Many commercially important plastics and fibers are addition polymers made from alkenes or substituted alkenes. They are called **addition polymers** because they are made by the sequential addition of the alkene monomer. The general formula for this addition reaction follows:

A Human Perspective

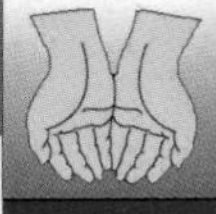

Life without Polymers?

What do Nike Air-Sole shoes, Saturn automobiles, disposable diapers, tires, shampoo, and artificial joints and skin share in common? These products and a great many other items we use every day are composed of synthetic or natural polymers. Indeed, the field of polymer chemistry has come a long way since the 1920s and 1930s when DuPont chemists invented nylon and Teflon.

Consider the disposable diaper. The outer, waterproof layer is composed of polyethylene. The polymerization reaction that produces polyethylene is shown in Section 12.4. The diapers have elastic to prevent leaking. The elastic is made of a natural polymer, rubber. The monomer from which natural rubber is formed is 2-methyl-1,3-butadiene. The common name of this monomer is *isoprene*. As we will see in coming chapters, isoprene is an important monomer in the synthesis of many natural polymers.

$$n\text{CH}_2{=}\underset{|}{\overset{\text{CH}_3}{\text{C}}}{-}\text{CH}{=}\text{CH}_2 \longrightarrow \sim\!\!\left[\text{CH}_2{-}\overset{\text{CH}_3}{\text{C}}{=}\text{CH}{-}\text{CH}_2\right]_n\!\!\sim$$

2-Methyl-1, 3,-butadiene (isoprene) — Rubber polymer

The diaper is filled with a synthetic polymer called poly(acrylic acid). This polymer has the remarkable ability to absorb many times its own weight in liquid. Polymers that have this ability are called superabsorbers, but polymer chemists have no idea why they have this property! The acrylate monomer and resulting poly(acrylic acid) polymer are shown here:

$$n\,\text{CH}_2{=}\text{CH}{-}\text{C}({=}\text{O}){-}\text{O}{-}\text{H} \longrightarrow \sim\!\!\left[\text{CH}_2{-}\text{CH}(\text{C}({=}\text{O}){-}\text{O}{-}\text{H})\right]_n\!\!\sim$$

Acrylate monomer — Poly(acrylic acid)

Another example of a useful polymer is Gore-Tex. This amazing polymer is made by stretching Teflon. Teflon is produced from the monomer tetrafluoroethene, as seen in the following reaction:

$$n\,\text{F}_2\text{C}{=}\text{CF}_2 \longrightarrow \sim\!\!\left[\text{CF}_2{-}\text{CF}_2\right]_n\!\!\sim$$

Tetrafluoroethene — Teflon

Clothing made from this fabric is used to protect firefighters because of its fire resistance. Because it also insulates, Gore-Tex clothing is used by military forces and by many amateur athletes, for protection during strenuous activity in the cold. In addition to its use in protective clothing, Gore-Tex has been used in millions of medical procedures for sutures, synthetic blood vessels, and tissue reconstruction.

To learn more about the fascinating topic of polymer chemistry, visit The Macrogalleria, www.psrc.usm.edu/macrog/index.html, an Internet site maintained by the Department of Polymer Science of the University of Southern Mississippi.

$$n\,\text{R}_2\text{C}{=}\text{CR}_2 \xrightarrow[\text{Pressure}]{\text{Heat}} \text{etc.}{-}\text{CR}_2{-}\text{CR}_2{-}\text{CR}_2{-}\text{CR}_2{-}\text{CR}_2{-}\text{CR}_2{-}\text{etc.}$$

Alkene monomer
R = H, X, or an alkyl group

Addition polymer

The product of the reaction is generally represented in a simplified manner:

$$\sim\!\!\left[\text{CR}_2{-}\text{CR}_2\right]_n\!\!\sim$$

An Environmental Perspective

Plastic Recycling

Plastics, first developed by British inventor Alexander Parkes in 1862, are amazing substances. Some serve as containers for many of our foods and drinks, keeping them fresh for long periods of time. Other plastics serve as containers for detergents and cleansers or are formed into pipes for our plumbing systems. We have learned to make strong, clear sheets of plastic that can be used as windows, and feather-light plastics that can be used as packaging materials. In the United States alone, seventy-five billion pounds of plastics are produced each year.

But plastics, amazing in their versatility, are a mixed blessing. One characteristic that makes them so useful, their stability, has created an environmental problem. It may take forty to fifty years for plastics discarded into landfill sites to degrade. Concern that we could soon be knee-deep in plastic worldwide has resulted in a creative new industry: plastic recycling.

Since there are so many types of plastics, it is necessary to identify, sort, and recycle them separately. To help with this sorting process, manufacturers place recycling symbols on their plastic wares. As you can see in the accompanying table, each symbol corresponds to a different type of plastic.

Polyethylene terephthalate, also known as PETE or simply #1, is a form of polyester often used to make bottles and jars to contain food. When collected, it is ground up into flakes and formed into pellets. The most common use for recycled PETE is the manufacture of polyester carpets. But it may also be spun into a cotton-candy-like form that can be used as a fiber filling for pillows or sleeping bags. It may also be rolled into thin sheets or ribbons and used as tapes for VCRs or tape decks. Reuse to produce bottles and jars is also common.

HDPE, or #2, is high-density polyethylene. Originally used for milk and detergent bottles, recycled HDPE is used to produce pipes, plastic lumber, trash cans, or bottles for storage of materials other than food. LDPE, #4, is identical to HDPE chemically, but it is produced in a less-dense, more flexible form. Originally used to produce plastic bags, recycled LDPE is also used to make trash bags, grocery bags, and plastic tubing and lumber.

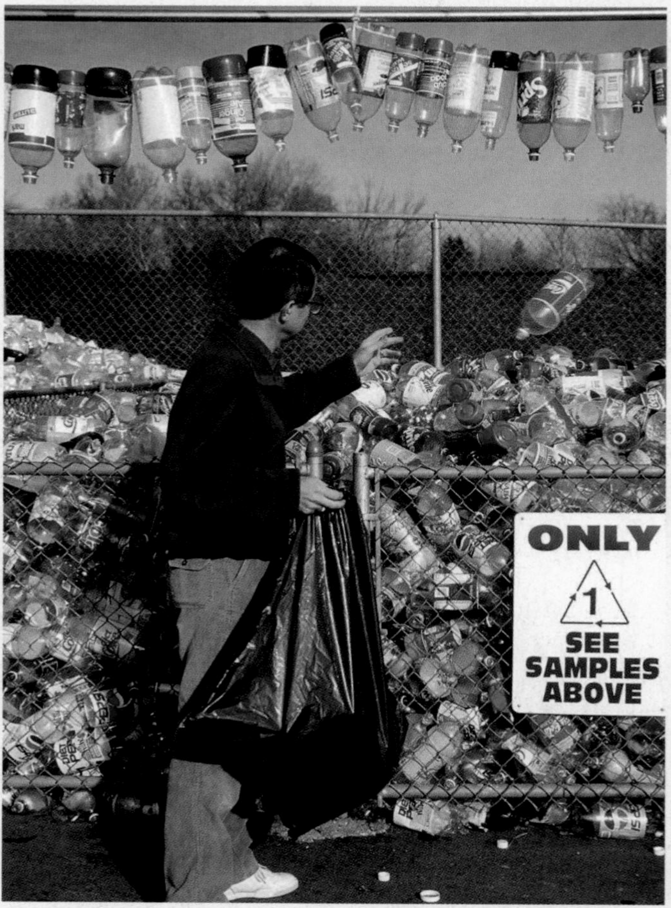

PVC, or #3, is one of the less commonly recycled plastics in the United States, although it is actively recycled in Europe. The recycled material is used to make non-food-bearing containers, shoe soles, flooring, sweaters, and pipes. Polypropylene, PP or

Polyethylene is a polymer made from the monomer ethylene (ethene):

$$n\mathrm{CH_2{=}CH_2} \longrightarrow \sim\!\sim\!\left[\mathrm{CH_2{-}CH_2}\right]_n\!\sim\!\sim$$

Ethene (ethylene) → Polyethylene

It is used to make bottles, injection-molded toys and housewares, and wire coverings.

Polypropylene is a plastic made from propylene (propene). It is used to make indoor-outdoor carpeting, packaging materials, toys, and housewares. When

#5, is found in margarine tubs, fabrics, and carpets. Recycled polypropylene has many uses, including fabrication of gardening implements.

You probably come into contact with polystyrene, PS or #6, almost every day. It is used to make foam egg cartons and meat trays, serving containers for fast food chains, CD "jewel boxes," and "peanuts" used as packing material. At the current time, polystyrene food containers are not recycled. PS from nonfood products can be melted down and converted into pellets that are used to manufacture office desktop accessories, hangers, video and audio cassette housings, and plastic trays used to hold plants.

Code	Type	Name	Formula	Description	Examples
PETE	♳ 1	Polyethylene terephthalate	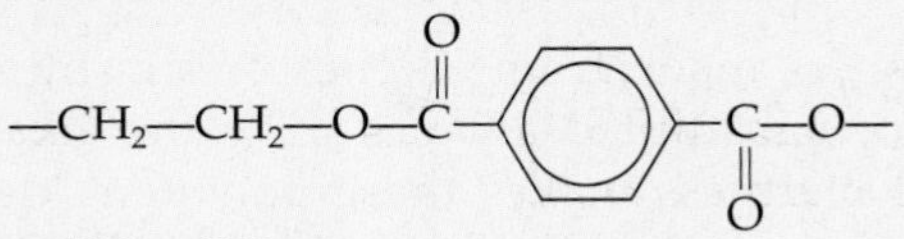 $-CH_2-CH_2-O-\overset{O}{\overset{\Vert}{C}}-C_6H_4-\underset{\underset{O}{\Vert}}{C}-O-$	Usually clear or green, rigid	Soda bottles, peanut butter jars, vegetable oil bottles
HDPE	♴ 2	High-density polyethylene	$-CH_2-CH_2-$	Semirigid	Milk and water jugs, juice and bleach bottles
PVC	♵ 3	Polyvinyl chloride	$-\underset{\underset{Cl}{\vert}}{CH}-CH_2-$	Semirigid	Detergent and cleanser bottles, pipes
LDPE	♶ 4	Low-density polyethylene	$-CH_2-CH_2-$	Flexible, not crinkly	Six-pack rings, bread bags, sandwich bags
PP	♷ 5	Polypropylene	$-\underset{\underset{CH_3}{\vert}}{CH}-CH_2-$	Semirigid	Margarine tubs, straws, screw-on lids
PS	♸ 6	Polystyrene	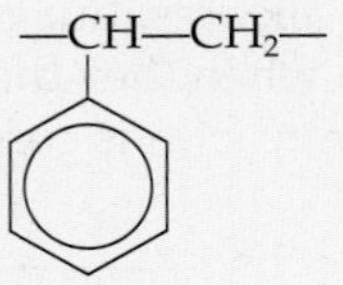 $-\underset{\underset{C_6H_5}{\vert}}{CH}-CH_2-$	Often brittle	Styrofoam, packing peanuts, egg cartons, foam cups
Other	♹ 7	Multilayer plastics	N/A	Squeezable	Ketchup and syrup bottles

propylene polymerizes, a methyl group is located on every other carbon of the main chain:

$$n\,CH_2{=}\overset{\overset{CH_3}{\vert}}{CH} \longrightarrow \sim\!\!\left[CH_2-\overset{\overset{CH_3}{\vert}}{CH}\right]_n\!\!\sim$$

or

$$\sim CH_2-\overset{\overset{CH_3}{\vert}}{CH}-CH_2-\overset{\overset{CH_3}{\vert}}{CH}-CH_2-\overset{\overset{CH_3}{\vert}}{CH}\sim$$

Table 12.1 Some Important Addition Polymers of Alkenes

Monomer name	Formula	Polymer	Uses
Styrene	$CH_2{=}CH{-}C_6H_5$ (benzene ring)	Polystyrene	Styrofoam containers
Acrylonitrile	$CH_2{=}CHCN$	Polyacrylonitrile (Orlon)	Clothing
Methyl methacrylate	$CH_2{=}C(CH_3){-}\overset{O}{\overset{\|}{C}}OCH_3$	Polymethyl methacrylate (Plexiglas, Lucite)	Basketball backboards
Vinyl chloride	$CH_2{=}CHCl$	Polyvinyl chloride (PVC)	Plastic pipe, credit cards
Tetrafluoroethene	$CF_2{=}CF_2$	Polytetrafluoroethylene (Teflon)	Nonstick surfaces

Polymers made from alkenes or substituted alkenes are simply very large alkanes or substituted alkanes. Like the alkanes, they are typically inert. This chemical inertness makes these polymers ideal for making containers to hold juices, chemicals, and fluids used medically. They are also used to make sutures, catheters, and other indwelling devices. A variety of polymers made from substituted alkenes are listed in Table 12.1.

12.6 Aromatic Hydrocarbons

Learning Goal

In the early part of the nineteenth century chemists began to discover organic compounds having chemical properties quite distinct from the alkanes, alkenes, and alkynes. They called these substances *aromatic compounds* because many of the first examples were isolated from the pleasant-smelling resins of tropical trees. The carbon:hydrogen ratio of these compounds suggested a very high degree of unsaturation, similar to the alkenes and alkynes. Imagine, then, how puzzled these early organic chemists must have been when they discovered that these compounds do not undergo the kinds of addition reactions common for the alkenes and alkynes.

$$CH_2{=}CH_2 + Br_2 \longrightarrow \underset{Br}{\underset{|}{C}}H_2{-}\underset{Br}{\underset{|}{C}}H_2$$

$$C_6H_6 \text{ (benzene, Kekulé structure)} + Br_2 \longrightarrow \text{No reaction}$$

We no longer define aromatic compounds as those having a pleasant aroma; in fact, many do not. We now recognize **aromatic hydrocarbons** as those that exhibit a much higher degree of chemical stability than their chemical composition would predict. The most common group of aromatic compounds is based on the six-member aromatic ring, the benzene ring. The structure of the benzene ring is represented in various ways in Figure 12.6.

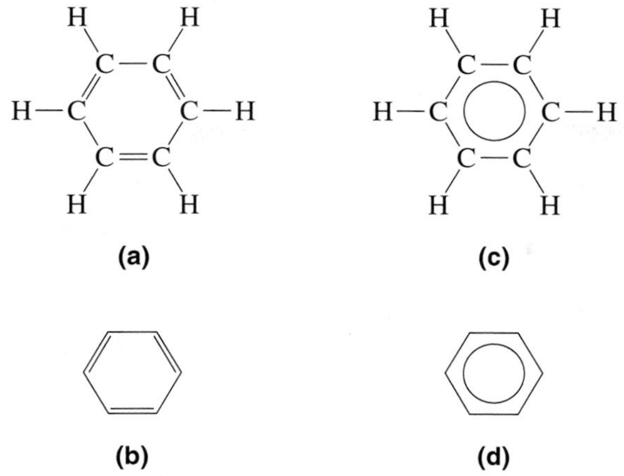

(a)

(c)

(b)

(d)

Figure 12.6
Four ways to represent the benzene molecule. Structure (b) is a simplified diagram of structure (a). Structure (d), a simplified diagram of structure (c), is the most commonly used representation.

Structure and Properties

The benzene ring consists of six carbon atoms joined in a planar hexagonal arrangement. Each carbon atom is bonded to one hydrogen atom. Friedrich Kekulè proposed a model for the structure of benzene in 1865. He proposed that single and double bonds alternated around the ring (a conjugated system of double bonds). To explain why benzene did not decolorize bromine—in other words, didn't react like an unsaturated compound—he suggested that the double and single bonds shift positions rapidly. We show this as a resonance model today.

Resonance models are described in Section 4.4.

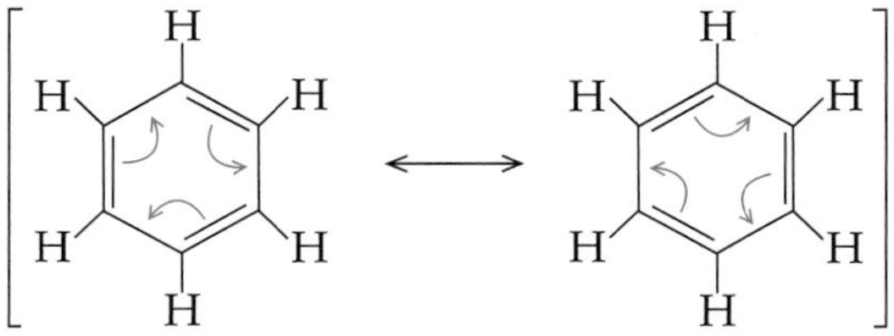

Benzene as a resonance hybrid

The current model of the structure of benzene is based on the idea of overlapping orbitals. Each carbon is bonded to two others by sharing a pair of electrons (σ bonds). Each carbon atom also shares a pair of electrons with a hydrogen atom. The remaining six electrons are located in p orbitals that are perpendicular to the plane of the ring. These p orbitals overlap laterally to form pi (π) orbitals that form a cloud of electrons above and below the ring. These π orbitals are shaped like doughnuts, as shown in Figure 12.7.

Two symbols are commonly used to represent the benzene ring. The representation in Figure 12.6b is the structure proposed by Kekulé. The structure in Figure 12.6d represents the π clouds.

The equal sharing of the six electrons of the p orbitals results in a rigid, flat ring structure, in contrast to the relatively flexible, nonaromatic cyclohexane ring. The model also explains the unusual chemical stability of benzene and its resistance to addition reactions. The electrons of the π cloud are said to be *delocalized.* That means they have much more space and freedom of movement than they would have if they were restricted to individual double bonds. Because electrons repel one another, the system is more stable when the electrons have more space to occupy. As a result, benzene is unusually stable and resists addition reactions typical of alkenes.

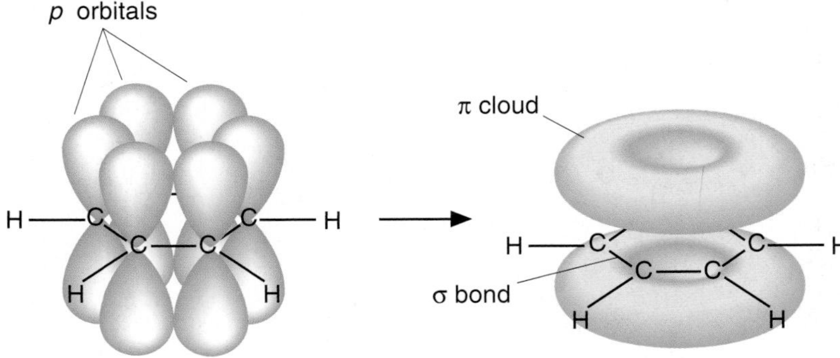

Figure 12.7
The current model of the bonding in benzene.

Nomenclature

Learning Goal

Most simple aromatic compounds are named as derivatives of benzene. Thus benzene is the parent compound, and the name of any atom or group bonded to benzene is used as a prefix, as in these examples:

NO_2 — Nitrobenzene; CH_2CH_3 — Ethylbenzene; Br — Bromobenzene

Other members of this family have unique names based on history rather than logic:

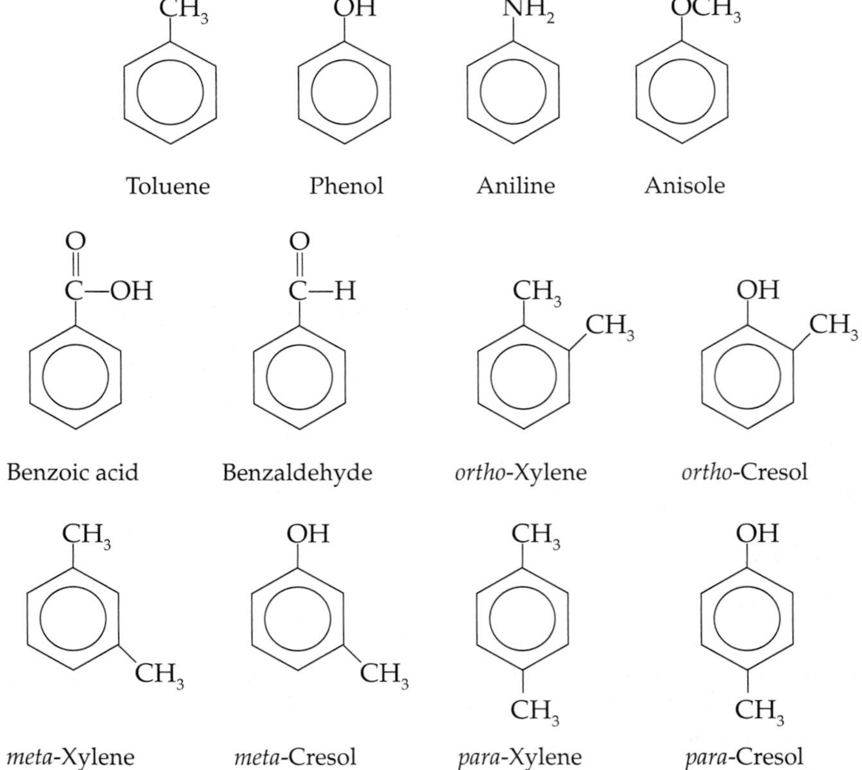

When two groups are present on the ring, three possible orientations exist, and they may be named by either the I.U.P.A.C. Nomenclature System or the common system of nomenclature. If the groups or atoms are located on two adjacent carbons, they are referred to as *ortho (o)* in the common system or with the prefix 1,2- in the I.U.P.A.C. system. If they are on carbons separated by one carbon atom, they are termed *meta (m)* in the common system or 1,3- in the I.U.P.A.C. system. Finally, if the substituents are on carbons separated by two carbon atoms, they are said to be *para (p)* in the common system or 1,4- in the I.U.P.A.C. system. The following examples demonstrate both of these systems:

Two groups 1,2 or *ortho* Two groups 1,3 or *meta* Two groups 1,4 or *para*
G = Any group

If three or more groups are attached to the benzene ring, numbers must be used to describe their location. The names of the substituents are given in alphabetical order.

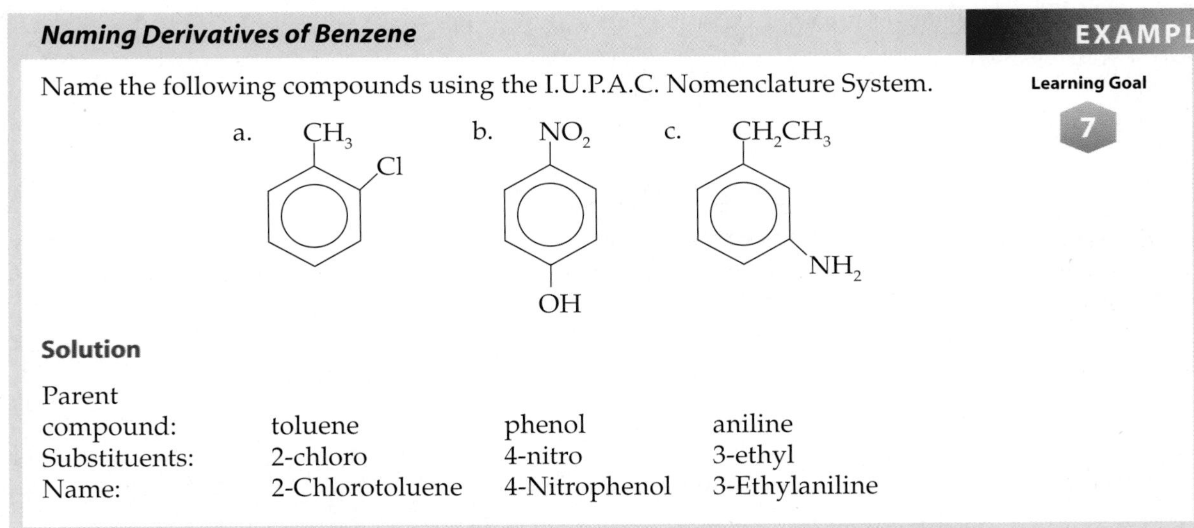

EXAMPLE 12.10

Naming Derivatives of Benzene

Learning Goal 7

Name the following compounds using the I.U.P.A.C. Nomenclature System.

Solution

	a.	b.	c.
Parent compound:	toluene	phenol	aniline
Substituents:	2-chloro	4-nitro	3-ethyl
Name:	2-Chlorotoluene	4-Nitrophenol	3-Ethylaniline

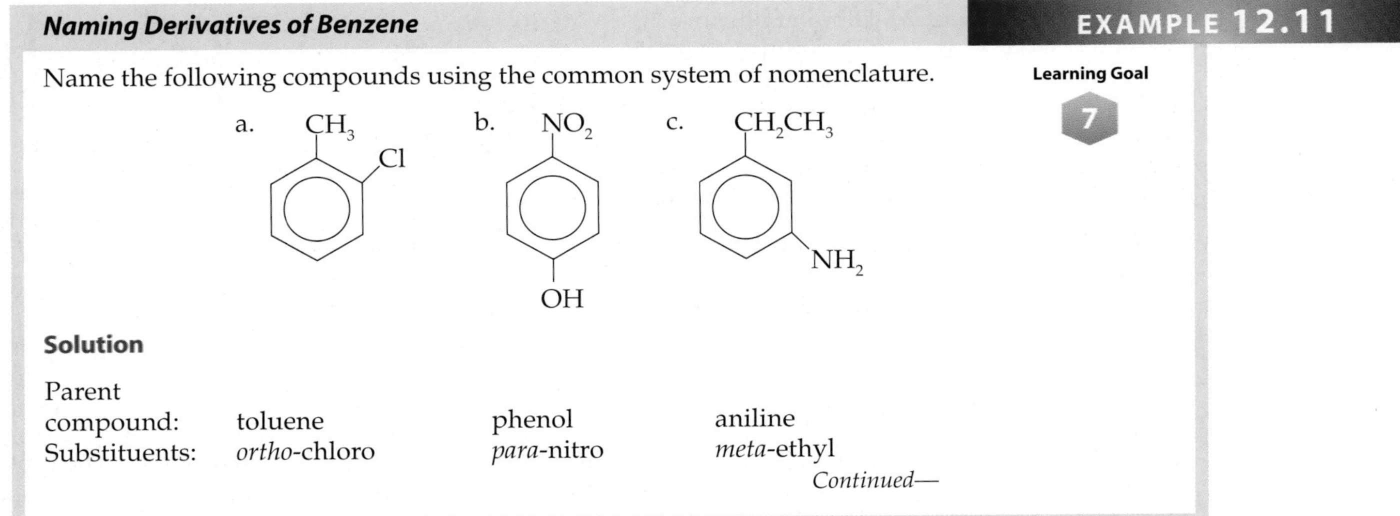

EXAMPLE 12.11

Naming Derivatives of Benzene

Learning Goal 7

Name the following compounds using the common system of nomenclature.

Solution

	a.	b.	c.
Parent compound:	toluene	phenol	aniline
Substituents:	*ortho*-chloro	*para*-nitro	*meta*-ethyl

Continued—

EXAMPLE 12.11 —*Continued*

Name:	*ortho*-Chlorotoluene	*para*-Nitrophenol	*meta*-Ethylaniline
Abbreviated Name:	*o*-Chlorotoluene	*p*-Nitrophenol	*m*-Ethylaniline

In I.U.P.A.C. nomenclature, the group—C_6H_5 derived by removing one hydrogen from benzene, is called the **phenyl group.** An aromatic hydrocarbon with an aliphatic side chain is named as a phenyl substituted hydrocarbon. For example:

$CH_3CHCH_2CH_3$

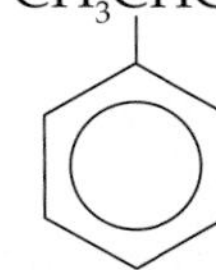

2-Phenylbutane

$CH_3CHCH{=}CH_2$

3-Phenyl-1-butene

One final special name that occurs frequently in aromatic compounds is the benzyl group:

$C_6H_5CH_2—$ or $—CH_2—$

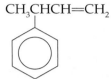

The use of this group name is illustrated by:

$—CH_2Cl$

Benzyl chloride

$—CH_2OH$

Benzyl alcohol

Question 12.13

Draw each of the following compounds:

a. 1,3,5-Trichlorobenzene
b. *ortho*-Cresol
c. 2,5-Dibromophenol
d. *para*-Dinitrobenzene
e. 2-Nitroaniline
f. *meta*-Nitrotoluene

Question 12.14

Draw each of the following compounds:

a. 2,3-Dichlorotoluene
b. 3-Bromoaniline
c. 1-Bromo-3-ethylbenzene
d. *o*-Nitrotoluene
e. *p*-Xylene
f. *o*-Dibromobenzene

Reactions Involving Benzene

Learning Goal

As we have noted, benzene does not readily undergo addition reactions. The typical reactions of benzene are **substitution reactions,** in which a hydrogen atom is replaced by another atom or group of atoms.

A HUMAN PERSPECTIVE

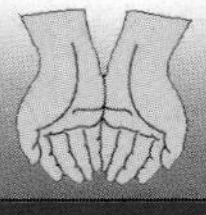

Aromatic Compounds and Carcinogenesis

We come into contact with many naturally occurring aromatic compounds each day. Originally, the name *aromatic* was given to these compounds because of the pleasant aromas of some members of this family. Indeed, many food flavorings and fragrances that we enjoy contain aromatic rings. Examples of other aromatic compounds include preservatives (such as BHT, butylated hydroxytoluene), insecticides (such as DDT), pharmaceutical drugs (such as aspirin), and toiletries.

The *polynuclear aromatic hydrocarbons* (PAH) are an important family of aromatic hydrocarbons that generally have toxic effects. They have also been shown to be carcinogenic, that is, they cause cancer. PAH are formed from the joining of the rings so that they share a common bond (edge). Three common examples are shown:

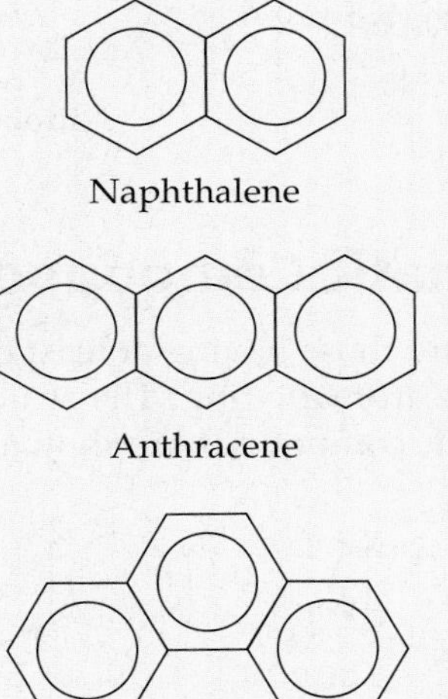

Naphthalene

Anthracene

Phenanthrene

The more complex members of this family (typically consisting of five or six rings at a minimum) are among the most potent carcinogens known. It has been shown that the carcinogenic nature of these chemicals results from their ability to bind to the nucleic acid (DNA) in cells. As we will see in Chapter 24, the ability of the DNA to guide the cell faithfully from generation to generation is dependent on the proper expression of the genetic information, a process called *transcription,* and the accurate copying or replication of the DNA. Accurate DNA replication is essential so that every new cell inherits a complete copy of all the genetic information that is identical to that of the original parent cell. If a mistake is made in the DNA replication process, the result is an error, or mutation, in the new DNA molecule. Some of these errors may cause the new cell to grow out of control, resulting in cancer.

Polynuclear aromatic hydrocarbons are thought to cause cancer by covalently binding to the DNA in cells and interfering with the correct replication of the DNA. Some of the mutations that result may cause a cell to begin to divide in an uncontrolled fashion, giving rise to a cancerous tumor.

Benzopyrene (shown below) is found in tobacco smoke, smokestack effluents, charcoal-grilled meat, and automobile exhaust. It is one of the strongest carcinogens known. It is estimated that a wide variety of all cancers are caused by chemical carcinogens, such as PAH, in the environment.

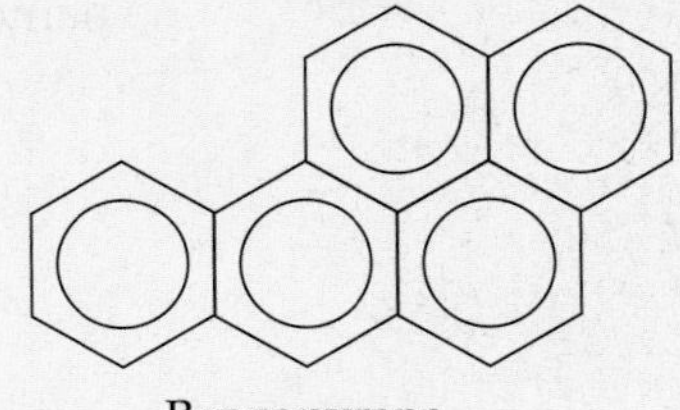

Benzopyrene

Benzene can react (by substitution) with Cl_2 or Br_2. These reactions require either iron or an iron halide as a catalyst. For example:

$$C_6H_6 + Cl_2 \xrightarrow{FeCl_3} C_6H_5Cl + HCl$$

Benzene Chlorine Chlorobenzene

$$\text{Benzene} + Br_2 \xrightarrow{FeBr_3} \text{Bromobenzene (}C_6H_5\text{—Br)} + HBr$$

Benzene Bromine Bromobenzene

When a second equivalent of the halogen is added, three isomers—*para, ortho,* and *meta*—are formed.

Benzene also reacts with sulfur trioxide by substitution. Concentrated sulfuric acid is required as the catalyst. Benzenesulfonic acid, a strong acid, is the product:

$$\text{Benzene} + SO_3 \xrightarrow{\text{Concentrated } H_2SO_4} C_6H_5\text{—}SO_2\text{—OH} + H_2O$$

Benzene Sulfur trioxide Benzenesulfonic acid

Benzene can also undergo nitration with concentrated nitric acid dissolved in concentrated sulfuric acid. This reaction requires temperatures in the range of 50–55°C.

$$\text{Benzene} + HNO_3 \xrightarrow[50\text{–}55^\circ C]{\text{Concentrated } H_2SO_4} C_6H_5\text{—}NO_2 + H_2O$$

Benzene Nitric acid Nitrobenzene

12.7 Heterocyclic Aromatic Compounds

Learning Goal

Heterocyclic aromatic compounds are those having at least one atom other than carbon as part of the structure of the aromatic ring. The structures and common names of several heterocyclic aromatic compounds are shown:

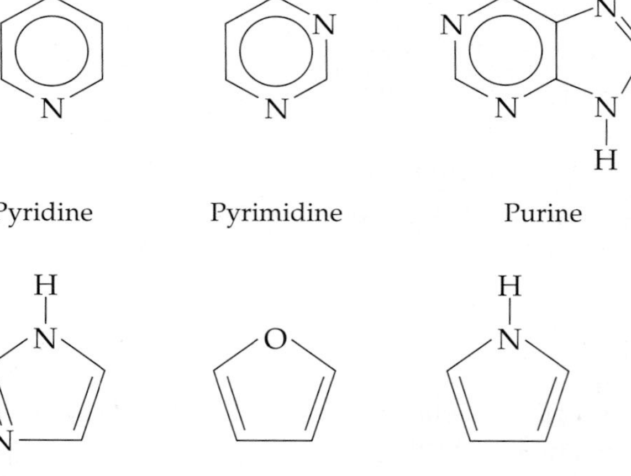

See the Chemistry Connection: The Nicotine Patch in Chapter 16.

All these compounds are more similar to benzene in stability and chemical behavior than they are to the alkenes. Many of these compounds are components of molecules that have significant effects on biological systems. For instance, the purines and pyrimidines are components of DNA (deoxyribonucleic acid) and RNA (ribonucleic acid). DNA and RNA are the molecules responsible for storing and expressing the genetic information of an organism. The pyridine ring is found in nicotine, the addictive compound in tobacco. The pyrrole ring is a component of the porphyrin ring found in hemoglobin and chlorophyll.

Fe^{3+}

Porphyrin

The imidazole ring is a component of cimetidine, a drug used in the treatment of stomach ulcers. The structure of cimetidine is shown below:

CH_3 $CH_2SCH_2CH_2NHC(=NCN)NHCH_3$

Cimetidine

We will discuss a subset of the heterocyclic aromatic compounds, the heterocyclic amines, in Chapter 16.

Summary of Reactions

Addition reactions of alkenes

Hydrogenation:

$$R_2C{=}CR_2 + H{-}H \xrightarrow[\text{Heat or pressure}]{\text{Pt, Pd, or Ni}} R_2CH{-}CHR_2$$

Alkene + Hydrogen → Alkane

Halogenation:

$$R_2C{=}CR_2 + X{-}X \longrightarrow R_2CX{-}CXR_2$$

Alkene + Halogen → Alkyl dihalide

Hydration:

$$R_2C{=}CR_2 + H{-}OH \xrightarrow{H^+} R_2CH{-}C(OH)R_2$$

Alkene + Water → Alcohol

Hydrohalogenation:

$$R_2C{=}CR_2 + H{-}X \longrightarrow R_2CH{-}CXR_2$$

Alkene + Hydrogen halide → Alkyl halide

Addition Polymers of Alkenes

$$n\ R_2C{=}CR_2 \longrightarrow \left[\begin{array}{c}R\ \ R\\ |\ \ \ |\\ \sim C{-}C\sim\\ |\ \ \ |\\ R\ \ R\end{array}\right]_n$$

Alkene monomer — Addition polymer

Reactions of Benzene

Halogenation:

$$\text{Benzene} + X_2 \xrightarrow{FeX_3} \text{C}_6\text{H}_5{-}X + HX$$

Benzene — Halogen — Halobenzene

Sulfonation:

$$\text{Benzene} + SO_3 \xrightarrow{\text{Concentrated } H_2SO_4} \text{C}_6\text{H}_5{-}SO_2{-}OH + H_2O$$

Benzene — Sulfur trioxide — Benzenesulfonic acid

Nitration:

$$\text{Benzene} + HNO_3 \xrightarrow[\text{50–55°C}]{\text{Concentrated } H_2SO_4} \text{C}_6\text{H}_5{-}NO_2 + H_2O$$

Benzene — Nitric acid — Nitrobenzene

Summary

12.1 Alkenes and Alkynes: Structure and Physical Properties

Alkenes and *alkynes* are *unsaturated hydrocarbons*. Alkenes are characterized by the presence of at least one carbon-carbon double bond and have the general molecular formula C_nH_{2n}. Alkynes are characterized by the presence of at least one carbon-carbon triple bond and have the general molecular formula C_nH_{2n-2}. The physical properties of the alkenes and alkynes are similar to those of alkanes, but their chemical properties are quite different.

12.2 Alkenes and Alkynes: Nomenclature

Alkenes and alkynes are named by identifying the parent compound and replacing the *-ane* ending of the alkane with *-ene* (for an alkene) or *-yne* (for an alkyne). The parent chain is numbered to give the lowest number to the first of the two carbons involved in the double bond (or triple bond). Finally, all groups are named and numbered.

12.3 Geometric Isomers: A Consequence of Unsaturation

The carbon-carbon double bond is rigid. This allows the formation of *geometric isomers*, or isomers that differ from one another depending on whether chemical groups are on the same or opposite sides of the rigid double bonds. When groups are on the same side of a double bond, the prefix *cis* is used to describe the compound. When groups are on opposite sides of a double bond, the prefix *trans* is used.

12.4 Alkenes in Nature

Alkenes and polyenes (alkenes with several carbon-carbon double bonds) are common in nature. Ethene, the simplest alkene, is a plant growth substance involved in fruit ripening, senescence and leaf fall, and responses to environmental stresses. Isoprenoids, or terpenes, are polyenes built from one or more isoprene units. Isoprenoids include steroids, chlorophyll and other photosynthetic pigments, and vitamins A, D, and K.

12.5 Reactions Involving Alkenes

Whereas alkanes undergo *substitution reactions*, alkenes and alkynes undergo *addition reactions*. The principal addition reactions of the unsaturated hydrocarbons are *halogenation*, *hydration*, *hydrohalogenation*, and *hydrogenation*. *Polymers* can be made from alkenes or substituted alkenes.

12.6 Aromatic Hydrocarbons

Aromatic hydrocarbons contain benzene rings. The rings can be represented as having alternating double and single bonds. However, it is more accurate to portray σ (sigma) bonds between carbons of the ring and a π (pi) cloud of electrons above and below the ring. Simple aromatic compounds are named as derivatives of benzene. Several members of this family have historical common names, such as aniline, phenol, and toluene. Aromatic compounds do not undergo addition reactions. The typical reactions of benzene are *substitution* reactions: halogenation, nitration, and sulfonation.

12.7 Heterocyclic Aromatic Compounds

Heterocyclic aromatic compounds are those having at least one atom other than carbon as part of the structure of the aromatic ring. They are more similar to benzene in stability and chemical behavior than they are to the alkenes. Many of these compounds are components of molecules that have significant effects on biological systems, including DNA, RNA, hemoglobin, and nicotine.

Key Terms

addition polymer (12.5)
addition reaction (12.5)
alkene (12.1)
alkyne (12.1)
aromatic hydrocarbon (12.6)
geometric isomers (12.3)
halogenation (12.5)
heterocyclic aromatic compound (12.7)
hydration (12.5)
hydrogenation (12.5)
hydrohalogenation (12.5)
Markovnikov's rule (12.5)
monomer (12.5)
phenyl group (12.6)
polymer (12.5)
substitution reaction (12.6)
unsaturated hydrocarbon (Intro)

Questions and Problems

Alkenes and Alkynes: Structure and Physical Properties

12.15 Write the general formulas for alkanes, alkenes, and alkynes.

12.16 What are the characteristic functional groups of alkenes and alkynes?

12.17 Describe the geometry of ethene.

12.18 What are the bond angles in ethene?

12.19 Describe the geometry of ethyne.

12.20 What are the bond angles in ethyne?

Alkenes and Alkynes: Nomenclature and Geometric Isomers

12.21 Draw a condensed formula for each of the following compounds:

a. 2-Methyl-2-hexene
b. *trans*-3-Heptene
c. *cis*-1-Chloro-2-pentene
d. *cis*-2-Chloro-2-methyl-3-heptene
e. *trans*-5-Bromo-2,6-dimethyl-3-octene

12.22 Draw a condensed formula for each of the following compounds:

a. 2-Hexyne
b. 4-Methyl-1-pentyne
c. 1-Chloro-4,4,5-trimethyl-2-heptyne
d. 2-Bromo-3-chloro-7,8-dimethyl-4-decyne

12.23 Name each of the following using the I.U.P.A.C. Nomenclature System:

a. $CH_3CH_2CHCH{=}CH_2$
   (with CH_3 on the third carbon)

b. $CH_2CH_2CH_2CH_2{-}Br$
   (with $CH_2CH{=}CH_2$ on the first carbon)

c. $CH_3CH_2CH{=}CHCHCH_2CH_3$
   (with Br on the fifth carbon)

d. 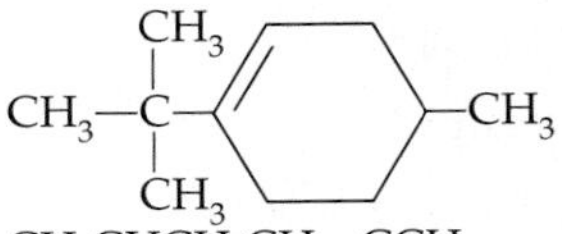 $CH_3{-}C(CH_3)_2{-}$ (cyclohexene ring) $-CH_3$

e. $CH_3CHCH_2CH{=}CCH_3$
   (with CH_3 on the second carbon and CH_3 on the fifth carbon)

f. $Cl{-}CH_2CHC{\equiv}C{-}H$
   (with CH_3 on the second carbon)

g. $CH_3CHCH_2CH_2CH_2{-}C{\equiv}C{-}H$
   (with Cl on the second carbon)

h. (cyclopentene ring with Br and Cl on adjacent carbons)

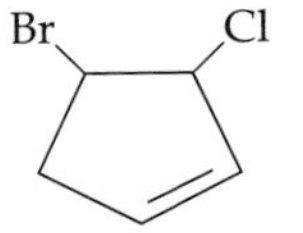

12.24 Draw each of the following compounds using condensed formulas:

a. 1,3,5-Trifluoropentane
b. *cis*-2-Octene
c. Dipropylacetylene
d. 3,3,5-Trimethyl-1-hexene
e. 1-Bromo-3-chloro-1-heptyne

12.25 Of the following compounds, which can exist as *cis-trans* geometric isomers? Draw the two geometric isomers.

a. 2,3-Dibromobutane
b. 2-Heptene
c. 2,3-Dibromo-2-butene
d. Propene
e. 1-Bromo-1-chloro-2-methylpropene
f. 1,1-Dichloroethene
g. 1,2-Dibromoethene
h. 3-Ethyl-2-methyl-2-hexene

12.26 Which of the following alkenes would not exhibit *cis-trans* geometric isomerism?

a. 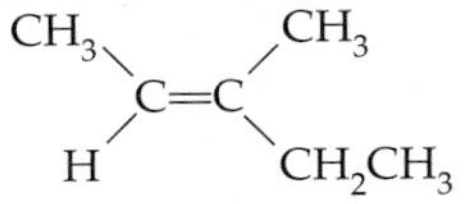 $(CH_3)(H)C{=}C(CH_3)(CH_2CH_3)$

b. $(CH_3)(CH_3)C{=}C(CH_3)(H)$

c. $(CH_3CH_2)(CH_3)C{=}C(CHCH_3)(CHCH_2CH_3)$, where the upper CH bears CH_2CH_3 and the lower CH bears CH_3

d. 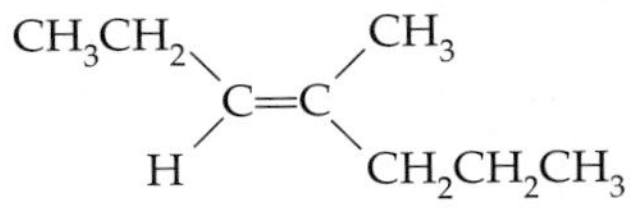 $(CH_3CH_2)(H)C{=}C(CH_3)(CH_2CH_2CH_3)$

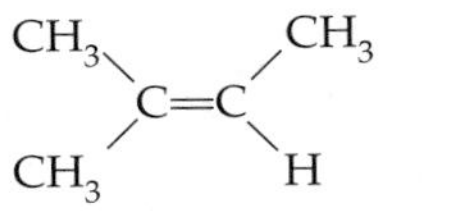

12.27 Which of the following structures have incorrect I.U.P.A.C. names? If incorrect, give the correct I.U.P.A.C. name.

a. $CH_3C{\equiv}C{-}CH_2CHCH_3$
   CH_3 (on CH)

2-Methyl-4-hexyne

b. CH_3CH_2, CH_2CH_3 / C=C / CH_3CH_2, H

3-Ethyl-3-hexyne

c. $CH_3CHCH_2{-}C{\equiv}C{-}CH_2CHCH_3$
   CH_3 (on C-2); CH_2CH_3 (on C-7)

2-Ethyl-7-methyl-4-octyne

d. CH_3CH_2, CH_2CHCH_3 (Cl on CH) / C=C / H, H

trans-6-Chloro-3-heptene

e. $ClCH_2$, H / C=C / H, $CHCH_2CH_3$ (CH_3 on CH)

1-Chloro-5-methyl-2-hexene

12.28 Which of the following can exist as *cis* and *trans* isomers?

a. $H_2C{=}CH_2$
b. $CH_3CH{=}CHCH_3$
c. $Cl_2C{=}CBr_2$
d. $ClBrC{=}CClBr$
e. $(CH_3)_2C{=}C(CH_3)_2$

12.29 Provide the I.U.P.A.C. name for each of the following molecules:

a. $CH_2{=}CHCH_2CH_2CH{=}CHCH_2CH_2CH_3$
b. $CH_2{=}CHCH_2CH{=}CHCH_2CH{=}CHCH_3$
c. $CH_3CH{=}CHCH_2CH{=}CHCH_2CH_3$
d. $CH_3CH{=}CHCHCH{=}CHCH_3$
   CH_3

12.30 Provide the I.U.P.A.C. name for each of the following molecules:

a. $CH_3C{=}CHCHCH{=}CCH_3$
   Br, CH_3, Br

b. $CH_2{=}CHCHCH{=}CHCHCH_2CH_3$
   CH_3, CH_2CH_3

c. $CH_2{=}CHCCH{=}CHCH_2CH{=}CHCHCH_3$
   CH_3, CH_3 (on C-3); CH_3 (on C-9)

d. $CH_3CHCHCH{=}CHCH_2CH{=}CHCHCH_3$
   CH_3 (on C-2), CH_2CH_3 (on C-3), CH_2CH_3 (on C-9)

Reactions Involving Alkenes

12.31 How could you distinguish between a sample of cyclohexane and a sample of hexene (both C_6H_{12}) using a simple chemical test? (*Hint:* Refer to the subsection entitled "Halogenation: Addition of X_2 to an Alkene.")

12.32 Quantitatively, 1 mol of Br_2 is consumed per mole of alkene, and 2 mol of Br_2 are consumed per mole of alkyne. How many moles of Br_2 would be consumed for 1 mol of each of the following:

a. 2-Hexyne
b. Cyclohexene
c. (cyclopentadiene ring)$-CH{=}CH_2$
d. (cyclohexene ring)$-C{\equiv}C{-}CH_3$

12.33 Complete each of the following reactions by supplying the missing reactant or product(s) as indicated by a question mark:

a. $CH_3CH_2CH{=}CHCH_2CH_3 + ? \longrightarrow CH_3CH_2CH_2CH_2CH_2CH_3$

b. $CH_3{-}\underset{\underset{CH_2}{\|}}{C}{-}CH_3 + ? \longrightarrow CH_3\overset{\overset{CH_3}{|}}{\underset{\underset{CH_3}{|}}{C}}{-}OH$

c. ? + (cyclohexene) $\longrightarrow$ (cyclohexane with Br and H)

d. $2CH_3CH_2CH_2CH_2CH_2CH_3 + ?O_2 \xrightarrow{Heat}$? + ? (complete combustion)

e. ? + (cyclohexane) $\longrightarrow$ Cl$-$(cyclohexane) + HCl

f. ? $\xrightarrow{H_2O,\ H^+}$ (cyclopentane with OH)

12.34 Draw and name the product in each of the following reactions:

a. Cyclopentene + H_2O (H^+)
b. Cyclopentene + HCl
c. Cyclopentene + H_2
d. Cyclopentene + HI

12.35 A hydrocarbon with a formula C_5H_{10} decolorized Br_2 and consumed 1 mol of hydrogen upon hydrogenation. Draw all the isomers of C_5H_{10} that are possible based on the above information.

12.36 Triple bonds react in a manner analogous to that of double bonds. The extra pair of electrons in the triple bond, however, generally allows 2 mol of a given reactant to *add* to the triple bond in contrast to 1 mol with the double bond. The "rich get richer" rule holds. Predict the major product in each of the following reactions:

a. Acetylene with 2 mol HCl
b. Propyne with 2 mol HBr
c. 2-Butyne with 2 mol HI

12.37 Complete each of the following by supplying the missing product indicated by the question mark:

a. 2-Butene $\xrightarrow{HBr}$?
b. 3-Methyl-2-hexene $\xrightarrow{HI}$?
c. (cyclopentene) $\xrightarrow{HCl}$?

12.38 Bromine is often used as a laboratory spot test for unsaturation in an aliphatic hydrocarbon. Bromine in CCl_4 is red. When bromine reacts with an alkene or alkyne, the alkyl halide formed is colorless; hence a disappearance of the red color is a positive test for unsaturation. A student tested the contents of two vials, A and B, both containing compounds with a molecular formula, C_6H_{12}. Vial A decolorized bromine, but vial B did not. How may the results for vial B be explained? What class of compound would account for this?

12.39 What is meant by the term *polymer?*

12.40 What is meant by the term *monomer?*

12.41 Write an equation representing the synthesis of Teflon from tetrafluoroethene. (*Hint:* Refer to Table 12.1.)

12.42 Write an equation representing the synthesis of polystyrene. (*Hint:* Refer to Table 12.1.)

12.43 Provide the I.U.P.A.C. name for each of the following molecules. Write a balanced equation for the hydration of each.

a. $CH_3CH{=}CHCH_2CH_3$

b. $CH_2CH{=}CH_2$ (Br on the first carbon)

c. (cyclohexene ring with CH_3 and CH_3 substituents)

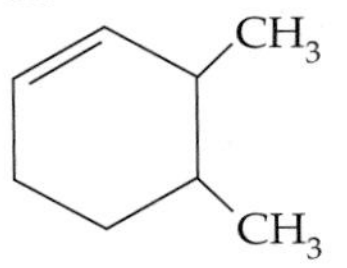

12.44 Provide the I.U.P.A.C. name for each of the following molecules. Write a balanced equation for the hydration of each.

a. (cyclopentadiene ring with $-CH_2CH_3$)

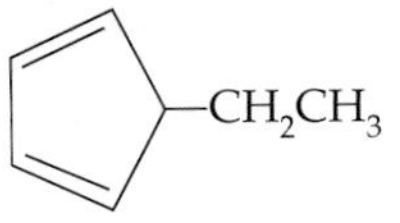

b. $CH_3CH{=}CHCH_2CH{=}CHCH_2CH{=}CHCH_3$

c. $CH_3CH{=}CHCCH_3$ (with CH_2CH_3 and CH_3 on the fourth carbon)

12.45 Write an equation for the addition reaction that produced each of the following molecules:

a. $CH_2CH_2CH_2CHCH_3$ (OH on C-1, CH_3 on C-4)

b. $CH_3CH_2CHCH_2CH_2CH_3$ (Br on C-3)

c. (cyclohexane ring with Br and CH_3)

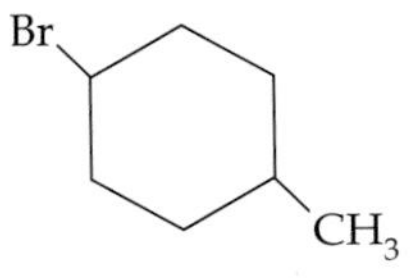

d. (cyclopentane ring with $-CH_2CH_3$ and OH)

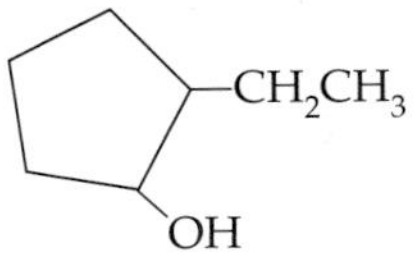

12.46 Write an equation for the addition reaction that produced each of the following molecules:

a. $CH_3CH_2CHCHCH_3$ (OH on C-3, CH_2CH_3 on C-4)

b. $CH_3CHCH_2CH_3$ (OH on C-2)

c. (cyclohexane ring with HO, CH_3, CH_3)

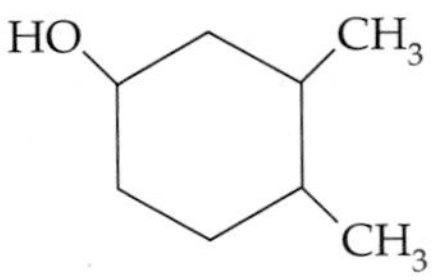

d. (cyclooctane ring with Br, OH, OH, Br)

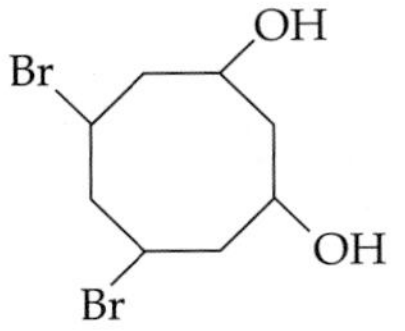

12.47 Draw the structure of each of the following compounds and write a balanced equation for the complete hydration of each:

a. 1,4-Hexadiene
b. 2,4,6-Octatriene
c. 1,3 Cyclohexadiene
d. 1,3,5-Cyclooctatriene

12.48 Draw the structure of each of the following compounds and write a balanced equation for the hydrobromination of each:

a. 3-Methyl-1,4-hexadiene
b. 4-Bromo-1,3-pentadiene
c. 3-Chloro-2,4-hexadiene
d. 3-Bromo-1,3-Cyclohexadiene

Aromatic Hydrocarbons

12.49 Draw the structure for each of the following compounds:

a. 2,4-Dibromotoluene
b. 1,2,4-Triethylbenzene
c. Isopropylbenzene
d. 2-Bromo-5-chlorotoluene

12.50 Name each of the following compounds, using the I.U.P.A.C. system.

a. (benzene ring with CH_3, CH_3)

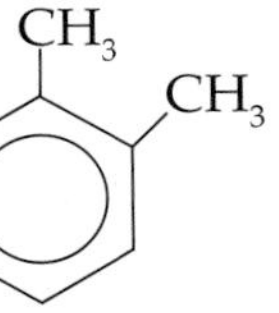

b. (benzene ring with NO_2, NO_2)

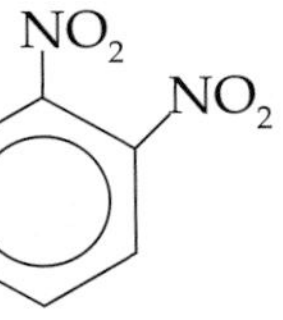

c. (benzene ring with CH_2CH_3 and CH_3)

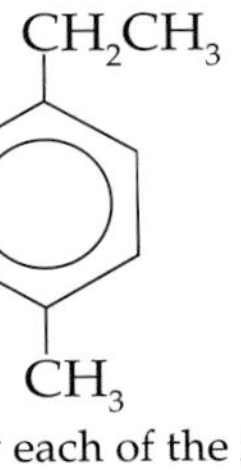

d. (benzene ring with Br, Cl, CH_3)

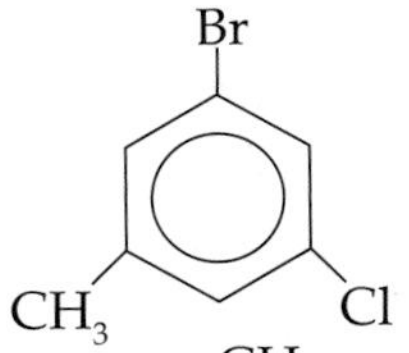

e. (benzene ring with CH_3, O_2N, NO_2, Br)

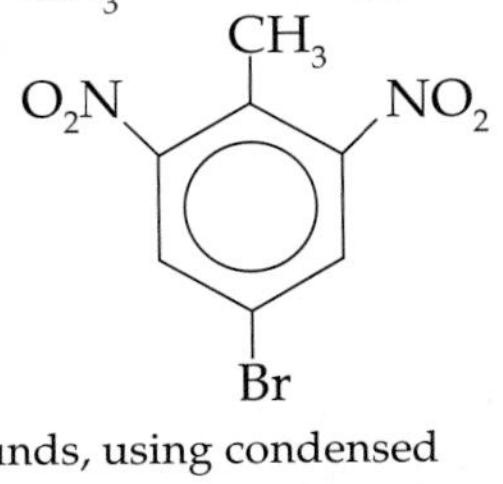

12.51 Draw each of the following compounds, using condensed formulas:

a. *meta*-Cresol
b. Propylbenzene
c. 1,3,5-Trinitrobenzene
d. *m*-Chlorotoluene

12.52 Draw each of the following compounds, using condensed formulas:

a. *p*-Xylene
b. Isopropylbenzene
c. *m*-Nitroanisole
d. *p*-Methylbenzaldehyde

12.53 Describe the Kekulé model for the structure of benzene.
12.54 Describe the current model for the structure of benzene.
12.55 How does a substitution reaction differ from an addition reaction?
12.56 Give an example of a substitution reaction and of an addition reaction.
12.57 Write an equation showing the reaction of benzene with Cl_2 and $FeCl_3$.
12.58 Write an equation showing the reaction of benzene with SO_3. Be sure to note the catalyst required.

Heterocyclic Aromatic Compounds

12.59 Draw the general structure of a pyrimidine.
12.60 What biological molecules contain pyrimidine rings?
12.61 Draw the general structure of a purine.
12.62 What biological molecules contain purine rings?

Critical Thinking Problems

1. There is a plastic polymer called polyvinylidene difluoride (PVDF) that can be used to sense a baby's breath and thus be used to prevent sudden infant death syndrome (SIDS). The secret is that this polymer can be specially processed so that it becomes piezoelectric (produces an electrical current when it is physically deformed) and pyroelectric (develops an electrical potential when its temperature changes). When a PVDF film is placed beside a sleeping baby, it will set off an alarm if the baby stops breathing. The structure of this polymer is shown here:

$$\left[\begin{array}{c} \text{F}\quad\text{H}\quad\text{F}\quad\text{H} \\ -\text{C}-\text{C}-\text{C}-\text{C}- \\ \text{F}\quad\text{H}\quad\text{F}\quad\text{H} \end{array}\right]$$

Go to the library and investigate some of the other amazing uses of PVDF. Draw the structure of the alkene from which this compound is produced.

2. Isoprene is the repeating unit of the natural polymer rubber. It is also the starting material for the synthesis of cholesterol and several of the lipid-soluble vitamins, including vitamin A and vitamin K. The structure of isoprene is seen below.

$$CH_2{=}\overset{\displaystyle CH_3}{\overset{|}{C}}{-}CH{=}CH_2$$

What is the I.U.P.A.C. name for isoprene?

3. When polyacrylonitrile is burned, toxic gases are released. In fact, in airplane fires, more passengers die from inhalation of toxic fumes than from burns. Refer to Table 12.1 for the structure of acrylonitrile. What toxic gas would you predict to be the product of the combustion of these polymers?
4. If a molecule of polystyrene consists of 25,000 monomers, what is the molar mass of the molecule?
5. A factory produces one million tons of polypropylene. How many moles of propene would be required to produce this amount? What is the volume of this amount of propene at 25°C and 1 atm?

Sugar-free jelly beans

13 Alcohols, Phenols, Thiols, and Ethers

ORGANIC CHEMISTRY

Outline

Learning Goals

1. Rank selected alcohols by relative water solubility, boiling points, or melting points.
2. Write the names and draw the structures for common alcohols.
3. Discuss the biological, medical, or environmental significance of several alcohols.
4. Classify alcohols as primary, secondary, or tertiary.
5. Write equations representing the preparation of alcohols by the hydration of an alkene.
6. Write equations representing the preparation of alcohols by hydrogenation (reduction) of aldehydes or ketones.
7. Write equations showing the dehydration of an alcohol.
8. Write equations representing the oxidation of alcohols.
9. Discuss the role of oxidation and reduction reactions in the chemistry of living systems.
10. Discuss the use of phenols as germicides.
11. Write names and draw structures for common ethers and discuss their use in medicine.
12. Write equations representing the dehydration reaction between two alcohol molecules.
13. Write names and draw structures for simple thiols and discuss their biological significance.

CHEMISTRY CONNECTION

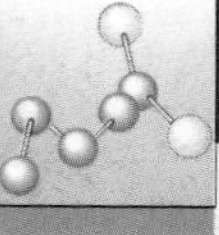

Polyols for the Sweet Tooth

Do you crave sweets, but worry about the empty calories in sugary treats? If so, you are not alone. Research tells us that, even as babies, we demonstrate preference for sweet tastes over all others. But there are many reasons to reduce our intake of refined sugars, in particular sucrose or table sugar. Too many people eat high-calorie, low-nutrition snacks rather than more nutritious foods. This can lead to obesity, a problem that is very common in our society. In addition, sucrose is responsible for tooth decay. Lactic acid, one of the products of the metabolism of sucrose by bacteria on our teeth, dissolves the tooth enamel, which results in a cavity. For those with diabetes, glucose intolerance, or hypoglycemia, sucrose in the diet makes it difficult to maintain a constant blood sugar level.

The food chemistry industry has invested billions of dollars in the synthesis of sugar substitutes. We recognize names such as aspartame (Equal or Nutrasweet) and saccharin (Sweet & Low) because they are the most common non-nutritive sweeteners worldwide. We also buy products, including candies, soft drinks, and gums, which are advertised to be "sugar-free." But you might be surprised to find that many of these products are not free of calories. A check of the nutritional label may reveal that these products contain sorbitol, mannitol, or one or more other members of a class of compounds called sugar alcohols, or *polyols (poly—*many; *ols—*alcohol or hydroxyl groups).

Sugar alcohols are found in many foods, including fruits, vegetables, and mushrooms. Others are made by hydrogenation or fermentation of carbohydrates from wheat or corn. But all are natural products. Compared to sucrose, they range in sweetness from about half to nearly the same; they also have fewer calories per gram than sucrose (about one-third to one-half the calories). Polyols also cause a cooling sensation in the mouth. This cooling is caused by a negative heat of solution (they must absorb heat from the surroundings in order to dissolve) and is used to advantage in breath freshening mints and gums.

Sorbitol, the most commonly used sugar alcohol, is about 0.6 times the sweetness of sucrose. While sucrose contains four calories per gram, sorbitol is only about 2.6 C/g. Discovered in 1872 in the berries of Mountain Ash trees, sorbitol has a smooth mouthfeel, which makes it ideal as a texturizing agent in foods. It also has a pleasant, cool, sweet flavor and acts as a humectant, keeping foods from losing moisture. No acceptable daily intake (ADI) has been specified for sorbitol, which is an indication that it is considered to be a very safe food additive. However, it has been observed that ingestion of more than 50–80 g/day may have a laxative effect.

Mannitol, a structural isomer of sorbitol, is found naturally in asparagus, olives, pineapple, and carrots. For use in the food industry, it is extracted from seaweed. Mannitol has about 0.7 the sweetness of sucrose and only about 1.6 C/g. As for sorbitol, no ADI has been specified; but ingestion of more than 20 g/day may cause diarrhea and bloating.

Xylitol was discovered in 1891 and has been used as a sweetener since the 1960s. Found in many fruits and vegetables, it has about the same sweetness as sucrose, but only one-third of the caloric value. Its high cooling effect, as well as sweetness, contribute to its popularity as a sweetening agent in hard candies and gums, as well as in oral health products. In fact, extensive studies suggest that use of xylitol-sweetened gum (7–10 g of xylitol per day) between meals results in a 30–60% decrease in dental cavities.

While polyols give us the sweetness that we enjoy without all of the calories, cavities, and blood sugar peaks of sucrose, they are not without a negative side. Some studies have reported weight gain by individuals who overeat these "sugar-free" foods. The American Diabetes Association has reported that these foods are "acceptable in moderate amounts but should not be eaten in excess." In fact, some diabetics have suffered elevated blood sugar after overeating foods containing polyols. Finally, as we noted above, when ingested in excess, sugar alcohols may cause bloating and diarrhea.

The use of these natural products continues to be investigated by the food and pharmaceutical industries. As we learn more about them, sugar alcohols continue to be versatile food additives. Using them in moderation, we can enjoy the benefits that they confer, without suffering uncomfortable side effects.

Sorbitol	Mannitol	Xylitol
CH_2OH	CH_2OH	CH_2OH
H—C—OH	HO—C—H	H—C—OH
HO—C—H	HO—C—H	HO—C—H
H—C—OH	H—C—OH	H—C—OH
H—C—OH	H—C—OH	CH_2OH
CH_2OH	CH_2OH	

Structures of three of the sugar alcohols used in the food industry.

Introduction

The characteristic functional group of the *alcohols* and *phenols* is the *hydroxyl group (—OH).* Alcohols have the general structure R—OH, in which R is any alkyl group. Phenols are similar in structure but contain an aryl group in place of the alkyl group. Both can be viewed as substituted water molecules in which one of the hydrogen atoms has been replaced by an alkyl or aryl group.

An aryl group is an aromatic ring with one hydrogen atom removed.

General formula: Example:

R–O–H (Alcohol) C$_6$H$_5$–O–H (Phenol) CH_3–O–H Methanol (methyl alcohol)

Ethers have two alkyl or aryl groups attached to the oxygen atom and may be thought of as substituted alcohols. The functional group characteristic of an ether is R—O—R. *Thiols* are a family of compounds that contain the sulfhydryl group (—SH). They, too, have a structure similar to that of alcohols.

R and R′ = alkyl or aryl group

R–O–R[1] Ethers; CH_3–O–CH_3 Methoxymethane (dimethyl ether); R—SH Thiol; CH_3—SH Methanethiol

Many important biological molecules, including sugars (carbohydrates), fats (lipids), and proteins, contain hydroxyl and/or thiol groups.

D-Glucose, *a sugar*: H–C=O, H–C–OH, HO–C–H, H–C–OH, H–C–OH, CH_2–OH

Lysine vasopressin (partial structure), *a protein*: O^-–C=O–CH_2–NH–$CHCH_2CH_2CH_2NH_2$ … (Portion of chain omitted for clarity) … CH–CH_2–C$_6$H$_4$–OH, NH, C=O, $CHCH_2SH$, NH_2

Monolaurin, *a lipid*: CH_2–O–C(=O)–$(CH_2)_{10}$–CH_3, CH–OH, CH_2–OH

In biological systems the hydroxyl group is often involved in a variety of reactions such as oxidation, reduction, hydration, and dehydration. In glycolysis (a metabolic pathway by which glucose is degraded and energy is harvested in the form of ATP), several steps center on the reactivity of the hydroxyl group. The majority of the consumable alcohol in the world (ethanol) is produced by fermentation reactions carried out by yeasts.

Glycolysis and fermentation are discussed in Chapter 21.

The thiol group is found in the structure of some amino acids and is essential for keeping proteins in the proper three-dimensional shape required for their biological function. Thus these functional groups play a central role in the structure and chemical properties of biological molecules. The thiol group of the amino acid cysteine is highlighted in blue in the structure of lysine vasopressin presented above.

Learning Goal 1

13.1 Alcohols: Structure and Physical Properties

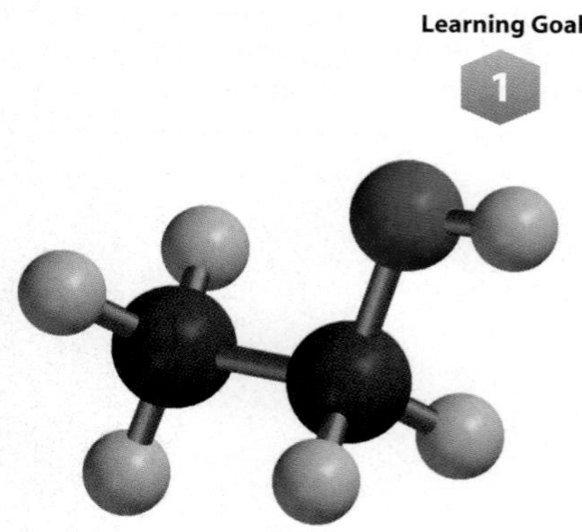

Figure 13.1
Ball-and-stick model of the simple alcohol ethanol.

An **alcohol** is an organic compound that contains a **hydroxyl group** (—OH) attached to an alkyl group (Figure 13.1). The R—O—H portion of an alcohol is similar to the structure of water. The oxygen and the two atoms bonded to it lie in the same plane, and the R—O—H bond angle is approximately 104°, which is very similar to the H—O—H bond angle of water.

The hydroxyl groups of alcohols are very polar because the oxygen and hydrogen atoms have significantly different electronegativities. Because the two atoms involved in this polar bond are oxygen and hydrogen, hydrogen bonds can form between alcohol molecules (Figure 13.2).

Electronegativity is discussed in Section 4.1. Hydrogen bonding is described in detail in Section 6.2.

As a result of this intermolecular hydrogen bonding, alcohols boil at much higher temperatures than hydrocarbons of similar molecular weight. These higher boiling points are caused by the large amount of heat needed to break the hydrogen bonds that attract the alcohol molecules to one another. Compare the boiling points of butane and propanol, which have similar molecular weights:

$CH_3CH_2CH_2CH_3$	$CH_3CH_2CH_2OH$
Butane	1-Propanol
M.W. = 58	M.W. = 60
b.p. = −0.4°C	b.p. = 97.2°C

Alcohols with fewer than four or five carbon atoms are very soluble in water, and those with five to eight carbons are moderately soluble in water. This is due to the ability of the alcohol to form intermolecular hydrogen bonds with water molecules (see Figure 13.2b). As the nonpolar, or hydrophobic, portion of an alcohol (the carbon chain) becomes larger relative to the polar, hydrophilic, region (the hydroxyl group), the water solubility of an alcohol decreases. As a result large alcohols are nearly insoluble in water. The term *hydrophobic,* which literally means "water fearing," is used to describe a molecule or a region of a molecule that is nonpolar and, thus, more soluble in nonpolar solvents than in water. Similarly, the term *hydrophilic,* meaning water loving, is used to describe a polar molecule or region of a molecule that is more soluble in the polar solvent water than in a nonpolar solvent.

***Intermolecular* hydrogen bonds are attractive forces between two molecules. *Intramolecular* hydrogen bonds are attractive forces between polar groups within the same molecule.**

An increase in the number of hydroxyl groups along a carbon chain will increase the influence of the polar hydroxyl group. It follows, then that diols and triols are more water soluble than alcohols with only a single hydroxyl group.

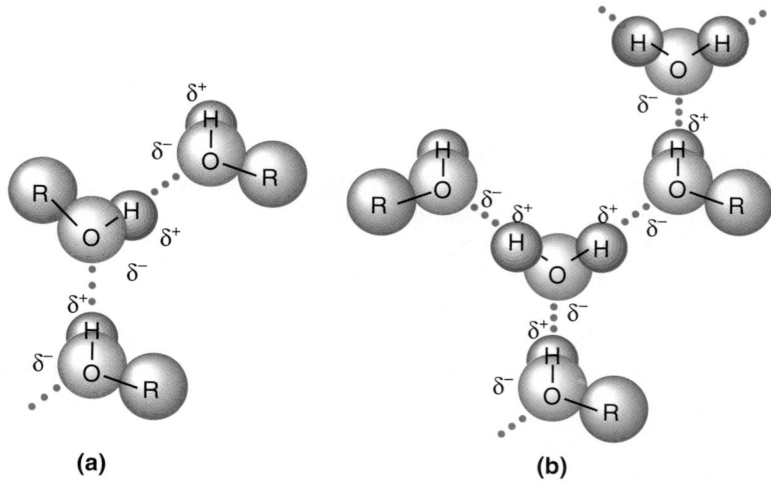

Figure 13.2
(a) Hydrogen bonding between alcohol molecules. (b) Hydrogen bonding between alcohol molecules and water molecules.

The presence of polar hydroxyl groups in large biological molecules—for instance, proteins and nucleic acids—allows intramolecular hydrogen bonding that keeps these molecules in the shapes needed for biological function.

13.2 Alcohols: Nomenclature

I.U.P.A.C. Names

Learning Goal

The way to determine the parent compound was described in Section 11.2.

In the I.U.P.A.C. Nomenclature System, alcohols are named according to the following steps:

- Determine the name of the *parent compound*, the longest continuous carbon chain containing the—OH group.
- Replace the *-e* ending of the alkane chain with the *-ol* ending of the alcohol. Following this pattern, an alkane becomes an alkanol. For instance, ethan*e* becomes ethan*ol*, and propan*e* becomes propan*ol*.
- Number the parent chain to give the carbon bearing the hydroxyl group the lowest possible number.
- Name and number all substituents, and add them as prefixes to the "alkanol" name.
- Alcohols containing two hydroxyl groups are named *-diols*. Those bearing three hydroxyl groups are called *-triols*. A number giving the position of each of the hydroxyl groups is needed in these cases.

Using I.U.P.A.C. Nomenclature to Name an Alcohol — **EXAMPLE 13.1**

Learning Goal

Name the following alcohol using I.U.P.A.C. nomenclature.

Solution

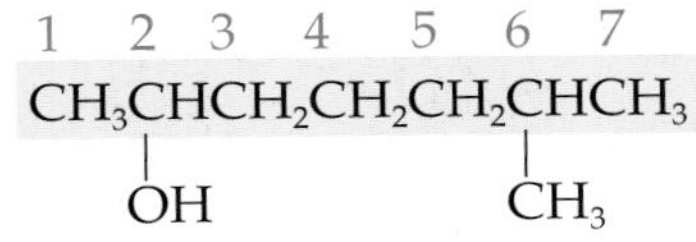

Parent compound: heptane (becomes heptanol)
Position of —OH: carbon-2 (*not* carbon-6)
Substituents: 6-methyl
Name: 6-Methyl-2-heptanol

Using I.U.P.A.C. Nomenclature to Name Alcohols — **EXAMPLE 13.2**

Learning Goal

Name the following cyclic alcohol using I.U.P.A.C. nomenclature.

Solution

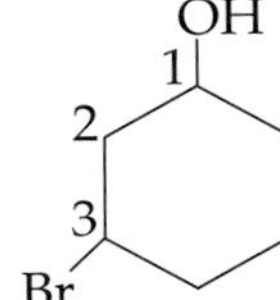

Remember that this line structure represents a cyclic molecule composed of six carbon atoms and associated hydrogen atoms, as follows:

Continued—

EXAMPLE 13.2 —*Continued*

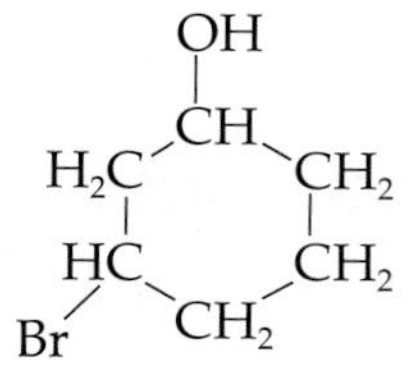

```
              OH
              |
             CH
         H2C/    \CH2
          |        |
         HC        CH2
        /  \     /
      Br     CH2
```

Parent compound: cyclohexane (becomes cyclohexanol)
Position of —OH: carbon-1 (*not* carbon-3)
Substituents: 3-bromo (*not* 5-bromo)
Name: 3-Bromocyclohexanol (it is assumed that the —OH is on carbon-1 in cyclic structures)

Common Names

See Section 11.2 for the names of the common alkyl groups.

The common names for alcohols are derived from the alkyl group corresponding to the parent compound. The name of the alkyl group is followed by the word *alcohol.* For some alcohols, such as ethylene glycol and glycerol, historical names are used. The following examples provide the I.U.P.A.C. and common names of several alcohols:

CH_3CHCH_3 (OH on C-2)	$HOCH_2CH_2OH$	CH_3CH_2OH
2-Propanol (isopropyl alcohol)	1,2-Ethanediol (ethylene glycol)	Ethanol (ethyl alcohol) (grain alcohol)

Question 13.1

Use the I.U.P.A.C. Nomenclature System to name each of the following compounds.

a. $CH_3CHCH_2CH_2CH_2OH$ with CH_3 on C-2

b. $CH_3CHCH_2CHCH_3$ with OH on C-2 and CH_2CH_3 on C-4

c. $CH_2—CH—CH_2$ with OH on each carbon
(Common name: Glycerol)

d. $CH_3CH_2CH—CHCH_2CH_2OH$ with Cl on C-3 and CH_3 on C-4

Question 13.2

Give the common name and the I.U.P.A.C. name for each of the following compounds.

a. $CH_3CH_2CH_2CH_2CH_2CH_2CH_2OH$
b. CH_3CHCH_3 with OH on C-2
c. CH_3CHCH_2OH with CH_3 on C-2

13.3 Medically Important Alcohols

Methanol

Learning Goal

Methanol (methyl alcohol), CH_3OH, is a colorless and odorless liquid that is used as a solvent and as the starting material for the synthesis of methanal (formaldehyde). Methanol is often called *wood alcohol* because it can be made by heating wood in the absence of air. Methanol is toxic and can cause blindness and perhaps death if ingested. Methanol may also be used as fuel, especially for "formula" racing cars.

Ethanol

Ethanol (ethyl alcohol), CH_3CH_2OH, is a colorless and odorless liquid and is the alcohol in alcoholic beverages. It is also widely used as a solvent and as a raw material for the preparation of other organic chemicals.

The ethanol used in alcoholic beverages comes from the **fermentation** of carbohydrates (sugars and starches). The beverage produced depends on the starting material and the fermentation process: scotch (grain), bourbon (corn), burgundy wine (grapes and grape skins), and chablis wine (grapes without red skins). The following equation summarizes the fermentation process:

Fermentation reactions are described in detail in Section 21.4 and in A Human Perspective: Fermentations: The Good, the Bad, and the Ugly.

$$\underset{\text{Sugar (glucose)}}{C_6H_{12}O_6} \xrightarrow[\text{enzyme action}]{\text{Several steps involving}} \underset{\text{Ethanol (ethyl alcohol)}}{2CH_3CH_2OH} + 2CO_2$$

The alcoholic beverages listed have quite different alcohol concentrations. Wines are generally 12–13% alcohol because the yeasts that produce the ethanol are killed by ethanol concentrations of 12–13%. To produce bourbon or scotch with an alcohol concentration of 40–45% ethanol (80 or 90 proof), the original fermentation products must be distilled.

Distillation is the separation of compounds in a mixture based on differences in boiling points.

The sale and use of pure ethanol (100% ethanol) are regulated by the federal government. To prevent illegal use of pure ethanol, it is *denatured* by the addition of a denaturing agent, which makes it unfit to drink but suitable for laboratory applications.

2-Propanol

2-Propanol (isopropyl alcohol),

$$\begin{array}{c} CH_3CHCH_3 \\ | \\ OH \end{array}$$

was commonly called *rubbing alcohol* because patients with high fevers were often given alcohol baths to reduce body temperature. Rapid evaporation of the alcohol results in skin cooling. This practice is no longer commonly used.

It is also used as a disinfectant, an astringent (skin-drying agent), an industrial solvent, and a raw material in the synthesis of organic chemicals. It is colorless, has a very slight odor, and is toxic when ingested.

1,2-Ethanediol

1,2-Ethanediol (ethylene glycol),

$$\begin{array}{cc} CH_2 - & CH_2 \\ | & | \\ OH & OH \end{array}$$

is used as automobile antifreeze. When added to water in the radiator, the ethylene glycol solute lowers the freezing point and raises the boiling point of the

The colligative properties of solutions are discussed in Section 7.6.

A MEDICAL PERSPECTIVE

Fetal Alcohol Syndrome

The first months of pregnancy are a time of great joy and anticipation but are not without moments of anxiety. On her first visit to the obstetrician the mother-to-be is tested for previous exposure to a number of infectious diseases that could damage the fetus. She is provided with information about diet, weight gain, and drugs that could harm the baby. Among the drugs that should be avoided are alcoholic beverages.

The use of alcoholic beverages by a pregnant woman can cause *fetal alcohol syndrome (FAS)*. A *syndrome* is a set of symptoms that occur together and are characteristic of a particular disease. In this case, physicians have observed that infants born to women with chronic alcoholism showed a reproducible set of abnormalities including mental retardation, poor growth before and after birth, and facial malformations.

Mothers who report only social drinking may have children with *fetal alcohol effects*, a less severe form of fetal alcohol syndrome. This milder form is characterized by a reduced birth weight, some learning disabilities, and behavioral problems.

How does alcohol consumption cause these varied symptoms? No one is exactly sure, but it is well known that the alcohol consumed by the mother crosses the placenta and enters the bloodstream of the fetus. Within about fifteen minutes the concentration of alcohol in the blood of the fetus is as high as that of the mother! However, the mother has enzymes to detoxify the alcohol in her blood; the fetus does not. Now consider that alcohol can cause cell division to stop or be radically altered. It is thought that even a single night on the town could be enough to cause FAS by blocking cell division during a critical developmental period.

This raises the question "How much alcohol can a pregnant woman safely drink?" As we have seen, the severity of the symptoms seems to increase with the amount of alcohol consumed by the mother. However, it is virtually impossible to do the scientific studies that would conclusively determine the risk to the fetus caused by different amounts of alcohol. There is some evidence that suggests that there is a risk associated with drinking even one ounce of absolute (100%) alcohol each day. Because of these facts and uncertainties, the American Medical Association and the U.S. Surgeon General recommend that pregnant women completely abstain from alcohol.

water. Ethylene glycol has a sweet taste but is extremely poisonous. For this reason, color additives are used in antifreeze to ensure that it is properly identified.

1,2,3-Propanetriol

1,2,3-Propanetriol (glycerol),

$$\underset{\text{OH}}{\text{CH}_2}-\underset{\text{OH}}{\text{CH}}-\underset{\text{OH}}{\text{CH}_2}$$

is a viscous, sweet-tasting, nontoxic liquid. It is very soluble in water and is used in cosmetics, pharmaceuticals, and lubricants. Glycerol is obtained as a by-product of the hydrolysis of fats.

13.4 Classification of Alcohols

Learning Goal

Alcohols are classified as **primary (1°)**, **secondary (2°)**, or **tertiary (3°)**, depending on the number of alkyl groups attached to the **carbinol carbon**, the carbon bearing the hydroxyl (—OH) group. If no alkyl groups are attached, the alcohol is methyl alcohol; if there is a single alkyl group, the alcohol is a primary alcohol; an alcohol with two alkyl groups bonded to the carbon bearing the hydroxyl group is a secondary alcohol, and if three alkyl groups are attached, the alcohol is a tertiary alcohol.

$$\text{H}-\overset{\text{OH}}{\underset{\text{H}}{\text{C}}}-\text{H} \qquad \text{R}-\overset{\text{OH}}{\underset{\text{H}}{\text{C}}}-\text{H} \qquad \text{R}-\overset{\text{OH}}{\underset{\text{H}}{\text{C}}}-\text{R} \qquad \text{R}-\overset{\text{OH}}{\underset{\text{R}}{\text{C}}}-\text{R}$$

Methyl alcohol — 1° Alcohol — 2° Alcohol — 3° Alcohol

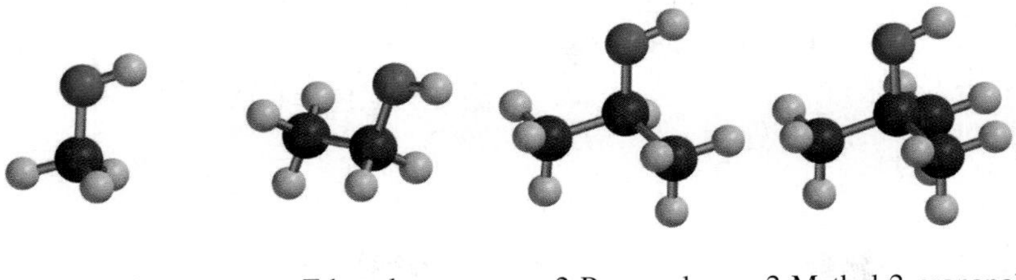

Methanol Ethanol 2-Propanol 2-Methyl-2-propanol

Classifying Alcohols

EXAMPLE 13.3

Learning Goal

Classify each of the following alcohols as primary, secondary, or tertiary.

Solution

In each of the structures shown below, the carbinol carbon is shown in red:

$$CH_3\underset{\displaystyle OH}{\underset{|}{C}}HCH_3$$

This alcohol, 2-propanol, is a secondary alcohol because there are two alkyl groups attached to the carbinol carbon.

$$CH_3\underset{\displaystyle OH}{\underset{|}{\overset{\displaystyle CH_3}{\overset{|}{C}}}}CH_3$$

This alcohol, 2-methyl-2-propanol, is a tertiary alcohol because there are three alkyl groups attached to the carbinol carbon.

$$CH_3CH_2\underset{\displaystyle CH_2CH_3}{\underset{|}{C}}H\overset{\displaystyle OH}{\overset{|}{C}}H_2$$

This alcohol, 2-ethyl-1-butanol, is a primary alcohol because there is only one alkyl group attached to the carbinol carbon.

Question 13.3

Classify each of the following alcohols as 1°, 2°, 3°, or aromatic (phenol).

a. $CH_3CH_2CH_2CH_2OH$

b. $CH_3CH_2\underset{\displaystyle OH}{\underset{|}{C}}HCH_2CH_3$

c. (cyclopentane ring bearing) CH_3 and OH on the same carbon

d. (benzene ring)—OH

e. HO (on cyclopentane ring)

Question 13.4

Classify each of the following alcohols as 1°, 2°, or 3°.

a. $CH_3CH_2CHCH_3$ (with OH on the third carbon)

b. $CH_3CH_2CH_2—C—CH_3$ (with CH_3 and OH on the central carbon)

c. $CH_3CH_2—OH$

d. $CH_2—CH—OH$ over $CH_2—CH_2$ (four-membered ring)

13.5 Reactions Involving Alcohols

Preparation of Alcohols

Learning Goal

Hydration of alkenes is described in Section 12.5.

As we saw in the last chapter, the most important reactions of alkenes are *addition reactions*. Addition of a water molecule to the carbon-carbon double bond of an alkene produces an alcohol. This reaction, called **hydration,** requires a trace of acid (H^+) as a catalyst, as shown in the following equation:

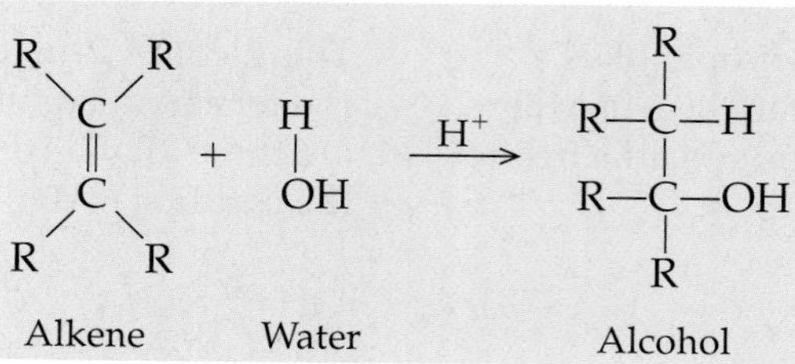

EXAMPLE 13.4

Writing an Equation Representing the Preparation of an Alcohol by the Hydration of an Alkene

Write an equation representing the preparation of cyclohexanol from cyclohexene.

Solution

Begin by writing the structure of cyclohexene. Recall that cyclohexene is a six-carbon cyclic alkene. Now add the water molecule to the equation.

Cyclohexene + H—OH ⟶

Cyclohexene Water

You will recognize that the hydration reaction involves the addition of a water molecule to the carbon-carbon double bond. Recall that the reaction requires a trace of acid as a catalyst. Complete the equation by adding the catalyst and product, cyclohexanol.

Continued—

EXAMPLE 13.4 —*Continued*

$$C_6H_{10} + H\text{—}OH \xrightarrow{H^+} C_6H_{11}OH$$

Cyclohexene Water Cyclohexanol

Alcohols may also be prepared via the hydrogenation (reduction) of aldehydes and ketones. This reaction, summarized as follows, is discussed in Section 14.4, and is similar to the hydrogenation of alkenes.

Learning Goal

In an aldehyde, R^1 and R^2 may be either alkyl groups or H. In ketones, R^1 and R^2 are both alkyl groups.

$$R^1\text{—}C(=O)\text{—}R^2 + H\text{—}H \xrightarrow{\text{Catalyst}} R^1\text{—}CH(OH)\text{—}R^2$$

Aldehyde or Ketone Hydrogen Alcohol

EXAMPLE 13.5

Writing an Equation Representing the Preparation of an Alcohol by the Hydrogenation (Reduction) of an Aldehyde

Write an equation representing the preparation of 1-propanol from propanal.

Solution

Begin by writing the structure of propanal. Propanal is a three-carbon aldehyde. Aldehydes are characterized by the presence of a carbonyl group (—C=O) attached to the end of the carbon chain of the molecule. After you have drawn the structure of propanal, add diatomic hydrogen to the equation.

$$CH_3\text{—}CH_2\text{—}CHO + H\text{—}H \xrightarrow{\text{catalyst}}$$

Propanol Hydrogen

Notice that the general equation reveals this reaction to be another example of a hydrogenation reaction. As the hydrogens are added to the carbon-oxygen double bond, it is converted to a carbon-oxygen single bond, as the carbonyl oxygen becomes a hydroxyl group.

$$CH_3\text{—}CH_2\text{—}CHO + H\text{—}H \xrightarrow{\text{catalyst}} CH_3\text{—}CH_2\text{—}CH_2\text{—}OH$$

Propanal Hydrogen 1-Propanol

EXAMPLE 13.6 Writing an Equation Representing the Preparation of an Alcohol by the Hydrogenation (Reduction) of a Ketone

Learning Goal

Write an equation representing the preparation of 2-propanol from propanone.

Solution

Begin by writing the structure of propanone. Propanone is a three-carbon ketone. Ketones are characterized by the presence of a carbonyl group (—C=O) located anywhere within the carbon chain of the molecule. In the structure of propanone, the carbonyl group must be associated with the center carbon. After you have drawn the structure of propanone, add diatomic hydrogen to the equation.

$$CH_3-\overset{O}{\overset{\|}{C}}-CH_3 + H-H \xrightarrow{\text{catalyst}}$$

Propanone Hydrogen

Notice that this reaction is another example of a hydrogenation reaction. As the hydrogens are added to the carbon-oxygen double bond, it is converted to a carbon-oxygen single bond, as the carbonyl oxygen becomes a hydroxyl group.

$$CH_3-\overset{O}{\overset{\|}{C}}-CH_3 + H-H \xrightarrow{\text{catalyst}} CH_3-\underset{H}{\overset{OH}{C}}-CH_3$$

Propanone Hydrogen 2-Propanol

Question 13.5

Write an equation representing the hydration of cyclopentene. Provide structures and names for the reactants and products.

Question 13.6

Write an equation representing the reduction of butanone. Provide the structures and names for the reactants and products. [*Hint:* butanone is a four-carbon ketone.]

Dehydration of Alcohols

Learning Goal

Alcohols undergo **dehydration** (lose water) when heated with concentrated sulfuric acid (H_2SO_4) or phosphoric acid (H_3PO_4). Dehydration is an example of an **elimination reaction,** that is, a reaction in which a molecule loses atoms or ions from its structure. In this case, the —OH and —H are "eliminated" from adjacent carbons in the alcohol to produce an alkene and water. We have just seen that alkenes can be hydrated to give alcohols. Dehydration is simply the reverse process: the conversion of an alcohol back to an alkene. This is seen in the following general reaction and the examples that follow:

$$\text{R}-\underset{\text{H}}{\overset{\text{H}}{\text{C}}}-\underset{\text{OH}}{\overset{\text{H}}{\text{C}}}-\text{H} \xrightarrow{\text{H}^+,\ \text{heat}} \text{R}-\text{CH}=\text{CH}_2 + \text{H}-\text{OH}$$

Alcohol | Alkene | Water

$$\text{H}-\underset{\text{H}}{\overset{\text{H}}{\text{C}}}-\underset{\text{OH}}{\overset{\text{H}}{\text{C}}}-\text{H} \xrightarrow{\text{H}^+,\ \text{heat}} \text{CH}_2=\text{CH}_2 + \text{H}-\text{OH}$$

Ethanol (ethyl alcohol) | Ethene (ethylene)

$$\text{CH}_3\text{CH}_2\text{CH}_2\text{OH} \xrightarrow{\text{H}^+,\ \text{heat}} \text{CH}_3\text{CH}=\text{CH}_2 + \text{H}_2\text{O}$$

1-Propanol (propyl alcohol) | Propene (propylene)

In some cases, dehydration of alcohols produces a mixture of products, as seen in the following example:

$$\text{CH}_3\text{CH}_2-\underset{\text{OH}}{\text{CH}}-\text{CH}_3 \xrightarrow[\text{heat}]{\text{H}^+} \text{CH}_3\text{CH}_2-\text{CH}=\text{CH}_2 + \text{CH}_3-\text{CH}=\text{CH}-\text{CH}_3 + \text{H}_2\text{O}$$

2-Butanol | 1-Butene (minor product) | 2-Butene (major product)

Notice in the equation shown above and in Example 13.7, the major product is the more highly substituted alkene. In 1875 the Russian chemist Alexander Zaitsev developed a rule to describe such reactions. **Zaitsev's rule** states that in an elimination reaction, the alkene with the greatest number of alkyl groups on the double bonded carbon (the more highly substituted alkene) is the major product of the reaction.

Predicting the Products of Alcohol Dehydration — EXAMPLE 13.7

Predict the products of the dehydration of 3-methyl-2-butanol.

Solution

Assuming that no rearrangement occurs, the product(s) of a dehydration of an alcohol will contain a double bond in which one of the carbons was the original carbinol carbon—the carbon to which the hydroxyl group is attached. Consider the following reaction:

Continued—

EXAMPLE 13.7 —*Continued*

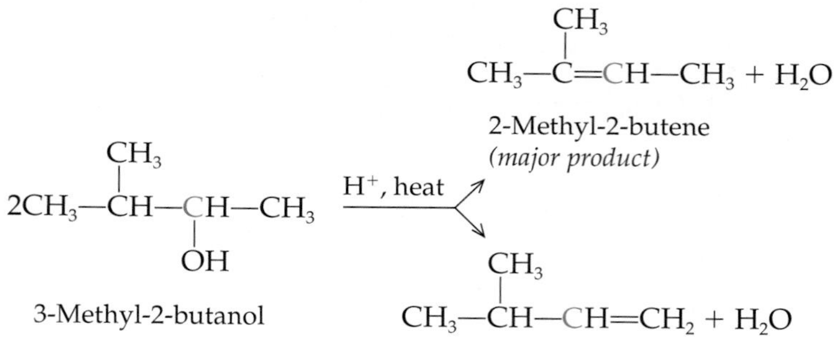

$$2CH_3{-}\underset{|}{\overset{CH_3}{\overset{|}{C}H}}{-}\underset{\underset{OH}{|}}{CH}{-}CH_3 \xrightarrow{H^+,\ heat} CH_3{-}\overset{CH_3}{\overset{|}{C}}{=}CH{-}CH_3 + H_2O$$

3-Methyl-2-butanol → 2-Methyl-2-butene (*major product*)

$$ \xrightarrow{H^+,\ heat} CH_3{-}\overset{CH_3}{\overset{|}{C}H}{-}CH{=}CH_2 + H_2O$$

3-Methyl-1-butene (*minor product*)

It is clear that both the major and minor products have a double bond to carbon number 2 in the original alcohol (this carbon is set off in color). Zaitsev's Rule tells us that in dehydration reactions with more than one product possible, the more highly branched alkene predominates. In the reaction shown, 2-methyl-2-butene has three alkyl groups at the double bond, whereas 3-methyl-1-butene has only two alkyl groups at the double bond. The more highly branched alkene is more stable and thus is the major product.

Question 13.7

Predict the products obtained on reacting each of the following alkenes with water and a trace of acid:

a. Ethene
b. Propene
c. 1-Butene
d. 2-Butene
e. 2-Methylpropene

Question 13.8

Draw the alkene products that would be produced on dehydration of each of the following alcohols:

a. 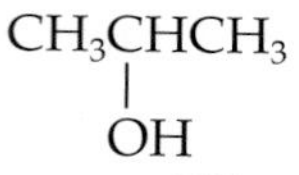 $CH_3\underset{\underset{OH}{|}}{C}HCH_3$

b. $CH_3CH_2\underset{\underset{OH}{|}}{C}HCH_3$

c. $CH_3{-}\underset{\underset{OH}{|}}{\overset{\overset{CH_3}{|}}{C}}{-}CH_2CH_3$

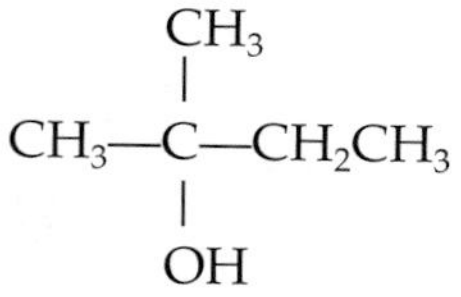

d. 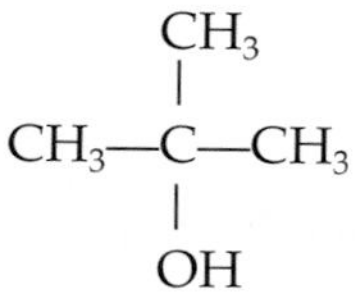 $CH_3{-}\underset{\underset{OH}{|}}{\overset{\overset{CH_3}{|}}{C}}{-}CH_3$

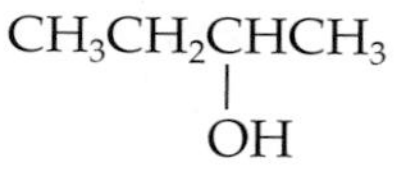

This reaction, and the other reactions of glycolysis, are considered in Section 21.3.

The dehydration of 2-phosphoglycerate to phosphoenolpyruvate is a critical step in the metabolism of the sugar glucose. In the following structures the circled P represents a phosphoryl group (PO_4^{2-}).

The squiggle (~) represents a high energy bond.

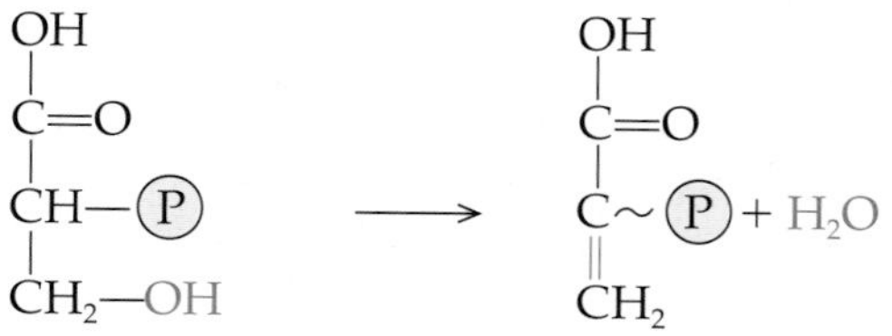

$$\begin{array}{c} OH \\ | \\ C{=}O \\ | \\ CH{-}(P) \\ | \\ CH_2{-}OH \end{array} \longrightarrow \begin{array}{c} OH \\ | \\ C{=}O \\ | \\ C{\sim}(P) + H_2O \\ \| \\ CH_2 \end{array}$$

2-Phosphoglycerate → Phosphoenolpyruvate

Oxidation Reactions

Learning Goal

Alcohols may be oxidized with a variety of oxidizing agents to aldehydes, ketones, and carboxylic acids. The most commonly used oxidizing agents are solutions of basic potassium permanganate ($KMnO_4/OH^-$) and chromic acid (H_2CrO_4). The symbol [O] over the reaction arrow is used throughout this book to designate any general oxidizing agent, as in the following reactions:

Oxidation of methanol produces the aldehyde methanal:

$$H-\underset{H}{\overset{OH}{C}}-H \xrightarrow{[O]} H-\overset{O}{\overset{\|}{C}}-H$$

Methanol (methyl alcohol) An alcohol — Methanal (formaldehyde) An aldehyde

An *oxidation* reaction involves a gain of oxygen or the loss of hydrogen. A *reduction* reaction involves the loss of oxygen or gain of hydrogen. If two hydrogens are gained or lost for every oxygen gained or lost, the reaction is neither an oxidation nor a reduction.

Note that the symbol [O] is used throughout this book to designate any oxidizing agent.

Oxidation of a primary alcohol produces an aldehyde:

$$R^1-\underset{H}{\overset{OH}{C}}-H \xrightarrow{[O]} R^1-\overset{O}{\overset{\|}{C}}-H$$

1° Alcohol — An aldehyde

As we will see in Section 14.4, aldehydes can undergo further oxidation to produce carboxylic acids.

Writing an Equation Representing the Oxidation of a Primary Alcohol — EXAMPLE 13.8

Write an equation showing the oxidation of 2,2-dimethyl-1-propanol to produce 2,2-dimethylpropanal.

Solution

Begin by writing the structure of the reactant, 2,2-dimethyl-1-propanol and indicate the need for an oxidizing agent by placing the designation [O] over the reaction arrow:

$$CH_3-\underset{CH_3}{\overset{CH_3}{C}}-\underset{H}{\overset{H}{C}}-OH \xrightarrow{[O]}$$

2,2-Dimethyl-1-propanol

Now show the oxidation of the hydroxyl group to the aldehyde carbonyl group.

$$CH_3-\underset{CH_3}{\overset{CH_3}{C}}-\underset{H}{\overset{H}{C}}-OH \xrightarrow{[O]} CH_3-\underset{CH_3}{\overset{CH_3}{C}}-\overset{O}{\overset{\|}{C}}-H$$

2,2-Dimethyl-1-propanol — 2,2-Dimethylpropanal

Oxidation of a secondary alcohol produces a ketone:

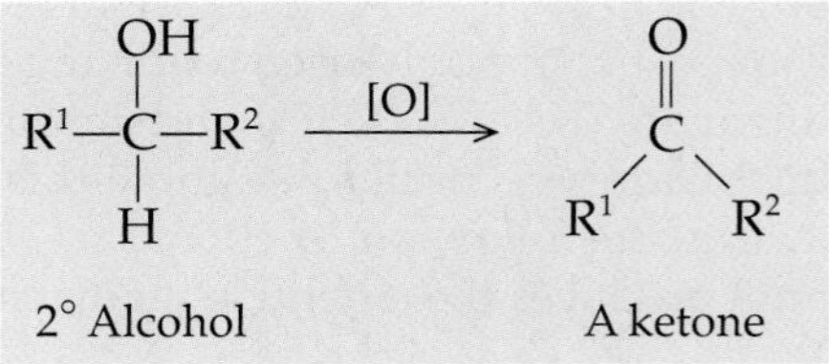

$$\underset{\text{2° Alcohol}}{R^1\text{—}\overset{\overset{\displaystyle OH}{|}}{\underset{\underset{\displaystyle H}{|}}{C}}\text{—}R^2} \xrightarrow{[O]} \underset{\text{A ketone}}{R^1\text{—}\overset{\overset{\displaystyle O}{\|}}{C}\text{—}R^2}$$

EXAMPLE 13.9 Writing an Equation Representing the Oxidation of a Secondary Alcohol

Learning Goal 8

Write an equation showing the oxidation of 2-propanol to produce propanone.

Solution

Begin by writing the structure of the reactant, 2-propanol, and indicate the need for an oxidizing agent by placing the designation [O] over the reaction arrow:

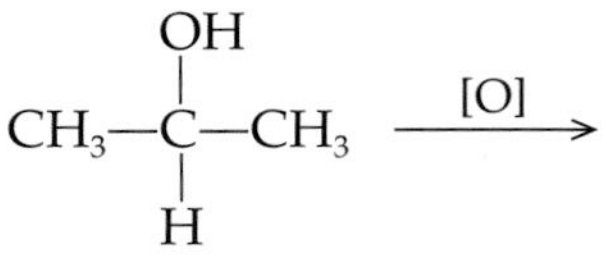

$$\underset{\text{2-Propanol}}{CH_3\text{—}\overset{\overset{\displaystyle OH}{|}}{\underset{\underset{\displaystyle H}{|}}{C}}\text{—}CH_3} \xrightarrow{[O]}$$

Now show the oxidation of the hydroxyl group to the ketone carbonyl group.

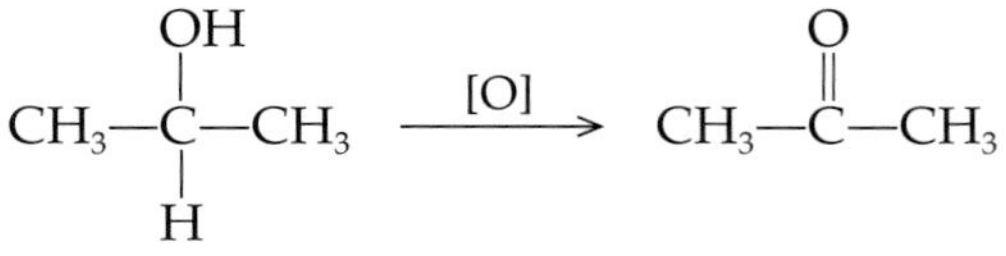

$$\underset{\text{2-Propanol}}{CH_3\text{—}\overset{\overset{\displaystyle OH}{|}}{\underset{\underset{\displaystyle H}{|}}{C}}\text{—}CH_3} \xrightarrow{[O]} \underset{\text{Propanone}}{CH_3\text{—}\overset{\overset{\displaystyle O}{\|}}{C}\text{—}CH_3}$$

Tertiary alcohols cannot be oxidized:

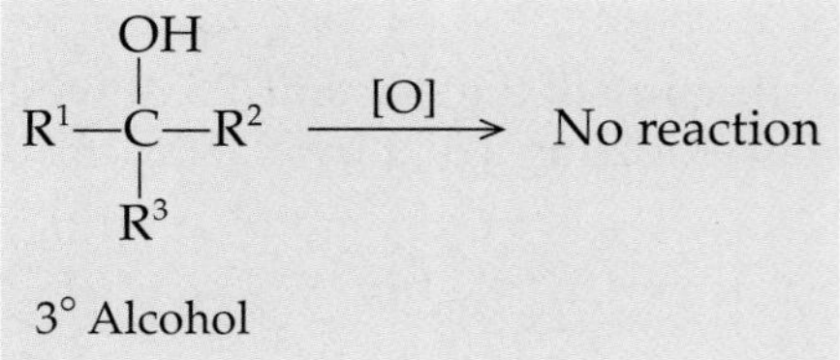

$$\underset{\text{3° Alcohol}}{R^1\text{—}\overset{\overset{\displaystyle OH}{|}}{\underset{\underset{\displaystyle R^3}{|}}{C}}\text{—}R^2} \xrightarrow{[O]} \text{No reaction}$$

For the oxidation reaction to occur, the carbon bearing the hydroxyl group must contain at least one C—H bond. Because tertiary alcohols contain three C—C bonds to the carbinol carbon, they cannot undergo oxidation.

EXAMPLE 13.10 Writing an Equation Representing the Oxidation of a Tertiary Alcohol

Write an equation showing the oxidation of 2-methyl-2-propanol.

Solution

Begin by writing the structure of the reactant, 2-methyl-2-propanol and indicate the need for an oxidizing agent by placing the designation [O] over the reaction arrow:

Continued—

EXAMPLE 13.10 —*Continued*

$$CH_3—\overset{\displaystyle OH}{\underset{\displaystyle CH_3}{C}}—CH_3 \xrightarrow{[O]}$$

2-Methyl-2-propanol

The structure of 2-methyl-2-propanol reveals that it is a tertiary alcohol. Therefore no oxidation reaction can occur because the carbon bearing the hydroxyl group is bonded to three other carbon atoms, not to a hydrogen atom.

$$CH_3—\overset{\displaystyle OH}{\underset{\displaystyle CH_3}{C}}—CH_3 \xrightarrow{[O]} \text{No reaction}$$

2-Methyl-2-propanol

When ethanol is metabolized in the liver, it is oxidized to ethanal (acetaldehyde). If too much ethanol is present in the body, an overabundance of ethanal is formed, which causes many of the adverse effects of the "morning-after hangover." Continued oxidation of ethanal produces ethanoic acid (acetic acid), which is used as an energy source by the cell and eventually oxidized to CO_2 and H_2O. These reactions, summarized as follows, are catalyzed by liver enzymes.

These reactions are discussed further in Sections 14.4 and 15.1.

$$CH_3CH_2—OH \longrightarrow CH_3\overset{\displaystyle O}{\overset{\|}{C}}—H \longrightarrow CH_3\overset{\displaystyle O}{\overset{\|}{C}}—OH \longrightarrow CO_2 + H_2O$$

Ethanol (ethyl alcohol) — Ethanal (acetaldehyde) — Ethanoic acid (acetic acid)

13.6 Oxidation and Reduction in Living Systems

Learning Goal

Before beginning a discussion of oxidation and reduction in living systems, we must understand how to recognize **oxidation** (loss of electrons) and **reduction** (gain of electrons) in organic compounds. It is easy to determine when an oxidation or a reduction occurs in inorganic compounds because the process is accompanied by a change in charge. For example,

$$Ag^0 \longrightarrow Ag^+ + 1e^-$$

With the loss of an electron, the neutral atom is converted to a positive ion, which is oxidation. In contrast,

$$:\ddot{\underset{..}{Br}}\cdot + e^- \longrightarrow :\ddot{\underset{..}{Br}}:^-$$

With the gain of one electron, the bromine atom is converted to a negative ion, which is reduction.

When organic compounds are involved, however, there may be no change in charge, and it is often difficult to determine whether oxidation or reduction has occurred. The following simplified view may help.

In organic systems, *oxidation* may be recognized as a gain of oxygen or a loss of hydrogen. A *reduction* reaction may involve a loss of oxygen or gain of hydrogen.

A HUMAN PERSPECTIVE

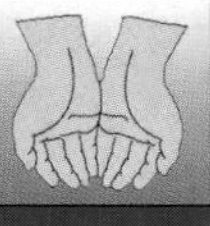

Alcohol Consumption and the Breathalyzer Test

Ethanol has been used widely as a beverage, a medicinal, and a solvent in numerous pharmaceutical preparations. Such common usage often overshadows the fact that ethanol is a toxic substance. Ethanol consumption is associated with a variety of long-term effects, including cirrhosis of the liver, death of brain cells, and alcoholism. Alcohol consumed by the mother can even affect the normal development of her unborn child and result in fetal alcohol syndrome. For these reasons, over-the-counter cough and cold medications that were once prepared in ethanol are now manufactured in alcohol-free form.

Short-term effects, linked to the social use of ethanol, center on its effects on behavior, reflexes, and coordination. Blood alcohol levels of 0.05–0.15% seriously inhibit coordination. Blood levels in excess of 0.10% are considered evidence of intoxication in most states. Blood alcohol levels in the range of 0.30–0.50% produce unconsciousness and the risk of death.

The loss of some coordination and reflex action is particularly serious when the affected individual attempts to operate a motor vehicle. Law enforcement has come to rely on the "breathalyzer" test to screen for individuals suspected of driving while intoxicated. Those with a positive breathalyzer test are then given a more accurate blood test to establish their guilt or innocence.

The suspect is required to exhale into a solution that will react with the unmetabolized alcohol in the breath. The partial pressure of the alcohol in the exhaled air has been demonstrated to be proportional to the blood alcohol level. The solution is an acidic solution of dichromate ion, which is yellow-orange. The alcohol reduces the chromium in the dichromate ion from +6 to +3, the Cr^{3+} ion, which is green. The intensity of the green color is measured, and it is proportional to the amount of ethanol that was oxidized. The reaction is:

$$\underset{\text{Yellow-orange}}{16H^+ + 2Cr_2O_7^{2-}} + 3CH_3CH_2OH \longrightarrow 3CH_3COOH + \underset{\text{Green}}{4Cr^{3+}} + 11H_2O$$

The breathalyzer test is a technological development based on a scientific understanding of the chemical reactions that ethanol may undergo—a further example of the dependence of technology on science.

Consider the following compounds. An alkane may be oxidized to an alcohol by gaining an oxygen. A primary or secondary alcohol may be oxidized to an aldehyde or ketone, respectively, by the loss of hydrogen. Finally, an aldehyde may be oxidized to a carboxylic acid by gaining an oxygen.

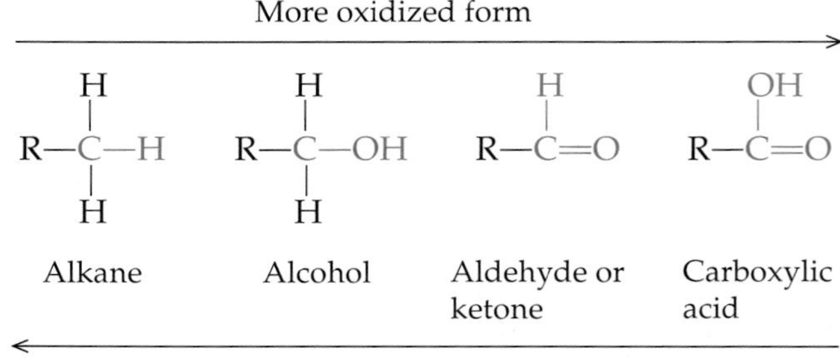

Thus the conversion of an alkane to an alcohol, an alcohol to a carbonyl compound, and a carbonyl compound (aldehyde) to a carboxylic acid are all examples of oxidations. Conversions in the opposite direction are reductions.

Oxidation and reduction reactions also play an important role in the chemistry of living systems. In living systems these reactions are catalyzed by the action of various enzymes called *oxidoreductases*. These enzymes require compounds called *coenzymes* to accept or donate hydrogen in the reactions that they catalyze.

Nicotinamide adenine dinucleotide, NAD^+, is a coenzyme commonly involved in biological oxidation–reduction reactions (Figure 13.3). Now consider the final reaction of the citric acid cycle, an energy-harvesting pathway essential to life. In this reaction, catalyzed by malate dehydrogenase, malate is oxidized to produce oxaloacetate:

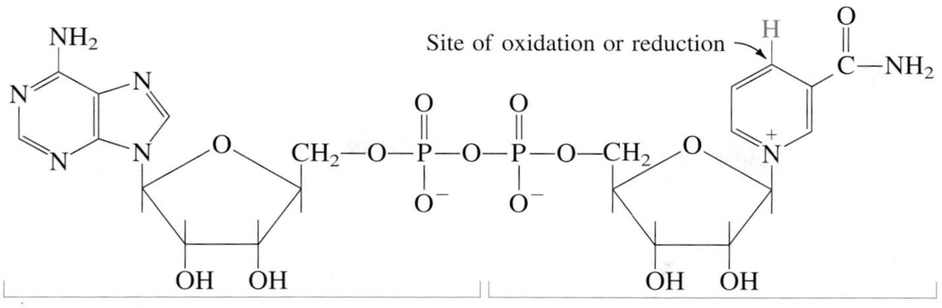

Site of oxidation or reduction

Adenine nucleotide | Nicotinamide nucleotide

Nicotinamide adenine dinucleotide (NAD^+)

Figure 13.3
Nicotinamide adenine dinucleotide.

$$\text{HO—CH(COO}^-\text{)—CH}_2\text{—COO}^- + NAD^+ \xrightarrow{\text{Malate dehydrogenase}} {}^-\text{OOC—C(=O)—CH}_2\text{—COO}^- + NADH + H^+$$

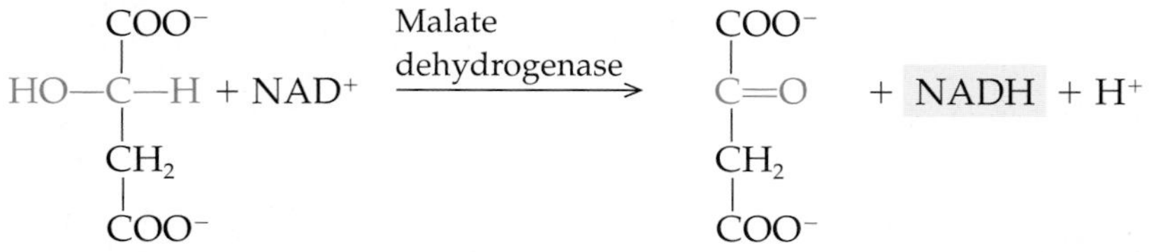

Malate → Oxaloacetate

NAD^+ participates by accepting hydrogen from the malate. As malate is oxidized, NAD^+ is reduced to NADH.

NAD^+ actually accepts a hydride anion, H^-, hydrogen with two electrons.

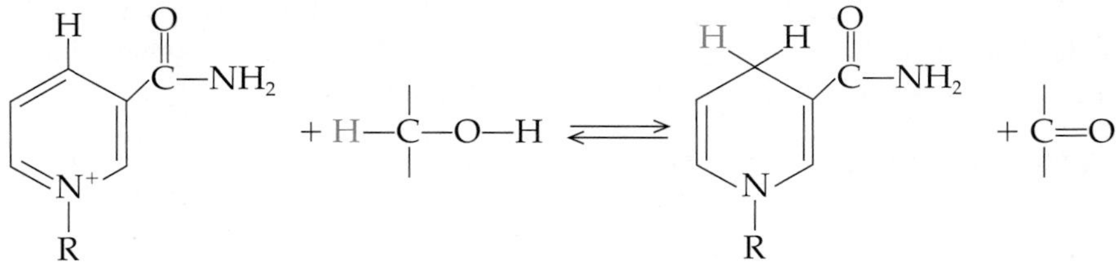

$$NAD^+ + \text{H—C—O—H} \rightleftharpoons NADH + \text{C=O}$$

NAD^+
Oxidized form

NADH
Reduced form

We will study many other biologically important oxidation-reduction reactions in upcoming chapters.

13.7 Phenols

Learning Goal 10

Phenols are compounds in which the hydroxyl group is attached to a benzene ring (Figure 13.4). Like alcohols, they are polar compounds because of the polar hydroxyl group. Thus the simpler phenols are somewhat soluble in water. They are found in flavorings and fragrances (mint and savory) and are used as preservatives (butylated hydroxytoluene, BHT). Examples include:

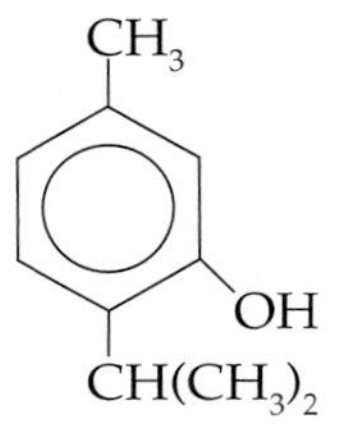

Thymol (mint)

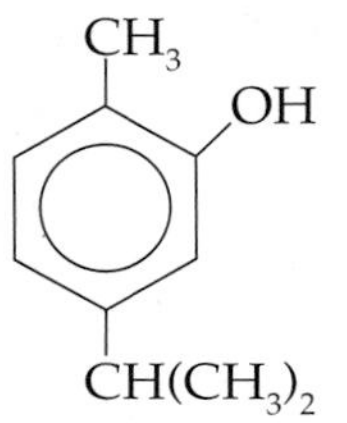

Carvacrol (savory)

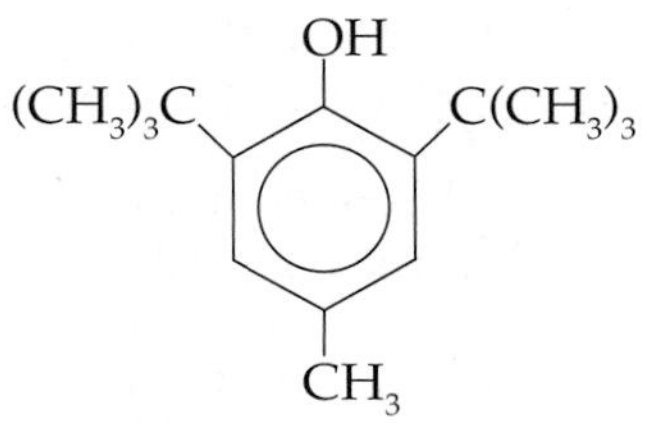

Butylated hydroxytoluene, BHT (food preservative)

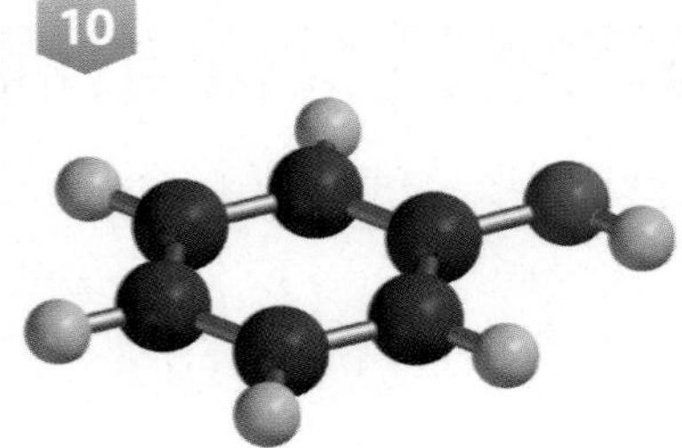

Figure 13.4
Ball-and-stick model of phenol.

A dilute solution of phenol must be used because concentrated phenol causes severe burns and because phenol is not highly soluble in water.

Phenols are also widely used in health care as germicides. In fact, carbolic acid, a dilute solution of phenol, was used as an antiseptic and disinfectant by Joseph Lister in his early work to decrease postsurgical infections. He used carbolic acid to bathe surgical wounds and to "sterilize" his instruments. Other derivatives of phenol that are used as antiseptics and disinfectants include hexachlorophene, hexylresorcinol, and *o*-phenylphenol. The structures of these compounds are shown below:

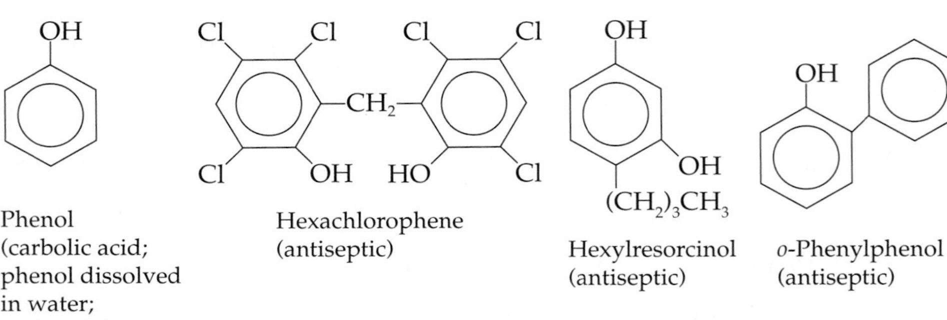

Phenol (carbolic acid; phenol dissolved in water; antiseptic)

Hexachlorophene (antiseptic)

Hexylresorcinol (antiseptic)

o-Phenylphenol (antiseptic)

13.8 Ethers

Learning Goal 11

Ethers have the general formula R—O—R, and thus they are structurally related to alcohols (R—O—H). The C—O bonds of ethers are polar, so ether molecules are polar (Figure 13.5). However, ethers do not form hydrogen bonds to one another because there is no —OH group. Therefore they have much lower boiling points than alcohols of similar molecular weight but higher boiling points than alkanes of similar molecular weight. Compare the following examples:

$CH_3CH_2CH_2CH_3$	$CH_3—O—CH_2CH_3$	$CH_3CH_2CH_2OH$
Butane	Methoxyethane	1-Propanol
(butane)	(ethyl methyl ether)	(propyl alcohol)
M.W. = 58	M.W. = 60	M.W. = 60
b.p. = −0.5°C	b.p. = 7.9°C	b.p. = 97.2°C

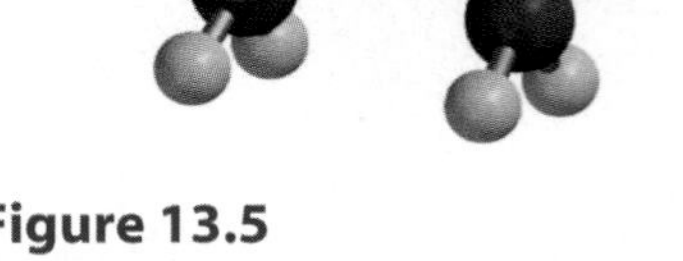

Figure 13.5
Ball-and-stick model of the ether, methoxymethane (dimethyl ether).

In the I.U.P.A.C. system of naming ethers the —OR substituent is named as an alkoxy group. This is analogous to the name *hydroxy* for the —OH group. Thus $CH_3—O—$ is methoxy, $CH_3CH_2—O—$ is ethoxy, and so on.

An alkoxy group is an alkyl group bonded to an oxygen atom (—OR).

EXAMPLE 13.11 ***Using I.U.P.A.C. Nomenclature to Name an Ether***

Name the following ether using I.U.P.A.C. nomenclature.

Solution

$$\begin{array}{c} \quad\; O—CH_3 \\ | \\ CH_3CH_2CHCH_2CH_2CH_2CH_2CH_2CH_3 \\ 1\;\;2\;\;3\;\;4\;\;5\;\;6\;\;7\;\;8\;\;9 \end{array}$$

Parent compound: nonane
Position of alkoxy group: carbon-3 (*not* carbon-7)
Substituents: 3-methoxy
Name: 3-Methoxynonane

In the common system of nomenclature, ethers are named by placing the names of the two alkyl groups attached to the ether oxygen as prefixes in front of the word *ether.* The names of the two groups can be placed either alphabetically or by size (smaller to larger), as seen in the following examples:

$CH_3—O—CH_3$	$CH_3—O—CH_2CH_3$	$CH_3CH_2—O—CH(CH_3)_2$
Dimethyl ether or methyl ether	Ethyl methyl ether or methyl ethyl ether	Ethyl isopropyl ether

Naming Ethers Using the Common Nomenclature System

EXAMPLE 13.12

Learning Goal 11

Write the common name for each of the following ethers.

Solution

	$CH_3CH_2—O—CH_2CH_3$	$CH_3—O—CH_2CH_2CH_3$
Alkyl Groups:	two ethyl groups	methyl and propyl
Name:	Diethyl ether	Methyl propyl ether

Notice that there is only one correct name for methyl propyl ether because the methyl group is smaller than the propyl group and it would be first in an alphabetical listing also.

Chemically ethers are moderately inert. They do not react with reducing agents or bases under normal conditions. However, they are extremely volatile and highly flammable (easily oxidized in air) and hence must always be treated with great care.

Ethers may be prepared by a dehydration reaction (removal of water) between two alcohol molecules, as shown in the following general reaction. The reaction requires heat and acid.

$$\underset{\text{Alcohol}}{R—OH} + \underset{\text{Alcohol}}{R'—OH} \xrightarrow{H^+} \underset{\text{Ether}}{R—O—R'} + \underset{\text{Water}}{H_2O}$$

Writing an Equation Representing the Synthesis of an Ether via a Dehydration Reaction

EXAMPLE 13.13

Learning Goal 12

Write an equation showing the synthesis of dimethyl ether.

Solution

The alkyl substituents of this ether are two methyl groups. Thus, the alcohol that must undergo dehydration to produce dimethyl ether is methanol.

$$\underset{\text{Methanol}}{CH_3OH} + \underset{\text{Methanol}}{CH_3OH} \xrightarrow{H^+} \underset{\text{Dimethyl ether}}{CH_3—O—CH_3} + \underset{\text{Water}}{H_2O}$$

Question 13.9

Write an equation showing the dehydration reaction that would produce diethyl ether. Provide structures and names for all reactants and products.

Question 13.10

Write an equation showing the dehydration reaction between two molecules of 2-propanol. Provide structures and names for all reactants and products.

Diethyl ether was the first general anesthetic used. The dentist Dr. William Morton is credited with its introduction in the 1800s. Diethyl ether functions as an anesthetic by interacting with the central nervous system. It appears that diethyl ether (and many other general anesthetics) functions by accumulating in the lipid material of the nerve cells, thereby interfering with nerve impulse transmission. This results in analgesia, a lessened perception of pain.

Halogenated ethers are also routinely used as general anesthetics. They are less flammable than diethyl ether and are therefore safer to store and work with. *Penthrane* and *enthrane* (trade names) are two of the more commonly used members of this family:

$$CH_3{-}O{-}\underset{F}{\overset{F}{C}}{-}\underset{Cl}{\overset{Cl}{C}}{-}H \qquad\qquad H{-}\underset{F}{\overset{F}{C}}{-}O{-}\underset{F}{\overset{F}{C}}{-}\underset{Cl}{\overset{F}{C}}{-}H$$

Penthrane Enthrane

13.9 Thiols

Compounds that contain the sulfhydryl group (—SH) are called **thiols.** They are similar to alcohols in structure, but the sulfur atom replaces the oxygen atom.

Thiols and many other sulfur compounds have nauseating aromas. They are found in substances as different as the defensive spray of the North American striped skunk, onions, and garlic.

$$HS{-}H_2C{-}CH{=}CH{-}CH_3 \quad (\text{H and } CH_3 \text{ trans})$$

trans-2-Butene-1-thiol
North American skunk
defense spray

Learning Goal

The I.U.P.A.C. rules for naming thiols are similar to those for naming alcohols, except that the full name of the alkane is retained. The suffix *-thiol* follows the name of the parent compound.

EXAMPLE 13.14 ***Naming Thiols Using the I.U.P.A.C. Nomenclature System***

Write the I.U.P.A.C. name for the thiols shown below.

Solution

Retain the full name of the parent compound and add the suffix *-thiol*.

Continued—

EXAMPLE 13.14 —*Continued*

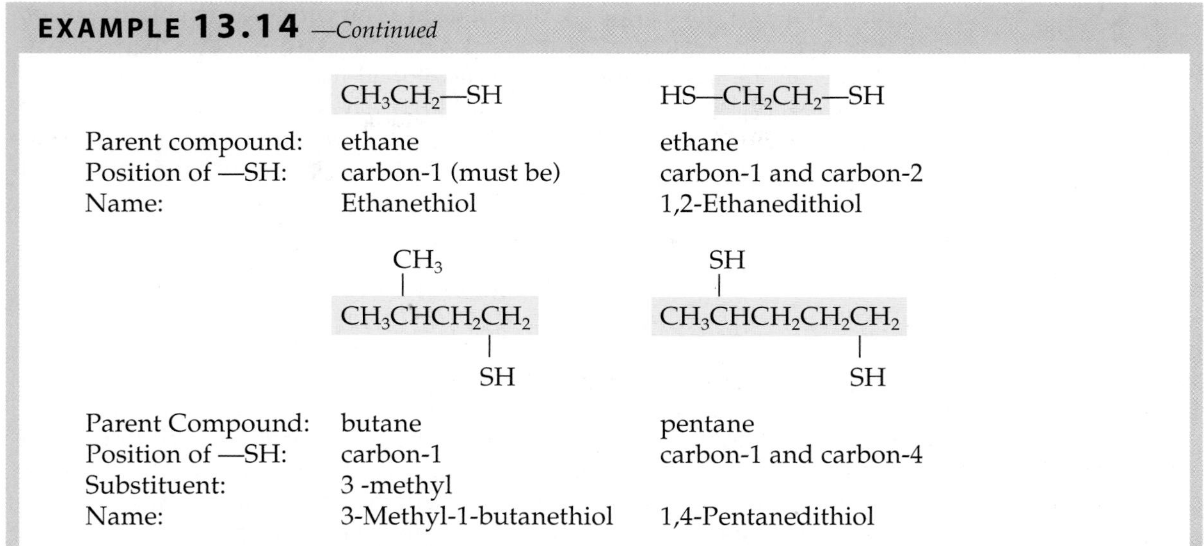

	CH_3CH_2—SH	HS—CH_2CH_2—SH
Parent compound:	ethane	ethane
Position of —SH:	carbon-1 (must be)	carbon-1 and carbon-2
Name:	Ethanethiol	1,2-Ethanedithiol

	$CH_3CHCH_2CH_2$ (with CH_3 on carbon-3, SH on carbon-1)	$CH_3CHCH_2CH_2CH_2$ (with SH on carbon-4 and carbon-1)
Parent Compound:	butane	pentane
Position of —SH:	carbon-1	carbon-1 and carbon-4
Substituent:	3 -methyl	
Name:	3-Methyl-1-butanethiol	1,4-Pentanedithiol

The amino acid cysteine is a thiol that has an important role to play in the structure and shape of many proteins. Two cysteine molecules can undergo oxidation to form cystine. The new bond formed is called a **disulfide** bond (—S—S—) bond.

Amino acids are the subunits from which proteins are made. A protein is a long polymer, or chain, of many amino acids bonded to one another.

$$2\ {}^{+}NH_3\text{—CH(COO}^-\text{)—CH}_2\text{—S—H} \underset{\text{Reduction}}{\overset{\text{Oxidation}}{\rightleftharpoons}} {}^{+}NH_3\text{—CH(COO}^-\text{)—CH}_2\text{—S—S—CH}_2\text{—CH(NH}_3^{+}\text{)—COO}^- + 2H^+ + 2e^-$$

2 Cysteine ⟶ Cystine (Disulfide Bond)

$${}^{+}NH_3\text{—CH(CH}_2\text{—SH)—C(=O)—O}^-$$

Cysteine

If the two cysteines are in different protein chains, the disulfide bond between them forms a bridge joining them together (Figure 13.6). If the two cysteines are in the same protein chain, a loop is formed.

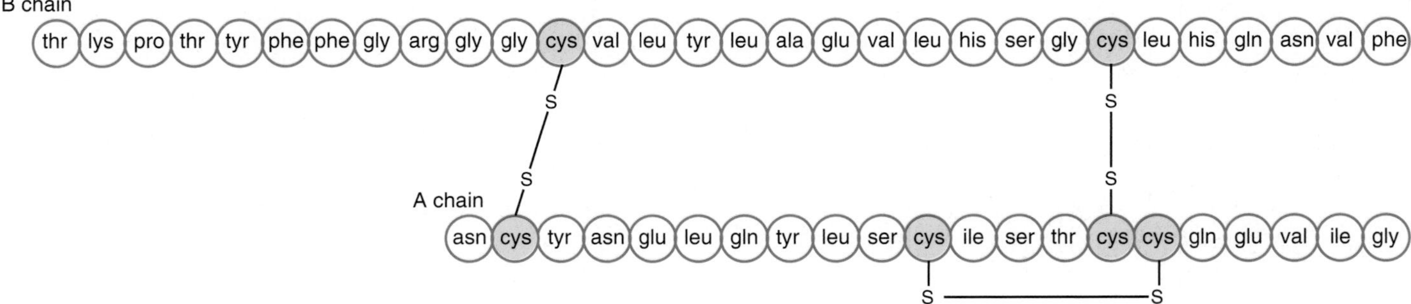

Figure 13.6
Human insulin, a hormone that controls blood sugar levels, is made up of two protein chains. Each of the circles represents an amino acid in the protein chains. Disulfide bonds hold the A and B chains of insulin together. Disulfide bonds also join sections of the A chain together.

British Anti-Lewisite (BAL) is a dithiol used as an antidote in mercury poisoning. It was originally developed as an antidote to a mustard-gas-like chemical warfare agent called Lewisite. Lewisite was developed near the end of World War I and never used. By the onset of World War II, Lewisite was considered to be obsolete because of the discovery of BAL, an effective, inexpensive antidote. The two thiol groups of BAL form a water-soluble complex with mercury (or with the arsenic in Lewisite) that is excreted from the body in the urine.

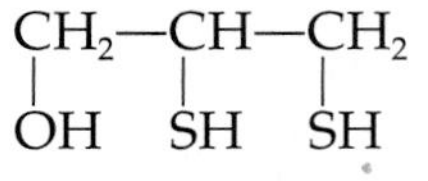

$$\underset{\text{OH}}{CH_2}-\underset{\text{SH}}{CH}-\underset{\text{SH}}{CH_2}$$

BAL

The reactions involving coenzyme A are discussed in detail in Chapters 21, 22, and 23.

A high-energy bond is one that releases a great deal of energy when it is broken.

Coenzyme A is a thiol that serves as a "carrier" of acetyl groups (CH_3CO—) in biochemical reactions. It plays a central role in metabolism by shuttling acetyl groups from one reaction to another. When the two-carbon acetate group is attached to coenzyme A, the product is acetyl coenzyme A (acetyl CoA). The bond between coenzyme A and the acetyl group is a high-energy *thioester bond.* In effect, formation of the high-energy thioester bond "energizes" the acetyl group so that it can participate in other biochemical reactions.

$$CH_3-\overset{O}{\overset{\|}{C}}\sim S-CH_2-CH_2-\underset{H}{N}-\overset{O}{\overset{\|}{C}}-CH_2-CH_2-\underset{H}{N}-\overset{O}{\overset{\|}{C}}-\underset{HO}{\overset{H}{C}}-\underset{CH_3}{\overset{CH_3}{C}}-CH_2-O-\underset{^-O}{\overset{O}{\overset{\|}{P}}}-O-\underset{^-O}{\overset{O}{\overset{\|}{P}}}OCH_2\text{-(ribose, 3'-}OPO_3^{2-}\text{)-adenine}$$

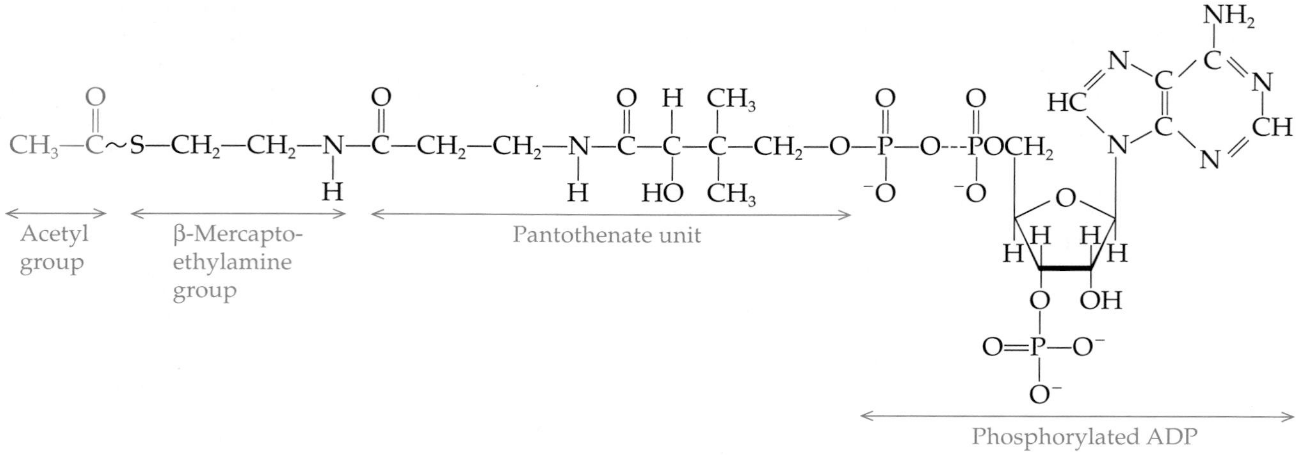

Acetyl group; β-Mercapto-ethylamine group; Pantothenate unit; Phosphorylated ADP

Acetyl coenzyme A
(acetyl CoA)

Acetyl CoA is made and used in the energy-producing reactions that provide most of the energy for life processes. It is also required for the biosynthesis of many biological molecules.

Summary of Reactions

Preparation of alcohols

Hydration of alkenes:

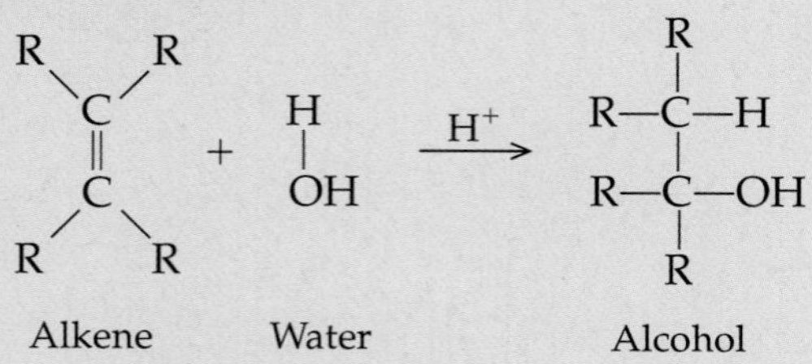

$$R_2C=CR_2 + H-OH \xrightarrow{H^+} R_2C(H)-C(OH)R_2$$

Alkene + Water → Alcohol

Reduction of an aldehyde or ketone:

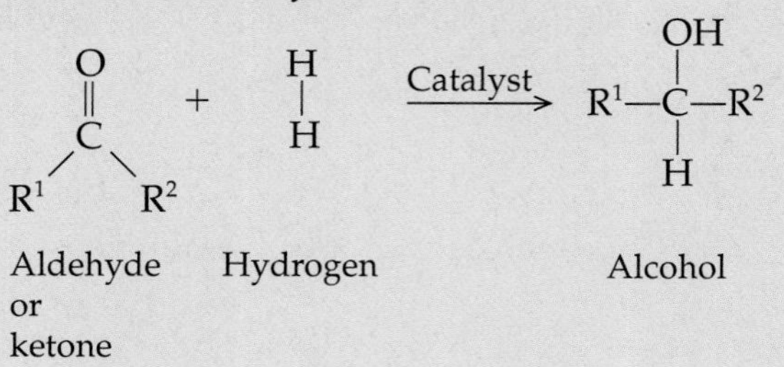

$$R^1-\overset{O}{\overset{\|}{C}}-R^2 + H-H \xrightarrow{\text{Catalyst}} R^1-\underset{H}{\overset{OH}{C}}-R^2$$

Aldehyde or ketone + Hydrogen → Alcohol

Dehydration of alcohols

$$R{-}\underset{H}{\overset{H}{C}}{-}\underset{OH}{\overset{H}{C}}{-}H \xrightarrow{H^+,\ \text{heat}} R{-}CH{=}CH_2 + HOH$$

Alcohol — Alkene — Water

Oxidation reactions

Oxidation of a primary alcohol:

$$R^1{-}\underset{H}{\overset{OH}{C}}{-}H \xrightarrow{[O]} R^1{-}\overset{O}{\overset{\|}{C}}{-}H$$

1° Alcohol — An aldehyde

Oxidation of a secondary alcohol:

$$R^1{-}\underset{H}{\overset{OH}{C}}{-}R^2 \xrightarrow{[O]} R^1{-}\overset{O}{\overset{\|}{C}}{-}R^2$$

2° Alcohol — A ketone

Oxidation of a tertiary alcohol:

$$R^1{-}\underset{R^2}{\overset{OH}{C}}{-}R^3 \xrightarrow{[O]} \text{No reaction}$$

3° Alcohol

Dehydration synthesis of an ether

$$R{-}OH + R'{-}OH \xrightarrow{H^+} R{-}O{-}R' + H_2O$$

Alcohol — Alcohol — Ether — Water

Summary

13.1 Alcohols: Structure and Physical Properties

Alcohols are characterized by the *hydroxyl group (—OH)* and have the general formula R—OH. They are very polar, owing to the polar hydroxyl group, and are able to form intermolecular hydrogen bonds. Because of hydrogen bonding between alcohol molecules, they have higher boiling points than hydrocarbons of comparable molecular weight. The smaller alcohols are very water soluble.

13.2 Alcohols: Nomenclature

In the I.U.P.A.C. system, alcohols are named by determining the parent compound and replacing the *-e* ending with *-ol*. The chain is numbered to give the hydroxyl group the lowest possible number. Common names are derived from the alkyl group corresponding to the parent compound.

13.3 Medically Important Alcohols

Methanol is a toxic alcohol that is used as a solvent. Ethanol is the alcohol consumed in beer, wine, and distilled liquors. Isopropanol is used as a disinfectant. Ethylene glycol (1,2-ethanediol) is used as antifreeze, and glycerol (1,2,3-propanetriol) is used in cosmetics and pharmaceuticals.

13.4 Classification of Alcohols

Alcohols may be classified as *primary, secondary,* or *tertiary,* depending on the number of alkyl groups attached to the *carbinol* carbon, the carbon bearing the hydroxyl group. A primary alcohol has a single alkyl group bonded to the *carbinol carbon*. Secondary and tertiary alcohols have two and three alkyl groups, respectively.

13.5 Reactions Involving Alcohols

Alcohols can be prepared by the *hydration* of alkenes or reduction of aldehydes and ketones. Alcohols can undergo *dehydration* to yield alkenes. Primary and secondary alcohols undergo oxidation reactions to yield aldehydes and ketones, respectively. Tertiary alcohols do not undergo oxidation.

13.6 Oxidation and Reduction in Living Systems

In organic and biological systems *oxidation* involves the gain of oxygen or loss of hydrogen. *Reduction* involves the loss of oxygen or gain of hydrogen. Nicotinamide adenine dinucleotide, NAD^+, is a coenzyme involved in many biological oxidation and reduction reactions.

13.7 Phenols

Phenols are compounds in which the hydroxyl group is attached to a benzene ring; they have the general formula

Ar—OH. Many phenols are important as antiseptics and disinfectants.

13.8 Ethers

Ethers are characterized by the R—O—R functional group. Ethers are generally nonreactive but are extremely flammable. Diethyl ether was the first general anesthetic used in medical practice. It has since been replaced by penthrane and enthrane, which are less flammable.

13.9 Thiols

Thiols are characterized by the sulfhydryl group (—SH). The amino acid cysteine is a thiol that is extremely important for maintaining the correct shapes of proteins. Coenzyme A is a thiol that serves as a "carrier" of acetyl groups in biochemical reactions.

Key Terms

alcohol (13.1)
carbinol carbon (13.4)
dehydration (13.5)
disulfide (13.9)
elimination reaction (13.5)
ether (13.8)
fermentation (13.3)
hydration (13.5)
hydroxyl group (13.1)
oxidation (13.6)
phenol (13.7)
primary (1°) alcohol (13.4)
reduction (13.6)
secondary (2°) alcohol (13.4)
tertiary (3°) alcohol (13.4)
thiol (13.9)
Zaitsev's rule (13.5)

Questions and Problems

Alcohols: Structure and Physical Properties

13.11 Arrange the following compounds in order of increasing boiling point, beginning with the lowest:

a. $CH_3CH_2CH_2CH_2CH_3$

b. $CH_3CHCH_2CHCH_3$
OH OH

c. $CH_3CHCH_2CH_2CH_3$
OH

d. $CH_3CH_2CH_2—O—CH_2CH_3$

13.12 Why do alcohols have higher boiling points than alkanes? Why are small alcohols readily soluble in water whereas large alcohols are much less soluble?

13.13 Which member of each of the following pairs is more soluble in water?

a. CH_3CH_2OH or $CH_3CH_2CH_2CH_2OH$

b. $CH_3CH_2CH_2CH_2CH_3$ or $CH_3CH_2CH_2CH_2—OH$

c. OH (cyclopentanol) or CH_3CHCH_3
OH

13.14 Arrange the three alcohols in each of the following sets in order of increasing solubility in water:

a. $CH_3CH_2CH_2CH_2CH_2OH$ $CH_3CHCH_2CHCH_2CH_3$
OH OH

$CH_3CHCH_2CHCH_2CH_2OH$
OH OH

b. Pentyl alcohol 1-Hexanol Ethylene glycol

Alcohols: Nomenclature

13.15 Give the I.U.P.A.C. name for each of the following compounds:

a. $CH_3CH_2CH_2CH_2CH_2CH_2CH_2OH$

b. CH_3CHCH_3
OH

c. CH_3
$CH_3—C—CH_3$
CH_2OH

d. Br
$CH_3CH_2CHCH_2CH_2CH_2OH$

e. CH_3
$CH_3CH—CCH_2CH_2CH_3$
OH CH_3

f. $CH_2CH_2CH_2CH_3$
$CH_3CH_2CCH_2CH_3$
OH

13.16 Draw each of the following, using complete structural formulas:

a. 3-Hexanol
b. 1,2,3-Pentanetriol
c. 2-Methyl-2-pentanol
d. Cyclohexanol
e. 3,4-Dimethyl-3-heptanol

13.17 Give the I.U.P.A.C. name for each of the following compounds:

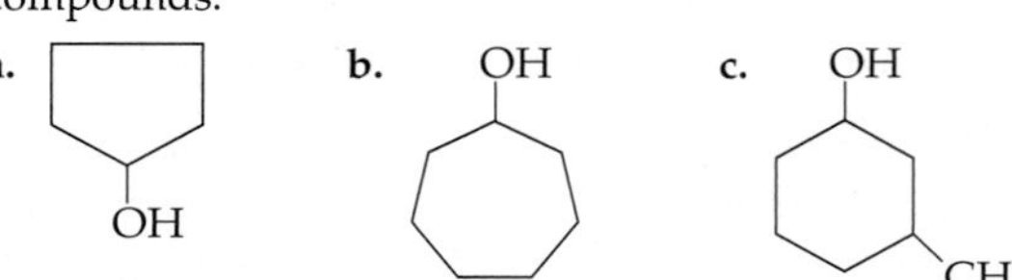

13.18 Draw each of the following alcohols:

a. 1-Iodo-2-butanol
b. 1,2-Butanediol
c. Cyclobutanol

13.19 Give the common name for each of the following compounds:

a. CH_3OH
b. CH_3CH_2OH
c. $CH_2—CH_2$
OH OH
d. $CH_3CH_2CH_2OH$

13.20 Draw the structure of each of the following compounds:

a. Pentyl alcohol
b. Isopropyl alcohol
c. Octyl alcohol
d. Propyl alcohol

13.21 Draw a complete structural formula for each of the following compounds:

a. 4-Methyl-2-hexanol
b. Isobutyl alcohol
c. 1,5-Pentanediol
d. 3-Nonanol
e. 1,3,5-Cyclohexanetriol

13.22 Name each of the following alcohols using the I.U.P.A.C. Nomenclature System:

a. OH
CH_3

b. OH
Br

c. $CH_3CHCH_2CHCH_2CHCH_3$
OH OH OH

d. OH
$CH_3CH_2CHCHCHCH_3$
OH OH

Medically Important Alcohols

13.23 What is denatured alcohol? Why is alcohol denatured?

13.24 What are the principal uses of methanol, ethanol, and isopropyl alcohol?

13.25 What is fermentation?

13.26 Why do wines typically have an alcohol concentration of 12–13%?

13.27 Why must fermentation products be distilled to produce liquors such as scotch?

13.28 If a bottle of distilled alcoholic spirits—for example, scotch whiskey—is labeled as 80 proof, what is the percentage of alcohol in the scotch?

Classification of Alcohols

13.29 Classify each of the following as a 1°, 2°, or 3° alcohol:

a. 3-Methyl-1-butanol
b. 2-Methylcyclopentanol
c. *t*-Butyl alcohol
d. 1-Methylcyclopentanol
e. 2-Methyl-2-pentanol

13.30 Classify each of the following as a 1°, 2°, or 3° alcohol:

a. $CH_3CH_2CH_2CH_2CH_2CH_2CH_2OH$

b. CH_3CHCH_3
OH

c. CH_3
$CH_3—C—CH_3$
CH_2OH

d. Br
$CH_3CH_2CHCH_2CH_2CH_2OH$

e. CH_3
$CH_3CH—CCH_2CH_2CH_3$
OH CH_3

13.31 Classify each of the following as a primary, secondary, or tertiary alcohol:

a. CH_2CH_3
$CH_3CH_2C—OH$
CH_3

b. $CH_3CHCH_2CHCH_3$
OH Br

c. $CH_3CH_2CH_2OH$

d. CH_3
$CH_3CCH_2CH_2CH_3$
OH

13.32 Classify each of the following as a primary, secondary, or tertiary alcohol:

a. 2-Methyl-2-butanol
b. 1,2-Dimethylcyclohexanol
c. 2,3,4-Trimethylcyclopentanol
d. 3,3-Dimethyl-2-pentanol

Reactions Involving Alcohols

13.33 Predict the products formed by the hydration of the following alkenes:

a. 1-Pentene
b. 2-Pentene
c. 3-Methyl-1-butene
d. 3,3-Dimethyl-1-butene

13.34 Draw the alkene products of the dehydration of the following alcohols:

a. 2-Pentanol
b. 3-Methyl-1-pentanol
c. 2-Butanol
d. 4-Chloro-2-pentanol
e. 1-Propanol

13.35 Write an equation showing the hydration of each of the following alkenes. Name each of the products using the I.U.P.A.C. Nomenclature System.

a. 2-Hexene
b. Cyclopentene
c. 1-Octene
d. 1-Methylcyclohexene

13.36 Write an equation showing the dehydration of each of the following alcohols. Name each of the reactants and products using the I.U.P.A.C. Nomenclature System.

a. $CH_3CHCH_2CH_3$
OH

b. $CH_2{=}CHCH_2CHCH_3$
OH

c. $CH_3CH_2CH_2OH$

d. OH
CH_3

13.37 What product(s) would result from the oxidation of each of the following alcohols with, for example, potassium permanganate? If no reaction occurs, write N.R.

a. 2-Butanol
b. 2-Methyl-2-hexanol
c. Cyclohexanol
d. 1-Methyl-1-cyclopentanol

13.38 We have seen that ethanol is metabolized to ethanal (acetaldehyde) in the liver. What would be the product formed, under the same conditions, from each of the following alcohols?

a. CH_3OH
b. $CH_3CH_2CH_2OH$
c. $CH_3CH_2CH_2CH_2OH$

13.39 Give the oxidation products of the following alcohols. If no reaction occurs, write N.R.

a. $CH_3CH_2CH(OH)CH_2CH_3$

b. $CH_3CH_2CH_2OH$

c. $CH_3CH(OH)CH_2CH(CH_3)CH_3$

d. $CH_3—C(OH)(CH_3)—CH_2CH_3$

e. $C_6H_5—CH_2CH_2CH_2OH$

13.40 Write an equation, using complete structural formulas, demonstrating each of the following chemical transformations:

a. Oxidation of an alcohol to an aldehyde

b. Oxidation of an alcohol to a ketone

c. Dehydration of a cyclic alcohol to a cycloalkene

d. Hydrogenation of an alkene to an alkane

13.41 Write the reaction, occurring in the liver, that causes the oxidation of ethanol. What is the product of this reaction and what symptoms are caused by the product?

13.42 Write the reaction, occurring in the liver, that causes the oxidation of methanol. What is the product of this reaction and what is the possible result of the accumulation of the product in the body?

13.43 Write an equation for the preparation of 2-butanol from 1-butene. What type of reaction is involved?

13.44 Write a general equation for the preparation of an alcohol from an aldehyde or ketone. What type of reaction is involved?

13.45 Show how acetone can be prepared from propene.

$CH_3C(=O)CH_3$

Acetone

13.46 Show how bromocyclopentane can be prepared from cyclopentanol.

13.47 Give the oxidation products for cholesterol.

Cholesterol

13.48 Why is a tertiary alcohol not oxidized?

Oxidation and Reduction in Living Systems

13.49 Define the terms *oxidation* and *reduction.*

13.50 How do we recognize oxidation and reduction in organic compounds?

13.51 Arrange the following compounds from the most reduced to the most oxidized:

$CH_3CH_2C(=O)—OH$ $CH_3CH_2CH_3$

$CH_3CH_2C(=O)—H$ $CH_3CH_2CH_2OH$

13.52 What is the role of the coenzyme nicotinamide adenine dinucleotide (NAD^+) in enzyme-catalyzed oxidation–reduction reactions?

Phenols

13.53 2,4,6-Trinitrophenol is known by the common name *picric acid.* Picric acid is a solid but is readily soluble in water. In solution it is used as a biological tissue stain. As a solid, it is also known to be unstable and may explode. In this way it is similar to 2,4,6-trinitrotoluene (TNT). Draw the structures of picric acid and TNT. Why is picric acid readily soluble in water whereas TNT is not?

13.54 Name the following aromatic compounds using the I.U.P.A.C. system:

a. benzene ring with OH and NO_2 (adjacent)

b. benzene ring with $CH(CH_3)_2$ and OH (para)

c. benzene ring with HO, Br, and Cl

d. benzene ring with OH, Br, and CH_3

13.55 List some phenol compounds that are commonly used as antiseptics or disinfectants.

13.56 Why must a dilute solution of phenol be used for disinfecting environmental surfaces?

Ethers

13.57 Draw all of the alcohols and ethers of molecular formula $C_4H_{10}O$.

13.58 Name each of the isomers drawn for Problem 13.57.

13.59 Give the I.U.P.A.C. names for penthrane and enthrane (see Section 13.8).

13.60 Why have penthrane and enthrane replaced diethyl ether as a general anesthetic?

13.61 Ethers may be prepared by the removal of water (dehydration) between two alcohols, as shown. Give the structure(s) of the ethers formed by the reaction of the following alcohol(s) under acidic conditions with heat.

Example: $CH_3OH + HOCH_3 \xrightarrow[\text{Heat}]{H^+,} CH_3OCH_3 + H_2O$

a. $2CH_3CH_2OH \longrightarrow ?$

b. $CH_3OH + CH_3CH_2OH \longrightarrow ?$

c. $(CH_3)_2CHOH + CH_3OH \longrightarrow ?$

d. 2 cyclopentyl$—CH_2OH \longrightarrow ?$

13.62 We have seen that alcohols are capable of hydrogen bonding to each other. Hydrogen bonding is also possible between alcohol molecules and water molecules or between alcohol molecules and ether molecules. Ether molecules *do not* hydrogen bond to each other, however. Explain.

13.63 Name each of the following ethers using the I.U.P.A.C. Nomenclature System:

a. $CH_3CH(OCH_2CH_3)CH_2CH_2CH_3$

b. $CH_3CH_2CH(OCH_3)CH_3$

c. $CH_3CH_2CH_2CH(OCH_2CH_3)$

d. 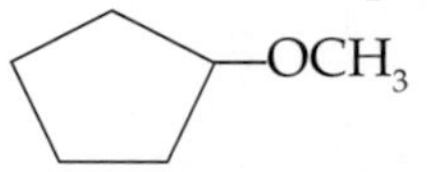 (cyclopentyl)$-OCH_3$

13.64 Draw the structural formula for each of the following ethers:
- a. Methyl propyl ether
- b. 2-Methoxyoctane
- c. Diisopropyl ether
- d. 3-Ethoxypentane

Thiols

13.65 Cystine is an amino acid formed from the oxidation of two cysteine molecules to form a disulfide bond. The molecular formula of cystine is $C_6H_{12}O_4N_2S_2$. Draw the structural formula of cystine. (*Hint:* For the structure of cysteine, see page 385.)

13.66 Explain the way in which British Anti-Lewisite acts as an antidote for mercury poisoning.

13.67 Give the I.U.P.A.C. name for each of the following thiols. (*Hint:* Use the rules for alcohol nomenclature and the suffix *-thiol.*)
- a. $CH_3CH_2CH_2{-}SH$
- b. $CH_3CH(SH)CH_2CH_3$
- c. $CH_3{-}C(CH_2CH_3)(SH){-}CH_3$
- d. 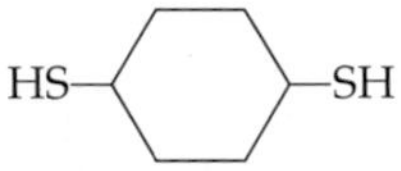 HS–(cyclohexane)–SH

13.68 Give the I.U.P.A.C. name for each of the following thiols. (*Hint:* Use the rules for alcohol nomenclature and the suffix *-thiol.*)
- a. $CH_2(SH)CH(SH)CH_3$
- b. 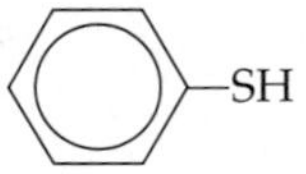 (benzene ring)–SH
- c. $CH_3CH(SH)CH_2CH_2CH_3$
- d. $CH_3CH_2CH_2CH_2CH_2CH_2CH_2SH$

Critical Thinking Problems

1. You are provided with two solvents: water (H_2O) and hexane ($CH_3CH_2CH_2CH_2CH_2CH_3$). You are also provided with two biological molecules whose structures are shown here:

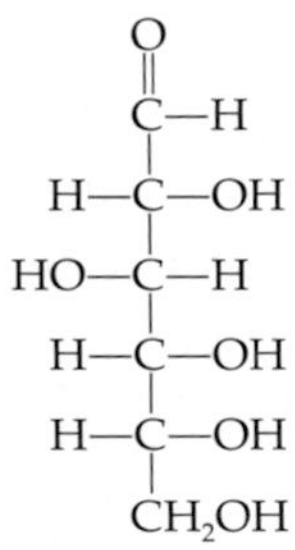

$$\begin{array}{c} O \\ \| \\ C{-}H \\ | \\ H{-}C{-}OH \\ | \\ HO{-}C{-}H \\ | \\ H{-}C{-}OH \\ | \\ H{-}C{-}OH \\ | \\ CH_2OH \end{array}$$

$$\begin{array}{l} H{-}\overset{H}{\overset{|}{C}}{-}O{-}\overset{O}{\overset{\|}{C}}{-}CH_2CH_2CH_2CH_2CH_2CH_2CH_2CH_2CH_2CH_2CH_2CH_2CH_2CH_2CH_3 \\ H{-}C{-}O{-}\overset{O}{\overset{\|}{C}}{-}CH_2CH_2CH_2CH_2CH_2CH_2CH_2CH_2CH_2CH_2CH_2CH_2CH_2CH_2CH_3 \\ H{-}\underset{H}{\underset{|}{C}}{-}O{-}\overset{O}{\overset{\|}{C}}{-}CH_2CH_2CH_2CH_2CH_2CH_2CH_2CH_2CH_2CH_2CH_2CH_2CH_2CH_2CH_3 \end{array}$$

Predict which biological molecule would be more soluble in water and which would be more soluble in hexane. Defend your prediction. Design a careful experiment to test your hypothesis.

Consider the digestion of dietary molecules in the digestive tract. Which of the two biological molecules shown in this problem would be more easily digested under the conditions present in the digestive tract?

2. Cholesterol is an alcohol and a steroid (Chapter 18). Diets that contain large amounts of cholesterol have been linked to heart disease and atherosclerosis, hardening of the arteries. The narrowing of the artery, caused by plaque buildup, is very apparent. Cholesterol is directly involved in this buildup. Describe the various functional groups and principal structural features of the cholesterol molecule. Would you use a polar or nonpolar solvent to dissolve cholesterol? Explain your reasoning.

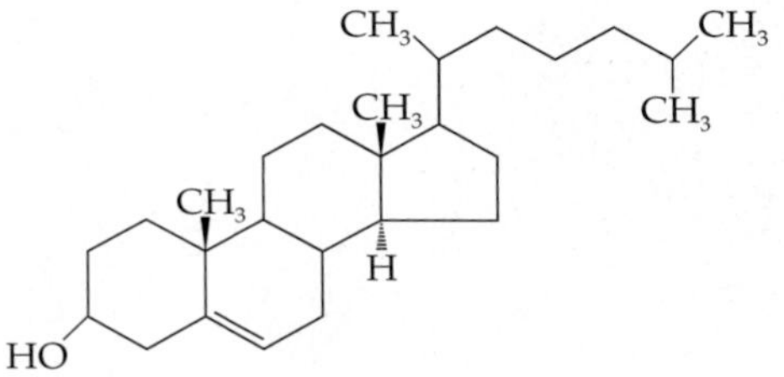

CH3 CH3 CH3 CH3 CH3 H HO

Cholesterol

3. An unknown compound A is known to be an alcohol with the molecular formula $C_4H_{10}O$. When dehydrated, compound A gave only one alkene product, C_4H_8, compound B. Compound A could not be oxidized. What are the identities of compound A and compound B?

4. Sulfides are the sulfur analogs of ethers, that is, ethers in which oxygen has been substituted by a sulfur atom. They are named in an analogous manner to the ethers with the term *sulfide* replacing *ether.* For example, $CH_3{-}S{-}CH_3$ is dimethyl sulfide. Draw the sulfides that correspond to the following ethers and name them:
   - a. diethyl ether
   - b. methyl propyl ether
   - c. dibutyl ether
   - d. ethyl phenyl ether

5. Dimethyl sulfoxide (DMSO) has been used by many sports enthusiasts as a linament for sore joints; it acts as an anti-inflammatory agent and a mild analgesic (pain killer). However, it is no longer recommended for this purpose because it carries toxic impurities into the blood. DMSO is a sulfoxide—it contains the S=O functional group. DMSO is prepared from dimethyl sulfide by mild oxidation, and it has the molecular formula C_2H_6SO. Draw the structure of DMSO.

Vanilla plant blossom and structure of vanillin.

14

Aldehydes and Ketones

ORGANIC CHEMISTRY

Outline

Learning Goals

1. Draw the structures and discuss the physical properties of aldehydes and ketones.
2. From the structures, write the common and I.U.P.A.C. names of aldehydes and ketones.
3. List several aldehydes and ketones that are of natural, commercial, health, and environmental interest and describe their significance.
4. Write equations for the preparation of aldehydes and ketones by the oxidation of alcohols.
5. Write equations representing the oxidation of carbonyl compounds.
6. Write equations representing the reduction of carbonyl compounds.
7. Write equations for the preparation of hemiacetals, hemiketals, acetals, and ketals.
8. Draw the keto and enol forms of aldehydes and ketones.
9. Write equations showing the aldol condensation.

CHEMISTRY CONNECTION

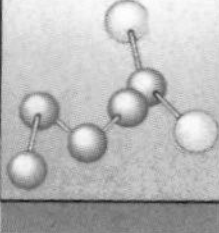

Genetic Complexity from Simple Molecules

Examine any cellular life-form and you will find the same basic genetic system. One or more deoxyribonucleic acid (DNA) molecules carry the genetic code for all the proteins needed by the cell as enzymes—biological catalysts that speed up energy harvesting and biosynthetic reactions—as essential structural elements, and much more. But DNA cannot be read directly to produce these critical proteins. Instead, the genetic information carried by the DNA is first used to produce a variety of ribonucleic acid (RNA) molecules. These RNA molecules work together, along with other molecules, to produce the proteins.

For decades scientists have been trying to figure out how this amazing genetic system could have evolved from a variety of small, simple molecules that were found in the shallow seas and atmosphere of earth perhaps four billion years ago. Which molecule might have formed first? After all, for the system to work all three molecules are needed: DNA to carry the information, RNA to carry and interpret it, and proteins to do all the cellular chores.

A startling discovery in the 1980s suggested an answer. It was discovered that some RNA molecules could act as biological catalysts. In other words, these RNA molecules could do two jobs: carry genetic information and catalyze chemical reactions. A hypothesis was developed: that RNA was the first biological molecule and that our genetic system evolved from it.

Could RNA have evolved from simple molecules in the "primordial soup"? In the 1960s two of the components of RNA, adenine and guanine, were synthesized in the laboratory from simple molecules and energy sources thought to be present on early earth. In 1995 researchers discovered that, by adding the carbonyl-group-containing molecule urea to their mixture, they could make large amounts of two other components of RNA, uracil and cytosine.

The remaining requirements for making an RNA molecule are phosphate groups and the sugar ribose. Phosphate would have been readily available, but what about ribose? Ribose, it turned out, could easily be produced from the simplest *aldehyde*, formaldehyde.

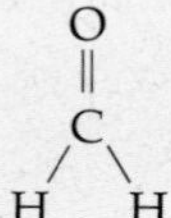

Formaldehyde would have been found in the shallow seas. We can find it today in comets, and many comets struck the early earth. In the laboratory, under conditions designed to imitate those on earth four billion years ago, formaldehyde molecules form chains. These chains twist into cyclic ring structures, including the sugar ribose.

These experiments suggest that all the precursors needed to make RNA could have formed spontaneously and thus that RNA may have been the information-carrying molecule from which our genetic system evolved. But other researchers argue that RNA is too fragile to have survived the conditions on early earth.

We may never know exactly how the first self-replicating genetic system formed. But it is intriguing to speculate about the origin of life and to consider the organic molecules and reactions that may have been involved. In this chapter we consider the aldehydes and ketones, two families of organic molecules containing the carbonyl group. As we will see in this and later chapters, the carbonyl group is a functional group that characterizes many biological molecules and affects their properties and reactivity.

Introduction

The **carbonyl group** consists of a carbon atom bonded to an oxygen atom by a double bond.

Carbonyl group

Compounds containing a carbonyl group are called *carbonyl compounds.* This group includes the aldehydes and ketones covered in this chapter, as well as the carboxylic acids and amides discussed in Chapters 15 and 16.

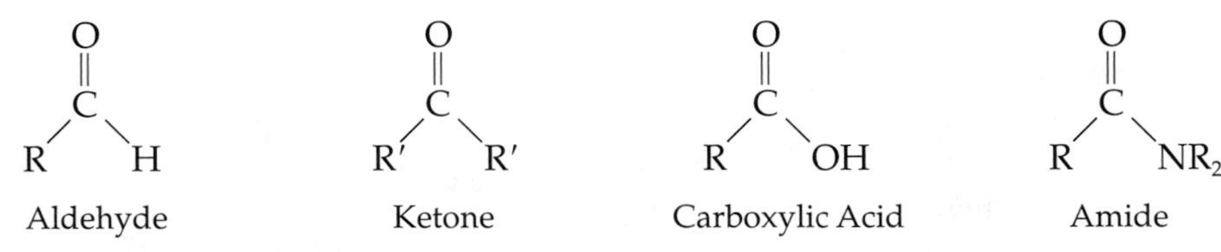

R=H, alkyl, or aryl group R′=alkyl or aryl group

We will study the aldehydes and ketones together because of their similar chemical and physical properties. They are distinguished by the location of the carbonyl group within the carbon chain. In *aldehydes* the carbonyl group is always located at the end of the carbon chain (carbon-1). In *ketones* the carbonyl group is located within the carbon chain of the molecule. Thus, in ketones the carbonyl carbon is attached to two other carbon atoms. However, in aldehydes the carbonyl carbon is attached to at least one hydrogen atom; the second atom attached to the carbonyl carbon of an aldehyde may be another hydrogen or a carbon atom (Figure 14.1).

14.1 Structure and Physical Properties

Learning Goal 1

Aldehydes and **ketones** are polar compounds because of the polar carbonyl group.

Because of the dipole-dipole attractions between molecules, they boil at higher temperatures than hydrocarbons or ethers that have the same number of carbon atoms or are of equivalent molecular weight. Because they cannot form intermolecular hydrogen bonds, their boiling points are lower than those of alcohols of comparable molecular weight. These trends are clearly demonstrated in the following examples:

Dipole-dipole attraction

$CH_3CH_2CH_2CH_3$	$CH_3—O—CH_2CH_3$	$CH_3CH_2CH_2—OH$	$CH_3CH_2—C(=O)—H$	$CH_3—C(=O)—CH_3$
Butane (butane)	Methoxyethane (ethyl methyl ether)	1-Propanol (propyl alcohol)	Propanal (propionaldehyde)	Propanone (acetone)
M.W. = 58	M.W. = 60	M.W. = 60	M.W. = 58	M.W. = 58
b.p. −0.4 °C	b.p. 7.0 °C	b.p. 97.2 °C	b.p. 49 °C	b.p. 56 °C

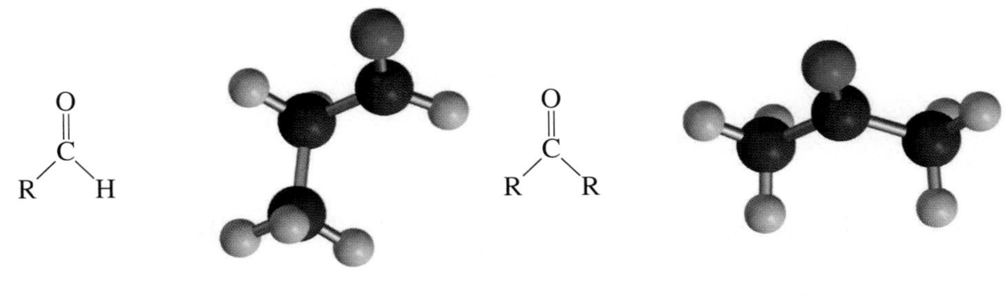

Figure 14.1
The structure of aldehydes and ketones. (a) The general structure of an aldehyde and a ball-and-stick model of the aldehyde propanal. (b) The general structure of a ketone and a ball-and-stick model of the ketone propanone.

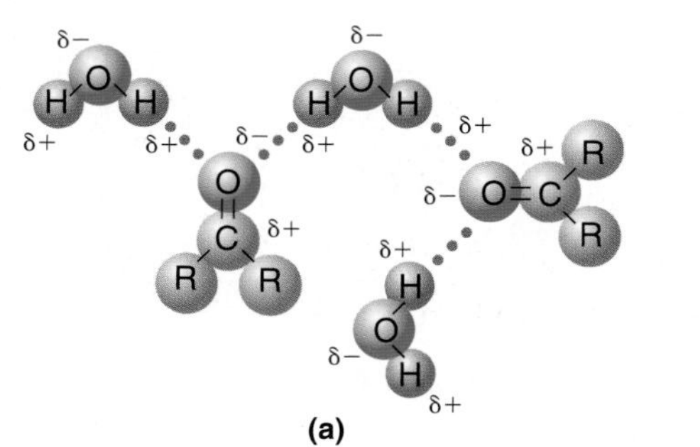

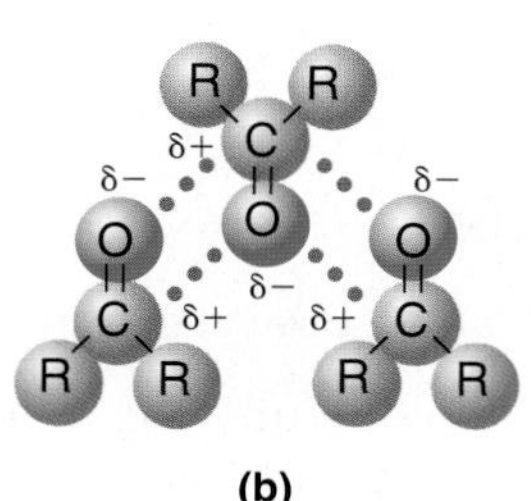

Figure 14.2
(a) Hydrogen bonding between the carbonyl group of an aldehyde or ketone and water. (b) Polar interactions between carbonyl groups of aldehydes or ketones.

Aldehydes and ketones can form intermolecular hydrogen bonds with water (Figure 14.2). As a result, the smaller members of the two families (five or fewer carbon atoms) are reasonably soluble in water. However, as the carbon chain length increases, the compounds become less polar and more hydrocarbonlike. These larger compounds are soluble in nonpolar organic solvents.

Question 14.1

Which member in each of the following pairs will be more water soluble?

a. $CH_3(CH_2)_2CH_3$ or $CH_3-\overset{\overset{O}{\|}}{C}-CH_3$

b. $CH_3-\underset{\underset{O}{\|}}{C}-CH_2CH_2CH_3$ or $CH_3-\underset{\underset{OH}{|}}{CH}-CH_2CH_2CH_3$

c. $\underset{|}{CH_3}$ (cyclopentane ring) or $H-\underset{|}{C}=O$ (cyclopentane ring)

d. $\underset{\underset{OH}{|}}{CH_2}-\underset{\underset{OH}{|}}{CH_2}$ or $H-\underset{\underset{O}{\|}}{C}-\underset{\underset{O}{\|}}{C}-H$

Question 14.2

Which member in each of the following pairs would have a higher boiling point?

a. $CH_3CH_2-\underset{\underset{O}{\|}}{C}-OH$ or $CH_3CH_2-\underset{\underset{O}{\|}}{C}-H$

b. $CH_3-\underset{\underset{O}{\|}}{C}-OH$ or $CH_3-\overset{\overset{O}{\|}}{C}-CH_3$

c. CH_3CH_2-OH or $CH_3-\overset{\overset{O}{\|}}{C}-H$

d. $CH_3(CH_2)_6CH_3$ or $CH_3(CH_2)_5-\underset{\underset{O}{\|}}{C}-H$

14.2 I.U.P.A.C. Nomenclature and Common Names

Naming Aldehydes

Learning Goal

In the I.U.P.A.C. system, aldehydes are named according to the following set of rules:

- Determine the parent compound, that is, the longest continuous carbon chain containing the carbonyl group.
- Replace the final *-e* of the parent alkane with *-al*.
- Number the chain beginning with the carbonyl carbon (or aldehyde group) as carbon-1.
- Number and name all substituents as usual. No number is used for the position of the carbonyl group because it is always at the end of the parent chain. Therefore, it must be carbon-1.

Several examples are provided here with common names given in parentheses:

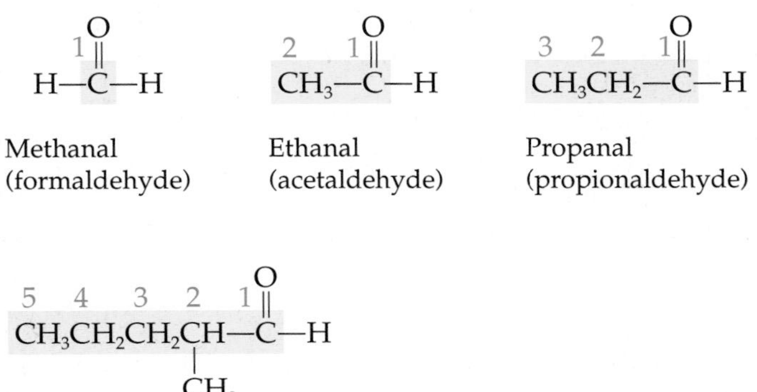

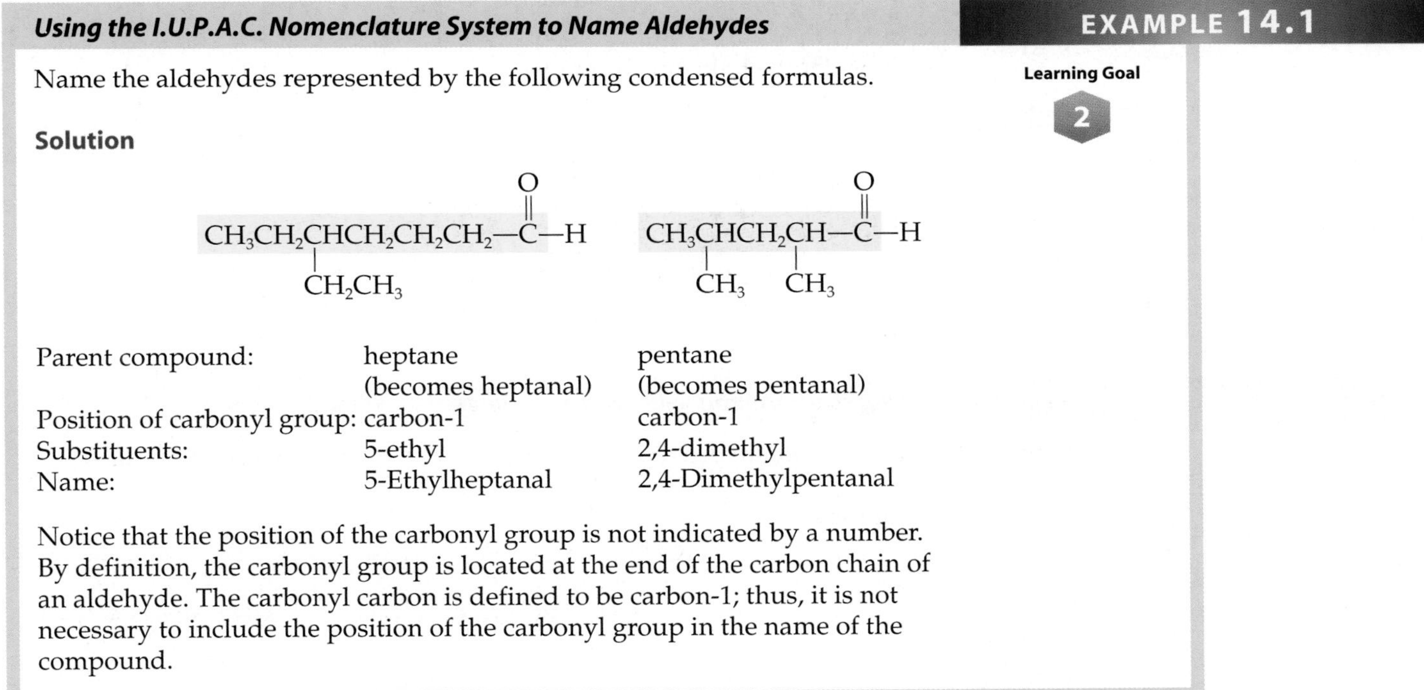

Using the I.U.P.A.C. Nomenclature System to Name Aldehydes
EXAMPLE 14.1
Learning Goal
2
Name the aldehydes represented by the following condensed formulas.
Solution
Parent compound: heptane (becomes heptanal) pentane (becomes pentanal)
Position of carbonyl group: carbon-1 carbon-1
Substituents: 5-ethyl 2,4-dimethyl
Name: 5-Ethylheptanal 2,4-Dimethylpentanal
Notice that the position of the carbonyl group is not indicated by a number. By definition, the carbonyl group is located at the end of the carbon chain of an aldehyde. The carbonyl carbon is defined to be carbon-1; thus, it is not necessary to include the position of the carbonyl group in the name of the compound.

Carboxylic acid nomenclature is described in Section 15.1.

The common names of the aldehydes are derived from the same Latin roots as the corresponding carboxylic acids. The common names of the first five aldehydes are presented in Table 14.1.

In the common system of nomenclature, substituted aldehydes are named as derivatives of the straight-chain parent compound (see Table 14.1). Greek letters

Table 14.1 I.U.P.A.C. and Common Names and Formulas for Several Aldehydes

I.U.P.A.C. Name	Common Name	Formula
Methanal	Formaldehyde	$H{-}C({=}O){-}H$
Ethanal	Acetaldehyde	$CH_3{-}C({=}O){-}H$
Propanal	Propionaldehyde	$CH_3CH_2{-}C({=}O){-}H$
Butanal	Butyraldehyde	$CH_3CH_2CH_2{-}C({=}O){-}H$
Pentanal	Valeraldehyde	$CH_3CH_2CH_2CH_2{-}C({=}O){-}H$

are used to indicate the position of the substituents. The carbon atom bonded to the carbonyl group is the α-carbon, the next is the β-carbon, and so on.

$$\overset{\delta}{-C}-\overset{\gamma}{C}-\overset{\beta}{C}-\overset{\alpha}{C}-\overset{O}{\overset{\|}{C}}-H$$

Consider the following examples:

$$CH_3CH_2CH_2\underset{\substack{|\\CH_3}}{CH}-\overset{\substack{O\\\|}}{C}-H \qquad CH_3CH_2\underset{\substack{|\\CH_3}}{C}HCH_2-\overset{\substack{O\\\|}}{C}-H$$

2-Methylpentanal (α-methylvaleraldehyde)

3-Methylpentanal (β-methylvaleraldehyde)

EXAMPLE 14.2 Using the Common Nomenclature System to Name Aldehydes

Learning Goal

Name the aldehydes represented by the following condensed formulas.

Solution

$$\overset{\delta}{CH_3}\underset{\substack{|\\Br}}{\overset{\gamma}{C}}H\overset{\beta}{C}H_2\overset{\alpha}{C}H_2-\overset{\substack{O\\\|}}{C}-H \qquad \overset{\gamma}{CH_3}\underset{\substack{|\\CH_3}}{\overset{\beta}{C}}H\overset{\alpha}{C}H_2-\overset{\substack{O\\\|}}{C}-H$$

Parent compound:	pentane (becomes valeraldehyde)	butane (becomes butyraldehyde)
Position of carbonyl group:	carbon-1 (must be!)	carbon-1 (must be!)
Substituents:	γ-bromo	β-methyl
Name:	γ-Bromovaleraldehyde	β-Methylbutyraldehyde

Continued—

EXAMPLE 14.2 —*Continued*

Notice that the substituents are designated by Greek letters, rather than by Arabic numbers. In the common system of nomenclature for aldehydes, the carbon atom bonded to the carbonyl group is called the α-carbon, the next is the β-carbon, etc. Remember to use these Greek letters to indicate the position of the substituents when naming aldehydes using the common system of nomenclature.

Also remember that by definition, the carbonyl group is located at the beginning of the carbon chain of an aldehyde. Thus, it is not necessary to include the position of the carbonyl group in the name of the compound.

Question 14.3

Use the I.U.P.A.C. and common nomenclature systems to name each of the following compounds.

a. $CH_3\overset{\overset{CH_3}{|}}{\underset{\underset{CH_3}{|}}{C}}HCHCH_2—\overset{\overset{O}{\|}}{C}—H$

b. $CH_3\underset{\underset{Cl}{|}}{C}H—\overset{\overset{O}{\|}}{C}—H$

c. $CH_3CH_2CH_2\underset{\underset{CH_2CH_3}{|}}{C}H—\overset{\overset{O}{\|}}{C}—H$

d. $CH_3\underset{\underset{OH}{|}}{C}HCH_2—\overset{\overset{O}{\|}}{C}—H$

Question 14.4

Write the condensed structural formula for each of the following compounds.

a. 3-Methylnonanal
b. β-Bromovaleraldehyde
c. 4-Fluorohexanal
d. α,β-Dimethylbutyraldehyde

Naming Ketones

Learning Goal 

The rules for naming ketones in the I.U.P.A.C. Nomenclature System are directly analogous to those for naming aldehydes. In ketones, however, the *-e* ending of the parent alkane is replaced with the *-one* suffix of the ketone family, and the location of the carbonyl carbon is indicated with a number. The longest carbon chain is numbered to give the carbonyl carbon the lowest possible number. For example,

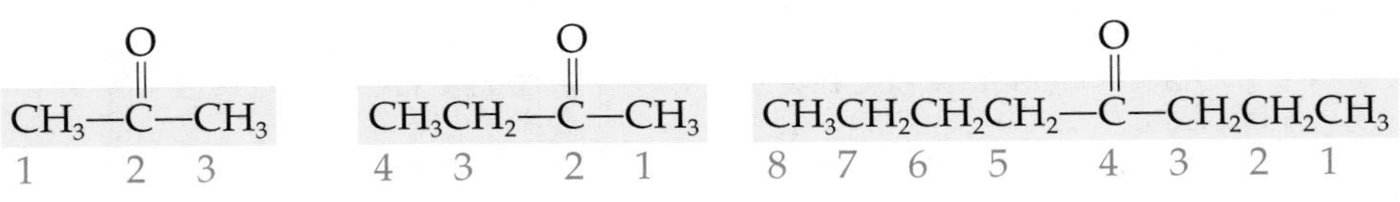

Propanone (no number necessary) (acetone)	Butanone (no number necessary) (methyl ethyl ketone)	4-Octanone (*not* 5-octanone) (butyl propyl ketone)

EXAMPLE 14.3 *Using the I.U.P.A.C Nomenclature System to Name Ketones*

Name the ketones represented by the following condensed formulas.

Continued—

EXAMPLE 14.3 —*Continued*

Solution

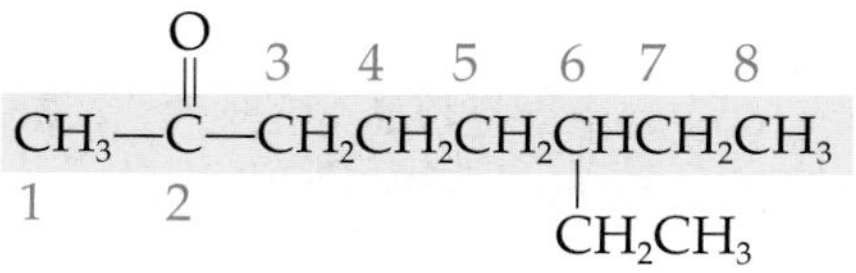

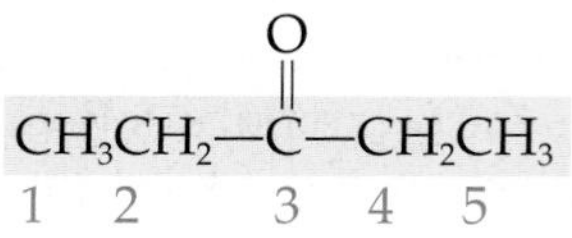

$$\overset{\overset{\text{O}}{\|}}{\text{CH}_3\text{—C—CH}_2\text{CH}_2\text{CH}_2\text{CHCH}_2\text{CH}_3}$$ with CH_2CH_3 on carbon-6 (numbered 1 2 3 4 5 6 7 8)

$$CH_3CH_2\text{—}\overset{\overset{O}{\|}}{C}\text{—}CH_2CH_3$$ (numbered 1 2 3 4 5)

Parent compound:	octane (becomes octanone)	pentane (becomes pentanone)
Position of carbonyl group:	carbon-2 (not carbon-7)	carbon-3
Substituents:	6-ethyl	none
Name:	6-Ethyl-2-octanone	3-Pentanone

The common names of ketones are derived by naming the alkyl groups that are bonded to the carbonyl carbon. These are used as prefixes followed by the word *ketone.* The alkyl groups may be arranged alphabetically or by size (smaller to larger).

EXAMPLE 14.4 *Using the Common Nomenclature System to Name Ketones*

Name the ketones represented by the following condensed formulas.

Solution

Identify the alkyl groups that are bonded to the carbonyl carbon.

$$CH_3CH_2CH_2\text{—}\overset{\overset{O}{\|}}{C}\text{—}CH_3 \qquad CH_3CH_2\text{—}\overset{\overset{O}{\|}}{C}\text{—}CH_2CH_2CH_2CH_2CH_3$$

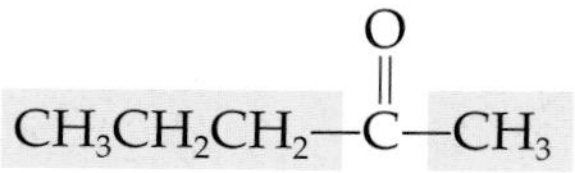

Alkyl groups:	propyl and methyl	ethyl and pentyl
Name:	Methyl propyl ketone	Ethyl pentyl ketone

Because the two *groups* bonded to the carbonyl carbon are named, a ketone is actually one carbon longer than an aldehyde with a *similar common name.* For example, methyl butyl ketone has six carbons, but β-methylbutyraldehyde has only five.

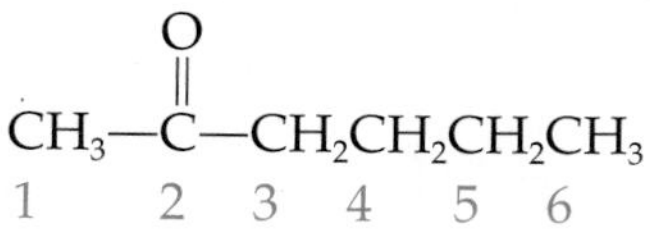

$$CH_3\text{—}\overset{\overset{O}{\|}}{C}\text{—}CH_2CH_2CH_2CH_3$$ (numbered 1 2 3 4 5 6)

Methyl butyl ketone

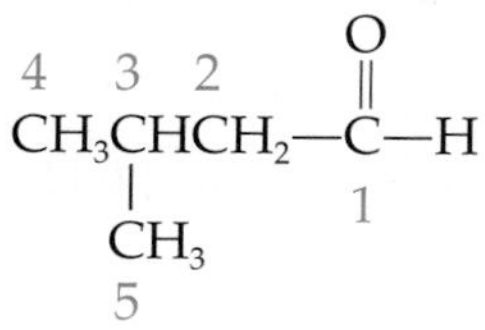

$$CH_3\underset{\underset{CH_3}{|}}{C}HCH_2\text{—}\overset{\overset{O}{\|}}{C}\text{—H}$$ (numbered 4 3 2 1, with CH_3 as 5)

β-Methylbutyraldehyde

This is because the aldehyde carbonyl carbon is included in the name of the parent chain, butyraldehyde. The carbonyl carbon of the ketone is not included in the name. It is treated only as the carbon to which the two alkyl or aryl groups are attached.

Question 14.5

Use the I.U.P.A.C. Nomenclature System to name each of the following compounds.

a. $CH_3\underset{\displaystyle I}{\underset{|}{C}}H-\overset{\displaystyle O}{\overset{\|}{C}}-CH_3$

b. $CH_3\underset{\displaystyle CH_3}{\underset{|}{C}}H-\overset{\displaystyle O}{\overset{\|}{C}}-CH_3$

c. $CH_3\underset{\displaystyle F}{\underset{|}{C}}H-\overset{\displaystyle O}{\overset{\|}{C}}-CH_2CH_3$

d. $CH_3\underset{\displaystyle CH_2CH_2CH_2CH_3}{\underset{|}{C}}HCH_2-\overset{\displaystyle O}{\overset{\|}{C}}-CH_3$

e. $CH_3\underset{\displaystyle CH_3}{\underset{|}{C}}H-\overset{\displaystyle O}{\overset{\|}{C}}-CH_2CH_3$

Question 14.6

Write the condensed formula for each of the following compounds.

a. Methyl isopropyl ketone (What is the I.U.P.A.C. name for this compound?)
b. 4-Heptanone
c. 2-Fluorocyclohexanone
d. Hexachloroacetone (What is the I.U.P.A.C. name of this compound?)

14.3 Important Aldehydes and Ketones

Learning Goal

Methanal (formaldehyde) is a gas (b.p. −21°C). It is available commercially as an aqueous solution called *formalin*. Formalin has been used as a preservative for tissues and as an embalming fluid. For other uses of formaldehyde, see A Clinical Perspective: Medical Applications of Aldehydes.

Ethanal (acetaldehyde) is produced from ethanol in the liver. Ethanol is oxidized in this reaction, which is catalyzed by the liver enzyme alcohol dehydrogenase. The ethanal that is produced in this reaction is responsible for the symptoms of a hangover.

Propanone (acetone), the simplest ketone, is an important and versatile solvent for organic compounds. It has the ability to dissolve organic compounds and is also miscible with water. As a result, it has a number of industrial applications and is used as a solvent in adhesives, paints, cleaning solvents, nail polish, and nail polish remover. Propanone is flammable and should therefore be treated with appropriate care. *Butanone,* a four-carbon ketone, is also an important industrial solvent.

Many aldehydes and ketones are produced industrially as food and fragrance chemicals, medicinals, and agricultural chemicals. They are particularly important to the food industry, in which they are used as artificial and/or natural additives to food. Vanillin, a principal component of natural vanilla, is shown in Figure 14.3. Artificial vanilla flavoring is a dilute solution of synthetic vanillin dissolved in ethanol. Figure 14.3 also shows other examples of important aldehydes and ketones.

$H-\overset{\displaystyle O}{\overset{\|}{C}}-H$

Methanal

$CH_3-\overset{\displaystyle O}{\overset{\|}{C}}-H$

Ethanal

$CH_3-\overset{\displaystyle O}{\overset{\|}{C}}-CH_3$

Propanone

$CH_3-\overset{\displaystyle O}{\overset{\|}{C}}-CH_2CH_3$

Butanone

Question 14.7

Draw the structure of the aldehyde synthesized from ethanol in the liver.

Question 14.8

Draw the structure of a ketone that is an important, versatile solvent for organic compounds.

$C_6H_5-\overset{O}{\overset{\|}{C}}-H$ Benzaldehyde—almonds

Vanillin—vanilla beans (HO–, CH_3O– substituted benzene ring with $-\overset{O}{\overset{\|}{C}}-H$)

$C_6H_5-CH=CH-\overset{O}{\overset{\|}{C}}-H$ Cinnamaldehyde—cinnamon

$CH_3-\overset{CH_3}{\overset{|}{C}}=CH-CH_2-CH_2-\overset{CH_3}{\overset{|}{C}}=CH-\overset{O}{\overset{\|}{C}}-H$ Citral—lemongrass

α-Demascone—berry flavoring

$CH_3CH_2CH_2CH_2CH_2CH_2-\overset{O}{\overset{\|}{C}}-CH_3$ 2-Octanone—mushroom flavoring

Figure 14.3
Important aldehydes and ketones.

A CLINICAL PERSPECTIVE

Medical Applications of Aldehydes

Most aldehydes have irritating, unpleasant odors. Formalin, an aqueous solution of formaldehyde, has often been used to preserve biological tissues and for embalming. It has also been used to disinfect environmental surfaces, body fluids, and feces. Under no circumstances is it used as an antiseptic on human tissue because of its noxious fumes and the skin irritation that it causes.

Formaldehyde is useful in the production of killed virus vaccines. A deadly virus, such as polio virus, can be treated with heat and formaldehyde. Formaldehyde reacts with the genetic information (RNA) of the virus, damaging it irreparably. It also reacts with the virus proteins but does not change their shape. Thus when you are injected with the Salk killed polio vaccine, the virus can't replicate and harm you. However, it will be recognized by your immune system, which will produce antibodies that will protect you against polio virus infection.

Formaldehyde can also be produced in the body! Drinking wood alcohol (methanol) causes blindness, respiratory failure, convulsions, and death. The liver enzyme *alcohol dehydrogenase,* whose function it is to detoxify alcohols, catalyzes the conversion of methanol to formaldehyde (methanal). The formaldehyde produced reacts with cellular proteins, causing the range of symptoms mentioned.

Acetaldehyde (ethanal) is produced from ethanol by the liver enzymes and is largely responsible for the symptoms of hangover experienced after a night of too much partying. This aldehyde is useful in treating alcoholics because of the unpleasant symptoms that it causes. When taken orally, in combination with alcohol, acetaldehyde quickly produces symptoms of a violent hangover with none of the *perceived* benefits of drinking alcohol.

The liver enzymes that oxidize ethanol to acetaldehyde are the same as those that oxidize methanol to formaldehyde. Physicians take advantage of this in the treatment of wood alcohol poisoning by trying to keep those enzymes busy with a reaction that produces a *less* toxic (not nontoxic) by-product. In cases of methanol poisoning, the patient receives ethanol intravenously. The ethanol should then be in greater concentration than methanol and should compete successfully for the liver enzymes and be converted to acetaldehyde. This gives the body time to excrete the methanol before it is oxidized to the potentially deadly formaldehyde.

14.4 Reactions Involving Aldehydes and Ketones

Preparation of Aldehydes and Ketones

Learning Goal

Aldehydes and ketones are prepared primarily by the **oxidation** of the corresponding alcohol. As we saw in Chapter 13, the oxidation of methyl alcohol gives methanal (formaldehyde). The oxidation of a primary alcohol produces an aldehyde, and the oxidation of a secondary alcohol yields a ketone. Tertiary alcohols do not undergo oxidation under the conditions normally used.

Differentiating the Oxidation of Primary, Secondary, and Tertiary Alcohols — EXAMPLE 14.5

Use specific examples to show the oxidation of a primary, a secondary, and a tertiary alcohol.

Solution

The oxidation of a primary alcohol to an aldehyde:

$$CH_3CH_2CH_2{-}\overset{\displaystyle H}{\underset{\displaystyle H}{\overset{|}{\underset{|}{C}}}}{-}OH \xrightarrow[\text{dichromate}]{\text{Pyridinium}} CH_3CH_2CH_2{-}\overset{\displaystyle O}{\overset{\|}{C}}{-}H$$

1-Butanol (butyl alcohol) → Butanal (butyraldehyde)

Continued—

EXAMPLE 14.5 —*Continued*

A mild oxidizing agent must be used in the oxidation of a primary alcohol to an aldehyde. Otherwise, the aldehyde will be further oxidized to a carboxylic acid.

The oxidation of a secondary alcohol to a ketone:

$$CH_3CH_2CH_2CH_2\text{—}\overset{\displaystyle CH_3}{\underset{\displaystyle H}{\overset{|}{\underset{|}{C}}}}\text{—}OH \xrightarrow[H_2O]{KMnO_4,\ OH^-} CH_3CH_2CH_2CH_2\text{—}\overset{\displaystyle O}{\overset{\|}{C}}\text{—}CH_3$$

2-Hexanol 2-Hexanone

Tertiary alcohols cannot undergo oxidation:

$$CH_3CH_2CH_2\text{—}\overset{\displaystyle CH_3}{\underset{\displaystyle CH_3}{\overset{|}{\underset{|}{C}}}}\text{—}OH \xrightarrow{H_2CrO_4} \text{No reaction}$$

2-Methyl-2-pentanol

Question 14.9 Write an equation showing the oxidation of 1-propanol.

Question 14.10 Write an equation showing the oxidation of 2-butanol.

Oxidation Reactions

Learning Goal

Aldehydes are easily oxidized further to carboxylic acids, whereas ketones do not generally undergo further oxidation. The reason is that an aldehydic carbon-hydrogen bond, present in the aldehyde but not in the ketone, is needed for the reaction to occur. In fact, aldehydes are so easily oxidized that it is often very difficult to prepare them because they continue to react to give the carboxylic acid rather than the desired aldehyde. Aldehydes are susceptible to air oxidation, even at room temperature, and cannot be stored for long periods. The following example shows a general equation for the oxidation of an aldehyde to a carboxylic acid:

$$R\text{—}\overset{\displaystyle O}{\overset{\|}{C}}\text{—}H \xrightarrow{[O]} R\text{—}\overset{\displaystyle O}{\overset{\|}{C}}\text{—}OH$$

Aldehyde Carboxylic acid

Many oxidizing agents can be used. Both basic potassium permanganate and chromic acid are good oxidizing agents, as the following specific example shows:

$$CH_3\text{—}\overset{\displaystyle O}{\overset{\|}{C}}\text{—}H \xrightarrow[H_2O, OH^-]{KMnO_4,} CH_3\text{—}\overset{\displaystyle O}{\overset{\|}{C}}\text{—}O^-$$

Ethanal (acetaldehyde) Ethanoate anion (acetate anion)

In basic solution, the product is the carboxylic acid anion:

$$CH_3\text{—}\overset{\displaystyle O}{\overset{\|}{C}}\text{—}O^-$$

The rules for naming carboxylic acid anions are described in Section 15.1.

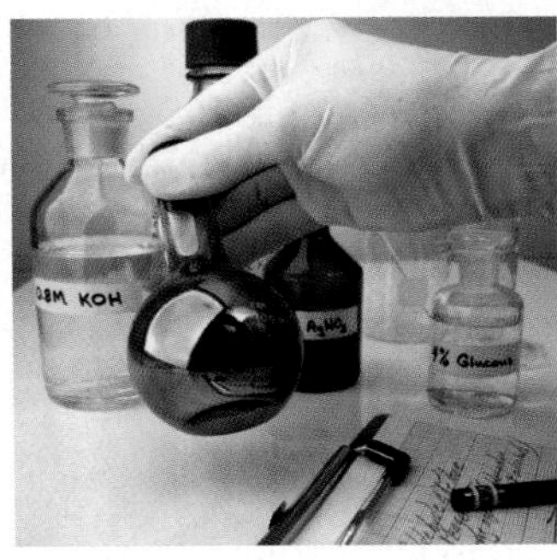

Figure 14.4
The silver precipitate produced by the Tollens' reaction is deposited on glass. Silver mirrors are made in a similar process.

The oxidation of benzaldehyde to benzoic acid is an example of the conversion of an *aromatic* aldehyde to the corresponding aromatic carboxylic acid:

$$C_6H_5\text{—}\overset{\text{O}}{\overset{\|}{\text{C}}}\text{—H} \xrightarrow{H_2CrO_4} C_6H_5\text{—}\overset{\text{O}}{\overset{\|}{\text{C}}}\text{—OH}$$

Benzaldehyde Benzoic acid

Silver ions are very mild oxidizing agents. They will oxidize aldehydes but not alcohols.

Aldehydes and ketones can be distinguished on the basis of differences in their reactivity. The most common laboratory test for aldehydes is the **Tollens' test.** When exposed to the Tollens' reagent, a basic solution of $Ag(NH_3)_2^+$, an aldehyde undergoes oxidation. The silver ion (Ag^+) is reduced to silver metal (Ag^0) as the aldehyde is oxidized to a carboxylic acid anion.

$$\text{R—}\overset{\text{O}}{\overset{\|}{\text{C}}}\text{—H} + Ag(NH_3)_2^+ \longrightarrow \text{R—}\overset{\text{O}}{\overset{\|}{\text{C}}}\text{—O}^- + Ag^0$$

Aldehyde — Silver ammonia complex—Tollens' reagent — Carboxylate anion — Silver metal mirror

Silver metal precipitates from solution and coats the flask, producing a smooth silver mirror, as seen in Figure 14.4. The test is therefore often called the Tollens' silver mirror test. The commercial manufacture of silver mirrors uses a similar process. Ketones cannot be oxidized to carboxylic acids and do not react with the Tollens' reagent.

EXAMPLE 14.6

Writing Equations for the Reaction of an Aldehyde and of a Ketone with Tollens' Reagent

Write equations for the reaction of propanal and 2-pentanone with Tollens' reagent.

Continued—

EXAMPLE 14.6 —*Continued*

Solution

$$CH_3CH_2{-}\overset{\overset{\Large O}{\|}}{C}{-}H + Ag(NH_3)_2^+ \longrightarrow CH_3CH_2{-}\overset{\overset{\Large O}{\|}}{C}{-}O^- + Ag^0$$

Propanal — Propanoate anion

$$CH_3CH_2CH_2{-}\overset{\overset{\Large O}{\|}}{C}{-}CH_3 + Ag(NH_3)_2^+ \longrightarrow \text{No reaction}$$

2-Pentanone

Question 14.11 Write an equation for the reaction of ethanal with Tollens' reagent.

Question 14.12 Write an equation for the reaction of propanone with Tollens' reagent.

Cu(II) is an even milder oxidizing agent than silver ion.

Another test that is used to distinguish between aldehydes and ketones is **Benedict's test.** Here, a buffered aqueous solution of copper(II) hydroxide and sodium citrate reacts to oxidize aldehydes but does not generally react with ketones. Cu^{2+} is reduced to Cu^+ in the process. Cu^{2+} is soluble and gives a blue solution, whereas the Cu^+ precipitates as the red solid copper(I) oxide, Cu_2O.

All simple sugars (monosaccharides) are either aldehydes or ketones. Glucose is an aldehyde sugar that is commonly called *blood sugar* because it is the sugar found transported in the blood and used for energy by many cells. In uncontrolled diabetes, glucose may be found in the urine. One means to determine the amount of glucose in the urine is to use Benedict's test and look for the color change. The amount of precipitate formed is directly proportional to the amount of glucose in the urine (Figure 14.5). The reaction of glucose with the Benedict's reagent is represented in the following equation:

$$\begin{array}{c} O{=}C{-}H \\ | \\ H{-}C{-}OH \\ | \\ HO{-}C{-}H \\ | \\ H{-}C{-}OH \\ | \\ H{-}C{-}OH \\ | \\ CH_2OH \end{array} + 2Cu^{2+} \xrightarrow{OH^-} \begin{array}{c} O{=}C{-}O^- \\ | \\ H{-}C{-}OH \\ | \\ HO{-}C{-}H \\ | \\ H{-}C{-}OH \\ | \\ H{-}C{-}OH \\ | \\ CH_2OH \end{array} + Cu_2O$$

Glucose

We should also note that when the carbonyl group of a ketone is bonded to a $—CH_2OH$ group, the molecule will give a positive Benedict's test. This occurs because such ketones are converted to aldehydes under basic conditions. In Chapter 17 we will see that this applies to the ketone sugars, as well. They are converted to aldehyde sugars and react with Benedict's reagent.

Figure 14.5
The amount of precipitate formed, and thus the color change observed, in the Benedict's test are directly proportional to the amount of reducing sugar in the sample.

Reduction Reactions

Aldehydes and ketones are both readily reduced to the corresponding alcohol by a variety of reducing agents. Throughout the text the symbol [H] over the reaction arrow represents a reducing agent.

Learning Goal

6

One way to recognize reduction, particularly in organic chemistry, is the gain of hydrogen. Oxidation and reduction are discussed in Section 13.6.

Hydrogenation was first discussed in Section 12.5 for the hydrogenation of alkenes.

The classical method of aldehyde or ketone reduction is **hydrogenation.** The carbonyl compound is reacted with hydrogen gas and a catalyst (nickel, platinum, or palladium metal) in a pressurized reaction vessel. Heating may also be necessary. The carbon-oxygen double bond (the carbonyl group) is reduced to a carbon-oxygen single bond. This is similar to the reduction of an alkene to an alkane (the reduction of a carbon-carbon double bond to a carbon-carbon single bond). The addition of hydrogen to a carbon-oxygen double bond is shown in the following example:

$$R^1-\overset{\overset{\large O}{\|}}{C}-R^2 + H-H \xrightarrow{Pt} R^1-\underset{\underset{\large R^2}{|}}{\overset{\overset{\large OH}{|}}{C}}-H$$

Aldehyde or ketone — Hydrogen — Alcohol

The hydrogenation (reduction) of a ketone produces a secondary alcohol, as seen in the following equation showing the reduction of the ketone, 3-octanone:

$$CH_3CH_2-\overset{\overset{\large O}{\|}}{C}-CH_2CH_2CH_2CH_2CH_3 + H_2 \xrightarrow{Ni} CH_3CH_2-\underset{\underset{\large H}{|}}{\overset{\overset{\large OH}{|}}{C}}-CH_2CH_2CH_2CH_2CH_3$$

3-Octanone (A ketone) — Hydrogen — 3-Octanol (A secondary alcohol)

A MEDICAL PERSPECTIVE

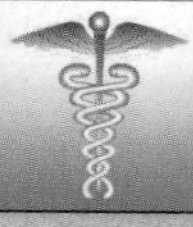

That Golden Tan Without the Fear of Skin Cancer

Self-tanning lotions have become very popular in recent years. This seems to be the result of our growing understanding of the link between exposure to the sun and skin cancer and to improvements in the quality of the tan produced by these self-tanners.

The active ingredient in most self-tanners is dihydroxyacetone (DHA).

$$\begin{array}{c} CH_2OH \\ | \\ C{=}O \\ | \\ CH_2OH \end{array}$$

Dihydroxyacetone

This simple, three-carbon molecule has two hydroxyl groups (−OH) and is, thus, an alcohol. It also has a carbonyl group at the center carbon:

$$\gt C{=}O$$

Carbonyl group

This makes the compound a ketone (Chapter 14). In fact, because it has both hydroxyl groups and a carbonyl group, DHA is a sugar, more precisely, a keto sugar. This sugar is actually a by-product of our own metabolism; we produce it in the metabolic pathway called glycolysis (Chapter 22).

A researcher by the name of Eva Wittgenstein discovered the tanning reaction while she was studying a human genetic disorder in children. These children were unable to store glycogen, a polysaccharide, or sugar polymer, which is our major energy storage molecule in the liver. She was trying to treat the disease by feeding large doses of DHA to the children. Sometimes, however, the children spit up some of the sickeningly sweet solution, which ended up on their clothes and skin. Dr. Wittgenstein noticed that the skin darkened at the site of these spills and decided to investigate the observation.

DHA works because of a reaction between its carbonyl group and a free amino group ($-NH_3^+$) of several amino acids in the skin protein keratin. Amino acids are the building blocks of the biological polymers called proteins (Chapter 19); keratin is just one such protein. The DHA produces brown-colored compounds called melanoids when it bonds to the keratins. These polymeric melanoids are chemically linked to cells of the stratum corneum, the dead, outermost layer of the skin. DHA does not penetrate this outer layer; so the chemical reaction that causes tanning only affects the stratum corneum. As this dead skin sloughs off, so does your tan!

Over the years research has improved the quality of the tan that is produced. Early self-tanning lotions produced an orange tan; the tans from today's lotions are much more natural. The DHA used today is in a much purer form and the other components of the lotion have been redesigned to promote greater penetration. Research has also taught us that the tanning reaction works best at acid pH; so newer formulations are buffered

The hydrogenation of an aldehyde results in the production of a primary alcohol, as seen in the following equation showing the reduction of the aldehyde, butanal:

$$CH_3CH_2CH_2{-}\overset{\overset{\displaystyle O}{\|}}{C}{-}H + H_2 \xrightarrow{Pt} CH_3CH_2CH_2{-}\underset{\underset{\displaystyle H}{|}}{\overset{\overset{\displaystyle OH}{|}}{C}}{-}H$$

Butanal (An aldehyde) Hydrogen 1-Butanol (A primary alcohol)

EXAMPLE 14.7 ***Writing an Equation Representing the Hydrogenation of a Ketone***

Write an equation showing the hydrogenation of 3-pentanone.

Solution

The product of the reduction of a ketone is a secondary alcohol, in this case, 3-pentanol.

Continued—

to pH 5. All of these changes have resulted in self-tanners that produce a longer lasting tan with a more natural, golden color.

We have also learned that it is important to exfoliate before using a self-tanner. Anywhere that the dead skin layer is thicker, there will be more keratin. From our study of chemistry we have learned that when we begin with more reactant, we often get more product. The greater the amount of product in this case, the darker the color! The resulting tan often looked splotchy or streaky. By gently removing some of the stratum corneum by a gentle exfoliation process, the surface of the skin, and hence the tanning reaction, becomes more uniform.

More recently some companies have added a new sugar, erythrulose, to the self-tanning lotions.

$$\begin{array}{c} CH_2OH \\ | \\ C{=}O \\ | \\ HO{-}CH \\ | \\ CH_2OH \end{array}$$

Erythrulose

Erythrulose is a four-carbon keto sugar that reacts in exactly the same way as DHA. However, since they are different compounds, they do produce different melanoids with slightly different properties, including color. The tan produced by erythrulose is less reddish in tone than that produced by DHA. However, while a DHA tan develops in 2–6 hours, an erythrulose tan requires 2 days. For this reason, erythrulose is usually not used alone, but only in combination with DHA.

People often ask whether the tan from a bottle can protect against burn, much as a natural tan does. The melanoids do absorb light of the same wavelengths absorbed by melanin (the substance formed by suntanning), so you might expect some protection against sunburn. However, the protection is minimal, rated at a sun protection factor (SPF) of only 2 or 3. Self-tanners offer an excellent substitute to a suntan. They produce the same golden tan without the danger of overexposure to the sun's harmful ultraviolet rays.

EXAMPLE 14.7 —*Continued*

$$CH_3CH_2{-}\overset{\overset{\large O}{\|}}{C}{-}CH_2CH_3 + H_2 \xrightarrow{Pt} CH_3CH_2{-}\overset{\overset{\large OH}{|}}{\underset{\underset{\large H}{|}}{C}}{-}CH_2CH_3$$

3-Pentanone 3-Pentanol

EXAMPLE 14.8

Writing an Equation Representing the Hydrogenation of an Aldehyde

Write an equation showing the hydrogenation of 3-methylbutanal.

Solution

Recall that the reduction of an aldehyde results in the production of a primary alcohol, in this case, 3-methyl-1-butanol.

Continued—

EXAMPLE 14.8 —*Continued*

$$CH_3CHCH_2{-}\overset{\overset{\large O}{\|}}{C}{-}H + H_2 \xrightarrow{Pt} CH_3CHCH_2{-}\overset{\overset{\large OH}{|}}{\underset{\underset{\large H}{|}}{C}}{-}H$$

(with a CH_3 branch on the second carbon of each chain)

3-Methylbutanal → 3-Methyl-1-butanol

Question 14.13 Write an equation for the hydrogenation of propanone.

Question 14.14 Write an equation for the hydrogenation of butanone.

Question 14.15 Label each of the following as an oxidation or a reduction reaction.

a. Ethanal to ethanol
b. Benzoic acid to benzaldehyde
c. Cyclohexanone to cyclohexanol
d. 2-Propanol to propanone
e. 2,3-Butanedione (found in butter) to 2,3-butanediol

Question 14.16 Write an equation for each of the reactions in Question 14.15.

Addition Reactions

Learning Goal

The principal reaction of the carbonyl group is the **addition reaction** across the polar carbon-oxygen bond. Such reactions require that a catalytic amount of acid be present in solution, as shown by the H^+ over the arrow for the reactions shown in the following examples.

An example of an addition reaction is the reaction of aldehydes with alcohols in the presence of catalytic amounts of acid. In this reaction, the hydrogen of the alcohol adds to the carbonyl oxygen. The alkoxyl group of the alcohol (—OR) adds to the carbonyl carbon. The predicted product, having the structure just described, is a **hemiacetal.**

$$R{-}\overset{\overset{\large OH}{|}}{\underset{\underset{\large H}{|}}{C}}{-}OR$$

General structure of a hemiacetal

However, this is not the product typically isolated from this reaction. Hemiacetals are quite reactive. In the presence of acid and excess alcohol, they undergo a substitution reaction in which the —OH group of the hemiacetal is exchanged for another —OR group from the alcohol. The product of this reaction is an **acetal.** Acetal formation is a reversible reaction, as the general equation shows:

$$R^1{-}\overset{\overset{\large O}{\|}}{C}{-}H + \underset{\underset{\large OR^2}{|}}{H} \underset{}{\overset{H^+}{\rightleftharpoons}} R^1{-}\overset{\overset{\large OH}{|}}{\underset{\underset{\large H}{|}}{C}}{-}OR^2 + \underset{\underset{\large OR^2}{|}}{H} \overset{H^+}{\rightleftharpoons} R^1{-}\overset{\overset{\large OR^2}{|}}{\underset{\underset{\large H}{|}}{C}}{-}OR^2 + H_2O$$

Aldehyde Alcohol Hemiacetal Acetal

Consider the acid-catalyzed reaction between propanal and methanol:

$$CH_3CH_2\text{—}\overset{\overset{\displaystyle O}{\|}}{C}\text{—}H + CH_3OH \underset{}{\overset{H^+}{\rightleftharpoons}} CH_3CH_2\text{—}\underset{\underset{\displaystyle H}{|}}{\overset{\overset{\displaystyle OH}{|}}{C}}\text{—}OCH_3 + CH_3OH \overset{H^+}{\rightleftharpoons} CH_3CH_2\text{—}\underset{\underset{\displaystyle H}{|}}{\overset{\overset{\displaystyle OCH_3}{|}}{C}}\text{—}OCH_3 + H_2O$$

Propanal Methanol Hemiacetal Propanal dimethyl acetal

Addition reactions will also occur between a ketone and an alcohol. In this case the more reactive intermediate is called a **hemiketal** and the product is called a **ketal.** The general equation for ketal formation is shown below:

$$\underset{R^1 \quad R^2}{\overset{\overset{\displaystyle O}{\|}}{C}} + \underset{\underset{\displaystyle OR^3}{|}}{H} \overset{H^+}{\rightleftharpoons} R^1\text{—}\underset{\underset{\displaystyle R^2}{|}}{\overset{\overset{\displaystyle OH}{|}}{C}}\text{—}OR^3 + \underset{\underset{\displaystyle OR^3}{|}}{H} \overset{H^+}{\rightleftharpoons} R^1\text{—}\underset{\underset{\displaystyle R^2}{|}}{\overset{\overset{\displaystyle OR^3}{|}}{C}}\text{—}OR^3 + H_2O$$

Ketone Alcohol Hemiketal Ketal

Recognizing Hemiacetals, Acetals, Hemiketals, and Ketals

EXAMPLE 14.9

Solution

A simple scheme is helpful in recognizing these four types of compounds. Begin by drawing a carbon atom with four bonds and follow the flow chart as additional groups are added that will identify the molecules.

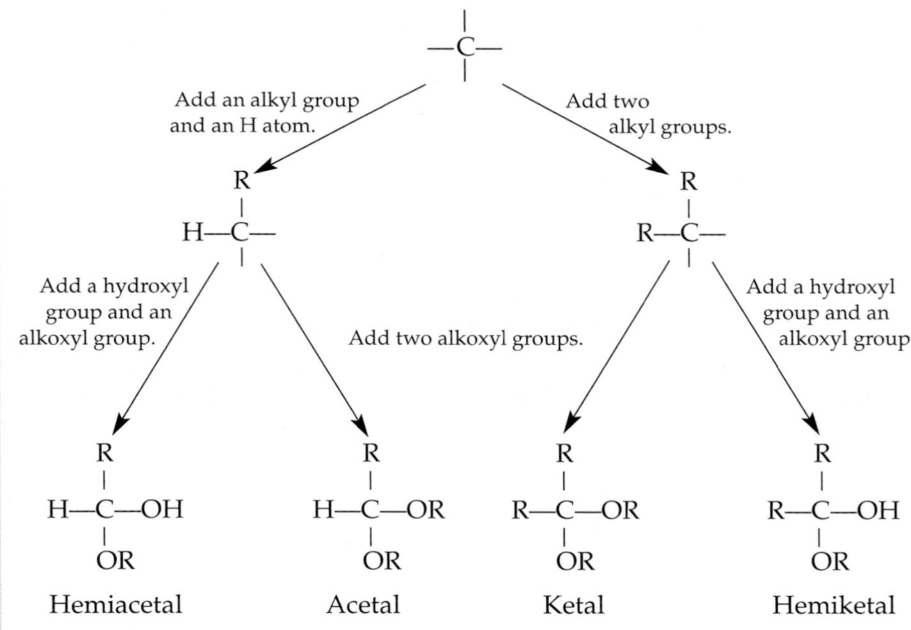

A ketal is the final product in the reaction between propanone and ethanol, seen in the following equation:

$$CH_3{-}\overset{\overset{\displaystyle O}{\|}}{C}{-}CH_3 + CH_3CH_2OH \underset{}{\overset{H^+}{\rightleftharpoons}} CH_3{-}\overset{\overset{\displaystyle OH}{|}}{\underset{\underset{\displaystyle CH_3}{|}}{C}}{-}OCH_2CH_3 + CH_3CH_2OH \overset{H^+}{\rightleftharpoons} CH_3{-}\overset{\overset{\displaystyle OCH_2CH_3}{|}}{\underset{\underset{\displaystyle CH_3}{|}}{C}}{-}OCH_2CH_3 + H_2O$$

Propanone Ethanol Hemiketal Ketal

Hemiacetals and hemiketals are readily formed in carbohydrates. Monosaccharides contain several hydroxyl groups and one carbonyl group. The linear form of a monosaccharide quickly undergoes an intramolecular reaction in solution to give a cyclic hemiacetal or hemiketal.

Earlier we noted that hemiacetals and hemiketals formed in *intermolecular* reactions were unstable and continued to react, forming acetals and ketals. This is not the case with the *intramolecular* reactions involving five- or six-carbon sugars. In these reactions the cyclic or ring form of the molecule is more stable than the linear form. This reaction is shown for the sugar glucose (blood sugar) in Figure 14.6 and is discussed in detail in Section 17.2.

When the hemiacetal or hemiketal of one monosaccharide reacts with a hydroxyl group of another monosaccharide, the product is an acetal or a ketal. A sugar molecule made up of two monosaccharides is called a *disaccharide.* The C—O—C bond between the two monosaccharides is called a *glycosidic bond* (Figure 14.7).

Keto-Enol Tautomers

Learning Goal

Many aldehydes and ketones may exist in an equilibrium mixture of two constitutional or structural isomers called *tautomers*. Tautomers differ from one another in the placement of a hydrogen atom and a double bond. One tautomer is the *keto form* (on the left in the equation below). The keto form has the structure typical of an aldehyde or ketone. The other form is called the *enol form* (on the right in the equation below). The enol form has a structure containing a carbon-carbon double bond (*en*) and a hydroxyl group, the functional group characteristic of alcohols (*ol*).

$$R^1{-}\overset{\overset{\displaystyle H}{|}}{\underset{\underset{\displaystyle R^2}{|}}{C}}{-}\overset{\overset{\displaystyle O}{\|}}{C}{-}R^3 \rightleftharpoons \begin{matrix} R^1 & & OH \\ & C{=}C & \\ R^2 & & R^3 \end{matrix}$$

(R^1, R^2, and R^3 = H or alkyl group)

Keto form Enol form

Because the keto form of most simple aldehydes and ketones is more stable, they exist mainly in that form.

EXAMPLE 14.10

Writing an Equation Representing the Equilibrium between the Keto and Enol Forms of a Simple Aldehyde

Draw the enol form of ethanal and write an equation representing the equilibrium between the keto and enol forms of this molecule.

Solution

$$H{-}\overset{\overset{\displaystyle H}{|}}{\underset{\underset{\displaystyle H}{|}}{C}}{-}\overset{\overset{\displaystyle O}{\|}}{C}{-}H \rightleftharpoons H{-}\underset{\underset{\displaystyle H}{|}}{C}{=}\overset{\overset{\displaystyle O{-}H}{|}}{C}{-}H$$

Ethanal
Keto form
More stable

Enol form
Less stable

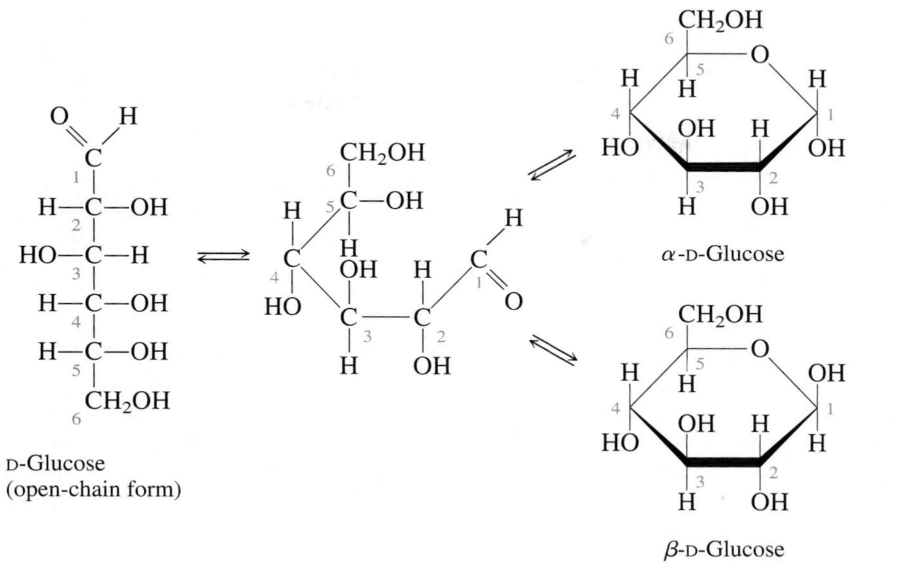

Figure 14.6
Hemiacetal formation in sugars, shown for the intramolecular reaction of D-glucose.

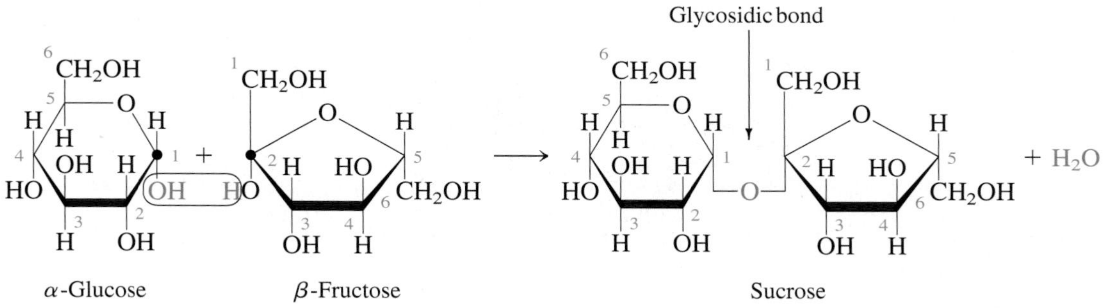

Figure 14.7
Acetal formation, demonstrated in the formation of the disaccharide sucrose, common table sugar. The reaction between the hydroxyl groups of the monosaccharides glucose and fructose produces the acetal sucrose. The bond between the two sugars is a glycosidic bond.

Phosphoenolpyruvate is a biologically important enol. In fact, it is the highest energy phosphorylated compound in living systems.

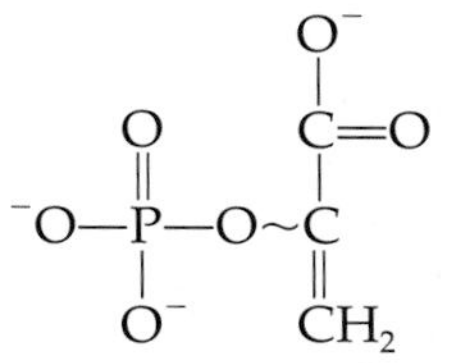

Phosphoenolpyruvate is produced in the next-to-last step in the metabolic pathway called *glycolysis,* which is the first stage of carbohydrate breakdown. In the final reaction of glycolysis the phosphoryl group from phosphoenolpyruvate is transferred to adenosine diphosphate (ADP). The reaction produces ATP, the major energy currency of the cell.

The glycolysis pathway is discussed in detail in Chapter 21.

Aldol Condensation

The **aldol condensation** is a reaction in which aldehydes or ketones react to form larger molecules. A new carbon-carbon bond is formed in the process:

Learning Goal

$$R^1{-}CH_2{-}\overset{\overset{O}{\|}}{C}{-}R + R^2{-}CH_2{-}\overset{\overset{O}{\|}}{C}{-}R \underset{\text{enzyme}}{\overset{OH^- \text{ or}}{\rightleftharpoons}} R^1{-}CH_2{-}\underset{R}{\overset{OH}{C}}{-}\underset{R^2}{CH}{-}\overset{\overset{O}{\|}}{C}{-}R$$

R = H, alkyl, or aryl group

Aldehyde or Ketone — Aldehyde or Ketone — Aldol

This is actually a very complex reaction that occurs in multiple steps. Here we focus on the end results of the reaction, using the example of the reaction between two molecules of ethanal. As shown in the equation below, the α-carbon (carbon-2) of one aldehyde forms a bond with the carbonyl carbon of a second aldehyde (shown in blue). A bond also forms between a hydrogen atom on that same α-carbon and the carbonyl oxygen (shown in red).

$$H{-}\underset{H}{\overset{H}{C}}{-}\overset{\overset{O}{\|}}{C}{-}H + H{-}\underset{H}{\overset{H}{C}}{-}\overset{\overset{O}{\|}}{C}{-}H \overset{OH^-}{\rightleftharpoons} H{-}\underset{H}{\overset{H}{C}}{-}\underset{H}{\overset{OH}{C}}{-}CH_2{-}\overset{\overset{O}{\|}}{C}{-}H$$

Ethanal — Ethanal — 3-Hydroxybutanal (β-hydroxybutyraldehyde)

The result is similar when two ketones react:

$$CH_3{-}\overset{\overset{O}{\|}}{C}{-}CH_3 \quad H{-}\underset{H}{\overset{H}{C}}{-}\overset{\overset{O}{\|}}{C}{-}CH_3 \overset{OH^-}{\rightleftharpoons} CH_3{-}\underset{CH_3}{\overset{OH}{C}}{-}CH_2{-}\overset{\overset{O}{\|}}{C}{-}CH_3$$

Propanone — Propanone — 4-Hydroxy-4-methyl-2-pentanone

In the laboratory the aldol condensation is catalyzed by dilute base. But the same reaction occurs in our cells, where it is catalyzed by an enzyme. This reaction is one of many in a pathway that makes the sugar glucose from smaller molecules. This pathway is called gluconeogenesis (gluco- [sugar], neo- [new], genesis [to make]), which simply means make new sugar. This pathway is critical during starvation or following strenuous exercise. Under those conditions blood glucose concentrations may fall dangerously low. Because the brain can use only glucose as an energy source, it is essential that the body be able to produce it quickly.

Gluconeogenesis is described in Chapter 21.

One of the steps in the pathway is an aldol condensation between the ketone dihydroxyacetone phosphate and the aldehyde glyceraldehyde-3-phosphate.

Dihydroxyacetonephosphate is a phosphorylated form of dihydroxyacetone (DHA), the active ingredient in self-tanning lotions. See *A Medical Perspective: That Golden Tan Without the Fear of Skin Cancer* on page 408.

$$\begin{array}{c} CH_2OPO_3^{2-} \\ | \\ C{=}O \\ | \\ H{-}COH \\ | \\ H \end{array} + \begin{array}{c} H\diagdown\ \diagup\!\!\!O \\ C \\ | \\ H{-}C{-}OH \\ | \\ CH_2OPO_3^{2-} \end{array} \overset{\text{Aldolase}}{\rightleftharpoons} \begin{array}{c} CH_2OPO_3^{2-} \\ | \\ C{=}O \\ | \\ HO{-}C{-}H \\ | \\ H{-}C{-}OH \\ | \\ H{-}C{-}OH \\ | \\ CH_2OPO_3^{2-} \end{array}$$

Dihydroxyacetone phosphate — Glyceraldehyde-3-phosphate — Fructose-1,6-bisphosphate

A HUMAN PERSPECTIVE

The Chemistry of Vision

β-Carotene is found in many yellow vegetables, as well as in tomatoes and spinach. When β-carotene is cleaved, two molecules of vitamin A are produced. In the body, vitamin A is converted to 11-*cis*-retinal, an unsaturated aldehyde that is a vital component in the photochemical transformations that make up the vision process.

CH_2OH

Vitamin A

Cleavage at this position gives two molecules of vitamin A.

β-Carotene

The retina of the eye contains two types of cells that are responsible for vision: *rods* and *cones*. Rods are primarily responsible for vision in dim light. Cones are responsible for vision in bright light and for the detection of color.

In the retina a protein called *opsin* combines with 11-*cis*-retinal to form a modified protein, *rhodopsin*. The 11-*cis*-retinal portion of rhodopsin is a prosthetic group (a nonprotein portion of a protein that is necessary for its action).

When light strikes the rods, the light energy is absorbed by the 11-*cis*-retinal, which is photochemically converted to 11-*trans*-retinal. This causes a change in the shape of rhodopsin itself, producing *metarhodopsin*. In the next step, 11-*trans*-retinal dissociates from opsin to begin the visual process. This dissociation causes ions to flow more freely into the rod cells. The influx of ions, in turn, stimulates nerve cells that send signals to the brain. Interpretation of those signals produces the visual image.

Following the initial light stimulus, retinal returns to the *cis* isomer and reassociates with opsin. The system is then ready for the next impulse of light. However, some retinal is lost in the process and must be replaced by conversion of dietary vitamin A to retinal. As you might expect, a deficiency of vitamin A can have terrible consequences. In children, lack of vitamin A causes *xerophthalmia*, an eye disease that results first in night blindness and eventually in total blindness. This can be prevented by an adequate dietary supply of this vitamin.

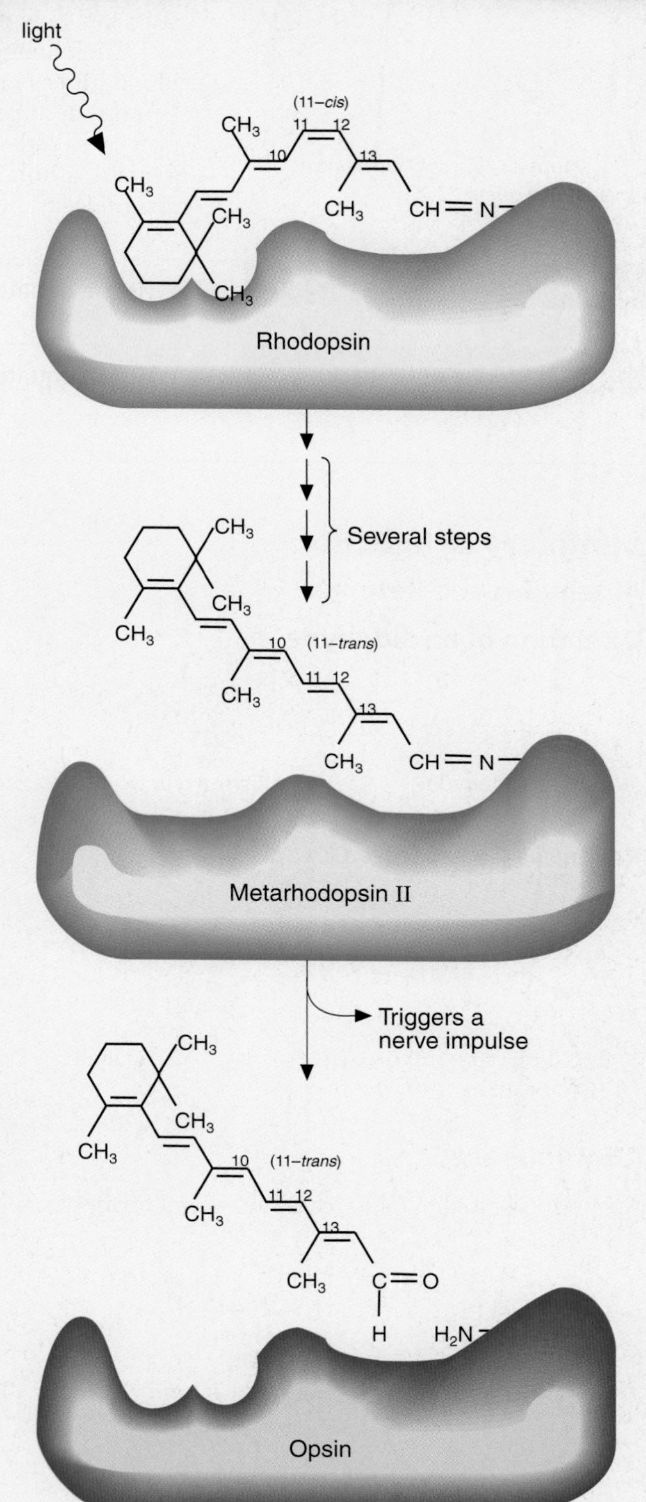

Light is absorbed by rhodopsin converting 11-*cis*-retinal to 11-*trans*-retinal. Metarhodopsin is converted to opsin and a nerve impulse is sent to the brain.

The speed and specificity of this reaction are ensured by the enzyme *aldolase*. The product is the sugar fructose-1,6-bisphosphate, which is converted by another enzyme into glucose-1,6-bisphosphate. Removal of the two phosphoryl groups results in a new molecule of glucose for use by the body as an energy source.

Gluconeogenesis occurs under starvation conditions to provide a supply of blood glucose to nourish the brain. However, when glucose is plentiful, it is broken down to provide ATP energy for the cell. The pathway for glucose degradation is called *glycolysis*. In glycolysis, the reaction just shown is reversed. In general, aldol condensation reactions are reversible. These reactions are called *reverse aldols*.

ATP, the universal energy currency, is discussed in Section 21.1.

Question 14.17 Write an equation for the aldol condensation of two molecules of propanal.

Question 14.18 Write an equation for the aldol condensation of two molecules of butanal.

Summary of Reactions

Aldehydes and Ketones

Oxidation of an aldehyde

$$\underset{\text{Aldehyde}}{\text{R—C(=O)—H}} \xrightarrow{[O]} \underset{\text{Carboxylic acid}}{\text{R—C(=O)—OH}}$$

Reduction of aldehydes and ketones

$$\underset{\substack{\text{Aldehyde}\\ \text{or Ketone}}}{R^1\text{—C(=O)—}R^2} + \underset{\text{Hydrogen}}{\text{H—H}} \xrightarrow{\text{Pt}} \underset{\text{Alcohol}}{R^1\text{—C(OH)(H)—}R^2}$$

Addition reactions

Addition of an alcohol to a ketone—ketal formation:

$$\underset{\text{Ketone}}{R^1\text{—C(=O)—}R^2} + \underset{\text{Alcohol}}{\text{H—O}R^3} \underset{}{\overset{\text{H}^+}{\rightleftharpoons}} \underset{\text{Hemiketal}}{R^1\text{—C(OH)(}R^2\text{)—O}R^3} + \text{H—O}R^3$$

$$\overset{\text{H}^+}{\rightleftharpoons} \underset{\text{Ketal}}{R^1\text{—C(O}R^3\text{)(}R^2\text{)—O}R^3} + H_2O$$

Addition of an alcohol to an aldehyde—acetal formation:

$$\underset{\text{Aldehyde}}{R^1\text{—C(=O)—H}} + \underset{\text{Alcohol}}{\text{H—O}R^2} \overset{\text{H}^+}{\rightleftharpoons} \underset{\text{Hemiacetal}}{R^1\text{—C(OH)(H)—O}R^2} + \text{H—O}R^2$$

$$\overset{\text{H}^+}{\rightleftharpoons} \underset{\text{Acetal}}{R^1\text{—C(O}R^2\text{)(H)—O}R^2} + H_2O$$

Keto-enol tautomerization

$$\underset{\text{Keto form}}{R^1\text{—C(H)(}R^2\text{)—C(=O)—}R^3} \rightleftharpoons \underset{\text{Enol form}}{(R^1)(R^2)\text{C=C(OH)(}R^3\text{)}}$$

Aldol condensation

$$R^1{-}CH_2{-}\overset{\overset{\displaystyle O}{\|}}{C}{-}R + R^2{-}CH_2{-}\overset{\overset{\displaystyle O}{\|}}{C}{-}R \xrightleftharpoons[\text{enzyme}]{OH^- \text{ or}} R^1{-}CH_2{-}\underset{\underset{\displaystyle R}{|}}{\overset{\overset{\displaystyle OH}{|}}{C}}{-}\underset{\underset{\displaystyle R^2}{|}}{CH}{-}\overset{\overset{\displaystyle O}{\|}}{C}{-}R$$

Aldehyde — Aldehyde — Aldol

R = H, alkyl, or aryl group

Summary

14.1 Structure and Physical Properties

The *carbonyl group* ($>C{=}O$) is characteristic of the *aldehydes* and *ketones*. The carbonyl group and the two groups attached to it are coplanar. In ketones the carbonyl carbon is attached to two carbon-containing groups, whereas in aldehydes the carbonyl carbon is attached to at least one hydrogen; the second group attached to the carbonyl carbon in aldehydes may be another hydrogen or a carbon atom. Owing to the polar carbonyl group, aldehydes and ketones are polar compounds. Their boiling points are higher than those of comparable hydrocarbons but lower than those of comparable alcohols. Small aldehydes and ketones are reasonably soluble in water because of the hydrogen bonding between the carbonyl group and water molecules. Larger carbonyl-containing compounds are less polar and thus are more soluble in nonpolar organic solvents.

14.2 I.U.P.A.C. Nomenclature and Common Names

In the I.U.P.A.C. Nomenclature System, aldehydes are named by determining the parent compound and replacing the final *-e* of the parent alkane with *-al*. The chain is numbered beginning with the carbonyl carbon as carbon-1. Ketones are named by determining the parent compound and replacing the *-e* ending of the parent alkane with the *-one* suffix of the ketone family. The longest carbon chain is numbered to give the carbonyl carbon the lowest possible number. In the common system of nomenclature, substituted aldehydes are named as derivatives of the parent compound. Greek letters indicate the position of substituents. Common names of ketones are derived by naming the alkyl groups bonded to the carbonyl carbon. These names are followed by the word *ketone*.

14.3 Important Aldehydes and Ketones

Many members of the aldehyde and ketone families are important as food and fragrance chemicals, medicinals, and agricultural chemicals. Methanal (formaldehyde) is used to preserve tissue. Ethanal causes the symptoms of a hangover and is oxidized to produce acetic acid commercially. Propanone is a useful and versatile solvent for organic compounds.

14.4 Reactions Involving Aldehydes and Ketones

In the laboratory, aldehydes and ketones are prepared by the oxidation of alcohols. Oxidation of a primary alcohol produces an aldehyde; oxidation of a secondary alcohol yields a ketone. Tertiary alcohols do not react under these conditions. Aldehydes and ketones can be distinguished from one another on the basis of their ability to undergo oxidation reactions. The *Tollens' test* and *Benedict's test* are the most common such tests. Aldehydes are easily oxidized to carboxylic acids. Ketones do not undergo further oxidation reactions. Aldehydes and ketones are readily reduced to alcohols by *hydrogenation*. The most common reaction of the carbonyl group is *addition* across the highly polar carbon-oxygen double bond. The addition of an alcohol to an aldehyde produces a *hemiacetal*. The hemiacetal reacts with a second alcohol molecule to form an *acetal*. The reaction of a ketone with an alcohol produces a *hemiketal*. A hemiketal reacts with a second alcohol molecule to form a *ketal*. Hemiacetals and hemiketals are readily formed in carbohydrates. Aldol condensation is a reaction in which aldehydes and ketones form larger molecules. Aldehydes and ketones may exist as an equilibrium mixture of keto and enol tautomers.

Key Terms

acetal (14.4)
addition reaction (14.4)
aldehyde (14.1)
aldol condensation (14.4)
Benedict's test (14.4)
carbonyl group (Intro)
hemiacetal (14.4)
hemiketal (14.4)
hydrogenation (14.4)
ketal (14.4)
ketone (14.1)
oxidation (14.4)
Tollens' test (14.4)

Questions and Problems

Structure and Physical Properties

14.19 Simple ketones (for example, acetone) are often used as industrial solvents for many organically based products such as adhesives and paints. They are often considered "universal solvents," because they dissolve so many diverse materials. Why are these chemicals such good solvents?

14.20 Explain briefly why simple (containing fewer than five carbon atoms) aldehydes and ketones exhibit appreciable solubility in water.

14.21 Draw intermolecular hydrogen bonding between ethanal and water.

14.22 Draw the polar interactions that occur between acetone molecules.

14.23 Why do alcohols have higher boiling points than aldehydes or ketones of comparable molecular weight?

14.24 Why do hydrocarbons have lower boiling points than aldehydes or ketones of comparable molecular weight?

Nomenclature

14.25 Draw each of the following using complete structural formulas:

a. Methanal
b. 7,8-Dibromooctanal
c. Acetone
d. Hydroxyethanal

14.26 Draw each of the following using condensed structural formulas:

a. 3-Chloro-2-pentanone
b. Benzaldehyde
c. 4-Bromo-3-hexanone
d. 2-Chlorocyclohexanone

14.27 Use the I.U.P.A.C. Nomenclature System to name each of the following compounds.

a. $CH_3-\overset{O}{\overset{\|}{C}}-CH_2CH_3$

b. $H-\overset{O}{\overset{\|}{C}}-\underset{CH_2CH_2CH_2CH_3}{\underset{|}{C}H}CH_2CH_3$

c. $Cl-\underset{Cl}{\underset{|}{\overset{Cl}{\overset{|}{C}}}}-\overset{O}{\overset{\|}{C}}-CH_3$

d. O, Cl

14.28 Name each of the following using the I.U.P.A.C. Nomenclature System:

a.

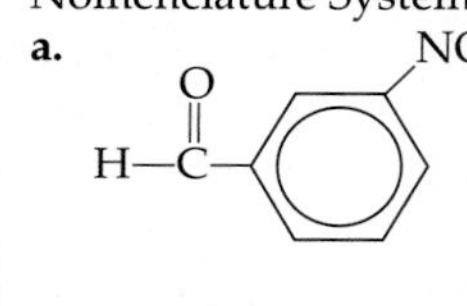

b. O, HO, OH

c. $CH_3CH_2CH_2-\overset{O}{\overset{\|}{C}}-H$

d. $CH_3\underset{CH_3}{\underset{|}{\overset{Br}{\overset{|}{C}}}}CH_2CH_2-\overset{O}{\overset{\|}{C}}-H$

14.29 List the rules for naming ketones under the I.U.P.A.C. Nomenclature System.

14.30 List the rules for naming aldehydes under the I.U.P.A.C. Nomenclature System.

14.31 Give the I.U.P.A.C. name for each of the following compounds:

a. $CH_3\underset{Br}{\underset{|}{C}H}CH_2-\overset{O}{\overset{\|}{C}}-H$

b. $CH_3\underset{Cl}{\underset{|}{\overset{CH_3}{\overset{|}{C}}}}CH_2-\overset{O}{\overset{\|}{C}}-CH_2CH_2CH_3$

c. $CH_3-\overset{O}{\overset{\|}{C}}-CH_2\underset{CH_2CH_3}{\underset{|}{\overset{CH_2CH_3}{\overset{|}{C}}}}CH_2CH_3$

d. $CH_3\overset{O}{\overset{\|}{C}}CH_2\underset{Cl}{\underset{|}{C}H}CH_2CH_3$

14.32 Give the I.U.P.A.C. name for each of the following compounds:

a. $CH_3\underset{CH_3}{\underset{|}{C}H}CH_2\underset{CH_3}{\underset{|}{C}H}-\overset{O}{\overset{\|}{C}}-CH_2CH_3$

b. CH_3, CH_3, O

c. $CH_3CH_2\overset{CH_3}{\overset{|}{C}H}CH_2\overset{O}{\overset{\|}{C}}-H$

d. O, Cl, Cl

14.33 Give the common name for each of the following compounds:

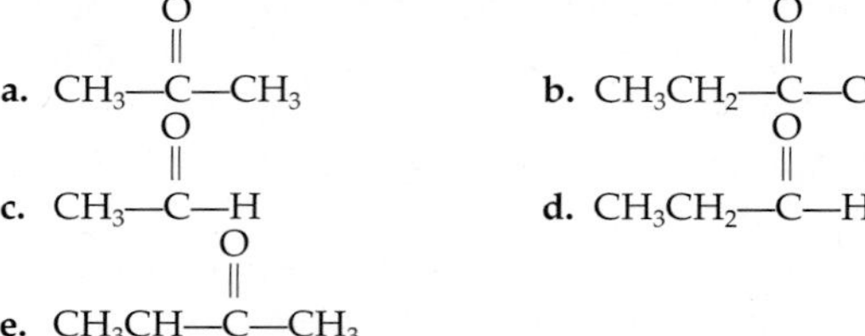

a. $CH_3-\overset{O}{\overset{\|}{C}}-CH_3$

b. $CH_3CH_2-\overset{O}{\overset{\|}{C}}-CH_3$

c. $CH_3-\overset{O}{\overset{\|}{C}}-H$

d. $CH_3CH_2-\overset{O}{\overset{\|}{C}}-H$

e. $CH_3\underset{CH_3}{\underset{|}{C}H}-\overset{O}{\overset{\|}{C}}-CH_3$

14.34 Give the common name for each of the following compounds:

a. $CH_3CH_2-\overset{O}{\overset{\|}{C}}-CH_2CH_3$

b. $CH_3CH_2CH_2\underset{CH_3}{\underset{|}{C}H}\overset{O}{\overset{\|}{C}}-H$

c. $CH_3\overset{O}{\overset{\|}{C}}CH_2CH_2CH_3$

d. $CH_3CH_2CH_2CH_2CH_2\overset{O}{\overset{\|}{C}}-H$

14.35 Draw the structure of each of the following compounds:

a. 3-Hydroxybutanal
b. 2-Methylpentanal
c. 4-Bromohexanal
d. 3-Iodopentanal
e. 2-Hydroxy-3-methylheptanal

14.36 Draw the structure of each of the following compounds:

a. Dimethyl ketone
b. Methyl propyl ketone
c. Ethyl butyl ketone
d. Diisopropyl ketone

Important Aldehydes and Ketones

14.37 Why is acetone a good solvent for many organic compounds?

14.38 List several uses for formaldehyde.

14.39 Ethanal is produced by the oxidation of ethanol. Where does this reaction occur in the body?

14.40 List several aldehydes and ketones that are used as food or fragrance chemicals.

Reactions Involving Aldehydes and Ketones

14.41 Draw the structures of each of the following compounds. Then draw and name the product that you would expect to produce by oxidizing each of these alcohols:

a. 2-Butanol
b. 2-Methyl-1-propanol
c. Cyclopentanol

14.42 Draw the structures of each of the following compounds. Then draw and name the product that you would expect to produce by oxidizing each of these alcohols.

a. 2-Methyl-2-propanol
b. 2-Nonanol
c. 1-Decanol

14.43 Draw the generalized equation for the oxidation of a primary alcohol.

14.44 Draw the generalized equation for the oxidation of a secondary alcohol.

14.45 Draw the structures of the reactants and products for each of the following reactions. Label each as an oxidation or a reduction reaction:

a. Ethanal to ethanol
b. Cyclohexanone to cyclohexanol
c. 2-Propanol to propanone

14.46 An unknown has been determined to be one of the following three compounds:

$CH_3CH_2—C(=O)—CH_2CH_3$ 3-Pentanone

$CH_3CH_2CH_2CH_2—C(=O)—H$ Pentanal

$CH_3CH_2CH_2CH_2CH_3$ Pentane

The unknown is fairly soluble in water and produces a silver mirror when treated with the silver ammonia complex. A red precipitate appears when it is treated with the Benedict's reagent. Which of the compounds is the correct structure for the unknown? Explain your reasoning.

14.47 Which of the following compounds would be expected to give a positive Tollens' test?

a. 3-Pentanone
b. Cyclohexanone
c. 3-Methylbutanal
d. Cyclopentanol
e. 2,2-Dimethyl-1-pentanol
f. Acetaldehyde

14.48 Write an equation representing the reaction of glucose with the Benedict's reagent. How was this test used in medicine?

14.49 Write an equation for the addition of one ethanol molecule to each of the following aldehydes and ketones:

a. $CH_3—C(=O)—CH_3$ b. $CH_3—C(=O)—H$

14.50 Write an equation for the addition of one ethanol molecule to each of the following aldehydes and ketones:

a. $CH_3CH_2—C(=O)—H$ b. $CH_3—C(=O)—CH_2CH_2CH_3$

14.51 What is the general name for the product that is formed when an aldehyde reacts with one molecule of alcohol?

14.52 What is the general name of the product that is formed when a ketone reacts with one molecule of alcohol?

14.53 What is the general name for the product that is formed when an aldehyde reacts with two molecules of alcohol?

14.54 What is the general name of the product that is formed when a ketone reacts with two molecules of alcohol?

14.55 Write an equation for the addition of two methanol molecules to each of the following aldehydes and ketones:

a. $CH_3—C(=O)—CH_3$ b. $CH_3—C(=O)—H$

14.56 Write an equation for the addition of two methanol molecules to each of the following aldehydes and ketones:

a. $CH_3CH_2—C(=O)—H$ b. $CH_3—C(=O)—CH_2CH_2CH_3$

14.57 An aldehyde can be oxidized to produce a carboxylic acid. Draw the carboxylic acid that would be produced by the oxidation of each of the following aldehydes:

a. Methanal
b. Ethanal
c. Propanal
d. Butanal

14.58 Indicate whether each of the following statements is true or false.

a. Aldehydes and ketones can be oxidized to produce carboxylic acids.
b. Oxidation of a primary alcohol produces an aldehyde.
c. Oxidation of a tertiary alcohol produces a ketone.
d. Alcohols can be produced by the oxidation of an aldehyde or ketone.

14.59 Indicate whether each of the following statements is true or false.

a. Ketones, but not aldehydes, react in the Tollens' silver mirror test.
b. Addition of one alcohol molecule to an aldehyde results in formation of a hemiacetal.
c. The cyclic forms of monosaccharides are intramolecular hemiacetals or intramolecular hemiketals.
d. Disaccharides (sugars composed of two covalently joined monosaccharides) are acetals, ketals, or both.

14.60 An alcohol can be oxidized to produce an aldehyde or a ketone. What aldehyde or ketone is produced by the oxidation of each of the following alcohols?

a. Methanol
b. 1-Propanol
c. 3-Pentanol
d. 2-Methyl-2-butanol

14.61 Write an equation for the aldol condensation of two molecules of ethanal.

14.62 Write an equation for the aldol condensation of two molecules of hexanal.

14.63 Draw the keto and enol forms of propanone.

14.64 Draw the keto and enol forms of 2-butanone.

14.65 Draw the hemiacetal or hemiketal that results from the reaction of each of the following aldehydes or ketones with ethanol:

a. $CH_3CH_2CH_2—C(=O)—CH_3$

b. 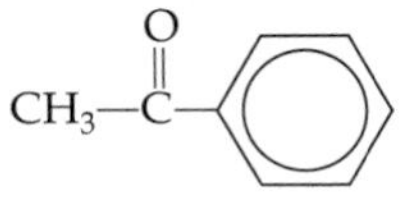 $CH_3—C(=O)—C_6H_5$ (benzene ring)

c. cyclopentanone (five-membered ring with =O)

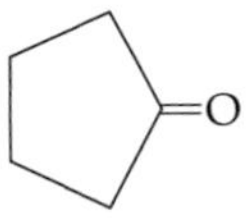

14.66 Identify each of the following compounds as a hemiacetal, hemiketal, acetal, or ketal:

a. 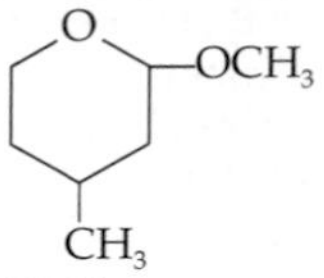 (tetrahydropyran ring with OCH_3 on carbon adjacent to O and CH_3 at position 4)

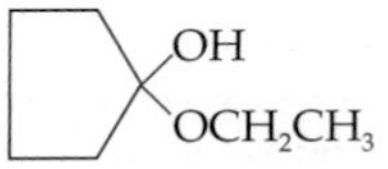

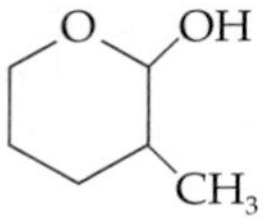

b. (cyclopentane ring with OH and OCH_2CH_3 on the same carbon)

c. CH_3CCH_3 with OH and OCH_2CH_3 on the central carbon

d. (tetrahydropyran ring with OH on carbon adjacent to O and CH_3 on the next carbon)

e. CH_3CCH_3 with OCH_3 and OCH_2CH_3 on the central carbon

f. $CH_3CH{=}CHCCH_3$ with OCH_3 and OH on the last substituted carbon

14.67 Complete the following synthesis by supplying the missing reactant(s), reagent(s), or product(s) indicated by the question marks:

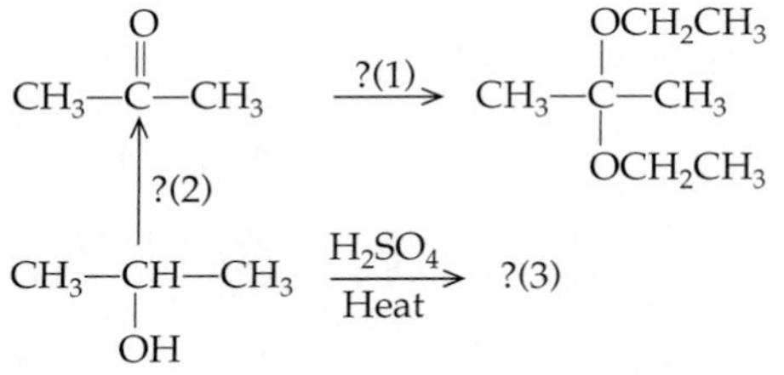

$$CH_3{-}\overset{O}{\overset{\|}{C}}{-}CH_3 \xrightarrow{?(1)} CH_3{-}C(OCH_2CH_3)_2{-}CH_3$$

$$CH_3{-}CH(OH){-}CH_3 \xrightarrow{?(2)} CH_3{-}\overset{O}{\overset{\|}{C}}{-}CH_3$$

$$CH_3{-}CH(OH){-}CH_3 \xrightarrow[\text{Heat}]{H_2SO_4} ?(3)$$

14.68 Which alcohol would you oxidize to produce each of the following compounds?

a. $CH_3\overset{CH_3}{CH}CH_2\overset{O}{\overset{\|}{C}}CH_3$

b. $H\overset{O}{\overset{\|}{C}}CH_2CH_2\overset{O}{\overset{\|}{C}}H$

c. 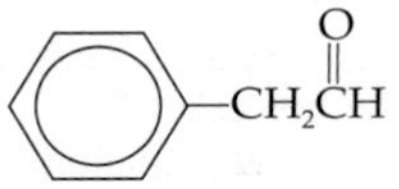 (benzene ring)$-CH_2\overset{O}{\overset{\|}{C}}H$

d. $H\overset{O}{\overset{\|}{C}}CH_2\overset{O}{\overset{\|}{C}}CH_3$

e. $CH_3C(CH_3)_2CH_2CH_2\overset{O}{\overset{\|}{C}}H$

f. 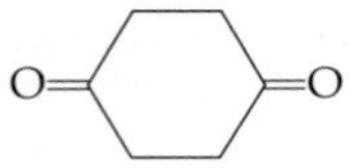 O=(cyclohexane ring)=O

Critical Thinking Problems

1. Review the material on the chemistry of vision and, with respect to the isomers of retinal, discuss the changes in structure that occur as the nerve impulses (that result in vision) are produced. Provide complete structural formulas of the retinal isomers that you discuss.

2. Classify the structure of β-D-fructose as a hemiacetal, hemiketal, acetal, or ketal. Explain your choice.

(structure of β-D-fructose: furanose ring with CH_2OH, O, OH, H, HO, CH_2OH, OH, H)

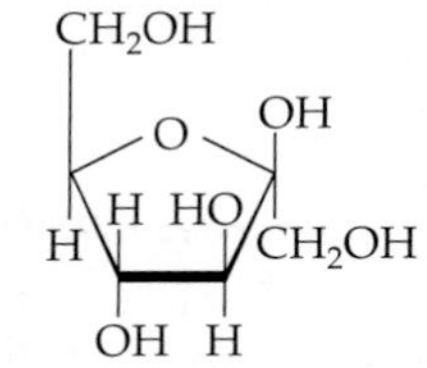

3. Design a synthesis for each of the following compounds, using any inorganic reagent of your choice and any hydrocarbon or alkyl halide of your choice:
   a. Octanal
   b. Cyclohexanone
   c. 2-Phenylethanoic acid

4. When alkenes react with ozone, O_3, the double bond is cleaved, and an aldehyde and/or a ketone is produced. The reaction, called *ozonolysis*, is shown in general as:

$$\rangle C{=}C\langle \; + O_3 \longrightarrow \; \rangle C{=}O + O{=}C\langle$$

Predict the ozonolysis products that are formed when each of the following alkenes is reacted with ozone:
   a. 1-Butene
   b. 2-Hexene
   c. *cis*-3,6-Dimethyl-3-heptene

5. Lactose is the major sugar found in mammalian milk. It is a dissacharide composed of the monosaccharides glucose and galactose:

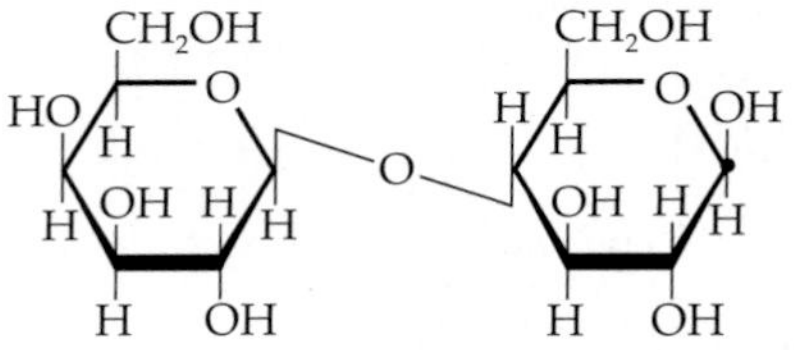

(structure of lactose: two pyranose rings with CH_2OH, HO, H, OH, O linkage, OH)

Is lactose a hemiacetal, hemiketal, acetal, or ketal? Explain your choice or choices.

6. The following are the keto and enol tautomers of phenol:

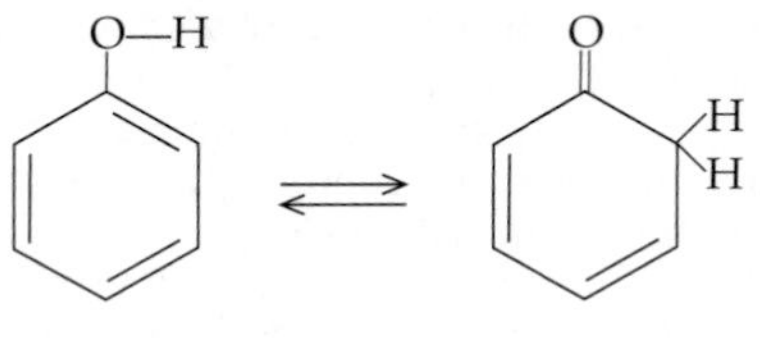

(structures: phenol with O–H ⇌ cyclohexadienone with O and H, H)

Enol form of phenol — Keto form of phenol

We have seen that most simple aldehydes and ketones exist mainly in the keto form because it is more stable. Phenol is an exception, existing primarily in the enol form. Propose a hypothesis to explain this.

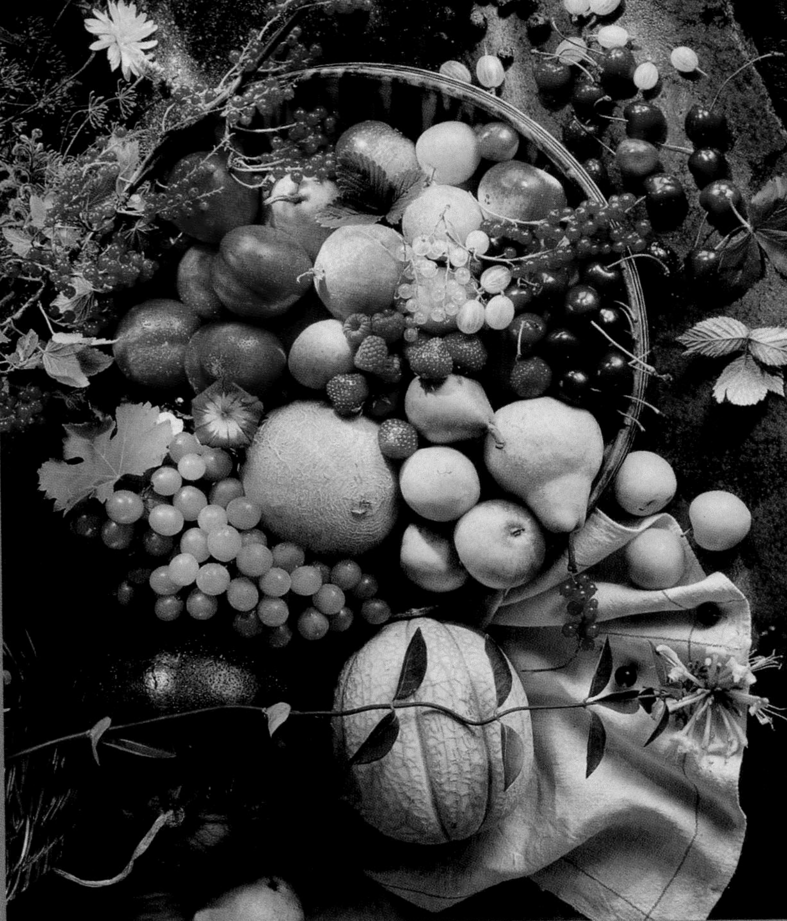

Natural fruit flavors are complex mixtures of esters and other organic compounds.

15

Carboxylic Acids and Carboxylic Acid Derivatives

Outline

ORGANIC CHEMISTRY

Learning Goals

1. Write structures and describe the physical properties of carboxylic acids.
2. Determine the common and I.U.P.A.C. names of carboxylic acids.
3. Describe the biological, medical, or environmental significance of several carboxylic acids.
4. Write equations that show the synthesis of a carboxylic acid.
5. Write equations representing acid–base reactions of carboxylic acids.
6. Write equations representing the preparation of an ester.
7. Write structures and describe the physical properties of esters.
8. Determine the common and I.U.P.A.C. names of esters.
9. Write equations representing the hydrolysis of an ester.
10. Define the term *saponification* and describe how soap works in the emulsification of grease and oil.
11. Determine the common and I.U.P.A.C. names of acid chlorides.
12. Write equations representing the synthesis of acid chlorides.
13. Determine the common and I.U.P.A.C. names of acid anhydrides.
14. Write equations representing the synthesis of acid anhydrides.
15. Discuss the significance of thioesters and phosphoesters in biological systems.

CHEMISTRY CONNECTION

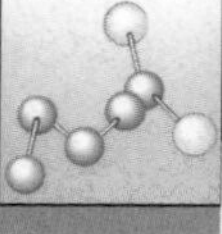

Wake Up, Sleeping Gene

A common carboxylic acid, butyric acid, holds the promise of being an effective treatment for two age-old human genetic diseases. Sickle cell anemia and β-thalassemia are human genetic diseases of the β-globin portion of hemoglobin, the protein that carries oxygen from the lungs to tissues throughout the body. Normal hemoglobin consists of two α-globin proteins, two β-globin proteins, and four heme groups. In sickle cell anemia the faulty β-globin gene calls for the synthesis of a sticky form of hemoglobin that forms long polymers. This distorts the red blood cells into elongated, sickled shapes that get stuck in capillaries and cannot provide the oxygen needed by the tissues. In β-thalassemia there may be no β-globin produced at all. This results in short-lived red blood cells and severe anemia.

These two genetic diseases do not affect the fetus because before birth and for several weeks after birth, a fetal globin is made, rather than the adult β-globin. Fetal hemoglobin has a stronger affinity for oxygen than the adult form, ensuring that the fetus gets enough oxygen from the mother's blood through the placenta.

Two observations have led to a possible treatment of these diseases. First, physicians found some sickle cell anemia patients who suffered only mild symptoms because they continued to make high levels of fetal hemoglobin. Second was the observation that some babies born to diabetic mothers continued to produce fetal hemoglobin for an unusually long time after birth. Coincidentally, there was an unexpectedly high concentration of aminobutyric acid, a modified carboxylic acid, in the blood of these infants.

Susan Perrine of the Children's Hospital Oakland Research Center decided to try to reawaken the dormant fetal globin gene. She and her colleagues injected a sodium butyrate solution (the sodium salt of butyric acid) into three sickle cell patients and three β-thalassemia patients. As a result of the two- to three-week treatment, fetal hemoglobin production was boosted as much as 45% in these individuals. One β-thalassemia patient even experienced a complete reversal of the symptoms. Moreover, this treatment had few adverse side effects.

Longer studies with larger numbers of patients will be needed before this treatment can be declared a total success. However, Perrine's results hold the promise of a full and active life for individuals who were previously limited in activity and expected a short life span.

In this chapter we study the properties and reactions of the carboxylic acids; their salts, such as the sodium butyrate used to treat hemoglobin disorders; and their derivatives, the esters. We will focus on the importance of these molecules in biological systems, medicine, and the food industry.

Introduction

Carboxylic acids (Figure 15.1a) have the following general structure:

$$\underset{\text{Aromatic carboxylic acid}}{\text{Ar}-\overset{\overset{\text{O}}{\|}}{\text{C}}-\text{OH}} \qquad \underset{\text{Aliphatic carboxylic acid}}{\text{R}-\overset{\overset{\text{O}}{\|}}{\text{C}}-\text{OH}}$$

They are characterized by the carboxyl group, shown in red, which may also be written in condensed form as —COOH or $—CO_2H$. The name *carboxylic acid* describes this family of compounds quite well. The term *carboxylic* is taken from the terms *carbonyl* and *hydroxyl*, the two structural units that make up the carboxyl group. The word *acid* in the name tells us one of the more important properties of these molecules: They dissociate in water to release protons. Thus they are acids.

In fact, carboxylic acids are weak acids because they partially dissociate in water.

In this chapter we will also study the esters (Figure 15.1b), which have the following general structures:

$$\text{R}-\overset{\overset{\text{O}}{\|}}{\text{C}}-\text{O}-\text{R} \qquad \text{Ar}-\overset{\overset{\text{O}}{\|}}{\text{C}}-\text{O}-\text{Ar} \qquad \text{Ar}-\overset{\overset{\text{O}}{\|}}{\text{C}}-\text{O}-\text{R}$$

Examples of aliphatic and aromatic esters

The group shown in red is called the **acyl group.** The acyl group is part of the functional group of the carboxylic acid derivatives, including the esters, acid chlorides, acid anhydrides, and amides.

Amides are discussed in Chapter 16.

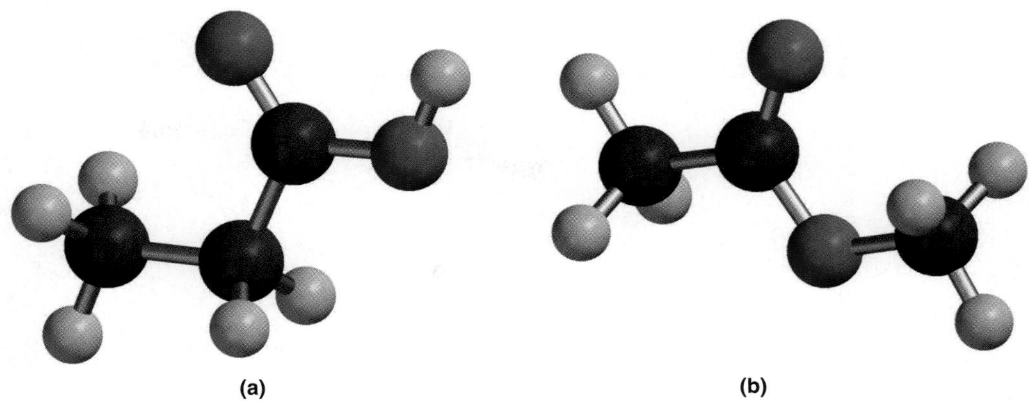

Figure 15.1
Ball-and-stick models of (a) a carboxylic acid, propanoic acid, and (b) an ester, methyl ethanoate.

15.1 Carboxylic Acids

Structure and Physical Properties

Learning Goal

The **carboxyl group** consists of two very polar functional groups, the carbonyl group and the hydroxyl group. Thus **carboxylic acids** are very polar compounds. In addition, they can hydrogen bond to one another and to molecules of a polar solvent such as water. As a result of intermolecular hydrogen bonding, they boil at higher temperatures than aldehydes, ketones, or alcohols of comparable molecular weight. A comparison of the boiling points of an alkane, alcohol, ether, aldehyde, ketone, and carboxylic acid of comparable molecular weight is shown below:

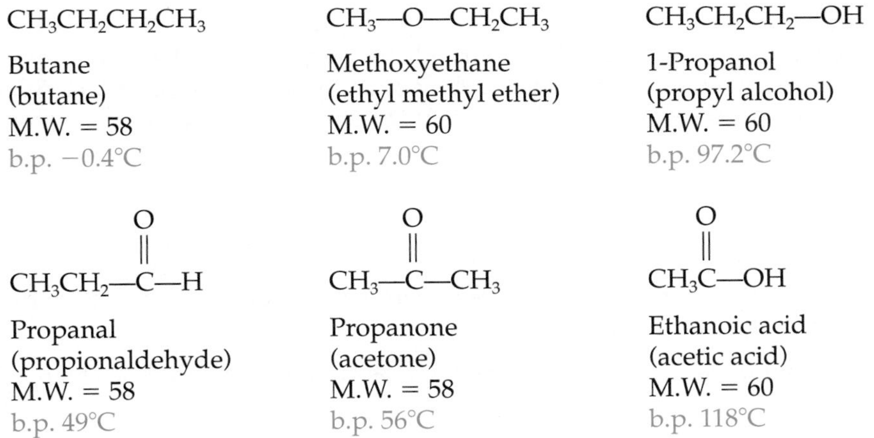

$CH_3CH_2CH_2CH_3$
Butane
(butane)
M.W. = 58
b.p. −0.4°C

$CH_3—O—CH_2CH_3$
Methoxyethane
(ethyl methyl ether)
M.W. = 60
b.p. 7.0°C

$CH_3CH_2CH_2—OH$
1-Propanol
(propyl alcohol)
M.W. = 60
b.p. 97.2°C

$CH_3CH_2—C(=O)—H$
Propanal
(propionaldehyde)
M.W. = 58
b.p. 49°C

$CH_3—C(=O)—CH_3$
Propanone
(acetone)
M.W. = 58
b.p. 56°C

$CH_3C(=O)—OH$
Ethanoic acid
(acetic acid)
M.W. = 60
b.p. 118°C

As with alcohols, the smaller carboxylic acids are soluble in water (Figure 15.2). However, solubility falls off dramatically as the carbon content of the carboxylic acid increases because the molecules become more hydrocarbonlike and less polar. For example, acetic acid (the carboxylic acid found in vinegar) is completely soluble in water, but hexadecanoic acid (a sixteen-carbon carboxylic acid found in palm oil) is insoluble in water.

The lower-molecular-weight carboxylic acids have sharp, sour tastes and unpleasant aromas. Formic acid, HCOOH, is used as a chemical defense by ants and causes the burning sensation of the ant bite. Acetic acid, CH_3COOH, is found in vinegar; propionic acid, CH_3CH_2COOH, is responsible for the tangy flavor of Swiss cheese; and butyric acid, $CH_3CH_2CH_2COOH$, causes the stench associated with rancid butter and gas gangrene.

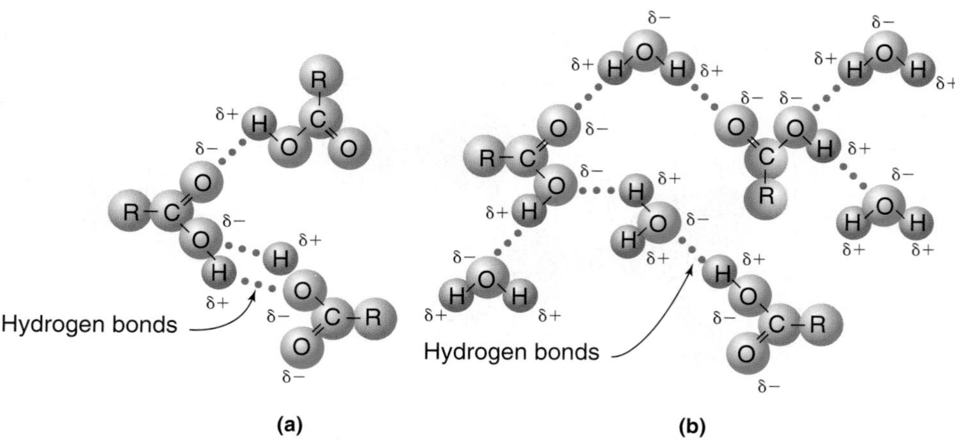

Figure 15.2
Hydrogen bonding (a) in carboxylic acids and (b) between carboxylic acids and water.

The longer-chain carboxylic acids are generally called **fatty acids** and are important components of biological membranes and triglycerides, the major lipid storage form in the body.

Question 15.1

Which member of each of the following pairs has the lower boiling point?

a. Hexanoic acid or 3-hexanone
b. 3-Hexanone or 3-hexanol
c. 3-Hexanol or hexane
d. Dipropyl ether or hexanal
e. Hexanal or hexanoic acid

Question 15.2

The functional group is largely responsible for the physical and chemical properties of the various chemical families. Why would you predict that a carboxylic acid would be more polar and have a higher boiling point than an alcohol of comparable molecular weight?

Nomenclature

Learning Goal

In the I.U.P.A.C. Nomenclature System, carboxylic acids are named according to the following set of rules:

- Determine the parent compound, the longest continuous carbon chain bearing the carboxyl group.
- Replace the *-e* ending of the parent alkane with the suffix *-oic acid.* If there are two carboxyl groups, the suffix *-dioic acid* is used.
- Number the chain so that the carboxyl carbon is carbon-1.
- Name and number substituents in the usual way.

The following examples illustrate the naming of carboxylic acids:

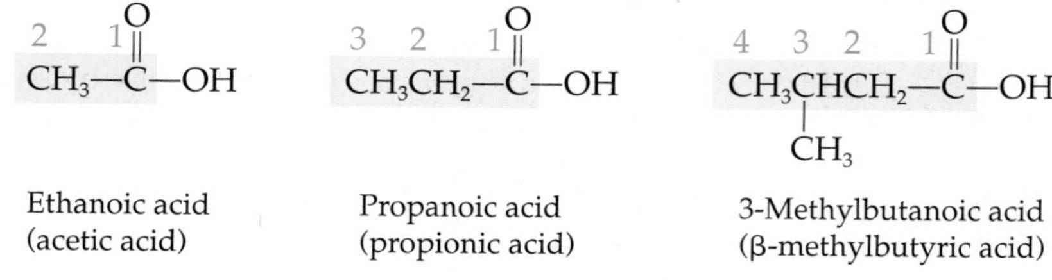

Ethanoic acid (acetic acid)

Propanoic acid (propionic acid)

3-Methylbutanoic acid (β-methylbutyric acid)

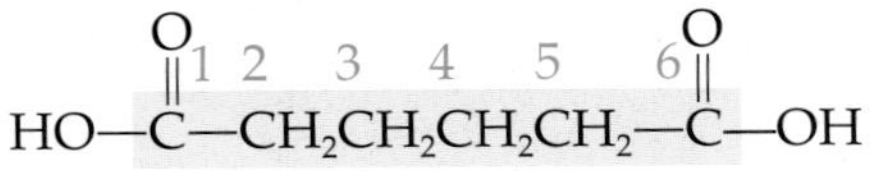

$$\overset{\displaystyle O}{\overset{\|}{\underset{1}{HO{-}C}}}{-}\underset{2\ \ 3\ \ 4\ \ 5}{CH_2CH_2CH_2CH_2}{-}\overset{\displaystyle O}{\overset{\|}{\underset{6}{C}}}{-}OH$$

Hexanedioic acid
(adipic acid)

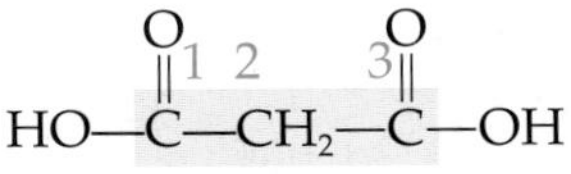

$$\overset{\displaystyle O}{\overset{\|}{\underset{1}{HO{-}C}}}{-}\underset{2}{CH_2}{-}\overset{\displaystyle O}{\overset{\|}{\underset{3}{C}}}{-}OH$$

Propanedioic acid
(malonic acid)

Use the I.U.P.A.C. Nomenclature System to Name a Carboxylic Acid — EXAMPLE 15.1

Name the following carboxylic acids using the I.U.P.A.C. Nomenclature System.

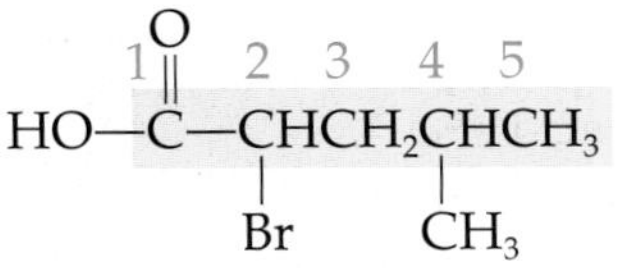

$$HO{-}\overset{\displaystyle O}{\overset{\|}{\underset{1}{C}}}{-}\underset{\underset{\displaystyle Br}{|}}{\overset{2}{C}H}\overset{3}{CH_2}\underset{\underset{\displaystyle CH_3}{|}}{\overset{4}{C}H}\overset{5}{CH_3}$$

Solution

Parent compound: pentane (becomes pentanoic acid)
Position of —COOH: carbon-1 (Must be!)
Substituents: 2-bromo and 4-methyl
Name: 2-Bromo-4-methylpentanoic acid

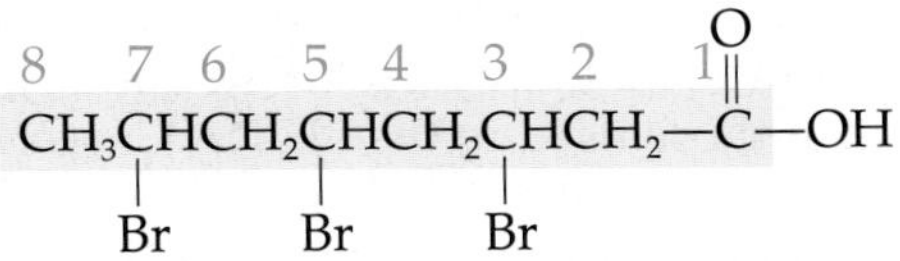

$$\overset{8}{CH_3}\underset{\underset{\displaystyle Br}{|}}{\overset{7}{C}H}\overset{6}{CH_2}\underset{\underset{\displaystyle Br}{|}}{\overset{5}{C}H}\overset{4}{CH_2}\underset{\underset{\displaystyle Br}{|}}{\overset{3}{C}H}\overset{2}{CH_2}{-}\overset{\displaystyle O}{\overset{\|}{\underset{1}{C}}}{-}OH$$

Solution

Parent compound: octane (becomes octanoic acid)
Position of —COOH: carbon-1 (Must be!)
Substituents: 3,5,7-bromo
Name: 3,5,7-Tribromooctanoic acid

The carboxylic acid derivatives of cycloalkanes are named by adding the suffix *carboxylic acid* to the name of the cycloalkane or substituted cycloalkane. The carboxyl group is always on carbon-1 and other substituents are named and numbered as usual.

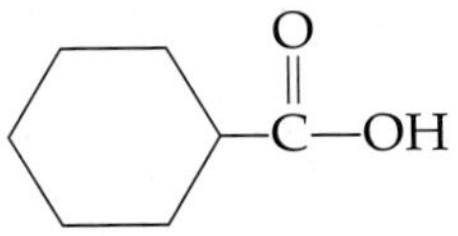

$$C_6H_{11}{-}\overset{\displaystyle O}{\overset{\|}{C}}{-}OH$$

Cyclohexanecarboxylic acid

Question 15.3

Determine the I.U.P.A.C. name for each of the following structures.

a. 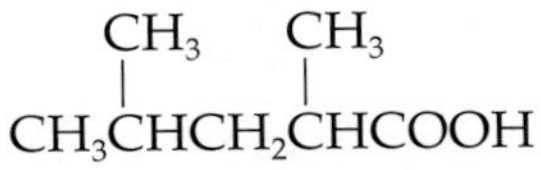 $CH_3\underset{}{\overset{\overset{\displaystyle CH_3}{|}}{C}H}CH_2\overset{\overset{\displaystyle CH_3}{|}}{C}HCOOH$

b. 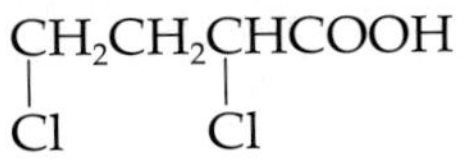 $\underset{\underset{\displaystyle Cl}{|}}{C}H_2CH_2\underset{\underset{\displaystyle Cl}{|}}{C}HCOOH$

c. COOH

CH_3

d. COOH

CH_2CH_3

Question 15.4

Write the structure for each of the following carboxylic acids.

a. 2,3-Dihydroxybutanoic acid
b. 2-Bromo-3-chloro-4-methylhexanoic acid
c. 1,4-Cyclohexanedicarboxylic acid
d. 4-Hydroxycyclohexanecarboxylic acid

Learning Goal

As we have seen so often, the use of common names, rather than systematic names, still persists. Often these names have evolved from the source of a given compound. This is certainly true of the carboxylic acids. Table 15.1 shows the I.U.P.A.C. and common names of several carboxylic acids, as well as their sources and the Latin or Greek words that gave rise to the common names. Not only are the prefixes different than those used in the I.U.P.A.C. system, the suffix is different as well. Common names end in *-ic acid* rather than *-oic acid.*

In the common system of nomenclature, substituted carboxylic acids are named as derivatives of the parent compound (see Table 15.1). Greek letters are

Table 15.1 Names and Sources of Some Common Carboxylic Acids

Name	Structure	Source	Root
Formic acid (methanoic acid)	HCOOH	Ants	L: *formica*, ant
Acetic acid (ethanoic acid)	CH_3COOH	Vinegar	L: *acetum*, vinegar
Propionic acid (propanoic acid)	CH_3CH_2COOH	Swiss cheese	Gk: *protos*, first *pion*, fat
Butyric acid (butanoic acid)	$CH_3(CH_2)_2COOH$	Rancid butter	L: *butyrum*, butter
Valeric acid (pentanoic acid)	$CH_3(CH_2)_3COOH$	Valerian root	
Caproic acid (hexanoic acid)	$CH_3(CH_2)_4COOH$	Goat fat	L: *caper*, goat
Caprylic acid (octanoic acid)	$CH_3(CH_2)_6COOH$	Goat fat	L: *caper*, goat
Capric acid (decanoic acid)	$CH_3(CH_2)_8COOH$	Goat fat	L: *caper*, goat
Palmitic acid (hexadecanoic acid)	$CH_3(CH_2)_{14}COOH$	Palm oil	
Stearic acid (octadecanoic acid)	$CH_3(CH_2)_{16}COOH$	Tallow (beef fat)	Gk: *stear*, tallow

Note: I.U.P.A.C. names are shown in parentheses.

used to indicate the position of the substituent. The carbon atom bonded to the carboxyl group is the α-carbon, the next is the β-carbon, and so on.

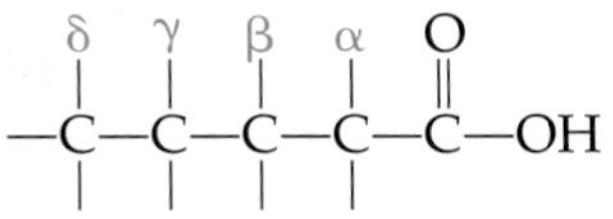

$$-\overset{\delta}{\underset{|}{\overset{|}{C}}}-\overset{\gamma}{\underset{|}{\overset{|}{C}}}-\overset{\beta}{\underset{|}{\overset{|}{C}}}-\overset{\alpha}{\underset{|}{\overset{|}{C}}}-\overset{O}{\overset{\|}{C}}-OH$$

Some examples of common names are

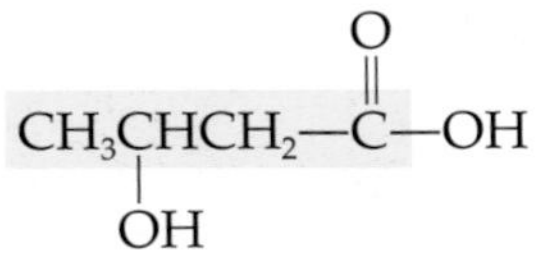

$$CH_3\underset{OH}{\underset{|}{C}}HCH_2-\overset{O}{\overset{\|}{C}}-OH \qquad CH_3\underset{OH}{\underset{|}{C}}H-\overset{O}{\overset{\|}{C}}-OH$$

β-Hydroxybutyric acid 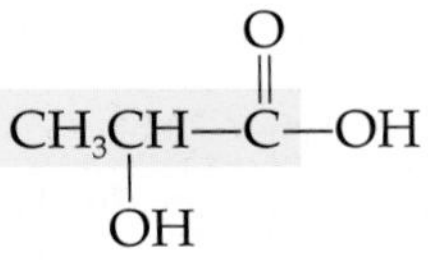 α-Hydroxypropionic acid

Naming Carboxylic Acids Using the Common System of Nomenclature — EXAMPLE 15.2

Write the common name for each of the following carboxylic acids.

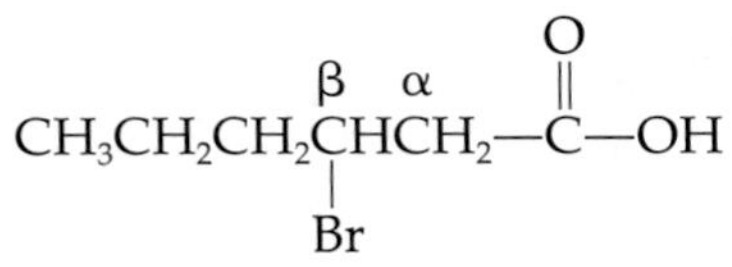

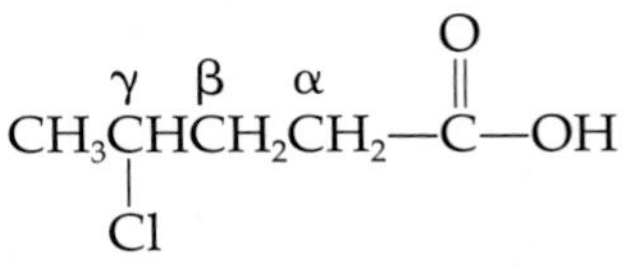

$$CH_3CH_2CH_2\underset{Br}{\underset{|}{\overset{\beta}{C}}}H\overset{\alpha}{C}H_2-\overset{O}{\overset{\|}{C}}-OH \qquad CH_3\underset{Cl}{\underset{|}{\overset{\gamma}{C}}}H\overset{\beta}{C}H_2\overset{\alpha}{C}H_2-\overset{O}{\overset{\|}{C}}-OH$$

Solution

Parent compound:	caproic acid	valeric acid
Substituents:	β-bromo	γ-chloro
Name:	β-Bromocaproic acid	γ-Chlorovaleric acid

Benzoic acid is the simplest aromatic carboxylic acid.

$$C_6H_5-\overset{O}{\overset{\|}{C}}-OH$$

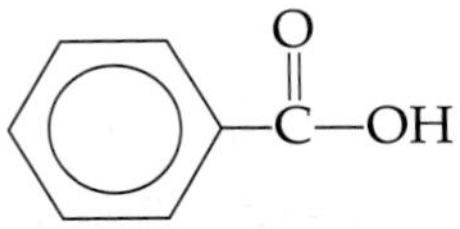

Benzoic acid

In many cases the aromatic carboxylic acids are named, in either system, as derivatives of benzoic acid. Generally, the *-oic acid* or *-ic acid* suffix is attached to the appropriate prefix. However, "common names" of substituted benzoic acids (for example, toluic acid and phthalic acid) are frequently used.

Nomenclature of aromatic compounds is described in Section 12.6.

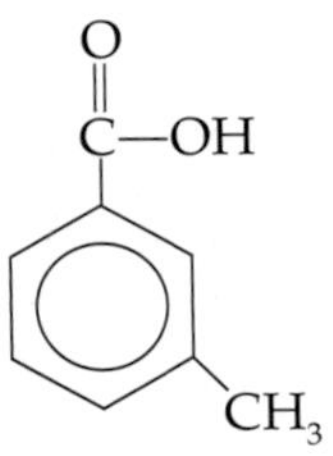

m-Toluic acid 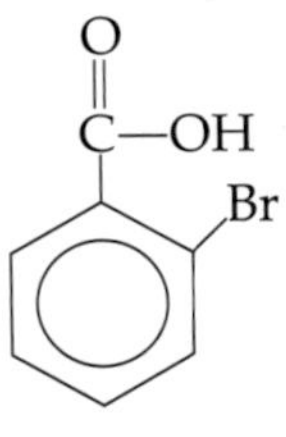 *o*-Bromobenzoic acid 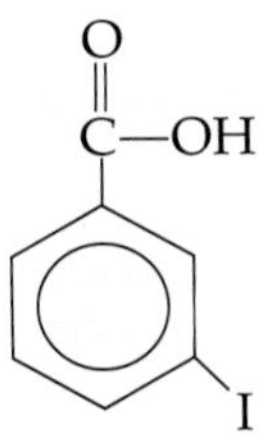 *m*-Iodobenzoic acid 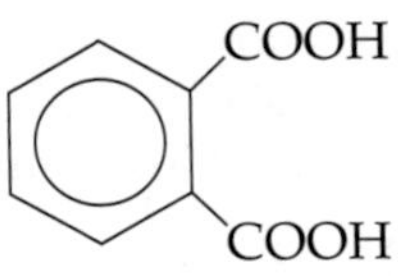 Phthalic acid

The phenyl group is benzene with one hydrogen removed.

Often the phenyl group is treated as a substituent, and the name is derived from the appropriate alkanoic acid parent chain. For example:

$C_6H_5CH_2COOH$

2-Phenylethanoic acid
(α-phenylacetic acid)

$C_6H_5CH_2CH_2COOH$

3-Phenylpropanoic acid
(β-phenylpropionic acid)

EXAMPLE 15.3 *Naming Aromatic Carboxylic Acids*

Name the following aromatic carboxylic acids.

$p\text{-}ClC_6H_4COOH$

Solution

It is simplest to name the compound as a derivative of benzoic acid. The substituent, Cl, is attached to carbon-4 of the benzene ring. This compound is 4-chlorobenzoic acid or *p*-chlorobenzoic acid.

$C_6H_5CH(CH_3)CH_2CH_2COOH$

Solution

This compound is most easily named by treating the phenyl group as a substituent. The phenyl group is bonded to carbon-4 (or the γ-carbon, in the common system of nomenclature). The parent compound is pentanoic acid (valeric acid in the common system). Hence the name of this compound is 4-phenylpentanoic acid or γ-phenylvaleric acid.

Question 15.5

Draw structures for each of the following compounds.

a. *o*-Toluic acid
b. 2,4,6-Tribromobenzoic acid
c. 2,2,2-Triphenylethanoic acid

Question 15.6

Draw structures for each of the following compounds.

a. *p*-Toluic acid
b. 3-Phenylhexanoic acid
c. 3-Phenylcyclohexanecarboxylic acid

Some Important Carboxylic Acids

Learning Goal

As Table 15.1 shows, many carboxylic acids occur in nature. Fatty acids can be isolated from a variety of sources including palm oil, coconut oil, butter, milk, lard, and tallow (beef fat). More complex carboxylic acids are also found in a variety of foodstuffs. For example, citric acid is found in citrus fruits and is often used to give the sharp taste to sour candies. It is also added to foods as a preservative and antioxidant. Adipic acid (hexanedioic acid) gives tartness to soft drinks and helps to retard spoilage.

Bacteria in milk produce lactic acid as a product of fermentation of sugars. Lactic acid contributes a tangy flavor to yogurt and buttermilk. It is also used as a food preservative to lower the pH to a level that retards microbial growth that causes food spoilage. Lactic acid is also produced in muscle cells when an individual is exercising strenuously. If the level of lactic acid in the muscle and bloodstream becomes high enough, the muscle can't continue to work.

$$\begin{array}{c} COOH \\ | \\ H{-}C{-}H \\ | \\ HO{-}C{-}COOH \\ | \\ H{-}C{-}H \\ | \\ COOH \end{array} \qquad \begin{array}{c} COOH \\ | \\ H{-}C{-}OH \\ | \\ CH_3 \end{array} \qquad \begin{array}{c} COOH \\ | \\ H{-}C{-}H \\ | \\ H{-}C{-}H \\ | \\ H{-}C{-}H \\ | \\ H{-}C{-}H \\ | \\ COOH \end{array}$$

Citric acid *(citrus fruit)*   Lactic acid *(yogurt)*   Adipic acid *(beet juice)*

Reactions Involving Carboxylic Acids

Preparation of Carboxylic Acids

Learning Goal 4

Many of the small carboxylic acids are prepared on a commercial scale. For example, ethanoic (acetic) acid, found in vinegar, is produced commercially by the **oxidation** of either ethanol or ethanal as shown here:

$$\underset{\text{Ethanol}}{CH_3CH_2OH} \quad \text{or} \quad \underset{\text{Ethanal}}{CH_3{-}\overset{\overset{\displaystyle O}{\|}}{C}{-}H} \xrightarrow{\text{Oxidizing agent}} \underset{\text{Ethanoic acid}}{CH_3{-}\overset{\overset{\displaystyle O}{\|}}{C}{-}OH}$$

A variety of oxidizing agents, including oxygen, can be used, and catalysts are often required to provide acceptable yields. Other simple carboxylic acids can be made by oxidation of the appropriate primary alcohol or aldehyde.

In the laboratory, carboxylic acids are prepared by the oxidation of aldehydes or primary alcohols. Most common oxidizing agents, such as chromic acid, can be used. The general reaction is

These reactions were discussed in Sections 13.5 and 14.4.

$$\underset{\text{Primary alcohol}}{R{-}CH_2OH} \xrightarrow{[O]} \underset{\text{Aldehyde}}{R{-}\overset{\overset{\displaystyle O}{\|}}{C}{-}H} \xrightarrow{[O]} \underset{\text{Carboxylic acid}}{R{-}\overset{\overset{\displaystyle O}{\|}}{C}{-}OH}$$

An Environmental Perspective

Garbage Bags from Potato Peels

One of the problems facing society is our enormous accumulation of trash. This has prompted intense efforts to recycle aluminum, paper, and plastics. But one of the problems that remains is the plastic trash bag. When garbage in plastic bags is buried in landfills, the soil bacteria are unable to degrade the plastic and thus can't get to the biodegradable materials inside. Imagine a twenty-fourth century archeologist excavating one of these monuments to our society!

Intensive research is underway to invent a truly biodegradable trash bag. One of the more creative methods is to make plastic sheets from lactic acid. Lactic acid is a natural carboxylic acid produced by fermentation of sugars, particularly in milk and working muscle. Many soil bacteria can degrade polymers of lactic acid. Thus these trash bags would be easily broken down in landfill soil.

To make plastics from lactic acid requires a large supply of this carboxylic acid. As it turns out, this supply can be obtained from garbage!

About ten billion pounds of potato waste are created each year from the process of making french fries. In fact, nearly half the mass of the potato is wasted. Several billion liters of whey, a carbohydrate-rich liquid left over from cheese making, are also dumped down the drain. Both of these waste products can be easily converted to glucose, which can be converted into lactic acid. Lactic acid molecules are then converted into long polymers. These polymers are used to make sheets that can be fashioned into trash bags.

$$H_3C-\underset{\displaystyle OH}{\overset{\displaystyle H}{\overset{|}{\underset{|}{C}}}}-\overset{\displaystyle O}{\overset{\|}{C}}-OH$$

$$-\overset{\displaystyle O}{\overset{\|}{C}}-\underset{\displaystyle H}{\overset{\displaystyle CH_3}{C}}-O-\overset{\displaystyle O}{\overset{\|}{C}}-\underset{\displaystyle CH_3}{\overset{\displaystyle H}{C}}-O-\overset{\displaystyle O}{\overset{\|}{C}}-\underset{\displaystyle H}{\overset{\displaystyle CH_3}{C}}-O-$$

Scientists are making biodegradable plastic from garbage. Lactic acid can be prepared from potato peels or whey. The lactic acid can then be polymerized to produce the plastic polylactic acid, which can be broken down by microorganisms.

Because the lactic acid polymers are biocompatible, they have already been applied to medical practice. For instance, some sutures are made from lactic acid plastics.

Several problems remain to be solved before these trash bags appear in the market. The chief problem is that the end product (polylactic acid) is currently too expensive for commercial production. However, future research and development promise to produce an "environmentally friendly" garbage bag.

EXAMPLE 15.4 ***Writing Equations for the Oxidation of a Primary Alcohol to a Carboxylic Acid***

Write an equation showing the oxidation of 1-propanol to propanoic acid.

Solution

$$CH_3CH_2-\underset{\displaystyle H}{\overset{\displaystyle H}{\overset{|}{\underset{|}{C}}}}-OH \xrightarrow{H_2CrO_4} CH_3CH_2-\overset{\displaystyle O}{\overset{\|}{C}}-H \xrightarrow[\text{oxidation}]{\text{Continued}} CH_3CH_2-\overset{\displaystyle O}{\overset{\|}{C}}-OH$$

1-Propanol (propyl alcohol) — Propanal (propionaldehyde) — Propanoic acid (propionic acid)

Learning Goal

Acid–Base Reactions

The carboxylic acids behave as acids because they are proton donors. They are weak acids that dissociate to form a carboxylate ion and a hydrogen ion, as in

$$\text{R—}\overset{\text{O}}{\overset{\|}{\text{C}}}\text{—OH} \rightleftharpoons \text{R—}\overset{\text{O}}{\overset{\|}{\text{C}}}\text{—O}^- + \text{H}^+$$

Carboxylic acid — Carboxylate anion — Hydrogen ion

The properties of weak acids are described in Sections 9.1 and 9.2.

Carboxylic acids are weak acids because they dissociate only slightly in solution. The majority of the acid remains in solution in the undissociated form. Typically, less than 5% of the acid is ionized (approximately five carboxylate ions to every ninety-five carboxylic acid molecules).

When strong bases are added to a carboxylic acid, neutralization occurs. The acid protons are removed by the OH^- to form water and the carboxylate ion. The equilibrium shown in the reaction above is shifted to the right, owing to removal of H^+. This is an illustration of LeChatelier's principle.

LeChatelier's principle is described in Section 8.4.

$$\text{R—}\overset{\text{O}}{\overset{\|}{\text{C}}}\text{—OH} + \text{NaOH} \longrightarrow \text{R—}\overset{\text{O}}{\overset{\|}{\text{C}}}\text{—O}^-\text{Na}^+ + \text{H}_2\text{O}$$

Carboxylic acid — Strong base — Carboxylic acid salt — Water

The following examples show the neutralization of acetic acid and benzoic acid in solutions of the strong base NaOH.

The carboxylate anion and the cation of the base form the carboxylic acid salt.

$$\text{CH}_3\text{—}\overset{\text{O}}{\overset{\|}{\text{C}}}\text{—OH} + \text{NaOH} \longrightarrow \text{CH}_3\text{—}\overset{\text{O}}{\overset{\|}{\text{C}}}\text{—O}^-\text{Na}^+ + \text{H}_2\text{O}$$

Acetic acid — Sodium hydroxide (strong base) — Sodium acetate — Water

$$\text{C}_6\text{H}_5\text{—}\overset{\text{O}}{\overset{\|}{\text{C}}}\text{—OH} + \text{NaOH} \longrightarrow \text{C}_6\text{H}_5\text{—}\overset{\text{O}}{\overset{\|}{\text{C}}}\text{—O}^-\text{Na}^+ + \text{H}_2\text{O}$$

Benzoic acid — Sodium hydroxide (strong base) — Sodium benzoate

Sodium benzoate is commonly used as a food preservative.

Note that the salt of a carboxylic acid is named by replacing the *-ic acid* suffix with *-ate*. Thus acetic acid becomes acetate, and benzoic acid becomes benzoate. This name is preceded by the name of the appropriate cation, sodium in the examples above.

EXAMPLE 15.5

Writing an Equation to Show the Neutralization of a Carboxylic Acid by a Strong Base

Write an equation showing the neutralization of propanoic acid by potassium hydroxide.

Solution

The protons of the acid are removed by the OH^- of the base. This produces water. The cation of the base, in this case potassium ion, forms the salt of the carboxylic acid.

Continued—

EXAMPLE 15.5 —*Continued*

$$\underset{\text{Propanoic acid}}{CH_3CH_2{-}\overset{\overset{\displaystyle O}{\|}}{C}{-}OH} + \underset{\text{Potassium hydroxide}}{KOH} \longrightarrow \underset{\text{Potassium salt of propanoic acid}}{CH_3CH_2{-}\overset{\overset{\displaystyle O}{\|}}{C}{-}O^-K^+} + \underset{\text{Water}}{H_2O}$$

EXAMPLE 15.6 ***Naming the Salt of a Carboxylic Acid***

Write the common and I.U.P.A.C. names of the salt produced in the reaction shown in Example 15.5.

Solution

$$CH_3CH_2{-}\overset{\overset{\displaystyle O}{\|}}{C}{-}O^-K^+$$

I.U.P.A.C. name of the parent carboxylic acid:	Propanoic acid
Replace the *-ic acid* ending with *-ate:*	Propanoate
Name of the cation of the base:	Potassium
Name of the carboxylic acid salt:	Potassium propanoate
Common name of the parent carboxylic acid:	Propionic acid
Replace the *-ic acid* ending with *-ate:*	Propionate
Name of the cation of the base:	Potassium
Name of the carboxylic acid salt:	Potassium propionate

Soaps are made by a process called saponification, which is the base-catalyzed hydrolysis of an ester. This is described in detail in Section 15.2.

Carboxylic acid salts are ionic substances. As a result, they are very soluble in water. The long-chain carboxylic acid salts (fatty acid salts) are called *soaps.*

Question 15.7

Write the formula of the organic product obtained through each of the following reactions.

a. $CH_3CH_2CH_2OH \xrightarrow{H_2CrO_4} ?$

b. $H\overset{\overset{\displaystyle O}{\|}}{C}CH_2CH_2CH_2CH_3 \xrightarrow{H_2CrO_4} ?$

c. $CH_3CH_2COOH + KOH \longrightarrow ?$

d. $CH_3CH_2CH_2COOH + Ba(OH)_2 \longrightarrow ?$

Question 15.8

Complete each of the following reactions by supplying the missing product(s).

a. $CH_3CH_2OH \xrightarrow{H_2CrO_4} ?$

b. $H\overset{\overset{\displaystyle O}{\|}}{C}CH_2CH_2\overset{\overset{\displaystyle O}{\|}}{C}H \xrightarrow{H_2CrO_4} ?$

c. $CH_3CH_2CH_2CH_2CH_2COOH + KOH \longrightarrow ?$

d. Benzoic acid + sodium hydroxide $\longrightarrow ?$

Esterification

Learning Goal 6

Carboxylic acids react with alcohols to form esters and water according to the general reaction:

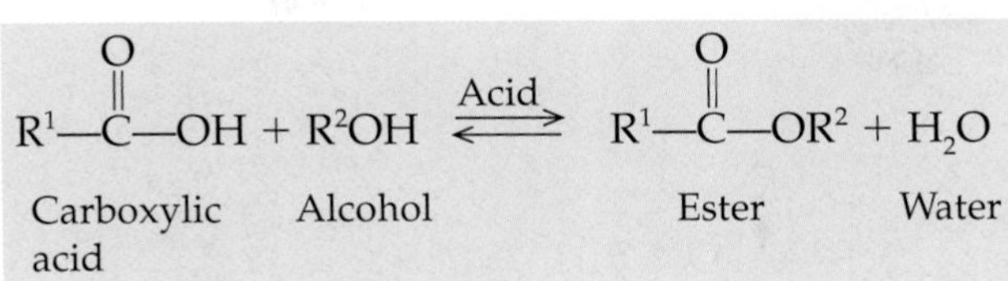

The details of these reactions will be examined in Section 15.2.

15.2 Esters

Structure and Physical Properties

Learning Goal 7

Esters are mildly polar and have pleasant aromas. Many esters are found in natural foodstuffs; banana oil (3-methylbutyl ethanoate; common name, isoamyl acetate), pineapples (ethyl butanoate; common name, ethyl butyrate), and raspberries (isobutyl methanoate; common name, isobutyl formate) are but a few examples.

Esters boil at approximately the same temperature as aldehydes or ketones of comparable molecular weight. The simpler ones are somewhat soluble in water.

Nomenclature

See A Human Perspective: The Chemistry of Flavor and Fragrance.

Esters are **carboxylic acid derivatives,** organic compounds derived from carboxylic acids. They are formed from the reaction of a carboxylic acid with an alcohol, and both of these reactants are reflected in the naming of the ester. They are named according to the following set of rules:

Learning Goal 8

- Use the *alkyl* or *aryl* portion of the alcohol name as the first name.
- The *-ic acid* ending of the name of the carboxylic acid is replaced with *-ate* and follows the first name.

For example, in the following reaction, ethanoic acid reacts with methanol to produce methyl ethanoate:

$$CH_3—\overset{\displaystyle O}{\overset{\|}{C}}—OH + CH_3OH \underset{}{\overset{H^+,\ heat}{\rightleftharpoons}} CH_3—\overset{\displaystyle O}{\overset{\|}{C}}—OCH_3 + H_2O$$

Ethanoic *acid* (acetic acid) | *Methanol* (methyl alcohol) | *Methyl* ethano*ate* (methyl acetate)

Similarly, acet*ic acid* and *ethanol* react to produce *ethyl* acet*ate,* and the product of the reaction between benzo*ic acid* and *isopropyl* alcohol is *isopropyl* benzo*ate.*

EXAMPLE 15.7

Naming Esters Using the I.U.P.A.C. and Common Nomenclature Systems

Write the I.U.P.A.C. and common names for each of the following esters.

$$CH_3CH_2CH_2—\overset{\displaystyle O}{\overset{\|}{C}}—OCH_2CH_3$$

Continued—

A HUMAN PERSPECTIVE

The Chemistry of Flavor and Fragrance

Carboxylic acids are often foul smelling. For instance, butyric acid is one of the worst smelling compounds imaginable.

$$CH_3CH_2CH_2—\overset{O}{\overset{\|}{C}}—OH$$

Butanoic acid
(butyric acid)

It is the smell you perceive in rancid butter. Butyric acid is also a product of fermentation reactions carried out by *Clostridium perfringens*. This organism is the most common cause of gas gangrene. Butyric acid contributes to the notable foul smell accompanying this infection.

By forming esters of butyric acid, one can generate compounds with pleasant smells. For instance, methyl butyrate is used in artificial fruit flavorings. Ethyl butyrate is the essence of pineapple oil.

$$CH_3CH_2CH_2—\overset{O}{\overset{\|}{C}}—OCH_3$$

Methyl butanoate
(methyl butyrate)

$$CH_3CH_2CH_2—\overset{O}{\overset{\|}{C}}—OCH_2CH_3$$

Ethyl butanoate
(ethyl butyrate)

Volatile esters are often pleasant in both aroma and flavor. Natural fruit flavors are complex mixtures of many esters and other organic compounds. Chemists can isolate these mixtures and identify the chemical components. With this information they are able to synthesize artificial fruit flavors, using just a few of the esters found in the natural fruit. As a result, the artificial flavors rarely have the full-bodied flavor of nature's original blend.

$$H—\overset{O}{\overset{\|}{C}}—OCH_2CH_3$$ Rum

Ethyl methanoate
(ethyl formate)

$$H—\overset{O}{\overset{\|}{C}}—OCH_2\underset{\underset{CH_3}{|}}{C}HCH_3$$ Raspberries

Isobutyl methanoate
(isobutyl formate)

$$CH_3—\overset{O}{\overset{\|}{C}}—OCH_2CH_2\overset{CH_3}{\overset{|}{C}}HCH_3$$ Bananas

3-Methylbutyl ethanoate
(isoamyl acetate)

$$CH_3—\overset{O}{\overset{\|}{C}}—OCH_2CH_2CH_2CH_2CH_2CH_2CH_2CH_3$$ Oranges

Octyl ethanoate
(octyl acetate)

$$CH_3CH_2CH_2—\overset{O}{\overset{\|}{C}}—OCH_3$$ Apples

Methyl butanoate
(methyl butyrate)

$$CH_3CH_2CH_2—\overset{O}{\overset{\|}{C}}—OCH_2CH_3$$ Pineapples

Ethyl butanoate
(ethyl butyrate)

$$CH_3CH_2CH_2—\overset{O}{\overset{\|}{C}}—OCH_2CH_2CH_2CH_2CH_3$$ Apricots

Pentyl butanoate
(pentyl butyrate)

$$C_6H_4(OH)—\overset{O}{\overset{\|}{C}}—OCH_3$$ Oil of wintergreen

Methyl salicylate

$$CH_3CH_2CH_2—\overset{O}{\overset{\|}{C}}—SCH_3$$ Strawberries

Methyl thiobutanoate
(methyl thiobutyrate)
(a thioester in which sulfur replaces oxygen)

EXAMPLE 15.7 —*Continued*

Solution

I.U.P.A.C. and common names of parent carboxylic acid:	butanoic acid	butyric acid
Replace the *ic acid* ending of the carboxylic acid with *-ate:*	butanoate	butyrate
Name of the alkyl portion of the alcohol:	ethyl	ethyl
I.U.P.A.C. and common names of the ester:	Ethyl butanoate	Ethyl butyrate

$$CH_3{-}\overset{\overset{\Large O}{\|}}{C}{-}OCH_2CH_2CH_3$$

Solution

I.U.P.A.C. and common names of parent carboxylic acid:	ethanoic acid	acetic acid
Replace the *ic acid* ending of the carboxylic acid with *-ate:*	ethanoate	acetate
Name of the alkyl portion of the alcohol:	propyl	propyl
I.U.P.A.C. and common names of the ester:	Propyl ethanoate	Propyl acetate

Naming esters is analogous to naming the salts of carboxylic acids. Consider the following comparison:

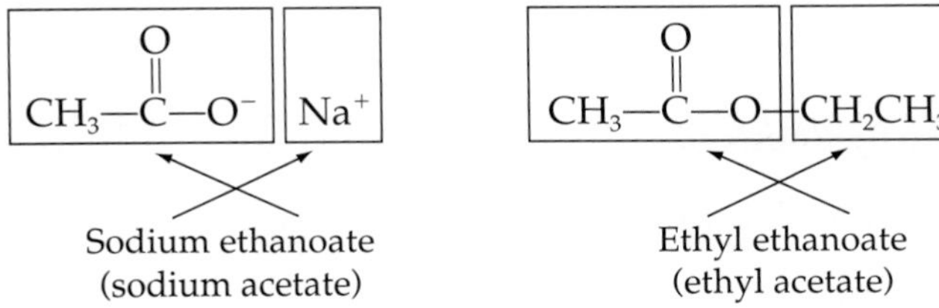

Sodium ethanoate (sodium acetate) — Ethyl ethanoate (ethyl acetate)

As shown in this example, the alkyl group of the alcohol, rather than Na^+, has displaced the acidic hydrogen of the carboxylic acid.

Reactions Involving Esters

Preparation of Esters

Learning Goal

The conversion of a carboxylic acid to an ester requires heat and is catalyzed by a trace of acid (H^+). When esters are prepared directly from a carboxylic acid and an alcohol, a water molecule is lost, as in the reaction:

$$\underset{\text{Carboxylic acid}}{R^1{-}\overset{\overset{O}{\|}}{C}{-}OH} + \underset{\text{Alcohol}}{R^2OH} \xrightleftharpoons{H^+,\ \text{heat}} \underset{\text{Ester}}{R^1{-}\overset{\overset{O}{\|}}{C}{-}OR^2} + \underset{\text{Water}}{H_2O}$$

Esterification is reversible. The direction of the reaction is determined by the conditions chosen. Excess alcohol favors ester formation. The carboxylic acid is favored when excess water is present.

$$\underset{\substack{\text{Propanoic acid}\\\text{(propionic acid)}}}{CH_3CH_2{-}\overset{\overset{O}{\|}}{C}{-}OH} + \underset{\substack{\text{Methanol}\\\text{(methyl alcohol)}}}{CH_3OH} \xrightleftharpoons{H^+,\ \text{heat}} \underset{\substack{\text{Methyl propanoate}\\\text{(methyl propionate)}}}{CH_3CH_2{-}\overset{\overset{O}{\|}}{C}{-}OCH_3} + H_2O$$

EXAMPLE 15.8 Writing Equations Representing Esterification Reactions

Write an equation showing the esterification reactions that would produce ethyl butanoate and propyl ethanoate.

Solution

The name, ethyl butanoate, tells us that the alcohol used in the reaction is ethanol and the carboxylic acid is butanoic acid. We must remember that a trace of acid and heat are required for the reaction and that the reaction is reversible. With this information, we can write the following equation representing the reaction:

$$\underset{\substack{\text{Butanoic acid}\\\text{(Butyric acid)}}}{CH_3CH_2CH_2{-}\overset{\overset{\displaystyle O}{\|}}{C}{-}OH} + \underset{\text{Ethanol}}{CH_3CH_2OH} \xrightleftharpoons{H^+,\ \text{heat}} \underset{\substack{\text{Ethyl butanoate}\\\text{(ethyl butyrate)}}}{CH_3CH_2CH_2{-}\overset{\overset{\displaystyle O}{\|}}{C}{-}OCH_2CH_3} + H_2O$$

Similarly, the name propyl ethanoate reveals that the alcohol used in this reaction is 1-propanol and the carboxylic acid must be ethanoic acid. Knowing that we must indicate that the reaction is reversible and that heat and a trace of acid are required, we can write the following equation:

$$\underset{\substack{\text{Ethanoic acid}\\\text{(Acetic acid)}}}{CH_3{-}\overset{\overset{\displaystyle O}{\|}}{C}{-}OH} + \underset{\text{1-Propanol}}{CH_3CH_2CH_2OH} \xrightleftharpoons{H^+,\ \text{heat}} \underset{\substack{\text{Propyl ethanoate}\\\text{(propyl acetate)}}}{CH_3{-}\overset{\overset{\displaystyle O}{\|}}{C}{-}OCH_2CH_2CH_3} + H_2O$$

EXAMPLE 15.9 Designing the Synthesis of an Ester

Design the synthesis of ethyl propanoate from organic alcohols.

Solution

The ease with which alcohols are oxidized to aldehydes, ketones, or carboxylic acids (depending on the alcohol that you start with and the conditions that you employ), coupled with the ready availability of alcohols, provides the pathway necessary to many successful synthetic transformations. For example, let's develop a method for synthesizing ethyl propanoate, using any inorganic reagent you wish but limiting yourself to organic alcohols that contain three or fewer carbon atoms:

$$\underset{\substack{\text{Ethyl propanoate}\\\text{(ethyl propionate)}}}{CH_3CH_2{-}\overset{\overset{\displaystyle O}{\|}}{C}{-}O{-}CH_2CH_3}$$

Ethyl propanoate can be made from propanoic acid and ethanol:

$$\underset{\substack{\text{Propanoic acid}\\\text{(propionic acid)}}}{CH_3CH_2{-}\overset{\overset{\displaystyle O}{\|}}{C}{-}OH} + \underset{\substack{\text{Ethanol}\\\text{(ethyl alcohol)}}}{CH_3CH_2OH} \xrightleftharpoons{H^+,\ \text{heat}} \underset{\substack{\text{Ethyl propanoate}\\\text{(ethyl propionate)}}}{CH_3CH_2{-}\overset{\overset{\displaystyle O}{\|}}{C}{-}O{-}CH_2CH_3} + H_2O$$

Continued—

EXAMPLE 15.9 —*Continued*

Ethanol is a two-carbon alcohol that is an allowed starting material, but propanoic acid is not. Can we now make propanoic acid from an alcohol of three or fewer carbons? Yes!

$$CH_3CH_2CH_2OH \xrightarrow{[O]} CH_3CH_2{-}\overset{\overset{\displaystyle O}{\|}}{C}{-}H \xrightarrow{[O]} CH_3CH_2{-}\overset{\overset{\displaystyle O}{\|}}{C}{-}OH$$

Propanol (propyl alcohol) → Propanoic acid (propionic acid)

Propanol is a three-carbon alcohol, an allowed starting material. The synthesis is now complete. By beginning with ethanol and propanol, ethyl propanoate can be synthesized easily.

Hydrolysis of Esters

Esters undergo **hydrolysis** reactions in water, as shown in the general reaction:

Learning Goal

$$R^1{-}\overset{\overset{\displaystyle O}{\|}}{C}{-}OR^2 + H_2O \underset{}{\overset{H^+,\ heat}{\rightleftharpoons}} R^1{-}\overset{\overset{\displaystyle O}{\|}}{C}{-}OH + R^2OH$$

Ester + Water ⇌ Carboxylic acid + Alcohol

In general, the term *hydrolysis* refers to cleavage of any bond by the addition of a water molecule.

This reaction requires heat. A small amount of acid (H^+) may be added to catalyze the reaction, as in the following example:

$$CH_3CH_2{-}\overset{\overset{\displaystyle O}{\|}}{C}{-}OCH_2CH_2CH_3 + H_2O \overset{H^+,\ heat}{\rightleftharpoons} CH_3CH_2{-}\overset{\overset{\displaystyle O}{\|}}{C}{-}OH + CH_3CH_2CH_2OH$$

Propyl propanoate (propyl propionate) + Water ⇌ Propanoic acid (propionic acid) + 1-Propanol (propanol)

The base-catalyzed hydrolysis of an ester is called **saponification.**

Learning Goal

$$R^1{-}\overset{\overset{\displaystyle O}{\|}}{C}{-}OR^2 + H_2O \xrightarrow{NaOH,\ heat} R^1{-}\overset{\overset{\displaystyle O}{\|}}{C}{-}O^-Na^+ + R^2OH$$

Ester + Water → Carboxylic acid salt + Alcohol

Under basic conditions the salt of the carboxylic acid is actually produced.

$$CH_3{-}\overset{\overset{\displaystyle O}{\|}}{C}{-}OCH_2CH_2CH_2CH_3 \xrightarrow{NaOH,\ heat} CH_3{-}\overset{\overset{\displaystyle O}{\|}}{C}{-}O^-Na^+ + CH_3CH_2CH_2OH$$

Butyl ethanoate (butyl acetate) → Sodium ethanoate (sodium acetate) + 1-Butanol (butyl alcohol)

The carboxylic acid is formed when the reaction mixture is neutralized with an acid such as HCl.

$$CH_3-\overset{\overset{\displaystyle O}{\|}}{C}-O^-Na^+ + HCl \longrightarrow CH_3-\overset{\overset{\displaystyle O}{\|}}{C}-OH + NaCl$$

Sodium ethanoate (sodium acetate) → Ethanoic acid (acetic acid)

Question 15.9

Complete each of the following reactions by supplying the missing products. Draw the structure and use the I.U.P.A.C. Nomenclature System to name the product.

a. $CH_3-\overset{\overset{\displaystyle O}{\|}}{C}-OCH_2CH_2CH_3 + H_2O \xrightleftharpoons{H^+,\ heat} ?$

b. $CH_3CH_2CH_2CH_2CH_2-\overset{\overset{\displaystyle O}{\|}}{C}-OCH_2CH_2CH_3 + H_2O \xrightarrow{KOH,\ heat} ?$

c. $CH_3CH_2CH_2CH_2-\overset{\overset{\displaystyle O}{\|}}{C}-OCH_3 + H_2O \xrightarrow{NaOH,\ heat} ?$

d. $CH_3CH_2CH_2CH_2CH_2-\overset{\overset{\displaystyle O}{\|}}{C}-O\underset{\underset{\displaystyle CH_3}{|}}{C}HCH_2CH_2CH_3 + H_2O \xrightleftharpoons{H^+,\ heat} ?$

Question 15.10

Draw the products that result from the saponification of methyl benzoate with sodium hydroxide followed by neutralization with hydrochloric acid.

Triesters of glycerol are more commonly referred to as triglycerides. We know them as solid fats, generally from animal sources, and liquid oils, typically from plants. We will study triglycerides in detail in Section 18.3.

Fats and oils are triesters of the alcohol glycerol. When they are hydrolyzed by saponification, the products are **soaps,** which are the salts of long-chain carboxylic acids (fatty acid salts). According to Roman legend, soap was discovered by washerwomen following a heavy rain on Mons Sapo ("Mount Soap"). An important sacrificial altar was located on the mountain. The rain mixed with the remains of previous animal sacrifices—wood ash and animal fat—at the base of the altar. Thus the three substances required to make soap accidentally came together—water, fat, and alkali (potassium carbonate and potassium hydroxide, called *potash,* leached from the wood ash). The soap mixture flowed down the mountain and into the Tiber River, where the washerwomen quickly realized its value.

We still use the old Roman recipe to make soap from water, a strong base, and natural fats and oils obtained from animals or plants. The carbon content of the fatty acid salts governs the solubility of a soap. The lower-molecular-weight carboxylic acid salts (up to twelve carbons) have greater solubility in water and give a lather containing large bubbles. The higher-molecular-weight carboxylic acid salts (fourteen to twenty carbons) are much less soluble in water and produce a lather with fine bubbles. The nature of the cation also affects the solubility of the soap. In general, the potassium salts of carboxylic acids are more soluble in water than the sodium salts. The synthesis of a soap is shown in Figure 15.3.

The role of soap in the removal of soil and grease is best understood by considering the functional groups in soap molecules and studying the way in which they interact with oil and water. The long, continuous hydrocarbon side chains of soap molecules resemble alkanes, and they dissolve other nonpolar compounds such as oils and greases ("like dissolves like"). The large nonpolar hydrocarbon part of the molecule is described as *hydrophobic,* which means "water-fearing." This part of the molecule is repelled by water. The highly polar carboxylate end of the molecule is called *hydrophilic,* which means "water-loving."

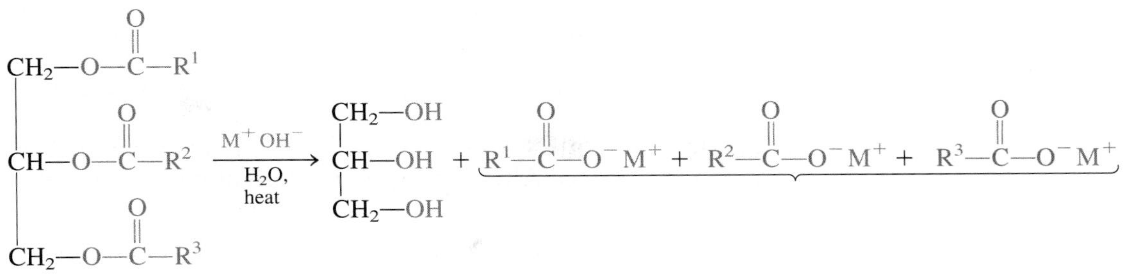

$$\begin{array}{l} \text{CH}_2\text{—O—}\overset{\displaystyle \text{O}}{\overset{\|}{\text{C}}}\text{—R}^1 \\ | \\ \text{CH—O—}\overset{\displaystyle \text{O}}{\overset{\|}{\text{C}}}\text{—R}^2 \\ | \\ \text{CH}_2\text{—O—}\overset{\displaystyle \text{O}}{\overset{\|}{\text{C}}}\text{—R}^3 \end{array} \xrightarrow[\text{H}_2\text{O, heat}]{\text{M}^+\text{OH}^-} \begin{array}{l} \text{CH}_2\text{—OH} \\ | \\ \text{CH—OH} \\ | \\ \text{CH}_2\text{—OH} \end{array} + \underbrace{\text{R}^1\text{—}\overset{\displaystyle \text{O}}{\overset{\|}{\text{C}}}\text{—O}^-\text{M}^+ + \text{R}^2\text{—}\overset{\displaystyle \text{O}}{\overset{\|}{\text{C}}}\text{—O}^-\text{M}^+ + \text{R}^3\text{—}\overset{\displaystyle \text{O}}{\overset{\|}{\text{C}}}\text{—O}^-\text{M}^+}$$

Fat or oil (triglyceride) — Glycerol — Soap (Mixture of carboxylic acid salts)

where $M^+ = Na^+$ or K^+

Figure 15.3
Saponification is the base-catalyzed hydrolysis of a glycerol triester.

When soap is dissolved in water, the carboxylate end actually dissolves. The hydrocarbon part is repelled by the water molecules so that a thin film of soap is formed on the surface of the water with the hydrocarbon chains protruding outward. When soap solution comes in contact with oil or grease, the hydrocarbon part dissolves in the oil or grease, but the polar carboxylate group remains dissolved in water. When particles of oil or grease are surrounded by soap molecules, the resulting "units" formed are called *micelles*. A simplified view of this phenomenon is shown in Figure 15.4.

A more detailed diagram of a micelle is found in Figure 23.1.

Micelles repel one another because they are surrounded on the surface by the negatively charged carboxylate ions. Mechanical action (for example, scrubbing or tumbling in a washing machine) causes oil or grease to be surrounded by soap molecules and broken into small droplets so that relatively small micelles are formed. These small micelles are then washed away. Careful examination of this solution shows that it is an *emulsion* containing suspended micelles.

An emulsion is a suspension of very fine droplets of one liquid in another. In this case it is oil in water.

Condensation Polymers

As we saw in Chapter 12, *polymers* are macromolecules, very large molecules. They result from the combination of many smaller molecules, usually in a repeating pattern, to give molecules whose molecular weight may be 10,000 g/mol or greater. The small molecules that make up the polymer are called *monomers*.

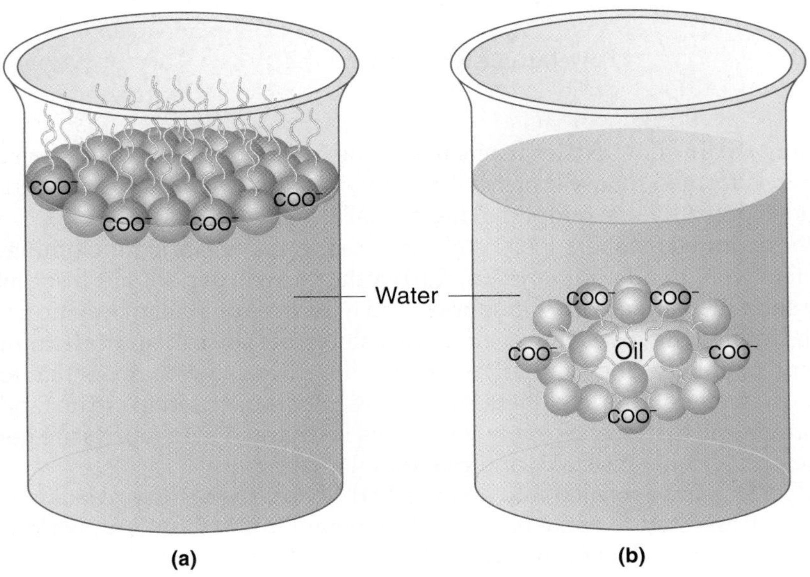

Figure 15.4
Simplified view of the action of a soap. The wiggly lines represent the long, continuous carbon chains of each soap molecule. (a) The thin film of soap molecules that forms at the water surface reduces surface tension. (b) Particles of oil and grease are surrounded by soap molecules to form a micelle.

A polymer may be made from a single type of monomer. Such a polymer would have the following general structure:

chain continues~A-A-A-A-A-A-A-A-A-A-A-A-A-A-A-A-A-A-A-A-A~chain continues

The addition polymers of alkenes that we studied in Chapter 12 are examples of this type of polymer. Alternatively, two different monomers may be copolymerized, producing a polymer with the following structure:

chain continues~A-B-A-B-A-B-A-B-A-B-A-B-A-B-A-B~chain continues

Polyesters are an example of this latter type of polymer. They are also known as condensation polymers. **Condensation polymers** are formed by the polymerization of monomers in a reaction that forms a small molecule such as water or an alcohol. Polyesters are synthesized by reacting a dicarboxylic acid and a dialcohol (diol). Notice that each of the combining molecules has two reactive functional groups.

n HOOC—C_6H_4—COOH + n $HOCH_2CH_2OH$

Terephthalic acid 1,2-Ethanediol

$\xrightarrow{H^+}$

$HOCH_2CH_2O$—C(=O)—C_6H_4—COOH + H_2O

Another molecule of terephthalic acid can react here.

Another molecule of 1,2-ethanediol can react here.

Reaction continues

$\left[—OCH_2CH_2O—\overset{O}{\overset{\|}{C}}—C_6H_4—\overset{O}{\overset{\|}{C}}—OCH_2CH_2O—\overset{O}{\overset{\|}{C}}—C_6H_4—\overset{O}{\overset{\|}{C}}—O— \right]_n$

Polyethylene terephthalate
PETE

Each time a pair of molecules reacts using one functional group from each, a new molecule is formed that still has two reactive groups. The product formed in this reaction is polyethylene terephthalate, or PETE.

When formed as fibers, polyesters are used to make fabric for clothing. These polyesters were trendy in the 1970s, during the "disco" period, but lost their popularity soon thereafter. Polyester fabrics, and a number of other synthetic polymers used in clothing, have become even more fashionable since the introduction of microfiber technology. The synthetic polymers are extruded into fibers that are only half the diameter of fine silk fibers. When these fibers are used to create fabrics, the result is a fabric that drapes freely yet retains its shape. These fabrics are generally lightweight, wrinkle resistant, and remarkably strong.

Polyester can be formed into a film called Mylar. These films, coated with aluminum foil, are used to make balloons that remain inflated for long periods. They are also used as the base for recording tapes and photographic film.

PETE can be used to make shatterproof plastic bottles, such as those used for soft drinks. However, these bottles cannot be recycled and reused directly because they cannot withstand the high temperatures required to sterilize them. PETE can't be used for any foods, such as jellies, that must be packaged at high temperatures. For these uses, a new plastic, PEN, or polyethylene naphthalate, is used.

$$\left[-\overset{O}{\overset{\|}{C}}-C_{10}H_6-\overset{O}{\overset{\|}{C}}-O-CH_2CH_2- \right]_n$$

Napthalate group   Ethylene group

15.3 Acid Chlorides and Acid Anhydrides

Acid Chlorides

Learning Goal 

Learning Goal 

Acid chlorides are carboxylic acid derivatives having the general formula

$$R-\overset{O}{\overset{\|}{C}}-Cl$$

They are named by replacing the *-ic* acid ending of the common name with *-yl chloride* or the *-oic acid* ending of the I.U.P.A.C. name of the carboxylic acid with *-oyl chloride*. For example,

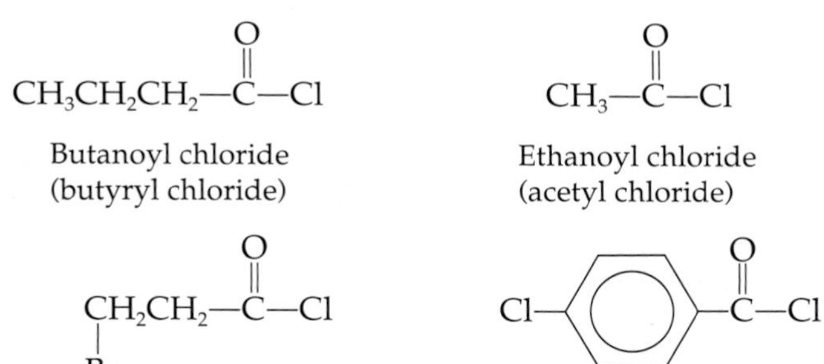

3-Bromopropanoyl chloride (β-bromopropionyl chloride)   4-Chlorobenzoyl chloride (*p*-chlorobenzoyl chloride)

Acid chlorides are noxious, irritating chemicals and must be handled with great care. They are slightly polar and boil at approximately the same temperature as the corresponding aldehyde or ketone of comparable molecular weight. They react violently with water and therefore cannot be dissolved in that solvent. Acid chlorides have little commercial value other than their utility in the synthesis of esters and amides, two of the other carboxylic acid derivatives.

Acid chlorides are prepared from the corresponding carboxylic acid by reaction with one of several inorganic acid chlorides, including PCl_3, PCl_5, or $SOCl_2$. The general reaction is summarized in the following equation:

$$\underset{\text{Carboxylic acid}}{R-\overset{O}{\overset{\|}{C}}-OH} \xrightarrow{\text{inorganic acid chloride}} \underset{\text{Acid chloride}}{R-\overset{O}{\overset{\|}{C}}-Cl} + \text{inorganic products}$$

The following equations show the synthesis of ethanoyl chloride and benzoyl chloride:

$$\underset{\text{Ethanoic acid (acetic acid)}}{CH_3{-}\overset{\overset{\displaystyle O}{\|}}{C}{-}OH} \xrightarrow[\text{(Phosphorus trichloride)}]{PCl_3} \underset{\text{Ethanoyl chloride (acetyl chloride)}}{CH_3{-}\overset{\overset{\displaystyle O}{\|}}{C}{-}Cl} + \text{inorganic products}$$

$$\underset{\text{Benzoic acid}}{C_6H_5{-}\overset{\overset{\displaystyle O}{\|}}{C}{-}OH} \xrightarrow[\text{(Thionyl chloride)}]{SOCl_2} \underset{\text{Benzoyl chloride}}{C_6H_5{-}\overset{\overset{\displaystyle O}{\|}}{C}{-}Cl} + \text{inorganic products}$$

EXAMPLE 15.10 *Writing Equations Representing the Synthesis of Acid Chlorides*

Write equations representing the following reactions:
(a) Butanoic acid with phosphorus trichloride
(b) 2-Bromobenzoic acid with thionyl chloride

Solution

(a) Begin by drawing the structure of the reactant, butanoic acid. The inorganic acid chloride (phosphorus trichloride) can be represented over the reaction arrow. The product is drawn by replacing the —OH of the carboxyl group with —Cl. This gives us the following equation:

$$CH_3CH_2CH_2{-}\overset{\overset{\displaystyle O}{\|}}{C}{-}OH \xrightarrow{PCl_3} CH_3CH_2CH_2{-}\overset{\overset{\displaystyle O}{\|}}{C}{-}Cl + \text{inorganic products}$$

(b) Begin by drawing the structure of the reactant, 2-bromobenzoic acid. The inorganic acid chloride (thionyl chloride) can be represented over the reaction arrow. The product is drawn by replacing the —OH of the carboxyl group with —Cl. This gives us the following equation:

$$2\text{-}Br{-}C_6H_4{-}\overset{\overset{\displaystyle O}{\|}}{C}{-}OH \xrightarrow{SOCl_2} 2\text{-}Br{-}C_6H_4{-}\overset{\overset{\displaystyle O}{\|}}{C}{-}Cl + \text{inorganic products}$$

EXAMPLE 15.11 *Naming Acid Chlorides*

Name the products of the reactions shown in Example 15.10.

Solution

$$\underset{\text{Butanoic acid (butyric acid)}}{CH_3CH_2CH_2{-}\overset{\overset{\displaystyle O}{\|}}{C}{-}OH} \xrightarrow{PCl_3} \underset{\text{Butanoyl chloride (butyryl chloride)}}{CH_3CH_2CH_2{-}\overset{\overset{\displaystyle O}{\|}}{C}{-}Cl} + \text{inorganic products}$$

Continued—

EXAMPLE 15.11 —*Continued*

The carboxylic acid that is the reactant in this reaction is butanoic acid (butyric acid). By dropping the *-oic acid* ending of the I.U.P.A.C. name (or the *-ic acid* ending of the common name) and replacing it with *-oyl chloride* (or *-yl chloride*), we can write the I.U.P.A.C. and common names of the product of this reaction. They are butanoyl chloride and butyryl chloride, respectively.

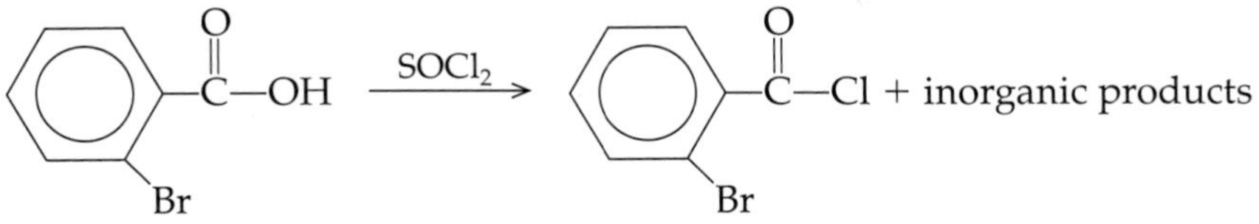

$$\text{BrC}_6\text{H}_4\text{—}\overset{\overset{\text{O}}{\|}}{\text{C}}\text{—OH} \xrightarrow{\text{SOCl}_2} \text{BrC}_6\text{H}_4\text{—}\overset{\overset{\text{O}}{\|}}{\text{C}}\text{—Cl} + \text{inorganic products}$$

2-Bromobenzoic acid (*o*-bromobenzoic acid) → 2-Bromobenzoyl chloride (*o*-bromobenzoyl chloride)

The carboxylic acid in this reaction is 2-bromobenzoic acid. By dropping the *-oic acid* ending of the I.U.P.A.C. name and replacing it with *-oyl chloride,* we can name the product of this reaction: 2-bromobenzoyl chloride. It is equally correct to name this compound *o*-bromobenzoyl chloride.

The reaction that occurs when acid chlorides react violently with water is hydrolysis. The products are the carboxylic acid and hydrochloric acid.

$$CH_3\text{—}\overset{\overset{O}{\|}}{C}\text{—}Cl + H_2O \longrightarrow CH_3\text{—}\overset{\overset{O}{\|}}{C}\text{—}OH + HCl$$

Ethanoyl chloride (acetyl chloride) → Ethanoic acid (acetic acid)

Question 15.11

Write an equation showing the synthesis of each of the following acid chlorides. Provide the I.U.P.A.C. names of the carboxylic acid reactants and the acid chloride products.

a. $CH_3\underset{\underset{CH_3}{|}}{CH}\text{—}\overset{\overset{O}{\|}}{C}\text{—}Cl$

b. $CH_3\text{—}\overset{\overset{O}{\|}}{C}\text{—}Cl$

c. $CH_3CH_2CH_2CH_2CH_2\text{—}\overset{\overset{O}{\|}}{C}\text{—}Cl$

d. $CH_3CH_2\underset{\underset{Br}{|}}{C}HCH_2CH_2\text{—}\overset{\overset{O}{\|}}{C}\text{—}Cl$

Question 15.12

Write an equation showing the synthesis of each of the following acid chlorides. Provide the common names of the carboxylic acid reactants and the acid chloride products.

a. $H\text{—}\overset{\overset{O}{\|}}{C}\text{—}Cl$

b. $CH_3CH_2\underset{\underset{CH_3}{|}}{CH}\text{—}\overset{\overset{O}{\|}}{C}\text{—}Cl$

c. $CH_3CH_2\text{—}\overset{\overset{O}{\|}}{C}\text{—}Cl$

d. $CH_3CH_2CH_2CH_2CH_2CH_2\text{—}\overset{\overset{O}{\|}}{C}\text{—}Cl$

Acid Anhydrides

Learning Goal 13

Learning Goal 14

Acid anhydrides are molecules with the following general formula:

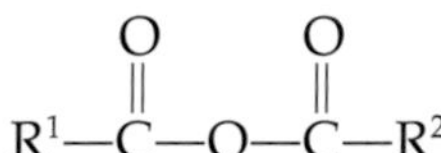

$$R^1{-}\overset{\overset{\large O}{\|}}{C}{-}O{-}\overset{\overset{\large O}{\|}}{C}{-}R^2$$

The name of the family is really quite fitting. The structure above reveals that acid anhydrides are actually two carboxylic acid molecules with a water molecule removed. The word *anhydride* means "without water."

$$R^1{-}\overset{\overset{\large O}{\|}}{C}{-}O{-}H + HO{-}\overset{\overset{\large O}{\|}}{C}{-}R^2$$

$$\downarrow$$

$$H{-}OH + R^1{-}\overset{\overset{\large O}{\|}}{C}{-}O{-}\overset{\overset{\large O}{\|}}{C}{-}R^2$$

Acid anhydrides are classified as *symmetrical* if both acyl groups are the same. Symmetrical acid anhydrides are named by replacing the *acid* ending of the carboxylic acid with the word *anhydride*. For example,

Learning Goal

$$CH_3{-}\overset{\overset{\large O}{\|}}{C}{-}O{-}\overset{\overset{\large O}{\|}}{C}{-}CH_3$$

Ethanoic anhydride
(acetic anhydride)

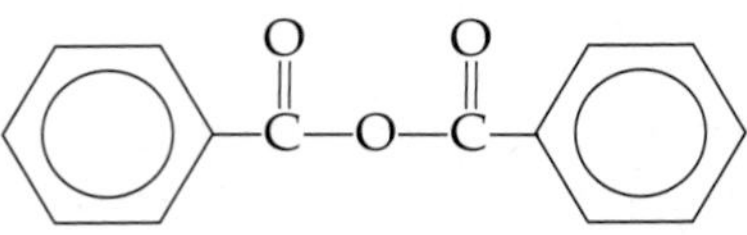

Benzoic anhydride

Unsymmetrical anhydrides are those having two different acyl groups. They are named by arranging the names of the two parent carboxylic acids and following them with the word *anhydride*. The names of the carboxylic acids may be arranged by size or alphabetically. For example:

$$CH_3{-}\overset{\overset{\large O}{\|}}{C}{-}O{-}\overset{\overset{\large O}{\|}}{C}{-}CH_2CH_3$$

Ethanoic propanoic anhydride
(acetic propionic anhydride)

$$CH_3{-}\overset{\overset{\large O}{\|}}{C}{-}O{-}\overset{\overset{\large O}{\|}}{C}{-}CH_2CH_2CH_2CH_3$$

Ethanoic pentanoic anhydride
(acetic valeric anhydride)

Most acid anhydrides cannot be formed in a reaction between the parent carboxylic acids. One typical pathway for the synthesis of an acid anhydride is the reaction between an acid chloride and a carboxylate anion. This general reaction is seen in the equation below:

$$\underset{\text{Acid chloride}}{R^1{-}\overset{\overset{\large O}{\|}}{C}{-}Cl} \xrightarrow[\text{carboxylate ion}]{R^2{-}\overset{\overset{O}{\|}}{C}{-}O^-} \underset{\text{Acid anhydride}}{R^1{-}\overset{\overset{\large O}{\|}}{C}{-}O{-}\overset{\overset{\large O}{\|}}{C}{-}R^2} + \underset{\text{Chloride ion}}{Cl^-}$$

The synthesis of ethanoic anhydride from ethanoic acid is

$$\underset{\substack{\text{Ethanoic acid}\\\text{(acetic acid)}}}{CH_3{-}\overset{\overset{\large O}{\|}}{C}{-}OH} \xrightarrow{SOCl_2} \underset{\substack{\text{Ethanoyl chloride}\\\text{(acetyl chloride)}}}{CH_3{-}\overset{\overset{\large O}{\|}}{C}{-}Cl} \xrightarrow{CH_3{-}\overset{\overset{O}{\|}}{C}{-}O^-} \underset{\substack{\text{Ethanoic anhydride}\\\text{(acetic anhydride)}}}{CH_3{-}\overset{\overset{\large O}{\|}}{C}{-}O{-}\overset{\overset{\large O}{\|}}{C}{-}CH_3}$$

Acid anhydrides readily undergo hydrolysis. The rate of the hydrolysis reaction may be increased by the addition of a trace of acid or hydroxide base to the solution.

$$CH_3CH_2-\overset{\overset{O}{\|}}{C}-O-\overset{\overset{O}{\|}}{C}-CH_2CH_3 + H_2O \xrightarrow{\text{Heat}} 2CH_3CH_2-\overset{\overset{O}{\|}}{C}-OH$$

Propanoic anhydride (propionic anhydride) → Propanoic acid (propionic acid)

EXAMPLE 15.12

Writing Equations Representing the Synthesis of Acid Anhydrides

Learning Goal 14

Write an equation representing the synthesis of propanoic anhydride.

Solution

Propanoic anhydride can be synthesized in a reaction between propanoyl chloride and the propanoate anion. This gives us the following equation:

$$CH_3CH_2-\overset{\overset{O}{\|}}{C}-Cl \xrightarrow[\text{Propanoate ion}]{CH_3CH_2-\overset{\overset{O}{\|}}{C}-O^-} CH_3CH_2-\overset{\overset{O}{\|}}{C}-O-\overset{\overset{O}{\|}}{C}-CH_2CH_3 + Cl^-$$

Propanoyl chloride → Propanoic anhydride + Chloride ion

EXAMPLE 15.13

Naming Acid Anhydrides

Learning Goal 13

Write the I.U.P.A.C. and common names for each of the following acid anhydrides.

$$CH_3CH_2CH_2-\overset{\overset{O}{\|}}{C}-O-\overset{\overset{O}{\|}}{C}-CH_2CH_2CH_3$$

Solution

This is a symmetrical acid anhydride. The I.U.P.A.C. name of the four-carbon parent carboxylic acid is butanoic acid (common name butyric acid). To name the anhydride, simply replace the word *acid* with the word *anhydride.* The I.U.P.A.C. name of this compound is butanoic anhydride (common name butyric anhydride).

$$CH_3-\overset{\overset{O}{\|}}{C}-O-\overset{\overset{O}{\|}}{C}-CH_2CH_2CH_2CH_2CH_3$$

Solution

This is an unsymmetrical anhydride. The I.U.P.A.C. names of the two parent carboxylic acids are ethanoic acid (two-carbon) and hexanoic acid (six-carbon). To name an unsymmetrical anhydride, the term *anhydride* is preceded by the names of the two parent acids. The I.U.P.A.C. name of this compound is ethanoic hexanoic anhydride. The common names of the two parent carboxylic acids are acetic acid and caproic acid. Thus, the common name of this compound is acetic caproic anhydride.

Question 15.13

Write an equation showing the synthesis of each of the following acid anhydrides. Provide the I.U.P.A.C. names of the carboxylic acid reactants and the acid anhydride products.

a. $CH_3CH(CH_3)CH_2-C(=O)-O-C(=O)-CH_2CH(CH_3)CH_3$

b. $H-C(=O)-O-C(=O)-CH_3$

Question 15.14

Write an equation showing the synthesis of each of the following acid anhydrides. Provide the common names of the carboxylic acid reactants and the acid anhydride products.

a. $CH_3CH(CH_2CH_3)CH_2-C(=O)-O-C(=O)-CH_2CH(CH_2CH_3)CH_3$

b. $CH_3-C(=O)-O-C(=O)-CH_2CH_2CH_3$

Acid anhydrides can also react with an alcohol. This reaction produces an ester and a carboxylic acid. This is an example of an acyl group transfer reaction, as shown in the following general reaction:

$$\underset{\text{Alcohol}}{R-OH} + \underset{\text{Acid anhydride}}{R-C(=O)-O-C(=O)-R} \longrightarrow \underset{\text{Ester}}{R-C(=O)-OR} + \underset{\text{Carboxylic acid}}{R-C(=O)-OH}$$

The acyl group of the acid anhydride is transferred to the oxygen of the alcohol in this reaction. The alcohol and anhydride reactants and ester product are described below. The carboxylic acid product is omitted.

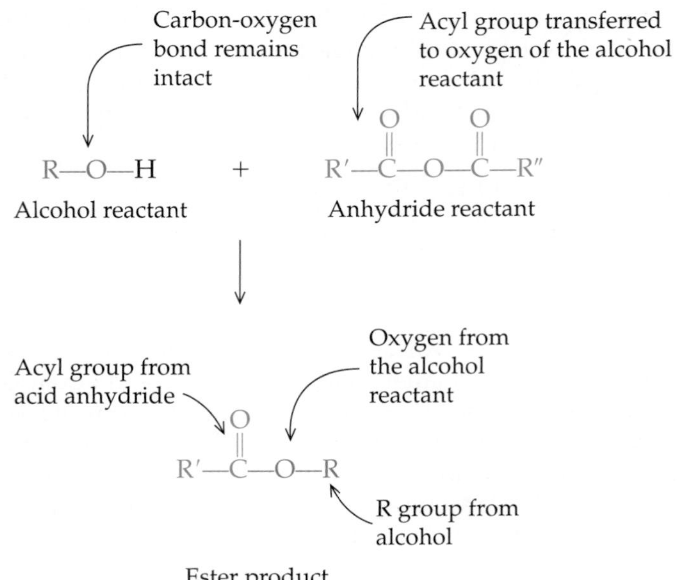

Other acyl group donors include thioesters and esters. As we will see in the final section of this chapter, acyl transfer reactions are very important in nature, particularly in the pathways responsible for breakdown of food molecules and harvesting cellular energy.

15.4 Nature's High-Energy Compounds: Phosphoesters and Thioesters

Learning Goal

An alcohol can react with phosphoric acid to produce a phosphate ester, or **phosphoester,** as in

$$\text{ROH} + \text{HO—}\overset{\overset{\text{O}}{\|}}{\underset{\underset{\text{OH}}{|}}{\text{P}}}\text{—OH} \longrightarrow \text{R—O—}\overset{\overset{\text{O}}{\|}}{\underset{\underset{\text{OH}}{|}}{\text{P}}}\text{—OH} + H_2O$$

Alcohol | Phosphoric acid | Phosphate ester | Water

Phosphate esters of simple sugars or monosaccharides are very important in the energy-harvesting biochemical pathways that provide energy for all life functions. One such pathway is *glycolysis.* This pathway is the first stage in the breakdown of sugars. The first reaction in this pathway is the formation of a phosphate ester of the six-carbon sugar, glucose. The phosphorylation of glucose to produce glucose-6-phosphate is represented in the following equation:

The many phosphorylated intermediates in the metabolism of sugars will be discussed in Chapter 21.

In fact the word *glycolysis* comes from two Greek words that mean "splitting sugars" (*glykos,* "sweet," and *lysis,* "to split"). In this pathway, the six-carbon sugar glucose is split, and then oxidized, to produce two three-carbon molecules, called *pyruvate.*

$$\beta\text{-D-Glucose} + \text{ATP} \xrightarrow{\text{Hexokinase}} \beta\text{-D-Glucose-6-phosphate} + \text{ADP}$$

In this reaction the source of the phosphoryl group is **adenosine triphosphate (ATP),** which is the universal energy currency for all living organisms. As such, ATP is used to store energy released in cellular metabolic reactions and provides the energy required for most of the reactions that occur in the cell. The transfer of a phosphoryl group from ATP to glucose "energizes" the glucose molecule in preparation for other reactions of the pathway.

ATP consists of a nitrogenous base (adenine) and a phosphate ester of the five-carbon sugar ribose (Figure 15.5). The triphosphate group attached to ribose is made up of three phosphate groups bonded to one another by phosphoric anhydride bonds. When two phosphate groups react with one another, a water molecule is lost. Because water is lost, the resulting bond is called a **phosphoric anhydride,** or *phosphoanhydride,* bond.

Phosphoryl **is the term used to describe the functional group derived from phosphoric acid that is part of another molecule.**

$$\text{RO—}\overset{\overset{\text{O}}{\|}}{\underset{\underset{\text{OH}}{|}}{\text{P}}}\text{—OH} + \text{HO—}\overset{\overset{\text{O}}{\|}}{\underset{\underset{\text{OH}}{|}}{\text{P}}}\text{—OH} \longrightarrow \text{RO—}\overset{\overset{\text{O}}{\|}}{\underset{\underset{\text{OH}}{|}}{\text{P}}}\text{—O—}\overset{\overset{\text{O}}{\|}}{\underset{\underset{\text{OH}}{|}}{\text{P}}}\text{—OH} + H_2O$$

Phosphate ester | Phosphate group | Phosphoric anhydride bond

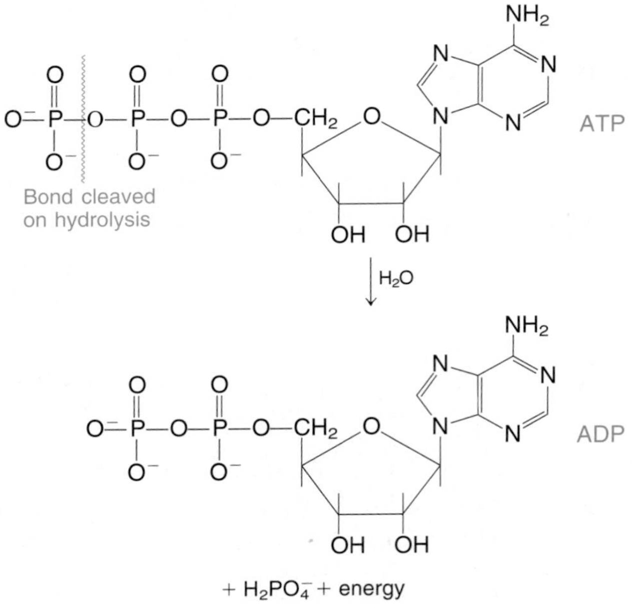

$$\mathrm{O^-{-}\overset{O}{\overset{\|}{\underset{O^-}{\underset{|}{P}}}}{-}O{-}\overset{O}{\overset{\|}{\underset{O^-}{\underset{|}{P}}}}{-}O{-}\overset{O}{\overset{\|}{\underset{O^-}{\underset{|}{P}}}}{-}O{-}CH_2{-}(ribose\text{-}adenine)}$$

Bond cleaved on hydrolysis

ATP

$\downarrow H_2O$

ADP

$+ H_2PO_4^- +$ energy

Figure 15.5
The hydrolysis of the phosphoric anhydride bond of ATP is accompanied by the release of energy that is used for biochemical reactions in the cell.

The functions and properties of ATP in energy metabolism are discussed in Section 21.1.

The energy of ATP is made available through hydrolysis of either of the two phosphoric anhydride bonds, as shown in Figure 15.5. This is an exothermic process; that is, energy is given off. When the phosphoryl group is transferred to another molecule—for instance, glucose—some of that energy resides in the phosphorylated sugar, thereby "energizing" it. The importance of ATP as an energy source becomes apparent when we realize that we synthesize and break down an amount of ATP equivalent to our body weight each day.

Thiols are described in Section 13.9.

Cellular enzymes can carry out a reaction between a thiol and a carboxylic acid to produce a **thioester:**

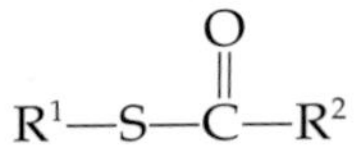

$$R^1{-}S{-}\overset{\overset{\displaystyle O}{\|}}{C}{-}R^2$$

Thioester

The reactions that produce thioesters are essential in energy-harvesting pathways as a means of "activating" acyl groups for subsequent breakdown reactions. The complex thiol coenzyme A is the most important acyl group activator in the cell. The detailed structure of coenzyme A appears in Section 13.9, but it is generally abbreviated CoA—SH to emphasize the importance of the sulfhydryl group. The most common thioester is the acetyl ester, called **acetyl coenzyme A** (acetyl CoA).

β-Oxidation is the pathway for the breakdown of fatty acids. Like glycolysis, it is an energy-harvesting pathway.

Acetyl CoA carries the acetyl group from glycolysis or β-oxidation of a fatty acid to an intermediate of the citric acid cycle. This reaction is an example of an acyl group transfer reaction. In this case, the acyl group donor is a thioester—acetyl CoA. The acyl group being transferred is the acetyl group, which is transferred to the carbonyl carbon of oxaloacetate. This reaction is shown here:

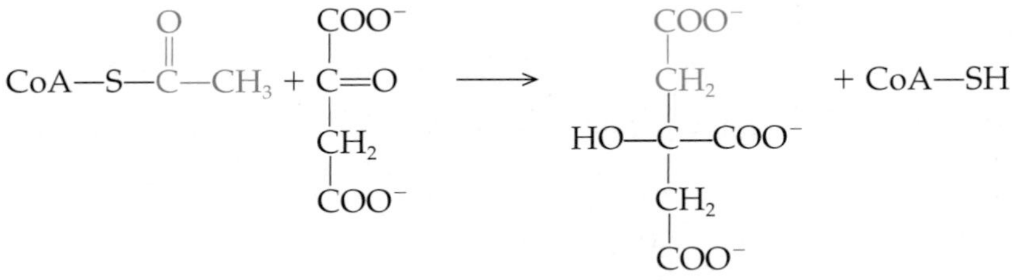

The acyl group of a carboxylic acid is named by replacing the *-oic acid* or *-ic* suffix with *-yl*. For instance, the acyl group of acetic acid is the acetyl group:

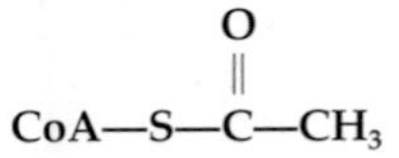

$$\mathrm{CoA{-}S{-}\overset{\overset{O}{\|}}{C}{-}CH_3}$$

Acetyl coenzyme A (acetyl CoA)

$$\mathrm{CoA{-}S{-}\overset{\overset{O}{\|}}{C}{-}CH_3 + C(=O)(COO^-)CH_2COO^- \longrightarrow {}^-OOC{-}CH_2{-}C(OH)(COO^-){-}CH_2{-}COO^- + CoA{-}SH}$$

Acetyl CoA Oxaloacetate Citrate Coenzyme A

A HUMAN PERSPECTIVE

Carboxylic Acid Derivatives of Special Interest

Analgesics (pain killers) and antipyretics (fever reducers)

A*spirin* (*acetylsalicylic acid*) is the most widely used drug in the world. Hundreds of millions of dollars are spent annually on this compound. It is used primarily as a pain reliever (analgesic) and in the reduction of fever (antipyretic). Aspirin is among the drugs often referred to as NSAIDS, or nonsteroidal anti-inflammatory drugs. These drugs inhibit the inflammatory response by inhibiting an enzyme called cyclooxygenase, which is the first enzyme in the pathway for the synthesis of prostaglandins. Prostaglandins are responsible, in part, for pain and fever. Thus, aspirin and other NSAIDS reduce pain and fever by decreasing prostaglandin synthesis. Aspirin's side effects are a problem for some individuals. Because aspirin inhibits clotting, its use is not recommended during pregnancy, nor should it be used by individuals with ulcers. In those instances, *acetaminophen,* found in the over-the-counter pain-reliever, Tylenol, is often prescribed.

The search for NSAIDS that are more effective and yet gentler on the stomach has provided two new analgesics for the over-the-counter market. These are ibuprofen (sold as Motrin, Advil, Nuprin) and naproxen (sold as Naprosyn, Naprelan, Anaprox, and Aleve).

$CH_3CH(CH_3)CH_2{-}C_6H_4{-}CH(CH_3){-}C(=O){-}OH$

Ibuprofen

$CH_3O{-}C_{10}H_6{-}CH(CH_3){-}C(=O){-}OH$

Naproxen

$C_6H_4(C(=O){-}OH)(O{-}C(=O){-}CH_3)$

Acetylsalicylic acid
(aspirin)

$HO{-}C_6H_4{-}NH{-}C(=O){-}CH_3$

Acetaminophen
(Tylenol)

Some common analgesics.

Pheromones

Pheromones, chemicals secreted by animals, influence the behavior of other members of the same species. They often represent the major means of communication among simpler animals. The term *pheromone* literally means "to carry" and "to excite" (Greek, *pherein,* to carry; Greek, *horman,* to excite). They are chemicals carried or shed by one member of the species and used to alert other members of the species.

Pheromones may be involved in sexual attraction, trail marking, aggregation or recruitment, territorial marking, or signaling alarm. Others may be involved in defense or in species socialization—for example, designating various classes within the species as a whole. Among all of the pheromones, insect pheromones have been the most intensely studied. Many of the insect pheromones are carboxylic acids or acid derivatives. Examples of members of this class of chemicals are provided in the accompanying figure along with the principal function of each compound.

$CH_3CH_2CH{=}CH(CH_2)_9CH_2OC(=O)CH_3$

Tetracecenyl acetate
(European corn borer sex pheromone)

$CH_3C(=O)CH_2CH_2CH_2CH_2CH_2{-}CH{=}CH{-}CO_2H$ (trans)

9-Keto-*trans*-2-decenoic acid
(queen bee socializing/royalty pheromone)

$CH_3(CH_2)_3{-}CH{=}CH{-}(CH_2)_5CH_2OC(=O)CH_3$ (cis)

cis-7-Dodecenyl acetate
(cabbage looper sex pheromone)

Glycolysis, β-oxidation, and the citric acid cycle are cellular energy-harvesting pathways that we will study in Chapters 21, 22, and 23.

As we will see in Chapter 22, the citric acid cycle is an energy-harvesting pathway that completely oxidizes the acetyl group to two CO_2 molecules. The electrons that are harvested in the process are used to produce large amounts of ATP. Coenzyme A also serves to activate the acyl group of fatty acids during β-oxidation, the pathway by which fatty acids are oxidized to produce ATP.

Summary of Reactions

Preparation of Carboxylic Acids

$$\text{R—CH}_2\text{OH} \xrightarrow{[\text{O}]} \text{R—}\overset{\text{O}}{\overset{\|}{\text{C}}}\text{—H} \xrightarrow{[\text{O}]} \text{R—}\overset{\text{O}}{\overset{\|}{\text{C}}}\text{—OH}$$

Primary alcohol — Aldehyde — Carboxylic acid

Dissociation of Carboxylic Acids

$$\text{R—}\overset{\text{O}}{\overset{\|}{\text{C}}}\text{—OH} \rightleftharpoons \text{R—}\overset{\text{O}}{\overset{\|}{\text{C}}}\text{—O}^- + \text{H}^+$$

Carboxylic acid — Carboxylate anion — Hydrogen ion

Neutralization of Carboxylic Acids

$$\text{R—}\overset{\text{O}}{\overset{\|}{\text{C}}}\text{—OH} + \text{NaOH} \longrightarrow \text{R—}\overset{\text{O}}{\overset{\|}{\text{C}}}\text{—O}^-\text{Na}^+ + \text{H}_2\text{O}$$

Carboxylic acid — Strong base — Carboxylic acid salt — Water

Esterification

$$\text{R}^1\text{—}\overset{\text{O}}{\overset{\|}{\text{C}}}\text{—OH} + \text{R}^2\text{OH} \underset{}{\overset{\text{H}^+,\ \text{heat}}{\rightleftharpoons}} \text{R}^1\text{—}\overset{\text{O}}{\overset{\|}{\text{C}}}\text{—OR}^2 + \text{H}_2\text{O}$$

Carboxylic acid — Alcohol — Ester — Water

Acid Hydrolysis of Esters

$$\text{R}^1\text{—}\overset{\text{O}}{\overset{\|}{\text{C}}}\text{—OR}^2 + \text{H}_2\text{O} \overset{\text{H}^+,\ \text{heat}}{\rightleftharpoons} \text{R}^1\text{—}\overset{\text{O}}{\overset{\|}{\text{C}}}\text{—OH} + \text{R}^2\text{OH}$$

Ester — Water — Carboxylic acid — Alcohol

Saponification

$$\text{R}^1\text{—}\overset{\text{O}}{\overset{\|}{\text{C}}}\text{—OR}^2 + \text{H}_2\text{O} \xrightarrow{\text{NaOH, heat}} \text{R}^1\text{—}\overset{\text{O}}{\overset{\|}{\text{C}}}\text{—O}^-\text{Na}^+ + \text{R}^2\text{OH}$$

Ester — Water — Carboxylic acid salt — Alcohol

Synthesis of Acid Chlorides

$$\text{R—}\overset{\text{O}}{\overset{\|}{\text{C}}}\text{—OH} \xrightarrow{\text{inorganic acid chloride}} \text{R—}\overset{\text{O}}{\overset{\|}{\text{C}}}\text{—Cl} + \text{inorganic products}$$

Carboxylic acid — Acid chloride

Synthesis of Acid Anhydrides

$$\text{R}^1\text{—}\overset{\text{O}}{\overset{\|}{\text{C}}}\text{—Cl} \xrightarrow[\text{carboxylate ion}]{\text{R}^2\text{—}\overset{\text{O}}{\overset{\|}{\text{C}}}\text{—O}^-} \text{R}^1\text{—}\overset{\text{O}}{\overset{\|}{\text{C}}}\text{—O—}\overset{\text{O}}{\overset{\|}{\text{C}}}\text{—R}^2 + \text{Cl}^-$$

Acid chloride — Acid anhydride — Chloride ion

Formation of a Phosphoester

$$\text{ROH} + \text{HO—}\underset{\text{OH}}{\underset{|}{\overset{\text{O}}{\overset{\|}{\text{P}}}}}\text{—OH} \longrightarrow \text{R—O—}\underset{\text{OH}}{\underset{|}{\overset{\text{O}}{\overset{\|}{\text{P}}}}}\text{—OH} + \text{H}_2\text{O}$$

Alcohol — Phosphoric acid — Phosphate ester — Water

Summary

15.1 Carboxylic Acids

The functional group of the *carboxylic acids* is the *carboxyl group* (—COOH). Because the carboxyl group is extremely polar and carboxylic acids can form intermolecular hydrogen bonds, they have higher boiling points and melting points than alcohols. The lower-molecular-weight carboxylic acids are water soluble and tend to taste sour and have unpleasant aromas. The longer-chain carboxylic acids are called *fatty acids*. Carboxylic acids are named (I.U.P.A.C.) by replacing the *-e* ending of the parent compound with *-oic acid*. Common names are often derived from the source of the carboxylic acid. They are synthesized by the oxidation of primary alcohols or aldehydes. Carboxylic acids are weak acids. They are neutralized by strong bases to form salts. *Soaps* are salts of long-chain carboxylic acids (fatty acids).

15.2 Esters

Esters are mildly polar and have pleasant aromas. The boiling points and melting points of esters are comparable to those of aldehydes and ketones. Esters are formed from the reaction between a carboxylic acid and an alcohol. They can undergo hydrolysis back to the parent carboxylic acid and alcohol. The base-catalyzed hydrolysis of an ester is called *saponification*.

15.3 Acid Chlorides and Acid Anhydrides

Acid chlorides are noxious chemicals formed in the reaction of a carboxylic acid and reagents such as PCl_3 or $SOCl_2$. *Acid anhydrides* are formed by the combination of an acid chloride and a carboxylate anion. Acid anhydrides can react with an alcohol to produce an ester and a carboxylic acid. This is an example of an acyl group transfer reaction.

15.4 Nature's High-Energy Compounds: Phosphoesters and Thioesters

An alcohol can react with phosphoric acid to produce a phosphate ester (*phosphoester*). When two phosphate groups are joined, the resulting bond is a *phosphoric anhydride* bond. These two functional groups are important to the structure and function of *adenosine triphosphate (ATP)*, the universal energy currency of all cells. Cellular enzymes can carry out a reaction between a thiol and a carboxylic acid to produce a *thioester*. This reaction is essential for the activation of acyl groups in carbohydrate and fatty acid metabolism. Coenzyme A is the most important thiol involved in these pathways.

Key Terms

acetyl coenzyme A (15.4)
acid anhydride (15.3)
acid chloride (15.3)
acyl group (Intro)
adenosine triphosphate (ATP) (15.4)
carboxyl group (15.1)
carboxylic acid (15.1)
carboxylic acid derivative (15.2)
condensation polymer (15.2)
ester (15.2)
fatty acid (15.1)
hydrolysis (15.2)
oxidation (15.1)
phosphoester (15.4)
phosphoric anhydride (15.4)
saponification (15.2)
soap (15.2)
thioester (15.4)

Questions and Problems

Carboxylic acids: Structure and Nomenclature

15.15 Write the complete structural formulas for each of the following carboxylic acids:

a. 2-Bromopentanoic acid
b. 2-Bromo-3-methylbutanoic acid
c. 2-Bromocyclohexanecarboxylic acid

15.16 Write the complete structural formulas for each of the following carboxylic acids:

a. 2,6-Dichlorocyclohexanecarboxylic acid
b. 2,4,6-Trimethylstearic acid
c. Propenoic acid

15.17 Name each of the following carboxylic acids, using both the common and the I.U.P.A.C. Nomenclature System:

a. $H-C(=O)-OH$
b. $CH_3CH(CH_3)CH_2-C(=O)-OH$
c. cyclopentyl$-C(=O)-OH$

15.18 Name each of the following carboxylic acids, using both the common and I.U.P.A.C. Nomenclature Systems:

a. $CH_3CH_2CH(Br)CH(CH_3)CH_2-C(=O)-OH$
b. $CH_3CH_2CH(CH_2CH_3)CH_2CH_2-C(=O)-OH$
c. 3-methylcyclohexyl$-C(=O)-OH$ (ring bearing CH_3)

15.19 Write a complete structural formula and determine the I.U.P.A.C. name for each of the carboxylic acids of molecular formula $C_4H_8O_2$.

15.20 Write the general structure of an aldehyde, a ketone, a carboxylic acid, and an ester. What similarities exist among these structures?

15.21 Write the condensed structure of each of the following carboxylic acids:
a. 4,4-Dimethylhexanoic acid
b. 3-Bromo-4-methylpentanoic acid
c. 2,3-Dinitrobenzoic acid
d. 3-Methylcyclohexanecarboxylic acid

15.22 Use I.U.P.A.C. nomenclature to write the names for each of the following carboxylic acids:

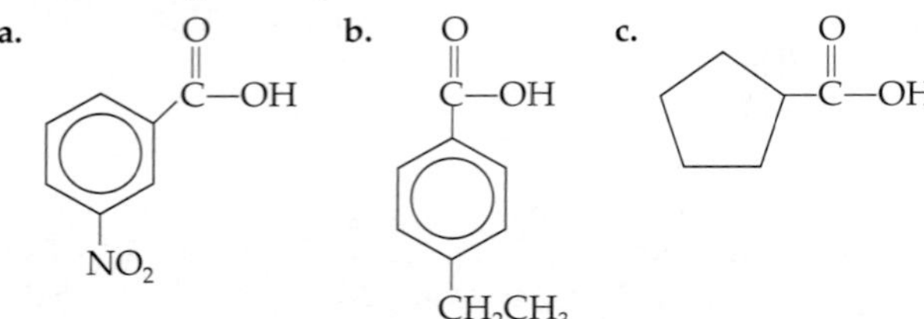

15.23 Provide the common and I.U.P.A.C. names for each of the following compounds:

a. $CH_3CH(OH)-C(=O)-OH$

b. $CH_3CH(OH)CH_2-C(=O)-OH$

c. $CH_3C(CH_3)_2CH_2CH_2-C(=O)-OH$

d. $CH_3CH_2CCl_2CH_2-C(=O)-OH$

15.24 Draw the structure of each of the following carboxylic acids:
a. β-Chlorobutyric acid
b. α, β-Dibromovaleric acid
c. β, γ-Dihydroxybutyric acid
d. δ-Bromo-γ-chloro-β-methylcaproic acid

Carboxylic Acids: Structure and Properties

15.25 Which member in each of the following pairs has the higher boiling point?
a. Heptanoic acid or 1-heptanol
b. Propanal or 1-propanol
c. Methyl pentanoate or pentanoic acid
d. 1-Butanol or butanoic acid

15.26 Which member in each of the following pairs is more soluble in water?

a. $CH_3CH_2CH_2CH_2CH_2-C(=O)-OH$ or $CH_3CH_2CH_2CH_2CH_2-C(=O)-O^-Na^+$

b. $CH_3CH_2CH_2CH_2CH_2CH_2CH_2CH_2CH_3$ or $CH_3CH_2CH_2CH_2CH_2CH_2CH_2CH_2OH$

c. $CH_3CH_2-O-CH_2CH_3$ or $CH_3CH_2-C(=O)-OCH_3$

d. $CH_3CH_2-O-CH_2CH_3$ or $CH_3CH_2CH_2CH_2CH_2CH_3$

e. Decanoic acid or ethanoic acid

15.27 Describe the properties of low-molecular-weight carboxylic acids.

15.28 What are some of the biological functions of the long chain carboxylic acids called fatty acids?

15.29 Why are citric acid and adipic acid added to some food products?

15.30 What is the function of lactic acid in food products? Of what significance is lactic acid in muscle metabolism?

Carboxylic Acids: Reactions

15.31 How are carboxylic acids produced commercially?

15.32 Carboxylic acids are described as weak acids. What is meant by that description?

15.33 How is a soap prepared?

15.34 How do soaps assist in the removal of oil and grease from clothing?

15.35 Complete each of the following reactions by supplying the missing portion indicated by a question mark:

a. $CH_3-C(=O)-H \xrightarrow{H_2CrO_4} ?$

b. $CH_3CH_2CH_2-C(=O)-OH + CH_3OH \xrightarrow{H^+,\ heat} ?$

c. $C_5H_9-C(=O)-OH + ? \longrightarrow C_5H_9-C(=O)-OCH_3$ (cyclopentyl)

15.36 Complete each of the following reactions by supplying the missing part(s) indicated by the question mark(s):

a. $CH_3CH_2CH_2OH \xrightarrow{?(1)} CH_3CH_2-C(=O)-OH \underset{?(3)}{\overset{NaOH}{\rightleftharpoons}} ?(4)$

$CH_3CH_2-C(=O)-OH \xrightarrow{?(2)} CH_3CH_2-C(=O)-OCH(CH_3)CH_3$

b. $CH_3COOH + NaOH \longrightarrow ?$

c. $CH_3CH_2CH_2CH_2CH_2COOH + NaOH \longrightarrow ?$

d. $? + CH_3CH_2CH(CH_3)OH \xrightarrow{H^+} CH_3-C(=O)-OCH(CH_3)CH_2CH_3$

15.37 How might $CH_3CH_2CH_2CH_2CH_2OH$ be converted to each of the following products?
a. $CH_3CH_2CH_2CH_2CHO$
b. $CH_3CH_2CH_2CH_2COOH$

15.38 Which of the following alcohols can be oxidized to a carboxylic acid? Name the carboxylic acid produced. For those alcohols that cannot be oxidized to a carboxylic acid, name the final product.
a. Ethanol
b. 2-Propanol
c. 1-Propanol
d. 3-Pentanol

Esters: Structure and Nomenclature

15.39 Write each of the following, using condensed formulas:
a. Methyl benzoate
b. Butyl decanoate
c. Methyl propionate
d. Ethyl propionate

15.40 Write each of the following using condensed formulas:
a. Ethyl *m*-nitrobenzoate
b. Isopropyl acetate
c. Methyl butyrate

15.41 Use the I.U.P.A.C. Nomenclature System to name each of the following esters:

a. $CH_3-\overset{\overset{O}{\|}}{C}-OCH_2CH_3$

b. $CH_3CH_2-\overset{\overset{O}{\|}}{C}-OCH_3$

c. $CH_3\underset{}{\overset{CH_3}{\overset{|}{C}}}HCH_2-\overset{\overset{O}{\|}}{C}-OCH_3$

d. $C_6H_5-\overset{\overset{O}{\|}}{C}-O-C_5H_9$ (benzene ring; cyclopentyl)

15.42 Use the I.U.P.A.C. Nomenclature System to name each of the following:

a. $C_6H_{11}-\overset{\overset{O}{\|}}{C}-OCH_2CH_2CH_3$ (cyclohexane ring)

b. $C_6H_5-\overset{\overset{O}{\|}}{C}-OCH_3$ (benzene ring)

c. $\underset{Br}{\underset{|}{C}}H_2\underset{Br}{\underset{|}{C}}HCH_2CH_2-\overset{\overset{O}{\|}}{C}-OCH_2CH_3$

Esters: Reactions

15.43 Complete each of the following reactions by supplying the missing portion indicated with a question mark:

a. $CH_3CH_2CH_2-\overset{\overset{O}{\|}}{C}-OH + CH_3CH_2OH \xrightarrow{H^+,\ heat} ?$

b. $CH_3CH_2-\overset{\overset{O}{\|}}{C}-OCH_2CH_3 + H_2O \xrightarrow{H^+,\ heat} ?$

c. $CH_3\overset{CH_3}{\overset{|}{C}}HCH_2CH_2-\overset{\overset{O}{\|}}{C}-OH + ? \xrightarrow{H^+,\ heat} CH_3\overset{CH_3}{\overset{|}{C}}HCH_2CH_2-\overset{\overset{O}{\|}}{C}-OCH_2CH_2CH_3$

d. $CH_3CH_2\overset{Br}{\overset{|}{C}}HCH_2-\overset{\overset{O}{\|}}{C}-OCH_2CH_3 + H_2O \xrightarrow{OH^-,\ heat} ?$

15.44 Complete each of the following reactions by supplying the missing portion indicated with a question mark:

a. $? + CH_3-\underset{CH_3}{\underset{|}{\overset{CH_3}{\overset{|}{C}}}}-OH \xrightarrow{?} CH_3CH_2-\overset{\overset{O}{\|}}{C}-O-\underset{CH_3}{\underset{|}{\overset{CH_3}{\overset{|}{C}}}}-CH_3$

b. $CH_3CH_2CH_2CH_2COOH + CH_3CH_2CH_2CH_2OH \xrightarrow{H^+,\ heat} ?$

c. $CH_3-\underset{CH_3}{\underset{|}{\overset{CH_3}{\overset{|}{C}}}}-CH_2-\overset{\overset{O}{\|}}{C}-OCH_2CH_2-\underset{CH_3}{\underset{|}{\overset{CH_3}{\overset{|}{C}}}}-CH_3 + H_2O \xrightarrow{H^+,\ heat} ?$

d. $CH_3CH_2-\overset{\overset{O}{\|}}{C}-OCH_3 + H_2O \xrightarrow{OH^-,\ heat} ?$

15.45 What is saponification? Give an example using specific molecules.

15.46 When the methyl ester of hexanoic acid is hydrolyzed in aqueous sodium hydroxide in the presence of heat, a homogeneous solution results. When the solution is acidified with dilute aqueous hydrochloric acid, a new product forms. What is the new product? Draw its structure.

15.47 The structure of salicylic acid is shown. If this acid reacts with methanol, the product is an ester, methyl salicylate. Methyl salicylate is known as oil of wintergreen and is often used as a flavoring agent. Draw the structure of the product of this reaction.

$o\text{-}HO-C_6H_4-\overset{\overset{O}{\|}}{C}-OH + CH_3OH \xrightarrow{H^+} ?$

15.48 When salicylic acid reacts with acetic anhydride, one of the products is an ester, acetylsalicylic acid. Acetylsalicylic acid is the active ingredient in aspirin. Complete the equation below by drawing the structure of acetylsalicylic acid. (*Hint:* Acid anhydrides are hydrolyzed by water.)

$o\text{-}HO-C_6H_4-\overset{\overset{O}{\|}}{C}-OH + CH_3\overset{\overset{O}{\|}}{C}-O-\overset{\overset{O}{\|}}{C}CH_3 \longrightarrow ?$

15.49 Compound A ($C_6H_{12}O_2$) reacts with water, acid, and heat to yield compound B ($C_5H_{10}O_2$) and compound C (CH_4O). Compound B is acidic. Deduce possible structures of compounds A, B, and C.

15.50 What products are formed when methyl *o*-bromobenzoate reacts with each of the following?
a. Aqueous acid and heat
b. Aqueous base and heat

Acid Chlorides and Acid Anhydrides

15.51 Supply the missing reagents (indicated by the question marks) necessary to complete each of the following transformations:

a. $CH_3-\overset{\overset{O}{\|}}{C}-OH \xrightarrow{?} CH_3-\overset{\overset{O}{\|}}{C}-Cl$

b. $C_5H_9-\overset{\overset{O}{\|}}{C}-Cl \xrightarrow{?} C_5H_9-\overset{\overset{O}{\|}}{C}-O-\overset{\overset{O}{\|}}{C}-CH_3$ (cyclopentane ring)

c. $C_6H_5-\overset{\overset{O}{\|}}{C}-Cl \xrightarrow{?} C_6H_{11}-\overset{\overset{O}{\|}}{C}-O-\overset{\overset{O}{\|}}{C}-C_6H_5$ (cyclohexane ring; benzene ring)

15.52 Supply the missing reagents (indicated by the question marks) necessary to complete each of the following transformations. All of the reactions may require more than one step to complete.

a. $CH_3CH_2CH_2CH_2{-}OH \xrightarrow{?} CH_3CH_2CH_2{-}\overset{O}{\overset{\|}{C}}{-}Cl$

b. $CH_3CH_2{-}OH \xrightarrow{?} CH_3{-}\overset{O}{\overset{\|}{C}}{-}O{-}\overset{O}{\overset{\|}{C}}{-}CH_2CH_3$

c. Ethanol $\xrightarrow{?}$ ethanoic anhydride

15.53 Complete each of the following reactions by supplying the missing product:

a. 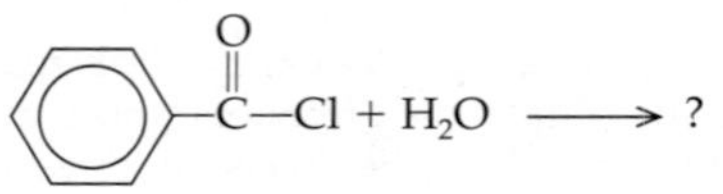 $C_6H_5{-}\overset{O}{\overset{\|}{C}}{-}Cl + H_2O \longrightarrow ?$

b. $CH_3{-}\overset{O}{\overset{\|}{C}}{-}O{-}\overset{O}{\overset{\|}{C}}{-}CH_3 + H_2O \xrightarrow{\text{Heat}} ?$

15.54 Use the I.U.P.A.C. Nomenclature System to name the products and reactants in Problem 15.53.

15.55 Write the condensed formula for each of the following compounds:
a. Decanoic anhydride
b. Acetic anhydride
c. Valeric anhydride
d. Benzoyl chloride

15.56 Write a condensed formula for each of the following compounds:
a. Propanoyl chloride
b. Heptanoyl chloride
c. Pentanoyl chloride

15.57 Describe the physical properties of acid chlorides.

15.58 Describe the physical properties of acid anhydrides.

15.59 Write an equation for the reaction of each of the following acid anhydrides with ethanol.
a. Propanoic anhydride
b. Ethanoic anhydride
c. Methanoic anhydride

15.60 Write an equation for the reaction of each of the following acid anhydrides with propanol. Name each of the products using the I.U.P.A.C. Nomenclature System.
a. Butanoic anhydride
b. Pentanoic anhydride
c. Methanoic anhydride

Phosphoesters and Thioesters

15.61 By reacting phosphoric acid with an excess of ethanol, it is possible to obtain the mono-, di-, and triesters of phosphoric acid. Draw all three of these products.

15.62 What is meant by a phosphoric anhydride bond?

15.63 We have described the molecule ATP as the body's energy storehouse. What do we mean by this designation? How does ATP actually store energy and provide it to the body as needed?

15.64 Write an equation for each of the following reactions:
a. Ribose + phosphoric acid
b. Methanol + phosphoric acid
c. Adenosine diphosphate + phosphoric acid

15.65 Draw the thioester bond between the acetyl group and coenzyme A.

15.66 Explain the significance of thioester formation in the metabolic pathways involved in fatty acid and carbohydrate breakdown.

15.67 It is also possible to form esters of other inorganic acids such as sulfuric acid and nitric acid. One particularly noteworthy product is nitroglycerine, which is both highly unstable (explosive) and widely used in the treatment of the heart condition known as angina, a constricting pain in the chest usually resulting from coronary heart disease. In the latter case its function is to alleviate the pain associated with angina. Nitroglycerine may be administered as a tablet (usually placed just beneath the tongue when needed) or as a salve or paste that can be applied to and absorbed through the skin. Nitroglycerine is the trinitroester of glycerol. Draw the structure of nitroglycerine, using the structure of glycerol.

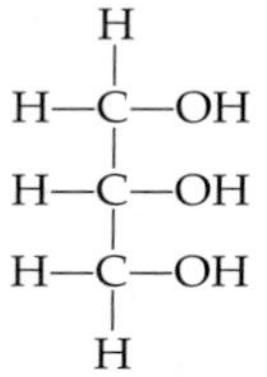

$$\begin{array}{c} H \\ | \\ H{-}C{-}OH \\ | \\ H{-}C{-}OH \\ | \\ H{-}C{-}OH \\ | \\ H \end{array}$$

Glycerol

15.68 Show the structure of the thioester that would be formed between coenzyme A and stearic acid.

Critical Thinking Problems

1. Radioactive isotopes of an element behave chemically in exactly the same manner as the nonradioactive isotopes. As a result, they can be used as tracers to investigate the details of chemical reactions. A scientist is curious about the origin of the bridging oxygen atom in an ester molecule. She has chosen to use the radioactive isotope oxygen-18 to study the following reaction:

$$CH_3CH_2OH + CH_3\overset{O}{\overset{\|}{C}}{-}OH \xrightarrow{H^+,\ \text{heat}} CH_3\overset{O}{\overset{\|}{C}}{-}O{-}CH_2CH_3 + H_2O$$

Design experiments using oxygen-18 that will demonstrate whether the oxygen in the water molecule came from the —OH of the alcohol or the —OH of the carboxylic acid.

2. Triglycerides are the major lipid storage form in the human body. They are formed in an esterification reaction between glycerol (1,2,3-propanetriol) and three fatty acids (long chain carboxylic acids). Write a balanced equation for the formation of a triglyceride formed in a reaction between glycerol and three molecules of decanoic acid.

3. Chloramphenicol is a very potent, broad-spectrum antibiotic. It is reserved for life-threatening bacterial infections because it is quite toxic. It is also a very bitter tasting chemical. As a result, children had great difficulty taking the antibiotic. A clever chemist found that the taste could be improved considerably by producing the palmitate ester. Intestinal enzymes hydrolyze the ester, producing chloramphenicol, which can then be absorbed. The following structure is the palmitate ester of chloramphenicol. Draw the structure of chloramphenicol.

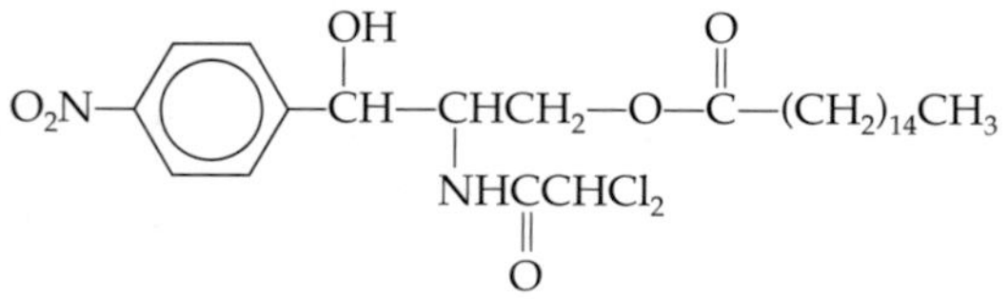

$$O_2N{-}C_6H_4{-}\overset{OH}{\overset{|}{C}H}{-}\underset{NH\overset{O}{\overset{\|}{C}}CHCl_2}{\underset{|}{C}H}CH_2{-}O{-}\overset{O}{\overset{\|}{C}}{-}(CH_2)_{14}CH_3$$

Chloramphenicol palmitate

4. Acetyl coenzyme A (acetyl CoA) can serve as a donor of acetate groups in biochemical reactions. One such reaction is the formation of acetylcholine, an important neurotransmitter involved in nerve signal transmission at neuromuscular junctions. The structure of choline is shown below. Draw the structure of acetylcholine.

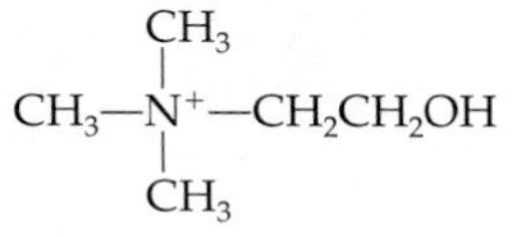

$$CH_3-N^+(CH_3)_2-CH_2CH_2OH$$

Choline

5. Hormones are chemical messengers that are produced in a specialized tissue of the body and travel through the bloodstream to reach receptors on cells of their target tissues. This specific binding to target tissues often stimulates a cascade of enzymatic reactions in the target cells. The work of Earl Sutherland and others led to the realization that there is a *second messenger* within the target cells. Binding of the hormone to the hormone receptor in the cell membrane triggers the enzyme adenyl cyclase to produce adenosine-3′,5′-monophosphate, which is also called *cyclic AMP,* from ATP. The reaction is summarized as follows:

$$\text{ATP} \xrightarrow{Mg^{+2},\ \text{adenyl cyclase}} \text{cyclic AMP} + PP_i + H^+$$

PP_i is the abbreviation for a pyrophosphate group, shown here:

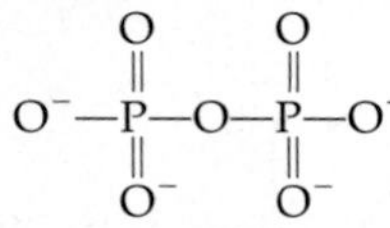

$$O^- -P(=O)(O^-)-O-P(=O)(O^-)-O^-$$

The structure of ATP is shown here with the carbon atoms of the sugar ribose numbered according to the convention used for nucleotides:

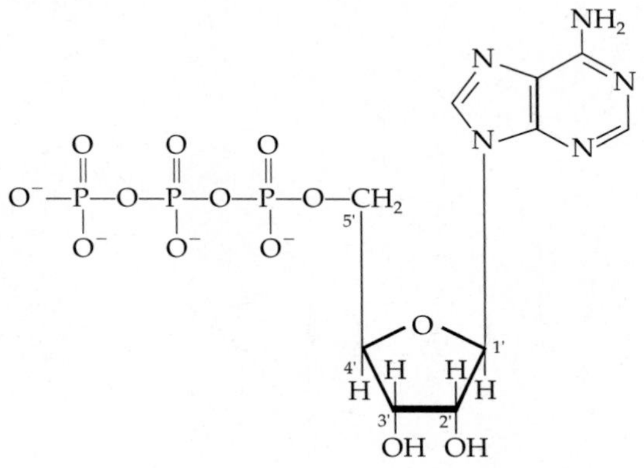

Adenosine-5'-triphosphate

Draw the structure of adenosine-3′,5′-monophosphate.

Ethnobotanists continue to search for medically active compounds from the rain forest.

16

Amines and Amides

Outline

ORGANIC CHEMISTRY

Learning Goals

1. Classify amines as primary, secondary, or tertiary.
2. Describe the physical properties of amines.
3. Draw and name simple amines using the Chemical Abstracts, common, and I.U.P.A.C. nomenclature system.
4. Write equations representing the synthesis of amines.
5. Write equations showing the basicity and neutralization of amines.
6. Describe the structure of quaternary ammonium salts and discuss their use as antiseptics and disinfectants.
7. Discuss the biological significance of heterocyclic amines.
8. Describe the physical properties of amides.
9. Draw the structure and write the common and I.U.P.A.C. names of amides.
10. Write equations representing the preparation of amides.
11. Write equations showing the hydrolysis of amides.
12. Draw the general structure of an amino acid.
13. Draw and discuss the structure of a peptide bond.
14. Describe the function of neurotransmitters.

CHEMISTRY CONNECTION

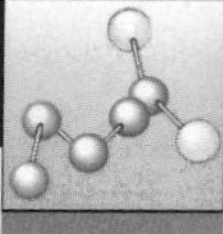

The Nicotine Patch

Smoking cigarettes is one of the most difficult habits to break—so difficult, in fact, that physicians now suspect that smoking is more than a habit: It's an addiction. The addictive drug in tobacco is *nicotine*.

Nicotine, one of the *heterocyclic amines* that we will study in this chapter (see Figure 16.3), is a highly toxic compound. In fact, it has been used as an insecticide! Small doses from cigarette smoking initially stimulate the autonomic (involuntary) nervous system. However, repeated small doses of nicotine obtained by smokers eventually depress the involuntary nervous system. As a result, the smoker *needs* another cigarette.

Some people have been able to quit smoking through behavioral modification programs, hypnosis, or sheer willpower. Others quit only after smoking has contributed to life-threatening illness, such as emphysema or a heart attack. Yet there are people who cannot quit even after a diagnosis of lung cancer.

One promising advance to help people quit smoking is the nicotine patch. The patch is applied to the smoker's skin, and nicotine from the patch slowly diffuses through the skin and into the bloodstream. Because the body receives a constant small dose of nicotine, the smoker no longer craves a cigarette. Of course, the long-range goal is to completely cure the addiction to nicotine. This is done by decreasing doses of nicotine in the patches as the treatment period continues. Eventually, after a period of about three months, the former smoker no longer needs the patches.

There are those who criticize this therapy because nicotine is a toxic chemical. However, the benefits seem to outweigh any negative aspects. A smoker inhales not only nicotine, but also dozens of other substances that have been shown to cause cancer. When someone successfully quits smoking, the body no longer suffers the risks associated with the substances in cigarette smoke.

In this chapter we study the structure and properties of amines and their derivatives, the amides. We will see that several are important pain killers, decongestants, and antibiotics, and others are addictive drugs and carcinogens.

Introduction

In this chapter we introduce an additional element into the structure of organic molecules. That element is nitrogen, the fourth most common atom in living systems. It is an important component of the structure of the nucleic acids, DNA and RNA, which are the molecules that carry the genetic information for living systems. It is also essential to the structure and function of proteins, molecules that carry out the majority of the work in biological systems. Some proteins serve as enzymes that catalyze the chemical reactions that allow life to exist. Other proteins, the antibodies, protect us against infection by a variety of infectious agents. Proteins are also structural components of the cell and of the body.

Excess nitrogen is removed from the body as urea, first synthesized by the father of organic chemistry, Friederich Wöhler (Chapter 11). Urea synthesis in the body is described in Chapter 22.

One class of organic molecules containing nitrogen is the amines. Amines are characterized by the presence of an amino group (—NH_2).

General structure of an amine

The nitrogen atom of the amino group may have one or more of its hydrogen atoms replaced by an organic group. General structures of these types of amines are shown below:

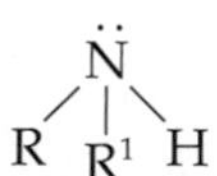

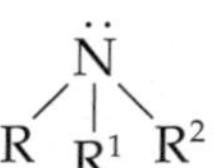

Amines are very common in biological systems and exhibit important physiological activity. Consider hist*amine*. Histamine contributes to the inflammatory response that causes the symptoms of colds and allergies, including swollen mucous membranes, congestion, and excessive nasal secretions. We take antihist*amines* to help relieve these symptoms. Ephedrine is an antihistamine that has been extracted from the leaves of the *ma-huang* plant in China for over two thousand years. Today it is the antihistamine found in many over-the-counter cold medications. This decongestant helps to shrink swollen mucous membranes and reduce nasal secretions. The structures of histamine and ephedrine are shown below.

$CH_2CH_2NH_2$ (on imidazole ring: HN, N^+H)

Histamine

OH / C_6H_5—CHCHNHCH$_3$ / CH_3

Ephedrine

The other group of nitrogen-containing organic compounds we will investigate in this chapter is the amides. Amides are the products of a reaction between an amine and a carboxylic acid derivative. They have the following general structure:

$$\mathrm{R{-}\overset{\overset{\large O}{\|}}{C}{-}NR_2}$$

From the carboxylic acid → R—C(=O); From the amine → NR_2; amide bond (C—N)

(R = H or an alkyl or aryl group)

General structure of an amide

The general structure of an amino acid is

$$\mathrm{H_2N{-}\underset{R}{\overset{H}{C}}{-}COOH}$$

The amino acids that are the subunits from which proteins are built are characterized by the presence of both an amino group and a carboxyl group. When amino acids are bonded to one another to produce a protein chain, the amino group of one amino acid reacts with the carboxyl group of another amino acid. The amide bond that results is called a *peptide bond.*

In this chapter we will explore the chemistry of the organic molecules that contain nitrogen. In upcoming chapters we will investigate the structure and properties of the nitrogen-containing biological molecules.

16.1 Amines

Structure and Physical Properties

Learning Goal

Amines are organic derivatives of ammonia and, like ammonia, they are basic. In fact, amines are the most important type of organic base found in nature. We can think of them as substituted ammonia molecules in which one, two, or three of the ammonia hydrogens have been replaced by an organic group:

The geometry of ammonia is described in Section 4.4.

$\ddot{N}$(H)(H)(H) —R substitutes for H→ $\ddot{N}$(H)(H)(R)

Ammonia — An amine

The structures drawn above and in Figure 16.1 reveal that like ammonia, amines are pyramidal. The nitrogen atom has three groups bonded to it and has a nonbonding pair of electrons.

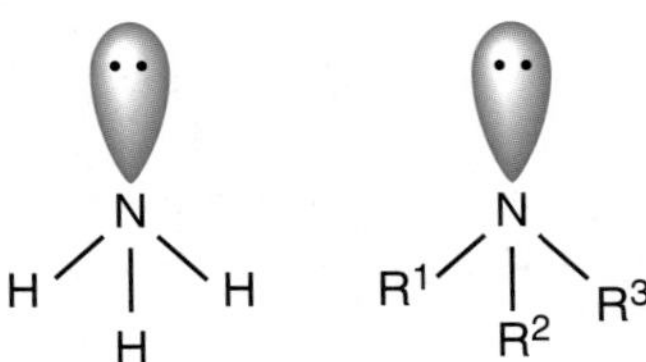

Figure 16.1
The pyramidal structure of amines. Note the similarities in structure between an amine and the ammonia molecule.

Amines are classified according to the number of alkyl or aryl groups attached to the nitrogen. In a **primary (1°) amine,** one of the hydrogens is replaced by an organic group. In a **secondary (2°) amine,** two hydrogens are replaced. In a **tertiary (3°) amine,** three organic groups replace the hydrogens:

$$\underset{\text{Ammonia}}{H{-}\overset{\overset{H}{|}}{N}{-}H} \qquad \underset{\text{1° amine (primary amine)}}{R{-}\overset{\overset{H}{|}}{N}{-}H} \qquad \underset{\text{2° amine (secondary amine)}}{R{-}\overset{\overset{H}{|}}{N}{-}R} \qquad \underset{\text{3° amine (tertiary amine)}}{R{-}\overset{\overset{R}{|}}{N}{-}R}$$

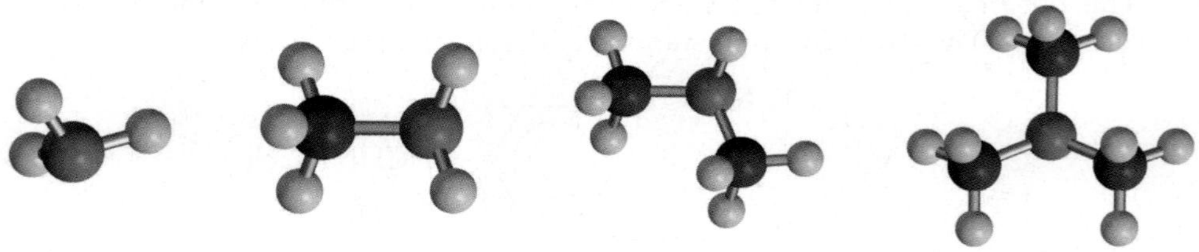

Ammonia

Methanamine (methylamine)

N-Methylmethanamine (dimethylamine)

N, N-Dimethylmethanamine (trimethylamine)

EXAMPLE 16.1 Classifying Amines as Primary, Secondary, or Tertiary

Classify each of the following compounds as a primary, secondary, or tertiary amine.

Solution

Compare the structure of the amine with that of ammonia.

$$H{-}\overset{\overset{CH_3}{|}}{N}{-}H \qquad H{-}\overset{\overset{H}{|}}{N}{-}H$$ 1° amine: one hydrogen replaced

$$CH_3{-}\overset{\overset{CH_3}{|}}{N}{-}H \qquad H{-}\overset{\overset{H}{|}}{N}{-}H$$ 2° amine: two hydrogens replaced

$$CH_3{-}\overset{\overset{CH_3}{|}}{N}{-}CH_3 \qquad H{-}\overset{\overset{H}{|}}{N}{-}H$$ 3° amine: three hydrogens replaced

Question 16.1

Determine whether each of the following amines is primary, secondary, or tertiary.

a. $CH_3CH_2{-}\overset{\overset{CH_2CH_3}{|}}{N}{-}CH_3$

b. $CH_3CH_2CH_2{-}NH_2$

c. $CH_3{-}\overset{\overset{H}{|}}{N}{-}CH_3$

Question 16.2

Classify each of the following amines as primary, secondary, or tertiary.

a. $CH_3{-}NH_2$

b. $CH_3CH_2CH_2CH_2{-}\overset{\overset{CH_3}{|}}{N}{-}CH_2CH_3$

c. $H{-}\overset{\overset{CH_2CH_2CH_3}{|}}{N}{-}CH_3$

Hydrogen bonding is described in Section 6.2.

The nitrogen atom is more electronegative than the hydrogen atoms in amines. As a result, the N—H bond is polar, and hydrogen bonding between amine molecules or between amine molecules and water can occur (Figure 16.2).

Question 16.3

Refer to Figure 16.2 and draw a similar figure showing the hydrogen bonding that occurs between water and a 2° amine.

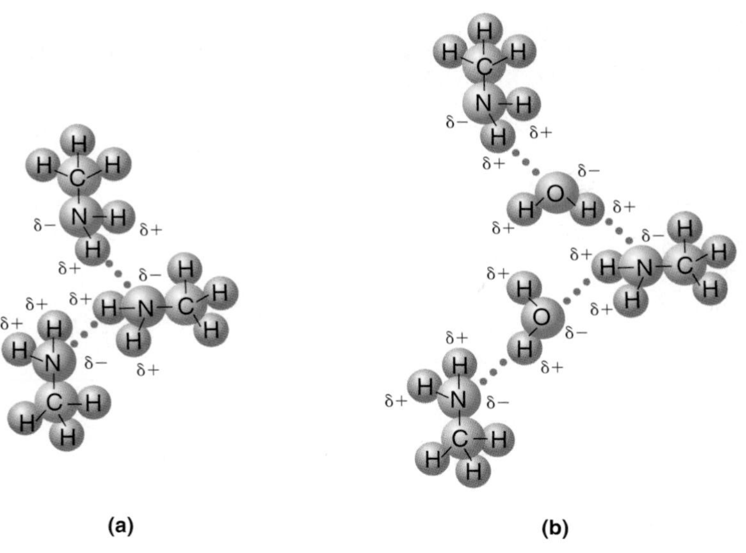

Figure 16.2
Hydrogen bonding (a) in methylamine and (b) between methylamine and water. Dotted lines represent hydrogen bonds.

Question 16.4

Refer to Figure 16.2 and draw hydrogen bonding between two primary amines.

Learning Goal 2

The ability of primary and secondary amines to form N—H···N hydrogen bonds is reflected in their boiling points (Table 16.1). Primary amines have boiling points well above those of alkanes of similar molecular weight but considerably lower than those of comparable alcohols. Consider the following examples:

$CH_3CH_2CH_3$	$CH_3CH_2NH_2$	CH_3CH_2OH
Propane	Ethanamine	Ethanol
M.W. = 44 g/mol	M.W. = 45 g/mol	M.W. = 46 g/mol
b.p. = −42.2° C	b.p. = 16.6° C	b.p. = 78.5° C

Tertiary amines do not have an N—H bond. As a result they cannot form intermolecular hydrogen bonds with other tertiary amines. Consequently, their boiling points are lower than those of primary or secondary amines of comparable molecular weight. This is seen in a comparison of the boiling points of propanamine (propylamine; M.W. = 59) and *N,N*-dimethylmethanamine (trimethylamine; M.W. = 59). Trimethylamine, the tertiary amine, has a boiling point of 2.9°C, whereas propylamine, the primary amine, has a boiling point of 48.7°C. Clearly the inability of trimethylamine molecules to form intermolecular hydrogen bonds results in a much lower boiling point.

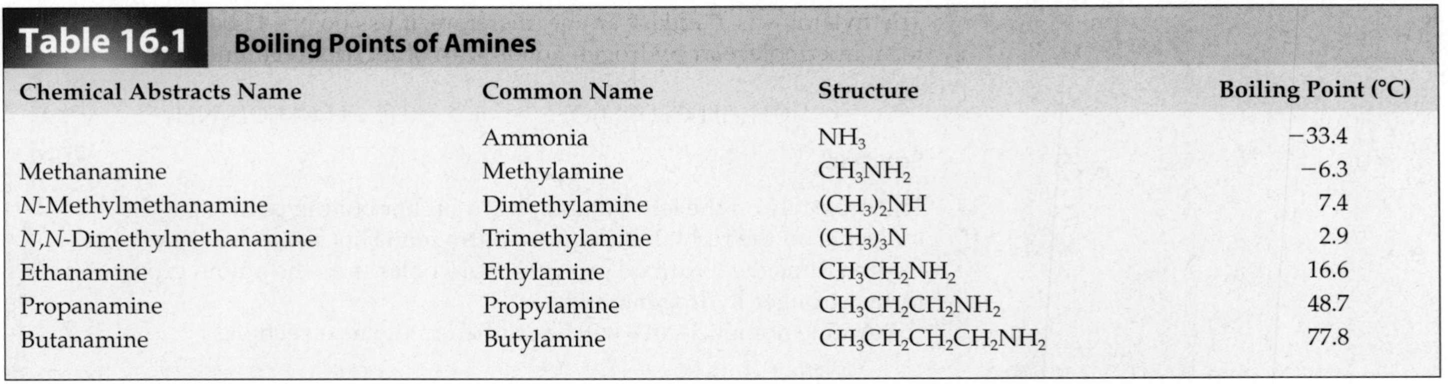

Table 16.1 Boiling Points of Amines

Chemical Abstracts Name	Common Name	Structure	Boiling Point (°C)
	Ammonia	NH_3	−33.4
Methanamine	Methylamine	CH_3NH_2	−6.3
N-Methylmethanamine	Dimethylamine	$(CH_3)_2NH$	7.4
N,N-Dimethylmethanamine	Trimethylamine	$(CH_3)_3N$	2.9
Ethanamine	Ethylamine	$CH_3CH_2NH_2$	16.6
Propanamine	Propylamine	$CH_3CH_2CH_2NH_2$	48.7
Butanamine	Butylamine	$CH_3CH_2CH_2CH_2NH_2$	77.8

Table 16.2 Comparison of the Boiling Points of Selected Alcohols and Amines

Name	Molecular Weight (g/mol)	Boiling Point (°C)
Methanol	32	64.5
Methanamine	31	−6.3
Ethanol	46	78.5
Ethanamine	45	16.6
Propanol	60	97.2
Propanamine	59	48.7

$CH_3CH_2CH_2—NH_2$
Propanamine
(propylamine)
M.W. = 59 g/mol
b.p. = 48.7° C

$CH_3CH_2—N(H)—CH_3$
N-Methylethanamine
(ethylmethylamine)
M.W. = 59 g/mol
b.p. = 36.7° C

$CH_3—N(CH_3)—CH_3$
N,N-Dimethylmethanamine
(trimethylamine)
M.W. = 59 g/mol
b.p. = 2.9° C

The intermolecular hydrogen bonds formed by primary and secondary amines are not as strong as the hydrogen bonds formed by alcohols because nitrogen is not as electronegative as oxygen. For this reason primary and secondary amines have lower boiling points than alcohols (Table 16.2).

All amines can form intermolecular hydrogen bonds with water (O—H···N). As a result, small amines (six or fewer carbons) are soluble in water. As we have noted for other families of organic molecules, water solubility decreases as the length of the hydrocarbon (hydrophobic) portion of the molecule increases.

EXAMPLE 16.2 *Predicting the Physical Properties of Amines*

Which member of each of the following pairs of molecules has the higher boiling point?

$CH_3CH_2N(CH_2CH_3)CH_2CH_3$ or $CH_3CH_2CH_2CH_2CH_2CH_2NH_2$

Solution

The molecule on the right, hexanamine, has a higher boiling point than the molecule on the left, *N,N*-dimethylmethanamine (triethylamine). Triethylamine is a tertiary amine; therefore, it has no N—H bond and cannot form intermolecular hydrogen bonds with other triethylamine molecules.

$CH_3CH_2CH_2CH_2OH$ or $CH_3CH_2CH_2CH_2NH_2$

Solution

The molecule on the left, butanol, has a higher boiling point than the molecule on the right, butanamine. Nitrogen is not as electronegative as oxygen, thus the hydroxyl group is more polar than the amino group and forms stronger hydrogen bonds.

Amine nomenclature will be studied in the next section.

Question 16.5

Which compound in each of the following pairs would you predict to have a higher boiling point? Explain your reasoning.

a. Methanol or methylamine
b. Dimethylamine or water
c. Methylamine or ethylamine
d. Propylamine or butane

Question 16.6

Compare the boiling points of methylamine, dimethylamine, and trimethylamine. Explain why they are different.

Nomenclature

Learning Goal

Several systems for naming amines have evolved, including, of course, the I.U.P.A.C. Nomenclature System. However, the I.U.P.A.C. names are not the most commonly used names for amines and will be mentioned only briefly in this text.

The nomenclature system that has gained great popularity is known as the *Chemical Abstracts* or CA system. This system has been approved for use by I.U.P.A.C. and will be presented here because it is logical and easy to use. In the CA system the final *-e* of the name of the parent compound is dropped, and the suffix *-amine* is added. For instance,

$CH_3—NH_2$ Methanamine

$CH_3CH_2CH_2—NH_2$ Propanamine

$CH_3CH_2CH_2CH(NH_2)CH_3$ 2-Pentanamine

For secondary or tertiary amines the prefix *N*-alkyl is added to the name of the parent compound. For example,

$CH_3—NH—CH_2CH_3$ *N*-Methylethanamine

$CH_3—N(CH_3)—CH_3$ *N,N*-Dimethylmethanamine

In the I.U.P.A.C. Nomenclature System, the parent chain is the longest continuous carbon chain to which the amino group is bonded. The name of the parent alkane is used as the suffix and is preceded by the word *amino* and a number that designates the position of the amino group on the chain. A few examples follow:

CH_3NH_2 Aminomethane

$CH_3CH_2CH_2NH_2$ 1-Aminopropane

$CH_3CH(NH_2)CH_3$ 2-Aminopropane

As with the *Chemical Abstracts* System, if a substituent is present on the nitrogen, it is designated by the prefix *N*.

$CH_3CH_2CH_2—NH—CH_3$ *N*-Methyl-1-aminopropane

$CH_3CH_2CH_2—N(CH_3)—CH_3$ *N,N*-Dimethyl-1-aminopropane

Several aromatic amines have special names that have also been approved for use by I.U.P.A.C. For example, the amine of benzene is given the name *aniline*. In the CA system, aniline is named *benzenamine*.

NH_2 on benzene ring — Aniline or benzenamine

NH_2 and CH_3 on benzene ring (meta) — *m*-Toluidine or *meta*-toluidine

NH_2 and CH_3 on benzene ring (ortho) — *o*-Toluidine or *ortho*-toluidine

NH_2 and CH_3 on benzene ring (para) — *p*-Toluidine or *para*-toluidine

If additional groups are attached to the nitrogen of an aromatic amine, they are indicated with the letter *N*- followed by the name of the group.

EXAMPLE 16.3 ***Using the Chemical Abstracts System to Name Amines***

Name the following amine using the CA system.

$$CH_3CH_2CH_2\text{—}NH\text{—}CH_3$$

Solution

Parent compound: propane (becomes propanamine)
Additional group on N: methyl (becomes *N*-methyl)
Name: *N*-Methylpropanamine

Common names are often used for the simple amines. The common names of the alkyl groups bonded to the amine nitrogen are followed by the ending *-amine*. Each group is listed alphabetically in one continuous word followed by the suffix *-amine:*

$CH_3\text{—}NH_2$ Methylamine

$CH_3\text{—}NH\text{—}CH_3$ Dimethylamine

$CH_3\text{—}N(CH_3)\text{—}CH_3$ Trimethylamine

$CH_3CH_2\text{—}NH_2$ Ethylamine

$CH_3CH_2\text{—}NH\text{—}CH_3$ Ethylmethylamine

Table 16.3 compares these systems of nomenclature for a number of simple amines.

Table 16.3 I.U.P.A.C., Common and Chemical Abstracts Names of Amines

Compound	CA Name	Common Name	I.U.P.A.C. Name
$R\text{—}NH_2$	Alkan*amine*	Alkyl*amine*	*Amino*alkane
$CH_3\text{—}NH_2$	Methanamine	Methylamine	Aminomethane
$CH_3CH_2\text{—}NH_2$	Ethanamine	Ethylamine	Aminoethane
$CH_3CH_2CH_2\text{—}NH_2$	Propanamine	Propylamine	Aminopropane
$CH_3\text{—}NH\text{—}CH_3$	*N*-Methylmethanamine	Dimethylamine	*N*-Methylaminomethane
$CH_3\text{—}NH\text{—}CH_2CH_3$	*N*-Methylethanamine	Ethylmethylamine	*N*-Methylaminoethane
$CH_3\text{—}N(CH_3)\text{—}CH_3$	*N,N*-Dimethylmethanamine	Trimethylamine	*N,N*-Dimethylaminomethane

A CLINICAL PERSPECTIVE

Medically Important Amines

Although amines play many different roles in our day-to-day lives, one important use is in medicine. A host of drugs derived from amines is responsible for improving the quality of life, whereas others, such as cocaine and heroin, are highly addictive.

Amphetamines, such as benzedrine and methedrine, stimulate the central nervous system. They elevate blood pressure and pulse rate and are often used to decrease fatigue. Medically, they have been used to treat depression and epilepsy. Amphetamines have also been prescribed as diet pills because they decrease the appetite. Their use is controlled by federal law because excess use of amphetamines can cause paranoia and mental illness.

$C_6H_5-CH_2-CH(CH_3)-NH_2$

2-Amino-1-phenylpropane (Amphetamine)

Benzedrine

$C_6H_5-CH_2-CH(CH_3)-NH-CH_3$

2-Methylamino-1-phenylpropane (Methamphetamine)

Methedrine

Many of the medicinal amines are *analgesics* (pain relievers) or *anesthetics* (pain blockers). Novocaine and related compounds, for example, are used as local anesthetics. Demerol is a very strong pain reliever.

$H_2N-C_6H_4-C(=O)-O-CH_2CH_2-\overset{\oplus}{N}H(CH_2CH_3)-CH_2CH_3\ Cl^{\ominus}$

Novocaine

Demerol: a piperidine ring bearing $N-CH_3$, with a phenyl group and $-C(=O)-O-CH_2CH_3$ on the same ring carbon

Demerol

Ephedrine and neosynephrine are used as *decongestants* in cough syrups and nasal sprays. They cause shrinking of the membranes that line the nasal passages. These compounds are related to two chemicals that are important to the functioning of the central nervous system, L-dopa and dopamine, which are described in Section 16.5.

$C_6H_5-CH(OH)-CH(CH_3)-NH-CH_3$

Ephedrine

$HO-C_6H_4-CH(OH)-CH_2-NH-CH_3$

Neosynephrine

The *sulfa drugs,* the first chemicals used to fight bacterial infections, are synthesized from amines.

$H_2N-C_6H_4-SO_2-NH_2$

Sulfanilamide—a sulfa drug

Question 16.7

Use the structure of aniline provided and draw the complete structural formula for each of the following amines.

a. *N*-Methylaniline
b. *N,N*-Dimethylaniline
c. *N*-Ethylaniline
d. *N*-Isopropylaniline

Question 16.8

Name each of the following amines using the CA and common nomenclature systems.

a. $CH_3CH(NH_2)CH_2CH_3$ (NH$_2$ on C-2)

b. $CH_3—C(NH_2)(CH_3)—CH_3$

c. $CH_3CH(NHCH_3)CH_2CH_3$

d. $CH_3CH(N(CH_3)CH_2CH_3)CH_3$

Question 16.9

Draw the complete structural formula for each of the following compounds:

a. 2-Propanamine
b. 3-Octanamine
c. *N*-Ethyl-2-heptanamine
d. 2-Methyl-2-pentanamine
e. 4-Chloro-5-iodo-1-nonanamine
f. *N,N*-Diethyl-1-pentanamine

Question 16.10

Draw the condensed formula for each of the following compounds:

a. Diethylmethylamine
b. 4-Methylpentylamine
c. *N*-Methylaniline
d. Triisopropylamine
e. Methyl-*t*-butylamine
f. Ethylhexylamine

Reactions Involving Amines

Preparation of Amines

Learning Goal 4

In the laboratory, amines are prepared by the reduction of amides and nitro compounds.

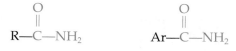

Ar—NO_2

Examples of amides

A nitro compound

As we will see in Section 16.3, amides are neutral nitrogen compounds that produce an amine and a carboxylic acid when hydrolyzed. Nitro compounds are prepared by the nitration of an aromatic compound.

Primary amines are readily produced by reduction of a nitro compound, as in the following reaction:

We are using the general symbol [H] to represent any reducing agent just as we used [O] to represent an oxidizing agent in previous chapters. Several different reducing agents may be used to effect the changes shown here; for example, metallic iron and acid may be used to reduce aromatic nitro compounds and $LiAlH_4$ in ether reduces amides.

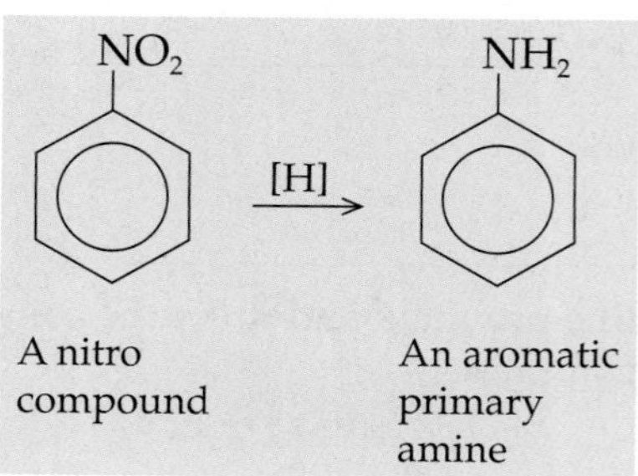

In this reaction the nitro compound is nitrobenzene and the product is aniline.

Amides may also be reduced to produce primary, secondary, or tertiary amines.

$$R^1{-}\overset{\overset{\large O}{\|}}{C}{-}N\begin{matrix}R^2\\ \\ R^3\end{matrix} \xrightarrow{[H]} R^1CH_2N\begin{matrix}R^2\\ \\ R^3\end{matrix}$$

Amide — Amine

(R^2 and R^3 may be a hydrogen atom or an organic group.)

If R^2 and R^3 are hydrogen atoms, the product will be a primary amine:

$$CH_3{-}\overset{\overset{\large O}{\|}}{C}{-}NH_2 \xrightarrow{[H]} CH_3{-}CH_2{-}NH_2$$

Ethanamide — Ethanamine (ethylamine)

If either R^2 or R^3 is an organic group, the product will be a secondary amine:

$$CH_3CH_2{-}\overset{\overset{\large O}{\|}}{C}{-}NH{-}CH_3 \xrightarrow{[H]} CH_3CH_2CH_2NH{-}CH_3$$

N-Methylpropanamide — *N*-Methylpropanamine (methylpropylamine)

If both R^2 and R^3 are organic groups, the product will be a tertiary amine:

$$CH_3{-}\overset{\overset{\large O}{\|}}{C}{-}\overset{\overset{\large CH_3}{|}}{N}{-}CH_3 \xrightarrow{[H]} CH_3CH_2{-}\overset{\overset{\large CH_3}{|}}{N}{-}CH_3$$

N,N-Dimethylethanamide — *N,N*-Dimethylethanamine

Designing the Synthesis of an Amine — EXAMPLE 16.4

Design a synthesis to produce 1-hexanamine.

Solution

How do we plan the synthesis of an amine? In the same way that we have approached all synthesis: by working in reverse. Consider the following amine:

$$CH_3CH_2CH_2CH_2CH_2CH_2{-}NH_2$$

1-Hexanamine

In planning the synthesis of this compound we try to determine the identity of the immediate precursor of the desired product. We know that two principal pathways are used in the synthesis of amines: the reduction of amides and the reduction of nitro compounds. We can plan accordingly. Because 1-hexanamine is an aliphatic amine, the immediate precursor must be hexanamide.

$$CH_3CH_2CH_2CH_2CH_2{-}\overset{\overset{\large O}{\|}}{C}{-}NH_2 \xrightarrow{[H]} CH_3CH_2CH_2CH_2CH_2CH_2{-}NH_2$$

Hexanamide — 1-Hexanamine

In designing a synthesis, remember to work in reverse through the various pathways that develop, always keeping in mind the end product that you desire and the types of starting materials that you prefer. As your knowledge of reactions and your experience increase, your ability to create synthetic routes will increase proportionately.

Basicity

Learning Goal 

Amines behave as weak bases when dissolved in water. The nonbonding pair (lone pair) of electrons of the nitrogen atom can be shared with a proton (H^+) from a water molecule, producing an **alkylammonium ion.** Hydroxide ions are also formed, so the resulting solution is basic.

$$R{-}\ddot{N}H_2 + H{-}OH \rightleftharpoons R{-}NH_3^+ + OH^-$$

Amine, Water, Alkylammonium ion, Hydroxide ion

$$CH_3{-}\ddot{N}H_2 + H{-}OH \rightleftharpoons CH_3{-}NH_3^+ + OH^-$$

Methylamine, Water, Methylammonium ion, Hydroxide ion

Neutralization

Because amines are bases, they react with acids to form alkylammonium salts.

$$R{-}\ddot{N}H_2 + HCl \longrightarrow R{-}NH_3^+\ Cl^-$$

Amine, Acid, Alkylammonium salt

Recall that the reaction of an acid and a base gives a salt (Section 9.3).

The reaction of methylamine with hydrochloric acid shown is typical of these reactions.

$$CH_3{-}\ddot{N}H_2 + HCl \longrightarrow CH_3{-}NH_3^+\ Cl^-$$

Methylamine, Hydrochloric acid, Methylammonium chloride

Alkylammonium salts are named by replacing the suffix *-amine* with *ammonium*. This is then followed by the name of the anion. The salts are ionic and hence are quite soluble in water.

A variety of important drugs are amines. They are usually administered as alkylammonium salts because the salts are much more soluble in aqueous solutions and in body fluids.

Question 16.11

Complete each of the following reactions by supplying the missing product(s).

a. NH_2–cyclopentane $+ HBr \longrightarrow ?$

b. $CH_3CH_2{-}NH{-}CH_3 + H_2O \longrightarrow ?$

c. $CH_3{-}NH_2 + H_2O \longrightarrow ?$

A Medical Perspective

Secondary Amines and Cancer

Many of the amines and amine derivatives have been linked to cancer in animals and, in several cases, humans. Little is known about the mode of action of many of these compounds, and a great deal of research is underway to determine how they are involved in the development of cancer. It appears that these *carcinogens* (cancer-causing chemicals) damage the DNA, causing mutations in genes that control cell division. This damage results in the formation of "outlaw" cancer cells. These compounds damage the DNA by alkylating (adding an alkyl group to) the DNA chain.

It is known that secondary amines can react with substances such as nitrous acid (HNO_2) to form nitrosamines. Nitrosamines, in turn, can react further to give diazocompounds. Diazocompounds act as the alkylating agents to alkylate the DNA. The reaction, in part, is as follows:

$$(CH_3)_2N\text{—}H \xrightarrow{HNO_2} (CH_3)_2N\text{—}N{=}O + H_2O$$

2° amine → *N*-nitrosamine

Many of the medicines and foods that we consume contain secondary amines. In addition, the nitrite ion is widely used as a preservative for bacon, ham, sausage, and other meat products. Nitrites give the meat a pleasant pink color and keep it from turning grey. They also inhibit the growth of some harmful bacteria. It is possible that nitrites in foods could react with the acid found in the saliva and stomach juices to form nitrous acid. This might, in turn, undergo the reactions shown below. Fortunately, the concentration of nitrite in foods is very low. However, research continues in an effort to determine whether the nitrites and secondary amines in the food we eat play a role in colon cancers.

$$CH_3\text{—}N{=}N\text{—}OH \xrightarrow{DNA} CH_3\text{—}DNA + N_2 + H_2O$$

Diazocompound → Alkylated—DNA

Question 16.12

Complete each of the following reactions by supplying the missing product(s).

a. $CH_3\text{—}NH_2 + HI \longrightarrow ?$
b. $CH_3CH_2\text{—}NH_2 + HBr \longrightarrow ?$
c. $(CH_3CH_2)_2NH + HCl \longrightarrow ?$

Alkylammonium salts can neutralize hydroxide ions. In this reaction, water is formed and the protonated amine cation is converted into an amine.

$$R\text{—}NH_3^+ + OH^- \longrightarrow R\text{—}\ddot{N}H_2 + H\text{—}OH$$

Alkylammonium salt + Hydroxide ion → Amine + Water

Thus, by adding a strong acid to a water-insoluble amine, a water soluble alkylammonium salt can be formed. The salt can just as easily be converted back to an amine by the addition of a strong base. The ability to manipulate the solubility of physiologically active amines through interconversion of the amine and its corresponding salt is extremely important in the development, manufacture, and administration of many important drugs.

The local anesthetic novocaine, which is often used in dentistry and for minor surgery, is injected as an amine salt. **See A Clinical Perspective: Medically Important Amines.**

Quaternary Ammonium Salts

Learning Goal

Phospholipids and biological membranes are discussed in Sections 18.3 and 18.6.

The function of acetylcholine is described in greater detail in Section 16.5.

Quaternary ammonium salts are ammonium salts that have four organic groups bonded to the nitrogen. They have the following general structure:

$$R_4N^+X^- \qquad (R = \text{any alkyl or aryl group}; \; X^- = \text{a halide anion, most commonly } Cl^-)$$

Quaternary ammonium salts that have a very long carbon chain, sometimes called "quats," are used as disinfectants and antiseptics because they have detergent activity. Two popular quats are benzalkonium chloride (Zephiran) and cetylpyridinium chloride, found in the mouthwash Cepacol.

$$\left[C_6H_5\text{—}CH_2\text{—}N^+(CH_3)_2\text{—}C_{18}H_{37}\right]Cl^-$$

Benzalkonium chloride

$$\left[C_5H_5N^+\text{—}(CH_2)_{15}CH_3\right]Cl^-$$

Cetylpyridinium chloride

Choline is an important quaternary ammonium salt in the body. It is part of the hydrophilic "head" of the membrane phospholipid lecithin. Choline is also a precursor for the synthesis of the neurotransmitter acetylcholine.

$$CH_3\text{—}N^+(CH_3)_2\text{—}CH_2CH_2OH \;\; Cl^-$$

Choline

16.2 Heterocyclic Amines

Learning Goal

Heterocyclic amines are cyclic compounds that have at least one nitrogen atom in the ring structure. The structures and common names of several heterocyclic amines important in nature are shown here. They are represented by their structural formulas and by abbreviated line formulas.

Pyrrole

Pyridine

Imidazole

Pyrimidine

The heterocyclic amines shown below are examples of fused ring structures. Each ring pair shares two carbon atoms in common. Thus, two fused rings share one or more common bonds as part of their ring backbones. Consider the structures of a purine, indole, and porphyrin, which are shown as structural formulas and as line diagrams.

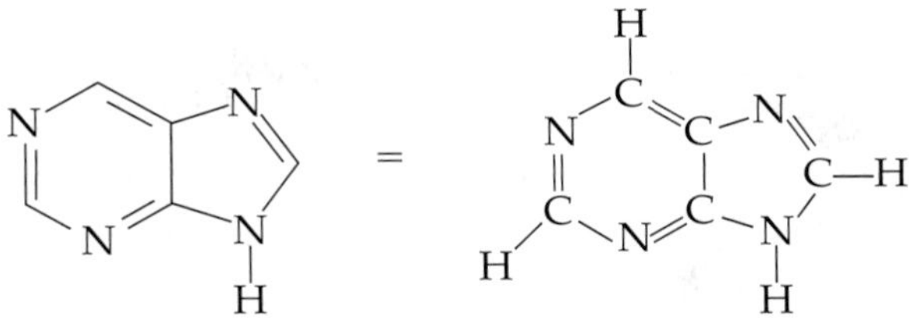

Purine

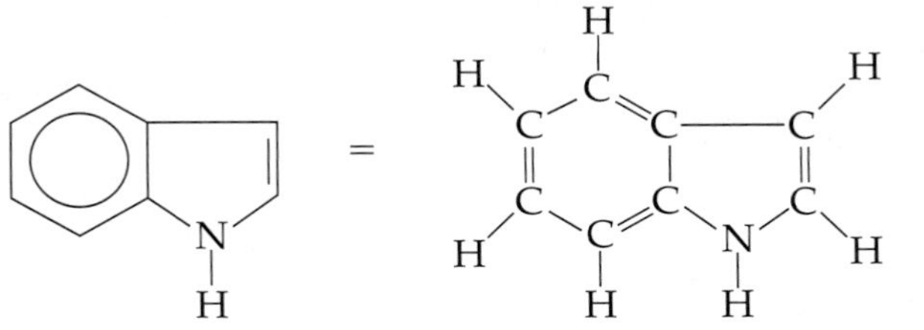

Indole

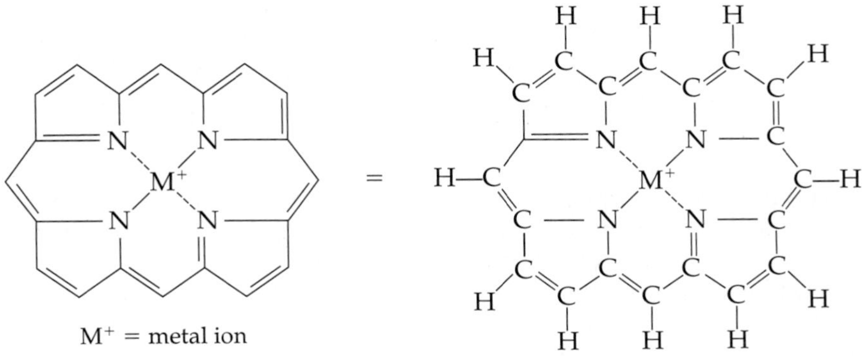

M^+ = metal ion

Porphyrin

The pyrimidine and purine rings are found in DNA and RNA. The porphyrin ring structure is found in hemoglobin (an oxygen-carrying blood protein), myoglobin (an oxygen-carrying protein found in muscle tissue), and chlorophyll (a photosynthetic plant pigment). The indole and pyridine rings are found in many **alkaloids,** which are naturally occurring compounds with one or more nitrogen-containing heterocyclic rings. The alkaloids include cocaine, nicotine, quinine, morphine, heroin, and LSD (Figure 16.3).

The structures of purines and pyrimidines are presented in Section 24.1.

The structure of the heme group found in hemoglobin and myoglobin is presented in Section 19.9.

Lysergic acid diethylamide (LSD) is a hallucinogenic compound that may cause severe mental disorders. Cocaine is produced by the coca plant. In small doses it is used as an anesthetic for the sinuses and eyes. An **anesthetic** is a drug that causes a lack of sensation in any part of the body (local anesthetic) or causes unconsciousness (general anesthetic). In higher doses, cocaine causes an intense feeling of euphoria followed by a deep depression. Cocaine is addictive because the user needs larger and larger amounts to overcome the depression. Nicotine is one of the simplest heterocyclic amines and appears to be the addictive component of cigarette smoke.

Morphine was the first alkaloid to be isolated from the sap of the opium poppy. Morphine is a strong **analgesic,** a drug that acts as a pain killer. However, it is a powerful and addictive narcotic. Codeine, also produced by the opium poppy, is a less powerful analgesic than morphine, but it is one of the most effective cough suppressants known. Heroin is produced in the laboratory by adding

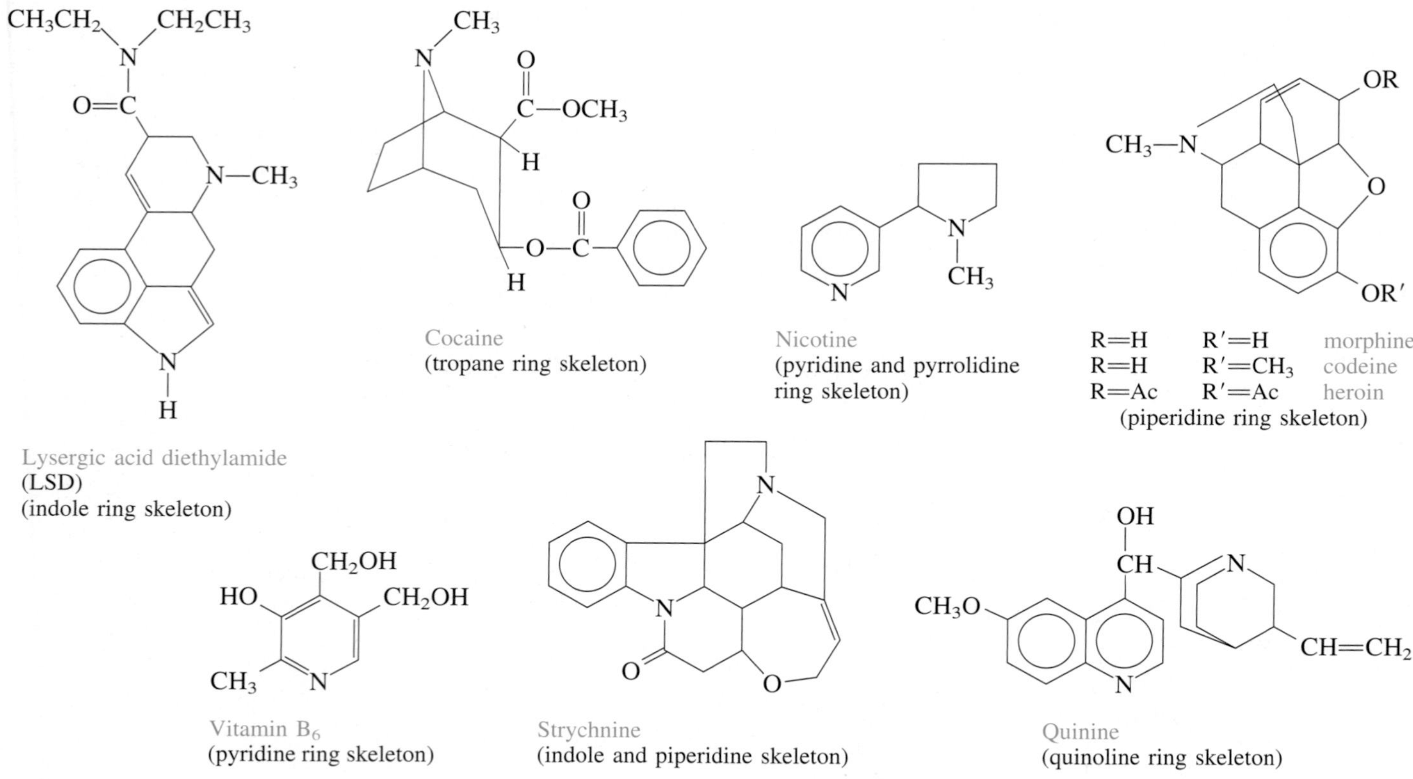

Figure 16.3
Structures of several heterocyclic amines with biological activity.

two acetyl groups to morphine. It was initially made in the hopes of producing a compound with the benefits of morphine but lacking the addictive qualities. However, heroin is even more addictive than morphine.

Strychnine is found in the seeds of an Asiatic tree. It is extremely toxic and was commonly used as a rat poison at one time. Quinine, isolated from the bark of South American trees, was the first effective treatment for malaria. Vitamin B_6 is one of the water-soluble vitamins required by the body.

16.3 Amides

Amides are the products formed in a reaction between a carboxylic acid derivative and ammonia or an amine. The general structure of an amide is shown here.

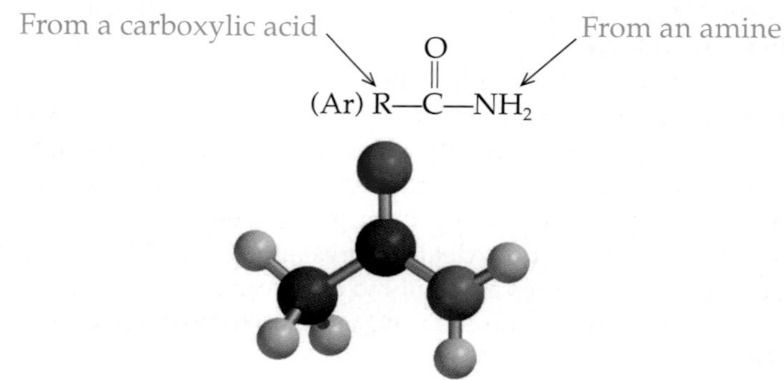

Ethanamide

The amide group is composed of two portions: the carbonyl group from a carboxylic acid and the amino group from ammonia or an amine. The bond between the carbonyl carbon and the nitrogen of the amine or ammonia is called the **amide bond.**

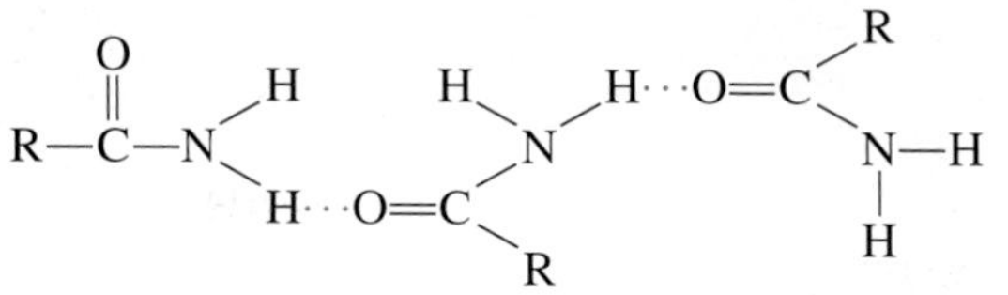

Figure 16.4
Hydrogen bonding in amides.

Structure and Physical Properties

Learning Goal

Most amides are solids at room temperature. They have very high boiling points, and the simpler ones are quite soluble in water. Both of these properties are a result of strong intermolecular hydrogen bonding between the N—H bond of one amide and the C=O group of a second amide, as shown in Figure 16.4.

Unlike amines, amides are not bases (proton acceptors). The reason is that the highly electronegative oxygen atom of the carbonyl group causes a very strong attraction between the lone pair of nitrogen electrons and the carbonyl group. As a result, the unshared pair of electrons cannot "hold" a proton.

Because of the attraction of the carbonyl group for the lone pair of nitrogen electrons, the structure of the C—N bond of an amide is a *resonance hybrid.*

Resonance hybrids are discussed in Section 4.4.

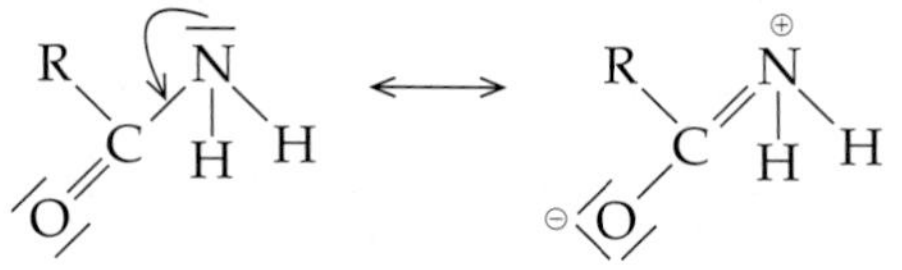

Nomenclature

Learning Goal

The common and I.U.P.A.C. names of the amides are derived from the common and I.U.P.A.C. names of the carboxylic acids from which they were made. Remove the *-ic acid* ending of the common name or the *-oic acid* ending of the I.U.P.A.C. name of the carboxylic acid, and replace it with the ending *-amide*. Several examples of the common and I.U.P.A.C. nomenclature are provided in Table 16.4 and in the following structures:

Nomenclature of carboxylic acid is described in Section 15.1.

Table 16.4 I.U.P.A.C. and Common Names of Simple Amides

Compound	I.U.P.A.C. Name	Common Name
$R{-}C({=}O){-}NH_2$	Alkan*amide* (*-amide* replaces the *-oic acid* ending of the I.U.P.A.C. name of carboxylic acid)	Alkan*amide* (*-amide* replaces the *-ic acid* ending of the common name of carboxylic acid)
$H{-}C({=}O){-}NH_2$	Methanamide	Formamide
$CH_3{-}C({=}O){-}NH_2$	Ethanamide	Acetamide
$CH_3CH_2{-}C({=}O){-}NH_2$	Propanamide	Propionamide
$H{-}C({=}O){-}NHCH_3$	*N*-Methylmethanamide	*N*-Methylformamide
$CH_3{-}C({=}O){-}NHCH_3$	*N*-Methylethanamide	*N*-Methylacetamide

A Medical Perspective

Semisynthetic Penicillins

The antibacterial properties of penicillin were discovered by Alexander Fleming in 1929. These natural penicillins produced by several species of the mold *Penicillium*, had a number of drawbacks. They were effective only against a type of bacteria referred to as Gram positive because of a staining reaction based on their cell wall structure. They were also very susceptible to destruction by bacterial enzymes called β-lactamases, and some were destroyed by stomach acid and had to be administered by injection.

To overcome these problems, chemists have produced semisynthetic penicillins by modifying the core structure. The core of penicillins is 6-aminopenicillanic acid, which consists of a thiazolidine ring fused to a β-lactam ring. In addition, there is an R group bonded via an amide bond to the core structure.

β-lactam ring; Site of amidase attack; thiazolidine ring

6-Aminopenicillanic acid

The β-lactam ring confers the antimicrobial properties. However, the R group determines the degree of antibacterial activity, the pharmacological properties, including the types of bacteria against which it is active, and the degree of resistance to the β-lactamases exhibited by any particular penicillin antibiotic. These are the properties that must be modified to produce penicillins that are acid resistant, effective with a broad spectrum of bacteria, and β-lactamase resistant.

Chemists simply remove the natural R group by cleaving the amide bond with an enzyme called an amidase. They then replace the R group and test the properties of the "new" antibiotic. Among the resulting semisynthetic penicillins are ampicillin, methicillin, and oxacillin.

Ampicillin

Methicillin

Oxacillin

$$CH_3-\overset{O}{\overset{\|}{C}}-NH_2$$

Ethan*oic acid* → Ethan*amide*
or
Acet*ic acid* → Acet*amide*

$$CH_3CH_2-\overset{O}{\overset{\|}{C}}-NH_2$$

Propan*oic acid* → Propan*amide*
or
Propion*ic acid* → Propion*amide*

Substituents on the nitrogen are placed as prefixes and are indicated by *N*- followed by the name of the substituent. There are no spaces between the prefix and the amide name. For example:

$$CH_3CH_2-\overset{O}{\overset{\|}{C}}-NH-CH_3$$

N-Methylpropanamide

$$CH_3CH_2CH_2CH_2CH_2-\overset{O}{\overset{\|}{C}}-NH-CH_2CH_2CH_3$$

N-Propylhexanamide

A CLINICAL PERSPECTIVE

Medically Important Amides

Barbiturates are often referred to as "downers." Barbiturates are derived from amides and are used as sedatives. They are also used as anticonvulsants for epileptics and for people suffering from a variety of brain disorders that manifest themselves in neurosis, anxiety, and tension.

Barbital—a barbiturate

Phenacetin and acetaminophen are also amides. Acetaminophen is an aromatic amide that is commonly used in place of aspirin, particularly by people who are allergic to aspirin or who suffer stomach bleeding from the use of aspirin. It was first synthesized in 1893 and is the active ingredient in Tylenol and Datril. Like aspirin, acetaminophen relieves pain and reduces fever. However, unlike aspirin, it is not an anti-inflammatory drug.

Phenacetin was synthesized in 1887 and used as an analgesic for almost a century. Its structure and properties are similar to those of acetaminophen. However, it was banned by the U.S. Food and Drug Administration in 1983 because of the kidney damage and blood disorders that it causes.

Phenacetin

Acetaminophen

Reactions Involving Amides

Preparation of Amides

Learning Goal

Amides are prepared from carboxylic acid derivatives, either acid chlorides or acid anhydrides. Recall that acid chlorides are made from carboxylic acids by reaction with reagents such as PCl_5.

Formation of acid chlorides is described in Section 15.3.

$$\underset{\text{Carboxylic acid}}{R{-}\overset{O}{\overset{\|}{C}}{-}OH} \xrightarrow{PCl_5} \underset{\text{Acid chloride}}{R{-}\overset{O}{\overset{\|}{C}}{-}Cl}$$

These acid chlorides rapidly react with either ammonia or amines, as in:

$$\underset{\text{Acid chloride}}{R{-}\overset{O}{\overset{\|}{C}}{-}Cl} + \underset{\text{Ammonia or amine}}{2NH_3} \longrightarrow \underset{\text{Amide}}{R{-}\overset{O}{\overset{\|}{C}}{-}NH_2} + \underset{\text{Ammonium chloride or alkylammonium chloride}}{NH_4^+Cl^-}$$

Note that two molar equivalents of ammonia or amine are required in this reaction. The **acyl group**

$$R{-}\overset{O}{\overset{\|}{C}}{-}$$

of the acid chloride is transferred from the Cl atom to the N atom of one of the ammonia or amine molecules. The second ammonia (or amine) reacts with the HCl formed in the transfer reaction to produce ammonium chloride or alkylammonium chloride.

The reaction between butanoyl chloride and methanamine to produce *N*-methylbutanamide is an example of an *acyl group transfer reaction.*

$$CH_3CH_2CH_2-\overset{\overset{O}{\|}}{C}-Cl + 2CH_3NH_2 \longrightarrow$$

Butanoyl chloride — Methanamine

$$CH_3CH_2CH_2-\overset{\overset{O}{\|}}{C}-NH-CH_3 + CH_3NH_3{}^+Cl^-$$

N-Methylbutanamide — Methylammonium chloride

The reaction between an amine and an acid anhydride is also an acyl group transfer. The general equation for the synthesis of an amide in the reaction between an acid anhydride and ammonia or an amine is

$$R-\overset{\overset{O}{\|}}{C}-O-\overset{\overset{O}{\|}}{C}-R + 2NH_3 \longrightarrow R-\overset{\overset{O}{\|}}{C}-NH_2 + R-\overset{\overset{O}{\|}}{C}-O^-NH_4{}^+$$

Acid anhydride — Ammonia or amine — Amide — Carboxylic acid salt

When subjected to heat, the ammonium salt loses a water molecule to produce a second amide molecule.

Amino acids have both a carboxyl group and an amino group and are discussed in detail in Sections 16.4 and 19.1.

A well-known commercial amide is the artificial sweetener aspartame or NutraSweet. Although the name suggests that it is a sugar, it is not a sugar at all. In fact, it is the methylester of a molecule composed of two amino acids, aspartic acid and phenylalanine, joined by an amide bond (Figure 16.5a).

Packages of aspartame carry the warning: "Phenylketonurics: Contains Phenylalanine." Digestion of aspartame and heating to high temperatures during cooking break both the ester bond and the amide bond, which releases the amino acid phenylalanine. People with the genetic disorder phenylketonuria (PKU) cannot break down this amino acid. As a result, it builds up to toxic levels that can cause mental retardation in an infant born with the condition. This no longer occurs because every child is tested for PKU at the time of birth and each is treated with a diet that limits the amount of phenylalanine to only the amount required for normal growth.

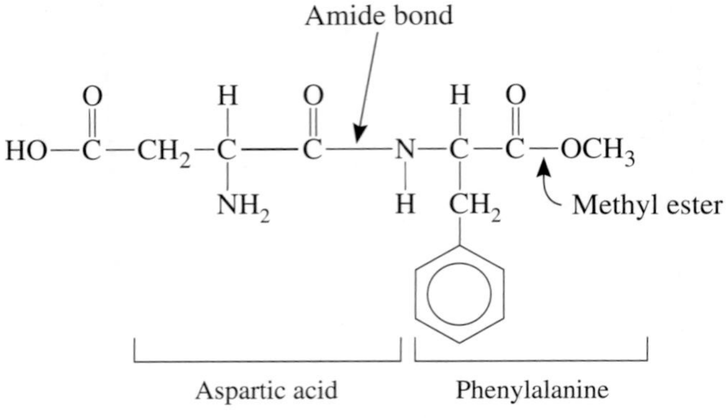

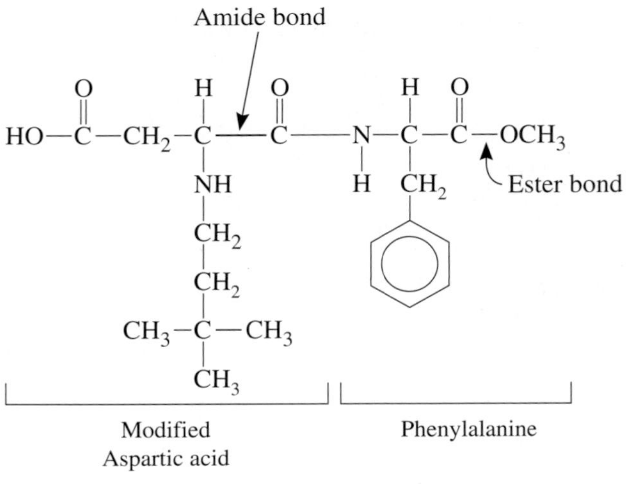

Figure 16.5
The amide bond. (a) NutraSweet, the dipeptide aspartame, is a molecule composed of two amino acids joined by an amide (peptide) bond. (b) Neotame, a newly approved sweetener, is also a dipeptide. One of the amino acids has been modified so that it is safe for use by phenylketonurics.

In July 2002 the Food and Drug Administration approved a new artificial sweetener that is related to aspartame. Called neotame, it has the same core structure as aspartame, but a 3,3-dimethylbutyl group has been added to the aspartic acid (Figure 16.5b). Digestion and heating still cause breakage of the ester bond, but the bulky 3,3-dimethylbutyl group blocks the breakage of the amide bond. Neotame can be used without risk by people with PKU and also retains its sweetness during cooking.

Question 16.13

What is the structure of the amine that, on reaction with the acid chlorides shown, will give each of the following products?

a. $? + CH_3{-}\overset{\overset{\displaystyle O}{\|}}{C}{-}Cl \longrightarrow CH_3{-}\overset{\overset{\displaystyle O}{\|}}{C}{-}NHCH_3 + CH_3NH_3{}^+Cl^-$

b. $? + CH_3CH_2CH_2CH_2\underset{\underset{\displaystyle CH_2CH_3}{|}}{CH}{-}\overset{\overset{\displaystyle O}{\|}}{C}{-}Cl \longrightarrow (CH_3)_2N{-}\overset{\overset{\displaystyle O}{\|}}{C}{-}\underset{\underset{\displaystyle CH_2CH_3}{|}}{C}HCH_2CH_2CH_2CH_3 + (CH_3)_2NH_2{}^+Cl^-$

Question 16.14

What are the structures of the acid chlorides and the amines that will react to give each of the following products?

a. *N*-Ethylhexanamide
b. *N*-Propylbutanamide

Hydrolysis of Amides

Learning Goal

Hydrolysis of an amide results in breaking the amide bond to produce a carboxylic acid and ammonia or an amine. It is very difficult to hydrolyze the amide bond. In fact, the reaction requires heating the amide in the presence of a strong acid or base.

$$R{-}\overset{\overset{\displaystyle O}{\|}}{C}{-}NH{-}R^1 + H_3O^+ \longrightarrow R{-}\overset{\overset{\displaystyle O}{\|}}{C}{-}OH + R^1{-}\overset{+}{N}H_3$$

Amide — Strong acid — Carboxylic acid — Alkylammonium ion or ammonium ion

$$CH_3CH_2CH_2{-}\overset{\overset{\displaystyle O}{\|}}{C}{-}NH_2 + H_3O^+ \longrightarrow CH_3CH_2CH_2{-}\overset{\overset{\displaystyle O}{\|}}{C}{-}OH + \overset{+}{N}H_4$$

Butanamide (butyramide) — Butanoic acid (butyric acid)

If a strong base is used, the products are the amine and the salt of the carboxylic acid:

$$R{-}\overset{\overset{\displaystyle O}{\|}}{C}{-}NH{-}R^1 + NaOH \longrightarrow R{-}\overset{\overset{\displaystyle O}{\|}}{C}{-}O^-Na^+ + R^1{-}NH_2$$

Amide — Strong base — Carboxylic acid salt — Amine or ammonia

$$\mathrm{CH_3CH_2{-}\overset{\overset{\displaystyle O}{\|}}{C}{-}NHCH_3 + NaOH \longrightarrow CH_3CH_2{-}\overset{\overset{\displaystyle O}{\|}}{C}{-}O^-Na^+ + CH_3NH_2}$$

N-Methylpropanamide (*N*-methylpropionamide) — Sodium propanoate (sodium propionate) — Methanamine (methylamine)

16.4 A Preview of Amino Acids, Proteins, and Protein Synthesis

Learning Goal 12

Learning Goal 13

In Chapter 19 we will describe the structure of proteins, the molecules that carry out the majority of the biological processes essential to life. Proteins are polymers of amino acids. As the name suggests, amino acids have two essential functional groups, an amino group (—NH_2) and a carboxyl group (—COOH). Typically amino acids have the following general structure:

$$\mathrm{H_2N{-}\underset{\underset{\displaystyle R}{|}}{\overset{\overset{\displaystyle H}{|}}{C}}{-}COOH}$$

(R may be a hydrogen atom or an organic group.)

In the cell the amino group is usually protonated and the carboxyl group is usually ionized to the carboxylate anion. In the future we will represent an amino acid in the following way:

$$\mathrm{H_3\overset{+}{N}{-}\underset{\underset{\displaystyle R}{|}}{\overset{\overset{\displaystyle H}{|}}{C}}{-}COO^-}$$

The amide bond that forms between the carboxyl group of one amino acid and the amino group of another is called the **peptide bond.** The joining of amino acids by amide bonds produces small *peptides* and larger *proteins.* Because protein structure and function are essential for life processes, it is fortunate indeed that the amide bonds that hold them together are not easily hydrolyzed at physiological pH and temperature.

The process of protein synthesis in the cell mimics amide formation in the laboratory; it involves acyl group transfer. There are several important differences between the chemistry in the laboratory and the chemistry in the cell. During protein synthesis, the **aminoacyl group** of the amino acid is transferred, rather than the acyl group of a carboxylic acid. In addition, the aminoacyl group is not transferred from a carboxylic acid derivative; it is transferred from a special carrier molecule called a **transfer RNA (tRNA).** When the aminoacyl group is covalently bonded to a tRNA, the resulting structure is called an *aminoacyl tRNA:*

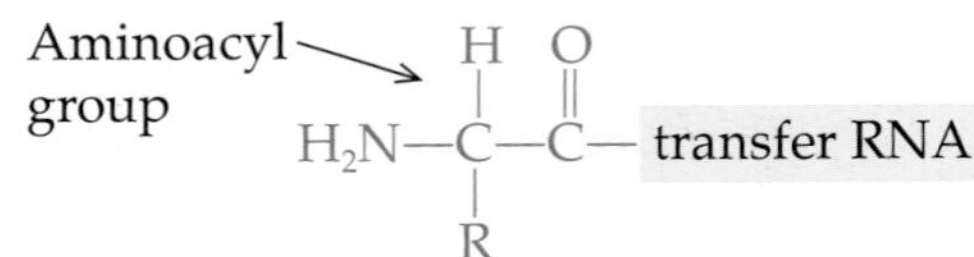

The aminoacyl group of the aminoacyl tRNA is transferred to the amino group nitrogen to form a peptide bond. The transfer RNA is recycled by binding to another of the same kind of aminoacyl group.

More than one hundred kinds of proteins, nucleotides, and RNA molecules participate in the incredibly intricate process of protein synthesis. In Chapter 19 we will study protein structure and learn about the many functions of proteins in the life of the cell. In Chapter 24 we will study the details of protein synthesis to see how these aminoacyl transfer reactions make us the individuals that we are.

16.5 Neurotransmitters

Learning Goal 14

Neurotransmitters are chemicals that carry messages, or signals, from a nerve cell to a target cell. The target cell may be another nerve cell or a muscle cell. Neurotransmitters are classified as being *excitatory,* stimulating their target cell, or inhibitory. One feature shared by the neurotransmitters is that they are all nitrogen-containing compounds. Some of them have rather complex structures and one, nitric oxide (NO), consists of only two atoms.

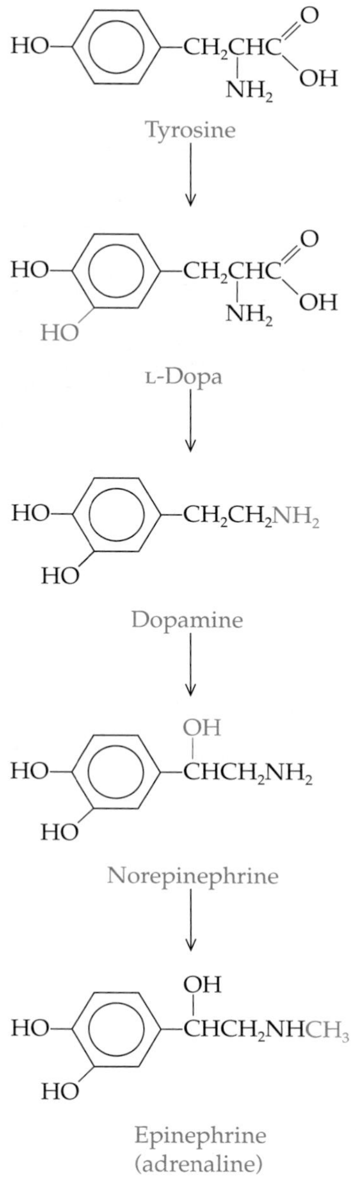

Catecholamines

All of these neurotransmitters are synthesized from the amino acid tyrosine (Figure 16.6). *Dopamine* is critical to good health. A deficiency in this neurotransmitter, for example, results in Parkinson's disease, a disorder characterized by tremors, monotonous speech, loss of memory and problem-solving ability, and loss of motor function. In the brain, dopamine is synthesized from L-dopa. It would seem logical to treat Parkinson's disease with dopamine. However, dopamine cannot cross the blood-brain barrier to enter brain cells. Therefore, L-dopa is used to treat this disorder. It is converted to dopamine upon entry into brain cells.

Just as too little dopamine causes Parkinson's disease, an excess is associated with schizophrenia. Dopamine also appears to play a role in addictive behavior. In proper amounts, it causes a pleasant, satisfied feeling. The greater the amount of dopamine, the more intense the sensation, the "high." Several drugs have been shown to increase the levels of dopamine. Among these are cocaine, heroin, amphetamines, alcohol, and nicotine. Marijuana also causes an increase in brain dopamine, raising the possibility that it, too, has the potential to produce addiction.

Both *epinephrine* (adrenaline) and *norepinephrine* are involved in the "fight or flight" response. Epinephrine stimulates the breakdown of glycogen to produce glucose. The glucose is metabolized to provide energy for the body. Norepinephrine is involved with the central nervous system in the stimulation of other glands and the constriction of blood vessels. All of these responses prepare the body to meet the stressful situation.

HO–C6H4–$CH_2CH(NH_2)COOH$

Tyrosine

↓

(HO)2C6H3–$CH_2CH(NH_2)COOH$

L-Dopa

↓

(HO)2C6H3–$CH_2CH_2NH_2$

Dopamine

↓

(HO)2C6H3–$CH(OH)CH_2NH_2$

Norepinephrine

↓

(HO)2C6H3–$CH(OH)CH_2NHCH_3$

Epinephrine (adrenaline)

Figure 16.6
The pathway for synthesis of dopamine, epinephrine, and norepinephrine.

Serotonin

Serotonin is synthesized from the amino acid tryptophan (Figure 16.7). It is a heterocyclic amine. A deficiency of serotonin has been associated with depression. It is also thought to be involved in bulimia and anorexia nervosa, as well as the carbohydrate-cravings that characterize seasonal affective disorder (SAD), a depression caused by a decrease in daylight during autumn and winter.

Serotonin also affects the perception of pain, thermoregulation, and sleep. There are those who believe that a glass of warm milk will help you fall asleep. We have all noticed how sleepy we become after that big Thanksgiving turkey dinner. Both milk protein and turkey are exceptionally high in tryptophan, the precursor of serotonin!

Prozac (fluoxetine), one of the newest generation of antidepressant drugs, is one of the most widely prescribed drugs in the United States.

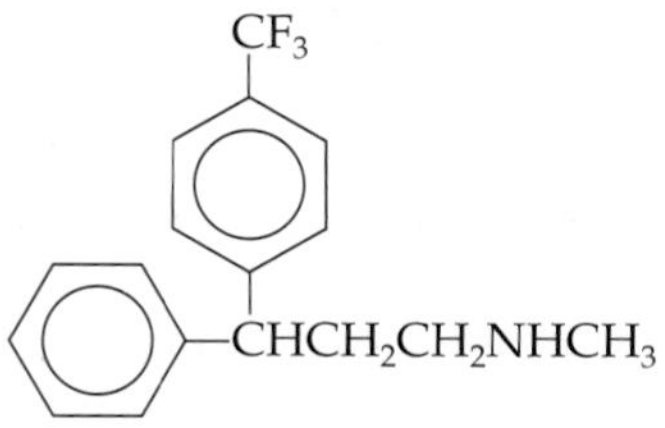

CF_3–C6H4– / C6H5–$CHCH_2CH_2NHCH_3$

Prozac (fluoxetine)

It is a member of a class of drugs called selective serotonin reuptake inhibitors (SSRI). By inhibiting the reuptake, Prozac effectively increases the level of serotonin, relieving the symptoms of depression.

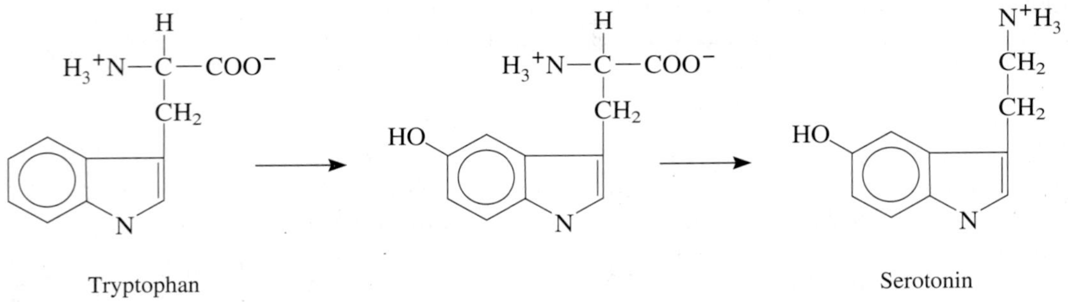

Tryptophan

Serotonin

Figure 16.7
Synthesis of serotonin from the amino acid tryptophan.

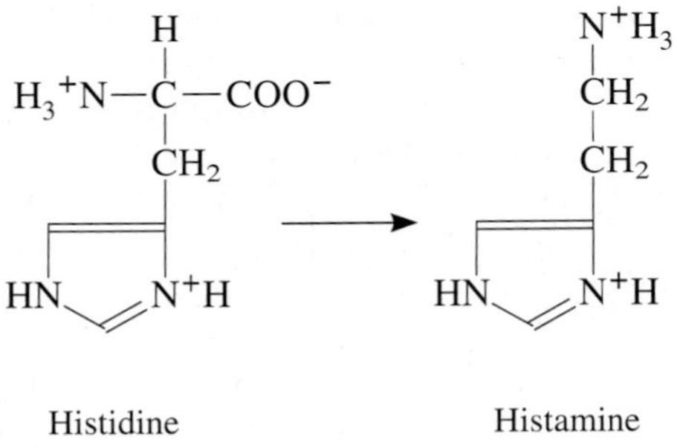

Histidine

Histamine

Figure 16.8
Synthesis of histamine from the amino acid histidine.

Histamine

Histamine, produced in many tissues, is synthesized by removing the carboxyl group from the amino acid histidine (Figure 16.8). It has many, often annoying, physiological roles. Histamine is released during the allergic response. It causes the itchy skin rash associated with poison ivy or insect bites. It also promotes the red, watery eyes and respiratory symptoms of hay fever.

Many antihistamines are available to counteract the symptoms of histamine release. These act by competing with histamine for binding to target cells. If histamine cannot bind to these target cells, the allergic response stops.

Benadryl is an antihistamine that is available as an ointment to inhibit the itchy rash response to allergens. It is also available as an oral medication to block the symptoms of systemic allergies. You need only visit the "colds and allergies" aisle of your grocery store to find dozens of medications containing antihistamines.

Histamines also stimulate secretion of stomach acid. When this response occurs frequently, the result can be chronic heartburn. The reflux of stomach acid into the esophagus can result in erosion of tissue and ulceration. The excess stomach acid may also contribute to development of stomach ulcers. The drug marketed as Tagamet (cimetidine) has proven to be an effective inhibitor of this histamine response, providing relief from chronic heartburn.

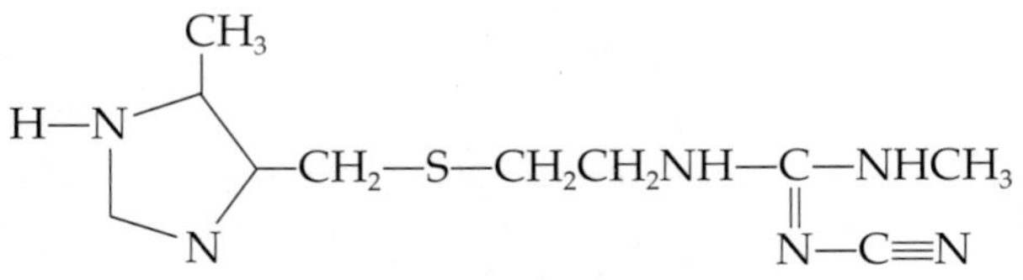

Tagamet (cimetidine)

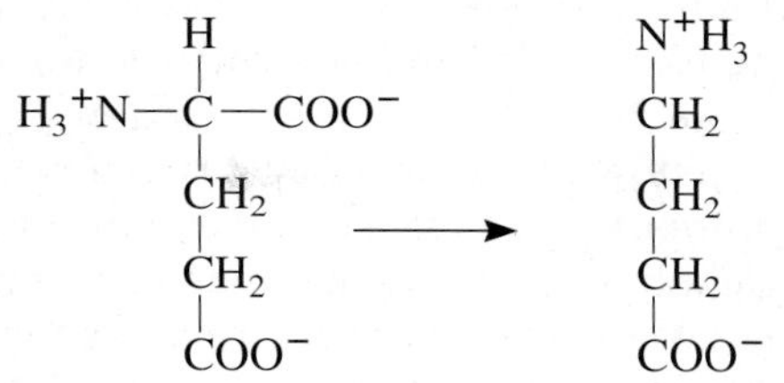

$$H_3{}^+N{-}\underset{\displaystyle CH_2CH_2COO^-}{\overset{H}{C}}{-}COO^- \longrightarrow {}^+H_3N{-}CH_2{-}CH_2{-}CH_2{-}COO^-$$

Glutamate γ-Aminobutyric acid

Figure 16.9
Synthesis of GABA from the amino acid glutamate.

$$HO{-}CH_2CH_2{-}N^+(CH_3)_3 + CH_3{-}\overset{O}{\overset{\|}{C}}{-}S{-}\text{Coenzyme A}$$

Choline Acetyl Coenzyme A

$$\downarrow$$

$$CH_3{-}\overset{O}{\overset{\|}{C}}{-}O{-}CH_2CH_2{-}N^+(CH_3)_3 + \text{Coenzyme A}$$

Acetylcholine

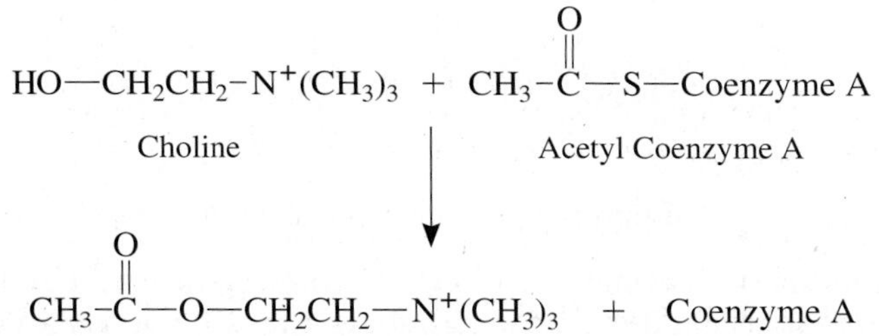

Figure 16.10
Synthesis of acetylcholine.

γ-Aminobutyric Acid and Glycine

γ-Aminobutyric acid (GABA) is produced by removal of a carboxyl group from the amino acid glutamate (Figure 16.9). Both GABA and the amino acid *glycine*

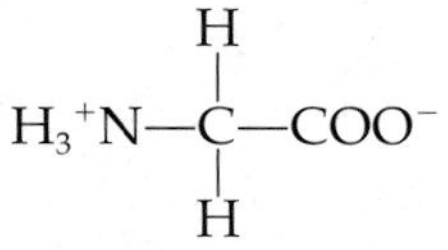

$$H_3{}^+N{-}\underset{H}{\overset{H}{C}}{-}COO^-$$

Glycine

are inhibitory neurotransmitters acting in the central nervous system. One class of tranquilizers, the benzodiazopines, relieves aggressive behavior and anxiety. These drugs have been shown to enhance the inhibitory activity of GABA, suggesting one of the roles played by this neurotransmitter.

Acetylcholine

Acetylcholine is a neurotransmitter that functions at the neuromuscular junction, carrying signals from the nerve to the muscle. It is synthesized in a reaction between the quaternary ammonium ion choline and acetyl coenzyme A (Figure 16.10). When it is released from the nerve cell, acetylcholine binds to receptors on the surface of muscle cells. This binding stimulates the muscle cell to contract. Acetylcholine is then broken down to choline and acetate ion.

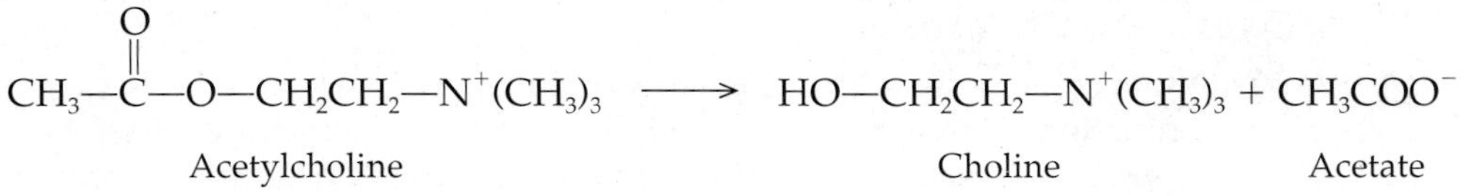

$$CH_3{-}\overset{O}{\overset{\|}{C}}{-}O{-}CH_2CH_2{-}N^+(CH_3)_3 \longrightarrow HO{-}CH_2CH_2{-}N^+(CH_3)_3 + CH_3COO^-$$

Acetylcholine Choline Acetate

These molecules are essentially recycled. They are taken up by the nerve cell where they are used to resynthesize acetylcholine, which is stored in the nerve cell until it is needed.

Nicotine is an agonist of acetylcholine. An agonist is a compound that binds to the receptor for another compound and causes or enhances the biological response. By binding to acetylcholine receptors, nicotine causes the sense of alertness and calm many smokers experience. Nerve cells that respond to nicotine may also signal nerve cells that produce dopamine. As noted above, the dopamine may be responsible for the addictive property of nicotine.

Inhibitors of acetylcholinesterase, the enzyme that catalyzes the breakdown of acetylcholine, are used both as poisons and as drugs. Among the most important poisons of acetylcholinesterase are a class of compounds known as organophosphates. One of these is *diisopropyl fluorophosphate* (DIFP). This molecule forms a covalently bonded intermediate with the enzyme, irreversibly inhibiting its activity.

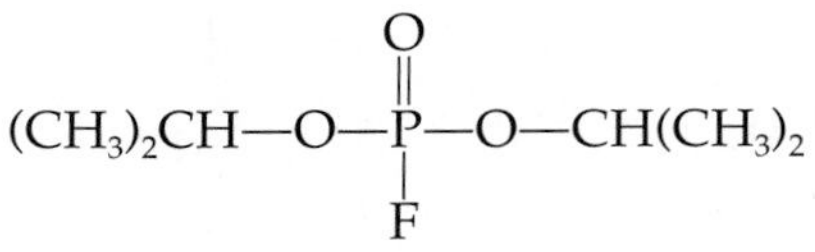

$$(CH_3)_2CH-O-\overset{\overset{\displaystyle O}{\|}}{\underset{\underset{\displaystyle F}{|}}{P}}-O-CH(CH_3)_2$$

Diisopropyl fluorophosphate (DIFP)

The covalent intermediate is stable, and acetylcholinesterase is, therefore, inactive. It is unable to break down the acetylcholine. As a result, nerve transmission continues, resulting in muscle spasm. Death may occur as a result of laryngeal spasm. Antidotes for poisoning by organophosphates, which include many insecticides and nerve gases, have been developed. The antidotes work by reversing the effects of the inhibitor. One of these antidotes is *pyridine aldoxime methiodide* (PAM). This molecule displaces the organophosphate group from the active site of the enzyme, alleviating the effects of the poison.

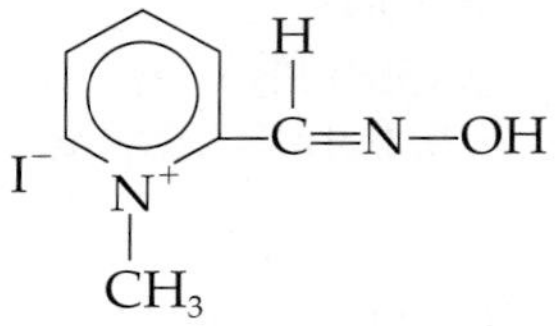

$$I^- \quad \text{(pyridinium ring, } N^+-CH_3\text{)}-\overset{\overset{\displaystyle H}{|}}{C}=N-OH$$

Pyridine aldoxime methiodide

A molecule called *succinylcholine* is a competitive inhibitor of acetylcholine binding to the receptor and can be used as a muscle relaxant in surgical procedures. This compound has a structure that resembles acetylcholine closely enough that it can bind to the acetylcholine receptor. However, it does not have the ability to stimulate muscle contraction. The result is that muscles relax. Normal muscle contraction resumes after infusion of the drug ceases.

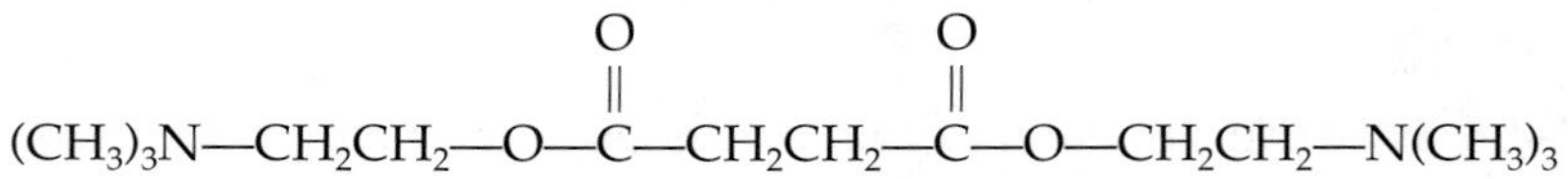

$$(CH_3)_3N-CH_2CH_2-O-\overset{\overset{\displaystyle O}{\|}}{C}-CH_2CH_2-\overset{\overset{\displaystyle O}{\|}}{C}-O-CH_2CH_2-N(CH_3)_3$$

Succinylcholine

Acetylcholine nerve transmission is discussed in further detail in A Clinical Perspective: Enzymes, Nerve Transmission, and Nerve Agents in Chapter 20.

Nitric Oxide and Glutamate

Nitric oxide (NO) is an amazing little molecule that has been shown to have many physiological functions. Among these is its ability to act as a neurotransmitter. NO is synthesized in many areas of the brain from the amino acid arginine. Research has suggested that NO works in conjunction with another neurotransmitter, the amino acid glutamate (see the structure of glutamate in Figure 16.9). Glutamate released from one nerve cell binds to receptors on its target cell. This triggers the target cell to produce NO, which then diffuses back to the original nerve cell. The NO signals the cell to release more glutamate, thus stimulating this neural pathway even further. This is a kind of positive feedback loop. It is thought that this NO-glutamate mechanism is involved in learning and the formation of memories.

Summary of Reactions

Preparation of Amines

$$C_6H_5NO_2 \xrightarrow{[H]} C_6H_5NH_2$$

A nitro compound → An aromatic primary amine

$$R^1\text{—}\overset{\displaystyle O}{\overset{\|}{C}}\text{—}NR^2R^3 \xrightarrow{[H]} R^1CH_2NR^2R^3$$

Amide → Amine

Basicity of Amines

$$R\text{—}NH_2 + H\text{—}OH \rightleftharpoons R\text{—}\overset{+}{N}H_3 + OH^-$$

Amine + Water ⇌ Alkylammonium ion + Hydroxide ion

Neutralization of Amines

$$R\text{—}NH_2 + HCl \longrightarrow R\text{—}\overset{+}{N}H_3\ Cl^-$$

Amine + Acid → Alkylammonium salt

Preparation of Amides

$$R\text{—}\overset{\displaystyle O}{\overset{\|}{C}}\text{—}Cl + 2NH_3 \longrightarrow R\text{—}\overset{\displaystyle O}{\overset{\|}{C}}\text{—}NH_2 + NH_4^+Cl^-$$

Acid chloride + Ammonia or amine → Amide + Ammonium chloride or alkylammonium chloride

$$R\text{—}\overset{\displaystyle O}{\overset{\|}{C}}\text{—}O\text{—}\overset{\displaystyle O}{\overset{\|}{C}}\text{—}R + 2NH_3 \longrightarrow R\text{—}\overset{\displaystyle O}{\overset{\|}{C}}\text{—}NH_2 + R\text{—}\overset{\displaystyle O}{\overset{\|}{C}}\text{—}O^-NH_4^+$$

Acid anhydride + Ammonia or amine → Amide + Carboxylic acid salt

Hydrolysis of Amides

$$R\text{—}\overset{\displaystyle O}{\overset{\|}{C}}\text{—}NH\text{—}R^1 + H_3O^+ \longrightarrow R\text{—}\overset{\displaystyle O}{\overset{\|}{C}}\text{—}OH + R^1\text{—}\overset{+}{N}H_3$$

Amide + Strong acid → Carboxylic acid + Alkylammonium ion

$$R\text{—}\overset{\displaystyle O}{\overset{\|}{C}}\text{—}NH\text{—}R^1 + NaOH \longrightarrow R\text{—}\overset{\displaystyle O}{\overset{\|}{C}}\text{—}O^-Na^+ + R^1\text{—}NH_2$$

Amide + Strong base → Carboxylic acid salt + Amine or ammonia

Summary

16.1 Amines

Amines are a family of organic compounds that contain an amino group or substituted amino group. A *primary amine* has the general formula RNH_2; a *secondary amine* has the general formula R_2NH; and a *tertiary amine* has the general formula R_3N. In the *Chemical Abstracts* nomenclature system, amines are named as *alkanamines*. In the I.U.P.A.C. system they are named as *aminoalkanes*. In the common system they are named as *alkylamines*. Amines behave as weak bases, forming *alkylammonium ions* in water and alkylammonium salts when they react with acids. *Quaternary ammonium salts* are ammonium salts that have four organic groups bonded to the nitrogen atom.

16.2 Heterocyclic Amines

Heterocyclic amines are cyclic compounds having at least one nitrogen atom in the ring structure. *Alkaloids* are natural plant products that contain at least one heterocyclic ring. Many alkaloids have powerful biological effects.

16.3 Amides

Amides are formed in a reaction between a carboxylic acid derivative and an amine (or ammonia). The *amide bond* is the bond between the carbonyl carbon of the *acyl group* and the nitrogen of the amine. In the I.U.P.A.C. Nomenclature System they are named by replacing the *-oic acid* ending of the carboxylic acid with the *-amide* ending. In the common system of nomenclature the *-ic acid* ending of the carboxylic acid is replaced by the *-amide* ending. Hydrolysis of an amide produces a carboxylic acid and an amine (or ammonia).

16.4 A Preview of Amino Acids, Proteins, and Protein Synthesis

Proteins are polymers of amino acids joined to one another by *amide bonds* called *peptide bonds*. During protein synthesis the *aminoacyl group* of one amino acid is transferred from a carrier molecule called a *transfer RNA* to the amino group nitrogen of another amino acid.

16.5 Neurotransmitters

Neurotransmitters are chemicals that carry messages, or signals, from a nerve cell to a target cell, which may be another nerve cell or a muscle cell. They may be inhibitory or excitatory and all are nitrogen-containing compounds. The catecholamines include dopamine, norepinephrine, and epinephrine. Too little dopamine results in Parkinson's disease. Too much is associated with schizophrenia. Dopamine is also associated with addictive behavior. A deficiency of serotonin is associated with depression and eating disorders. Serotonin is involved in pain perception, regulation of body temperature, and sleep. Histamine contributes to allergy symptoms. Antihistamines block histamines and provide relief from allergies. γ-Aminobutyric acid (GABA) and glycine are inhibitory neurotransmitters. It is believed that GABA is involved in control of aggressive behavior. Acetylcholine is a neurotransmitter that functions at the neuromuscular junction, carrying signals from the nerve to the muscle. Nitric oxide and glutamate function in a positive feedback loop that is thought to be involved in learning and the formation of memories.

Key Terms

acyl group (16.3)
alkaloid (16.2)
alkylammonium ion (16.1)
amide (16.3)
amide bond (16.3)
amine (16.1)
aminoacyl group (16.4)
analgesic (16.2)
anesthetic (16.2)
heterocyclic amine (16.2)
neurotransmitter (16.5)
peptide bond (16.4)
primary (1°) amine (16.1)
quaternary ammonium salt (16.1)
secondary (2°) amine (16.1)
tertiary (3°) amine (16.1)
transfer RNA (tRNA) (16.4)

Questions and Problems

Amines

16.15 For each pair of compounds predict which would have greater solubility in water. Explain your reasoning.
- a. Pentane or 1-butanamine
- b. Cyclohexane or 2-pentanamine

16.16 For each pair of compounds predict which would have the higher boiling point. Explain your reasoning.
- a. Ethanamine or ethanol
- b. Butane or 1-propanamine
- c. Methanamine or water
- d. Ethylmethylamine or butane

16.17 Explain why a tertiary amine such as triethylamine has a significantly lower boiling point than its primary amine isomer, 1-hexanamine.

16.18 Draw a diagram to illustrate your answer to Problem 16.17.

16.19 Use the *Chemical Abstracts* system of nomenclature to name each of the following amines:
- a. $CH_3CH_2CH(CH_3)—NH_2$ (with CH_3 on the CH)
- b. $CH_3CH_2CH_2CHCH_2CH_3$ with NH_2 on the CH
- c. Cyclopentane ring—NH_2
- d. $(CH_3)_3C—NH_2$

16.20 Use the CA and common nomenclature systems to name each of the following amines:
- a. $CH_3CH_2CH_2CH_2CH_2CH_2CH_2CH_2—NH_2$
- b. Cl—(benzene ring)—NH_2

c. 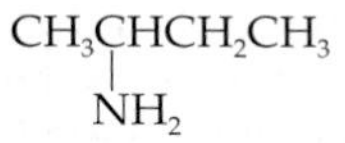 $CH_3CHCH_2CH_3$ with NH_2 on C-2

$$CH_3CHCH_2CH_3 \\ \quad | \\ \quad NH_2$$

d. 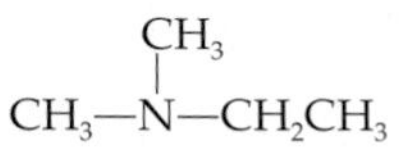 $CH_3—N(CH_3)—CH_2CH_3$

16.21 Draw the structure of each of the following compounds:
a. Diethylamine
b. Butylamine
c. 3-Decanamine
d. 3-Bromo-2-pentanamine
e. Triphenylamine

16.22 Draw the structure of each of the following compounds:
a. *N,N*-Dipropylaniline
b. Cyclohexanamine
c. 2-Bromocyclopentanamine
d. Tetraethylammonium iodide
e. 3-Bromobenzenamine

16.23 Draw each of the following compounds with condensed formulas:
a. 2-Pentanamine
b. 2-Bromo-1-butanamine
c. Ethylisopropylamine
d. Cyclopentanamine

16.24 Draw each of the following compounds with condensed formulas:
a. Dipentylamine
b. 3,4-Dinitroaniline
c. 4-Methyl-3-heptanamine
d. *t*-Butylpentylamine
e. 3-Methyl-3-hexanamine
f. Trimethylammonium iodide

16.25 Draw structural formulas for the eight isomeric amines that have the molecular formula $C_4H_{11}N$. Name each of the isomers using the CA system, and determine whether each isomer is a 1°, 2°, or 3° amine.

16.26 Draw all of the isomeric amines of molecular formula C_3H_9N. Name each of the isomers, using the CA system, and determine whether each isomer is a primary, secondary, or tertiary amine.

16.27 Classify each of the following amines as 1°, 2°, or 3°:
a. Cyclohexanamine
b. Dibutylamine
c. 2-Methyl-2-heptanamine
d. Tripentylamine

16.28 Classify each of the following amines as primary, secondary, or tertiary:
a. Benzenamine
b. *N*-Ethyl-2-pentanamine
c. Ethylmethylamine
d. Tripropylamine
e. *m*-Chloroaniline

16.29 Write an equation to show a reaction that would produce each of the following products:

a. 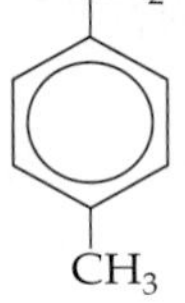 Benzene ring with NH_2 at top and CH_3 at bottom (para)

b. 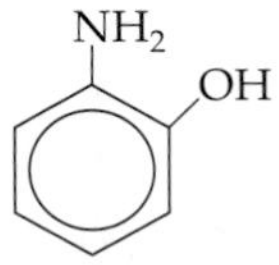 Benzene ring with NH_2 and OH on adjacent carbons (ortho)

c. Benzene ring with NH_2

d. 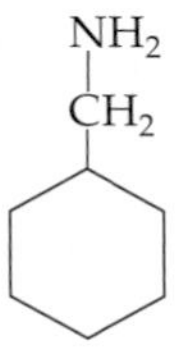 Cyclohexane ring with $CH_2—NH_2$

16.30 Write an equation to show a reaction that would produce each of the following amines:
a. 1-Pentanamine
b. *N,N*-Dimethylethanamine
c. *N*-Ethylpropanamine

16.31 Complete each of the following reactions by supplying the missing reactant or product indicated by a question mark:

a. 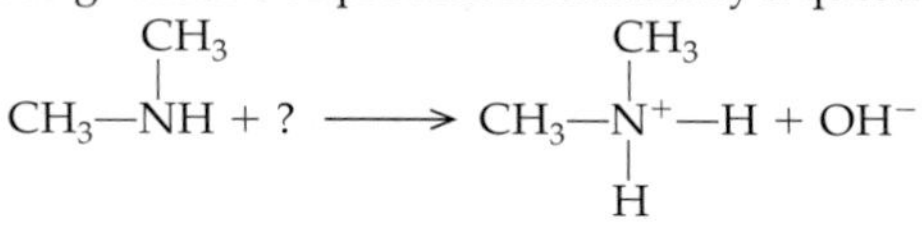 $CH_3—NH(CH_3) + ? \longrightarrow CH_3—N^+H(CH_3)—H + OH^-$

b. 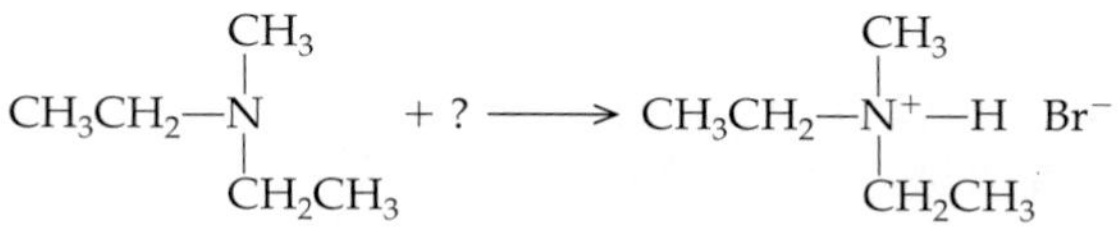 $CH_3CH_2—N(CH_3)(CH_2CH_3) + ? \longrightarrow CH_3CH_2—N^+(CH_3)(CH_2CH_3)—H\ Br^-$

c. $CH_3CH_2CH_2—NH_2 + H_2O \longrightarrow ? + OH^-$

d. 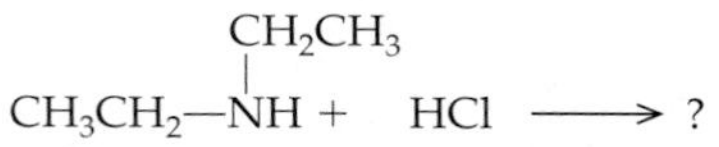 $CH_3CH_2—NH(CH_2CH_3) + HCl \longrightarrow ?$

16.32 Complete each of the following reactions by supplying the missing reactant or product indicated by a question mark:

a. $CH_3CH_2—NH_2 + H_2O \longrightarrow ? + OH^-$

b.  $? + HCl \longrightarrow CH_3CH_2CH_2—N^+H(CH_2CH_2CH_3)—H\ Cl^-$

c. 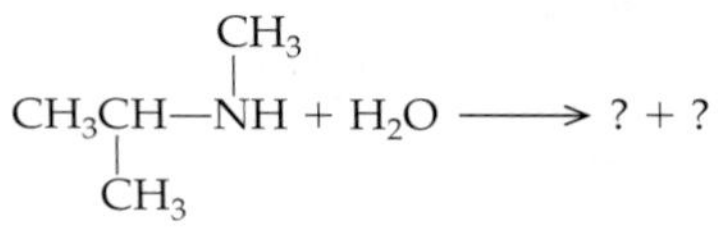 $CH_3CH(CH_3)—NH(CH_3) + H_2O \longrightarrow ? + ?$

d. $NH_3 + HBr \longrightarrow ?$

16.33 Briefly explain why the lower-molecular-weight amines (fewer than five carbons) exhibit appreciable solubility in water.

16.34 Why is the salt of an amine appreciably more soluble in water than the amine from which it was formed?

16.35 Most drugs containing amine groups are not administered as the amine but rather as the ammonium salt. Can you suggest a reason why?

16.36 Why does aspirin upset the stomach, whereas acetaminophen (Tylenol) does not?

16.37 Putrescine and cadaverine are two odoriferous amines that are produced by decaying flesh. Putrescine is 1,4-diaminobutane, and cadaverine is 1,5-diaminopentane. Draw the structures of these two compounds.

16.38 How would you quickly convert an alkylammonium salt into a water-insoluble amine? Explain the rationale for your answer.

Heterocyclic Amines

16.39 Indole and pyridine rings are found in alkaloids.
a. Sketch each ring.
b. Name one compound containing each of the ring structures and indicate its use.

16.40 What is an alkaloid?

16.41 List some heterocyclic amines that are used in medicine.

16.42 Distinguish between the terms *analgesic* and *anesthetic*.

Amides

16.43 Use the I.U.P.A.C. and common systems of nomenclature to name the following amides:

a. 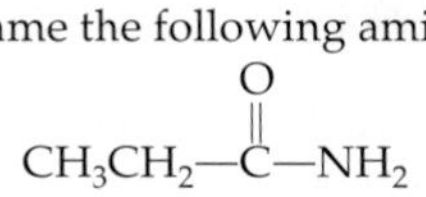 $CH_3CH_2—\overset{O}{\overset{\|}{C}}—NH_2$

b. $CH_3CH_2CH_2CH_2{-}\overset{O}{\overset{\|}{C}}{-}NH_2$

c. $CH_3{-}\overset{O}{\overset{\|}{C}}{-}N(CH_3)_2$

16.44 Use the I.U.P.A.C. Nomenclature System to name each of the following amides:

a. $CH_3CH_2\underset{Br}{\underset{|}{C}}HCH_2{-}\overset{O}{\overset{\|}{C}}{-}NH_2$

b. 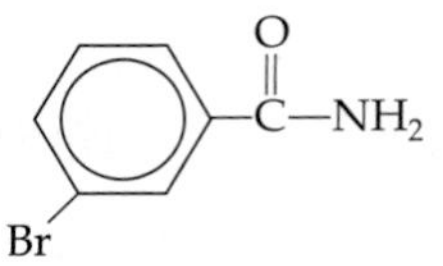 (benzene ring with Br in the meta position) $-\overset{O}{\overset{\|}{C}}{-}NH_2$

c. $CH_3\underset{CH_3}{\underset{|}{C}}H{-}\overset{O}{\overset{\|}{C}}{-}NH_2$

16.45 Draw the condensed structural formula of each of the following amides:
a. Ethanamide
b. *N*-Methylpropanamide
c. *N,N*-Diethylbenzamide
d. 3-Bromo-4-methylhexanamide
e. *N,N*-Dimethylacetamide

16.46 Draw the condensed formula of each of the following amides:
a. Acetamide
b. 4-Methylpentanamide
c. *N,N*-Dimethylpropanamide
d. Formamide
e. *N*-Ethylpropionamide

16.47 The active ingredient in many insect repellents is *N,N*-diethyl-*m*-toluamide. Draw the structure of this compound. Which carboxylic acid and amine would be released by hydrolysis of this compound?

16.48 When an acid anhydride and an amine are combined, an amide is formed. This approach may be used to synthesize acetaminophen, the active ingredient in Tylenol. Complete the following reaction to determine the structure of acetaminophen:

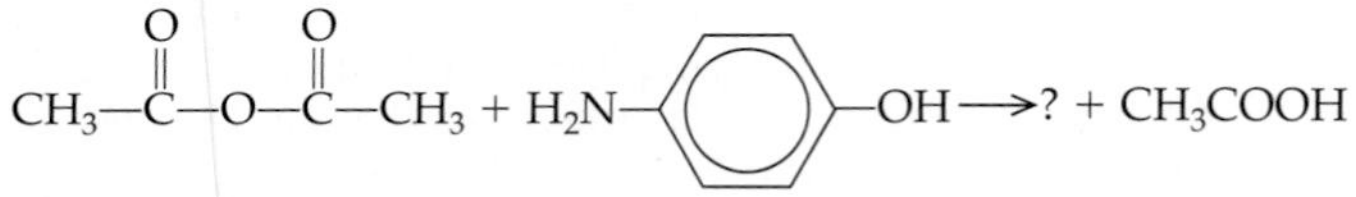

$$CH_3{-}\overset{O}{\overset{\|}{C}}{-}O{-}\overset{O}{\overset{\|}{C}}{-}CH_3 + H_2N{-}C_6H_4{-}OH \longrightarrow ? + CH_3COOH$$

16.49 Explain why amides are neutral in the acid-base sense.

16.50 The amide bond is stabilized by resonance. Draw the contributing resonance forms of the amide bond.

16.51 Lidocaine is often used as a local anesthetic. For medicinal purposes it is often used in the form of its hydrochloride salt because the salt is water soluble. In the structure of lidocaine hydrochloride shown, locate the amide functional group.

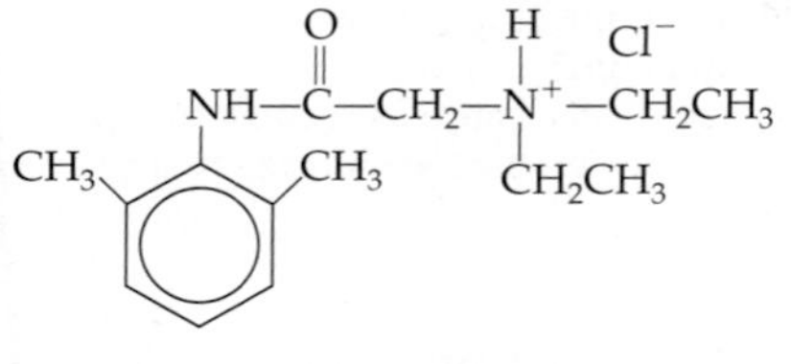

$$(2,6\text{-}(CH_3)_2C_6H_3){-}NH{-}\overset{O}{\overset{\|}{C}}{-}CH_2{-}\overset{H}{\overset{|}{N^+}}(CH_2CH_3){-}CH_2CH_3 \quad Cl^-$$

Lidocaine hydrochloride

16.52 Locate the amine functional group in the structure of lidocaine. Is lidocaine a primary, secondary, or tertiary amine?

16.53 The antibiotic penicillin BT contains functional groups discussed in this chapter. In the structure of penicillin BT shown, locate and name as many functional groups as you can.

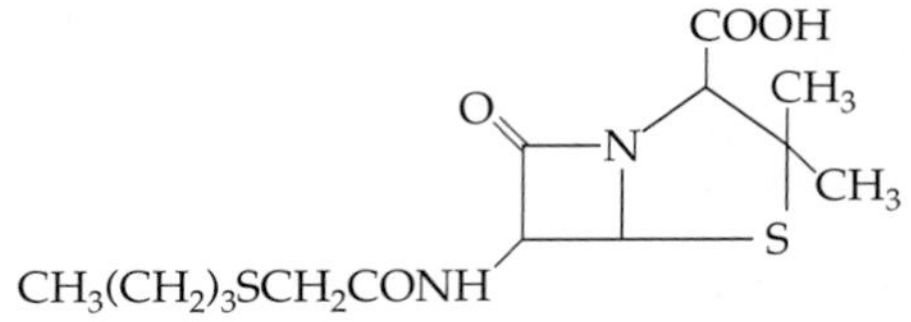

(structure: β-lactam ring fused to a thiazolidine ring bearing COOH, two CH_3 groups and S; side chain $CH_3(CH_2)_3SCH_2CONH$)

Penicillin BT

16.54 The structure of saccharin, an artificial sweetener, is shown. Circle the amide group.

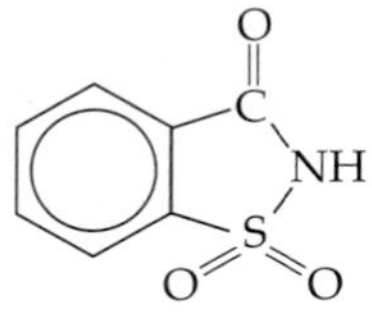

(structure: benzene ring fused to a five-membered ring containing C=O, NH, and SO_2)

Saccharin

16.55 Complete each of the following reactions by supplying the missing reactant(s) or product(s) indicated by a question mark. Provide the systematic name for all the reactants and products.

a. $CH_3{-}\overset{O}{\overset{\|}{C}}{-}NHCH_3 + H_3O^+ \longrightarrow ? + ?$

b. $? + H_3O^+ \longrightarrow CH_3CH_2CH_2{-}\overset{O}{\overset{\|}{C}}{-}OH + CH_3NH_3^+$

c. $CH_3\underset{CH_3}{\underset{|}{C}}HCH_2{-}\overset{O}{\overset{\|}{C}}{-}NHCH_2CH_3 + ? \longrightarrow CH_3\underset{CH_3}{\underset{|}{C}}HCH_2{-}\overset{O}{\overset{\|}{C}}{-}OH + ?$

16.56 Complete each of the following by supplying the missing reagents. Draw the structures of each of the reactants and products.
a. *N*-Methylpropanamide + ? ⟶ methanamine + ?
b. *N,N*-Dimethylacetamide + strong acid ⟶ ? + ?
c. Formamide + strong acid ⟶ ? + ?

16.57 Complete each of the following reactions by supplying the missing reactant(s) or product(s) indicated by a question mark.

a. $? + 2CH_3CH_2CH_2{-}NH_2 \longrightarrow CH_3CH_2CH_2{-}NH{-}\overset{O}{\overset{\|}{C}}{-}CH_2CH_3 + CH_3CH_2{-}\overset{O}{\overset{\|}{C}}{-}O^-\ \overset{+}{N}H_3{-}CH_2CH_2CH_3$

b. $CH_3CH_2{-}\overset{O}{\overset{\|}{C}}{-}Cl + 2NH_3 \longrightarrow ? + ?$

c. $? + ? \longrightarrow CH_3CH_2CH_2{-}\overset{O}{\overset{\|}{C}}{-}NH{-}CH_2CH_3 + CH_3CH_2{-}NH_3^+Cl^-$

16.58 Write two equations for the synthesis of each of the following amides. In one equation use an acid chloride as a reactant. In the second equation use an acid anhydride.
  a. Ethanamide
  b. *N*-Propylpentanamide
  c. Propionamide

A Preview of Amino Acids, Proteins, and Protein Synthesis

16.59 Draw the general structure of an amino acid.
16.60 The amino acid glycine has a hydrogen atom as its R group, and the amino acid alanine has a methyl group. Draw these two amino acids.
16.61 Draw a dipeptide composed of glycine and alanine. Begin by drawing glycine with its amino group on the left. Circle the amide bond.
16.62 What is the name of the amide bond formed between two amino acids?
16.63 Draw the amino acid alanine (see Problem 16.60). Place a star by the chiral carbon.
16.64 Does glycine have a chiral carbon? Explain your reasoning.
16.65 Describe acyl group transfer.
16.66 Describe the relationship between acyl group transfer and the process of protein synthesis.

Neurotransmitters

16.67 Define the term *neurotransmitter.*
16.68 What are the two general classes of neurotransmitters? What distinguishes them from one another?
16.69 a. What symptoms result from a deficiency of dopamine?
  b. What is the name of this condition?
  c. What symptoms result from an excess of dopamine?
16.70 What is the starting material in the synthesis of dopamine, epinephrine, and norepinephrine?
16.71 Explain the connection between addictive behavior and dopamine.
16.72 Why is L-dopa used to treat Parkinson's disease rather than dopamine?
16.73 What is the function of epinephrine?
16.74 What is the function of norepinephrine?
16.75 What is the starting material from which serotonin is made?
16.76 What symptoms are associated with a deficiency of serotonin?
16.77 What physiological processes are affected by serotonin?
16.78 How does Prozac relieve the symptoms of depression?
16.79 What are the physiological roles of histamine?
16.80 How do antihistamines function to control the allergic response?
16.81 What type of neurotransmitters are γ-aminobutyric acid and glycine?
16.82 Explain the evidence for a relationship between γ-aminobutyric acid and aggressive behavior.
16.83 Explain the function of acetylcholine at the neuromuscular junction.
16.84 Explain why organophosphates are considered to be poisons.
16.85 How does pyridine aldoxime methiodide function as an antidote for organophosphate poisoning?
16.86 Explain the mechanism by which glutamate and NO may function to promote development of memories and learning.

Critical Thinking Problems

1. Histamine is made and stored in blood cells called *mast cells.* Mast cells are involved in the allergic response. Release of histamine in response to an allergen causes dilation of capillaries. This, in turn, allows fluid to leak out of the capillary resulting in local swelling. It also causes an increase in the volume of the vascular system. If this increase is great enough, a severe drop in blood pressure may cause shock. Histamine is produced by decarboxylation (removal of the carboxylate group as CO_2) of the amino acid histidine shown below. Draw the structure of histamine.

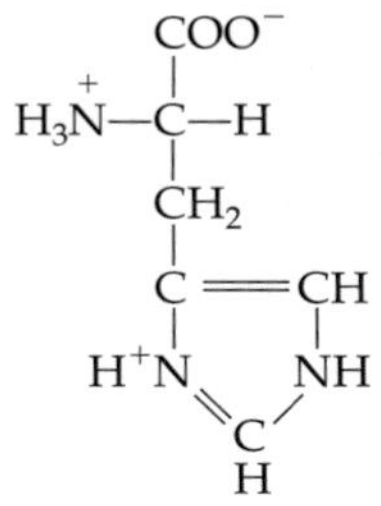

COO^-
$H_3\overset{+}{N}$—C—H
CH_2
C═CH
H^+N NH
C
H

2. Carnitine tablets are sold in health food stores. It is claimed that carnitine will enhance the breakdown of body fat. Carnitine is a tertiary amine found in mitochondria, cell organelles in which food molecules are completely oxidized and ATP is produced. Carnitine is involved in transporting the acyl groups of fatty acids from the cytoplasm into the mitochondria. The fatty acyl group is transferred from a fatty acyl CoA molecule and esterified to carnitine. Inside the mitochondria the reaction is reversed and the fatty acid is completely oxidized. The structure of carnitine is shown here:

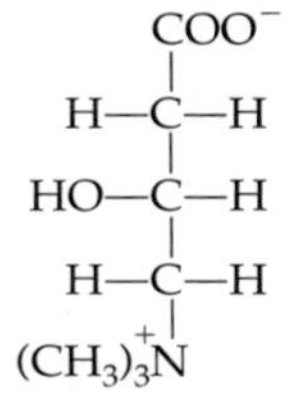

COO^-
H—C—H
HO—C—H
H—C—H
$(CH_3)_3\overset{+}{N}$

Draw the acyl carnitine molecule that is formed by esterification of palmitic acid with carnitine.

3. The amino acid proline has a structure that is unusual among amino acids. Compare the general structure of an amino acid with that of proline, shown here:

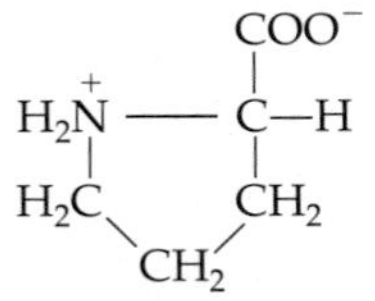

COO^-
$H_2\overset{+}{N}$ — C—H
H_2C CH_2
CH_2

What is the major difference between proline and the other amino acids? Draw the structure of a dipeptide in which the amino group of proline forms a peptide bond with the carboxyl group of alanine.

4. Bulletproof vests are made of the polymer called Kevlar. It is produced by the copolymerization of the following two monomers:

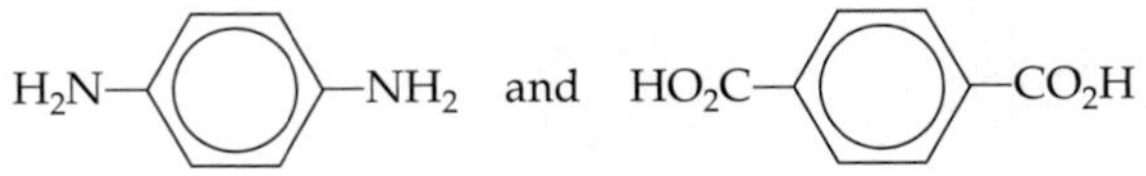

H_2N—C₆H₄—NH_2 and HO_2C—C₆H₄—CO_2H

Draw the structure of a portion of Kevlar polymer.

Would "looking-glass milk" be nutritious?

BIOCHEMISTRY

17

Carbohydrates

Outline

Learning Goals

1. Explain the difference between complex and simple carbohydrates and know the amounts of each recommended in the daily diet.
2. Apply the systems of classifying and naming monosaccharides according to the functional group and number of carbons in the chain.
3. Determine whether a molecule has a chiral center.
4. Explain stereoisomerism.
5. Identify monosaccharides as either D- or L-.
6. Draw and name the common monosaccharides using structural formulas.
7. Given the linear structure of a monosaccharide, draw the Haworth projection of its α- and β-cyclic forms and vice versa.
8. By inspection of the structure, predict whether a sugar is a reducing or a nonreducing sugar.
9. Discuss the use of the Benedict's reagent to measure the level of glucose in urine.
10. Draw and name the common disaccharides and discuss their significance in biological systems.
11. Describe the difference between galactosemia and lactose intolerance.
12. Discuss the structural, chemical, and biochemical properties of starch, glycogen, and cellulose.

CHEMISTRY CONNECTION

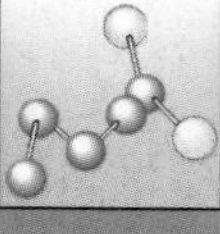

Chemistry Through the Looking Glass

In his children's story *Through the Looking Glass,* Lewis Carroll's heroine Alice wonders whether "looking-glass milk" would be good to drink. As we will see in this chapter, many biological molecules, such as the sugars, exist as two stereoisomers, *enantiomers,* that are mirror images of one another. Because two mirror-image forms occur, it is rather remarkable that in our bodies, and in most of the biological world, only one of the two is found. For instance, the common sugars are members of the D-family, whereas all the common amino acids that make up our proteins are members of the L-family. It is not too surprising, then, that the enzymes in our bodies that break down the sugars and proteins we eat are *stereospecific,* that is, they recognize only one mirror-image isomer. Knowing this, we can make an educated guess that "looking-glass milk" could not be digested by our enzymes and therefore would not be a good source of food for us. It is even possible that it might be toxic to us!

Pharmaceutical chemists are becoming more and more concerned with the stereochemical purity of the drugs that we take. Consider a few examples. In 1960 the drug thalidomide was commonly prescribed in Europe as a sedative. However, during that year, hundreds of women who took thalidomide during pregnancy gave birth to babies with severe birth defects. Thalidomide, it turned out, was a mixture of two enantiomers. One is a sedative; the other is a teratogen, a chemical that causes birth defects.

One of the common side effects of taking antihistamines for colds or allergies is drowsiness. Again, this is the result of the fact that antihistamines are mixtures of enantiomers. One causes drowsiness; the other is a good decongestant.

One enantiomer of the compound carvone is associated with the smell of spearmint; the other produces the aroma of caraway seeds or dill. One mirror-image form of limonene smells like lemons; the other has the aroma of oranges.

The pain reliever ibuprofen is currently sold as a mixture of enantiomers, but one is a much more effective analgesic than the other.

Taste, smell, and the biological effects of drugs in the body all depend on the stereochemical form of compounds and their interactions with cellular enzymes or receptors. As a result, chemists are actively working to devise methods of separating the isomers in pure form. Alternatively, methods of conducting stereospecific syntheses that produce only one stereoisomer are being sought. By preparing pure stereoisomers, the biological activity of a compound can be much more carefully controlled. This will lead to safer medications.

In this chapter we will begin our study of stereochemistry, the spatial arrangement of atoms in molecules, with the carbohydrates. Later, we will examine the stereochemistry of the amino acids that make up our proteins and consider the stereochemical specificity of the metabolic reactions that are essential to life. A more complete treatment of stereochemistry is found in Appendix D: Stereochemistry and Stereoisomers Revisited.

Introduction

Emil Fischer's father, a wealthy businessman, once said that Emil was too stupid to be a businessman and had better be a student. Lucky for the field of biochemistry, Emil did just that. Originally he wanted to study physics, but his cousin Otto Fischer convinced him to study chemistry. Beginning as an organic chemist, Fischer launched a career that eventually led to groundbreaking research in biochemistry. By following his career, we get a glimpse of the entire field.

Early on, Fischer discovered the active ingredients in coffee and tea, caffeine and theobromine. Eventually he discovered their structures and synthesized them in the laboratory. In work that he carried out between 1882 and 1906, Fischer demonstrated that adenine and guanine, along with some other compounds found in plants and animals, all belonged to one family of compounds. He called these the purines. All the purines have the same core structure and differ from one another by the functional groups attached to the ring.

N N N N H

Purine core structure

We will study the purines in Chapter 24 where we will learn that these are two of the essential components of the genetic molecules DNA and RNA. We will see that DNA is a double helix. Each strand of the helix is made up of a backbone of alternating sugars (ribose in RNA and deoxyribose in DNA) and phosphoryl groups. The purines are one of the two types of nitrogenous bases that we will study. They project into the helix. By reading the order of these nitrogenous bases, we can decipher the genetic code of an organism.

The structure of the purines and pyrimidines was first described in Section 16.2. It is discussed in much more detail in Sections 24.1 and 24.2.

In 1884 Fischer began his monumental work on sugars. In 1890 he established the stereochemical nature of all sugars, and between 1891 and 1894 he worked out the stereochemical configuration of all the known sugars and predicted all the possible stereoisomers. Stereochemistry is the study of molecules that have two mirror-image isomers. We will find, as Fischer did, that nature has "selected" only one of the two mirror-image forms for common use in biological systems. Fischer studied virtually all the sugars, but one of his greatest successes was the synthesis of glucose, fructose, and mannose, three six-carbon sugars. We will learn much more about the structure and function of carbohydrates, as well as about stereochemistry in this chapter.

The structure of simple sugars and the function in biological systems are discussed in Sections 17.2 and 17.4. Stereochemistry is discussed in Section 17.3.

Between 1899 and 1908, Fischer turned his attention to proteins. He developed methods to separate and identify individual amino acids and discovered an entirely new class, the cyclic amino acids (those with ring structures). Fischer also worked on protein synthesis from the amino acids. He demonstrated the nature of the peptide bond and discovered how to make small peptides in the laboratory.

As we will learn in Chapter 19, amino acids are all characterized by a common core structure, having both a carboxyl group and an amino group.

$$\begin{array}{c} \quad\ \ \text{H}\ \ \text{O} \\ \quad\ \ |\ \ \ \| \\ \text{H}_3{}^{+}\text{N}\text{—}\text{C}\text{—}\text{C}\text{—}\text{O}^{-} \\ \quad\ \ | \\ \quad\ \ \text{R} \end{array}$$

Amino acid core structure

We will study the amino acids and the structure of proteins in Chapter 19.

The peptide bond that forms between amino acids is actually an amide bond between these two groups.

See also Section 16.4, A Preview of Amino Acids, Proteins, and Protein Synthesis.

While each of the amino acids has a common core, they differ from one another in the nature of a side chain, or R group. The amino acids are classified based on the properties they acquire from these R groups.

Quite late in his career, Fischer even got around to studying the fats, a heterogenous group of substances characterized by their hydrophobic nature. In Chapter 18, we will study the incredibly diverse family of fats, or lipids.

Fischer's personal life was not as happy as his professional life. His wife died after only seven years of marriage, leaving Fischer with three sons. One son died in World War I and another committed suicide at the age of twenty-five. However, his third son, Hermann Otto Laurenz Fischer, followed in his father's footsteps, becoming a professor of biochemistry at the University of California at Berkeley.

In 1902, Fischer was awarded the Nobel Prize for his work on sugar and purine synthesis. But a glance at all of his accomplishments helps us to realize that this great man, who began his career as an organic chemist, established the field of biochemistry through his extraordinary studies of the molecules of life.

17.1 Types of Carbohydrates

Learning Goal

We begin our study of biochemistry with the carbohydrates. Carbohydrates are produced in plants by photosynthesis (Figure 17.1). Natural carbohydrate sources such as grains and cereals, breads, sugar cane, fruits, milk, and honey are an important source of energy for animals. **Carbohydrates,** especially glucose, are the

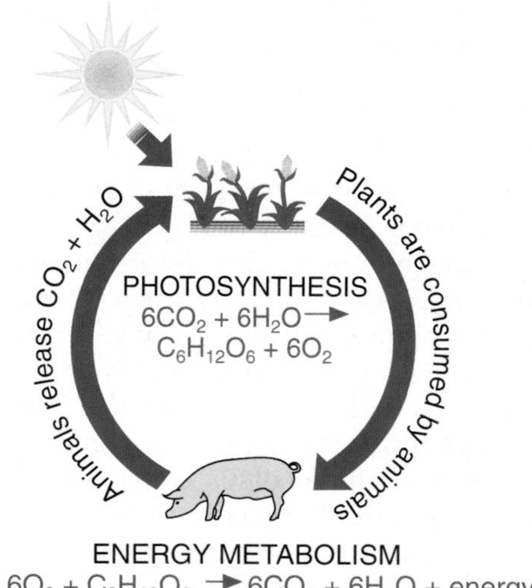

Figure 17.1
Carbohydrates are produced by the process of photosynthesis, which uses the energy of sunlight to produce hexoses from CO_2 and H_2O. The plants use these hexoses to harvest energy and produce ATP to fuel cellular function and to produce macromolecules, including starch, cellulose, fats, nucleic acids, and proteins. Animals depend on plants as a source of organic carbon. The hexoses are metabolized to generate ATP and are used as precursors for the biosynthesis of carbohydrates, fats, proteins, and nucleic acids.

A kilocalorie is the same as the calorie referred to in the "count-your-calories" books and on nutrition labels.

See A Human Perspective: Tooth Decay and Simple Sugars.

primary energy source for the brain and nervous system and can be used by many other tissues. When "burned" by cells for energy, each gram of carbohydrate releases approximately four kilocalories of energy.

A healthy diet should contain both complex carbohydrates, such as starches and cellulose, and simple sugars, such as fructose and sucrose. However, the quantity of simple sugars, especially sucrose, should be minimized because large quantities of sucrose in the diet promote obesity and tooth decay.

Complex carbohydrates are better for us than the simple sugars. Starch, found in rice, potatoes, breads, and cereals, is an excellent energy source. In addition, the complex carbohydrates, such as cellulose, provide us with an important supply of dietary fiber.

It is hard to determine exactly what percentage of the daily diet *should* consist of carbohydrates. The *actual* percentage varies widely throughout the world, from 80% in the Far East, where rice is the main component of the diet, to 40–50% in the United States. Currently, it is recommended that about 58% of the calories in the diet should come from carbohydrates and that no more than 10% of the daily caloric intake should be sucrose. The U.S. Department of Agriculture has adopted a food pyramid to show the recommended amounts of various foods in the diet (Figure 17.2). The foods at the bottom of the pyramid, grains (breads, cereals, rice, pasta), and at the next level, fruits and vegetables, should be the most abundant in our diet. These food groups are our major sources of dietary carbohydrates.

Question 17.1

What is the current recommendation for the amount of carbohydrates that should be included in the diet? Of the daily intake of carbohydrates, what percentage should be simple sugar?

Question 17.2

Distinguish between simple and complex sugars. What are some sources of complex carbohydrates?

Monosaccharides such as glucose and fructose are the simplest carbohydrates because they contain a single (*mono-*) sugar (*saccharide*) unit. **Disaccharides,** including sucrose and lactose, consist of two monosaccharide units joined through

A HUMAN PERSPECTIVE

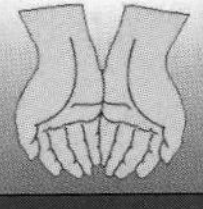

Tooth Decay and Simple Sugars

How many times have you heard the lecture from parents or your dentist about brushing your teeth after a sugary snack? Annoying as this lecture might be, it is based on sound scientific data that demonstrate that the cause of tooth decay is plaque and acid formed by the bacterium *Streptococcus mutans* using sucrose as its substrate.

Saliva is teeming with bacteria in concentrations up to one hundred million (10^8) per milliliter of saliva! Within minutes after you brush your teeth, sticky glycoproteins in the saliva adhere to tooth surfaces. Then millions of oral bacteria immediately bind to this surface.

Although all oral bacteria adhere to the tooth surface, only *S. mutans* causes dental caries, or cavities. Why does this bacterium cause cavities when all the others do not? The answer lies in the special enzyme called *glucosyl transferase* that is found on the surface of *S. mutans* cells.

Glucosyl transferase is a very specific enzyme. It can act only on the disaccharide sucrose, which it breaks down into glucose and fructose. As the accompanying diagram shows, the enzyme then adds the glucose to a growing polysaccharide chain called *dextran,* which adheres tightly to both the tooth enamel and the bacteria. *Plaque* is made up of huge masses of bacteria, embedded in dextran, adhering to the tooth surface.

This is just the first stage of cavity formation. Note in the accompanying figure that the second sugar released by the cleavage of sucrose is fructose. The bacteria use the fructose in the energy-harvesting pathways of glycolysis and lactic acid fermentation. Production of lactic acid decreases the pH on the tooth surface and begins to dissolve calcium from the tooth enamel.

Why is the acid not washed away from the tooth surface? After all, we produce about 1 L of saliva each day, which should dilute the acid and remove it from the tooth surface. The problem is the dextran plaque; it is not permeable to saliva, and thus plaque keeps the bacteria and the lactic acid localized on the enamel.

What measures can we take to prevent tooth decay? Practice good oral hygiene: brushing after each meal and flossing regularly reduce plaque buildup. Eat a diet rich in calcium; this helps to build strong tooth enamel. Include many complex carbohydrates in the diet; these cannot be used by glucosyl transferase and will not lead to the formation of acid. Further, the complex carbohydrates from fruits and vegetables help prevent decay by mechanically removing plaque from tooth surfaces. Avoid sucrose-containing snacks between meals. Studies have shown that the consumption of a sucrose-rich dessert with a meal, followed by brushing, does not produce many cavities. However, even small amounts of sugar ingested between meals are very cariogenic.

Researchers have developed a vaccine that prevents tooth decay in rats. Such a vaccine may one day be available for humans.

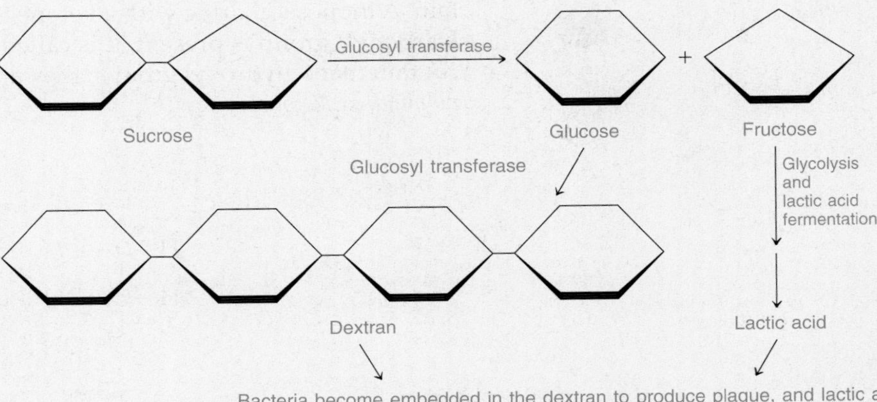

Action of the glucosyl transferase of *Streptococcus mutans*, which is responsible for tooth decay.

bridging oxygen atoms. Such a bond is called a **glycosidic bond. Oligosaccharides** consist of three to ten monosaccharide units joined by glycosidic bonds. The largest and most complex carbohydrates are the **polysaccharides,** which are long, often highly branched, chains of monosaccharides. Starch, glycogen, and cellulose are all examples of polysaccharides.

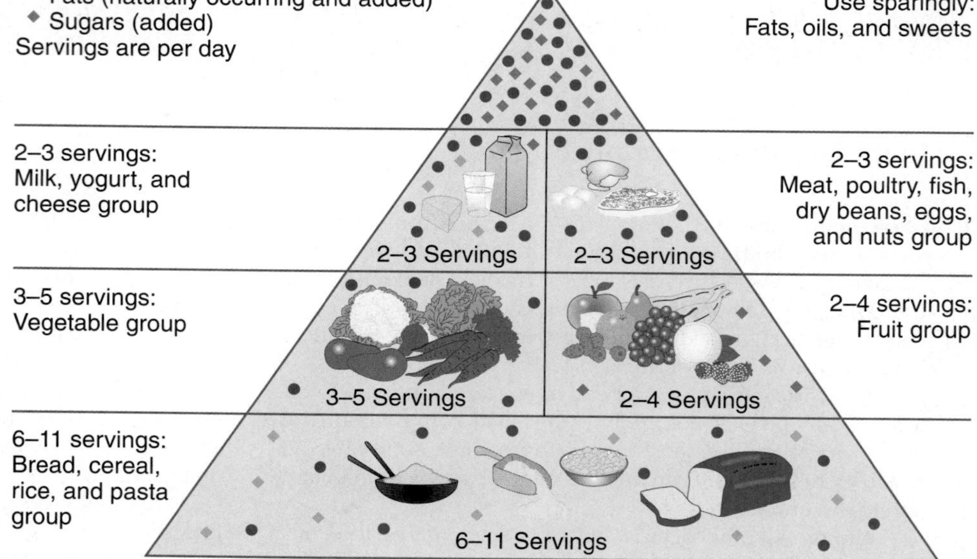

Figure 17.2
The U.S. Department of Agriculture has adopted a food pyramid to explain that carbohydrates in the form of cereals, grains, fruits, and vegetables should make up the majority of our diet. Fats and sweets should be consumed sparingly. Source: U.S. Department of Agriculture.

17.2 Monosaccharides

The importance of phosphorylated sugars in metabolic reactions is discussed in Sections 15.4 and 21.3.

Monosaccharides are composed of carbon, hydrogen, and oxygen, and most are characterized by the general formula $(CH_2O)_n$, in which n is any integer from 3 to 7. As we will see, this general formula is an oversimplification because several biologically important monosaccharides are chemically modified. For instance, several blood group antigen and bacterial cell wall monosaccharides are substituted with amino groups. Many of the intermediates in carbohydrate metabolism carry phosphate groups. Deoxyribose, the monosaccharide found in DNA, has one fewer oxygen atom than the general formula above would predict.

Learning Goal

Monosaccharides can be named on the basis of the functional groups they contain. A monosaccharide with a ketone (carbonyl) group is a **ketose.** If an aldehyde (carbonyl) group is present, it is called an **aldose.** Because monosaccharides also contain many hydroxyl groups, they are sometimes called *polyhydroxyaldehydes* or *polyhydroxyketones*.

```
Aldehyde            H                    CH2OH        Ketone
functional          |                    |            functional
group               C=O                  C=O          group
                    |                    |
                 H—C—OH               H—C—OH
                    |                    |
                 H—C—OH              HO—C—H
                    |                    |
                    CH2OH                CH2OH

               An aldose             A ketose
```

Another system of nomenclature tells us the number of carbon atoms in the main skeleton. A three-carbon monosaccharide is a ***tri*ose,** a four-carbon sugar is a ***tetr*ose,** a five-carbon sugar is a ***pent*ose,** a six-carbon sugar is a ***hex*ose,** and so on. Combining the two naming systems gives even more information about the structure and composition of a sugar. For example, an aldotetrose is a four-carbon sugar that is also an aldehyde.

In addition to these general names, each monosaccharide has a unique name. These names are shown in blue for the following structures. Because the monosaccharides can exist in several different isomeric forms, it is important to provide the complete name. Thus the complete names of the following structures are

D-glyceraldehyde, D-glucose, and D-fructose. These names tell us that the structure represents one particular sugar and also identifies the sugar as one of two possible isomeric forms (D- or L-).

H	H	CH_2OH
|	|	|
C=O	C=O	C=O
|	|	|
H—C—OH	H—C—OH	HO—C—H
|	|	|
CH_2OH	HO—C—H	H—C—OH
	|	|
	H—C—OH	H—C—OH
	|	|
	H—C—OH	CH_2OH
	|	
	CH_2OH	
Aldose	Aldose	Ketose
Triose	Hexose	Hexose
Aldotriose	Aldohexose	Ketohexose
D-Glyceraldehyde	D-Glucose	D-Fructose

Question 17.3

What is the structural difference between an aldose and a ketose?

Question 17.4

Explain the difference between:

a. A ketohexose and an aldohexose
b. A triose and a pentose

17.3 Stereoisomers and Stereochemistry

Stereoisomers

Learning Goal 3 Learning Goal 4 Learning Goal 5

The prefixes D- and L- found in the complete name of a monosaccharide are used to identify one of two possible isomeric forms called **stereoisomers.** By definition, each member of a pair of stereoisomers must have the same molecular formula and the same bonding. How then do isomers of the D-family differ from those of the L-family? D- and L-isomers differ in the spatial arrangements of atoms in the molecule.

Stereochemistry is the study of the different spatial arrangements of atoms. A general example of a pair of stereoisomers is shown in Figure 17.3. In this example the general molecule C-abcd is formed from the bonding of a central carbon to four different groups: a, b, c, and d. This results in two molecules rather than one. Each isomer is bonded together through the exact *same* bonding pattern, yet they are *not* identical. If they were identical, they would be superimposable one upon the other; *they are not.* They are therefore stereoisomers. These two stereoisomers have a mirror-image relationship that is analogous to the mirror-image relationship of the left and right hands (see Figure 17.3b).

Build models of these compounds using toothpicks and gumdrops of five different colors to prove this to yourself.

For a more detailed discussion of stereochemistry, see Appendix D, Stereochemistry and Stereoisomers Revisited.

Two stereoisomers that are nonsuperimposable mirror images of one another are called a pair of **enantiomers.** Molecules that can exist in enantiomeric forms are called **chiral molecules.** The term simply means that as a result of different three-dimensional arrangements of atoms, the molecule can exist in two mirror-image forms. For any pair of nonsuperimposable mirror-image forms (enantiomers), one is always designated D- and the other L-.

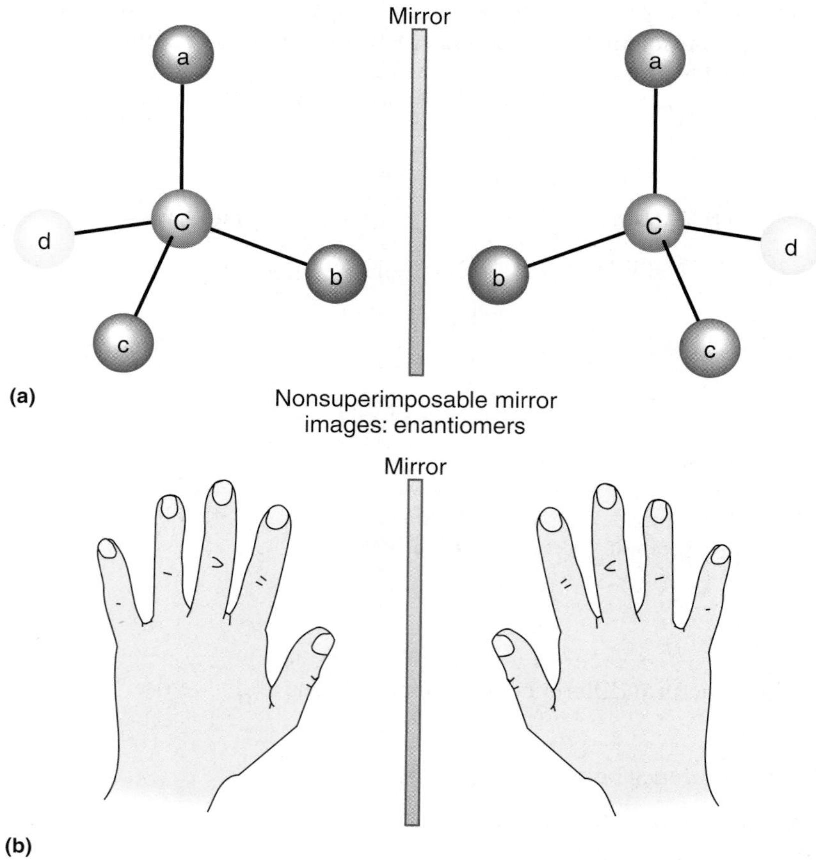

Figure 17.3
(a) A pair of enantiomers for the general molecule C-abcd. (b) Mirror-image right and left hands.

A carbon atom that has four different groups bonded to it is called an **asymmetric** (or chiral) **carbon** atom. Any molecule containing a chiral carbon can exist as a pair of enantiomers. Consider the simplest carbohydrate, **glyceraldehyde,** which is shown in Figure 17.4. Note that the second carbon is bonded to four different groups. It is therefore a chiral carbon. As a result, we can draw two enantiomers of glyceraldehyde that are nonsuperimposable mirror images of one another. Larger biological molecules typically have more than one chiral carbon.

Rotation of Plane-Polarized Light

Stereoisomers can be distinguished from one another by their different optical properties. Each member of a pair of stereoisomers will rotate plane-polarized light in different directions.

The polarimeter, measurement of the rotation of plane-polarized light, and the calculation of specific rotation are discussed in detail in Appendix D, Stereochemistry and Stereoisomers Revisited.

As we learned in Chapter 3, white light is a form of electromagnetic radiation that consists of many different wavelengths (colors) vibrating in *planes* that are all perpendicular to the direction of the light beam (Figure 17.5a). To measure optical properties of enantiomers, scientists use special light sources to produce *monochromatic light*, that is, light of a single wavelength. The monochromatic light is passed through a polarizing material, like a Polaroid lens, so that only waves in one plane can pass through. The light that emerges from the lens is *plane-polarized light* (Figure 17.5b).

Applying these principles, scientists have developed the *polarimeter* to measure the ability of a compound to change the angle of the plane of plane-polarized light (Figure 17.5c). The polarimeter allows the determination of the specific rotation of a compound, that is, the measure of its ability to rotate plane-polarized light.

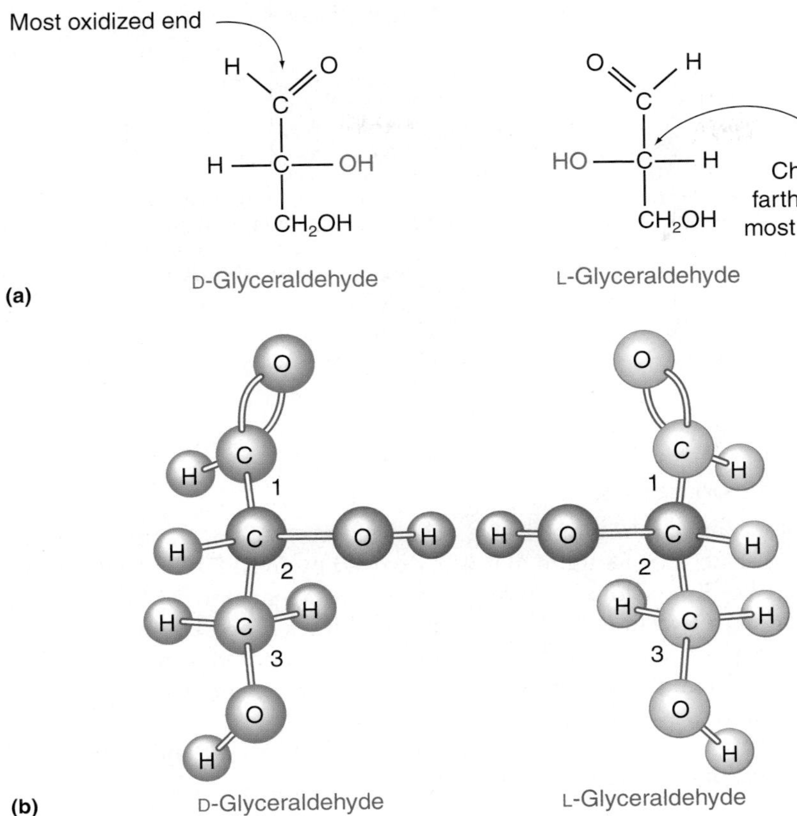

Figure 17.4
(a) Structural formulas of D- and L-glyceraldehyde. The end of the molecule with the carbonyl group is the most oxidized end. The D- or L-configuration of a monosaccharide is determined by the orientation of the functional groups attached to the chiral carbon farthest from the oxidized end. In the D-enantiomer the —OH is to the right. In the L-enantiomer, the —OH is to the left. (b) A three-dimensional representation of D- and L-glyceraldehyde.

Some compounds rotate light in a clockwise direction. These are said to be *dextrorotatory* and are designated by a plus sign (+) before the specific rotation value. Other substances rotate light in a counterclockwise direction. These are called *levorotatory* and are indicated by a minus sign (−) before the specific rotation value.

The Relationship between Molecular Structure and Optical Activity

In 1848, Louis Pasteur was the first to see a relationship between the structure of a compound and the effect of that compound on plane-polarized light. In his studies of winemaking, Pasteur noticed that salts of tartaric acid were formed as a by-product. It is a tribute to his extraordinary powers of observation that he noticed that two types of crystals were formed and that they were mirror images of one another. Using a magnifying glass and forceps, Pasteur separated the left-handed and right-handed crystals into separate piles. When he measured the optical activity of each of the mirror-image forms and of the original mixed sample, he obtained the following results:

- A solution of the original mixture of crystals was optically inactive.
- But both of the mirror-image crystals were optically active. In fact, the specific rotation produced by each was identical in magnitude but was of opposite sign.

Although Pasteur's work opened the door to understanding the relationship between structure and optical activity, it was not until 1874 that the Dutch chemist van't Hoff and the French chemist LeBel independently came up with a basis for the observed optical activity: tetrahedral carbon atoms bonded to four different atoms or groups of atoms. Thus, two enantiomers, which are identical to one another in all other chemical and physical properties, will rotate plane-polarized light to the same degree, but in opposite directions.

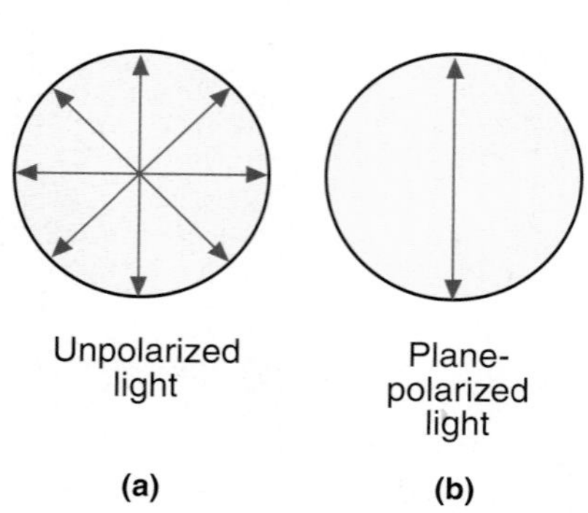

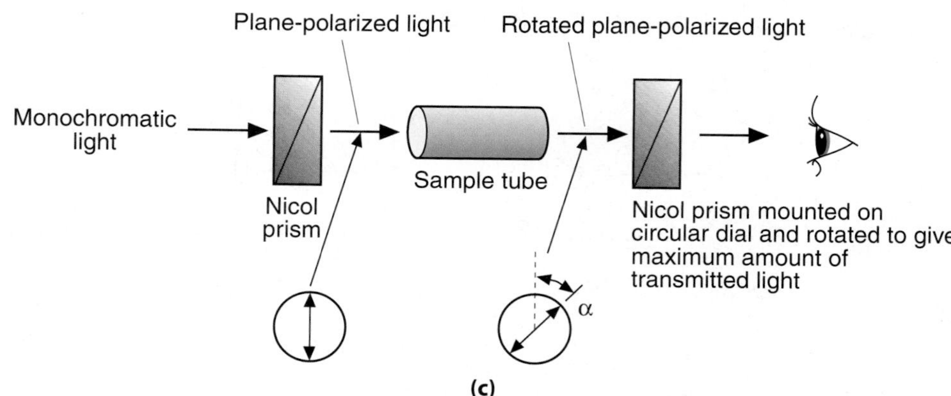

Figure 17.5
Light as wave motion. (a) Viewed from the end of the axis of propagation, light contains waves traveling in many planes. (b) Plane-polarized light contains light traveling in only one plane. (c) Schematic drawing of a polarimeter.

Fischer Projection Formulas

Let's take another look at the aldotriose glyceraldehyde to learn a convenient way to represent the structure of sugars that was devised by Emil Fischer. The **Fischer Projection** is a two-dimensional drawing of a molecule that shows a chiral carbon at the intersection of two lines and with horizontal lines representing bonds projecting out of the page and vertical lines representing bonds that project into the page. For sugars, the aldehyde or ketone group, the most oxidized carbon, is always represented at the "top."

Let's look again at Figure 17.4b. We can represent these ball-and-stick models with three-dimensional wedge drawings:

D-Glyceraldehyde	L-Glyceraldehyde
CHO H—C—OH CH_2OH	CHO HO—C—H CH_2OH

Remember that in the wedge diagram, the solid wedges represent bonds directed toward the reader. The dashed wedges represent bonds directed away from the reader and into the page. In these molecules, the center carbon in the representation is the only chiral carbon in the structure. To convert these wedge representations to a Fischer Projection, simply use a horizontal line in place of each solid wedge and use a vertical line to represent each dashed wedge. The chiral carbon is represented by the point at which the vertical and horizontal lines cross, as shown below.

D-Glyceraldehyde		L-Glyceraldehyde	
CHO H—C—OH CH_2OH	CHO H—┼—OH CH_2OH	CHO HO—C—H CH_2OH	CHO HO—┼—H CH_2OH

The D- and L- System of Nomenclature

In 1891 Emil Fischer devised a nomenclature system that would allow scientists to distinguish between enantiomers. Fischer knew that there are two enantiomers of glyceraldehyde that rotated plane-polarized light in opposite directions. He did not have the sophisticated tools needed to make an absolute connection between the structure and the direction of rotation of plane-polarized light. He simply decided that the (+) enantiomer would be the one with the hydroxyl group of the chiral carbon on the right:

```
      CHO
       |
  H ———+——— OH
       |
     CH2OH
```

D-Glyceraldehyde

It was not until 1952 that researchers were able to demonstrate that Fischer had guessed correctly when he proposed the structures of the (+) and (−) enantiomers of glyceraldehyde.

The enantiomer that rotated plane-polarized light in the (−) or levorotatory direction, he called L-glyceraldehyde (Figure 17.4).

While specific rotation is an experimental value that must be measured, the D- and L- designations of all other monosaccharides are determined by comparison of their structures with D- and L-glyceraldehyde. Sugars with more than three carbons will have more than one chiral carbon. *By convention*, it is the position of the hydroxyl group on the chiral carbon farthest from the carbonyl group (the most oxidized end of the molecule) that determines whether a monosaccharide is in the D- or L- configuration. If the —OH group is on the right, the molecule is in the D-configuration. If the —OH group is on the left, the molecule is in the L-configuration. Almost all carbohydrates in living systems are members of the D-family.

```
      H              H              CH2OH
      |              |                |
     C=O            C=O              C=O
      |              |                |
  H—C—OH         H—C—OH          HO—C—H
      |              |                |
    CH2OH        HO—C—H           H—C—OH
                     |                |
                 H—C—OH           H—C—OH
                     |                |
                 H—C—OH             CH2OH
                     |
                   CH2OH
```

D-Glyceraldehyde D-Glucose D-Fructose

The structures and designations of D- and L-glyceraldehyde are defined by convention. In fact, the D- and L- terminology is generally applied only to carbohydrates and amino acids. For organic molecules the D- and L- convention has been replaced by a new system that provides the absolute configuration of a chiral carbon. This system, called the (R) and (S) system, is described in Appendix D, Stereochemistry and Stereoisomers Revisited.

Question 17.5

Indicate whether each of the following molecules is an aldose or a ketose.

```
a.    CH3            b.    H             c.    CH2OH
      |                    |                   |
     C=O                  C=O                 C=O
      |                    |                   |
  H—C—OH              H—C—OH              HO—C—H
      |                    |                   |
    CH2OH             H—C—OH               H—C—OH
                           |                   |
                      HO—C—H               H—C—OH
                           |                   |
                         CH2OH               CH2OH

d.    H              e.    CH3           f.    H
      |                    |                   |
     C=O                  C=O                 C=O
      |                    |                   |
  H—C—OH              H—C—OH              HO—C—H
      |                    |                   |
    CH2OH             H—C—OH               H—C—OH
                           |                   |
                      H—C—OH              HO—C—H
                           |                   |
                         CH2OH               CH2OH
```

Question 17.6

Determine the configuration (D- or L-) for each of the molecules in Question 17.5.

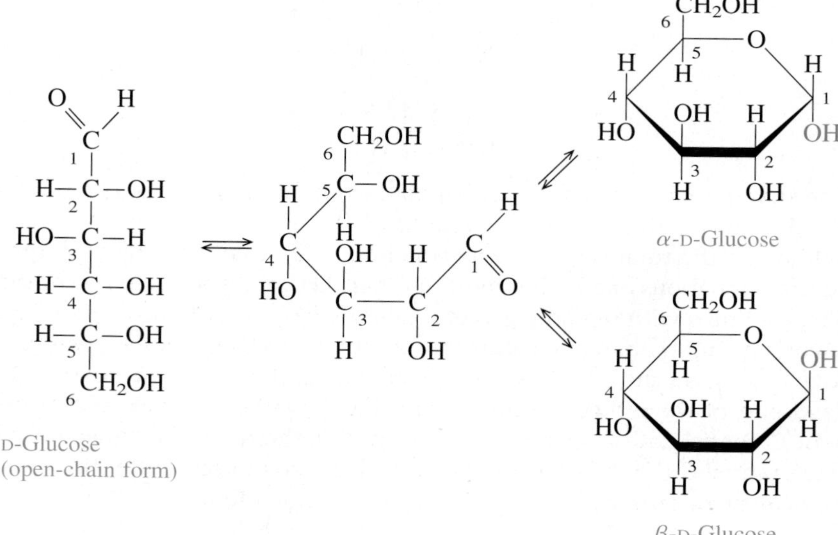

Figure 17.6
Cyclization of glucose to give α- and β-D-glucose. Note that the carbonyl carbon (C-1) becomes chiral in this process, yielding the α- and β- forms of glucose.

17.4 Biologically Important Monosaccharides

Learning Goal

Monosaccharides, the simplest carbohydrates, have backbones of from three to seven carbons. There are many monosaccharides, but we will focus on those that are most common in biological systems. These include the five- and six-carbon sugars: glucose, fructose, galactose, ribose, and deoxyribose.

Glucose

Glucose is the most important sugar in the human body. It is found in numerous foods and has several common names, including dextrose, grape sugar, and blood sugar. Glucose is broken down in glycolysis and other pathways to release energy for body functions.

The concentration of glucose in the blood is critical to normal body function. As a result, it is carefully controlled by the hormones insulin and glucagon. Normal blood glucose levels are 100–120 mg/100 mL, with the highest concentrations appearing after a meal. Insulin stimulates the uptake of the excess glucose by most cells of the body, and after 1–2 hr, levels return to normal. If blood glucose concentrations drop too low, the individual feels lightheaded and shaky. When this happens, glucagon stimulates the liver to release glucose into the blood, reestablishing normal levels. We will take a closer look at this delicate balancing act in Section 23.6.

The molecular formula of glucose, an aldohexose, is $C_6H_{12}O_6$. The structure of glucose is shown in Figure 17.6, and the method used to draw this structure is described in Example 17.1.

EXAMPLE 17.1 ***Drawing the Structure of a Monosaccharide***

Learning Goal

Draw the structure for D-glucose.

Solution

Glucose is an aldohexose.

Step 1. Draw six carbons in a straight vertical line; each carbon is separated from the ones above and below it by a bond:

Continued—

EXAMPLE 17.1 —*Continued*

```
1C
 |
2C
 |
3C
 |
4C
 |
5C
 |
6C
```

Step 2. The most highly oxidized carbon is, by convention, drawn as the uppermost carbon (carbon-1). In this case, carbon-1 is an aldehyde carbon:

```
      H
      |
1     C=O
      |
2   —C—
      |
3   —C—
      |
4   —C—
      |
5   —C—
      |
6   —C—
      |
```

Most oxidized end of carbon chain; aldehyde

Step 3. The atoms are added to the next to the last carbon atom, at the bottom of the chain, to give either the D- or L-configuration as desired. Remember, when the —OH group is to the right, you have D-glucose. When in doubt, compare your structure to D-glyceraldehyde!

```
      H
      |
      C=O
      |
    —C—
      |
    —C—                 H
      |                 |
    —C—                 C=O
      |                 |
  H—C—OH            H—C—OH
      |                 |
      CH2OH             CH2OH

  D-Isomer        D-Glyceraldehyde
```

Compare chiral carbons farthest from the carbonyl group

Step 4. All the remaining atoms are then added to give the desired carbohydrate. For example, one would draw the following structure for D-glucose.

Continued—

EXAMPLE 17.1 —*Continued*

```
      H
      |
      C=O
      |
   H—C—OH
      |
  HO—C—H
      |
   H—C—OH
      |
   H—C—OH
      |
      CH₂OH
```

D-Glucose

The positions for the hydrogen atoms and the hydroxyl groups on the remaining carbons must be learned for each sugar. For instance, the complete structures of D-fructose and D-galactose are shown later in this section.

Hemiacetal structure,

```
       OH
       |
  R¹—C—OR²
       |
       H
```

and formation are described in Section 14.4.

The term *intramolecular* tells us that the reacting carbonyl and hydroxyl groups are part of the same molecule.

In actuality the open-chain form of glucose is present in very small concentrations in cells. It exists in cyclic form under physiological conditions because the carbonyl group at C-1 of glucose reacts with the hydroxyl group at C-5 to give a six-membered ring. In the discussion of aldehydes, we noted that the reaction between an aldehyde and an alcohol yields a **hemiacetal.** When the aldehyde portion of the glucose molecule reacts with the C-5 hydroxyl group, the product is a cyclic *intramolecular hemiacetal*. For D-glucose, two isomers can be formed in this reaction (see Figure 17.6). These isomers are called α- and β-D-glucose. The two isomers formed differ from one another in the location of the —OH attached to the hemiacetal carbon, C-1. Such isomers, differing in the arrangement of bonds around the hemiacetal carbon, are called **anomers.** In the α-anomers, the C-1 (anomeric carbon) hydroxyl group is below the ring, and in the β-anomers, the C-1 hydroxyl group is above the ring. Like the stereoisomers discussed previously, the α and β forms can be distinguished from one another because they rotate plane-polarized light differently.

In Figure 17.6 a new type of structural formula, called a **Haworth projection,** is presented. Although on first inspection it appears complicated, it is quite simple to derive a Haworth projection from a structural formula, as Example 17.2 shows.

EXAMPLE 17.2 Drawing the Haworth Projection of a Monosaccharide from the Structural Formula

Learning Goal

Draw the Haworth projections of α- and β-D-glucose.

Solution

1. Before attempting to draw a Haworth projection, look at the first steps of ring formation shown here:

Continued—

EXAMPLE 17.2 —*Continued*

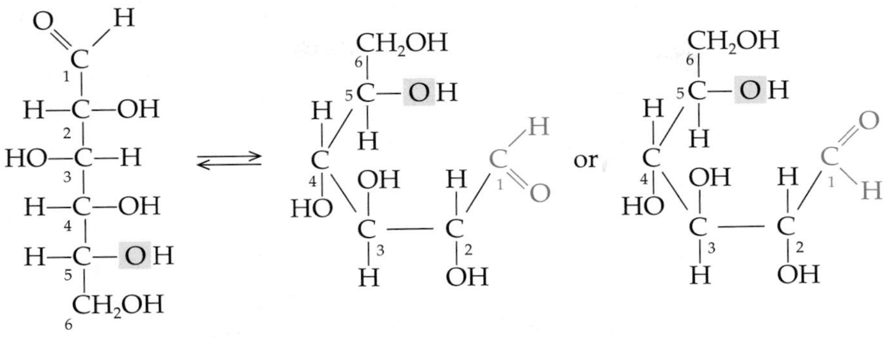

Glucose (open chain)

Glucose (intermediates in ring formation)

Try to imagine that you are seeing the molecules shown above in three dimensions. Some of the substituent groups on the molecule will be above the ring, and some will be beneath it. The question then becomes: How do you determine which groups to place above the ring and which to place beneath the ring?

2. Look at the two-dimensional structural formula. Note the groups (drawn in blue) to the left of the carbon chain. These are placed above the ring in the Haworth projection.

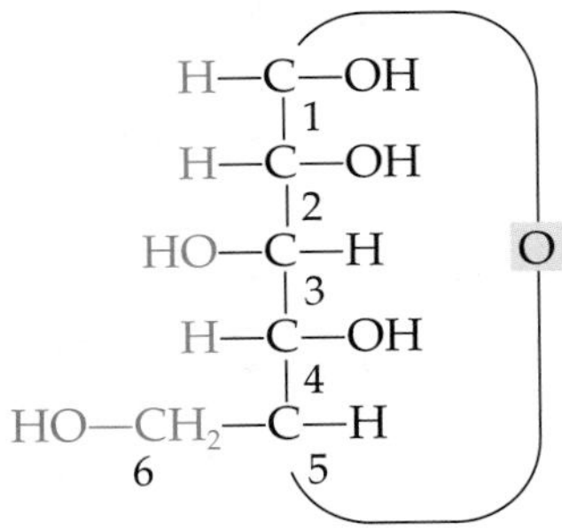

α-D-Glucose 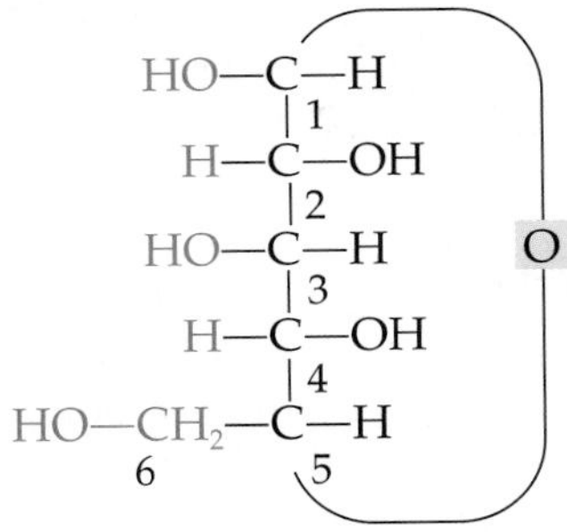 β-D-Glucose

3. Now note the groups (drawn in red) to the right of the carbon chain. These will be located beneath the carbon ring in the Haworth projection.

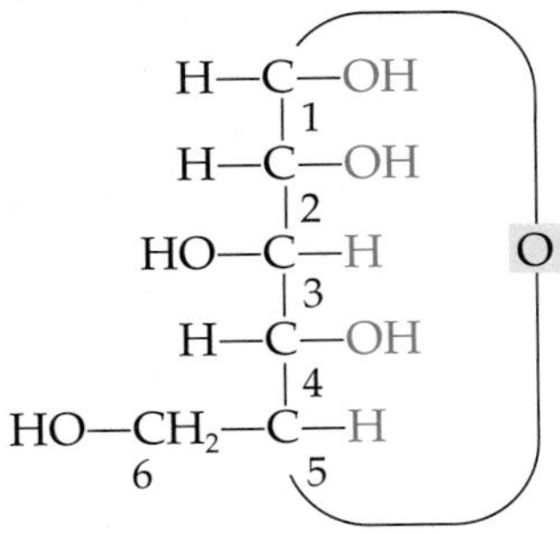

α-D-Glucose 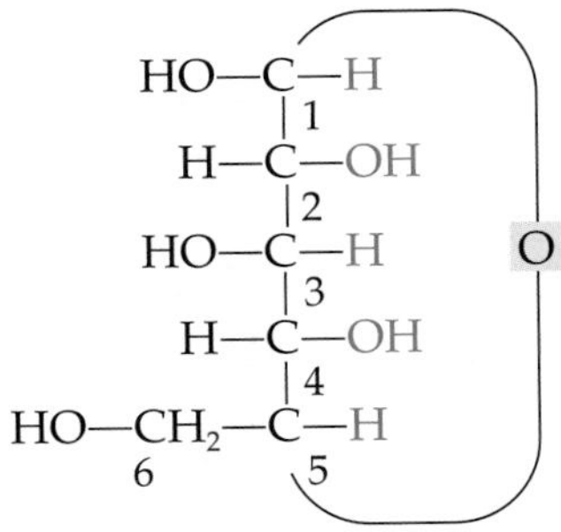 β-D-Glucose

Continued—

EXAMPLE 17.2 —*Continued*

4. Thus in the Haworth projection of the cyclic form of any D-sugar the —CH_2OH group is always "up." When the —OH group at C-1 is also "up," *cis* to the —CH_2OH group, the sugar is β-D-glucose. When the —OH group at C-1 is "down," *trans* to the —CH_2OH group, the sugar is α-D-glucose.

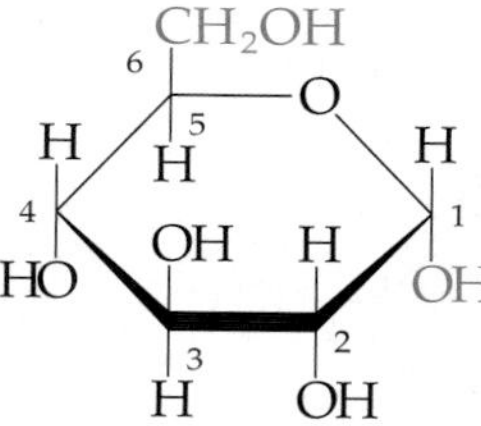

Haworth projection
α-D-Glucose

Haworth projection
β-D-Glucose

Question 17.7 Refer to the linear structure of D-galactose. Draw the Haworth projections of α- and β-D-galactose.

Question 17.8 Refer to the linear structure of D-ribose. Draw the Haworth projections of α- and β-D-ribose. Note that D-ribose is a pentose.

Fructose

Fructose, also called levulose and fruit sugar, is the sweetest of all sugars. It is found in large amounts in honey, corn syrup, and sweet fruits. The structure of fructose is similar to that of glucose. When there is a —CH_2OH group instead of a —CHO group at carbon-1 and a —C=O group instead of CHOH at carbon-2, the sugar is a ketose. In this case it is D-fructose.

Cyclization of fructose produces α- and β-D-fructose:

D-Fructose

α-D-Fructose

β-D-Fructose

Fructose is a ketose, or ketone sugar. Recall that the reaction between an alcohol and a ketone yields a **hemiketal.** Thus the reaction between the C-2 keto group and the C-5 hydroxyl group in the fructose molecule produces an *intramolecular hemiketal.* Fructose forms a five-membered ring structure.

Hemiketal formation is described in Section 14.4.

Galactose

Another important hexose is **galactose.** The linear structure of D-galactose and the Haworth projections of α-D-galactose and β-D-galactose are shown here:

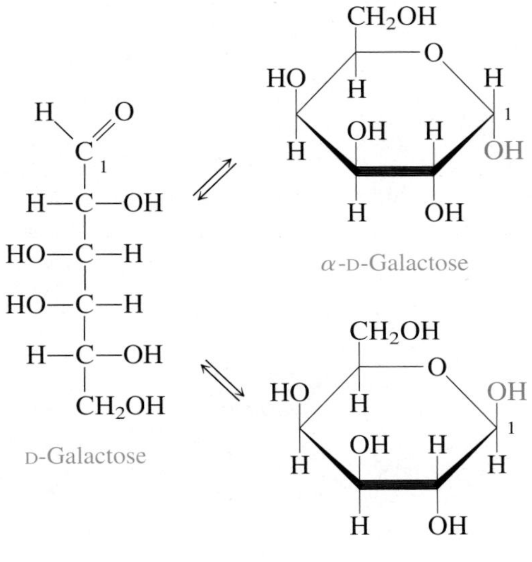

D-Galactose

α-D-Galactose

β-D-Galactose

Galactose is found in biological systems as a component of the disaccharide lactose, or milk sugar. This is the principal sugar found in the milk of most mammals. β-D-Galactose and a modified form, β-D-*N*-acetylgalactosamine, are also components of the blood group antigens.

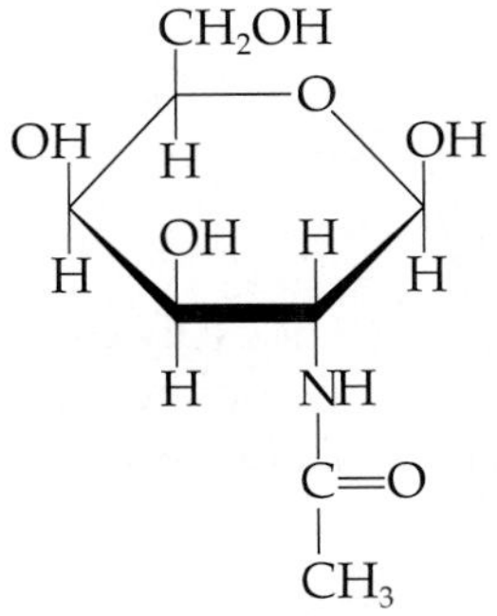

β-D-*N*-Acetylgalactosamine

Ribose and Deoxyribose, Five-Carbon Sugars

Ribose is a component of many biologically important molecules, including RNA, and various coenzymes, a group of compounds required by many of the enzymes that carry out biochemical reactions in the body. The structure of the five-carbon sugar D-ribose is shown in its open-chain form and in the α- and β-cyclic forms.

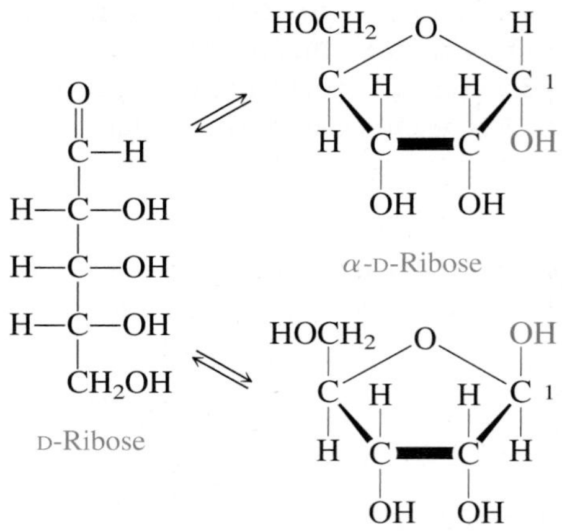

D-Ribose

α-D-Ribose

β-D-Ribose

DNA, the molecule that carries the genetic information of the cell, contains 2-deoxyribose. In this molecule the —OH group at C-2 has been replaced by a hydrogen, hence the designation "2-deoxy," indicating the absence of an oxygen.

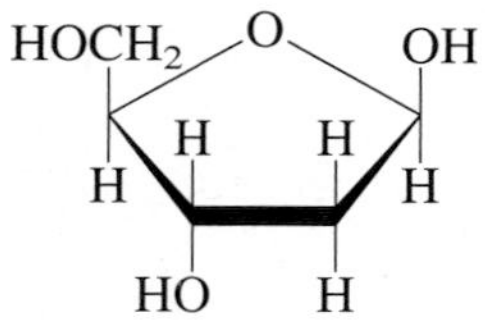

β-D-2-Deoxyribose

Reducing Sugars

Learning Goal 8

Learning Goal 9

The aldehyde group of aldoses is readily oxidized by the Benedict's reagent. Recall that the **Benedict's reagent** is a basic buffer solution that contains Cu^{2+} ions. The Cu^{2+} ions are reduced to Cu^{+} ions, which, in basic solution, precipitate as brick-red Cu_2O. The aldehyde group of the aldose is oxidized to a carboxylic acid, which undergoes an acid-base reaction to produce a carboxylate anion.

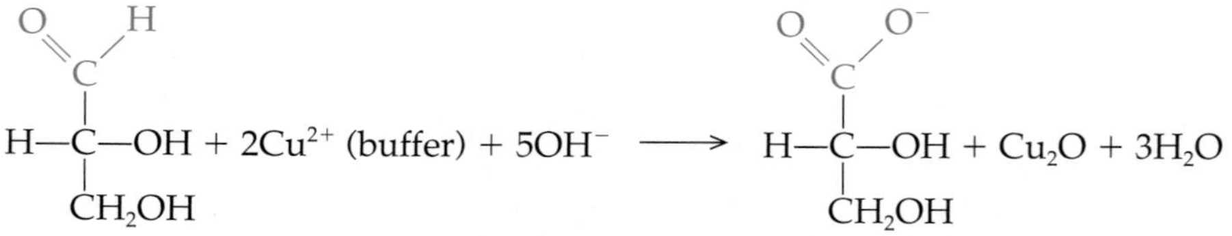

$$\text{HOCH}_2\text{CH(OH)CHO} + 2Cu^{2+} \text{ (buffer)} + 5OH^- \longrightarrow \text{HOCH}_2\text{CH(OH)COO}^- + Cu_2O + 3H_2O$$

Although ketones generally are not easily oxidized, ketoses are an exception to that rule. Because of the —OH group on the carbon next to the carbonyl group, ketoses can be converted to aldoses, under basic conditions, via an *enediol reaction:*

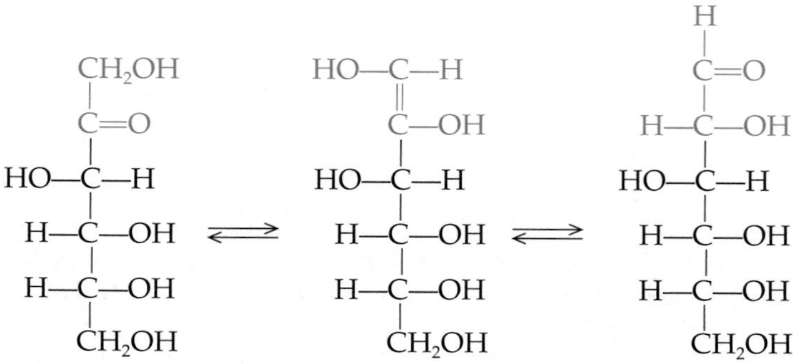

D-Fructose ⇌ Enediol ⇌ D-Glucose

The name of the enediol reaction is derived from the structure of the intermediate through which the ketose is converted to the aldose: It has a double bond (ene), and it has two hydroxyl groups (diol). Because of this enediol reaction, ketoses are also able to react with Benedict's reagent, which is basic. Because the metal ions in the solution are reduced, the sugars are serving as reducing agents and are called **reducing sugars.** All monosaccharides and all the common disaccharides, except sucrose, are reducing sugars.

For many years the Benedict's reagent was used to test for *glucosuria,* the presence of excess glucose in the urine. Individuals suffering from *Type I insulin-dependent diabetes mellitus* do not produce the hormone insulin, which controls the uptake of glucose from the blood. When the blood glucose level rises above 160 mg/100 mL, the kidney is unable to reabsorb the excess, and glucose is found in the urine. Although the level of blood glucose could be controlled by the injection of insulin, urine glucose levels were monitored to ensure that the amount of insulin injected was correct. The Benedict's reagent was a useful tool because the amount of Cu_2O formed, and hence the degree of color change in the reaction, is directly proportional to the amount of reducing sugar in the urine. A brick-red color indicates a very high concentration of glucose in the urine. Yellow, green, and blue-green solutions indicate decreasing amounts of glucose in the urine, and a blue solution indicates an insignificant concentration.

See A Clinical Perspective: Diabetes Mellitus and Ketone Bodies in Chapter 23.

Use of Benedict's reagent to test urine glucose levels has largely been replaced by chemical tests that provide more accurate results. The most common technology is based on a test strip that is impregnated with the enzyme glucose oxidase and other agents that will cause a measurable color change. In one such kit, the compounds that result in color development include the enzyme peroxidase, a compound called orthotolidine, and a yellow dye. When a drop of blood is placed on the strip, the glucose oxidase catalyzes the conversion of glucose into gluconic acid and hydrogen peroxide.

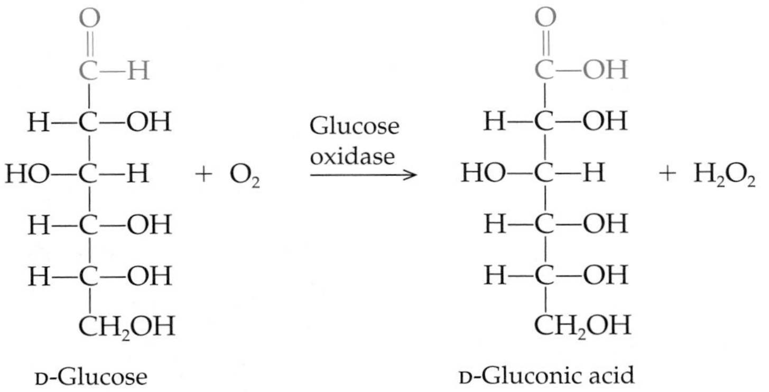

$$
\begin{array}{c}
\mathrm{O} \\
\| \\
\mathrm{C{-}H} \\
| \\
\mathrm{H{-}C{-}OH} \\
| \\
\mathrm{HO{-}C{-}H} \\
| \\
\mathrm{H{-}C{-}OH} \\
| \\
\mathrm{H{-}C{-}OH} \\
| \\
\mathrm{CH_2OH}
\end{array}
+ O_2 \xrightarrow[\text{oxidase}]{\text{Glucose}}
\begin{array}{c}
\mathrm{O} \\
\| \\
\mathrm{C{-}OH} \\
| \\
\mathrm{H{-}C{-}OH} \\
| \\
\mathrm{HO{-}C{-}H} \\
| \\
\mathrm{H{-}C{-}OH} \\
| \\
\mathrm{H{-}C{-}OH} \\
| \\
\mathrm{CH_2OH}
\end{array}
+ H_2O_2
$$

D-Glucose D-Gluconic acid

The enzyme peroxidase catalyzes a reaction between the hydrogen peroxide and orthotolidine. This produces a blue product. The yellow dye on the test strip simply serves to "dilute" the blue end product, thereby allowing greater accuracy of the test over a wider range of glucose concentrations. The test strip remains yellow if there is no glucose in the sample. It will vary from a pale green to a dark blue, depending on the concentration of glucose in the urine sample.

Frequently, doctors recommend that diabetics monitor their *blood* glucose levels multiple times each day because this provides a more accurate indication of how well the diabetic is controlling his or her diet. Many small, inexpensive glucose meters are available that couple the oxidation of glucose by glucose oxidase with an appropriate color change system. As with the urine test, the intensity of the color change is proportional to the amount of glucose in the blood. A photometer within the device reads the color change and displays the glucose concentration. An even newer technology uses a device that detects the electrical charge

Actually, glucose oxidase can only oxidize β-D-glucose. However, in the blood there is an equilibrium mixture of the α and β anomers of glucose. Fortunately, α-D-glucose is very quickly converted to β-D-glucose.

generated by the oxidation of glucose. In this case, it is the amount of electrical charge that is proportional to the glucose concentration.

17.5 Biologically Important Disaccharides

Learning Goal 10

Acetals and ketals are described in Section 14.4:

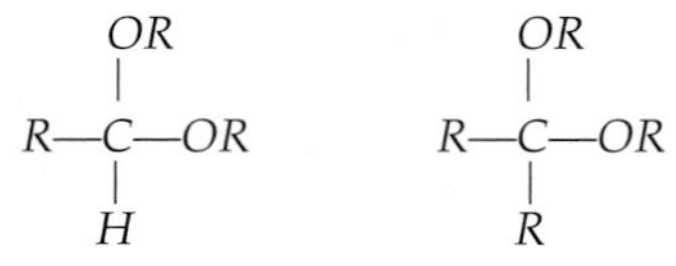

Acetal: R—C(OR)(H)—OR *Ketal*: R—C(OR)(R)—OR

Recall that disaccharides consist of two monosaccharides joined through an "oxygen bridge." In biological systems, monosaccharides exist in the cyclic form and, as we have seen, they are actually hemiacetals or hemiketals. Recall that when a hemiacetal reacts with an alcohol, the product is an *acetal,* and when a hemiketal reacts with an alcohol, the product is a *ketal.* In the case of disaccharides, the alcohol comes from a second monosaccharide. The acetals or ketals formed are given the general name *glycosides,* and the carbon-oxygen bonds are called *glycosidic bonds.*

Glycosidic bond formation is nonspecific; that is, it can occur between a hemiacetal or hemiketal and any of the hydroxyl groups on the second monosaccharide. However, in biological systems, we commonly see only particular disaccharides, such as maltose (Figure 17.7), lactose (see Figure 17.9), or sucrose (see Figure 17.10). These specific disaccharides are produced in cells because the reactions are catalyzed by enzymes. Each enzyme catalyzes the synthesis of one specific disaccharide, ensuring that one particular pair of hydroxyl groups on the reacting monosaccharides participates in glycosidic bond formation.

Maltose

If an α-D-glucose and a second glucose are linked, as shown in Figure 17.7, the disaccharide is **maltose,** or malt sugar. This is one of the intermediates in the hydrolysis of starch. Because the C-1 hydroxyl group of α-D-glucose is attached to C-4 of another glucose molecule, the disaccharide is linked by an α $(1 \rightarrow 4)$ glycosidic bond.

Maltose is a reducing sugar. Any disaccharide that has a hemiacetal hydroxyl group (a free —OH group at C-1) is a reducing sugar. This is because the cyclic structure can open at this position to form a free aldehyde. Disaccharides that do not contain a hemiacetal group on C-1 do not react with the Benedict's reagent and are called **nonreducing sugars.**

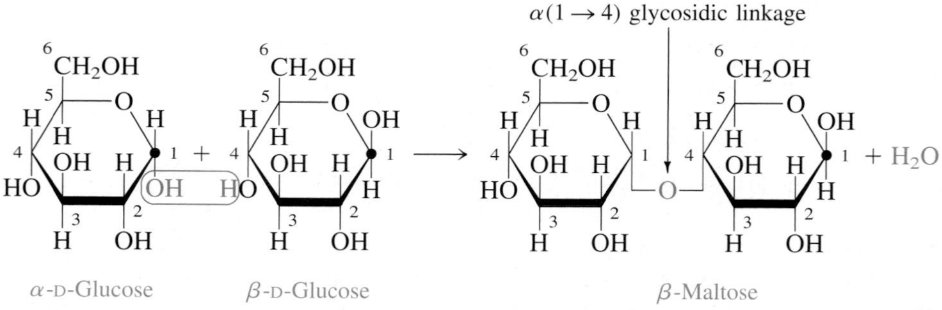

α-D-Glucose + β-D-Glucose → β-Maltose + H_2O [α(1 → 4) glycosidic linkage]

Figure 17.7
Glycosidic bond formed between the C-1 hydroxyl group of α-D-glucose and the C-4 hydroxyl group of β-D-glucose. The disaccharide is called β-maltose because the hydroxyl group at the reducing end of the disaccharide has the β-configuration.

β-D-Glucose

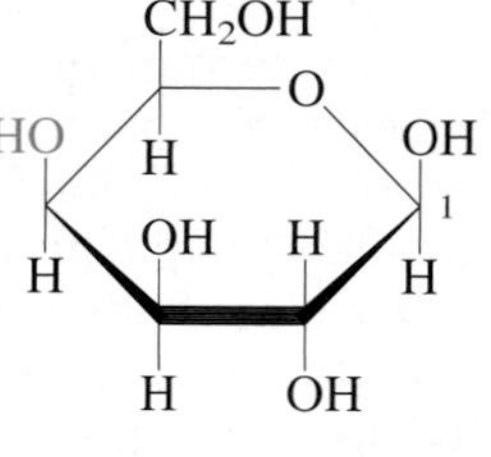
β-D-Galactose

Figure 17.8
Comparison of the cyclic forms of glucose and galactose. Note that galactose is identical to glucose except in the position of the C-4 hydroxyl group.

Lactose

Learning Goal

Milk sugar, or **lactose,** is a disaccharide made up of one molecule of β-D-galactose and one of either α- or β-D-glucose. Galactose differs from glucose only in the configuration of the hydroxyl group at C-4 (Figure 17.8). In the cyclic form of glucose the C-4 hydroxyl group is "down," and in galactose it is "up." In lactose the C-1 hydroxyl group of β-D-galactose is bonded to the C-4 hydroxyl group of either an α- or β-D-glucose. The bond between the two monosaccharides is therefore a β(1 → 4) glycosidic bond (Figure 17.9).

Lactose is the principal sugar in the milk of most mammals. To be used by the body as an energy source, lactose must be hydrolyzed to produce glucose and galactose. Note that this is simply the reverse of the reaction shown in Figure 17.9. Glucose liberated by the hydrolysis of lactose is used directly in the energy-harvesting reactions of glycolysis. However, a series of reactions is necessary to convert galactose into a phosphorylated form of glucose that can be used in cellular metabolic reactions. In humans the genetic disease **galactosemia** is caused by the absence of one or more of the enzymes needed for this conversion. A toxic compound formed from galactose accumulates in people who suffer from galactosemia. If the condition is not treated, galactosemia leads to severe mental retardation, cataracts, and early death. However, the effects of this disease can be avoided entirely by providing galactosemic infants with a diet that does not contain galactose. Such a diet, of course, cannot contain lactose and therefore must contain no milk or milk products.

Glycolysis is discussed in Chapter 21.

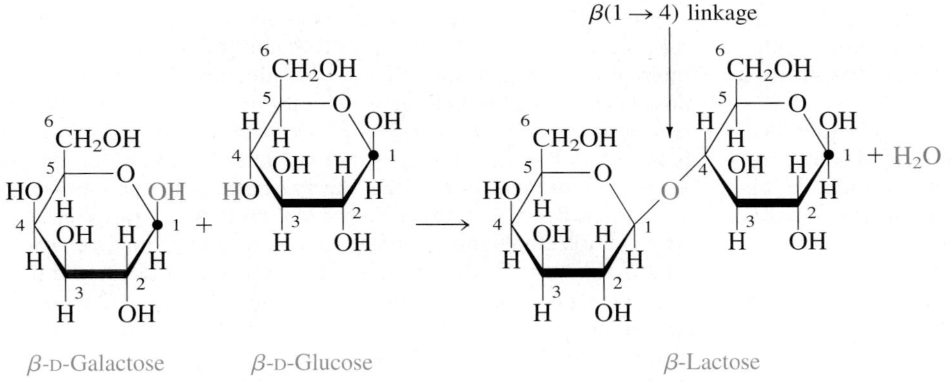

β(1 → 4) linkage

β-D-Galactose + β-D-Glucose → β-Lactose + H_2O

Figure 17.9
Glycosidic bond formed between the C-1 hydroxyl group of β-D-galactose and the C-4 hydroxyl group of β-D-glucose. The disaccharide is called β-lactose because the hydroxyl group at the reducing end of the disaccharide has the β-configuration.

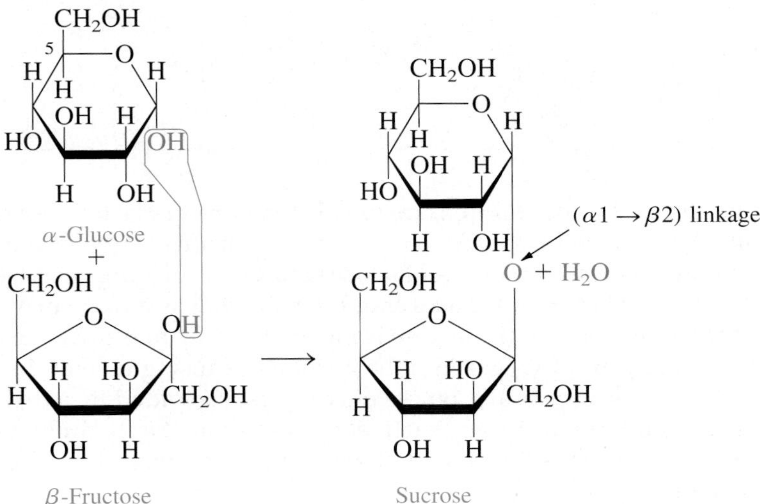

Figure 17.10
Glycosidic bond formed between the C-1 hydroxyl of α-D-glucose and the C-2 hydroxyl of β-D-fructose. This bond is called an (α1 → β2) glycosidic linkage. The disaccharide formed in this reaction is sucrose.

A HUMAN PERSPECTIVE

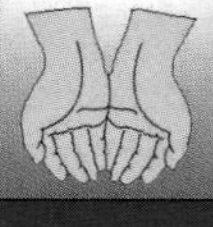

Blood Transfusions and the Blood Group Antigens

The first blood transfusions were tried in the seventeenth century, when physicians used animal blood to replace human blood lost by hemorrhages. Unfortunately, many people died as a result of this attempted cure, and transfusions were banned in much of Europe. Transfusions from human donors were somewhat less lethal, but violent reactions often led to the death of the recipient, and by the nineteenth century, transfusions had been abandoned as a medical failure.

In 1904, Dr. Karl Landsteiner performed a series of experiments on the blood of workers in his laboratory. His results explained the mysterious transfusion fatalities, and blood transfusions were reinstated as a lifesaving clinical tool. Landsteiner took blood samples from his coworkers. He separated the blood cells from the serum, the liquid component of the blood, and mixed these samples in test tubes. When he mixed serum from one individual with blood cells of another, Landsteiner observed that, in some instances, the serum samples caused clumping, or *agglutination,* of red blood cells (RBC). The agglutination reaction always indicated that the two bloods were incompatible and transfusion could lead to life-threatening reactions. As a result of many such experiments, Landsteiner showed that there are four human blood groups, designated A, B, AB, and O.

We now know that differences among blood groups reflect differences among oligosaccharides attached to the proteins and lipids of the RBC membranes. The oligosaccharides on the RBC surface have a common core, as shown in the accompanying figure, consisting of β-D-*N*-acetylgalactosamine, galactose, *N*-acetylneuraminic acid (sialic acid), and L-fucose. It is the terminal monosaccharide of this oligosaccharide that distinguishes the cells of different blood types and governs the compatibility of the blood types.

The A blood group antigen has β-D-*N*-acetylgalactosamine at its end, whereas the B blood group antigen has α-D-galactose. In type O blood, neither of these sugars is found on the cell surface; only the core oligosaccharide is present. Some of the oligosaccharides on type AB blood cells have a terminal β-D-*N*-acetylgalactosamine, whereas others have a terminal α-D-galactose.

Why does agglutination occur? The clumping reaction that occurs when incompatible bloods are mixed is an antigen-antibody reaction. Antigens are large molecules, often portions of bacteria or viruses, that stimulate the immune defenses of the body to produce protective antibodies. Antibodies bind to the foreign antigens and help to destroy them.

People with type A blood also have antibodies against type B blood (anti-B antibodies) in the blood serum. If the person with type A blood receives a transfusion of type B blood, the anti-B antibodies bind to the type B blood cells, causing clumping and destruction of those cells that can result in death. Individuals with type B blood also produce anti-A antibodies and therefore cannot receive a transfusion from a type A individual. Those with type AB blood are considered to be *universal recipients* because they have neither anti-A nor anti-B antibodies in their blood. (If they did, they would destroy their own red blood cells!) Thus in emergency situations a patient with type AB blood can receive blood from an individual of any blood type without serious transfusion reactions. Type O blood has no A or B antigens on the RBC but has both anti-A and anti-B antibodies. Because of the presence of both types of antibodies, type O individuals can receive transfusions only from a person who is also type O. On the other hand, the absence of A and B antigens on the red blood cell surface means that type O blood can be safely transfused into patients of any blood type. Hence type O individuals are *universal donors*.

Many adults, and some children, are unable to hydrolyze lactose because they do not make the enzyme *lactase*. This condition, which affects 20% of the population of the United States, is known as **lactose intolerance.** Undigested lactose remains in the intestinal tract and causes cramping and diarrhea that can eventually lead to dehydration. Some of the lactose is metabolized by intestinal bacteria that release organic acids and CO_2 gas into the intestines, causing further discomfort. Lactose intolerance is unpleasant, but its effects can be avoided by a diet that excludes milk and milk products. Alternatively, the enzyme that hydrolyzes lactose is available in tablet form. When ingested with dairy products it breaks down the lactose, preventing symptoms.

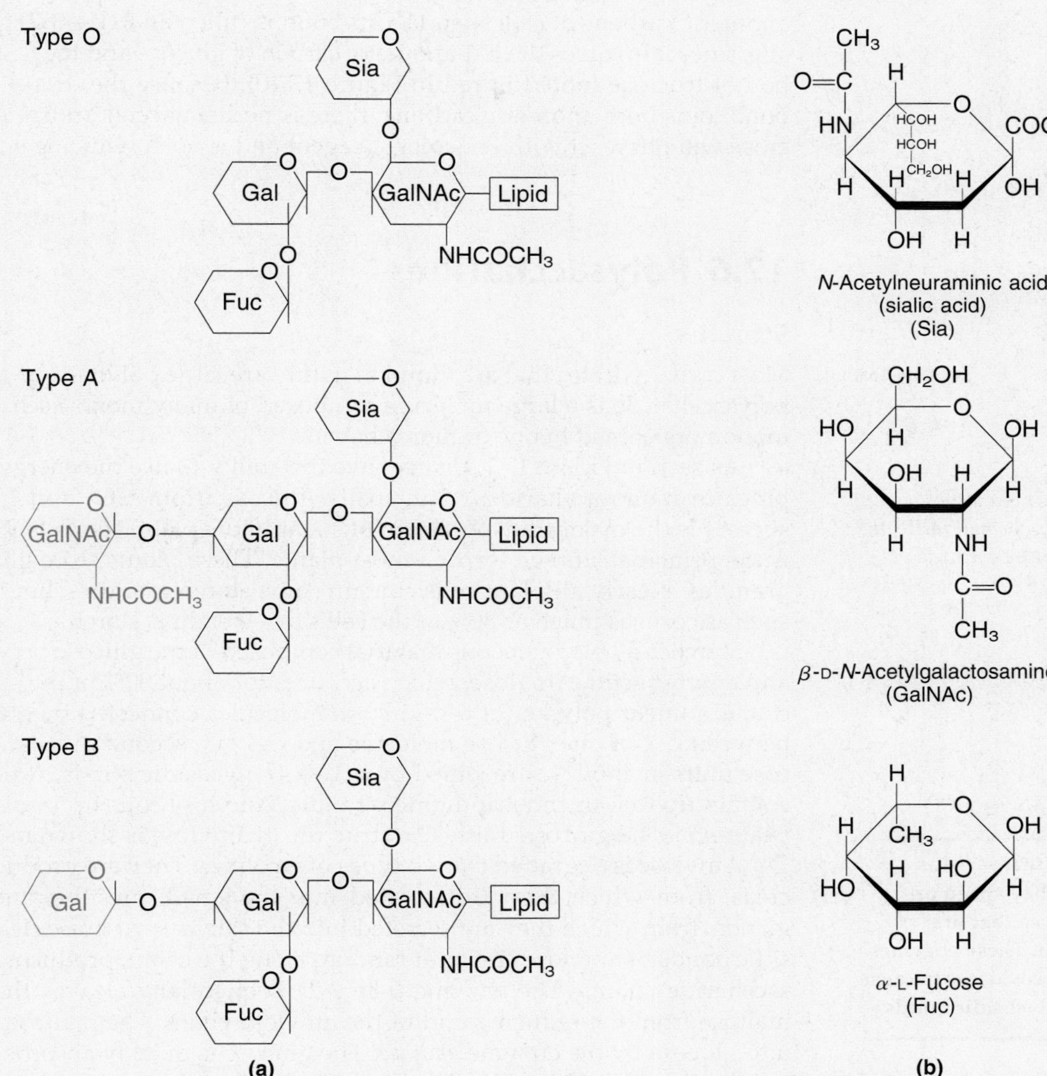

Schematic diagram of the blood group oligosaccharides. (a) Only the core oligosaccharide is found on the surface of type O red blood cells. On type A red blood cells, β-D-*N*-acetylgalactosamine is linked to the galactose (Gal) of the core oligosaccharide. On type B red blood cells, a galactose molecule is found attached to the galactose of the core oligosaccharide. (b) The structures of some of the unusual monosaccharides found in the blood group oligosaccharides.

Sucrose

Sucrose is also called table sugar, cane sugar, or beet sugar. Sucrose is an important carbohydrate in plants. It is water soluble and can easily be transported through the circulatory system of the plant. It cannot be synthesized by animals. High concentrations of sucrose produce a high osmotic pressure, which inhibits the growth of microorganisms, so it is used as a preservative. Of course, it is also widely used as a sweetener. In fact, it is estimated that the average American consumes 100–125 pounds of sucrose each year. It has been suggested that sucrose in the diet is undesirable because it represents a source of empty calories; that is, it contains no vitamins or minerals. However, the only negative association that has been scientifically verified is the link between sucrose in the diet and dental caries, or cavities (see A Human Perspective: Tooth Decay and Simple Sugars on p. 493).

Sucrose is a disaccharide of α-D-glucose joined to β-D-fructose (Figure 17.10). The glycosidic linkage between α-D-glucose and β-D-fructose is quite different from those that we have examined for lactose and maltose. This bond involves the anomeric carbons of *both* sugars! This bond is called an (α1 → β2) glycosidic linkage, since it involves the C-1 anomeric carbon of glucose and the C-2 anomeric carbon of fructose (noted in red in Figure 17.10). Because the (α1 → β2) glycosidic bond joins both anomeric carbons, there is no hemiacetal group. As a result, sucrose will not react with Benedict's reagent and is not a reducing sugar.

17.6 Polysaccharides

Starch

Learning Goal

A polymer (Section 12.5) is a large molecule made up of many small units, the monomers, held together by chemical bonds.

Most carbohydrates that are found in nature are large polymers of glucose. Thus a polysaccharide is a large molecule composed of many monosaccharide units (the monomers) joined in one or more chains.

As seen in Figure 17.1, plants have the ability to use the energy of sunlight to produce monosaccharides, principally glucose, from CO_2 and H_2O. Although sucrose is the major transport form of sugar in the plant, starch (a polysaccharide) is the principal storage form in most plants. These plants store glucose in starch granules. Nearly all plant cells contain some starch granules, but in some seeds, such as corn, as much as 80% of the cell's dry weight is starch.

Starch is a heterogeneous material composed of the glucose polymers **amylose** and **amylopectin.** Amylose, which accounts for about 80% of the starch of a plant cell, is a linear polymer of α-D-glucose molecules connected by glycosidic bonds between C-1 of one glucose molecule and C-4 of a second glucose. Thus the glucose units in amylose are joined by α(1 → 4) glycosidic bonds. A single chain can contain up to four thousand glucose units. Amylose coils up into a helix that repeats every six glucose units. The structure of amylose is shown in Figure 17.11.

Enzymes are proteins that serve as biological catalysts. They speed up biochemical reactions so that life processes can function. These enzymes are called α(1 → 4) glycosidases because they cleave α(1 → 4) glycosidic bonds.

Amylose is degraded by two types of enzymes. They are produced in the pancreas, from which they are secreted into the small intestine, and the salivary glands, from which they are secreted into the saliva. α-*Amylase* cleaves the glycosidic bonds of amylose chains at random along the chain, producing shorter polysaccharide chains. The enzyme β-*amylase* sequentially cleaves the disaccharide maltose from the reducing end of the amylose chain. The maltose is hydrolyzed into glucose by the enzyme *maltase.* The glucose is quickly absorbed by intestinal cells and used by the cells of the body as a source of energy.

Amylopectin is a highly branched amylose in which the branches are attached to the C-6 hydroxyl groups by α(1 → 6) glycosidic bonds (Figure 17.12). The main chains consist of α(1 → 4) glycosidic bonds. Each branch contains 20–25 glucose units, and there are so many branches that the main chain can scarcely be distinguished.

Glycogen

Glycogen is the major glucose storage molecule in animals. The structure of glycogen is similar to that of amylopectin. The "main chain" is linked by α(1 → 4) glycosidic bonds, and it has numerous α(1 → 6) glycosidic bonds, which provide many branch points along the chain. Glycogen differs from amylopectin only by having more and shorter branches. Otherwise, the two molecules are virtually identical. The structure of glycogen is shown in Figure 17.12.

(a) (b)

Figure 17.11
Structure of amylose. (a) A linear chain of α-D-glucose joined in α(1 → 4) glycosidic linkage makes up the primary structure of amylose. (b) Owing to hydrogen bonding, the amylose chain forms a left-handed helix that contains six glucose units per turn.

(a)

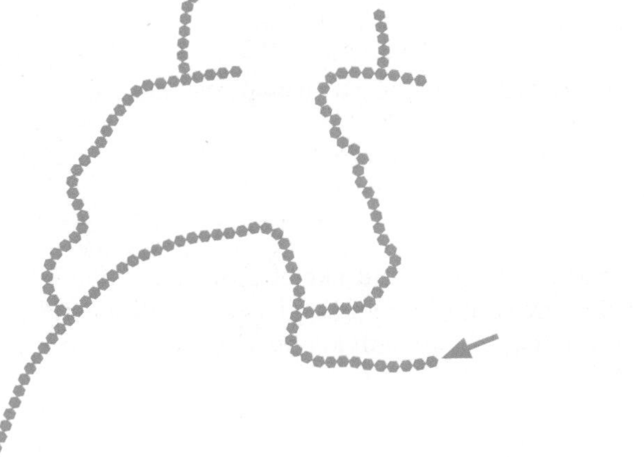

(b)

(c)

Figure 17.12
Structure of amylopectin and glycogen. (a) Both amylopectin and glycogen consist of chains of α-D-glucose molecules joined in α(1 → 4) glycosidic linkages. Branching from these chains are other chains of the same structure. Branching occurs by formation of α(1 → 6) glycosidic bonds between glucose units. (b) A representation of the branched-chain structure of amylopectin. (c) A representation of the branched-chain structure of glycogen. Glycogen differs from amylopectin only in that the branches are shorter and there are more of them.

A CLINICAL PERSPECTIVE

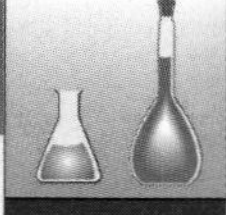

The Bacterial Cell Wall

The major component of bacterial cell walls is a complex polysaccharide known as a *peptidoglycan*. The name tells us that this structure consists of sugar molecules (-glycan) and peptides (peptido-; short polymers of amino acids).

As the accompanying structure shows, the carbohydrate portion of the peptidoglycan is a polymer of alternating units of two modified glucose molecules called *N-acetylglucosamine* and *N-acetylmuramic acid*. These two unusual monosaccharides

N-acetylmuramic acid *N*-acetylglucosamine

L-Alanine

D-Glutamic acid

L-Lysine

D-Alanine

Connected to pentapeptide interbridge

Structures of *N*-acetylglucosamine and *N*-acetylmuramic acid in β(1 → 4) glycosidic linkage. Note the tetrapeptide bridge linked to the *N*-acetylmuramic acid.

Glycogen is stored in the liver and skeletal muscle. Glycogen synthesis and degradation in the liver are carefully regulated. As we will see in Section 21.7, these two processes are intimately involved in keeping blood glucose levels constant.

Cellulose

The most abundant polysaccharide, indeed the most abundant organic molecule in the world, is **cellulose,** a polymer of β-D-glucose units linked by β(1 → 4) glycosidic bonds (Figure 17.13). A molecule of cellulose typically contains about 3000 glucose

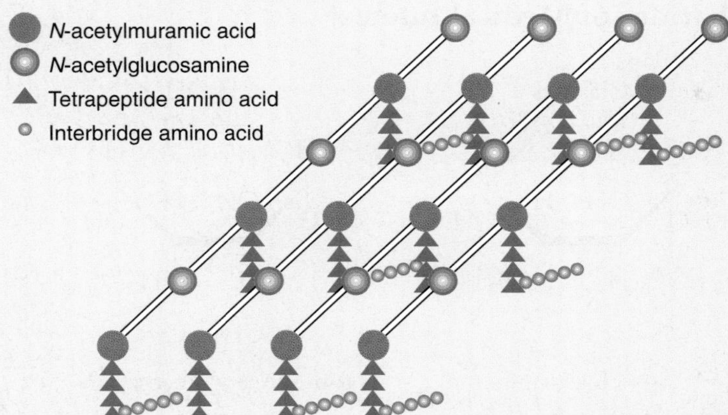

The three-dimensional structure of one layer of peptidoglycan.

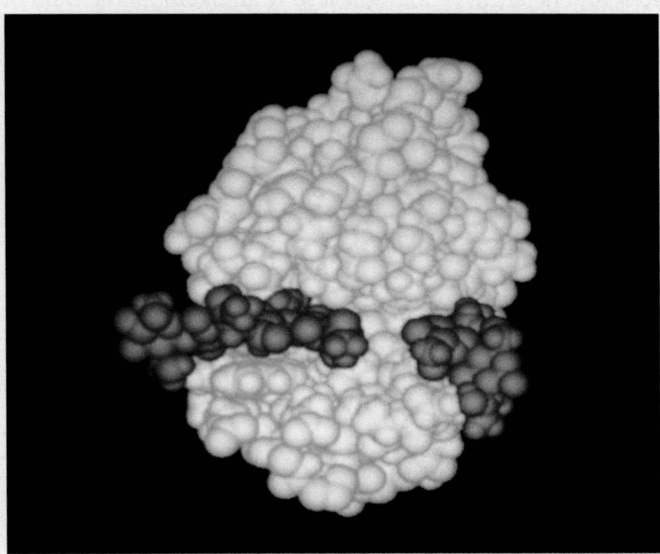

Conformation of lysozyme bound to its substrate. The enzyme binds with a six-sugar portion of the bacterial cell wall and cleaves it. The substrate fits into a deep crevice on the surface of the enzyme.

are joined by a $\beta(1 \rightarrow 4)$ glycosidic bond. In addition, each N-acetylmuramic acid is bonded to a tetrapeptide, a chain of four amino acids.

The structural strength of the cell wall is a result of pentapeptide cross-bridges that link the repeat units to one another (see the figure at the left). Millions of such cross-linkages produce an enormous peptidoglycan molecule, dozens of layers thick, around the bacterium. This thick wall is very rigid. It allows the bacterium to maintain its shape and protects it from bursting if the salt concentration of the environment is too low (hypotonic conditions).

Our bodies are constantly being assaulted by a variety of bacteria, and as you might expect, we have evolved protective mechanisms to minimize the damage. For instance, the enzyme *lysozyme,* found in tears and saliva, catalyzes the hydrolysis of the $\beta(1 \rightarrow 4)$ glycosidic bonds of peptidoglycan. As the accompanying figure shows, the enzyme has a deep groove on the surface (the active site) that a six-sugar unit of the cell wall can slip into like a bank card into the slot of an automatic teller machine. Lysozyme then catalyzes bond breakage and destroys the cell wall of the bacterium.

The penicillins are antibiotics that interfere with bacterial cell wall synthesis. The human body has no structures similar to the bacterial cell wall, so treatment with penicillins selectively destroys the bacteria, causing no harm to the patient. In practice, however, it must always be remembered that some individuals may develop an allergy to penicillins.

Penicillin G

The penicillins inhibit the enzyme that catalyzes the formation of the cross-linkage between the tetrapeptides. The antibiotic binds irreversibly to the active site of that enzyme so that it cannot bind to the tetrapeptide tail. Thus no cross-linkage can be made. Without the rigid, highly cross-linked peptidoglycan, the bacterial cells rupture and die.

units, but the largest known cellulose, produced by the alga *Valonia,* contains 26,000 glucose molecules.

Cellulose is a structural component of the plant cell wall. The unbranched structure of the cellulose polymer and the $\beta(1 \rightarrow 4)$ glycosidic linkages allow cellulose molecules to form long, straight chains of parallel cellulose molecules called *fibrils.* These fibrils are quite rigid and are held together tightly by hydrogen bonds; thus it is not surprising that cellulose is a cell wall structural element.

In contrast to glycogen, amylose, and amylopectin, cellulose *cannot* be digested by humans. The reason is that we cannot synthesize the enzyme *cellulase,* which can

A MEDICAL PERSPECTIVE

Monosaccharide Derivatives and Heteropolysaccharides of Medical Interest

Many of the carbohydrates with important functions in the human body are either derivatives of simple monosaccharides or are complex polymers of monosaccharide derivatives. One type of monosaccharide derivatives, the uronates, is formed when the terminal—CH_2OH group of a monosaccharide is oxidized to a carboxylate group. α-D-Glucuronate is a uronate of glucose:

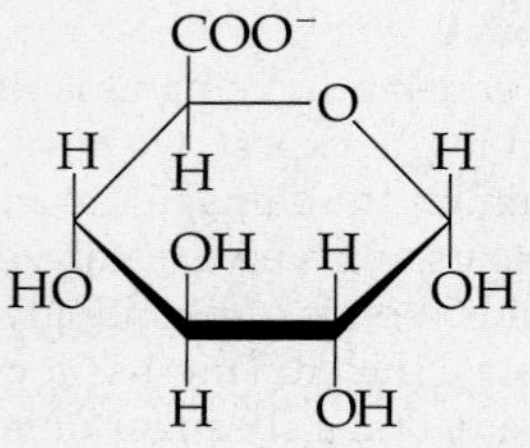

α-D-Glucuronate

In liver cells, α-D-glucuronate is bonded to hydrophobic molecules, such as steroids, to increase their solubility in water. When bonded to the modified sugar, steroids are more readily removed from the body.

Amino sugars are a second important group of monosaccharide derivatives. In amino sugars one of the hydroxyl groups (usually on carbon-2) is replaced by an amino group. Often these are found in complex oligosaccharides that are attached to cellular proteins and lipids. The most common amino sugars, D-glucosamine and D-galactosamine, are often found in the *N*-acetyl form. *N*-acetylglucosamine (see A Clinical Perspective: The Bacterial Cell Wall on pp. 514–515) is a component of bacterial cell walls and *N*-acetylgalactosamine is a component of the A, B, O blood group antigens (see preceding, A Human Perspective: Blood Transfusions and the Blood Group Antigens).

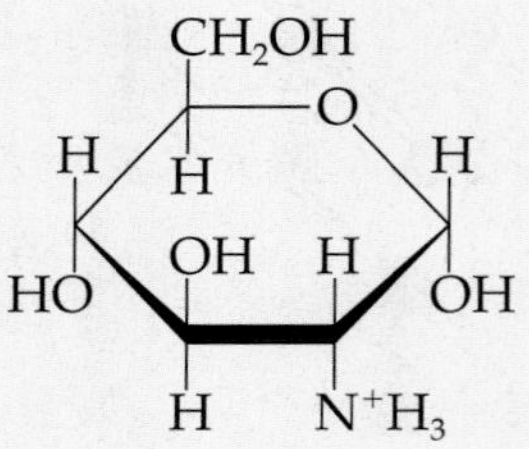

α-D-Glucosamine

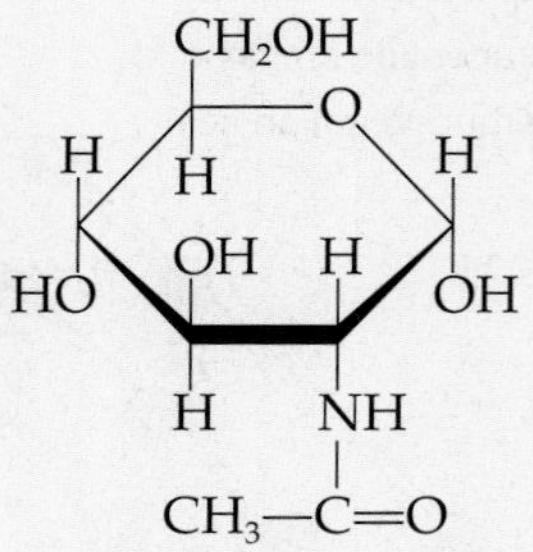

α-D-*N*-Acetylglucosamine

Heteropolysaccharides are long-chain polymers that contain more than one type of monosaccharide, many of which are amino sugars. As a result, they are often referred to as *glycosaminoglycans,* which include chondroitin sulfate, hyaluronic acid, and heparin. Hyaluronic acid is abundant in the fluid of joints and in the vitreous humor of the eye. Chondroitin sulfate is an important component of cartilage; and heparin has anticoagulant function. The structures of the repeat units of these polymers are shown below.

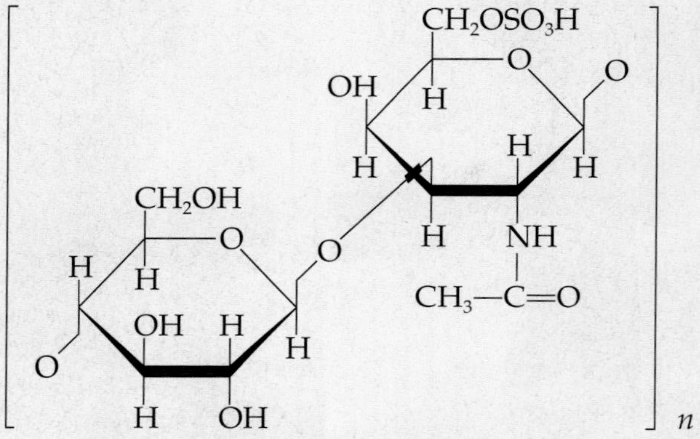

Repeat unit of chondroitin sulfate

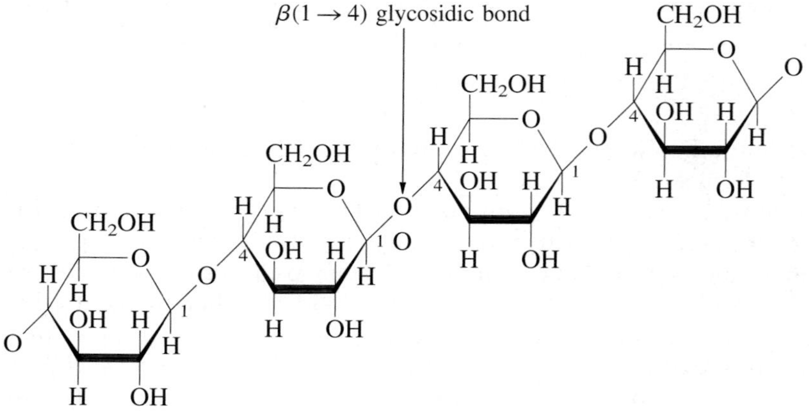

$\beta(1 \rightarrow 4)$ glycosidic bond

Figure 17.13
The structure of cellulose.

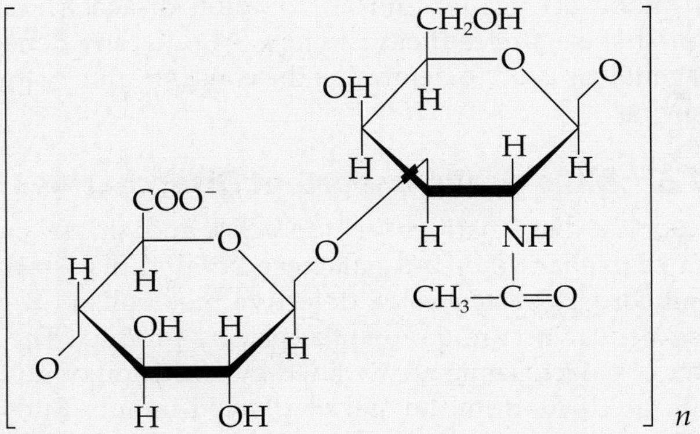

Repeat unit of hyaluronic acid

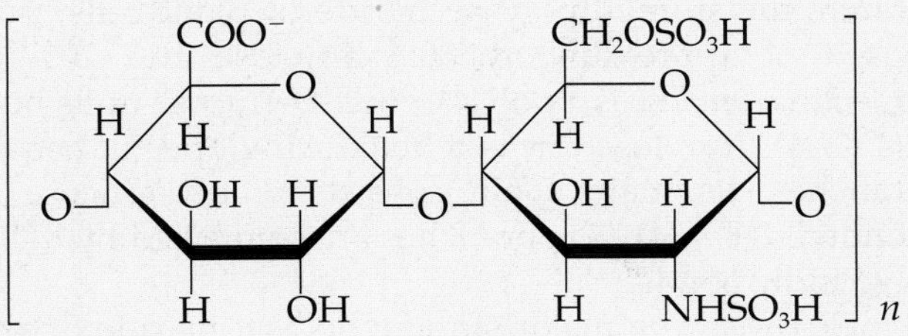

Repeat unit of heparin

Two of these molecules have been studied as potential treatments for osteoarthritis, a painful, degenerative disease of the joints. The amino sugar D-glucosamine is thought to stimulate the production of collagen. Collagen is one of the main components of articular cartilage, which is the shock-absorbing cushion within the joints. With aging, some of the D-glucosamine is lost, leading to a reduced cartilage layer and to the onset and progression of arthritis. It has been suggested that ingestion of D-glucosamine can actually "jump-start" production of cartilage and help repair eroded cartilage in arthritic joints.

It has also been suggested that chondroitin sulfate can protect existing cartilage from premature breakdown. It absorbs large amounts of water, which is thought to facilitate diffusion of nutrients into the cartilage, providing precursors for the synthesis of new cartilage. The increased fluid also acts as a shock absorber.

Studies continue on the effect that D-glucosamine and chondroitin sulfate have on degenerative joint disease. To date the studies are inconclusive because a large placebo effect is observed with sufferers of osteoarthritis. Many people in the control groups of these studies also experience relief of symptoms when they receive treatment with a placebo, such as a sugar pill.

Capsules containing D-glucosamine and chondroitin sulfate are available over the counter, and many sufferers of osteoarthritis prefer to take this nutritional supplement as an alternative to any nonsteroidal anti-inflammatory drugs (NSAID), such as ibuprofen. Although NSAID can reduce inflammation and pain, long-term use of NSAID can result in stomach ulcers, damage to auditory nerves, and kidney damage.

hydrolyze the $\beta(1 \rightarrow 4)$ glycosidic linkages of the cellulose polymer. Indeed, only a few animals, such as termites, cows, and goats, are able to digest cellulose. These animals have, within their digestive tracts, microorganisms that produce the enzyme cellulase. The sugars released by this microbial digestion can then be absorbed and used by these animals. In humans, cellulose from fruits and vegetables serves as fiber in the diet.

Question 17.9

What chemical reactions are catalyzed by α-amylase and β-amylase?

Question 17.10

What is the function of cellulose in the human diet? How does this relate to the structure of cellulose?

Summary

17.1 Types of Carbohydrates

Carbohydrates are found in a wide variety of naturally occurring substances and serve as principal energy sources for the body. Dietary carbohydrates include complex carbohydrates, such as starch in potatoes, and simple carbohydrates, such as sucrose.

Carbohydrates are classified as *monosaccharides* (one sugar unit), *disaccharides* (two sugar units), *oligosaccharides* (three to ten sugar units), or *polysaccharides* (many sugar units).

17.2 Monosaccharides

Monosaccharides that have an aldehyde as their most oxidized functional group are *aldoses,* and those having a ketone group as their most oxidized functional group are *ketoses.* They may be classified as *trioses, tetroses, pentoses,* and so forth, depending on the number of carbon atoms in the carbohydrate.

17.3 Stereoisomers and Stereochemistry

Stereoisomers of monosaccharides exist because of the presence of *chiral* carbon atoms. They are classified as D- or L- depending on the arrangement of the atoms on the chiral carbon farthest from the aldehyde or ketone group. If the —OH on this carbon is to the right, the stereoisomer is of the D-family. If the —OH group is to the left, the stereoisomer is of the L-family.

Each member of a pair of stereoisomers will rotate plane-polarized light in different directions. A polarimeter is used to measure the direction of rotation of plane-polarized light. Compounds that rotate light in a clockwise direction are termed dextrorotatory and are designated by a plus sign (+). Compounds that rotate light in a counterclockwise direction are called levorotatory and are indicated by a minus sign (−).

The *Fischer Projection* is a two-dimensional drawing of a molecule that shows a chiral carbon at the intersection of two lines. Horizontal lines represent bonds projecting out of the page and vertical lines represent bonds that project into the page. The most oxidized carbon is always represented at the "top" of the structure.

17.4 Biologically Important Monosaccharides

Important monosaccharides include *glyceraldehyde, glucose, fructose,* and *ribose.* Monosaccharides containing five or six carbon atoms can exist as five-membered or six-membered rings. Formation of a ring produces a new chiral carbon at the original carbonyl carbon, which is designated either α or β depending on the orientation of the groups. The cyclization of an aldose produces an intramolecular *hemiacetal,* and the cyclization of a ketose yields an intramolecular *hemiketal.*

Reducing sugars are oxidized by the *Benedict's reagent.* All monosaccharides and all common disaccharides, except sucrose, are reducing sugars. At one time Benedict's reagent was used to determine the concentration of glucose in urine.

17.5 Biologically Important Disaccharides

Important disaccharides include *lactose* and *sucrose.* Lactose is a disaccharide of β-D-galactose bonded β(1 → 4) with D-glucose. In *galactosemia,* defective metabolism of galactose leads to accumulation of a toxic by-product. The ill effects of galactosemia are avoided by exclusion of milk and milk products from the diet of affected infants. Sucrose is a dimer composed of α-D-glucose bonded (α1 → β2) with β-D-fructose.

17.6 Polysaccharides

Starch, the storage polysaccharide of plant cells, is composed of approximately 80% amylose and 20% amylopectin. *Amylose* is a polymer of α-D-glucose units bonded α(1 → 4). Amylose forms a helix. Amylopectin has many branches. Its main chain consists of α-D-glucose units bonded α(1 → 4). The branches are connected by α(1 → 6) glycosidic bonds.

Glycogen, the major storage polysaccharide of animal cells, resembles amylopectin, but it has more, shorter branches. The liver reserve of glycogen is used to regulate blood glucose levels.

Cellulose is a major structural molecule of plants. It is a β(1 → 4) polymer of D-glucose that can contain thousands of glucose monomers. Cellulose cannot be digested by animals because they do not produce an enzyme capable of cleaving the β(1 → 4) glycosidic linkage.

Key Terms

aldose (17.2)
amylopectin (17.6)
amylose (17.6)
anomer (17.4)
asymmetric carbon (17.3)
Benedict's reagent (17.4)
carbohydrate (17.1)
cellulose (17.6)
chiral molecule (17.3)
disaccharide (17.1)
enantiomers (17.3)
Fischer Projection (17.3)
fructose (17.4)
galactose (17.4)
galactosemia (17.5)
glucose (17.4)
glyceraldehyde (17.3)
glycogen (17.6)
glycosidic bond (17.1)
Haworth projection (17.4)
hemiacetal (17.4)
hemiketal (17.4)
hexose (17.2)
ketose (17.2)
lactose (17.5)
lactose intolerance (17.5)
maltose (17.5)
monosaccharide (17.1)
nonreducing sugar (17.5)
oligosaccharide (17.1)

pentose (17.2)
polysaccharide (17.1)
reducing sugar (17.4)
ribose (17.4)
saccharide (17.1)
stereochemistry (17.3)
stereoisomers (17.3)
sucrose (17.5)
tetrose (17.2)
triose (17.2)

Questions and Problems

Types of Carbohydrates

17.11 What is the difference between a monosaccharide and a disaccharide?

17.12 What is a polysaccharide?

17.13 Read the labels on some of the foods in your kitchen, and see how many products you can find that list one or more carbohydrates among the ingredients in the package. Make a list of these compounds, and attempt to classify them according to parent structure (e.g., monosaccharides, disaccharides, polysaccharides).

17.14 Some disaccharides are often referred to by their common names. What are the chemical names of (a) milk sugar, (b) beet sugar, and (c) cane sugar?

17.15 How many kilocalories of energy are released when 1 g of carbohydrate is "burned" or oxidized?

17.16 List some natural sources of carbohydrates.

17.17 Draw and provide the names of an aldohexose and a ketohexose.

17.18 Draw and provide the name of an aldotriose.

Monosaccharides

17.19 Identify each of the following sugars. Label each as either a hemiacetal or a hemiketal:

a.

CH_2OH
H, O, OH
H
HO, OH, H, H
H, OH

c.

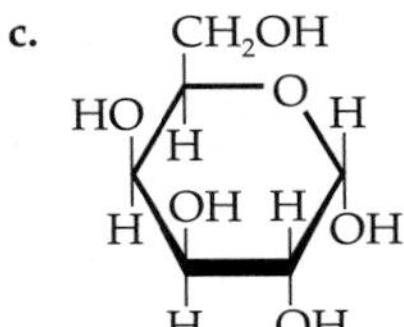

b. $HOCH_2$ O OH
H HO
H CH_2OH
OH H

17.20 Draw the open-chain form of the sugars in Problem 17.19.

17.21 Draw all of the different possible aldotrioses of molecular formula $C_3H_6O_3$.

17.22 Draw all of the different possible aldotetroses of molecular formula $C_4H_8O_4$.

Stereoisomers and Stereochemistry

17.23 Is there any difference between dextrose and D-glucose?

17.24 The structure of D-glucose is shown. Draw its mirror image.

H
C=O
H—C—OH
HO—C—H
H—C—OH
H—C—OH
CH_2OH

17.25 How are D- and L-glyceraldehyde related?

17.26 Determine whether each of the following is a D- or L-sugar:

a.
O
CH
H—C—OH
H—C—OH
CH_2OH

b.
O
CH
H—C—OH
H—C—OH
HO—C—H
CH_2OH

c.
O
CH
H—C—OH
H—C—OH
HO—C—H
CH_2OH

d.
O
CH
H—C—OH
HO—C—H
H—C—OH
CH_2OH

17.27 Define the term *stereoisomer.*

17.28 Define the term *enantiomer.*

17.29 Define the term *chiral carbon.*

17.30 Draw an example of a molecule with one or more chiral carbons. Note the carbon or carbons that are chiral with asterisks (*).

17.31 Explain how a polarimeter works.

17.32 What is plane-polarized light?

17.33 Draw a Fischer Projection formula for each of the following compounds. Indicate each of the chiral carbons with asterisks (*).

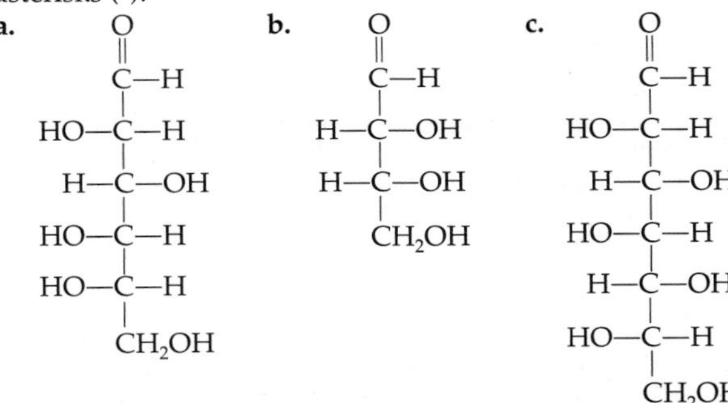

17.34 Draw a Fischer Projection formula for each of the following compounds. Indicate each of the chiral carbons with asterisks (*).

a.
O
C—H
H—C—H
HO—C—H
HO—C—H
HO—C—H
CH_2OH

b.
O
C—H
H—C—H
H—C—OH
CH_2OH

c.
O
C—H
HO—C—H
HO—C—H
HO—C—H
H—C—OH
HO—C—H
CH_2OH

Biologically Important Monosaccharides

17.35 Why does cyclization of D-glucose give two isomers, α- and β-D-glucose?

17.36 Draw the structure of the open chain form of D-fructose, and show how it cyclizes to form α- and β-D-fructose.

17.37 Which of the following would give a positive Benedict's Test?
a. Sucrose
b. Glycogen
c. β-Maltose
d. α-Lactose

17.38 Why was the Benedict's reagent useful for determining the amount of glucose in the urine?

17.39 Describe what is meant by a pair of enantiomers. Draw an example of a pair of enantiomers.

17.40 What is a chiral carbon atom?

17.41 When discussing sugars, what do we mean by an intramolecular hemiacetal?

17.42 When discussing sugars, what do we mean by an intramolecular hemiketal?

Biologically Important Disaccharides

17.43 Maltose is a disaccharide isolated from amylose that consists of two glucose units linked $\alpha(1 \rightarrow 4)$. Draw the structure of this molecule.

17.44 Sucrose is a disaccharide formed by linking α-D-glucose and β-D-fructose by an $(\alpha 1 \rightarrow \beta 2)$ bond. Draw the structure of this disaccharide.

17.45 What is the major biological source of lactose?

17.46 What metabolic defect causes galactosemia?

17.47 What simple treatment prevents most of the ill effects of galactosemia?

17.48 What are the major physiological effects of galactosemia?

17.49 What is lactose intolerance?

17.50 What is the difference between lactose intolerance and galactosemia?

Polysaccharides

17.51 What is the difference between the structure of cellulose and the structure of amylose?

17.52 How does the structure of amylose differ from that of amylopectin and glycogen?

17.53 What is the major physiological purpose of glycogen?

17.54 Where in the body do you find glycogen stored?

17.55 Where are α-amylase and β-amylase produced?

17.56 Where do α-amylase and β-amylase carry out their enzymatic functions?

Critical Thinking Problems

1. The six-member glucose ring structure is not a flat ring. Like cyclohexane, it can exist in the chair conformation. Build models of the chair conformation of α- and β-D-glucose. Draw each of these structures. Which would you predict to be the more stable isomer? Explain your reasoning.
2. The following is the structure of salicin, a bitter-tasting compound found in the bark of willow trees:

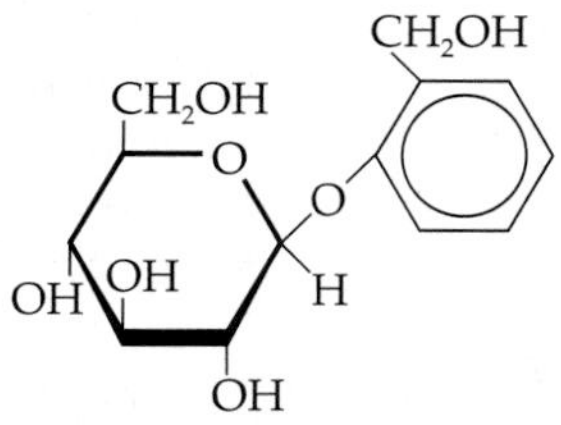

   CH_2OH CH_2OH O   O   OH   OH   H   OH

   Salicin

   The aromatic ring portion of this structure is quite insoluble in water. How would forming a glycosidic bond between the aromatic ring and β-D-glucose alter the solubility? Explain your answer.
3. Ancient peoples used salicin to reduce fevers. Write an equation for the acid-catalyzed hydrolysis of the O-glycosidic bond of salicin. Compare the aromatic product with the structure of acetylsalicylic acid (aspirin). Use this information to develop a hypothesis explaining why ancient peoples used salicin to reduce fevers.
4. Chitin is a modified cellulose in which the C-2 hydroxyl group of each glucose is replaced by

   $$—NH\overset{\overset{O}{\|}}{C}CH_3$$

   This nitrogen-containing polysaccharide makes up the shells of lobsters, crabs, and the exoskeletons of insects. Draw a portion of a chitin polymer consisting of four monomers.
5. Pectins are polysaccharides obtained from fruits and berries and used to thicken jellies and jams. Pectins are $\alpha(1 \rightarrow 4)$ linked D-galacturonic acid. D-Galacturonic acid is D-galactose in which the C-6 hydroxyl group has been oxidized to a carboxyl group. Draw a portion of a pectin polymer consisting of four monomers.
6. Peonin is a red pigment found in the petals of peony flowers. Consider the structure of peonin:

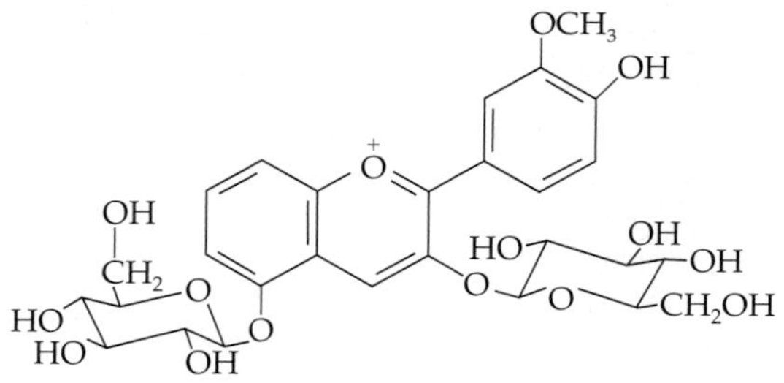

   OCH_3 OH $\overset{+}{O}$ OH CH_2 O   HO   HO   O   OH   HO   O   O   OH   OH CH_2OH

   Why do you think peonin is bonded to two hexoses? What monosaccharide(s) would be produced by acid-catalyzed hydrolysis of peonin?

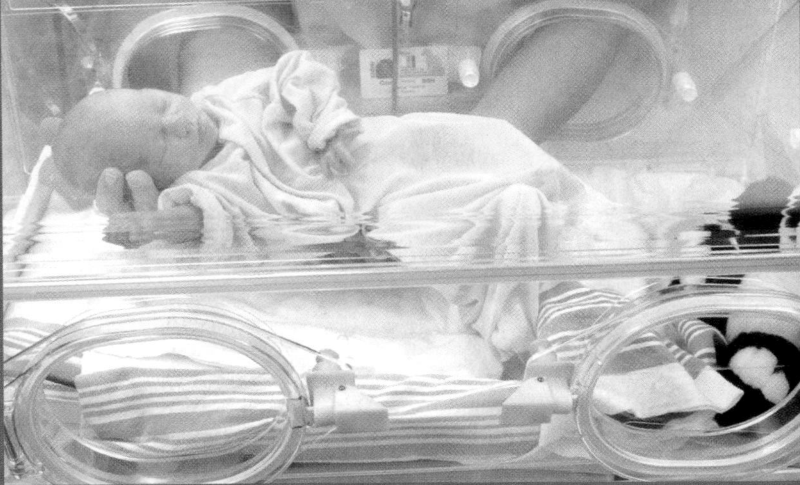

A premature baby in an incubator.

18

Lipids and Their Functions in Biochemical Systems

BIOCHEMISTRY

Outline

Learning Goals

1. Discuss the physical and chemical properties and biological function of each of the families of lipids.
2. Write the structures of saturated and unsaturated fatty acids.
3. Compare and contrast the structure and properties of saturated and unsaturated fatty acids.
4. Write equations representing the reactions that fatty acids undergo.
5. Describe the functions of prostaglandins.
6. Discuss the mechanism by which aspirin reduces pain.
7. Draw the structure of a phospholipid and discuss its amphipathic nature.
8. Discuss the general classes of sphingolipids and their functions.
9. Draw the structure of the steroid nucleus and discuss the functions of steroid hormones.
10. Describe the function of lipoproteins in triglyceride and cholesterol transport in the body.
11. Draw the structure of the cell membrane and discuss its functions.
12. Discuss passive and facilitated diffusion of materials through a cell membrane.
13. Explain the process of osmosis.
14. Describe the mechanism of action of a Na^{+}-K^{+} ATPase.

CHEMISTRY CONNECTION

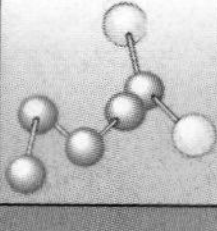

Lifesaving Lipids

In the intensive-care nursery the premature infant struggles for life. Born three and a half months early, the baby weighs only 1.6 pounds, and the lungs labor to provide enough oxygen to keep the tiny body alive. Premature infants often have respiratory difficulties because they have not yet begun to produce *pulmonary surfactant*.

Pulmonary surfactant is a combination of phospholipids and proteins that reduces surface tension in the alveoli of the lungs. (Alveoli are the small, thin-walled air sacs in the lungs.) This allows efficient gas exchange across the membranes of the alveolar cells; oxygen can more easily diffuse from the air into the tissues and carbon dioxide can easily diffuse from the tissues into the air.

Without pulmonary surfactant, gas exchange in the lungs is very poor. Pulmonary surfactant is not produced until early in the sixth month of pregnancy. Premature babies born before they have begun secretion of natural surfactant suffer from *respiratory distress syndrome (RDS)*, which is caused by the severe difficulty they have obtaining enough oxygen from the air that they breathe.

Until recently, RDS was a major cause of death among premature infants, but now a lifesaving treatment is available. A fine aerosol of an artificial surfactant is administered directly into the trachea. The Glaxo-Wellcome Company product EXOSURF Neonatal contains the phospholipid lecithin to reduce surface tension; 1-hexadecanol, which spreads the lecithin; and a polymer called tyloxapol, which disperses both the lecithin and the 1-hexadecanol.

Artificial pulmonary surfactant therapy has dramatically reduced premature infant death caused by RDS and appears to have reduced overall mortality for all babies born weighing less than 700 g (about 1.5 pounds). Advances such as this have come about as a result of research on the makeup of body tissues and secretions in both healthy and diseased individuals. Often, such basic research provides the information needed to develop effective therapies.

In this chapter we will study the chemistry of lipids with a wide variety of structures and biological functions. Among these are the triglycerides that stock our adipose tissue, pain-producing prostaglandins, and steroids that determine our secondary sexual characteristics.

Introduction

Lipids seem to be the most controversial group of biological molecules, particularly in the fields of medicine and nutrition. One concern is the use of anabolic steroids by athletes. Although these hormones increase muscle mass and enhance performance, we are just beginning to understand the damage they cause to the body.

We are concerned about what types of dietary fat we should consume. We hear frequently about the amounts of saturated fats and cholesterol in our diets because a strong correlation has been found between these lipids and heart disease. Large quantities of dietary saturated fats may also predispose an individual to colon, esophageal, stomach, and breast cancers. As a result, we are advised to reduce our intake of cholesterol and saturated fats.

Nonetheless, lipids serve a wide variety of functions essential to living systems and are required in our diet. Standards of fat intake have not been experimentally determined. However, the most recent U.S. Dietary Guidelines recommend that dietary fat not exceed 30% of the daily caloric intake, and no more than 10% should be saturated fats.

18.1 Biological Functions of Lipids

Learning Goal

The term **lipids** actually refers to a collection of organic molecules of varying chemical composition. They are grouped together on the basis of their solubility in nonpolar solvents. Lipids are commonly subdivided into four main groups:

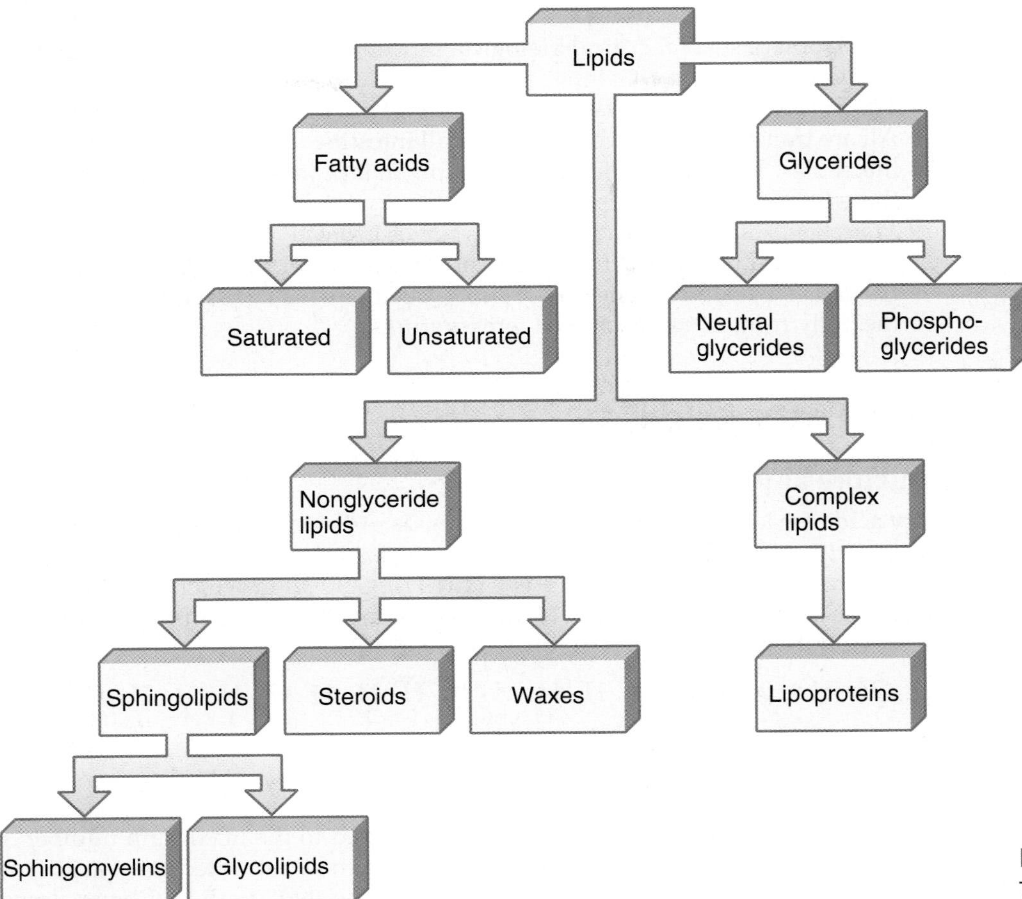

Figure 18.1
The classification of lipids.

1. *Fatty acids* (saturated and unsaturated)
2. *Glycerides* (glycerol-containing lipids)
3. *Nonglyceride lipids* (sphingolipids, steroids, waxes)
4. *Complex lipids* (lipoproteins)

In this chapter we examine the structure, properties, chemical reactions, and biological functions of each of the lipid groups shown in Figure 18.1.

As a result of differences in their structures, lipids serve many different functions in the human body. The following brief list will give you an idea of the importance of lipids in biological processes:

- *Energy source.* Like carbohydrates, lipids are an excellent source of energy for the body. When oxidized, each gram of fat releases 9 kcal of energy, or more than twice the energy released by oxidation of a gram of carbohydrate.
- *Energy storage.* Most of the energy stored in the body is in the form of lipids (triglycerides). Stored in fat cells called *adipocytes*, these fats are a particularly rich source of energy for the body.
- *Cell membrane structural components.* Phosphoglycerides, sphingolipids, and steroids make up the basic structure of all cell membranes. These membranes control the flow of molecules into and out of cells and allow cell-to-cell communication.
- *Hormones.* The steroid hormones are critical chemical messengers that allow tissues of the body to communicate with one another. The prostaglandins exert strong biological effects on both the cells that produce them and other cells of the body.

Lipid-soluble vitamins are discussed in detail in Appendix E.

- *Vitamins.* The lipid-soluble vitamins, A, D, E, and K, play a major role in the regulation of several critical biological processes, including blood clotting and vision.
- *Vitamin absorption.* Dietary fat serves as a carrier of the lipid-soluble vitamins. All are transported into cells of the small intestine in association with fat molecules. Therefore a diet that is too low in fat can result in a deficiency of these four vitamins.
- *Protection.* Fats serve as a shock absorber, or protective layer, for the vital organs. About 4% of the total body fat is reserved for this critical function.
- *Insulation.* Fat stored beneath the skin (subcutaneous fat) serves to insulate the body from extremes of cold temperatures.

18.2 Fatty Acids

Structure and Properties

Learning Goal

Fatty acids are long-chain monocarboxylic acids. As a consequence of their biosynthesis, fatty acids generally contain an *even number* of carbon atoms. The general formula for a **saturated fatty acid** is $CH_3(CH_2)_nCOOH$, in which n in biological systems is an even integer. If $n = 16$, the result is an 18-carbon saturated fatty acid, stearic acid, having the following structural formula:

```
   H  H  H  H  H  H  H  H  H  H  H  H  H  H  H  H  H  O
   |  |  |  |  |  |  |  |  |  |  |  |  |  |  |  |  |  ||
H—C—C—C—C—C—C—C—C—C—C—C—C—C—C—C—C—C—C—OH
   |  |  |  |  |  |  |  |  |  |  |  |  |  |  |  |  |
   H  H  H  H  H  H  H  H  H  H  H  H  H  H  H  H  H
```

The saturated fatty acids may be thought of as derivatives of alkanes, the saturated hydrocarbons described in Chapter 11.

Note that each of the carbons in the chain is bonded to the maximum number of hydrogen atoms. To help remember the structure of a saturated fatty acid, you might think of each carbon in the chain being "saturated" with hydrogen atoms. Examples of common saturated fatty acids are given in Table 18.1. An example of an **unsaturated fatty acid** is the 18-carbon unsaturated fatty acid oleic acid, which has the following structural formula:

```
   H  H  H  H  H  H  H  H               H  H  H  H  H  H  H  O
   |  |  |  |  |  |  |  |               |  |  |  |  |  |  |  ||
H—C—C—C—C—C—C—C—C               C—C—C—C—C—C—C—C—OH
   |  |  |  |  |  |  |  |  \          /  |  |  |  |  |  |  |
   H  H  H  H  H  H  H  H    C═C    H  H  H  H  H  H  H
                            /     \
                           H       H
```

The unsaturated fatty acids may be thought of as derivatives of the alkenes, the unsaturated hydrocarbons discussed in Chapter 12.

In the case of unsaturated fatty acids there is at least one carbon-to-carbon double bond. Because of the double bonds, the carbon atoms involved in these bonds are not "saturated" with hydrogen atoms. The double bonds found in almost all naturally occurring unsaturated fatty acids are in the *cis* configuration. In addition, the double bonds are not randomly located in the hydrocarbon chain. Both the placement and the geometric configuration of the double bonds are dictated by the enzymes that catalyze the biosynthesis of unsaturated fatty acids. Examples of common unsaturated fatty acids are also given in Table 18.1.

EXAMPLE 18.1 ***Writing the Structural Formula of an Unsaturated Fatty Acid***

Draw the structural formula for palmitoleic acid.

Solution

The I.U.P.A.C. name of palmitoleic acid is *cis*-9-hexadecenoic acid. The name tells us that this is a 16-carbon fatty acid having a carbon-to-carbon

Continued—

Table 18.1 **Common Saturated and Unsaturated Fatty Acids**

Common Saturated Fatty Acids

Common Name	I.U.P.A.C. Name	Melting Point (°C)	RCOOH Formula	Condensed Formula
Capric	Decanoic	32	$C_9H_{19}COOH$	$CH_3(CH_2)_8COOH$
Lauric	Dodecanoic	44	$C_{11}H_{23}COOH$	$CH_3(CH_2)_{10}COOH$
Myristic	Tetradecanoic	54	$C_{13}H_{27}COOH$	$CH_3(CH_2)_{12}COOH$
Palmitic	Hexadecanoic	63	$C_{15}H_{31}COOH$	$CH_3(CH_2)_{14}COOH$
Stearic	Octadecanoic	70	$C_{17}H_{35}COOH$	$CH_3(CH_2)_{16}COOH$
Arachidic	Eicosanoic	77	$C_{19}H_{39}COOH$	$CH_3(CH_2)_{18}COOH$

Common Unsaturated Fatty Acids

Common Name	I.U.P.A.C. Name	Melting Point (°C)	RCOOH Formula	Number of Double Bonds	Position of Double Bonds
Palmitoleic	*cis*-9-Hexadecenoic	0	$C_{15}H_{29}COOH$	1	9
Oleic	*cis*-9-Octadecenoic	16	$C_{17}H_{33}COOH$	1	9
Linoleic	*cis*,*cis*-9,12-Octadecadienoic	5	$C_{17}H_{31}COOH$	2	9, 12
Linolenic	All *cis*-9,12,15-Octadecatrienoic	−11	$C_{17}H_{29}COOH$	3	9, 12, 15
Arachidonic	All *cis*-5,8,11,14-Eicosatetraenoic	−50	$C_{19}H_{31}COOH$	4	5, 8, 11, 14

Condensed Formula

Palmitoleic	$CH_3(CH_2)_5CH{=}CH(CH_2)_7COOH$
Oleic	$CH_3(CH_2)_7CH{=}CH(CH_2)_7COOH$
Linoleic	$CH_3(CH_2)_4CH{=}CH{-}CH_2{-}CH{=}CH(CH_2)_7COOH$
Linolenic	$CH_3CH_2CH{=}CH{-}CH_2{-}CH{=}CH{-}CH_2{-}CH{=}CH(CH_2)_7COOH$
Arachidonic	$CH_3(CH_2)_4CH{=}CH{-}CH_2{-}CH{=}CH{-}CH_2{-}CH{=}CH{-}CH_2{-}CH{=}CH{-}(CH_2)_3COOH$

EXAMPLE 18.1 —*Continued*

double bond between carbons 9 and 10. The name also reveals that this is the *cis* isomer.

```
   H  H  H  H  H  H              H  H  H  H  H  H  H
   |  |  |  |  |  |              |  |  |  |  |  |  |
H—C—C—C—C—C—C               C—C—C—C—C—C—C—COOH
   |  |  |  |  |  | \          / |  |  |  |  |  |  |
   H  H  H  H  H  H  C=C      H  H  H  H  H  H  H
                    /    \
                   H      H

  16 15 14 13 12 11  10  9   8  7  6  5  4  3  2  1
```

Examination of Table 18.1 and Figure 18.2 reveals several interesting and important points about the physical properties of fatty acids.

- The melting points of saturated fatty acids increase with increasing carbon number, as is the case with alkanes. Saturated fatty acids containing ten or more carbons are solids at room temperature.
- The melting point of a saturated fatty acid is greater than that of an unsaturated fatty acid of the same chain length. The reason is that saturated fatty acid chains tend to be fully extended and to stack in a regular structure, thereby causing increased intermolecular attraction. Introduction of a *cis*

Learning Goal

The relationship between alkane chain length and melting point is described in Section 11.2.

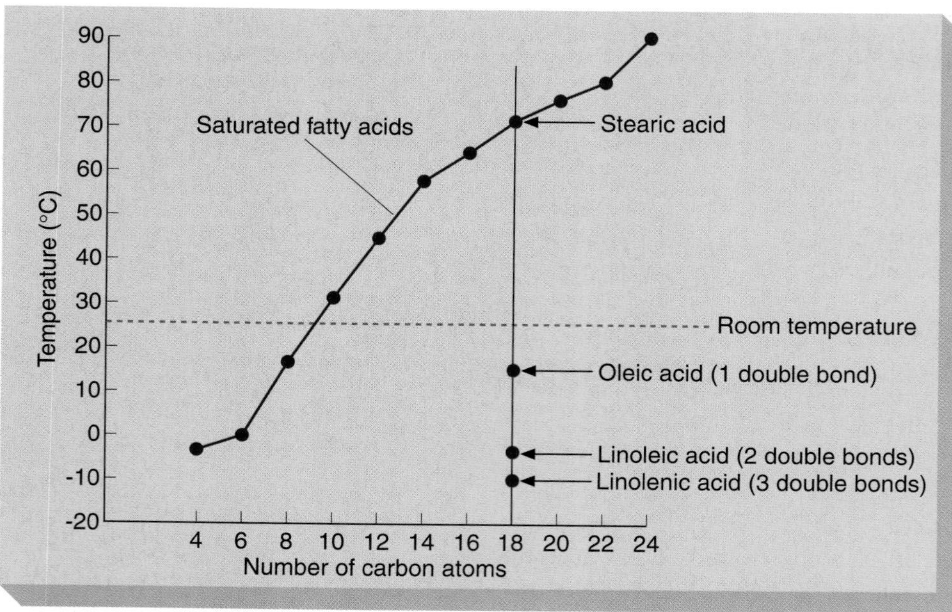

Figure 18.2

The melting points of fatty acids. Melting points of both saturated and unsaturated fatty acids increase as the number of carbon atoms in the chain increases. The melting points of unsaturated fatty acids are lower than those of the corresponding saturated fatty acid with the same number of carbon atoms. Also, as the number of double bonds in the chain increases, the melting points decrease.

double bond into the hydrocarbon chain produces a rigid 30° bend. Such "kinked" molecules cannot stack in an organized arrangement and thus have lower intermolecular attractions and lower melting points.

- As in the case for saturated fatty acids, the melting points of unsaturated fatty acids increase with increasing hydrocarbon chain length.

The relationship between alkene chain length and melting point is described in Section 12.1.

EXAMPLE 18.2

Examining the Similarities and Differences between Saturated and Unsaturated Fatty Acids

Construct a table comparing the structure and properties of saturated and unsaturated fatty acids.

Solution

Property	**Saturated Fatty Acid**	**Unsaturated Fatty Acid**
Chemical composition	Carbon, hydrogen, oxygen	Carbon, hydrogen, oxygen
Chemical structure	Hydrocarbon chain with a terminal carboxyl group	Hydrocarbon chain with a terminal carboxyl group
Carbon-carbon bonds within the hydrocarbon chain	Only C—C single bonds	At least one C—C double bond
Hydrocarbon chains are characteristic of what group of hydrocarbons	Alkanes	Alkenes
"Shape" of hydrocarbon chain	Linear, fully extended	Bend in carbon chain at site of C—C double bond
Physical state at room temperature	Solid	Liquid
Melting point for two fatty acids of the same hydrocarbon chain length	Higher	Lower
Relationship between melting point and chain length	Longer chain length, higher melting point	Longer chain length, higher melting point

Question 18.1

Draw formulas for each of the following fatty acids:

a. Oleic acid
b. Lauric acid
c. Linoleic acid
d. Stearic acid

Question 18.2

What is the I.U.P.A.C. name for each of the fatty acids in Question 18.1? (*Hint:* Review the naming of carboxylic acids in Section 15.1 and Table 18.1.)

Chemical Reactions of Fatty Acids

Learning Goal

The reactions of fatty acids are identical to those of short-chain carboxylic acids. The major reactions that they undergo include esterification, acid hydrolysis of esters, saponification, and addition at the double bond.

Esterification

Esterification is described in Sections 15.1 and 15.2.

In **esterification,** fatty acids react with alcohols to form esters and water according to the following general equation:

$$\underset{\text{Fatty acid}}{R^1{-}\overset{\overset{\large O}{\|}}{C}{-}OH} + \underset{\text{Alcohol}}{HOR^2} \xrightarrow{H^+,\ heat} \underset{\text{Ester}}{R^1{-}\overset{\overset{\large O}{\|}}{C}{-}OR^2} + \underset{\text{Water}}{H{-}OH}$$

Acid Hydrolysis

Acid hydrolysis is discussed in Section 15.2.

Recall that hydrolysis is the reverse of esterification, producing fatty acids from esters:

$$\underset{\text{Ester}}{R^1{-}\overset{\overset{\large O}{\|}}{C}{-}OR^2} + \underset{\text{Water}}{HO{-}H} \xrightarrow{H^+,\ heat} \underset{\text{Fatty acid}}{R^1{-}\overset{\overset{\large O}{\|}}{C}{-}OH} + \underset{\text{Alcohol}}{R^2OH}$$

Saponification

Saponification is described in Section 15.2.

Saponification is the base-catalyzed hydrolysis of an ester:

$$\underset{\text{Ester}}{R^1{-}\overset{\overset{\large O}{\|}}{C}{-}OR^2} + \underset{\text{Base}}{NaOH} \longrightarrow \underset{\text{Salt}}{R^1{-}\overset{\overset{\large O}{\|}}{C}{-}O^-Na^+} + \underset{\text{Alcohol}}{R^2OH}$$

The role of soaps in removal of dirt and grease is described in Section 15.2.

Examples of micelles are shown in Figures 15.4 and 23.1.

The product of this reaction, an ionized salt, is a soap. Because soaps have a long uncharged hydrocarbon tail and a negatively charged terminus (the carboxylate group), they form micelles that dissolve oil and dirt particles. Thus the dirt is emulsified and broken into small particles, and can be rinsed away.

Problems can arise when "hard" water is used for cleaning because the high concentrations of Ca^{2+} and Mg^{2+} in such water cause fatty acid salts to precipitate. Not only does this interfere with the emulsifying action of the soap, it also leaves a hard scum on the surface of sinks and tubs.

$$2R{-}\overset{\overset{\large O}{\|}}{C}{-}O^- + Ca^{2+} \longrightarrow (R{-}\overset{\overset{\large O}{\|}}{C}{-}O^-)_2Ca^{2+}(s)$$

Reaction at the Double Bond (Unsaturated Fatty Acids)

Hydrogenation is discussed in Section 12.5.

Hydrogenation is an example of an addition reaction. The following is a typical example of the addition of hydrogen to the double bonds of a fatty acid:

A HUMAN PERSPECTIVE

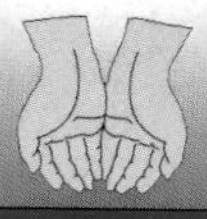

Mummies Made of Soap

In the Smithsonian Museum of Natural History in Washington, D.C., one can find a great many wonders of the natural world. None is quite as macabre as the corpse made of soap. The man in question died of yellow fever in the eighteenth century and was buried near Boston. Actually, he was buried alongside a woman, perhaps the love of his life, who has been dubbed "Soap Woman." She, too, died of yellow fever. However, the couple has been separated for quite some time, because Soap Woman has been on display at the Mutter Museum at the College of Physicians in Philadelphia since 1874.

Recently Soap Woman became a television celebrity when a CT Scan done to examine the body was filmed for "The Mummy Road Show," a presentation of the National Geographic Channel. One reason for the examination was to try to understand the conditions that caused this chemical conversion. At the present time, no one is precisely sure how these two people turned to soap. One clue resides in the environment of the burial site. Apparently the groundwater running through the graves was very basic. Another clue to the mystery is that our soap couple was overweight. Certainly these two factors contributed to the saponification reactions that converted this chubby couple into blocks of soap.

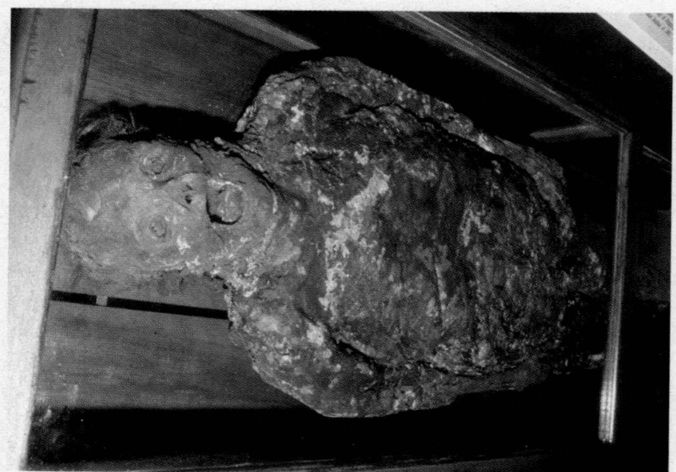

Soap Woman.

$$\underset{\text{Linoleic acid}}{CH_3(CH_2)_4CH{=}CHCH_2CH{=}CH(CH_2)_7COOH} \xrightarrow{2H_2,\ Ni} \underset{\text{Stearic acid}}{CH_3(CH_2)_{16}COOH}$$

Hydrogenation is used in the food industry to convert polyunsaturated vegetable oils into saturated solid fats. *Partial hydrogenation* is carried out to add hydrogen to some, but not all, double bonds in polyunsaturated oils. In this way liquid vegetable oils are converted into solid form. Crisco is one example of a hydrogenated vegetable oil.

Hydrogenation of vegetable oils produces a mixture of *cis* and *trans* unsaturated fatty acids. The *trans* unsaturated fatty acids are thought to contribute to atherosclerosis (hardening of the arteries).

Margarine is also produced by partial hydrogenation of vegetable oils, such as corn oil or soybean oil. The extent of hydrogenation is carefully controlled so that the solid fat will be spreadable and have the consistency of butter when eaten. If too many double bonds were hydrogenated, the resulting product would have the undesirable consistency of animal fat. Artificial color is added to the product, and it may be mixed with milk to produce a butterlike appearance and flavor.

EXAMPLE 18.3 *Writing Equations Representing the Chemical Reactions of Fatty Acids*

Write an equation for each of the following reactions and indicate the I.U.P.A.C. names of each of the organic reactants and products:

a. The esterification of capric acid with propyl alcohol

Continued—

EXAMPLE 18.3 —*Continued*

Solution

$$CH_3(CH_2)_8{-}\overset{\overset{\displaystyle O}{\|}}{C}{-}OH + CH_3CH_2CH_2OH \xrightarrow{H^+,\ heat} CH_3(CH_2)_8{-}\overset{\overset{\displaystyle O}{\|}}{C}{-}O{-}CH_2CH_2CH_3 + H_2O$$

Decanoic acid — Propanol — Propyl decanoate

b. The acid hydrolysis of methyl decanoate

Solution

$$CH_3(CH_2)_8{-}\overset{\overset{\displaystyle O}{\|}}{C}{-}O{-}CH_3 + H_2O \xrightarrow{H^+,\ heat} CH_3OH + CH_3(CH_2)_8{-}\overset{\overset{\displaystyle O}{\|}}{C}{-}OH$$

Methyl decanoate — Methanol — Decanoic acid

c. The base-catalyzed hydrolysis of ethyl dodecanoate

Solution

$$CH_3(CH_2)_{10}{-}\overset{\overset{\displaystyle O}{\|}}{C}{-}O{-}CH_2CH_3 + NaOH \longrightarrow CH_3(CH_2)_{10}{-}\overset{\overset{\displaystyle O}{\|}}{C}{-}O^-Na^+ + CH_3CH_2OH$$

Ethyl dodecanoate — Sodium dodecanoate — Ethanol

d. Hydrogenation of oleic acid

Solution

$$CH_3(CH_2)_7CH{=}CH(CH_2)_7{-}\overset{\overset{\displaystyle O}{\|}}{C}{-}OH \xrightarrow{H_2,\ Ni} CH_3(CH_2)_{16}{-}\overset{\overset{\displaystyle O}{\|}}{C}{-}OH$$

cis-9-Octadecenoic acid — Octadecanoic acid

Question 18.3

Write the complete equation for each of the following reactions:

a. Esterification of lauric acid and ethanol
b. Reaction of oleic acid with NaOH
c. Hydrogenation of arachidonic acid

Question 18.4

Write the complete equation for each of the following reactions:

a. Esterification of capric acid and 2-pentanol
b. Reaction of lauric acid with KOH
c. Hydrogenation of palmitoleic acid

A hormone is a chemical signal that is produced by a specialized tissue and is carried by the bloodstream to target tissues. Eicosanoids are referred to as hormonelike because they affect the cells that produce them, as well as other target tissues.

Eicosanoids: Prostaglandins, Leukotrienes, and Thromboxanes

Some of the unsaturated fatty acids containing more than one double bond cannot be synthesized by the body. For many years it has been known that linolenic acid and linoleic acid, called the **essential fatty acids,** are necessary for specific biochemical functions and must be supplied in the diet (see Table 18.1). The function of linoleic acid became clear in the 1960s when it was discovered that linoleic acid is required for the biosynthesis of **arachidonic acid,** the precursor of a class of hormonelike molecules known as **eicosanoids.** The name is derived from the Greek word *eikos,* meaning "twenty," because they are all derivatives of twenty-carbon fatty acids. The eicosanoids include three groups of structurally related compounds: the prostaglandins, the leukotrienes, and the thromboxanes.

Prostaglandins are extremely potent biological molecules with hormonelike activity. They received the name *prostaglandins* because they were originally isolated from seminal fluid produced in the prostate gland, but more recently they also have been isolated from most animal tissues. Prostaglandins are unsaturated carboxylic acids consisting of a twenty-carbon skeleton that contains a five-carbon ring.

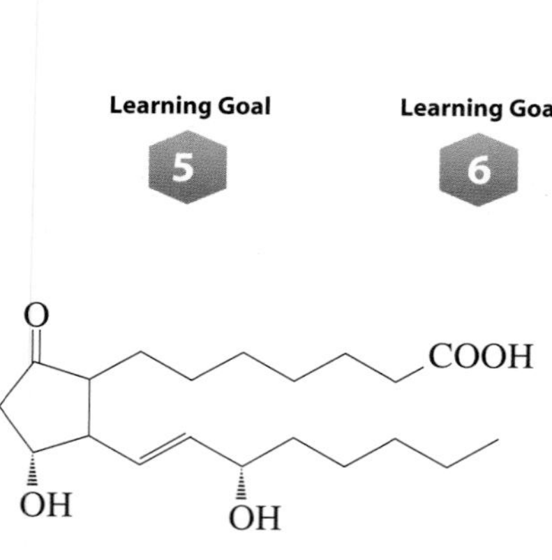

Prostaglandin E_1

Prostaglandin F_1

Prostaglandin E_2

Prostaglandin F_2

Figure 18.3
The structures of four prostaglandins.

Several general classes of prostaglandins are grouped under the designations A, B, E, and F, among others. The nomenclature of prostaglandins is based on the arrangement of the carbon skeleton and the number and orientation of double bonds, hydroxyl groups, and ketone groups. For example, in the name PGF_2, PG stands for prostaglandin, F indicates a particular group of prostaglandins with a hydroxyl group bonded to carbon-9, and 2 indicates that there are two carbon-carbon double bonds in the compound. The examples in Figure 18.3 illustrate the general structure of prostaglandins and the current nomenclature system.

Prostaglandins are made in most tissues, and exert their biological effects on the cells that produce them and on other cells in the immediate vicinity. The extraordinary range of prostaglandin functions includes

- stimulation of smooth muscle,
- regulation of steroid biosynthesis,
- inhibition of gastric secretion,
- inhibition of hormone-sensitive lipases,
- inhibition of platelet aggregation,
- stimulation of platelet aggregation,
- regulation of nerve transmission,
- sensitization to pain, and
- mediation of the inflammatory response.

Because the prostaglandins and the closely related leukotrienes and thromboxanes affect so many body processes and because they often cause opposing effects in different tissues, it can be difficult to keep track of their many regulatory functions. The following is a brief summary of some of the biological processes that are thought to be regulated by the prostaglandins, leukotrienes, and thromboxanes.

1. **Blood clotting.** Blood clots form when a blood vessel is damaged, yet such clotting along the walls of undamaged vessels could result in heart attack or stroke. *Thromboxane* A_2 (Figure 18.4) is produced by platelets in the blood and stimulates constriction of the blood vessels and aggregation of the platelets. Conversely, PGI_2 (prostacyclin) is produced by the cells lining the blood vessels and has precisely the opposite effect of thromboxane A_2. Prostacyclin inhibits platelet aggregation and causes dilation of blood vessels and thus prevents the untimely production of blood clots.
2. **The inflammatory response.** The inflammatory response is another of the body's protective mechanisms. When tissue is damaged by mechanical injury, burns, or invasion by microorganisms, a variety of white blood cells descend on the damaged site to try to minimize the tissue destruction. The result of this response is swelling, redness, fever, and pain. Prostaglandins

are thought to promote certain aspects of the inflammatory response, especially pain and fever. Drugs such as aspirin block prostaglandin synthesis and help to relieve the symptoms. We will examine the mechanism of action of these drugs later in this section.

3. **Reproductive system.** PGE_2 stimulates smooth muscle contraction, particularly uterine contractions. An increase in the level of prostaglandins has been noted immediately before the onset of labor. PGE_2 has also been used to induce second trimester abortions. There is strong evidence that dysmenorrhea (painful menstruation) suffered by many women may be the result of an excess of two prostaglandins. Indeed, drugs, such as ibuprofen, that inhibit prostaglandin synthesis have been approved by the FDA and are found to provide relief from these symptoms.
4. **Gastrointestinal tract.** Prostaglandins have been shown to both inhibit the secretion of acid and increase the secretion of a protective mucus layer into the stomach. In this way, prostaglandins help to protect the stomach lining. Consider for a moment the possible side effect that prolonged use of a drug such as aspirin might have on the stomach—ulceration of the stomach lining. Because aspirin inhibits prostaglandin synthesis, it may actually encourage stomach ulcers by inhibiting the formation of the normal protective mucus layer, while simultaneously allowing increased secretion of stomach acid.
5. **Kidneys.** Prostaglandins produced in the kidneys cause the renal blood vessels to dilate. The greater flow of blood through the kidney results in increased water and electrolyte excretion.
6. **Respiratory tract.** Eicosanoids produced by certain white blood cells, the *leukotrienes* (see Figure 18.4), promote the constriction of the bronchi associated with asthma. Other prostaglandins promote bronchodilation.

As this brief survey suggests, the prostaglandins have numerous, often antagonistic effects. Although they do not fit the formal definition of a hormone (a substance produced in a specialized tissue and transported by the circulatory system to target tissues *elsewhere* in the body), the prostaglandins are clearly strong biological regulators with far-reaching effects.

As mentioned, prostaglandins stimulate the inflammatory response and, as a result, are partially responsible for the cascade of events that cause pain. Aspirin has long been known to alleviate such pain, and we now know that it does so by inhibiting the synthesis of prostaglandins (Figure 18.5).

The first two steps of prostaglandin synthesis (Figure 18.6), the release of arachidonic acid from the membrane and its conversion to PGH_2 by the enzyme cyclooxygenase, occur in all tissues that are able to produce prostaglandins. The conversion of PGH_2 into the other biologically active forms is tissue specific and requires the appropriate enzymes, which are found only in certain tissues.

Aspirin works by inhibiting the cyclooxygenase, which catalyzes the first step in the pathway leading from arachidonic acid to PGH_2. The acetyl group of aspirin becomes covalently bound to the enzyme, thereby inactivating it (Figure 18.5). Because the reaction catalyzed by cyclooxygenase occurs in all cells, aspirin effectively inhibits synthesis of all of the prostaglandins.

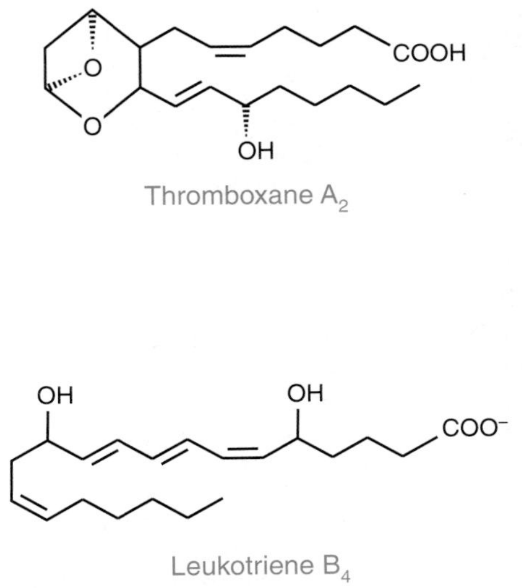

Thromboxane A_2

Leukotriene B_4

Figure 18.4
The structures of thromboxane A_2 and leukotriene B_4.

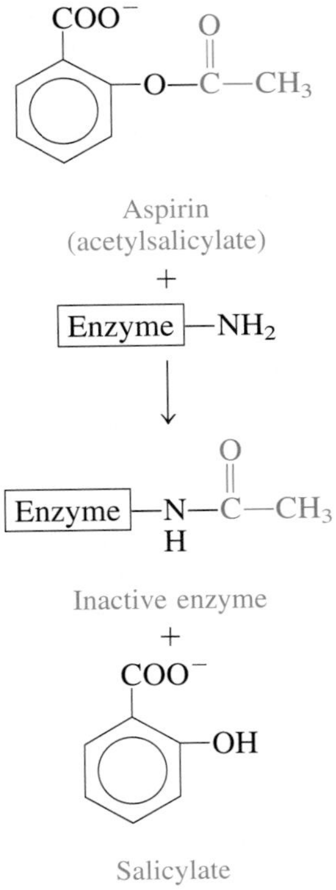

Aspirin (acetylsalicylate)

+

Enzyme—NH_2

Inactive enzyme

+

Salicylate

Figure 18.5
Aspirin inhibits the synthesis of prostaglandins by acetylating the enzyme cyclooxygenase. The acetylated enzyme is no longer functional.

18.3 Glycerides

Neutral Glycerides

Glycerides are lipid esters that contain the glycerol molecule and fatty acids. They may be subdivided into two classes: neutral glycerides and phosphoglycerides. Neutral glycerides are nonionic and nonpolar. Phosphoglyceride molecules have a polar region, the phosphoryl group, in addition to the nonpolar fatty acid tails. The structures of each of these types of glycerides are critical to their function.

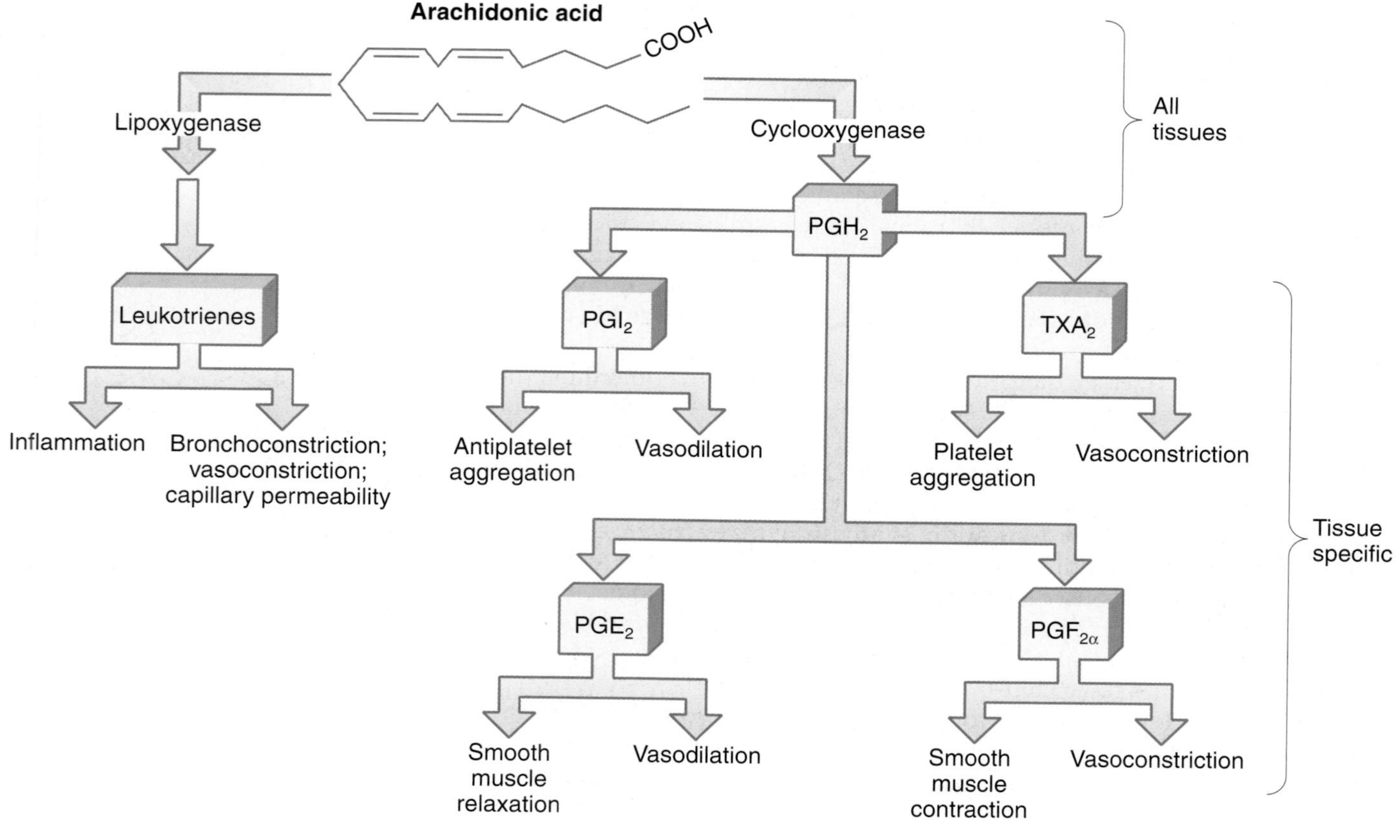

Figure 18.6
A summary of the synthesis of several prostaglandins from arachidonic acid.

The esterification of glycerol with a fatty acid produces a **neutral glyceride.** Esterification may occur at one, two, or all three positions, producing **monoglycerides, diglycerides,** or **triglycerides.** You will also see these referred to as *mono-, di-,* or *triacylglycerols.*

EXAMPLE 18.4 *Writing an Equation for the Synthesis of a Monoglyceride*

Write a general equation for the esterification of glycerol and one fatty acid.

Solution

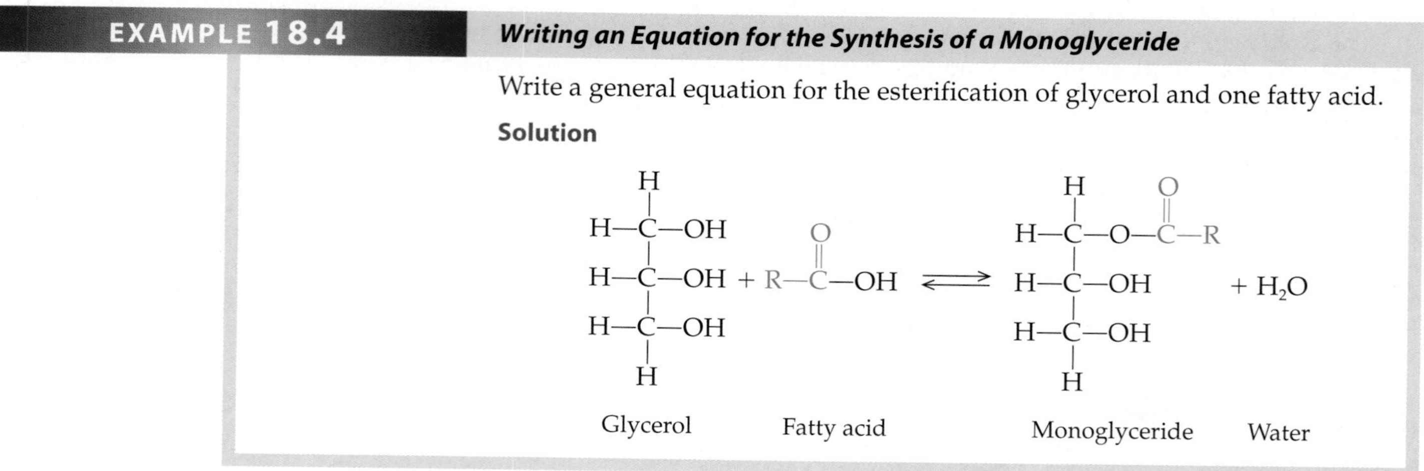

Although monoglycerides and diglycerides are present in nature, the most important neutral glycerides are the triglycerides, the major component of fat cells. The triglyceride consists of a glycerol backbone (shown in black) joined to three fatty acid units through ester bonds (shown in red). The formation of a triglyceride is shown in the following equation:

$$
\begin{array}{l}
\quad\ \ \mathrm{H} \\
\quad\ \ | \\
\mathrm{H{-}C{-}OH} \\
\quad\ \ | \\
\mathrm{H{-}C{-}OH} \\
\quad\ \ | \\
\mathrm{H{-}C{-}OH} \\
\quad\ \ | \\
\quad\ \ \mathrm{H}
\end{array}
+ 3\mathrm{R{-}\overset{\overset{\displaystyle O}{\|}}{C}{-}OH}
\rightleftharpoons
\begin{array}{l}
\quad\ \ \mathrm{H} \quad\quad\ \mathrm{O} \\
\quad\ \ | \quad\quad\quad\ \| \\
\mathrm{H{-}C{-}O{-}C{-}R} \\
\quad\ \ | \quad\quad\quad\ \mathrm{O} \\
\quad\ \ | \quad\quad\quad\ \| \\
\mathrm{H{-}C{-}O{-}C{-}R} \\
\quad\ \ | \quad\quad\quad\ \mathrm{O} \\
\quad\ \ | \quad\quad\quad\ \| \\
\mathrm{H{-}C{-}O{-}C{-}R} \\
\quad\ \ | \\
\quad\ \ \mathrm{H}
\end{array}
+ 3H_2O
$$

Glycerol Fatty acids Triglyceride Water

Triglycerides (triacylglycerols) are named by using the "backbone" name, *glycerol,* as the suffix. The name(s) of the fatty acyl group(s) are placed before it. The fatty acyl group is named by dropping the ending *-ic acid* and replacing it with the ending *-oyl.* In the examples below, the name of the triglyceride on the left is tristearoylglycerol. The prefix *tri-* tells us that there are three stearic acyl groups attached to the glycerol backbone. The triglyceride on the right is a *mixed triglyceride;* that is, there are three different fatty acyl groups attached to the glycerol backbone. They are listed according to their placement along the glycerol backbone.

$$
\begin{array}{l}
\quad\ \ \mathrm{H} \quad\quad\ \mathrm{O} \\
\quad\ \ | \quad\quad\quad\ \| \\
\mathrm{H{-}\underset{1}{C}{-}O{-}C{-}(CH_2)_{16}CH_3} \\
\quad\ \ | \quad\quad\quad\ \mathrm{O} \\
\quad\ \ | \quad\quad\quad\ \| \\
\mathrm{H{-}\underset{2}{C}{-}O{-}C{-}(CH_2)_{16}CH_3} \\
\quad\ \ | \quad\quad\quad\ \mathrm{O} \\
\quad\ \ | \quad\quad\quad\ \| \\
\mathrm{H{-}\underset{3}{C}{-}O{-}C{-}(CH_2)_{16}CH_3} \\
\quad\ \ | \\
\quad\ \ \mathrm{H}
\end{array}
\qquad\qquad
\begin{array}{l}
\quad\ \ \mathrm{H} \quad\quad\ \mathrm{O} \\
\quad\ \ | \quad\quad\quad\ \| \\
\mathrm{H{-}\underset{1}{C}{-}O{-}C{-}(CH_2)_{14}CH_3} \\
\quad\ \ | \quad\quad\quad\ \mathrm{O} \\
\quad\ \ | \quad\quad\quad\ \| \\
\mathrm{H{-}\underset{2}{C}{-}O{-}C{-}(CH_2)_7CH{=}CH(CH_2)_7CH_3} \\
\quad\ \ | \quad\quad\quad\ \mathrm{O} \\
\quad\ \ | \quad\quad\quad\ \| \\
\mathrm{H{-}\underset{3}{C}{-}O{-}C{-}(CH_2)_{16}CH_3} \\
\quad\ \ | \\
\quad\ \ \mathrm{H}
\end{array}
$$

Tristearoylglycerol 1-Palmitoyl-2-oleoyl-3-stearoylglycerol

Because there are no charges (+ or −) on these molecules, they are called *neutral glycerides.* These long molecules readily stack with one another and constitute the majority of the lipids stored in the body's fat cells.

Lipid metabolism is discussed in Chapter 23.

The principal function of triglycerides in biochemical systems is the storage of energy. If more energy-rich nutrients are consumed than are required for metabolic processes, much of the excess is converted to neutral glycerides and stored as triglycerides in fat cells of *adipose tissue.* When energy is needed, the triglycerides are metabolized by the body, and energy is released. For this reason, exercise, along with moderate reduction in caloric intake, is recommended for overweight individuals. Exercise, an energy-demanding process, increases the rate of metabolism of fats and results in weight loss.

See A Human Perspective: Losing Those Unwanted Pounds of Adipose Tissue in Chapter 23.

Phosphoglycerides

Learning Goal

Phospholipids are a group of lipids that are phosphate esters. The presence of the phosphoryl group results in a molecule with a polar head (the phosphoryl group) and a nonpolar tail (the alkyl chain of the fatty acid). Because the phosphoryl group ionizes in solution, a charged lipid results.

Phosphoesters are described in Section 15.4.

The most abundant membrane lipids are derived from glycerol-3-phosphate and are known as **phosphoglycerides.** Phosphoglycerides contain acyl groups derived from long-chain fatty acids at C-1 and C-2 of glycerol-3-phosphate. At C-3 the phosphoryl group is joined to glycerol by a phosphoester bond. The simplest phosphoglyceride contains a free phosphoryl group and is known as a **phosphatidate** (Figure 18.7). When the phosphoryl group is attached to another hydrophilic molecule, a more complex phosphoglyceride is formed. For example, *phosphatidylcholine*

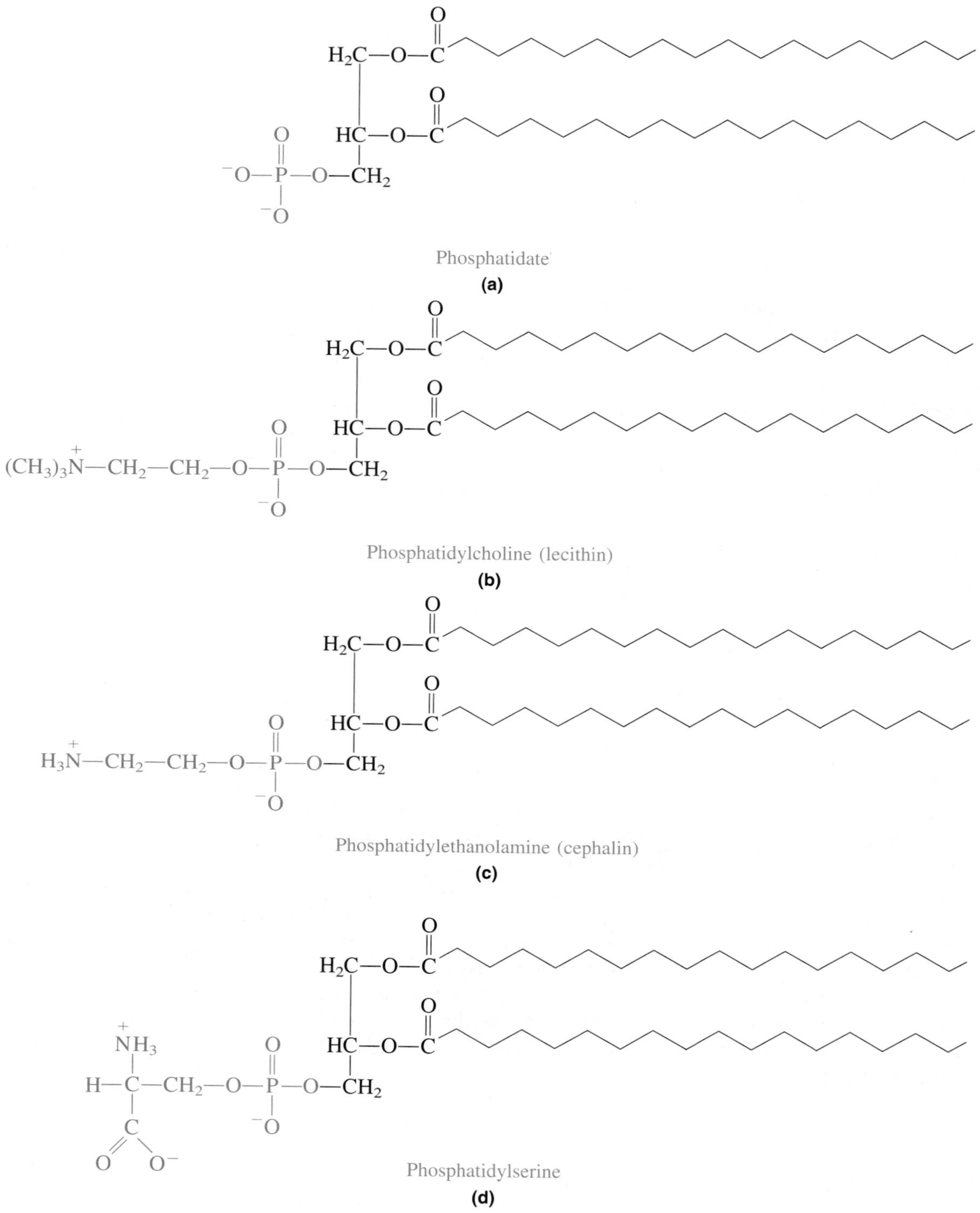

Phosphatidate

(a)

Phosphatidylcholine (lecithin)

(b)

Phosphatidylethanolamine (cephalin)

(c)

Phosphatidylserine

(d)

Figure 18.7
The structures of (a) phosphatidate and the common membrane phospholipids, (b) phosphatidylcholine (lecithin), (c) phosphatidylethanolamine (cephalin), and (d) phosphatidyl serine.

(lecithin) and *phosphatidylethanolamine (cephalin)* are found in the membranes of most cells (Figure 18.7).

Lecithin possesses a polar "head" and a nonpolar "tail." Thus, it is an *amphipathic* molecule. This structure is similar to that of soap and detergent molecules, discussed earlier. The ionic "head" is hydrophilic and interacts with water molecules, whereas the nonpolar "tail" is hydrophobic and interacts with nonpolar molecules. This amphipathic nature is central to the structure and function of cell membranes.

In addition to being a component of cell membranes, lecithin is the major phospholipid in pulmonary surfactant. It is also found in egg yolks and soybeans and is used as an emulsifying agent in ice cream. An **emulsifying agent** aids in the suspension of triglycerides in water. The amphipathic lecithin serves as a bridge, holding together the highly polar water molecules and the nonpolar triglycerides. Emulsification occurs because the hydrophilic head of lecithin dissolves in water and its hydrophobic tail dissolves in the triglycerides.

Cephalin is similar in general structure to lecithin; the amine group bonded to the phosphoryl group is the only difference.

See the Chemistry Connection: Lifesaving Lipids, at the beginning of this chapter.

Question 18.5

Using condensed formulas, draw the mono-, di-, and triglycerides that would result from the esterification of glycerol with each of the following acids.

a. Oleic acid
b. Capric acid
c. Palmitic acid
d. Lauric acid

Question 18.6

Name the triglycerides that are produced in the reactions discussed in Question 18.5.

18.4 Nonglyceride Lipids

Sphingolipids

Learning Goal 8

Sphingolipids are lipids that are not derived from glycerol. Like phospholipids, sphingolipids are amphipathic, having a polar head group and two nonpolar fatty acid tails, and are structural components of cellular membranes. They are derived from sphingosine, a long-chain, nitrogen-containing (amino) alcohol:

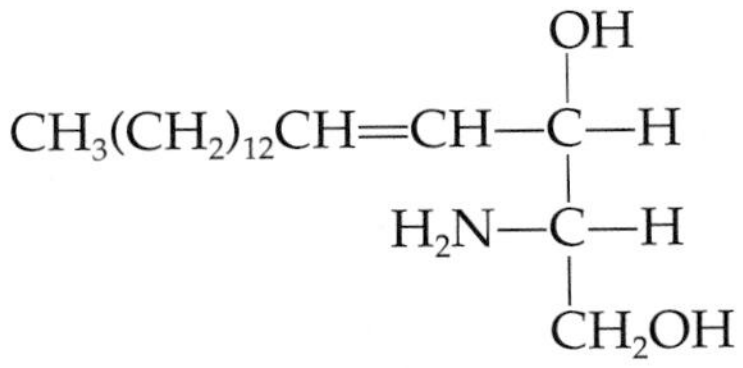

$$
\begin{array}{r}
OH \\
CH_3(CH_2)_{12}CH{=}CH{-}\overset{|}{\underset{|}{C}}{-}H \\
H_2N{-}\overset{}{\underset{|}{C}}{-}H \\
CH_2OH
\end{array}
$$

Sphingosine

The sphingolipids include the sphingomyelins and the glycosphingolipids. The **sphingomyelins** are the only class of sphingolipids that are also phospholipids:

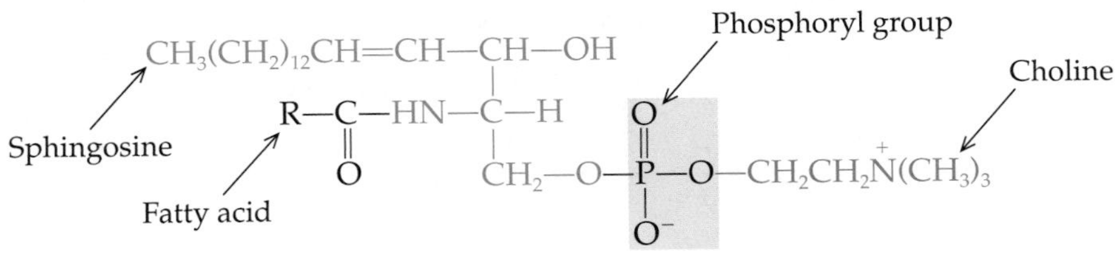

$$
\begin{array}{l}
CH_3(CH_2)_{12}CH{=}CH{-}CH{-}OH \\
\qquad R{-}\underset{\|}{C}{-}HN{-}\underset{|}{C}{-}H \\
\qquad\quad O \qquad\quad CH_2{-}O{-}\overset{O}{\overset{\|}{\underset{\underset{O^-}{|}}{P}}}{-}O{-}CH_2CH_2\overset{+}{N}(CH_3)_3
\end{array}
$$

Sphingomyelin

Sphingomyelins are located throughout the body, but are particularly important structural lipid components of nerve cell membranes. They are found in abundance in the myelin sheath that surrounds and insulates cells of the central nervous system. In humans, about 25% of the lipids of the myelin sheath are sphingomyelins. Their role is essential to proper cerebral function and nerve transmission.

Glycosphingolipids, or *glycolipids,* include the cerebrosides, sulfatides, and gangliosides and are built on a ceramide backbone structure, which is a fatty acid amide derivative of sphingosine:

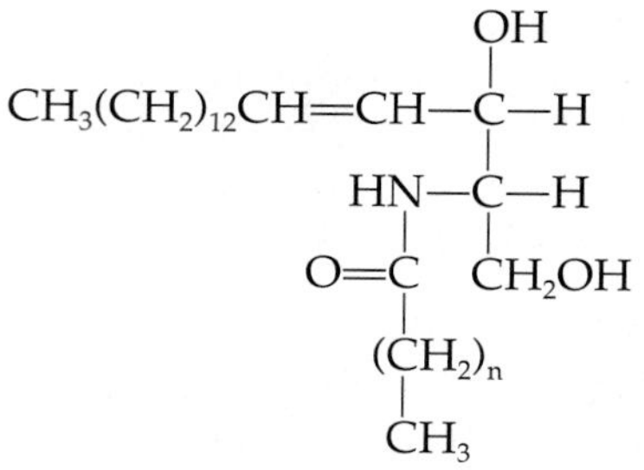

OH

$CH_3(CH_2)_{12}CH{=}CH{-}C{-}H$

HN—C—H

O=C CH_2OH

$(CH_2)_n$

CH_3

Ceramide

The *cerebrosides* are characterized by the presence of a single monosaccharide head group. Two common cerebrosides are glucocerebroside, found in the membranes of macrophages (cells that protect the body by ingesting and destroying foreign microorganisms) and galactocerebroside, found almost exclusively in the membranes of brain cells. Glucocerebroside consists of ceramide bonded to the hexose glucose; galactocerebroside consists of ceramide joined to the monosaccharide galactose.

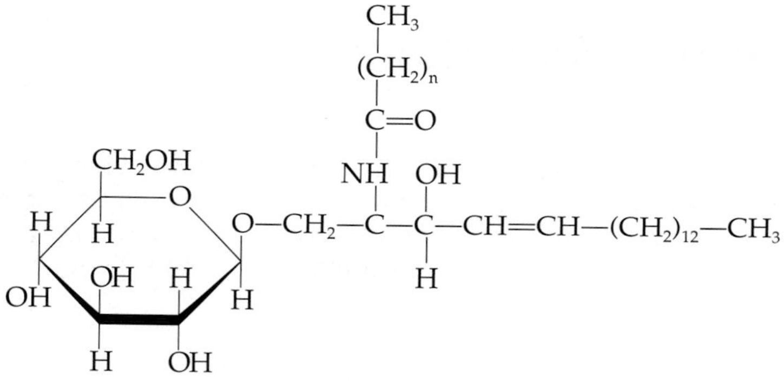

CH_3

$(CH_2)_n$

C=O

CH_2OH NH OH

O

H H $O{-}CH_2{-}C{-}C{-}CH{=}CH{-}(CH_2)_{12}{-}CH_3$

OH H H

OH H

H OH

Glucocerebroside

CH_3

$(CH_2)_n$

C=O

CH_2OH NH OH

O

OH H $O{-}CH_2{-}C{-}C{-}CH{=}CH{-}(CH_2)_{12}{-}CH_3$

OH H H

H H

H OH

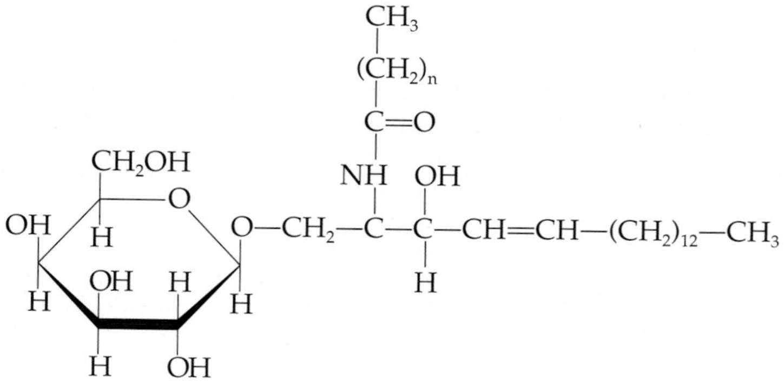

Galactocerebroside

Sulfatides are derivatives of galactocerebroside that contain a sulfate group. Notice that they carry a negative charge at physiological pH.

A CLINICAL PERSPECTIVE

Disorders of Sphingolipid Metabolism

There are a number of human genetic disorders that are caused by a deficiency in one of the enzymes responsible for the breakdown of sphingolipids. In general, the symptoms are caused by the accumulation of abnormally large amounts of these lipids within particular cells. It is interesting to note that three of these diseases, Niemann-Pick disease, Gaucher's disease, and Tay-Sachs disease are found much more frequently among Ashkenazi Jews of Northern European heritage than among other ethnic groups.

Of the four subtypes of Niemann-Pick disease, type A is the most severe. It is inherited as a recessive disorder (i.e., a defective copy of the gene must be inherited from each parent) that results in an absence of the enzyme sphingomyelinase. The absence of this enzyme causes the storage of large amounts of sphingomyelin and cholesterol in the brain, bone marrow, liver, and spleen.

Symptoms may begin when a baby is only a few months old. The parents may notice a delay in motor development and/or problems with feeding. Although the infants may develop some motor skills, they quickly begin to regress as they lose muscle strength and tone, as well as vision and hearing. The disease progresses rapidly and the children typically die within the first few years of life.

Tay-Sachs disease is a lipid storage disease caused by an absence of the enzyme hexosaminidase, which functions in ganglioside metabolism. As a result of the enzyme deficiency, the ganglioside, shown in Section 18.4, accumulates in the cells of the brain causing neurological deterioration. Like Niemann-Pick disease, it is an autosomal recessive genetic trait that becomes apparent in the first few months of the life of an infant and rapidly progresses to death within a few years. Symptoms include listlessness, irritability, seizures, paralysis, loss of muscle tone and function, blindness, deafness, and delayed mental and social skills.

Gaucher's disease is an autosomal recessive genetic disorder resulting in a deficiency of the enzyme glucocerebrosidase. In the normal situation, this enzyme breaks down glucocerebroside, which is an intermediate in the synthesis and degradation of complex glycosphingolipids found in cellular membranes. In Gaucher's disease, glucocerebroside builds up in macrophages found in the liver, spleen, and bone marrow. These cells become engorged with excess lipid and displace healthy, normal cells in bone marrow. The symptoms of Gaucher's disease include severe anemia, thrombocytopenia (reduction in the number of platelets), and hepatosplenomegaly (enlargement of the spleen and liver). There can also be skeletal problems including bone deterioration and secondary fractures.

Fabry's disease is an X-linked inherited disorder caused by the deficiency of the enzyme α-galactosidase A. This disease afflicts as many as fifty thousand people worldwide. Typically, symptoms, including pain in the fingers and toes and a red rash around the waist, begin to appear when individuals reach their early twenties. A preliminary diagnosis can be confirmed by determining the concentration of the enzyme α-galactosidase A. Patients with Fabry's disease have an increased risk of kidney and heart disease, and a reduced life expectancy. Because this is an X-linked disorder, it is more common among males than females.

CH_3
$(CH_2)_n$
$C{=}O$
CH_2OH NH OH
OH H O $O{-}CH_2{-}C{-}C{-}CH{=}CH{-}(CH_2)_{12}{-}CH_3$
OSO_3^- H H
H H
H OH

A sulfatide of galactocerebroside

Gangliosides are glycolipids that possess oligosaccharide groups, including one or more molecules of *N*-acetylneuraminic acid (sialic acid). First isolated from membranes of nerve tissue, gangliosides are found in most tissues of the body.

A ganglioside associated with Tay-Sachs disease

Steroids

Learning Goal

Lipid digestion is described in Section 23.1.

Steroids are a naturally occurring family of organic molecules of biochemical and medical interest. A great deal of controversy has surrounded various steroids. We worry about the amount of cholesterol in the diet and the possible health effects. We are concerned about the use of anabolic steroids by athletes wishing to build muscle mass and improve their performance. However, members of this family of molecules derived from cholesterol have many important functions in the body. The bile salts that aid in the emulsification and digestion of lipids are steroid molecules, as are the sex hormones testosterone and estrone.

$$CH_2{=}\underset{|}{\overset{CH_3}{C}}{-}CH{=}CH_2$$

Isoprene

The structure and function of the lipid-soluble vitamins are found in Appendix E, Lipid-Soluble Vitamins.

The steroids are members of a large, diverse collection of lipids called the *isoprenoids.* All of these compounds are built from one or more five-carbon units called *isoprene.*

Terpene is the general term for lipids that are synthesized from isoprene units. Examples of terpenes include the steroids and bile salts, the lipid-soluble vitamins, chlorophyll, and certain plant hormones.

All steroids contain the steroid nucleus (steroid carbon skeleton) as shown here:

Carbon skeleton of the steroid nucleus

Steroid nucleus

The steroid carbon skeleton consists of four fused rings. Each ring pair has two carbons in common. Thus two fused rings share one or more common bonds as part of their ring backbones. For example, rings A and B, B and C, and C and D are all fused in the preceding structure. Many steroids have methyl groups attached to carbons 10 and 13, as well as an alkyl, alcohol, or ketone group attached to carbon-17.

Cholesterol, a common steroid, is found in the membranes of most animal cells. It is an amphipathic molecule and is readily soluble in the hydrophobic re-

A HUMAN PERSPECTIVE

Anabolic Steroids and Athletics

In the 1988 Summer Olympics, Ben Johnson of Canada ran the fastest 100-meter race in history, 9.79 seconds, and was awarded the Gold Medal. Little more than two days later, Michele Verdier of the International Olympic Committee stood at a press conference and read the following statement: "The urine sample of Ben Johnson, Canada, Athletics, 100 meters, collected Saturday, 24th September 1988, was found to contain metabolites of a banned substance, namely stanozolol, an anabolic steroid." Johnson was disqualified, and Carl Lewis of the United States became the Olympic Gold Medalist in the 100-meter race.

Why do athletes competing in power sports take anabolic steroids? Use of anabolic steroids has a number of desirable effects for the athlete. First, they help build the muscle mass needed to succeed in sprints or weight lifting. They hasten the healing of muscle damage caused by the intense training of the competitive athlete. Finally, anabolic steroids may help the athlete maintain an aggressive attitude, not just during the competition but also throughout training.

If these hormones have such beneficial effects, why not allow all athletes to use them? Unfortunately, the beneficial effects are far outweighed by the negative side effects. These include kidney and liver damage, stroke, impotence and infertility, an increase in cardiovascular disease, and extremely aggressive behavior.

Stanozolol

Even though these life-threatening side effects are well known, athletes continue to use anabolic steroids. The temptation must have been too great for Ben Johnson. After a period of suspension from amateur athletics, Johnson again entered competition. In March 1993, testing before a track meet revealed that he had used anabolic steroids to enhance his performance. As a result, he was forever banned from amateur competition.

gion of membranes. It is involved in regulation of the fluidity of the membrane as a result of the nonpolar fused ring. However, the hydroxyl group is polar and functions like the polar heads of sphingolipids and phospholipids. There is a strong correlation between the concentration of cholesterol found in the blood plasma and heart disease, particularly **atherosclerosis** (hardening of the arteries). Cholesterol, in combination with other substances, contributes to a narrowing of the artery passageway. As narrowing increases, more pressure is necessary to ensure adequate blood flow, and high blood pressure (*hypertension*) develops. Hypertension is also linked to heart disease.

Cholesterol

Egg yolks contain a high concentration of cholesterol, as do many dairy products and animal fats. As a result, it has been recommended that the amounts of these products in the diet be regulated to moderate the dietary intake of cholesterol.

Bile salts are amphipathic derivatives of cholesterol that are synthesized in the liver and stored in the gallbladder. The principal bile salts in humans are cholate and chenodeoxycholate.

Bile salts are described in greater detail in Section 23.1.

Cholate

Chenodeoxycholate

Bile salts are emulsifying agents whose polar hydroxyl groups interact with water and whose hydrophobic regions bind to lipids. Following a meal, bile flows from the gallbladder to the duodenum (the uppermost region of the small intestine). Here the bile salts emulsify dietary fats into small droplets that can be more readily digested by lipases (lipid digesting enzymes) also found in the small intestine.

Steroids play a role in the reproductive cycle. In a series of chemical reactions, cholesterol is converted to the steroid *progesterone,* the most important hormone associated with pregnancy. Produced in the ovaries and in the placenta, progesterone is responsible for both the successful initiation and the successful completion of pregnancy. It prepares the lining of the uterus to accept the fertilized egg. Once the egg is attached, progesterone is involved in the development of the fetus and plays a role in the suppression of further ovulation during pregnancy.

Progesterone

Testosterone

Estrone

Testosterone, a male sex hormone found in the testes, and *estrone,* a female sex hormone, are both produced by the chemical modification of progesterone. These hormones are involved in the development of male and female sex characteristics.

Many steroids, including progesterone, have played important roles in the development of birth control agents. 19-Norprogesterone was one of the first synthetic birth control agents. It is approximately ten times as effective as progesterone in providing birth control. However, its utility was severely limited because this compound could not be administered orally and had to be taken by injection. A related compound, norlutin (chemical name: 17-α-ethynyl-19-nortestosterone), was found to provide both the strength and the effectiveness of 19-norprogesterone and could be taken orally.

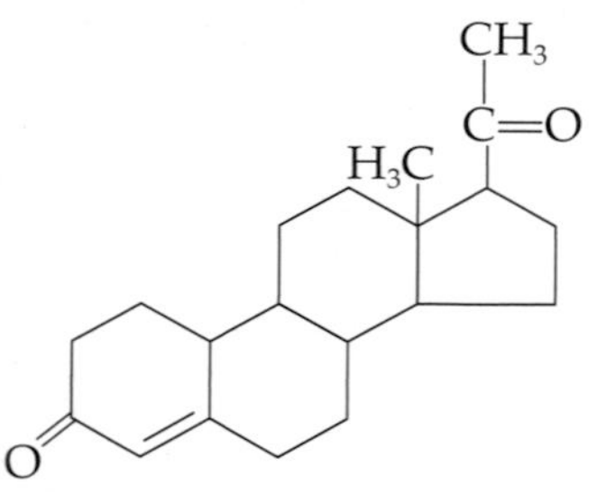

19-Norprogesterone

Norlutin

Currently "combination" oral contraceptives are prescribed most frequently. These include a progesterone and an estrogen. These newer products confer better contraceptive protection than either agent administered individually. They are also used to regulate menstruation in patients with heavy menstrual bleeding. First investigated in the late 1950s and approved by the FDA in 1961, there are at least thirty combination pills currently available. In addition, a transdermal patch for the treatment of postmenopausal osteoporosis is being investigated.

All of these compounds act by inducing a false pregnancy, which prevents ovulation. When oral contraception is discontinued, ovulation usually returns within

A CLINICAL PERSPECTIVE

Steroids and the Treatment of Heart Disease

The foxglove plant (*Digitalis purpurea*) is an herb that produces one of the most powerful known stimulants of heart muscle. The active ingredients of the foxglove plant (digitalis) are the so-called cardiac glycosides, or *cardiotonic steroids*, which include digitoxin, digosin, and gitalin.

Digitoxin

The structure of digitoxin, one of the cardiotonic steroids produced by the foxglove plant.

Digitalis purpurea, the foxglove plant.

These drugs are used clinically in the treatment of congestive heart failure, which results when the heart is not beating with strong, efficient strokes. When the blood is not propelled through the cardiovascular system efficiently, fluid builds up in the lungs and lower extremities (edema). The major symptoms of congestive heart failure are an enlarged heart, weakness, edema, shortness of breath, and fluid accumulation in the lungs.

This condition was originally described in 1785 by a physician, William Withering, who found a peasant woman whose folk medicine was famous as a treatment for chronic heart problems. Her potion contained a mixture of more than twenty herbs, but Dr. Withering, a botanist as well as physician, quickly discovered that foxglove was the active ingredient in the mixture. Withering used *Digitalis purpurea* successfully to treat congestive heart failure and even described some cautions in its use.

The cardiotonic steroids are extremely strong heart stimulants. A dose as low as 1 mg increases the stroke volume of the heart (volume of blood per contraction), increases the strength of the contraction, and reduces the heart rate. When the heart is pumping more efficiently because of stimulation by digitalis, the edema disappears.

Digitalis can be used to control congestive heart failure, but the dose must be carefully determined and monitored because the therapeutic dose is close to the dose that causes toxicity. The symptoms that result from high body levels of cardiotonic steroids include vomiting, blurred vision and lightheadedness, increased water loss, convulsions, and death. Only a physician can determine the initial dose and maintenance schedule for an individual to control congestive heart failure and yet avoid the toxic side effects.

three menstrual cycles. Although there have been problems associated with "the pill," it appears to be an effective and safe method of family planning for much of the population.

Cortisone is also important to the proper regulation of a number of biochemical processes. For example, it is involved in the metabolism of carbohydrates. Cortisone is also used in the treatment of rheumatoid arthritis, asthma, gastrointestinal disorders, many skin conditions, and a variety of other diseases. However, treatment with cortisone is not without risk. Some of the possible side effects of cortisone therapy include fluid retention, sodium retention, and potassium loss that can lead to congestive heart failure. Other side effects include muscle weakness, osteoporosis, gastrointestinal upsets including peptic ulcers, and neurological symptoms, including vertigo, headaches, and convulsions.

Cortisone

Aldosterone

Aldosterone is a steroid hormone produced by the adrenal cortex and secreted into the bloodstream when blood sodium ion levels are too low. Upon reaching its target tissues in the kidney, aldosterone activates a set of reactions that cause sodium ions and water to be returned to the blood. If sodium levels are elevated, aldosterone is not secreted from the adrenal cortex and the sodium ions filtered out of the blood by the kidney will be excreted.

Question 18.7

Draw the structure of the steroid nucleus. Note the locations of the A, B, C, and D steroid rings.

Question 18.8

What is meant by the term *fused ring?*

Waxes

$$CH_3(CH_2)_{14}-\overset{O}{\overset{\|}{C}}-O-(CH_2)_{29}CH_3$$

Myricyl palmitate (beeswax)

$$CH_3(CH_2)_{14}-\overset{O}{\overset{\|}{C}}-O-(CH_2)_{15}CH_3$$

Cetyl palmitate (whale oil)

Waxes are derived from many different sources and have a variety of chemical compositions, depending on the source. Paraffin wax, for example, is composed of a mixture of solid hydrocarbons (usually straight-chain compounds). The natural waxes generally are composed of a long-chain fatty acid esterified to a long-chain alcohol. Because the long hydrocarbon tails are extremely hydrophobic, waxes are completely insoluble in water. Waxes are also solid at room temperature, owing to their high molecular weights. Two examples of waxes are myricyl palmitate, a major component of beeswax, and whale oil (spermaceti wax), from the head of the sperm whale, which is composed of cetyl palmitate.

Naturally occurring waxes have a variety of uses. Lanolin, which serves as a protective coating for hair and skin, is used in skin creams and ointments. Carnauba wax is used in automobile polish. Whale oil was once used as a fuel, in ointments, and in candles. However, synthetic waxes have replaced whale oil to a large extent, because of efforts to ban the hunting of whales.

18.5 Complex Lipids

Learning Goal

Complex lipids are lipids that are bonded to other types of molecules. The most common and important complex lipids are plasma lipoproteins, which are responsible for the transport of other lipids in the body.

Lipids are only sparingly soluble in water, and the movement of lipids from one organ to another through the bloodstream requires a transport system that uses **plasma lipoproteins.** Lipoprotein particles consist of a core of hydrophobic lipids surrounded by amphipathic proteins, phospholipids, and cholesterol (Figure 18.8).

There are four major classes of human plasma lipoproteins:

- **Chylomicrons,** which have a density of less than 0.95 g/mL, carry dietary triglycerides from the intestine to other tissues. The remaining lipoproteins are classified by their densities.

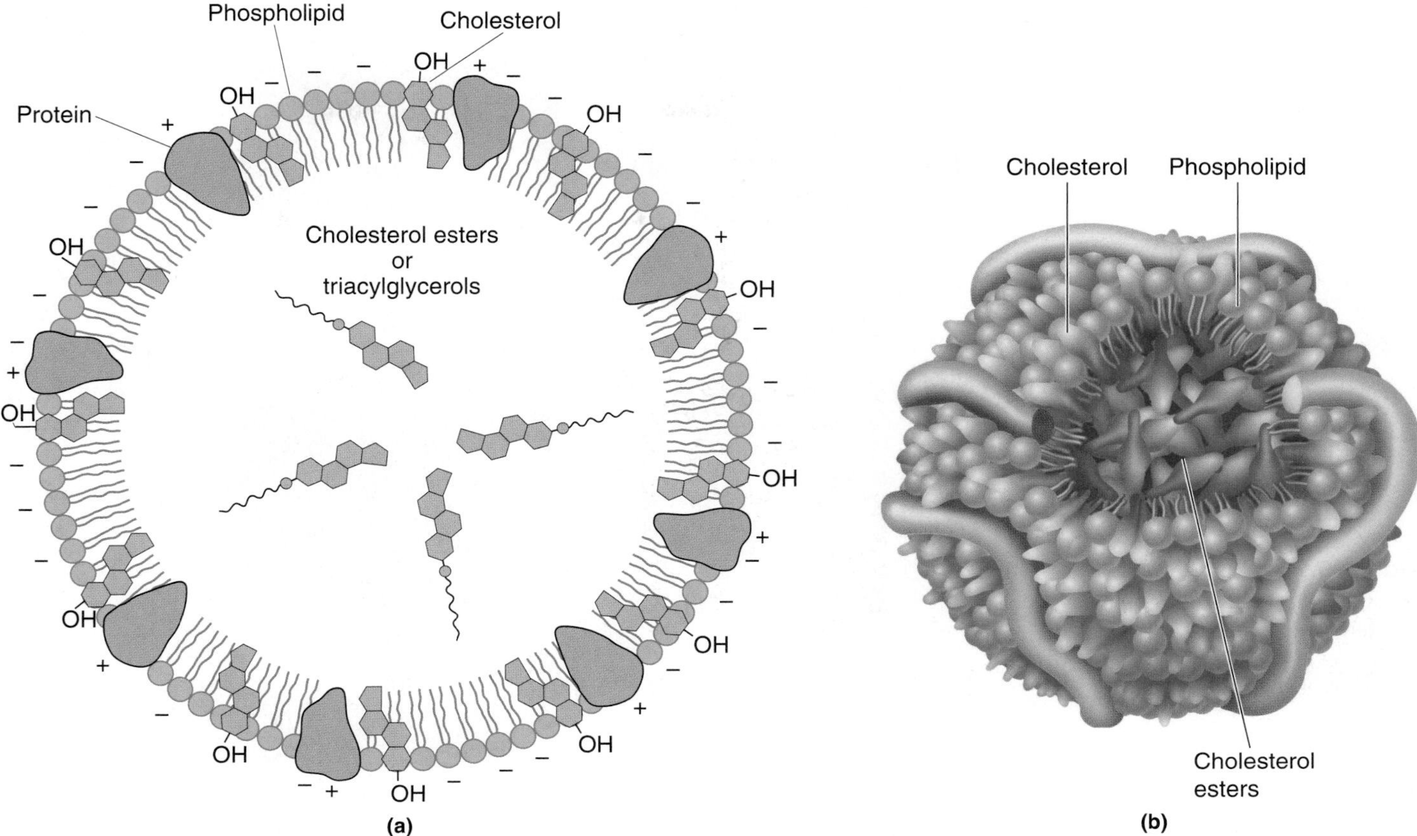

Figure 18.8
A model for the structure of a plasma lipoprotein. The various lipoproteins are composed of a shell of protein, cholesterol, and phospholipids surrounding more hydrophobic molecules such as triglycerides or cholesterol esters (cholesterol esterified to fatty acids). (a) Cross section, (b) three-dimensional view.

- **Very low density lipoproteins (VLDL)** have a density of 0.95–1.019 g/mL. They bind triglycerides synthesized in the liver and carry them to adipose and other tissues for storage.
- **Low-density lipoproteins (LDL)** are characterized by a density of 1.019–1.063 g/mL. They carry cholesterol to peripheral tissues and help regulate cholesterol levels in those tissues. These are richest in cholesterol, frequently carrying 40% of the plasma cholesterol.
- **High-density lipoproteins (HDL)** have a density of 1.063–1.210 g/mL. They are bound to plasma cholesterol; however, they transport cholesterol from peripheral tissues to the liver.

A summary of the composition of each of the plasma lipoproteins is presented in Figure 18.9.

Chylomicrons are aggregates of triglycerides and protein that transport dietary triglycerides to cells throughout the body. Not all lipids in the blood are derived directly from the diet. Triglycerides and cholesterol are also synthesized in the liver and also are transported through the blood in lipoprotein packages. Triglycerides are assembled into VLDL particles that carry the energy-rich lipid molecules either to tissues requiring an energy source or to adipose tissue for storage. Similarly, cholesterol is assembled into LDL particles for transport from the liver to peripheral tissues.

Entry of LDL particles into the cell is dependent on a specific recognition event and binding between the LDL particle and a protein receptor embedded within the membrane. Low-density lipoprotein receptors (LDL receptors) are found in the membranes of cells outside the liver and are responsible for the uptake of cholesterol

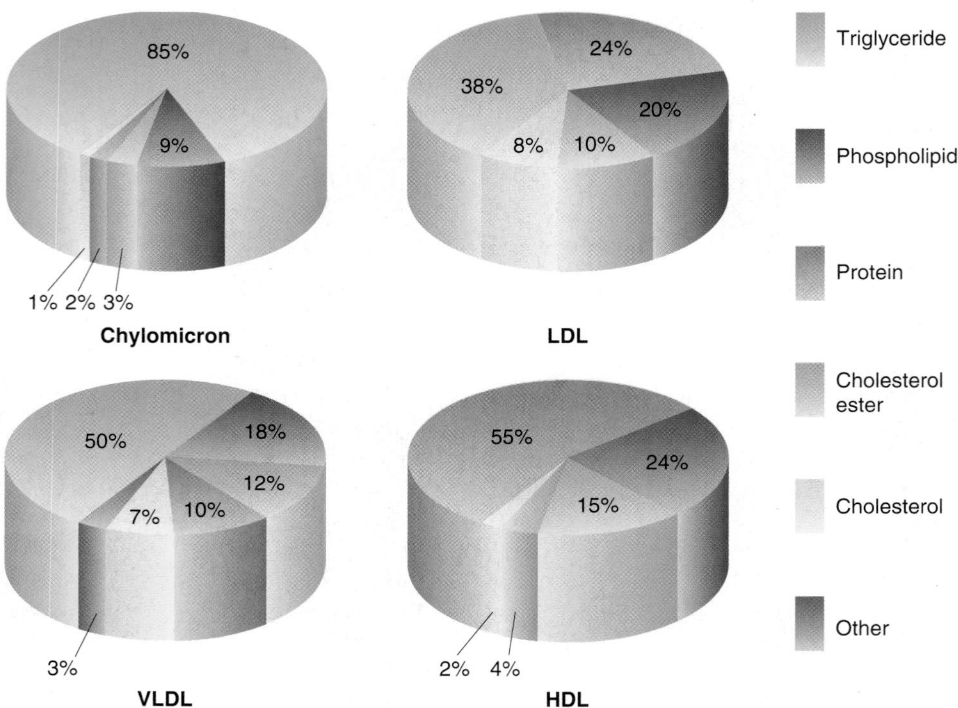

Figure 18.9
A summary of the relative amounts of cholesterol, phospholipid, protein, triglycerides, and cholesterol esters in the four classes of lipoproteins.

by the cells of various tissues. LDL (lipoprotein bound to cholesterol) binds specifically to the LDL receptor, and the complex is taken into the cell by a process called *receptor-mediated endocytosis* (Figure 18.10). The membrane begins to be pulled into the cell at the site of the LDL receptor complexes. This draws the entire LDL particle into the cell. Eventually, the portion of the membrane surrounding the LDL particles pinches away from the cell membrane and forms a membrane around the LDL particles. As we will see in Section 18.6, membranes are fluid and readily flow. Thus they can form a vesicle or endosome containing the LDL particles.

Cellular digestive organelles known as *lysosomes* fuse with the endosomes. This fusion is accomplished when the membranes of the endosome and the lysosome flow together to create one larger membrane-bound body or vesicle. Hydrolytic enzymes from the lysosome then digest the entire complex to release cholesterol into the cytoplasm of the cell. There, cholesterol inhibits its own biosynthesis and activates an enzyme that stores cholesterol in cholesterol ester droplets. High concentrations of cholesterol inside the cell also inhibit the synthesis of LDL receptors to ensure that the cell will not take up too much cholesterol. People who have a genetic defect in the gene coding for the LDL receptor do not take up as much cholesterol. As a result they accumulate LDL cholesterol in the plasma. This excess plasma cholesterol is then deposited on the artery walls, causing atherosclerosis. This disease is called *hypercholesterolemia.*

Liver lipoprotein receptors enable large amounts of cholesterol to be removed from the blood, thus ensuring low concentrations of cholesterol in the blood plasma. Other factors being equal, the person with the most lipoprotein receptors will be the least vulnerable to a high-cholesterol diet and have the least likelihood of developing atherosclerosis.

There is also evidence that high levels of HDL in the blood help reduce the incidence of atherosclerosis, perhaps because HDL carries cholesterol from the peripheral tissues back to the liver. In the liver, some of the cholesterol is used for bile synthesis and secreted into the intestines, from which it is excreted.

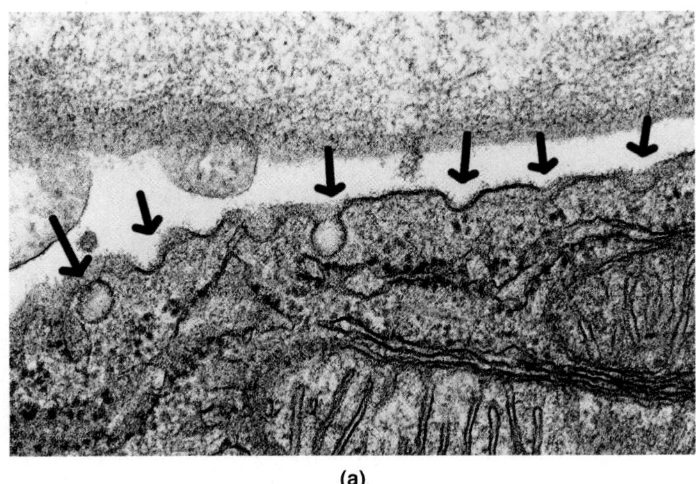

(a)

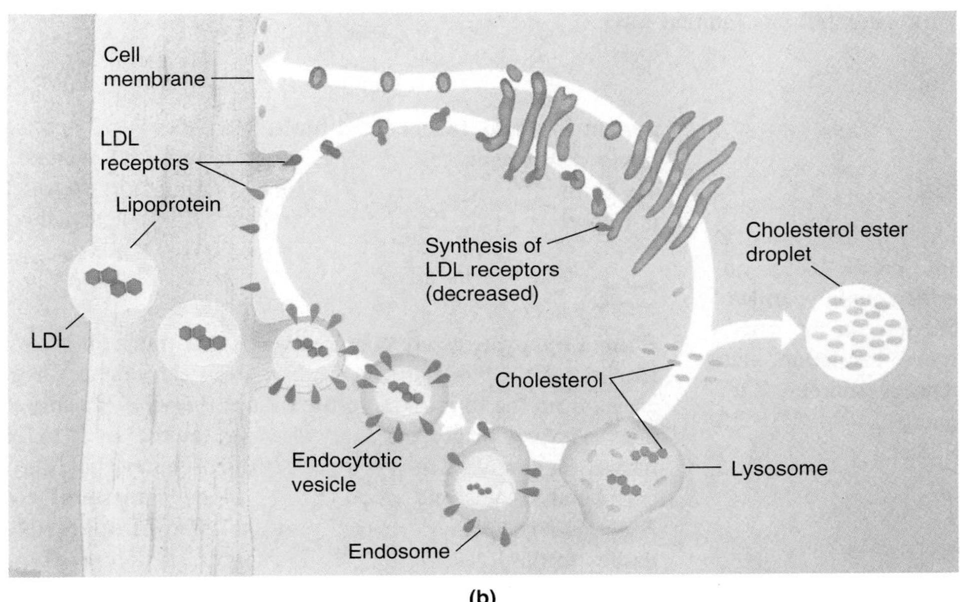

(b)

Figure 18.10
Receptor-mediated endocytosis.
(a) Electron micrographs of the process of receptor-mediated endocytosis.
(b) Summary of the events of receptor-mediated endocytosis of LDL.

A final correlation has been made between diet and atherosclerosis. People whose diet is high in saturated fats tend to have high levels of cholesterol in the blood. Although the relationship between saturated fatty acids and cholesterol metabolism is unclear, it is known that a diet rich in unsaturated fats results in decreased cholesterol levels. In fact, the use of unsaturated fat in the diet results in a decrease in the level of LDL and an increase in the level of HDL. With the positive correlation between heart disease and high cholesterol levels, the current dietary recommendations include a diet that is low in fat and the substitution of unsaturated fats (vegetable oils) for saturated fats (animal fats).

Recently an inflammatory protein, the C-reactive protein, has been implicated in atherosclerosis. A test for the level of this protein in the blood is being suggested as a way to predict the risk of heart attack. It is hoped that a sensitive test will be available within a year.

Question 18.9

What is the mechanism of uptake of cholesterol from plasma?

Question 18.10

What is the role of lysosomes in the metabolism of plasma lipoproteins?

Figure 18.11
(a) Cartoon drawing of a phospholipid. (b) Space-filling model of a phospholipid. (c) Cartoon of a phospholipid bilayer membrane. (d) Line formula representation of a bilayer membrane composed of phospholipids, cholesterol, and sphingolipids.

18.6 The Structure of Biological Membranes

Learning Goal

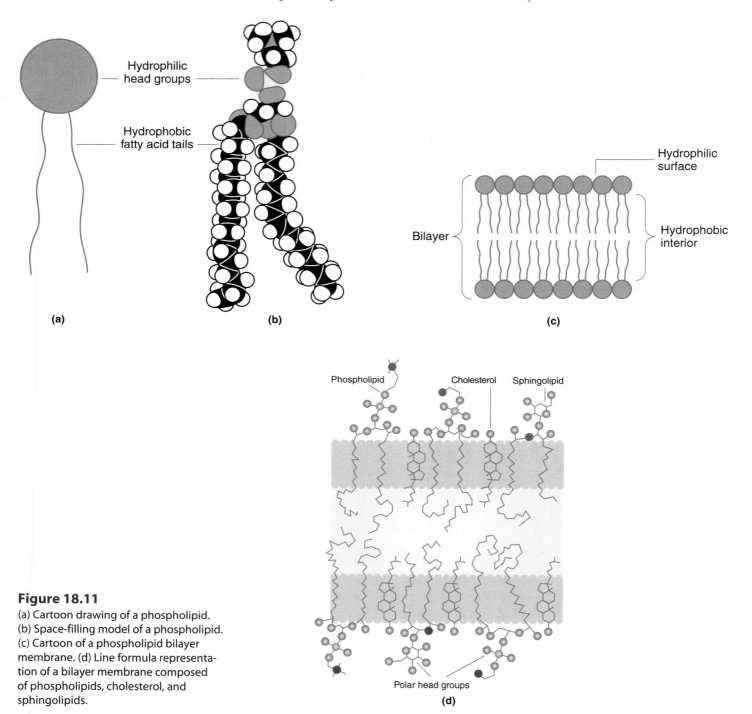

Biological membranes are *lipid bilayers* in which the hydrophobic hydrocarbon tails are packed in the center of the bilayer and the ionic head groups are exposed on the surface to interact with water (Figure 18.11). The hydrocarbon tails of membrane phospholipids provide a thin shell of nonpolar material that prevents mixing of molecules on either side. The nonpolar tails of membrane phospholipids thus provide a barrier between the interior of the cell and its surroundings. The polar heads of lipids are exposed to water, and they are highly solvated. Little exchange, known colloquially as "flip-flop," occurs between lipids on the outer and inner halves of the bilayers (Figure 18.12). The movement of a lipid molecule within one sheath of the bilayer, by contrast, is rapid. A bacterial cell is about 2 μm long, and a lipid molecule diffuses from one end of the cell to the other in a second.

The two layers of the phospholipid bilayer membrane are not identical in composition. For instance, in human red blood cells, approximately 80% of the phospholipids in the outer layer of the membrane are phosphatidylcholine and sphingomyelin; whereas phosphatidylethanolamine and phosphatidylserine make up approximately 80% of the inner layer. In addition, carbohydrate groups are found attached only to those phospholipids found on the outer layer of a membrane. Here they participate in receptor and recognition functions.

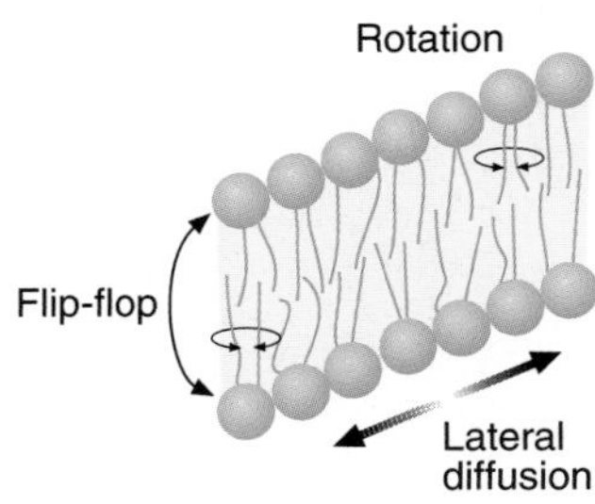

Figure 18.12
Lateral diffusion in a biological membrane is rapid, but "flip-flop" across the membrane almost never occurs.

Fluid Mosaic Structure of Biological Membranes

As we have just noted, membranes are not static; they are composed of molecules in motion. The fluidity of biological membranes is determined by the proportions of saturated and unsaturated fatty acid groups in the membrane phospholipids. About half of the fatty acids that are isolated from membrane lipids from all sources are unsaturated.

The unsaturated fatty acid tails of the phospholipids contribute to membrane fluidity because of the bends introduced into the hydrocarbon chain by the double bonds. Because of these "kinks," the fatty acid tails do not pack together tightly.

We also find that the percentage of unsaturated fatty acid groups in membrane lipids is inversely proportional to the temperature of the environment. Bacteria, for example, have different ratios of saturated and unsaturated fatty acids in their membrane lipids, depending on the temperatures of their surroundings. For instance, the membranes of bacteria that grow in the Arctic Ocean have high levels of unsaturated fatty acids so that their membranes remain fluid even at these frigid temperatures. Conversely, the organisms that live in the hot springs of Yellowstone National Park, with temperatures near the boiling point of water, have membranes with high levels of saturated fatty acids. This flexibility in fatty acid content enables the bacteria to maintain the same membrane fluidity over a temperature range of almost 100°C.

Generally, the body temperatures of mammals are quite constant, and the fatty acid composition of their membrane lipids is therefore usually very uniform. One interesting exception is the reindeer. Much of the year the reindeer must travel through ice and snow. Thus the hooves and lower legs must continue to function at much colder temperatures than the rest of the body. Because of this, the percentage of unsaturation in the membranes varies along the length of the reindeer leg. We find that the proportion of unsaturated fatty acid groups increases closer to the hoof. The lower freezing points and greater fluidity of lipids that contain a high proportion of unsaturated fatty acid groups permit the membranes to function in the low temperatures of ice and snow to which the lower leg is exposed.

Thus membranes are fluid, regardless of the environmental temperature conditions. In fact, it has been estimated that membranes have the consistency of olive oil.

Although the hydrophobic barrier created by the fluid lipid bilayer is an important feature of membranes, the proteins embedded within the lipid bilayer are equally important and are responsible for critical cellular functions. The presence of these membrane proteins was revealed by an electron microscopic technique called *freeze-fracture.* Cells are frozen to very cold temperatures and then fractured with a very fine diamond knife. Some of the cells are fractured between the two layers of the lipid bilayer. When viewed with the electron microscope, the membrane appeared to be a mosaic, studded with proteins. Because of the fluidity of membranes and the appearance of the proteins seen by electron microscopy, our concept of membrane structure is called the **fluid mosaic model** (Figure 18.13).

Some of the observed proteins, called **peripheral membrane proteins,** are bound only to one of the surfaces of the membrane by interactions between ionic

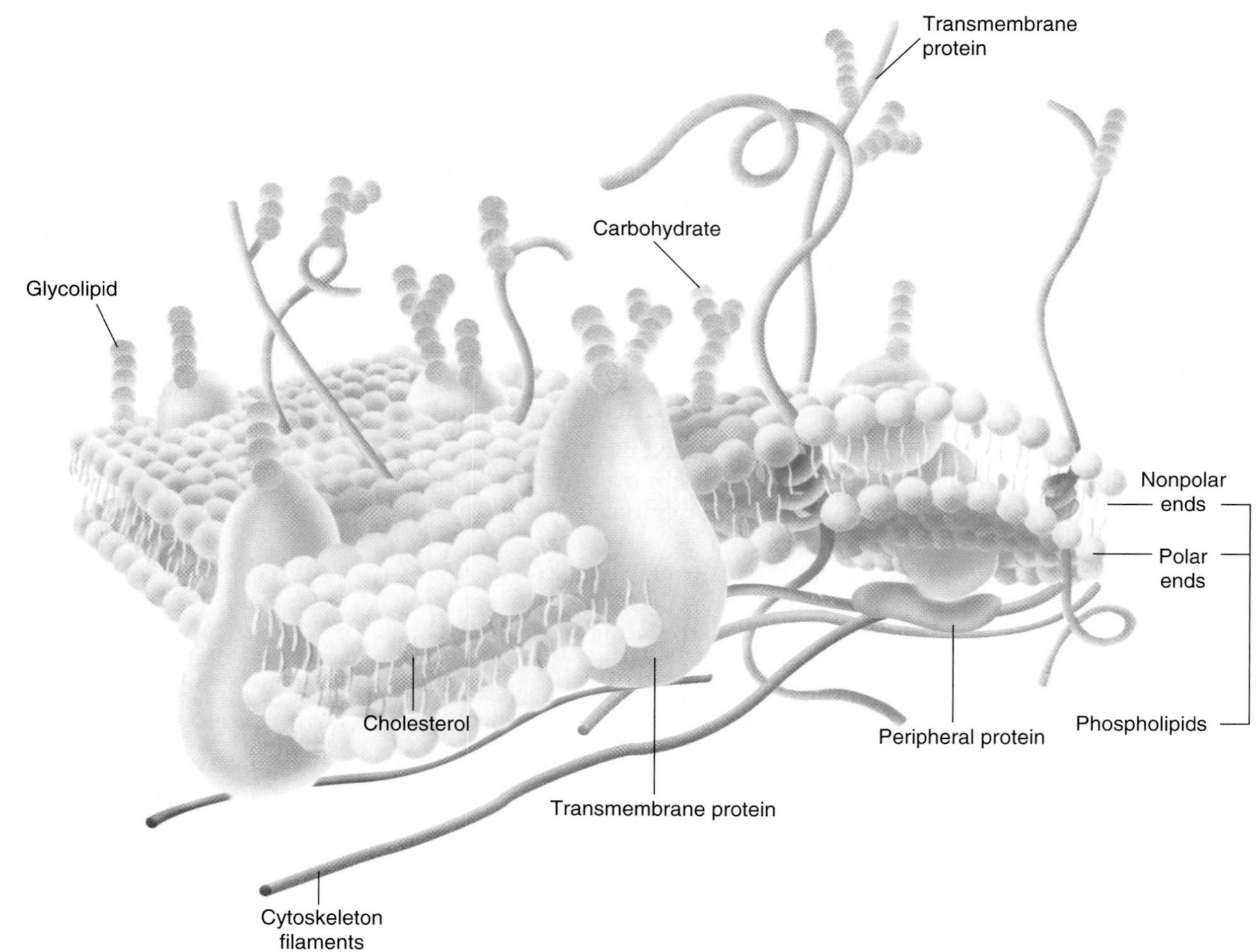

Figure 18.13
The fluid mosaic model of membrane structure.

head groups of the membrane lipids and ionic amino acids on the surface of the peripheral protein. Other membrane proteins, called **transmembrane proteins,** are embedded within the membrane and extend completely through it, being exposed both inside and outside the cell.

Just as the phospholipid composition of the membrane is asymmetric, so too is the orientation of transmembrane proteins. Each transmembrane protein has hydrophobic regions that associate with the fatty acid tails of membrane phospholipids. Each also has a unique hydrophilic domain that is always found associated with the outer layer of the membrane and is located on the outside of the cell. This region of the protein typically has oligosaccharides covalently attached. Hence these proteins are *glycoproteins.* Similarly, each transmembrane protein has a second hydrophilic domain that is always found associated with the inner layer of the membrane and projects into the cytoplasm of the cell. Typically this region of the transmembrane protein is attached to filaments of the cytoplasmic skeleton.

Because the lipid bilayer is fluid, there can be rapid lateral diffusion of membrane proteins through the lipid bilayer; but membrane proteins, like membrane lipids, do not "flip-flop" across the membrane or turn in the membrane like a revolving door of a department store.

Membranes are dynamic structures, as we may infer from our knowledge about the mobility of membrane proteins and lipids. The mobility of proteins embedded in biological membranes was studied by labeling certain proteins in human and mouse cell membranes with red and green fluorescent dyes. The human and mouse cells were fused; in other words, special techniques were used to cause the membranes of

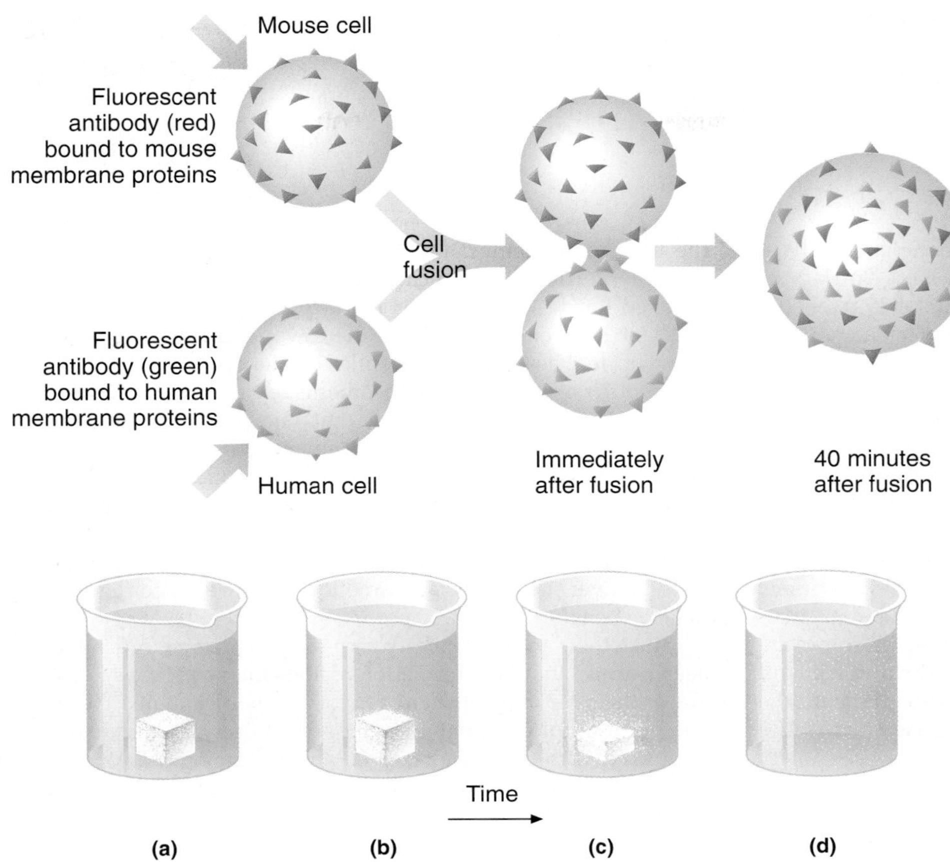

Figure 18.14
Demonstration that membranes are fluid and that proteins move freely in the plane of the lipid bilayer.

Figure 18.15
Diffusion results in the *net* movement of sugar and water molecules from the area of high concentration to the area of low concentration. Eventually, the concentrations of sugar and water throughout the beaker will be equal.

the mouse and human cell to flow together to create a single cell. The new cell was observed through a special ultraviolet or fluorescence microscope. The red and green patches were localized within regions of their original cell membranes when the experiment began. An hour later the color patches were uniformly distributed in the fused cellular membrane (Figure 18.14). This experiment suggests that we can think of the fluid mosaic membrane as an ocean filled with mobile, floating icebergs.

Membrane Transport

Learning Goal

The cell membrane mediates the interaction of the cell with its environment and is responsible for the controlled passage of material into and out of the cell. The external cell membrane controls the entrance of fuel and the exit of waste products. Internal cellular membranes partition metabolites among cell organelles. Most of these transport processes are controlled by transmembrane transport proteins. These transport proteins are the cellular gatekeepers, whose function in membrane transport is analogous to the function of enzymes in carrying out cellular chemical reactions. However, some molecules pass through the membranes unassisted, by the **passive transport** processes of diffusion and osmosis. These are referred to as passive processes because they do not require any energy expenditure by the cell.

Passive Diffusion: The Simplest Form of Membrane Transport

If we put a teaspoon of instant iced tea on the surface of a glass of water, the molecules soon spread throughout the solution. The molecules of both the solute (tea) and the solvent (water) are propelled by random molecular motion. The initially concentrated tea becomes more and more dilute. This process of the *net* movement of a solute with the gradient (from an area of high concentration to an area of low concentration) is called *diffusion* (Figure 18.15).

A CLINICAL PERSPECTIVE

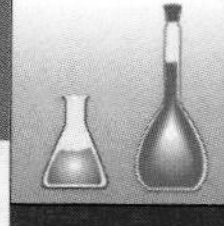

Liposome Delivery Systems

Liposomes were discovered by Dr. Alec Bangham in 1961. During his studies on phospholipids and blood clotting, he found that if he mixed phospholipids and water, tiny phospholipid bilayer sacs, called liposomes, would form spontaneously. Since that first observation, liposomes have been developed as efficient delivery systems for everything from antitumor and antiviral drugs, to the hair-loss therapy minoxidil!

If a drug is included in the solution during formation of liposomes, the phospholipids will form a sac around the solution. In this way the drug becomes encapsulated within the phospholipid sphere. These liposomes can be injected intravenously or applied to body surfaces. Sometimes scientists include hydrophilic molecules in the surface of the liposome. This increases the length of time that they will remain in circulation in the bloodstream. These so-called stealth liposomes are being used to carry anticancer drugs, such as doxorubicin and mitoxantrone. Liposomes are also being used as carriers for the antiviral drugs, such as AZT and ddC, that are used to treat human immunodeficiency virus infection.

A clever trick to help target the drug-carrying liposome is to include an antibody on the surface of the liposome. These antibodies are proteins designed to bind specifically to the surface of a tumor cell. Upon attaching to the surface of the tumor cell, the liposome "membrane" fuses with the cell membrane. In this way the deadly chemicals are delivered only to those cells targeted for destruction. This helps to avoid many of the unpleasant side effects of chemotherapy treatment that occur when normal healthy cells are killed by the drug.

Another application of liposomes is in the cosmetics industry. Liposomes can be formed that encapsulate a vitamin, herbal agent, or other nutritional element. When applied to the skin, the liposomes pass easily through the outer layer of dead skin, delivering their contents to the living skin cells beneath. As with the pharmaceutical liposomes, these liposomes, sometimes called cosmeceuticals, fuse with skin cells. Thus, they directly deliver the beneficial cosmetic agent directly to the cells that can benefit the most.

Since their accidental discovery forty years ago, much has been learned about the formation of liposomes and ways to engineer them for more efficient delivery of their contents. This is another example of the marriage of serendipity, an accidental discovery, with scientific research and technological application. As the development of new types of liposomes continues, we can expect that even more ways will be found to improve the human condition.

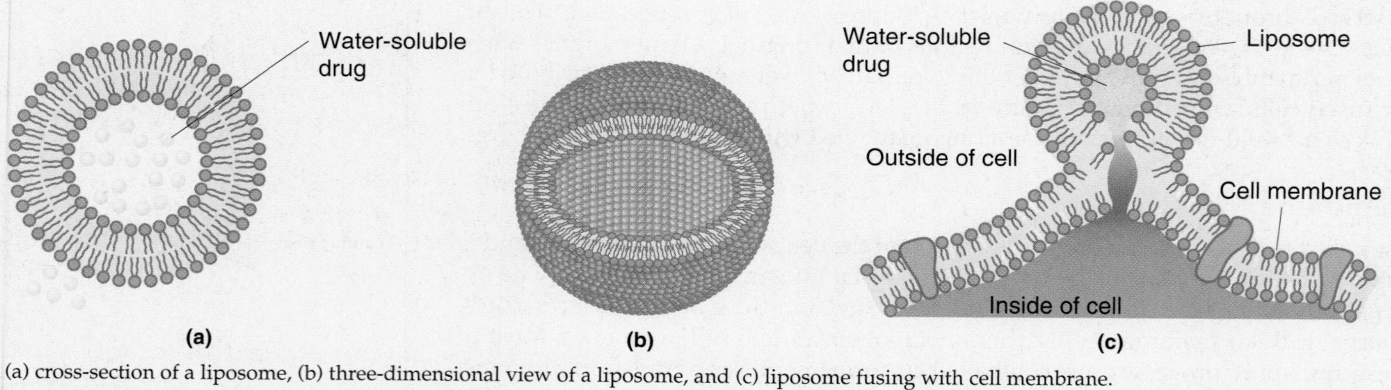

(a) cross-section of a liposome, (b) three-dimensional view of a liposome, and (c) liposome fusing with cell membrane.

Diffusion is one means of passive transport across membranes. Let's now suppose that a biological membrane is present and that a substance is found at one concentration outside the membrane and at half that concentration inside the membrane. If the solute can pass through the membrane, diffusion will occur with net transport of material from the region of initial high concentration to the region of initial low concentration, and the substance will equilibrate across the cell membrane (Figure 18.16). After a while, the concentration of the substance will be the same on both sides of the membrane; the system will be at equilibrium, and no more net change will occur.

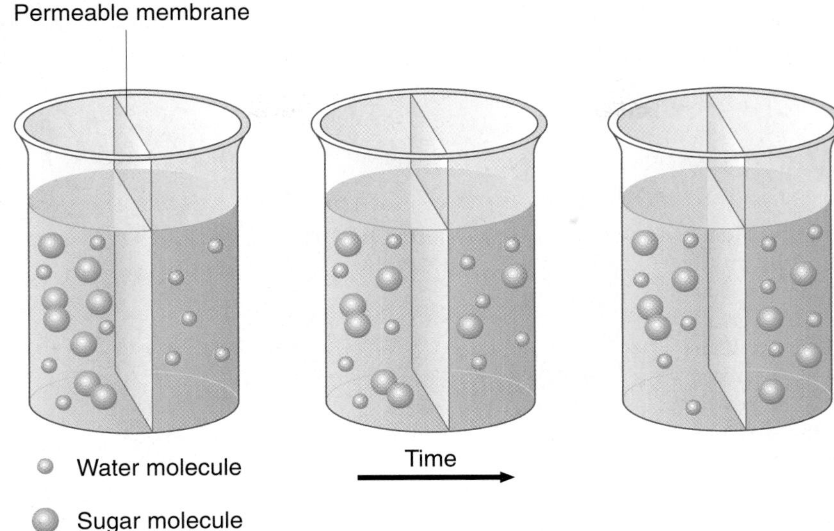

Figure 18.16
Diffusion of a solute through a membrane.

Of what practical value is the process of diffusion to the cell? Certainly, diffusion is able to distribute metabolites effectively throughout the interior of the cell. But what about the movement of molecules through the membrane? Because of the lipid bilayer structure of the membrane, only a few molecules are able to diffuse freely across a membrane. These include small molecules such as O_2 and CO_2. Any large or highly charged molecules or ions are not able to pass through the lipid bilayer directly. Such molecules require an assist from cell membrane proteins. Any membrane that allows the diffusion of some molecules but not others is said to be *selectively permeable.*

Facilitated Diffusion: Specificity of Molecular Transport

Most molecules are transported across biological membranes by specific protein carriers known as *permeases.* When a solute diffuses through a membrane from an area of high concentration to an area of low concentration by passing through a channel within a permease, the process is known as **facilitated diffusion.** No energy is consumed by facilitated diffusion; thus it is another means of passive transport, and the direction of transport depends upon the concentrations of metabolite on each side of the membrane.

Transport across cell membranes by facilitated diffusion occurs through pores within the permease that have conformations, or shapes, that are complementary to those of the transported molecules. The charge and conformation of the pore define the specificity of the carrier (Figure 18.17). Only molecules that have the correct shape can enter the pore. As a result, the rate of diffusion for any molecule is limited by the number of carrier permease molecules in the membrane that are responsible for the passage of that molecule.

The transport of glucose illustrates the specificity of carrier permease proteins. D-Glucose is transported by the glucose carrier, but its enantiomer, L-glucose, is not. Thus the glucose permease exhibits stereospecificity. In other words, the solute to be brought into the cell must "fit" precisely, like a hand in a glove, into a recognition site within the structure of the permease. In Chapter 20 we will see that enzymes that catalyze the biochemical reactions within cells show this same type of specificity.

Enzyme specificity is discussed in Section 20.5.

The rate of transport of metabolites into the cell has profound effects on the net metabolic rate of many cells. Insulin, a polypeptide hormone synthesized by the islet β-cells of the pancreas, increases the maximum rate of glucose transport by a factor of three to four. The result is that the metabolic activity of the cell is greatly increased.

The metabolic effects of insulin are described in Sections 21.7 and 23.6.

A CLINICAL PERSPECTIVE

Antibiotics That Destroy Membrane Integrity

The "age of antibiotics" began in 1927 when Alexander Fleming discovered, quite by accident, that a product of the mold *Penicillium* can kill susceptible bacteria. We now know that penicillin inhibits bacterial growth by interfering with cell wall synthesis. Since Fleming's time, hundreds of antibiotics, which are microbial products that either kill or inhibit the growth of susceptible bacteria or fungi, have been discovered. The key to antibiotic therapy is to find a "target" in the microbe, a metabolic process or structure that the human does not have. In this way the antibiotic will selectively inhibit the disease-causing organism without harming the patient.

Many antibiotics disrupt cell membranes. The cell membrane is not an ideal target for antibiotic therapy because all cells, human and bacterial, have membranes. Therefore both types of cells are damaged. Because these antibiotics exhibit a wide range of toxic side effects when ingested, they are usually used to combat infections topically (on body surfaces). In this way, damage to the host is minimized but the inhibitory effect on the microbe is maximized.

Polymyxins are antibiotics produced by the bacterium *Bacillus polymyxa*. They are protein derivatives having one end that is hydrophobic because of an attached fatty acid. The opposite end is hydrophilic. Because of these properties, the polymyxins bind to membranes with the hydrophobic end embedded within the membrane, while the hydrophilic end remains outside the cell. As a result, the integrity of the membrane is disrupted, and leakage of cellular constituents occurs, causing cell death.

Amphotericin B

Nystatin

The structures of amphotericin B and nystatin, two antifungal antibiotics.

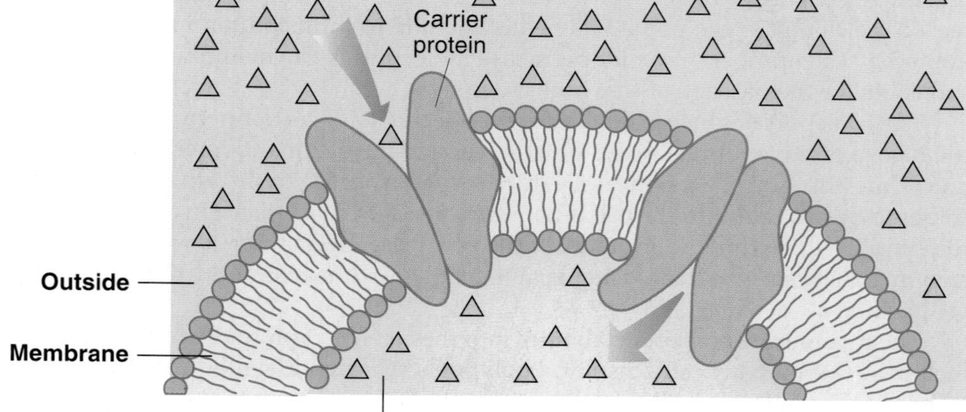

Figure 18.17
Transport by facilitated diffusion occurs through pores in the transport protein whose size and shape are complementary to those of the transported molecule.

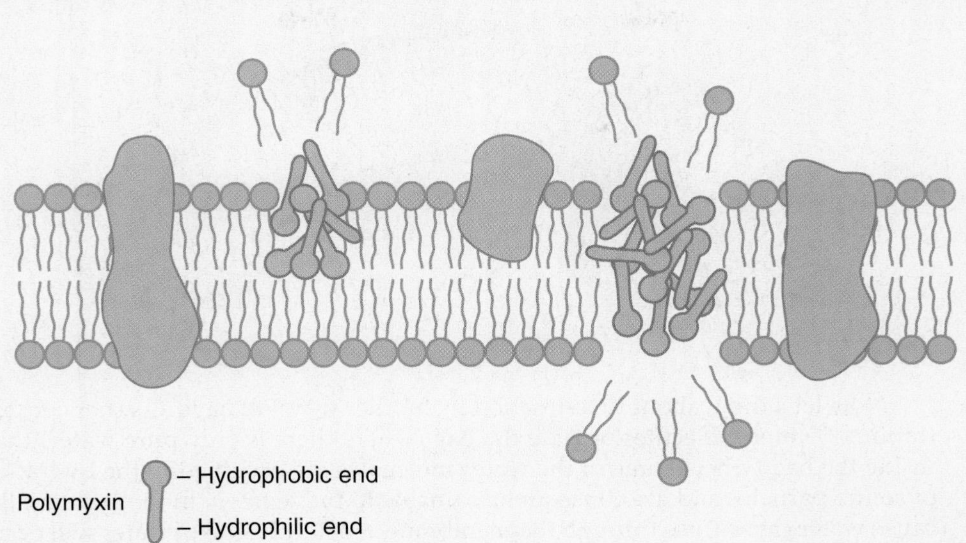

Polymyxins act like detergents, disrupting membrane integrity and killing the cell.

Although the polymyxins have been found to be useful in treating some urinary tract infections, pneumonias, and infections of burn patients, other antibiotics are now favored because of the toxic effects of the polymyxins on the kidney and central nervous system. Polymyxin B is still used topically and is available as an over-the-counter ointment in combination with two other antibiotics, neomycin and bacitracin.

Two other antibiotics that destroy membranes, amphotericin B and nystatin, are large ring structures that are used in treating serious systemic fungal infections. These antibiotics form complexes with ergosterol in the fungal cell membrane, and they disrupt the membrane permeability and cause leakage of cellular constituents. Neither is useful in treating bacterial infections because most bacteria have no ergosterol in their membranes. Both amphotericin B and nystatin are extremely toxic and cause symptoms that include nausea and vomiting, fever and chills, anemia, and renal failure. It is easy to understand why the use of these drugs is restricted to treatment of life-threatening fungal diseases.

Red blood cells use the same anion channel to transport Cl^- into the cell in exchange for HCO_3^- ions (Figure 18.18). This two-way transport is known as *antiport.* This transport occurs by facilitated diffusion so that Cl^- flows from a high exterior concentration to a low interior one, and bicarbonate flows from a high interior concentration to a low exterior one.

Osmosis: Passive Movement of a Solvent Across a Membrane

Learning Goal

Because a cell membrane is selectively permeable, it is not always possible for solutes to pass through it in response to a concentration gradient. In such cases the solvent diffuses through the membrane. Such membranes, permeable to solvent but not to solute, are specifically called **semipermeable membranes. Osmosis** is the diffusion of a solvent (water in biological systems) through a semipermeable membrane in response to a water concentration gradient.

Suppose that we place a 0.5 *M* glucose solution in a dialysis bag that is composed of a membrane with pores that allow the passage of water molecules but not glucose molecules. Consider what will happen when we place this bag into a beaker of pure water. We have created a gradient in which there is a higher concentration of glucose inside the bag than outside, but the glucose cannot diffuse through the bag to achieve equal concentration on both sides of the membrane.

Recall that there is an inverse relationship between the osmotic (solute) concentration of a solution and the water concentration of that solution, as discussed in Chapter 7.

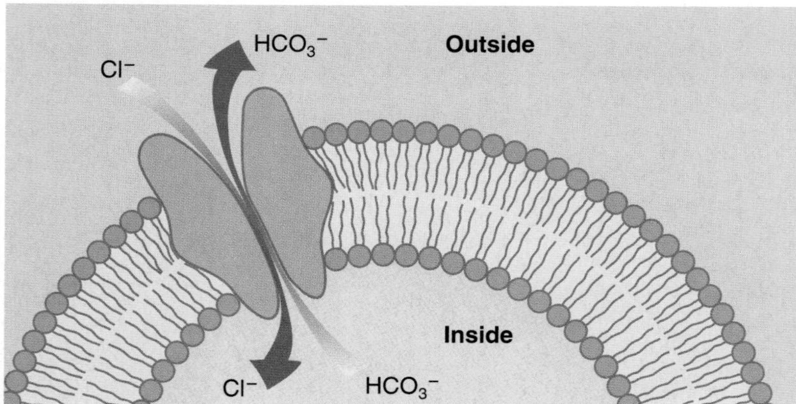

Figure 18.18
Transport of Cl^- and HCO_3^- ions in opposite directions across the red blood cell membrane.

Now let's think about this situation in another way. We have a higher concentration of water molecules outside the bag (where there is only pure water) than inside the bag (where some of the water molecules are occupied in the hydration of solute particles and are consequently unable to move freely in the system). Because water can diffuse through the membrane, a net diffusion of water will occur through the membrane into the bag. This is the process of osmosis (Figure 18.19).

As you have probably already guessed, this system can never reach equilibrium (equal concentrations inside and outside the bag) because regardless of how much water diffuses into the bag, diluting the glucose solution, the concentration of glucose will always be higher inside the bag (and the accompanying free water concentration will always be lower).

What happens when the bag has taken in as much water as it can, when it has expanded as much as possible? Now the walls of the bag exert a force that will stop the *net* flow of water into the bag. **Osmotic pressure** is the pressure that must be exerted to stop the flow of water across a selectively permeable membrane by osmosis. Stated more precisely, the osmotic pressure of a solution is the net pressure with which water enters it by osmosis from a pure water compartment when the two compartments are separated by a semipermeable membrane.

Osmotic concentration or *osmolarity* is the term used to describe the osmotic strength of a solution. It depends only on the ratio of the number of solute particles to the number of solvent particles. Thus the chemical nature and size of the solute are not important, only the concentration, expressed in molarity. For instance, a 2 *M* solution of glucose (a sugar of molecular weight 180) has the same osmolarity as a 2 *M* solution of albumin (a protein of molecular weight 60,000).

Blood plasma has an osmolarity equivalent to a 0.30 *M* glucose solution or a 0.15 *M* NaCl solution. This latter is true because in solution NaCl dissociates into Na^+ and Cl^- and thus contributes twice the number of solute particles as a molecule that does not ionize. If red blood cells, which have an osmolarity equal to blood plasma, are placed in a 0.30 *M* glucose solution, no net osmosis will occur because the osmolarity and water concentration inside the red blood cell are equal to those of the 0.30 *M* glucose solution. The solutions inside and outside the red blood cell are said to be **isotonic** (*iso* means "same," and *tonic* means "strength") **solutions.** Because the osmolarity is the same inside and outside, the red blood cell will remain the same size (Figure 18.20b).

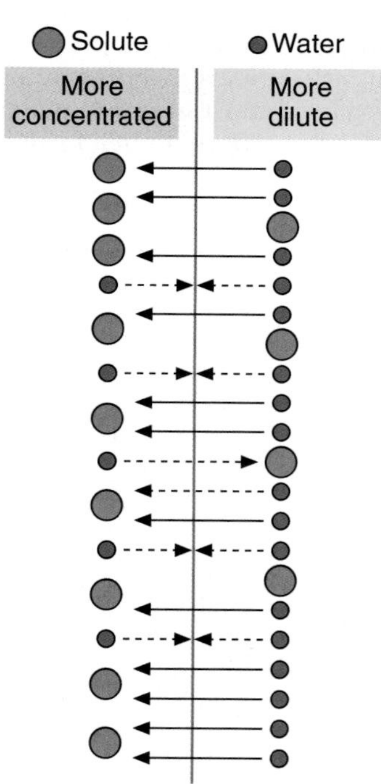

Figure 18.19
Osmosis across a membrane. The solvent, water, diffuses from an area of lower solute concentration to an area of higher solute concentration.

What happens if we now place the red blood cells into a **hypotonic solution,** in other words, a solution having a lower osmolarity than the cytoplasm of the cell? In this situation there will be a net movement of water into the cell as water diffuses down its concentration gradient. The membrane of the red blood cell does not have the strength to exert a sufficient pressure to stop this flow of water, and

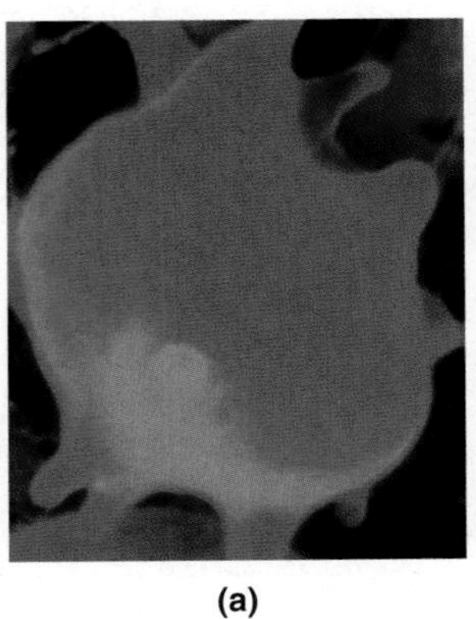
(a)

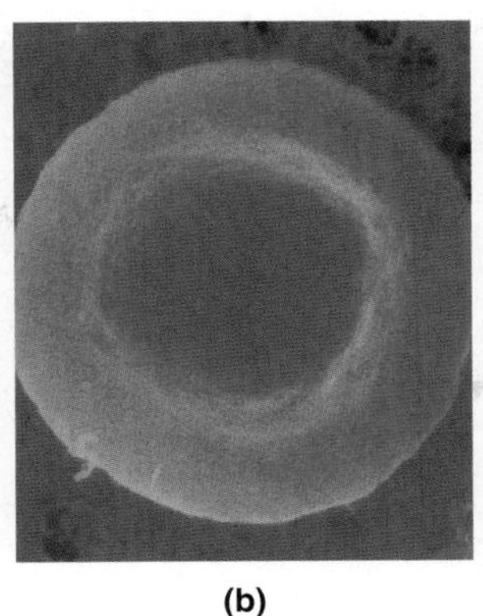
(b)

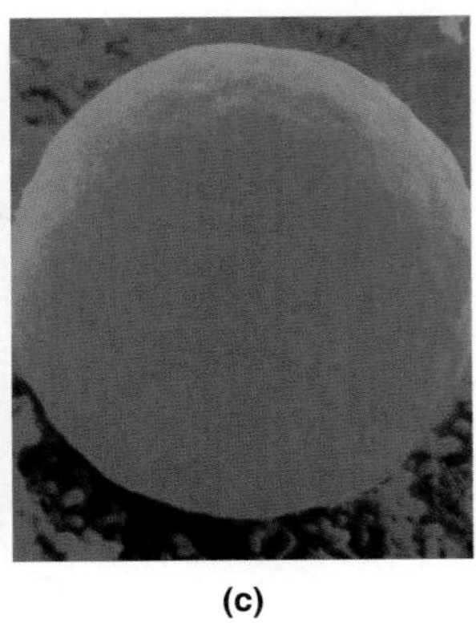
(c)

Figure 18.20
Scanning electron micrographs of red blood cells exposed to (a) hypertonic, (b) isotonic, and (c) hypotonic solutions.

the cell will swell and burst (Figure 18.20c). Alternatively, if we place the red blood cells into a **hypertonic solution** (one with a greater osmolarity than the cell), water will pass out of the cells, and they will shrink dramatically in size (Figure 18.20a).

These principles have important applications in the delivery of intravenous (IV) solutions into an individual. Normally, any fluids infused intravenously must have the correct osmolarity; they must be isotonic with the blood cells and the blood plasma. Such infusions are frequently either 5.5% dextrose (glucose) or "normal saline." The first solution is composed of 5.5 g of glucose per 100 mL of solution (0.30 *M*), and the latter of 9.0 g of NaCl per 100 mL of solution (0.15 *M*). In either case they have the same osmotic pressure and osmolarity as the plasma and blood cells and can therefore be safely administered without upsetting the osmotic balance between the blood and the blood cells.

Energy Requirements for Transport

Learning Goal 14

Simple diffusion, facilitated diffusion, and osmosis involve the spontaneous flow of materials from a region of higher concentration to an area of lower concentration (a concentration gradient). To survive, cells often must move substances "uphill," against a concentration gradient. This phenomenon, called **active transport,** requires energy. Many ions and food molecules are imported through the cell membrane by active transport. The energy used for this process may consume more than half of the total energy harvested by cellular metabolism.

A good example of active transport is the Na^+-K^+ ATPase, which moves these ions into and out of the cell against their gradients (Figure 18.21). Cells must maintain a high concentration of Na^+ outside the cell and a high concentration of K^+ inside the cell. This requires a continuous supply of cellular energy in the form of adenosine triphosphate (ATP). Over one-third of the total ATP produced by the cell is used to maintain these Na^+ and K^+ concentration gradients across the cell membrane. Thus, the name *Na^+-K^+ ATPase* refers to the enzymatic activity that hydrolyzes ATP. The hydrolysis of ATP releases the energy needed to move Na^+ and K^+ ions across the cell membrane. For each ATP molecule hydrolyzed, three Na^+ are moved out of the cell and two K^+ are transported into the cell.

Question 18.11

How does membrane transport resemble enzyme catalysis?

Question 18.12

Why is D-glucose transported by the glucose transport protein whereas L-glucose is not?

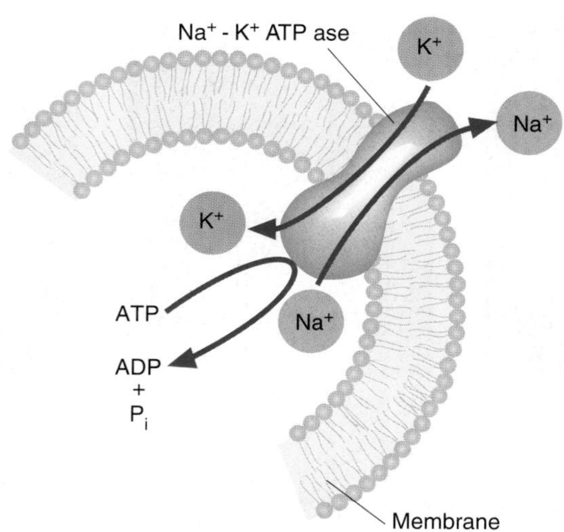

Figure 18.21
Schematic diagram of the operation of Na^+-K^+ ATPase.

Summary

18.1 Biological Functions of Lipids

Lipids are organic molecules characterized by their solubility in nonpolar solvents. Lipids are subdivided into classes based on structural characteristics: *fatty acids, glycerides,* nonglycerides, and *complex lipids.* Lipids serve many functions in the body, including energy storage, protection of organs, insulation, and absorption of vitamins. Other lipids are energy sources, hormones, or vitamins. Cells store chemical energy in the form of lipids, and the cell membrane is a lipid bilayer.

18.2 Fatty Acids

Fatty acids are *saturated* and *unsaturated* carboxylic acids containing between twelve and twenty-four carbon atoms. Fatty acids with even numbers of carbon atoms occur most frequently in nature. The reactions of fatty acids are identical to those of carboxylic acids. They include esterification, production by acid hydrolysis of esters, saponification, and addition at the double bond. *Prostaglandins,* thromboxanes, and leukotrienes are derivatives of twenty-carbon fatty acids that have a variety of physiological effects.

18.3 Glycerides

Glycerides are the most abundant lipids. The triesters of glycerol (*triglycerides*) are of greatest importance. Neutral triglycerides are important because of their ability to store energy. The ionic *phospholipids* are important components of all biological membranes.

18.4 Nonglyceride Lipids

Nonglyceride lipids consist of *sphingolipids, steroids,* and *waxes. Sphingomyelin* is a component of the myelin sheath around cells of the central nervous system. The *steroids* are important for many biochemical functions: *Cholesterol* is a membrane component; testosterone, progesterone, and estrone are sex hormones; and cortisone is an anti-inflammatory steroid that is important in the regulation of many biochemical pathways.

18.5 Complex Lipids

Plasma lipoproteins are complex lipids that transport other lipids through the bloodstream. *Chylomicrons* carry dietary triglycerides from the intestine to other tissues. *Very low density lipoproteins* carry triglycerides synthesized in the liver to other tissues for storage. *Low-density lipoproteins* carry cholesterol to peripheral tissues and help regulate blood cholesterol levels. *High-density lipoproteins* transport cholesterol from peripheral tissues to the liver.

18.6 The Structure of Biological Membranes

The *fluid mosaic model* of membrane structure pictures biological membranes that are composed of lipid bilayers in which proteins are embedded. Membrane lipids contain polar head groups and nonpolar hydrocarbon tails. The hydrocarbon tails of phospholipids are derived from saturated and unsaturated long-chain fatty acids containing an even number of carbon atoms. The lipids and proteins diffuse rapidly in the lipid bilayer but seldom cross from one side to the other.

The simplest type of membrane transport is passive diffusion of a substance across the lipid bilayer from the region of higher concentration to that of lower concentration. Many metabolites are transported across biological membranes by permeases that form pores through the membrane. The conformation of the pore is complementary to that of the substrate to be transported. Cells use energy to transport molecules across the plasma membrane against their concentration gradients, a process known as *active transport*.

The Na^+-K^+ ATPase hydrolyzes one molecule of ATP to provide the driving force for pumping three Na^+ out of the cell in exchange for two K^+.

Key Terms

active transport (18.6)
arachidonic acid (18.2)
atherosclerosis (18.4)
cholesterol (18.4)
chylomicron (18.5)
complex lipid (18.5)
diglyceride (18.3)
eicosanoid (18.2)
emulsifying agent (18.3)
essential fatty acid (18.2)
esterification (18.2)
facilitated diffusion (18.6)
fatty acid (18.2)
fluid mosaic model (18.6)
glyceride (18.3)
high-density lipoprotein (HDL) (18.5)
hydrogenation (18.2)
hypertonic solution (18.6)
hypotonic solution (18.6)
isotonic solution (18.6)
lipid (18.1)
low-density lipoprotein (LDL) (18.5)
monoglyceride (18.3)
neutral glyceride (18.3)
osmosis (18.6)
osmotic pressure (18.6)
passive transport (18.6)
peripheral membrane protein (18.6)
phosphatidate (18.3)
phosphoglyceride (18.3)
phospholipid (18.3)
plasma lipoprotein (18.5)
prostaglandin (18.2)
saponification (18.2)
saturated fatty acid (18.2)
semipermeable membrane (18.6)
sphingolipid (18.4)
sphingomyelin (18.4)
steroid (18.4)
terpene (18.4)
transmembrane protein (18.6)
triglyceride (18.3)
unsaturated fatty acid (18.2)
very low density lipoprotein (VLDL) (18.5)
wax (18.4)

Questions and Problems

Biological Functions of Lipids

18.13 List the four main groups of lipids.
18.14 List the biological functions of lipids.

Fatty Acids

18.15 What is the difference between a saturated and an unsaturated fatty acid?
18.16 Write the structure for a saturated and an unsaturated fatty acid.
18.17 As the length of the hydrocarbon chain of saturated fatty acids increases, what is the effect on the melting points?
18.18 As the number of carbon-carbon double bonds in fatty acids increases, what is the effect on the melting points?
18.19 Draw the structures of each of the following fatty acids:
- **a.** Decanoic acid
- **b.** Stearic acid
- **c.** *trans*-5-Decenoic acid
- **d.** *cis*-5-Decenoic acid

18.20 What are the common and I.U.P.A.C. names of each of the following fatty acids?
- **a.** $C_{15}H_{31}COOH$
- **b.** $C_{11}H_{23}COOH$
- **c.** $CH_3(CH_2)_5CH{=}CH(CH_2)_7COOH$
- **d.** $CH_3(CH_2)_7CH{=}CH(CH_2)_7COOH$

18.21 Write an equation for each of the following reactions:
- **a.** Esterification of glycerol with three molecules of myristic acid
- **b.** Acid hydrolysis of tristearoyl glycerol
- **c.** Reaction of decanoic acid with KOH
- **d.** Hydrogenation of linoleic acid

18.22 Write an equation for each of the following reactions:
- **a.** Esterification of glycerol with three molecules of palmitic acid
- **b.** Acid hydrolysis of trioleoyl glycerol
- **c.** Reaction of stearic acid with KOH
- **d.** Hydrogenation of oleic acid

18.23 What is the function of the essential fatty acids?
18.24 What molecules are formed from arachidonic acid?
18.25 What is the biochemical basis for the effectiveness of aspirin in decreasing the inflammatory response?
18.26 What is the role of prostaglandins in the inflammatory response?
18.27 List four effects of prostaglandins.
18.28 What are the functions of thromboxane A_2 and leukotrienes?

Glycerides

18.29 Draw the structure of the triglyceride molecule formed by esterification at C-1, C-2, and C-3 with hexadecanoic acid, *trans*-9-hexadecenoic acid, and *cis*-9-hexadecenoic acid, respectively.
18.30 Draw one possible structure of a triglyceride that contains the three fatty acids stearic acid, palmitic acid, and oleic acid.
18.31 Draw the structure of the phosphatidate formed between glycerol-3-phosphate that is esterified at C-1 and C-2 with capric and lauric acids, respectively.
18.32 Draw the structure of a lecithin molecule in which the fatty acyl groups are derived from stearic acid.
18.33 What are the structural differences between triglycerides (triacylglycerols) and phospholipids?
18.34 How are the structural differences between triglycerides and phospholipids reflected in their different biological functions?

Nonglyceride Lipids

18.35 What is a sphingolipid?
18.36 What is the function of sphingomyelin?
18.37 What is the role of cholesterol in biological membranes?
18.38 How does cholesterol contribute to atherosclerosis?
18.39 What are the biological functions of progesterone, testosterone, and estrone?
18.40 How has our understanding of the steroid sex hormones contributed to the development of oral contraceptives?
18.41 What is the medical application of cortisone?
18.42 What are the possible side effects of cortisone treatment?
18.43 A wax found in beeswax is myricyl palmitate. What fatty acid and what alcohol are used to form this compound?
18.44 A wax found in the head of sperm whales is cetyl palmitate. What fatty acid and what alcohol are used to form this compound?
18.45 What are isoprenoids?
18.46 What is a terpene?
18.47 List some important biological molecules that are terpenes.
18.48 Draw the five-carbon isoprene unit.

Complex Lipids

18.49 What are the four major types of plasma lipoproteins?
18.50 What is the function of each of the four types of plasma lipoproteins?

18.51 What is the relationship between atherosclerosis and high blood pressure?
18.52 How is LDL taken into cells?
18.53 How does a genetic defect in the LDL receptor contribute to atherosclerosis?
18.54 What is the correlation between saturated fats in the diet and atherosclerosis?

The Structure of Biological Membranes

18.55 How will the properties of a biological membrane change if the fatty acid tails of the phospholipids are converted from saturated to unsaturated chains?
18.56 What is the function of unsaturation in the hydrocarbon tails of membrane lipids?
18.57 What is the basic structure of a biological membrane?
18.58 Describe the fluid mosaic model of membrane structure.
18.59 Describe peripheral membrane proteins.
18.60 Describe transmembrane proteins and list some of their functions.
18.61 What is the major effect of cholesterol on the properties of biological membranes?
18.62 Why do the hydrocarbon tails of membrane phospholipids provide a barrier between the inside and outside of the cell?
18.63 What experimental observation shows that proteins diffuse within the lipid bilayers of biological membranes?
18.64 Why don't proteins turn around in biological membranes like revolving doors?

Biological Membranes: Transport

18.65 Explain the difference between simple diffusion across a membrane and facilitated diffusion.
18.66 Explain what would happen to a red blood cell placed in each of the following solutions:
a. hypotonic
b. isotonic
c. hypertonic
18.67 How does active transport differ from facilitated diffusion?
18.68 By what mechanism are Cl^- and HCO_3^- ions transported across the red blood cell membrane?
18.69 What is the meaning of the term *antiport?*
18.70 How does insulin affect the transport of glucose?
18.71 What properties of a transport protein (permease) determine its specificity?
18.72 Why is the function of the Na^+-K^+ ATPase an example of active transport?
18.73 What is the stoichiometry of the Na^+-K^+ ATPase?
18.74 How will the Na^+ and K^+ concentrations of a cell change if the Na^+-K^+ ATPase is inhibited?
18.75 What is meant by the term *active transport?*

Critical Thinking Problems

1. Olestra is a fat substitute that provides no calories, yet has all the properties of a naturally occurring fat. It has a creamy, tongue-pleasing consistency. Unlike other fat substitutes, olestra can withstand heating. Thus, it can be used to prepare foods such as potato chips and crackers. Olestra is a sucrose polyester and is produced by esterification of six, seven, or eight fatty acids to molecules of sucrose. Draw the structure of one such molecule having eight stearic acid acyl groups attached.
2. Liposomes can be made by vigorously mixing phospholipids (like phosphatidylcholine) in water. When the mixture is allowed to settle, spherical vesicles form that are surrounded by a phospholipid bilayer "membrane." Pharmaceutical chemists are trying to develop liposomes as a targeted drug delivery system. By adding the drug of choice to the mixture described above, liposomes form around the solution of drug. Specific proteins can be incorporated into the mixture that will end up within the phospholipid bilayers of the liposomes. These proteins are able to bind to targets on the surface of particular kinds of cells in the body. Explain why injection of liposome encapsulated pharmaceuticals might be a good drug delivery system.
3. "Cholesterol is bad and should be eliminated from the diet." Do you agree or disagree? Defend your answer.
4. Why would a phospholipid such as lecithin be a good emulsifying agent for ice cream?
5. When a plant becomes cold-adapted, the composition of the membranes changes. What changes in fatty acid and cholesterol composition would you predict? Explain your reasoning.
6. In terms of osmosis, explain why it would be preferable for a cell to store 10,000 molecules of glycogen each composed of 10^5 molecules of glucose rather than to store 10^9 individual molecules of glucose.

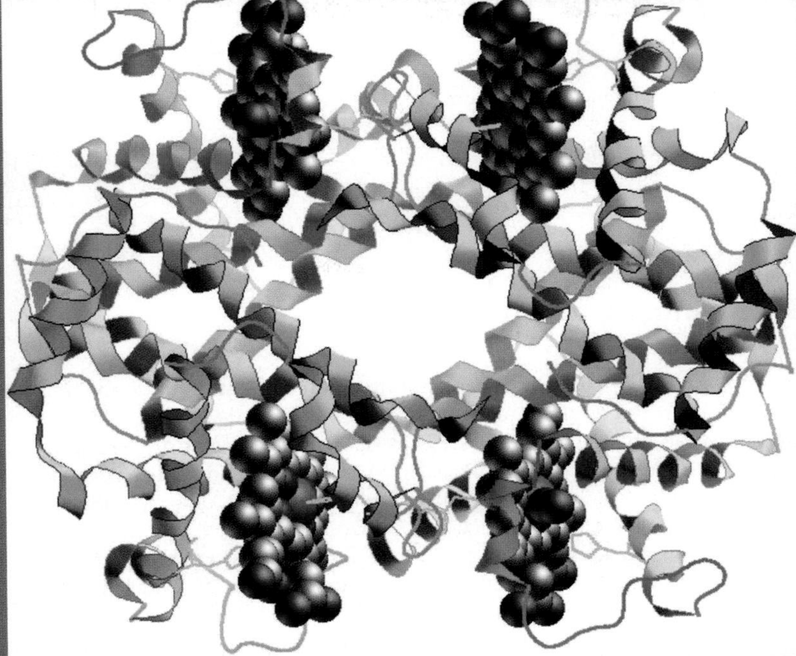

Computer-generated model of the structure of a protein.

19

Protein Structure and Function

BIOCHEMISTRY

Outline

Learning Goals

1. List the functions of proteins.
2. Draw the general structure of an amino acid and classify amino acids based on their R groups.
3. Describe the primary structure of proteins and draw the structure of the peptide bond.
4. Draw the structure of small peptides and name them.
5. Describe the types of secondary structure of a protein.
6. Discuss the forces that maintain secondary structure.
7. Describe the structure and functions of fibrous proteins.
8. Describe the tertiary and quaternary structure of a protein.
9. List the R group interactions that maintain protein conformation.
10. List examples of proteins that require prosthetic groups and explain the way in which they function.
11. Discuss the importance of the three-dimensional structure of a protein to its function.
12. Describe the roles of hemoglobin and myoglobin.
13. Describe how extremes of pH and temperature cause denaturation of proteins.
14. Explain the difference between essential and nonessential amino acids.

CHEMISTRY CONNECTION

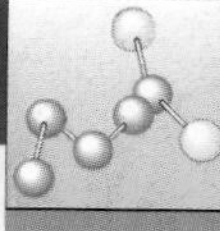

Angiogenesis Inhibitors: Proteins That Inhibit Tumor Growth

Cancer researchers have long known that solid tumors cannot grow larger than the size of a pinhead unless they stimulate the formation of new blood vessels that provide the growing tumor with nutrients and oxygen and remove the waste products of cellular metabolism. Studies of *angiogenesis,* the formation of new blood vessels, in normal tissues have provided new weapons in the arsenal of anticancer drugs.

Angiogenesis occurs through a carefully controlled sequence of steps. Consider the process of tissue repair. One of several protein growth factors stimulates the endothelial cells that form the lining of an existing blood vessel to begin growing, dividing, and migrating into the tissue to be repaired. Threads of new endothelial cells organize themselves into hollow cylinders, or tubules. These tubules become a new network of blood vessels throughout the damaged tissue. These new blood vessels bring the needed nutrients, oxygen, and other factors to the site of damage, allowing the tissue to be repaired and healing to occur.

In addition to the growth factors that stimulate this process, there are several other proteins that inhibit the formation of new blood vessels. In fact, the normal process of angiogenesis is dependent on the appropriate balance of the stimulatory growth factors and the inhibitory proteins.

The normal events of angiogenesis are duplicated at a critical moment in the growth of a tumor. Cells of the tumor secrete one or more of the growth factors known to stimulate angiogenesis. The newly formed blood vessels provide the cells of the growing tumor with everything needed to continue growing and dividing.

Metastasis, the spreading of tumor cells to other sites in the body, also requires angiogenesis. Typically, those tumors having more blood vessels are more likely to metastasize. Clinically, treatment of these tumors has a poorer outcome.

Researchers considered all of this information known about angiogenesis and its impact on tumor formation and metastasis. They developed the hypothesis that proteins that inhibit blood vessel formation might be effective weapons against developing tumors. If this hypothesis turned out to be supported by experimental data, there would be a number of advantages to the use of angiogenesis inhibitors. Because these proteins are normally produced by the human body, they should not have the toxic side effects caused by so many anticancer drugs. In addition, angiogenesis inhibitors can overcome the problem of cancer cell drug resistance. Most cancer cells are prone to mutations and mutant cells resistant to the anticancer drugs develop. The angiogenesis inhibitors target normal endothelial cells, which are genetically stable. As a result, drug resistance is much less likely to occur.

Endostatin is one of the anti-angiogenesis proteins. Discovered in 1997, it was found to be a protein of 20,000 g/mol, which is a fragment of the C-terminus of collagen XVIII. Experimentally, endostatin is a potent inhibitor of tumor growth. It binds to the heparin sulfate proteoglycans of the cell surface and interferes with growth factor signaling. As a result, the growth and division of endothelial cells is inhibited and new blood vessels are not formed.

Angiostatin is another anti-angiogenesis protein normally found in the human body. Discovered in 1994, it is a protein fragment of human plasminogen and has a molecular weight of 50,000 g/mol. The role of angiostatin in the human body is to block the growth of diseased tissue by inhibiting the formation of blood vessels. Like endostatin, it is hoped that angiostatin will block the growth of tumors by depriving them of their blood supply.

Currently, there are about twenty angiogenesis inhibitors being tested in clinical trials involving humans. Most are in phase I or II trials, which allow scientists to determine a safe dosage and assess the severity of any side effects. Only a small number of people are involved in phase I or II trials. In phase III trials, a large number of patients are divided into two groups. One group receives standard anticancer treatment plus a placebo. The other group receives standard treatment and the new drug.

As we await the results of the clinical trials involving the proteins endostatin and angiostatin, scientists explore alternative methods to attack cancer cells. Some of these involve a class of proteins called *antibodies* that can bind specifically to cancer cells and help to inhibit or destroy them.

As we will discover, there are many different classes of proteins that carry out a variety of functions for the body. Endostatin and angiostatin serve as regulatory proteins; the antibodies serve as the body's defense system against infectious diseases. These and many other proteins are the focus of this chapter.

Introduction

In the 1800s, Johannes Mulder came up with the name **protein,** a term derived from a Greek word that means "of first importance." Indeed, proteins are a very important class of food molecules because they provide an organism not only with carbon and hydrogen, but also with nitrogen and sulfur. These latter two elements are unavailable from fats and carbohydrates, the other major classes of food molecules.

In addition to their dietary importance, the proteins are the most abundant macromolecules in the cell, and they carry out most of the work in a cell. Protection of the body from infection, mechanical support and strength, and catalysis of metabolic reactions—all are functions of proteins that are essential to life.

19.1 Cellular Functions of Proteins

Learning Goal

Proteins have many biological functions, as the following short list suggests.

- **Enzymes** are biological catalysts. The majority of the enzymes that have been studied are proteins. Reactions that would take days or weeks or require extremely high temperatures without enzymes are completed in an instant. For example, the digestive enzymes *pepsin, trypsin,* and *chymotrypsin* break down proteins in our diet so that subunits can be absorbed for use by our cells.
- **Antibodies** (also called *immunoglobulins*) are specific protein molecules produced by specialized cells of the immune system in response to foreign **antigens.** These foreign invaders include bacteria and viruses that infect the body. Each antibody has regions that precisely fit and bind to a single antigen. It helps to end the infection by binding to the antigen and helping to destroy it or remove it from the body.

In the broadest sense, an antigen is any substance that stimulates an immune response.

- **Transport proteins** carry materials from one place to another in the body. The protein *transferrin* transports iron from the liver to the bone marrow, where it is used to synthesize the heme group for hemoglobin. The proteins *hemoglobin* and *myoglobin* are responsible for transport and storage of oxygen in higher organisms, respectively.
- **Regulatory proteins** control many aspects of cell function, including metabolism and reproduction. We can function only within a limited set of conditions. For life to exist, body temperature, the pH of the blood, and blood glucose levels must be carefully regulated. Many of the hormones that regulate body function, such as *insulin* and *glucagon,* are proteins.
- **Structural proteins** provide mechanical support to large animals and provide them with their outer coverings. Our hair and fingernails are largely composed of the protein *keratin.* Other proteins provide mechanical strength for our bones, tendons, and skin. Without such support, large, multicellular organisms like ourselves could not exist.
- **Movement proteins** are necessary for all forms of movement. Our muscles, including that most important muscle, the heart, contract and expand through the interaction of *actin* and *myosin* proteins. Sperm can swim because they have long flagella made up of proteins.
- **Nutrient proteins** serve as sources of amino acids for embryos or infants. Egg *albumin* and *casein* in milk are examples of nutrient storage proteins.

19.2 The α-Amino Acids

Learning Goal

The proteins of the body are made up of some combination of twenty different subunits called **α-amino acids.** The general structure of an α-amino acid is shown in Figure 19.1. We find that nineteen of the twenty amino acids that are commonly isolated from proteins have this same general structure; they are primary amines on the α-carbon. The remaining amino acid, proline, is a secondary amine.

The α-carbon in the general structure is attached to a carboxylate group (a carboxyl group that has lost a proton, $—COO^-$) and a protonated amino group (an amino group that has gained a proton, $—NH_3^+$). In aqueous solution of approximately pH 7, conditions required for life functions, amino acids in which the carboxylate group is protonated (—COOH) and the amino group is unprotonated

A CLINICAL PERSPECTIVE

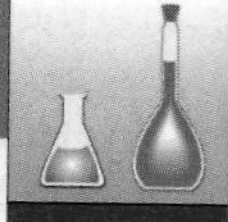

Proteins in the Blood

The blood plasma of a healthy individual typically contains 60–80 g/L of protein. This protein can be separated into five classes designated α through γ. The separation is based on the overall surface charge on each of the types of protein. (See A Clinical Perspective: Enzymes, Isoenzymes, and Myocardial Infarction in Chapter 20 for a discussion of the separation of proteins based on surface charge.)

The most abundant protein in the blood is albumin, making up about 55% of the blood protein. Albumin contributes to the osmotic pressure of the blood simply because it is a dissolved molecule. It also serves as a nonspecific transport molecule for important metabolites that are otherwise poorly soluble in water. Among the molecules transported through the blood by albumin are bilirubin (a waste product of the breakdown of hemoglobin), Ca^{2+}, and fatty acids (organic anions).

The α-globulins (α_1 and α_2) make up 13% of the plasma proteins. They include glycoproteins (proteins with sugar groups attached), high-density lipoproteins, haptoglobin (a transport protein for free hemoglobin), ceruloplasmin (a copper transport protein), prothrombin (a protein involved in blood clotting), and very low density lipoproteins. The most abundant is α_1-globulin α_1-antitrypsin. Although the name leads us to believe that this protein inhibits a digestive enzyme, trypsin, the primary function of α_1-antitrypsin is the inactivation of an enzyme that causes damage in the lungs (see also, A Medical Perspective: α_1-Antitrypsin and Familial Emphysema in Chapter 20). α_1-Antichymotrypsin is another inhibitor found in the bloodstream. This protein, along with amyloid proteins, is found in the amyloid plaques characteristic of Alzheimer's disease (AD). As a result, it has been suggested that an overproduction of this protein may contribute to AD. In the blood, α_1-antichymotrypsin is also found complexed to prostate specific antigen (PSA), the protein antigen that is measured as an indicator of prostate cancer. Elevated PSA levels are observed in those with the disease. It is interesting to note that PSA is a chymotrypsin-like proteolytic enzyme.

The β-globulins represent 13% of the blood plasma proteins and include transferrin (an iron transport protein) and low-density lipoprotein. Fibrinogen, a protein involved in coagulation of blood, comprises 7% of the plasma protein. Finally, the γ-globulins, IgG, IgM, IgA, IgD, and IgE, make up the remaining 11% of the plasma proteins. The γ-globulins are synthesized by B lymphocytes, but most of the remaining plasma proteins are synthesized in the liver. In fact, a frequent hallmark of liver disease is reduced amounts of one or more of the plasma proteins.

(—NH_2), do not exist. Under these conditions, the carboxyl group ionizes, and the basic amino group picks up the proton that is released. Any neutral molecule with equal numbers of positive and negative charges is called a *zwitterion*. Thus, amino acids in water exist as dipolar ions called zwitterions.

The α-carbon of each amino acid is also bonded to a hydrogen atom and a side chain, or R group. In a protein, the R groups interact with one another through a variety of weak attractive forces. These interactions participate in folding the protein chain into a precise three-dimensional shape that determines its ultimate function. They also serve to maintain that three-dimensional conformation.

The α-carbon is attached to four different groups in all amino acids except glycine. The α-carbon of α-amino acids is therefore chiral. That is, an α-amino acid isolated from a protein cannot be superimposed on its mirror image. Glycine has two hydrogen atoms attached to the α-carbon and is the only amino acid commonly found in proteins that is not chiral.

Stereochemistry is discussed in Section 17.3 and in Appendix D, Stereochemistry and Stereoisomers Revisited.

The configuration of α-amino acids isolated from proteins is L-. This is based on comparison of amino acids with D-glyceraldehyde (Figure 19.2). In Figure 19.2 we see a comparison of D- and L-glyceraldehyde with D- and L-alanine. Notice that the most oxidized end of the molecule, in each case the carbonyl group, is drawn at the top of the molecule. In the D-isomer of glyceraldehyde, the —OH group is on the right. Similarly, in the D-isomer of alanine, the —N^+H_3 is on the right. In the L-isomers of the two compounds, the —OH and —N^+H_3 groups are on the left. By this comparison with the enantiomers of glyceraldehyde, we can define the D- and L-enantiomers of the amino acids.

In Chapter 17 we learned that almost all of the monosaccharides found in nature are in the D-family. Just the opposite is true of the α-amino acids. Almost all of

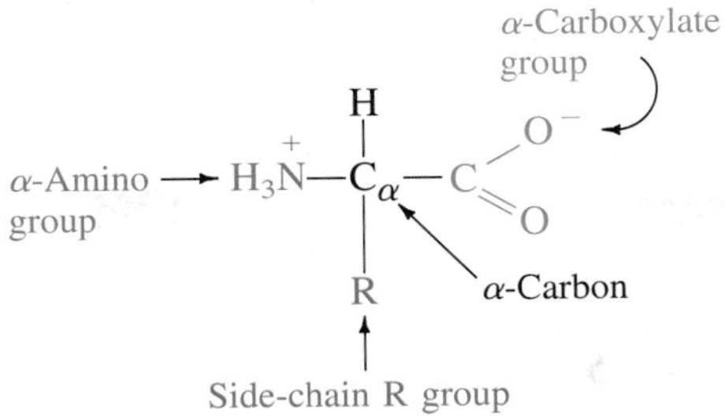

α-Carboxylate group
α-Amino group → $H_3\overset{+}{N}—C_\alpha(H)(R)—C(=O)O^-$
α-Carbon
Side-chain R group

Figure 19.1
General structure of an α-amino acid. All amino acids isolated from proteins, with the exception of proline, have this general structure.

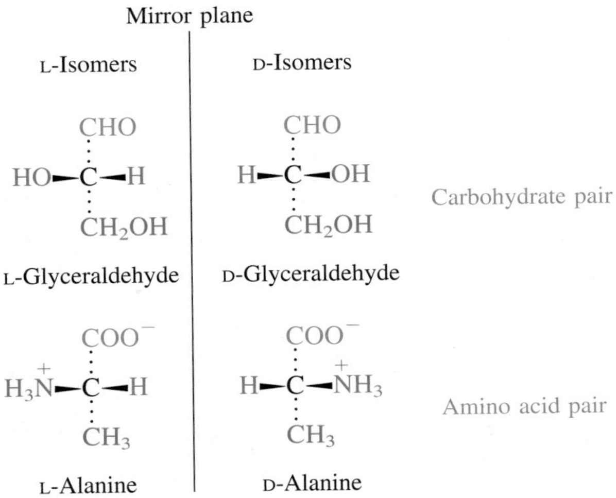

Mirror plane

L-Isomers | D-Isomers

L-Glyceraldehyde: CHO, HO—C—H, CH_2OH | D-Glyceraldehyde: CHO, H—C—OH, CH_2OH — Carbohydrate pair

L-Alanine: COO^-, $H_3\overset{+}{N}$—C—H, CH_3 | D-Alanine: COO^-, H—C—$\overset{+}{N}H_3$, CH_3 — Amino acid pair

Figure 19.2
Structure of D- and L-glyceraldehyde and their relationship to D- and L-alanine. (The student should build models of these compounds, from which it will be immediately apparent that the members of each pair are nonsuperimposable mirror images.)

the α-amino acids isolated from proteins in nature are members of the L-family. In other words, the orientation of the four groups around the chiral carbon of these α-amino acids resembles the orientation of the four groups around the chiral carbon of L-glyceraldehyde.

Because all of the amino acids have a carboxyl group and an amino group, all differences between amino acids depend upon their side-chain R groups. The amino acids are grouped in Figure 19.3 according to the polarity of their side chains.

The side chains of some amino acids are nonpolar. They prefer contact with one another over contact with water and are said to be **hydrophobic** ("water-fearing") **amino acids.** They are generally found buried in the interior of proteins, where they can associate with one another and remain isolated from water. Nine amino acids fall into this category: alanine, valine, leucine, isoleucine, proline, glycine, methionine, phenylalanine, and tryptophan. The R group of proline is unique; it is actually bonded to the α-amino group, forming a secondary amine.

The hydrophobic interaction between nonpolar R groups is one of the forces that helps maintain the proper three-dimensional shape of a protein.

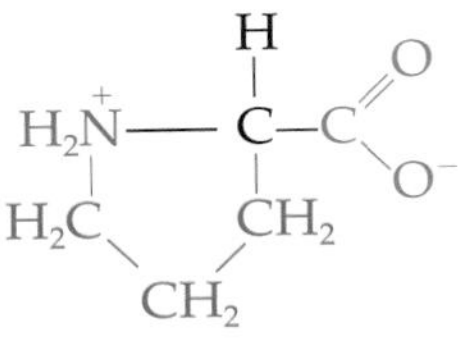

$H_2\overset{+}{N}$—C(H)—$C(=O)O^-$, H_2C, CH_2, CH_2

Proline (Pro)

The side chains of the remaining amino acids are polar. Because they are attracted to polar water molecules, they are said to be **hydrophilic** ("water-loving") **amino acids.** The hydrophilic side chains are often found on the surfaces of proteins. The polar amino acids can be subdivided into three classes.

- *Polar, neutral amino acids* have R groups that have a high affinity for water but that are not ionic at pH 7. Serine, threonine, tyrosine, cysteine, asparagine, and glutamine fall into this category. Most of these amino acids associate with one another by hydrogen bonding; but cysteine molecules form disulfide bonds with one another, as we will discuss in Section 19.6.
- *Negatively charged amino acids* have ionized carboxyl groups in their side chains. At pH 7 these amino acids have a net charge of −1. Aspartate and

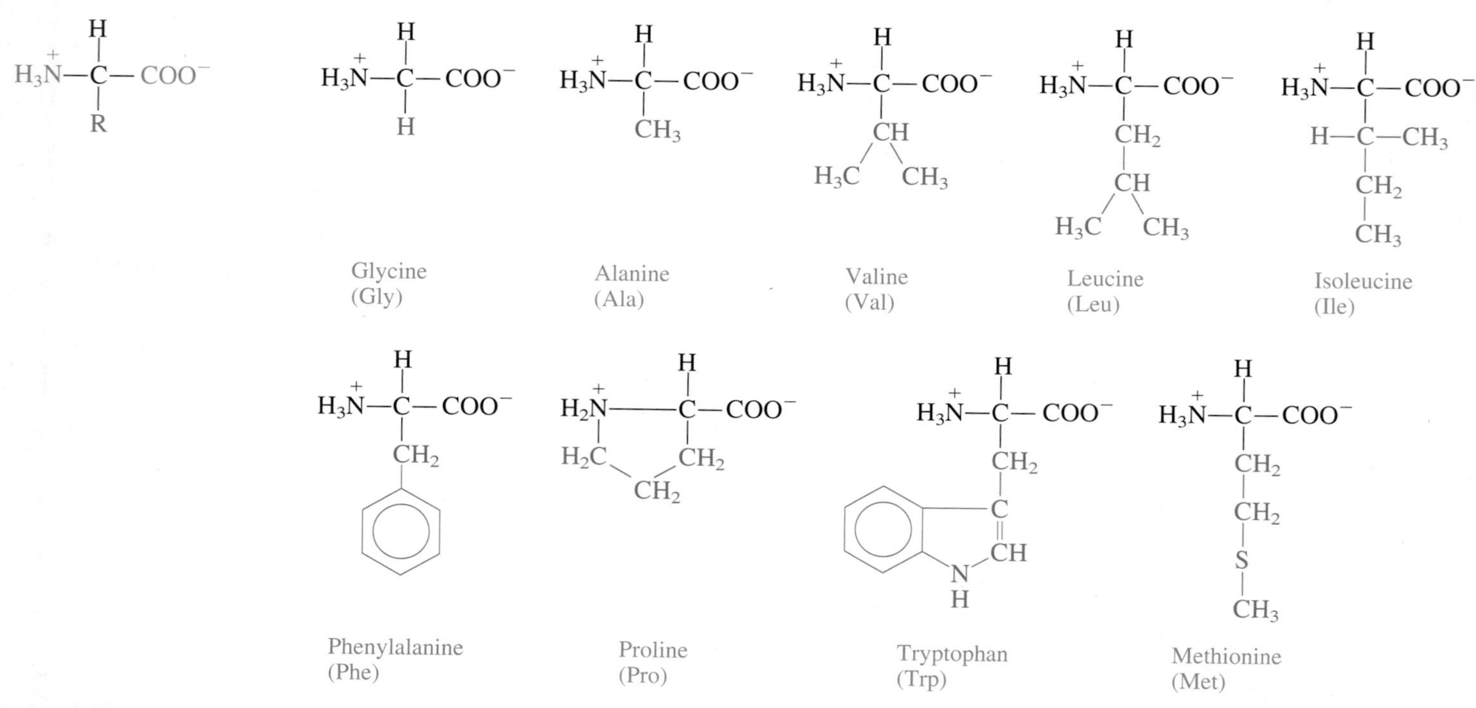

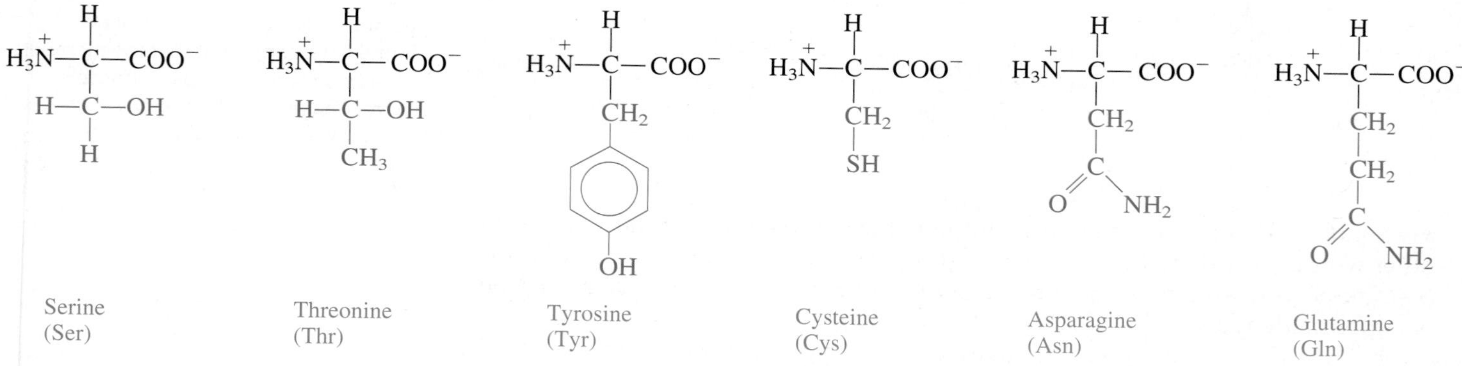

(d) Negatively charged amino acids

Aspartate (Asp)

Glutamate (Glu)

(e) Positively charged amino acids

Lysine (Lys)

Arginine (Arg)

Histidine (His)

Figure 19.3

Structures of the amino acids at pH 7.0. (a) The general structure of an amino acid. Structures of (b) the hydrophobic; (c) polar, neutral; (d) negatively charged; and (e) positively charged amino acids.

Table 19.1 Names and Three-Letter Abbreviations of the α-Amino Acids

Amino Acid	Three-Letter Abbreviation
Alanine	ala
Arginine	arg
Asparagine	asn
Aspartate	asp
Cysteine	cys
Glutamic acid	glu
Glutamine	gln
Glycine	gly
Histidine	his
Isoleucine	ile
Leucine	leu
Lysine	lys
Methionine	met
Phenylalanine	phe
Proline	pro
Serine	ser
Threonine	thr
Tryptophan	trp
Tyrosine	tyr
Valine	val

glutamate are the two amino acids in this category. They are acidic amino acids because ionization of the carboxylic acid releases a proton.

- *Positively charged amino acids.* At pH 7, lysine, arginine, and histidine have a net positive charge because their side chains contain positive groups. These amino groups are basic because the side chain reacts with water, picking up a proton and releasing a hydroxide anion.

The names of the amino acids can be abbreviated by a three-letter code. These abbreviations are shown in Table 19.1.

Hydrogen bonding (Section 6.2) is another weak interaction that helps maintain the proper three-dimensional structure of a protein. The positively and negatively charged amino acids within a protein can interact with one another to form ionic bridges. This is yet another attractive force that helps to keep the protein chain folded in a precise way.

Question 19.1

Write the three-letter abbreviation and draw the structure of each of the following amino acids.

a. Glycine
b. Proline
c. Threonine
d. Aspartate
e. Lysine

Question 19.2

Indicate whether each of the amino acids listed in Question 19.1 is polar, nonpolar, basic, or acidic.

19.3 The Peptide Bond

Learning Goal

Proteins are linear polymers of L-α-amino acids. The carboxyl group of one amino acid is linked to the amino group of another amino acid. The amide bond formed in the reaction is called a **peptide bond** (Figure 19.4).

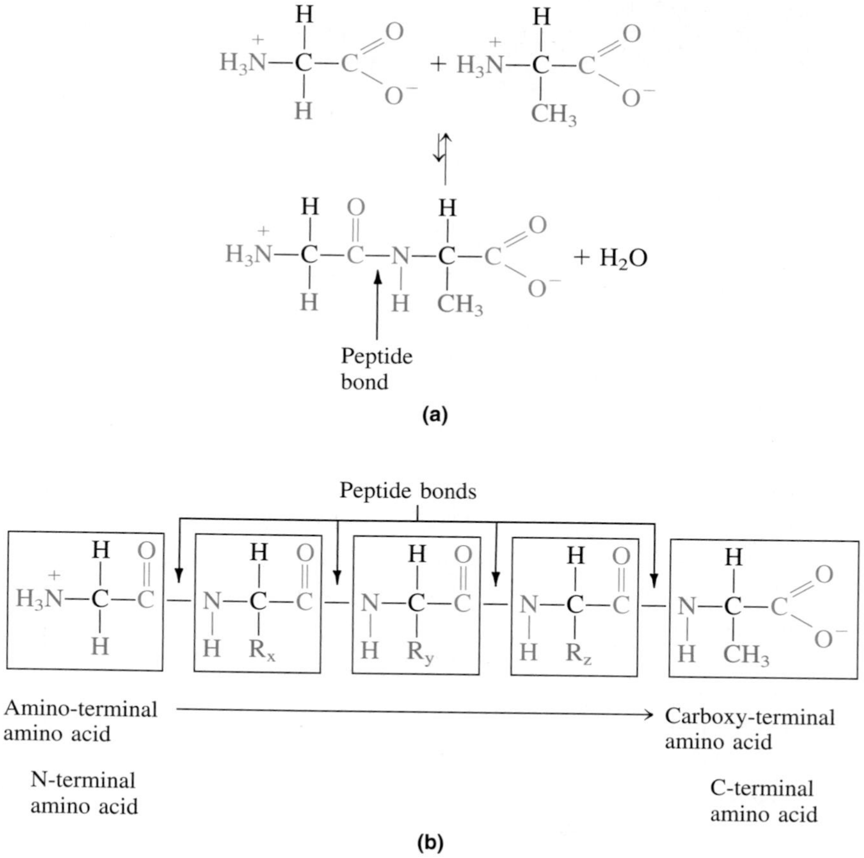

Figure 19.4

(a) Condensation of two α-amino acids to give a dipeptide. The two amino acids shown are glycine and alanine. (b) Structure of a pentapeptide. Amino acid residues are enclosed in boxes. Glycine is the amino-terminal amino acid, and alanine is the carboxy-terminal amino acid.

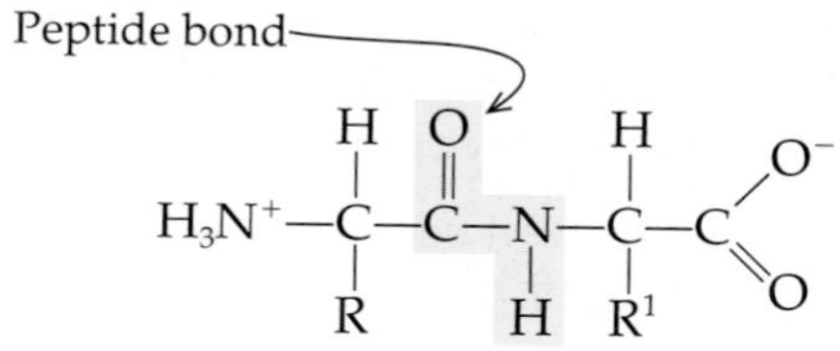

To understand why the N-terminal amino acid is placed first and the C-terminal amino acid is placed last, we need to look at the process of protein synthesis. As we will see in Section 24.6, the N-terminal amino acid is the first amino acid of the protein. It forms a peptide bond involving its carboxyl group and the amino group of the second amino acid in the protein. Thus a free amino group literally projects from the "left" end of the protein. Similarly, the C-terminal amino acid is the last amino acid added to the protein during protein synthesis. Because the peptide bond is formed between the amino group of this amino acid and the carboxyl group of the previous amino acid, a free carboxyl group projects from the "right" end of the protein chain.

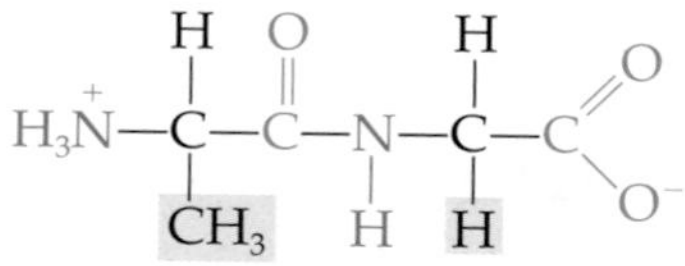

The molecule formed by condensing two amino acids is called a *dipeptide.* The amino acid with a free α-N^+H_3 group is known as the amino terminal, or simply the **N-terminal amino acid,** and the amino acid with a free —COO^- group is known as the carboxyl, or **C-terminal amino acid.** Structures of proteins are conventionally written with their N-terminal amino acid on the left.

The number of amino acids in small peptides is indicated by the prefixes *di-* (two units), *tri-* (three units), *tetra-* (four units), and so forth. Peptides are named as derivatives of the C-terminal amino acid, which receives its entire name. For all other amino acids, the ending *-ine* is changed to *-yl.* Thus the dipeptide alanyl-glycine has glycine as its C-terminal amino acid, as indicated by its full name, *glycine:*

Alanyl-glycine
(ala-gly)

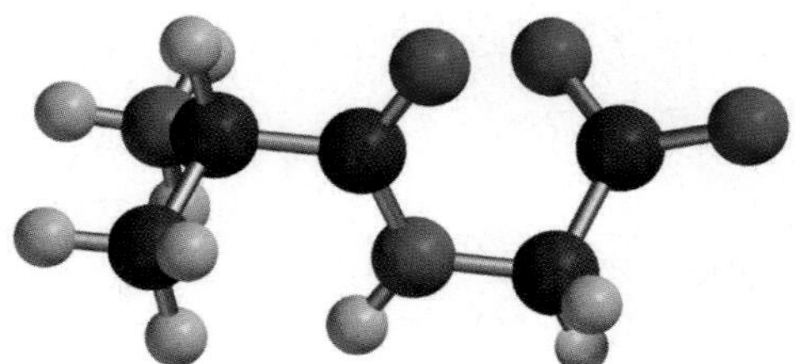

Alanyl-glycine

The dipeptide formed from alanine and glycine that has alanine as its C-terminal amino acid is glycyl-alanine:

```
       H  O     H
       |  ||    |    //O
 +     |  ||    |   //
H3N—C—C—N—C—C
       |     |  |   \
       H     H  CH3  O⁻
```

Glycyl-alanine
(gly-ala)

These two dipeptides have the same amino acid composition, but different amino acid sequences.

The structures of small peptides can easily be drawn with practice if certain rules are followed. First note that the backbone of the peptide contains the repeating sequence

```
N—C—C—N—C—C—N—C—C
  1  2     1  2     1  2
```

in which N is the α-amino group, carbon-1 is the α-carbon, and carbon-2 is the carboxyl group. Carbon-1 is always bonded to a hydrogen atom and to the R group side chain that is unique to each amino acid. Continue drawing as outlined in Example 19.1.

EXAMPLE 19.1

Writing the Structure of a Tripeptide

Learning Goal

4

Draw the structure of the tripeptide alanyl-glycyl-valine.

Solution

Step 1. Write the backbone for a tripeptide. It will contain three sets of three atoms, or nine atoms in all. Remember that the N-terminal amino acid is written to the left.

```
N—C—C      N—C—C      N—C—C
 Set 1      Set 2      Set 3
```

Step 2. Add oxygens to the carboxyl carbons and hydrogens to the amino nitrogens:

```
   H          O          O          O
   |          ||         ||         ||
H—N⁺—C—C—N—C—C—N—C—C—O⁻
   |          |          |
   H          H          H
```

Step 3. Add hydrogens to the α-carbons:

Continued—

A HUMAN PERSPECTIVE

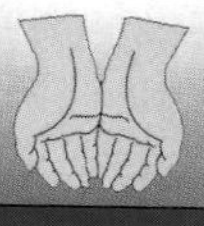

The Opium Poppy and Peptides in the Brain

The seeds of the oriental poppy contain morphine. *Morphine* is a narcotic that has a variety of effects on the body and the brain, including drowsiness, euphoria, mental confusion, and chronic constipation. Although morphine was first isolated in 1805, not until the 1850s and the advent of the hypodermic was it effectively used as a painkiller. During the American Civil War, morphine was used extensively to relieve the pain of wounds and amputations. It was at this time that the addictive properties were noticed. By the end of the Civil War, over 100,000 soldiers were addicted to morphine.

As a result of the Harrison Act (1914), morphine came under government control and was made available only by prescription. Although morphine is addictive, *heroin*, a derivative of morphine, is much more addictive and induces a greater sense of euphoria that lasts for a longer time.

Why do heroin and morphine have such powerful effects on the brain? Both drugs have been found to bind to *receptors* on the surface of the cells of the brain. The function of these receptors is to bind specific chemical signals and to direct the brain cells to respond. Yet it seemed odd that the cells of our brain should have receptors for a plant chemical. This mystery was solved in 1975, when John Hughes discovered that the brain itself synthesizes small peptide hormones with a morphinelike structure. Two of these opiate peptides are called *methionine enkephalin*, or met-enkephalin, and *leucine enkephalin*, or leu-enkephalin.

These neuropeptide hormones have a variety of effects. They inhibit intestinal motility and blood flow to the gastrointestinal tract. This explains the chronic constipation of morphine users. In addition, it is thought that these *enkephalins* play a role in pain perception, perhaps serving as a pain blockade. This is supported by the observation that they are found in higher concentrations in the bloodstream following painful stimulation. It is further suspected that they may play a role in mood and mental health. The so-called runner's high is thought to be a euphoria brought about by an excessively long or strenuous run!

Unlike morphine, the action of enkephalins is short-lived. They bind to the cellular receptor and thereby induce the cells to respond. Then they are quickly destroyed by enzymes in the brain that hydrolyze the peptide bonds of the enkephalin. Once destroyed, they are no longer able to elicit a cellular response. Morphine and heroin bind to these same receptors and induce the cells to respond. However, these drugs are not destroyed and therefore persist in the brain for long periods at concentrations high enough to continue to cause biological effects.

Many researchers are working to understand why drugs like morphine and heroin are addictive. Studies with cells in culture have suggested one mechanism for morphine tolerance and addiction. Normally, when the cell receptors bind to enkephalins, this signals the cell to decrease the production of a chemical messenger called *cyclic AMP*, or simply cAMP. (This compound is very closely related to the nucleotide adenosine-5′-monophosphate.) The decrease in cAMP level helps to block pain and elevate one's mood. When morphine is applied to these cells they initially respond by decreasing cAMP levels. However, with chronic use of morphine the cells become

CH_3 N CH_3—C(=O)—O ... O ... O—C(=O)—CH_3

Heroin

CH_3 N OH O OH

Morphine

The structures of heroin and morphine.

EXAMPLE 19.1 —*Continued*

$$H_3N^+—CH—C(=O)—NH—CH—C(=O)—NH—CH—C(=O)—O^-$$

Continued—

$$H_3\overset{+}{N}-CH(CH_2C_6H_4OH)-CO-NH-CH_2-CO-NH-CH_2-CO-NH-CH(CH_2C_6H_5)-CO-NH-CH(CH_2CH_2SCH_3)-CO_2^-$$

Tyr-Gly-Gly-Phe-Met
Methionine enkephalin

$$H_3\overset{+}{N}-CH(CH_2C_6H_4OH)-CO-NH-CH_2-CO-NH-CH_2-CO-NH-CH(CH_2C_6H_5)-CO-NH-CH(CH_2CH(CH_3)_2)-CO_2^-$$

Tyr-Gly-Gly-Phe-Leu
Leucine enkephalin

Structures of the peptide opiates leucine enkephalin and methionine enkephalin. These are the body's own opiates.

desensitized; that is, they do not decrease cAMP production and thus behave as though no morphine were present. However, a greater amount of morphine will once again cause the decrease in cAMP levels. Thus addiction and the progressive need for more of the drug seem to result from biochemical reactions in the cells.

This logic can be extended to understand withdrawal symptoms. When an addict stops using the drug, he or she exhibits withdrawal symptoms that include excessive sweating, anxiety, and tremors. The cause may be that the high levels of morphine were keeping the cAMP levels low, thus reducing pain and causing euphoria. When morphine is removed completely, the cells overreact and produce huge quantities of cAMP. The result is all of the unpleasant symptoms known collectively as the *withdrawal syndrome*.

Clearly, morphine and heroin have demonstrated the potential for misuse and are a problem for society in several respects. Often, the money needed to support a drug habit is acquired by illegal means such as robbery, theft, and prostitution. More recently, it has become apparent that the use of shared needles for the injection of drugs is resulting in the alarming spread of the virus responsible for acquired immune deficiency syndrome (AIDS). Nonetheless, morphine remains one of the most effective painkillers known. Certainly, for people suffering from cancer, painful burns, or serious injuries, the risk of addiction is far outweighed by the benefits of relief from excruciating pain.

EXAMPLE 19.1 *—Continued*

Step 4. Add the side chains. In this example (ala-gly-val) they are, from left to right, $-CH_3$, H, and $-CH(CH_3)_2$:

$$H_2N^+H-CH(CH_3)-CO-NH-CH(H)-CO-NH-CH(CH(CH_3)_2)-CO-O^-$$

Question 19.3

Write the structure of each of the following peptides at pH 7:

a. Alanyl-phenylalanine
b. Lysyl-alanine
c. Phenylalanyl-tyrosyl-leucine

Question 19.4

Write the structure of each of the following peptides at pH 7:

a. Glycyl-valyl-serine
b. Threonyl-cysteine
c. Isoleucyl-methionyl-aspartate

Although you might expect free rotation about the peptide bond, this is not the case. Because the lone pair of electrons of the nitrogen atom interacts with the carbon and oxygen of the carbonyl group, the molecule exhibits resonance. This gives the peptide bond a partially double bond character:

$$-C_{\alpha}-C(=O)-\ddot{N}(H)-C_{\alpha}- \quad \longleftrightarrow \quad -C_{\alpha}-C(-O^{-})=\overset{+}{N}(H)-C_{\alpha}-$$

As a result, the peptide bond is planar and somewhat rigid. This is quite important physiologically because it makes protein structures relatively rigid. If they could not hold their shapes, they could not function.

19.4 The Primary Structure of Proteins

Learning Goal

The **primary structure** of a protein is the amino acid sequence of the protein chain. It results from the covalent bonding between the amino acids in the chain (peptide bonds). The primary structures of proteins are translations of information contained in genes. Each protein has a different primary structure with different amino acids in different places along the chain.

The genetic code and the process of protein synthesis are described in Sections 24.5 and 24.6.

Mutations and their effect on protein synthesis are discussed in Section 24.7.

Genes can change by the process of mutation during the course of evolution. A mutation in a gene can result in a change in the primary amino acid sequence of a protein. Over longer periods, more of these changes will occur. If two species of organisms diverged (became new species) very recently, the differences in the amino acid sequences of their proteins will be few. On the other hand, if they diverged millions of years ago, there will be many more differences in the amino acid sequences of their proteins. As a result, we can compare evolutionary relationships between species by comparing the primary structures of proteins present in both species.

19.5 The Secondary Structure of Proteins

Learning Goal

Learning Goal

The primary sequence of a protein, the chain of covalently linked amino acids, folds into regularly repeating structures that resemble designs in a tapestry. These repeating structures define the **secondary structure** of the protein. The secondary structure is the result of hydrogen bonding between the amide hydrogens and carbonyl oxygens of the peptide bonds. Many hydrogen bonds are needed to maintain the secondary structure and thereby the overall structure of the protein. Different regions of a protein chain may have different types of secondary structure. Some regions of a protein chain may have a random or nonregular structure; however, the two most common types of secondary structure are the α-helix and the β-pleated sheet because they maximize hydrogen bonding in the backbone.

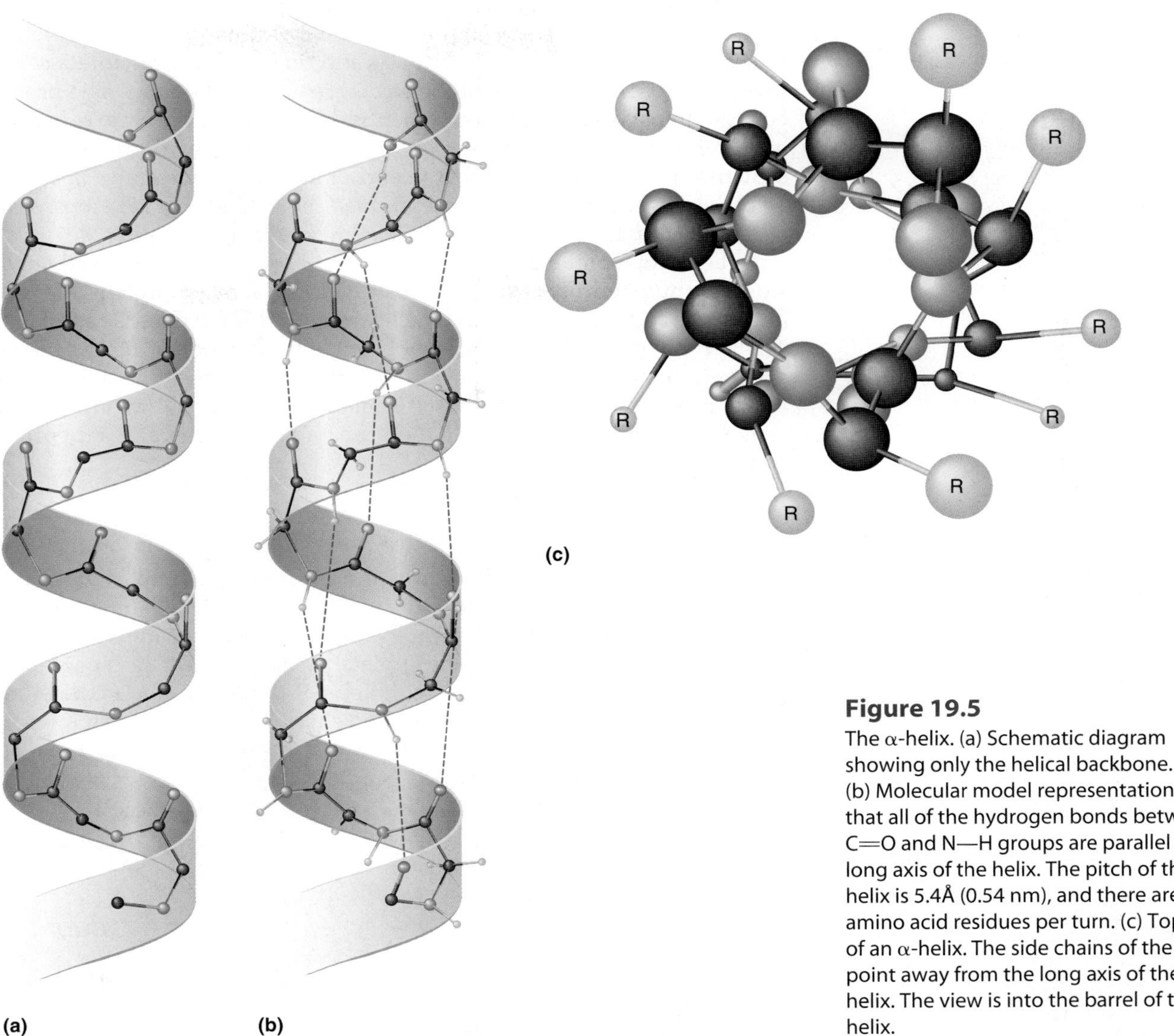

Figure 19.5
The α-helix. (a) Schematic diagram showing only the helical backbone. (b) Molecular model representation. Note that all of the hydrogen bonds between C═O and N—H groups are parallel to the long axis of the helix. The pitch of the helix is 5.4Å (0.54 nm), and there are 3.6 amino acid residues per turn. (c) Top view of an α-helix. The side chains of the helix point away from the long axis of the helix. The view is into the barrel of the helix.

α-Helix

The most common type of secondary structure is a coiled, helical conformation known as the **α-helix** (Figure 19.5). The α-helix has several important features.

- Every amide hydrogen and carbonyl oxygen associated with the peptide backbone is involved in a hydrogen bond when the chain coils into an α-helix. These hydrogen bonds lock the α-helix into place.
- Every carbonyl oxygen is hydrogen-bonded to an amide hydrogen four amino acids away in the chain.
- The hydrogen bonds of the α-helix are parallel to the long axis of the helix (see Figure 19.5).
- The polypeptide chain in an α-helix is right-handed. It is oriented like a normal screw. If you turn a screw clockwise it goes into the wall; turned counterclockwise, it comes out of the wall.
- The repeat distance of the helix, or its pitch, is 5.4 Å, and there are 3.6 amino acids per turn of the helix.

Learning Goal 7

Fibrous proteins are structural proteins arranged in fibers or sheets that have only one type of secondary structure. The **α-keratins** are fibrous proteins that form the covering (hair, wool, nails, hooves, and fur) of most land animals. Human hair provides a typical example of the structure of the α-keratins. The proteins of hair consist almost exclusively of polypeptide chains coiled up into α-helices. A single α-helix is coiled in a bundle with two other helices to give a three-stranded superstructure called a *protofibril* that is part of an array known as a *microfibril* (Figure 19.6). These structures, which resemble "molecular pigtails," possess great mechanical strength, and they are virtually insoluble in water.

The fibrous proteins of muscle are also composed of proteins that contain considerable numbers of α-helices. **Myosin,** one of the major proteins of muscle, for example, is a rodlike structure in which two α-helices form a coiled coil (Figure 19.7).

The major structural property of a coiled coil superstructure of α-helices is its great mechanical strength. This property is applied very efficiently in both the fibrous proteins of skin and those of muscle. As you can imagine, these proteins must be very strong to carry out their functions of mechanical support and muscle contraction.

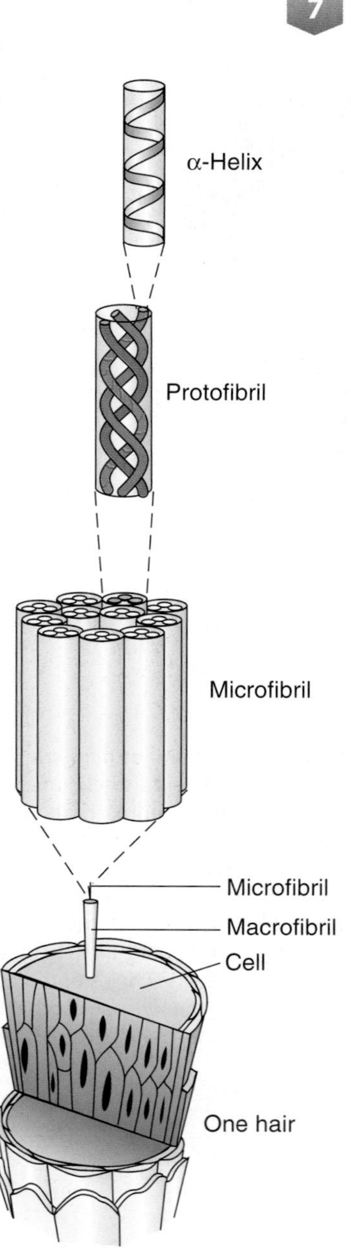

Figure 19.6
Structure of the α-keratins. These proteins are assemblies of triple-helical protofibrils that are assembled in an array known as a *microfibril.* These in turn are assembled into macrofibrils. Hair is a collection of macrofibrils and hair cells.

β-Pleated Sheet

The second common secondary structure in proteins resembles the pleated folds of drapery and is known as **β-pleated sheet** (Figure 19.8). All of the carbonyl oxygens and amide hydrogens in a β-pleated sheet are involved in hydrogen bonds, and the polypeptide chain is nearly completely extended. The polypeptide chains in a β-pleated sheet can have two orientations. If the N-termini are head to head, the structure is known as a *parallel* β-pleated sheet. And if the N-terminus of one chain is aligned with the C-terminus of a second chain (head to tail), the structure is known as an *antiparallel* β-pleated sheet.

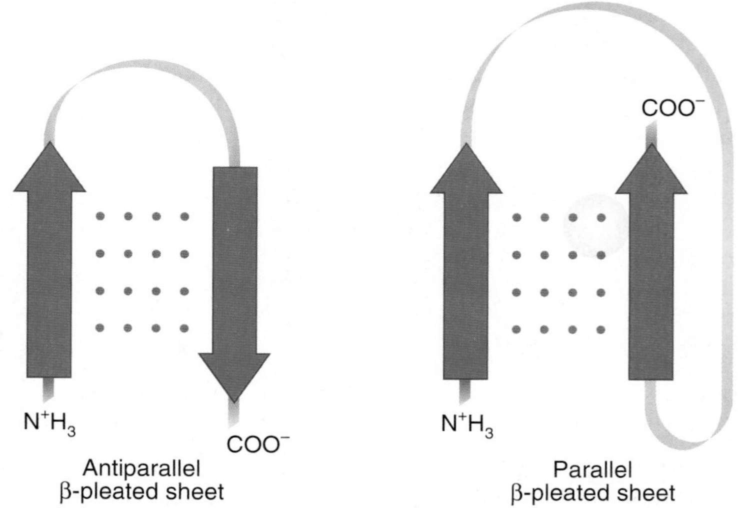

Some fibrous proteins are composed of β-pleated sheets. For example, the silkworm produces *silk fibroin,* a protein whose structure is an antiparallel β-pleated sheet (Figure 19.9). The polypeptide chains of a β-pleated sheet are almost completely extended, and silk does not stretch easily. Glycine accounts for nearly half of the amino acids of silk fibroin. Alanine and serine account for most of the others. The methyl groups of alanines and the hydroxymethyl groups of serines lie on opposite sides of the sheet. Thus the stacked sheets nestle comfortably, like sheets of corrugated cardboard, because the R groups are small enough to allow the stacked-sheet superstructure.

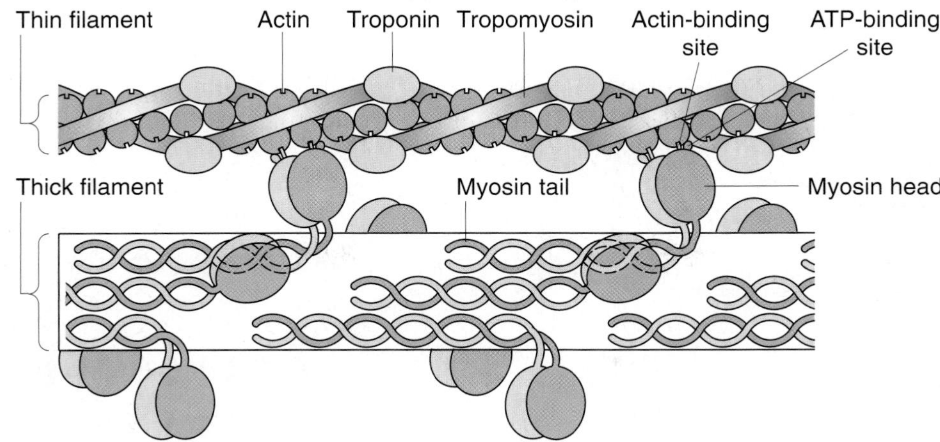

Figure 19.7
Schematic diagram of the structure of myosin. This muscle protein consists of a rodlike coil of α-helices with two globular heads, also composed of protein, attached to myosin at its C-terminus. In muscle, myosin molecules are assembled into thick filaments that alternate with thin filaments composed of the proteins actin, troponin, and tropomyosin. Working together, these filaments allow muscles to contract and relax.

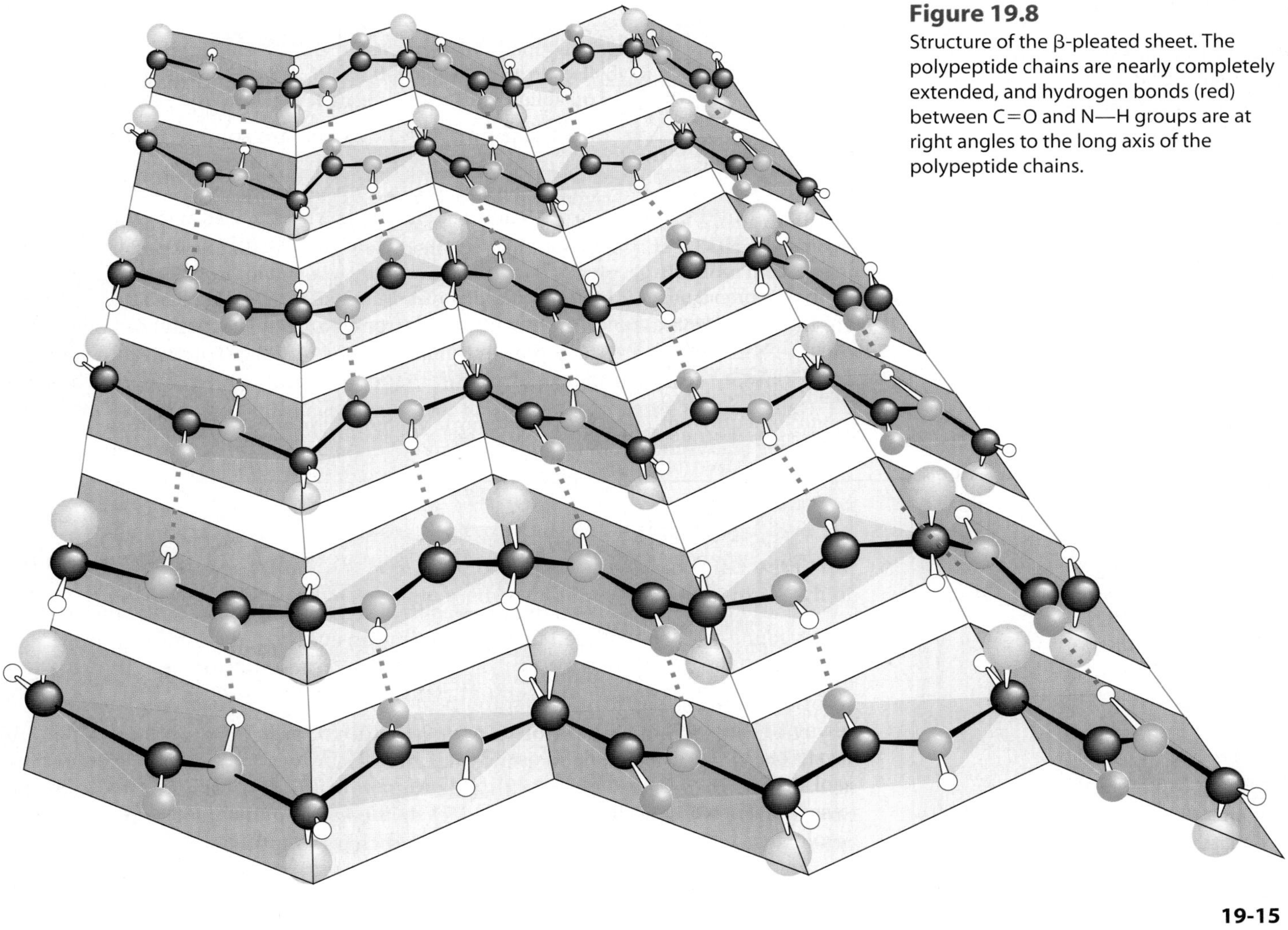

Figure 19.8
Structure of the β-pleated sheet. The polypeptide chains are nearly completely extended, and hydrogen bonds (red) between C=O and N—H groups are at right angles to the long axis of the polypeptide chains.

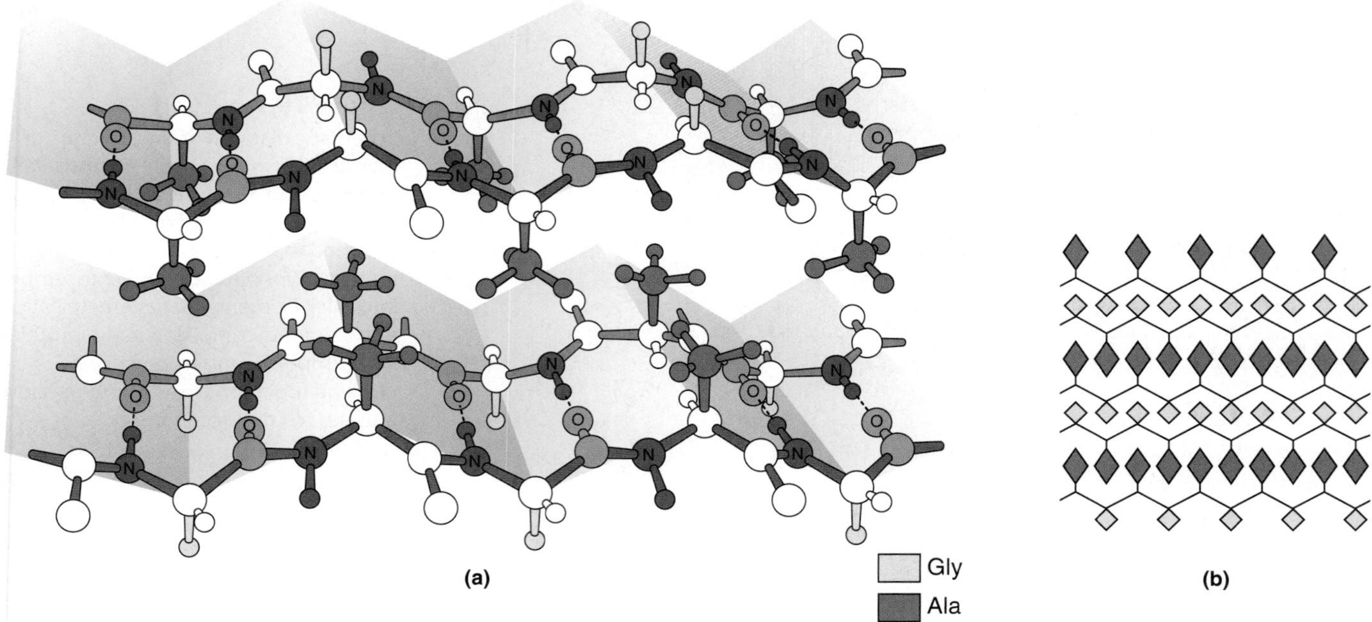

Figure 19.9
The structure of silk fibroin is almost entirely antiparallel β-pleated sheet. (a) The molecular structure of a portion of the silk fibroin protein. (b) A schematic representation of the antiparallel β-pleated sheet with the nestled R groups.

19.6 The Tertiary Structure of Proteins

Learning Goal

Most fibrous proteins, such as silk, collagen, and the α-keratins, are almost completely insoluble in water. (Our skin would do us very little good if it dissolved in the rain.) The majority of cellular proteins, however, are soluble in the cell cytoplasm. Soluble proteins are usually **globular proteins.** Globular proteins have three-dimensional structures called the **tertiary structure** of the protein, which are distinct from their secondary structure. The polypeptide chain with its regions of secondary structure, α-helix and β-pleated sheet, further folds on itself to achieve the tertiary structure.

We have seen that the forces that maintain the secondary structure of a protein are hydrogen bonds between the amide hydrogen and the carbonyl oxygen of the peptide bond. What are the forces that maintain the tertiary structure of a protein? The globular tertiary structure forms spontaneously and is maintained as a result of interactions among the side chains, the R groups, of the amino acids. The structure is maintained by the following molecular interactions:

Learning Goal

- Van der Waals forces between the R groups of nonpolar amino acids that are hydrophobic
- Hydrogen bonds between the polar R groups of the polar amino acids
- Ionic bonds (salt bridges) between the R groups of oppositely charged amino acids
- Covalent bonds between the thiol-containing amino acids. Two of the polar cysteines can be oxidized to a dimeric amino acid called *cystine* (Figure 19.10). The disulfide bond of cystine can be a cross-link between different proteins, or it can tie two segments within a protein together.

In the next section we will see that some proteins have an additional level of structure, quaternary structure, that also influences function.

The bonds that maintain the tertiary structure of proteins are shown in Figure 19.11. The importance of these bonds becomes clear when we realize that it is the tertiary structure of the protein that defines its biological function. Most of the time, nonpolar amino acids are buried, closely packed, in the interior of a globular protein, out of contact with water. Polar and charged amino acids lie on the surfaces of globular proteins. Globular proteins are extremely compact. The tertiary

A HUMAN PERSPECTIVE 

Collagen: A Protein That Holds Us Together

Collagen is the most abundant protein in the human body, making up about one-third of the total protein content. It provides mechanical strength to bone, tendon, skin, and blood vessels. Collagen fibers in bone provide a scaffolding around which *hydroxyapatite* (a calcium phosphate polymer) crystals are arranged. Skin contains loosely woven collagen fibers that can expand in all directions. The corneas of the eyes are composed of collagen. As we consider these tissues, we realize that they have quite different properties, ranging from tensile strength (tendons) and flexibility (blood vessels) to transparency (cornea).

How could such diverse structures be composed of a single protein? The answer lies in the fact that collagen is actually a family of twenty genetically distinct, but closely related proteins. Although the differences in the amino acid sequence of these different collagen proteins allow them to carry out a variety of functions in the body, they all have a similar three-dimensional structure. Collagen is composed of three left-handed polypeptide helices that are twisted around one another to form a "superhelix" called a *triple helix.* Each of the individual peptide chains of collagen is a left-handed helix, but they are wrapped around one another in the right-handed sense.

Structure of the collagen triple helix.

Every third amino acid in the collagen chain is glycine. It is important to the structure because the triple-stranded helix forms as a result of interchain hydrogen bonding involving glycine. Thus, every third amino acid on one strand is in very close contact with the other two strands. Glycine has another advantage; it is the only amino acid with an R group small enough for the space allowed by the triple-stranded structure.

Two unusual, hydroxylated amino acids account for nearly one-fourth of the amino acids in collagen. These amino acids are 4-hydroxyproline and 5-hydroxylysine.

4-Hydroxyproline

5-Hydroxylysine

Structures of 4-hydroxyproline and 5-hydroxylysine, two amino acids found only in collagen.

These amino acids are an important component of the structure of collagen because they form covalent cross-linkages between adjacent molecules within the triple strand. They can also participate in interstrand hydrogen bonding to further strengthen the structure.

When collagen is synthesized, the amino acids proline and lysine are incorporated into the chain of amino acids. These are later modified by two enzymes to form 4-hydroxyproline and 5-hydroxylysine. Both of these enzymes require vitamin C to carry out these reactions. In fact, this is the major known physiological function of vitamin C. Without hydroxylation, hydrogen bonds cannot form and the triple helix is weak, resulting in fragile blood vessels

Vitamin C (ascorbic acid)

People who are deprived of vitamin C, as were sailors on long voyages before the eighteenth century, develop *scurvy,* a disease of collagen metabolism. The symptoms of scurvy include skin lesions, fragile blood vessels, and bleeding gums. The British Navy provided the antidote to scurvy by including limes, which are rich in vitamin C, in the diets of its sailors. The epithet *limey,* a slang term for *British,* entered the English language as a result.

$$2\ \text{---NH---CH(CH}_2\text{---SH)---C(=O)---} \underset{\text{Reduction}}{\overset{\text{Oxidation}}{\rightleftharpoons}} \text{---NH---CH(CH}_2\text{---S---S---CH}_2\text{)CH---NH---} + 2H^+ + 2e^-$$

Cysteine → Cystine

Figure 19.10
Oxidation of two cysteines to give the dimer cystine. This reaction occurs in cells and is readily reversible.

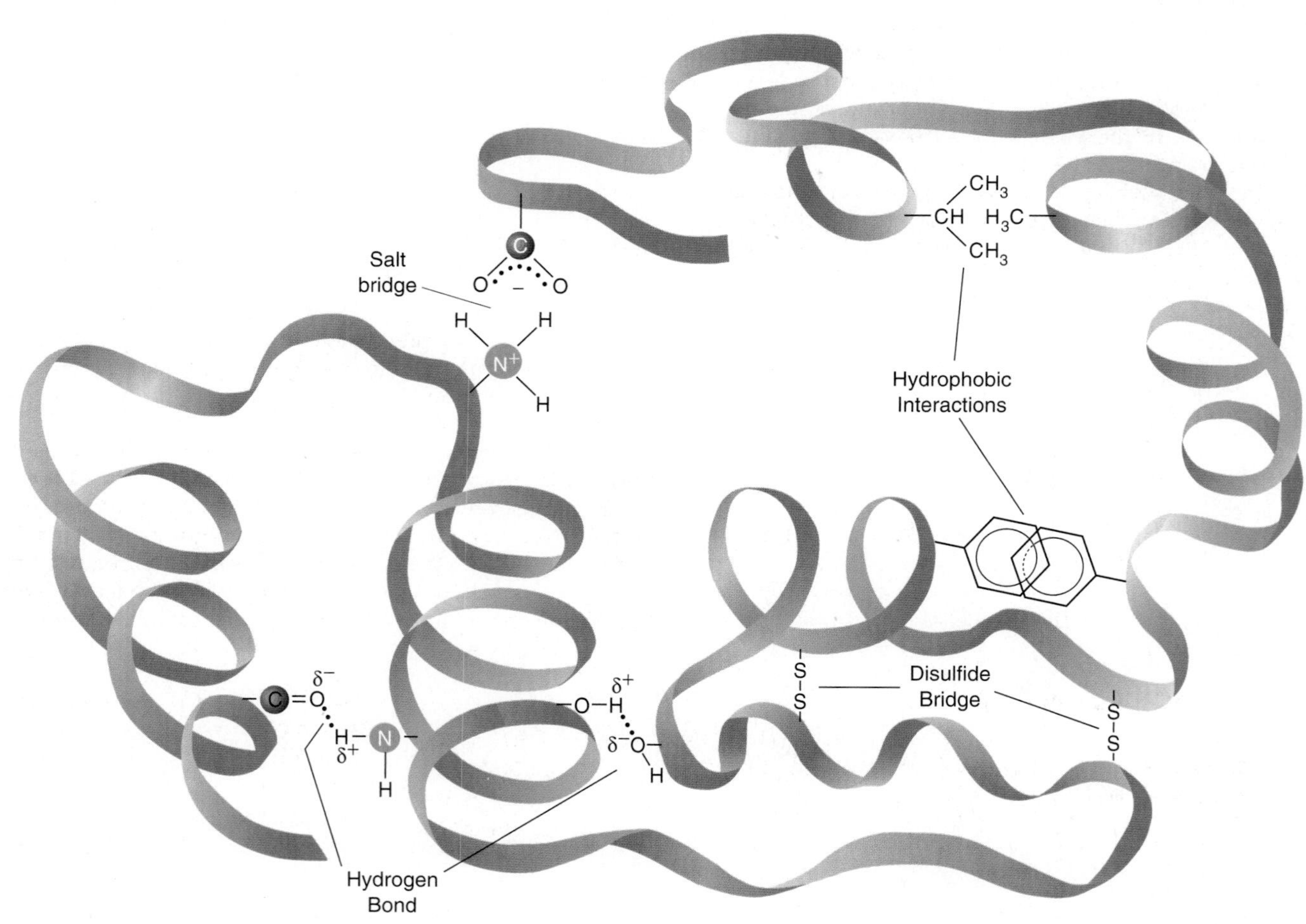

Figure 19.11
Summary of the weak interactions that help maintain the tertiary structure of a protein.

structure can contain regions of α-helix and regions of β-pleated sheet. "Hinge" regions of random coil connect regions of α-helix and β-pleated sheet. Because of its cyclic structure, proline disrupts an α-helix. As a result, proline is often found in these hinge regions. The exact amount of each type of secondary structure varies from one protein to the next.

19.7 The Quaternary Structure of Proteins

Learning Goal

For many proteins the functional form is not composed of a single peptide but is rather an aggregate of smaller globular peptides. For instance, the protein hemoglobin is composed of four individual globular peptide subunits: two identical α-subunits and two identical β-subunits. Only when the four peptides are bound to one another is the protein molecule functional. The association of several polypeptides to produce a functional protein defines the **quaternary structure** of a protein.

The designations α- and β- used to describe the subunits of hemoglobin do not refer to types of secondary structure.

The forces that hold the quaternary structure of a protein are the same as those that hold the tertiary structure. These include hydrogen bonds between polar amino acids, ionic bridges between oppositely charged amino acids, van der Waals forces between nonpolar amino acids, and disulfide bridges.

Learning Goal

In some cases the quaternary structure of a functional protein involves binding to a nonprotein group. This additional group is called a **prosthetic group.** For example, many of the receptor proteins on cell surfaces are **glycoproteins.** These are proteins with sugar groups covalently attached. Each of the subunits of hemoglobin is bound to an iron-containing heme group. The heme group is a large, unsaturated organic cyclic amine with an iron ion coordinated within it. As in the case of hemoglobin, the prosthetic group often determines the function of a protein. For instance, in hemoglobin it is the iron-containing heme groups that have the ability to bind reversibly to oxygen.

Question 19.5

Describe the four levels of protein structure.

Question 19.6

What are the weak interactions that maintain the tertiary structure of a protein?

19.8 An Overview of Protein Structure and Function

Let's summarize the various types of protein structure and their relationship to one another (Figure 19.12).

- *Primary Structure:* The primary structure of the protein is the amino acid sequence of the protein. The primary structure results from the formation of covalent peptide bonds between amino acids. Peptide bonds are amide bonds formed between the α-carboxylate group of one amino acid and the α-amino group of another.
- *Secondary Structure:* As the protein chain grows, numerous opportunities for noncovalent interactions in the backbone of the polypeptide chain become available. These cause the chain to fold and orient itself in a variety of conformational arrangements. The secondary level of structure includes the α-helix and the β-pleated sheet, which are the result of hydrogen bonding between the amide hydrogens and carbonyl oxygens of the peptide bonds. Different portions of the chain may be involved in different types of secondary structure arrangements; some regions might be α-helix and others might be a β-pleated sheet.
- *Tertiary Structure:* When we discuss tertiary structure, we are interested in the overall folding of the entire chain. In other words, we are concerned with the further folding of the secondary structure. Are the two ends of the chain close together or far apart? What general shape is involved? Both noncovalent interactions between the R groups of the amino acids and covalent disulfide bridges play a role in determining the tertiary structure. The noncovalent interactions include hydrogen bonding, ionic bonding, and van der Waals forces.

The three-dimensional structure of a protein is the feature that allows it to carry out its specific biological function. However, we must always remember that it is the primary structure, the order of the R groups, that determines how the protein will fold and what the ultimate shape will be.

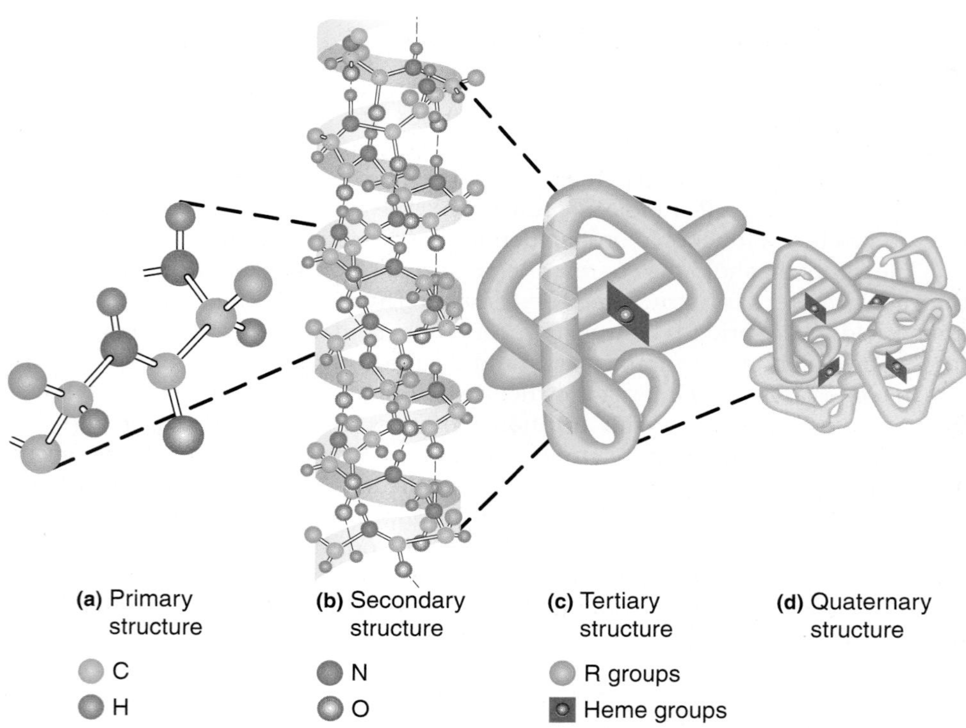

Figure 19.12
Summary of the four levels of protein structure, using hemoglobin as an example.

- *Quaternary Structure:* Like tertiary structure, quaternary structure is concerned with the topological, spatial arrangements of two or more peptide chains with respect to each other. How is one chain oriented with respect to another? What is the overall shape of the final functional protein?

Learning Goal

The quaternary structure is maintained by the same forces that are responsible for the tertiary structure. It is the tertiary and quaternary structures of the protein that ultimately define its function. Some have a fibrous structure with great mechanical strength. These make up the major structural components of the cell and the organism. Often they are also responsible for the movement of the organism. Others fold into globular shapes. Most of the transport proteins, regulatory proteins, and enzymes are globular proteins. The very precise three-dimensional structure of the transport proteins allows them to recognize a particular molecule and facilitate its entry into the cell. Similarly, it is the specific three-dimensional shape of regulatory proteins that allows them to bind to their receptors on the surfaces of the target cell. In this way they can communicate with the cell, instructing it to take some course of action. In Chapter 20 we will see that the three-dimensional structure of enzyme active sites allows them to bind to their specific reactants and speed up biochemical reactions.

As we will see with the example of sickle cell hemoglobin in the next section, an alteration of just a single amino acid within the primary structure of a protein can have far-reaching implications. When an amino acid replaces another in a peptide, there is a change in the R group at that position in the protein chain. This leads to different tertiary and perhaps quaternary structure because the nature of the noncovalent interactions is altered by changing the R group that is available for that bonding. Similarly, replacement of another amino acid with proline can disrupt important regions of secondary structure. Thus changes in the primary amino acid sequence can change the three-dimensional structure of a protein in ways that cause it to be nonfunctional. In the case of sickle cell hemoglobin, this protein malfunction can lead to death.

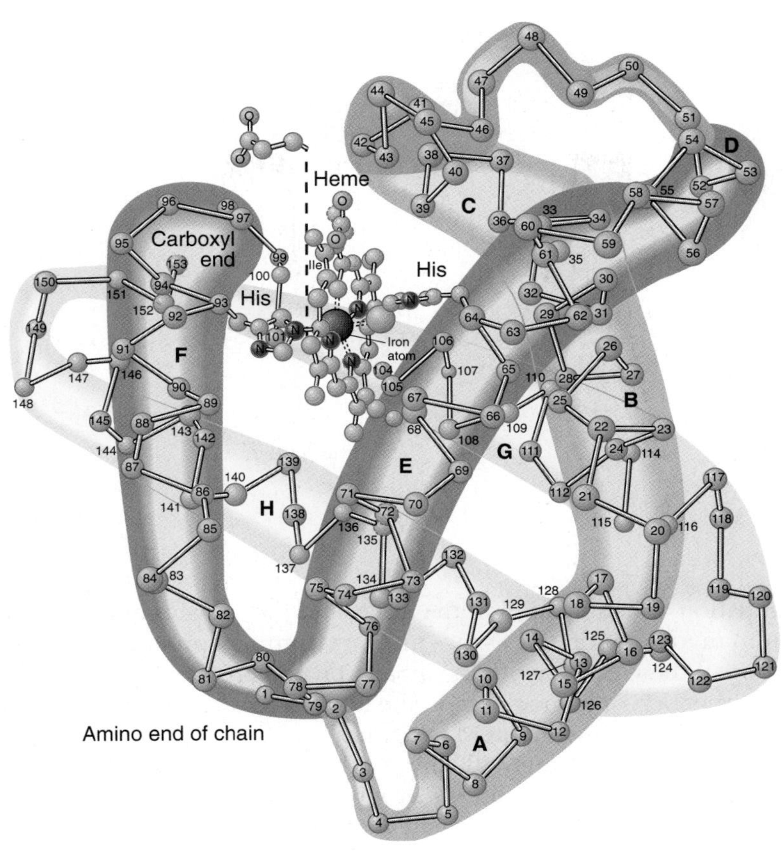

Figure 19.13
Myoglobin. The heme group has an iron atom to which oxygen binds.

19.9 Myoglobin and Hemoglobin

Myoglobin and Oxygen Storage

Learning Goal

Most of the cells of our bodies are buried in the interior of the body and cannot directly get food molecules or eliminate waste. The circulatory system solves this problem by delivering nutrients and oxygen to body cells and carrying away wastes. Our cells require a steady supply of oxygen, but oxygen is only slightly soluble in aqueous solutions. To overcome this solubility problem, we have an oxygen transport protein, **hemoglobin.** Hemoglobin is found in red blood cells and is the oxygen transport protein of higher animals. **Myoglobin** is the oxygen storage protein of skeletal muscle.

The structure of myoglobin (Mb) is shown in Figure 19.13. The **heme group** (Figure 19.14) is also an essential component of this protein. The Fe^{2+} ion in the heme group is the binding site for oxygen in both myoglobin and hemoglobin. Fortunately, myoglobin has a greater attraction for oxygen than does hemoglobin. Thus there is efficient transfer of oxygen from the bloodstream to the cells of the body.

Hemoglobin and Oxygen Transport

Hemoglobin (Hb) is a tetramer composed of four polypeptide subunits: two α-subunits and two β-subunits (Figure 19.15). Because each subunit of hemoglobin contains a heme group, a hemoglobin molecule can bind four molecules of oxygen:

$$\underset{\text{Deoxyhemoglobin}}{\text{Hb}} + 4O_2 \longrightarrow \underset{\text{Oxyhemoglobin}}{\text{Hb}(O_2)_4}$$

Figure 19.14
Structure of the heme prosthetic group, which binds to myoglobin and hemoglobin.

The oxygenation of hemoglobin in the lungs and the transfer of oxygen from hemoglobin to myoglobin in the tissues are very complex processes. We begin our investigation of these events with the inhalation of a breath of air.

The oxygenation of hemoglobin in the lungs is greatly favored by differences in the oxygen partial pressure (pO_2) in the lungs and in the blood. The pO_2 in the air in the lungs is approximately 100 mm Hg; the pO_2 in oxygen-depleted blood is only about 40 mm Hg. Oxygen diffuses from the region of high pO_2 in the lungs to the region of low pO_2 in the blood. There it enters red blood cells and binds to the Fe^{2+} ions of the heme groups of deoxyhemoglobin, forming oxyhemoglobin. This binding actually helps bring more O_2 into the blood.

The events of oxygen binding are somewhat complex. Deoxyhemoglobin has a space in the center where the organic anion 2,3-bisphosphoglycerate (BPG) binds. When one of the four deoxyhemoglobin subunits binds O_2, a shape change in the protein expels the BPG. This initiates a cascade of events in which the remaining three hemoglobin subunits sequentially undergo a shape change that increases their ability to accept an oxygen molecule. Thus, once the first subunit accepts an O_2, the remaining three quickly follow suit.

Hemoglobin
α-chains
β-chains
Heme groups

Figure 19.15
Structure of hemoglobin. The protein contains four subunits, designated α and β. The α- and β-subunits face each other across a central cavity. Each subunit in the tetramer contains a heme group that binds oxygen.

The H^+ concentration also affects the ability of deoxyhemoglobin to bind oxygen. When deoxyhemoglobin binds oxygen, it releases protons. These are quickly removed in a reaction catalyzed by the enzyme *carbonic anhydrase.*

$$H^+ + HCO_3^- \xrightarrow{\text{Carbonic anhydrase}} CO_2 + H_2O$$

This is a readily reversible reaction. In fact, the HCO_3^- involved in this reaction was originally formed by the reverse reaction when CO_2, the waste product of aerobic metabolism, entered the blood from actively metabolizing tissue.

$$H^+ + HCO_3^- \xleftarrow{\text{Carbonic anhydrase}} CO_2 + H_2O$$

In the lungs the protons released by the deoxyhemoglobin "push" the reaction to the right, releasing the waste product, CO_2. This diffuses from the region of higher pCO_2 in the blood to the region of lower pCO_2 in the air in the lungs. It is then removed from the body by exhalation.

The forward reaction also helps the oxygenation of hemoglobin because removal of H^+ from the blood stimulates oxygen binding by deoxyhemoglobin. Thus, the release of H^+ during the oxygenation of deoxyhemoglobin stimulates both the removal of waste CO_2 from the body and the binding of O_2 by hemoglobin.

The Danish physiologist Christian Bohr, father of nuclear physicist Niels Bohr, first observed that a decrease of H^+ stimulates the oxygenation of hemoglobin. This phenomenon is called the *Bohr effect*.

When oxygenated blood reaches actively metabolizing tissue, waste CO_2 diffuses into the blood. This initiates the reversal of the processes just described. CO_2 enters the red blood cell where carbonic anhydrase catalyzes its combination with H_2O to produce HCO_3^- and H^+. Just as removal of H^+ from the blood in the lungs stimulated oxygenation of hemoglobin, the increase in H^+ in the blood in contact with metabolizing tissues enhances the release of oxygen from hemoglobin. The released oxygen can then diffuse from the blood, the region of higher pO_2, into the tissues that require it.

BPG further accelerates the release of oxygen from hemoglobin. As soon as O_2 is released from one of the four hemoglobin subunits, BPG begins to work its way back into the space in the hemoglobin molecule. This causes shape changes in the subunits that allow them to immediately release their bound oxygen.

Myoglobin in the tissues has a higher affinity for oxygen than hemoglobin does because it does not bind BPG. Myoglobin quickly binds the O_2 that diffuses into the tissues and releases it when it is required for aerobic respiration.

Oxygen Transport from Mother to Fetus

A fetus receives its oxygen from its mother by simple diffusion across the placenta. If both the fetus and the mother had the same type of hemoglobin, this transfer process would not be efficient, because the hemoglobin of the fetus and the mother would have the same affinity for oxygen. The fetus, however, has a unique type of hemoglobin, called *fetal hemoglobin*. This unique hemoglobin molecule has a greater affinity for oxygen than does the mother's hemoglobin because it does not bind BPG as well as adult hemoglobin. Oxygen is therefore efficiently transported, via the circulatory system, from the lungs of the mother to the fetus. The biosynthesis of fetal hemoglobin stops shortly after birth when the genes encoding fetal hemoglobin are switched "off" and the genes coding for adult hemoglobin are switched "on."

See Chemistry Connection: Wake Up, Sleeping Gene, in Chapter 15.

Question 19.7

Why is oxygen efficiently transferred from hemoglobin in the blood to myoglobin in the muscles?

Question 19.8

How is oxygen efficiently transferred from mother to fetus?

Sickle Cell Anemia

Sickle cell anemia is a human genetic disease that first appeared in tropical west and central Africa. It afflicts about 0.4% of African Americans. These individuals produce a mutant hemoglobin known as sickle cell hemoglobin (Hb S). Sickle cell anemia receives its name from the sickled appearance of the red blood cells that form in this condition (Figure 19.16). The sickled cells are unable to pass through the small capillaries of the circulatory system, and circulation is hindered. This results in damage to many organs, especially bone and kidney, and can lead to death at an early age.

The genetic basis of this alteration is discussed in Chapter 24.

Sickle cell hemoglobin differs from normal hemoglobin by a single amino acid. In the β-chain of sickle cell hemoglobin, a valine (a hydrophobic amino acid) has replaced a glutamic acid (a negatively charged amino acid). This substitution provides a basis for the binding of hemoglobin S molecules to one another. When oxyhemoglobin S unloads its oxygen, individual deoxyhemoglobin S molecules bind to one another as long polymeric fibers. This occurs because the valine fits into a hydrophobic pocket on the surface of a second deoxyhemoglobin S molecule. The fibers generated in this way radically alter the shape of the red blood cell, resulting in the sickling effect.

When hemoglobin is carrying O_2, it is called oxyhemoglobin. When it is not bound to O_2, it is called deoxyhemoglobin.

A MEDICAL PERSPECTIVE

Immunoglobulins: Proteins That Defend the Body

A living organism is subjected to a constant barrage of bacterial, viral, parasitic, and fungal diseases. Without a defense against such perils we would soon perish. All vertebrates possess an *immune system.* In humans the immune system is composed of about 10^{12} cells, about as many as the brain or liver, which protect us from foreign invaders. This immune system has three important characteristics.

1. **It is highly specific.** The immune response to each infection is specific to, or directed against, only one disease organism or similar, related organisms.
2. **It has a memory.** Once the immune system has responded to an infection, the body is protected against reinfection by the same organism. This is the reason that we seldom suffer from the same disease more than once. Most of the diseases that we suffer recurrently, such as the common cold and flu, are actually caused by many different strains of the same virus. Each of these strains is "new" to the immune system.
3. **It can recognize "self" from "nonself."** When we are born, our immune system is already aware of all the antigens of our bodies. These it recognizes as "self" and will not attack. Every antigen that is not classified as "self" will be attacked by the immune system when it is encountered. Some individuals suffer from a defect of the immune response that allows it to attack the cells of one's own body. The result is an *autoimmune reaction* that can be fatal.

One facet of the immune response is the synthesis of *immunoglobulins,* or *antibodies,* that specifically bind a single macromolecule called an *antigen.* These antibodies are produced by specialized white blood cells called *B lymphocytes.* We are born with a variety of B lymphocytes that are capable of producing antibodies against perhaps a million different antigens. When a foreign antigen enters the body, it binds to the B lymphocyte that was preprogrammed to produce antibodies to destroy it. This stimulates the B cell to grow and divide. Then all of these new B cells produce antibodies that will bind to the disease agent and facilitate its destruction. Each B cell produces only one type of antibody with an absolute specificity for its target antigen. Many different B cells respond to each infection because the disease-causing agent is made up of many different antigens. Antibodies are made that bind to many of the antigens of the invader. This primary immune response is rather slow. It can take a week or two before there are enough B cells to produce a high enough level of antibodies in the blood to combat an infection.

Because the immune response has a memory, the second time we encounter a disease-causing agent the antibody response is immediate. This is why it is extremely rare to suffer from mumps, measles, or chickenpox a second time. We take advantage of this property of the immune system to protect ourselves against many diseases. In the process of *vaccination* a person can be immunized against an infectious disease by injection of a small amount of the antigens of the virus or microorganism (the vaccine). The B lymphocytes of the body then manufacture antibodies against the antigens of the infectious agent. If the individual comes into contact with the disease-causing microorganism at some later time, the sensitized B lymphocytes "remember" the antigen and very quickly produce a large amount of specific antibody to overwhelm the microorganism or virus before it can cause overt disease.

Immunoglobulin molecules contain four peptide chains that are connected by disulfide bonds and arranged in a Y-shaped quaternary structure.

Each immunoglobulin has two identical antigen-binding sites located at the tips of the Y and is therefore bivalent. Because most antigens have three or more antibody-binding sites, immunoglobulins can form large cross-linked antigen-antibody complexes that precipitate from solution.

Immunoglobulin G (IgG) is the major serum immunoglobulin. Some immunoglobulin G molecules can cross cell membranes and thus can pass between mother and fetus through the placenta, before birth. This is important because the immune system of a fetus is immature and cannot provide adequate protection from disease. Fortunately, the IgG acquired from the mother protects the fetus against most bacterial and viral infections that it might encounter before birth.

There are four additional types of antibody molecules that vary in their protein composition, but all have the same general Y shape. One of these is IgM, which is the first antibody produced in response to an infection. Secondarily, the B cell produces IgG molecules with the same antigen-binding region

Sickle cell anemia occurs in individuals who have inherited the gene for sickle cell hemoglobin from both parents. Afflicted individuals produce 90–100% defective β-chains. Individuals who inherit one normal gene and one defective gene produce both normal and altered β-chains. About 10% of African Americans carry a single copy of the defective gene, a condition known as *sickle cell trait.* Although not severely affected, they have a 50% chance of passing the gene to each of their children.

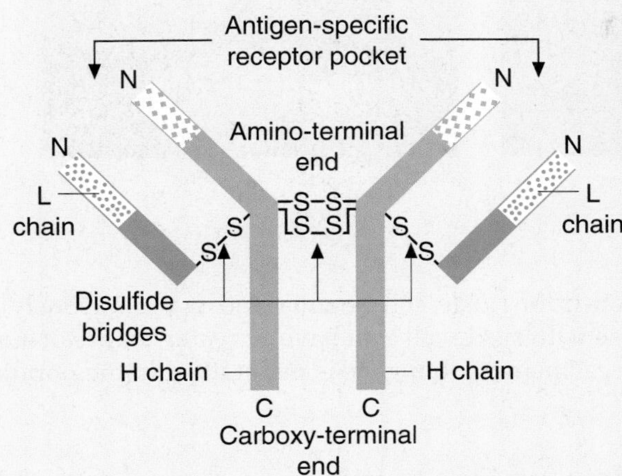

Schematic diagram of a Y-shaped immunoglobulin molecule. The binding sites for antigens are at the tips of the Y.

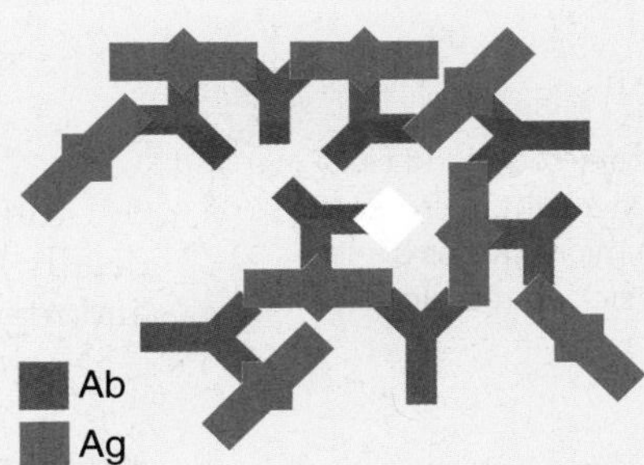

Schematic diagram of cross-linked immunoglobulin–antigen lattice.

but a different protein composition in the rest of the molecule. IgA is the immunoglobulin responsible for protecting the body surfaces, such as the mucous membranes of the gut, the oral cavity, and the genitourinary tract. IgA is also found in mother's milk, protecting the newborn against diseases during the first few weeks of life. IgD is found in very small amounts and is thought to be involved in the regulation of antibody synthesis. The last type of immunoglobulin is IgE. For many years the function of IgE was unknown. It is found in large quantities in the blood of people suffering from allergies and is therefore thought to be responsible for this "overblown" immunological reaction to dust particles and pollen grains.

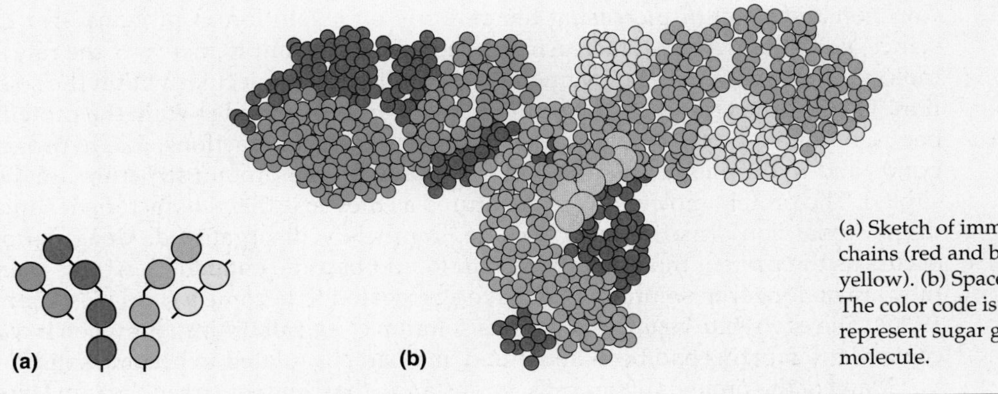

(a) Sketch of immunoglobulin G showing the two heavy chains (red and blue) and the two light chains (green and yellow). (b) Space-filling model of immunoglobulin G. The color code is the same as in (a). The gray balls represent sugar groups attached to the immunoglobulin molecule.

An interesting relationship exists between sickle cell trait and resistance to malaria. In some parts of Africa, up to 20% of the population has sickle cell trait. In those same parts of Africa, one of the leading causes of death is malaria. The presence of sickle cell trait is linked to an increased resistance to malaria because the malarial parasite cannot feed efficiently on sickled red blood cells. People who have sickle cell disease die young; those without sickle cell trait have a high probability of succumbing to malaria. Occupying the middle ground, people who have sickle

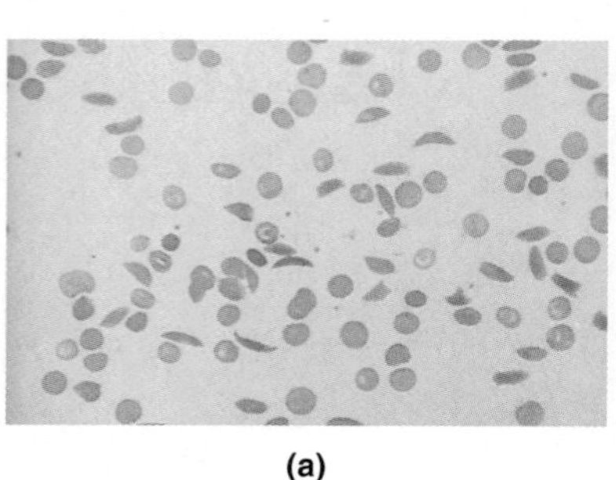
(a)

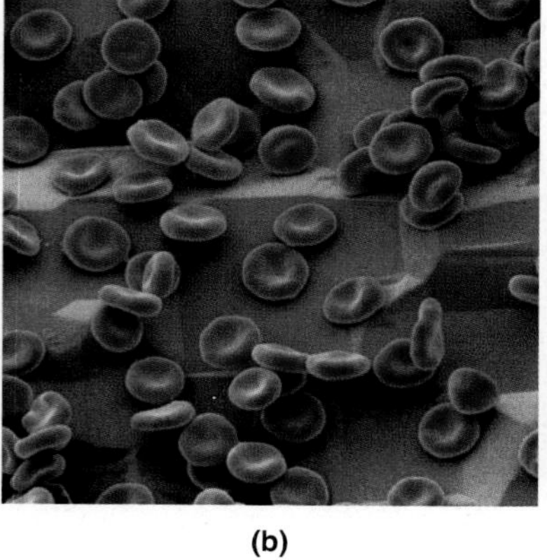
(b)

(c)

Figure 19.16
(a) Sickled and normal red blood cells photographed with a light microscope. Scanning electron micrographs of (b) normal and (c) sickled red blood cells.

cell trait do not suffer much from sickle cell anemia and simultaneously resist deadly malaria. Because those with sickle cell trait have a greater chance of survival and reproduction, the sickle cell hemoglobin gene is maintained in the population.

19.10 Denaturation of Proteins

Learning Goal

13

We have shown that the shape of a protein is absolutely essential to its function. We have also mentioned that life can exist only within a rather narrow range of temperature and pH. How are these two concepts related? As we will see, extremes of pH or temperature have a drastic effect on protein conformation, causing the molecules to lose their characteristic three-dimensional shape. **Denaturation** occurs when the organized structures of a globular protein, the α-helix, the β-pleated sheet, and tertiary folds become completely disorganized. However, it does not alter the primary structure. Denaturation of an α-helical protein is shown in Figure 19.17.

Temperature

Consider the effect of increasing temperature on a solution of proteins—for instance, egg white. At first, increasing the temperature simply increases the rate of molecular movement, the movement of the individual molecules within the solution. Then, as the temperature continues to increase, the bonds within the proteins begin to vibrate more violently. Eventually, the weak interactions, like hydrogen bonds and hydrophobic interactions, that maintain the protein structure are disrupted. The protein molecules are denatured as they lose their characteristic three-dimensional conformation and become completely disorganized. **Coagulation** occurs as the protein molecules then unfold and become entangled. At this point they are no longer in solution; they have aggregated to become a solid (see Figure 19.17). The egg white began as a viscous solution of egg albumins; but when it was cooked, the proteins had been denatured and had coagulated to become solid.

Many of the proteins of our cells, for instance, the enzymes, are in the same kind of viscous solution within the cytoplasm. To continue to function properly, they must remain in solution and maintain the correct three-dimensional configuration. If the body temperature becomes too high, or if local regions of the body are subjected to very high temperatures, as when you touch a hot cookie sheet, cellular proteins become denatured. They lose their function, and the cell or the organism dies.

pH

Because of the R groups of the amino acids, all proteins have a characteristic electric charge. Because every protein has a different amino acid composition, each will have a characteristic net electric charge on its surface. The positively and negatively

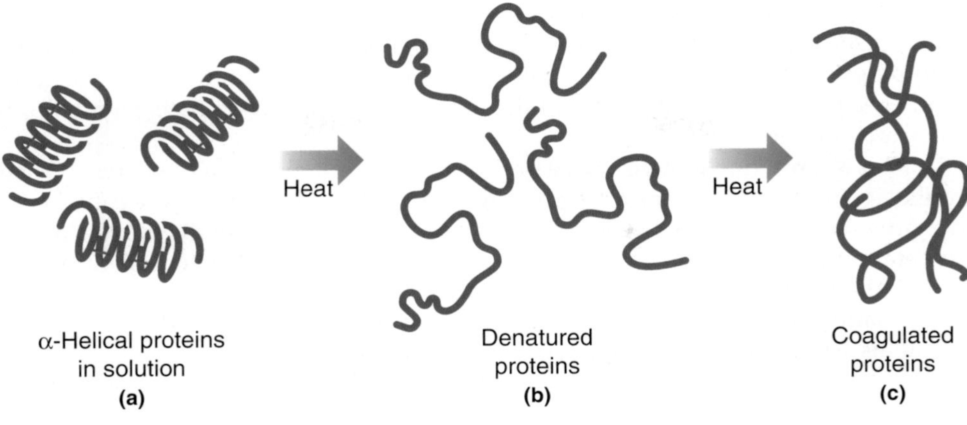

Figure 19.17
The denaturation of proteins by heat. (a) The α-helical proteins are in solution. (b) As heat is applied, the hydrogen bonds maintaining the secondary structure are disrupted, and the protein structure becomes disorganized. The protein is denatured. (c) The denatured proteins clump together, or coagulate, and are now in an insoluble form.

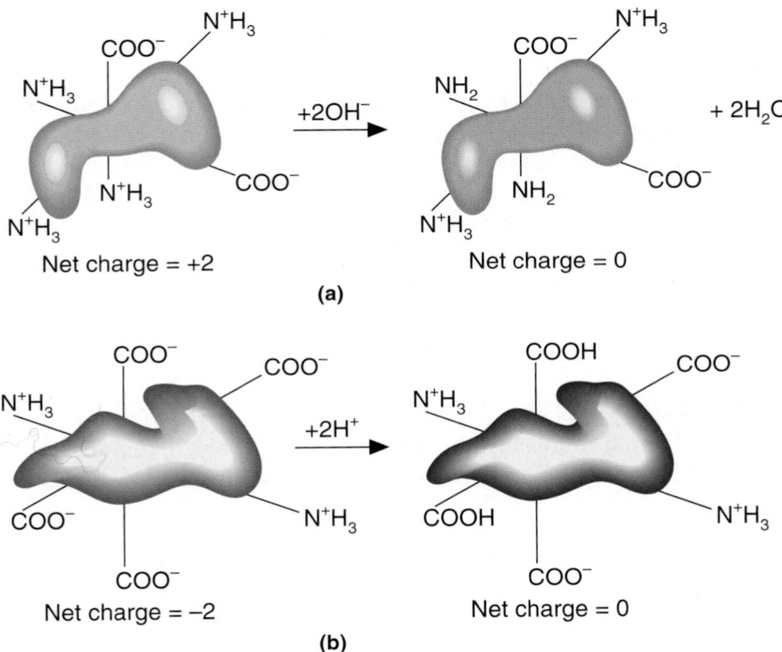

Figure 19.18
The effect of pH on proteins. (a) This protein has an overall charge of 2+. When a base is added, some of the protonated amino groups lose their protons. Now the protein is isoelectric; it has an equal number of positive and negative charges. (b) This protein has an overall charge of 2−. As acid is added, some of the carboxylate groups are protonated. The result is that the protein becomes isoelectric.

charged R groups on the surface of the molecule interact with ions and water molecules, and these interactions keep the protein in solution within the cytoplasm.

The protein shown in Figure 19.18a has a net charge of 2+ because it has two extra —N^+H_3 groups. If we add 2 moles of base, such as NaOH, the protonated amino groups lose their protons and thus become electrically neutral. Now the net charge of the protein is zero. The pH at which a protein has an equal number of positive and negative charges, that is, a net charge of zero, is called the **isoelectric point.** The protein shown in Figure 19.18b has a net charge of 2− because of two additional carboxylate groups. When 2 moles of acid are added, the carboxylate groups become protonated. They are now electrically neutral, and the net charge on the protein is zero. As in the preceding example, the protein solution is at the isoelectric point.

When the pH of a protein solution is above the isoelectric point, all the protein molecules will have a net negative surface charge. Below the isoelectric point, they will have a net positive charge. In either case, these like-charged molecules repel one another, and this repulsion helps keep these very large molecules in solution.

At the isoelectric point the protein molecules no longer have a net surface charge. As a result they no longer strongly repel one another and are at their least soluble. Under these conditions, there is a tendency for them to clump together and precipitate out of solution. In this case, proteins may coagulate even though they are not denatured.

Lactate fermentation is discussed in Section 21.4.

This is a reaction that you have probably observed in your own kitchen. When milk sits in the refrigerator for a prolonged period, the bacteria in the milk begin to grow. They use the milk sugar, lactose, as an energy source in the process of fermentation and produce lactic acid as a by-product. As the bacteria continue to grow, the concentration of lactic acid increases. The additional acid results in the protonation of exposed carboxylate groups on the surface of the dissolved milk proteins. They become isoelectric and coagulate into a solid curd.

See A Clinical Perspective: Proteins in the Blood, earlier in this chapter.

Imagine for a moment what would happen if the pH of the blood were to become too acidic or too basic. Blood is a fluid that contains water and dissolved electrolytes, a variety of cells, including the red blood cells responsible for oxygen transport, and many different proteins. These proteins include fibrinogen, which is involved in the clotting reaction; immunoglobulins, which protect us from disease; and albumins, which carry hydrophobic molecules in the blood.

When the blood pH drops too low, blood proteins become polycations. Similarly, when the blood pH rises too high, the proteins become polyanions. In either case, the proteins will unfold because of charge repulsion and loss of stabilizing ionic interactions. Under these extreme conditions, the denatured blood proteins would no longer be able to carry out their required functions. The blood cells would also die as their critical enzymes were denatured. The hemoglobin in the red blood cells would become denatured and would no longer be able to transport oxygen. Fortunately, the body has a number of mechanisms, such as the carbonate buffer system discussed earlier, to avoid the radical changes in the blood pH that can occur as a result of metabolic or respiratory difficulties.

Organic Solvents

Polar organic solvents, such as rubbing alcohol (2-propanol), denature proteins by disrupting hydrogen bonds within the protein, in addition to forming hydrogen bonds with the solvent, water. The nonpolar regions of these solvents interfere with hydrophobic interactions in the interior of the protein molecule, thereby disrupting the conformation. Traditionally, a 70% solution of rubbing alcohol was often used as a disinfectant or antiseptic. However, recent evidence suggests that it is not an effective agent in this capacity.

Detergents

Detergents have both a hydrophobic region (the fatty acid tail) and a polar or hydrophilic region. When detergents interact with proteins, they disrupt hydrophobic interactions, causing the protein chain to unfold.

Heavy Metals

Heavy metals such as mercury (Hg^{2+}) or lead (Pb^{2+}) may form bonds with negatively charged side chain groups. This interferes with the salt bridges formed between amino acid R groups of the protein chain, resulting in loss of conformation. Heavy metals may also bind to sulfhydryl groups of a protein. This may cause a profound change in the three-dimensional structure of the protein, accompanied by loss of function.

Mechanical Stress

Stirring, whipping, or shaking can disrupt the weak interactions that maintain protein conformation. This is the reason that whipping egg whites produces a stiff meringue.

How does high temperature denature proteins?

Question 19.9

How does extremely low pH cause proteins to coagulate?

Question 19.10

19.11 Dietary Protein and Protein Digestion

Learning Goal

Proteins, as well as carbohydrates and fats, are a major type of energy source in the diet. As do carbohydrates and fats, proteins serve several dietary purposes. They can be oxidized to provide energy. In addition, the amino acids liberated by the hydrolysis of proteins are used directly in biosynthesis. The protein synthetic machinery of the cell can incorporate amino acids, released by the digestion of dietary protein, directly into new cellular proteins. Amino acids are also used in the biosynthesis of a large number of important molecules called the *nitrogen compounds*. This group includes some hormones, the heme groups of hemoglobin and myoglobin, and the nitrogen-containing bases found in DNA and RNA.

Digestion of dietary protein begins in the stomach. The stomach enzyme *pepsin* begins the digestion by hydrolyzing some of the peptide bonds of the protein. This breaks the protein down into smaller peptides.

Production of pepsin and other proteolytic digestive enzymes must be carefully controlled because the active enzymes would digest and destroy the cell that produces them. Thus, the stomach lining cells that make pepsin actually synthesize and secrete an inactive form called *pepsinogen*. Pepsinogen has an additional forty-two amino acids in its primary structure. These are removed in the stomach to produce active pepsin.

The inactive form of a proteolytic enzyme is called a zymogen or proenzyme. These are discussed in Section 20.9.

Protein digestion continues in the small intestine where the enzymes trypsin, chymotrypsin, elastase, and others catalyze the hydrolysis of peptide bonds at different sites in the protein. For instance, chymotrypsin cleaves peptide bonds on the carbonyl side of aromatic amino acids and trypsin cleaves peptide bonds on the carbonyl side of basic amino acids. Together these proteolytic enzymes degrade large dietary proteins into amino acids that can be absorbed by cells of the small intestine.

The specificity of proteolytic enzymes is described in Section 20.11.

See Section 20.11 and Figure 20.12 for a more detailed picture of the action of digestive proteases.

Amino acids can be divided into two major nutritional classes. **Essential amino acids** are those that cannot be synthesized by the body and are required in the diet. **Nonessential amino acids** are those amino acids that can be synthesized by the body and need not be included in the diet. Table 19.2 lists the essential and nonessential amino acids.

Proteins are also classified as *complete* or *incomplete*. Protein derived from animal sources is generally **complete protein.** That is, it provides all of the essential and nonessential amino acids in approximately the correct amounts for biosynthesis. In contrast, protein derived from vegetable sources is generally **incomplete protein** because it lacks a sufficient amount of one or more essential amino acids. People who want to maintain a strictly vegetarian diet or for whom animal protein is often not available have the problem that no single high-protein vegetable has all of the essential amino acids to ensure a sufficient daily intake. For example, the major protein of beans contains abundant lysine and tryptophan but very little methionine, whereas corn contains considerable methionine but very little tryptophan or lysine. A mixture of corn and beans, however, satisfies both requirements. This combination, called *succotash*, was a staple of the diet of Native Americans for centuries.

Eating a few vegetarian meals each week can provide all the required amino acids and simultaneously help reduce the amount of saturated fats in the diet. Many ethnic foods apply the principle of mixing protein sources. Mexican foods such as tortillas and refried beans, Cajun dishes of spicy beans and rice, Indian cuisine of rice and lentils, and even the traditional American peanut butter sandwich are examples of ways to mix foods to provide complete protein.

Table 19.2 The Essential and Nonessential Amino Acids

Essential Amino Acids	Nonessential Amino Acids
Isoleucine	Alanine
Leucine	Arginine[1]
Lysine	Asparagine
Methionine	Aspartate
Phenylalanine	Cysteine[2]
Threonine	Glutamate
Tryptophan	Glutamine
Valine	Glycine
	Histidine[1]
	Proline
	Serine
	Tyrosine[2]

[1]Histidine and arginine are essential amino acids for infants but not for healthy adults.
[2]Cysteine and tyrosine are considered to be semiessential amino acids. They are required by premature infants and adults who are ill.

Question 19.11 Why must vegetable sources of protein be mixed to provide an adequate diet?

Question 19.12 What are some common sources of dietary protein?

Summary

19.1 Cellular Functions of Proteins

Proteins serve as biological catalysts (enzymes) and protective *antibodies*. *Transport proteins* carry materials throughout the body. Protein hormones regulate conditions in the body. Proteins also provide mechanical support and are needed for movement.

19.2 The α-Amino Acids

Proteins are made from twenty different *amino acids*, each having an α-COO^- group and an α-N^+H_3 group. They differ only in terms of side-chain R groups. All *α-amino acids* are chiral except glycine. Naturally occurring amino acids have the same chirality, designated L. The amino acids are grouped according to the polarity of their R groups.

19.3 The Peptide Bond

Amino acids are joined by *peptide bonds* to produce peptides and proteins. The peptide bond is an amide bond formed in the reaction between the carboxyl group of one amino acid and the amino group of another. The peptide bond is planar and relatively rigid.

19.4 The Primary Structure of Proteins

Proteins are linear polymers of amino acids. The linear sequence of amino acids defines the *primary structure* of the protein. Evolutionary relationships between species of organisms can be deduced by comparing the primary structures of their proteins.

19.5 The Secondary Structure of Proteins

The *secondary structure* of a protein is the folding of the primary sequence into an *α-helix* or *β-pleated sheet*. These structures are maintained by hydrogen bonds between the amide hydrogen and the carbonyl oxygen of the peptide bond. Usually *structural proteins*, such as the *α-keratins* and silk fibroin, are composed entirely of α-helix or β-pleated sheet.

19.6 The Tertiary Structure of Proteins

Globular proteins contain varying amounts of α-helix and β-pleated sheet folded into higher levels of structure called the *tertiary structure*. The tertiary structure of a protein is maintained by attractive forces between the R groups of amino acids. These forces include hydrophobic interactions, hydrogen bonds, ionic bridges, and disulfide bonds.

19.7 The Quaternary Structure of Proteins

Some proteins are composed of more than one peptide. They are said to have *quaternary structure.* Weak attractions between amino acid R groups hold the peptide subunits of the protein together. Some proteins require an attached, nonprotein *prosthetic group.*

19.8 An Overview of Protein Structure and Function

The primary structure of a protein dictates the way in which it folds into secondary and tertiary levels of structure. It also determines the way in which a protein may associate with other peptide subunits in the quaternary structure. An amino acid change in the primary structure may drastically affect protein folding. If the protein does not fold properly, and assume its correct three-dimensional shape, it will not be able to carry out its cellular function.

19.9 Myoglobin and Hemoglobin

Myoglobin, the oxygen storage protein of skeletal muscle, has a prosthetic group called the *heme group.* The heme group is the site of oxygen binding. *Hemoglobin* consists of four peptides. It transports oxygen from the lungs to the tissues. Myoglobin has a greater affinity for oxygen than does hemoglobin, and so oxygen is efficiently transferred from hemoglobin in the blood to myoglobin in tissues. Fetal hemoglobin has a greater affinity for oxygen than does maternal hemoglobin, and oxygen transfer occurs efficiently across the placenta from the mother to the fetus. A mutant hemoglobin is responsible for the genetic disease *sickle cell anemia.*

19.10 Denaturation of Proteins

Heat disrupts the hydrogen bonds and hydrophobic interactions that maintain protein structure. As a result, the protein unfolds and the organized structure is lost. The protein is said to be *denatured. Coagulation,* or clumping, occurs when the protein chains unfold and become entangled. When this occurs, proteins are no longer soluble. Changes in pH may cause proteins to become *isoelectric* (equal numbers of positive and negative charges). Isoelectric proteins coagulate because they no longer repel each other. If pH drops very low, proteins become polycations; if pH rises very high, proteins become polyanions. In either case, proteins become denatured owing to charge repulsion.

19.11 Dietary Protein and Protein Digestion

Essential amino acids must be acquired in the diet; *nonessential amino acids* can be synthesized by the body. *Complete proteins* contain all the essential and nonessential amino acids. *Incomplete proteins* are missing one or more essential amino acids. Protein digestion begins in the stomach, where proteins are degraded by the enzyme pepsin. Further digestion occurs in the small intestine by enzymes such as trypsin and chymotrypsin.

Key Terms

α-amino acid (19.2)
α-helix (19.5)
α-keratin (19.5)
antibody (19.1)
antigen (19.1)
β-pleated sheet (19.5)
coagulation (19.10)
complete protein (19.11)
C-terminal amino acid (19.3)
denaturation (19.10)
enzyme (19.1)
essential amino acid (19.11)
fibrous protein (19.5)
globular protein (19.6)
glycoprotein (19.7)
heme group (19.9)
hemoglobin (19.9)
hydrophilic amino acid (19.2)
hydrophobic amino acid (19.2)
incomplete protein (19.11)
isoelectric point (19.10)
movement protein (19.1)
myoglobin (19.9)
myosin (19.5)
nonessential amino acid (19.11)
N-terminal amino acid (19.3)
nutrient protein (19.1)
peptide bond (19.3)
primary structure (of a protein) (19.4)
prosthetic group (19.7)
protein (Intro)
quaternary structure (of a protein) (19.7)
regulatory protein (19.1)
secondary structure (of a protein) (19.5)
sickle cell anemia (19.9)
structural protein (19.1)
tertiary structure (of a protein) (19.6)
transport protein (19.1)

Questions and Problems

Cellular Functions of Proteins

19.13 List five biological functions of proteins.
19.14 Provide an example of a protein that carries out each of the functions listed in the answer to Problem 19.13.

The α-Amino Acids

19.15 Write the basic general structure of an L-α-amino acid.
19.16 Why are all of the α-amino acids except glycine chiral?
19.17 What is the importance of the R groups of the amino acids?
19.18 Describe the classification of the R groups of the amino acids, and provide an example of each class.
19.19 Write the structures of the nine amino acids that have hydrophobic side chains.
19.20 Write the structures of the aromatic amino acids. Indicate whether you would expect to find each on the surface or buried in a globular protein.

The Peptide Bond

19.21 Write the structure of each of the following peptides:
- **a.** His-trp-cys
- **b.** Gly-leu-ser
- **c.** Arg-ile-val

19.22 Write the structure of each of the following peptides:
- **a.** Ile-leu-phe
- **b.** His-arg-lys
- **c.** Asp-glu-ser

19.23 What properties of the peptide bond are responsible for its geometry?
19.24 Explain why the peptide bond is relatively rigid.

The Primary Structure of Proteins

19.25 Explain the relationship between the primary structure of a protein and the gene for that protein.
19.26 Explain how comparison of the primary structure of a protein from different organisms can be used to deduce evolutionary relationships between them.
19.27 What is the primary structure of a protein?
19.28 What type of bond joins the amino acids to one another in the primary structure of a protein?

The Secondary Structure of Proteins

19.29 Define the secondary structure of a protein.
19.30 What are the two most common types of secondary structure?
19.31 What type of secondary structure is characteristic of:
 a. The α-keratins?
 b. Silk fibroin?
19.32 Describe the forces that maintain the two types of secondary structure: α-helix and β-pleated sheet.
19.33 Define fibrous proteins.
19.34 What is the relationship between the structure of fibrous proteins and their functions?
19.35 Describe a parallel β-pleated sheet.
19.36 Compare a parallel β-pleated sheet to an antiparallel β-pleated sheet.

The Tertiary Structure of Proteins

19.37 Define the tertiary structure of a protein.
19.38 Use examples of specific amino acids to show the variety of weak interactions that maintain tertiary protein structure.
19.39 Write the structure of the amino acid produced by the oxidation of cysteine.
19.40 What is the role of cystine in maintaining protein structure?
19.41 Explain the relationship between the secondary and tertiary protein structures.
19.42 Why is the amino acid proline often found in the random coil hinge regions of the tertiary structure?

The Quaternary Structure of Proteins

19.43 Describe the quaternary structure of proteins.
19.44 What weak interactions are responsible for maintaining quaternary protein structure?
19.45 What is a glycoprotein?
19.46 What is a prosthetic group?

An Overview of Protein Structure and Function

19.47 Why is hydrogen bonding so important to protein structure?
19.48 Explain why α-keratins that have many disulfide bonds between adjacent polypeptide chains are much less elastic and much harder than those without disulfide bonds.
19.49 How does the structure of the peptide bond make the structure of proteins relatively rigid?
19.50 The primary structure of a protein known as histone H4, which tightly binds DNA, is identical in all mammals and differs by only one amino acid between the calf and pea seedlings. What does this extraordinary conservation of primary structure imply about the importance of that one amino acid?
19.51 What does it mean to say that the structure of proteins is genetically determined?
19.52 Explain why genetic mutations that result in the replacement of one amino acid with another can lead to the formation of a protein that cannot carry out its biological function.

Myoglobin and Hemoglobin

19.53 What is the function of hemoglobin?
19.54 What is the function of myoglobin?
19.55 Describe the structure of hemoglobin.
19.56 Describe the structure of myoglobin.
19.57 What is the function of heme in hemoglobin and myoglobin?
19.58 Write an equation representing the binding to and release of oxygen from hemoglobin.
19.59 Carbon monoxide binds tightly to the heme groups of hemoglobin and myoglobin. How does this affinity reflect the toxicity of carbon monoxide?
19.60 The blood of the horseshoe crab is blue because of the presence of a protein called hemocyanin. What is the function of hemocyanin?
19.61 Why does replacement of glutamic acid with valine alter hemoglobin and ultimately result in sickle cell anemia?
19.62 How do sickled red blood cells hinder circulation?
19.63 What is the difference between sickle cell disease and sickle cell trait?
19.64 How is it possible for sickle cell trait to confer a survival benefit on the person who possesses it?

Denaturation of Proteins

19.65 Define the term *denaturation*.
19.66 What is the difference between denaturation and coagulation?
19.67 Why is heat an effective means of sterilization?
19.68 As you increase the temperature of an enzyme-catalyzed reaction, the rate of the reaction initially increases. It then reaches a maximum rate and finally dramatically declines. Keeping in mind that enzymes are proteins, how do you explain these changes in reaction rate?
19.69 Why is it important that blood have several buffering mechanisms to avoid radical pH changes?
19.70 Define the term *isoelectric*.
19.71 Why do proteins become polycations at extremely low pH?
19.72 Why do proteins become polyanions at very high pH?
19.73 Yogurt is produced from milk by the action of dairy bacteria. These bacteria produce lactic acid as a by-product of their metabolism. The pH decrease causes the milk proteins to coagulate. Why are food preservatives not required to inhibit the growth of bacteria in yogurt?
19.74 Wine is made from the juice of grapes by varieties of yeast. The yeast cells produce ethanol as a by-product of their fermentation. However, when the ethanol concentration reaches 12–13%, all the yeast die. Explain this observation.

Dietary Protein and Protein Digestion

19.75 Why is it necessary to mix vegetable proteins to provide an adequate vegetarian diet?
19.76 Name some ethnic foods that apply the principle of mixing vegetable proteins to provide all of the essential amino acids.
19.77 What is the difference between essential and nonessential amino acids?
19.78 What is the difference between a complete protein and an incomplete protein?
19.79 Why must synthesis of digestive enzymes be carefully controlled?
19.80 What is the relationship between pepsin and pepsinogen?

Critical Thinking Problems

1. Calculate the length of an α-helical polypeptide that is twenty amino acids long. Calculate the length of a region of antiparallel β-pleated sheet that is forty amino acids long.
2. Proteins involved in transport of molecules or ions into or out of cells are found in the membranes of all cells. They are classified as transmembrane proteins because some regions are embedded within the lipid bilayer, whereas other regions protrude into the cytoplasm or outside the cell. Review the classification of amino acids based on the properties of their R groups. What type of amino acids would you expect to find in the regions of the proteins embedded within the membrane? What type of amino acids would you expect to find on the surface of the regions in the cytoplasm or that protrude outside the cell?
3. A biochemist is trying to purify the enzyme hexokinase from a bacterium that normally grows in the Arctic Ocean at 5°C. In the next lab, a graduate student is trying to purify the same protein from a bacterium that grows in the vent of a volcano at 98°C. To maintain the structure of the protein from the Arctic bacterium, the first biochemist must carry out all her purification procedures at refrigerator temperatures. The second biochemist must perform all his experiments in a warm room incubator. In molecular terms, explain why the same kind of enzyme from organisms with different optimal temperatures for growth can have such different thermal properties.
4. The α-keratin of hair is rich in the amino acid cysteine. The location of these cysteines in the protein chain is genetically determined; as a result of the location of the cysteines in the protein, a person may have curly, wavy, or straight hair. How can the location of cysteines in α-keratin result in these different styles of hair? Propose a hypothesis to explain how a "perm" causes straight hair to become curly.
5. Calculate the number of different pentapeptides you can make in which the amino acids phenylalanine, glycine, serine, leucine, and histidine are each found. Imagine how many proteins could be made from the twenty amino acids commonly found in proteins.

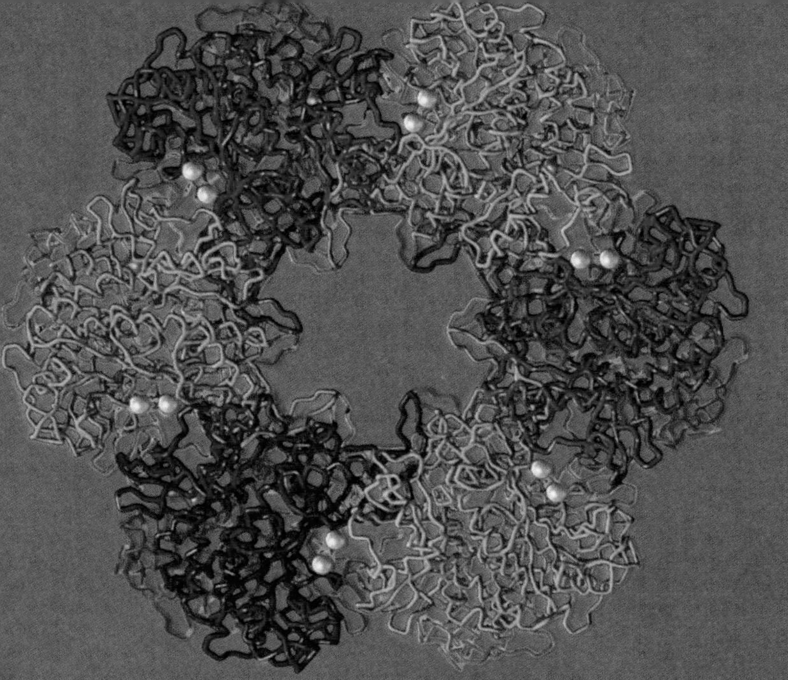

Molecular model of glutamine synthetase.

BIOCHEMISTRY

20

Enzymes

Outline

Learning Goals

1. Classify enzymes according to the type of reaction catalyzed and the type of specificity.
2. Give examples of the correlation between an enzyme's common name and its function.
3. Describe the effect that enzymes have on the activation energy of a reaction.
4. Explain the effect of substrate concentration on enzyme-catalyzed reactions.
5. Discuss the role of the active site and the importance of enzyme specificity.
6. Describe the difference between the lock-and-key model and the induced fit model of enzyme-substrate complex formation.
7. Discuss the roles of cofactors and coenzymes in enzyme activity.
8. Explain how pH and temperature affect the rate of an enzyme-catalyzed reaction.
9. Describe the mechanisms used by cells to regulate enzyme activity.
10. Discuss the mechanisms by which certain chemicals inhibit enzyme activity.
11. Discuss the role of the enzyme chymotrypsin and other serine proteases.
12. Provide examples of medical uses of enzymes.

CHEMISTRY CONNECTION

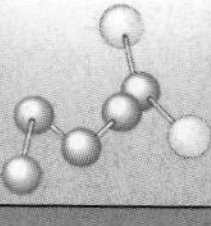

Super Hot Enzymes and the Origin of Life

Imagine the earth about four billion years ago: It was young then, not even a billion years old. Beginning as a red-hot molten sphere, slowly the earth's surface had cooled and become solid rock. But the interior, still extremely hot, erupted through the crust spewing hot gases and lava. Eventually these eruptions produced craggy land masses and an atmosphere composed of gases like hydrogen, carbon dioxide, ammonia, and water vapor. As the water vapor cooled, it condensed into liquid water, forming ponds and shallow seas.

At the dawn of biological life, the surface of the earth was still very hot and covered with rocky peaks and hot shallow oceans. The atmosphere was not very inviting either—filled with noxious gases and containing no molecular oxygen. Yet this is the environment that fostered the beginnings of life on this planet.

Some scientists think that they have found bacteria—living fossils—that may be very closely related to the first inhabitants of earth. The bacteria thrive at temperatures higher than the boiling point of water. Some need only H_2, CO_2, and H_2O for their metabolic processes and they quickly die in the presence of molecular oxygen.

But this lifestyle raises some uncomfortable questions. For instance, how do these bacteria survive at these extreme temperatures that would cook the life-forms with which we are more familiar? Researcher Mike Adams of the University of Georgia has found some of the answers. Adams and his students have studied the structure of an enzyme, a protein that acts as a biological catalyst, from one of these extraordinary bacteria. He compared the structure of the super hot enzyme with that of the same enzyme purified from an organism that grows at "normal" temperatures. The overall three-dimensional structures of the two enzymes were very similar. This makes sense because they both catalyze the same reaction.

The question, then, is why is the super hot enzyme so stable at very high temperatures, while its low temperature counterpart is not. The answer lay in the tertiary structure of the enzyme. Adams observed that the three-dimensional structure of the super hot enzyme is held together by many more R group interactions than are found in the low-temperature version. These R group interactions, along with other differences, keep the protein stable and functional even at temperatures above 100°C.

In Chapter 19 we studied the structure and properties of proteins. We are now going to apply that knowledge to the study of a group of proteins that do the majority of the work for the cell. These special proteins, the enzymes, catalyze the biochemical reactions that break down food molecules to allow the cell to harvest energy. They also catalyze the biosynthetic reactions that produce the molecules required for cellular life. In this chapter we will study the properties of this extraordinary group of proteins and learn how they dramatically speed up biochemical reactions.

Introduction

The enzymes discussed in this chapter are proteins; however, several ribonucleic acid (RNA) molecules have been demonstrated to have the ability to catalyze biological reactions. These are *ribozymes*.

An **enzyme** is a biological molecule that serves as a catalyst for a biochemical reaction. The majority of enzymes are proteins. Without enzymes to speed up biochemical reactions, life could not exist. The life of the cell depends on the simultaneous occurrence of hundreds of chemical reactions that must take place rapidly under mild conditions. It is possible, for example, to add water to an alkene. However, this reaction is usually carried out at a temperature of 100°C in aqueous sulfuric acid. Such conditions would kill a cell. The fragile cell must carry out its chemical reactions at body temperature (37°C) and in the absence of any strong acids or bases. How can this be accomplished? In Section 8.3 we saw that catalysts lower the energy of activation of a chemical reaction and thereby increase the rate of the reaction. This allows reactions to occur under milder conditions. The cell uses enzymes to solve the problem of chemical reactions that must occur rapidly under the mild conditions found within the cell. The enzyme *facilitates* a biochemical reaction, lowering the energy of activation and increasing the rate of the reaction. The efficient functioning of enzymes is essential for the life of the cell and of the organism.

The twin phenomena of high specificity and rapid reaction rates are the cornerstones of enzyme activity and the topic of this chapter. A typical cell contains thousands of different molecules, each of which is important to the chemistry of life processes. Each enzyme "recognizes" only one, or occasionally a few, of these molecules. One of the most remarkable features of enzymes is this *specificity*. Each can recognize and bind to a single type of *substrate* or reactant. The molecular size, shape, and charge distribution of both the enzyme and substrate must be compatible for this selective binding process to occur. The enzyme then transforms the substrate into the *product* with lightning speed. In fact, enzyme-catalyzed reactions often occur from one million to one hundred million times faster than the corresponding uncatalyzed reaction.

The enzyme *catalase* provides one of the most spectacular examples of the increase in reaction rates brought about by enzymes. This enzyme is required for life in an oxygen-containing environment. In this environment the process of the aerobic (oxygen-requiring) breakdown of food molecules produces hydrogen peroxide (H_2O_2). Because H_2O_2 is toxic to the cell, it must be destroyed. One molecule of catalase converts *forty million* molecules of hydrogen peroxide to harmless water and oxygen every second:

$$2H_2O_2 \xrightarrow[\text{(an enzyme)}]{\text{Catalase}} 2H_2O + O_2$$

Reaction occurs forty million times every second!

This is the same reaction that you witness when you pour hydrogen peroxide on a wound. The catalase released from injured cells rapidly breaks down the hydrogen peroxide. The bubbles that you see are oxygen gas released as a product of the reaction.

20.1 Nomenclature and Classification

Classification of Enzymes

Learning Goal

Enzymes may be classified according to the type of reaction that they catalyze. The six classes are as follows.

Oxidoreductases

Recall that redox reactions involve electron transfer from one substance to another (Section 9.5).

Oxidoreductases are enzymes that catalyze oxidation–reduction (redox) reactions. *Lactate dehydrogenase* is an oxidoreductase that removes hydrogen from a molecule of lactate. Other subclasses of the oxidoreductases include oxidases and reductases.

$$\text{HO—C(COO}^-\text{)(CH}_3\text{)—H} + NAD^+ \underset{}{\overset{\text{Lactate dehydrogenase}}{\rightleftharpoons}} \text{C(COO}^-\text{)(CH}_3\text{)=O} + NADH + H^+$$

Lactate — Pyruvate

Transferases

The significance of phosphate group transfers in energy metabolism is discussed in Sections 21.1 and 21.3.

Transferases are enzymes that catalyze the transfer of functional groups from one molecule to another. For example, a *transaminase* catalyzes the transfer of an amino functional group, and a *kinase* catalyzes the transfer of a phosphate group. Kinases play a major role in energy-harvesting processes involving ATP. In the adrenal glands, norepinephrine is converted to epinephrine by the enzyme *phenylethanolamine-N-methyltransferase* (PNMT), a *transmethylase*.

$$\text{Methyl group donor} + \text{(HO)}_2\text{C}_6\text{H}_3\text{—CH(OH)CH}_2\text{NH}_2 \overset{\text{PNMT}}{\rightleftharpoons} \text{(HO)}_2\text{C}_6\text{H}_3\text{—CH(OH)CH}_2\text{NH—CH}_3$$

Norepinephrine — Epinephrine

Hydrolysis of esters is described in Section 15.2. The action of lipases in digestion is discussed in Section 23.1.

Hydrolases

Hydrolases catalyze hydrolysis reactions, that is, the addition of a water molecule to a bond resulting in bond breakage. These reactions are important in the digestive process. For example, *lipases* catalyze the hydrolysis of the ester bonds in triglycerides:

$$\begin{array}{l} CH_2-O-\overset{O}{\overset{\|}{C}}(CH_2)_nCH_3 \\ | \\ CH-O-\overset{O}{\overset{\|}{C}}(CH_2)_nCH_3 \\ | \\ CH_2-O-\overset{O}{\overset{\|}{C}}(CH_2)_nCH_3 \end{array} + 3H_2O \xrightarrow{\text{Lipase}} \begin{array}{l} CH_2OH \\ | \\ CHOH \\ | \\ CH_2OH \end{array} + 3CH_3(CH_2)_nCOOH$$

Triglyceride Glycerol Fatty acids

Lyases

Lyases catalyze the addition of a group to a double bond or the removal of a group to form a double bond. *Carbonic anhydrase* is an example of a lyase. This reaction, which we studied in Section 19.9, occurs in the blood and is one of the body's mechanisms for buffering body fluids. In this reaction we see the addition of a group to a double bond.

$$O{=}C{=}O + H-OH \underset{}{\overset{\text{Carbonic anhydrase}}{\rightleftharpoons}} \begin{array}{c} O{=}C-OH \\ | \\ OH \end{array}$$

Carbon dioxide Water Carbonic acid

Citrate lyase catalyzes a far more complicated reaction in which we see the removal of a group and formation of a double bond. Specifically, citrate lyase catalyzes the removal of an acetyl group from a molecule of citrate. The products of this reaction include oxaloacetate, acetyl CoA, ADP, and an inorganic phosphate group (P_i):

$$\begin{array}{c} COO^- \\ | \\ CH_2 \\ | \\ ^-OOC-C-OH \\ | \\ CH_2 \\ | \\ COO^- \end{array} + ATP + \text{Coenzyme A} + H_2O \xrightarrow{\text{Citrate lyase}}$$

Citrate

$$\begin{array}{c} COO^- \\ | \\ CH_2 \\ | \\ C{=}O \\ | \\ COO^- \end{array} + CH_3-\overset{O}{\overset{\|}{C}}\sim S-CoA + ADP + P_i$$

Oxaloacetate Acetyl CoA

Recall that the squiggle (~) represents a high-energy bond.

Isomerases

Isomerases rearrange the functional groups within a molecule and catalyze the conversion of one isomer into another. For example, *phosphoglycerate mutase* converts one structural isomer, 3-phosphoglycerate, into another, 2-phosphoglycerate:

$$\mathrm{COO^-{-}CH(OH){-}CH_2{-}O{-}PO_3^{2-}} \xrightleftharpoons{\text{Phosphoglycerate mutase}} \mathrm{COO^-{-}CH(O{-}PO_3^{2-}){-}CH_2OH}$$

3-Phosphoglycerate 2-Phosphoglycerate

Ligases

Ligases are enzymes that catalyze a reaction in which a C—C, C—S, C—O, or C—N bond is made or broken. This is accompanied by an ATP-ADP interconversion. For example, *DNA ligase* catalyzes the joining of the hydroxyl group of a nucleotide in a DNA strand with the phosphoryl group of the adjacent nucleotide to form a phosphoester bond:

The use of DNA ligase in recombinant DNA studies is detailed in Section 24.8.

ATP, adenosine triphosphate, is the universal energy currency for all life-forms we have studied. It is formed by the addition of a phosphoryl group to a molecule of ADP, adenosine diphosphate. Formation of ATP requires energy. Thus, when a ligase breaks a bond, the energy released is used to form ATP. When a ligase forms a bond, energy is required. That energy is provided by the hydrolysis of the terminal phosphoric anhydride bond in ATP.

$$\text{DNA strand}{-}3'{-}\mathrm{OH} + {}^-\mathrm{O{-}P(=O)(O^-){-}O}{-}5'{-}\text{DNA strand} \xrightarrow{\text{DNA ligase}} \text{DNA strand}{-}3'{-}\mathrm{O{-}P(=O)(O^-){-}O}{-}5'{-}\text{DNA strand}$$

Classifying Enzymes According to the Type of Reaction That They Catalyze — EXAMPLE 20.1

Classify the enzyme that catalyzes each of the following reactions, and explain your reasoning.

$$\mathrm{H_3N^+{-}CH(CH_3){-}C(=O){-}NH{-}CH_2{-}COO^-} + \mathrm{H_2O} \longrightarrow \mathrm{H_3N^+{-}CH(CH_3){-}COO^-} + \mathrm{H_3N^+{-}CH_2{-}COO^-}$$

Alanylglycine Alanine Glycine

Solution

The reaction occurring here involves breaking a bond, in this case a peptide bond, by the addition of a water molecule. The enzyme is classified as a hydrolase, specifically a peptidase.

Continued—

EXAMPLE 20.1 —*Continued*

$$\text{Glucose} + \text{ATP (Adenosine triphosphate)} \longrightarrow \text{Glucose-6-phosphate} + \text{ADP (Adenosine diphosphate)}$$

Solution

This is the first reaction in the biochemical pathway called *glycolysis*. A phosphoryl group is transferred from a donor molecule, adenosine triphosphate, to the recipient molecule, glucose. The products are glucose-6-phosphate and adenosine diphosphate. This enzyme, called *hexokinase*, is an example of a transferase.

$$\underset{\text{Malate}}{^{-}OOC-CH_2-CH(OH)-COO^{-}} + NAD^{+} \longrightarrow \underset{\text{Oxaloacetate}}{^{-}OOC-CH_2-C(=O)-COO^{-}} + NADH$$

Solution

In this reaction, the reactant malate is oxidized and the coenzyme NAD^{+} is reduced. The enzyme that catalyzes this reaction, malate dehydrogenase, is an oxidoreductase.

$$\underset{\text{Dihydroxyacetone phosphate}}{HOCH_2-C(=O)-CH_2-O-PO_3^{2-}} \rightleftharpoons \underset{\text{Glyceraldehyde-3-phosphate}}{OHC-CH(OH)-CH_2-O-PO_3^{2-}}$$

Solution

Careful inspection of the structure of the reactant and the product reveals that they each have the same number of carbon, hydrogen, oxygen, and phosphorus atoms; thus, they must be structural isomers. The enzyme must be an isomerase. Its name is triose phosphate isomerase.

Question 20.1

To which class of enzymes does each of the following belong?

a. Pyruvate kinase
b. RNA ligase
c. Triose isomerase
d. Pyruvate dehydrogenase
e. Phosphoglucoisomerase

Question 20.2

To which class of enzymes does each of the following belong?

a. Phosphofructokinase
b. Lipase
c. Acetoacetate decarboxylase
d. Succinate dehydrogenase
e. Alanine transaminase

Nomenclature of Enzymes

Learning Goal 2

The common name for a hydrolase is derived from the name of the **substrate,** the reactant that binds to the enzyme and is converted to product. Names of other enzymes reflect the type of reactions that they catalyze. Because of this, the function of the enzyme is generally conveyed directly by its common name.

Let's look at a few examples of this simple concept. *Urea* is the substrate acted on by the enzyme *urease:*

Urea − a + ase = Urease

Substrate Enzyme

Note that the name of this enzyme is simply the name of the substrate with the ending *-ase* added. With the exception of some historical common names, the general ending for the name of an enzyme is *-ase.* For instance, *lactose* is the substrate of *lactase:*

Lactose − ose + ase = Lactase

Substrate Enzyme

Other enzymes may be named for the reactions they catalyze. For example,

Dehydrogenases remove hydrogen atoms, transferring them to a coenzyme.

Decarboxylases remove carboxyl groups.

Coenzymes are molecules required by some enzymes to serve as donors or acceptors of electrons, hydrogen atoms, or other functional groups during a chemical reaction. Coenzymes are discussed in Section 20.7.

The prefix *de-* indicates that a functional group is being removed. Hydrogenases and carboxylases, on the other hand, add hydrogen or carboxyl groups. Some enzyme names include *both* the names of the substrate and of the reaction type. For example, *lactate dehydrogenase* removes hydrogen atoms from lactate ions, and *pyruvate decarboxylase* removes the carboxyl group from pyruvate.

The complete name for lactate dehydrogenase is lactate: NAD oxidoreductase. This systematic name tells us the substrate, coenzyme, and type of reaction catalyzed.

As in other areas of chemistry, historical names, having no relationship to either substrate or reaction, continue to be used. In these cases the substrates and reactions must simply be memorized. Examples of some historical common names include catalase, pepsin, chymotrypsin, and trypsin.

Question 20.3

What is the substrate for each of the following enzymes?

a. Sucrase
b. Pyruvate decarboxylase
c. Succinate dehydrogenase

Question 20.4 What chemical reaction is mediated by each of the enzymes in Question 20.3?

20.2 The Effect of Enzymes on the Activation Energy of a Reaction

Learning Goal

How does an enzyme speed up a chemical reaction? It changes the path by which the reaction occurs, providing a lower energy route for the conversion of the substrate into the **product,** the chemical species that results from the enzyme-catalyzed reaction. Thus enzymes speed up reactions by lowering the activation energy of the reaction.

Recall that every chemical reaction is characterized by an equilibrium constant. Consider, for example, the simple equilibrium

The activation energy (Section 8.3) of a reaction is the threshold energy that must be overcome to produce a chemical reaction.

Equilibrium constants are described in Section 8.4.

Energy, rate, and equilibrium are described in Chapter 8.

$$aA \rightleftharpoons bB$$

The equilibrium constant for this reaction, K_{eq}, is defined as

$$K_{eq} = \frac{[B]^b}{[A]^a} = \frac{[\text{product}]^b}{[\text{reactant}]^a}$$

This equilibrium constant is actually a reflection of the difference in energy between reactants and products. It is a measure of the relative stabilities of the reactants and products. No matter how the chemical reaction occurs (which path it follows), the difference in energy between the reactants and the products is always the same. **An enzyme cannot therefore alter the equilibrium constant for the reaction that it catalyzes.** An enzyme does, however, change the path by which the process occurs, providing a lower energy route for the conversion of the substrate into the product. An enzyme increases the rate of a chemical reaction by lowering the activation energy for the reaction (Figure 20.1). An enzyme thus increases the rate at which the reaction it catalyzes reaches equilibrium.

20.3 The Effect of Substrate Concentration on Enzyme-Catalyzed Reactions

Learning Goal

The rates of uncatalyzed chemical reactions often double every time the substrate concentration is doubled (Figure 20.2a). Therefore as long as the substrate concentration increases, there is a direct increase in the rate of the reaction. For enzyme-

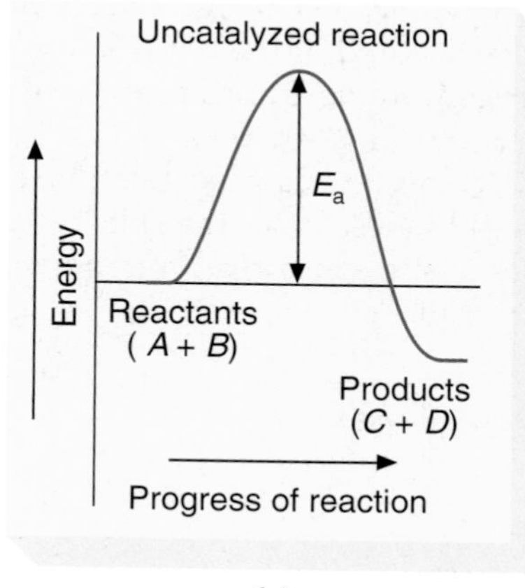

(a)

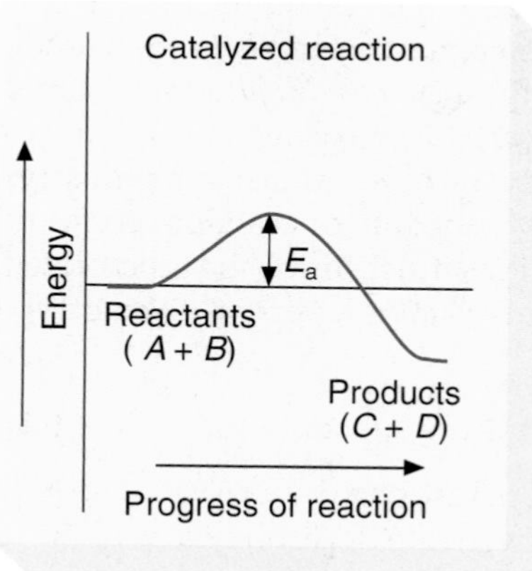

(b)

Figure 20.1
Diagram of the difference in energy between the reactants (*A* and *B*) and products (*C* and *D*) for a reaction. Enzymes cannot change this energy difference but act by lowering the activation energy (E_a) for the reaction, thereby speeding up the reaction.

catalyzed reactions, however, this is not the case. Although the rate of the reaction is initially responsive to the substrate concentration, at a certain concentration of substrate the rate of the reaction reaches a maximum value. A graph of the rate of reaction, V, versus the substrate concentration, $[S]$, is shown in Figure 20.2b. We see that the rate of the reaction initially increases rapidly as the substrate concentration is increased but that the rate levels off at a maximum value. At its maximum rate the active sites of all the enzyme molecules in solution are occupied by a substrate molecule. The active site is the region of the enzyme that specifically binds the substrate and catalyzes the reaction. A new molecule of substrate cannot bind to a given enzyme molecule until the substrate molecule already bound to the enzyme is converted to product. Thus it appears that the enzyme-catalyzed reaction occurs in two stages. The first, rapid step is the formation of the *enzyme-substrate complex.* The second step is slower and thus controls the rate at which the reaction can occur. It is said to be *rate-limiting* and involves conversion of the substrate to product and the release of the product and enzyme from the resulting enzyme-product complex. In effect, the rate of the reaction is limited by the speed with which the substrate is converted into product and the product is released. Thus the reaction rate is dependent on the availability of the enzyme.

20.4 The Enzyme-Substrate Complex

The following series of reversible reactions represents the steps in an enzyme-catalyzed reaction. The first step (highlighted in blue) involves the encounter of the enzyme with its substrate and the formation of an **enzyme-substrate complex.**

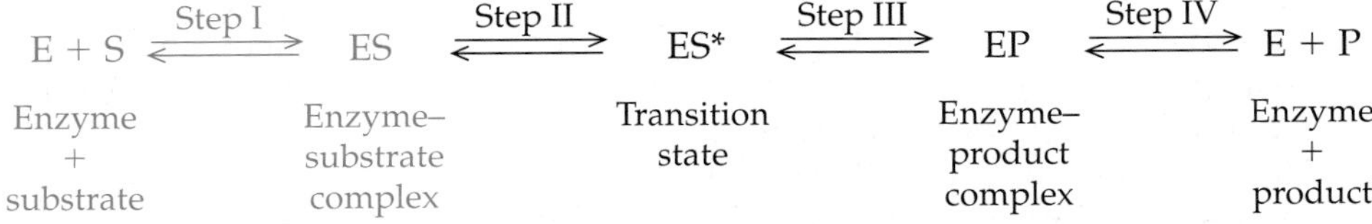

The part of the enzyme that binds with the substrate is called the **active site.** We find that the properties of the active site are crucial to the function of the enzyme and have the following general characteristics:

- Enzyme active sites are pockets or clefts in the surface of the enzyme. The R groups in the active site that are involved in catalysis are called *catalytic groups.*
- The shape of the active site is complementary to the shape of the substrate. That is, the substrate fits neatly into the active site of the enzyme.

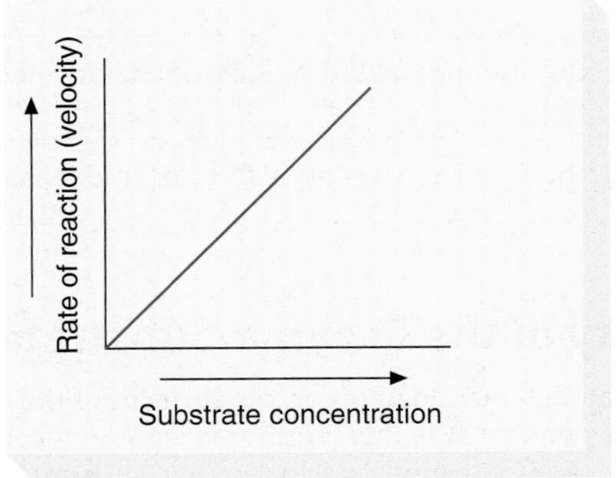

(a)

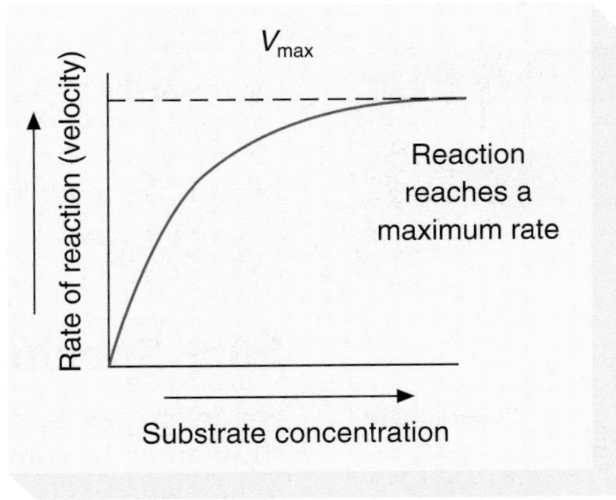

(b)

Figure 20.2
Plot of the rate or velocity, V, of a reaction versus the concentration of substrate, $[S]$, for (a) an uncatalyzed reaction and (b) an enzyme-catalyzed reaction. For an enzyme-catalyzed reaction the rate is at a maximum when all of the enzyme molecules are bound to the substrate. Beyond this concentration of substrate, further increases in substrate concentration have no effect on the rate of the reaction.

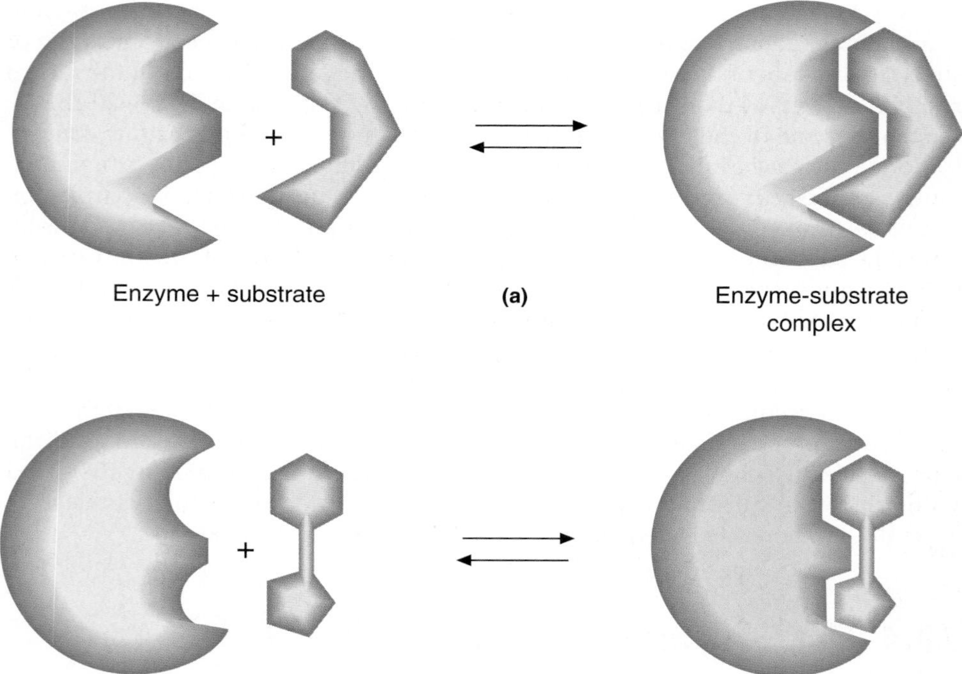

Figure 20.3
(a) The lock-and-key model of enzyme-substrate binding assumes that the enzyme active site has a rigid structure that is precisely complementary in shape and charge distribution to the substrate. (b) The induced fit model of enzyme-substrate binding. As the enzyme binds to the substrate, the shape of the active site conforms precisely to the shape of the substrate. The shape of the substrate may also change.

The overall shape of a protein is maintained by many weak interactions. At any time a few of these weak interactions may be broken by heat energy or a local change in pH. If only a few bonds are broken, they will re-form very quickly. The overall result is that there is a brief change in the shape of the enzyme. Thus the protein or enzyme can be viewed as a flexible molecule, changing shape slightly in response to minor local changes.

- An enzyme attracts and holds its substrate by weak, noncovalent interactions. The R groups involved in substrate binding, and not necessarily catalysis, make up the *binding site.*
- The conformation of the active site determines the specificity of the enzyme because only the substrate that fits into the active site will be used in a reaction.

Learning Goal

6

The **lock-and-key model** of enzyme activity, shown in Figure 20.3a, was devised by Emil Fischer in 1894. At that time it was thought that the substrate simply snapped into place like a piece of a jigsaw puzzle or a key into a lock.

Today we know that proteins are flexible molecules. This led Daniel E. Koshland, Jr., to propose a more sophisticated model of the way enzymes and substrates interact. This model, proposed in 1958, is called the **induced fit model** (Figure 20.3b). In this model the active site of the enzyme is not a rigid pocket into which the substrate fits precisely; rather, it is a flexible pocket that *approximates* the shape of the substrate. When the substrate enters the pocket, the active site "molds" itself around the substrate. This produces the perfect enzyme-substrate "fit."

Question 20.5 Compare the lock-and-key and induced fit models of enzyme-substrate binding.

Question 20.6 What is the relationship between an enzyme active site and its substrate?

20.5 Specificity of the Enzyme-Substrate Complex

Learning Goal 1 Learning Goal 5

For an enzyme-substrate interaction to occur, the surfaces of the enzyme and substrate must be complementary. It is this requirement for a specific fit that determines whether an enzyme will bind to a particular substrate and carry out a chemical reaction.

Enzyme specificity is the ability of an enzyme to bind only one, or a very few, substrates and thus catalyze only a single reaction. To illustrate the specificity of enzymes, consider the following reactions.

The enzyme urease catalyzes the decomposition of urea to carbon dioxide and ammonia as follows:

$$\underset{\text{Urea}}{H_2N{-}\overset{\overset{\displaystyle O}{\|}}{C}{-}NH_2} + H_2O \xrightarrow{\text{Urease}} CO_2 + 2NH_3$$

This reaction is one that you may have observed if you have a cat. Urea is a waste product of the breakdown of proteins and is removed from the body in urine. Bacteria in kitty litter produce urease. As the urease breaks down the urea in the cat urine, the ammonia released produces the distinctive odor of an untended litter box.

Methylurea, in contrast, though structurally similar to urea, is not affected by urease:

$$\underset{\text{Methylurea}}{H_2N{-}\overset{\overset{\displaystyle O}{\|}}{C}{-}NHCH_3} + H_2O \xrightarrow{\text{Urease}} \text{no reaction}$$

Not all enzymes exhibit the same degree of specificity. Four classes of enzyme specificity have been observed.

- **Absolute specificity:** An enzyme that catalyzes the reaction of only one substrate has **absolute specificity.** Aminoacyl tRNA synthetases exhibit absolute specificity. Each must attach the correct amino acid to the correct transfer RNA molecule. If the wrong amino acid is attached to the transfer RNA, it will mistakenly be added to a peptide chain, producing a nonfunctional protein.

Aminoacyl tRNA synthetases are discussed in Section 24.6. Aminoacyl group transfer reactions were described in Section 16.4.

- **Group specificity:** An enzyme that catalyzes processes involving similar molecules containing the same functional group has **group specificity.** Hexokinase is a group-specific enzyme that catalyzes the addition of a phosphoryl group to the hexose sugar glucose in the first step of glycolysis. Hexokinase can also add a phosphoryl group to several other six-carbon sugars.

Hexokinase activity is described in Section 21.3.

- **Linkage specificity:** An enzyme that catalyzes the formation or breakage of only certain bonds in a molecule has **linkage specificity.** Proteases, such as trypsin, chymotrypsin, and elastase, are enzymes that selectively hydrolyze peptide bonds. Thus these enzymes are linkage specific.

Proteolytic enzymes are discussed in Section 20.11.

- **Stereochemical specificity:** An enzyme that can distinguish one enantiomer from the other has **stereochemical specificity.** Most of the enzymes of the human body show stereochemical specificity. Because we use only D-sugars and L-amino acids, the enzymes involved in digestion and metabolism recognize only those particular stereoisomers.

20.6 The Transition State and Product Formation

How does enzyme-substrate binding result in a faster chemical reaction? The precise answer to this question is probably different for each enzyme-substrate pair, and, indeed, we understand the exact mechanism of catalysis for very few enzymes. Nonetheless, we can look at the general features of enzyme-substrate interactions that result in enhanced reaction rate and product formation. To do this, we must once again look at the steps of an enzyme-catalyzed reaction, focusing on the remaining steps highlighted in blue:

$$\underset{\substack{\text{Enzyme}\\+\\\text{substrate}}}{E + S} \underset{}{\overset{\text{Step I}}{\rightleftharpoons}} \underset{\substack{\text{Enzyme–}\\\text{substrate}\\\text{complex}}}{ES} \overset{\text{Step II}}{\rightleftharpoons} \underset{\substack{\text{Transition}\\\text{state}}}{ES^*} \overset{\text{Step III}}{\rightleftharpoons} \underset{\substack{\text{Enzyme–}\\\text{product}\\\text{complex}}}{EP} \overset{\text{Step IV}}{\rightleftharpoons} \underset{\substack{\text{Enzyme}\\+\\\text{product}}}{E + P}$$

A MEDICAL PERSPECTIVE

HIV Protease Inhibitors and Pharmaceutical Drug Design

In 1981 the Centers for Disease Control in Atlanta, Georgia, recognized a new disease syndrome, acquired immune deficiency syndrome (AIDS). The syndrome is characterized by an impaired immune system, a variety of opportunistic infections and cancer, and brain damage that results in dementia. It soon became apparent that the disease was being transmitted by blood and blood products, as well as by sexual conduct.

The earliest drugs that proved effective in the treatment of HIV infections were all inhibitors of replication of the genetic material of the virus. While these treatments were initially effective, prolonging the lives of many, it was not long before viral mutants resistant to these drugs began to appear.

In 1989 a group of scientists revealed the three-dimensional structure of the HIV protease. This structure is shown in the accompanying figure. This enzyme is necessary for viral replication because the virus has an unusual strategy for making all of its proteins. Rather than make each protein individually, it makes large "polyproteins" that must then be cut by the HIV protease to form the final proteins required for viral replication.

Since scientists realized that this enzyme was essential for HIV replication, they decided to engineer a substance that would inhibit the enzyme by binding irreversibly to the active site, in essence plugging it up. The challenge, then, was to design a molecule that would be the plug. Researchers knew the primary structure (amino acid sequence) of the HIV protease from earlier nucleic acid sequencing studies. By 1989 they also had a very complete picture of the three-dimensional nature of the molecule, which they had obtained by X-ray crystallography.

Putting all of this information into a sophisticated computer modeling program, they could look at the protease from any angle. They could see the location of each of the R groups of each of the amino acids in the active site. This kind of information allowed the scientists to design molecules that would be complementary to the shape and charge distribution of the enzyme active site—in other words, structural analogs of the normal protease substrate. It was not long before the scientists had produced several candidates for the HIV protease inhibitor.

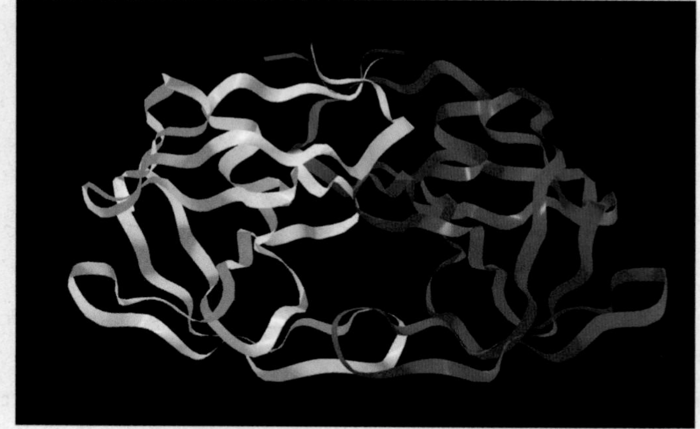

The human immunodeficiency virus protease.

But, there are many tests that a drug candidate must pass before it can be introduced into the market as safe and effective. Scientists had to show that the candidate drugs would bind effectively to the HIV protease and block its function, thereby inhibiting virus replication. Properties such as the solubility, the efficiency of absorption by the body, the period of activity in the body, and the toxicity of the drug candidates all had to be determined.

By 1996 there were three protease inhibitors available to those with HIV infection. There are currently seven of these drugs on the market. In many cases development and testing of a drug candidate can take up to fifteen years. In the case of the first HIV protease inhibitors, the first three drugs were on the market in less than eight years. This is a testament both to the urgent need for HIV treatments and to the technology available to attack the problem.

In Section 20.4 we examined the events of step I by which the enzyme and substrate interact to form the enzyme-substrate complex. In this section we will look at the events that lead to product formation. We have already described the enzyme as a flexible molecule; the substrate also has a degree of flexibility. The continued interaction between the enzyme and substrate changes the shape or position of the substrate in such a way that the molecular configuration is no longer energetically stable (step II). In this state, the **transition state,** the shape of the substrate is altered, because of its interaction with the enzyme, into an intermediate form having features of both the substrate and the final product. This transition state, in turn, favors the conversion of the substrate into product (step III). The product remains bound to the enzyme for a very brief time, then in step IV the product and enzyme dissociate from one another, leaving the enzyme completely unchanged.

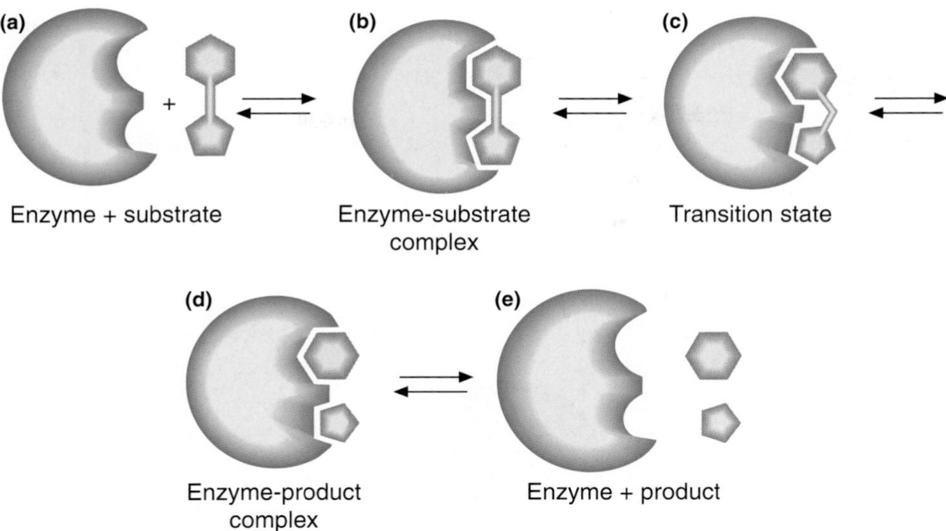

Figure 20.4
Bond breakage is facilitated by the enzyme as a result of stress on a bond. (a, b) The enzyme-substrate complex is formed. (c) In the transition state, the enzyme changes shape and thereby puts stress on the O-glycosidic linkage holding the two monosaccharides together. This lowers the energy of activation of this reaction. (d, e) The bond is broken, and the products are released.

What kinds of transition state changes might occur in the substrate that would make a reaction proceed more rapidly?

1. The enzyme might put "stress" on a bond and thereby facilitate bond breakage. Consider the hydrolysis of the sugar sucrose by the enzyme sucrase. The enzyme catalyzes the hydrolysis of the disaccharide sucrose into the monosaccharides glucose and fructose. The formation of the enzyme-substrate complex (Figures 20.4a and 20.4b) results in a change in the shape of the enzyme. This, in turn, may stretch or put pressure on one of the bonds of the substrate. Such a stress weakens the bond, allowing it to be broken much more easily than in the absence of the enzyme. This is represented in Figure 20.4c as the bending of the O-glycosidic bond between the fructose and the glucose. In the transition state the substrate has a molecular form resembling both the disaccharide, the original substrate, and the two monosaccharides, the eventual products. Clearly, the stress placed on the bond weakens it, and much less energy is required to break the bond to form products (Figures 20.4d and 20.4e). This also has the effect of speeding up the reaction.
2. An enzyme may facilitate a reaction by bringing two reactants into close proximity and in the proper orientation for reaction to occur. Consider now the condensation reaction between glucose and fructose to produce sucrose (Figure 20.5a). Each of the sugars has five hydroxyl groups that could undergo condensation to produce a disaccharide. But the purpose is to produce sucrose, not some other disaccharide. By random molecular collision there is a one in twenty-five chance that the two molecules will collide in the proper orientation to produce sucrose. The probability that the two will react is actually much less than that because of a variety of conditions in addition to orientation that must be satisfied for the reaction to occur. For example, at body temperature, most molecular collisions will not have a sufficient amount of energy to overcome the energy of activation, even if the molecules are in the proper orientation. The enzyme can facilitate the reaction by bringing the two molecules close together in the correct alignment (Figure 20.5b), thereby forcing the desired reactive groups of the two molecules together in the transition state and greatly speeding up the reaction.
3. The active site of an enzyme may modify the pH of the microenvironment surrounding the substrate. To accomplish this, the enzyme may, for example, serve as a donor or an acceptor of H^+. As a result, there would be a change in the pH in the vicinity of the substrate without disturbing the normal pH elsewhere in the cell.

Glucose + Fructose $\rightleftharpoons$ Sucrose + H_2O

(a)

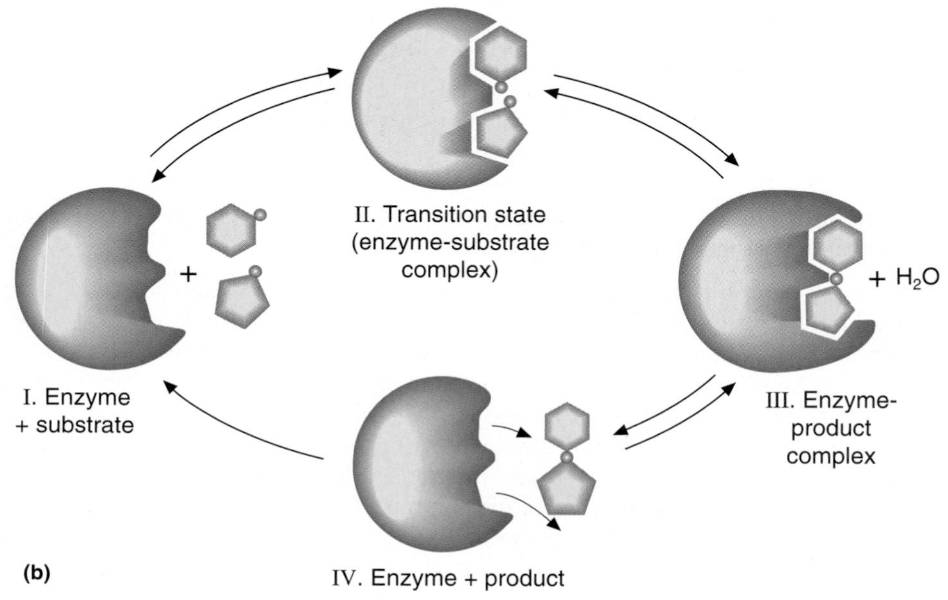

Figure 20.5
An enzyme may lower the energy of activation required for a reaction by holding the substrates in close proximity and in the correct orientation. (a) A condensation reaction in which glucose and fructose are joined in O-glycosidic linkage to produce sucrose. (b) The enzyme-substrate complex forms, bringing the two monosaccharides together with the hydroxyl groups involved in the linkage extended toward one another.

Question 20.7

Summarize three ways in which an enzyme might lower the energy of activation of a reaction.

Question 20.8

What is the transition state in an enzyme-catalyzed reaction?

20.7 Cofactors and Coenzymes

Learning Goal

In Section 19.7 we saw that some proteins require an additional nonprotein prosthetic group to function. The same is true of some enzymes. The polypeptide portion of such an enzyme is called the **apoenzyme,** and the nonprotein prosthetic group is called the **cofactor.** Together they form the active enzyme called the **holoenzyme.** Cofactors may be metal ions, organic compounds, or organometallic compounds. They must be bound to the enzyme to maintain the correct configuration of the enzyme active site (Figure 20.6). When the cofactor is bound and the active site is in the proper conformation, the enzyme can bind the substrate and catalyze the reaction.

Other enzymes require the temporary binding of a **coenzyme.** Such binding is generally mediated by weak interactions like hydrogen bonds. The coenzymes are organic molecules that generally serve as carriers of electrons or chemical groups.

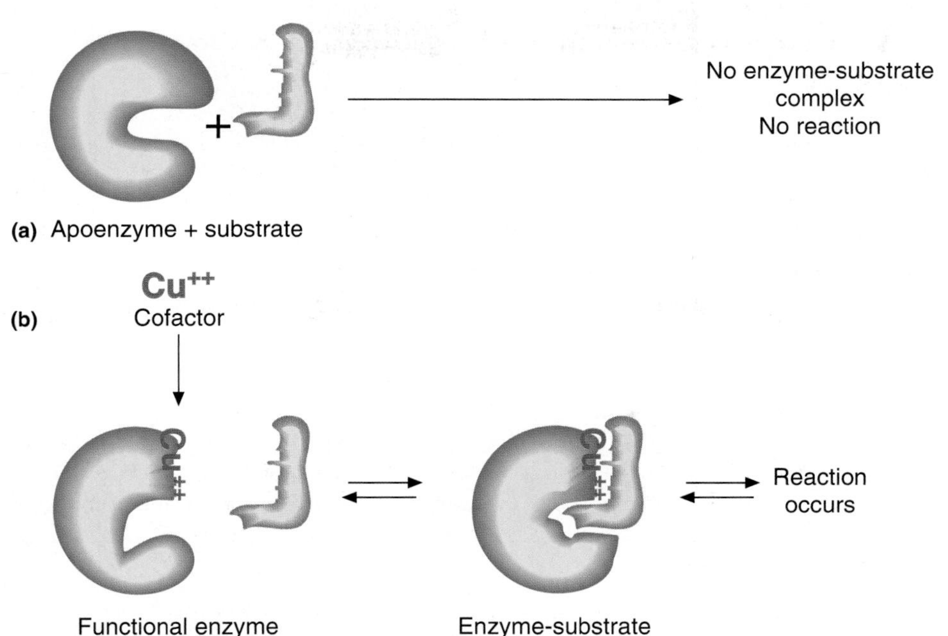

Figure 20.6
(a) The apoenzyme is unable to bind to its substrate. (b) When the required cofactor, in this case a copper ion, Cu^{2+}, is available, it binds to the apoenzyme. Now the active site takes on the correct configuration, the enzyme-substrate complex forms, and the reaction occurs.

Holoenzyme

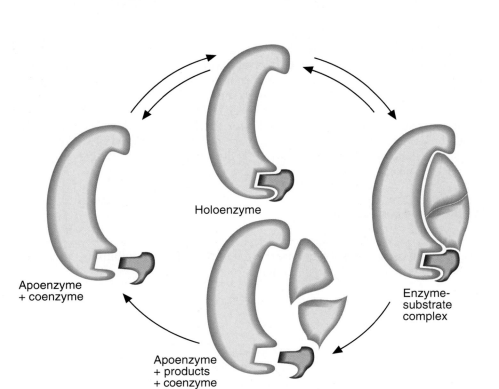

Apoenzyme
+ coenzyme

Enzyme-
substrate
complex

Apoenzyme
+ products
+ coenzyme

Figure 20.7
Some enzymes require a coenzyme to facilitate the reaction. The apoenzyme binds the coenzyme and then the substrate. The coenzyme is a part of the catalytic domain and will either donate or accept functional groups, allowing the reaction to occur. Once the product is formed, both the product and the coenzyme are released.

In chemical reactions they may either donate groups to the substrate or serve as recipients of groups that are removed from the substrate (Figure 20.7).

Often coenzymes contain modified vitamins as part of their structure. A **vitamin** is an organic substance that is required in the diet in only small amounts. Of the water-soluble vitamins, only vitamin C has not been associated with a coenzyme. Table 20.1 is a summary of some coenzymes and the water-soluble vitamins from which they are made.

Water soluble vitamins are discussed in greater detail in Appendix F.

Table 20.1 The Water-Soluble Vitamins and the Coenzymes of Which They Are Structural Components

Vitamin	Coenzyme	Function
Thiamine (B_1)	Thiamine pyrophosphate	Decarboxylation reactions
Riboflavin (B_2)	Flavin mononucleotide (FMN)	Carrier of H atoms
	Flavin adenine dinucleotide (FAD)	
Niacin (B_3)	Nicotinamide adenine dinucleotide (NAD^+)	Carrier of hydride ions
	Nicotinamide adenine dinucleotide phosphate ($NADP^+$)	
Pyridoxine (B_6)	Pyridoxal phosphate	Carriers of amino and carboxyl groups
	Pyridoxamine phosphate	
Cyanocobalamin (B_{12})	Deoxyadenosyl cobalamin	Coenzyme in amino acid metabolism
Folic acid	Tetrahydrofolic acid	Coenzyme for 1-C transfer
Pantothenic acid	Coenzyme A	Acyl group carrier
Biotin	Biocytin	Coenzyme in CO_2 fixation
Ascorbic acid	Unknown	Hydroxylation of proline and lysine in collagen

Nicotinamide adenine dinucleotide (NAD^+), shown in Figure 20.8, is an example of a coenzyme that is of critical importance in the oxidation reactions of the cellular energy-harvesting processes. The NAD^+ molecule has the ability to accept a hydride ion, a hydrogen atom with two electrons, from the substrate of the energy-harvesting reactions. The substrate is oxidized, and the portion of NAD^+ that is derived from the vitamin *niacin* is reduced to produce NADH. The NADH subsequently yields the hydride ion to the first acceptor in an electron transport chain. This regenerates the NAD^+ and provides electrons for the generation of ATP, the chemical energy required by the cell. Also shown in Figure 20.8 is the hydride carrier $NADP^+$ and the hydrogen atom carrier FAD. Both are used in the oxidation-reduction reactions that harvest energy for the cell. Unlike NADH and $FADH_2$, NADPH serves as "reducing power" for the cell by donating hydride ions in biochemical reactions. Like NAD^+, $NADP^+$ is derived from niacin. FAD is made from the vitamin *riboflavin*.

Question 20.9

Why does the body require the water-soluble vitamins?

Question 20.10

What are the coenzymes formed from each of the following vitamins? What are the functions of each of these coenzymes?

a. Pantothenic acid
b. Niacin
c. Riboflavin

20.8 Environmental Effects

Effect of pH

Learning Goal

Most enzymes are active only within a very narrow pH range. The cellular cytoplasm has a pH of 7, and most cytoplasmic enzymes function at a maximum efficiency at this pH. A plot of the relative rate at which a typical cytoplasmic enzyme catalyzes its specific reaction versus pH is provided in Figure 20.9.

The pH at which an enzyme functions optimally is called the **pH optimum.** Making the solution more basic or more acidic sharply decreases the rate of the

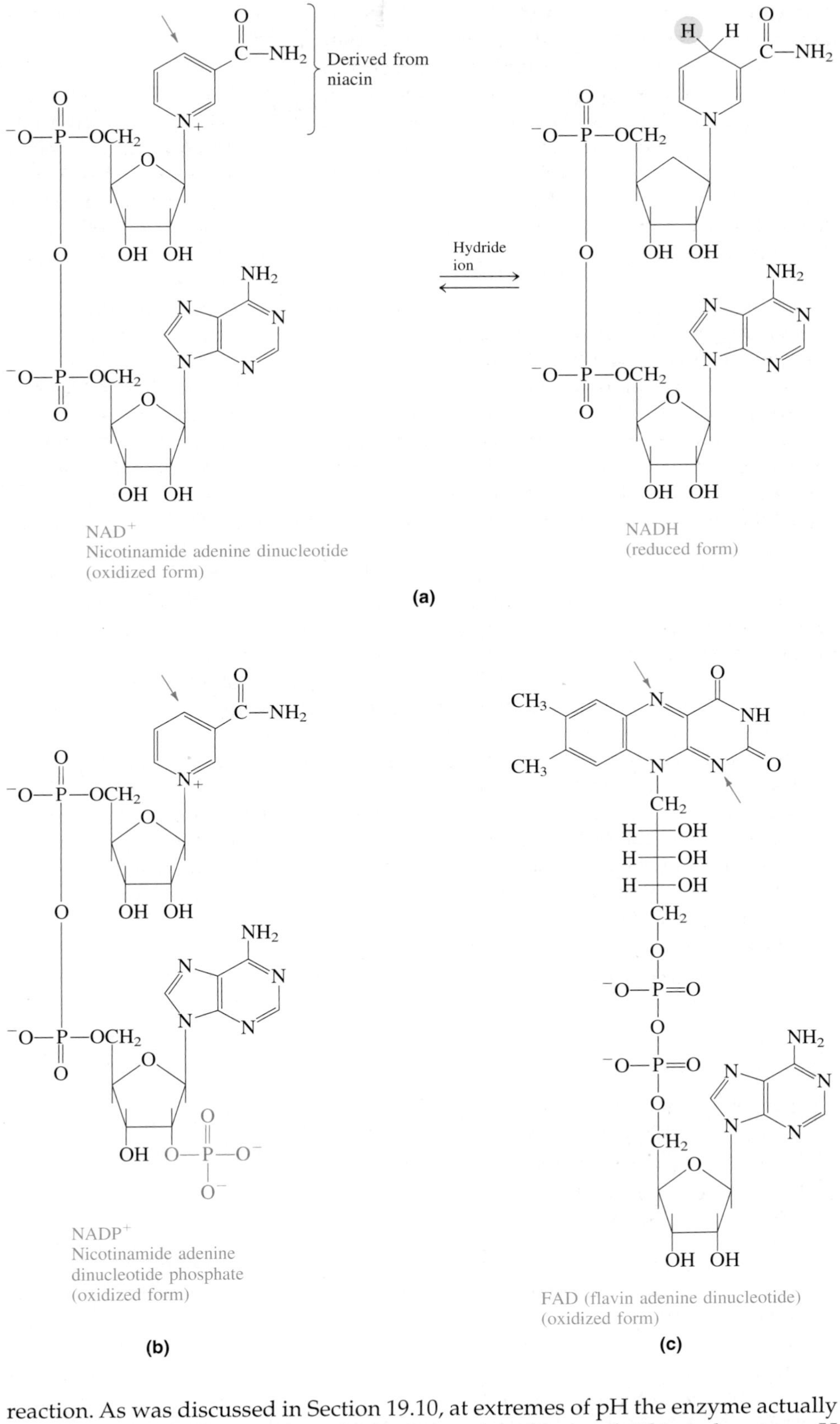

NAD$^+$
Nicotinamide adenine dinucleotide
(oxidized form)

Derived from niacin

Hydride ion

NADH
(reduced form)

(a)

NADP$^+$
Nicotinamide adenine dinucleotide phosphate
(oxidized form)

(b)

FAD (flavin adenine dinucleotide)
(oxidized form)

(c)

Figure 20.8
The structure of three coenzymes. (a) The oxidized and reduced forms of nicotinamide adenine dinucleotide. (b) The oxidized form of the closely related hydride ion carrier, nicotinamide adenine dinucleotide phosphate ($NADP^+$), which accepts hydride ions at the same position as NAD^+ (colored arrow). (c) The oxidized form of flavin adenine dinucleotide (FAD) accepts hydrogen atoms at the positions indicated by the colored arrows.

reaction. As was discussed in Section 19.10, at extremes of pH the enzyme actually loses its biologically active conformation and is *denatured*. This is because pH changes alter the degree of ionization of the R groups of the amino acids within the protein chain, as well as the extent to which they can hydrogen bond. Just as these

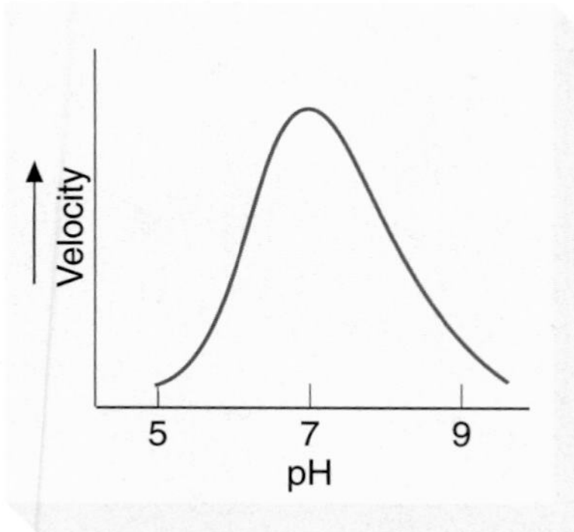

Figure 20.9
Effect of pH on the rate of an enzyme-catalyzed reaction. The enzyme functions most efficiently at pH 7. The rate of the reaction falls rapidly as the solution is made either more acidic or more basic.

interactions can drastically alter the overall configuration of a protein, so too can less drastic changes in the R groups of an enzyme active site destroy the ability to form the enzyme-substrate complex.

Although the cytoplasm of the cell and the fluids that bathe the cells have a pH that is carefully controlled so that it remains at about pH 7, there are environments within the body in which enzymes must function at a pH far from 7. Protein sequences have evolved that can maintain the proper three-dimensional structure under extreme conditions of pH. For instance, the pH of the stomach is approximately 2 as a result of the secretion of hydrochloric acid by specialized cells of the stomach lining. The proteolytic digestive enzyme *pepsin* must effectively degrade proteins at this extreme pH. In the case of pepsin the enzyme has evolved in such a way that it can maintain a stable tertiary structure at a pH of 2 and is catalytically most active in the hydrolysis of peptides that have been denatured by very low pH. Thus pepsin has a pH optimum of 2.

In a similar fashion, another proteolytic enzyme, *trypsin,* functions under the conditions of higher pH found in the intestine. Both pepsin and trypsin cleave peptide bonds by virtually identical mechanisms, yet their amino acid sequences have evolved so that they are stable and active in very different environments.

The body has used adaptation of enzymes to different environments to protect itself against one of its own destructive defense mechanisms. Within the cytoplasm of a cell are organelles called *lysosomes.* Christian de Duve, who discovered lyosomes in 1956, called them "suicide bags" because they are membrane-bound vesicles containing about fifty different kinds of hydrolases or hydrolytic enzymes. The purpose of these enzymes is to degrade large molecules into small molecules that are useful for energy-harvesting reactions. For instance, some of the enzymes in the lysosomes can degrade proteins to amino acids, and others hydrolyze polysaccharides into monosaccharides. Certain cells of the immune defense system engulf foreign invaders, such as bacteria and viruses. They then use the hydrolytic enzymes in the lysosomes to degrade and destroy the invaders and use the simple sugars, amino acids, and lipids that are produced as energy sources.

What would happen if the hydrolytic enzymes of the lysosome were accidentally released into the cytoplasm of the cell? Certainly, the result would be the destruction of cellular macromolecules and death of the cell. Because of this danger, the cell invests a great deal of energy in maintaining the integrity of the lysosomal membranes. An additional protective mechanism relies on the fact that lysosomal enzymes function optimally at an acid pH (pH 4.8). Should some of these enzymes leak out of the lysosome or should a lysosome accidentally rupture, the cytoplasmic pH of 7.0–7.3 renders them inactive.

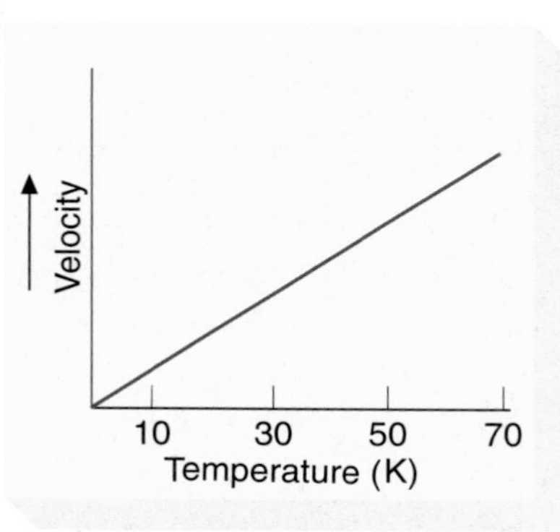

(a)

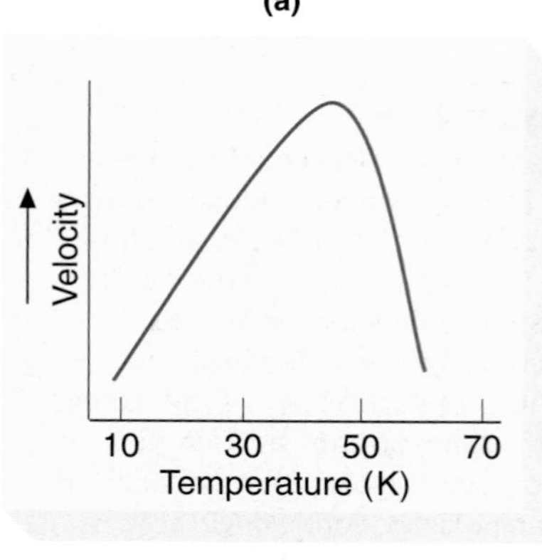

(b)

Figure 20.10
Effect of temperature on (a) uncatalyzed reactions and (b) enzyme-catalyzed reactions.

Effect of Temperature

Enzymes are rapidly destroyed if the temperature of the solution rises much above 37°C, but they remain stable at much lower temperatures. It is for this reason that solutions of enzymes used for clinical assays are stored in refrigerators or freezers before use. Figure 20.10 shows the effects of temperature on enzyme-catalyzed and uncatalyzed reactions. The rate of the uncatalyzed reaction steadily increases with increasing temperature because more collisions occur with sufficient energy to overcome the energy barrier for the reaction. The rate of an enzyme-catalyzed reaction also increases with modest increases in temperature because there are increasing numbers of collisions between the enzyme and the substrate. At the **temperature optimum** the enzyme is functioning optimally and the rate of the reaction is maximal. Above the temperature optimum, increasing temperature begins to increase the vibrational energy of the bonds within the enzyme. Eventually, so many bonds and weak interactions are disrupted that the enzyme becomes denatured, and the reaction stops.

A Medical Perspective

α_1-Antitrypsin and Familial Emphysema

Nearly two million people in the United States suffer from emphysema. Emphysema is a respiratory disease caused by destruction of the alveoli, the tiny, elastic air sacs of the lung. This damage results from the irreversible destruction of a protein called elastin, which is needed for the strength and flexibility of the walls of the alveoli. When elastin is destroyed, the small air passages in the lungs, called bronchioles, become narrower or may even collapse. This severely limits the flow of air into and out of the lung, causing respiratory distress, and in extreme conditions, death.

Some people have a genetic predisposition to emphysema. This is called familial emphysema. These individuals have been found to have a genetic defect in the gene that encodes the human plasma protein α_1-antitrypsin. As the name suggests, α_1-antitrypsin is an inhibitor of the proteolytic enzyme trypsin. But, as we have seen in this chapter, trypsin is just one member of a large family of proteolytic enzymes called the serine proteases. In the case of the α_1-antitrypsin activity in the lung, it is the inhibition of the enzyme elastase that is the critical event.

Elastase damages or destroys elastin, which in turn promotes the development of emphysema. People with normal levels of α_1-antitrypsin are protected from familial emphysema because their α_1-antitrypsin inhibits elastase and, thus, protects the elastin. The result is healthy alveoli in the lungs. However, individuals with a genetic predisposition to emphysema have very low levels of α_1-antitrypsin. This is due to a mutation that causes a single amino acid substitution in the protein chain. Because elastase in the lungs is not effectively controlled, severe lung damage characteristic of emphysema occurs.

Emphysema is also caused by cigarette smoking. Is there a link between these two forms of emphysema? The answer is yes; research has revealed that components of cigarette smoke cause the oxidation of a methionine near the amino terminus of the α_1-antitrypsin. This chemical damage destroys α_1-antitrypsin activity. There are enzymes in the lung that reduce the methionine, converting it back to its original chemical form and restoring α_1-antitrypsin activity. However, it is obvious that over a long period, smoking seriously reduces the level of α_1-antitrypsin activity. The accumulated lung damage results in emphysema in many chronic smokers.

At the current time the standard treatment of emphysema is the use of inhaled oxygen. It has been found that α_1-antitrypsin can be isolated from blood. Studies have shown that introduction of this material by intravenous infusion is both safe and effective. However, the level of α_1-antitrypsin in the blood must be maintained by repeated administration.

The α_1-antitrypsin gene has been cloned. In experiments with sheep it was shown that the protein remains stable when administered as an aerosol. It is still functional after it has passed through the pulmonary epithelium. This research offers hope of an effective treatment for this frightful disease.

Because heating enzymes and other proteins destroys their characteristic three-dimensional structure, and hence their activity, a cell cannot survive very high temperatures. Because cells cannot function without proper enzyme activity, heat is an effective means of sterilizing medical instruments and solutions for transfusion or clinical tests. Although instruments can be sterilized by dry heat (160°C) applied for at least two hours in a dry air oven, autoclaving is a quicker, more reliable procedure. The autoclave works on the principle of the pressure cooker. Air is pumped out of the chamber, and steam under pressure is pumped into the chamber until a pressure of two atmospheres is achieved. The pressure causes the temperature of the steam, which would be 100°C at atmospheric pressure, to rise to 121°C. Within twenty minutes, all the bacteria and viruses are killed. This is the most effective means of destroying the very heat-resistant endospores that are formed by many bacteria of clinical interest. These bacteria include the genera *Bacillus* and *Clostridium*, which are responsible for such unpleasant and deadly diseases as anthrax, gas gangrene, tetanus, and botulism food poisoning.

However, not all enzymes are inactivated by heating, even to rather high temperatures. Certain bacteria live in such out-of-the-way places as coal slag heaps, which are actually burning. Others live in deep vents on the ocean floor where temperatures and pressures are extremely high. Still others grow in the hot springs of Yellowstone National Park, where some bacteria thrive at temperatures near the boiling point of water. These organisms, along with their enzymes, survive under such incredible conditions because the amino acid sequences of their proteins dictate structures that are stable at such seemingly impossible temperature extremes.

Question 20.11 How does a decrease in pH alter the activity of an enzyme?

Question 20.12 Heating is an effective mechanism for killing bacteria on surgical instruments. How does elevated temperature result in cellular death?

20.9 Regulation of Enzyme Activity

Learning Goal

One of the major ways in which enzymes differ from nonbiological catalysts is that the activity of the enzyme is often regulated by the cell. There are many reasons for this control of enzyme function. Some involve energy considerations. If the cell runs out of chemical energy, it will die; therefore many mechanisms exist to conserve cellular energy. For instance, it is a great waste of energy to produce an enzyme if the substrate is not available. Similarly, if the product of an enzyme-catalyzed reaction is present in excess, it is a waste of energy for the enzyme to continue to catalyze the reaction, thereby producing more of the unwanted product.

Just as there are many reasons for regulation of enzyme activity, there are many mechanisms for such regulation. The simplest mechanism is to produce the enzyme only when the substrate is present. This mechanism is used by bacteria to regulate the enzymes needed to break down various sugars to yield ATP for cellular work. The bacteria have no control over their environment or over what food sources, if any, might be available. It would be an enormous waste of energy to produce all of the enzymes that are needed to break down all the possible sugars. Thus the bacteria save energy by producing the enzymes only when a specific sugar substrate is available. Other mechanisms for regulating enzyme activity include use of allosteric enzymes, feedback inhibition, production of zymogens, and protein modification. Let's take a look at these regulatory mechanisms in some detail.

Allosteric Enzymes

One type of enzyme regulation involves enzymes that have more than a single binding site. These enzymes, called **allosteric enzymes,** meaning "other forms," are enzymes whose active sites can be altered by binding of small molecules called *effector molecules.* As shown in Figure 20.11, the effector binding to its binding site alters the shape of the active site of the enzyme. The result can be to convert the active site to an inactive configuration, **negative allosterism,** or to convert the active site to the active configuration, **positive allosterism.** In either case, binding of the effector molecule regulates enzyme activity by determining whether it will be active or inactive.

In upcoming chapters we will begin our study of metabolic pathways. A metabolic pathway is a series of biochemical reactions that accomplishes the breakdown or synthesis of one or more biological molecules. Consider the pathway glycolysis, which is the first stage of the breakdown of carbohydrates to release energy that can be used by the cell (adenosine triphosphate or ATP). Such a pathway must be attuned to the demands of the body. When more energy is required, the reactions of the pathway should occur more quickly, releasing more energy. However, if the energy demand is low, the reactions should slow down, producing less ATP.

The third reaction in glycolysis is the transfer of a phosphoryl group from an ATP molecule to a molecule of fructose-6-phosphate. This reaction, shown here, is catalyzed by an enzyme called *phosphofructokinase:*

$$\text{Fructose-6-phosphate} + \text{ATP} \xrightarrow{\text{Phosphofructokinase}} \text{Fructose-1,6-bisphosphate} + \text{ADP}$$

Phosphofructokinase activity is sensitive to both positive and negative allosterism. For instance, when ATP is present in abundance, a signal that the body has sufficient energy, it binds to an effector binding site on phosphofructokinase. This inhibits the activity of the enzyme and, thus, slows the entire pathway. An abundance of AMP (adenosine monophosphate), which is a precursor of ATP, is evidence that the body needs to make ATP to have a sufficient energy supply. When AMP binds to an effector binding site on phosphofructokinase, enzyme activity is increased, speeding up the reaction and the entire pathway.

Feedback Inhibition

Allosteric enzymes are the basis for **feedback inhibition** of biochemical pathways. This system functions on the same principle as the thermostat on your furnace. You set the thermostat at 70°F; the furnace turns on and produces heat until the sensor in the thermostat registers a room temperature of 70°F. It then signals the furnace to shut off.

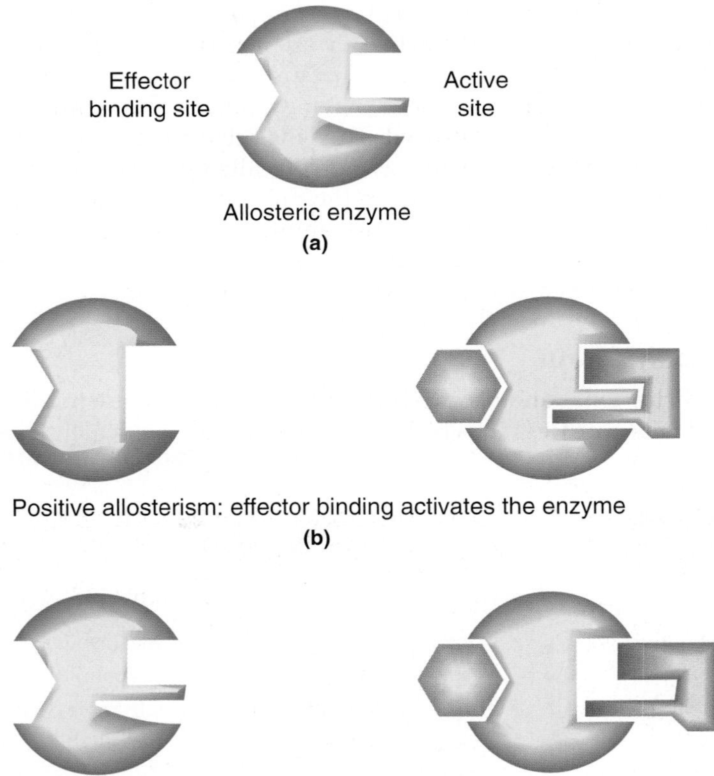

Figure 20.11
(a) Allosteric enzymes regulate a great many biochemical pathways. The allosteric enzyme has an active site and an effector binding site. (b) Positive allosterism. (c) Negative allosterism.

Feedback inhibition usually regulates pathways of enzymes involved in the synthesis of a biological molecule. Such a pathway can be shown schematically as follows:

$$A \xrightarrow{E_1} B \xrightarrow{E_2} C \xrightarrow{E_3} D \xrightarrow{E_4} E \xrightarrow{E_5} F$$

In this pathway the starting material, A, is converted to B by the enzyme E_1. Enzyme E_2 immediately converts B to C, and so on until the final product, F, has been synthesized. If F is no longer needed, it is a waste of cellular energy to continue to produce it.

To avoid this waste of energy, the cell uses feedback inhibition, in which the product can shut off the entire pathway for its own synthesis. This is the result of the fact that the product, F, acts as a negative allosteric effector on one of the early enzymes of the pathway. For instance, enzyme E_1 may have an effector-binding site for F in addition to the active site that binds to A. When F is present in excess, it binds to the effector-binding site. This binding causes the active site to close so that it cannot bind to substrate A. Thus A is not converted to B. If no B is produced, there is no substrate for enzyme E_2, and the entire pathway ceases to operate. The product, F, has turned off all the steps involved in its own synthesis, just as the heat produced by the furnace is ultimately responsible for turning off the furnace itself.

Zymogens

Another means of regulating enzyme activity involves the production of the enzyme in an inactive form called a **zymogen,** or a *proenzyme.* It is then converted, usually by proteolysis (hydrolysis of the protein), to the active form when it has reached the site of its activity. What is the purpose of this type of mechanism? On first examination it seems wasteful to add a step to the synthesis of an enzyme. But consider for a moment the very destructive nature of some of the enzymes that are necessary for life. The enzymes pepsin, trypsin, and chymotrypsin are all proteolytic enzymes of the digestive tract. They are necessary to life because they degrade dietary proteins into amino acids that are used by the cell. But what would happen to the cells that produce these enzymes if they were synthesized in active form? Those cells would be destroyed. Thus the cells of the stomach that produce pepsin actually produce an inactive zymogen, called *pepsinogen.* Pepsinogen has an additional forty-two amino acids. In the presence of stomach acid and previously activated pepsin, the extra forty-two amino acids are cleaved off, and the zymogen is transformed into the active enzyme. Table 20.2 lists several other zymogens and the enzymes that convert them to active form.

Protein Modification

Protein modification is another mechanism that the cell can use to turn an enzyme on or off. This is a process in which a chemical group is covalently added to or

Table 20.2 Zymogens of the Digestive Tract

Zymogen	Activator	Enzyme
Proelastase	Trypsin	Elastase
Trypsinogen	Trypsin	Trypsin
Chymotrypsinogen A	Trypsin + chymotrypsin	Chymotrypsin
Pepsinogen	Acid pH + pepsin	Pepsin
Procarboxypeptidases	Trypsin	Carboxypeptidase A, carboxypeptidase B

removed from the protein. This covalent modification either activates the enzyme or turns it off.

The most common type of protein modification is phosphorylation or dephosphorylation of an enzyme. Typically, the phosphoryl group is added to (or removed from) the R group of a serine, tyrosine, or threonine in the protein chain of the enzyme. Notice that these three amino acids have a free —OH in their R group, which serves as the site for the addition of the phosphoryl group.

The covalent modification of an enzyme's structure is catalyzed by other enzymes. *Protein kinases* add phosphoryl groups to a target enzyme, while *phosphatases* remove them. For some enzymes it is the phosphorylated form that is active. For instance, in adipose tissue, phosphorylation activates the enzyme triacylglycerol lipase, an enzyme that breaks triglycerides down to fatty acids and glycerol. Glycogen phosphorylase, an enzyme involved in the breakdown of glycogen, is also activated by the addition of a phosphoryl group. However, for some enzymes phosphorylation inactivates the enzyme. This is true for glycogen synthase, an enzyme involved in the synthesis of glycogen. When this enzyme is phosphorylated, it becomes inactive.

The convenient aspect of this type of regulation is the reversibility. An enzyme can quickly be turned on or off in response to environmental or physiological conditions.

20.10 Inhibition of Enzyme Activity

Learning Goal

Many chemicals can bind to enzymes and either eliminate or drastically reduce their catalytic ability. These chemicals, called *enzyme inhibitors,* have been used for hundreds of years. When she poisoned her victims with arsenic, Lucretia Borgia was unaware that it was binding to the thiol groups of cysteine amino acids in the proteins of her victims and thus interfering with the formation of disulfide bonds needed to stabilize the tertiary structure of enzymes. However, she was well aware of the deadly toxicity of heavy metal salts like arsenic and mercury. When you take penicillin for a bacterial infection, you are taking another enzyme inhibitor. Penicillin inhibits several enzymes that are involved in the synthesis of bacterial cell walls.

The effect of penicillin on bacterial cell wall biosynthesis is discussed in A Clinical Perspective: The Bacterial Cell Wall in Chapter 17.

Enzyme inhibitors are classified on the basis of whether the inhibition is reversible or irreversible, competitive or noncompetitive. Reversibility deals with whether the inhibitor will eventually dissociate from the enzyme, releasing it in the active form. Competition refers to whether the inhibitor is a structural analog, or look-alike, of the natural substrate. If so, the inhibitor and substrate will compete for the enzyme active site.

Irreversible Inhibitors

Irreversible enzyme inhibitors, such as arsenic, usually bind very tightly, sometimes even covalently, to the enzyme. This generally involves binding of the inhibitor to one of the R groups of an amino acid in the active site. Inhibitor binding may block the active site binding groups so that the enzyme-substrate complex cannot form. Alternatively, an inhibitor may interfere with the catalytic groups of the active site, thereby effectively eliminating catalysis. Irreversible inhibitors generally inhibit many different enzymes. Examples include snake venoms and nerve gases.

Reversible, Competitive Inhibitors

Reversible, competitive enzyme inhibitors are often referred to as **structural analogs,** that is, they are molecules that resemble the structure and charge distribution of the natural substrate for a particular enzyme. Because of this resemblance, the inhibitor can occupy the enzyme active site. However, no reaction can occur, and enzyme activity is inhibited. This inhibition is said to be competitive because

See A Human Perspective: Fooling the AIDS Virus with "Look-Alike" Nucleotides in Chapter 24.

A CLINICAL PERSPECTIVE

Enzymes, Nerve Transmission, and Nerve Agents

The transmission of nerve impulses at the *neuromuscular junction* involves many steps, one of which is the activity of a critical enzyme, called *acetylcholinesterase,* which catalyzes the hydrolysis of the chemical messenger, *acetylcholine,* that initiated the nerve impulse. The need for this enzyme activity becomes clear when we consider the events that begin with a message from the nerve cell and end in the appropriate response by the muscle cell. Acetylcholine is a chemical messenger that transmits a message from the nerve cell to the muscle cell. Such a molecule is known as a *neurotransmitter.* Acetylcholine is stored in membrane-bound bags, called *synaptic vesicles,* in the nerve cell ending.

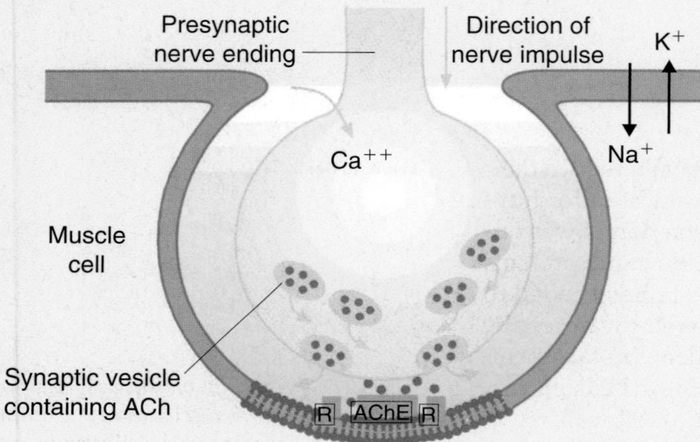

Schematic diagram of the synapse at the neuromuscular junction. The nerve impulse causes acetylcholine (ACh) to be released from synaptic vesicles. Acetylcholine diffuses across the synaptic cleft and binds to a specific receptor protein (R) on the postsynaptic membrane. A channel opens that allows Na^+ ions to flow into the cell and K^+ ions to flow out of the cell. This results in muscle contraction. Any ACh remaining in the synaptic cleft is destroyed by acetylcholinesterase (AChE) to terminate the stimulation of the muscle cell.

Acetylcholinesterase comes into play in the following way. The arrival of a nerve impulse at the end plate of the nerve axon causes an influx of Ca^{2+}. This causes the acetylcholine-containing vesicles to migrate to the nerve cell membrane that is in contact with the muscle cell. This is called the *presynaptic membrane.* The vesicles fuse with the *presynaptic membrane* and release the neurotransmitter. The acetylcholine then diffuses across the *nerve synapse* (the space between the nerve and muscle cells) and binds to the acetylcholine receptor protein in the *postsynaptic membrane* of the muscle cell. This receptor then opens pores in the membrane through which Na^+ and K^+ ions flow into and out of the cell, respectively. This generates the nerve impulse and causes the muscle to contract. If acetylcholine remains at the neuromuscular junction, it will continue to stimulate the muscle contraction. To stop this continued stimulation, acetylcholine is hydrolyzed, and hence, destroyed by acetylcholinesterase. When this happens, choline is no longer able to bind to the acetylcholine receptor and nerve stimulation ceases.

$$\underset{\text{Acetylcholine}}{H_3C{-}\overset{\overset{\displaystyle O}{\|}}{C}{-}O{-}CH_2CH_2{-}N^+{-}(CH_3)_3} + H_2O$$

$$\Big\downarrow\!\!\uparrow \text{ Acetylcholinesterase}$$

$$\underset{\text{Acetate}}{H_3C{-}C{\genfrac{}{}{0pt}{}{\nearrow O}{\searrow O^-}}} + \underset{\text{Choline}}{HO{-}CH_2CH_2{-}N^+{-}(CH_3)_3} + H^+$$

the inhibitor and the substrate compete for binding to the enzyme active site. Thus, the degree of inhibition depends on their relative concentrations. If the inhibitor is in excess or binds more strongly to the active site, it will occupy the active site more frequently, and enzyme activity will be greatly decreased. On the other hand, if the natural substrate is present in excess, it will more frequently occupy the active site, and there will be little inhibition.

The sulfa drugs, the first antimicrobics to be discovered, are **competitive inhibitors** of a bacterial enzyme needed for the synthesis of the vitamin folic acid.

Inhibitors of acetylcholinesterase are used both as poisons and as drugs. Among the most important inhibitors of acetylcholinesterase are a class of compounds known as *organophosphates.* One of these is the nerve agent Sarin (isopropylmethylfluorophosphate). Sarin forms a covalently bonded intermediate with the active site of acetylcholinesterase. Thus, it acts as an irreversible, noncompetitive inhibitor.

Pyridine aldoxime methiodide (PAM)

Sarin

Serine in the acetylcholinesterase active site

HF

Sarin is covalently bonded to the serine in the active site.

Sarin is covalently bonded to the serine in the active site.

Complex formed between sarin and PAM

Regenerated enzyme

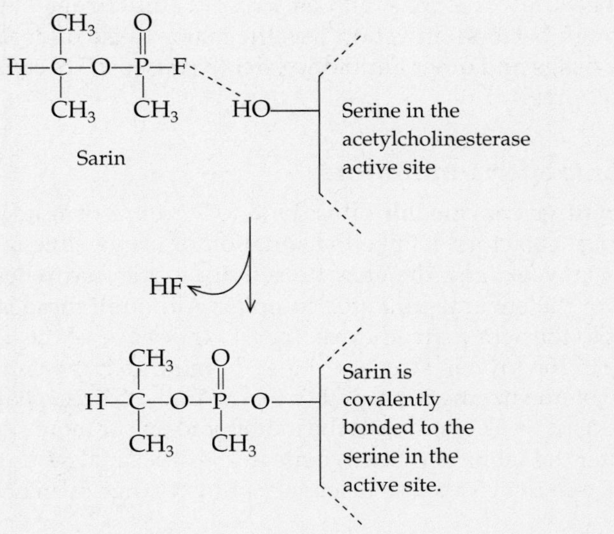

The covalent intermediate is stable, and acetylcholinesterase is therefore inactive. It is unable to react with other substrates. Nerve transmission continues, resulting in muscle spasm. Death may occur as a result of laryngeal spasm. Antidotes for poisoning by organophosphates, which include many insecticides and nerve gases, have been developed. The antidotes work by reversing the effects of the inhibitor. One of these antidotes is known as *PAM,* an acronym for *pyridine aldoxime methiodide.* This molecule displaces the organophosphate group from the active site of the enzyme, alleviating the effects of the poison.

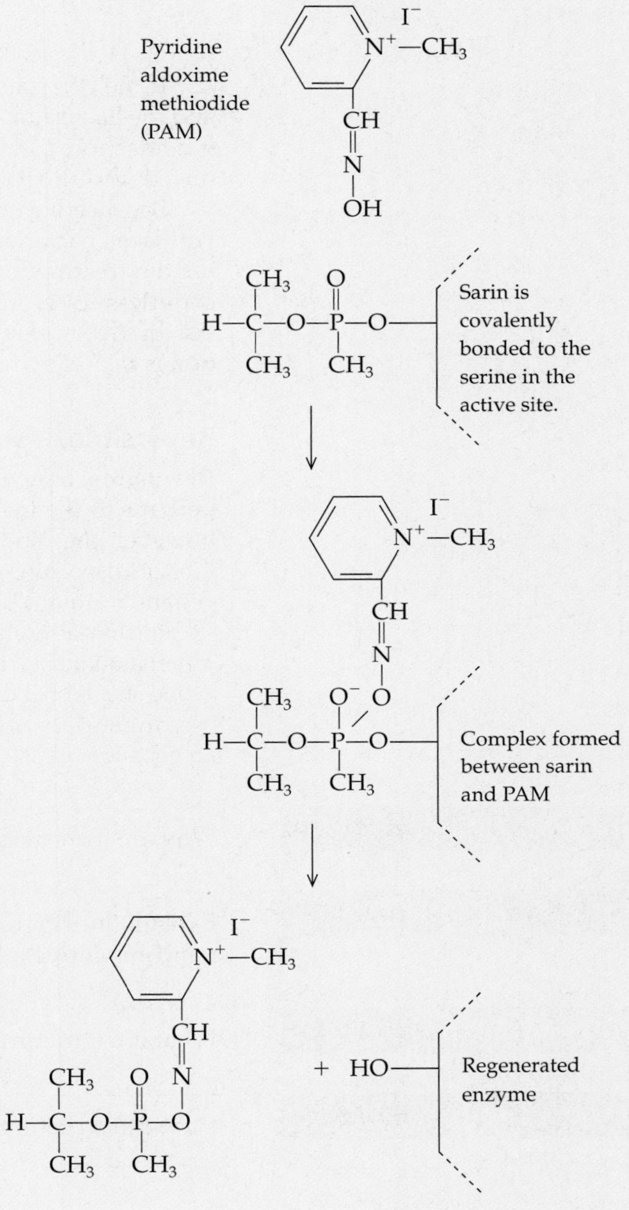

Folic acid is a vitamin required for the transfer of methyl groups in the biosynthesis of methionine and nitrogenous bases. Humans cannot synthesize folic acid and must obtain it from the diet. Bacteria, on the other hand, must make folic acid because they cannot take it in from the environment.

In addition to the folic acid supplied in the diet, we obtain folic acid from our intestinal bacteria.

para-Aminobenzoic acid (PABA) is the substrate for an early step in folic acid synthesis. The sulfa drugs, the prototype for which was discovered by Gerhard Domagk in the 1930s, are structural analogs of PABA and thus competitive inhibitors of the enzyme that uses PABA as its normal substrate.

$H_2N-C_6H_4-\overset{O}{\overset{\|}{C}}-OH$ $H_2N-C_6H_4-\underset{O}{\underset{\|}{\overset{O}{\overset{\|}{S}}}}-NH_2$

p-Aminobenzoic acid Sulfanilamide

If the correct substrate (PABA) is bound by the enzyme, the reaction occurs, and the bacterium lives. However, if the sulfa drug is present in excess over PABA, it binds more frequently to the active site of the enzyme. No folic acid will be produced, and the bacterial cell will die.

Because we obtain our folic acid from our diets, sulfa drugs do not harm us. However, bacteria are selectively killed. Luckily, we can capitalize on this property for the treatment of bacterial infections, and as a result, sulfa drugs have saved countless lives. Although bacterial infection was the major cause of death before the discovery of sulfa drugs and other antibiotics, death caused by bacterial infection is relatively rare at present.

Reversible, Noncompetitive Inhibitors

Reversible, noncompetitive enzyme inhibitors bind to R groups of amino acids or perhaps to the metal ion cofactors. Unlike the situation of irreversible inhibition, however, the binding is weak, and the enzyme activity is restored when the inhibitor dissociates from the enzyme-inhibitor complex. Although these inhibitors generally do not bind to the active site, they do modify the shape of the active site by binding elsewhere in the protein structure. Keep in mind that the entire three-dimensional structure of an enzyme is needed to maintain the correct shape of the active site. Ionic bonding or weak bonding of the inhibitor to one or more sites on the enzyme surface can alter the shape of the active site in a fashion analogous to that of an allosteric effector. These inhibitors also inactivate a broad range of enzymes.

Question 20.13 Why are irreversible inhibitors considered to be poisons?

Question 20.14 Explain the difference between an irreversible inhibitor and a reversible, noncompetitive inhibitor.

Question 20.15 What is a structural analog?

Question 20.16 How can structural analogs serve as enzyme inhibitors?

20.11 Proteolytic Enzymes

Learning Goal

Proteolytic enzymes are protein-cleaving enzymes, that is, they break the peptide bonds that maintain the primary protein structure. *Chymotrypsin*, for example, is an enzyme that hydrolyzes dietary proteins in the small intestine. It acts specifically at peptide bonds on the carbonyl side of the peptide bond. The C-terminal amino acids of the peptides released by bond cleavage are methionine, tyrosine, tryptophan, and phenylalanine. The specificity of chymotrypsin depends upon the presence of a *hydrophobic pocket*, a cluster of hydrophobic amino acids brought together by the three-dimensional folding of the protein chain. The flat aromatic side chains of certain amino acids (tyrosine, tryptophan, phenylalanine) slide into this pocket, providing the binding specificity required for catalysis (Figure 20.12).

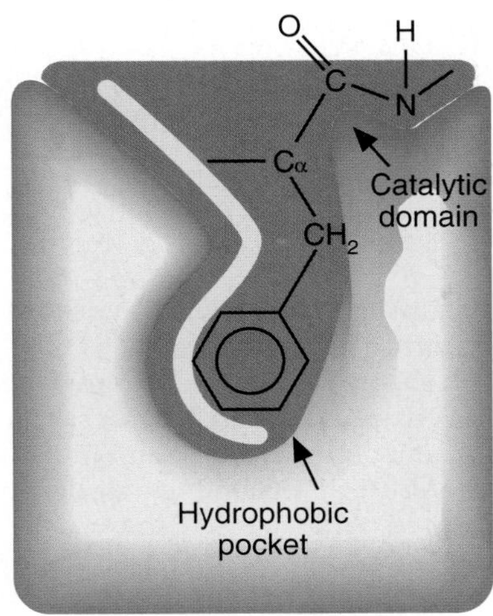

Figure 20.12
The specificity of chymotrypsin is determined by a hydrophobic pocket that holds the aromatic side chain of the substrate. This brings the peptide bond to be cleaved into the catalytic domain of the active site.

How can we determine which bond is cleaved by a protease such as chymotrypsin? To know which bond is cleaved, we must write out the sequence of amino acids in the region of the peptide that is being cleaved. This can be determined experimentally by amino acid sequencing techniques. Remember that the N-terminal amino acid is written to the left and the C-terminal amino acid to the right. Consider a protein having within it the sequence —Ala-Phe-Gly—. A reaction is set up in which the enzyme, chymotrypsin, is mixed with the protein substrate. After the reaction has occurred, the products are purified, and their amino acid sequences are determined. Experiments of this sort show that chymotrypsin cleaves the bond between phenylalanine and glycine, which is the peptide bond on the carbonyl side of amino acids having an aromatic side chain.

$$H_3^+N-CH(CH_3)-C(=O)-NH-CH(CH_2C_6H_5)-C(=O)-NH-CH_2-C(=O)-O^-$$

Peptide bond cleaved

Ala —— Phe —— Gly

> **These enzymes are called *serine proteases* because they have the amino acid serine in the catalytic region of the active site that is essential for hydrolysis of the peptide bond.**

The **pancreatic serine proteases** trypsin, chymotrypsin, and elastase all hydrolyze peptide bonds. These enzymes are the result of *divergent evolution* in which a single ancestral gene first duplicated and then each copy evolved individually. They have similar primary structures, similar tertiary structures (Figure 20.13), and virtually identical mechanisms of action. However, as a result of evolution, these enzymes all have different specificities:

- Chymotrypsin cleaves peptide bonds on the carbonyl side of aromatic amino acids and large, hydrophobic amino acids such as methionine.
- Trypsin cleaves peptide bonds on the carbonyl side of basic amino acids.
- Elastase cleaves peptide bonds on the carbonyl side of glycine and alanine.

These enzymes have different pockets for the side chains of their substrates; *different keys fit different locks*. This difference manifests itself in the substrate specificity alluded to above. For example, the binding pocket of trypsin is long, narrow, and negatively charged to accommodate lysine or arginine R groups. Yet although the binding pockets have undergone divergent evolution, the catalytic sites have

A CLINICAL PERSPECTIVE

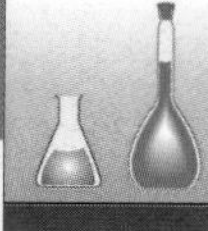

Enzymes, Isoenzymes, and Myocardial Infarction

A patient is brought into the emergency room with acute, squeezing chest pains; shallow, irregular breathing; and pale, clammy skin. The immediate diagnosis is myocardial infarction, a heart attack. The first thoughts of the attending nurses and physicians concern the series of treatments and procedures that will save the patient's life. It is a short time later, when the patient's condition has stabilized, that the doctor begins to consider the battery of enzyme assays that will confirm the diagnosis.

Myocardial infarction occurs when the blood supply to the heart muscle is blocked for an extended time. If this lack of blood supply, called *ischemia,* is prolonged, the myocardium suffers irreversible cell damage and muscle death, or infarction. When this happens, the concentration of cardiac enzymes in the blood rises dramatically as the dead cells release their contents into the bloodstream. Although many enzymes are liberated, three are of prime importance. These three enzymes, creatine phosphokinase (CPK), lactate dehydrogenase (LDH), and aspartate aminotransferase/serum glutamate–oxaloacetate transaminase (AST/SGOT), show a characteristic sequential rise in blood serum level following myocardial infarction and then return to normal. This enzyme profile, shown in the accompanying figure, is characteristic of and the basis for the diagnosis of a heart attack.

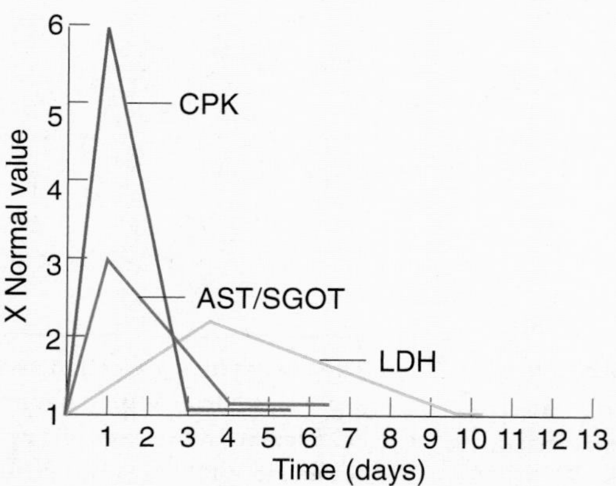

Characteristic pattern of serum cardiac enzyme concentrations following a myocardial infarction.

To ensure against misdiagnosis caused by tissue damage in other organs, the levels of other serum enzymes, including alanine aminotransferase/serum glutamate–pyruvate transaminase (ALT/SGPT) and isocitrate dehydrogenase (ICD), are also measured. ALT and AST are usually determined simultaneously to differentiate between cardiac and hepatic disease. The concentration of ALT is higher in liver disease, whereas the serum concentration of AST is higher following acute myocardial infarction. ICD is found primarily in the liver, and serum levels would not be elevated after a heart attack.

The use of LDH and CPK levels alone can also lead to a misdiagnosis because these enzymes are produced by many tissues. How can a clinician diagnose heart disease with confidence when the elevated serum enzyme levels could indicate coexisting disease in another tissue? The physician is able to make such a decision because of the presence of *isoenzymes,* which provide diagnostic accuracy because they reveal the tissue of origin.

Isoenzymes are forms of the same enzyme with slightly different amino acid sequences. The binding and catalytic sites are the same, but there are differences in the scaffolding sequences of the enzyme that maintain the three-dimensional structure of the protein. Each of the cells of the body contains the genes that could direct the production of all the different forms of these enzymes, yet the expression is *tissue-specific.* This means that the genes for certain isoenzymes are expressed preferentially in different types of tissue.

It is not clear why a certain isoenzyme is "turned on" in the liver whereas another predominates in the heart, but it is known that we can distinguish among the different forms in the laboratory on the basis of their migration through a gel placed in an electric field. This process is called *gel electrophoresis.* This test is based on the fact that each protein has a characteristic surface charge resulting from the R groups of the amino acids. If these proteins are placed in a gel matrix and an electrical current is applied, the proteins migrate as a function of that charge. The figure on the next page shows the position of the five isoenzymes of LDH following electrophoresis.

Imagine a mixture of serum proteins, each with a different overall charge, subjected to an electric field. The proteins with the greatest negative charge will be most strongly attracted to the positive pole and will migrate rapidly toward it, whereas those with a lesser negative charge will migrate much more slowly. Thus, the proteins will be distributed throughout the gel based on their characteristic overall charge.

Once electrophoresis is terminated, the location of the bands of each isoenzyme must be determined and the amount of each must be measured. To do this an enzyme assay is carried out. The substrate is added to the gel in a solution that provides proper conditions for the enzyme. The product is a colored substance that can be seen visually. By inspecting the positions of the stained bands on the gel, one can determine which tissue isoenzymes are present.

To measure the amount of each, the intensity of the color can be measured with a spectrophotometer. The intensity of the color is directly proportional to the amount of the product, and thus to the amount of enzyme in the original sample. This allows the laboratory to calculate the concentration of each isoenzyme in the sample.

These data give the clinician an accurate picture of the nature of the diseased tissue. For a heart attack victim, only creatine phosphokinase isoenzyme 2 (CPK II, the predominant heart isoenzyme) will be elevated in the three days following the heart attack. CPK I (brain) and CPK III (skeletal muscle) levels will remain unchanged. Similarly, only LDH 1, the lactate dehydrogenase isoenzyme made in heart muscle, will be elevated. The levels of LDH 2–5 will remain within normal values.

The physician also has enzymes available to treat a heart attack patient. Most myocardial infarctions are the result of a *thrombus,* or clot, within a coronary blood vessel. The clot restricts blood flow to the heart muscle. One technique that shows promise for treatment following a coronary thrombosis, a heart attack caused by the formation of a clot, is destruction of the clot by intravenous or intracoronary injection of an enzyme called *streptokinase.* This enzyme, formerly purified from the pathogenic bacterium *Streptococcus pyogenes* but now available through recombinant DNA techniques, catalyzes the production of the proteolytic enzyme plasmin from its zymogen, plasminogen. Plasmin has the ability to degrade a fibrin clot into subunits. Of course, this has the effect of dissolving the clot that is responsible for restricted blood flow to the heart, but there is an additional protective function as well. The subunits produced by plasmin degradation of fibrin clots are able to inhibit further clot formation by inhibiting thrombin.

Recombinant DNA technology has provided medical science with yet another, perhaps more promising, clot-dissolving enzyme. *Tissue-type plasminogen activator (TPA)* is a proteolytic enzyme that occurs naturally in the body as a part of the anticlotting mechanisms. TPA converts the zymogen, plasminogen, into the active enzyme, plasmin. Injection of TPA within two hours of the initial chest pain can significantly improve the circulation to the heart and greatly improve the patient's chances of survival.

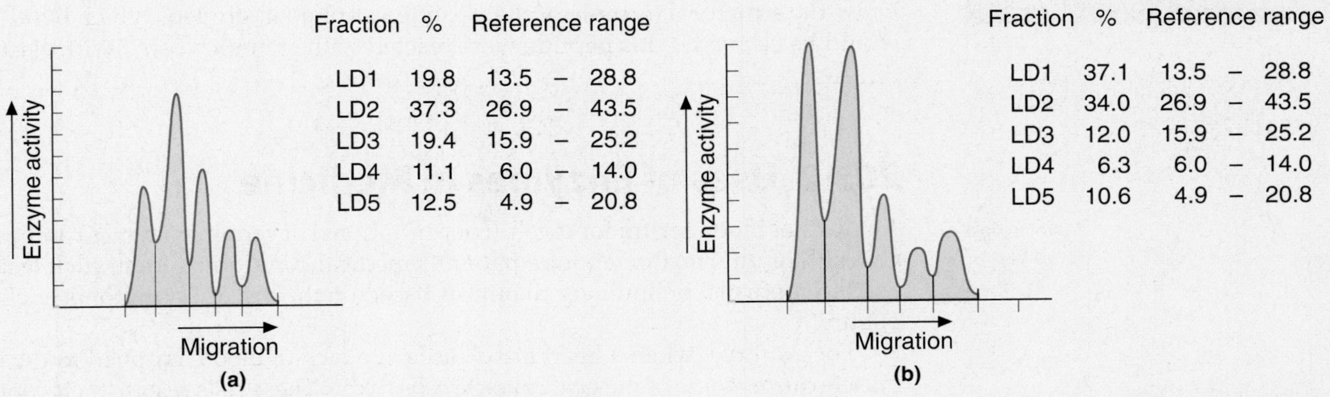

(a)

Fraction	%	Reference range
LD1	19.8	13.5 – 28.8
LD2	37.3	26.9 – 43.5
LD3	19.4	15.9 – 25.2
LD4	11.1	6.0 – 14.0
LD5	12.5	4.9 – 20.8

(b)

Fraction	%	Reference range
LD1	37.1	13.5 – 28.8
LD2	34.0	26.9 – 43.5
LD3	12.0	15.9 – 25.2
LD4	6.3	6.0 – 14.0
LD5	10.6	4.9 – 20.8

A profile of the serum isoenzymes of lactate dehydrogenase. (a) The pattern of LDH isoenzymes from a normal individual. (b) The pattern of LDH isoenzymes of an individual suffering from a myocardial infarction.

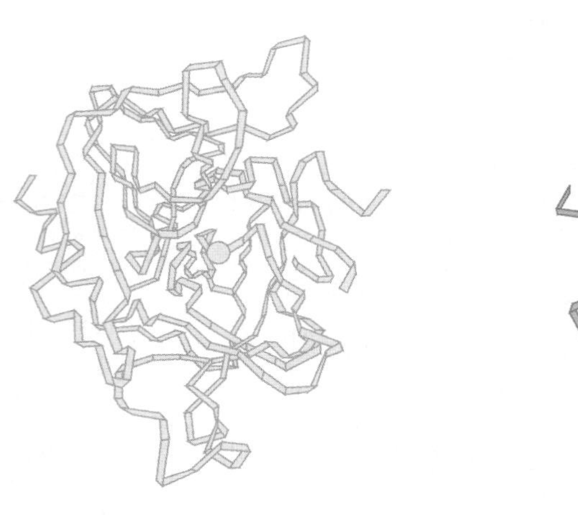

Figure 20.13
Structures of chymotrypsin and elastase are virtually identical, suggesting that these enzymes have evolved from a common ancestral protease.

remained unchanged, and the mechanism of proteolytic action is the same for all the serine proteases. In each case, the mechanism involves a serine R group.

Question 20.17

Draw the structural formulas of the following peptides and show which bond would be cleaved by chymotrypsin.

a. ala-phe-ala
b. tyr-ala-tyr
c. trp-val-gly
d. phe-ala-pro

Question 20.18

Draw the structural formula of the peptide val-phe-ala-gly-leu. Which bond would be cleaved if this peptide were reacted with chymotrypsin? With elastase?

20.12 Uses of Enzymes in Medicine

Learning Goal

12

Analysis of blood serum for levels (concentrations) of certain enzymes can provide a wealth of information about a patient's medical condition. Often, such tests are used to confirm a preliminary diagnosis based on the disease symptoms or clinical picture.

For example, when a heart attack occurs, a lack of blood supplied to the heart muscle causes some of the heart muscle cells to die. These cells release their contents, including their enzymes, into the bloodstream. Simple tests can be done to measure the amounts of certain enzymes in the blood. Such tests, called *enzyme assays,* are very precise and specific because they are based on the specificity of the enzyme-substrate complex. If you wish to test for the enzyme lactate dehydrogenase (LDH), you need only to add the appropriate substrate, in this case pyruvate and NADH. The reaction that occurs is the oxidation of NADH to NAD^+ and the reduction of pyruvate to lactate. To measure the rate of the chemical reaction, one can measure the disappearance of the substrate or the accumulation of one of the products. In the case of LDH, spectrophotometric methods (based on the light-absorbing properties of a substrate or product) are available to measure the rate of production of NAD^+. The choice of substrate determines what enzyme activity is to be measured.

The role of LDH and other enzymes in disease diagnosis is discussed in A Clinical Perspective: Enzymes, Isoenzymes, and Myocardial Infarction.

Elevated blood serum concentrations of the enzymes amylase and lipase are indications of pancreatitis, an inflammation of the pancreas. Liver diseases such as cirrhosis and hepatitis result in elevated levels of one of the isoenzymes of lactate dehydrogenase (LDH_5), and elevated levels of alanine aminotransferase/serum glutamate–pyruvate transaminase (ALT/SGPT) and aspartate aminotransferase/serum glutamate–oxaloacetate transaminase (AST/SGOT) in blood serum. In fact, these latter two enzymes also increase in concentration following heart attack, but the physician can differentiate between these two conditions by considering the relative increase in the two enzymes. If ALT/SGPT is elevated to a greater extent than AST/SGOT, it can be concluded that the problem is liver dysfunction.

Enzymes are also used as analytical reagents in the clinical laboratory owing to their specificity. They often selectively react with one substance of interest, producing a product that is easily measured. An example of this is the clinical analysis of urea in blood. The measurement of urea levels in blood is difficult because of the complexity of blood. However, if urea is converted to ammonia using the enzyme urease, the ammonia becomes an *indicator* of urea, because it is produced from urea, and it is easily measured. This test, called the *blood urea nitrogen (BUN) test,* is useful in the diagnosis of kidney malfunction and serves as one example of the utility of enzymes in clinical chemistry.

Enzyme replacement therapy can also be used in the treatment of certain diseases. One such disease, Gaucher's disease, is a genetic disorder resulting in a deficiency of the enzyme *glucocerebrosidase.* In the normal situation, this enzyme breaks down a glycolipid called *glucocerebroside,* which is an intermediate in the synthesis and degradation of complex glycosphingolipids found in cellular membranes. Glucocerebrosidase is found in the lysosomes, where it hydrolyzes glucocerebroside into glucose and ceramide.

See A Clinical Perspective: Disorders of Sphingolipid Metabolism in Chapter 18.

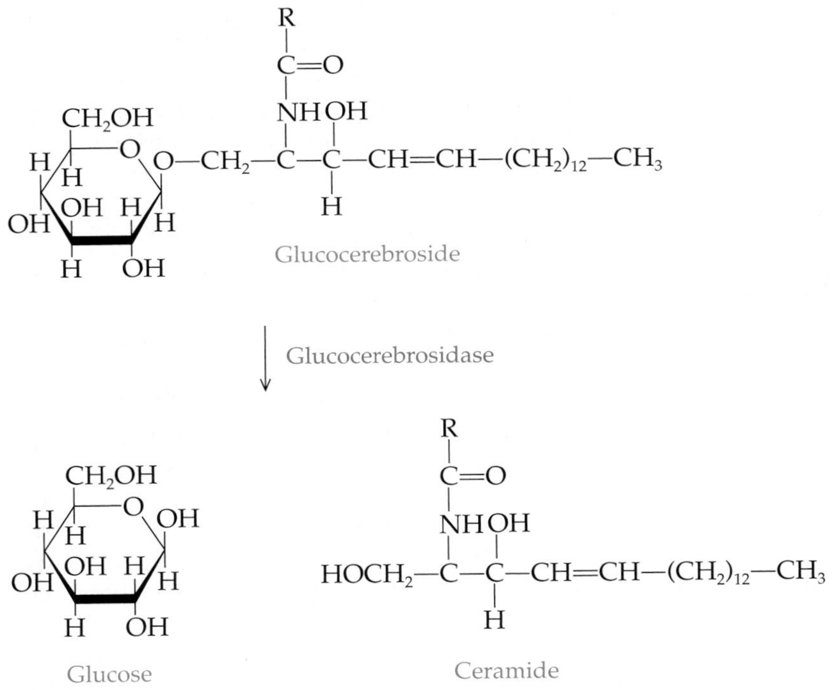

$$\text{Glucose ring}-O-CH_2-\underset{\underset{\underset{\underset{R}{|}}{C=O}}{NH}}{C}-\underset{H}{\overset{OH}{C}}-CH=CH-(CH_2)_{12}-CH_3$$

Glucocerebroside

↓ Glucocerebrosidase

$$\text{Glucose} \quad + \quad HOCH_2-\underset{\underset{\underset{\underset{R}{|}}{C=O}}{NH}}{C}-\underset{H}{\overset{OH}{C}}-CH=CH-(CH_2)_{12}-CH_3$$

Glucose | Ceramide

In Gaucher's disease, the enzyme is not present and glucocerebroside builds up in macrophages found in the liver, spleen, and bone marrow. These cells become engorged with excess lipid that cannot be metabolized and then displace healthy, normal cells in bone marrow. The symptoms of Gaucher's disease include severe anemia, thrombocytopenia (reduction in the number of platelets), and hepatosplenomegaly (enlargement of the spleen and liver). There can also be skeletal problems including bone deterioration and secondary fractures.

Recombinant DNA technology has been used by the Genzyme Corporation to produce the human lysosomal enzyme β-glucocerebrosidase. Given the trade name *Cerezyme*, the enzyme hydrolyzes glucocerebroside into glucose and ceramide so that the products can be metabolized normally. Patients receive Cerezyme intravenously over the course of one to two hours. The dosage and treatment schedule can be tailored to the individual. The results of testing are very encouraging. Patients experience improved red blood cell and platelet counts and reduced hepatosplenomegaly.

Summary

20.1 Nomenclature and Classification

Enzymes are most frequently named by using the common system of nomenclature. The names are useful because they are often derived from the name of the substrate and/or the reaction of the substrate that is catalyzed by the enzyme. Enzymes are classified according to function. The six general classes are *oxidoreductases, transferases, hydrolases, lyases, isomerases,* and *ligases.*

20.2 The Effect of Enzymes on the Activation Energy of a Reaction

Enzymes are the biological catalysts of cells. They lower the activation energies but do not alter the equilibrium constants of the reactions they catalyze.

20.3 The Effect of Substrate Concentration on Enzyme-Catalyzed Reactions

With uncatalyzed reactions, increases in substrate concentration result in an increase in reaction rate. For enzyme-catalyzed reactions, an increase in substrate concentration initially causes an increase in reaction rate, but at a particular concentration the reaction rate reaches a maximum. At this concentration all enzyme active sites are filled with substrate.

20.4 The Enzyme-Substrate Complex

Formation of an *enzyme-substrate complex* is the first step of an enzyme-catalyzed reaction. This involves the binding of the substrate to the active site of the enzyme. The lock-and-key model of substrate binding describes the enzyme as a rigid structure into which the substrate fits precisely. The newer induced fit model describes the enzyme as a flexible molecule. The shape of the active site approximates the shape of the substrate and then "molds" itself around the substrate.

20.5 Specificity of the Enzyme-Substrate Complex

Enzymes are also classified on the basis of their specificity. The four classifications of specificity are *absolute, group, linkage,* and *stereochemical specificity.* An enzyme with *absolute specificity* catalyzes the reaction of only a single substrate. An enzyme with *group specificity* catalyzes reactions involving similar substrates with the same functional group. An enzyme with *linkage specificity* catalyzes reactions involving a single kind of bond. An enzyme with *stereochemical specificity* catalyzes reactions involving only one enantiomer.

20.6 The Transition State and Product Formation

An enzyme-catalyzed reaction is mediated through an unstable *transition state.* This may involve the enzyme putting "stress" on a bond, bringing reactants into close proximity and in the correct orientation, or altering the local pH.

20.7 Cofactors and Coenzymes

Cofactors are metal ions, organic compounds, or organometallic compounds that bind to an enzyme and help maintain the correct configuration of the active site. The term *coenzyme* refers specifically to an organic group that binds transiently to the enzyme during the reaction. It accepts or donates chemical groups.

20.8 Environmental Effects

Enzymes are sensitive to pH and temperature. High temperatures or extremes of pH rapidly inactivate most enzymes by denaturing them.

20.9 Regulation of Enzyme Activity

Enzymes differ from inorganic catalysts in that they are regulated by the cell. Some of the means of enzyme regulation include allosteric regulation, *feedback inhibition,* production of inactive forms, or *zymogens,* and *protein modification. Allosteric enzymes* have an effector binding site, as well as an active site. Effector binding renders the enzyme active (*positive allosterism*) or inactive (*negative allosterism*). In feedback inhibition the product of a biosynthetic pathway turns off the entire pathway via negative allosterism. In protein modification, adding or removing a covalently bound group either activates or inactivates an enzyme.

20.10 Inhibition of Enzyme Activity

Enzyme activity can be destroyed by a variety of inhibitors. *Irreversible inhibitors,* or poisons, bind tightly to enzymes and destroy their activity permanently. *Competitive inhibitors* are generally *structural analogs* of the natural substrate for the enzyme. They compete with the normal substrate for binding to the active site. When the competitive inhibitor is bound by the active site, the reaction cannot occur, and no product is produced.

20.11 Proteolytic Enzymes

Proteolytic enzymes (proteases) catalyze the hydrolysis of peptide bonds. The *pancreatic serine proteases* chymotrypsin, trypsin, and elastase have similar structures and mechanisms of action, but different substrate specificities. It is thought that they evolved from a common ancestral protease.

20.12 Uses of Enzymes in Medicine

Analysis of blood serum for unusually high levels of certain enzymes provides valuable information on a patient's condition. Such analysis is used to diagnose heart attack, liver disease, and pancreatitis. Enzymes are also used as analytical reagents, as in the blood urea nitrogen (BUN) test, and in the treatment of disease.

Key Terms

absolute specificity (20.5)
active site (20.4)
allosteric enzyme (20.9)
apoenzyme (20.7)
coenzyme (20.7)
cofactor (20.7)
competitive inhibitor (20.10)
enzyme (Intro)
enzyme specificity (20.5)
enzyme-substrate complex (20.4)
feedback inhibition (20.9)
group specificity (20.5)
holoenzyme (20.7)
hydrolase (20.1)
induced fit model (20.4)
irreversible enzyme inhibitor (20.10)
isomerase (20.1)
ligase (20.1)
linkage specificity (20.5)
lock-and-key model (20.4)
lyase (20.1)
negative allosterism (20.9)
oxidoreductase (20.1)
pancreatic serine protease (20.11)
pH optimum (20.8)
positive allosterism (20.9)
product (20.2)
protein modification (20.9)
proteolytic enzyme (20.11)
reversible, competitive enzyme inhibitor (20.10)
reversible, noncompetitive enzyme inhibitor (20.10)
stereochemical specificity (20.5)
structural analog (20.10)
substrate (20.1)
temperature optimum (20.8)
transferase (20.1)
transition state (20.6)
vitamin (20.7)
zymogen (20.9)

Questions and Problems

Nomenclature and Classification

20.19 Match each of the following substrates with its corresponding enzyme:

1. Urea	**a.** Lipase
2. Hydrogen peroxide	**b.** Glucose-6-phosphatase
3. Lipid	**c.** Peroxidase
4. Aspartic acid	**d.** Sucrase
5. Glucose-6-phosphate	**e.** Urease
6. Sucrose	**f.** Aspartase

20.20 Give a systematic name for the enzyme that would act on each of the following substrates:
a. Alanine
b. Citrate
c. Ampicillin
d. Ribose
e. Methylamine

20.21 Describe the function implied by the name of each of the following enzymes:
a. Citrate decarboxylase
b. Adenosine diphosphate phosphorylase
c. Oxalate reductase
d. Nitrite oxidase
e. *cis-trans* Isomerase

20.22 List the six classes of enzymes based on the type of reaction catalyzed. Briefly describe the function of each class, and provide an example of each.

The Effect of Enzymes on the Activation Energy of a Reaction

20.23 What is the activation energy of a reaction?
20.24 What is the effect of an enzyme on the activation energy of a reaction?
20.25 Write and explain the equation for the equilibrium constant of an enzyme-mediated reaction. Does the enzyme alter the K_{eq}?
20.26 If an enzyme does not alter the equilibrium constant of a reaction, how does it speed up the reaction?

The Effect of Substrate Concentration on Enzyme-Catalyzed Reactions

20.27 What is the effect of doubling the substrate concentration on the rate of a chemical reaction?
20.28 Why doesn't the rate of an enzyme-catalyzed reaction increase indefinitely when the substrate concentration is made very large?
20.29 Draw a graph that describes the effect of increasing the concentration of the substrate on the rate of an enzyme-catalyzed reaction.
20.30 What does a graph of enzyme activity versus substrate concentration tell us about the nature of enzyme-catalyzed reactions?

The Enzyme-Substrate Complex

20.31 Name three major properties of enzyme active sites.
20.32 If enzyme active sites are small, why are enzymes so large?
20.33 What is the lock-and-key model of enzyme-substrate binding?
20.34 Why is the induced fit model of enzyme-substrate binding a much more accurate model than the lock-and-key model?

Specificity of the Enzyme-Substrate Complex

20.35 List and define four classes of enzyme specificities.
20.36 Give an example of an enzyme that is representative of each class of enzyme specificity.

The Transition State and Product Formation

20.37 Outline the four general stages in an enzyme-catalyzed reaction.
20.38 Describe the transition state.

20.39 What types of transition states might be envisioned that would decrease the energy of activation of an enzyme?

20.40 If an enzyme catalyzed a reaction by modifying the local pH, what kind of amino acid R groups would you expect to find in the active site?

Cofactors and Coenzymes

20.41 What is the role of a cofactor in enzyme activity?

20.42 How does a coenzyme function in an enzyme-catalyzed reaction?

20.43 What is the function of NAD^+? What class of enzymes would require a coenzyme of this sort?

20.44 What is the function of FAD? What class of enzymes would require this coenzyme?

Environmental Effects

20.45 List the factors that affect enzyme activity.

20.46 How will each of the following changes in conditions alter the rate of an enzyme-catalyzed reaction?

a. Decreasing the temperature from 37°C to 10°C
b. Increasing the pH of the solution from 7 to 11
c. Heating the enzyme from 37°C to 100°C

20.47 Why does an enzyme lose activity when the pH is drastically changed from optimum pH?

20.48 Define the optimum pH for enzyme activity.

20.49 High temperature is an effective mechanism for killing bacteria on surgical instruments. How does high temperature result in cellular death?

20.50 An increase in temperature will increase the rate of a reaction if a nonenzymatic catalyst is used; however, an increase in temperature will eventually *decrease* the rate of a reaction when an enzyme catalyst is used. Explain the apparent contradiction of these two statements.

20.51 What is a lysosome?

20.52 Of what significance is it that lysosomal enzymes have a pH optimum of 4.8?

20.53 Why are enzymes that are used for clinical assays in hospitals stored in refrigerators?

20.54 Why do extremes of pH inactivate enzymes?

Regulation of Enzyme Activity

20.55 a. Why is it important for cells to regulate the level of enzyme activity?
b. Why must synthesis of digestive enzymes be carefully controlled?

20.56 What is an allosteric enzyme?

20.57 What is the difference between positive and negative allosterism?

20.58 a. Define feedback inhibition.
b. Describe the role of allosteric enzymes in feedback inhibition.
c. Is this positive or negative allosterism?

20.59 What is a zymogen?

20.60 Three zymogens that are involved in digestion of proteins in the stomach and intestines are pepsinogen, chymotrypsinogen, and trypsinogen. What is the advantage of producing these enzymes as inactive peptides?

Inhibition of Enzyme Activity

20.61 Define *competitive enzyme inhibition.*

20.62 How do the sulfa drugs selectively kill bacteria while causing no harm to humans?

20.63 Describe the structure of a structural analog.

20.64 How can structural analogs serve as enzyme inhibitors?

20.65 Define *irreversible enzyme inhibition.*

20.66 Why are irreversible enzyme inhibitors often called *poisons?*

20.67 Suppose that a certain drug company manufactured a compound that had nearly the same structure as a substrate for a certain enzyme but that could not be acted upon chemically by the enzyme. What type of interaction would the compound have with the enzyme?

20.68 The addition of phenylthiourea to a preparation of the enzyme polyphenoloxidase completely inhibits the activity of the enzyme.

a. Knowing that phenylthiourea binds all copper ions, what conclusion can you draw about whether polyphenoloxidase requires a cofactor?
b. What kind of inhibitor is phenylthiourea?

Proteolytic Enzymes

20.69 What do the similar structures of chymotrypsin, trypsin, and elastase suggest about their evolutionary relationship?

20.70 What properties are shared by chymotrypsin, trypsin, and elastase?

20.71 Draw the complete structural formula for the peptide tyr-lys-ala-phe. Show which bond would be broken when this peptide is reacted with chymotrypsin.

20.72 Repeat Question 20.71 for the peptide trp-pro-gly-tyr.

20.73 The sequence of a peptide that contains ten amino acid residues is as follows:

ala-gly-val-leu-trp-lys-ser-phe-arg-pro

Indicate with arrows and label the peptide bond(s) that are cleaved by elastase, trypsin, and chymotrypsin.

20.74 What structural features of trypsin, chymotrypsin, and elastase account for their different specificities?

Uses of Enzymes in Medicine

20.75 List the enzymes whose levels are elevated in blood serum following a myocardial infarction.

20.76 List the enzymes whose levels are elevated as a result of hepatitis or cirrhosis of the liver.

Critical Thinking Problems

1. Ethylene glycol is a poison that causes about fifty deaths a year in the United States. Treating people who have drunk ethylene glycol with massive doses of ethanol can save their lives. Suggest a reason for the effect of ethanol.

2. Generally speaking, feedback inhibition involves regulation of the first step in a pathway. Consider the following hypothetical pathway:

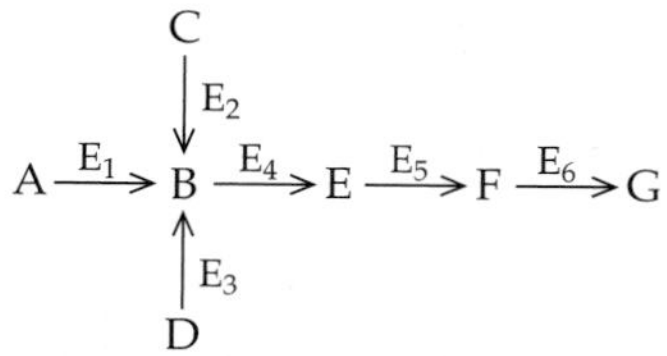

Which step in this pathway do you think should be regulated? Explain your reasoning.

3. In an amplification cascade, each step greatly increases the amount of substrate available for the next step, so that a very large amount of the final product is made. Consider the following hypothetical amplification cascade:

$$\begin{array}{l} A_{active} \\ \downarrow \\ B_{inactive} \longrightarrow B_{active} \\ \qquad\qquad\downarrow \\ \qquad C_{inactive} \longrightarrow C_{active} \\ \qquad\qquad\qquad\qquad\downarrow \\ \qquad\qquad D_{inactive} \longrightarrow D_{active} \end{array}$$

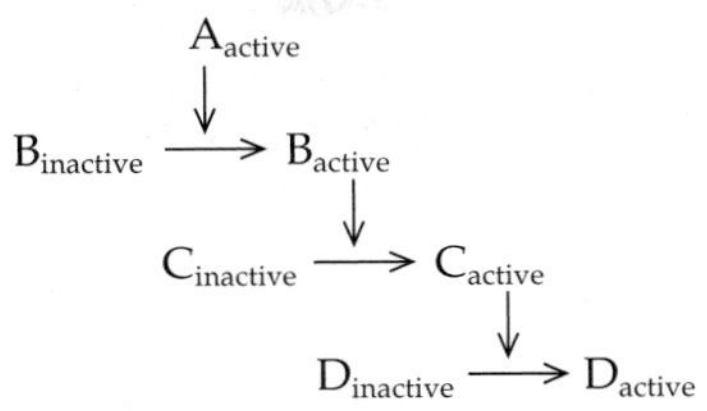

If each active enzyme in the pathway converts one hundred molecules of its substrate to active form, how many molecules of D will be produced if the pathway begins with one molecule of A?

4. L-1-(*p*-toluenesulfonyl)-amido-2-phenylethylchloromethyl ketone (TPCK, shown below) inhibits chymotrypsin, but not trypsin. Propose a hypothesis to explain this observation.

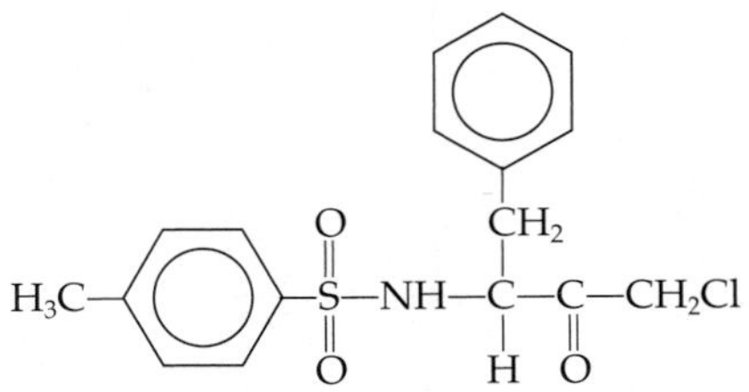

$$H_3C-C_6H_4-SO_2-NH-CH(CH_2C_6H_5)-C(=O)-CH_2Cl$$

5. A graduate student is trying to make a "map" of a short peptide so that she can eventually determine the amino acid sequence. She digested the peptide with several proteases and determined the sizes of the resultant digestion products.

Enzyme	M.W. of Digestion Products
Trypsin	2000, 3000
Chymotrypsin	500, 1000, 3500
Elastase	500, 1000, 1500, 2000

Suggest experiments that would allow the student to map the order of the enzyme digestion sites along the peptide.

Some familiar fermentation products.

21

Carbohydrate Metabolism

Outline

BIOCHEMISTRY

Learning Goals

1. Discuss the importance of ATP in cellular energy transfer processes.
2. Describe the three stages of catabolism of dietary proteins, carbohydrates, and lipids.
3. Discuss glycolysis in terms of its two major segments.
4. Looking at an equation representing any of the chemical reactions that occur in glycolysis, describe the kind of reaction that is occurring and the significance of that reaction to the pathway.
5. Describe the mechanism of regulation of the rate of glycolysis. Discuss particular examples of that regulation.
6. Discuss the practical and metabolic roles of fermentation reactions.
7. List several products of the pentose phosphate pathway that are required for biosynthesis.
8. Compare glycolysis and gluconeogenesis.
9. Summarize the regulation of blood glucose levels by glycogenesis and glycogenolysis.

CHEMISTRY CONNECTION

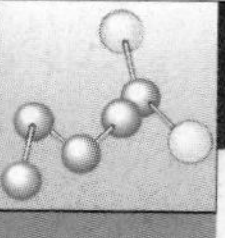

The Man Who Got Tipsy from Eating Pasta

Imagine becoming drunk after eating a plate of spaghetti or a bag of potato chips. That is exactly what happened to Charles Swaart while he was stationed in Tokyo after World War II. Suddenly, he would be completely drunk—without having swallowed a drop of alcohol.

During the next two decades, Swaart continued to have unexplainable bouts of drunkenness and horrible hangovers. The problem was so serious that his liver was being destroyed. But in 1964, Swaart heard of a man in Japan who suffered from the same mysterious—and embarrassing—symptoms. After twenty-five years, physicians diagnosed the problem. There was a mutant strain of the yeast *Candida albicans* living in the gastrointestinal tract of the Japanese man. These yeast cells were using carbohydrates from the man's diet to make ethanol. The metabolic pathways used by these yeast cells were glycolysis and alcohol fermentation, two of the pathways that we will study in this chapter.

Swaart took advantage of the therapy used in Japan. He had to try several antibiotics over the years. But finally, in 1975, all of the mutant yeast cells in his intestine were killed, and his life returned to normal.

Why was it so difficult for physicians to solve this medical mystery? Nonfermenting *Candida albicans* is a regular inhabitant of the human gut. It took some very clever scientific detective work to find this mutant ethanol-producing strain. The scientists even have a hypothesis about where the mutant yeast came from. They think that the radiation released in one of the atomic bomb blasts at Nagasaki or Hiroshima may have caused the mutation.

In this chapter we begin our study of the chemical reactions used by all organisms to provide energy for cellular work. In Chapter 24 we will look at the kinds of DNA damage (mutations) that can produce changes in the structure or function of an organism.

Introduction

Just as we need energy to run, jump, and think, the cell needs a ready supply of cellular energy for the many functions that support these activities. Cells need energy for *active transport*, to move molecules between the environment and the cell. Energy is also needed for *biosynthesis* of small metabolic molecules and production of macromolecules from these intermediates. Finally, energy is required for *mechanical work*, including muscle contraction and motility of sperm cells. Table 21.1 lists some examples of each of these energy-requiring processes.

We need a supply of energy-rich food molecules that can be degraded, or *oxidized*, to provide this needed cellular energy. Our diet includes three major sources of energy: carbohydrates, fats, and proteins. Each of these types of large biological molecules must be broken down into its basic subunits—simple sugars, fatty acids and glycerol, and amino acids—before they can be taken into the cell and used to produce cellular energy. Of these classes of food molecules, carbohydrates are the most readily used. The pathway for the first stages of carbohydrate breakdown is called *glycolysis*. We find the same pathway in organisms as different as the simple bacterium and humans.

Recall that the potential energy of a compound is the bond energy of that compound.

In this chapter we are going to examine the steps of this ancient energy-harvesting pathway. We will see that it is responsible for the capture of some of the bond energy of carbohydrates and the storage of that energy in the molecular form of *adenosine triphosphate (ATP)*. Glycolysis actually releases and stores very little (2.2%) of the potential energy of glucose, but the pathway also serves as a source of biosynthetic building blocks. It also modifies the carbohydrates in such a way that other pathways are able to release as much as 40% of the potential energy.

Table 21.1 The Types of Cellular Work That Require Energy

Biosynthesis: Synthesis of Metabolic Intermediates and Macromolecules
Synthesis of glucose from CO_2 and H_2O in the process of photosynthesis in plants
Synthesis of amino acids
Synthesis of nucleotides
Synthesis of lipids
Protein synthesis from amino acids
Synthesis of nucleic acids
Synthesis of organelles and membranes
Active Transport: Movement of Ions and Molecules
Transport of H^+ to maintain constant pH
Transport of food molecules into the cell
Transport of K^+ and Na^+ into and out of nerve cells for transmission of nerve impulses
Secretion of HCl from parietal cells into the stomach
Transport of waste from the blood into the urine in the kidneys
Transport of amino acids and most hexose sugars into the blood from the intestine
Accumulation of calcium ions in the mitochondria
Motility
Contraction and flexion of muscle cells
Separation of chromosomes during cell division
Ability of sperm to swim via flagella
Movement of foreign substances out of the respiratory tract by cilia on the epithelial lining of the trachea
Translocation of eggs into the fallopian tubes by cilia in the female reproductive tract

21.1 ATP: The Cellular Energy Currency

Learning Goal

The degradation of fuel molecules, called **catabolism,** provides the energy for cellular energy-requiring functions, including **anabolism,** or biosynthesis. Actually, the energy of a food source can be released in one of two ways: as heat or, more important to the cell, as chemical bond energy. We can envision two alternative modes of aerobic degradation of the simple sugar glucose. We can, metaphorically, simply set the glucose afire. This would result in its complete oxidation to CO_2 and H_2O and would release 686 kcal/mol of glucose. Yet in terms of a cell, what would be accomplished? Nothing. All of the potential energy of the bonds of glucose is lost as heat and light.

The cell uses a different strategy. With a series of enzymes, biochemical pathways in the cell carry out a step-by-step oxidation of glucose. Small amounts of energy are released at several points in the pathway and that energy is harvested and saved in the bonds of a molecule that has been called the *universal energy currency.* This molecule is **adenosine triphosphate (ATP).**

ATP serves as a "go-between" molecule that couples the *exergonic* (energy releasing) reactions of catabolism and the endergonic (energy requiring) reactions of anabolism. To understand how this molecule harvests the energy and releases it for energy-requiring reactions, we must take a look at the structure of this amazing molecule (Figure 21.1). ATP is a **nucleotide,** which means that it is a molecule composed of a nitrogenous base; a five-carbon sugar; and one, two, or three phosphoryl groups.

Nitrogenous bases are heterocyclic amines. Their structure and functions are discussed in Section 16.2. The five-carbon sugars are discussed in Section 17.4. More information on the structure of nucleotides is found in Section 24.1.

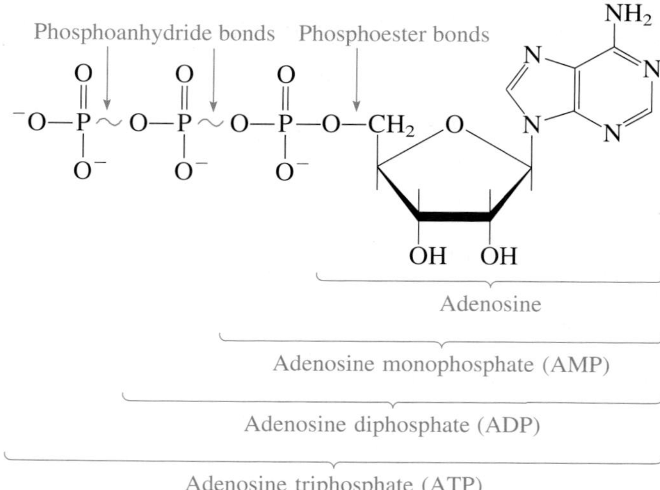

Figure 21.1
The structure of the universal energy currency, ATP.

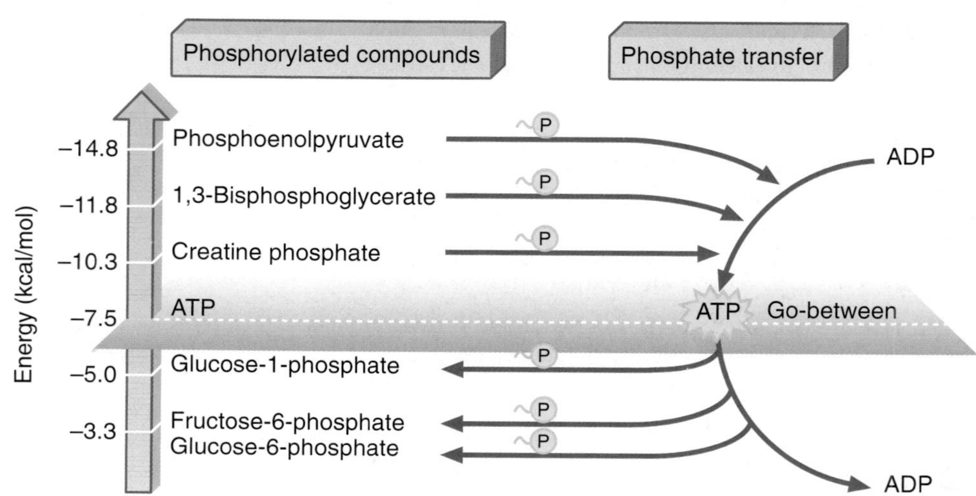

Figure 21.2
An energy comparison of several phosphorylated compounds that are important in cellular metabolic processes.

In ATP, a phosphoester bond joins the first phosphoryl group to the five-carbon sugar ribose. The next two phosphoryl groups are joined to one another by phosphoanhydride bonds (Figure 21.1). Recall that the phosphoanhydride bond is a *high-energy bond.* When it is broken or hydrolyzed, a large amount of energy is released. When the phosphoanhydride bond of ATP is broken, the energy that is released can be used for cellular work. These high-energy bonds are indicated as squiggles (~) in Figure 21.1.

Nature's high energy bonds, including phosphoanhydride and phosphoester bonds, are discussed in Section 15.4.

The structure of ATP is only a part of the reason that the molecule is a good go-between in energy transformations in the cell. ATP must have a higher energy content than the compounds to which it will donate energy, but it must also contain less energy than the compounds that are involved in forming it. In this way, all the reactions are favored because both the reactions that produce ATP and the hydrolysis of ATP to provide energy for cellular work are exergonic. Figure 21.2 shows the relative energies of some phosphorylated compounds, including ATP, that are involved in energy metabolism.

Hydrolysis of ATP yields adenosine diphosphate (ADP), an inorganic phosphate group (P_i), and energy (Figure 21.3). The energy released by this hydrolysis of ATP is then used to drive biological processes, for instance, the phosphorylation of glucose or fructose.

See Sections 15.4 and 24.1 for further information on the structure of ATP and hydrolysis of the phosphoanhydride bonds.

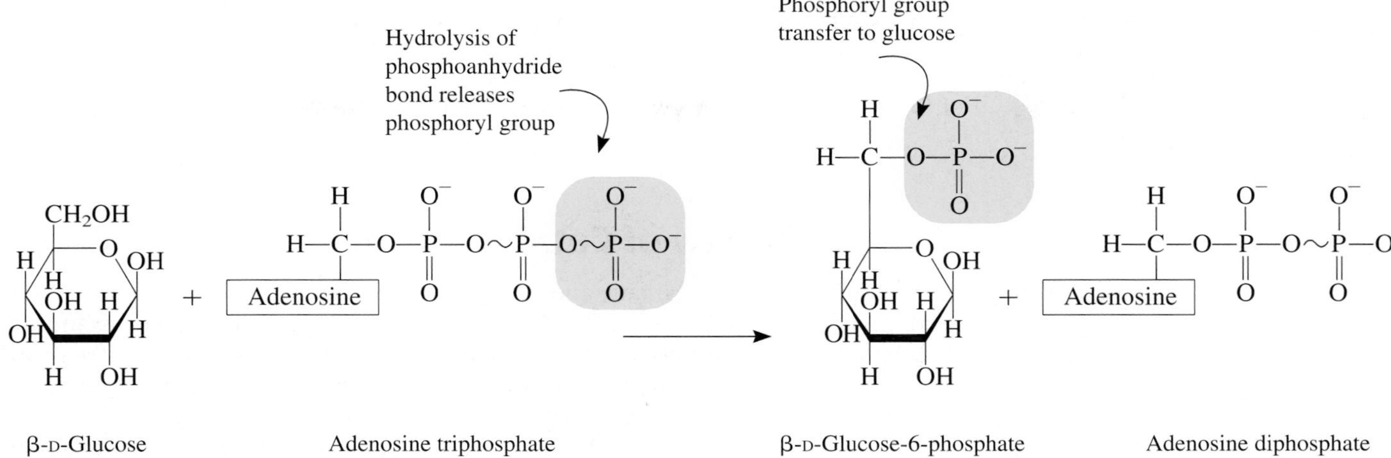

Figure 21.3
The hydrolysis of the phosphoanhydride bond of ATP releases inorganic phosphate and energy. In this coupled reaction catalyzed by an enzyme, the phosphoryl group and some of the released energy are transferred to β-D-glucose.

An example of the way in which the energy of ATP is used can be seen in the first step of glycolysis, the anaerobic degradation of glucose to produce chemical energy. The first step involves the transfer of a phosphoryl group, $—PO_3^{2-}$, from ATP to the C-6 hydroxyl group of glucose (Figure 21.3). This reaction is catalyzed by the enzyme hexokinase.

This reaction can be dissected to reveal the role of ATP as a source of energy. Although this is a coupled reaction, we can think of it as a two-step process. The first step is the hydrolysis of ATP to ADP and phosphate, abbreviated P_i. This is an exergonic reaction that *releases* about 7 kcal/mol of energy:

$$\text{ATP} + \text{H}_2\text{O} \longrightarrow \text{ADP} + \text{P}_i + 7\ \text{kcal/mol}$$

The second step, the synthesis of glucose-6-phosphate from glucose and phosphate, is an endergonic reaction that *requires* 3.0 kcal/mol:

$$3.0\ \text{kcal/mol} + \text{glucose} + \text{P}_i \longrightarrow \text{glucose-6-phosphate} + \text{H}_2\text{O}$$

These two chemical reactions can then be added to give the equation showing the way in which ATP hydrolysis is *coupled* to the phosphorylation of glucose:

$$\begin{array}{r} \text{ATP} + \text{H}_2\text{O} \longrightarrow \text{ADP} + \text{P}_i + 7\ \text{kcal/mol} \\ 3.0\ \text{kcal/mol} + \text{glucose} + \text{P}_i \longrightarrow \text{glucose-6-phosphate} + \text{H}_2\text{O} \\ \hline \text{Net: ATP} + \text{glucose} \longrightarrow \text{glucose-6-phosphate} + \text{ADP} + 4\ \text{kcal/mol} \end{array}$$

Because the hydrolysis of ATP releases more energy than is required to synthesize glucose-6-phosphate from glucose and phosphate, there is an overall energy release in this process and the reaction proceeds spontaneously to the right. The product, glucose-6-phosphate, has more energy than the reactant, glucose, because it now carries some of the energy from the original phosphoanhydride bond of ATP.

The primary function of all catabolic pathways is to harvest the chemical energy of fuel molecules and to store that energy by the production of ATP. This continuous production of ATP is what provides the stored potential energy that is used to power most cellular functions.

Question 21.1 Why is ATP called the universal energy currency?

Question 21.2 List five biological activities that require ATP.

21.2 Overview of Catabolic Processes

Learning Goal

Although carbohydrates, fats, and proteins can all be degraded to release energy, carbohydrates are the most readily used energy source. We will begin by examining the oxidation of the hexose glucose. In Chapters 22 and 23 we will see how the pathways of glucose oxidation are also used for the degradation of fats and proteins.

Any catabolic process must begin with a supply of nutrients. When we eat a meal, we are eating quantities of carbohydrates, fats, and proteins. From this point the catabolic processes can be broken down into a series of stages. The three stages of catabolism are summarized in Figure 21.4.

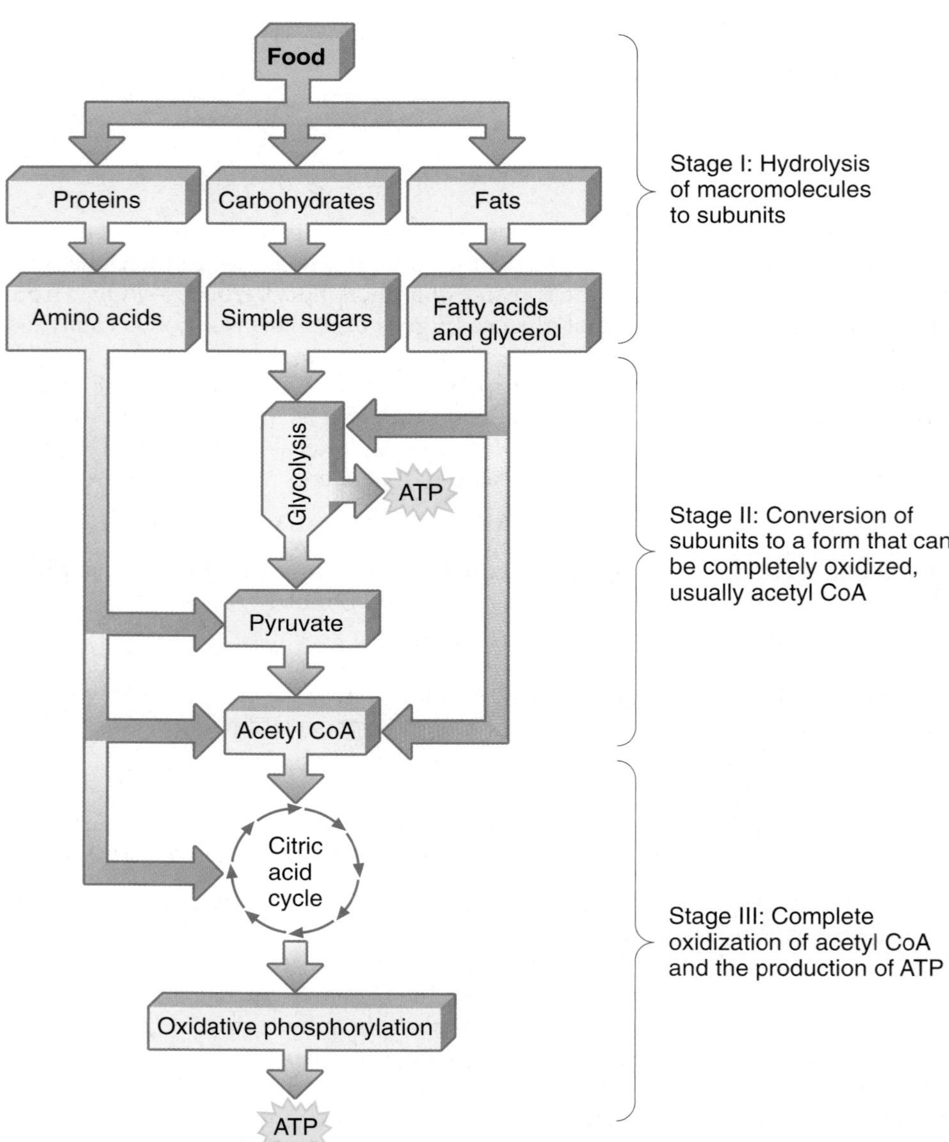

Figure 21.4
The three stages of the conversion of food into cellular energy in the form of ATP.

Stage I: Hydrolysis of Dietary Macromolecules into Small Subunits

The purpose of the first stage of catabolism is to degrade large food molecules into their component subunits. These subunits—simple sugars, amino acids, fatty acids, and glycerol—are then taken into the cells of the body for use as an energy source. The process of digestion is summarized in Figure 21.5.

Polysaccharides are hydrolyzed to monosaccharides. This process begins in the mouth, where the enzyme amylase begins the hydrolysis of starch. Digestion continues in the small intestine, where pancreatic amylase further hydrolyzes the starch into maltose (a disaccharide of glucose). Maltase catalyzes the hydrolysis of maltose, producing two glucose molecules. Similarly, sucrose is hydrolyzed to glucose and fructose by the enzyme sucrase, and lactose (milk sugar) is degraded into the monosaccharides glucose and galactose by the enzyme lactase in the small intestine. The monosaccharides are taken up by the epithelial cells of the intestine in an energy-requiring process called *active transport*.

In the laboratory, a strong acid or base and high temperatures are required for hydrolysis of amide bonds (Section 16.3). However, this reaction proceeds quickly under physiological conditions when catalyzed by enzymes (Section 20.11).

The digestion of proteins begins in the stomach, where the low pH denatures the proteins so that they are more easily hydrolyzed by the enzyme pepsin. They are further degraded in the small intestine by trypsin, chymotrypsin, elastase, and other proteases. The products of protein digestion—amino acids and short oligopeptides—are taken up by the cells lining the intestine. This uptake also involves an active transport mechanism.

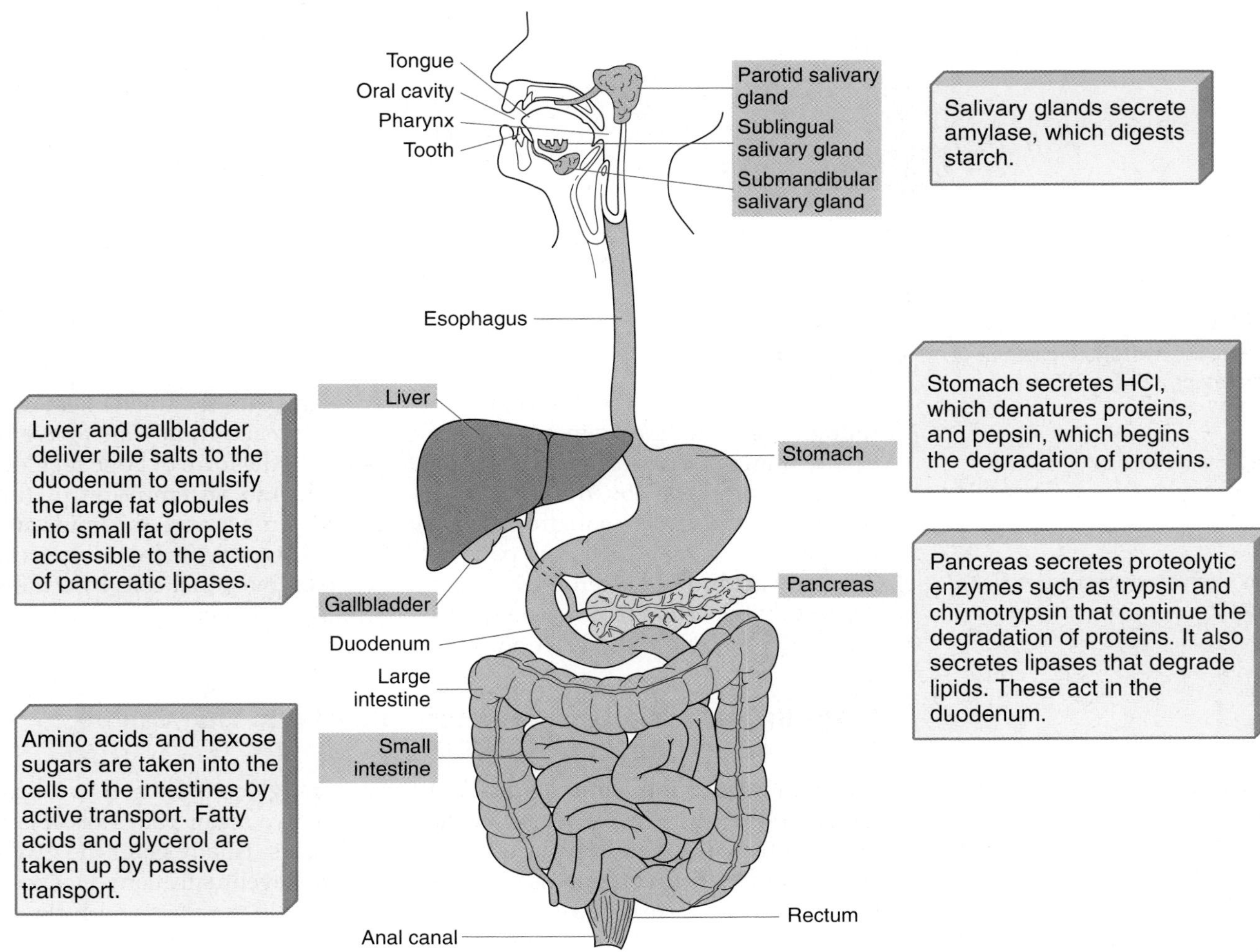
Tongue
Oral cavity
Pharynx
Tooth
Parotid salivary gland
Sublingual salivary gland
Submandibular salivary gland
Salivary glands secrete amylase, which digests starch.
Esophagus
Liver
Stomach
Stomach secretes HCl, which denatures proteins, and pepsin, which begins the degradation of proteins.
Liver and gallbladder deliver bile salts to the duodenum to emulsify the large fat globules into small fat droplets accessible to the action of pancreatic lipases.
Pancreas
Pancreas secretes proteolytic enzymes such as trypsin and chymotrypsin that continue the degradation of proteins. It also secretes lipases that degrade lipids. These act in the duodenum.
Gallbladder
Duodenum
Large intestine
Small intestine
Amino acids and hexose sugars are taken into the cells of the intestines by active transport. Fatty acids and glycerol are taken up by passive transport.
Rectum
Anal canal

Figure 21.5
An overview of the digestive processes that hydrolyze carbohydrates, proteins, and fats.

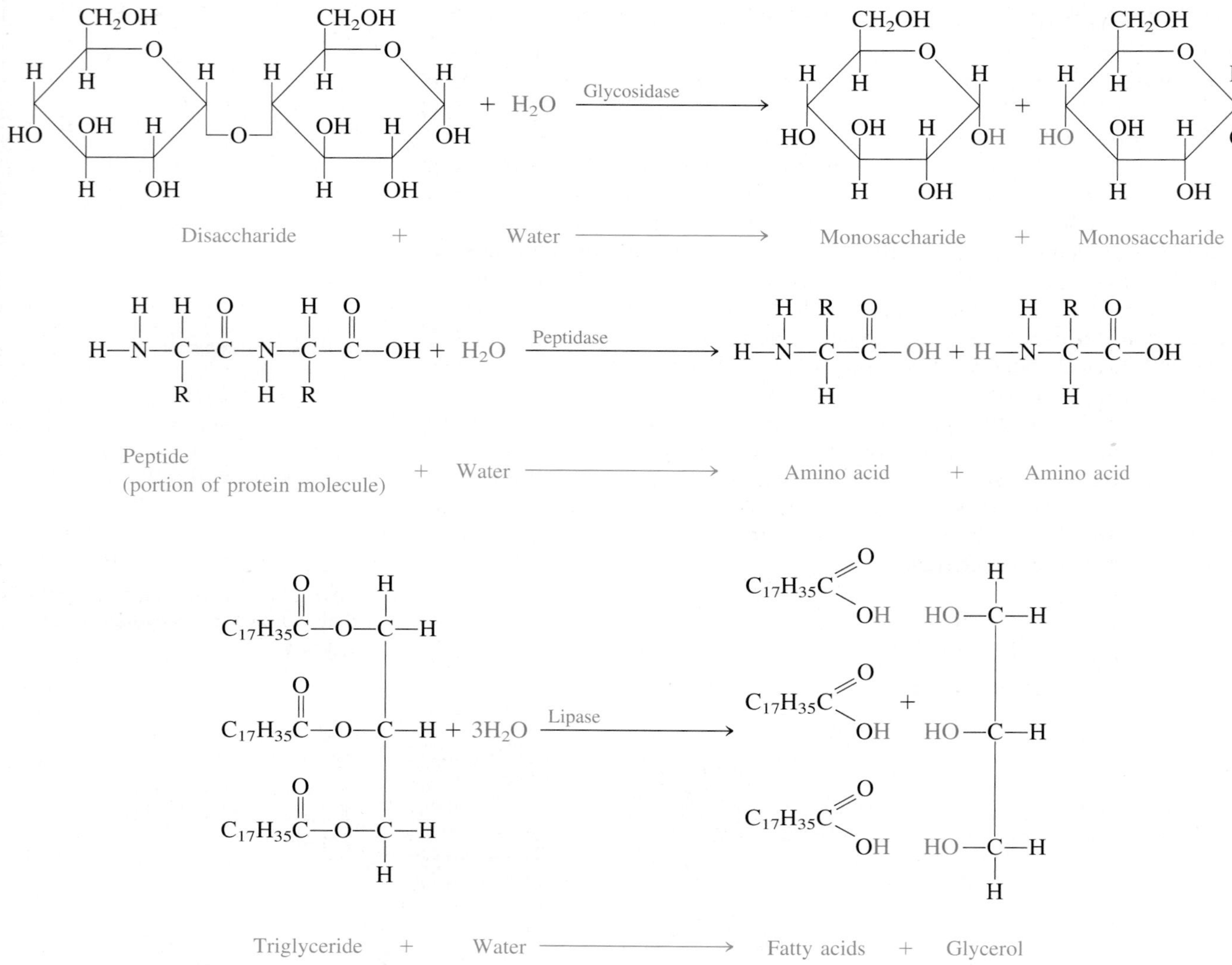

Disaccharide + Water $\xrightarrow{\text{Glycosidase}}$ Monosaccharide + Monosaccharide

$$\text{H-N(H)-C(H)(R)-C(=O)-N(H)-C(H)(R)-C(=O)-OH} + H_2O \xrightarrow{\text{Peptidase}} \text{H-N(H)-C(R)(H)-C(=O)-OH} + \text{H-N(H)-C(R)(H)-C(=O)-OH}$$

Peptide (portion of protein molecule) + Water ⟶ Amino acid + Amino acid

$$\text{Triglyceride } (C_{17}H_{35}C(=O)-O-CH_2,\ C_{17}H_{35}C(=O)-O-CH,\ C_{17}H_{35}C(=O)-O-CH_2) + 3H_2O \xrightarrow{\text{Lipase}} 3\,C_{17}H_{35}COOH + HOCH_2-CH(OH)-CH_2OH$$

Triglyceride + Water ⟶ Fatty acids + Glycerol

Figure 21.6
A summary of the hydrolysis reactions of carbohydrates, proteins, and fats.

The digestion and transport of fats are considered in greater detail in Chapter 23.

Active and passive transport are discussed in Section 18.6.

Digestion of fats does not begin until the food reaches the small intestine, even though there are lipases in both the saliva and stomach fluid. Fats arrive in the duodenum, the first portion of the small intestine, in the form of large fat globules. Bile salts produced by the liver break these up into an emulsion of tiny fat droplets. Because the small droplets have a greater surface area, the lipids are now more accessible to the action of pancreatic lipase. This enzyme hydrolyzes the fats into fatty acids and glycerol, which are taken up by intestinal cells by a transport process that does not require energy. This process is called *passive transport*. A summary of these hydrolysis reactions is shown in Figure 21.6.

Stage II: Conversion of Monomers into a Form That Can Be Completely Oxidized

The citric acid cycle is considered in detail in Section 22.4.

The monosaccharides, amino acids, fatty acids, and glycerol must now be assimilated into the pathways of energy metabolism. The two major pathways are glycolysis and the citric acid cycle (see Figure 21.4). Sugars usually enter the glycolysis pathway in the form of glucose or fructose. They are eventually converted to acetyl CoA, which is a form that can be completely oxidized in the citric acid cycle. Amino groups are removed from amino acids, and the remaining carbon skeletons enter the catabolic processes at many steps of the citric acid cycle. Fatty acids are converted to acetyl CoA and enter the citric acid cycle in that form. Glycerol, produced by the hydrolysis of fats, is converted to glyceraldehyde-3-phosphate, one of the intermediates of glycolysis, and enters energy metabolism at that level.

Stage III: The Complete Oxidation of Nutrients and the Production of ATP

Acetyl CoA carries two-carbon remnants of the nutrients, acetyl groups, to the citric acid cycle. Acetyl CoA enters the cycle, and electrons and hydrogen atoms are harvested during the complete oxidation of the acetyl group to CO_2. Coenzyme A is released (recycled) to carry additional acetyl groups to the pathway. The electrons and hydrogen atoms that are harvested are used in the process of oxidative phosphorylation to produce ATP.

Oxidative phosphorylation is described in Section 22.6.

Question 21.3

Briefly describe the three stages of catabolism.

Question 21.4

Discuss the digestion of dietary carbohydrates, lipids, and proteins.

21.3 Glycolysis

An Overview

Glycolysis, also known as the Embden-Meyerhof Pathway, is a pathway for carbohydrate catabolism that begins with the substrate D-glucose. The very fact that all organisms can use glucose as an energy source for glycolysis suggests that glycolysis was the first successful energy-harvesting pathway that evolved on the earth. The pathway evolved at a time when the earth's atmosphere was *anaerobic;* no free oxygen was available. As a result, glycolysis requires no oxygen; it is an anaerobic process. Further, it must have evolved in very simple, single-celled organisms, much like bacteria. These organisms did not have complex organelles in the cytoplasm to carry out specific cellular functions. Thus glycolysis was a process carried out by enzymes that were free in the cytoplasm. To this day, glycolysis remains an anaerobic process carried out by cytoplasmic enzymes, even in cells as complex as our own.

The ten steps of glycolysis are outlined in Figure 21.7. The first substrate in the pathway is the hexose sugar glucose. Ten enzymes are needed to carry out the reactions of the pathway. The first reactions of glycolysis involve an energy investment. ATP molecules are hydrolyzed, energy is released, and phosphoryl groups are added to the hexose sugars. In the remaining steps of glycolysis, energy is harvested to produce a net gain of ATP.

The three major products of glycolysis are seen in Figure 21.7. These are chemical energy in the form of ATP, chemical energy in the form of NADH, and two three-carbon pyruvate molecules. Each of these products is considered below:

- **Chemical energy as ATP.** Four ATP molecules are formed by the process of **substrate-level phosphorylation.** This means that a high-energy phosphoryl group from one of the substrates in glycolysis is transferred to ADP to form ATP. The two substrates involved in these transfer reactions are 1,3-bisphosphoglycerate and phosphoenolpyruvate (see Figure 21.7, steps 7 and 10). Although four ATP molecules are produced during glycolysis, the *net* gain is only two ATP molecules because two ATP molecules are used early in glycolysis (Figure 21.7, steps 1 and 3). The two ATP molecules produced

A MEDICAL PERSPECTIVE

Genetic Disorders of Glycolysis

Imagine always having difficulty with physical exercise. Imagine the coach telling you to get tough and run that lap again, since you were last! Imagine being accused of being lazy because you didn't carry your share of the camping gear. Imagine all that and not knowing why it is that you can't keep up with your friends or the others in your physical education class. This has been the fate of thousands of people who suffer from a metabolic myopathy—a muscle (*myo-*) disorder (*-pathy*) caused by an inability to extract the energy from the food that you eat.

The onset of fatigue during exercise is called *exercise intolerance*. It is one of the major symptoms of a metabolic myopathy. But simple fatigue is just the mildest of the symptoms. Overexertion may cause episodes of muscle breakdown (rhabdomyolysis) in which the muscle cells, unable to provide enough ATP energy for themselves, begin to die. The muscle breakdown causes greatly elevated blood levels of creatine kinase. Creatine kinase is an abundant enzyme in muscle that is critical in energy metabolism (see A Human Perspective: Exercise and Energy Metabolism, in Chapter 22). When muscle cells die, this enzyme is released into the bloodstream. Another symptom is myoglobinuria (myoglobin in the urine). Recall that myoglobin is the oxygen storage protein in muscle. When muscles die, myoglobin ends up in the urine, turning it the color of cola soft drinks. Myoglobinuria may even cause kidney damage. Accompanying these clinical symptoms, many people describe intense muscle pain. They describe it as a cramp, but it is not a cramp, since the muscle is not able to contract because of the lack of energy. Rather, the pain is caused by cell death and tissue damage that result from an inability to produce enough ATP.

There are three glycolytic enzyme deficiencies that lead to metabolic myopathy. The first is phosphofructokinase deficiency or Tarui's disease. Although this is not a sex-linked disorder, the great majority of sufferers are males (nine males to one female). The disorder is most frequently found in U.S. Ashkenazi Jews and Italian families. Onset of symptoms typically occurs between the ages of twenty and forty, although some severe cases have been reported in infants and young children. Patients experiencing the late-onset form of Tarui's disease typically experienced exercise intolerance when they were younger. Vigorous exercise results in myoglobinuria and severe muscle pain. Meals high in carbohydrates worsen the exercise intolerance. Early-onset disease is often associated with respiratory failure, cardiomyopathy (heart muscle disease), seizures, and cortical blindness.

Phosphoglycerate kinase deficiency is a sex-linked genetic disorder (located on the X chromosome). As a result, far more males than females suffer from this disease. There are many clinical features associated with this deficiency, although only rarely are they all found in the same patient. These symptoms range from mental retardation and seizures to a slowly progressive myopathy.

Phosphoglycerate mutase deficiency has been mapped on chromosome 7. The disorder is found predominantly in U.S. African American, Italian, and Japanese families. The clinical features include exercise intolerance, muscle pain, and myoglobinuria following more intense exercise.

Since each of these disorders is caused by the lack of an enzyme, scientists are trying to design a way to replace the lost activity. Oral medication will not work because enzymes are proteins. They would simply be digested, like any other dietary protein. Enzyme replacement therapy is one approach that is being studied. This would involve periodic injections of the enzyme into the bloodstream, a treatment just like the injection of insulin by diabetics. Enzyme replacement therapy would require a large supply of the enzyme. Following the model of insulin, the gene for the enzyme could be cloned into bacteria. The bacteria would then produce the protein, which would be purified for use by humans.

Another strategy is to introduce the gene for the missing enzyme into the patient's cells. This is called gene therapy. This method would also require that the gene for the enzyme be cloned. The DNA would then have to be introduced into the body using a safe procedure that would promote entry into target cells. There are still many obstacles to overcome before this type of treatment will be a reality for sufferers of metabolic myopathy.

A number of physicians are using a commonsense approach to the management of these disorders. Logic tells us that if a person cannot harvest energy from carbohydrates in the diet, perhaps a diet high in protein and lipids might be beneficial. As with any condition of this sort, it is important to consult a physician who understands the metabolic disorder and who will design and supervise a customized diet.

represent only 2.2% of the potential energy of the glucose molecule. Thus glycolysis is not a very efficient energy-harvesting process.

The structure of NAD^+ and the way it functions as a hydride anion carrier are shown in Figure 20.8 and described in Section 20.7.

- **Chemical energy in the form of reduced NAD^+, NADH. Nicotinamide adenine dinucleotide (NAD^+)** is a coenzyme derived from the vitamin niacin. The reduced form of NAD^+, NADH, carries hydride anions, hydrogen atoms with two electrons ($H:^-$), removed during the oxidation of one of the substrates, glyceraldehyde-3-phosphate (see Figure 21.7, step 6). Under aerobic conditions the electrons and hydrogen atom are transported

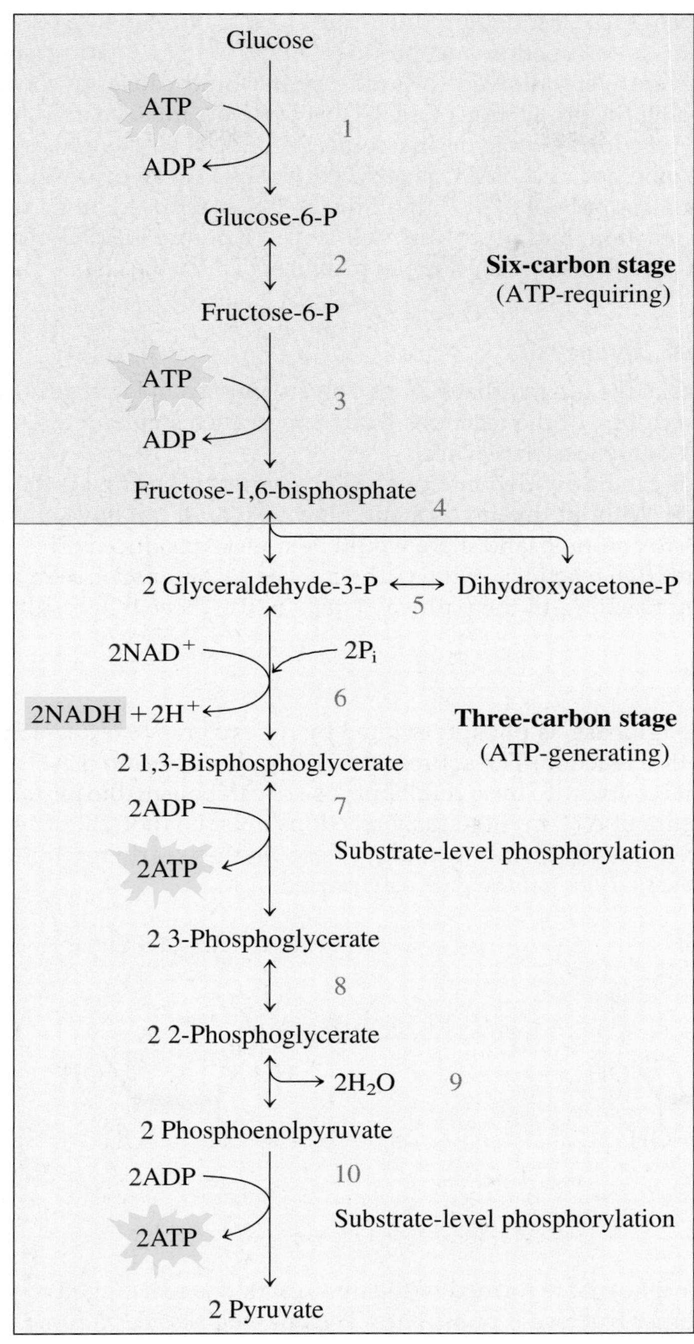

Figure 21.7
A summary of the reactions of glycolysis. These reactions occur in the cell cytoplasm.

from the cytoplasm into the mitochondria. Here they enter an electron transport system for the generation of ATP by **oxidative phosphorylation.** Under anaerobic conditions, NADH is used as a source of electrons in fermentation reactions.

- **Two pyruvate molecules.** At the end of glycolysis the six-carbon glucose molecule has been converted into two three-carbon pyruvate molecules. The fate of the pyruvate also depends on whether the reactions are occurring in the presence or absence of oxygen. Under aerobic conditions it is used to produce acetyl CoA destined for the citric acid cycle and complete oxidation. Under anaerobic conditions it is used as an electron acceptor in fermentation reactions.

In any event these last two products must be used in some way so that glycolysis can continue to function and produce ATP. There are two reasons for this. First, if pyruvate were allowed to build up, it would cause glycolysis to stop, thereby stopping the production of ATP. Thus pyruvate must be used in some kind of follow-up reaction, aerobic or anaerobic. Second, in step 6, glyceraldehyde-3-phosphate is oxidized and NAD^+ is reduced (accepts the hydride anion). The cell has only a small supply of NAD^+. If all the NAD^+ is reduced, none will be available for this reaction, and glycolysis will stop. Therefore NADH must be reoxidized so that glycolysis can continue to produce ATP for the cell.

Reactions of Glycolysis

Learning Goal

Learning Goal

The structures of the intermediates of glycolysis are seen in Figure 21.8, along with a concise description of the reactions that occur at each step and the names of the enzymes that catalyze each reaction.

Glycolysis can be divided into two major segments. The first is the investment of ATP energy. Without this investment, glucose would not have enough energy for glycolysis to continue, and there would be no ATP produced. This segment includes the first five reactions of the pathway. The second major segment involves the remaining reactions of the pathway (steps 6–10), those that result in a net energy yield.

The enzyme name can tell us a lot about the reaction (see Section 20.1). The suffix *-kinase* tells us that the enzyme is a transferase that will transfer a phosphoryl group, in this case from an ATP molecule to the substrate. The prefix *hexo-* gives us a hint that the substrate is a six-carbon sugar. Hexokinase predominantly phosphorylates the six-carbon sugar glucose.

Reaction 1

The substrate, glucose, is phosphorylated by the enzyme *hexokinase* in a coupled phosphorylation reaction. The source of the phosphoryl group is ATP. At first this reaction seems contrary to the overall purpose of catabolism, the *production* of ATP. The expenditure of ATP in these early reactions must be thought of as an "investment." The cell actually goes into energy "debt" in these early reactions, but this is absolutely necessary to get the pathway started.

Glucose + ATP $\xrightarrow{\text{Hexokinase}}$ Glucose-6-phosphate ($CH_2OPO_3^{2-}$) + ADP + H^+

Glucose Glucose-6-phosphate

The enzyme name, phosphoglucose isomerase, provides clues to the reaction that is being catalyzed (Section 20.1). *Isomerase* tells us that the enzyme will catalyze the interconversion of one isomer into another. *Phosphoglucose* suggests that the substrate is a phosphorylated form of glucose.

Reaction 2

The glucose-6-phosphate formed in the first reaction is rearranged to produce the structural isomer fructose-6-phosphate. The enzyme *phosphoglucose isomerase* catalyzes this isomerization. The result is that the C-1 carbon of the six-carbon sugar is exposed; it is no longer part of the ring structure. Examination of the open-chain structures reveals that this isomerization converts an aldose into a ketose.

Glucose-6-phosphate ($CH_2OPO_3^{2-}$) $\xrightleftharpoons{\text{Phosphoglucose isomerase}}$ Fructose-6-phosphate ($^{2-}O_3POH_2C$, CH_2OH)

Glucose-6-phosphate Fructose-6-phosphate

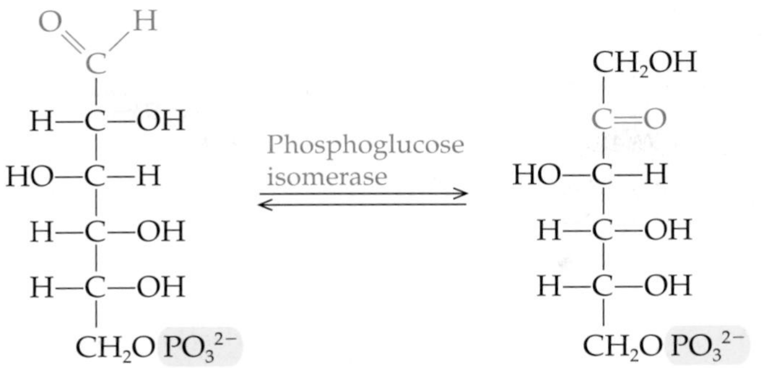

Glucose-6-phosphate (an aldose) ⇌ (Phosphoglucose isomerase) Fructose-6-phosphate (a ketose)

This is an enediol reaction. It occurs through exactly the same steps as the conversion of fructose to glucose that we discussed in Section 17.4.

Reaction 3

A second energy "investment" is catalyzed by the enzyme *phosphofructokinase*. The phosphoanhydride bond in ATP is hydrolyzed, and a phosphoester linkage between the phosphoryl group and the C-1 hydroxyl group of fructose-6-phosphate is formed. The product is fructose-1,6-bisphosphate.

The suffix *-kinase* in the name of the enzyme once again tells us that this is a coupled reaction in which ATP is hydrolyzed and a phosphoryl group is released and transferred to another molecule. The prefix *phosphofructo-* indicates that the substrate that will be phosphorylated is a phosphorylated form of fructose.

Fructose-6-phosphate + ATP → (Phosphofructokinase) Fructose-1,6-bisphosphate + ADP + H^+

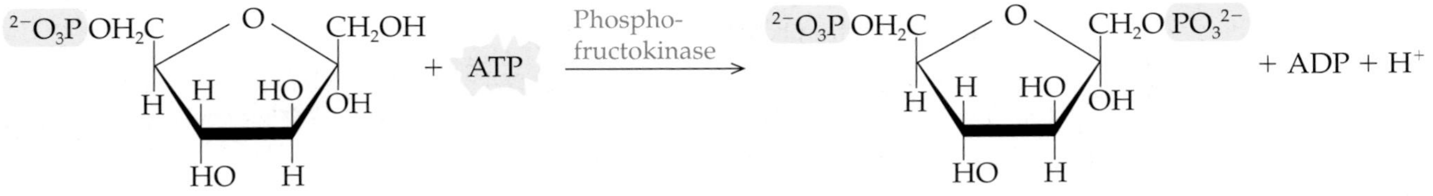

Reaction 4

Fructose-1,6-bisphosphate is split into two three-carbon intermediates in a reaction catalyzed by the enzyme *aldolase*. The products are glyceraldehyde-3-phosphate (G3P) and dihydroxyacetone phosphate.

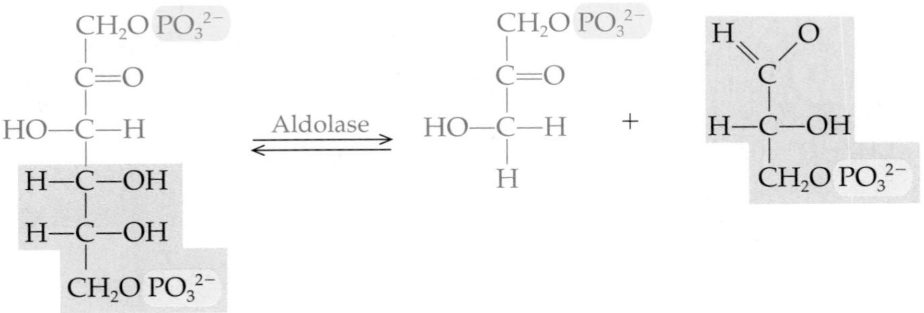

Fructose-1,6-bisphosphate ⇌ (Aldolase) Dihydroxyacetone phosphate + Glyceraldehyde-3-phosphate

In aldol condensation, aldehydes and ketones react to form a larger molecule (Section 14.4). This reaction is a reverse aldol condensation. The large ketone sugar fructose-1,6-bisphosphate is broken down into dihydroxyacetone phosphate (a ketone) and glyceraldehyde-3-phosphate (an aldehyde).

The double reaction arrows tell us that this is a reversible reaction. The reverse reaction is an aldol condensation (Section 14.4) that we will study in the pathway for glucose synthesis called gluconeogenesis (Section 21.6).

Reaction 5

Because G3P is the only substrate that can be used by the next enzyme in the pathway, the dihydroxyacetone phosphate is rearranged to become a second molecule of G3P. The enzyme that mediates this isomerization is *triose phosphate isomerase*.

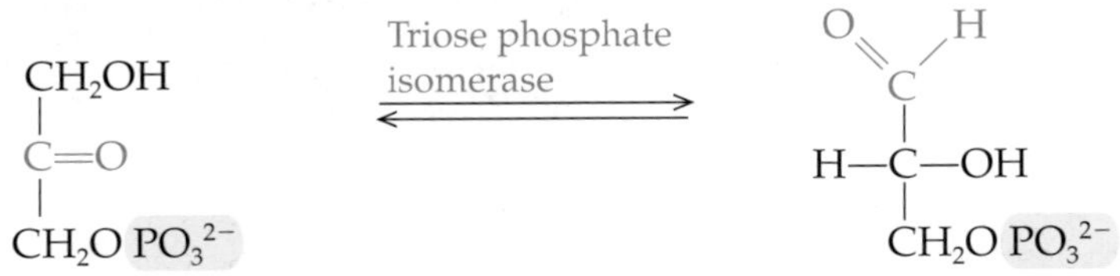

Dihydroxyacetone phosphate ⇌ (Triose phosphate isomerase) Glyceraldehyde-3-phosphate

The enzyme name hints that two isomers of a phosphorylated three-carbon sugar are going to be interconverted (Section 20.1). The ketone dihydroxyacetone phosphate and its isomeric aldehyde, phosphoglyceraldehyde-3-phosphate are interconverted through an enediol intermediate.

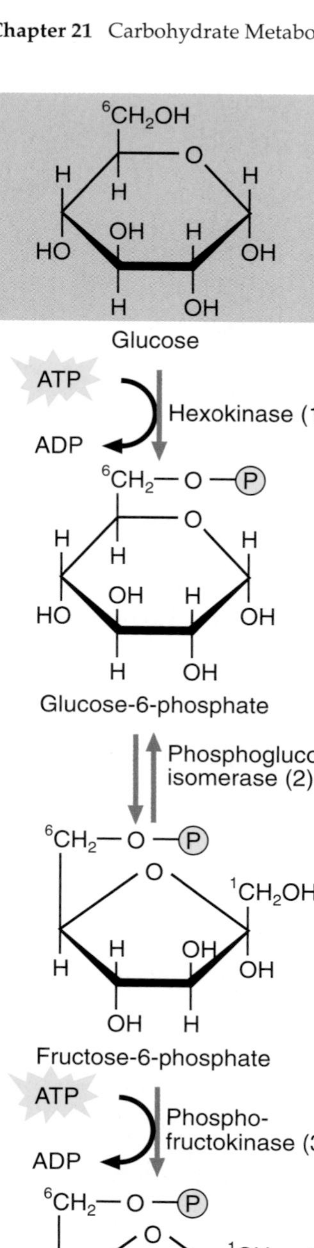

1. Glucose is phosphorylated at the expense of ATP to produce glucose-6-phosphate.

2. Glucose-6-phosphate is rearranged to produce fructose-6-phosphate.

3. Fructose-6-phosphate is phosphorylated to produce fructose-1,6-bisphosphate at the expense of another ATP. The expenditure of 2 ATP represents an energy investment to "activate" the glucose for its eventual oxidation.

4 and 5. Aldolase cleaves the six-carbon fructose-1,6-bisphosphate into two nonidentical three-carbon molecules, dihydroxyacetone phosphate and glyceraldehyde-3-phosphate. The dihydroxyacetone phosphate is converted to glyceraldehyde-3-phosphate by the enzyme triose phosphate isomerase.

Figure 21.8
The intermediates and enzymes of glycolysis.

2 Glyceraldehyde-3-phosphate

2Pi, 2 NAD^+ → 2 $NADH + H^+$

Glyceraldehyde-3-phosphate dehydrogenase (6)

6. Glyceraldehyde-3-phosphate is oxidized and NADH is produced. An inorganic phosphate group is transferred to the carboxylate group to produce 1,3-bisphosphoglycerate.

2 $^3CH_2—O—Ⓟ$ / ^2CHOH / $^1C=O$ / $O \sim Ⓟ$

1,3-Bisphosphoglycerate

2 ADP → 2 ATP

Phosphoglycerate kinase (7)

7. ATP is produced in the first substrate level phosphorylation in the pathway. The phosphoryl group is transferred from the substrate to ADP to produce ATP.

2 $^3CH_2—O—Ⓟ$ / ^2CHOH / $^1C=O$ / O^-

3-Phosphoglycerate

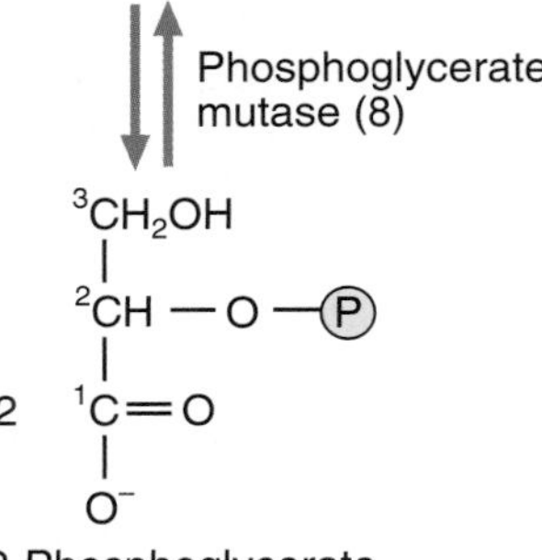

8. The C-3 phosphoryl group of 3-phosphoglycerate is transferred to the second carbon.

2-Phosphoglycerate

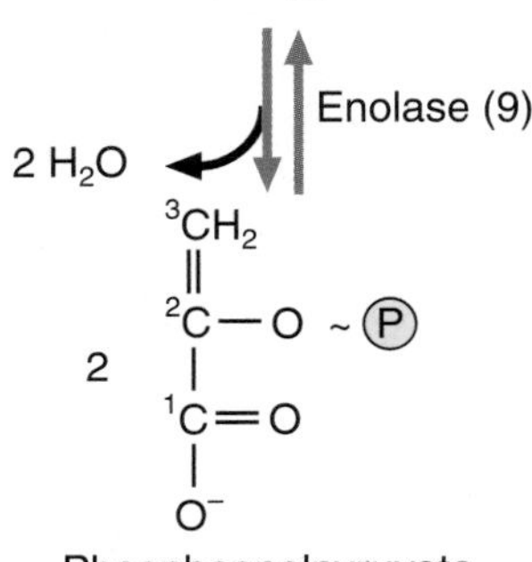

9. Dehydration of 2-phosphoglycerate generates the energy-rich molecule phosphoenolpyruvate.

Phosphoenolpyruvate

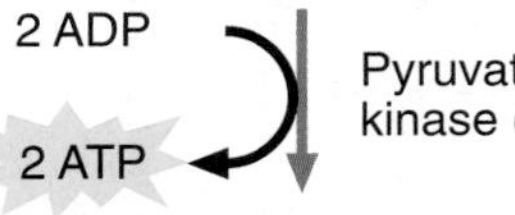

Pyruvate kinase (10)

10. The final substrate level phosphorylation produces ATP and pyruvate.

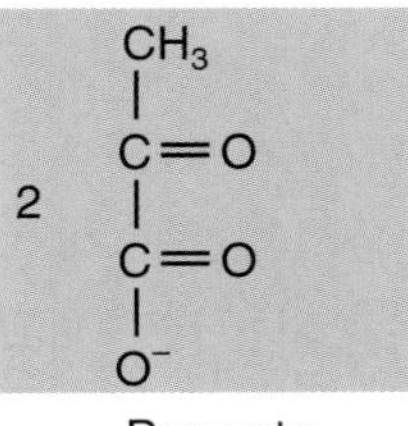

Pyruvate

The enzyme is glyceraldehyde-3-phosphate dehydrogenase. This tells us that the substrate glyceraldehyde-3-phosphate is going to be oxidized. Recall that in organic (and biochemical) reactions, oxidation is typically recognized as a gain of oxygen or a loss of hydrogen (Section 13.6). In this reaction, we see that the aldehyde group has been oxidized to a carboxylate group (Section 14.4).

Actually the intermediate of the oxidation reaction is a high-energy thioester formed between the enzyme and the substrate (Section 15.4). When this bond is hydrolyzed, enough energy is released to allow the formation of a bond between an oxygen atom of an inorganic phosphate group and the substrate.

Reaction 6

In this reaction the aldehyde glyceraldehyde-3-phosphate is oxidized to a carboxylic acid in a reaction catalyzed by *glyceraldehyde-3-phosphate dehydrogenase.* This is the first step in glycolysis that harvests energy, and it involves the reduction of the coenzyme nicotinamide adenine dinucleotide (NAD^+). This reaction occurs in two steps. First, NAD^+ is reduced to NADH as the aldehyde group of glyceraldehyde-3-phosphate is oxidized to a carboxyl group. Second, an inorganic phosphate group is transferred to the carboxyl group to give 1,3-bisphosphoglycerate. Notice that the new bond is denoted with a squiggle (~), indicating that this is a high-energy bond. This, and all remaining reactions of glycolysis, occur twice for each glucose because each glucose has been converted into two molecules of glyceraldehyde-3-phosphate.

$$\text{Glyceraldehyde-3-phosphate (O=C(H)–CH(OH)–}CH_2O\,PO_3^{2-}) + NAD^+ + P_i \xrightleftharpoons{\text{Glyceraldehyde-3-phosphate dehydrogenase}} \text{1,3-Bisphosphoglycerate (O=C(O}\sim PO_3^{2-})\text{–CH(OH)–}CH_2O\,PO_3^{2-}) + NADH + H^+$$

Glyceraldehyde-3-phosphate

1,3-Bisphosphoglycerate

Once again, the enzyme name reveals a great deal about the reaction. The suffix *-kinase* tells us that a phosphoryl group will be transferred. In this case, a phosphoester bond in the substrate 1,3-bisphosphoglycerate is hydrolyzed and ADP is phosphorylated. Note that this is a reversible reaction.

Reaction 7

In this reaction, energy is harvested in the form of *ATP.* The enzyme *phosphoglycerate kinase* catalyzes the transfer of the phosphoryl group of 1,3-bisphosphoglycerate to ADP. This is the first substrate-level phosphorylation of glycolysis, and it produces ATP and 3-phosphoglycerate. It is a coupled reaction in which the high-energy bond is hydrolyzed and the energy released is used to drive the synthesis of ATP.

$$\text{1,3-Bisphosphoglycerate (O=C(O}\sim PO_3^{2-})\text{–CH(OH)–}CH_2O\,PO_3^{2-}) + ADP + H^+ \xrightleftharpoons{\text{Phosphoglycerate kinase}} \text{3-Phosphoglycerate (O=C(O}^-)\text{–CH(OH)–}CH_2O\,PO_3^{2-}) + ATP$$

1,3-Bisphosphoglycerate

3-Phosphoglycerate

The suffix *-mutase* indicates another type of isomerase. Notice that the chemical formulas of the substrate and reactant are the same. The only difference is in the location of the phosphoryl group.

Reaction 8

3-Phosphoglycerate is isomerized to produce 2-phosphoglycerate in a reaction catalyzed by the enzyme *phosphoglycerate mutase.* The phosphoryl group attached to the third carbon of 3-phosphoglycerate is transferred to the second carbon.

$$\text{3-Phosphoglycerate (O=C(O}^-)\text{–CH(OH)–CH(H)–O}\,PO_3^{2-}) \xrightleftharpoons{\text{Phosphoglycerate mutase}} \text{2-Phosphoglycerate (O=C(O}^-)\text{–CH(O}\,PO_3^{2-})\text{–CH(H)–OH)}$$

3-Phosphoglycerate

2-Phosphoglycerate

Reaction 9

In this step the enzyme *enolase* catalyzes the dehydration (removal of a water molecule) of 2-phosphoglycerate. The energy-rich product is phosphoenolpyruvate, the highest energy phosphorylated compound in metabolism.

$$\text{H—}\overset{\overset{\text{O}\quad\text{O}^-}{\text{C}}}{\text{C}}\text{—O PO}_3^{2-} \;/\; \text{H—C—OH} \;/\; \text{H} \quad \xrightleftharpoons{\text{Enolase}} \quad \text{C(COO}^-\text{)—O}\sim\text{PO}_3^{2-} = \text{CH}_2 + \text{H}_2\text{O}$$

2-Phosphoglycerate → Phosphoenolpyruvate

In Section 14.4 we learned that aldehydes and ketones exist in an equilibrium mixture of two tautomers called the keto and enol forms. The dehydration of 2-phosphoglycerate produces the molecule phosphoenolpyruvate, which is in the enol form. In this case, the enol is extremely unstable. Because of this instability, the phosphoester bond in the product is a high-energy bond; in other words, a great deal of energy is released when this bond is broken.

Reaction 10

Here we see the final substrate-level phosphorylation in the pathway, which is catalyzed by *pyruvate kinase*. Phosphoenolpyruvate serves as a donor of the phosphoryl group that is transferred to ADP to produce ATP. This is another coupled reaction in which hydrolysis of the phosphoester bond in phosphoenolpyruvate provides energy for the formation of the phosphoanhydride bond of ATP. The final product of glycolysis is pyruvate.

The enzyme name indicates that a phosphoryl group will be transferred (*kinase*) and that the product will be pyruvate. Pyruvate is a keto tautomer and is much more stable than the enol (Section 14.4).

$$\text{C(COO}^-\text{)—O}\sim\text{PO}_3^{2-}(=\text{CH}_2) + \text{ADP} + \text{H}^+ \xrightarrow{\text{Pyruvate kinase}} \text{C(COO}^-\text{)=O (CH}_3\text{)} + \text{ATP}$$

Phosphoenolpyruvate → Pyruvate

It should be noted that reactions 6 through 10 occur twice per glucose molecule, because the starting six-carbon sugar is split into two three-carbon molecules. Thus in reaction 6, two NADH molecules are generated, and a total of four ATP molecules are made (steps 7 and 10). The net ATP gain from this pathway is, however, only two ATP molecules because there was an energy investment of two ATP molecules in the early steps of the pathway. This investment was paid back by the two ATP molecules produced by substrate-level phosphorylation in step 7. The actual energy yield is produced by substrate-level phosphorylation in reaction 10.

Question 21.5

What is substrate-level phosphorylation?

Question 21.6

What are the major products of glycolysis?

Question 21.7

Describe an overview of the reactions of glycolysis.

Question 21.8

How do the names of the first three enzymes of the glycolytic pathway relate to the reactions they catalyze?

Regulation of Glycolysis

Learning Goal

Energy-harvesting pathways, such as glycolysis, are responsive to the energy needs of the cell. Reactions of the pathway speed up when there is a demand for ATP. They slow down when there is abundant ATP to meet the energy requirements of the cell.

One of the major mechanisms for the control of the rate of the glycolytic pathway is the use of *allosteric enzymes*. In addition to the active site, which binds the substrate, allosteric enzymes have an effector binding site, which binds a chemical signal that alters the rate at which the enzyme catalyzes the reaction. Effector binding may increase (positive allosterism) or decrease the rate of reaction (negative allosterism).

The chemical signals, or effectors, that indicate the energy needs of the cell include molecules such as ATP. If the ATP concentration is high, the cell must have sufficient energy. Similarly, ADP and AMP, which are precursors of ATP, are indicators that the cell is in need of ATP. In fact, all of these molecules are allosteric effectors that alter the rate of irreversible reactions catalyzed by enzymes in the glycolytic pathway.

The enzyme hexokinase, which catalyzes the phosphorylation of glucose, is allosterically inhibited by the product of the reaction it catalyzes, glucose-6-phosphate. A buildup of this product indicates that the reactions of glycolysis are not proceeding at a rapid rate, presumably because the energy demands of the cell are being met.

Phosphofructokinase, the enzyme that catalyzes the third reaction in glycolysis, is a key regulatory enzyme in the pathway. ATP is an allosteric inhibitor of phosphofructokinase, whereas AMP and ADP are allosteric activators. Another allosteric inhibitor of phosphofructokinase is citrate. As we will see in the next chapter, citrate is the first intermediate in the citric acid cycle. The citric acid cycle is a pathway that results in the complete oxidation of the pyruvate produced by glycolysis. A high concentration of citrate signals that sufficient substrate is entering the citric acid cycle. The inhibition of phosphofructokinase by citrate is an example of *feedback inhibition:* the product, citrate, allosterically inhibits the activity of an enzyme early in the pathway.

There are additional mechanisms that regulate the rate of glycolysis, but we will focus on those that involve principles studied previously (Section 20.9).

The last enzyme in glycolysis, pyruvate kinase, is also subject to allosteric regulation. In this case, fructose-1,6-bisphosphate is the allosteric activator. It is interesting that fructose-1,6-bisphosphate is the product of the reaction catalyzed by phosphofructokinase. Thus, activation of phosphofructokinase results in the activation of pyruvate kinase. This is an example of *feedforward activation* because the product of an earlier reaction causes activation of an enzyme later in the pathway.

21.4 Fermentations

Learning Goal

Aerobic respiration is discussed in Chapter 22.

In the overview of glycolysis we noted that the pyruvate produced must be used up in some way so that the pathway will continue to produce ATP. Similarly, the NADH produced by glycolysis in step 6 (see Figure 21.8) must be reoxidized at a later time, or glycolysis will grind to a halt as the available NAD^+ is used up. If the cell is functioning under aerobic conditions, NADH will be reoxidized, and pyruvate will be completely oxidized by aerobic respiration. Under anaerobic conditions, however, different types of fermentation reactions accomplish these purposes. **Fermentations** are catabolic reactions that occur with no net oxidation. Pyruvate or an organic compound produced from pyruvate is reduced as NADH is oxidized. We will examine two types of fermentation pathways in detail: lactate fermentation and alcohol fermentation.

Lactate Fermentation

Lactate fermentation is familiar to anyone who has performed strenuous exercise. If you exercise so hard that your lungs and circulatory system can't deliver enough oxygen to the working muscles, your aerobic (oxygen-requiring) energy-harvesting pathways are not able to supply enough ATP to your muscles. But the muscles still demand energy. Under these anaerobic conditions, lactate fermentation begins. In this reaction the enzyme *lactate dehydrogenase* reduces pyruvate to lactate. NADH is the reducing agent for this process (Figure 21.9). As pyruvate is reduced, NADH is oxidized, and NAD^+ is again available, permitting glycolysis to continue.

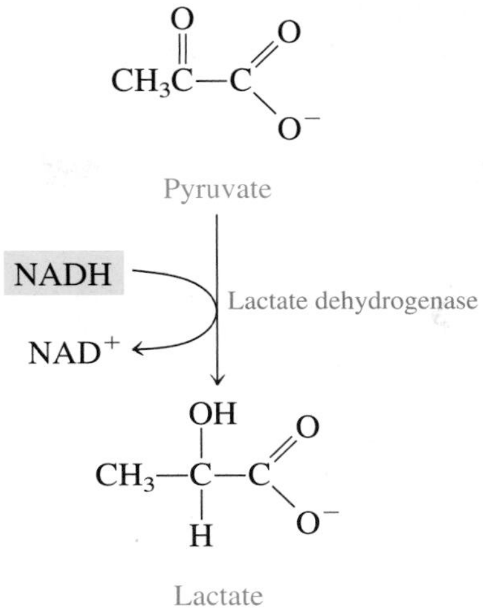

$$CH_3\overset{O}{\overset{\|}{C}}-C\begin{matrix}\nearrow O \\ \searrow O^-\end{matrix}$$

Pyruvate

NADH → NAD$^+$ (Lactate dehydrogenase)

$$CH_3-\underset{H}{\underset{|}{\overset{OH}{\overset{|}{C}}}}-C\begin{matrix}\nearrow O \\ \searrow O^-\end{matrix}$$

Lactate

Figure 21.9
The final reaction of lactate fermentation.

The lactate produced in the working muscle passes into the blood. Eventually, if strenuous exercise is continued, the concentration of lactate becomes so high that this fermentation can no longer continue. Glycolysis, and thus ATP production, stops. The muscle, deprived of energy, can no longer function. This point of exhaustion is called the **anaerobic threshold.**

Of course, most of us do not exercise to this point. When exercise is finished, the body begins the process of reclaiming all of the potential energy that was lost in the form of lactate. The liver takes up the lactate from the blood and converts it back to pyruvate. Now that a sufficient supply of oxygen is available, the pyruvate can be completely oxidized in the much more efficient aerobic energy-harvesting reactions to replenish the store of ATP. Alternatively, the pyruvate may be converted to glucose and used to restore the supply of liver and muscle glycogen. This exchange of metabolites between the muscles and liver is called the *Cori Cycle.*

The Cori Cycle is described in Section 21.6 and shown in Figure 21.13.

A variety of bacteria are able to carry out lactate fermentation under anaerobic conditions. This is of great importance in the dairy industry, because these organisms are used to produce yogurt and some cheeses. The tangy flavor of yogurt is contributed by the lactate produced by these bacteria. Unfortunately, similar organisms also cause milk to spoil.

As we saw in A Human Perspective: Tooth Decay and Simple Sugars (Chapter 17), the lactate produced by oral bacteria is responsible for the gradual removal of calcium from tooth enamel and the resulting dental cavities.

Alcohol Fermentation

Alcohol fermentation has been appreciated, if not understood, since the dawn of civilization. The fermentation process itself was discovered by Louis Pasteur during his studies of the chemistry of winemaking and "diseases of wines." Under anaerobic conditions, yeast are able to ferment the sugars produced by fruit and grains. The sugars are broken down to pyruvate by glycolysis. This is followed by the two reactions of alcohol fermentation. First, *pyruvate decarboxylase* removes CO_2 from the pyruvate producing acetaldehyde (Figure 21.10). Second, *alcohol dehydrogenase* catalyzes the reduction of acetaldehyde to ethanol but, more important, reoxidizes NADH in the process. The regeneration of NAD$^+$ allows glycolysis to continue, just as in the case of lactate fermentation.

These applications and other fermentations are described in A Human Perspective: Fermentations: The Good, the Bad, and the Ugly.

The two products of alcohol fermentation, then, are ethanol and CO_2. We take advantage of this fermentation in the production of wines and other alcoholic beverages and in the process of breadmaking.

Question 21.9

How is the alcohol fermentation in yeast similar to lactate production in skeletal muscle?

A HUMAN PERSPECTIVE 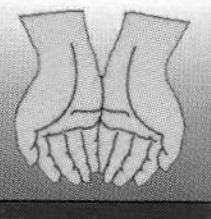

Fermentations: The Good, the Bad, and the Ugly

In this chapter we have seen that fermentation is an anaerobic, cytoplasmic process that allows continued ATP generation by glycolysis. ATP production can continue because the pyruvate produced by the pathway is utilized in the fermentation and because NAD^+ is regenerated.

The stable end products of alcohol fermentation are CO_2 and ethanol. These have been used by humankind in a variety of ways, including the production of alcoholic beverages, bread making, and alternative fuel sources.

The production of bread, wine, and cheese depends on fermentation processes.

If alcohol fermentation is carried out by using fruit juices in a vented vat, the CO_2 will escape, and the result will be a still wine (not bubbly). But conditions must remain anaerobic; otherwise, fermentation will stop, and aerobic energy-harvesting reactions will ruin the wine. Fortunately for vintners (wine makers), when a vat is fermenting actively, enough CO_2 is produced to create a layer that keeps the oxygen-containing air away from the fermenting juice, thus maintaining an anaerobic atmosphere.

Now suppose we want to make a sparkling wine, such as champagne. To do this, we simply have to trap the CO_2 produced. In this case the fermentation proceeds in a sealed bottle, a very strong bottle. Both the fermentation products, CO_2 and ethanol, accumulate. Under pressure within the sealed bottle the CO_2 remains in solution. When the top is "popped," the pressure is released, and the CO_2 comes out of solution in the form of bubbles.

In either case the fermentation continues until the alcohol concentration reaches 12–13%. At that point the yeast "stews in its own juices"! That is, 12–13% ethanol kills the yeast cells that produce it. This points out a last generalization about fermentations. The stable fermentation end product, whether it is lactate or ethanol, eventually accumulates to a concentration that is toxic to the organism. Muscle fatigue is the early effect of lactate buildup in the working muscle. In the same way, continued accumulation of the fermentation product can lead to concentrations that are fatal if there is no means of getting rid of the

Question 21.10

Why must pyruvate be used and NADH be reoxidized so that glycolysis can continue?

$$CH_3\overset{\overset{O}{\|}}{C}{-}CO_2^- \quad \text{Pyruvate}$$

$$\downarrow \text{Pyruvate decarboxylase}$$

$$CH_3CHO + CO_2$$

Acetaldehyde

$$\downarrow \text{NADH} \rightarrow \text{Alcohol dehydrogenase} \rightarrow NAD^+$$

$$CH_3CH_2OH \quad \text{Ethanol}$$

Figure 21.10
The final two reactions of alcohol fermentation.

toxic product or of getting away from it. For single-celled organisms the result is generally death. Our bodies have evolved in such a way that lactate buildup contributes to muscle fatigue that causes the exerciser to stop the exercise. Then the lactate is removed from the blood and converted to glucose by the process of gluconeogenesis.

Another application of alcohol fermentation is the use of yeast in bread making. When we mix the water, sugar, and dried yeast, the yeast cells begin to grow and carry out the process of fermentation. This mixture is then added to the flour, milk, shortening, and salt, and the dough is placed in a warm place to rise. The yeast continues to grow and ferment the sugar, producing CO_2 that causes the bread to rise. Of course, when we bake the bread, the yeast cells are killed, and the ethanol evaporates, but we are left with a light and airy loaf of bread.

Today, alcohol produced by fermentation is being considered as an alternative fuel to replace the use of some fossil fuels. Geneticists and bioengineers are trying to develop strains of yeast that can survive higher alcohol concentrations and thus convert more of the sugar of corn and other grains into alcohol.

Bacteria perform a variety of other fermentations. The propionibacteria produce propionic acid and CO_2. The acid gives Swiss cheese its characteristic flavor, and the CO_2 gas produces the characteristic holes in the cheese. Other bacteria, the clostridia, perform a fermentation that is responsible in part for the horrible symptoms of gas gangrene. When these bacteria are inadvertently introduced into deep tissues by a puncture wound, they find a nice anaerobic environment in which to grow. In fact, these organisms are *obligate anaerobes;* that is, they are killed by even a small amount of oxygen. As they grow, they perform a fermentation called the *butyric acid, butanol, acetone fermentation*. This results in the formation of CO_2, the gas associated with gas gangrene. The CO_2 infiltrates the local tissues and helps to maintain an anaerobic environment because oxygen from the local blood supply cannot enter the area of the wound. Now able to grow well, these bacteria produce a variety of toxins and enzymes that cause extensive tissue death and necrosis. In addition, the fermentation produces acetic acid, ethanol, acetone, isopropanol, butanol, and butyric acid (which is responsible, along with the necrosis, for the characteristic foul smell of gas gangrene). Certainly, the presence of these organic chemicals in the wound enhances tissue death.

Gas gangrene is very difficult to treat. Because the bacteria establish an anaerobic region of cell death and cut off the local circulation, systemic antibiotics do not infiltrate the wound and kill the bacteria. Even our immune response is stymied. Treatment usually involves surgical removal of the necrotic tissue accompanied by antibiotic therapy. In some cases a hyperbaric oxygen chamber is employed. The infected extremity is placed in an environment with a very high partial pressure of oxygen. The oxygen forced into the tissues is poisonous to the bacteria, and they die.

These are but a few examples of the fermentations that have an effect on humans. Regardless of the specific chemical reactions, all fermentations share the following traits:

- They use pyruvate produced in glycolysis.
- They reoxidize the NADH produced in glycolysis.
- They are self-limiting because the accumulated stable fermentation end product eventually kills the cell that produces it.

21.5 The Pentose Phosphate Pathway

Learning Goal

The **pentose phosphate pathway** is an alternative pathway for glucose oxidation. It provides the cell with energy in the form of reducing power for biosynthesis. Specifically, NADPH is produced in the oxidative stage of this pathway. NADPH is the reducing agent required for many biosynthetic pathways.

The details of the pentose phosphate pathway will not be covered in this text. But an overview of the key reactions will allow us to understand the importance of the pathway (Figure 21.11). We can consider the pathway in three stages. The first is the oxidative stage, which can be summarized as

$$\text{glucose-6-phosphate} + 2\text{NADP}^+ + H_2O \longrightarrow \text{ribulose-5-phosphate} + 2\text{NADPH} + CO_2$$

These reactions provide the NADPH required for biosynthesis.

The second stage involves isomerization reactions that convert ribulose-5-phosphate into ribose-5-phosphate or xylulose-5-phosphate. The pathway's name reflects the production of these phosphorylated five-carbon sugars (pentose phosphates).

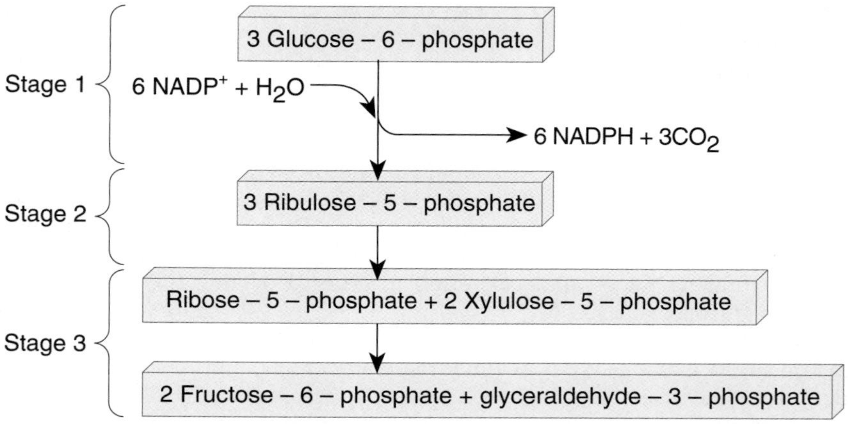

Stage 1: 3 Glucose – 6 – phosphate; $6\ NADP^+ + H_2O$ → $6\ NADPH + 3CO_2$

Stage 2: 3 Ribulose – 5 – phosphate

Stage 3: Ribose – 5 – phosphate + 2 Xylulose – 5 – phosphate → 2 Fructose – 6 – phosphate + glyceraldehyde – 3 – phosphate

Figure 21.11
Summary of the major stages of the pentose phosphate pathway.

The third stage is a complex series of reactions involving C—C bond breakage and formation. The result of these reactions is the formation of two molecules of fructose-6-phosphate and one molecule of glyceraldehyde-3-phosphate from three molecules of pentose phosphate.

In addition to providing reducing power (NADPH), the pentose phosphate pathway provides sugar phosphates that are required for biosynthesis. For instance, ribose-5-phosphate is used for the synthesis of nucleotides such as ATP. The four-carbon sugar phosphate, erythrose-4-phosphate, produced in the third stage of the pentose phosphate pathway is a precursor of the amino acids phenylalanine, tyrosine, and tryptophan.

The pathway for fatty acid biosynthesis is discussed in Section 23.4.

The pentose phosphate pathway is most active in tissues involved in cholesterol and fatty acid biosynthesis. These two processes require abundant NADPH. Thus the liver, which is the site of cholesterol synthesis and a major site for fatty acid biosynthesis, and adipose (fat) tissue, where active fatty acid synthesis also occurs, have very high levels of pentose phosphate pathway enzymes.

21.6 Gluconeogenesis: The Synthesis of Glucose

Learning Goal

Under normal conditions we have enough glucose to satisfy our needs. However, under some conditions the body must make glucose. This is necessary following strenuous exercise to replenish the liver and muscle supplies of glycogen. It also occurs during starvation so that the body can maintain adequate blood glucose levels to supply the brain cells and red blood cells. Under normal conditions these two tissues use only glucose for energy.

Under extreme conditions of starvation the brain eventually switches to the use of ketone bodies. Ketone bodies are produced, under certain circumstances, from the breakdown of lipids (Section 23.3).

Glucose is produced by the process of **gluconeogenesis,** the production of glucose from noncarbohydrate starting materials (Figure 21.12). Gluconeogenesis, an anabolic pathway, occurs primarily in the liver. Lactate, all the amino acids except leucine and lysine, and glycerol from fats can all be used to make glucose. However, the amino acids and glycerol are generally used only under starvation conditions.

At first glance, gluconeogenesis appears to be simply the reverse of glycolysis (compare Figures 21.12 and 21.7), because the intermediates of the two pathways are identical. But this is not the case, because steps 1, 3, and 10 of glycolysis are irreversible, and therefore the reverse reactions must be carried out by other enzymes. In step 1 of glycolysis, hexokinase catalyzes the phosphorylation of glucose. In gluconeogenesis the dephosphorylation of glucose-6-phosphate is carried out by the enzyme *glucose-6-phosphatase,* which is found in the liver but not in muscle. Similarly, reaction 3, the phosphorylation of fructose-6-phosphate catalyzed by phosphofructokinase, is irreversible. That step is bypassed in gluconeogenesis by using the enzyme *fructose-1,6-bisphosphatase.* Finally, the phosphorylation of ADP

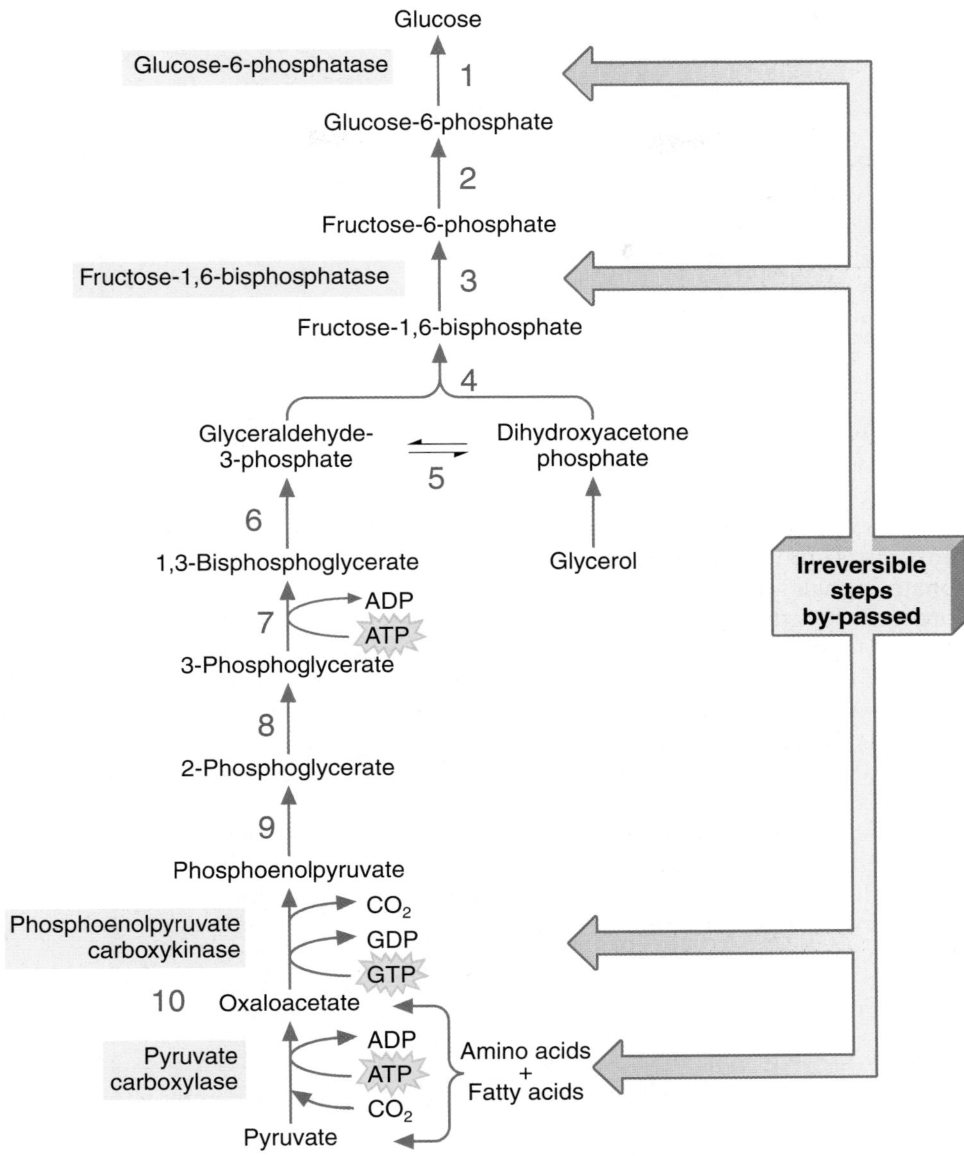

Figure 21.12
Comparison of the reactions of glycolysis and gluconeogenesis. All the reactions of glycolysis occur in the cytoplasm of the cell. However, in many human cells, pyruvate carboxylase is found in the mitochondria and phosphoenolpyruvate carboxykinase is located in the cytoplasm. Oxaloacetate, the product of the reaction catalyzed by pyruvate carboxylase, is shuttled out of the mitochondria and into the cytoplasm by a complex set of reactions.

catalyzed by pyruvate kinase, step 10 of glycolysis, cannot be reversed. The conversion of pyruvate to phosphoenolpyruvate actually involves two enzymes and some unusual reactions. First, the enzyme *pyruvate carboxylase* adds CO_2 to pyruvate. The product is the four-carbon compound oxaloacetate. Then *phosphoenolpyruvate carboxykinase* removes the CO_2 and adds a phosphoryl group. The donor of the phosphoryl group in this unusual reaction is **guanosine triphosphate (GTP).** This is a nucleotide like ATP, except that the nitrogenous base is guanine.

This last pair of reactions is complicated by the fact that pyruvate carboxylase is found in the mitochondria, whereas phosphoenolpyruvate carboxykinase is found in the cytoplasm. As we will see in Chapters 22 and 23, mitochondria are organelles in which the final oxidation of food molecules occurs and large amounts of ATP are produced. A complicated shuttle system transports the oxaloacetate produced in the mitochondria through the two mitochondrial membranes and into the cytoplasm. There, phosphoenolpyruvate carboxykinase catalyzes its conversion to phosphoenolpyruvate.

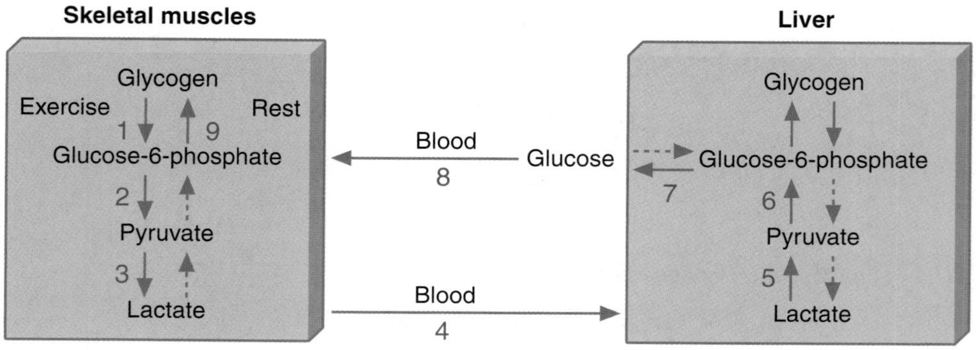

Skeletal muscles
Glycogen
Exercise
Rest
1
9
Glucose-6-phosphate
2
Pyruvate
3
Lactate
Blood
8
Glucose
Blood
4
Liver
Glycogen
Glucose-6-phosphate
7
6
Pyruvate
5
Lactate

Figure 21.13
The Cori Cycle.

If glycolysis and gluconeogenesis were not regulated in some fashion, the two pathways would occur simultaneously, with the disastrous effect that nothing would get done. Three convenient sites for this regulation are the three bypass reactions. Step 3 of glycolysis is catalyzed by the enzyme phosphofructokinase. This enzyme is stimulated by high concentrations of AMP, ADP, and inorganic phosphate, signals that the cell needs energy. When the enzyme is active, glycolysis proceeds. On the other hand, when ATP is plentiful, phosphofructokinase is inhibited, and fructose-1,6-bisphosphatase is stimulated. The net result is that in times of energy excess (high concentrations of ATP), gluconeogenesis will occur.

As we have seen, the conversion of lactate into glucose is important in mammals. As the muscles work, they produce lactate, which is converted back to glucose in the liver. The glucose is transported into the blood and from there back to the muscle. In the muscle it can be catabolized to produce ATP, or it can be used to replenish the muscle stores of glycogen. This cyclic process between the liver and skeletal muscles is called the **Cori Cycle** and is shown in Figure 21.13. Through this cycle, gluconeogenesis produces enough glucose to restore the depleted muscle glycogen reservoir within forty-eight hours.

Question 21.11

What are the major differences between gluconeogenesis and glycolysis?

Question 21.12

What do the three irreversible reactions of glycolysis have in common?

21.7 Glycogen Synthesis and Degradation

Learning Goal

Glucose is the sole source of energy of mammalian red blood cells and the major source of energy for the brain. Neither red blood cells nor the brain can store glucose; thus a constant supply must be available as blood glucose. This is provided by dietary glucose and by the production of glucose either by gluconeogenesis or by **glycogenolysis,** the degradation of glycogen. Glycogen is a long-branched-chain polymer of glucose. Stored in the liver and skeletal muscles, it is the principal storage form of glucose.

The total amount of glucose in the blood of a 70-kg (approximately 150-lb) adult is about 20 g, but the brain alone consumes 5–6 g of glucose per hour. Breakdown of glycogen in the liver mobilizes the glucose when hormonal signals register a need for increased levels of blood glucose. Skeletal muscle also contains substantial stores of glycogen, which provide energy for rapid muscle contraction. However, this glycogen is not able to contribute to blood glucose because muscle cells do not have the enzyme glucose-6-phosphatase. Because glucose cannot be formed from the glucose-6-phosphate, it cannot be released into the bloodstream.

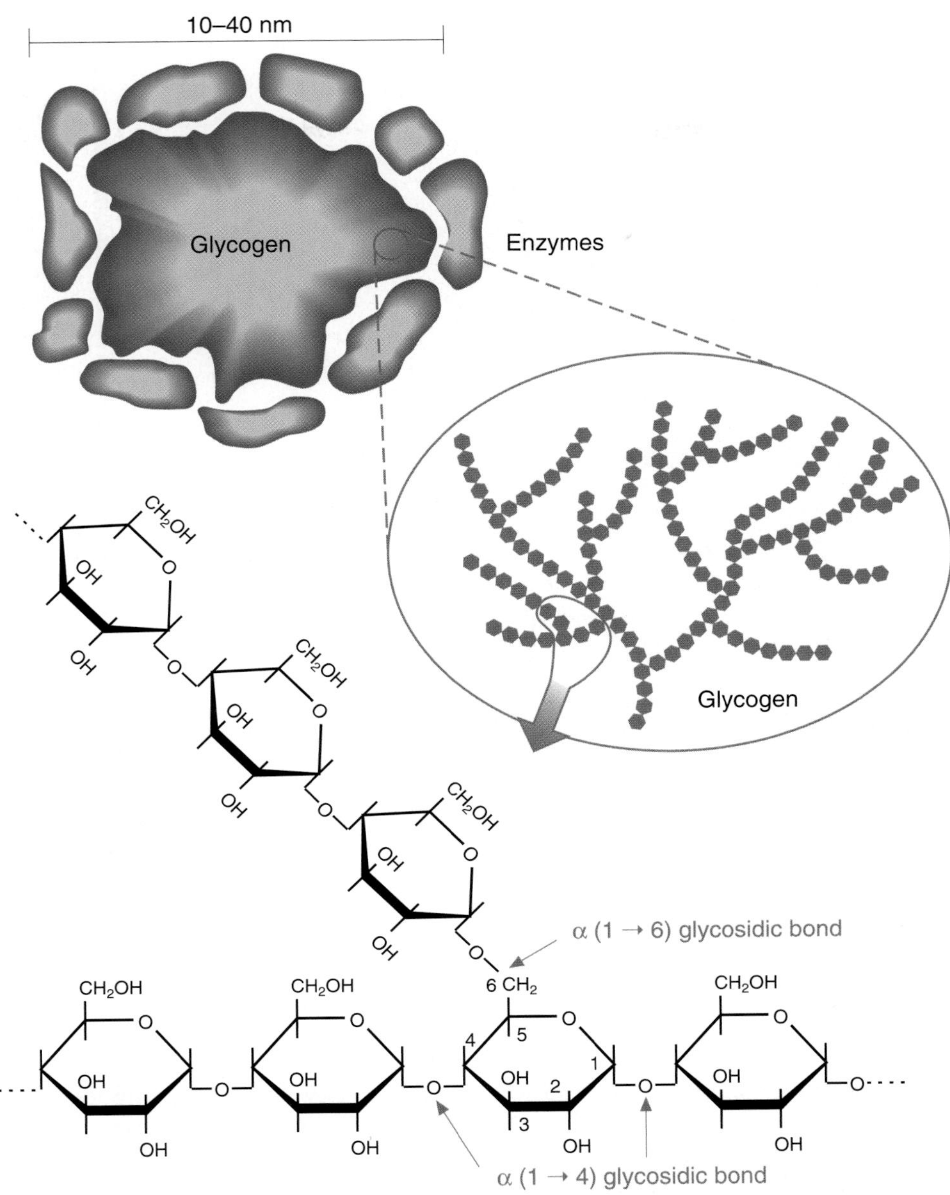

Figure 21.14
The structure of glycogen and a glycogen granule.

The Structure of Glycogen

Glycogen is a highly branched glucose polymer in which the "main chain" is linked by α (1 → 4) glycosidic bonds. The polymer also has numerous α (1 → 6) glycosidic bonds, which provide many branch points along the chain. This structure is shown schematically in Figure 21.14. **Glycogen granules** with a diameter of 10–40 nm are found in the cytoplasm of liver and muscle cells. These granules exist in complexes with the enzymes that are responsible for glycogen synthesis and degradation. The structure of such a granule is also shown in Figure 21.14.

Glycogenolysis: Glycogen Degradation

Two hormones control glycogenolysis, the degradation of glycogen. These are **glucagon,** a peptide hormone synthesized in the pancreas, and *epinephrine,* produced in the adrenal glands. Glucagon is released from the pancreas in response to low blood glucose, and epinephrine is released from the adrenal glands in response to a threat or a stress. Both situations require an increase in blood glucose, and both

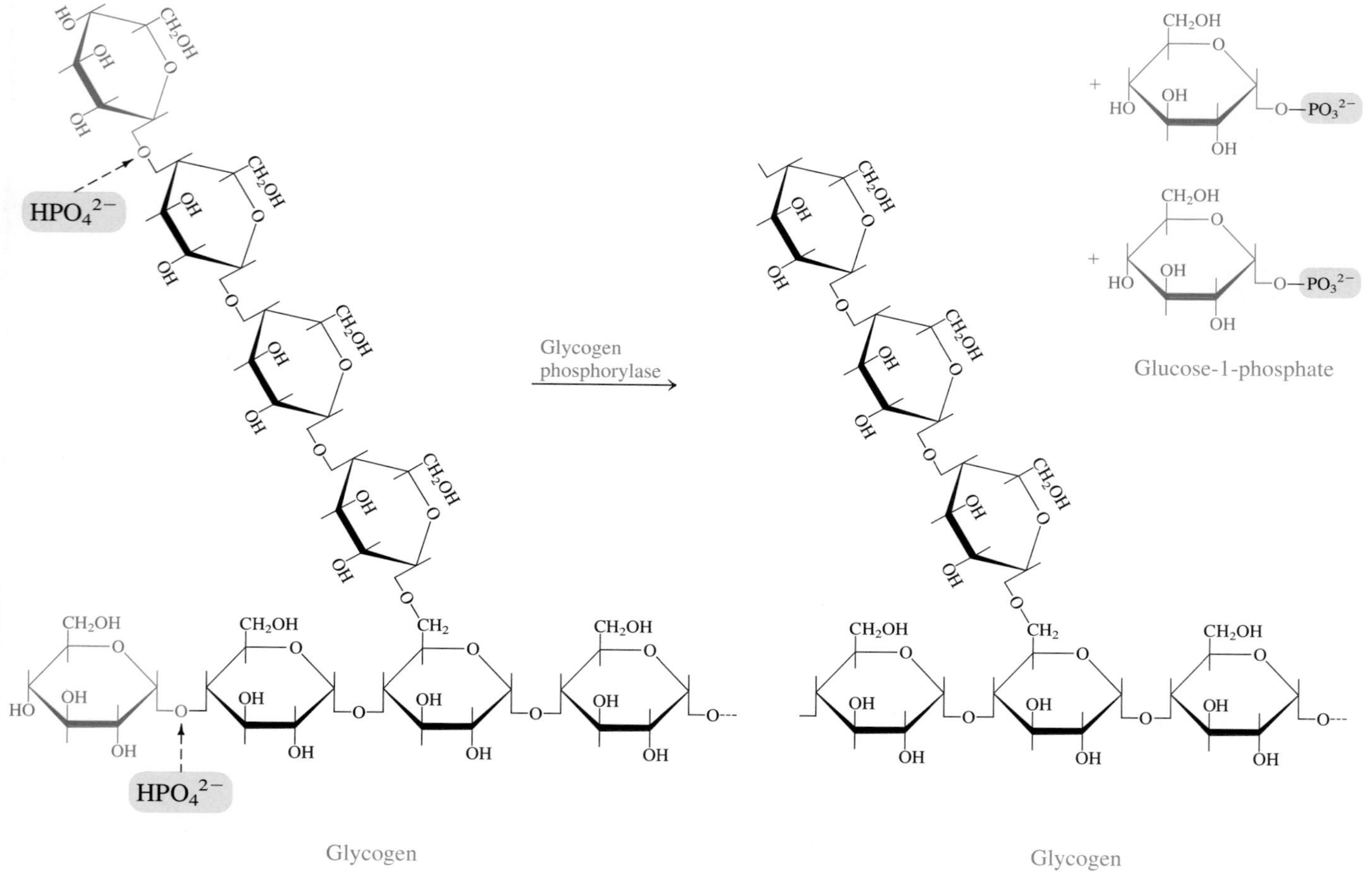

General reaction:

$$\text{Glycogen (glucose)}_x + n\ HPO_4^{2-} \xrightarrow[\text{phosphorylase}]{\text{Glycogen}} \text{(glucose)}_{x-n} + n\ \text{glucose-1-phosphate}$$

Figure 21.15
The action of glycogen phosphorylase in glycogenolysis.

hormones function by altering the activity of two enzymes, glycogen phosphorylase and glycogen synthase. *Glycogen phosphorylase* is involved in glycogen degradation and is activated; *glycogen synthase* is involved in glycogen synthesis and is inactivated. The steps in glycogen degradation are summarized as follows.

Step 1. The enzyme glycogen phosphorylase catalyzes *phosphorolysis* of a glucose at one end of a glycogen polymer (Figure 21.15). The reaction involves the displacement of a glucose unit of glycogen by a phosphate group. As a result of phosphorolysis, glucose-1-phosphate is produced without using ATP as the phosphoryl group donor.

Step 2. Glycogen contains many branches bound to the α (1 → 4) backbone by α (1 → 6) glycosidic bonds. These branches must be removed to allow the complete degradation of glycogen. The extensive action of glycogen phosphorylase produces a smaller polysaccharide with a single glucose bound by an α (1 → 6) glycosidic bond to the main chain. The enzyme *α (1 → 6) glycosidase,* also called the *debranching enzyme,* hydrolyzes the α (1 → 6) glycosidic bond at a branch point and frees one molecule of glucose (Figure 21.16). This molecule of glucose can be phosphorylated and utilized in glycolysis, or it may be released into the bloodstream for use elsewhere. Hydrolysis of the branch bond liberates another stretch of α(1 → 4)-linked glucose for the action of glycogen phosphorylase.

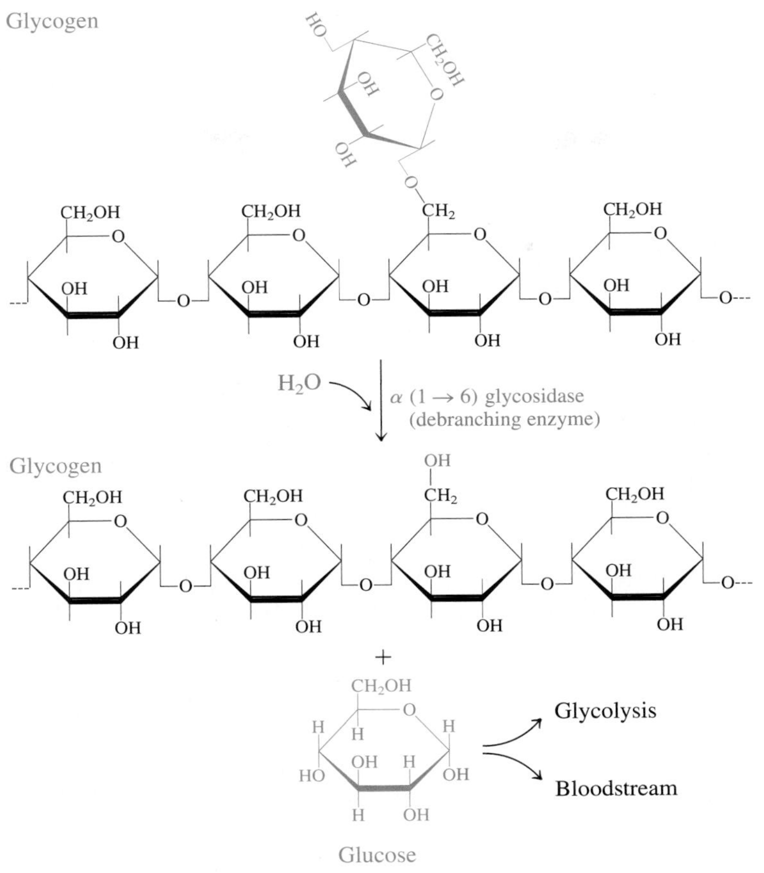

Figure 21.16
The action of α (1 → 6) glycosidase (debranching enzyme) in glycogen degradation.

Step 3. Glucose-1-phosphate is converted to glucose-6-phosphate by *phosphoglucomutase* (Figure 21.17). Glucose originally stored in glycogen enters glycolysis through the action of phosphoglucomutase. Alternatively, in the liver and kidneys it may be dephosphorylated for transport into the bloodstream.

Question 21.13

Explain the role of glycogen phosphorylase in glycogenolysis.

Question 21.14

How does the action of glycogen phosphorylase and phosphoglucomutase result in an energy savings for the cell if the product, glucose-6-phosphate, is used directly in glycolysis?

Glycogenesis: Glycogen Synthesis

The hormone **insulin,** produced by the pancreas in response to high blood glucose levels, stimulates the synthesis of glycogen, **glycogenesis.** Insulin is perhaps one of the most influential hormones in the body because it directly alters the metabolism and uptake of glucose in all but a few cells of the body.

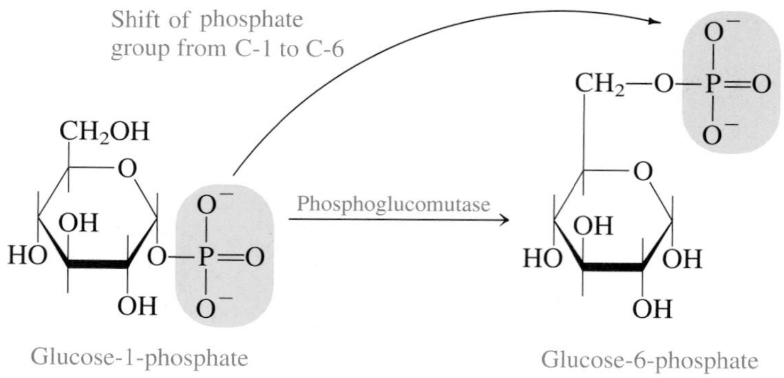

Figure 21.17
The action of phosphoglucomutase in glycogen degradation.

When blood glucose rises, as after a meal, the beta cells of the pancreas secrete insulin. It immediately accelerates the uptake of glucose by all the cells of the body except the brain and certain blood cells. In these cells the uptake of glucose is insulin-independent. The increased uptake of glucose is especially marked in the liver, heart, skeletal muscle, and adipose tissue.

In the liver, insulin promotes glycogen synthesis and storage by inhibiting glycogen phosphorylase, thus inhibiting glycogen degradation. It also stimulates glycogen synthase and glucokinase, two enzymes that are involved in glycogen synthesis.

Although glycogenesis and glycogenolysis share some reactions in common, the two pathways are not simply the reverse of one another. Glycogenesis involves some very unusual reactions, which we will now examine in detail.

The first reaction of glycogen synthesis in the liver traps glucose within the cell by phosphorylating it. In this reaction, catalyzed by the enzyme *glucokinase,* ATP serves as a phosphoryl donor, and glucose-6-phosphate is formed:

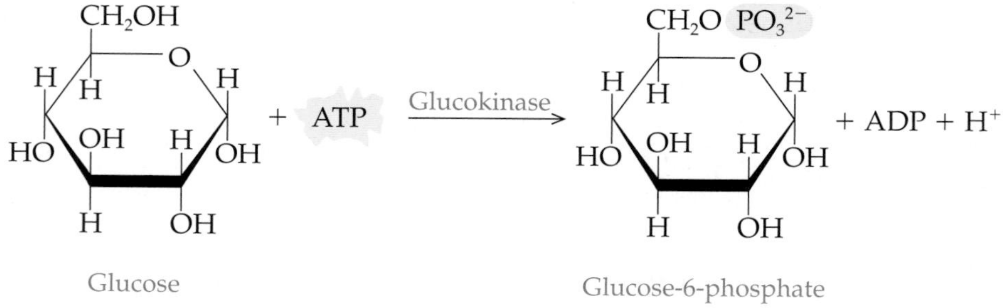

Glucose + ATP $\xrightarrow{\text{Glucokinase}}$ Glucose-6-phosphate + ADP + H^+

The second reaction of glycogenesis is the reverse of one of the reactions of glycogenolysis. The glucose-6-phosphate formed in the first step is isomerized to glucose-1-phosphate. The enzyme that catalyzes this step is phosphoglucomutase:

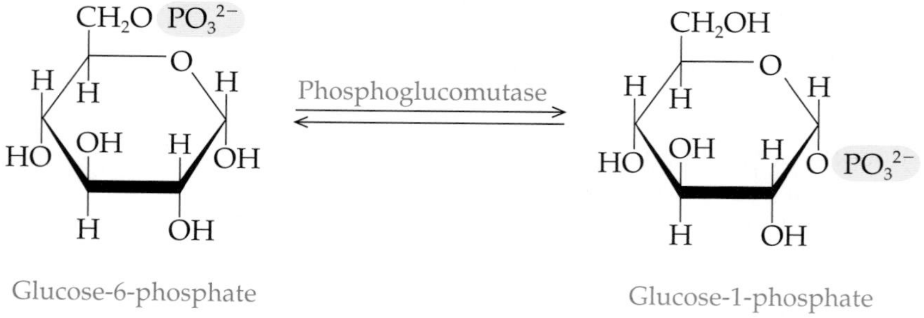

Glucose-6-phosphate $\xrightleftharpoons{\text{Phosphoglucomutase}}$ Glucose-1-phosphate

The glucose-1-phosphate must now be activated before it can be added to the growing glycogen chain. The high-energy compound that accomplishes this is the nucleotide **uridine triphosphate (UTP).** In this reaction, mediated by the enzyme *pyrophosphorylase,* the C-1 phosphoryl group of glucose is linked to the α phosphoryl group of UTP to produce UDP-glucose:

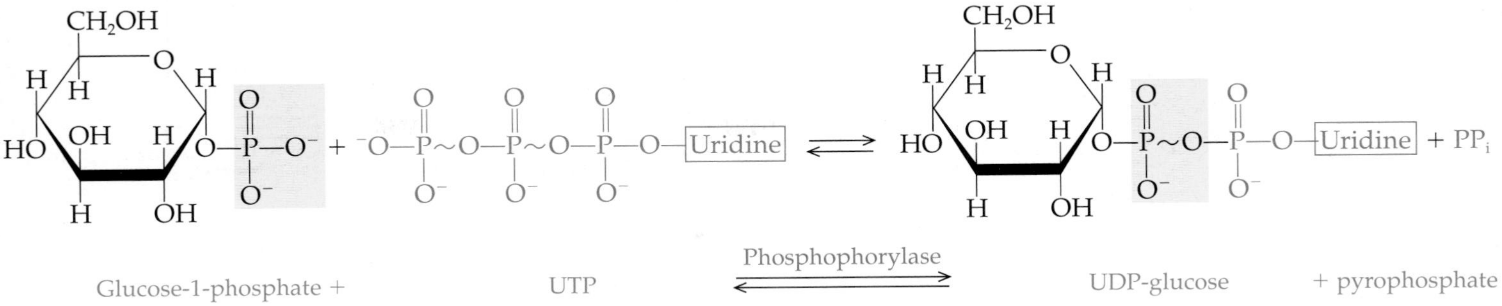

Glucose-1-phosphate + UTP $\rightleftharpoons$ (Phosphophorylase) UDP-glucose + pyrophosphate

This is accompanied by the release of a pyrophosphate group (PP_i). The structure of UDP-glucose is seen in Figure 21.18.

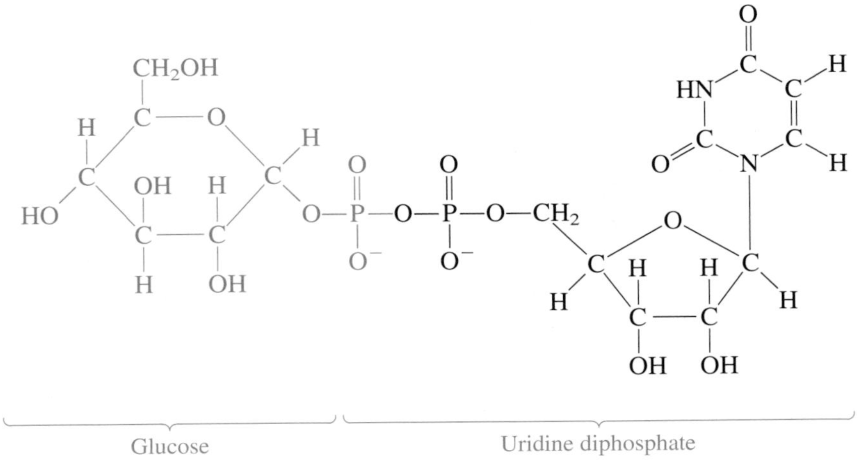

Glucose | Uridine diphosphate

Figure 21.18
The structure of UDP-glucose.

The UDP-glucose can now be used to extend glycogen chains. The enzyme glycogen synthase breaks the phosphoester linkage of UDP-glucose and forms an α $(1 \rightarrow 4)$ glycosidic bond between the glucose and the growing glycogen chain. UDP is released in the process.

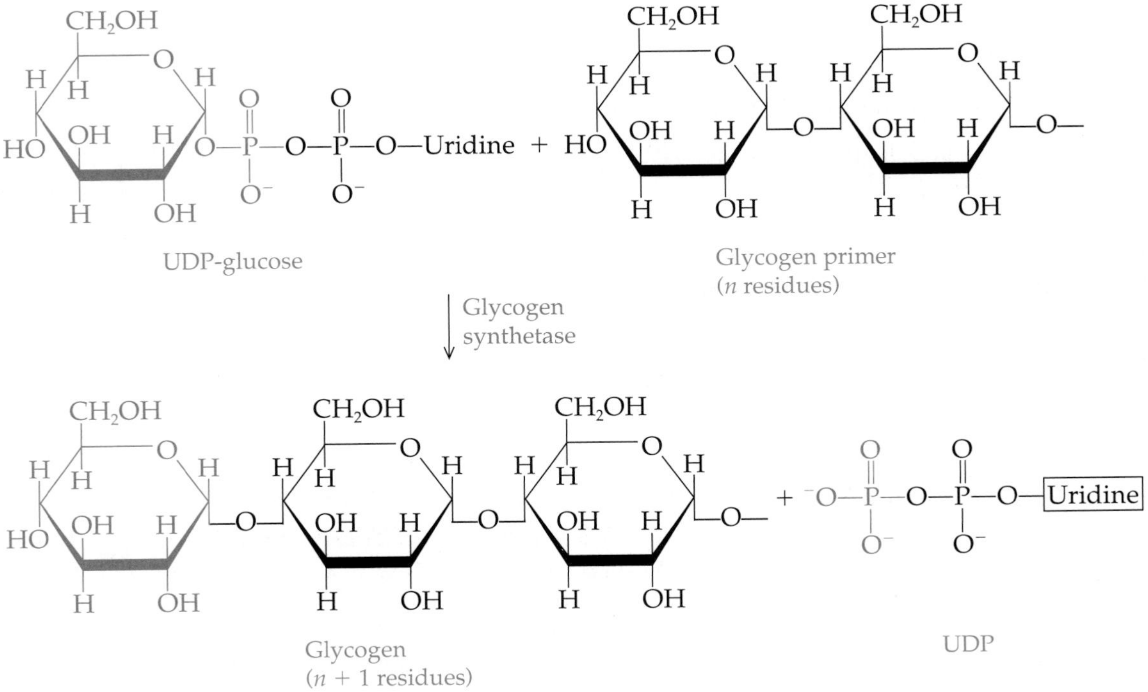

UDP-glucose + Glycogen primer (n residues)

Glycogen synthetase

Glycogen ($n + 1$ residues) + UDP

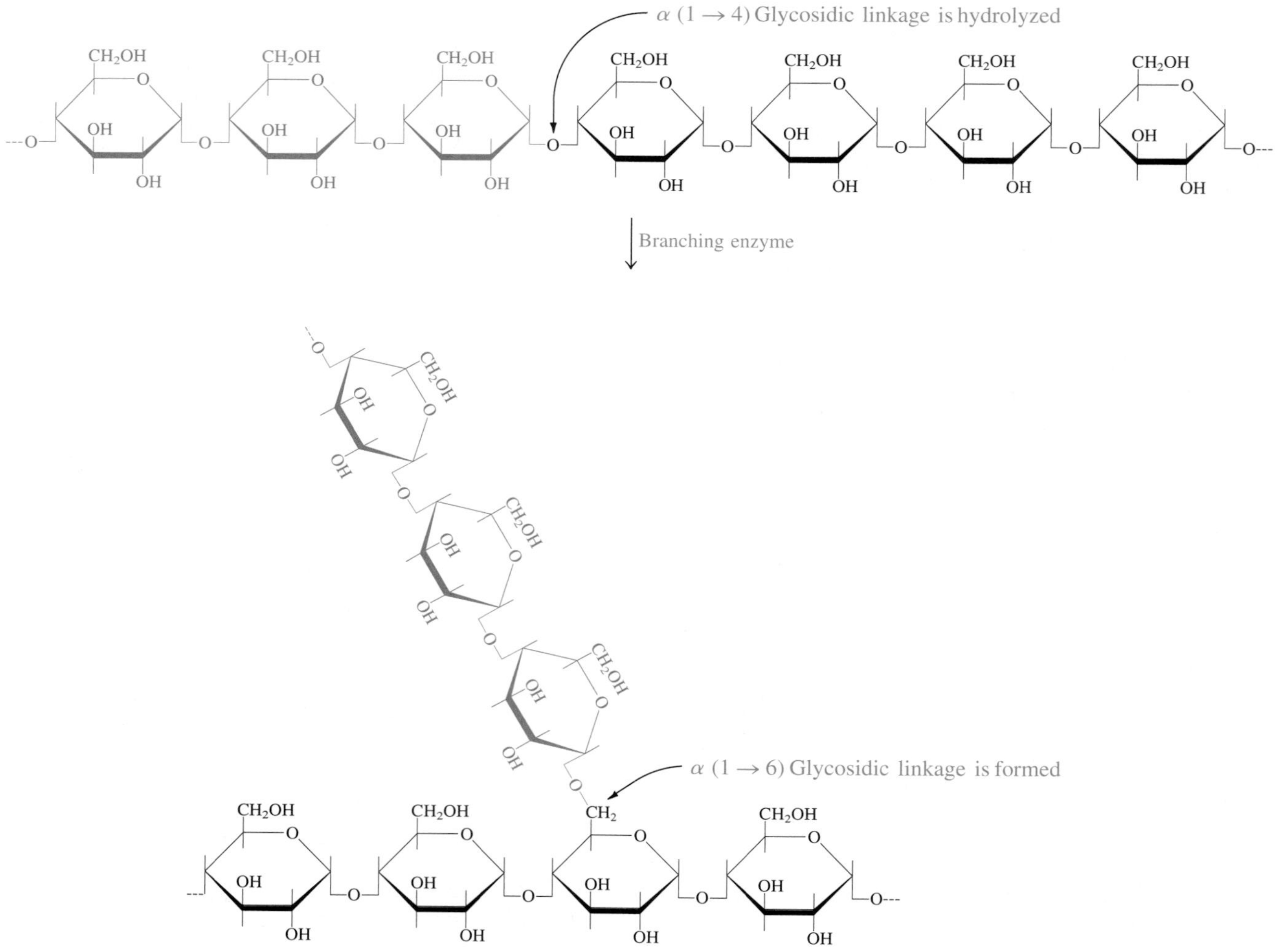

Figure 21.19
The action of the branching enzyme in glycogen synthesis.

Finally, we must introduce the α (1 → 6) glycosidic linkages to form the branches. The branches are quite important to proper glycogen utilization. As Figure 21.19 shows, the *branching enzyme* removes a section of the linear α (1 → 4) linked glycogen and reattaches it in α (1 → 6) glycosidic linkage elsewhere in the chain.

Question 21.15 Describe the way in which glucokinase traps glucose inside liver cells.

Question 21.16 Describe the reaction catalyzed by the branching enzyme.

Compatibility of Glycogenesis and Glycogenolysis

As was the case with glycolysis and gluconeogenesis, it would be futile for the cell to carry out glycogen synthesis and degradation simultaneously. The results achieved by the action of one pathway would be undone by the other. This problem is avoided by a series of hormonal controls that activate the enzymes of one pathway while inactivating the enzymes of the other pathway.

A CLINICAL PERSPECTIVE

Diagnosing Diabetes

When diagnosing diabetes, doctors take many factors and symptoms into consideration. However, there are two primary tests that are performed to determine whether an individual is properly regulating blood glucose levels. First and foremost is the fasting blood glucose test. A person who has fasted since midnight should have a blood glucose level between 70 and 110 mg/dL in the morning. If the level is 140 mg/dL on at least two occasions, a diagnosis of diabetes is generally made.

The second commonly used test is the glucose tolerance test. For this test the subject must fast for at least ten hours. A beginning blood sample is drawn to determine the fasting blood glucose level. This will serve as the background level for the test. The subject ingests 50–100 g of glucose (40 g/m^2 body surface), and the blood glucose level is measured at thirty minutes, and at one, two, and three hours after ingesting the glucose.

A graph is made of the blood glucose levels over time. For a person who does not have diabetes, the curve will show a peak of blood glucose at approximately one hour. There will be a reduction in the level, and perhaps a slight hypoglycemia (low blood glucose level) over the next hour. Thereafter, the blood glucose level stabilizes at normal levels.

An individual is said to have impaired glucose tolerance if the blood glucose level remains between 140 and 200 mg/dL two hours after ingestion of the glucose solution. This suggests that there is a risk of the individual developing diabetes and is reason to prescribe periodic testing to allow early intervention.

If the blood glucose level remains at or above 200 mg/dL after two hours, a tentative diagnosis of diabetes is made. However, this result warrants further testing on subsequent days to rule out transient problems, such as the effect of medications on blood glucose levels.

It was recently suggested that the upper blood glucose level of 200 mg/dL should be lowered to 180 mg/dL as the standard to diagnose impaired glucose tolerance and diabetes. This would allow earlier detection and intervention. Considering the grave nature of long-term diabetic complications, it is thought to be very beneficial to begin treatment at an early stage to maintain constant blood glucose levels. For more information on diabetes, see A Clinical Perspective: Diabetes Mellitus and Ketone Bodies, in Chapter 23.

When the blood glucose level is too high, a condition known as **hyperglycemia,** insulin stimulates the uptake of glucose via a transport mechanism. It further stimulates the trapping of the glucose by the elevated activity of glucokinase. Finally, it activates glycogen synthase, the last enzyme in the synthesis of glycogen chains. To further accelerate storage, insulin *inhibits* the first enzyme in glycogen degradation, glycogen phosphorylase. The net effect, seen in Figure 21.20, is that glucose is removed from the bloodstream and converted into glycogen in the liver. When the glycogen stores are filled, excess glucose is converted to fat and stored in adipose tissue.

Glucagon is produced in response to low blood glucose levels, a condition known as **hypoglycemia,** and has an effect opposite to that of insulin. It stimulates glycogen phosphorylase, which catalyzes the first stage of glycogen degradation. This accelerates glycogenolysis and release of glucose into the bloodstream. The effect is further enhanced because glucagon inhibits glycogen synthetase. The opposing effects of insulin and glucagon are summarized in Figure 21.20.

This elegant system of hormonal control ensures that the reactions involved in glycogen degradation and synthesis do not compete with one another. In this way they provide glucose when the blood level is too low, and they cause the storage of glucose in times of excess.

Question 21.17

Explain how glucagon affects the synthesis and degradation of glycogen.

Question 21.18

How does insulin affect the storage and degradation of glycogen?

A HUMAN PERSPECTIVE

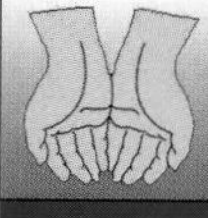

Glycogen Storage Diseases

Glycogen metabolism is important for the proper function of many aspects of cellular metabolism. Many diseases of glycogen metabolism have been discovered. Generally, these are diseases that result in the excessive accumulation of glycogen in the liver, muscle, and tubules of the kidneys. Often they are caused by defects in one of the enzymes involved in the degradation of glycogen.

One example is an inherited defect of glycogen metabolism known as *von Gierke's disease*. This disease results from a defective gene for glucose-6-phosphatase, which catalyzes the final step of gluconeogenesis and glycogenolysis. People who lack glucose-6-phosphatase cannot convert glucose-6-phosphate to glucose. As we have seen, the liver is the primary source of blood glucose, and much of this glucose is produced by gluconeogenesis. Glucose-6-phosphate, unlike glucose, cannot cross the cell membrane, and the liver of a person suffering from von Gierke's disease cannot provide him or her with glucose. The blood sugar level falls precipitously low between meals. In addition, the lack of glucose-6-phosphatase also affects glycogen metabolism. Because glucose-6-phosphatase is absent, the supply of glucose-6-phosphate in the liver is large. This glucose-6-phosphate can also be converted to glycogen. A person suffering from von Gierke's disease has a massively enlarged liver as a result of enormously increased stores of glycogen.

Defects in other enzymes of glycogen metabolism also exist. *Cori's disease* is caused by a genetic defect in the debranching enzyme. As a result, individuals who have this disease cannot completely degrade glycogen and thus use their glycogen stores very inefficiently.

On the other side of the coin, *Andersen's disease* results from a genetic defect in the branching enzyme. Individuals who have this disease produce very long, unbranched glycogen chains. This genetic disorder results in decreased efficiency of glycogen storage.

A final example of a glycogen storage disease is *McArdle's disease*. In this syndrome the muscle cells lack the enzyme glycogen phosphorylase and cannot degrade glycogen to glucose. Individuals who have this disease have little tolerance for physical exercise because their muscles cannot provide enough glucose for the necessary energy-harvesting processes. It is interesting to note that the liver enzyme glycogen phosphorylase is perfectly normal, and these people respond appropriately with a rise of blood glucose levels under the influence of glucagon or epinephrine.

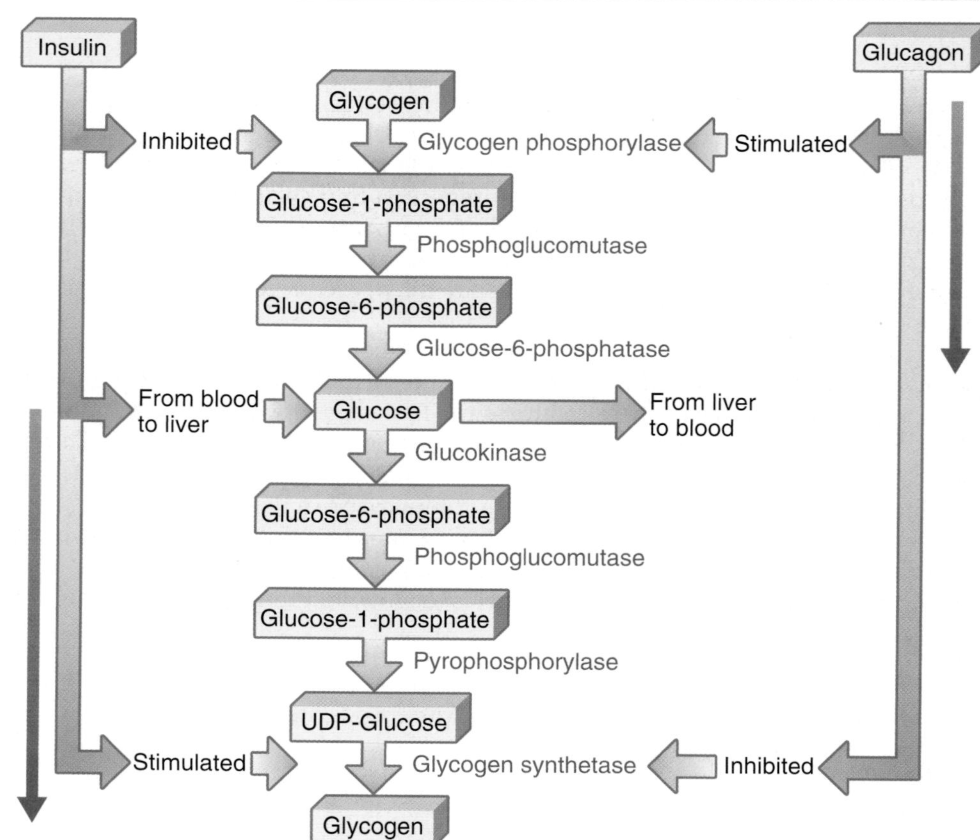

Figure 21.20
The opposing effects of the hormones insulin and glucagon on glycogen metabolism.

Summary

21.1 ATP: The Cellular Energy Currency

Adenosine triphosphate, ATP, is a *nucleotide* composed of adenine, the sugar ribose, and a triphosphate group. The energy released by the hydrolysis of the phosphoanhydride bond between the second and third phosphoryl groups provides the energy for most cellular work.

21.2 Overview of Catabolic Processes

The body needs a supply of ATP to carry out life processes. To provide this ATP, we consume a variety of energy-rich food molecules: carbohydrates, lipids, and proteins. In the digestive tract these large molecules are degraded into smaller molecules (monosaccharides, glycerol, fatty acids, and amino acids) that are absorbed by our cells. These molecules are further broken down to generate ATP.

21.3 Glycolysis

Glycolysis is the pathway for the *catabolism* of glucose that leads to pyruvate. It is an anaerobic process carried out by enzymes in the cytoplasm of the cell. The net harvest of ATP during glycolysis is two molecules of ATP per molecule of glucose. Two molecules of NADH are also produced. The rate of glycolysis responds to the energy demands of the cell. The regulation of glycolysis occurs through the allosteric enzymes hexokinase, phosphofructokinase, and pyruvate kinase.

21.4 Fermentations

Under anaerobic conditions the NADH produced by glycolysis is used to reduce pyruvate to lactate in skeletal muscle (lactate *fermentation*) or to convert acetaldehyde to ethanol in yeast (alcohol fermentation).

21.5 The Pentose Phosphate Pathway

The *pentose phosphate pathway* is an alternative pathway for glucose degradation that is particularly abundant in the liver and adipose tissue. It provides the cell with a source of NADPH to serve as a reducing agent for biosynthetic reactions. It also provides ribose-5-phosphate for nucleotide synthesis and erythrose-4-phosphate for biosynthesis of the amino acids tryptophan, tyrosine, and phenylalanine.

21.6 Gluconeogenesis: The Synthesis of Glucose

Gluconeogenesis is the pathway for glucose synthesis from noncarbohydrate starting materials. It occurs in mammalian liver. Glucose can be made from lactate, all the amino acids except lysine and leucine, and glycerol. Gluconeogenesis is not simply the reversal of glycolysis. Three steps in glycolysis in which ATP is produced or consumed are bypassed by different enzymes in gluconeogenesis. All other enzymes in gluconeogenesis are shared with glycolysis.

21.7 Glycogen Synthesis and Degradation

Glycogenesis is the pathway for the synthesis of glycogen, and *glycogenolysis* is the pathway for the degradation of glycogen. The concentration of blood glucose is controlled by the liver. A high blood glucose level causes secretion of *insulin*. This hormone stimulates glycogenesis and inhibits glycogenolysis. When blood glucose levels are too low, the hormone *glucagon* stimulates gluconeogenesis and glycogen degradation in the liver.

Key Terms

adenosine triphosphate (ATP) (21.1)
anabolism (21.1)
anaerobic threshold (21.4)
catabolism (21.1)
Cori Cycle (21.6)
fermentation (21.4)
glucagon (21.7)
gluconeogenesis (21.6)
glycogen (21.7)
glycogenesis (21.7)
glycogen granule (21.7)
glycogenolysis (21.7)
glycolysis (21.3)
guanosine triphosphate (GTP) (21.6)
hyperglycemia (21.7)
hypoglycemia (21.7)
insulin (21.7)
nicotinamide adenine dinucleotide (NAD^+) (21.3)
nucleotide (21.1)
oxidative phosphorylation (21.3)
pentose phosphate pathway (21.5)
substrate-level phosphorylation (21.3)
uridine triphosphate (UTP) (21.7)

Questions and Problems

ATP: The Cellular Energy Currency

21.19 What molecule is primarily responsible for conserving the energy released in catabolism?
21.20 Describe the structure of ATP.
21.21 Write a reaction showing the hydrolysis of the terminal phosphoanhydride bond of ATP.
21.22 What is meant by the term *high-energy bond?*

Glycolysis

21.23 Why does glycolysis require a supply of NAD^+ to function?
21.24 Why must the NADH produced in glycolysis be reoxidized to NAD^+?
21.25 What is the net energy yield of ATP in glycolysis?
21.26 How many molecules of ATP are produced by substrate-level phosphorylation during glycolysis?
21.27 Explain how muscle is able to carry out rapid contraction for prolonged periods even though its supply of ATP is sufficient only for a fraction of a second of rapid contraction.
21.28 Where in the muscle cell does glycolysis occur?
21.29 Write the balanced chemical equation for glycolysis.

21.30 Write a chemical equation for the transfer of a phosphoryl group from ATP to fructose-6-phosphate.

21.31 Which glycolysis reactions are catalyzed by each of the following enzymes?
- **a.** Hexokinase
- **b.** Pyruvate kinase
- **c.** Phosphoglycerate mutase
- **d.** Glyceraldehyde-3-phosphate dehydrogenase

21.32 Fill in the blanks:
- **a.** ________ molecules of ATP are produced per molecule of glucose that is converted to pyruvate.
- **b.** Two molecules of ATP are consumed in the conversion of ________ to fructose-1,6-bisphosphate.
- **c.** NAD^+ is ________ to NADH in the first energy-releasing step of glycolysis.
- **d.** The second substrate-level phosphorylation in glycolysis is phosphoryl group transfer from phosphoenolpyruvate to ________.

21.33 Examine the following pair of reactions and use them to answer Questions 21.33–21.36. What type of enzyme would catalyze each of these reactions?

```
(a)        O                     (b)      CH2OH
           ‖                               |
           C—H                             C=O
           |                               |
        H—C—OH                          HO—C—H
           |                               |
       HO—C—H                            H—C—OH
           |                               |
        H—C—OH          ⇌                H—C—OH
           |                               |
        H—C—OH                           H—C—H
           |                               |
        H—C—H                              O
           |                               |
           O                          O⁻—P—O⁻
           |                               ‖
      O⁻—P—O⁻                              O
           ‖
           O
```

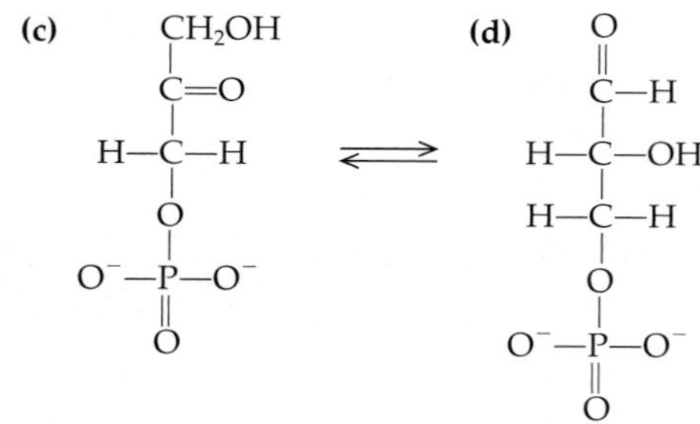

21.34 To which family of organic molecules do (a) and (d) belong? To which family of organic molecules do (b) and (c) belong?

21.35 What is the name of the type of intermediate formed in each of these reactions?

21.36 Draw the intermediate that would be formed in each of these reactions.

21.37 When an enzyme has the term *kinase* in the name, what type of reaction do you expect it to catalyze?

21.38 What features do the reactions catalyzed by hexokinase and phosphofructokinase share in common?

21.39 What is the role of NAD^+ in a biochemical oxidation reaction?

21.40 Write the equation for the reaction catalyzed by glyceraldehyde-3-phosphate dehydrogenase. Highlight the chemical changes that show this to be an oxidation reaction.

21.41 The enzyme that catalyzes step 9 of glycolysis is called enolase. What is the significance of that name?

21.42 Draw the enol tautomer of pyruvate.

21.43 What is the importance of the regulation of glycolysis?

21.44 Explain the role of allosteric enzymes in control of glycolysis.

21.45 What molecules serve as allosteric effectors of phosphofructokinase?

21.46 What molecule serves as an allosteric inhibitor of hexokinase?

21.47 Explain the role of citrate in the feedback inhibition of glycolysis.

21.48 Explain the feedforward activation mechanism that results in the activation of pyruvate kinase.

Fermentations

21.49 Write a balanced chemical equation for the conversion of acetaldehyde to ethanol.

21.50 Write a balanced chemical equation for the conversion of pyruvate to lactate.

21.51 After running a 100-m dash, a sprinter had a high concentration of muscle lactate. What process is responsible for production of lactate?

21.52 If the muscle of an organism had no lactate dehydrogenase, could anaerobic glycolysis occur in those muscle cells? Explain your answer.

21.53 What food products are the result of lactate fermentation?

21.54 Explain the value of alcohol fermentation in bread making.

21.55 What enzyme catalyzes the reduction of pyruvate to lactate?

21.56 What enzymes catalyze the conversion of pyruvate to ethanol and carbon dioxide?

21.57 A child was brought to the doctor's office suffering from a strange set of symptoms. When the child exercised hard, she became giddy and behaved as though drunk. What do you think is the metabolic basis of these symptoms?

21.58 A family started a batch of wine by adding yeast to grape juice and placing the mixture in a sealed bottle. Two weeks later, the bottle exploded. What metabolic reactions—and specifically, what product of those reactions—caused the bottle to explode?

The Pentose Phosphate Pathway

21.59 Describe the three stages of the pentose phosphate pathway.

21.60 Write an equation to summarize the pentose phosphate pathway.

21.61 Of what value are the ribose-5-phosphate and erythrose-4-phosphate that are produced in the pentose phosphate pathway?

21.62 Of what value is the NADPH that is produced in the pentose phosphate pathway?

Gluconeogenesis

21.63 What organ is primarily responsible for gluconeogenesis?

21.64 What is the physiological function of gluconeogenesis?

21.65 Lactate can be converted to glucose by gluconeogenesis. To what metabolic intermediate must lactate be converted so that it can be a substrate for the enzymes of gluconeogenesis?

21.66 L-Alanine can be converted to pyruvate. Can L-alanine also be converted to glucose? Explain your answer.

21.67 Explain why gluconeogenesis is not simply the reversal of glycolysis.

21.68 In step 10 of glycolysis, phosphoenolpyruvate is converted to pyruvate, and ATP is produced by substrate-level phosphorylation. How is this reaction bypassed in gluconeogenesis?

21.69 Which steps in the glycolysis pathway are irreversible?

21.70 What enzymatic reactions of gluconeogenesis bypass the irreversible steps of glycolysis?

Glycogen Synthesis and Degradation

21.71 What organs are primarily responsible for maintaining the proper blood glucose level?

21.72 Why must the blood glucose level be carefully regulated?

21.73 What does the term *hypoglycemia* mean?

21.74 What does the term *hyperglycemia* mean?

21.75 **a.** What enzymes involved in glycogen metabolism are stimulated by insulin?
b. What effect does this have on glycogen metabolism?
c. What effect does this have on blood glucose levels?

21.76 **a.** What enzyme is stimulated by glucagon?
b. What effect does this have on glycogen metabolism?
c. What effect does this have on blood glucose levels?

21.77 Explain how a defect in glycogen metabolism can cause hypoglycemia.

21.78 What defects of glycogen metabolism would lead to a large increase in the concentration of liver glycogen?

Critical Thinking Problems

1. An enzyme that hydrolyzes ATP (an ATPase) bound to the plasma membrane of certain tumor cells has an abnormally high activity. How will this activity affect the rate of glycolysis?
2. Explain why no net oxidation occurs during anaerobic glycolysis followed by lactate fermentation.
3. A certain person was found to have a defect in glycogen metabolism. The liver of this person could (a) make glucose-6-phosphate from lactate and (b) synthesize glucose-6-phosphate from glycogen but (c) could not synthesize glycogen from glucose-6-phosphate. What enzyme is defective?
4. A scientist added phosphate labeled with radioactive phosphorus (^{32}P) to a bacterial culture growing anaerobically (without O_2). She then purified all the compounds produced during glycolysis. Look carefully at the steps of the pathway. Predict which of the intermediates of the pathway would be the first one to contain radioactive phosphate. On which carbon of this compound would you expect to find the radioactive phosphate?
5. A two-month-old baby was brought to the hospital suffering from seizures. He deteriorated progressively over time, showing psychomotor retardation. Blood tests revealed a high concentration of lactate and pyruvate. Although blood levels of alanine were high, they did not stimulate gluconeogenesis. The doctor measured the activity of pyruvate carboxylase in the baby and found it to be only 1% of the normal level. What reaction is catalyzed by pyruvate carboxylase? How could this deficiency cause the baby's symptoms and test results?

Downhill skiing demands a great deal of energy.

BIOCHEMISTRY

22 Aerobic Respiration and Energy Production

Outline

Learning Goals

1. Name the regions of the mitochondria and the function of each region.
2. Describe the reaction that results in the conversion of pyruvate to acetyl CoA, describing the location of the reaction and the components of the pyruvate dehydrogenase complex.
3. Summarize the reactions of aerobic respiration.
4. Looking at an equation representing any of the chemical reactions that occur in the citric acid cycle, describe the kind of reaction that is occurring and the significance of that reaction to the pathway.
5. Explain the mechanisms for the control of the citric acid cycle.
6. Describe the process of oxidative phosphorylation.
7. Describe the conversion of amino acids to molecules that can enter the citric acid cycle.
8. Explain the importance of the urea cycle and describe its essential steps.
9. Discuss the cause and effect of hyperammonemia.
10. Summarize the role of the citric acid cycle in catabolism and anabolism.

CHEMISTRY CONNECTION

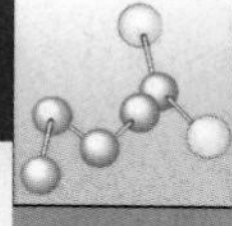

Mitochondria from Mom

In this chapter we will be studying the amazing, intricate set of reactions that allow us to completely degrade fuel molecules such as sugars and amino acids. These oxygen-requiring reactions occur in cellular organelles called *mitochondria.*

We are used to thinking of the organelles as a collection of membrane-bound structures that are synthesized under the direction of the genetic information in the nucleus of the cell. Not so with the mitochondria. These organelles have their own genetic information and are able to make some of their own proteins. They grow and multiply in a way very similar to the simple bacteria. This, along with other information on the structure and activities of mitochondria, has led researchers to conclude that the mitochondria are actually the descendants of bacteria captured by eukaryotic cells millions of years ago.

Recent studies of the mitochondrial genetic information (DNA) have revealed fascinating new information. For instance, although each of us inherited half our genetic information from our mothers and half from our fathers, each of us inherited all of our mitochondria from our mothers. The reason for this is that when the sperm fertilizes the egg, only the sperm nucleus enters the cell.

The observation that all of our mitochondria are inherited from our mothers led Dr. A. Wilson to study the mitochondrial DNA of thousands of women around the world. He thought that by looking for similarities and differences in the mitochondrial DNA he would be able to identify a "Mitochondrial Eve"—the mother of all humanity. He didn't really think that he could identify a single woman who would have lived tens of thousands of years ago. But he hoped to determine the location of the first population of human women to help answer questions about the origin of humankind. Although the idea was a good one, the study had a number of experimental flaws. Currently, a hot debate is going on among hundreds of scientists about the Mitochondrial Eve. This controversy should encourage better experiments and analysis to help us identify our origins and to better understand the workings of the mitochondria.

Like the mitochondria themselves, some genetic diseases of energy metabolism are maternally inherited. One such disease, Leber's hereditary optic neuropathy (LHON), causes blindness and heart problems. People with LHON have a reduced ability to make ATP. As a result, sensitive tissues that demand a great deal of energy eventually die. LHON sufferers eventually lose their sight because the optic nerve dies from lack of energy.

Researchers have identified and cloned a mutant mitochondrial gene that is responsible for LHON. The defect is a mutant form of *NADH dehydrogenase.* NADH dehydrogenase is a huge, complex enzyme that accepts electrons from NADH and sends them on through an electron transport system. Passage of electrons through the electron transport system allows the synthesis of ATP. If NADH dehydrogenase is defective, passage of electrons through the electron transport system is less efficient, and less ATP is made. In LHON sufferers, the result is eventual blindness.

In this chapter and the next, we will study some of the important biochemical reactions that occur in the mitochondria. A better understanding of the function of healthy mitochondria will eventually allow us to help those suffering from LHON and other mitochondrial genetic diseases.

Introduction

As we have seen, the anaerobic glycolysis pathway begins the breakdown of glucose and produces a small amount of ATP and NADH. But it is aerobic catabolic pathways that complete the oxidation of glucose to CO_2 and H_2O and provide most of the ATP needed by the body. In fact, this process, called *aerobic respiration,* produces thirty-six ATP molecules using the energy harvested from each glucose molecule that enters glycolysis. These reactions occur in metabolic pathways located in *mitochondria,* the cellular "power plants." Mitochondria are a type of membrane-enclosed cell *organelle.*

An organelle is a compartment within the cytoplasm that has a specialized function.

Here, in the mitochondria, the final oxidations of carbohydrates, lipids, and proteins occur. Here, also, the electrons that are harvested in these oxidation reactions are used to make ATP. In these remarkably efficient reactions, nearly 40% of the potential energy of glucose is stored as ATP.

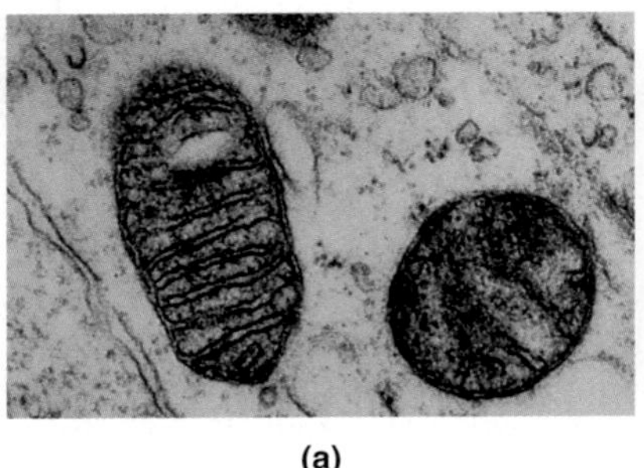
(a)

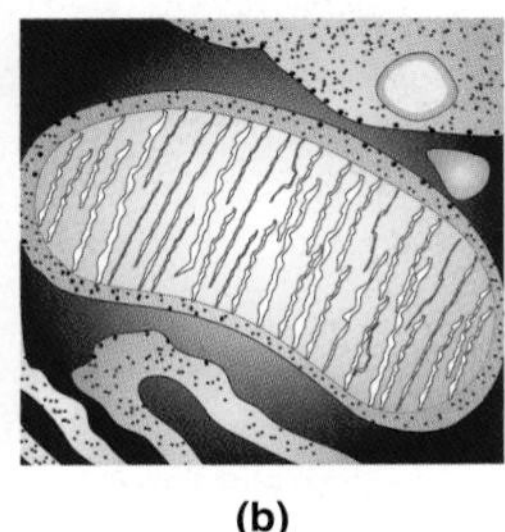
(b)

Figure 22.1
Structure of the mitochondrion.
(a) Electron micrograph of mitochondria.
(b) Schematic drawing of the mitochondrion.

22.1 The Mitochondria

Learning Goal

Mitochondria are football-shaped organelles that are roughly the size of a bacterial cell. They are surrounded by an **outer mitochondrial membrane** and an **inner mitochondrial membrane** (Figure 22.1). The space between the two membranes is the **intermembrane space,** and the space inside of the inner membrane is the **matrix space.** The enzymes of the citric acid cycle, of the β-oxidation pathway for the breakdown of fatty acids, and for the degradation of amino acids are all found in the mitochondrial matrix space.

Structure and Function

The outer mitochondrial membrane has many small pores through which small molecules (less than 10,000 g/mol) can pass. Thus, the small molecules to be oxidized for the production of ATP can easily enter the mitochondrial intermembrane space.

The inner membrane is highly folded to create a large surface area. The folded membranes are known as **cristae.** The inner mitochondrial membrane is almost completely impermeable to most substances. For this reason it has many transport proteins to bring particular fuel molecules into the matrix space. Also embedded within the inner mitochondrial membrane are the protein electron carriers of the *electron transport system* and *ATP synthase.* ATP synthase is a large complex of many proteins that catalyzes the synthesis of ATP.

Origin of the Mitochondria

Not only are mitochondria roughly the size of bacteria, they have several other features that have led researchers to suspect that they may once have been free-living bacteria that were "captured" by eukaryotic cells. They have their own genetic information (DNA). They also make their own ribosomes that are very similar to those of bacteria. These ribosomes allow the mitochondria to synthesize some of their own proteins. Finally, mitochondria are actually self-replicating; they grow in size and divide to produce new mitochondria. All of these characteristics suggest that the mitochondria that produce the majority of the ATP for our cells evolved from bacteria "captured" perhaps as long as 1.5×10^9 years ago.

As we will see in Chapter 24, ribosomes are complexes of protein and RNA that serve as small platforms for protein synthesis.

Question 22.1

What is the function of the mitochondria?

Question 22.2

How do the mitochondria differ from the other components of eukaryotic cells?

Question 22.3

Draw a schematic diagram of a mitochondrion, and label the parts of this organelle.

A HUMAN PERSPECTIVE

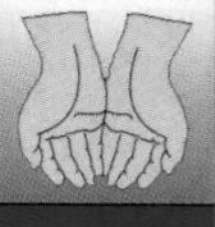

Exercise and Energy Metabolism

The Olympic sprinters get set in the blocks. The gun goes off, and roughly ten seconds later the 100-m dash is over. Elsewhere, the marathoners line up. They will run 26 miles and 385 yards in a little over two hours. Both sports involve running, but they utilize very different sources of energy.

Let's look at the sprinter first. The immediate source of energy for the sprinter is stored ATP. But the quantity of stored ATP is very small, only about three ounces. This allows the sprinter to run as fast as he or she can for about three seconds. Obviously, another source of stored energy must be tapped, and that energy store is *creatine phosphate:*

$$^{-}O-\overset{\overset{O}{\|}}{\underset{\underset{O^{-}}{|}}{P}}-\overset{\overset{H}{|}}{N}-\overset{\overset{NH}{\|}}{C}-\underset{\underset{CH_3}{|}}{N}-CH_2-C\begin{matrix}\diagup O^{-}\\ \diagdown\!\!\diagdown O\end{matrix}$$

The structure of creatine phosphate.

Creatine phosphate, stored in the muscle, donates its high-energy phosphate to ADP to produce new supplies of ATP. This will keep our runner in motion for another five or six seconds before the store of creatine phosphate is also depleted. This is almost enough energy to finish the 100-m dash, but in reality, all the runners are slowing down, owing to energy depletion, and the winner is the sprinter who is slowing down the least!

Consider a longer race, the 400-m or the 800-m. These runners run at maximum capacity for much longer. When they have depleted their ATP and creatine phosphate stores, they must synthesize more ATP. Of course, the cells have been making ATP all the time, but now the demand for energy is much greater. To supply this increased demand, the anaerobic energy-generating reactions (glycolysis and lactate fermentation, Chapter 21) and aerobic processes (citric acid cycle and oxidative phosphorylation) begin to function much more rapidly. Often, however, these athletes are running so strenuously that they cannot provide enough oxygen to the exercising muscle to allow oxidative phosphorylation to function efficiently. When this happens, the muscles must rely on glycolysis and lactate fermentation to provide *most* of the energy requirement. The chemical by-product of these anaerobic processes, lactate,

$$^{-}O-\overset{\overset{O}{\|}}{\underset{\underset{O^{-}}{|}}{P}}-\overset{\overset{H}{|}}{N}-\overset{\overset{NH}{\|}}{C}-\underset{\underset{CH_3}{|}}{N}-CH_2-C\begin{matrix}\diagup O^{-}\\ \diagdown\!\!\diagdown O\end{matrix} + ADP \xrightarrow{\text{Creatine kinase}} H_2N-\overset{\overset{NH}{\|}}{C}-\underset{\underset{CH_3}{|}}{N}-CH_2-C\begin{matrix}\diagup O^{-}\\ \diagdown\!\!\diagdown O\end{matrix} + ATP$$

Phosphoryl group transfer from creatine phosphate to ADP is catalyzed by the enzyme creatine kinase.

Question 22.4

Describe the evidence that suggests that mitochondria evolved from free-living bacteria.

22.2 Conversion of Pyruvate to Acetyl CoA

Learning Goal

As we saw in Chapter 21, under *anaerobic* conditions, glucose is broken down into two pyruvate molecules that are then converted to a stable fermentation product. This limited degradation of glucose releases very little of the potential energy of glucose. Under *aerobic* conditions the cells can use oxygen and completely oxidize glucose to CO_2 in a metabolic pathway called the *citric acid cycle.*

This pathway is often referred to as the *Krebs cycle* in honor of Sir Hans Krebs who worked out the steps of this cyclic pathway from his own experimental data and that of other researchers. It is also called the *tricarboxylic acid (TCA) cycle* because several of the early intermediates in the pathway have three carboxyl groups.

builds up in the muscle and diffuses into the bloodstream. However, the concentration of lactate inevitably builds up in the working muscle and causes muscle fatigue and, eventually, muscle failure. Thus, exercise that depends primarily on anaerobic ATP production cannot continue for very long.

The marathoner presents us with a different scenario. This runner will deplete his or her stores of ATP and creatine phosphate as quickly as a short-distance runner. The anaerobic glycolytic pathway will begin to degrade glucose provided by the blood at a more rapid rate, as will the citric acid cycle and oxidative phosphorylation. The major difference in ATP production between the long-distance runner and the short- or middle-distance runner is that the muscles of the long-distance runner derive almost all the energy through aerobic pathways. These individuals continue to run long distances at a pace that allows them to supply virtually all the oxygen needed by the exercising muscle. In fact, only aerobic pathways can provide a constant supply of ATP for exercise that goes on for hours. Theoretically, under such conditions our runner could run indefinitely, utilizing first his or her stored glycogen and eventually stored lipids. Of course, in reality, other factors such as dehydration and fatigue place limits on the athlete's ability to continue.

We have seen, then, that long-distance runners must have a great capacity to produce ATP aerobically, in the mitochondria, whereas short- and middle-distance runners need a great capacity to produce energy anaerobically, in the cytoplasm of the muscle cells. It is interesting to note that the muscles of these runners reflect these diverse needs.

When one examines muscle tissue that has been surgically removed, one finds two predominant types of muscle fibers. *Fast twitch muscle fibers* are large, relatively plump, pale cells. They have only a few mitochondria but contain a large reserve of glycogen and high concentrations of the enzymes that are needed for glycolysis and lactate fermentation. These muscle fibers fatigue rather quickly because fermentation is inefficient, quickly depleting the cell's glycogen store and causing the accumulation of lactate.

Slow twitch muscle fiber cells are about half the diameter of fast twitch muscle cells and are red. The red color is a result of the high concentrations of myoglobin in these cells. Recall that myoglobin stores oxygen for the cell (Section 19.9) and facilitates rapid diffusion of oxygen throughout the cell. In addition, slow twitch muscle fiber cells are packed with mitochondria. With this abundance of oxygen and mitochondria these cells have the capacity for extended ATP production via aerobic pathways—ideal for endurance sports like marathon racing.

It is not surprising, then, that researchers have found that the muscles of sprinters have many more fast twitch muscle fibers and those of endurance athletes have many more slow twitch muscle fibers. One question that many researchers are trying to answer is whether the type of muscle fibers an individual has is a function of genetic makeup or training. Is a marathon runner born to be a long-distance runner, or are his or her abilities due to the type of training the runner undergoes? There is no doubt that the training regimen for an endurance runner does indeed increase the number of slow twitch muscle fibers and that of a sprinter increases the number of fast twitch muscle fibers. But there is intriguing new evidence to suggest that the muscles of endurance athletes have a greater proportion of slow twitch muscle fibers before they ever begin training. It appears that some of us truly were born to run.

Once pyruvate enters the mitochondria, it must be converted to a two-carbon acetyl group. This acetyl group must be "activated" to enter the reactions of the citric acid cycle. Activation occurs when the acetyl group is bonded to the thiol group of coenzyme A. **Coenzyme A** is a large thiol derived from ATP and the vitamin pantothenic acid (Figure 22.2). It is an acceptor of acetyl groups (in red in Figure 22.2), which are bonded to it through a high-energy thioester bond. The acetyl coenzyme A (**acetyl CoA**) formed is the "activated" form of the acetyl group.

Coenzyme A is described in Sections 13.9 and 15.4.

Thioester bonds are discussed in Section 15.4.

The structure and function of pantothenic acid are discussed in Appendix F, Water-Soluble Vitamins.

Figure 22.3 shows us the reaction that converts pyruvate to acetyl CoA. First, pyruvate is decarboxylated, which means that it loses a carboxyl group that is released as CO_2. Next it is oxidized, and the hydride anion that is removed is accepted by NAD^+. Finally, the remaining acetyl group, CH_3CO—, is linked to coenzyme A by a thioester bond. This very complex reaction is carried out by three enzymes and five coenzymes that are organized together in a single bundle called the **pyruvate dehydrogenase complex** (see Figure 22.3). This organization allows the substrate to be passed from one enzyme to the next as each chemical reaction occurs. A schematic representation of this "disassembly line" is shown in Figure 22.3b.

This single reaction requires four coenzymes made from four different vitamins, in addition to the coenzyme lipoamide. These are thiamine pyrophosphate, derived from thiamine (Vitamin B_1); FAD, derived from riboflavin (Vitamin B_2);

These vitamins are discussed in Appendix F, Water-Soluble Vitamins.

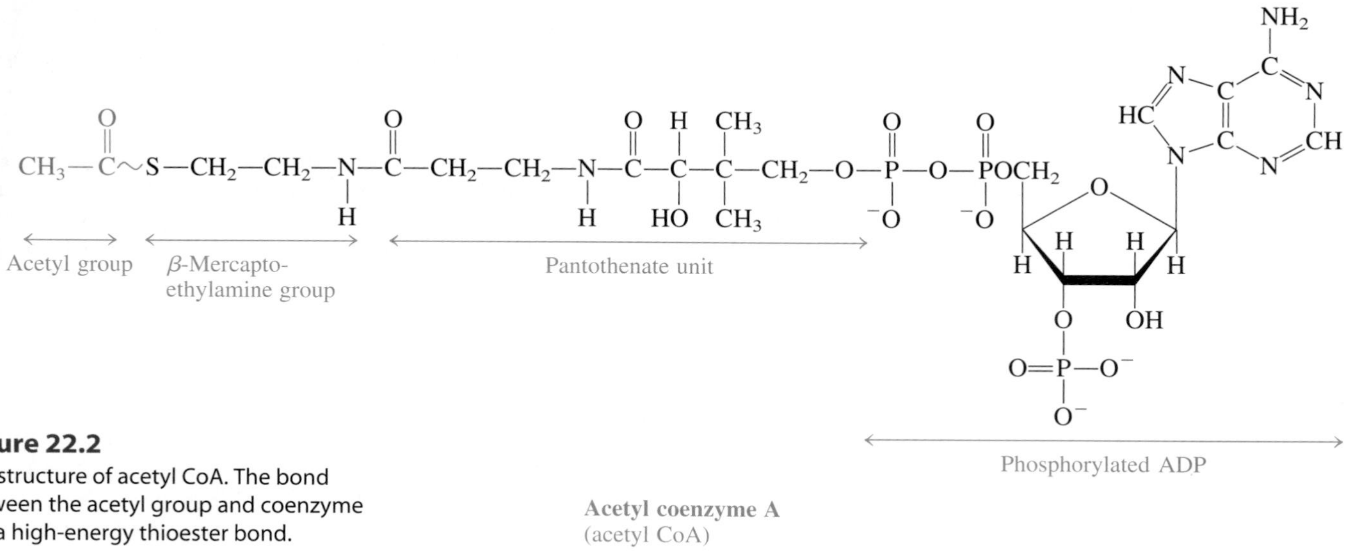

Figure 22.2
The structure of acetyl CoA. The bond between the acetyl group and coenzyme A is a high-energy thioester bond.

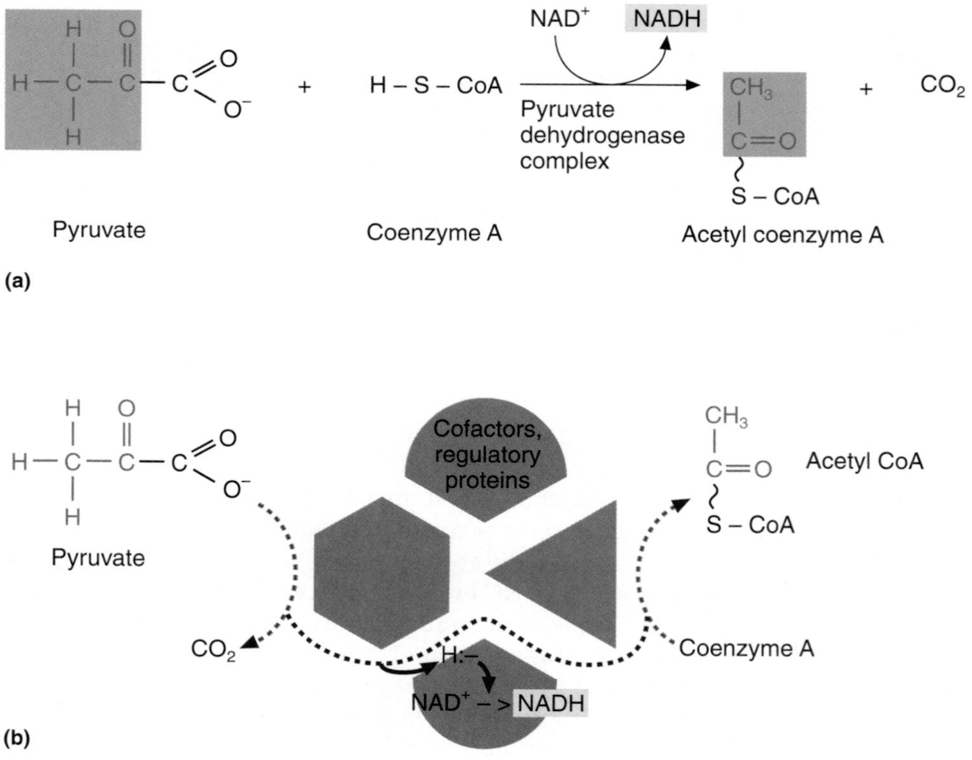

Figure 22.3
The decarboxylation and oxidation of pyruvate to produce acetyl CoA. (a) The overall reaction in which CO_2 and an $H:^-$ are removed from pyruvate and the remaining acetyl group is attached to coenzyme A. This requires the concerted action of three enzymes and five coenzymes. (b) The pyruvate dehydrogenase complex that carries out this reaction is actually a cluster of enzymes and coenzymes. The substrate is passed from one enzyme to the next as the reaction occurs.

NAD^+, derived from niacin; and coenzyme A, derived from pantothenic acid. Obviously, a deficiency in any of these vitamins would seriously reduce the amount of acetyl CoA that our cells could produce. This, in turn, would limit the amount of ATP that the body could make and would contribute to vitamin-deficiency diseases. Fortunately, a well-balanced diet provides an adequate supply of these and other vitamins.

In Figure 22.4 we see that acetyl CoA is a central character in cellular metabolism. It is produced by the degradation of glucose, fatty acids, and some amino

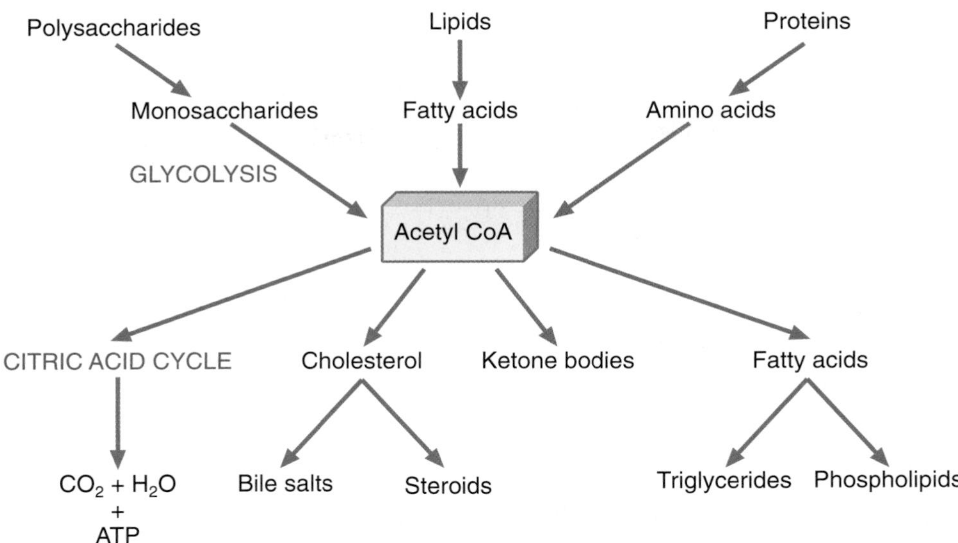

Figure 22.4
The central role of acetyl CoA in cellular metabolism.

acids. The major function of acetyl CoA in energy-harvesting pathways is to carry the acetyl group to the citric acid cycle, in which it will be used to produce large amounts of ATP. In addition to these catabolic duties, the acetyl group of acetyl CoA can also be used for *anabolic* or biosynthetic reactions to produce cholesterol and fatty acids. It is through this intermediate, acetyl CoA, that all the energy sources (fats, proteins, and carbohydrates) are interconvertible.

Question 22.5

What vitamins are required for acetyl CoA production from pyruvate?

Question 22.6

What is the major role of coenzyme A in catabolic reactions?

22.3 An Overview of Aerobic Respiration

Learning Goal

Aerobic respiration is the oxygen-requiring breakdown of food molecules and production of ATP. The different steps of aerobic respiration occur in different compartments of the mitochondria.

The enzymes for the citric acid cycle are found in the mitochondrial matrix space. The first enzyme catalyzes a reaction that joins the acetyl group of acetyl CoA (two carbons) to a four-carbon molecule (oxaloacetate) to produce citrate (six carbons). The remaining enzymes catalyze a series of rearrangements, decarboxylations (removal of CO_2), and oxidation–reduction reactions. The eventual products of this cyclic pathway are two CO_2 molecules and oxaloacetate—the molecule we began with.

Remember (Section 20.7) that it is really the hydride anion with its pair of electrons ($H:^-$) that is transferred to NAD^+ to produce NADH. Similarly, a pair of hydrogen atoms (and thus two electrons) are transferred to FAD to produce $FADH_2$.

At several steps in the citric acid cycle a substrate is oxidized. In three of these steps a pair of electrons is transferred from the substrate to NAD^+, producing NADH (three NADH molecules per turn of the cycle). At another step a pair of electrons is transferred from a substrate to FAD, producing $FADH_2$ (one $FADH_2$ molecule per turn of the cycle).

The electrons are passed from NADH or $FADH_2$, through an electron transport system located in the inner mitochondrial membrane, and finally to the terminal electron acceptor, molecular oxygen (O_2). The transfer of electrons through the electron transport system causes protons (H^+) to be pumped from the mitochondrial matrix into the intermembrane compartment. The result is a high-energy H^+ reservoir.

In the final step the energy of the H^+ reservoir is used to make ATP. This last step is carried out by the enzyme complex ATP synthase. As protons flow back into the mitochondrial matrix through a pore in the ATP synthase complex, the enzyme catalyzes the synthesis of ATP.

This long, involved process is called *oxidative phosphorylation,* because the energy of electrons from the *oxidation* of substrates in the citric acid cycle is used to *phosphorylate* ADP and produce ATP. The details of each of these steps will be examined in upcoming sections.

Question 22.7 What is meant by the term *oxidative phosphorylation?*

Question 22.8 What does the term *aerobic respiration* mean?

22.4 The Citric Acid Cycle (The Krebs Cycle)

Reactions of the Citric Acid Cycle

Learning Goal

The formation of acetyl CoA was described in Section 22.2.

Aldol condensation reactions are reactions between aldehydes and ketones to form larger molecules (Section 14.4).

The **citric acid cycle** is sometimes called the *Krebs cycle,* in honor of its discoverer, Sir Hans Krebs. It is the final stage of the breakdown of carbohydrates, fats, and amino acids released from dietary proteins (Figure 22.5).

To understand this important cycle, let's follow the fate of the acetyl group of an acetyl CoA as it passes through the citric acid cycle. The numbered steps listed below correspond to the steps in the citric acid cycle that are summarized in Figure 22.5.

Reaction 1. This is a condensation reaction between the acetyl group of acetyl CoA and oxaloacetate. Actually, this is another biological example of an aldol condensation reaction. It is catalyzed by the enzyme *citrate synthase.* The product that is formed is citrate:

$$\underset{\text{Oxaloacetate}}{^{-}OOC\text{—}C(=O)\text{—}CH_2\text{—}COO^-} + \underset{\text{Acetyl CoA}}{H_3C\text{—}C(=O)\sim S\text{—}CoA} + H_2O \xrightarrow{\text{Citrate synthase}} \underset{\text{Citrate}}{HO\text{—}C(COO^-)(CH_2COO^-)\text{—}CH_2\text{—}COO^-} + \underset{\text{Coenzyme A}}{HS\text{—}CoA} + H^+$$

Notice that the conversion of citrate to cis-aconitate is a biological example of the dehydration of an alcohol to produce an alkene (Section 13.5). The conversion of cis-aconitate to isocitrate is a biochemical example of the hydration of an alkene to produce an alcohol (Sections 12.5 and 13.5).

Reaction 2. The enzyme *aconitase* catalyzes the dehydration of citrate, producing *cis*-aconitate. The same enzyme, aconitase, then catalyzes addition of a water molecule to the *cis*-aconitate, converting it to isocitrate. The net effect of these two steps is the isomerization of citrate to isocitrate:

$$\underset{\text{Citrate}}{^{-}OOC\text{—}CH_2\text{—}C(OH)(COO^-)\text{—}CH_2\text{—}COO^-} \xrightarrow{\text{Aconitase}} \underset{\textit{cis}\text{-Aconitate}}{^{-}OOC\text{—}CH_2\text{—}C(COO^-)\text{=}CH\text{—}COO^-} + H_2O \xrightarrow{\text{Aconitase}} \underset{\text{Isocitrate}}{^{-}OOC\text{—}CH_2\text{—}CH(COO^-)\text{—}CH(OH)\text{—}COO^-}$$

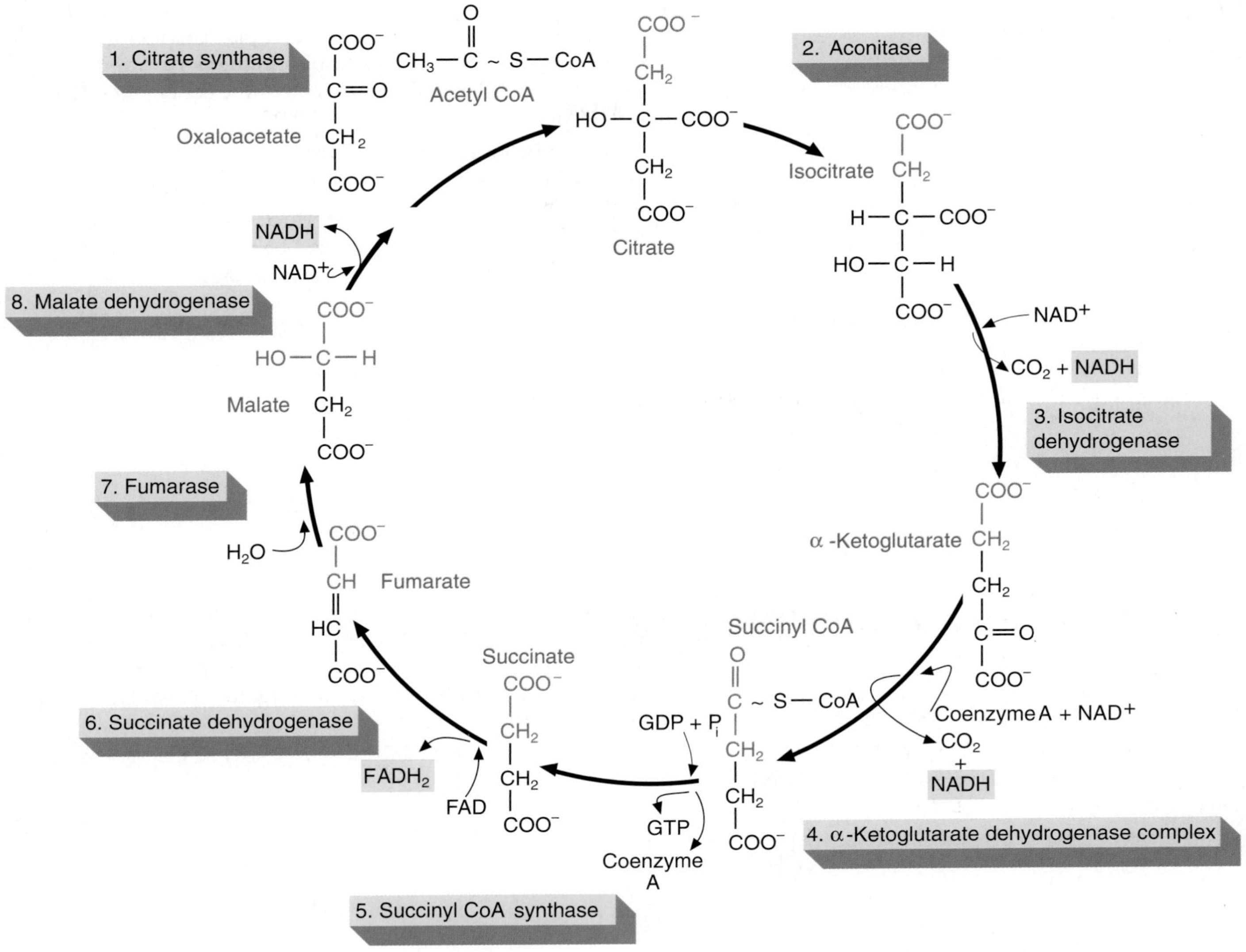

Figure 22.5
The reactions of the citric acid cycle.

Reaction 3. The first oxidative step of the citric acid cycle is catalyzed by *isocitrate dehydrogenase.* It is a complex reaction in which three things happen:

a. the hydroxyl group of isocitrate is oxidized to a ketone,
b. carbon dioxide is released, and
c. NAD^+ is reduced to NADH.

The product of this oxidative decarboxylation reaction is α-ketoglutarate:

$$\underset{\text{Isocitrate}}{\begin{matrix} COO^- \\ | \\ CH_2 \\ | \\ H{-}C{-}COO^- \\ | \\ HO{-}C{-}H \\ | \\ COO^- \end{matrix}} + NAD^+ \xrightarrow{\text{Isocitrate dehydrogenase}} \underset{\alpha\text{-Ketoglutarate}}{\begin{matrix} COO^- \\ | \\ CH_2 \\ | \\ CH_2 \\ | \\ C{=}O \\ | \\ COO^- \end{matrix}} + CO_2 + NADH$$

The oxidation of a secondary alcohol produces a ketone (Sections 13.5 and 14.4).

The structure of NAD^+ and its reduction to NADH are shown in Figure 20.8.

Remember, in organic (and thus biochemical) reactions, oxidation can be recognized as a gain of oxygen or loss of hydrogen (Section 13.6).

The pyruvate dehydrogenase complex was described in Section 22.2 and shown in Figure 22.3.

Reaction 4. Coenzyme A enters the picture again in this step of the citric acid cycle. The α-*ketoglutarate dehydrogenase* complex carries out this series of reactions. This complex is very similar to the pyruvate dehydrogenase complex and requires the same coenzymes. Once again, three chemical events occur:

a. α-ketoglutarate loses a carboxylate group as CO_2,
b. it is oxidized and NAD^+ is reduced to NADH, and
c. coenzyme A combines with the product, succinate, to form succinyl CoA. The bond thus formed between succinate and coenzyme A is a high-energy thioester linkage.

$$\begin{array}{l} COO^- \\ | \\ CH_2 \\ | \\ CH_2 \\ | \\ C{=}O \\ | \\ COO^- \end{array} + NAD^+ + \text{Coenzyme A} \xrightarrow[\text{complex}]{\alpha\text{-Ketoglutarate dehydrogenase}} \begin{array}{l} COO^- \\ | \\ CH_2 \\ | \\ CH_2 \\ | \\ C{\sim}S{-}CoA \\ \| \\ O \end{array} + CO_2 + NADH$$

α-Ketoglutarate　　　　Succinyl CoA

Reaction 5. Succinyl CoA is converted to succinate in this step, which once more is chemically very involved. The enzyme *succinyl CoA synthase* catalyzes a coupled reaction in which the high-energy thioester bond of succinyl CoA is hydrolyzed and an inorganic phosphate group is added to GDP to make GTP:

$$\begin{array}{l} COO^- \\ | \\ CH_2 \\ | \\ CH_2 \\ | \\ C{\sim}S{-}CoA \\ \| \\ O \end{array} + GDP + P_i \xrightarrow{\text{Succinyl CoA synthase}} \begin{array}{l} COO^- \\ | \\ CH_2 \\ | \\ CH_2 \\ | \\ COO^- \end{array} + GTP + \text{Coenzyme A}$$

Succinyl CoA　　　　Succinate

Another enzyme, *dinucleotide diphosphokinase*, then catalyzes the transfer of a phosphoryl group from GTP to ADP to make ATP:

$$GTP + ADP \xrightarrow{\text{Dinucleotide diphosphokinase}} GDP + ATP$$

The structure of FAD was shown in Figure 20.8.

Reaction 6. *Succinate dehydrogenase* then catalyzes the oxidation of succinate to fumarate in the next step. The oxidizing agent, *flavin adenine dinucleotide (FAD)*, is reduced in this step:

We studied hydrogenation of alkenes to produce alkanes in Section 12.5. This is simply the reverse.

$$\begin{array}{l} COO^- \\ | \\ CH_2 \\ | \\ CH_2 \\ | \\ COO^- \end{array} + FAD \xrightarrow{\text{Succinate dehydrogenase}} \begin{array}{r} COO^- \\ | \\ C{-}H \\ \| \\ H{-}C \\ | \\ COO^- \end{array} + FADH_2$$

Succinate　　　　Fumarate

Reaction 7. Addition of H_2O to the double bond of fumarate gives malate. The enzyme *fumarase* catalyzes this reaction:

$$\underset{\text{Fumarate}}{\begin{array}{c} COO^- \\ | \\ C{-}H \\ \| \\ H{-}C \\ | \\ COO^- \end{array}} + H_2O \xrightarrow{\text{Fumarase}} \underset{\text{Malate}}{\begin{array}{c} COO^- \\ | \\ HO{-}C{-}H \\ | \\ H{-}C{-}H \\ | \\ COO^- \end{array}}$$

This reaction is a biological example of the hydration of an alkene to produce an alcohol (Sections 12.5 and 13.5).

Reaction 8. In the final step of the citric acid cycle, *malate dehydrogenase* catalyzes the reduction of NAD^+ to NADH and the oxidation of malate to oxaloacetate. Because the citric acid cycle "began" with the addition of an acetyl group to oxaloacetate, we have come full circle.

$$\underset{\text{Malate}}{\begin{array}{c} COO^- \\ | \\ HO{-}C{-}H \\ | \\ CH_2 \\ | \\ COO^- \end{array}} + NAD^+ \xrightarrow{\text{Malate dehydrogenase}} \underset{\text{Oxaloacetate}}{\begin{array}{c} COO^- \\ | \\ C{=}O \\ | \\ CH_2 \\ | \\ COO^- \end{array}} + NADH$$

This reaction is a biochemical example of the oxidation of a secondary alcohol to a ketone, which we studied in Sections 13.5 and 15.4.

22.5 Control of the Citric Acid Cycle

Learning Goal

Just like glycolysis, the citric acid cycle is responsive to the energy needs of the cell. The pathway speeds up when there is a greater demand for ATP, and it slows down when ATP energy is in excess. In the last chapter we saw that several of the enzymes that catalyze the reactions of glycolysis are *allosteric enzymes.* Similarly, four enzymes or enzyme complexes involved in the complete oxidation of pyruvate via the citric acid cycle also are allosteric enzymes. They are able to bind to *effectors,* such as ATP or ADP, that alter the shape of the enzyme active site, either stimulating the rate of the reaction (positive allosterism) or inhibiting the reaction (negative allosterism).

Because the control of the pathway must be precise, there are several enzymatic steps that are regulated. These are summarized in Figure 22.6 and below:

1. *Conversion of pyruvate to acetyl CoA.* The pyruvate dehydrogenase complex is inhibited by high concentrations of ATP, acetyl CoA, and NADH. Of course, the presence of these compounds in abundance signals that the cell has an adequate supply of energy, and thus energy metabolism is slowed.
2. *Synthesis of citrate from oxaloacetate and acetyl CoA.* The enzyme citrate synthase is an allosteric enzyme. In this case the negative effector is ATP. Again, this is logical because an excess of ATP indicates that the cell has an abundance of energy.
3. *Oxidation and decarboxylation of isocitrate to α-ketoglutarate.* Isocitrate dehydrogenase is also an allosteric enzyme; however, the enzyme is controlled by the positive allosteric effector, ADP. ADP is a signal that the levels of ATP must be low, and therefore the rate of the citric acid cycle should be increased. Interestingly, isocitrate dehydrogenase is also *inhibited* by high levels of NADH and ATP.
4. *Conversion of α-ketoglutarate to succinyl CoA.* The α-ketoglutarate dehydrogenase complex is inhibited by high levels of the products of the reactions that it catalyzes, namely, NADH and succinyl CoA. It is further inhibited by high concentrations of ATP.

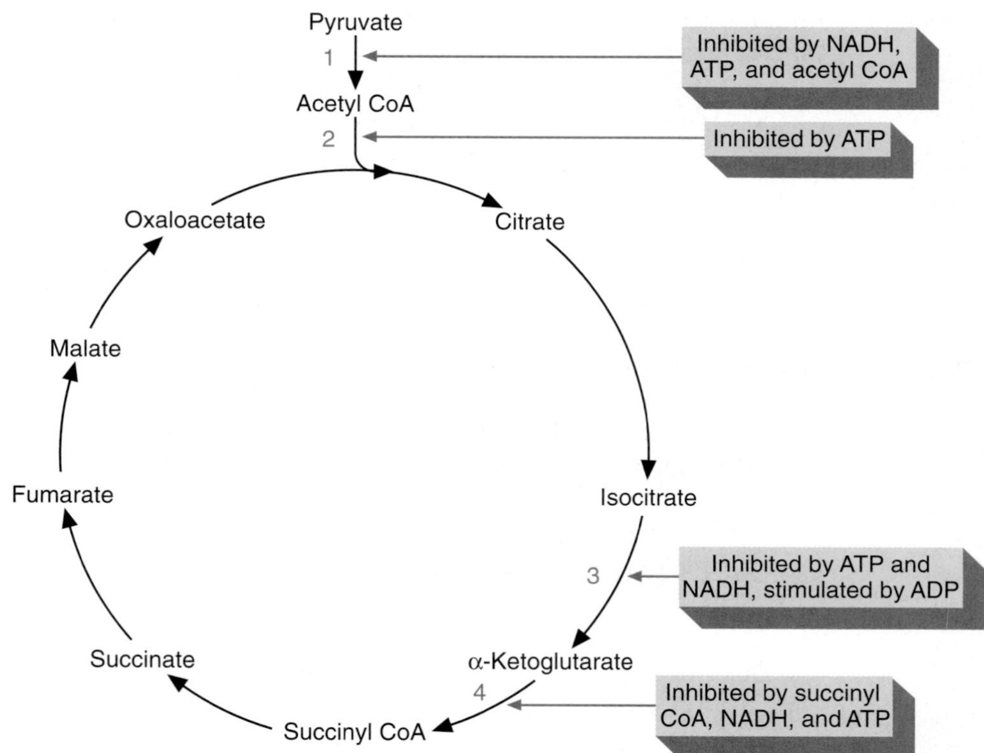

Figure 22.6
Regulation of the pyruvate dehydrogenase complex and the citric acid cycle.

22.6 Oxidative Phosphorylation

Learning Goal

In Section 22.3 we noted that the electrons carried by NADH can be used to produce three ATP molecules, and those carried by $FADH_2$ can be used to produce two ATP molecules. We turn now to the process by which the energy of electrons carried by these coenzymes is converted to ATP energy. It is a series of reactions called **oxidative phosphorylation,** which couples the oxidation of NADH and $FADH_2$ to the phosphorylation of ADP to generate ATP.

Electron Transport Systems and the Hydrogen Ion Gradient

Before we try to understand the mechanism of oxidative phosphorylation, let's first look at the molecules that carry out this complex process. Embedded within the mitochondrial inner membrane are **electron transport systems.** These are made up of a series of electron carriers, including coenzymes and cytochromes. All these molecules are located within the membrane in an arrangement that allows them to pass electrons from one to the next. This array of electron carriers is called the *respiratory electron transport system* (Figure 22.7). As you would expect in such sequential oxidation-reduction reactions, the electrons lose some energy with each transfer. Some of this energy is used to make ATP.

At three sites in the electron transport system, protons (H^+) can be pumped from the mitochondrial matrix to the intermembrane space. These H^+ contribute to a high-energy H^+ reservoir. At each of the three sites, enough H^+ are pumped into the H^+ reservoir to produce one ATP molecule. The first site is NADH dehydrogenase. Because electrons from NADH enter the electron transport system by being transferred to NADH dehydrogenase, all three sites actively pump H^+, and three ATP molecules are made (see Figure 22.7). $FADH_2$ is a less "powerful" electron donor. It transfers its electrons to an electron carrier that follows NADH dehydrogenase. As a result, when $FADH_2$ is oxidized, only the second and third sites pump H^+, and only two ATP molecules are made.

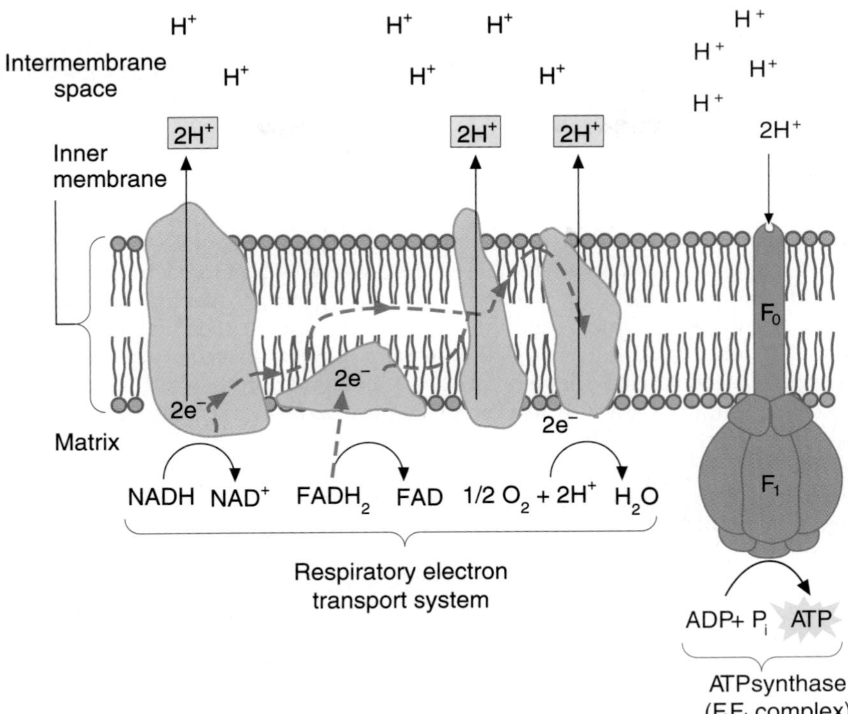

Figure 22.7
Electrons flow from NADH to molecular oxygen through a series of electron carriers embedded in the inner mitochondrial membrane. Protons are pumped from the mitochondrial matrix space into the intermembrane space. This results in a hydrogen ion reservoir in the intermembrane space. As protons pass through the channel in ATP synthase, their energy is used to phosphorylate ADP and produce ATP.

The last component needed for oxidative phosphorylation is a multiprotein complex called **ATP synthase,** also called the **F_0F_1 complex** (see Figure 22.7). The F_0 portion of the molecule is a channel through which H^+ pass. It spans the inner mitochondrial membrane, as shown in Figure 22.7. The F_1 part of the molecule is an enzyme that catalyzes the phosphorylation of ADP to produce ATP.

ATP Synthase and the Production of ATP

How does all this complicated machinery actually function? NADH carries electrons, originally from glucose, to the first carrier of the electron transport system, NADH dehydrogenase (see Figure 22.7). There, NADH is oxidized to NAD^+, which returns to the site of the citric acid cycle to be reduced again. As Figure 22.7 shows (dashed red line), the pair of electrons is passed to the next electron carrier, and H^+ are pumped to the intermembrane compartment. The electrons are passed sequentially through the electron transport system, and at two additional sites, H^+ from the matrix are pumped into the intermembrane compartment. With each transfer the electrons lose some of their potential energy. It is this energy that is used to transport H^+ across the inner mitochondrial membrane and into the H^+ reservoir. As mentioned above, $FADH_2$ donates its electrons to a carrier of lower energy and fewer H^+ are pumped into the reservoir.

The importance of keeping the electron transport system functioning becomes obvious when we consider what occurs in cyanide poisoning. Cyanide binds to the heme group iron of cytochrome oxidase, instantly stopping electron transfers and causing death within minutes!

Finally, the electrons arrive at the last carrier. They now have too little energy to accomplish any more work, but they *must* be donated to some final electron acceptor so that the electron transport system can continue to function. In aerobic organisms the **terminal electron acceptor** is molecular oxygen, O_2, and the product is water.

As the electron transport system continues to function, a high concentration of protons builds up in the intermembrane space. This creates an H^+ gradient across the inner mitochondrial membrane. Such a gradient is an enormous energy source, like water stored behind a dam. The mitochondria make use of the potential energy of the gradient to synthesize ATP energy.

A HUMAN PERSPECTIVE

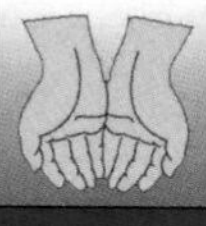

Brown Fat: The Fat That Makes You Thin?

Humans have two types of fat, or adipose, tissue. *White fat* is distributed throughout the body and is composed of aggregations of cells having membranous vacuoles containing stored triglycerides. The size and number of these storage vacuoles determine whether a person is overweight or not. The other type of fat is *brown fat*. Brown fat is a specialized tissue for heat production, called *nonshivering thermogenesis*. As the name suggests, this is a means of generating heat in the absence of the shivering response. The cells of brown fat look nothing like those of white fat. They do contain small fat vacuoles; however, the distinguishing feature of brown fat is the huge number of mitochondria within the cytoplasm. In addition, brown fat tissue contains a great many blood vessels. These provide oxygen for the thermogenic metabolic reactions.

Brown fat is most pronounced in newborns, cold-adapted mammals, and hibernators. One major difficulty faced by a newborn is temperature regulation. The baby leaves an environment in which he or she was bathed in fluid of a constant 37°C, body temperature. Suddenly, the child is thrust into a world that is much colder and in which he or she must generate his or her own warmth internally. By having a good reserve of active brown fat to generate that heat, the newborn is protected against cold shock at the time of birth. However, this thermogenesis literally burns up most of the brown fat tissue, and adults typically have so little brown fat that it can be found only by using a special technique called *thermography*, which detects temperature differences throughout a body. However, in some individuals, brown fat is very highly developed. For instance, the Korean diving women who spend 6–7 hours every day diving for pearls in cold water have a massive amount of brown fat to warm them by nonshivering thermogenesis. Thus development of brown fat is a mechanism of cold adaptation.

When it was noticed that such cold-adapted individuals were seldom overweight, a correlation was made between the amount of brown fat in the body and the tendency to become overweight. Studies done with rats suggest that, to some degree, fatness is genetically determined. In other words, you are as lean as your genes allow you to be. In these studies, cold-adapted and non-cold-adapted rats were fed cafeteria food—as much as they wanted—and their weight gain was monitored. In every case the cold-adapted rats, with their greater quantity of brown fat, gained significantly less weight than their non-cold-adapted counterparts, despite the fact that they ate as much as the non-cold-adapted rats. This and other studies led researchers to conclude that brown fat burns excess fat in a highly caloric diet.

How does brown fat generate heat and burn excess calories? For the answer we must turn to the mitochondrion. In addition to the ATP synthase and the electron transport system proteins that are found in all mitochondria, there is a protein in the inner mitochondrial membrane of brown fat tissue called *thermogenin*. This protein has a channel in the center through which the protons (H^+) of the intermembrane space could pass back into the mitochondrial matrix. Under normal conditions this channel is plugged by a GDP molecule so that it remains closed and the proton gradient can continue to drive ATP synthesis by oxidative phosphorylation.

When brown fat is "turned on," by cold exposure or in response to certain hormones, there is an immediate increase in the rate of glycolysis and β-oxidation of the stored fat (Chapter 23). These reactions produce acetyl CoA, which then fuels the citric acid cycle. The citric acid cycle, of course, produces NADH and $FADH_2$, which carry electrons to the electron transport system. Finally, the electron transport system pumps protons into the intermembrane space. Under usual conditions the energy of the proton gradient would be used to synthesize ATP. However, when brown fat is stimulated, the GDP that had plugged the pore in thermogenin is lost. Now protons pass freely back into the matrix space, and the proton gradient is dissipated. The energy of the gradient, no longer useful for generating ATP, is released as *heat*, the heat that warms and protects newborns and cold-adapted individuals.

Brown fat is just one of the body's many systems for maintaining a constant internal environment regardless of the conditions in the external environment. Such mechanisms, called *homeostatic mechanisms*, are absolutely essential to allow the body to adapt to and survive in an ever-changing environment.

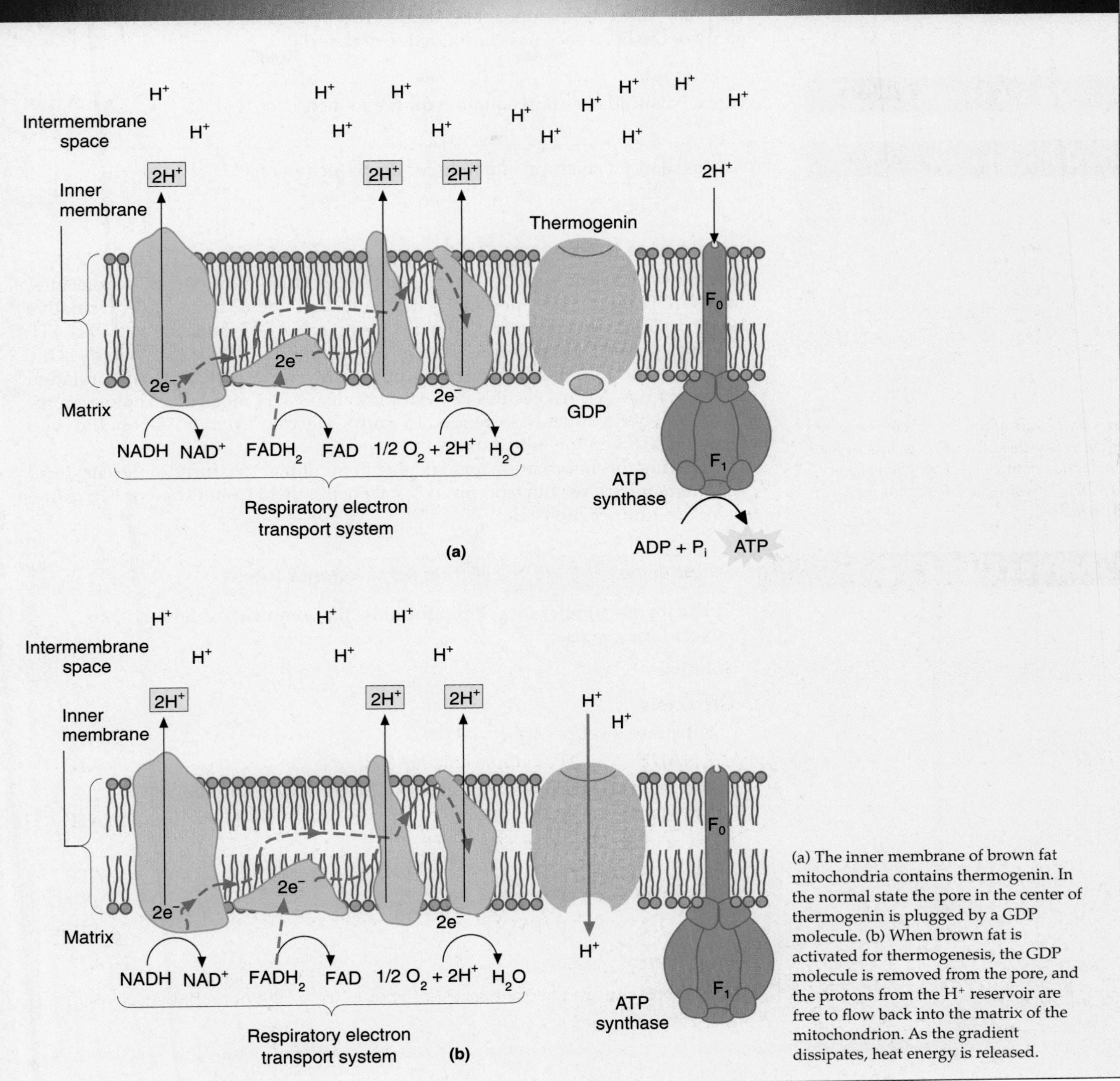

(a) The inner membrane of brown fat mitochondria contains thermogenin. In the normal state the pore in the center of thermogenin is plugged by a GDP molecule. (b) When brown fat is activated for thermogenesis, the GDP molecule is removed from the pore, and the protons from the H^+ reservoir are free to flow back into the matrix of the mitochondrion. As the gradient dissipates, heat energy is released.

ATP synthase harvests the energy of this gradient by making ATP. H^+ pass through the F_0 channel back into the matrix. This causes F_1 to become an active enzyme that catalyzes the phosphorylation of ADP to produce ATP. In this way the energy of the H^+ reservoir is harvested to make ATP.

Question 22.9 Write a balanced chemical equation for the reduction of NAD^+.

Question 22.10 Write a balanced chemical equation for the reduction of FAD.

Summary of the Energy Yield

One turn of the citric acid cycle results in the production of two CO_2 molecules, three NADH molecules, one $FADH_2$ molecule, and one ATP molecule. Oxidative phosphorylation yields three ATP molecules per NADH molecule and two ATP molecules per $FADH_2$ molecule. The only exception to these energy yields is the NADH produced in the cytoplasm during glycolysis. Oxidative phosphorylation yields only two ATP molecules per cytoplasmic NADH molecule. The reason for this is that energy must be expended to shuttle electrons from NADH in the cytoplasm to $FADH_2$ in the mitochondrion.

In some tissues of the body there is a more efficient shuttle system that results in the production of three ATP per cytoplasmic NADH. This system is described in Appendix G.

Knowing this information and keeping in mind that two turns of the citric acid cycle are required, we can sum up the total energy yield from the complete oxidation of one glucose molecule.

EXAMPLE 22.1 ***Determining the Yield of ATP from Aerobic Respiration***

Calculate the number of ATP produced by the complete oxidation of one molecule of glucose.

Solution

Glycolysis:	
Substrate-level phosphorylation	2 ATP
2 NADH × 2 ATP/cytoplasmic NADH	4 ATP
Conversion of 2 pyruvate molecules to 2 acetyl CoA molecules:	
2 NADH × 3 ATP/NADH	6 ATP
Citric acid cycle (two turns):	
2 GTP × 1 ATP/GTP	2 ATP
6 NADH × 3 ATP/NADH	18 ATP
2 $FADH_2$ × 2 ATP/$FADH_2$	4 ATP
	36 ATP

This represents an energy harvest of about 40% of the potential energy of glucose.

Aerobic metabolism is very much more efficient than anaerobic metabolism. The abundant energy harvested by aerobic metabolism has had enormous consequences for the biological world. Much of the energy released by the oxidation of fuels is not lost as heat but conserved in the form of ATP. Organisms that possess abundant energy have evolved into multicellular organisms and developed specialized functions. As a consequence of their energy requirements, all multicellular organisms are aerobic.

22.7 The Degradation of Amino Acids

Learning Goal

Carbohydrates are not our only source of energy. As we saw in Chapter 21, dietary protein is digested to amino acids that can also be used as an energy source, although this is not their major metabolic function. Most of the amino acids used for energy come from the diet. In fact, it is only under starvation conditions, when stored glycogen has been depleted, that the body begins to burn its own protein, for instance from muscle, as a fuel.

The fate of the mixture of amino acids provided by digestion of protein depends upon a balance between the need for amino acids for biosynthesis and the need for cellular energy. Only those amino acids that are not needed for protein synthesis are eventually converted into citric acid cycle intermediates and used as fuel.

The degradation of amino acids occurs primarily in the liver and takes place in two stages. The first stage is the removal of the α-amino group, and the second is the degradation of the carbon skeleton. In land mammals the amino group generally ends up in urea, which is excreted in the urine. The carbon skeletons can be converted into a variety of compounds, including citric acid cycle intermediates, pyruvate, acetyl CoA, or acetoacetyl CoA. The degradation of the carbon skeletons is summarized in Figure 22.8. Deamination reactions and the fate of the carbon skeletons of amino acids are the focus of this section.

Removal of α-Amino Groups: Transamination

The first stage of amino acid degradation, the removal of the α-amino group, is usually accomplished by a **transamination** reaction. **Transaminases** catalyze the transfer of the α-amino group from an α-amino acid to an α-keto acid:

$$\underset{\substack{\text{Donor}\\\text{amino}\\\text{acid}}}{H{-}C(NH_3^+)(R^1){-}COO^-} + \underset{\substack{\text{Acceptor}\\\text{keto}\\\text{acid}}}{R^2{-}C(=O){-}COO^-} \xrightleftharpoons{\text{Transaminase}} \underset{\substack{\alpha\text{-Keto acid}\\\text{of amino}\\\text{acid}}}{R^1{-}C(=O){-}COO^-} + \underset{\substack{\text{New}\\\text{amino}\\\text{acid}}}{H{-}C(NH_3^+)(R^2){-}COO^-}$$

The α-amino group of a great many amino acids is transferred to α-ketoglutarate to produce the amino acid glutamate and a new keto acid. This glutamate family of transaminases is especially important because the α-keto acid corresponding to glutamate is α-ketoglutarate, a citric acid cycle intermediate. The glutamate transaminases thus provide a direct link between amino acid degradation and the citric acid cycle.

Aspartate transaminase, catalyzes the transfer of the α-amino group of aspartate to α-ketoglutarate, producing oxaloacetate and glutamate:

$$\underset{\text{Aspartate}}{^-OOC{-}CH_2{-}CH(NH_3^+){-}COO^-} + \underset{\alpha\text{-Ketoglutarate}}{^-OOC{-}CH_2{-}CH_2{-}C(=O){-}COO^-} \rightleftharpoons \underset{\text{Oxaloacetate}}{^-OOC{-}CH_2{-}C(=O){-}COO^-} + \underset{\text{Glutamate}}{^-OOC{-}CH_2{-}CH_2{-}CH(NH_3^+){-}COO^-}$$

Another important transaminase in mammalian tissues is *alanine transaminase,* which catalyzes the transfer of the α-amino group of alanine to α-ketoglutarate and produces pyruvate and glutamate:

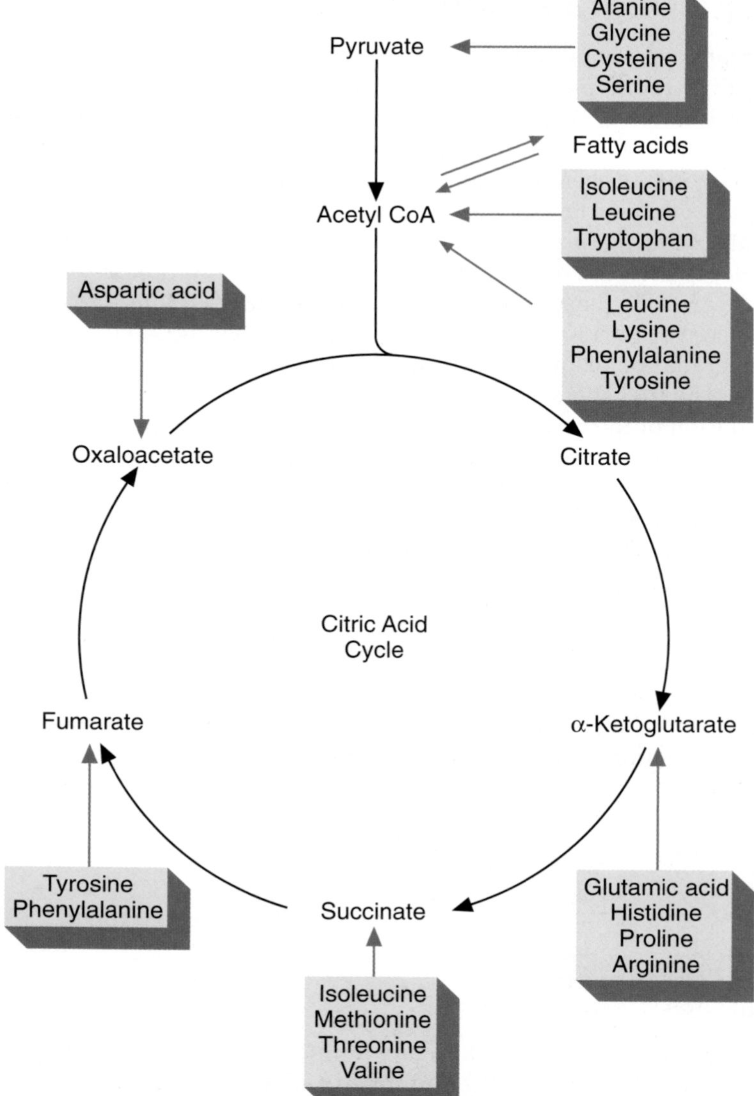

Figure 22.8
The carbon skeletons of amino acids can be converted to citric acid cycle intermediates and completely oxidized to produce ATP energy.

$$\underset{\text{Alanine}}{\begin{array}{c} \overset{+}{N}H_3 \\ | \\ H-C-COO^- \\ | \\ H-C-H \\ | \\ H \end{array}} + \underset{\alpha\text{-Ketoglutarate}}{\begin{array}{c} O \\ \| \\ C-COO^- \\ | \\ H-C-H \\ | \\ H-C-H \\ | \\ COO^- \end{array}} \rightleftharpoons \underset{\text{Pyruvate}}{\begin{array}{c} O \\ \| \\ C-COO^- \\ | \\ H-C-H \\ | \\ H \end{array}} + \underset{\text{Glutamate}}{\begin{array}{c} \overset{+}{N}H_3 \\ | \\ H-C-COO^- \\ | \\ H-C-H \\ | \\ H-C-H \\ | \\ COO^- \end{array}}$$

See Appendix F, Water-Soluble Vitamins, for more information on these vitamins and the coenzymes that are made from them.

All of the more than fifty transaminases that have been discovered require the coenzyme **pyridoxal phosphate.** This coenzyme is derived from vitamin B_6 (pyridoxine, Figure 22.9).

The transamination reactions shown above appear to be a simple transfer, but in reality, the reaction is much more complex. The transaminase binds the amino acid (aspartate in Figure 22.10a) in its active site. Then the α-amino group of aspartate is transferred to pyridoxal phosphate, producing pyridoxamine phosphate

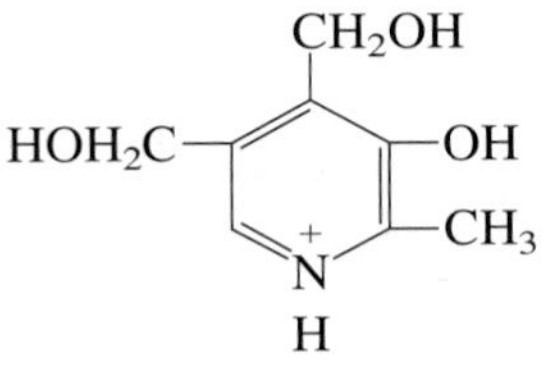

$$\text{HOH}_2\text{C} \text{—(pyridinium ring, } \overset{+}{\text{N}}\text{–H)— } \text{CH}_2\text{OH},\ \text{OH},\ \text{CH}_3$$

Pyridoxine
(vitamin B_6)

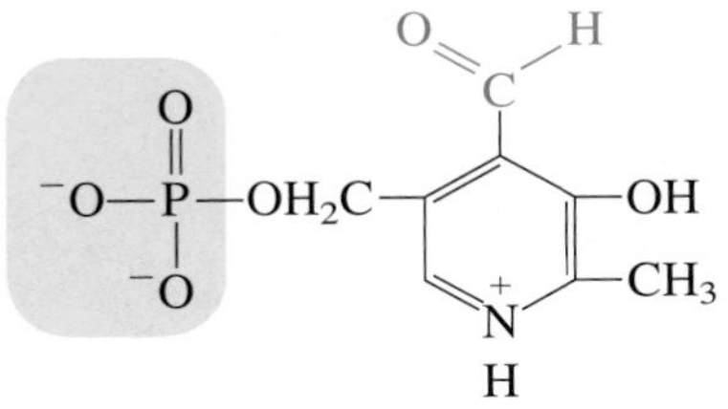

$$^{-}\text{O—}\underset{\underset{-}{\text{O}}}{\overset{\overset{\text{O}}{\|}}{\text{P}}}\text{—OH}_2\text{C— (pyridinium ring, } \overset{+}{\text{N}}\text{–H)— CHO, OH, CH}_3$$

Pyridoxal phosphate

Figure 22.9
The structure of pyridoxal phosphate, the coenzyme required for all transamination reactions, and pyridoxine, vitamin B_6, the vitamin from which it is derived.

and oxaloacetate (Figure 22.10b). The amino group is then transferred to an α-keto acid, in this case, α-ketoglutarate (Figure 22.10c), to produce the amino acid glutamate (Figure 22.10d). Next we will examine the fate of the amino group that has been transferred to α-ketoglutarate to produce glutamate.

Question 22.11

What is the role of pyridoxal phosphate in transamination reactions?

Question 22.12

What is the function of a transaminase?

Removal of α-Amino Groups: Oxidative Deamination

In the next stage of amino acid degradation, ammonium ion is liberated from the glutamate formed by the transaminase. This breakdown of glutamate, catalyzed by the enzyme *glutamate dehydrogenase*, occurs as follows:

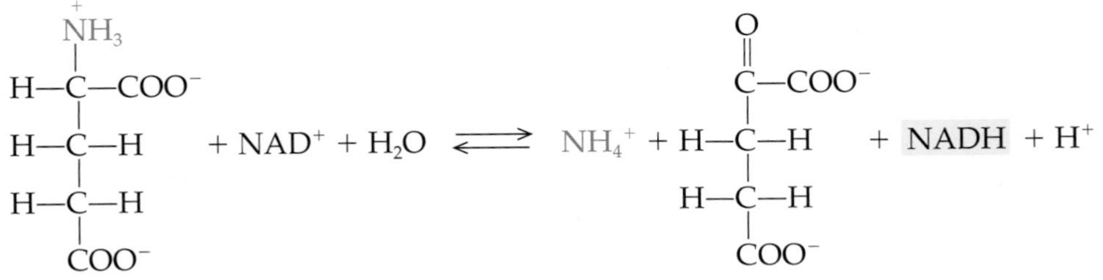

$$\underset{\text{Glutamate}}{^{-}\text{OOC—CH}_2\text{—CH}_2\text{—CH}(\overset{+}{\text{N}}\text{H}_3)\text{—COO}^{-}} + \text{NAD}^{+} + \text{H}_2\text{O} \rightleftharpoons \text{NH}_4^{+} + \underset{\alpha\text{-Ketoglutarate}}{^{-}\text{OOC—CH}_2\text{—CH}_2\text{—C(=O)—COO}^{-}} + \text{NADH} + \text{H}^{+}$$

This is an example of an **oxidative deamination,** an oxidation-reduction process in which NAD^+ is reduced to NADH and the amino acid is deaminated (the amino group is removed). A summary of the deamination reactions described is shown in Figure 22.11.

The Fate of Amino Acid Carbon Skeletons

The carbon skeletons produced by these and other deamination reactions enter glycolysis or the citric acid cycle at many steps. For instance, we have seen that

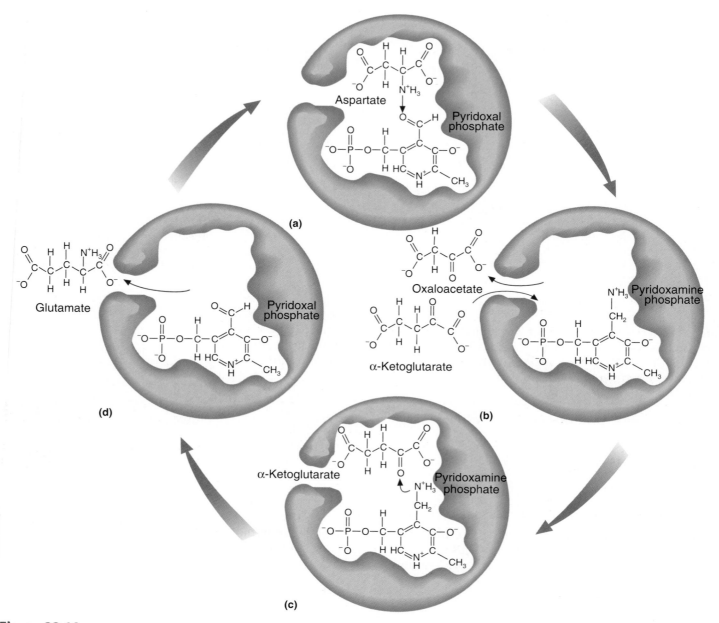

Figure 22.10
The mechanism of transamination.

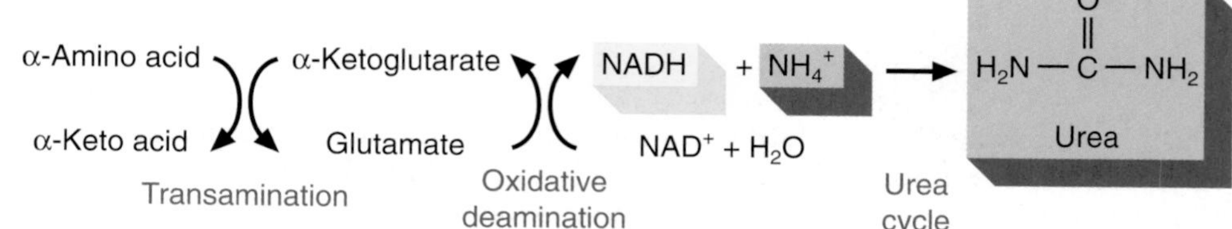

Figure 22.11
Summary of the deamination of an α-amino acid and the fate of the ammonium ion (NH_4^+).

transamination converts aspartate to oxaloacetate and alanine to pyruvate. The positions at which the carbon skeletons of various amino acids enter the energy-harvesting pathways are summarized in Figure 22.8.

22.8 The Urea Cycle

Learning Goal 

Oxidative deamination produces large amounts of ammonium ion. Because ammonium ions are extremely toxic, they must be removed from the body, regardless of the energy expenditure required. In humans, they are detoxified in the liver by converting the ammonium ions into urea. This pathway, called the **urea cycle,** is the method by which toxic ammonium ions are kept out of the blood. The excess ammonium ions incorporated in urea are excreted in the urine (Figure 22.12).

Reactions of the Urea Cycle

The five reactions of the urea cycle are shown in Figure 22.12, and details of the reactions are summarized as follows.

Step 1. The first step of the cycle is a reaction in which CO_2 and NH_4^+ form carbamoyl phosphate. This reaction, which also requires ATP and H_2O, occurs in the mitochondria and is catalyzed by the enzyme *carbamoyl phosphate synthetase.*

$$CO_2 + NH_4^+ + 2ATP + H_2O \longrightarrow H_2N{-}\overset{\displaystyle O}{\overset{\|}{C}}{-}O{-}\overset{\displaystyle O}{\overset{\|}{\underset{\underset{\displaystyle O^-}{|}}{P}}}{-}O^- + 2ADP + P_i + 3H^+$$

Carbamoyl phosphate

Step 2. The carbamoyl phosphate thus produced condenses with the amino acid ornithine to produce the amino acid citrulline. This reaction also occurs in the mitochondria and is catalyzed by the enzyme *ornithine transcarbamoylase.*

The urea cycle involves several unusual amino acids that are not found in polypeptides.

$$\begin{array}{c} \overset{+}{N}H_3 \\ | \\ H{-}C{-}H \\ | \\ H{-}C{-}H \\ | \\ H{-}C{-}H \\ | \\ H{-}C{-}\overset{+}{N}H_3 \\ | \\ COO^- \end{array} + H_2N{-}\overset{\displaystyle O}{\overset{\|}{C}}{-}O{-}\overset{\displaystyle O}{\overset{\|}{\underset{\underset{\displaystyle O^-}{|}}{P}}}{-}O^- \longrightarrow \begin{array}{c} H{-}N{-}\overset{\displaystyle O}{\overset{\|}{C}}{-}NH_2 \\ | \\ H{-}C{-}H \\ | \\ H{-}C{-}H \\ | \\ H{-}C{-}H \\ | \\ H{-}C{-}\overset{+}{N}H_3 \\ | \\ COO^- \end{array} + P_i$$

Ornithine Carbamoyl phosphate Citrulline

Step 3. Citrulline is transported into the cytoplasm and now condenses with aspartate to produce argininosuccinate. This reaction, which requires energy released by the hydrolysis of ATP, is catalyzed by the enzyme *argininosuccinate synthase.*

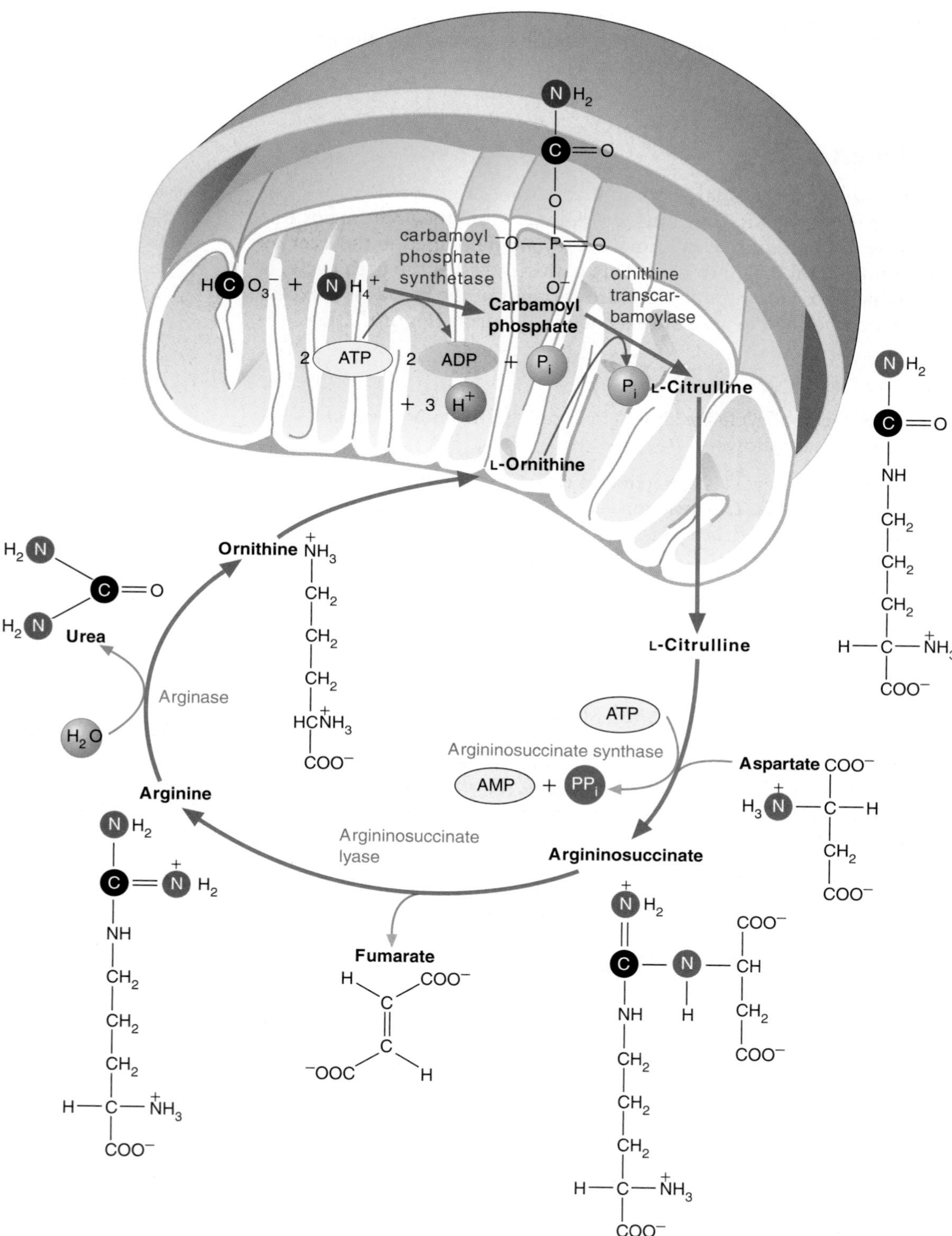

Figure 22.12
The urea cycle converts ammonium ions into urea, which is less toxic. The sources of the atoms are shown in color and the intracellular locations of the reactions are indicated. Citrulline, formed in the reaction between ornithine and carbamoyl phosphate, is transported out of the mitochondrion and into the cytoplasm. Ornithine, a substrate for the formation of citrulline, is transported from the cytoplasm into the mitochondrion.

$$\text{Citrulline} + \text{Aspartate} \xrightarrow{\text{ATP} \rightarrow \text{AMP} + \text{PP}_i} \text{Argininosuccinate}$$

Citrulline Aspartate Argininosuccinate

Step 4. Now the argininosuccinate is cleaved to produce the amino acid arginine and the citric acid cycle intermediate fumarate. This reaction is catalyzed by the enzyme *argininosuccinate lyase.*

$$\text{Argininosuccinate} \longrightarrow \text{Arginine} + \text{Fumarate}$$

Argininosuccinate Arginine Fumarate

Step 5. Finally, arginine is hydrolyzed to generate urea, the product of the reaction to be excreted, and ornithine, the original reactant in the cycle. *Arginase* is the enzyme that catalyzes this reaction.

$$\text{Arginine} + H_2O \longrightarrow H_2N\text{—}\overset{\overset{\displaystyle O}{\|}}{C}\text{—}NH_2 + \text{Ornithine}$$

Arginine Water Urea Ornithine

Note that one of the amino groups in urea is derived from the ammonium ion and the second is derived from the amino acid aspartate.

The abbreviation PP_i represents the pyrophosphate group, which consists of two phosphate groups joined by a phosphoanhydride bond:

$$^{-}O\text{—}\underset{O^-}{\overset{O}{P}}\text{—}O\text{—}\underset{O^-}{\overset{O}{P}}\text{—}O^{-}$$

Learning Goal

There are genetically transmitted diseases that result from a deficiency of one of the enzymes of the urea cycle. The importance of the urea cycle is apparent when we consider the terrible symptoms suffered by afflicted individuals. A deficiency of urea cycle enzymes causes an elevation of the concentration of NH_4^+, a condition known as **hyperammonemia.** If there is a complete deficiency of one of

A MEDICAL PERSPECTIVE

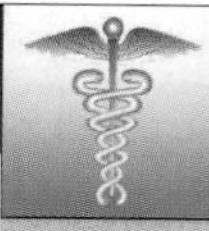

Pyruvate Carboxylase Deficiency

Pyruvate carboxylase is the enzyme that converts pyruvate to oxaloacetate.

$$\text{Pyruvate} + CO_2 + \text{ATP} + H_2O \longrightarrow \text{Oxaloacetate} + \text{ADP} + 2H^+$$

This reaction is important because it provides oxaloacetate for the citric acid cycle when the supplies have run low because of the demands of biosynthesis. It is also the enzyme that catalyzes the first step in gluconeogenesis, the pathway that provides the body with needed glucose in times of starvation or periods of exercise that deplete glycogen stores. But somehow these descriptions don't fill us with a sense of the importance of this enzyme and its jobs. It is not until we investigate a case study of a child born with pyruvate carboxylase deficiency that we see the full impact of this enzyme.

Pyruvate carboxylase deficiency is found in about 1 in 250,000 births; however, there is an increased incidence in native North American Indians who speak the Algonquin dialect and in the French. There are two types of genetic disorders that have been described. In the neonatal form of the disease, there is a complete absence of the enzyme. Symptoms are apparent at birth and the child is born with brain abnormalities. In the infantile form, the patient develops symptoms early in infancy. Again, it is neurological symptoms that draw attention to the condition. The infants do not develop mental or psychomotor skills. They may develop seizures and/or respiratory depression. In both cases, it is the brain that suffers the greatest damage. In fact, this is the case in most of the disorders that reduce energy metabolism because the brain has such high energy requirements.

Biochemically, patients exhibit quite a variety of symptoms. They show acidosis (low blood pH) due to accumulations of lactate and extremely high pyruvate concentrations in the blood. Blood levels of alanine are also high and large doses of alanine do not stimulate gluconeogenesis. Furthermore, a patient's cells accumulate lipid.

We can understand each of these symptoms by considering the pathways affected by the absence of this single enzyme. Lactic acidosis results from the fact that the body must rely on glycolysis and lactate fermentation for most of its energy needs. Alanine levels are high because it isn't being transaminated to pyruvate efficiently, because pyruvate levels are so high. In addition, alanine can't be converted to glucose by gluconeogenesis. Although the excess alanine is taken up by the liver and converted to pyruvate, the pyruvate can't be converted to glucose. Lipids accumulate because a great deal of pyruvate is converted to acetyl CoA. However, the acetyl CoA is not used to produce citrate as a result of the absence of oxaloacetate. So, the acetyl CoA is thus used to synthesize fatty acids, which are stored as triglycerides.

Dietary intervention has been tried. One such regimen is to supplement with aspartic acid and glutamic acid. The theory behind this treatment is as complex as the many symptoms of the disorder. Both amino acids can be aminated (amino groups added) in non-nervous tissue. This produces asparagine and glutamine, both of which are able to cross the blood-brain barrier. Glutamine is deaminated to glutamate, which is then transaminated to α-ketoglutarate, indirectly replenishing oxaloacetate. Asparagine can be deaminated to aspartate, which can be converted to oxaloacetate. This serves as a second supply of oxaloacetate. To date, these attempts at dietary intervention have not proved successful. Perhaps in time research will provide the tools for enzyme replacement therapy or gene therapy that could alleviate the symptoms.

the enzymes of the urea cycle, the result is death in early infancy. If there is a partial deficiency of one of the enzymes of the urea cycle, the result may be retardation, convulsions, and vomiting. In these milder forms of hyperammonemia, a low-protein diet leads to a lower concentration of NH_4^+ in blood and less severe clinical symptoms.

Question 22.13 What is the purpose of the urea cycle?

Question 22.14 Where do the reactions of the urea cycle occur?

22.9 Overview of Anabolism: The Citric Acid Cycle as a Source of Biosynthetic Intermediates

Learning Goal

So far, we have talked about the citric acid cycle only as an energy-harvesting mechanism. We have seen that dietary carbohydrates and amino acids enter the pathway at various stages and are oxidized to generate NADH and $FADH_2$, which, by means of oxidative phosphorylation, are used to make ATP.

However, the role of the citric acid cycle in cellular metabolism involves more than just **catabolism.** It plays a key role in **anabolism,** or biosynthesis, as well. Figure 22.13 shows the central role of glycolysis and the citric acid cycle as energy-harvesting reactions, as well as their role as a source of biosynthetic precursors.

As you may already suspect from the fact that amino acids can be converted into citric acid cycle intermediates, these same citric acid cycle intermediates can also be used as starting materials for the synthesis of amino acids. Oxaloacetate

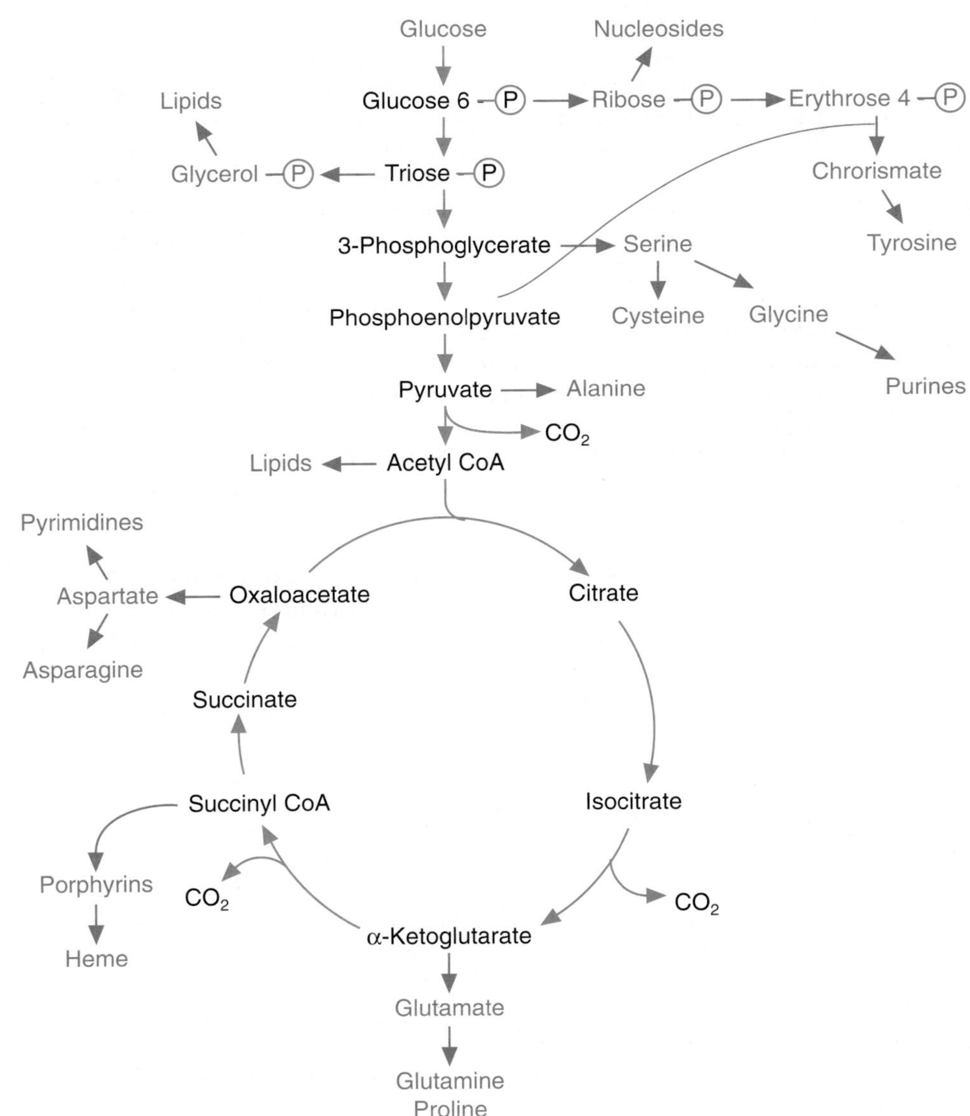

Figure 22.13
Glycolysis, the pentose phosphate pathway, and the citric acid cycle also provide a variety of precursors for the biosynthesis of amino acids, nitrogenous bases, and porphyrins.

provides the carbon skeleton for the one-step synthesis of the amino acid aspartate by the transamination reaction:

$$\text{oxaloacetate} + \text{glutamate} \rightleftharpoons \text{aspartate} + \alpha\text{-ketoglutarate}$$

Aside from providing aspartate for protein synthesis, this reaction provides aspartate for the urea cycle.

Asparagine is made from aspartate by the amination reaction

$$\text{aspartate} + NH_4^+ + \text{ATP} \longrightarrow \text{asparagine} + \text{AMP} + PP_i + H^+$$

The nine amino acids not shown in Figure 22.13 (histidine, isoleucine, leucine, lysine, methionine, phenylalanine, threonine, tryptophan, and valine) are called the essential amino acids (Section 19.11) because they cannot be synthesized by humans. Arginine is an essential amino acid for infants and adults under physical stress.

α-Ketoglutarate serves as the starting carbon chain for the family of amino acids including glutamate, glutamine, proline, and arginine. Glutamate is especially important because it serves as the donor of the α-amino group of almost all other amino acids. It is synthesized from NH_4^+ and α-ketoglutarate in a reaction mediated by glutamate dehydrogenase. This is the reverse of the reaction shown in Figure 22.11 and previously described. In this case the coenzyme that serves as the reducing agent is NADPH.

$$NH_4^+ + \alpha\text{-ketoglutarate} + \text{NADPH} \rightleftharpoons \text{L-glutamate} + \text{NADP}^+ + H_2O$$

Glutamine, proline, and arginine are synthesized from glutamate.

Actually, in humans tyrosine is made from the essential amino acid phenylalanine.

Examination of Figure 22.13 reveals that serine, glycine, and cysteine are synthesized from 3-phosphoglycerate; alanine is synthesized from pyruvate; and tyrosine is produced from phosphoenolpyruvate and the four-carbon sugar erythrose-4-phosphate, which, in turn, is synthesized from glucose-6-phosphate in the pentose phosphate pathway. In addition to the amino acid precursors, glycolysis and the citric acid cycle also provide precursors for lipids and the nitrogenous bases that are required to make DNA, the molecule that carries the genetic information. They also generate precursors for heme, the prosthetic group that is required for hemoglobin, myoglobin, and the cytochromes.

Clearly, the reactions of glycolysis and the citric acid cycle are central to both anabolic and catabolic cellular activities. Metabolic pathways that function in both anabolism and catabolism are called **amphibolic pathways.** Consider for a moment the difficulties that the dual nature of these pathways could present to the cell. When the cell is actively growing, there is a great demand for biosynthetic precursors to build new cell structures. A close look at Figure 22.13 shows us that periods of active cell growth and biosynthesis may deplete the supply of citric acid cycle intermediates. The problem is, the processes of growth and biosynthesis also require a great deal of ATP!

Carboxylation of pyruvate during gluconeogenesis is discussed in Section 21.6.

The solution to this problem is to have an alternative pathway for oxaloacetate synthesis that can produce enough oxaloacetate to supply the anabolic and catabolic requirements of the cell. Although bacteria and plants have several mechanisms, the only way that mammalian cells can produce more oxaloacetate is by the carboxylation of pyruvate, a reaction that is also important in gluconeogenesis. This reaction is

$$\text{pyruvate} + CO_2 + \text{ATP} \longrightarrow \text{oxaloacetate} + \text{ADP} + P_i$$

The enzyme that catalyzes this reaction is *pyruvate carboxylase.* It is a conjugated protein having as its covalently linked prosthetic group the vitamin *biotin*. This enzyme is "turned on" by high levels of acetyl CoA. A high concentration of acetyl CoA is a signal that the cell requires high levels of the citric acid cycle intermediates, particularly oxaloacetate, the beginning substrate.

The reaction catalyzed by pyruvate carboxylase is called an **anaplerotic reaction.** The term *anaplerotic* means to fill up. Indeed, this critical enzyme must constantly replenish the oxaloacetate and thus indirectly all the citric acid cycle intermediates that are withdrawn as biosynthetic precursors for the reactions summarized in Figure 22.13.

Question 22.15

Explain how the citric acid cycle serves as an amphibolic pathway.

Question 22.16

What is the function of an anaplerotic reaction?

Summary

22.1 The Mitochondria

The *mitochondria* are aerobic cell organelles that are responsible for most of the ATP production in eukaryotic cells. They are enclosed by a double membrane. The outer membrane permits low-molecular-weight molecules to pass through. The *inner mitochondrial membrane,* by contrast, is almost completely impermeable to most molecules. The inner mitochondrial membrane is the site where *oxidative phosphorylation* occurs. The enzymes of the *citric acid cycle,* of amino acid catabolism, and of fatty acid oxidation are located in the *matrix space* of the mitochondrion.

22.2 Conversion of Pyruvate to Acetyl CoA

Under aerobic conditions, pyruvate is oxidized by the pyruvate dehydrogenase complex. Acetyl CoA, formed in this reaction, is a central molecule in both catabolism and anabolism.

22.3 An Overview of Aerobic Respiration

Aerobic respiration is the oxygen-requiring degradation of food molecules and production of ATP. *Oxidative phosphorylation* is the process that uses the high-energy electrons harvested by oxidation of substrates of the citric acid cycle to produce ATP.

22.4 The Citric Acid Cycle (The Krebs Cycle)

The *citric acid cycle* is the final pathway for the degradation of carbohydrates, amino acids, and fatty acids. The citric acid cycle occurs in the matrix of the mitochondria. It is a cyclic series of biochemical reactions that accomplishes the complete oxidation of the carbon skeletons of food molecules.

22.5 Control of the Citric Acid Cycle

Because the rate of ATP production by the cell must vary with the amount of available oxygen and the energy requirements of the body at any particular time, the citric acid cycle is regulated at several steps. This allows the cell to generate more energy when needed, as for exercise, and less energy when the body is at rest.

22.6 Oxidative Phosphorylation

Oxidative phosphorylation is the process by which NADH and $FADH_2$ are oxidized and ATP is produced. Two molecules of ATP are produced when $FADH_2$ is oxidized, and three molecules of ATP are produced when NADH is oxidized. The complete oxidation of one glucose molecule by glycolysis, the citric acid cycle, and oxidative phosphorylation yields thirty-six molecules of ATP versus two molecules of ATP for anaerobic degradation of glucose by glycolysis and fermentation.

22.7 The Degradation of Amino Acids

Amino acids are oxidized in the mitochondria. The first step of amino acid catabolism is deamination, the removal of the amino group. The carbon skeletons of amino acids are converted into molecules that can enter the citric acid cycle.

22.8 The Urea Cycle

In the *urea cycle* the toxic ammonium ions released by deamination of amino acids are incorporated in urea, which is excreted in the urine.

22.9 Overview of Anabolism: The Citric Acid Cycle as a Source of Biosynthetic Intermediates

In addition to its role in *catabolism,* the citric acid cycle also plays an important role in cellular *anabolism,* or biosynthetic reactions. Many of the citric acid cycle intermediates are precursors for the synthesis of amino acids and macromolecules required by the cell. A pathway that functions in both catabolic and anabolic reactions is called an *amphibolic pathway.*

Key Terms

acetyl CoA (22.2)
aerobic respiration (22.3)
amphibolic pathway (22.9)
anabolism (22.9)
anaplerotic reaction (22.9)
ATP synthase (22.6)
catabolism (22.9)
citric acid cycle (22.4)
coenzyme A (22.2)
cristae (22.1)

electron transport system (22.6)
F_0F_1 complex (22.6)
hyperammonemia (22.8)
inner mitochondrial membrane (22.1)
intermembrane space (22.1)
matrix space (22.1)
mitochondria (22.1)
outer mitochondrial membrane (22.1)
oxidative deamination (22.7)
oxidative phosphorylation (22.6)
pyridoxal phosphate (22.7)
pyruvate dehydrogenase complex (22.2)
terminal electron acceptor (22.6)
transaminase (22.7)
transamination (22.7)
urea cycle (22.8)

Questions and Problems

The Mitochondria

22.17 What is the function of the intermembrane compartment of the mitochondria?

22.18 What biochemical processes occur in the matrix space of the mitochondria?

22.19 In what important way do the inner and outer mitochondrial membranes differ?

22.20 What kinds of proteins are found in the inner mitochondrial membrane?

Conversion of Pyruvate to Acetyl CoA

22.21 Under what metabolic conditions is pyruvate converted to acetyl CoA?

22.22 Write a chemical equation for the production of acetyl CoA from pyruvate. Under what conditions does this reaction occur?

22.23 How could a deficiency of riboflavin, thiamine, niacin, or pantothenic acid reduce the amount of ATP the body can produce?

22.24 In what form are the vitamins riboflavin, thiamine, niacin, and pantothenic acid needed by the pyruvate dehydrogenase complex?

The Citric Acid Cycle

22.25 Label each of the following statements as true or false:

a. Both glycolysis and the citric acid cycle are aerobic processes.

b. Both glycolysis and the citric acid cycle are anaerobic processes.

c. Glycolysis occurs in the cytoplasm, and the citric acid cycle occurs in the mitochondria.

d. The inner membrane of the mitochondria is virtually impermeable to most substances.

22.26 Fill in the blanks:

a. The proteins of the electron transport system are found in the ________, the enzymes of the citric acid cycle are found in the ________, and the hydrogen ion reservoir is found in the ________ of the mitochondria.

b. The infoldings of the inner mitochondrial membrane are called ________.

c. Energy released by oxidation in the citric acid cycle is conserved in the form of phosphoanhydride bonds in ________.

d. The purpose of the citric acid cycle is the ________ of the acetyl group.

22.27 **a.** To what metabolic intermediate is the acetyl group of acetyl CoA transferred in the citric acid cycle?

b. What is the product of this reaction?

22.28 To what final products is the acetyl group of acetyl CoA converted during oxidation in the citric acid cycle?

22.29 How many ions of NAD^+ are reduced to molecules of NADH during one turn of the citric acid cycle?

22.30 How many molecules of FAD are converted to $FADH_2$ during one turn of the citric acid cycle?

22.31 What is the net yield of ATP for anaerobic glycolysis?

22.32 How many molecules of ATP are produced by the complete degradation of glucose via glycolysis, the citric acid cycle, and oxidative phosphorylation?

22.33 What is the function of acetyl CoA in the citric acid cycle?

22.34 What is the function of oxaloacetate in the citric acid cycle?

22.35 GTP is formed in one step of the citric acid cycle. How is this GTP converted into ATP?

22.36 What is the chemical meaning of the term *decarboxylation?*

22.37 The first reaction in the citric acid cycle is an aldol condensation. Write the equation for this reaction and explain its significance.

22.38 A bacterial culture is given ^{14}C-labeled pyruvate as its sole source of carbon and energy. The following is the structure of the radiolabeled pyruvate.

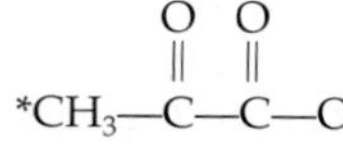

Follow the fate of the radioactive carbon through the reactions of the citric acid cycle.

22.39 The enzyme aconitase catalyzes the isomerization of citrate into isocitrate. Discuss the two reactions catalyzed by aconitase in terms of the chemistry of alcohols and alkenes.

22.40 A bacterial culture is given ^{14}C-labeled pyruvate as its sole source of carbon and energy. The following is the structure of the radiolabeled pyruvate.

$$CH_3{-}\overset{\overset{O}{\|}}{C^*}{-}\overset{\overset{O}{\|}}{C}{-}O^-$$

Follow the fate of the radioactive carbon through the reactions of the citric acid cycle.

22.41 In the oxidation of malate to oxaloacetate, what is the structural evidence that an oxidation reaction has occurred? What functional groups are involved?

22.42 In the oxidation of succinate to fumarate, what is the structural evidence that an oxidation reaction has occurred? What functional groups are involved?

22.43 To what class of enzymes does dinucleotide diphosphokinase belong? Explain your answer.

22.44 To what class of enzymes does succinate dehydrogenase belong? Explain your answer.

Control of the Citric Acid Cycle

22.45 What is the importance of the regulation of the citric acid cycle?

22.46 Explain the role of allosteric enzymes in control of the citric acid cycle.

22.47 What molecule serves as a signal to increase the rate of the reactions of the citric acid cycle?

22.48 What molecules serve as signals to decrease the rate of the reactions of the citric acid cycle?

Oxidative Phosphorylation

22.49 How many molecules of ATP are produced when one molecule of NADH is oxidized by oxidative phosphorylation?

22.50 How many molecules of ATP are produced when one molecule of $FADH_2$ is oxidized by oxidative phosphorylation?
22.51 What is the source of energy for the synthesis of ATP in mitochondria?
22.52 What is the name of the enzyme that catalyzes ATP synthesis in mitochondria?
22.53 What is the function of the electron transport systems of the mitochondria?
22.54 What is the cellular location of the electron transport systems?
22.55 a. Compare the number of molecules of ATP produced by glycolysis to the number of ATP molecules produced by oxidation of glucose by aerobic respiration.
b. Which pathway produces more ATP? Explain.
22.56 At which steps in the citric acid cycle do oxidation–reduction reactions occur?

The Degradation of Amino Acids

22.57 What chemical transformation is carried out by transaminases?
22.58 Write a chemical equation for the transfer of an amino group from alanine to α-ketoglutarate, catalyzed by a transaminase.
22.59 Why is the glutamate family of transaminases so important?
22.60 What biochemical reaction is catalyzed by glutamate dehydrogenase?
22.61 Into which citric acid cycle intermediate is each of the following amino acids converted?
a. Alanine
b. Glutamate
c. Aspartate
d. Phenylalanine
e. Threonine
f. Arginine
22.62 What is the net ATP yield for degradation of each of the amino acids listed in problem 22.62?
22.63 Write a balanced equation for the synthesis of glutamate that is mediated by the enzyme glutamate dehydrogenase.
22.64 Write a balanced equation for the transamination of aspartate.

The Urea Cycle

22.65 What metabolic condition is produced if the urea cycle does not function properly?
22.66 What is hyperammonemia? How are mild forms of this disease treated?
22.67 The structure of urea is

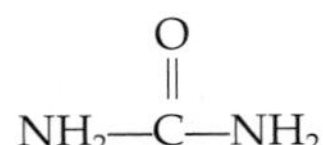

a. What substances are the sources of each of the amino groups in the urea molecule?
b. What substance is the source of the carbonyl group?
22.68 What is the energy source used for the urea cycle?

Overview of Anabolism: The Citric Acid Cycle as a Source of Biosynthetic Intermediates

22.69 From which citric acid cycle intermediate is the amino acid glutamate synthesized?
22.70 What amino acids are synthesized from α-ketoglutarate?
22.71 What is the role of the citric acid cycle in biosynthesis?
22.72 How are citric acid cycle intermediates replenished when they are in demand for biosynthesis?
22.73 What is meant by the term *essential amino acid?*
22.74 What are the nine essential amino acids?
22.75 Write a balanced equation for the reaction catalyzed by pyruvate carboxylase.
22.76 How does the reaction described in Problem 22.75 allow the citric acid cycle to fulfill its roles in both catabolism and anabolism?

Critical Thinking Problems

1. A one-month-old baby boy was brought to the hospital showing severely delayed development and cerebral atrophy. Blood tests showed high levels of lactate and pyruvate. By three months of age, very high levels of succinate and fumarate were found in the urine. Fumarase activity was absent in the liver and muscle tissue. The baby died at five months of age. This was the first reported case of fumarase deficiency and the defect was recognized too late for effective therapy to be administered. What reaction is catalyzed by fumarase? How would a deficiency of this mitochondrial enzyme account for the baby's symptoms and test results?
2. A certain bacterium can grow with ethanol as its only source of energy and carbon. Propose a pathway to describe how ethanol can enter a pathway that would allow ATP production and synthesis of precursors for biosynthesis.
3. Fluoroacetate has been used as a rat poison and can be fatal when eaten by humans. Patients with fluoroacetate poisoning accumulate citrate and fluorocitrate within the cells. What enzyme is inhibited by fluoroacetate? Explain your reasoning.
4. The pyruvate dehydrogenase complex is activated by removal of a phosphoryl group from pyruvate dehydrogenase. This reaction is catalyzed by the enzyme pyruvate dehydrogenase phosphate phosphatase. A baby is born with a defect in this enzyme. What effects would this defect have on the rate of each of the following pathways: aerobic respiration, glycolysis, lactate fermentation? Explain your reasoning.
5. Pyruvate dehydrogenase phosphate phosphatase is stimulated by Ca^{2+}. In muscles, the Ca^{2+} concentration increases dramatically during muscle contraction. How would the elevated Ca^{2+} concentration affect the rate of glycolysis and the citric acid cycle?
6. Liver contains high levels of nucleic acids. When excess nucleic acids are degraded, ribose-5-phosphate is one of the degradation products that accumulate in the cell. Can this substance be used as a source of energy? What pathway would be used?
7. In birds, arginine is an essential amino acid. Can birds produce urea as a means of removing ammonium ions from the blood? Explain your reasoning.

A tasty source of energy.

23

Fatty Acid Metabolism

Outline

BIOCHEMISTRY

Learning Goals

1. Summarize the digestion and storage of lipids.
2. Describe the degradation of fatty acids by β-oxidation.
3. Explain the role of acetyl CoA in fatty acid metabolism.
4. Understand the role of ketone body production in β-oxidation.
5. Compare β-oxidation of fatty acids and fatty acid biosynthesis.
6. Describe the regulation of lipid and carbohydrate metabolism in relation to the liver, adipose tissue, muscle tissue, and the brain.
7. Summarize the antagonistic effects of glucagon and insulin.

CHEMISTRY CONNECTION

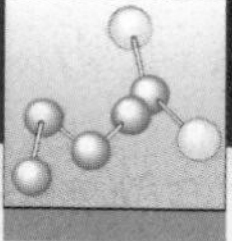

Obesity: A Genetic Disorder?

Approximately a third of all Americans are obese; that is, they are more than 20% overweight. One million are morbidly obese; they carry so much extra weight that it threatens their health. Many obese people simply eat too much and exercise too little, but others actually gain weight even though they eat fewer calories than people of normal weight. This observation led many researchers to the hypothesis that obesity in some people is a genetic disorder.

This hypothesis was supported by the 1950 discovery of an obesity mutation in mice. Selective breeding produced a strain of genetically obese mice from the original mutant mouse. The hypothesis was further strengthened by the results of experiments performed in the 1970s by Douglas Coleman. Coleman connected the circulatory systems of a genetically obese mouse and a normal mouse. The obese mouse started eating less and lost weight. Coleman concluded that there was a substance in the blood of normal mice that signals the brain to decrease the appetite. Obese mice, he hypothesized, can't produce this "satiety factor," and thus they continue to eat and gain weight.

In 1987, Jeffrey Friedman assembled a team of researchers to map and then clone the obesity gene that was responsible for appetite control. In 1994, after seven years of intense effort, the scientists achieved their goal, but they still had to demonstrate that the protein encoded by the cloned obesity gene did, indeed, have a metabolic effect. The gene was modified to be compatible with the genetic system of bacteria so that they could be used to manufacture the protein. When the engineered gene was then introduced into bacteria, they produced an abundance of the protein product. The protein was then purified in preparation for animal testing.

The researchers calculated that a normal mouse has about 12.5 mg of the protein in its blood. They injected that amount into each of ten mice that were so fat they couldn't squeeze into the feeding tunnels used for normal mice. The day after the first injection, graduate student Jeff Halaas observed that the mice had eaten less food. Injections were given daily, and each day the obese mice ate less. After two weeks of treatment, each of the ten mice had lost about 30% of its weight. In addition, the mice had become more active and their metabolisms had speeded up.

When normal mice underwent similar treatment, their body fat fell from 12.2% to 0.67%, which meant that these mice had no extra fat tissue. The 0.67% of their body weight represented by fat was accounted for by the membranes that surround each of the cells of their bodies! Because of the dramatic results, Friedman and his colleagues called the protein leptin, from the Greek word *leptos,* meaning slender.

The leptin protein is a hormone, and ongoing research is aimed at understanding how leptin works to control metabolism and food intake. Friedman has hypothesized that it is a signal in a metabolic thermostat. Fat cells produce leptin and secrete it into the bloodstream. As a result, the leptin concentration in a normal person is proportional to the amount of body fat. The blood concentration of the hormone is monitored by a center in the brain, probably the hypothalamus, a region known to control appetite and set metabolic rates. When the concentration reaches a certain level, it triggers the hypothalamus to suppress the appetite. If a genetically obese person, or mouse, produces no leptin or only small amounts of it, the hypothalamus "thinks" that the individual has too little body fat or is starving. Under these circumstances it does not send a signal to suppress hunger and the individual continues to eat.

The human leptin gene also has been cloned and shown to correct genetic obesity in mice. But what about obesity in humans? Leptin has been tested in a small number of obese individuals. Unfortunately, the dramatic results achieved with mice were *not* observed with humans. Why? It seems that nearly all of the obese volunteers already produced an abundance of leptin. In fact, fewer than ten people have been found, to date, who do not produce leptin. For these individuals, leptin injections do, indeed, reduce their appetites and lead to significant weight loss.

For the majority of obese humans, who produce leptin, perhaps a genetic defect exists in the leptin receptor. Or perhaps the genetics of obesity in humans is more complex than in mice. Clearly lipid metabolism in animals is a complex process and is not yet fully understood. The discovery of the leptin gene, and the hormone it produces, is just one part of the story. In this chapter we will study other aspects of lipid metabolism: the pathways for fatty acid degradation and biosynthesis and the processes by which dietary lipids are digested and excess lipids are stored.

Introduction

The metabolism of fatty acids and lipids revolves around the fate of acetyl CoA. We saw in Chapter 22 that, under aerobic conditions, pyruvate is converted to acetyl CoA, which feeds into the citric acid cycle. Fatty acids are also degraded to acetyl CoA and oxidized by the citric acid cycle, as are certain amino acids. Moreover, acetyl CoA is itself the starting material for the biosynthesis of fatty acids, cholesterol, and steroid hormones. Acetyl CoA is thus a key intermediary in lipid metabolism.

23.1 Lipid Metabolism in Animals

Digestion and Absorption of Dietary Triglycerides

Learning Goal

Triglycerides are highly hydrophobic ("water fearing"). Because of this they must be processed before they can be digested, absorbed, and metabolized. Because processing of dietary lipids occurs in the small intestine, the water soluble **lipases,** enzymes that hydrolyze triglycerides, that are found in the stomach and in the saliva are not very effective. In fact, most dietary fat arrives in the duodenum, the first part of the small intestine, in the form of fat globules. These fat globules stimulate the secretion of bile from the gallbladder. **Bile** is composed of micelles of lecithin, cholesterol, protein, bile salts, inorganic ions, and bile pigments. **Micelles** (Figure 23.1) are aggregations of molecules having a polar region and a nonpolar region. The nonpolar ends of bile salts tend to bunch together when placed in water. The hydrophilic ("water loving") regions of these molecules interact with water. Bile salts are made in the liver and stored in the gallbladder, awaiting the stimulus to be secreted into the duodenum. The major bile salts in humans are cholate and chenodeoxycholate (Figure 23.2).

See Sections 15.1 and 18.2 for a discussion of micelles.

Cholesterol is almost completely insoluble in water, but the conversion of cholesterol to bile salts creates *detergents* whose polar heads make them soluble in the aqueous phase and whose hydrophobic tails bind triglycerides. After a meal is eaten, bile flows through the common bile duct into the duodenum, where bile salts emulsify the fat globules into tiny droplets. This increases the surface area of the lipid molecules, allowing them to be more easily hydrolyzed by lipases (Figure 23.3).

Much of the lipid in these droplets is in the form of **triglycerides,** or triacylglycerols, which are fatty acid esters of glycerol. A protein called **colipase** binds to the surface of the lipid droplets and helps pancreatic lipases to stick to the surface and hydrolyze the ester bonds between the glycerol and fatty acids of the triglycerides (Figure 23.4). In this process, two of the three fatty acids are liberated, and the monoglycerides and free fatty acids produced mix freely with the micelles of bile. These micelles are readily absorbed through the membranes of the intestinal epithelial cells (Figure 23.5).

Triglycerides are described in Section 18.3.

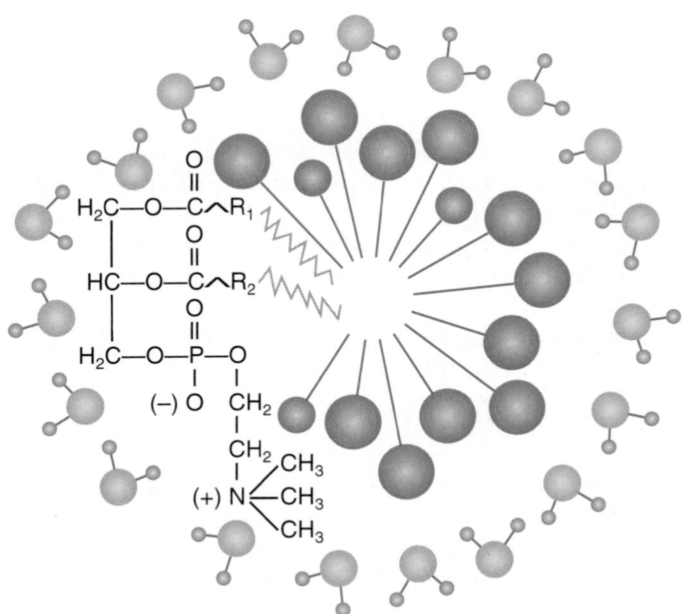

Figure 23.1
The structure of a micelle formed from the phospholipid lecithin. The straight lines represent the long hydrophobic fatty acid tails, and the spheres represent the hydrophilic heads of the phospholipid.

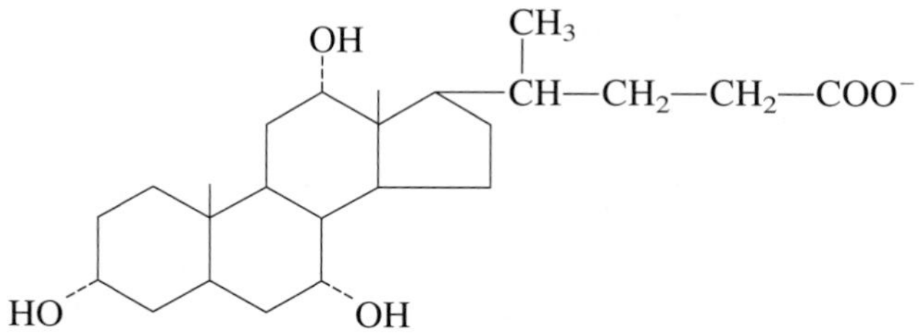

Cholate

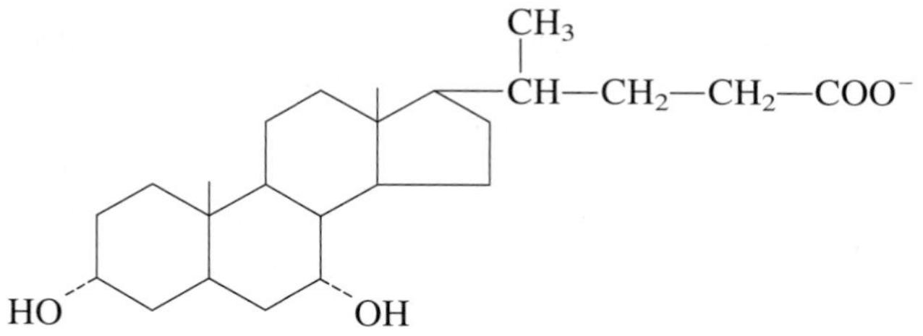

Chenodeoxycholate

Figure 23.2
Structures of the most common bile acids in human bile: cholate and chenodeoxycholate.

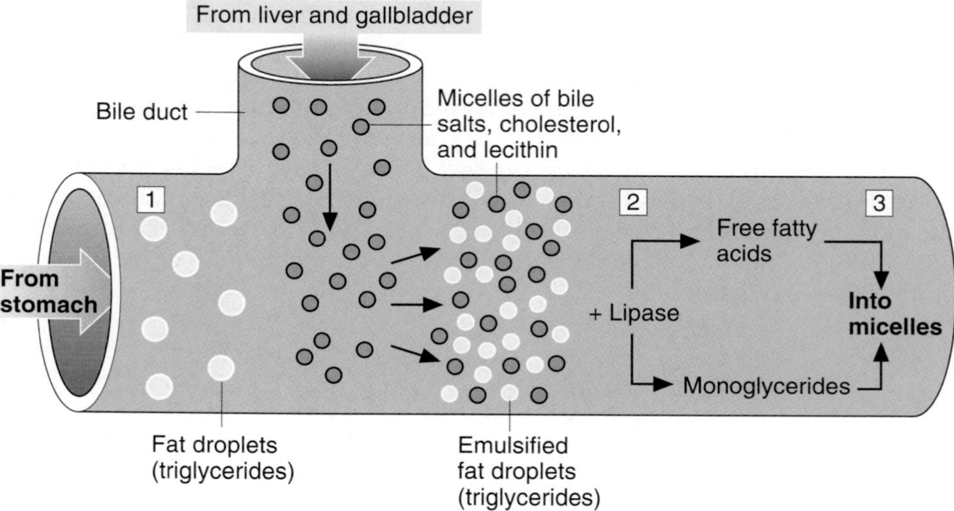

Figure 23.3
Stages of lipid digestion in the intestinal tract. Step 1 is the emulsification of fat droplets by bile salts. Step 2 is the hydrolysis of triglycerides in emulsified fat droplets into fatty acids and monoglycerides. Step 3 involves dissolving fatty acids and monoglycerides into micelles to produce "mixed micelles."

Plasma lipoproteins are described in Section 18.5.

Surprisingly, the monoglycerides and fatty acids are then reassembled into triglycerides that are combined with protein to produce the class of plasma lipoproteins called **chylomicrons** (Figure 23.5). These collections of lipid and protein are secreted into small lymphatic vessels and eventually arrive in the bloodstream. In the bloodstream the triglycerides are once again hydrolyzed to produce glycerol and free fatty acids that are then absorbed by the cells. If the body needs energy, these molecules are degraded to produce ATP. If the body does not need energy, these energy-rich molecules are stored.

Lipid Storage

Fatty acids are stored in the form of triglycerides. Most of the body's triglyceride molecules are stored as fat droplets in the cytoplasm of **adipocytes** (fat cells) that make up **adipose tissue.** Each adipocyte contains a large fat droplet that accounts for nearly the entire volume of the cell. Other cells, such as those of cardiac muscle, contain a few small fat droplets. In these cells the fat droplets are surrounded

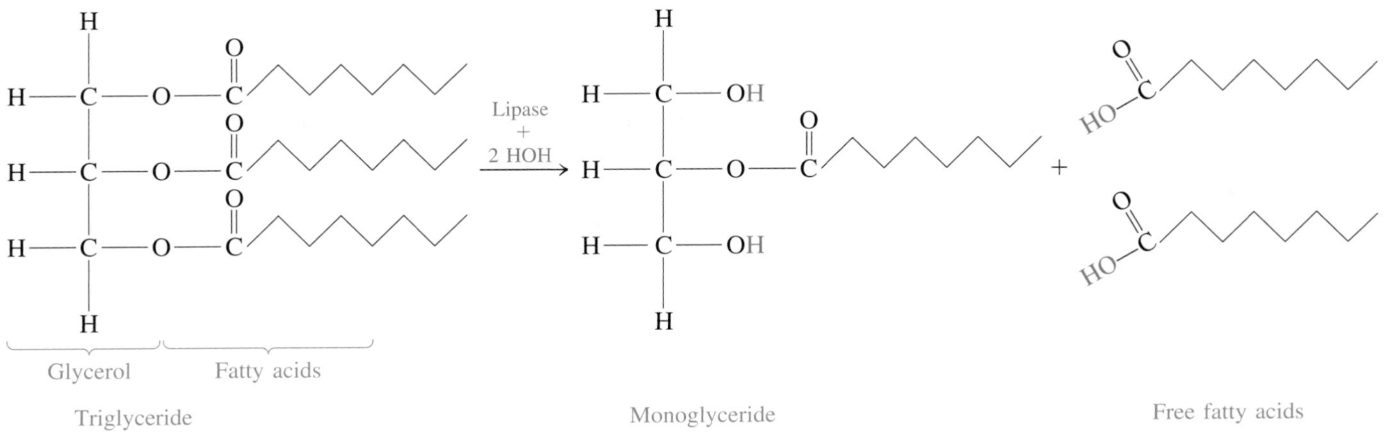

$$\text{Triglyceride (Glycerol + Fatty acids)} \xrightarrow[\text{2 HOH}]{\text{Lipase +}} \text{Monoglyceride} + \text{Free fatty acids}$$

Glycerol Fatty acids

Triglyceride

Monoglyceride

Free fatty acids

Figure 23.4
The action of pancreatic lipase in the hydrolysis of dietary lipids.

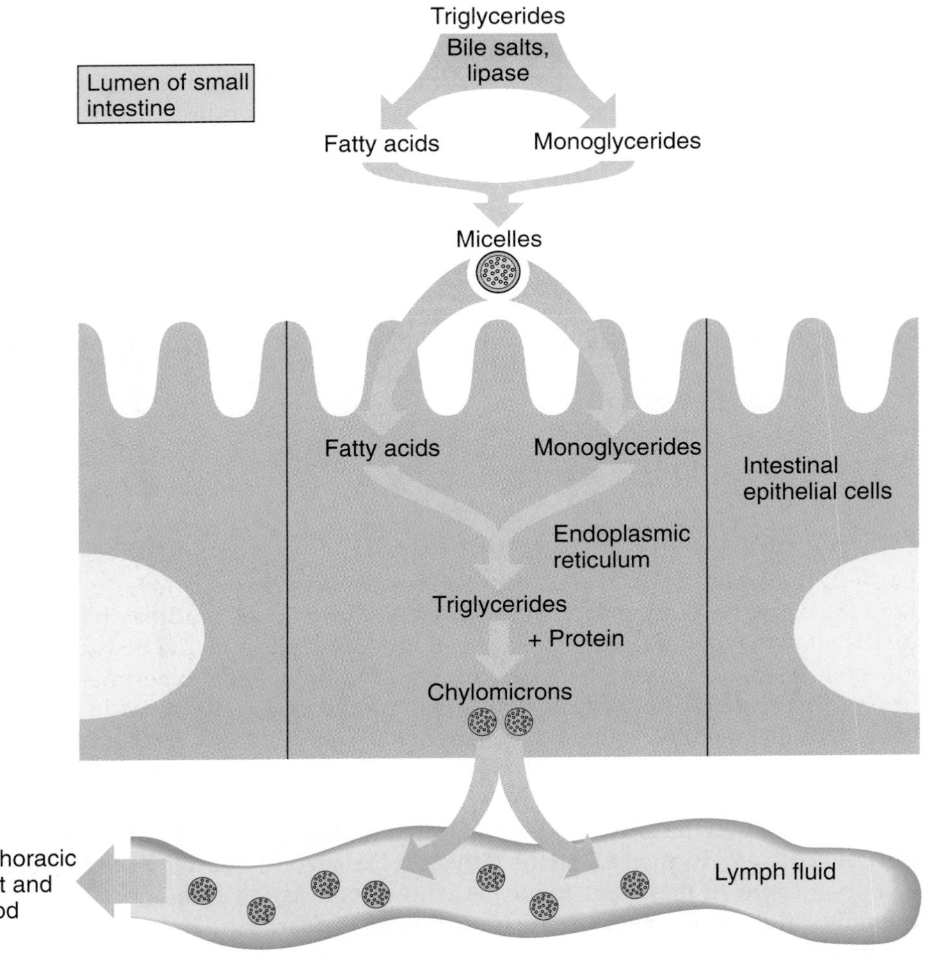

Figure 23.5
Passage of triglycerides in micelles into the cells of the intestinal epithelium.

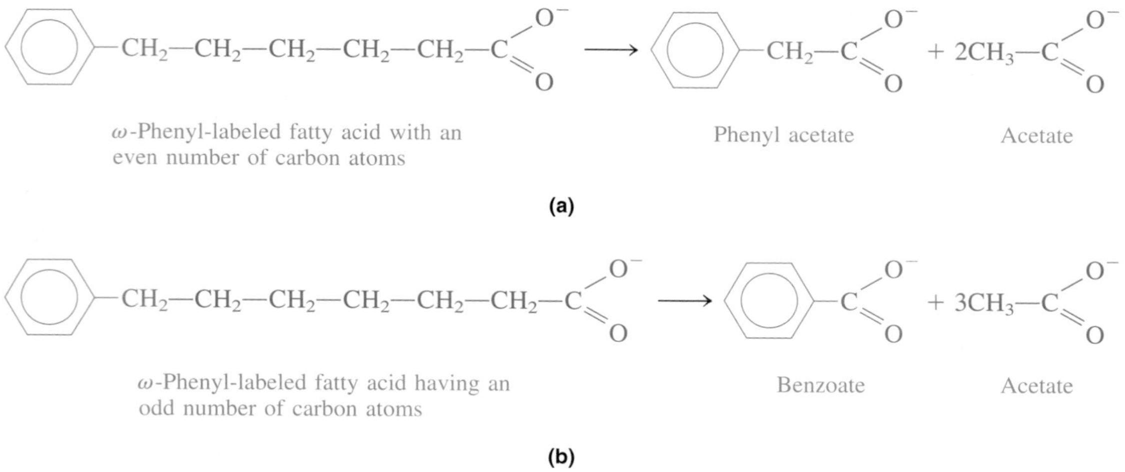

$$C_6H_5{-}CH_2{-}CH_2{-}CH_2{-}CH_2{-}CH_2{-}COO^- \longrightarrow C_6H_5{-}CH_2{-}COO^- + 2CH_3{-}COO^-$$

ω-Phenyl-labeled fatty acid with an even number of carbon atoms — Phenyl acetate — Acetate

(a)

$$C_6H_5{-}CH_2{-}CH_2{-}CH_2{-}CH_2{-}CH_2{-}CH_2{-}COO^- \longrightarrow C_6H_5{-}COO^- + 3CH_3{-}COO^-$$

ω-Phenyl-labeled fatty acid having an odd number of carbon atoms — Benzoate — Acetate

(b)

Figure 23.6
The last carbon of the chain is called the ω-carbon (omega-carbon), so the attached phenyl group is an ω-phenyl group. (a) Oxidation of ω-phenyl-labeled fatty acids occurs two carbons at a time. Fatty acids having an even number of carbon atoms are degraded to phenyl acetate and "acetate." (b) Oxidation of ω-phenyl-labeled fatty acids that contain an odd number of carbon atoms yields benzoate and "acetate."

by mitochondria. When the cells need energy, triglycerides are hydrolyzed to release fatty acids that are transported into the matrix space of the mitochondria. There the fatty acids are completely oxidized, and ATP is produced.

The fatty acids provided by the hydrolysis of triglycerides are a very rich energy source for the body. The complete oxidation of fatty acids releases much more energy than the oxidation of a comparable amount of glycogen.

Question 23.1 How do bile salts aid in the digestion of dietary lipids?

Question 23.2 Why must dietary lipids be processed before enzymatic digestion can be effective?

23.2 Fatty Acid Degradation

An Overview of Fatty Acid Degradation

Learning Goal 2

Learning Goal

Early in the twentieth century, a very clever experiment was done to determine how fatty acids are degraded. Recall from Chapter 10 that radioactive elements can be attached to biological molecules and followed through the body. A German biochemist, Franz Knoop, devised a similar kind of labeling experiment long before radioactive tracers were available. Knoop fed dogs fatty acids in which the usual terminal methyl group had a phenyl group attached to it. Such molecules are called ω-labeled (omega-labeled) fatty acids (Figure 23.6). When he isolated the metabolized fatty acids from the urine of the dogs, he found that phenyl acetate was formed when the fatty acid had an even number of carbon atoms in the chain. But benzoate was formed when the fatty acid had an odd number of carbon atoms. Knoop interpreted these data to mean that the degradation of fatty acids occurs by the removal of two-carbon acetate groups from the carboxyl end of the fatty acid. We now know that the two-carbon fragments produced by the degradation of fatty acids are not acetate, but acetyl CoA. The pathway for the breakdown of fatty acids into acetyl CoA is called **β-oxidation.**

This pathway is called β-oxidation because it involves the stepwise oxidation of the β-carbon of the fatty acid.

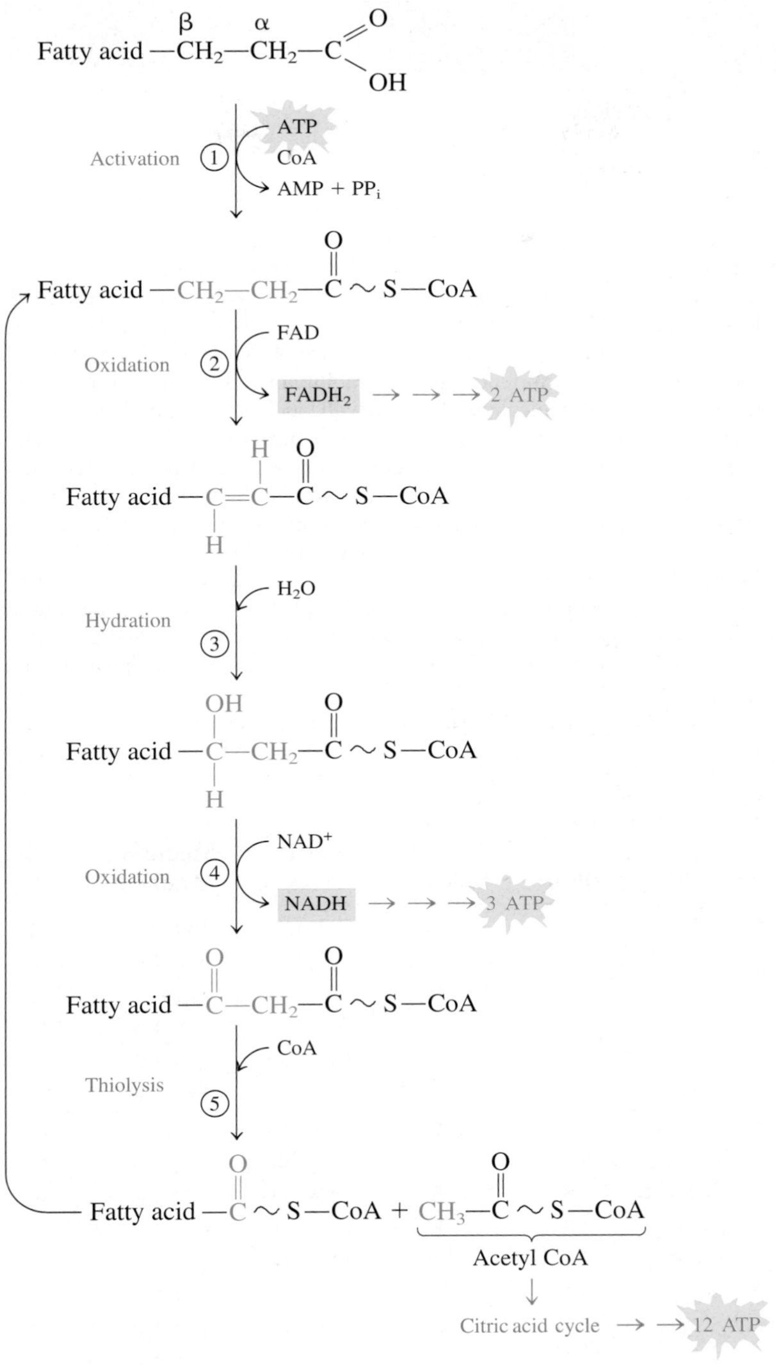

$$\text{Fatty acid}-\overset{\beta}{C}H_2-\overset{\alpha}{C}H_2-C(=O)OH$$

Activation ① ATP, CoA → AMP + PP_i

$$\text{Fatty acid}-CH_2-CH_2-\overset{O}{\overset{\|}{C}}\sim S-CoA$$

Oxidation ② FAD → $FADH_2$ → → → 2 ATP

$$\text{Fatty acid}-\underset{H}{\underset{|}{C}}=\overset{H}{\overset{|}{C}}-\overset{O}{\overset{\|}{C}}\sim S-CoA$$

Hydration ③ H_2O

$$\text{Fatty acid}-\overset{OH}{\overset{|}{\underset{H}{\underset{|}{C}}}}-CH_2-\overset{O}{\overset{\|}{C}}\sim S-CoA$$

Oxidation ④ NAD^+ → NADH → → → 3 ATP

$$\text{Fatty acid}-\overset{O}{\overset{\|}{C}}-CH_2-\overset{O}{\overset{\|}{C}}\sim S-CoA$$

Thiolysis ⑤ CoA

$$\text{Fatty acid}-\overset{O}{\overset{\|}{C}}\sim S-CoA + \underbrace{CH_3-\overset{O}{\overset{\|}{C}}\sim S-CoA}_{\text{Acetyl CoA}}$$

Acetyl CoA ↓ Citric acid cycle → → 12 ATP

Figure 23.7
The reactions in β-oxidation of fatty acids.

The β-oxidation cycle (steps 2–5, Figure 23.7) consists of a set of four reactions whose overall form is similar to the last four reactions of the citric acid cycle. Each trip through the sequence of reactions releases acetyl CoA and returns a fatty acyl CoA molecule that has two fewer carbons. One molecule of $FADH_2$, equivalent to two ATP molecules, and one molecule of NADH, equivalent to three ATP molecules, are produced for each cycle of β-oxidation.

Review Section 22.6 for the ATP yields that result from oxidation of $FADH_2$ and NADH.

A HUMAN PERSPECTIVE

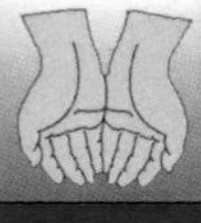

Losing Those Unwanted Pounds of Adipose Tissue

Weight, or overweight, is a topic of great concern to the American populace. A glance through almost any popular magazine quickly informs us that by today's standards, "beautiful" is synonymous with "thin." The models in all these magazines are extremely thin, and there are literally dozens of ads for weight-loss programs. Americans spend millions of dollars each year trying to attain this slim ideal of the fashion models.

Studies have revealed that this slim ideal is often below a desirable, healthy body weight. In fact, the suggested weight for a 6-foot tall male between 18 and 39 years of age is 179 pounds. For a 5′6″ female in the same age range, the desired weight is 142 pounds. For a 5′1″ female, 126 pounds is recommended. Just as being too thin can cause health problems, so too can obesity.

What is obesity, and does it have disadvantages beyond aesthetics? An individual is considered to be obese if his or her body weight is more than 20% above the ideal weight for his or her height. The accompanying table lists desirable body weights, according to sex, age, height, and body frame.

Overweight carries with it a wide range of physical problems, including elevated blood cholesterol levels; high blood pressure; increased incidence of diabetes, cancer, and heart disease; and increased probability of early death. It often causes psychological problems as well, such as guilt and low self-esteem.

Many factors may contribute to obesity. These include genetic factors, a sedentary lifestyle, and a preference for high-calorie, high-fat foods. However, the real concern is how to lose weight. How can we lose weight wisely and safely and keep the weight off for the rest of our lives? Unfortunately, the answer is *not* the answer that most people want to hear. The prevalence and financial success of the quick-weight-loss programs suggest that the majority of people want a program that is rapid and effortless. Unfortunately, most programs that promise dramatic

*Men**					*Women***				
Height					**Height**				
Feet	**Inches**	**Small Frame**	**Medium Frame**	**Large Frame**	**Feet**	**Inches**	**Small Frame**	**Medium Frame**	**Large Frame**
5	2	128–134	131–141	138–150	4	10	102–111	109–121	118–131
5	3	130–136	133–143	140–153	4	11	103–113	111–123	120–134
5	4	132–138	135–145	142–156	5	0	104–115	113–126	122–137
5	5	134–140	137–148	144–160	5	1	106–118	115–129	125–140
5	6	136–142	139–151	146–164	5	2	108–121	118–132	128–143
5	7	138–145	142–154	149–168	5	3	111–124	121–135	131–147
5	8	140–148	145–157	152–172	5	4	114–127	124–138	134–151
5	9	142–151	148–160	155–176	5	5	117–130	127–141	137–155
5	10	144–154	151–163	158–180	5	6	120–133	130–144	140–159
5	11	146–157	154–166	161–184	5	7	123–136	133–147	143–163
6	0	149–160	157–170	164–188	5	8	126–139	136–150	146–167
6	1	152–164	160–174	168–192	5	9	129–142	139–153	149–170
6	2	155–168	164–178	172–197	5	10	132–145	142–156	152–173
6	3	158–172	167–182	176–202	5	11	135–148	145–159	155–176
6	4	162–176	171–187	181–207	6	0	138–151	148–162	158–179

*Weights at ages 25–59 based on lowest mortality. Weight in pounds according to frame (in indoor clothing weighing 5 lb, shoes with 1″ heels).
**Weights at ages 25–59 based on lowest mortality. Weight in pounds according to frame (in indoor clothing weighing 3 lb, shoes with 1″ heels).
Reprinted with permission of the Metropolitan Life Insurance Companies *Statistical Bulletin*.

Activity	Kilocalories per hour*
Badminton, competitive singles	480
Basketball	360–660
Bicycling	
10 mph	420
11 mph	480
12 mph	600
13 mph	660
Calisthenics, heavy	600
Handball, competitive	660
Rope skipping, vigorous	800
Rowing machine	840
Running	
5 mph	600
6 mph	750
7 mph	870
8 mph	1,020
9 mph	1,130
10 mph	1,285
Skating, ice or roller, rapid	700
Skiing, downhill, vigorous	600
Skiing, cross-country	
2.5 mph	560
4 mph	600
5 mph	700
8 mph	1,020
Swimming, 25–50 yards per min.	360–750
Walking	
Level road, 4 mph (fast)	420
Upstairs	600–1,080
Uphill, 3.5 mph	480–900
Gardening, much lifting, stooping, digging	500
Mowing, pushing hand mower	450
Sawing hardwood	600
Shoveling, heavy	660
Wood chopping	560

*Caloric expenditure is based on a 150-lb person. There is a 10% increase in caloric expenditure for each 15 lb over this weight and a 10% decrease for each 15 lb under.
From E. L. Wynder, *The Book of Health: The American Health Foundation*. © 1981 Franklin Watts, Inc., New York. Used with permission.

weight reduction with little effort are usually ineffective or, worse, unsafe. The truth is that weight loss and management are best obtained by a program involving three elements.

1. *Reduced caloric intake.* A pound of body fat is equivalent to 3500 Calories (kilocalories). So if you want to lose 2 pounds each week, a reasonable goal, you must reduce your caloric intake by 1000 Calories per day. Remember that diets recommending fewer than 1200 Calories per day are difficult to maintain because they are not very satisfying and may be unsafe because they don't provide all the required vitamins and minerals. The best way to decrease Calories is to reduce fat and increase complex carbohydrates in the diet.
2. *Exercise.* Increase energy expenditures by 200–400 Calories each day. You may choose walking, running, or mowing the lawn; the type of activity doesn't matter, as long as you get moving. Exercise has additional benefits. It increases cardiovascular fitness, provides a psychological lift, and may increase the base rate at which you burn calories after exercise is finished. The accompanying table summarizes the caloric expenditure of several activities.
3. *Behavior modification.* For some people overweight is as much a psychological problem as it is a physical problem, and half the battle is learning to recognize the triggers that cause overeating. Several principles of behavior modification have been found to be very helpful.
   a. Keep a diary. Record the amount of foods eaten and the circumstances—for instance, a meal at the kitchen table or a bag of chips in the car on the way home.
   b. Identify your eating triggers. Do you eat when you feel stress, boredom, fatigue, joy?
   c. Develop a plan for avoiding or coping with your trigger situations or emotions. You might exercise when you feel that stress-at-the-end-of-the-day trigger or carry a bag of carrot sticks for the midmorning-boredom trigger.
   d. Set realistic goals, and reward yourself when you reach them. The reward should not be food related.

As you can see, there is no "quick fix" for safe, effective weight control. A commitment must be made to modify existing diet and exercise habits. Most important, those habits must be avoided forever and replaced by new, healthier behaviors and attitudes.

EXAMPLE 23.1 ***Predicting the Products of β-Oxidation of a Fatty Acid***

What products would be produced by the β-oxidation of 10-phenyldecanoic acid?

Solution

This ten-carbon fatty acid would be broken down into four acetyl CoA molecules and one phenyl acetate molecule. Because four cycles through β-oxidation are required to break down a ten-carbon fatty acid, four NADH molecules and four $FADH_2$ molecules would also be produced.

Question 23.3

What products would be formed by β-oxidation of each of the following fatty acids? (*Hint:* Refer to Example 23.1.)

a. 9-Phenylnonanoic acid
b. 8-Phenyloctanoic acid
c. 7-Phenylheptanoic acid
d. 12-Phenyldodecanoic acid

Question 23.4

What does ω refer to in the naming of ω-phenyl-labeled fatty acids?

The Reactions of β-Oxidation

Learning Goal

The enzymes that catalyze the β-oxidation of fatty acids are located in the matrix space of the mitochondria. Special transport mechanisms are required to bring fatty acid molecules into the mitochondrial matrix. Once inside, the fatty acids are degraded by the reactions of β-oxidation. As we will see, these reactions interact with oxidative phosphorylation and the citric acid cycle to produce ATP.

Reaction 1. The first step is an *activation* reaction that results in the production of a fatty acyl CoA molecule. A thioester bond is formed between coenzyme A and the fatty acid:

$$CH_3-(CH_2)_n-CH_2-CH_2-\overset{\overset{\Large O}{\|}}{\underset{\underset{\Large OH}{|}}{C}} \xrightarrow[\text{Coenzyme A}]{\text{ATP} \quad \text{AMP} + \text{PP}_i}$$

Fatty acid

$$CH_3-(CH_2)_n-CH_2-CH_2-\overset{\overset{\Large O}{\|}}{C}\sim S-CoA$$

thioester bond

Fatty acyl CoA

This reaction requires energy in the form of ATP, which is cleaved to AMP and pyrophosphate. This involves hydrolysis of two phosphoanhydride bonds. Here again we see the need to invest a small amount of energy so that a much greater amount of energy can be harvested later in the pathway. Coenzyme A is also required for this step. The product, a fatty acyl CoA, has a *high-energy* thioester bond between the fatty acid and coenzyme A. *Acyl-CoA ligase,* which

Acyl group transfer reactions are described in Section 15.4.

catalyzes this reaction, is located in the outer membrane of the mitochondria. The mechanism that brings the fatty acyl CoA into the mitochondrial matrix involves a carrier molecule called *carnitine.* The first step, catalyzed by the enzyme *carnitine acyltransferase I,* is the transfer of the fatty acyl group to carnitine, producing acylcarnitine and coenzyme A. Next a carrier protein located in the mitochondrial inner membrane transfers the acylcarnitine into the mitochondrial matrix. There *carnitine acyltransferase II* catalyzes the regeneration of fatty acyl CoA, which now becomes involved in the remaining reactions of β-oxidation.

Reaction 2. The next reaction is an *oxidation* reaction that removes a pair of hydrogen atoms from the fatty acid. These are used to reduce FAD to produce $FADH_2$. This *dehydrogenation* reaction is catalyzed by the enzyme *acyl-CoA dehydrogenase* and results in the formation of a carbon-carbon double bond:

$$CH_3{-}(CH_2)_n{-}CH_2{-}CH_2{-}\overset{\overset{\displaystyle O}{\|}}{C}{\sim}S{-}CoA \xrightarrow{FAD \;\to\; FADH_2} CH_3{-}(CH_2)_n{-}\underset{\underset{\displaystyle H}{|}}{C}{=}\overset{\overset{\displaystyle H}{|}}{C}{-}\overset{\overset{\displaystyle O}{\|}}{C}{\sim}S{-}CoA$$

Oxidative phosphorylation yields two ATP molecules for each molecule of $FADH_2$ produced by this oxidation–reduction reaction.

Reaction 3. The third reaction involves the *hydration* of the double bond produced in reaction 2. As a result the β-carbon is hydroxylated. This reaction is catalyzed by the enzyme *enoyl-CoA hydrase.*

$$CH_3{-}(CH_2)_n{-}\underset{\underset{\displaystyle H}{|}}{C}{=}\overset{\overset{\displaystyle H}{|}}{C}{-}\overset{\overset{\displaystyle O}{\|}}{C}{\sim}S{-}CoA \xrightarrow{H_2O} CH_3{-}(CH_2)_n{-}\underset{\underset{\displaystyle H}{|}}{\overset{\overset{\displaystyle OH}{|}}{C}}{-}CH_2{-}\overset{\overset{\displaystyle O}{\|}}{C}{\sim}S{-}CoA$$

Reaction 4. In this *oxidation* reaction the hydroxyl group of the β-carbon is now dehydrogenated. NAD^+ is reduced to form NADH that is subsequently used to produce three ATP molecules by oxidative phosphorylation. L-β-*Hydroxyacyl-CoA dehydrogenase* catalyzes this reaction.

$$CH_3{-}(CH_2)_n{-}\underset{\underset{\displaystyle H}{|}}{\overset{\overset{\displaystyle OH}{|}}{C}}{-}CH_2{-}\overset{\overset{\displaystyle O}{\|}}{C}{\sim}S{-}CoA \xrightarrow{NAD^+ \;\to\; NADH} CH_3{-}(CH_2)_n{-}\overset{\overset{\displaystyle O}{\|}}{C}{-}CH_2{-}\overset{\overset{\displaystyle O}{\|}}{C}{\sim}S{-}CoA$$

Reaction 5. The final reaction, catalyzed by the enzyme *thiolase,* is the cleavage that releases acetyl CoA. This is accomplished by *thiolysis,* attack of a molecule of coenzyme A on the β-carbon. The result is the release of acetyl CoA and a fatty acyl CoA that is two carbons shorter than the beginning fatty acid:

$$CH_3{-}(CH_2)_n{-}\overset{O}{\overset{\|}{C}}{-}CH_2{-}\overset{O}{\overset{\|}{C}}{\sim}S{-}CoA \xrightarrow{CoA} CH_3{-}(CH_2)_{n-2}{-}CH_2{-}CH_2{-}\overset{O}{\overset{\|}{C}}{\sim}S{-}CoA + CH_3{-}\overset{O}{\overset{\|}{C}}{\sim}S{-}CoA$$

The shortened fatty acyl CoA is further oxidized by cycling through reactions 2–5 until the fatty acid carbon chain is completely degraded to acetyl CoA. The acetyl CoA produced by β-oxidation of fatty acids then enters the reactions of the citric acid cycle. Of course, this eventually results in the production of 12 ATP molecules per molecule of acetyl CoA released during β-oxidation.

As an example of the energy yield from β-oxidation, the balance sheet for ATP production when the sixteen-carbon-fatty acid palmitic acid is degraded by β-oxidation is summarized in Figure 23.8. Complete oxidation of palmitate results in production of 129 molecules of ATP, *three and one half times more energy than results from the complete oxidation of an equivalent amount of glucose.*

EXAMPLE 23.2

Calculating the Amount of ATP Produced in Complete Oxidation of a Fatty Acid

How many molecules of ATP are produced in the complete oxidation of stearic acid, an eighteen-carbon saturated fatty acid?

Solution

Step 1 (activation)	−2 ATP
Steps 2–5:	
8 $FADH_2$ × 2 ATP/$FADH_2$	16 ATP
8 NADH × 3 ATP/NADH	24 ATP
9 acetyl CoA (to citric acid cycle):	
9 × 1 GTP × 1 ATP/GTP	9 ATP
9 × 3 NADH × 3 ATP/NADH	81 ATP
9 × 1 $FADH_2$ × 2 ATP/$FADH_2$	18 ATP
	146 ATP

Question 23.5 Write out the sequence of steps for β-oxidation of butyryl CoA.

Question 23.6 What is the energy yield from the complete degradation of butyryl CoA via β-oxidation, the citric acid cycle, and oxidative phosphorylation?

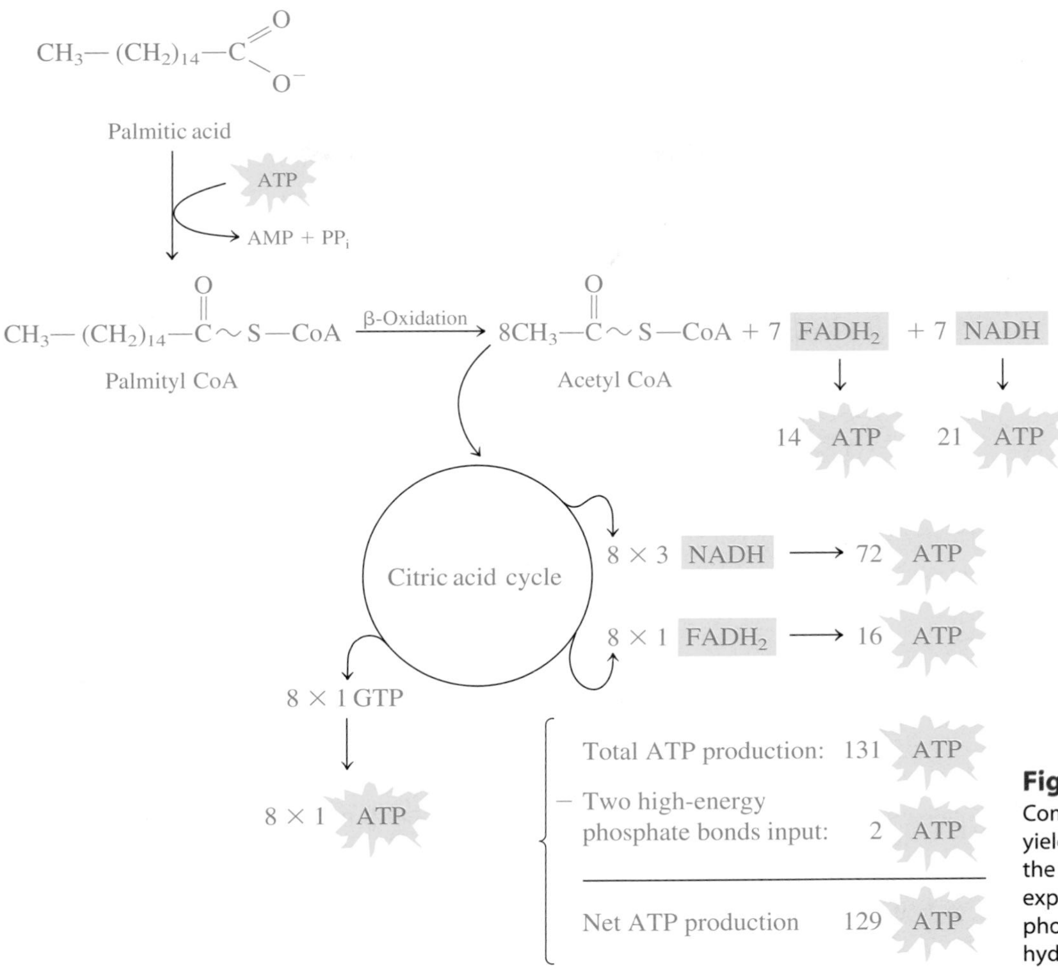

$$CH_3-(CH_2)_{14}-COO^-$$

Palmitic acid

ATP → AMP + PP_i

$$CH_3-(CH_2)_{14}-\overset{O}{\overset{\|}{C}}\sim S-CoA \xrightarrow{\beta\text{-Oxidation}} 8CH_3-\overset{O}{\overset{\|}{C}}\sim S-CoA + 7\ FADH_2 + 7\ NADH$$

Palmityl CoA — Acetyl CoA

7 $FADH_2$ → 14 ATP; 7 NADH → 21 ATP

Citric acid cycle

8 × 3 NADH ⟶ 72 ATP

8 × 1 $FADH_2$ ⟶ 16 ATP

8 × 1 GTP → 8 × 1 ATP

Total ATP production: 131 ATP

− Two high-energy phosphate bonds input: 2 ATP

Net ATP production 129 ATP

Figure 23.8
Complete oxidation of palmitic acid yields 129 molecules of ATP. Note that the activation step is considered to be an expenditure of two high-energy phosphoanhydride bonds because ATP is hydrolyzed to AMP + PP_i.

23.3 Ketone Bodies

Learning Goal

For the acetyl CoA produced by the β-oxidation of fatty acids to efficiently enter the citric acid cycle, there must be an adequate supply of oxaloacetate. If glycolysis and β-oxidation are occurring at the same rate, there will be a steady supply of pyruvate (from glycolysis) that can be converted to oxaloacetate. But what happens if the supply of oxaloacetate is too low to allow all of the acetyl CoA to enter the citric acid cycle? Under these conditions, acetyl CoA is converted to the so-called **ketone bodies:** β-hydroxybutyrate, acetone, and acetoacetate (Figure 23.9).

See Section 22.9 for a review of the reactions that provide oxaloacetate.

Ketosis

Ketosis, abnormally high levels of blood ketone bodies, is a situation that arises under some pathological conditions, such as starvation, a diet that is extremely low in carbohydrates (as with the high-protein liquid diets), or uncontrolled **diabetes mellitus.** The carbohydrate intake of a diabetic is normal, but the carbohydrates cannot get into the cell to be used as fuel. Thus diabetes amounts to starvation in the midst of plenty. In diabetes the very high concentration of ketone acids in the blood leads to **ketoacidosis.** The ketone acids are relatively strong acids and therefore readily dissociate to release H^+. Under these conditions the blood pH becomes acidic, which can lead to death.

Diabetes mellitus is a disease characterized by the appearance of glucose in the urine as a result of high blood glucose levels. The disease is usually caused by the inability to produce the hormone insulin.

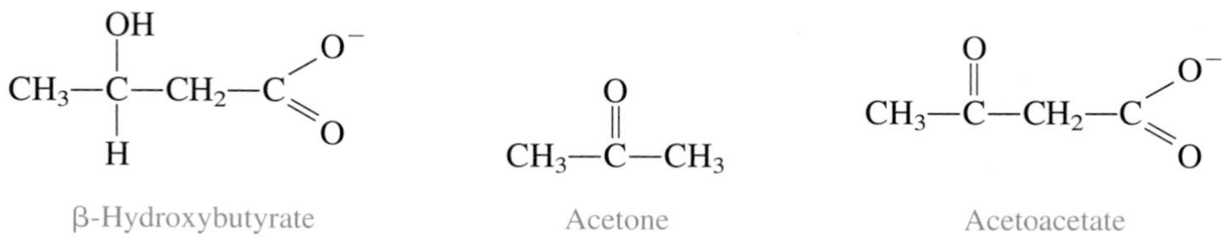

$$CH_3-\underset{H}{\overset{OH}{C}}-CH_2-C(=O)O^- \qquad CH_3-\overset{O}{\overset{\|}{C}}-CH_3 \qquad CH_3-\overset{O}{\overset{\|}{C}}-CH_2-C(=O)O^-$$

β-Hydroxybutyrate — Acetone — Acetoacetate

Figure 23.9
Structures of ketone bodies.

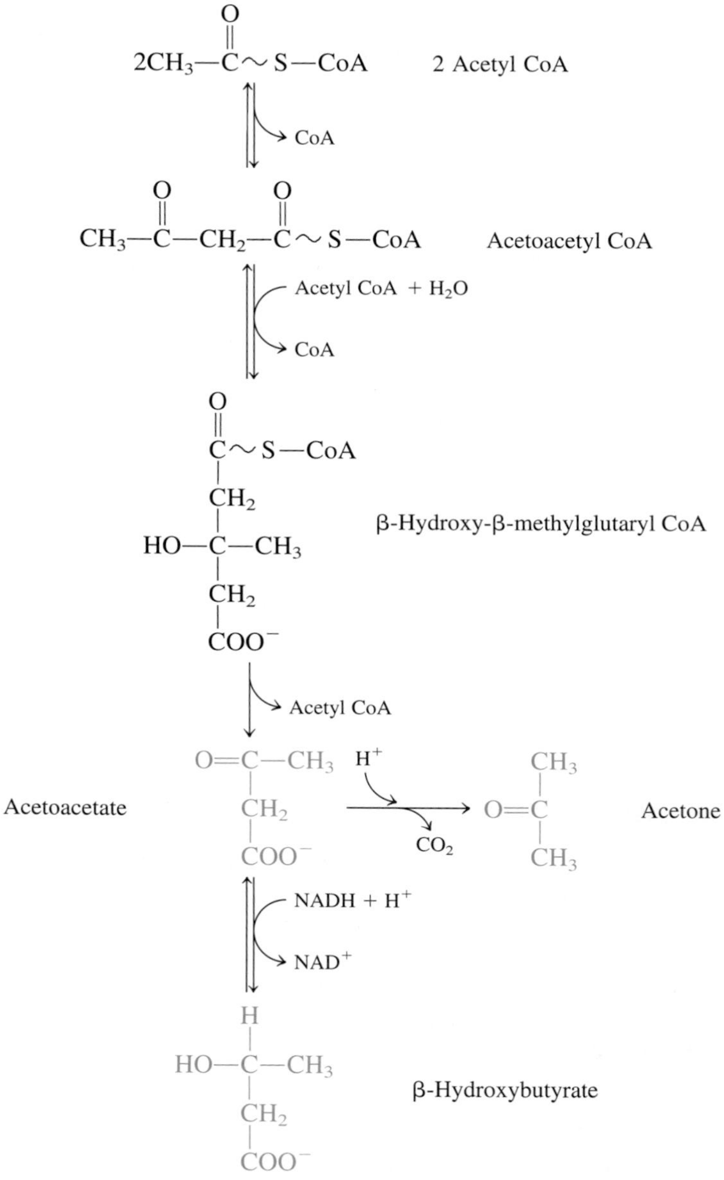

$$2CH_3-\overset{O}{\overset{\|}{C}}\sim S-CoA$$ 2 Acetyl CoA

⇅ → CoA

$$CH_3-\overset{O}{\overset{\|}{C}}-CH_2-\overset{O}{\overset{\|}{C}}\sim S-CoA$$ Acetoacetyl CoA

⇅ Acetyl CoA + H_2O → CoA

$$^{-}OOC-CH_2-\underset{CH_3}{\overset{OH}{C}}-CH_2-\overset{O}{\overset{\|}{C}}\sim S-CoA$$ β-Hydroxy-β-methylglutaryl CoA

↓ → Acetyl CoA

Acetoacetate $O=C(CH_3)-CH_2-COO^-$ —(H^+ → CO_2)→ $O=C(CH_3)_2$ Acetone

⇅ NADH + H^+ → NAD^+

$$HO-\overset{H}{C}(CH_3)-CH_2-COO^-$$ β-Hydroxybutyrate

Figure 23.10
Summary of the reactions involved in ketogenesis.

Ketogenesis

The pathway for the production of ketone bodies (Figure 23.10) begins with a "reversal" of the last step of β-oxidation. When oxaloacetate levels are low, the enzyme that normally carries out the last reaction of β-oxidation now catalyzes the fusion of two acetyl CoA molecules to produce acetoacetyl CoA:

$$2CH_3-\overset{\overset{\large O}{\|}}{C}\sim S-CoA \rightleftharpoons CH_3-\overset{\overset{\large O}{\|}}{C}-CH_2-\overset{\overset{\large O}{\|}}{C}\sim S-CoA + CoA$$

Acetyl CoA → Acetoacetyl CoA

Acetoacetyl CoA can react with a third acetyl CoA molecule to yield β-hydroxy-β-methylglutaryl CoA (HMG-CoA):

$$CH_3-\overset{\overset{\large O}{\|}}{C}-CH_2-\overset{\overset{\large O}{\|}}{C}\sim S-CoA + CH_3-\overset{\overset{\large O}{\|}}{C}\sim S-CoA + H_2O \rightleftharpoons$$

Acetoacetyl CoA, Acetyl CoA

$$^{-}OOC-CH_2-\overset{\overset{\large OH}{|}}{\underset{\underset{\large CH_3}{|}}{C}}-CH_2-\overset{\overset{\large O}{\|}}{C}\sim S-CoA + CoA + H^+$$

HMG-CoA

If HMG-CoA were formed in the cytoplasm, it would serve as a precursor for cholesterol biosynthesis. But ketogenesis, like β-oxidation, occurs in the mitochondrial matrix, and here HMG-CoA is cleaved to yield acetoacetate and acetyl CoA:

$$^{-}OOC-CH_2-\overset{\overset{\large OH}{|}}{\underset{\underset{\large CH_3}{|}}{C}}-CH_2-\overset{\overset{\large O}{\|}}{C}\sim S-CoA \longrightarrow {}^{-}OOC-CH_2-\overset{\overset{\large O}{\|}}{\underset{\underset{\large CH_3}{|}}{C}} + CH_3-\overset{\overset{\large O}{\|}}{C}\sim S-CoA$$

HMG-CoA → Acetoacetate + Acetyl CoA

In very small amounts, acetoacetate spontaneously loses carbon dioxide to give acetone. This is the reaction that causes the "acetone breath" that is often associated with uncontrolled diabetes mellitus. More frequently, it undergoes NADH-dependent reduction to produce β-hydroxybutyrate:

$$^{-}OOC-CH_2-\overset{\overset{\large O}{\|}}{\underset{\underset{\large CH_3}{|}}{C}} + H^+ \longrightarrow CH_3-\overset{\overset{\large O}{\|}}{C}-CH_3 + CO_2$$

Acetoacetate → Acetone

$$^{-}OOC-CH_2-\overset{\overset{\large O}{\|}}{\underset{\underset{\large CH_3}{|}}{C}} \xrightarrow{NADH \quad NAD^+} {}^{-}OOC-CH_2-\overset{\overset{\large OH}{|}}{\underset{\underset{\large H}{|}}{C}}-CH_3$$

Acetoacetate → β-Hydroxybutyrate

Acetoacetate and β-hydroxybutyrate are produced primarily in the liver. These metabolites diffuse into the blood and are circulated to other tissues, where they may be reconverted to acetyl CoA and used to produce ATP. In fact, the heart muscle derives most of its metabolic energy from the oxidation of ketone bodies, not from the oxidation of glucose. Other tissues that are best adapted to the use of glucose will increasingly rely on ketone bodies for energy when glucose becomes unavailable or limited. This is particularly true of the brain.

A CLINICAL PERSPECTIVE

Diabetes Mellitus and Ketone Bodies

More than one person, found unconscious on the streets of some metropolis, has been carted to jail only to die of complications arising from uncontrolled diabetes mellitus. Others are fortunate enough to arrive in hospital emergency rooms. A quick test for diabetes mellitus–induced coma is the odor of acetone on the breath of the afflicted person. Acetone is one of several metabolites produced by diabetics that are known collectively as *ketone bodies*.

The term *diabetes* was used by the ancient Greeks to designate diseases in which excess urine is produced. Two thousand years later, in the eighteenth century, the urine of certain individuals was found to contain sugar, and the name *diabetes mellitus* (Latin: *mellitus*, sweetened with honey) was given to this disease. People suffering from diabetes mellitus waste away as they excrete large amounts of sugar-containing urine.

The cause of insulin-dependent diabetes mellitus is an inadequate production of insulin by the body. Insulin is secreted in response to high blood glucose levels. It binds to the membrane receptor protein on its target cells. Binding increases the rate of transport of glucose across the membrane and stimulates glycogen synthesis, lipid biosynthesis, and protein synthesis. As a result, the blood glucose level is reduced. Clearly, the inability to produce sufficient insulin seriously impairs the body's ability to regulate metabolism.

Individuals suffering from diabetes mellitus do not produce enough insulin to properly regulate blood glucose levels. This generally results from the destruction of the β-cells of the islets of Langerhans. One theory to explain the mysterious disappearance of these cells is that a virus infection stimulates the immune system to produce antibodies that cause the destruction of the β-cells.

In the absence of insulin the uptake of glucose into the tissues is not stimulated, and a great deal of glucose is eliminated in the urine. Without insulin, then, adipose cells are unable to take up the glucose required to synthesize triglycerides. As a result, the rate of fat hydrolysis is much greater than the rate of fat resynthesis, and large quantities of free fatty acids are liberated into the bloodstream. Because glucose is not being efficiently taken into cells, carbohydrate metabolism slows, and there is an increase in the rate of lipid catabolism. In the liver this lipid catabolism results in the production of ketone bodies: acetone, acetoacetate, and β-hydroxybutyrate.

A similar situation can develop from improper eating, fasting, or dieting—any situation in which the body is not provided with sufficient energy in the form of carbohydrates. These ketone bodies cannot all be oxidized by the citric acid cycle, which is limited by the supply of oxaloacetate. The acetone concentration in blood rises to levels so high that acetone can be detected in the breath of untreated diabetics. The elevated concentration of ketones in the blood can overwhelm the buffering capacity of the blood, resulting in ketoacidosis. Ketones, too, will be excreted through the kidney. In fact, the presence of excess ketones in the urine can raise the osmotic concentration of the urine so that it behaves as an "osmotic diuretic," causing the excretion of enormous amounts of water. As a result, the patient may become severely dehydrated. In extreme cases the combination of dehydration and ketoacidosis may lead to coma and death.

It has been observed that diabetics also have a higher than normal level of glucagon in the blood. As we have seen, glucagon stimulates lipid catabolism and ketogenesis. It may be that the symptoms previously described result from both the deficiency of insulin and the elevated glucagon levels. The absence of insulin may cause the elevated blood glucose and fatty acid levels, whereas the glucagon, by stimulating ketogenesis, may be responsible for the ketoacidosis and dehydration.

There is no cure for diabetes. However, when the problem is the result of the inability to produce active insulin, blood glucose levels can be controlled moderately well by the injection of either animal insulin or human insulin produced from the cloned insulin gene. Unfortunately, one or even a few injections of insulin each day cannot mimic the precise control of blood glucose accomplished by the pancreas.

As a result, diabetics suffer progressive tissue degeneration that leads to early death. One primary cause of this degeneration is atherosclerosis, the deposition of plaque on the walls of blood vessels. This causes a high frequency of strokes, heart attack, and gangrene of the feet and lower extremities, often necessitating amputation. Kidney failure causes the death of about 20% of diabetics under forty years of age, and diabetic retinopathy (various kinds of damage to the retina of the eye) ranks fourth among the leading causes of blindness in the United States. Nerves are also damaged, resulting in neuropathies that can cause pain or numbness, particularly of the feet.

There is no doubt that insulin injections prolong the life of diabetics, but only the presence of a fully functioning pancreas can allow a diabetic to live a life free of the complications noted here. At present, pancreas transplants do not have a good track record. Only about 50% of the transplants are functioning after one year. It is hoped that improved transplantation techniques will be developed so that diabetics can live a normal life span, free of debilitating disease.

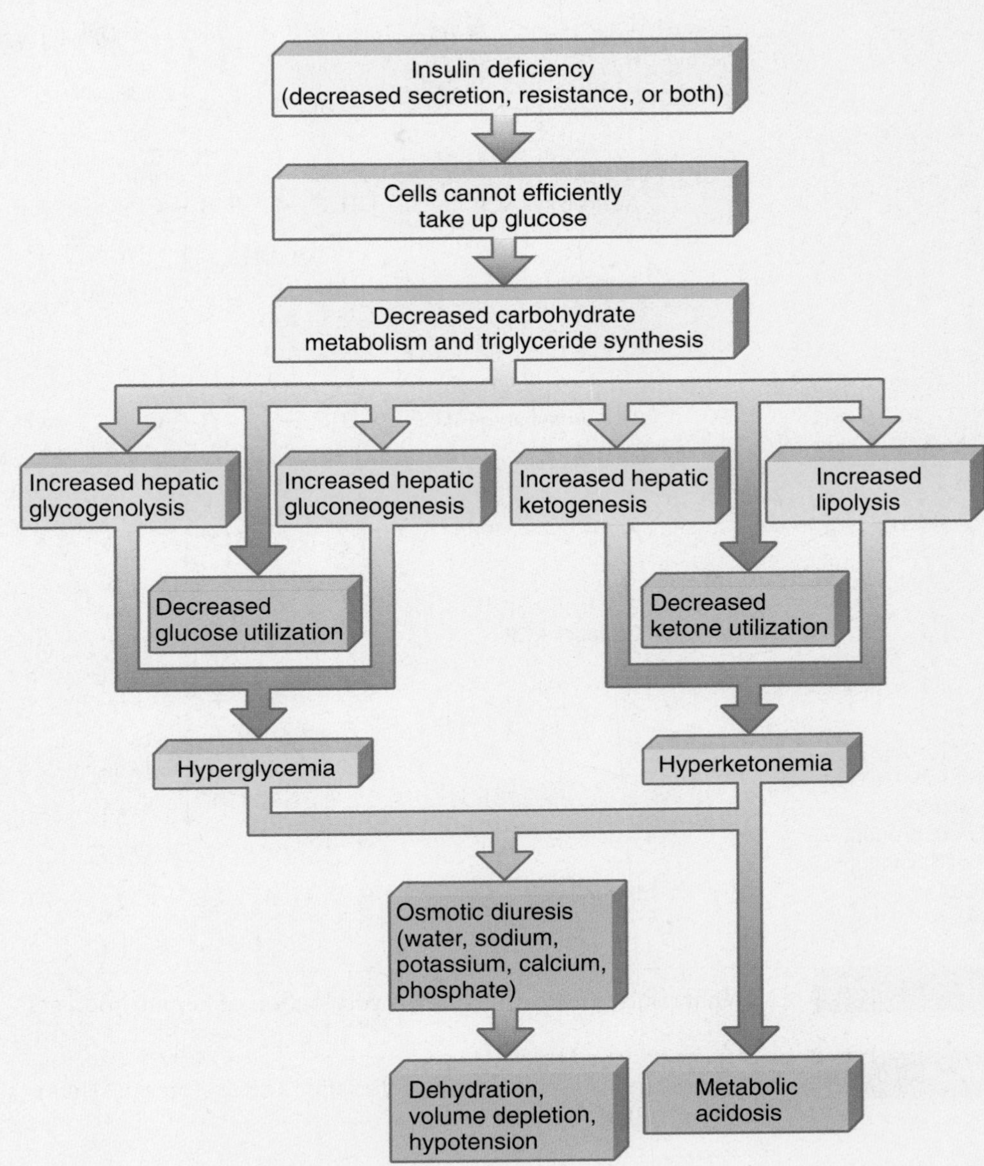

The metabolic events that occur in uncontrolled diabetes and that can lead to coma and death.

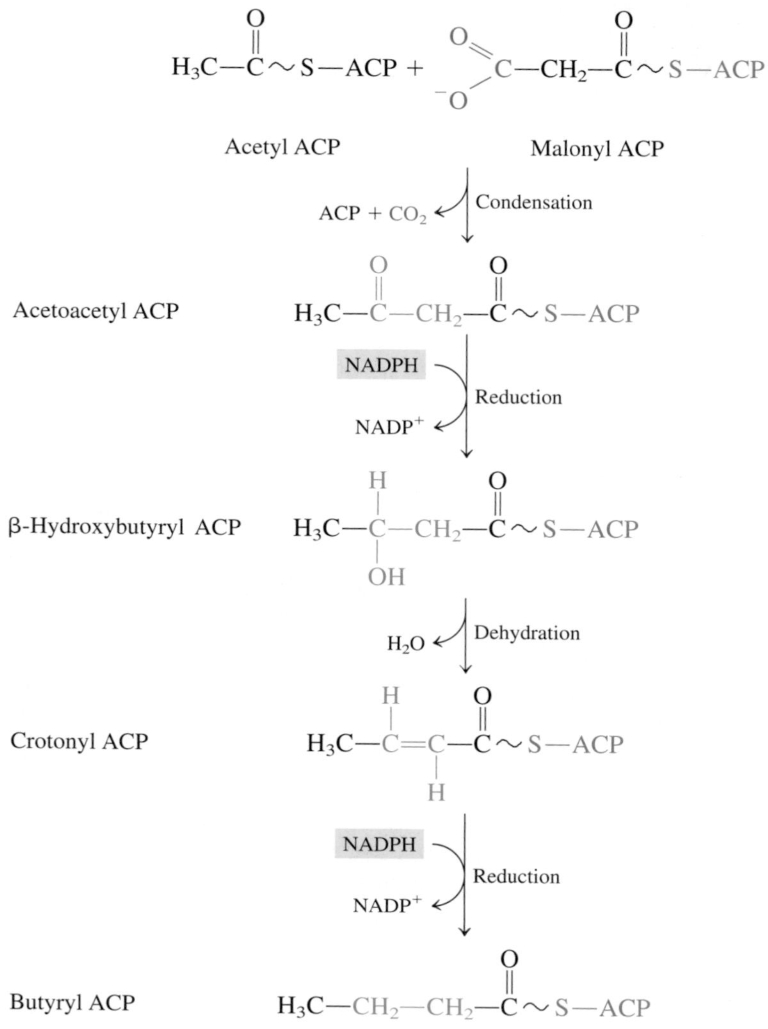

$$H_3C-\overset{O}{\overset{\|}{C}}\sim S-ACP \; + \; {}^{-}O-\overset{O}{\overset{\|}{C}}-CH_2-\overset{O}{\overset{\|}{C}}\sim S-ACP$$

Acetyl ACP — Malonyl ACP

↓ Condensation ($ACP + CO_2$)

Acetoacetyl ACP $H_3C-\overset{O}{\overset{\|}{C}}-CH_2-\overset{O}{\overset{\|}{C}}\sim S-ACP$

↓ Reduction (NADPH → $NADP^+$)

β-Hydroxybutyryl ACP $H_3C-\underset{OH}{\overset{H}{C}}-CH_2-\overset{O}{\overset{\|}{C}}\sim S-ACP$

↓ Dehydration (H_2O)

Crotonyl ACP $H_3C-\overset{H}{C}=\underset{H}{C}-\overset{O}{\overset{\|}{C}}\sim S-ACP$

↓ Reduction (NADPH → $NADP^+$)

Butyryl ACP $H_3C-CH_2-CH_2-\overset{O}{\overset{\|}{C}}\sim S-ACP$

Figure 23.11
Summary of fatty acid synthesis. Malonyl ACP is produced in two reactions: carboxylation of acetyl CoA to produce malonyl CoA and transfer of the malonyl acyl group from malonyl CoA to ACP.

Question 23.7

What conditions lead to excess production of ketone bodies?

Question 23.8

What is the cause of the characteristic "acetone breath" that is associated with uncontrolled diabetes mellitus?

23.4 Fatty Acid Synthesis

Learning Goal

All organisms possess the ability to synthesize fatty acids. In humans the excess acetyl CoA produced by carbohydrate degradation is used to make fatty acids that are then stored as triglycerides.

A Comparison of Fatty Acid Synthesis and Degradation

On first examination, fatty acid synthesis appears to be simply the reverse of β-oxidation. Specifically, the fatty acid chain is constructed by the sequential addition of two-carbon acetyl groups (Figure 23.11). Although the chemistry of fatty acid synthesis and breakdown are similar, there are several major differences between β-oxidation and fatty acid biosynthesis. These are summarized as follows.

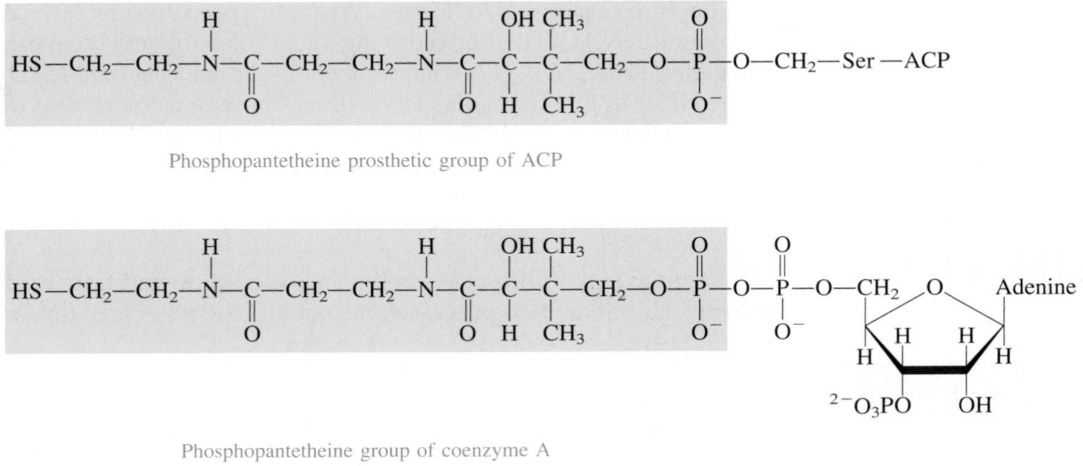

Figure 23.12
The structure of the phosphopantetheine group, the reactive group common to coenzyme A and acyl carrier protein, is highlighted in yellow.

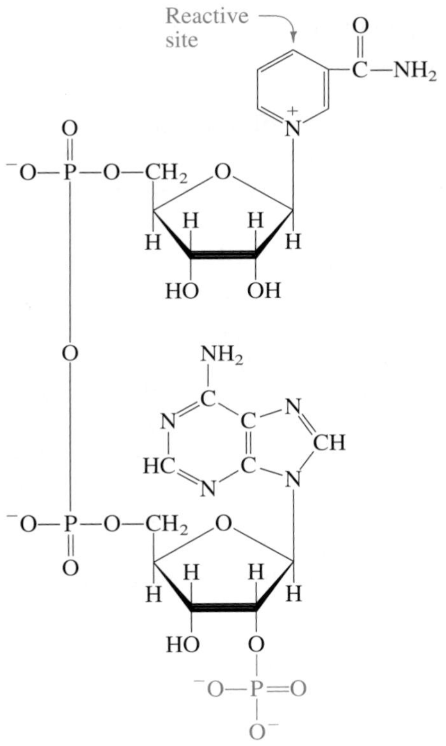

Figure 23.13
Structure of NADPH. The phosphate group shown in red is the structural feature that distinguishes NADPH from NADH.

- **Intracellular location.** The enzymes responsible for fatty acid biosynthesis are located in the cytoplasm of the cell, whereas those responsible for the degradation of fatty acids are in the mitochondria.
- **Acyl group carriers.** The activated intermediates of fatty acid biosynthesis are bound to a carrier molecule called the **acyl carrier protein (ACP)** (Figure 23.12). In β-oxidation the acyl group carrier was coenzyme A. However, there are important similarities between these two carriers. Both contain the **phosphopantetheine** group, which is made from the vitamin pantothenic acid. In both cases the fatty acyl group is bound by a thioester bond to the phosphopantetheine group.
- **Enzymes involved.** Fatty acid biosynthesis is carried out by a multienzyme complex known as *fatty acid synthase.* The enzymes responsible for fatty acid degradation are not physically associated in such complexes.

- **Electron carriers.** NADH and $FADH_2$ are produced by fatty acid oxidation, whereas NADPH is the reducing agent for fatty acid biosynthesis. As a general rule, *NADH is produced by catabolic reactions, and NADPH is the reducing agent of biosynthetic reactions.* These two coenzymes differ only by the presence of a phosphate group bound to the ribose ring of NADPH (Figure 23.13). The enzymes that use these coenzymes, however, are easily able to distinguish them on this basis.

Question 23.9

List the four major differences between β-oxidation and fatty acid biosynthesis that reveal that the two processes are not just the reverse of one another.

Question 23.10

What chemical group is part of coenzyme A and acyl carrier protein and allows both molecules to form thioester bonds to fatty acids?

23.5 The Regulation of Lipid and Carbohydrate Metabolism

Learning Goal

The metabolism of fatty acids and carbohydrates occurs to a different extent in different organs. As we will see in this section, the regulation of these two related aspects of metabolism is of great physiological importance.

The Liver

Glycogenesis is described in Section 21.7.

The liver provides a steady supply of glucose for muscle and brain and plays a major role in the regulation of blood glucose concentration. This regulation is under hormonal control. Recall that the hormone insulin causes blood glucose to be taken up by the liver and stored as glycogen (*glycogenesis*). In this way the liver reduces the blood glucose levels when they are too high.

The hormone glucagon, on the other hand, stimulates the breakdown of glycogen and the release of glucose into the bloodstream. Lactate produced by muscles under anaerobic conditions is also taken up by liver cells and is converted to glucose by gluconeogenesis. Both glycogen degradation (*glycogenolysis*) and *gluconeogenesis* are pathways that produce glucose for export to other organs when energy is needed (Figure 23.14).

Figure 23.14
The liver controls the concentration of blood glucose.

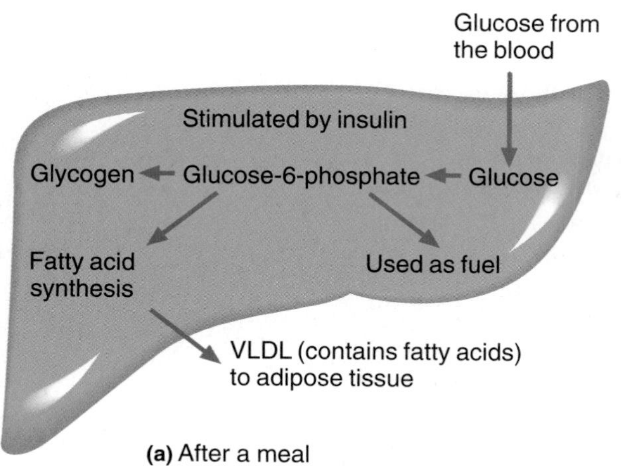

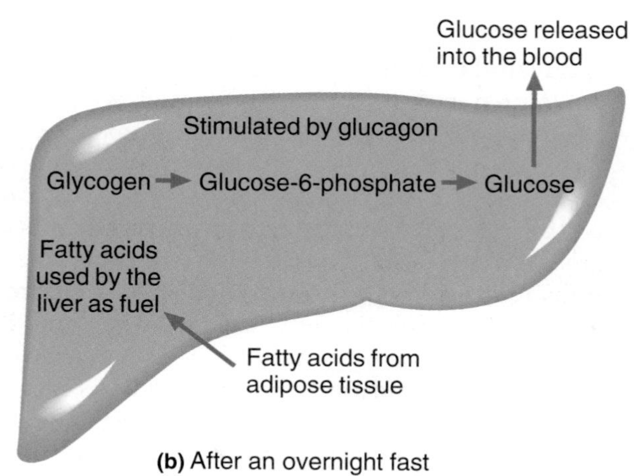

The liver also plays a central role in lipid metabolism. When excess fuel is available, the liver synthesizes fatty acids. These are used to produce triglycerides that are transported from the liver to adipose tissues by very low density lipoprotein (VLDL) complexes. In fact, VLDL complexes provide adipose tissue with its major source of fatty acids. This transport is particularly active when more calories are eaten than are burned! During fasting or starvation conditions, however, the liver converts fatty acids to acetoacetate and other ketone bodies. The liver cannot use these ketone bodies because it lacks an enzyme for the conversion of acetoacetate to acetyl CoA. Therefore the ketone bodies produced by the liver are exported to other organs where they are oxidized to make ATP.

VLDL is described in Section 18.5.

Adipose Tissue

Adipose tissue is the major storage depot of fatty acids. Triglycerides produced by the liver are transported through the bloodstream as components of VLDL complexes. The triglycerides are hydrolyzed by the same lipases that act on chylomicrons, and the fatty acids are absorbed by adipose tissue. The synthesis of triglycerides in adipose tissue requires glycerol-3-phosphate. However, adipose tissue is unable to make glycerol-3-phosphate and depends on glycolysis for its supply of this molecule. Thus adipose cells must have a ready source of glucose to synthesize and store triglycerides.

Triglycerides are constantly being hydrolyzed and resynthesized in the cells of adipose tissue. Lipases that are under hormonal control determine the rate of hydrolysis of triglycerides into fatty acids and glycerol. If glucose is in limited supply, there will not be sufficient glycerol-3-phosphate for the resynthesis of triglycerides, and the fatty acids and glycerol are exported to the liver for further processing (Figure 23.15).

H
H—C—OH
H—C—OH O
H—C—O—P—O^-
H O^-

Glycerol-3-phosphate

Muscle Tissue

The energy demand of *resting* muscle is generally supplied by the β-oxidation of fatty acids. The heart muscle actually prefers ketone bodies over glucose.

Working muscle, however, obtains energy by degradation of its own supply of glycogen. Glycogen degradation produces glucose-6-phosphate, which is directly funneled into glycolysis. If the muscle is working so hard that it doesn't get enough oxygen, it produces large amounts of lactate. This fermentation end product, as

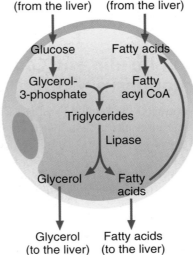

Figure 23.15
Synthesis and degradation of triglycerides in adipose tissue.

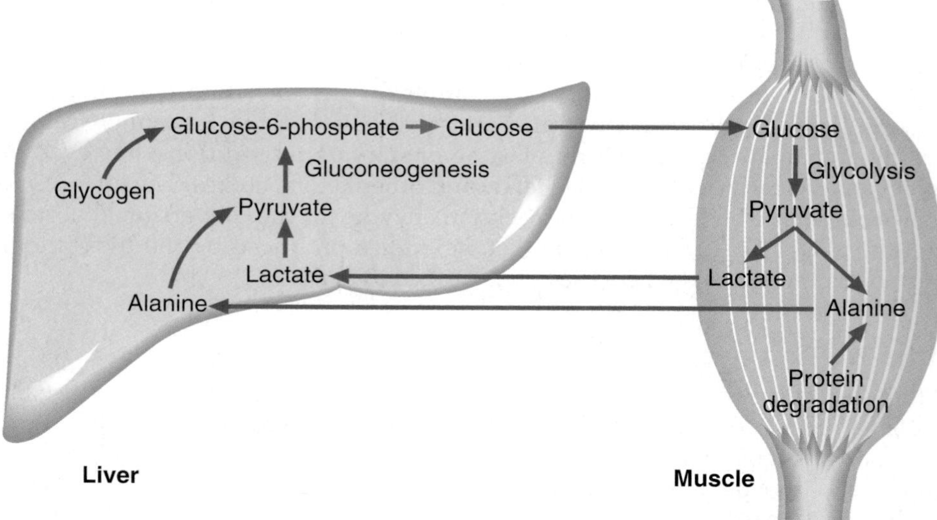

Figure 23.16
Metabolic relationships between liver and muscle.

Gluconeogenesis is described in Section 21.6.

well as alanine (from catabolism of proteins and transamination of pyruvate), is exported to the liver. Here they are converted to glucose by gluconeogenesis (Figure 23.16).

The Brain

Under normal conditions the brain uses glucose as its sole source of metabolic energy. When the body is in the resting state, about 60% of the free glucose of the body is used by the brain. Starvation depletes glycogen stores, and the amount of glucose available to the brain drops sharply. The ketone bodies acetoacetate and β-hydroxybutyrate are then used by the brain as an alternative energy source. Fatty acids are transported in the blood in complexes with proteins and cannot cross the blood-brain barrier to be used by brain cells as an energy source. But ketone bodies, which have a free carboxyl group, are soluble in blood and can enter the brain.

Question 23.11

How does the liver regulate blood glucose levels?

Question 23.12

Why is regulation of blood glucose levels important to the efficient function of the brain?

23.6 The Effects of Insulin and Glucagon on Cellular Metabolism

Learning Goal

The hormone **insulin** is produced by the β-cells of the islets of Langerhans in the pancreas. It is secreted from these cells in response to an increase in the blood glucose level. Insulin lowers the concentration of blood glucose by causing a number of changes in metabolism (Table 23.1).

The simplest way to lower blood glucose levels is to stimulate storage of glucose, both as glycogen and as triglycerides. *Insulin therefore activates biosynthetic processes and inhibits catabolic processes.*

The effect of insulin on glycogen metabolism is described in Section 21.7.

Insulin acts only on those cells, known as *target cells,* that possess a specific insulin receptor protein in their plasma membranes. The major target cells for insulin are liver, adipose, and muscle cells.

Table 23.1 Comparison of the Metabolic Effects of Insulin and Glucagon

Actions	Insulin	Glucagon
Cellular glucose transport	Increased	No effect
Glycogen synthesis	Increased	Decreased
Glycogenolysis in liver	Decreased	Increased
Gluconeogenesis	Decreased	Increased
Amino acid uptake and protein synthesis	Increased	No effect
Inhibition of amino acid release and protein degradation	Decreased	No effect
Lipogenesis	Increased	No effect
Lipolysis	Decreased	Increased
Ketogenesis	Decreased	Increased

The blood glucose level is normally about 10 mM. However, a substantial meal increases the concentration of blood glucose considerably and stimulates insulin secretion. Subsequent binding of insulin to the plasma membrane insulin receptor increases the rate of transport of glucose across the membrane and into cells.

Insulin exerts a variety of effects on all aspects of cellular metabolism:

- **Carbohydrate metabolism.** Insulin stimulates glycogen synthesis. At the same time it inhibits glycogenolysis and gluconeogenesis. The overall result of these activities is the storage of excess glucose.
- **Protein metabolism.** Insulin stimulates transport and uptake of amino acids, as well as the incorporation of amino acids into proteins.
- **Lipid metabolism.** Insulin stimulates uptake of glucose by adipose cells, as well as the synthesis and storage of triglycerides. As we have seen, storage of lipids requires a source of glucose, and insulin helps the process by increasing the available glucose. At the same time, insulin inhibits the breakdown of stored triglycerides.

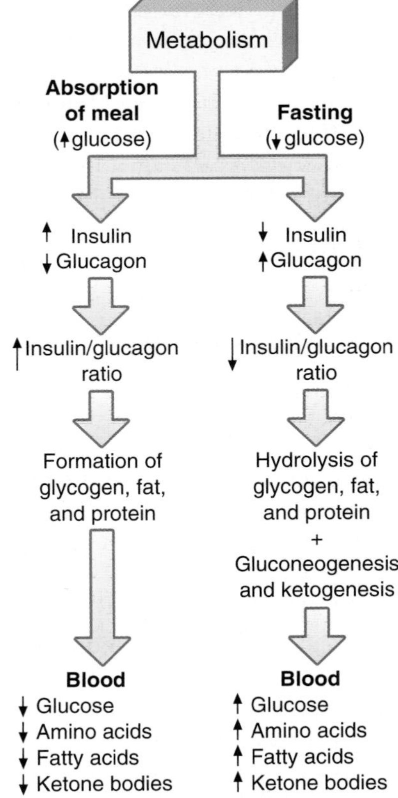

Figure 23.17
A summary of the antagonistic effects of insulin and glucagon.

As you may have already guessed, insulin is only part of the overall regulation of cellular metabolism in the body. A second hormone, **glucagon,** is secreted by the α-cells of the islets of Langerhans in response to decreased blood glucose levels. The effects of glucagon, generally the opposite of the effects of insulin, are summarized in Table 23.1. Although it has no direct effect on glucose uptake, glucagon inhibits glycogen synthesis and stimulates glycogenolysis and gluconeogenesis. It also stimulates the breakdown of fats and ketogenesis.

The antagonistic effects of these two hormones, seen in Figure 23.17, are critical for the maintenance of adequate blood glucose levels. During fasting, low blood glucose levels stimulate production of glucagon, which increases blood glucose by stimulating the breakdown of glycogen and the production of glucose by gluconeogenesis. This ensures a ready supply of glucose for the tissues, especially the brain. On the other hand, when blood glucose levels are too high, insulin is secreted. It stimulates the removal of the excess glucose by enhancing uptake and inducing pathways for storage.

Question 23.13

Summarize the effects of the hormone insulin on carbohydrate, lipid, and amino acid metabolism.

Question 23.14

Summarize the effects of the hormone glucagon on carbohydrate and lipid metabolism.

Summary

23.1 Lipid Metabolism in Animals

Dietary lipids (*triglycerides*) are emulsified into tiny fat droplets in the intestine by the action of *bile* salts. Pancreatic *lipase* catalyzes the hydrolysis of triglycerides into monoglycerides and fatty acids. These are absorbed by intestinal epithelial cells, reassembled into triglycerides, and combined with protein to form *chylomicrons*. Chylomicrons are transported to the cells of the body through the bloodstream. Fatty acids are stored as triglycerides (triacylglycerols) in fat droplets in the cytoplasm of *adipocytes*.

23.2 Fatty Acid Degradation

Fatty acids are degraded to acetyl CoA in the mitochondria by the β-*oxidation* pathway, which involves five steps: (1) the production of a fatty acyl CoA molecule, (2) oxidation of the fatty acid by an FAD-dependent dehydrogenase, (3) hydration, (4) oxidation by an NAD^+-dependent dehydrogenase, and (5) cleavage of the chain with release of acetyl CoA and a fatty acyl CoA that is two carbons shorter than the beginning fatty acid. The last four reactions are repeated until the fatty acid is completely degraded to acetyl CoA.

23.3 Ketone Bodies

Under some conditions, fatty acid degradation occurs more rapidly than glycolysis. As a result, a large amount of acetyl CoA is produced from fatty acids, but little oxaloacetate is generated from pyruvate. When oxaloacetate levels are too low, the excess acetyl CoA is converted to the *ketone bodies* acetone, acetoacetate, and β-hydroxybutyrate.

23.4 Fatty Acid Synthesis

Fatty acid biosynthesis occurs by the sequential addition of acetyl groups and, on first inspection, appears to be a simple reversal of the β-oxidation pathway. Although the biochemical reactions are similar, fatty acid synthesis differs from β-oxidation in the following ways: It occurs in the cytoplasm, utilizes *acyl carrier protein* and NADPH, and is carried out by a multienzyme complex, fatty acid synthase.

23.5 The Regulation of Lipid and Carbohydrate Metabolism

Lipid and carbohydrate metabolism occur to different extents in different organs. The liver regulates the flow of metabolites to brain, muscle, and adipose tissue and ultimately controls the concentration of blood glucose. *Adipose tissue* is the major storage depot for fatty acids. Triglycerides are constantly hydrolyzed and resynthesized in adipose tissue. Muscle oxidizes glucose, fatty acids, and ketone bodies. The brain uses glucose as a fuel except in prolonged fasting or starvation, when it will use ketone bodies as an energy source.

23.6 The Effects of Insulin and Glucagon on Cellular Metabolism

Insulin stimulates biosynthetic processes and inhibits catabolism in liver, muscle, and adipose tissue. Insulin is synthesized in the β-cells of the pancreas and is secreted when the blood glucose levels become too high. The insulin receptor protein binds to the insulin. This binding mediates a variety of responses in target tissues, including the storage of glucose and lipids. *Glucagon* is secreted when blood glucose levels are too low. It has the opposite effects on metabolism, including the breakdown of lipids and glycogen.

Key Terms

acyl carrier protein (ACP) (23.4)
adipocyte (23.1)
adipose tissue (23.1)
bile (23.1)
chylomicron (23.1)
colipase (23.1)
diabetes mellitus (23.3)
glucagon (23.6)
insulin (23.6)
ketoacidosis (23.3)
ketone bodies (23.3)
ketosis (23.3)
lipase (23.1)
micelle (23.1)
β-oxidation (23.2)
phosphopantetheine (23.4)
triglyceride (23.1)

Questions and Problems

Lipid Metabolism in Animals

23.15 What is the major storage form of fatty acids?
23.16 What tissue is the major storage depot for lipids?
23.17 What is the outstanding structural feature of an adipocyte?
23.18 What is the major metabolic function of adipose tissue?
23.19 What is the general reaction catalyzed by lipases?
23.20 Why are the lipases that are found in saliva and in the stomach not very effective at digesting triglycerides?
23.21 List three major biological molecules for which acetyl CoA is a precursor.
23.22 Why are triglycerides more efficient energy-storage molecules than glycogen?
23.23 What are chylomicrons, and what is their function?
23.24 **a.** What are very low density lipoproteins?
b. Compare the function of VLDLs with that of chylomicrons.
23.25 What is the function of the bile salts in the digestion of dietary lipids?

23.26 What is the function of colipase in the digestion of dietary lipids?
23.27 Describe the stages of lipid digestion.
23.28 Describe the transport of lipids digested in the lumen of the intestines to the cells of the body.

Fatty Acid Degradation

23.29 What products are formed when the ω-phenyl-labeled carboxylic acid 14-phenyltetradecanoic acid is degraded by β-oxidation?
23.30 What products are formed when the ω-phenyl-labeled carboxylic acid 5-phenylpentanoic acid is degraded by β-oxidation?
23.31 Calculate the number of ATP molecules produced by complete β-oxidation of the fourteen-carbon saturated fatty acid tetradecanoic acid (common name: myristic acid).
23.32 **a.** Write the sequence of steps that would be followed for one round of β-oxidation of hexanoic acid.
b. Calculate the number of ATP molecules produced by complete β-oxidation of hexanoic acid.
23.33 How many molecules of ATP are produced for each molecule of $FADH_2$ that is generated by β-oxidation?
23.34 How many molecules of ATP are produced for each molecule of NADH generated by β-oxidation?
23.35 What is the fate of the acetyl CoA produced by β-oxidation?
23.36 How many ATP molecules are produced from each acetyl CoA molecule generated in β-oxidation that enters the citric acid cycle?

Ketone Bodies

23.37 Draw the structures of acetoacetate and β-hydroxybutyrate.
23.38 Describe the relationship between the formation of ketone bodies and β-oxidation.
23.39 Why do uncontrolled diabetics produce large amounts of ketone bodies?
23.40 How does the presence of ketone bodies in the blood lead to ketoacidosis?
23.41 When does the heart use ketone bodies?
23.42 When does the brain use ketone bodies?

Fatty Acid Synthesis

23.43 **a.** What is the role of the phosphopantetheine group in fatty acid biosynthesis?
b. From what molecule is phosphopantetheine made?
23.44 What molecules involved in fatty acid degradation and fatty acid biosynthesis contain the phosphopantetheine group?
23.45 How does the structure of fatty acid synthase differ from that of the enzymes that carry out β-oxidation?
23.46 In what cellular compartments do fatty acid biosynthesis and β-oxidation occur?

The Regulation of Lipid and Carbohydrate Metabolism

23.47 What is the major metabolic function of the liver?
23.48 What is the fate of lactate produced in skeletal muscle during rapid contraction?
23.49 What are the major fuels of the heart, brain, and liver?
23.50 Why can't the brain use fatty acids as fuel?
23.51 Briefly describe triglyceride metabolism in an adipocyte.
23.52 What is the source of the glycerol molecule that is used in the synthesis of triglycerides?

The Effects of Insulin and Glucagon on Cellular Metabolism

23.53 Where is insulin produced?
23.54 Where is glucagon produced?
23.55 How does insulin affect carbohydrate metabolism?
23.56 How does glucagon affect carbohydrate metabolism?
23.57 How does insulin affect lipid metabolism?
23.58 How does glucagon affect lipid metabolism?
23.59 Why is it said that diabetes mellitus amounts to starvation in the midst of plenty?
23.60 What is the role of the insulin receptor in controlling blood glucose levels?

Critical Thinking Problems

1. Suppose that fatty acids were degraded by sequential oxidation of the α-carbon. What product(s) would Knoop have obtained with fatty acids with even numbers of carbon atoms? What product(s) would he have obtained with fatty acids with odd numbers of carbon atoms?
2. Oil-eating bacteria can oxidize long-chain alkanes. In the first step of the pathway, the enzyme monooxygenase catalyzes a reaction that converts the long-chain alkane into a primary alcohol. Data from research studies indicate that three more reactions are required to allow the primary alcohol to enter the β-oxidation pathway. Propose a pathway that would convert the long-chain alcohol into a product that could enter the β-oxidation pathway.
3. A young woman sought the advice of her physician because she was 30 pounds overweight. The excess weight was in the form of triglycerides carried in adipose tissue. Yet when the woman described her diet, it became obvious that she actually ate very moderate amounts of fatty foods. Most of her caloric intake was in the form of carbohydrates. This included candy, cake, beer, and soft drinks. Explain how the excess calories consumed in the form of carbohydrates ended up being stored as triglycerides in adipose tissue.
4. Olestra is a fat substitute that provides no calories, yet has a creamy, tongue-pleasing consistency. Because it can withstand heating, it can be used to prepare foods such as potato chips and crackers. Recently the Food and Drug Administration approved olestra for use in prepared foods. Olestra is a sucrose polyester produced by esterification of six, seven, or eight fatty acids to molecules of sucrose. Develop a hypothesis to explain why olestra is not a source of dietary calories.
5. Carnitine is a tertiary amine found in mitochondria that is involved in transporting the acyl groups of fatty acids from the cytoplasm into the mitochondria. The fatty acyl group is transferred from a fatty acyl CoA molecule and esterified to carnitine. Inside the mitochondria the reaction is reversed and the fatty acid enters the β-oxidation pathway.

   A seventeen-year-old male went to a university medical center complaining of fatigue and poor exercise tolerance. Muscle biopsies revealed droplets of triglycerides in his muscle cells. Biochemical analysis showed that he had only one-fifth of the normal amount of carnitine in his muscle cells.

   What effect will carnitine deficiency have on β-oxidation? What effect will carnitine deficiency have on glucose metabolism?

6. Acetyl CoA carboxylase catalyzes the formation of malonyl CoA from acetyl CoA and the bicarbonate anion, a reaction that requires the hydrolysis of ATP. Write a balanced equation showing this reaction.

The reaction catalyzed by acetyl CoA carboxylase is the rate-limiting step in fatty acid biosynthesis. The malonyl group is transferred from coenzyme A to acyl carrier protein; similarly, the acetyl group is transferred from coenzyme A to acyl carrier protein. This provides the two beginning substrates of fatty acid biosynthesis shown in Figure 23.11.

Consider the following case study. A baby boy was brought to the emergency room with severe respiratory distress. Examination revealed muscle pathology, poor growth, and severe brain damage. A liver biopsy revealed that the child didn't make acetyl CoA carboxylase. What metabolic pathway is defective in this child? How is this defect related to the respiratory distress suffered by the baby?

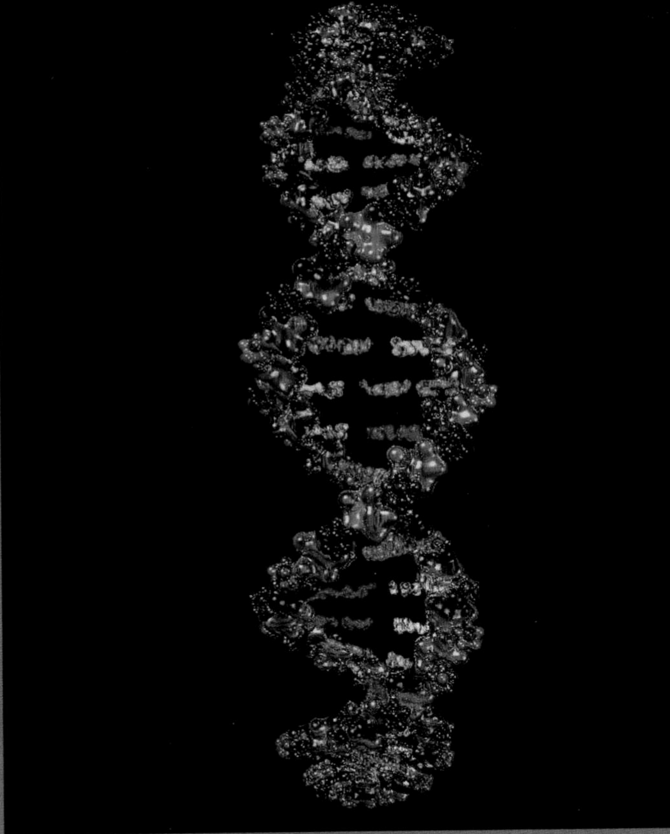

Computer-generated model of DNA.

24 Introduction to Molecular Genetics

Outline

Learning Goals

1. Draw the general structure of DNA and RNA nucleotides.
2. Describe the structure of DNA and compare it with RNA.
3. Explain DNA replication.
4. List three classes of RNA molecules and describe their functions.
5. Explain the process of transcription.
6. List and explain the three types of post-transcriptional modifications of eukaryotic mRNA.
7. Describe the essential elements of the genetic code, and develop a "feel" for its elegance.
8. Describe the process of translation.
9. Define mutation and understand how mutations cause cancer and cell death.
10. Describe the tools used in the study of DNA and in genetic engineering.
11. Describe the process of polymerase chain reaction and discuss potential uses of the process.
12. Discuss strategies for genome analysis and DNA sequencing.

CHEMISTRY CONNECTION

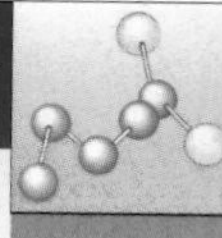

Molecular Genetics and Detection of Human Genetic Disease

It is estimated that 3–5% of the human population suffers from a serious genetic defect. That's 200 million people! But what if genetic disease could be detected and "cured"? Two new technologies, *gene therapy* and *preimplantation diagnosis*, may help us realize this dream.

For a couple with a history of a genetic disease in the family, pregnancy is a time of anxiety. Through *genetic counseling* these couples can learn the probability that their child has the disease. For about 200 genetic diseases the uncertainty can be eliminated. *Amniocentesis* (removal of 10–20 mL of fluid from the sac around the fetus) and *chorionic villus* sampling (removal of cells from a fetal membrane) are two procedures that are used to obtain fetal cells for genetic testing. Fetal cells are cultured and tested by enzyme assays and DNA tests to look for genetic diseases. If a genetic disease is diagnosed, the parents must make a difficult decision: to abort the fetus or to carry the child to term and deal with the effects of the genetic disease.

The power of modern molecular genetics is obvious in our ability to find a "bad" gene from just a few cells. But scientists have developed an even more impressive way to test for genetic disease before the embryo implants into the uterine lining. This technique, called *preimplantation diagnosis*, involves fertilizing a human egg and allowing the resulting conceptus to divide in a sterile petri dish. When the conceptus consists of 8–16 cells, *one* cell is removed for genetic testing. Only genetically normal embryos are implanted in the mother. Thus the genetic diseases that we can detect could be eliminated from the population by preimplantation diagnosis because only a conceptus with "good" genes is used.

Gene therapy is a second way in which genetic diseases may one day be eliminated. Foreign genes, including growth hormone, have been introduced into fertilized mouse eggs and the conceptuses implanted in female mice. The baby mice born with the foreign growth hormone gene were about three times larger than their normal littermates! One day, this kind of technology may be used to introduce normal genes into human fertilized eggs carrying a defective gene, thereby replacing the defective gene with a normal one.

In this chapter we will examine the molecules that carry and express our genetic information, DNA and RNA. Only by understanding the structure and function of these molecules have we been able to develop the amazing array of genetic tools that currently exists. We hope that as we continue to learn more about human genetics, we will be able to detect and one day correct most of the known genetic diseases.

Introduction

Look around at the students in your chemistry class. They all share many traits: upright stance, a head with two eyes, a nose, and a mouth facing forward, one ear on each side of the head, and so on. You would have no difficulty listing the similarities that define you and your classmates as *Homo sapiens*.

As you look more closely at the individuals you begin to notice many differences. Eye color, hair color, skin color, the shape of the nose, height, body build: all these traits, and many more, show amazing variety from one person to the next. Even within one family, in which the similarities may be more pronounced, each individual has a unique appearance. In fact, only identical twins look exactly alike—well, most of the time.

The molecule responsible for all these similarities and differences is deoxyribonucleic acid (DNA). Tightly wound up in structures called *chromosomes* in the nucleus of the cell, DNA carries the genetic code to produce the thousands of different proteins that make us who we are. These proteins include enzymes that are responsible for production of the pigment melanin. The more melanin we are genetically programmed to make, the darker our hair, eyes, and skin will be. Others are structural proteins. The gene for α-keratin that makes up hair determines whether our hair will be wavy, straight, or curly. Thousands of genes carry the genetic information for thousands of proteins that dictate our form and, some believe, our behavior.

Genetic traits are passed from one generation to the next. When a sperm fertilizes an ovum, a conceptus is created from a single set of maternal chromosomes and a single set of paternal chromosomes. As this fertilized egg divides, each daughter cell will receive one copy of each of these chromosomes. The genes on these chromosomes will direct fetal development, from that fertilized cell to a newborn with all the characteristics we recognize as human.

In this chapter we will explore the structure of DNA and the molecular events that translate the genetic information of a gene into the structure of a protein.

24.1 The Structure of the Nucleotide

Chemical Composition of DNA and RNA

Learning Goal

DNA was discovered by Friedrich Meischer in 1869. However, it was not recognized as the genetic information until 1950. In 1953 James Watson and Francis Crick published a paper describing the structure of the DNA molecule. From that paper and many other studies we know that **deoxyribonucleic acid (DNA)** and **ribonucleic acid (RNA)** are long polymers of **nucleotides.** Every nucleotide is composed of three units: a five-carbon sugar, a nitrogenous base, and either one, two, or three phosphoryl groups. Nitrogenous bases are heterocyclic ring structures having backbones consisting of carbon and nitrogen atoms. There are two classes of nitrogenous bases. The **purine** nitrogenous bases consist of a six-member ring fused to a five-member ring. The **pyrimidine** nitrogenous bases consist of a single six-member ring. The structures are seen in Figure 24.1.

The five-carbon sugar in RNA is ribose, and the sugar in DNA is 2′-deoxyribose. The only difference between these two sugars is the absence of an hydroxyl group on the 2′ carbon of 2′-deoxyribose. The purines in both DNA and RNA are adenine and guanine. Both DNA and RNA contain the pyrimidine cytosine; however, the fourth base is thymine in DNA and uracil in RNA. The chemical compositions of DNA and RNA are summarized in Table 24.1.

The carbon atoms of the sugars found in nucleic acids are indicated with a prime: 1′, 2′, 3′, and so on. This is to distinguish them from the ring atoms of the nitrogenous bases.

In addition, the nucleotides that make up DNA and RNA contain phosphoryl groups. Nucleotides may be *mono-*, *di-*, or *tri*phosphates.

Nucleotide Structure

Nucleotides are produced by the combination of a sugar, a nitrogenous base, and at least one phosphoryl group, as shown in Figure 24.2. Because this large structure contains two cyclic molecules, the sugar and the base, we must have an easy way to describe the ring atoms of each. For this reason the ring atoms of the sugar are designated with a prime to distinguish them from atoms in the base. The covalent bond between the sugar and the phosphoryl group is a phosphoester linkage formed by a condensation reaction between the 5′-OH of the sugar and an —OH of the phosphoryl group. The bond between the base and the sugar is a β-*N*-glycosidic linkage, and it joins the 1′-carbon of the sugar and a nitrogen atom of the nitrogenous base (N-9 of purines and N-1 of pyrimidines).

To name a nucleotide, simply begin with the name of the nitrogenous base, and apply the following simple rules:

- Remove the *-ine* ending, and replace it with either *-osine* for purines or *-idine* for pyrimidines. Uracil is the one exception to this rule. In this case the *-acil* ending is replaced with *-idine*, producing the name *uridine.*
- Nucleotides with the sugar ribose are **ribonucleotides,** and those having the sugar 2′-deoxyribose are **deoxyribonucleotides.** For a deoxyribonucleotide, the prefix *deoxy-* is placed before the modified nitrogenous base name. No prefix is required for ribonucleotides, or for thymidine, which is found only in DNA.
- Add a prefix to indicate the number of phosphoryl groups that are attached. A *mono*phosphate carries one phosphoryl group; a *di*phosphate carries two phosphoryl groups; and a *tri*phosphate carries three phosphoryl groups.

In diphosphates and triphosphates the phosphoryl groups are bonded to one another through phosphoanhydride bonds (Section 15.4).

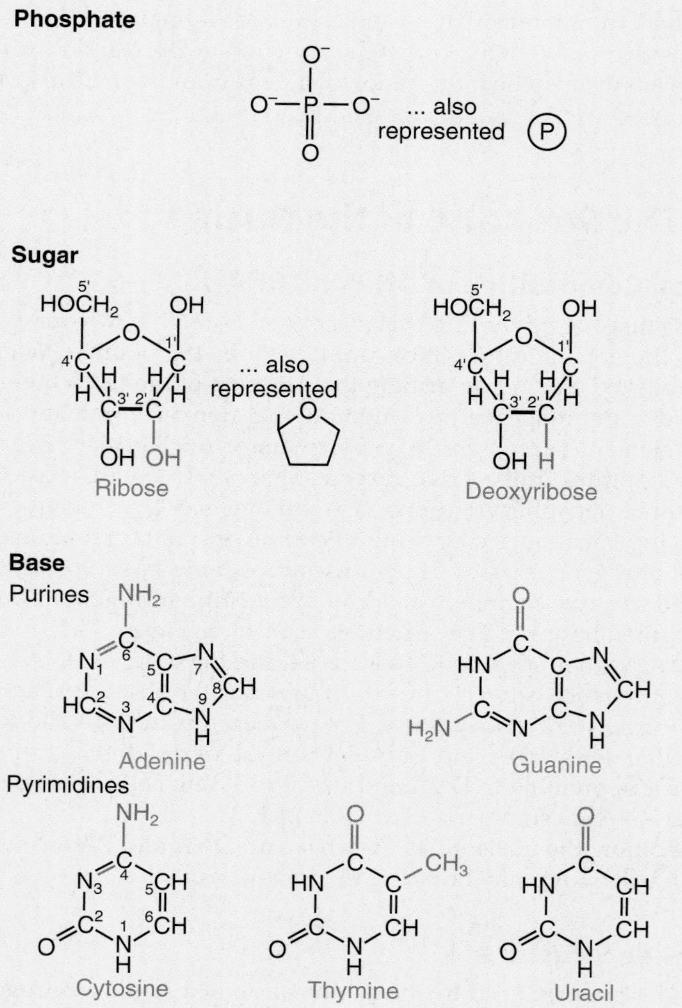

Phosphate

... also represented (P)

Sugar

Ribose

Deoxyribose

... also represented

Base

Purines

Adenine

Guanine

Pyrimidines

Cytosine

Thymine

Uracil

Figure 24.1
The components of nucleic acids include phosphate groups, the five-carbon sugars ribose and deoxyribose, and purine and pyrimidine nitrogenous bases. The ring positions of the sugars are designated with primes (′) to distinguish them from the ring positions of the bases.

Table 24.1 Chemical Composition of Nucleic Acids

	DNA	RNA
Sugar	2′-Deoxyribose	Ribose
Purine nitrogenous bases	Adenine (A)	Adenine (A)
	Guanine (G)	Guanine (G)
Pyrimidine nitrogenous bases	Cytosine (C)	Cytosine (C)
	Thymine (T)	Uracil (U)

Because the full names of the nucleotides are so cumbersome, a simple abbreviation is generally used. These abbreviations are summarized in Table 24.2.

Question 24.1

Referring to the structures in Figures 24.1 and 24.2, draw the structures and write the names for nucleotides consisting of the following units.

a. Ribose, adenine, two phosphoryl groups
b. 2′-Deoxyribose, guanine, three phosphoryl groups

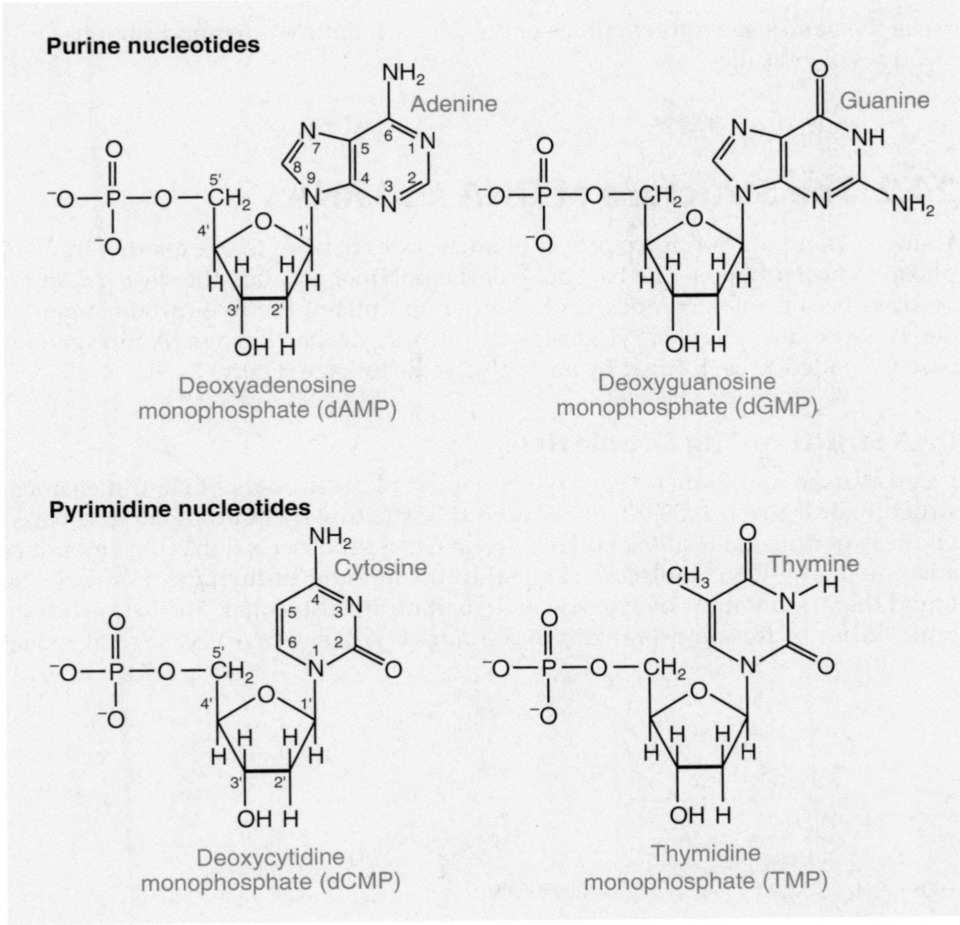

Figure 24.2
Structures, names, and common abbreviations of four deoxyribonucleotides or deoxyribonucleotide-5′-phosphates.

Table 24.2 Names and Abbreviations of the Ribonucleotides and Deoxyribonucleotides Containing Adenine

Nucleotide	Abbreviation
Deoxyadenosine monophosphate	dAMP
Deoxyadenosine diphosphate	dADP
Deoxyadenosine triphosphate	dATP
Adenosine monophosphate	AMP
Adenosine diphosphate	ADP
Adenosine triphosphate	ATP

Question 24.2

Referring to the structures in Figures 24.1 and 24.2, draw the structures and write the names for nucleotides consisting of the following units:

a. 2′-Deoxyribose, thymine, one phosphoryl group
b. Ribose, cytosine, three phosphoryl groups
c. Ribose, uracil, one phosphoryl group

Question 24.3

Write the names and abbreviations of the deoxyribonucleotides and ribonucleotides of guanine.

Question 24.4

Write the names and abbreviations of the deoxyribonucleotides and ribonucleotides of cytosine.

24.2 The Structure of DNA and RNA

Learning Goal

A single strand of DNA is a polymer of nucleotides bonded to one another by 3′–5′ phosphodiester bonds. The backbone of the polymer is called the *sugar-phosphate backbone* because it is composed of alternating units of the five-carbon sugar 2′-deoxyribose and phosphoryl groups in phosphodiester linkage. A nitrogenous base is bonded to each sugar by an *N*-glycosidic linkage (Figure 24.3).

DNA Structure: The Double Helix

James Watson and Francis Crick were the first to describe the three-dimensional structure of DNA in 1953. They deduced the structure by building models based on the experimental results of others. Irwin Chargaff observed that the amount of adenine in any DNA molecule is equal to the amount of thymine. Similarly, he found that the amounts of cytosine and guanine are also equal. The X-ray diffraction studies of Rosalind Franklin and Maurice Wilkens revealed several repeat

Figure 24.3
The covalent, primary structure of DNA. (a) The esterification reaction by which two nucleotides become linked by a phosphodiester bond. (b) A series of three covalently linked deoxyribonucleotides.

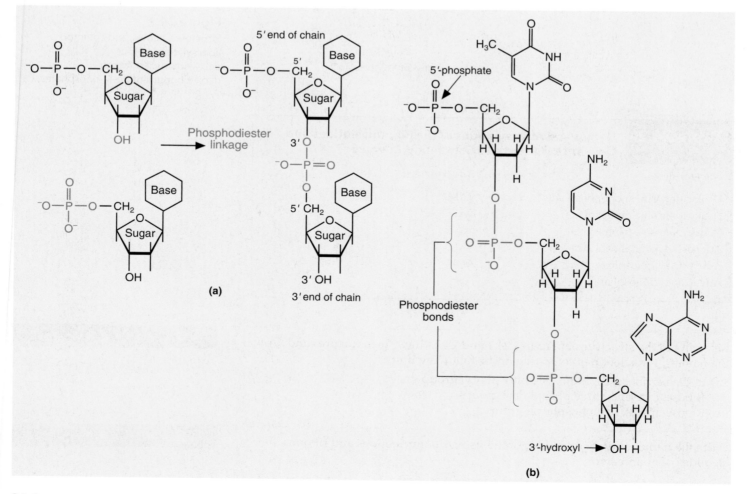

distances that characterize the structure of DNA: 0.34 nm, 3.4 nm, and 2 nm. (Look at the structure of DNA in Figure 24.4 to see the significance of these measurements.) With this information Watson and Crick concluded that DNA is a **double helix** of two strands of DNA wound around one another. It is useful to compare the structure of the double helix to a spiral staircase. The sugar-phosphate backbones of the two strands of DNA spiral around the outside of the helix like the handrails on a spiral staircase. The nitrogenous bases extend into the center at right angles to the axis of the helix. You can imagine the nitrogenous bases forming the steps of the staircase. The structure of this elegant molecule is shown in Figure 24.4.

The two strands of DNA are held together by hydrogen bonds between the nitrogenous bases in the center of the helix. Adenine forms two hydrogen bonds with

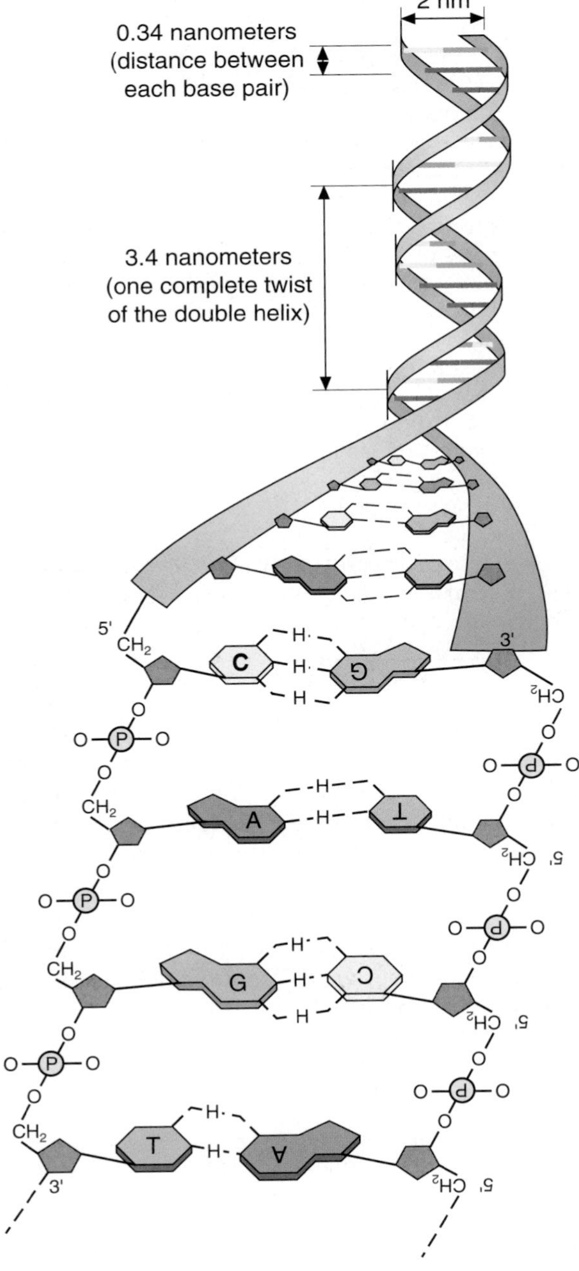

Figure 24.4
Schematic ribbon diagram of the DNA double helix showing the dimensions of the DNA molecule and the antiparallel orientation of the two strands.

A CLINICAL PERSPECTIVE

Fooling the AIDS Virus with "Look-Alike" Nucleotides

The virus that is responsible for the acquired immune deficiency syndrome (AIDS) is called the *human immunodeficiency virus,* or *HIV.* HIV is a member of a family of viruses called *retroviruses,* all of which have single-stranded RNA as their genetic material. The RNA is copied by a viral enzyme called *reverse transcriptase* into a double-stranded DNA molecule. This process is the opposite of the central dogma, which states that the flow of genetic information is from DNA to RNA. But these viruses reverse that flow, RNA to DNA. For this reason these viruses are called retroviruses, which literally means "backward viruses." The process of producing a DNA copy of the RNA is called *reverse transcription.*

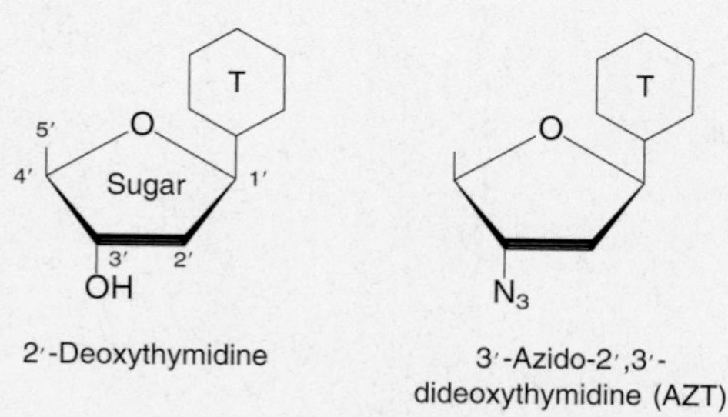

Comparison of the structures of the normal nucleoside, 2′-deoxythymidine, and the nucleoside analog, 3′-azido-2′, 3′-dideoxythymidine.

Because our genetic information is DNA and it is expressed by the classical DNA → RNA → protein pathway, our cells have no need for a reverse transcriptase enzyme. Thus the HIV reverse transcriptase is a good target for antiviral chemotherapy because inhibition of reverse transcription should kill the virus but have no effect on the human host. Many drugs have been tested for the ability to selectively inhibit HIV reverse transcription. Among these is the DNA chain terminator 3′-azido-2′, 3′-dideoxythymidine, commonly called *AZT* or *zidovudine.*

How does AZT work? It is one of many drugs that looks like one of the normal nucleosides. These are called *nucleoside analogs.* A nucleoside is just a nucleotide without any phosphate groups attached. The analog is phosphorylated by the cell and then tricks a polymerase, in this case viral reverse transcriptase, into incorporating it into the growing DNA chain in place of the normal phosphorylated nucleoside. AZT is a nucleoside analog that looks like the nucleoside thymidine except that in the 3′ position of the deoxyribose sugar there is an azido group (—N_3) rather than the 3′-OH group. Compare the structures of thymidine and AZT shown in the accompanying figure. The 3′-OH group is necessary for further DNA polymerization because it is there that the phosphoester linkage must be made between the growing DNA strand and the next nucleotide. If an azido group or some other group is present at the 3′ position, the nucleotide analog can be incorporated into the growing DNA strand, but further chain elongation is blocked, as shown in the following figure. If the viral RNA cannot be reverse transcribed into the DNA form, the virus will not be able to replicate and can be considered to be dead.

AZT is particularly effective because the HIV reverse transcriptase actually prefers it over the normal nucleotide, thymidine. Nonetheless, AZT is not a cure. At best it prolongs the life

thymine, and cytosine forms three hydrogen bonds with guanine (Figures 24.4 and 24.5). These are called **base pairs.** The two strands of DNA are **complementary strands** because the sequence of bases on one automatically determines the sequence of bases on the other. When there is an adenine on one strand, there will always be a thymine in the same location on the opposite strand.

The diameter of the double helix is 2.0 nm. This is dictated by the dimensions of the purine-pyrimidine base pairs. The helix completes one turn every ten base pairs. One complete turn is 3.4 nm. Thus each base pair advances the helix by 0.34 nm.

One last important feature of the DNA double helix is that the two strands are **antiparallel strands,** as this example shows:

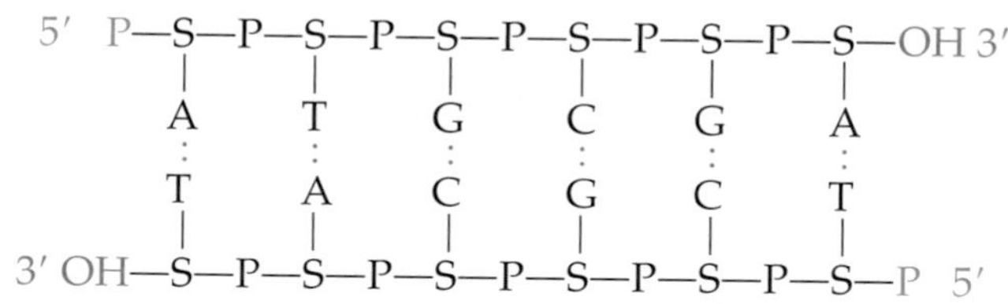

of a person with AIDS for a year or two. Eventually, however, AZT has a negative effect on the body. The cells of our bone marrow are constantly dividing to produce new blood cells: red blood cells to carry oxygen to the tissues, white blood cells of the immune system, and platelets for blood clotting. For cells to divide, they must replicate their DNA. The DNA polymerases of these dividing cells also accidentally incorporate AZT into the growing DNA chains with the result that cells of the bone marrow begin to die. This can result in anemia and even further depression of the immune response.

Another problem that has arisen with prolonged use of AZT is that AZT-resistant mutants of the virus appear. It is well known that HIV is a virus that mutates rapidly. Some of these mutant forms of the virus have an altered reverse transcriptase that will no longer use AZT. When these mutants appear, AZT is no longer useful in treating the infection.

It is hoped that research with other nucleoside analogs, alternative types of antiviral treatments, and combinations of drugs will provide a means of effectively treating HIV infection. Such a therapy must have fewer toxic side effects while stopping the replication of the virus and the progress of the disease.

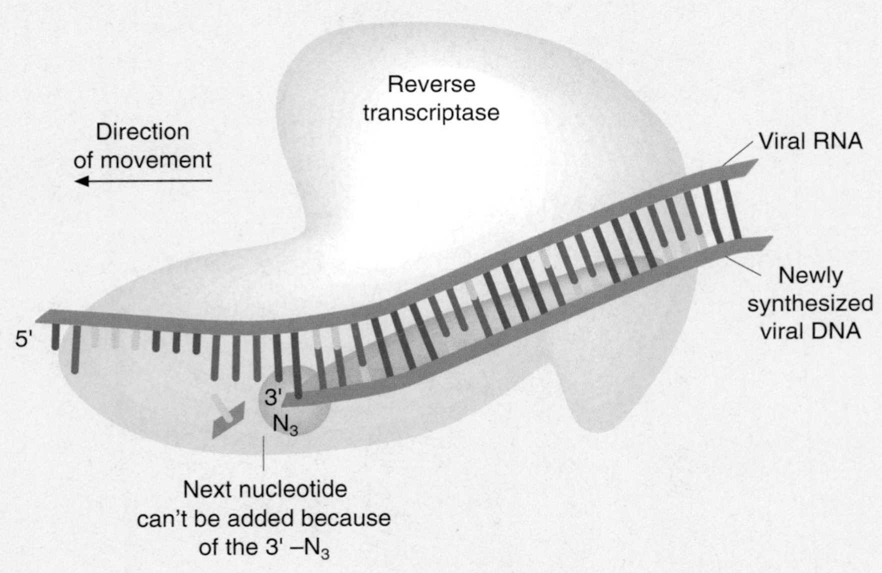

The mechanism by which AZT inhibits HIV reverse transcriptase. Incorporation of AZT into the growing HIV DNA strand in place of deoxythymidine results in DNA chain termination; the azido group on the 3′ carbon of the sugar cannot react to produce the phosphoester linkage required to add the next nucleotide.

In other words, the two strands of the helix run in opposite directions (see Figure 24.4). Only when the two strands are antiparallel can the base pairs form the hydrogen bonds that hold the two strands together.

Chromosomes

Chromosomes are pieces of DNA that carry the genetic instructions, or genes, of an organism. Organisms such as the prokaryotes have only a single chromosome and its structure is relatively simple. Others, the eukaryotes, have many chromosomes, each of which has many different levels of structure. The complete set of genetic information in all the chromosomes of an organism is called the **genome.**

Prokaryotes are organisms with a simple cellular structure in which there is no true nucleus surrounded by a nuclear membrane and there are no true membrane-bound organelles. This group includes all of the bacteria. In these organisms the chromosome is a circular DNA molecule that is supercoiled, which means that the helix is coiled on itself. The supercoiled DNA molecule is attached to a complex of proteins at roughly forty sites along its length, forming a series of loops. This structure, called the nucleoid, can be seen in Figure 24.6.

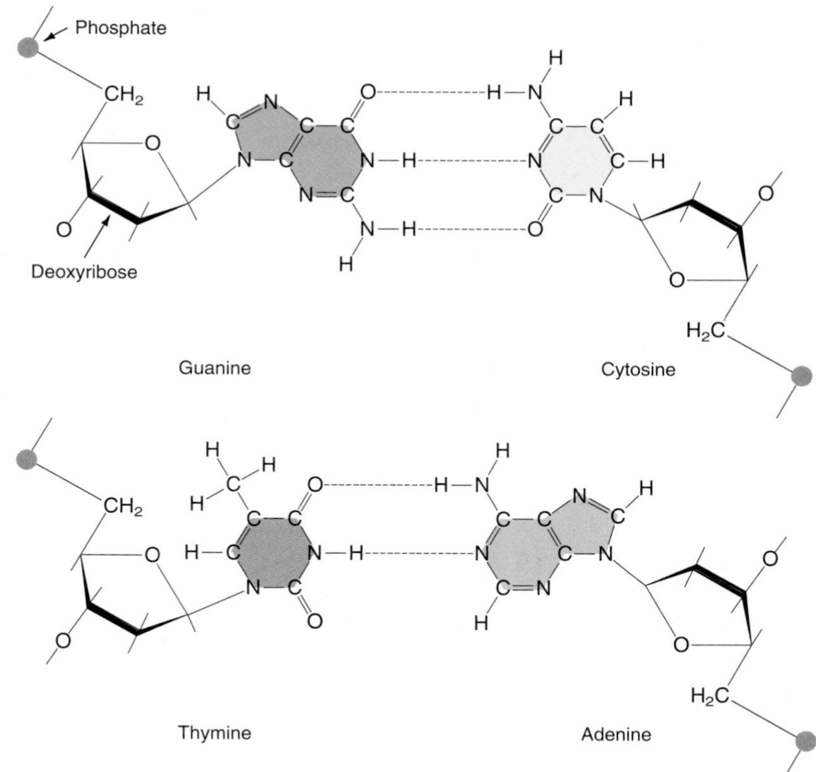

Figure 24.5
Base pairing in DNA. Adenine is always paired with thymine (A—T), and guanine is paired with cytosine (G—C).

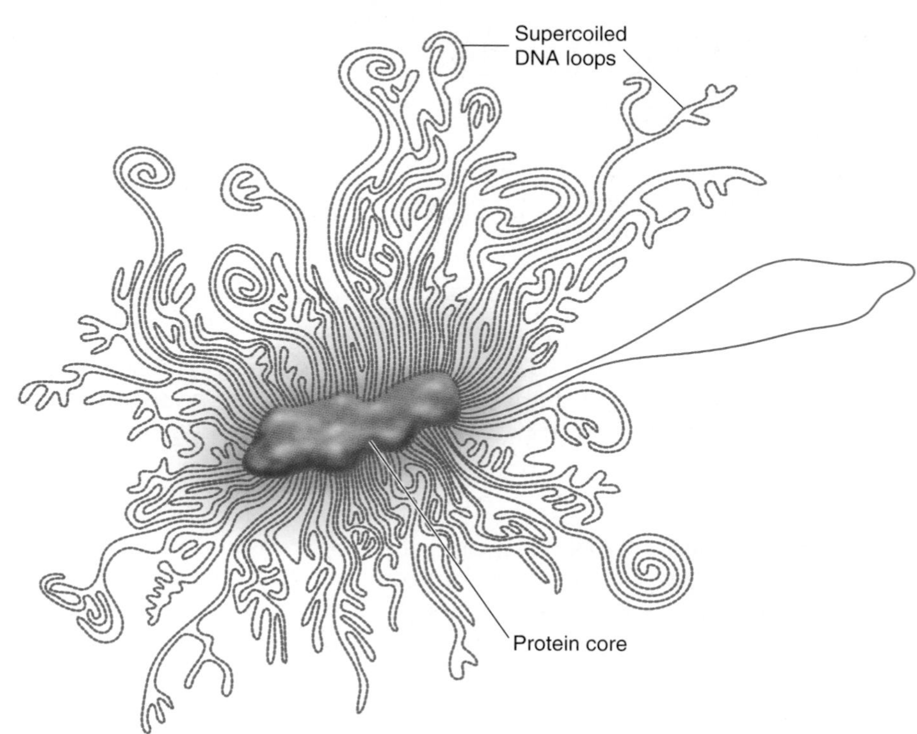

Figure 24.6
Structure of a bacterial nucleoid. The nucleoid is made up of the supercoiled, circular chromosome attached to a protein core.

Eukaryotes are organisms that have cells containing a true nucleus enclosed by a nuclear membrane. They also have a variety of membrane-bound organelles that segregate different cellular functions into different compartments. As an example, the processes of aerobic respiration are located within the mitochondria.

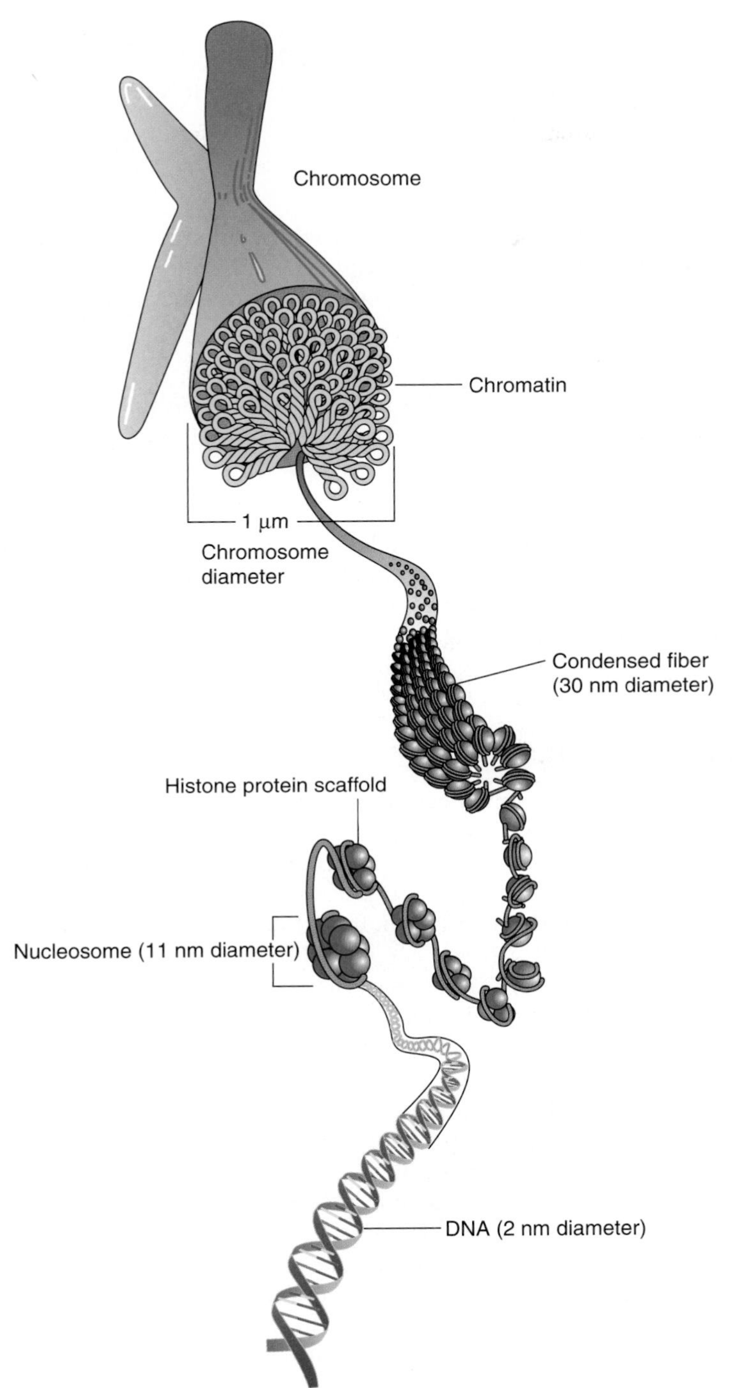

Figure 24.7
The eukaryotic chromosome has many levels of structure.

All animals, plants, and fungi are eukaryotes. The number and size of the chromosomes of eukaryotes vary from one species to the next. For instance, humans have 23 pairs of chromosomes, while the Adder's Tongue fern has 631 pairs of chromosomes. But the chromosome structure is the same for all those organisms that have been studied.

Eukaryotic chromosomes are very complex structures (Figure 24.7). The first level of structure is the **nucleosome,** which consists of a strand of DNA wrapped around a small disk made up of histone proteins. At this level the DNA looks like beads along a string. The string of beads then coils into a larger structure called the

30 nm fiber. These, in turn, are further coiled into a *200 nm fiber.* Other proteins are probably involved in the organization of the 200 nm fiber. The full complexities of the eukaryotic chromosome are not yet understood, but there are probably many such levels of coiled structures.

RNA Structure

The sugar-phosphate backbone of RNA consists of ribonucleotides, also linked by 3′–5′ phosphodiester bonds. These phosphodiester bonds are identical to those found in DNA. However, RNA molecules differ from DNA molecules in three basic properties.

- RNA molecules are usually single-stranded.
- The sugar-phosphate backbone of RNA consists of *ribonucleotides* linked by 3′–5′ phosphodiester bonds. Thus the sugar ribose is found in place of 2′-deoxyribose.
- The nitrogenous base uracil (U) replaces thymine (T).

Although RNA molecules are single-stranded, base pairing between uracil and adenine and between guanine and cytosine can still occur. We will show the importance of this property as we examine the way in which RNA molecules are involved in the expression of the genetic information in DNA.

24.3 DNA Replication

Learning Goal

DNA must be replicated before a cell divides so that each daughter cell inherits a copy of each gene. A cell that is missing a critical gene will die, just as an individual with a genetic disease, a defect in an important gene, may die early in life. Thus it is essential that the process of DNA replication produces an absolutely accurate copy of the original genetic information. If mistakes are made in critical genes, the result may be lethal mutations.

The first step in DNA replication is the separation of the strands of DNA. Proteins do this by breaking the hydrogen bonds between the base pairs. Then the enzyme **DNA polymerase** "reads" each parental strand, also called the *template,* and catalyzes the polymerization of a complementary daughter strand. Deoxyribonucleotide triphosphate molecules are the precursors for DNA replication. However, the last two phosphoryl groups are cleaved away in the process. This cleavage releases the energy needed by DNA polymerase to form the phosphoester linkage between the 3′-OH of 2′-deoxyribose and the 5′-phosphoryl group of the deoxyribonucleotide monophosphate to be added to the DNA chain.

Because DNA polymerase "reads" each parental strand and produces a new complementary daughter DNA strand, each new DNA molecule consists of one parental strand and one newly synthesized daughter strand. This mode of DNA replication is called **semiconservative replication** (Figure 24.8).

DNA polymerase can catalyze a reaction only between the 5′-phosphoryl group on a nucleotide to be added to the growing chain and the hydroxyl group on the growing daughter strand. Thus, the parental DNA strands can be copied in only one direction. This is a problem, since the two strands of a DNA molecule are antiparallel. As a result, only one daughter strand can be produced continuously from the **replication fork,** the point at which new nucleotides are added to the growing daughter strand.

The other strand must grow in the opposite direction of the moving replication fork. For this reason, it must be synthesized in short segments. Each of these is started at the replication fork as it moves in the opposite direction. The new pieces of DNA are covalently linked to one another by the enzyme DNA ligase (Figure 24.9).

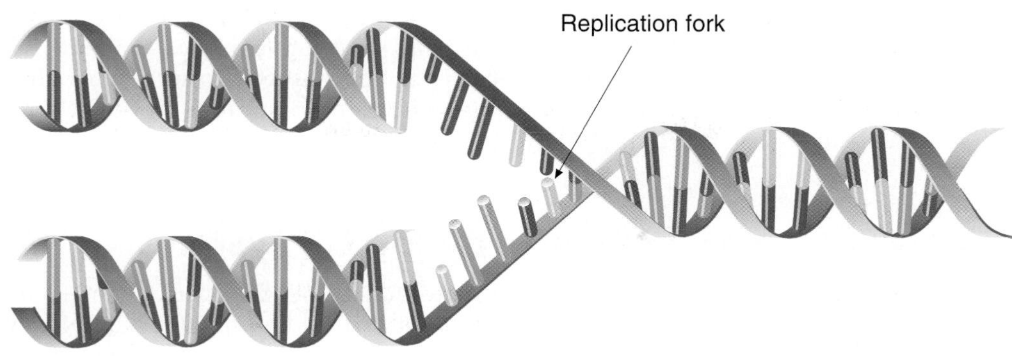

Figure 24.8
In semiconservative DNA replication, each parent strand serves as a template for the synthesis of a new daughter strand.

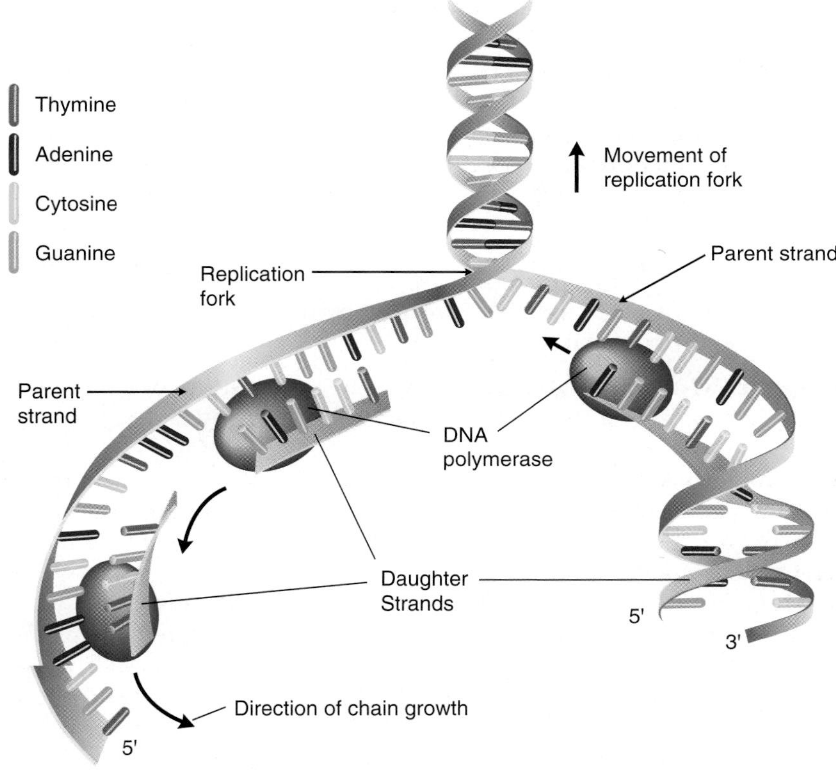

Figure 24.9
Because the two strands of DNA are antiparallel and DNA polymerase can only catalyze $5' \rightarrow 3'$ replication, only one of the two DNA strands (on the right) can be read continuously to produce a daughter strand. The other must be synthesized in segments that are extended away from the direction of movement of the replication fork (on the left). These discontinuous segments are later covalently joined together by DNA ligase.

Because it is critical to produce an accurate copy of the parental DNA, it is very important to avoid errors in the replication process. In addition to catalyzing the replication of new DNA strands, DNA polymerase is able to *proofread* the newly synthesized strand. If the wrong nucleotide has been added to the growing DNA strand, it is removed and replaced with the correct one. In this way a faithful copy of the parental DNA is ensured.

All the genetic information of bacteria such as *E. coli* is contained on a single circular piece of DNA made up of about three million nucleotide pairs and called the *chromosome.* DNA replication in *E. coli* begins at a unique sequence on the circular chromosome known as the **replication origin.** Replication occurs bidirectionally at the rate of about five hundred new nucleotides every second! Because DNA synthesis occurs bidirectionally, there are two replication forks moving in opposite directions. Replication is complete when the two replication forks meet halfway around the circular chromosome.

DNA replication in eukaryotes is more complex. The human genome consists of approximately three billion nucleotide pairs. Just one chromosome may be nearly one hundred times longer than a bacterial chromosome. To accomplish this huge job, DNA replication begins at many replication origins and proceeds bidirectionally along each chromosome.

24.4 Information Flow in Biological Systems

The **central dogma** of molecular biology states that in cells the flow of genetic information contained in DNA is a one-way street that leads from DNA to RNA to protein. The process by which a single strand of DNA serves as a template for the synthesis of an RNA molecule is called **transcription.** The word *transcription* is derived from the Latin word *transcribere* and simply means "to make a copy." Thus in this process, part of the information in the DNA is copied into a strand of RNA. The process by which the message is converted into protein is called **translation.** Unlike transcription the process of translation involves converting the information from one language to another. In this case the genetic information in the linear sequence of nucleotides is being translated into a protein, a linear sequence of amino acids. The expression of the information contained in DNA is fundamental to the growth, development, and maintenance of all organisms.

Classes of RNA Molecules

Learning Goal

Three classes of RNA molecules are produced by transcription: messenger RNA, transfer RNA, and ribosomal RNA.

1. **Messenger RNA (mRNA)** carries the genetic information for a protein from DNA to the ribosomes. It is a complementary RNA copy of a gene on the DNA.
2. **Ribosomal RNA (rRNA)** is a structural and functional component of the ribosomes, which are "platforms" on which protein synthesis occurs. There are three types of rRNA molecules in bacterial ribosomes and four in the ribosomes of eukaryotes.
3. **Transfer RNA (tRNA)** translates the genetic code of the mRNA into the primary sequence of amino acids in the protein. In addition to the primary structure, tRNA molecules have a cloverleaf-shaped secondary structure resulting from base pair hydrogen bonding (A—U and G—C) and a roughly L-shaped tertiary structure (Figure 24.10). The sequence CCA is found at the 3′ end of the tRNA. The 3′–OH group of the terminal nucleotide, adenosine, can be covalently attached to an amino acid. Three nucleotides at the base of the cloverleaf structure form the **anticodon.** As we will discuss in more detail in Section 24.6, this triplet of bases forms hydrogen bonds to a **codon** (complementary sequence of bases) on a messenger RNA (mRNA) molecule on the surface of a ribosome during protein synthesis. This hydrogen bonding of codon and anticodon brings the correct amino acid to the site of protein synthesis at the appropriate location in the growing peptide chain (Figure 24.11).

Transcription

Learning Goal

Transcription, shown in Figure 24.12, is catalyzed by the enzyme **RNA polymerase.** The process occurs in three stages. The first, called *initiation,* involves binding of RNA polymerase to a specific nucleotide sequence, the **promoter,** at the beginning of a gene. This interaction of RNA polymerase with specific promoter DNA sequences allows RNA polymerase to recognize the start point for transcription. It also determines which DNA strand will be transcribed. Unlike DNA repli-

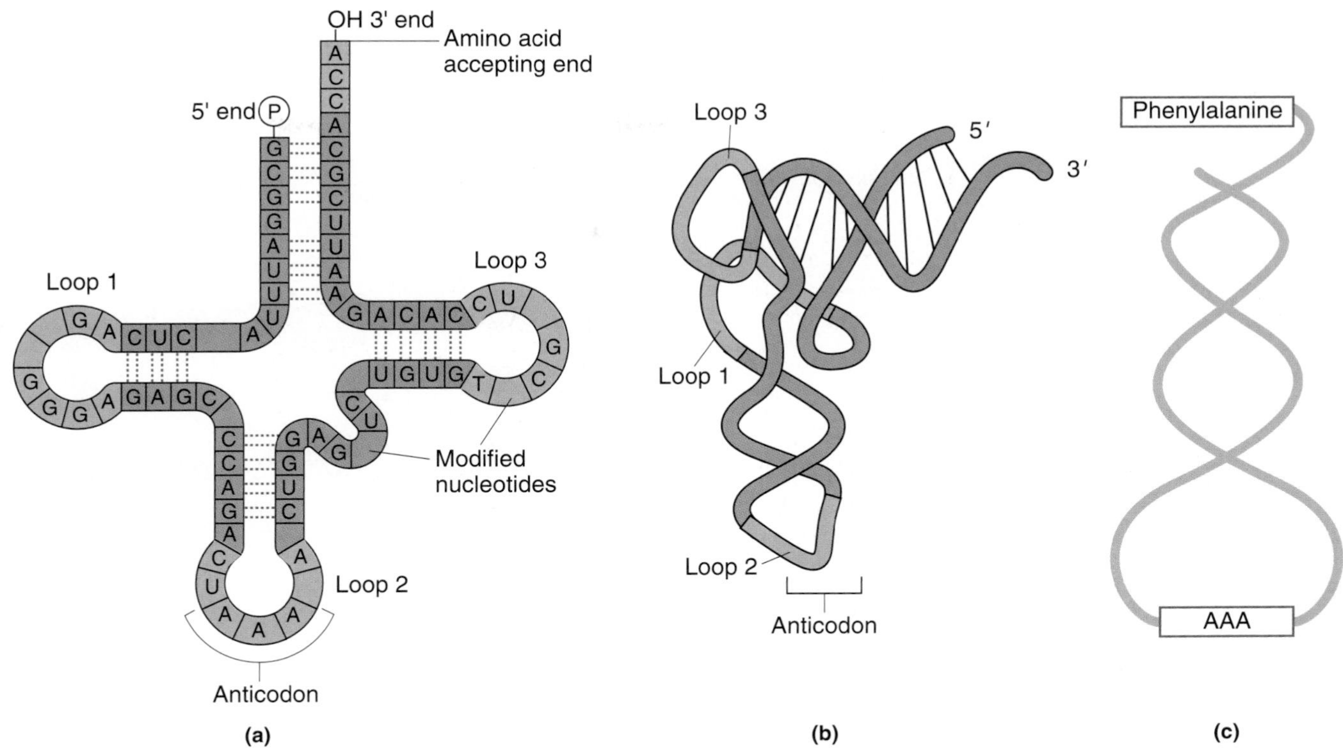

Figure 24.10
Structure of tRNA. (a) The primary structure of a tRNA is the linear sequence of ribonucleotides. Here we see the hydrogen-bonded secondary structure of a tRNA showing the three loops and the amino acid–accepting end. (b) The three-dimensional structure of a tRNA. (c) A schematic diagram that will be used to represent a tRNA throughout the chapter.

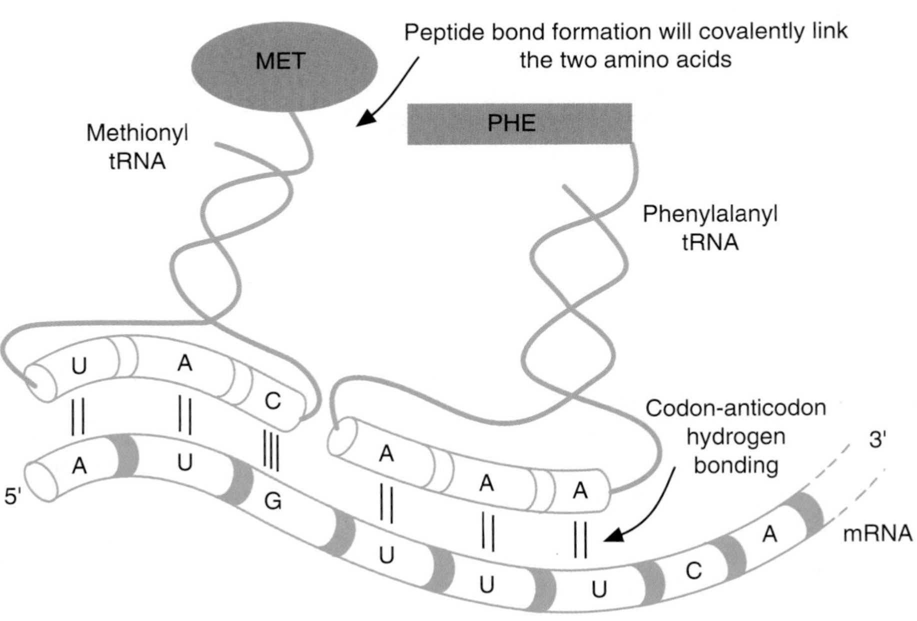

Figure 24.11
Codon-anticodon binding.

cation, transcription produces a complementary copy of only one of the two strands of DNA. As it binds to the DNA, RNA polymerase separates the two strands of DNA so that it can "read" the base sequence of the DNA.

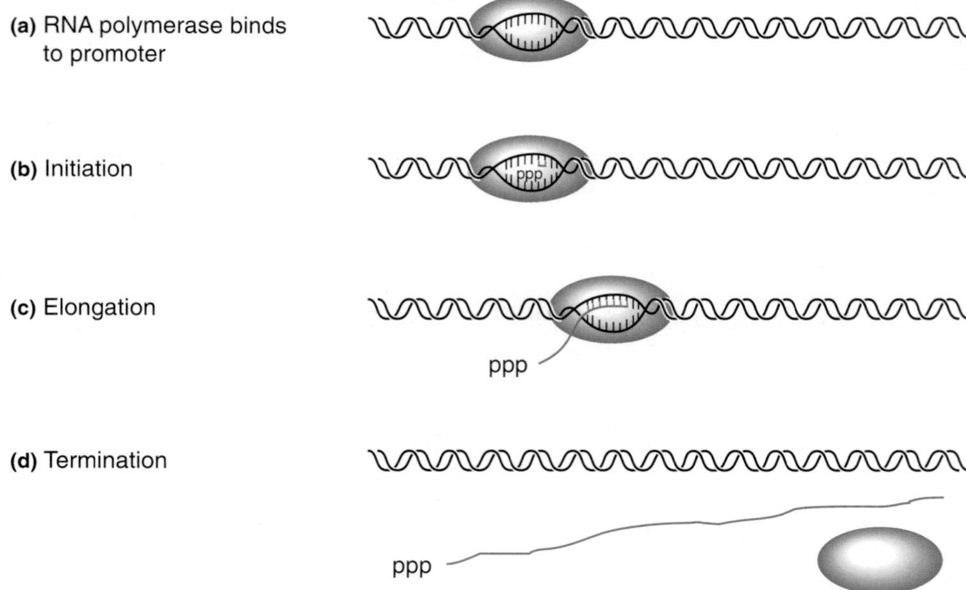

Figure 24.12
The stages of transcription.

The second stage, chain elongation, begins as the RNA polymerase "reads" the DNA template strand and catalyzes the polymerization of a complementary RNA copy. With each catalytic step, RNA polymerase transfers a complementary ribonucleotide to the end of the growing RNA chain and catalyzes the formation of a 3′–5′ phosphodiester bond between the 5′ phosphoryl group of the incoming ribonucleotide and the 3′ hydroxyl group of the last ribonucleotide of the growing RNA chain. This reaction is shown in Figure 24.13.

The final stage of transcription is termination. The RNA polymerase finds a termination sequence at the end of the gene and releases the newly formed RNA molecule.

Question 24.5

What is the function of RNA polymerase in the process of transcription?

Question 24.6

What is the function of the promoter sequence in the process of transcription?

Post-transcriptional Processing of RNA

Learning Goal

In bacteria, which are prokaryotes, termination releases a mature mRNA for translation. In fact, because prokaryotes have no nuclear membrane separating the DNA from the cytoplasm, translation begins long before the mRNA is completed. In eukaryotes, transcription produces a **primary transcript** that must undergo extensive **post-transcriptional modification** before it is exported out of the nucleus for translation in the cytoplasm.

Eukaryotic primary transcripts undergo three post-transcriptional modifications. These are the addition of a 5′ cap structure and a 3′ poly(A) tail, and RNA splicing.

In the first modification, a **cap structure** is enzymatically added to the 5′ end of the primary transcript. The cap structure (Figure 24.14) consists of 7-methylguanosine attached to the 5′ end of the RNA by a 5′–5′ triphosphate bridge. The first two nucleotides of the mRNA are also methylated. The cap structure is required for efficient translation of the final mature mRNA.

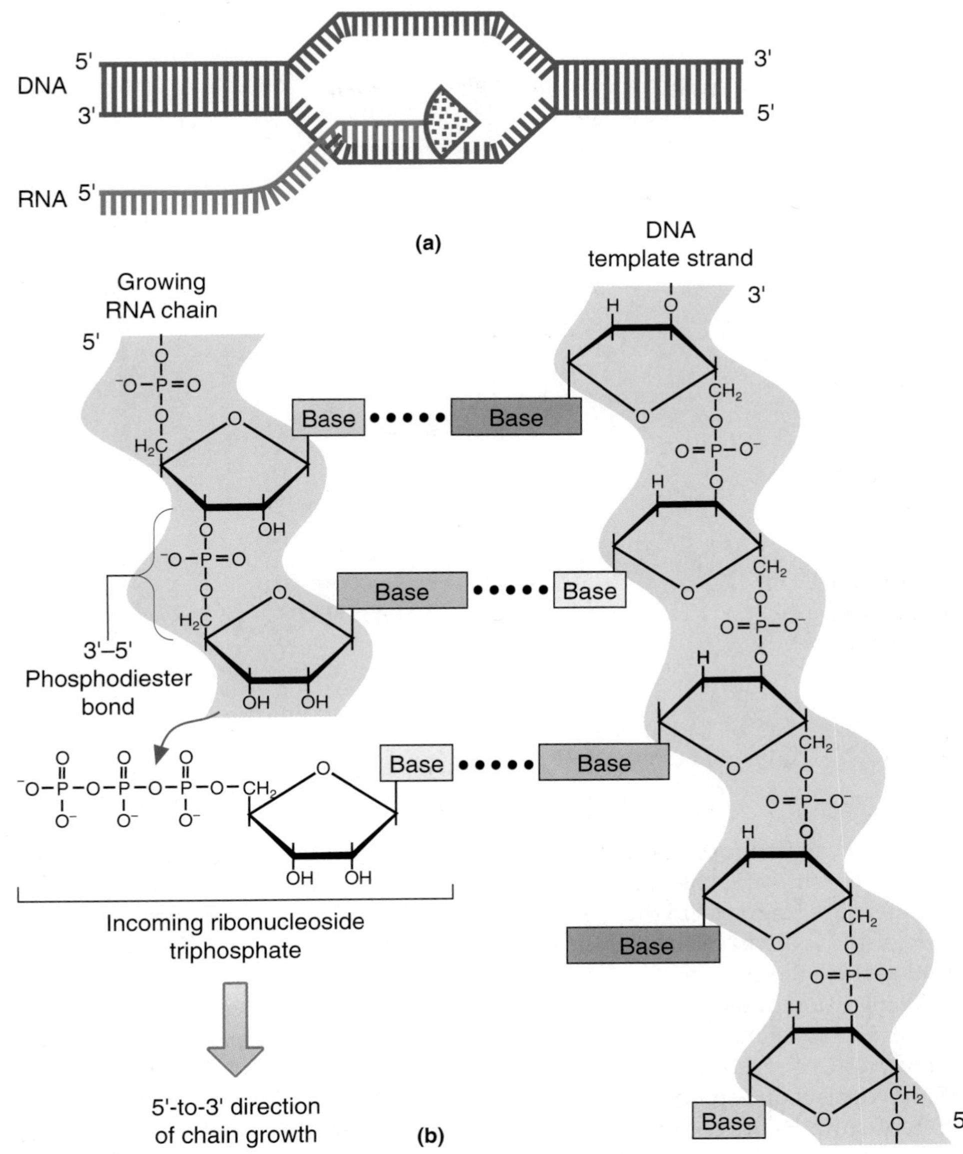

Figure 24.13
The reaction catalyzed by RNA polymerase. (a) RNA polymerase separates the two strands of DNA and produces an RNA copy of one of the two DNA strands. (b) Phosphodiester bond formation occurs as a nucleotide is added to the growing RNA chain.

The second modification is the enzymatic addition of a **poly(A) tail** to the 3′ end of the transcript. *Poly(A) polymerase* uses ATP and catalyzes the stepwise polymerization of 100–200 adenosine nucleotides on the 3′ end of the RNA. The poly(A) tail protects the 3′ end of the mRNA from enzymatic degradation and thus prolongs the lifetime of the mRNA.

The third modification, **RNA splicing,** involves the removal of portions of the primary transcript that are not protein coding. Bacterial genes are continuous; all the nucleotide sequences of the gene are found in the mRNA. However, study of the gene structure of eukaryotes revealed a fascinating difference. Eukaryotic genes are discontinuous; there are *extra* DNA sequences within these genes that do not encode any amino acid sequences for the protein. These sequences are called *intervening sequences* or **introns.** The primary transcript contains both the introns and the protein coding sequences, called **exons.** The presence of introns in the mRNA would make it impossible for the process of translation to synthesize the correct protein. Therefore they must be removed, which is done by the process of RNA splicing.

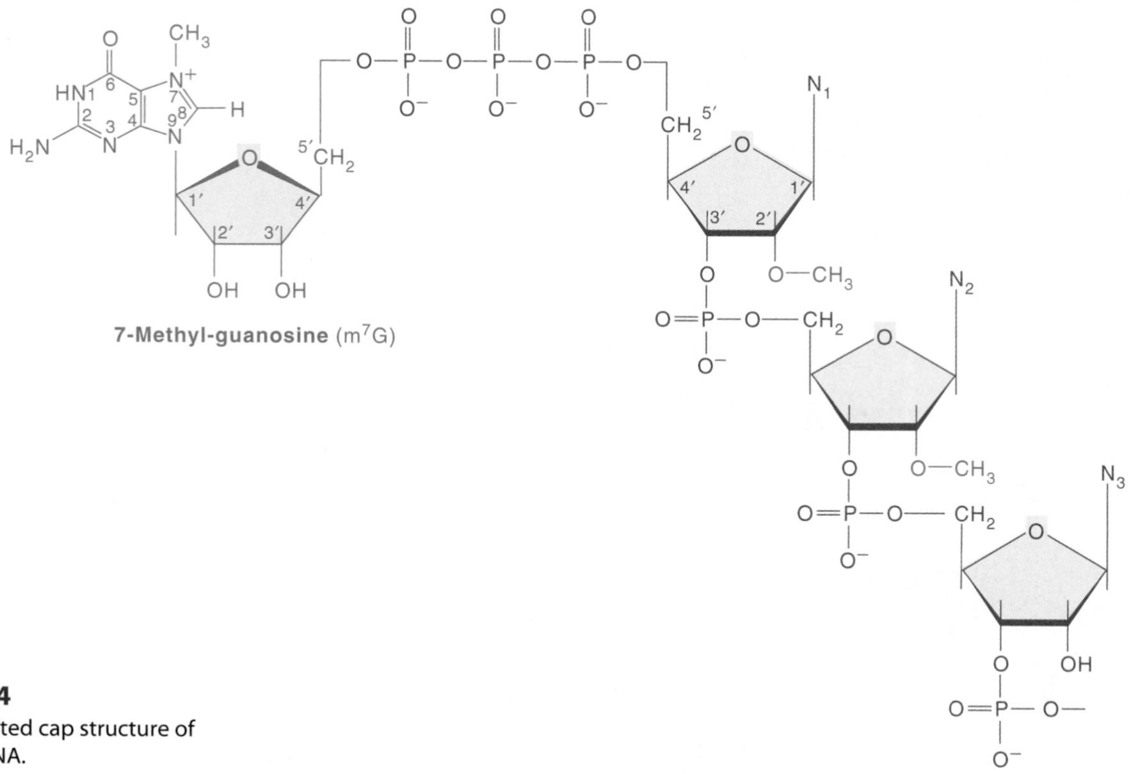

7-Methyl-guanosine (m^7G)

Figure 24.14
The 5′-methylated cap structure of eukaryotic mRNA.

As you can imagine, RNA splicing must be very precise. If too much, or too little, RNA is removed, the mRNA will not carry the correct code for the protein. Thus there are "signals" in the DNA to mark the boundaries of the introns. The sequence GpU is always found at the intron's 5′ boundary and the sequence ApG is found at the 3′ boundary.

Recognition of the splice boundaries and stabilization of the splicing complex requires the assistance of particles called *spliceosomes*. Spliceosomes are composed of a variety of *small nuclear ribonucleoproteins* (snRNPs, read "snurps"). Each snRNP consists of a small RNA and associated proteins. The RNA components of different snRNPs are complementary to different sequences involved in splicing. By hydrogen bonding to a splice boundary or intron sequences the snRNPs recognize and bring together the sequences involved in the splicing reactions.

One of the first eukaryotic genes shown to contain introns was the gene for the β subunit of adult hemoglobin (Figure 24.15). On the DNA the gene for β-hemoglobin is 1200 nucleotides long, but only 438 nucleotides carry the genetic information for protein. The remaining sequences are found in two introns of 116 and 646 nucleotides that are removed by splicing before translation. It is interesting that the larger intron is longer than the final β-hemoglobin mRNA! In the genes that have been studied, introns have been found to range in size from 50 to 20,000 nucleotides in length, and there may be many throughout a gene. Thus a typical human gene might be 10–30 times longer than the final mRNA.

24.5 The Genetic Code

Learning Goal

The mRNA carries the genetic code for a protein. But what is the nature of this code? In 1954, George Gamow proposed that because there are only four "letters" in the DNA alphabet (A, T, G, and C) and because there are twenty amino acids, the genetic code must contain words made of at least three letters taken from the

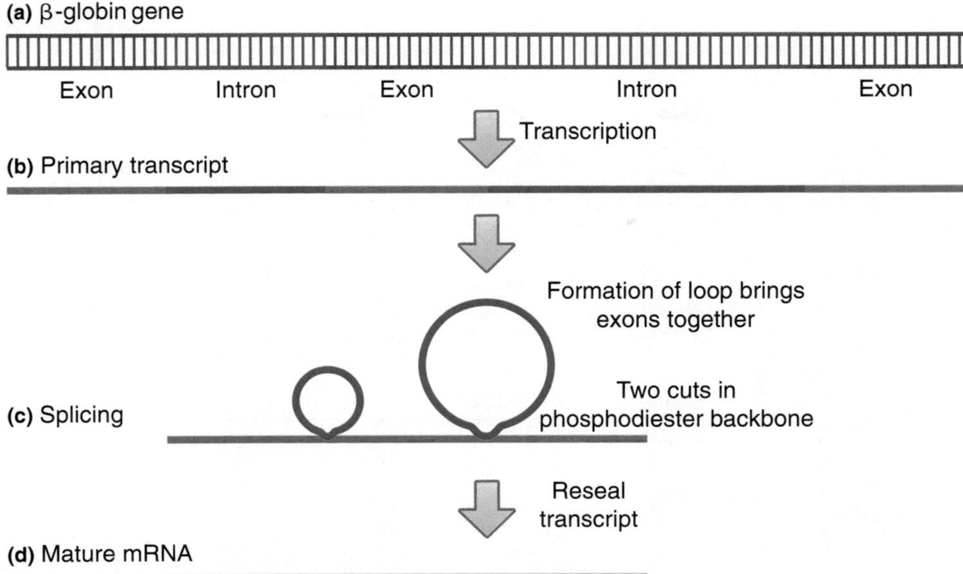

Figure 24.15
Schematic diagram of mRNA splicing. (a) The β-globin gene contains protein coding exons, as well as noncoding sequences called *introns.* (b) The primary transcript of the DNA carries both the introns and the exons. (c) The introns are looped out, the phosphodiester backbone of the mRNA is cut twice, and the pieces are tied together. (d) The final mature mRNA now carries only the coding sequences (exons) of the gene.

four letters in the DNA alphabet. How did he come to this conclusion? He reasoned that a code of two-letter words constructed from any combination of the four letters has a "vocabulary" of only sixteen words (4^2). In other words, there are only sixteen different ways to put A, T, C, and G together two bases at a time (AA, AT, AC, AG, TT, TA, etc.). That is not enough to encode all twenty amino acids. A code of four-letter words gives 256 words (4^4), far more than are needed. A code of three-letter words, however, has a possible vocabulary of sixty-four words (4^3), sufficient to encode the twenty amino acids but not too excessive.

A series of elegant experiments proved that Gamow was correct by demonstrating that the genetic code is, indeed, a triplet code. Mutations were introduced into the DNA of a bacterial virus. These mutations inserted (or deleted) one, two, or three nucleotides into a gene. The researchers then looked for the protein encoded by that gene. When one or two nucleotides were inserted, no protein was produced. However, when a third base was inserted, the sense of the mRNA was restored, and the protein was made. You can imagine this experiment by using a sentence composed of only three-letter words. For instance,

THE CAT RAN OUT

What happens to the "sense" of the sentence if we insert one letter?

THE FCA TRA NOU T

The reading frame of the sentence has been altered, and the sentence is now nonsense. Can we now restore the sense of the sentence by inserting a second letter?

THE FAC ATR ANO UT

No, we have not restored the sense of the sentence. Once again, we have altered the reading frame, but because our code has only three-letter words, the sentence is still nonsense. If we now insert a third letter, it should restore the correct reading frame:

THE FAT CAT RAN OUT

Indeed, by inserting three new letters we have restored the sense of the message by restoring the reading frame. This is exactly the way in which the message of the mRNA is interpreted. Each group of three nucleotides in the sequence of the mRNA is called a *codon,* and each codes for a single amino acid. If the sequence is

FIRST BASE	SECOND BASE				THIRD BASE
	U	C	A	G	
U	UUU Phenylalanine	UCU Serine	UAU Tyrosine	UGU Cysteine	U
U	UUC Phenylalanine	UCC Serine	UAC Tyrosine	UGC Cysteine	C
U	UUA Leucine	UCA Serine	UAA STOP	UGA STOP	A
U	UUG Leucine	UCG Serine	UAG STOP	UGG Tryptophan	G
C	CUU Leucine	CCU Proline	CAU Histidine	CGU Arginine	U
C	CUC Leucine	CCC Proline	CAC Histidine	CGC Arginine	C
C	CUA Leucine	CCA Proline	CAA Glutamine	CGA Arginine	A
C	CUG Leucine	CCG Proline	CAG Glutamine	CGG Arginine	G
A	AUU Isoleucine	ACU Threonine	AAU Asparagine	AGU Serine	U
A	AUC Isoleucine	ACC Threonine	AAC Asparagine	AGC Serine	C
A	AUA Isoleucine	ACA Threonine	AAA Lysine	AGA Arginine	A
A	AUG *(START)* Methionine	ACG Threonine	AAG Lysine	AGG Arginine	G
G	GUU Valine	GCU Alanine	GAU Aspartic acid	GGU Glycine	U
G	GUC Valine	GCC Alanine	GAC Aspartic acid	GGC Glycine	C
G	GUA Valine	GCA Alanine	GAA Glutamic acid	GGA Glycine	A
G	GUG Valine	GCG Alanine	GAG Glutamic acid	GGG Glycine	G

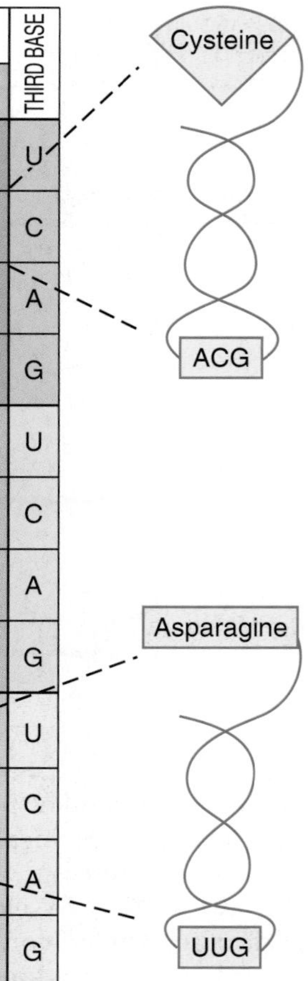

Cysteine
ACG
Asparagine
UUG

Figure 24.16
The genetic code. The table shows the possible codons found in mRNA. To read the universal biological language from this chart, find the first base in the column on the left, the second base from the row across the top, and the third base from the column to the right. This will direct you to one of the sixty-four squares in the matrix. Within that square you will find the codon and the amino acid that it specifies. In the cell this message is decoded by tRNA molecules like those shown to the right of the table.

interrupted or changed, it can change the amino acid composition of the protein that is produced or even result in the production of no protein at all.

As we noted, a three-letter genetic code contains sixty-four words, called *codons,* but there are only twenty amino acids. Thus there are forty-four more codons than are required to specify all of the amino acids found in proteins. Three of the codons—UAA, UAG, and UGA—specify termination signals for the process of translation. But this still leaves us with forty-one additional codons. What is the function of the "extra" code words? Francis Crick (recall Watson and Crick and the double helix) proposed that the genetic code is a **degenerate code.** The term *degenerate* is used to indicate that different triplet codons may serve as code words for the same amino acid.

The complete genetic code is shown in Figure 24.16. We can make several observations about the genetic code. First, methionine and tryptophan are the only amino acids that have a single codon. All others have at least two codons, and serine and leucine have six codons each. The genetic code is also somewhat mutation-resistant. For those amino acids that have multiple codons the first two bases are

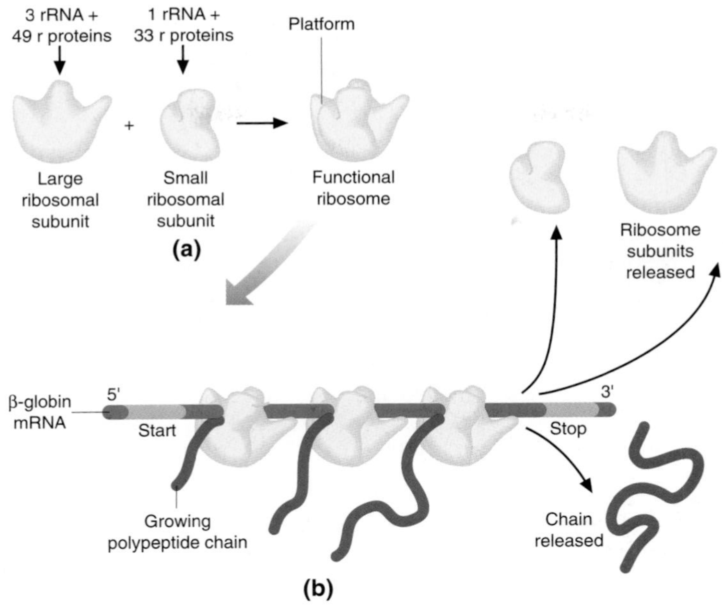

Figure 24.17
Structure of the ribosome. (a) The large and small subunits form the functional complex in association with an mRNA molecule. (b) A polyribosome translating the mRNA for a β-globin chain of hemoglobin.

often identical and thus identify the amino acid, and only the third position is variable. Mutations—changes in the nucleotide sequence—in the third position therefore often have no effect on the amino acid that is incorporated into a protein.

Question 24.7

Why is the genetic code said to be degenerate?

Question 24.8

Why is the genetic code said to be mutation-resistant?

24.6 Protein Synthesis

Learning Goal

The process of protein synthesis is called *translation*. It involves translating the genetic information from the sequence of nucleotides into the sequence of amino acids in the primary structure of a protein. Translation is carried out on **ribosomes,** which are complexes of ribosomal RNA (rRNA) and proteins. Each ribosome is made up of two subunits: a small and a large ribosomal subunit (Figure 24.17a). In eukaryotic cells the small ribosomal subunit contains one rRNA molecule and thirty-three different ribosomal proteins, and the large subunit contains three rRNA molecules and about forty-nine different proteins.

Protein synthesis involves the simultaneous action of many ribosomes on a single mRNA molecule. These complexes of many ribosomes along a single mRNA are known as *polyribosomes* or **polysomes** (Figure 24.17b). Each ribosome is synthesizing one copy of the protein molecule encoded by the mRNA. Thus many copies of a protein are simultaneously produced.

The Role of Transfer RNA

The codons of mRNA must be read if the genetic message is to be translated into protein. The molecule that decodes the information in the mRNA molecule into the primary structure of a protein is transfer RNA (tRNA). To decode the genetic message into the primary sequence of a protein, the tRNA must faithfully perform two functions.

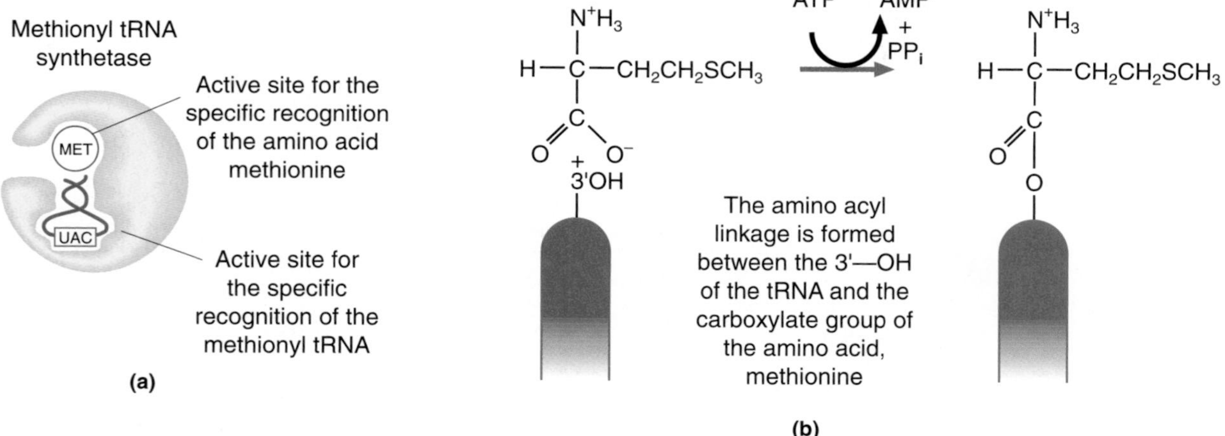

Figure 24.18
Methionyl tRNA synthetase. (a) The enzyme specifically recognizes the amino acid methionine in one region of the active site and the methionyl tRNA in another. (b) The acylation reaction that results in a covalent linkage of the amino acid to the tRNA.

First, the tRNA must covalently bind one, and only one, specific amino acid. There is at least one transfer RNA for each amino acid. As shown in Figure 24.10, all tRNA molecules have the sequence CCA at their 3′ ends. This is the site where the amino acid will be covalently attached to the tRNA molecule. Each tRNA is specifically recognized by the active site of an enzyme called an **aminoacyl tRNA synthetase.** This enzyme also recognizes the correct amino acid and covalently links the amino acid to the 3′ end of the tRNA molecule. Figure 24.18a shows the recognition of the amino acid methionine and its tRNA by the methionyl tRNA synthetase. The resulting structure is called an **aminoacyl tRNA,** in this case methionyl tRNA. In Figure 24.18b the reaction that results in the attachment of the aminoacyl group to the tRNA is shown. The covalently bound amino acid will be transferred from the tRNA to a growing polypeptide chain during protein synthesis.

Second, the tRNA must be able to recognize the appropriate codon on the mRNA that calls for that amino acid. This is mediated through a sequence of three bases called the *anticodon,* which is located at the bottom of the tRNA cloverleaf (refer to Figure 24.10). The anticodon sequence for each tRNA is complementary to the codon on the mRNA that specifies a particular amino acid. As you can see in Figure 24.11, the anticodon-codon complementary hydrogen bonding will bring the correct amino acid to the site of protein synthesis.

Question 24.9

How are codons related to anticodons?

Question 24.10

If the sequence of a codon on the mRNA is 5′-AUG-3′, what will the sequence of the anticodon be? Remember that the hydrogen bonding rules require antiparallel strands. It is easiest to write the anticodon first $3' \rightarrow 5'$ and then reverse it to the $5' \rightarrow 3'$ order.

The Process of Translation

Initiation

Learning Goal

The first stage of protein synthesis is *initiation.* Proteins called **initiation factors** are required to mediate the formation of a translation complex composed of an mRNA molecule, the small and large ribosomal subunits, and the initiator tRNA. This initiator tRNA recognizes the codon AUG and carries the amino acid methionine.

The ribosome has two sites for binding tRNA molecules. The first site, called the **peptidyl tRNA binding site (P-site),** holds the peptidyl tRNA, the growing peptide bound to a tRNA molecule. The second site, called the **aminoacyl tRNA**

binding site (A-site), holds the aminoacyl tRNA carrying the next amino acid to be added to the peptide chain. Each of the tRNA molecules is hydrogen bonded to the mRNA molecule by codon-anticodon complementarity. The entire complex is further stabilized by the fact that the mRNA is also bound to the ribosome. Figure 24.19a shows the series of events that result in the formation of the initiation complex. The initiator methionyl tRNA occupies the P-site in this complex.

Chain Elongation

The second stage of translation is *chain elongation.* This occurs in three steps that are repeated until protein synthesis is complete. We enter the action after a tetrapeptide has already been assembled, and a peptidyl tRNA occupies the P-site (Figure 24.19b).

The first event is binding of an aminoacyl-tRNA molecule to the empty A-site. Next, peptide bond formation occurs. This is catalyzed by an enzyme on the ribosome called *peptidyl transferase.* Now the peptide chain is shifted to the tRNA that occupies the A-site. Finally, the tRNA in the P-site falls away, and the ribosome changes positions so that the next codon on the mRNA occupies the A-site. This movement of the ribosome is called **translocation.** The process shifts the new peptidyl tRNA from the A-site to the P-site. The chain elongation stage of translation requires the hydrolysis of GTP to GDP and P_i. Several **elongation factors** are also involved in this process.

Recent evidence indicates that the peptidyl transferase is a catalytic region of the 28S ribosomal RNA.

Termination

The last stage of translation is *termination.* There are three **termination codons**—UAA, UAG, and UGA—for which there are no corresponding tRNA molecules. When one of these "stop" codons is encountered, translation is terminated. A **release factor** binds the empty A-site. The peptidyl transferase that had previously catalyzed peptide bond formation hydrolyzes the ester bond between the peptidyl tRNA and the last amino acid of the newly synthesized protein (Figure 24.19c). At this point the tRNA, the newly synthesized peptide, and the two ribosomal subunits are released.

Question 24.11

What is the function of the ribosomal P-site in protein synthesis?

Question 24.12

What is the function of the ribosomal A-site in protein synthesis?

The peptide that is released following translation is not necessarily in its final functional form. In some cases the peptide is proteolytically cleaved before it becomes functional. Synthesis of digestive enzymes uses this strategy. Sometimes the protein must associate with other peptides to form a functional protein, as in the case of hemoglobin. Cellular enzymes add carbohydrate or lipid groups to some proteins, especially those that will end up on the cell surface. These final modifications are specific for particular proteins and, like the sequence of the protein itself, are directed by the cellular genetic information.

Post-translational proteolytic cleavage of digestive enzymes is discussed in Section 20.11.

The quaternary structure of hemoglobin is described in Section 19.9.

24.7 Mutation, Ultraviolet Light, and DNA Repair

The Nature of Mutations

Learning Goal

Changes can occur in the nucleotide sequence of a DNA molecule. Such a genetic change is called a **mutation.** Mutations can arise from mistakes made by DNA polymerase during DNA replication. They also result from the action of chemicals, called **mutagens,** that damage the DNA.

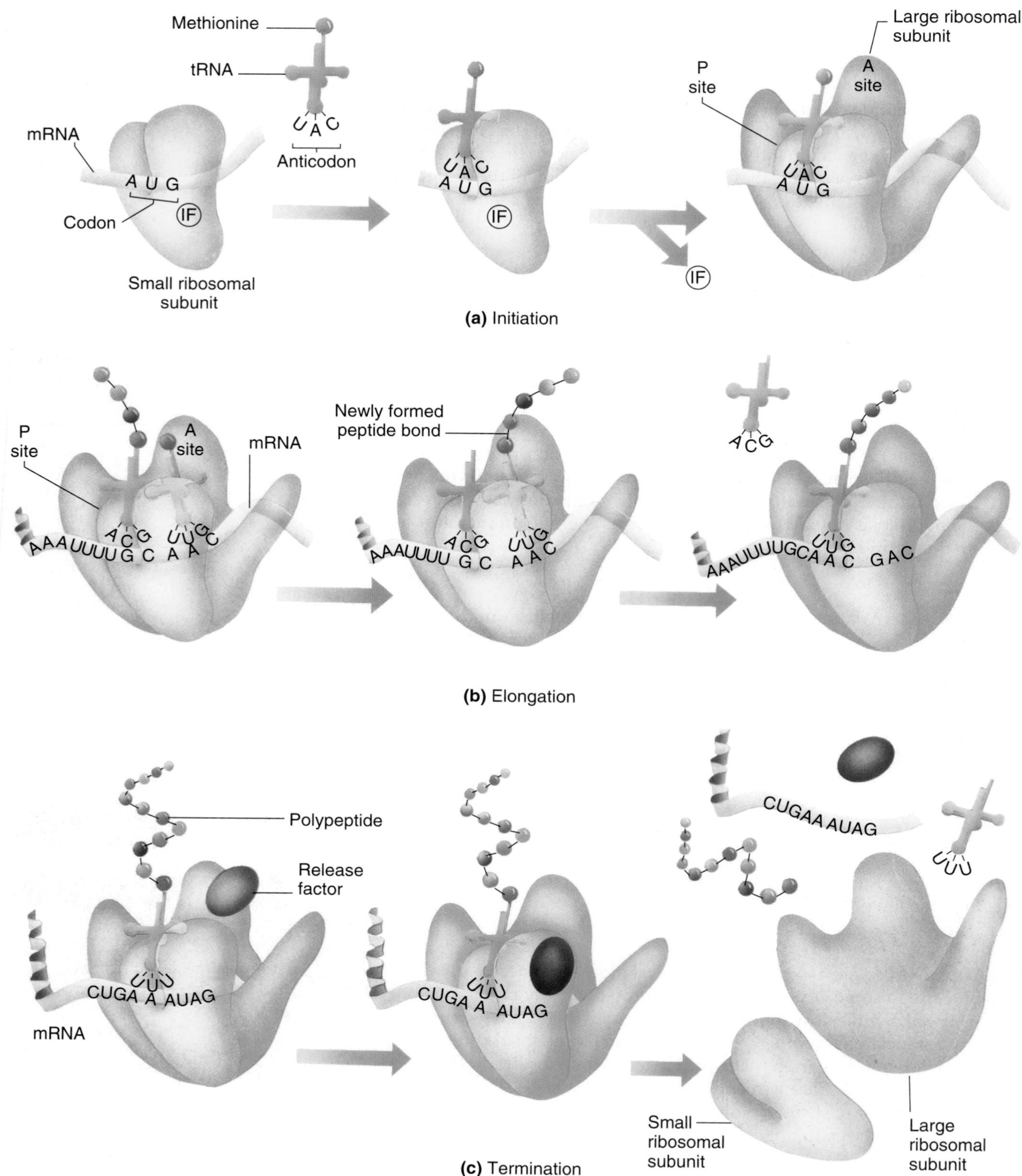

Figure 24.19

(a) Formation of an initiation complex sets protein synthesis in motion. The mRNA and proteins called *initiation factors* bind to the small ribosomal subunit. Next, a charged methionyl tRNA molecule binds, and finally, the initiation factors are released, and the large subunit binds. (b) The elongation phase of protein synthesis involves addition of new amino acids to the C-terminus of the growing peptide. An aminoacyl tRNA molecule binds at the empty A-site, and the peptide bond is formed. The uncharged tRNA molecule is released, and the peptidyl tRNA is shifted to the P-site as the ribosome moves along the mRNA. (c) Termination of protein synthesis occurs when a release factor binds the stop codon on mRNA. This leads to the hydrolysis of the ester bond linking the peptide to the peptidyl tRNA molecule in the P-site. The ribosome then dissociates into its two subunits, releasing the mRNA and the newly synthesized peptide.

Mutations are classified by the kind of change that occurs in the DNA. The substitution of a single nucleotide for another is called a **point mutation:**

ATGGACTTC:	normal DNA sequence
ATGCACTTC:	point mutation

Sometimes a single nucleotide or even large sections of DNA are lost. These are called **deletion mutations:**

ATGGACTTC:	normal DNA sequence
ATGTTC:	deletion mutation

Occasionally, one or more nucleotides are added to a DNA sequence. These are called **insertion mutations:**

ATGGACTTC:	normal DNA sequence
ATGCTCGACTTC:	insertion mutation

The Results of Mutations

Some mutations are **silent mutations;** that is, they cause no change in the protein. Often, however, a mutation has a negative effect on the health of the organism. The effect of a mutation depends on how it alters the genetic code for a protein. Consider the two codons for glutamic acid: GAA and GAG. A point mutation that alters the third nucleotide of GAA to GAG will still result in the incorporation of glutamic acid at the correct position in the protein. Similarly, a GAG to GAA mutation will also be silent.

There are approximately four thousand human genetic diseases that result from mutations. These occur because the mutation in the DNA changes the codon and results in incorporation of the wrong amino acid into the protein. This causes the protein to be nonfunctional or to function improperly.

Consider the human genetic disease sickle cell anemia. In the normal β-chain of hemoglobin, the sixth amino acid is glutamic acid. In the β-chain of sickle cell hemoglobin, the sixth amino acid is valine. How did this amino acid substitution arise? The answer lies in examination of the codons for glutamic acid and valine:

Glutamic acid:	GAA or GAG
Valine:	GUG, GUC, GUA, or GUU

A point mutation of A $\rightarrow$ U in the second nucleotide changes some codons for glutamic acid into codons for valine:

Glutamic acid codon		Valine codon
GAA	$\longrightarrow$	GUA
GAG	$\longrightarrow$	GUG

This mutation in a single codon leads to the change in amino acid sequence at position 6 in the β-chain of human hemoglobin from glutamic acid to valine. The result of this seemingly minor change is sickle cell anemia in individuals who inherit two copies of the mutant gene.

Question 24.13

The sequence of a gene on the mRNA is normally AUGCCCGACUUU. A point mutation in the gene results in the mRNA sequence AUGCGCGACUUU. What are the amino acid sequences of the normal and mutant proteins? Would you expect this to be a silent mutation?

A CLINICAL PERSPECTIVE

The Ames Test for Carcinogens

Each day we come into contact with a variety of chemicals, including insecticides, food additives, hair dyes, automobile emissions, and cigarette smoke. Some of these chemicals have the potential to cause cancer. How do we determine whether these agents are harmful? More particularly, how do we determine whether they cause cancer?

If we consider the example of cigarette smoke, we see that it can be years, even centuries, before a relationship is seen between a chemical and cancer. Europeans and Americans have been smoking since Sir Walter Raleigh introduced tobacco into England in the seventeenth century. However, it was not until three centuries later that physicians and scientists demonstrated the link between smoking and lung cancer. Obviously, this epidemiological approach takes too long, and too many people die. Alternatively, we can test chemicals by treating laboratory animals, such as mice, and observing them for various kinds of cancer. However, this, too, can take years, is expensive, and requires the sacrifice of many laboratory animals. How, then, can chemicals be tested for carcinogenicity (the ability to cause cancer) quickly and inexpensively? In the 1970s it was recognized that most carcinogens are also mutagens. That is, they cause cancer by causing mutations in the DNA, and the mutations cause the cells of the body to lose growth control. Bruce Ames, a biochemist and bacterial geneticist, developed a test using mutants of the bacterium *Salmonella typhimurium* that can demonstrate in 48–72 hours whether a chemical is a mutagen and thus a suspected carcinogen.

Ames chose several mutants of *S. typhimurium* that cannot grow unless the amino acid histidine is added to the growth medium. The Ames test involves subjecting these bacteria to a chemical and determining whether the chemical causes reversion of the mutation. In other words, the researcher is looking for a mutation that reverses the original mutation. When a reversion occurs, the bacteria will be able to grow in the absence of histidine.

The details of the Ames test are shown in the accompanying figure. Both an experimental and a control test are done. The control test contains no carcinogen and will show the number of spontaneous revertants that occur in the culture. If there are many colonies on the surface of the experimental plate and only a few colonies on the negative control plate, it can be concluded that the chemical tested is a mutagen. It is therefore possible that the chemical is also a carcinogen.

The Ames test has greatly accelerated our ability to test new compounds for mutagenic and possibly carcinogenic effects. However, once the Ames test identifies a mutagenic compound, testing in animals must be done to show conclusively that the compound also causes cancer.

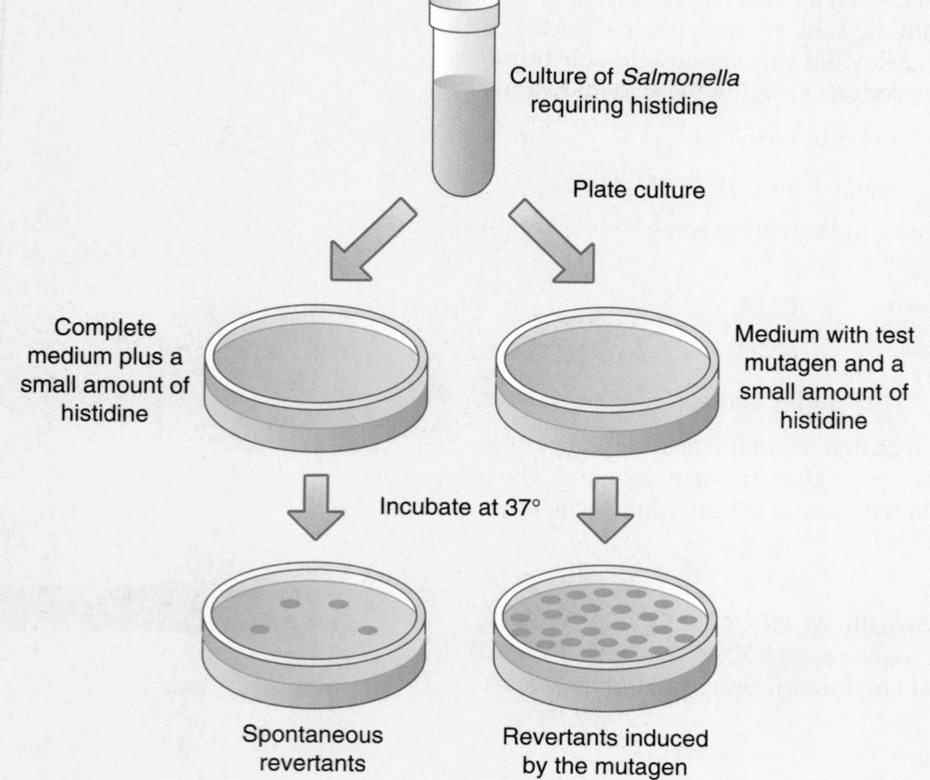

The Ames test for carcinogenic compounds.

Question 24.14

The sequence of a gene on the mRNA is normally AUGCCCGACUUU. A point mutation in the gene results in the mRNA sequence AUGCCGGACUUU. What are the amino acid sequences of the normal and mutant proteins? Would you expect this to be a silent mutation?

Mutagens and Carcinogens

Any chemical that causes a change in the DNA sequence is called a *mutagen*. Often, mutagens are also **carcinogens,** cancer-causing chemicals. Most cancers result from mutations in a single normal cell. These mutations result in the loss of normal growth control, causing the abnormal cell to proliferate. If that growth is not controlled or destroyed, it will result in the death of the individual. We are exposed to many carcinogens in the course of our lives. Sometimes we are exposed to a carcinogen by accident, but in some cases it is by choice. There are about three thousand chemical components in cigarette smoke, and several are potent mutagens. As a result, people who smoke have a much greater chance of lung cancer than those who don't.

Ultraviolet Light Damage and DNA Repair

Ultraviolet (UV) light is another agent that causes damage to DNA. Absorption of UV light by DNA causes adjacent pyrimidine bases to become covalently linked. The product (Figure 24.20) is called a **pyrimidine dimer.** As a result of pyrimidine dimer formation, there is no hydrogen bonding between these pyrimidine molecules and the complementary bases on the other DNA strand. This stretch of DNA cannot be replicated or transcribed!

Bacteria such as *Escherichia coli* have four different mechanisms to repair ultraviolet light damage. However, even a repair process can make a mistake. Mutations occur when the UV damage repair system makes an error and causes a change in the nucleotide sequence of the DNA.

In medicine the pyrimidine dimerization reaction is used to advantage in hospitals where germicidal (UV) light is used to kill bacteria in the air and on environmental surfaces, such as in a vacant operating room. This cell death is caused by pyrimidine dimer formation on a massive scale. The repair systems of the bacteria are overwhelmed, and the cells die.

Of course, the same type of pyrimidine dimer formation can occur in human cells as well. Lying out in the sun all day to acquire a fashionable tan exposes the

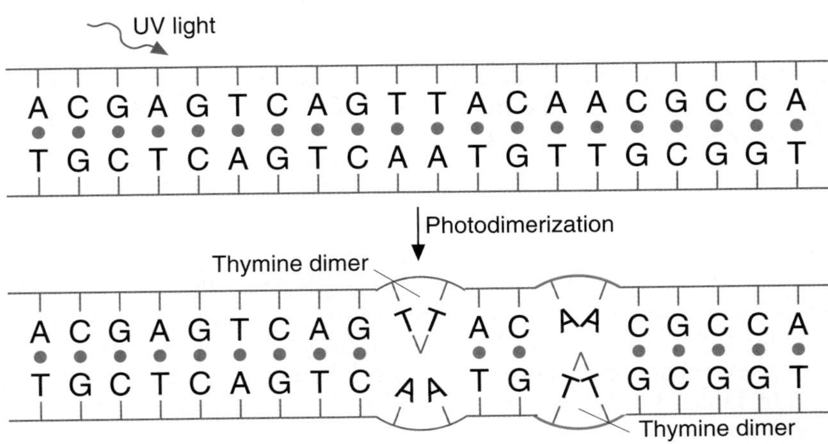

Figure 24.20
Photodimerization of adjacent pyrimidines (thymines in this example) results from the absorption of ultraviolet light.

skin to large amounts of UV light. This damages the skin by formation of many pyrimidine dimers. Exposure to high levels of UV from sunlight or tanning booths has been linked to a rising incidence of skin cancer in human populations.

Consequences of Defects in DNA Repair

The human repair system for pyrimidine dimers is quite complex, requiring at least five enzymes. The first step in repair of the pyrimidine dimer is the cleavage of the sugar-phosphate backbone of the DNA near the site of the damage. The enzyme that performs this cleavage of the sugar-phosphate backbone is called a *repair endonuclease.* If the gene encoding this enzyme is defective, pyrimidine dimers cannot be repaired. The accumulation of mutations combined with a simultaneous decrease in the efficiency of DNA repair mechanisms leads to an increased incidence of cancer. For example, a mutation in the repair endonuclease gene and in other genes in the repair pathway results in the genetic skin disorder called *xeroderma pigmentosum.* People who suffer from xeroderma pigmentosum are extremely sensitive to the ultraviolet rays of sunlight and develop multiple skin cancers, usually before the age of twenty.

24.8 Recombinant DNA

Tools Used in the Study of DNA

Learning Goal

Scientists are often asked why they study such seemingly unimportant subjects as bacterial DNA replication. One very good reason is that such studies often lend insight into the workings of human genetic systems. A second is that such research often produces the tools that allow great leaps into new technologies. Nowhere is this more true than in the development of recombinant DNA technology. Many of the techniques and tools used in recombinant DNA studies were developed or discovered during basic studies on bacterial DNA replication and gene expression. These include many enzymes that catalyze reactions of DNA molecules, gel electrophoresis, cloning vectors, and hybridization techniques.

Restriction Enzymes

Restriction enzymes are bacterial enzymes that "cut" the sugar-phosphate backbone of DNA molecules at specific nucleotide sequences. The first of these enzymes to be purified and studied was called EcoR1. The name is derived from the genus and species name of the bacteria from which it was isolated, in this case *E. coli.* The following is the specific nucleotide sequence recognized by EcoR1:

When EcoR1 cuts the DNA at this site, it does so in a staggered fashion. Specifically, it cuts between the G and the first A on both strands. Cutting produces two DNA fragments with the following structure:

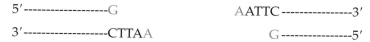

These staggered termini are called *sticky ends* because they can reassociate with one another by hydrogen bonding. This is a property of the DNA fragments generated by restriction enzymes that is very important to gene cloning.

Examples of other restriction enzymes and their specific recognition sequences are listed in Table 24.3. The sites on the sugar-phosphate backbone that are cut by the enzymes are indicated by slashes.

Table 24.3 Some Common Restriction Enzymes and Their Recognition Sequences

Restriction Enzyme	Recognition Sequence
BamHI	5′-G/GATCC-3′ 3′-CCTAG/G-5′
HindIII	5′-A/AGCTT-3′ 3′-TTCGA/A-5′
SalI	5′-G/TCGAC-3′ 3′-CAGCT/G-5′
BglII	5′-A/GATCT-3′ 3′-TCTAG/A-5′
PstI	5′-CTGCA/G-3′ 3′-G/ACGTC-5′

These enzymes are used to digest large DNA molecules into smaller fragments of specific size. Because a restriction enzyme always cuts at the same site, DNA from a particular individual generates a reproducible set of DNA fragments. This is convenient for the study or cloning of DNA from any source.

Agarose Gel Electrophoresis

One means of studying the DNA fragments produced by restriction enzyme digestion is agarose gel electrophoresis. The digested DNA sample is placed in a sample well in the gel, and an electrical current is applied. The negative charge of the phosphoryl groups in the sugar-phosphate backbone causes the DNA fragment to move through the gel away from the negative electrode (cathode) and toward the positive electrode (anode). The smaller DNA fragments move more rapidly than the larger ones, and as a result the DNA fragments end up distributed throughout the gel according to their size. The sizes of each fragment can be determined by comparison with the migration pattern of DNA fragments of known size.

Hybridization

Agarose gel electrophoresis allows the determination of the size of a DNA fragment. However, in recombinant DNA research it is also important to identify what gene is carried by a particular DNA fragment.

Hybridization is a technique used to identify the presence of a gene on a particular DNA fragment. This technique is based on the fact that complementary DNA sequences will hydrogen bond, or hybridize, to one another. In fact, even RNA can be used in hybridization studies. RNA can hybridize to DNA molecules or to other RNA molecules.

One technique, called Southern blotting, involves hybridization of DNA fragments from an agarose gel (Figure 24.21). DNA digested by a restriction enzyme is run on an agarose gel. Next the DNA fragments are transferred by blotting onto a special membrane filter. The DNA molecules on the filter are "melted" into single DNA strands so that they are ready for hybridization. The filter is then bathed in a solution containing a radioactive DNA or RNA molecule. This probe will hybridize to any DNA fragments on the filter that are complementary to it. X-ray film is used to detect any bands where the radioactive probe hybridized, thus locating the gene of interest.

DNA Cloning Vectors

DNA cloning experiments combine these technologies with a few additional tricks to isolate single copies of a gene and then produce billions of copies. To produce

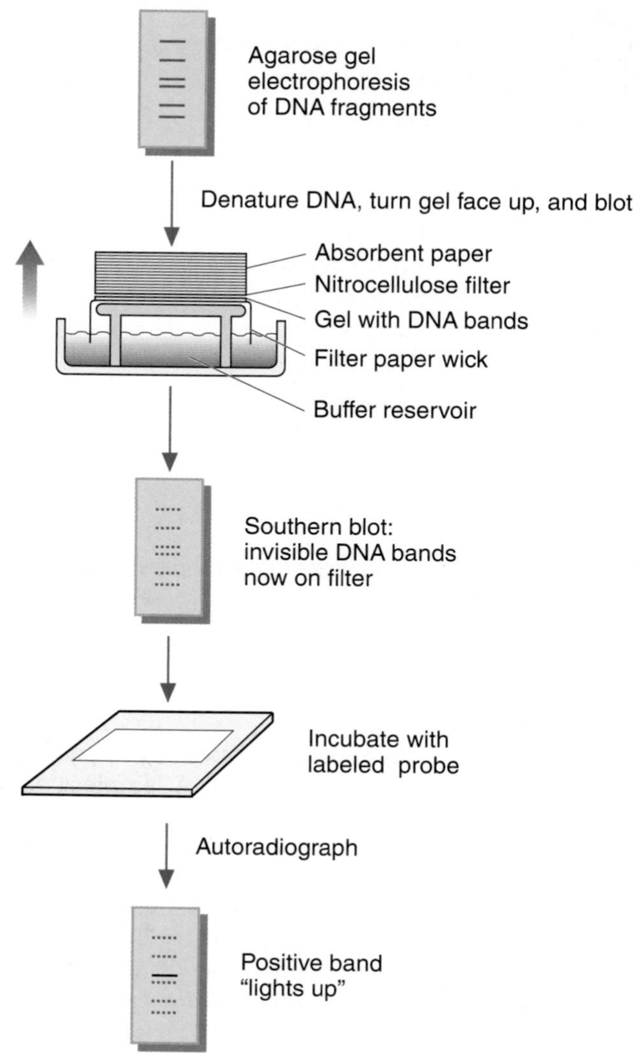

Figure 24.21
Southern blot hybridization.

multiple copies of a gene, it may be joined to a **cloning vector.** A cloning vector is a piece of DNA having its own replication origin so that it can be replicated inside a host cell. Often the bacterium *E. coli* serves as the host cell in which the vector carrying the cloned DNA is replicated in abundance.

There are two major kinds of cloning vectors. The first are bacterial virus or phage vectors. These are bacterial viruses that have been genetically altered to allow the addition of cloned DNA fragments. These viruses have all the genes required to replicate 100–200 copies of the virus (and cloned fragment) per infected cell.

Antibiotic resistance causes countless problems in the treatment of bacterial infections.

The second commonly used vector is a plasmid vector. Plasmids are extra pieces of circular DNA found in most kinds of bacteria. The plasmids that are used as cloning vectors often contain antibiotic resistance genes that are useful in the selection of cells containing a plasmid.

One such plasmid is pBR322. The restriction enzyme map of pBR322 is illustrated in Figure 24.22. This plasmid has two antibiotic resistance genes: one for ampicillin and one for tetracycline. Within the antibiotic resistance genes are several restriction enzyme sites that are convenient for cloning.

Genetic Engineering

Now that we have assembled most of the tools needed for a cloning experiment, we must decide which gene to clone. The example that we will use is the cloning

A HUMAN PERSPECTIVE

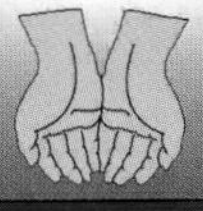

DNA Fingerprinting

DNA fingerprinting has become a valuable tool in many investigations. It can be used by forensic scientists to identify criminals. The courts use DNA fingerprinting in lawsuits to determine the paternity of a child. Wildlife conservationists use the technique to investigate whether elephant tusks that they confiscate were illegally obtained from endangered elephant species.

DNA fingerprinting was developed in the 1980s by Alec Jeffries of the University of Leicester in England. The idea grew out of basic molecular genetic studies of the human genome. Scientists observed that some DNA sequences varied greatly from one person to the next. Such hypervariable regions are made up of variable numbers of repeats of short DNA sequences. They are located at many sites on different chromosomes. Each person has a different number of repeats and when his or her DNA is digested with restriction enzymes, a unique set of DNA fragments is generated. Jeffries invented DNA fingerprinting by developing a set of DNA probes that detect these variable number tandem repeats (VNTRs) when used in hybridization with Southern blots.

Although several variations of DNA fingerprinting exist, the basic technique is quite simple. DNA is digested with restriction enzymes, producing a set of DNA fragments. These are separated by electrophoresis through an agarose gel. The DNA fragments are then transferred to membrane filters and hybridized with the radioactive probe DNA. The bands that hybridize the radioactive probe are visualized by exposing the membrane to X-ray film and developing a "picture" of the gel. The result is what Jeffries calls a *DNA fingerprint,* a set of twenty-five to sixty DNA bands that are unique to an individual.

It was originally estimated that the odds of two individuals (excluding identical twins) having the same DNA fingerprint were less than one in one hundred million or even one in a billion. At first blush these odds suggest that DNA fingerprinting is a very powerful tool for the identification of criminals using tissue samples.

However, recently, the technique has come under intense scrutiny. Some scientists have questioned the statistical methods used to calculate the odds of the individuality of a DNA fingerprint, concerned that within particular ethnic groups the variability may be much lower than was previously claimed. Others argued that slightly different methodologies used in different labs could lead to different results. Still others were concerned that the DNA obtained from a crime scene sampling may be partially degraded and thus yield misleading results.

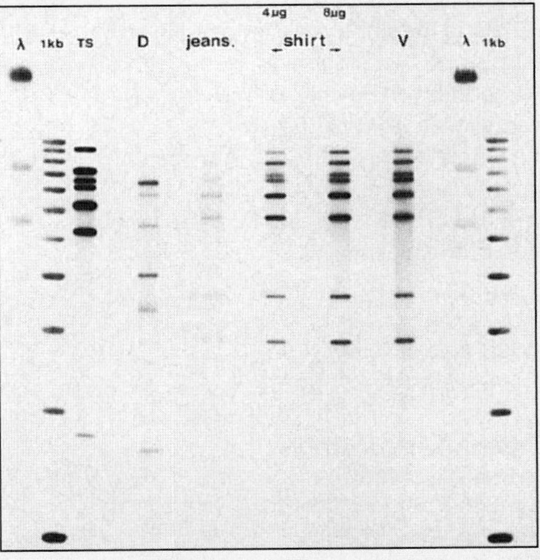

An example of a DNA fingerprint used in a criminal case. The DNA sample designated *V* is that of the victim and the sample designated *D* is that of the defendant. The samples labeled *jeans* and *shirt* were taken from the clothing of the defendant. The DNA bands from the defendant's clothing clearly match the DNA bands of the victim, providing evidence of the guilt of the defendant.

In 1992, a committee established by the National Academy of Sciences (NAS) supported the use of DNA fingerprint evidence in legal proceedings. However, they stated that there must be stringent quality control. The committee also established a more conservative means for calculating probability estimates. The probability estimates based on this new means of statistical analysis are in the range of one in several hundred thousand to one in a million—still impressive odds! Finally, the committee recommended an extensive study of the degree of variability of these VNTRs in various ethnic groups and populations. Implementation of these recommendations will improve DNA fingerprint technology and ensure that it is performed properly and that the data are interpreted correctly.

of the β-globin genes for normal and sickle cell hemoglobin. DNA from an individual with normal hemoglobin is digested with a restriction enzyme, perhaps BamHI. This is the target DNA. The vector DNA must be digested with the same enzyme (Figure 24.23). In our example we will use the plasmid vector pBR322.

The digested vector and target DNA are mixed together under conditions that encourage the BamHI sticky ends of the target and vector DNA to hybridize with one another. The sticky ends are then covalently linked by the enzyme DNA ligase.

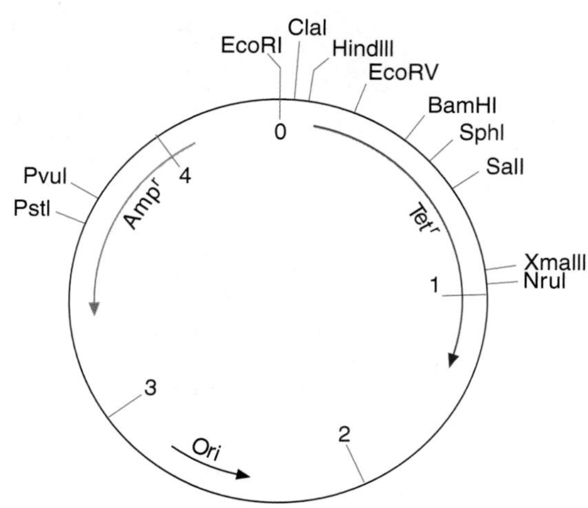

Figure 24.22
A restriction enzyme map of the plasmid cloning vector pBR322. Restriction enzyme cleavage sites for eleven enzymes are shown, along with the location of the ampicillin resistance gene (*red arrow*) and the tetracycline resistance gene (*blue arrow*). "Ori" is the origin of DNA replication of the plasmid.

1. Plant cell

or

Animal (human) cell

2. Bacterium

Major chromosome

Plasmid

DNA isolated from donor cell

3. Donor DNA and plasmid DNA are fragmented by same specific restriction enzyme

Plasmid isolated from bacterium

Donor and plasmid DNA mixed; a donor DNA fragment anneals to plasmid DNA by hydrogen-bonding of complementary bases; ring is restored and sealed by specific enzyme (DNA ligase)

Each plasmid incorporates a *different* DNA fragment

a b c

Recombinant plasmids

4. Engineered plasmid (recombinant DNA) introduced into a bacterium

a

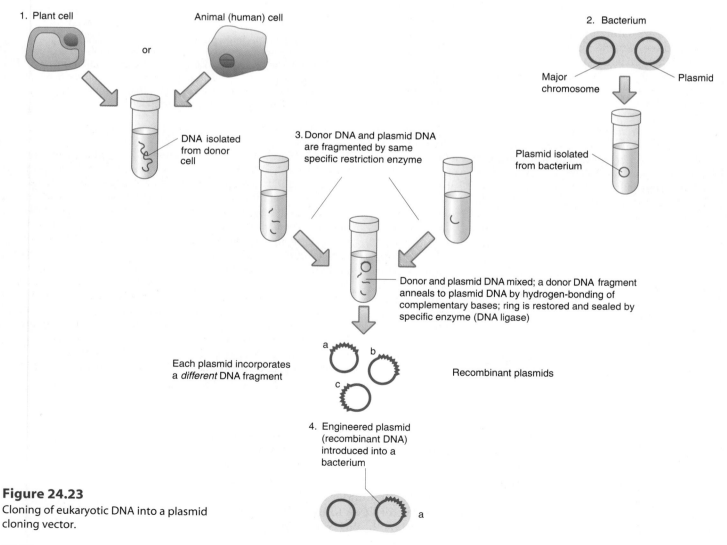

Figure 24.23
Cloning of eukaryotic DNA into a plasmid cloning vector.

This enzyme catalyzes the formation of phosphoester bonds between the two pieces of DNA.

Now the recombinant DNA molecules are introduced into bacterial cells by a process called transformation. Next, the cells of the transformation mixture are plated on a solid nutrient agar medium containing the antibiotic ampicillin. Only those cells containing the antibiotic resistance gene will survive and grow into bacterial colonies.

Now hybridization can be used to detect the clones that carry the β-globin gene. A replica of the experimental plate is made by transferring some cells from each colony onto a membrane filter. These cells are gently broken open so that the released DNA becomes attached to the membrane. When hybridization is carried out on these filters, the radioactive probe will hybridize only to the complementary sequences of the β-globin gene. When the membrane filter is exposed to X-ray film, a "spot" will appear on the developed film only at the site of a colony carrying the desired clone. By going back to the original plate, we can select cells from that colony and grow the cells for further study (Figure 24.24).

The same procedure can be used to clone the β-chain gene of sickle cell hemoglobin. Then the two can be studied and compared to determine the nature of the genetic defect.

This simple example makes it appear that all gene cloning is very easy and straightforward. This has proved to be far from the truth. Genetic engineers have had to overcome many obstacles to clone eukaryotic genes of particular medical interest. One of the first obstacles encountered was the presence of introns within eukaryotic genes. Bacteria that are used for cloning lack the enzymatic machinery to splice out introns. Molecular biologists found that a DNA copy of a eukaryotic mRNA could be made by using the enzyme reverse transcriptase from a family of viruses called retroviruses (see A Clinical Perspective: Fooling the AIDS Virus with "Look-Alike" Nucleotides, pp. 728–729). Such a DNA copy of the mRNA carries all the protein-coding sequences of a gene but none of the intron sequences. Thus bacteria are able to transcribe and translate the cloned DNA and produce valuable products for use in medicine and other applications.

This is only one of the many technical problems that have been overcome by the amazing developments in recombinant DNA technology. A brief but impressive list of medically important products of genetic engineering is presented in Table 24.4.

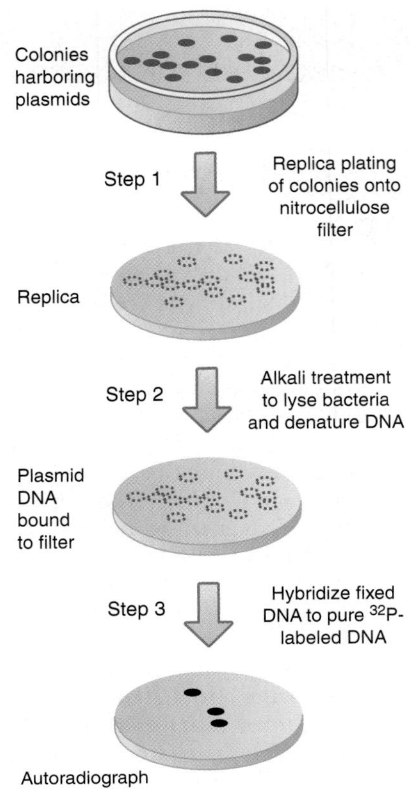

Figure 24.24
Colony blot hybridization for detection of cells carrying a plasmid clone of the β-chain gene of hemoglobin.

Table 24.4 A Brief List of Medically Important Proteins Produced by Genetic Engineering

Protein	Medical Condition Treated
Insulin	Insulin-dependent diabetes
Human growth hormone	Pituitary dwarfism
Factor VIII	Type A hemophilia
Factor IX	Type B hemophilia
Tissue plasminogen factor	Stroke, myocardial infarction
Streptokinase	Myocardial infarction
Interferon	Cancer, some virus infection
Interleukin-2	Cancer
Tumor necrosis factor	Cancer
Atrial natriuretic factor	Hypertension
Erythropoietin	Anemia
Thymosin α-1	Stimulate immune system
Hepatitis B virus (HBV) vaccine	Prevent HBV viral hepatitis
Influenza vaccine	Prevent influenza infection

A MEDICAL PERSPECTIVE

A Genetic Approach to Familial Emphysema

Familial emphysema is a human genetic disease resulting from the inability to produce the protein α_1-antitrypsin. See also A Medical Perspective: α_1-Antitrypsin and Familial Emphysema, in Chapter 20. In individuals who have inherited one or two copies of the α_1-antitrypsin gene, this serum protein protects the lungs from the enzyme elastase. Normally, elastase fights bacteria and helps in the destruction and removal of dead lung tissue. However, the enzyme can also cause lung damage. By inhibiting elastase, α_1-antitrypsin prevents lung damage. Individuals who have inherited two defective α_1-antitrypsin genes do not produce this protein and suffer from familial, or A1AD, emphysema. In the absence of α_1-antitrypsin, the elastase and other proteases cause the severe lung damage characteristic of emphysema.

A1AD is the second most common genetic disorder in Caucasians. It is estimated that there are 100,000 sufferers in the United States and that one in five Americans carries the gene. The disorder, discovered in 1963, is often misdiagnosed as asthma or chronic obstructive pulmonary disease. In fact, it is estimated that fewer than 5% of the sufferers are diagnosed with A1AD.

The α_1-antitrypsin gene has been cloned. Early experiments with sheep showed that the protein remains stable when administered as an aerosol and remains functional after it has passed through the pulmonary epithelium. This research offers hope of an effective treatment for this disease.

The current treatment involves weekly IV injections of α_1-antitrypsin. The supply of the protein, purified from human plasma that has been demonstrated to be virus free, is rather limited. Thus, the injections are expensive. In addition, they are painful. These two factors cause some sufferers to refuse the treatment.

Recently Dr. Terry Flotte and his colleagues at the University of Florida have taken a new approach. They have cloned the gene for α_1-antitrypsin into the DNA of adeno-associated virus. This virus is an ideal vector for human gene replacement therapy because it replicates only in cells that are not dividing and it does not stimulate a strong immune or inflammatory response. The researchers injected the virus carrying the cloned α_1-antitrypsin gene into the muscle tissue of mice, then tested for the level of α_1-antitrypsin in the blood. The results were very promising. Effective levels of α_1-antitrypsin were produced in the muscle cells of the mice and secreted into the bloodstream. Furthermore, the level of α_1-antitrypsin remained at therapeutic levels for more than four months. The research team is planning tests with larger animals and eventually will confirm their results in human trials.

24.9 Polymerase Chain Reaction

A bacterium originally isolated from a hot spring in Yellowstone National Park provides the key to a powerful molecular tool for the study of DNA. Polymerase chain reaction (PCR) allows scientists to produce unlimited amounts of any gene of interest and the bacterium *Thermus aquaticus* produces a heat-stable DNA polymerase (Taq polymerase) that allows the process to work.

The human genome consists of approximately three billion base pairs of DNA. But suppose you are interested in studying only one gene, perhaps the gene responsible for muscular dystrophy or cystic fibrosis. It's like looking for a needle in a haystack. Using PCR, a scientist can make millions of copies of the gene you want to study, while ignoring the thousands of other genes on human chromosomes.

The secret to this specificity is the synthesis of a DNA primer, a short piece of single-stranded DNA that will specifically hybridize to the beginning of a particular gene. DNA polymerases require a primer for initiation of DNA synthesis because they act by adding new nucleotides to the 3′—OH of the last nucleotide of the primer.

Learning Goal

11

To perform PCR, a small amount of DNA is mixed with Taq polymerase, the primers, and the four DNA nucleotide triphosphates. The mixture is then placed in an instrument called a *thermocycler.* The temperature in the thermocycler is raised to 94–96°C for several minutes to separate the two strands of DNA. The temperature is then dropped to 50–56°C to allow the primers to hybridize to the target DNA. Finally, the temperature is raised to 72°C to allow Taq polymerase to act, reading the template DNA strand and polymerizing a daughter strand

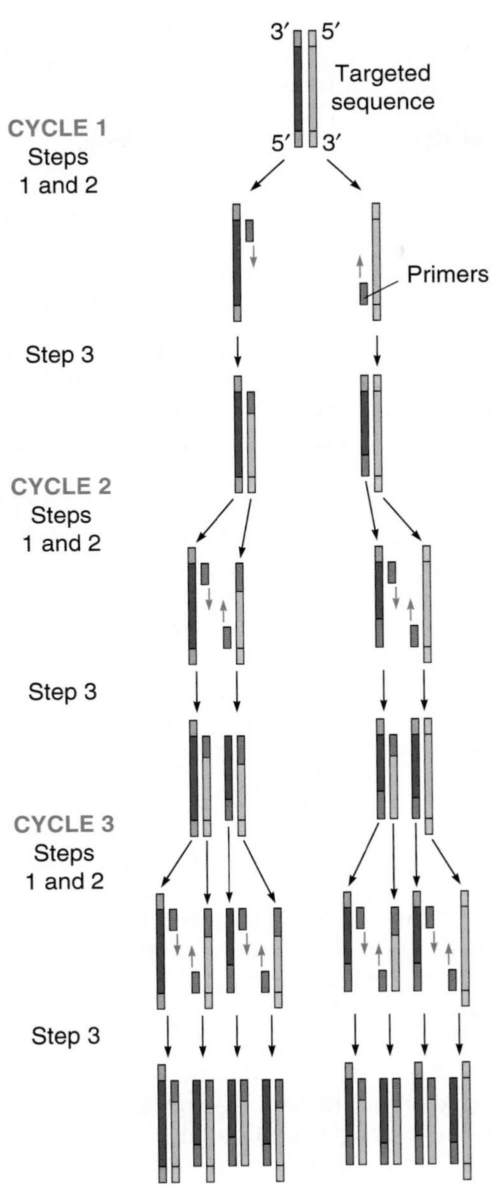

Figure 24.25
Polymerase chain reaction.

extended from the primer. At the end of this step, the amount of the gene has doubled (Figure 24.25).

Now the three steps are repeated. With each cycle the amount of the gene is doubled. Theoretically after thirty cycles, you have one billion times more DNA than you started with!

PCR can be used in genetic screening to detect the gene responsible for muscular dystrophy. It can also be used to diagnose disease. For instance, it can be used to amplify small amounts of HIV in the blood of an AIDS victim. It can also be used by forensic scientists to amplify DNA from a single hair follicle or a tiny drop of blood at a crime scene.

24.10 The Human Genome Project

In 1990 the Department of Energy and the National Institutes of Health began the Human Genome Project (HGP), a multinational project that would extend into the next millennium. The goals of the HGP were to identify all of the genes in human

DNA and to sequence the entire three billion nucleotide pairs of the genome. In order to accomplish these goals, enormous computer databases had to be developed to store the information and computer software had to be designed to analyze it.

Initially, the HGP planned to complete the work by the year 2005. However, as a result of technological advances made by those in the project, a working draft of the human genome was published in February 2001. Because of the accelerated pace of the project, the new completion date is predicted to be 2003.

Genetic Strategies for Genome Analysis

The strategy for HGP was rather straightforward. In order to determine the DNA sequence of the human genome, genomic libraries had to be produced. A *genomic library* is a set of clones representing the entire genome. The DNA sequences of each of these clones could then be determined. Of course, once the sequence of each of these clones is determined, there is no way to know how they are arranged along the chromosomes.

A second technique, called *chromosome walking*, provided both DNA sequence information, as well as a method for identifying the DNA sequences next to it on the chromosome. This method requires clones that are overlapping. To accomplish this, libraries of clones were made using many different restriction enzymes. The DNA sequence of a fragment is determined. Then that information is used to develop a probe for any clones in the library that are overlapping. Each time a DNA fragment is sequenced, the information is used to identify overlapping clones. This process continues, allowing scientists to walk along the chromosome in two directions until the entire sequence is cloned, mapped, and sequenced.

DNA Sequencing

The method of DNA sequencing that is used is based on a technique developed by Frederick Sanger. A cloned piece of DNA is separated into its two strands. Each of these will serve as a template strand to carry out DNA replication in test tubes. A primer strand is also needed. This is a short piece of DNA that will hybridize to the template strand. The primer is the starting point for addition of new nucleotides during DNA synthesis.

The DNA is then placed in four test tubes with all of the enzymes and nucleotides required for DNA synthesis. In addition, each tube contains an unusual nucleotide, called a dideoxynucleotide. These nucleotides differ from the standard nucleotides by having a hydrogen atom at the 3′ position of the deoxyribose, rather than a hydroxyl group. When a dideoxynucleotide is incorporated into a growing DNA chain, it acts as a chain terminator. Because it does not have a 3′-hydroxyl group, no phosphoester bond can be formed with another nucleotide and no further polymerization can occur.

Each of the four tubes containing the DNA, enzymes, and an excess of the nucleotides required for replication will also have a small amount of one of the four dideoxynucleotides. In the tube that receives dideoxyadenosine triphosphate (ddA), for example, DNA synthesis will begin. As replication proceeds, either the standard nucleotide or ddA will be incorporated into the growing strand. Since the standard nucleotide is present in excess, the dideoxynucleotide will be incorporated infrequently and randomly. This produces a family of DNA fragments that terminate at the location of one of the deoxyadenosines in the molecule.

The same reaction is done with each of the dideoxynucleotides. The DNA fragments are then separated by gel electrophoresis on a DNA sequencing gel. The four reactions are placed in four wells, side by side, on the gel. Following electrophoresis, the DNA sequence can be read directly from the gel, as shown in Figure 24.26.

When chain termination DNA sequencing was first done, radioactive isotopes were used to label the DNA strands. However, new technology has resulted in

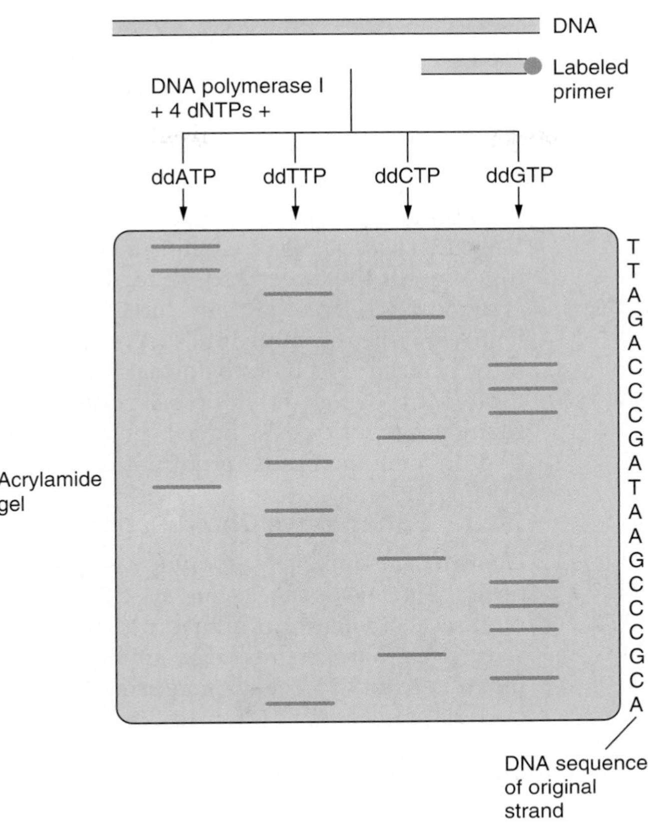

Figure 24.26
DNA sequencing by chain termination requires a template DNA strand and a radioactive primer. These are placed into each of four reaction mixtures that contain DNA polymerase, the four DNA nucleotides (dATP, dCTP, dGTP, and TTP), as well as one of the four dideoxynucleotides. Following the reaction, the products are separated on a DNA sequencing gel. The sequence is read from an autoradiograph of the gel.

automated systems that employ dideoxynucleotides that are labeled with fluorescent dyes, a different color for each dideoxynucleotide. Because each reaction (A, G, C, and T) will be a different color, all the reactions can be done in a single reaction mixture and the products separated on a single lane of a sequencing gel. A computer then "reads" the gel by distinguishing the color of each DNA band. The sequence information is directly stored into a databank for later analysis.

There is currently a vast amount of DNA information available on the Internet. The complete genomes of many bacteria have been reported, as well as the sequence information generated by the Human Genome Project. Because we know the genetic code, we can predict the amino acid sequence of proteins encoded by the genes. Researchers can also compare the sequences of normal genes with those of people suffering from genetic disorders. The enormity of the DNA information available, as well as the many types of analysis that need to be carried out, have given rise to an entirely new branch of science. The field of **bioinformatics** is a marriage of computer information sciences and DNA technology that is helping to devise methods for understanding, analyzing, and applying the DNA sequence information that we are gathering.

Summary

24.1 The Structure of the Nucleotide

DNA and RNA are polymers of *nucleotides,* which are composed of a five-carbon sugar (ribose in RNA and 2′-deoxyribose in DNA), a nitrogenous base, and one, two, or three phosphoryl groups. There are two kinds of nitrogenous bases, the *purines* (adenine and guanine) and the *pyrimidines* (cytosine, thymine, and uracil). *Deoxyribonucleotides* are the subunits of DNA. *Ribonucleotides* are the subunits of RNA.

24.2 The Structure of DNA and RNA

Nucleotides are joined by 3′–5′ phosphodiester bonds in both DNA and RNA. DNA is a *double helix,* two strands of

DNA wound around one another. The sugar-phosphate backbone is on the outside of the helix, and complementary pairs of bases extend into the center of the helix. The *base pairs* are held together by hydrogen bonds. Adenine base pairs with thymine, and cytosine base pairs with guanine. The two strands of DNA in the helix are antiparallel to one another. RNA is single stranded.

24.3 DNA Replication

DNA replication involves synthesis of a faithful copy of the DNA molecule. It is *semiconservative;* each daughter molecule consists of one parental strand and one newly synthesized strand. *DNA polymerase* "reads" each parental strand and synthesizes the complementary daughter strand according to the rules of base pairing.

24.4 Information Flow in Biological Systems

The *central dogma* states that the flow of biological information in cells is DNA → RNA → protein. There are three classes of RNA: *messenger RNA, transfer RNA,* and *ribosomal RNA. Transcription* is the process by which RNA molecules are synthesized. *RNA polymerase* catalyzes the synthesis of RNA. Transcription occurs in three stages: initiation, elongation, and termination. Eukaryotic genes contain *introns,* sequences that do not encode protein. These are removed from the primary transcript by the process of *RNA splicing.* The final mRNA contains only the protein coding sequences or *exons.* This final mRNA also has an added 5′ cap structure and 3′ poly(A) tail.

24.5 The Genetic Code

The genetic code is a triplet code. Each code word is called a *codon* and consists of three nucleotides. There are sixty-four codons in the genetic code. Of these, three are *termination codons* (UAA, UAG, and UGA), and the remaining sixty-one specify an amino acid. Most amino acids have several codons. As a result, the genetic code is said to be *degenerate.*

24.6 Protein Synthesis

The process of protein synthesis is called *translation.* The genetic code words on the mRNA are decoded by tRNA. Each tRNA has an *anticodon* that is complementary to a codon on the mRNA. In addition the tRNA is covalently linked to its correct amino acid. Thus hydrogen bonding between codon and anticodon brings the correct amino acid to the site of protein synthesis. Translation also occurs in three stages called initiation, chain elongation, and termination.

24.7 Mutation, Ultraviolet Light, and DNA Repair

Any change in a DNA sequence is a *mutation.* Mutations are classified according to the type of DNA alteration, including *point mutations, deletion mutations,* and *insertion mutations.* Ultraviolet light (UV) causes formation of *pyrimidine dimers.* Mistakes can be made during pyrimidine dimer repair, causing UV-induced mutations. Germicidal (UV) lamps are used to kill bacteria on environmental surfaces. UV damage to skin can result in skin cancer.

24.8 Recombinant DNA

Several tools are required for genetic engineering, including *restriction enzymes, agarose gel electrophoresis, hybridization,* and *cloning vectors.* Cloning a DNA fragment involves digestion of the target and vector DNA with a restriction enzyme. DNA ligase joins the target and vector DNA covalently, and the recombinant DNA molecules are introduced into bacterial cells by transformation. The desired clone is located by using antibiotic selection and hybridization. Many eukaryotic genes have been cloned for the purpose of producing medically important proteins.

24.9 Polymerase Chain Reaction

Using a heat-stable DNA polymerase produced by the bacterium *Thermus aquaticus* and specific DNA primers, polymerase chain reaction allows the amplification of DNA sequences that are present in small quantities. This technique is useful in genetic screening, diagnosis of viral or bacterial disease, and forensic science.

24.10 The Human Genome Project

The goals of the Human Genome Project are to identify and map the genes of the human genome and to determine the complete DNA sequence of each of the chromosomes. To do this, genomic libraries are generated and the DNA sequences of the clones are determined. To map the sequences along each chromosome, chromosome walking is used. DNA sequencing involves reactions in which DNA polymerase copies specific DNA sequences. Nucleotide analogues that cause chain termination (dideoxynucleotides) are incorporated randomly into the growing DNA chain. This generates a family of DNA fragments that differ in size by one nucleotide. DNA sequencing gels separate these fragments and provide DNA sequence data.

Key Terms

aminoacyl tRNA (24.6)
aminoacyl tRNA binding site of ribosome (A-site) (24.6)
aminoacyl tRNA synthetase (24.6)
anticodon (24.4)
antiparallel strands (24.2)
base pairs (24.2)
bioinformatics (24.10)
cap structure (24.4)
carcinogen (24.7)
central dogma (24.4)
chromosome (24.2)
cloning vector (24.8)
codon (24.4)
complementary strands (24.2)
degenerate code (24.5)
deletion mutation (24.7)
deoxyribonucleic acid (DNA) (24.1)
deoxyribonucleotide (24.1)
DNA polymerase (24.3)
double helix (24.2)
elongation factor (24.6)

eukaryote (24.2)
exon (24.4)
genome (24.2)
hybridization (24.8)
initiation factor (24.6)
insertion mutation (24.7)
intron (24.4)
messenger RNA (mRNA) (24.4)
mutagen (24.7)
mutation (24.7)
nucleosome (24.2)
nucleotide (24.1)
peptidyl tRNA binding site of ribosome (P-site) (24.6)
point mutation (24.7)
poly(A) tail (24.4)
polysome (24.6)
post-transcriptional modification (24.4)
primary transcript (24.4)
prokaryote (24.2)
promoter (24.4)
purine (24.1)
pyrimidine (24.1)
pyrimidine dimer (24.7)
release factor (24.6)
replication fork (24.3)
replication origin (24.3)
restriction enzyme (24.8)
ribonucleic acid (RNA) (24.1)
ribonucleotide (24.1)
ribosomal RNA (rRNA) (24.4)
ribosome (24.6)
RNA polymerase (24.4)
RNA splicing (24.4)
semiconservative replication (24.3)
silent mutation (24.7)
termination codon (24.6)
transcription (24.4)
transfer RNA (tRNA) (24.4)
translation (24.4)
translocation (24.6)

Questions and Problems

The Structure of the Nucleotide

24.15 Draw the structure of the purine ring, and indicate the nitrogen that is bonded to sugars in nucleotides.
24.16 a. Draw the ring structure of the pyrimidines.
b. In a nucleotide, which nitrogen atom of pyrimidine rings is bonded to the sugar?
24.17 ATP is the universal energy currency of the cell. What components make up the ATP nucleotide?
24.18 One of the energy-harvesting steps of the citric acid cycle results in the production of GTP. What is the structure of the GTP nucleotide?

The Structure of DNA and RNA

24.19 The two strands of a DNA molecule are antiparallel. What is meant by this description?
24.20 List three differences between DNA and RNA.
24.21 How many hydrogen bonds link the adenine-thymine base pair?
24.22 How many hydrogen bonds link the guanine-cytosine base pair?
24.23 Write the structure that results when deoxycytosine-5′-monophosphate is linked by a 3′ → 5′ phosphodiester bond to thymidine-5′-monophosphate.
24.24 Write the structure that results when adenosine-5′-monophosphate is linked by a 3′ → 5′ phosphodiester bond to uridine-5′-monophosphate.

DNA Replication

24.25 What is meant by semiconservative DNA replication?
24.26 Draw a diagram illustrating semiconservative DNA replication.
24.27 What are the two primary functions of DNA polymerase?
24.28 a. Why is DNA polymerase said to be template-directed?
b. Why is DNA replication a self-correcting process?
24.29 If a DNA strand had the nucleotide sequence

5′-ATGCGGCTAGAATATTCCA-3′

what would the sequence of the complementary daughter strand be?
24.30 If the sequence of a double-stranded DNA is

5′-G A A T T C C T T A A G G A T C G A T C-3′
| | | | | | | | | | | | | | | | | | | |
3′-C T T A A G G A A T T C C T A G C T A G-5′

what would the sequence of the two daughter DNA molecules be after DNA replication? Indicate which strands are newly synthesized and which are parental.
24.31 What is the replication origin of a DNA molecule?
24.32 What is occurring at the replication fork?

Information Flow in Biological Systems

24.33 What is the central dogma of molecular biology?
24.34 What are the roles of DNA, RNA, and protein in information flow in biological systems?
24.35 On what molecule is the anticodon found?
24.36 On what molecule is the codon found?
24.37 If a gene had the nucleotide sequence

5′-TACCTAGCTCTGGTCATTAAGGCAGTA-3′

what would the sequence of the mRNA be?
24.38 If a mRNA had the nucleotide sequence

5′-AUGCCCUUUCAUUACCCGGUA-3′

what was the sequence of the DNA strand that was transcribed?
24.39 What is meant by the term *RNA splicing?*
24.40 The following is the unspliced transcript of a eukaryotic gene:

What would the structure of the final mature mRNA look like, and which of the above sequences would be found in the mature mRNA?
24.41 List the three classes of RNA molecules.
24.42 What is the function of each of the classes of RNA molecules?
24.43 What is the function of the spliceosome?
24.44 What are snRNPs? How do they facilitate RNA splicing?
24.45 What is a poly(A) tail?
24.46 What is the purpose of the poly(A) tail on eukaryotic mRNA?
24.47 What is the cap structure?
24.48 What is the function of the cap structure on eukaryotic mRNA?

The Genetic Code

24.49 How many codons constitute the genetic code?
24.50 What is meant by a triplet code?
24.51 What is meant by the reading frame of a gene?
24.52 What happens to the reading frame of a gene if a nucleotide is deleted?
24.53 Which two amino acids are encoded by only one codon?
24.54 Which amino acids are encoded by six codons?
24.55 An essential gene has the codon 5′-UUU-3′ in a critical position. If this codon is mutated to the sequence 5′-UUA-3′, what is the expected consequence for the cell?
24.56 An essential gene has the codon 5′-UUA-3′ in a critical position. If this codon is mutated to the sequence 5′-UUG-3′, what is the expected consequence for the cell?

Protein Synthesis

24.57 What is the function of ribosomes?
24.58 What are the two tRNA binding sites on the ribosome?
24.59 Briefly describe the three stages of translation: initiation, elongation, and termination.
24.60 What peptide sequence would be formed from the mRNA 5′-AUGUGUAGUGACCAACCGAUUUCACUGUGA-3′?
24.61 By what type of bond is an amino acid linked to a tRNA molecule in an aminoacyl tRNA molecule?
24.62 Draw the structure of an alanine residue bound to the 3′ position of adenine at the 3′ end of alanyl tRNA.

Mutation, Ultraviolet Light, and DNA Repair

24.63 What damage does UV light cause in DNA, and how does this lead to mutations?
24.64 Explain why UV lights are effective germicides on environmental surfaces.
24.65 What is a carcinogen? Why are carcinogens also mutagens?
24.66 a. What causes the genetic disease xeroderma pigmentosum?
b. Why are people who suffer from xeroderma pigmentosum prone to cancer?

Recombinant DNA

24.67 What is a restriction enzyme?
24.68 Of what value are restriction enzymes in recombinant DNA research?
24.69 Describe the molecular basis of hybridization.
24.70 What is a cloning vector?
24.71 Name three products of recombinant DNA that are of value in the field of medicine.
24.72 a. What is the ultimate goal of genetic engineering?
b. What ethical issues does this goal raise?

Polymerase Chain Reaction

24.73 After ten cycles of polymerase chain reaction, how many copies of target DNA would you have for each original molecule in the mixture?
24.74 List several practical applications of polymerase chain reaction.

The Human Genome Project

24.75 What are the major goals of the Human Genome Project?
24.76 What are the potential benefits of the information gained in the Human Genome Project?
24.77 What is a genome library?
24.78 What is meant by the term *chromosome walking?*
24.79 What is a dideoxynucleotide?
24.80 How does a dideoxynucleotide cause chain termination in DNA replication?
24.81 A researcher has determined the sequence of the following five pieces of DNA. Using this sequence information, map the location of these pieces relative to one another.
a. 5′ AGCTCCTGATTTCATACAGTTTCTACTACCTACTA 3′
b. 5′ AGACATTCTATCTACCTAGACTATGTTCAGAA 3′
c. 5′ TTCAGAACTCATTCAGACCTACTACTATACCTTGGG AGCTCCT 3′
d. 5′ ACCTACTAGACTATACTACTACTAAGGGGACTATT CCAGACTT 3′

24.82 Draw a DNA sequencing gel that would represent the sequence shown below. Be sure to label which lanes of the gel represent each of the four dideoxynucleotides in the chain termination reaction mixture.

5′ GACTATCCTAG 3′

Critical Thinking Problems

1. It has been suggested that the triplet genetic code evolved from a two-nucleotide code. Perhaps there were fewer amino acids in the ancient proteins. Examine the genetic code in Figure 24.16. What features of the code support this hypothesis?
2. The strands of DNA can be separated by heating the DNA sample. The input heat energy breaks the hydrogen bonds between base pairs, allowing the strands to separate from one another. Suppose that you are given two DNA samples. One has a G + C content of 70% and the other has a G + C content of 45%. Which of these samples will require a higher temperature to separate the strands? Explain your answer.
3. A mutation produces a tRNA with a new anticodon. Originally the anticodon was 5′-CCA-3′; the mutant anticodon is 5′-UCA-3′. What effect will this mutant tRNA have on cellular translation?
4. You have just cloned an EcoR1 fragment that is 1650 base pairs (bp) and contains the gene for the hormone leptin. Your first job is to prepare a restriction enzyme map of the recombinant plasmid. You know that you have cloned into a plasmid vector that is 805 bp and that has only one EcoR1 site (the one into which you cloned). There are no other restriction enzyme sites in the plasmid. The following table shows the restriction enzymes used and the DNA fragment sizes that result. Draw a map of the circular recombinant plasmid and a representation of the gel from which the fragment sizes were obtained.

Restriction Enzymes	DNA Fragment Sizes (bp)
EcoR1	805, 1650
EcoR1 + BamHI	450, 805, 1200
EcoR1 + SalI	200, 805, 1450
BamHI + SalI	200, 250, 805, 1200

5. A scientist is interested in cloning the gene for blood clotting factor VIII into bacteria so that large amounts of the protein can be produced to treat hemophiliacs. Knowing that bacterial cells cannot carry out RNA splicing, she clones a complementary DNA copy of the factor VIII mRNA and introduces this into bacteria. However, there is no transcription of the cloned factor VIII gene. How could the scientist engineer the gene so that the bacterial cell RNA polymerase will transcribe it?

Appendix A

A Review of Mathematics Applied to Problem Solving in Chemistry

A.1 Algebraic Equations

Many of the problems and examples discussed in the text involve one or more of a limited number of algebraic equations. If you are having difficulty with the mathematics rather than the chemistry, this algebra review may prove useful. Let's consider a variety of common algebraic relationships and their application to chemistry problem solving.

One common algebraic equation that is often applied in chemical problem solving is

$$a = \frac{b}{c}$$

The expression for density,

$$d = \frac{m}{V}$$

is a typical example of this type of relationship.

Let's compare the similarity between the pure algebraic manipulation (left column) and the relationship between density (d), mass (m), and volume (V) (right column) in the following table.

Mathematical Function	Chemistry Connection
To solve for a, divide b by c.	To solve for d, divide m by V.
To solve for b, multiply both sides of the equation by c:	To solve for m, multiply both sides of the equation by V:
$a \times c = \frac{b}{\cancel{c}} \times \cancel{c}$	$d \times V = \frac{m}{\cancel{V}} \times \cancel{V}$
$b = a \times c$	$m = d \times V$

This method may be applied to problems such as those in Examples 1.18 and 1.19 in the text. Example A.1 presents a typical problem involving the concept of density.

Calculating Mass from Volume by Using Density — EXAMPLE A.1

Pure oxygen has a density of 0.00140 g/mL at 273K. What is the mass of an 8.00-L sample of oxygen?

Solution

Recall that a number such as 0.00140 g/mL can be expressed as

Continued—

EXAMPLE A.1 —*Continued*

$$1.40 \times 10^{-3}\ \text{g/mL}$$

(The decimal point is moved three positions to the right.) Also, using a conversion factor to make the volume units consistent, we have

$$8.00\ \cancel{\text{L oxygen}} \times \frac{10^3\ \text{mL oxygen}}{1\ \cancel{\text{L oxygen}}} = 8.00 \times 10^3\ \text{mL oxygen}$$

This problem can be solved by rearranging the expression for density, $d = m/V$, to the form

$$m = d \times V$$

Substituting the data that we have for density and volume gives

$$\frac{1.40 \times 10^{-3}\ \text{g oxygen}}{\cancel{\text{mL oxygen}}} \times 8.00 \times 10^3\ \cancel{\text{mL oxygen}} = 11.2\ \text{g oxygen}$$

We can use the density to calculate the mass of a liquid as well.

EXAMPLE A.2 ***Calculating Mass from Volume by Using Density***

A certain thermometer contains 0.500 mL of mercury. Calculate the mass, in grams, of the mercury in the thermometer. The density of mercury is 13.5 g/mL.

Solution

Using the density as a conversion factor from volume to mass, we have

$$\begin{aligned} m &= d \times V \\ &= (0.500\ \cancel{\text{mL mercury}})\left(\frac{13.5\ \text{g mercury}}{\cancel{\text{mL mercury}}}\right) \\ &= 6.75\ \text{g mercury} \end{aligned}$$

If we want to solve the same general expression,

$$a = \frac{b}{c}$$

for c or in the specific case of the density expression,

$$d = \frac{m}{V}$$

for V, we may use the following approach.

Mathematical Function	Chemistry Connection
To solve for c, multiply both sides of the equation by $\frac{c}{a}$:	To solve for V, multiply both sides of the equation by $\frac{V}{d}$:
$\cancel{a} \times \frac{c}{\cancel{a}} = \frac{b}{\cancel{c}} \times \frac{\cancel{c}}{a}$	$\cancel{d} \times \frac{V}{\cancel{d}} = \frac{m}{\cancel{V}} \times \frac{\cancel{V}}{d}$
$c = \frac{b}{a}$	$V = \frac{m}{d}$

This approach is useful in solving problems such as the one in Example 1.20. For instance, we can use the density to calculate the volume of a liquid.

Calculating Volume from Mass by Using Density

EXAMPLE A.3

Calculate the volume, in milliliters, of a liquid that has a density of 1.30 g/mL and a mass of 9.00 g.

Solution

Using the density as a conversion factor from mass to volume and following the algebraic solution yields

$$c = \frac{b}{a}$$

Then

$$V = \frac{m}{d}$$

and

$$V = \frac{9.00 \text{ g liquid}}{1.30 \text{ g/mL liquid}}$$

$$= 6.92 \text{ mL liquid}$$

Another frequently used algebraic expression is

$$\frac{a}{b} = \frac{c}{d}$$

The expression for Charles's law, relating initial and final volumes, V_i and V_f, and temperatures, T_i and T_f, depends on this algebraic form for its solution. One form of Charles's law follows this algebraic form:

$$\frac{V_i}{T_i} = \frac{V_f}{T_f}$$

Mathematical Function	Chemistry Connection
To solve for a, multiply both sides of the equation by b: $\frac{a}{\cancel{b}} \times \cancel{b} = \frac{c}{d} \times b$; $a = \frac{c \times b}{d}$	To solve for V_i, multiply both sides of the equation by T_i: $\frac{V_i}{\cancel{T_i}} \times \cancel{T_i} = \frac{V_f}{T_f} \times T_i$; $V_i = \frac{V_f \times T_i}{T_f}$
To solve for b, multiply both sides of the equation by $\frac{b \times d}{c}$: $\frac{a}{\cancel{b}} \times \frac{\cancel{b} \times d}{c} = \frac{\cancel{c}}{\cancel{d}} \times \frac{b \times \cancel{d}}{\cancel{c}}$; $\frac{a \times d}{c} = b$ or $b = \frac{a \times d}{c}$	To solve for T_i, multiply both sides of the equation by $\frac{T_i \times T_f}{V_f}$: $\frac{V_i}{\cancel{T_i}} \times \frac{\cancel{T_i} \times T_f}{V_f} = \frac{\cancel{V_f}}{\cancel{T_f}} \times \frac{T_i \times \cancel{T_f}}{\cancel{V_f}}$; $\frac{V_i \times T_f}{V_f} = T_i$ or $T_i = \frac{V_i \times T_f}{V_f}$

To solve for c, multiply both sides of the equation by d: $$\frac{a}{b} \times d = \frac{c}{d} \times d$$ $$\frac{a \times d}{b} = c \text{ or } c = \frac{a \times d}{b}$$	To solve for V_f, multiply both sides of the equation by T_f: $$\frac{V_i}{T_i} \times T_f = \frac{V_f}{T_f} \times T_f$$ $$\frac{V_i \times T_f}{T_i} = V_f \text{ or } V_f = \frac{V_i \times T_f}{T_i}$$

This equation is used in solving the problem in Example 6.2. We can use this equation to calculate the final volume of a balloon after it has undergone a change in temperature.

EXAMPLE A.4 ***Calculating a New Volume After a Temperature Change***

A balloon filled with helium has a volume of 10.0×10^3 L at 298 K. What would be the balloon's volume, at 255 K, if the pressure surrounding the balloon remained constant?

Solution

We use the Charles's law relationship,

$$\frac{V_i}{T_i} = \frac{V_f}{T_f}$$

and substitute the data into the Charles's law expression, rearranged as

$$c = \frac{a \times d}{b}$$

$$V_f = \frac{V_i \times T_f}{T_i}$$

$$= \frac{(10.0 \times 10^3 \text{ L})(255 \text{ K})}{(298 \text{ K})}$$

$$= 8.56 \times 10^3 \text{ L}$$

Mathematical Function	Chemistry Connection
To solve for d, multiply both sides of the equation by $\frac{b \times d}{a}$: $$\frac{a}{b} \times \frac{b \times d}{a} = \frac{c}{d} \times \frac{b \times d}{a}$$ $$d = \frac{b \times c}{a}$$	To solve for T_f, multiply both sides of the equation by $\frac{T_i \times T_f}{V_i}$: $$\frac{V_i}{T_i} \times \frac{T_i \times T_f}{V_i} = \frac{V_f}{T_f} \times \frac{T_i \times T_f}{V_i}$$ $$T_f = \frac{V_f \times T_i}{V_i}$$

This expression is useful in the calculation of temperature-volume problems, as illustrated in Example A.5.

Calculating a New Temperature After a Volume Change

EXAMPLE A.5

Calculate the final temperature of the gas in a balloon that was observed to expand from 3.00 L to 6.00 L as the balloon was heated from an initial temperature of 325 K.

Solution

As we have shown,

$$T_f = \frac{V_f \times T_i}{V_i}$$

$$= \frac{(6.00\ \cancel{L})\ (325\ K)}{3.00\ \cancel{L}}$$

$$= 650\ K$$

Another useful algebraic expression takes the form

$$a \times b = c \times d$$

The expression for Boyle's law, relating pressure to the initial and final volumes of a gas, depends on this algebraic form for its solution:

$$P_iV_i = P_fV_f$$

Mathematical Function	Chemistry Connection
To solve for a, divide both sides of the equation by b:	To solve for P_i, divide both sides of the equation by V_i:
$\frac{a \times \cancel{b}}{\cancel{b}} = \frac{c \times d}{b}$	$\frac{P_i \times \cancel{V_i}}{\cancel{V_i}} = \frac{P_f \times V_f}{V_i}$
$a = \frac{c \times d}{b}$	$P_i = \frac{P_f \times V_f}{V_i}$
Similarly, expressions for b, c, or d may be derived:	Similarly, expressions for V_i, P_f, and V_f, may be derived:
$b = \frac{c \times d}{a}$	$V_i = \frac{P_f \times V_f}{P_i}$
$c = \frac{a \times b}{d}$	$P_f = \frac{P_i \times V_i}{V_f}$
$d = \frac{a \times b}{c}$	$V_f = \frac{P_i \times V_i}{P_f}$

This method may be applied to the solution of problems such as those in Questions 6.5 and 6.6. The determination of a final pressure is illustrated in Example A.6.

Calculating the Pressure Needed to Compress a Gas

EXAMPLE A.6

A certain mass of air, at 25°C, occupies a volume of 8.00×10^2 mL at 2.75 atm pressure. What pressure must be applied to compress the gas to a volume of 1.00×10^2 mL if we assume no temperature change?

Continued—

EXAMPLE A.6 —*Continued*

Solution

Boyle's law applies directly, because there is no change in temperature or number of moles (the mass remains constant). Begin by identifying each term in the Boyle's law expression:

$$P_i = 2.75 \text{ atm}$$

$$V_i = 8.00 \times 10^2 \text{ mL}$$

$$V_f = 1.00 \times 10^2 \text{ mL}$$

$$P_iV_i = P_fV_f$$

and solve by using the algebraic form $c = \dfrac{a \times b}{d}$:

$$P_f = \frac{P_i \times V_i}{V_f}$$

$$= \frac{(2.75 \text{ atm})(8.00 \times 10^2 \cancel{\text{mL air}})}{(1.00 \times 10^2 \cancel{\text{mL air}})}$$

$$= 22.0 \text{ atm}$$

(*Note:* The calculation can be done with volume units of milliliters or liters. However, the units must be the *same* on both sides of the equation.)

The last algebraic form that we consider is

$$a = \frac{(b - c)}{d}$$

The expression relating temperature to Celsius and Fahrenheit units is a practical example of this algebraic form:

$$°\text{C} = \frac{(°\text{F} - 32)}{1.8}$$

Mathematical Function	Chemistry Connection
To solve for b, multiply both sides of the equation by d:	To solve for °F, multiply both sides of the equation by 1.8:
$d \times a = \cancel{d} \times \dfrac{(b - c)}{\cancel{d}}$	$1.8 \times °\text{C} = \cancel{1.8} \times \dfrac{°\text{F} - 32}{\cancel{1.8}}$
$b - c = (d \times a)$	$°\text{F} - 32 = 1.8 \times °\text{C}$
Then add c to both sides of the equation:	Then add 32 to both sides of the equation:
$b - c + c = (d \times a) + c$ $b = (d \times a) + c$	$°\text{F} - 32 + 32 = (1.8 \times °\text{C}) + 32$ $°\text{F} = (1.8 \times °\text{C}) + 32$

Consider Example A.7.

Converting Temperature Units **EXAMPLE A.7**

A patient has a temperature of 40°C. Calculate the corresponding temperature in units of degrees Fahrenheit.

Solution

By following the steps outlined in the preceding table, we can solve the equation

$$°C = \frac{(°F - 32)}{1.8}$$

for °F:

$$°F = (1.8 \times °C) + 32$$

Substituting, we have

$$°F = 1.8(40) + 32$$
$$= 104$$

A.2 Exponential Notation

In scientific calculations, data, results, and physical constants are often either very large or very small numbers. For example, Avogadro's number is

602,200,000,000,000,000,000,000 particles

and the atomic mass unit (amu), expressed in units of grams, is

0.000000000000000000000001661 g

Chapter 1 describes the use of scientific notation (exponential notation) to represent such numbers more conveniently. The rules for scientific notation, as summarized there, are as follows:

1. To convert a number greater than 1 to scientific notation, move the original decimal point x places to the left, and multiply the resulting number by 10^x. The exponent (x) is a *positive* number equal to the number of places the original decimal point was moved.

Avogadro's number is represented as 6.022×10^{23}. The decimal point is moved 23 places to the *left*.

2. To convert a number less than 1 to scientific notation, move the original decimal point x places to the right, and multiply the resulting number by 10^{-x}. The exponent ($-x$) is a *negative* number equal to the number of places the original decimal point was moved.

The atomic mass unit is represented as 1.661×10^{-24} g. The decimal point is moved 24 places to the *right*.

Addition and Subtraction Using Exponents

Consider the following mathematical operation:

$$(6.52 \times 10^{-3}) - (4.87 \times 10^{-4}) = ?$$

Note that the exponents are not identical. To use a calculator to solve this problem, simply enter the data in a format that is consistent with the way that your calculator is designed to function. The answer is displayed with the correct numerical value and exponent.

If you do not have access to a calculator, you must make both exponents identical before attempting to subtract the numbers. It makes no difference which exponent you choose to change; in this example, let's change the first exponent to agree with the second:

$$6.52 \times 10^{-3} \quad \text{is also} \quad 65.2 \times 10^{-4}$$

Moving the decimal point one place to the right makes the exponent one unit more negative.

Subtract (or add) the numerical portions of each and carry the exponential term:

$$(65.2 \times 10^{-4}) - (4.87 \times 10^{-4}) = 60.3 \times 10^{-4}$$

$$60.3 \times 10^{-4} \quad \text{is also} \quad 6.03 \times 10^{-3}$$

Moving the decimal point one place to the left makes the exponent one unit more positive.

Perhaps the most common error made in these calculations is one of omission; after subtraction of the nonexponential terms we may forget to carry along the exponent. This small oversight can have great consequences. In the preceding example, omitting the power of ten would produce a thousandfold error.

Multiplication and Division Using Exponents

To multiply numbers with exponents, carry out the following steps.

- Change all numbers to powers of 10.
- To determine the numerical value, multiply the numerical portion.
- To determine the exponent, algebraically add the exponents of 10.

EXAMPLE A.8 ***Multiplying Numbers with Positive Exponents***

$$220 \times 3500 = ?$$

Solution

$$2.2 \times 10^2 \times 3.5 \times 10^3 = 7.7 \times 10^{(2+3)} = 7.7 \times 10^5$$

EXAMPLE A.9 ***Multiplying Numbers with Negative Exponents***

$$220 \times 0.000035 = ?$$

Solution

$$2.2 \times 10^2 \times 3.5 \times 10^{-5} = 7.7 \times 10^{[2+(-5)]} = 7.7 \times 10^{-3}$$

To divide numbers with exponents, carry out the following steps.

- Change all numbers to powers of 10.
- To determine the numerical value, divide the numerical portion.
- To determine the exponent, algebraically subtract the exponents of 10. Note that the exponents in the denominator are subtracted from the exponents in the numerator.

EXAMPLE A.10

Dividing Numbers with Positive Exponents

$$\frac{770{,}000}{220} = ?$$

Solution

$$\frac{7.7 \times 10^5}{2.2 \times 10^2} = ?$$

$$= 3.5 \times 10^{(5-2)} = 3.5 \times 10^3$$

EXAMPLE A.11

Dividing Numbers with Negative Exponents

$$\frac{0.0077}{220} = ?$$

Solution

$$\frac{7.7 \times 10^{-3}}{2.20 \times 10^2} = ?$$

$$= 3.5 \times 10^{[-3-(+2)]} = 3.5 \times 10^{-5}$$

A.3 Proportionality

Experiments in chemistry, and in science in general, often look for relationships between two or more variable quantities. For example, Charles's law resulted from the observation that gas volumes increase as the temperature of the gas increases when the pressure and number of moles of the gas remain constant.

Direct Proportionality

The following data illustrate Charles's law.

Experiment	T (K)	V of Helium (L)
1	300	1.00
2	600	2.00
3	900	3.00

Doubling the temperature (from 300 K to 600 K) causes the volume of helium to double. Tripling the temperature (from 300 K to 900 K) causes the volume of helium to triple. If we begin with a temperature of 600 K, decreasing the temperature by one-half also decreases the volume by one-half.

Hence the temperature and volume of a gas are *directly proportional.* This relationship is expressed as

$$V \propto T$$

in which the symbol $\propto$ is shorthand for the words *proportional to;* it reads: "Volume is proportional to temperature." Use of a *proportionality constant, k,* and an *equals sign* to replace $\propto$ results in a valid mathematical equation:

$$V = kT$$

In this example, k is the Charles's law constant. Graphical representation of a direct proportion results in a straight-line (linear) relationship between variables:

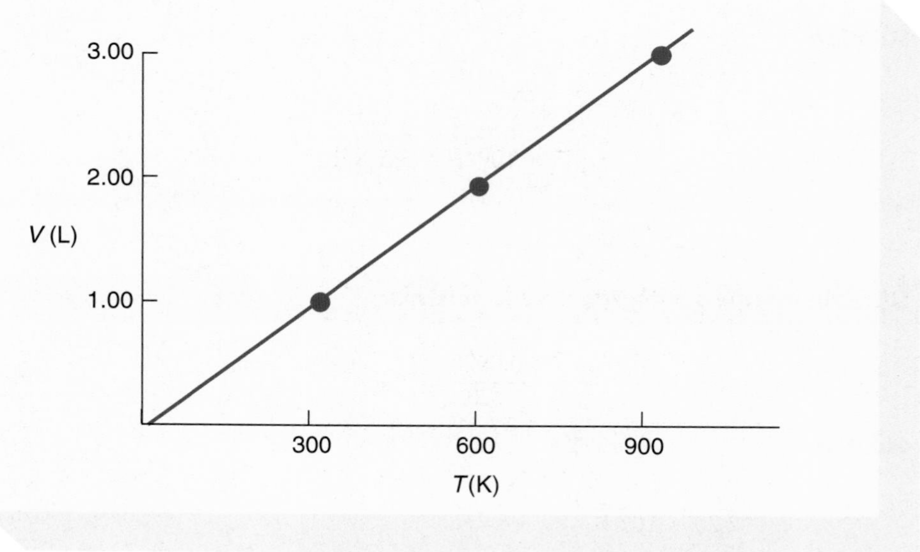

Graphs of this type allow prediction of the volume at *any* temperature within the range in which Charles's law applies.

Inverse Proportionality

Boyle's law resulted from the observation that gas volumes decrease as the pressure of the gas increases when the temperature and number of moles of the gas remain constant. The following data illustrate Boyle's law.

Experiment	P (atm)	V of Helium (L)
1	2.00	6.00
2	4.00	3.00
3	6.00	2.00

Doubling the pressure (from 2 atm to 4 atm) causes the volume of helium to decrease by a factor of one-half. Tripling the pressure (from 2 atm to 6 atm) decreases the volume to one-third of the original value (from 6 L to 2 L).

Pressure and volume are *inversely proportional*. This relationship is expressed as

$$V \propto \frac{1}{P}$$

in which again the symbol $\propto$ is short for the words *proportional to;* it reads: "Volume is inversely proportional to pressure." Use of a *proportionality constant*, k, and an *equals sign* to replace $\propto$ results in a valid mathematical equation:

$$V = \frac{k}{P}$$

The proportionality constant, k, is the Boyle's law constant. Graphically we may represent the data as

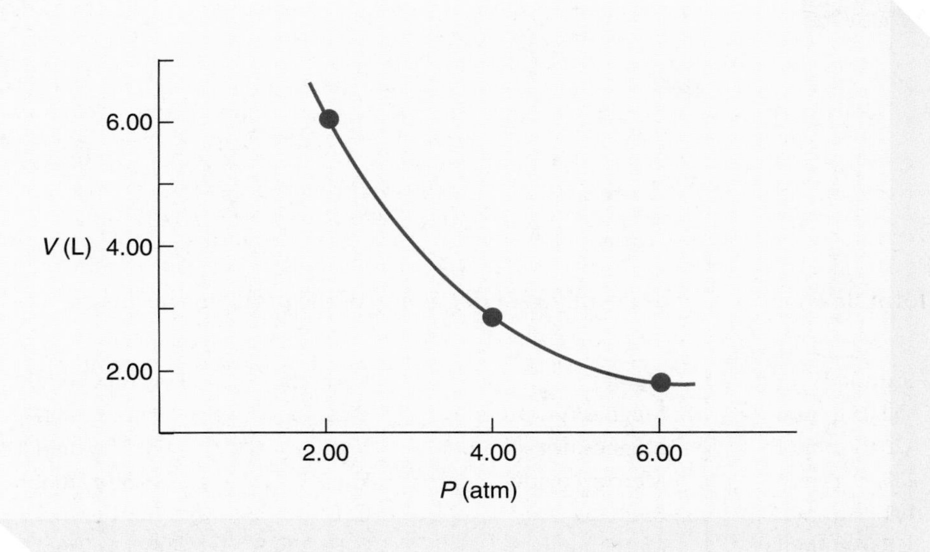

A curved relationship, such as that shown, is not ideal for predicting other pairs of values from the graph. However, regraphing the data as

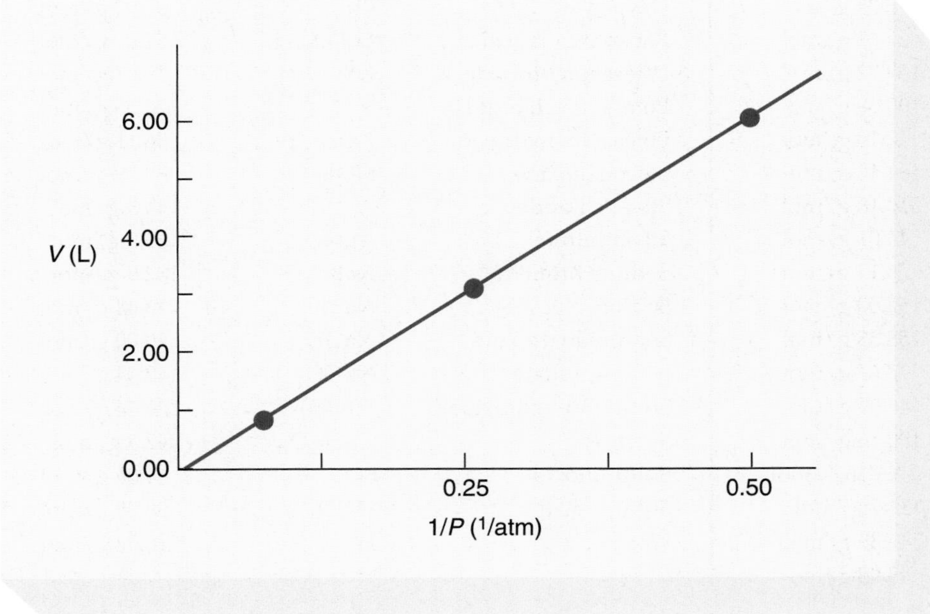

results in a linear relationship that is useful for predicting the volume at *any* pressure within the limits of Boyle's law.

Appendix B

Table of Formula Weights

The following list of names, formulas, and formula weights may be useful in solving many of the problems in Chapters 4–7.

Name	Formula	Formula Weight
Acetic acid	CH_3COOH	60.05 g/mol
Acetylene	C_2H_2	26.04 g/mol
Aluminum carbonate	$Al_2(CO_3)_3$	233.99 g/mol
Aluminum oxide	Al_2O_3	101.96 g/mol
Ammonia	NH_3	17.03 g/mol
Ammonium chloride	NH_4Cl	53.49 g/mol
Ammonium nitrate	NH_4NO_3	80.05 g/mol
Aspirin	$C_9H_8O_4$	180.15 g/mol
Barium carbonate	$BaCO_3$	197.35 g/mol
Boron oxide	B_2O_3	69.62 g/mol
Bromine	Br_2	159.82 g/mol
Calcium carbonate	$CaCO_3$	100.09 g/mol
Calcium hydride	CaH_2	42.10 g/mol
Calcium nitrate	$Ca(NO_3)_2$	164.10 g/mol
Calcium phosphate	$Ca_3(PO_4)_2$	310.18 g/mol
Carbon dioxide	CO_2	44.01 g/mol
Carbon disulfide	CS_2	76.13 g/mol
Chromium(III) oxide	Cr_2O_3	152.00 g/mol
Chromium(III) chloride	$CrCl_3$	158.35 g/mol
Diborane	B_2H_6	27.67 g/mol
Ethyl alcohol (Ethanol)	C_2H_5OH	46.07 g/mol
Glucose	$C_6H_{12}O_6$	180.16 g/mol
Hydrogen	H_2	2.016 g/mol
Iron(III) oxide	Fe_2O_3	159.7 g/mol
Lithium chloride	$LiCl$	42.39 g/mol
Lithium nitrate	$LiNO_3$	68.95 g/mol

Name	Formula	Formula Weight
Magnesium chloride	$MgCl_2$	95.21 g/mol
Magnesium sulfate	$MgSO_4$	120.37 g/mol
Mercury oxide	HgO	216.59 g/mol
Methane	CH_4	16.04 g/mol
Methionine	$C_5H_{11}NO_2S$	149.21 g/mol
Nitrogen	N_2	28.02 g/mol
Nitrous oxide	N_2O	44.02 g/mol
Octane	C_8H_{18}	114.2 g/mol
Oxygen	O_2	32.00 g/mol
Phosphorous acid	H_3PO_3	82.00 g/mol
Potassium bromide	KBr	119.01 g/mol
Potassium chloride	KCl	74.55 g/mol
Potassium hydroxide	KOH	56.11 g/mol
Silicon dioxide	SiO_2	60.09 g/mol
Silver chloride	$AgCl$	143.3 g/mol
Silver nitrate	$AgNO_3$	169.9 g/mol
Sodium bromide	$NaBr$	102.9 g/mol
Sodium chloride	$NaCl$	58.44 g/mol
Sodium hydroxide	$NaOH$	40.00 g/mol
Sodium sulfate	Na_2SO_4	142.04 g/mol
Strontium hydroxide	$Sr(OH)_2$	121.64 g/mol
Sucrose	$C_{12}H_{22}O_{11}$	342.3 g/mol
Sulfur dioxide	SO_2	64.06 g/mol
Sulfuric acid	H_2SO_4	98.08 g/mol
Water	H_2O	18.02 g/mol

Appendix C

Determination of Composition and Formulas of Compounds

C.1 Percentage Composition of Compounds

The chemical formula provides information about the composition of a compound in terms of moles. For example, 1 mol of glucose, $C_6H_{12}O_6$, contains 6 mol each of both carbon and oxygen atoms and 12 mol of hydrogen atoms.

Percentage composition of a compound, on the other hand, provides us with the composition of the compound in terms of mass. If we calculate the percentage composition of glucose, it will tell us the relative masses of C, H, and O that are present in any amount of the compound. The relative masses are expressed as percentages of the whole.

Calculating Percentage Composition

EXAMPLE C.1

Calculate the percentage composition of glucose ($C_6H_{12}O_6$).

Solution

First of all, calculate the molar mass of glucose:

$$6\,\cancel{\text{mol C}} \times \frac{12.01\text{ g C}}{1\,\cancel{\text{mol C}}} = 72.06\text{ g C}$$

$$12\,\cancel{\text{mol H}} \times \frac{1.008\text{ g H}}{1\,\cancel{\text{mol H}}} = 12.10\text{ g H}$$

$$6\,\cancel{\text{mol O}} \times \frac{16.00\text{ g O}}{1\,\cancel{\text{mol O}}} = 96.00\text{ g O}$$

and

$$72.06\text{ g} + 12.10\text{ g} + 96.00\text{ g} = 180.16\text{ g}$$

The percentage of carbon in glucose is the mass of carbon in glucose divided by the mass of the compound (in terms of grams in one mole of glucose). This quantity is multiplied by 100% to express the answer as a percentage:

$$\frac{72.06\text{ g C}}{180.16\text{ g glucose}} \times 100\% = 40.00\%\text{ C}$$

Similarly,

$$\frac{12.10\text{ g H}}{180.16\text{ g glucose}} \times 100\% = 6.72\%\text{ H}$$

and

Continued—

EXAMPLE C.1 —*Continued*

$$\frac{96.00 \text{ g O}}{180.16 \text{ g glucose}} \times 100\% = 53.29\% \text{ O}$$

The percentage composition of glucose is therefore 40.00% C, 6.72% H, and 53.29% O.

Helpful Hint: Note that the sum of the percentages to four significant figures should be 100%!

Let's now see how we can use the results of a percentage composition calculation to determine the mass of each element in any amount of the compound. Example C.2 presents such a case.

EXAMPLE C.2 ***Using Percentage Composition***

Calculate the mass of carbon, in grams, contained in 1.000 kg of glucose.

Solution

Using the information from Example C.1, convert the percentage of carbon in glucose to a decimal fraction:

40.00% C means $\frac{40.00 \text{ g C}}{100.0 \text{ g glucose}}$ which corresponds to 0.4000 g C/g glucose

Then

$$1.000 \text{ } \cancel{\text{kg glucose}} \times \frac{10^3 \text{ } \cancel{\text{g glucose}}}{1 \text{ } \cancel{\text{kg glucose}}} \times \frac{0.4000 \text{ g C}}{1 \text{ } \cancel{\text{g glucose}}} = 400.0 \text{ g C}$$

400.0 g of C is contained in 1.000 kg of glucose.

C.2 Determining Percentage Composition from Experimental Data

We have just shown that a knowledge of the chemical formula allows us to calculate the elemental percentage composition of a compound. In the laboratory it is more common to analyze a compound and determine its percentage composition by experimental means. This information is then used to calculate the simplest formula and molecular formula of the compound.

Calculation of the percentage composition of a compound is illustrated in Example C.3.

EXAMPLE C.3 ***Calculating Percentage Composition from Experimental Data***

A compound known to contain the elements carbon and hydrogen was analyzed quantitatively to determine the mass of each element present in a measured amount of the compound. A 0.800-g sample was taken for analysis and was found to contain 0.662 g C and 0.138 g H. Calculate the percentage composition.

Continued—

EXAMPLE C.3 *—Continued*

Solution

$$\% \text{ C} = \frac{\text{mass of carbon}}{\text{mass of sample}} \times 100\%$$

$$= \frac{0.662 \text{ g C}}{0.800 \text{ g sample}} \times 100\%$$

$$= 82.8\% \text{ C}$$

and

$$\% \text{ H} = \frac{\text{mass of hydrogen}}{\text{mass of sample}} \times 100\%$$

$$= \frac{0.138 \text{ g H}}{0.800 \text{ g sample}} \times 100\%$$

$$= 17.2\% \text{ H}$$

As a check of the results, note that the sum of the individual results, 82.8% and 17.2%, equals 100.0%, to four significant figures.

C.3 Determining the Simplest Formula of a Compound from Experimental Data

We now know how to calculate the percentage composition of a compound, using information from a quantitative determination of its individual elements. This information may also be used to deduce the simplest formula for the compound. Example C.4 illustrates the approach that should be used to arrive at the simplest formula.

Calculating the Simplest Formula **EXAMPLE C.4**

A compound containing only carbon and hydrogen was found (Example C.3) to have the following percentage composition:

C: 82.8% H: 17.2%

Calculate the simplest formula for this compound.

Solution

When the experimental result is given as a percentage rather than a mass, it is easiest to *assume* a 1.00×10^2-g sample and convert the percentages directly to mass, in grams.

For carbon:

$$82.8\% \text{ C} = \frac{82.8 \text{ g C}}{1.00 \times 10^2 \text{ g sample}}$$

For hydrogen:

$$17.2\% \text{ H} = \frac{17.2 \text{ g H}}{1.00 \times 10^2 \text{ g sample}}$$

Continued—

EXAMPLE C.4 —*Continued*

The number of moles of each element is then determined. For carbon:

$$82.8\ \cancel{\text{g C}} \times \frac{1\ \text{mol C}}{12.01\ \cancel{\text{g C}}} = 6.89\ \text{mol C}$$

For hydrogen:

$$17.2\ \cancel{\text{g H}} \times \frac{1\ \text{mol H}}{1.008\ \cancel{\text{g H}}} = 17.1\ \text{mol H}$$

The *ratio* of the number of moles is then calculated:

$$\frac{17.1\ \text{mol H}}{6.89\ \text{mol C}} = \frac{2.48\ \text{mol H}}{1\ \text{mol C}}$$

The simplest formula, which must be a *whole-number ratio* of atoms, 2.48:1, or $CH_{2.48}$, is inappropriate. However, multiplication of the numerator and denominator by 2 in this case results in a ratio of 5:2, or

$$C_2H_5$$

which is the simplest formula.

C.4 Calculating the Molecular Formula from the Simplest Formula

The simplest formula may also be the molecular formula, but not necessarily. The true molecular formula may be determined from the molar mass and the simplest formula. Consider Example C.5.

EXAMPLE C.5 ***Determining the Molecular Formula from the Simplest Formula***

If the molar mass of the compound in Example C.4 is 58.0 g/mol, what is the molecular formula? Recall that the simplest formula is C_2H_5.

Solution

Calculate the molar mass of the simplest formula:

2(atomic mass of carbon) + 5(atomic mass of hydrogen) =

$$2(12.0\ \text{g/mol}) + 5(1.008\ \text{g/mol}) = 29.0\ \text{g/mol}$$

Divide the molecular molar mass by the simplest molar mass to determine the number of simplest formula units contained in the molecular formula:

$$\frac{58.0\ \text{g/mol}}{29.0\ \text{g/mol}} = 2.00$$

Multiply all subscripts in the simplest formula by that ratio:

$$C_{2\times2}H_{2\times5}\ \text{or}\ C_4H_{10}$$

The molecular formula is

$$C_4H_{10}$$

which represents the compound butane, a common fuel.

Appendix D

Stereochemistry and Stereoisomers Revisited

D.1 Introduction to Isomers

Compounds with identical molecular formulas but different structures are called *isomers*. Many types of isomers exist, and several of them are discussed throughout this text. *Constitutional isomers* differ from one another in configuration; that is, they differ in terms of which atoms are bonded to one another. Constitutional or structural isomers can be interconverted only by breaking bonds within the molecule and forming new bonds. *Functional group isomers* are molecules having the same molecular formula but different functional groups. For instance, alcohols and ethers having the same number of carbon atoms are functional group isomers, such as:

$CH_3CH_2CH_2OH$	$CH_3CH_2OCH_3$
1-Propanol (C_3H_8O)	Ethylmethyl ether (C_3H_8O)

Similarly, carboxylic acids and esters having the same number of carbon atoms are also functional group isomers, as are aldehydes and ketones. *Geometric isomers,* also called *cis-trans* isomers, differ from one another in the placement of substituents on a double bond or ring.

Stereoisomers are the major focus of this appendix. By definition, stereoisomers are molecules that have the same structural formulas but differ in the arrangement of the atoms in space. Stereoisomers may be distinguished from one another by their different optical properties. They rotate plane-polarized light in different directions.

D.2 Rotation of Plane-Polarized Light

White light is a form of electromagnetic (EM) radiation and thus consists of waves in motion. In fact, white light is made up of many different wavelengths (colors) of light. The light waves vibrate in all directions, or *planes,* but are always perpendicular to the direction of the light beam (refer to Figure 17.5a and b). Special light sources, such as sodium or mercury lamps, and filters can be used to produce *monochromatic light,* light consisting of only a single wavelength.

When monochromatic light is passed through a polarizing material, such as a polaroid lens, only light waves in one plane can pass through; all others are filtered out. The light that emerges from the lens is called *plane-polarized light*. Polaroid lenses, like those found in polaroid sunglasses, consist of parallel arrays of crystals that can be imagined to look like the slats of Venetian blinds. When light interacts with this material, the emerging light beam is plane-polarized by the regular crystalline structure.

Applying these principles, scientists have developed an instrument called a *polarimeter* that is used to measure the optical activity of molecules. Specifically, the polarimeter measures the ability of a compound to change the angle of the plane of plane-polarized light (refer to Figure 17.5c).

The monochromatic light source of a polarimeter is generally a sodium lamp. The light waves are directed through a *polarizer,* and the emerging plane-polarized light passes through the sample. Finally, the light passes through an *analyzer.* If the plane of the light is not altered by the sample, the compound is *optically inactive.* However, if the plane of light is rotated either clockwise or counterclockwise, the sample is *optically active.*

The angle and direction of rotation are determined by rotating the analyzer, which is attached to a round dial graduated in degrees. First, the zero point is determined by passing light through the polarimeter without the sample present. The position that allows the maximum amount of light to pass through is the zero point. Next, the sample is placed in the polarimeter, and the analyzer is again rotated to allow the maximum amount of light to pass through. The *angle of rotation,* or *optical rotation,* is the difference between the zero point and the new angle obtained with the sample in place.

The observed angles of rotation are proportional to the number of optically active molecules in the sample that interact with light. Thus optical rotation is proportional to the concentration of the sample and to the length of the sample tube, because both affect the total number of molecules in the light path. To compare values from different laboratories that use different concentrations and apparatus, a standard reference, the specific rotation, was developed. Chemists have defined *specific rotation* [α], as the amount of rotation produced by 1.00 g of substance in 1.00 mL of solution and in a sample tube 1.00 decimeter (dm) in length. Because rotation is also a function of the temperature, the wavelength of monochromatic light, and the solvent used (if any), these experimental variables must also be reported. The following equation is used to calculate and express specific rotation:

$$[\alpha]^{t}_{\mathrm{D}} = \frac{[\alpha_{\mathrm{obs}}]}{l \times \mathrm{c}}$$

in which

$[\alpha]$ = specific rotation

$[\alpha_{\mathrm{obs}}]$ = observed rotation

l = sample tube length (dm)

c = concentration of sample (expressed as g/mL)

t = temperature (°C)

D = the most intense line of the Na spectrum (589.3 nm)

D.3 The Relationship between Molecular Structure and Optical Activity

As noted in Section 17.3 of the text, some optically active compounds rotate plane-polarized light clockwise. These are said to be *dextrorotatory* and are designated by a plus sign (+) before the specific rotation value. Substances that rotate plane-polarized light counterclockwise are called *levorotatory* and are designated by a minus sign (−) before the specific rotation value.

It was the experimental work of Louis Pasteur that first revealed a relationship between structure and optical activity. However, it was not until 1874 that the Dutch chemist van't Hoff and the French chemist LeBel independently came up with a basis for the observed optical activity: tetrahedral carbon atoms bonded to four different atoms or groups of atoms.

In Section 11.2 we saw that a carbon atom involved in four single bonds has tetrahedral geometry. If the carbon atom is bonded to two identical substituents and

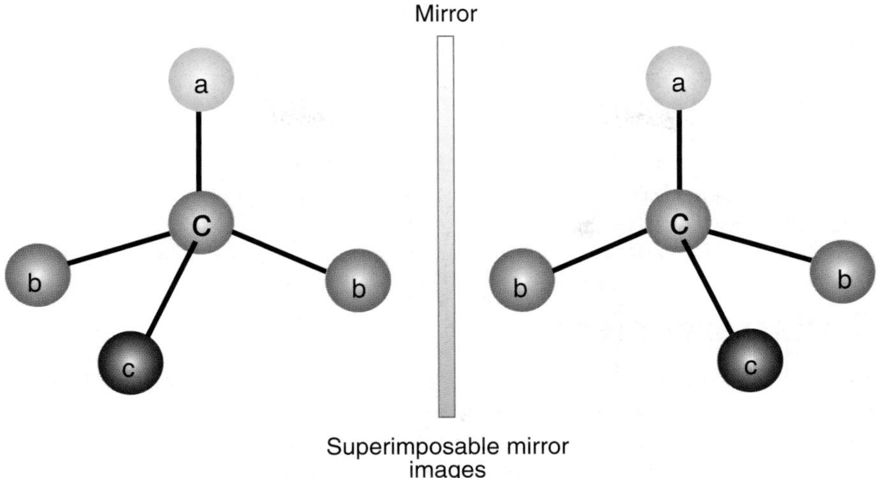

Figure D.1
A pair of superimposable mirror images. These molecules have an internal plane of symmetry that can be drawn through atoms a—C—c.

two nonidentical substituents, the resulting molecule is *symmetrical* (Figure D.1). In other words, a plane of symmetry can be drawn through this molecule. Furthermore, this molecule is superimposable on its mirror image. (Prove this to yourself by building the molecules with molecular models or toothpicks and gumdrops.)

Compare the structure in Figure D.1 with that shown in Figure 17.3 of the text. In that molecule the tetrahedral carbon is bonded to four nonidentical groups. The resulting molecule is *asymmetric*. No plane of symmetry can be drawn through the molecule, nor can the molecule be superimposed on its mirror image. (Build the molecules to demonstrate these characteristics.)

As discussed in Section 17.3, the analogy can be made between these mirror-image molecules and your left and right hands. Your hands are, indeed, mirror images of one another; you cannot draw a plane of symmetry through your hand, nor can you superimpose your left and right hands on one another.

A molecule that cannot be superimposed on its mirror image is said to be *chiral*. When a carbon atom is bonded to four different atoms or groups of atoms, it is called a *chiral carbon*. Two stereoisomers that are nonsuperimposable mirror images of one another are a pair of *enantiomers*. As mentioned in Section 17.3, the chemical and physical properties of enantiomers are identical, with the exception that they rotate plane-polarized light to the same degree but in opposite directions. This is exactly the phenomenon that Pasteur observed with the mirror-image crystals of tartaric acid salts.

Refer to Figure 17.4 in the text for the structures of the enantiomers of glyceraldehyde. Note that when you are comparing two structures to determine whether two molecules are enantiomers, you may rotate the structures as much as 180°, but you may never "flip" the structure out of the plane of the page. Always remember: If you are in doubt about the three-dimensional structure of a molecule, build it with a molecular model kit. This is particularly useful as you begin your study of organic chemistry, and it will help you in your future study of biochemistry.

D.4 Racemic Mixtures

When Louis Pasteur measured the specific rotation of the mixture of left- and right-handed tartaric acid salt crystals, he observed that it was optically inactive. The reason was that the mixture contained equal amounts of the (+) enantiomer and the (−) enantiomer. A mixture of equal amounts of a pair of enantiomers is called a *racemic mixture*, or simply a *racemate*. The prefix (±) is used to designate a racemic mixture. Consider the following situation:

50% (+) tartrate	+ 50% (−) tartrate	= (±) tartrate
$[\alpha]_D^{20} = +4.7$	$[\alpha]_D^{20} = -4.7$	$[\alpha]_D^{20} = 0$

50% (+) enantiomer + 50% (−) enantiomer = racemic mixture

In this situation the specific rotation is zero because the rotation caused by one enantiomer is canceled by the opposite rotation caused by the mirror-image enantiomer.

D.5 Diastereomers

So far, we have looked only at molecules containing a single carbon. In this case only two enantiomers are possible. However, it is quite common to find molecules with two or more chiral carbons. For a molecule of n chiral carbons the maximum possible number of different configurations is 2^n. Note that this formula predicts the *maximum* number of configurations. As we will see, there may actually be fewer.

EXAMPLE D.1 Drawing Stereoisomers for Compounds with More Than One Chiral Carbon

Draw all the possible stereoisomers of 2,3,4-trichlorobutanal.

Solution

1. There are two chiral carbons in this molecule, C-2 and C-3. Thus there are 2^2 or 4 possible stereoisomers.
2. There are two possible configurations for each of the chiral carbons (Cl on the left or on the right). Begin by drawing an isomer with both Cl atoms on the right (a). Now draw the mirror image (b). You have now generated the first pair of enantiomers, (a) and (b).

```
      CHO              CHO
       |                |
   H—C—Cl          Cl—C—H
       |                |
   H—C—Cl          Cl—C—H
       |                |
     CH2Cl            CH2Cl
      (a)              (b)
```

Enantiomers

3. Next, change the location of one of the two Cl atoms bonded to a chiral carbon to produce another possible isomer (c). Finally, draw the mirror image of (c) to produce the second set of enantiomers, (c) and (d).

```
      CHO              CHO
       |                |
   H—C—Cl          Cl—C—H
       |                |
  Cl—C—H            H—C—Cl
       |                |
     CH2Cl            CH2Cl
      (c)              (d)
```

Enantiomers

4. By this systematic procedure we have drawn the four possible isomers of 2,3,4-trichlorobutanal.

In Example D.1, structures (a) and (b) are clearly enantiomers, as are (c) and (d). But how do we describe the relationship between structure (a) and (c) or any of the pairs of stereoisomers that are *not* enantiomers? The term *diastereomers* is used to describe a pair of stereoisomers that are not enantiomers.

Although enantiomers differ from one another only in the direction of rotation of plane-polarized light, diastereomers are different in their chemical and physical properties.

D.6 Meso Compounds

As mentioned previously, the maximum number of configurations for a molecule with two chiral carbons is 2^2, or 4. However, if each of the two chiral carbons is bonded to the same four nonidentical groups, fewer than four stereoisomers exist. The example of tartaric acid, studied by Pasteur, helps to explain this phenomenon.

Drawing Stereoisomers of Compounds with More Than One Chiral Carbon — EXAMPLE D.2

Draw all the possible stereoisomers of tartaric acid, HOOC—CHOH—CHOH—COOH.

Solution

1. Proceeding as in Example D.1, you will generate the following four structures:

```
    COOH            COOH            COOH            COOH
     |               |               |               |
 H—C—OH         HO—C—H          H—C—OH         HO—C—H
     |               |               |               |
 H—C—OH         HO—C—H         HO—C—H           H—C—OH
     |               |               |               |
    COOH            COOH            COOH            COOH
    (a)             (b)             (c)             (d)

           Identical                      Enantiomers
```

2. Careful examination of pair (c) and (d) reveals that these molecules are nonsuperimposable mirror images. Thus they are enantiomers.
3. Similar inspection of structures (a) and (b) reveals that, although they are mirror images, they are identical. Structure (b) can simply be rotated 180° to produce structure (a); therefore they are identical.

Note that if you draw a line between chiral carbon-2 and chiral carbon-3 of structure (a) or (b) in Example D.2, the top half of the molecule is the mirror image of the bottom half. There is a plane of symmetry within the molecule:

```
    COOH
     |
 H—C—OH
-----+-----
 H—C—OH
     |
    COOH
```

As a result, structure (a) is optically inactive. Even though there are two chiral carbons, the rotation of plane-polarized light by chiral carbon-2 is canceled by the opposite rotation of plane-polarized light caused by chiral carbon-3. This molecule is *achiral* and is termed *meso*-tartaric acid. Any compound with an internal plane of symmetry (i.e., that can be superimposed on its mirror image) is optically inactive and is termed a *meso*-compound.

D.7 Assignment of Absolute Configuration: The (R) and (S) System

Absolute configuration is the actual arrangement of the four groups around a chiral carbon atom. The *(R) and (S) System* indicates the absolute configuration for any chiral carbon. In this system, (R) stands for a right-handed configuration (Latin, *rectus*), and (S) stands for a left-handed configuration (Latin, *sinister*).

To assign an (R) or (S) configuration to a chiral carbon, the following set of rules is used:

1. Priority rank the atoms or groups of atoms attached to the chiral carbon according to the sequence rules listed in Table D.1.
2. Draw the molecule with the lowest priority group projecting to the rear.
3. Draw a circular arrow from the group of highest priority to the group with the next highest priority.
4. If the arrow points in a clockwise direction (right), the configuration of the chiral carbon is (R); if the arrow points counterclockwise (left), it is (S).

Table D.1 Sequence Rules for Order of Priority

Rule	Example
1. For atoms, those with the highest atomic number are given the highest priority.	F < Cl < Br < I
2. If two isotopes of an element are present, the isotope of higher mass is given the higher priority.	$^{1}H < ^{2}H < ^{3}H$
3. If two atoms are identical, the atomic numbers of the next atoms are used to assign priority.	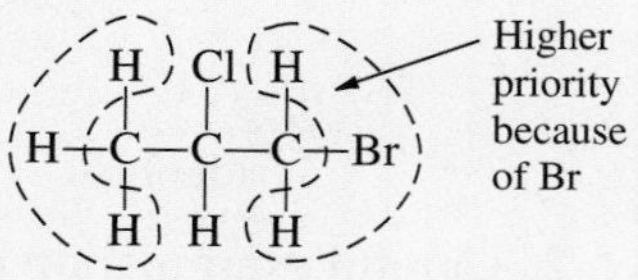 H–C(H)(H)–C(Cl)(H)–C(H)(H)–Br; Higher priority because of Br
4. Atoms attached by double or triple bonds are assigned single bond equivalences. Every double-bonded atom is duplicated, and every triple-bonded atom is triplicated.	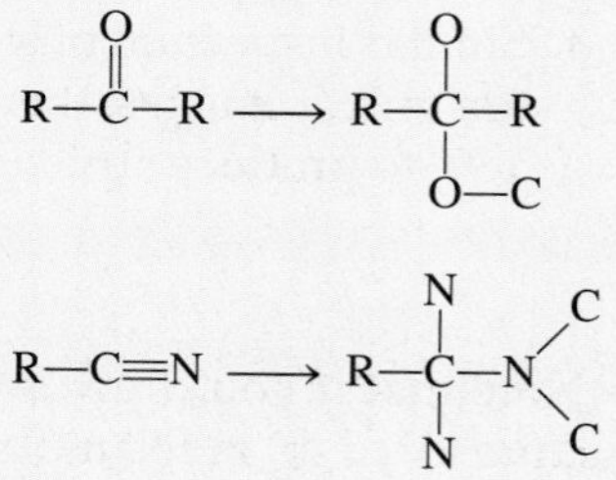 R–C(=O)–R ⟶ R–C(O)(O–C)–R R–C≡N ⟶ R–C(N)(N)–N(C)(C)

Appendix E Lipid-Soluble Vitamins

Vitamins are organic substances, required in the diet, that promote a variety of essential reactions in cells. Because they are not an energy source, they are required only in small amounts. However, if a vitamin is absent from the diet, the results are often catastrophic.

Vitamins A, D, E, and K are soluble in lipids and in biological membranes. Many of the functions of the lipid-soluble vitamins are intimately involved in metabolic processes that occur in membranes. The common sources and functions of the lipid-soluble vitamins are summarized in Table E.1.

The Food and Nutrition Board of the National Research Council of the National Academy of Sciences publishes information on the quantities of vitamins and minerals that are required in the diet. These are called *recommended dietary allowances (RDA)* and are defined as "the levels of intake of essential nutrients considered adequate to meet the known nutritional needs of practically all healthy persons." The RDA is determined by obtaining an estimate of the range of normal human needs. The value at the high end of the range is chosen, and an additional safety factor is added. Thus the RDA is by no means a minimum value, but rather a high estimate of daily requirements. It is important to remember that serious physical problems can follow ingestion of megadoses of many minerals or vitamins.

In some cases the RDA of a mineral or a vitamin cannot be determined owing to insufficient information. In those cases the Food and Nutrition Board expresses the suggested daily dose as the *estimated safe and adequate daily dietary intake (ESADDI)*.

Recently, a great deal of emphasis has been placed on vitamin supplements to combat stress, prevent the common cold, protect against various kinds of cancer and heart disease, offset the symptoms of premenstrual syndrome, delay the aging process, and improve one's sex life! Most nutritionists believe that a well-balanced diet provides all the nutrients, including the vitamins, required by the body. Indeed, when associations such as the American Cancer Society suggest that certain vitamins might help prevent cancers, they recommend that they be obtained from the natural food sources, rather than from vitamin supplements.

Table E.1 **Nutritional Sources, Functions, and Symptoms of Deficiency of the Lipid-Soluble Vitamins**

Vitamin	Source	Function	Symptoms of Deficiency
A, carotene	Egg yolk, liver, green and yellow vegetables, fruits	Synthesis of visual pigments	Night blindness and blindness in children
D_3, calciferol	Milk, action of sunlight on the skin	Regulation of calcium metabolism	Rickets (malformation of the bones)
E	Vegetable oil	Antioxidant, protection of cell membranes	Fragile red blood cells
K	Leafy vegetables, intestinal bacteria	Required for the carboxylation of prothrombin and other blood-clotting factors	Blood-clotting disorders

E.1 Vitamin A

Vitamin A is obtained in the active form called *retinol,* from animal sources such as liver and egg yolks. It is also acquired in the precursor form, *provitamin A* or *carotene,* from plant foods. Green and yellow vegetables and fruits are good sources of vitamin A. Carrots are especially rich in this vitamin.

$$H_3C\quad CH_3 \quad\ldots\quad C-(CH{=}CH-\overset{CH_3}{\overset{|}{C}}{=}CH)_2-CH{=}CH-(CH{=}\overset{CH_3}{\overset{|}{C}}-CH{=}CH)_2-C \quad\ldots\quad H_3C\quad CH_3$$

β-carotene

$$C-(CH{=}CH-\overset{CH_3}{\overset{|}{C}}{=}CH)_2-\overset{O}{\overset{\|}{C}}-H$$

Retinal

$$C-(CH{=}CH-\overset{CH_3}{\overset{|}{C}}{=}CH)_2-\underset{H}{\underset{|}{\overset{H}{\overset{|}{C}}}}-OH$$

Retinol (vitamin A)

Vitamin A helps maintain the skin and mucous membranes of the oral cavity and the digestive, respiratory, reproductive, and urinary tracts. Vitamin A is also critical for vision. The aldehyde form of Vitamin A, called *retinal,* binds to a protein called *opsin* to form the visual pigment *rhodopsin.* This pigment is found in the *rod* cells of the retina of the eye. These cells are responsible for black-and-white vision. As you might expect, a deficiency of vitamin A can have terrible consequences. In children, lack of vitamin A leads to *xerophthalmia,* an eye disease that results first in night blindness and eventually in total blindness. The disease can be prevented by an adequate dietary or supplementary supply of this vitamin. Because vitamin A is stored in the liver, a dose of 0.03 mg will protect a child for six months. Yet in countries that have suffered from cruel famines, even this amount of vitamin A is unavailable, and the burdens of malnutrition and disease lead to total blindness in thousands of children.

The current recommended dietary allowance (RDA) for vitamin A is expressed in international units (I.U.). The RDA for vitamin A is 5000 I.U., which is equal to 1 μg of retinol or 6 μg of β-carotene. Because vitamin A is a lipid-soluble vitamin that is stored in the liver, it is dangerous to ingest quantities larger than the RDA. Symptoms of vitamin A poisoning include elevated pressure of the spinal fluid and the fluid around the brain, as well as swelling around the optic nerve. These result in severe headaches. Other symptoms include anorexia, swelling of the spleen and liver, irritability, hair loss, and scaly dermatitis. It is interesting to note that early Arctic explorers suffered from vitamin A poisoning. Later it was found that this was the result of eating polar bear liver, which has an unusually high concentration of vitamin A.

E.2 Vitamin K

The formation of a blood clot in response to a wound is an intricate process that involves at least a dozen proteins in the blood serum. *Vitamin K* is involved in blood clotting.

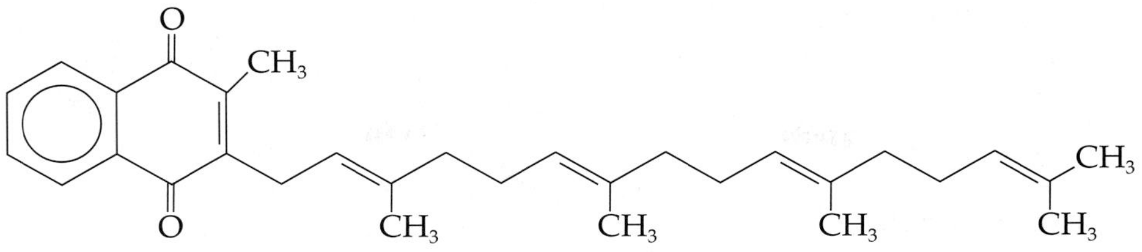

Vitamin K

In clot formation, molecules of the serum protein *prothrombin* must be activated to produce *thrombin,* which then initiates the final stages of clot formation. This requires binding of Ca^{2+} ions to the unusual amino acid γ-carboxyglutamate in prothrombin. Vitamin K is required as a coenzyme by the enzyme that adds the carboxyl groups to the normal amino acid glutamate to form γ-carboxyglutamate. Thus a deficiency of vitamin K in the diet leads to poor blood clotting.

The estimated safe and adequate daily dietary intake (ESADDI) for vitamin K is 70–140 μg/day. This is easily obtained in the diet by eating leafy vegetables. In addition, vitamin K is manufactured by our normal intestinal bacteria. It is extremely rare for adults to suffer from vitamin K deficiency, but it is observed in some individuals on antibiotic therapy or with fat absorption problems. However, newborns frequently suffer from vitamin K deficiency because they lack intestinal bacteria. They are often administered injections of vitamin K to prevent excessive bleeding in the early days of their lives.

Because vitamin K is another lipid-soluble vitamin, it is possible to suffer from hypervitaminosis K. The symptoms include gastrointestinal disturbances and anemia.

E.3 Vitamin D

Vitamin D plays a major role in the regulation of calcium levels and therefore is required for the proper formation of bone and teeth.

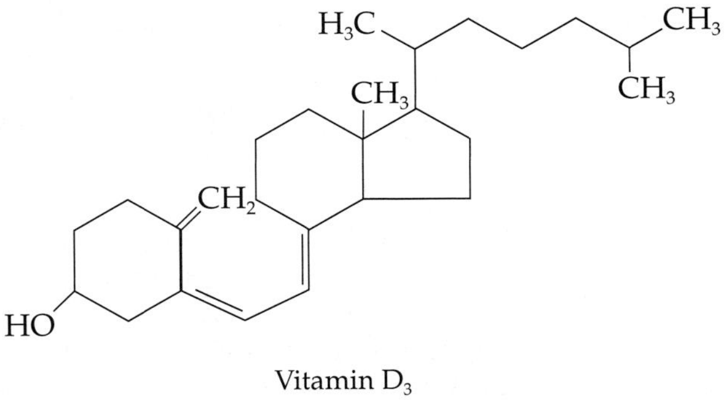

Vitamin D_3

The RDA for vitamin D is 10 μg/day for children and 5 μg/day for adults. Milk, liver, and fish oils are rich in this vitamin. It is also produced by the action of sunlight on the skin.

In children, vitamin D deficiency causes *rickets,* a disease that results in soft, deformed, and poorly calcified bones. Vitamin D deficiency is almost totally confined to children. However, the home-bound elderly who are not able to get outside and drink little milk may suffer from vitamin D deficiency.

The vitamin D produced in the skin results from the action of ultraviolet light on 7-dehydrocholesterol. This series of reactions is diagrammed in Figure E.1. Ultraviolet light causes the conversion of 7-dehydrocholesterol to vitamin D_3 (cholecalciferol). This alcohol is then hydroxylated in the liver to produce 25-hydroxyvitamin

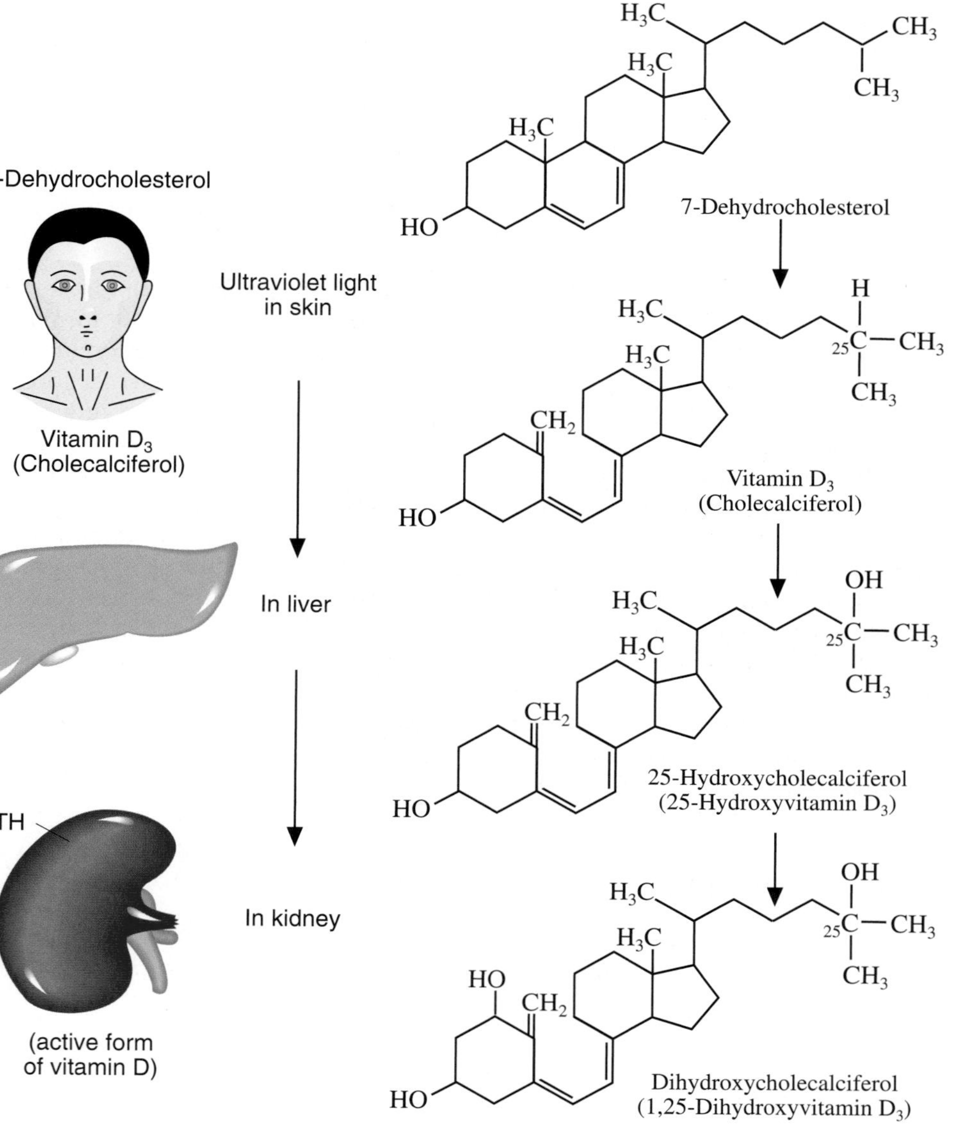

Figure E.1
The pathway of the synthesis of vitamin D_3 in the skin and its conversion to the hormone that is active in calcium metabolism.

D_3, also called *25-hydroxycholecalciferol*. In the kidney a final hydroxylation produces the hormone 1,25-dihydroxyvitamin D_3 (1,25-dihydroxycholecalciferol), the active form of the vitamin. 1,25-Dihydroxyvitamin D_3 is classified as a hormone because it is synthesized in one part of the body but exerts its effects elsewhere. Only about 1% of the body's calcium exists outside of bone, but regulating the concentration of calcium ions in the blood is critical because these soluble calcium ions are involved in many physiological processes from blood clotting to muscle contraction. When the level of calcium in the blood is low, 1,25-dihydroxyvitamin D_3 stimulates the uptake of calcium from the intestine and its transport into the blood. In the kidneys this hormone, along with parathyroid hormone, stimulates the reabsorption of calcium so that it is not lost in the urine. If the blood level of calcium is low enough, 1,25-dihydroxyvitamin D_3 even stimulates the removal of calcium from the bone. This, of course, leads to the weakening of the bone that can result in *osteomalacia*—very brittle, decalcified bones—later in life.

Ingestion of excess vitamin D can result in hypervitaminosis with the presentation of the following symptoms: renal failure, weight loss, and calcification of soft tissues of the body.

E.4 Vitamin E

Vitamin E is the least well understood of the lipid-soluble vitamins. In fact, the term vitamin E actually refers to a family of eight compounds called the *tocopherols*.

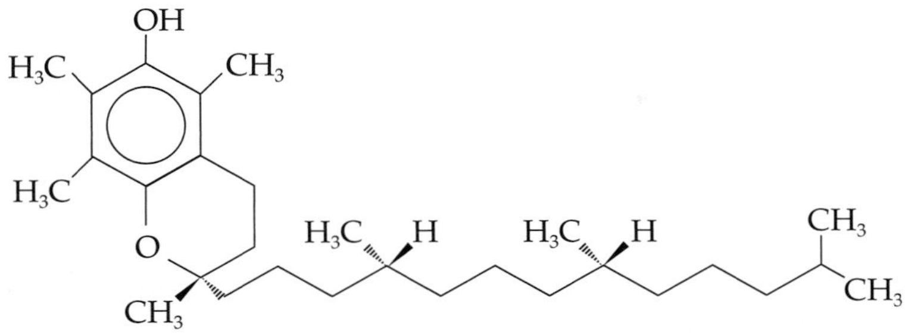

Vitamin E

Rats that are deprived of vitamin E become infertile, but the reasons for this effect are unknown. Vitamin E is known to prevent the oxidation of double bonds in the hydrocarbon tails of membrane lipids, and this may be its major function. Because oxidation reactions accelerate aging, some researchers believe that vitamin E may help to retard the aging process. The RDA for vitamin E is expressed in α-tocopherol equivalents (α-TE) because this is the most active form of vitamin E. The recommended daily intake is 10 α-TE for males and 8 α-TE for females. This is roughly the amount of vitamin E in a tablespoon of vegetable oil.

Compared to vitamins A and D, vitamin E is relatively nontoxic at high levels. However, it is unwise to drastically exceed the RDA because high levels of vitamin E may cause diarrhea, nausea, headache, and fatigue.

Appendix F Water-Soluble Vitamins

The *water-soluble vitamins* are organic substances needed in small amounts in the diet because they are required for a variety of essential enzymatic reactions in cells. The water-soluble vitamins are components of many coenzymes that are required by enzymes to carry out a variety of important biochemical reactions. Once ingested, these vitamins undergo chemical modifications that convert them into coenzymes. However, it serves no purpose to consume vast quantities of water-soluble vitamins by taking enormous doses of vitamin tablets because they are not stored in the body. Because they are soluble in water, the excess is simply excreted in the urine. Table 20.1 lists the coenzymes derived from the water-soluble vitamins and their chemical functions. Table F.1 provides the major nutritional sources of the water-soluble vitamins and the clinical conditions that result from their deficiency.

F.1 Pantothenic Acid

Pantothenic acid is essential for the normal metabolism of fats and carbohydrates. Like many other vitamins, pantothenic acid is abundant in meat, fish, poultry, whole-grain cereals, and legumes. The recommended daily allowance (RDA) of pantothenic acid is 4–7 mg per day. Pantothenic acid deficiency, which is rather rare in the United States except among alcoholics, manifests itself as gastrointestinal, neuromotor, and cardiovascular disorders. Pantothenic acid is converted to its biologically functional form, known as *coenzyme A,* in the body. Coenzyme A is

Table F.1 Major Nutritional Sources of Water-Soluble Vitamins Required by Humans and Some Physiological Effects of Deficiencies

Vitamin	Source	Symptoms of Deficiency
Thiamine (B_1)	Brain, liver, heart, whole grains	Beriberi, neuritis, mental disturbance
Riboflavin (B_2)	Milk, eggs, liver	Photophobia, dermatitis
Niacin (B_3)	Whole grains, liver	Pellagra, dermatitis, digestive problems
Pyridoxine (B_6)	Whole grains, liver, fish, kidney	Dermatitis, nervous disorders
Cobalamin (B_{12})	Liver, kidney, brain	Pernicious anemia
Folic acid	Liver, leafy vegetables, intestinal bacteria	Anemia
Pantothenic acid	Most foods	Neuromotor and cardiovascular disorders
Biotin	Egg yolk, intestinal bacteria	Scaly dermatitis, muscle pains, weakness
Ascorbic acid (C)	Citrus fruits, green leafy vegetables, tomatoes	Scurvy, failure to form collagen

important for the transfer of acyl groups in the metabolism of fatty acids and carbohydrates.

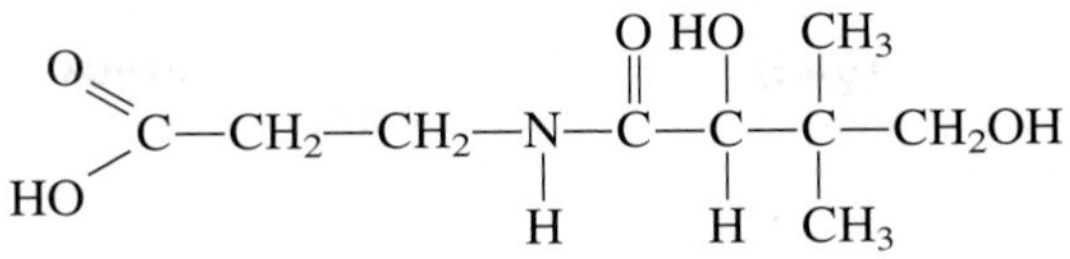

Pantothenic acid

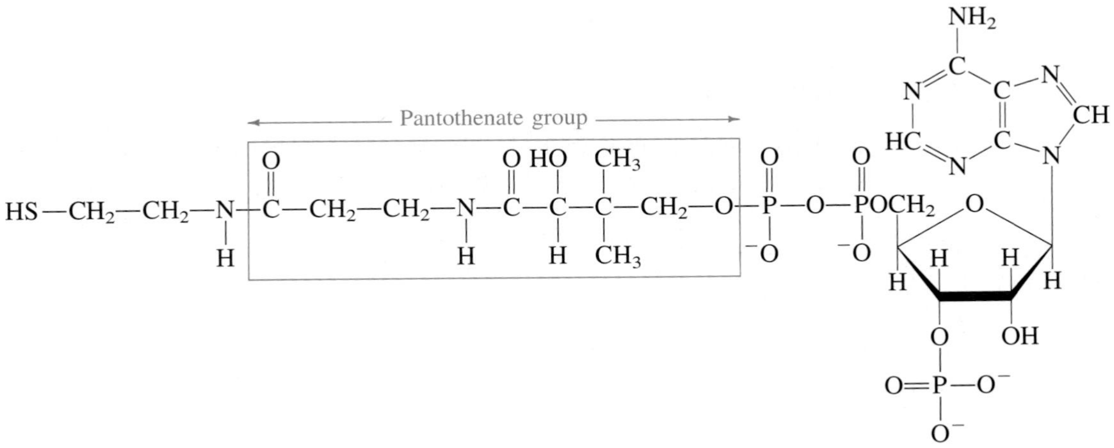

Pantothenate group

Coenzyme A
(CoA)

F.2 Niacin

Niacin (vitamin B_3) refers to both nicotinic acid and nicotinamide. Nicotinamide is an essential precursor for the coenzyme nicotinamide adenine dinucleotide (NAD^+).

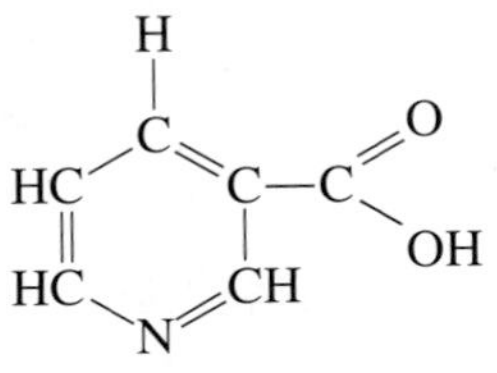

Nicotinic acid

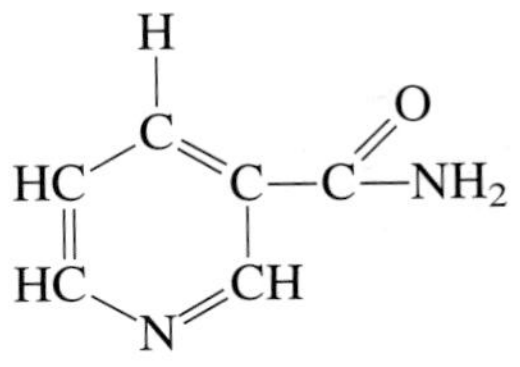

Nicotinamide

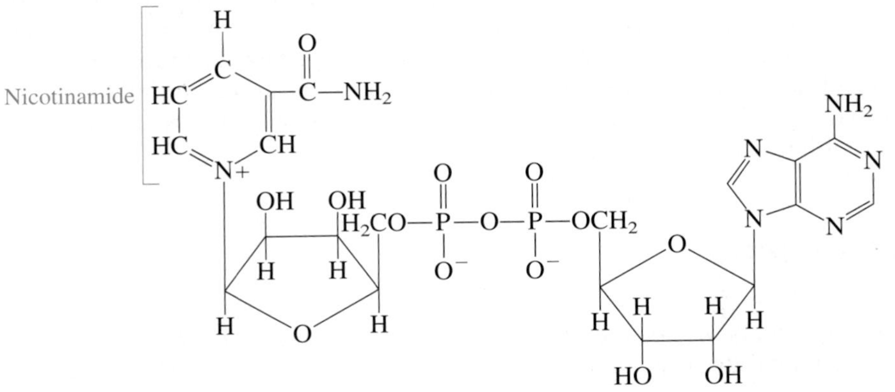

Nicotinamide

NAD^+ or nicotinamide adenine dinucleotide

Niacin is found in fish, lean meat, legumes, milk, and whole-grain and enriched cereals. The RDA for niacin is 20 mg per day. Niacin deficiency leads to dermatitis, diarrhea, dementia, and death. The most common illness that develops from niacin deficiency is *pellagra,* a form of dermatitis. This nutritional disease is found where corn is abundant in the diet and meat is scarce.

Corn actually contains a rather large amount of niacin, but it is present in a form that is not made available to the body simply by cooking this grain. American and South American Indians discovered centuries ago that soaking cornmeal in lime water (dilute calcium hydroxide) proved to be beneficial. The lime water releases nicotinamide in a form that can then be absorbed through the intestine.

F.3 Riboflavin

Riboflavin, or *vitamin* B_2, is abundant in milk, eggs, and dark green leafy vegetables.

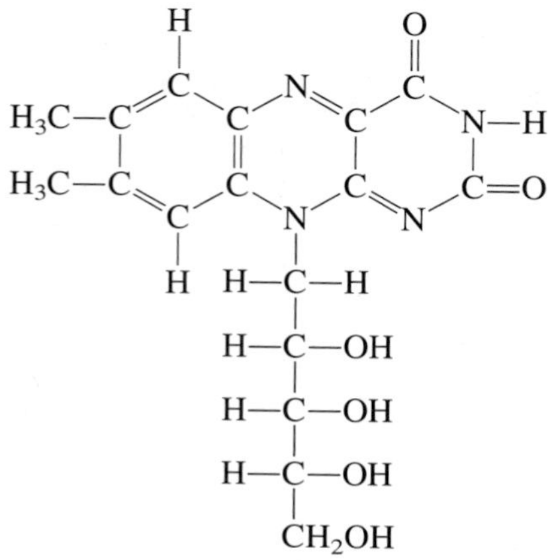

Riboflavin
(Vitamin B_2)

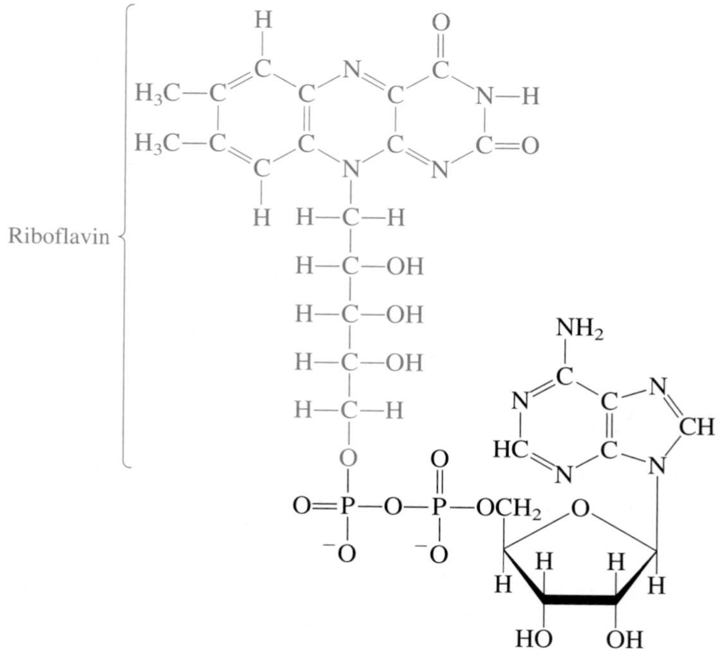

Riboflavin

Flavin adenine dinucleotide
(FAD)

As a component of the coenzyme flavin adenine dinucleotide (FAD), riboflavin is essential for the energy-releasing reactions of the cell. The RDA for riboflavin is about 1.7 mg. Severe riboflavin deficiency is rare in most parts of the world, but a marginal deficiency of this vitamin is common even in the United States. Mild riboflavin deficiency leads to dry, cracked lips and other mild forms of dermatitis. In severe cases, however, riboflavin deficiency leads to extreme sensitivity to sunlight and retarded growth in children.

F.4 Thiamine

Thiamine, also known as *vitamin* B_1, is required in the diet of all animals. Its biologically active form is the coenzyme thiamine pyrophosphate. This coenzyme is required for many decarboxylation reactions, including the decarboxylation of pyruvate to form CO_2 and acetyl coenzyme A in the transition reaction between glycolysis and the citric acid cycle.

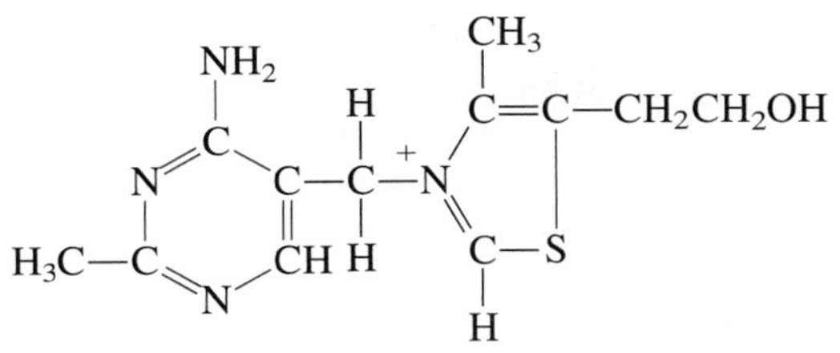

Thiamine (vitamin B_1)

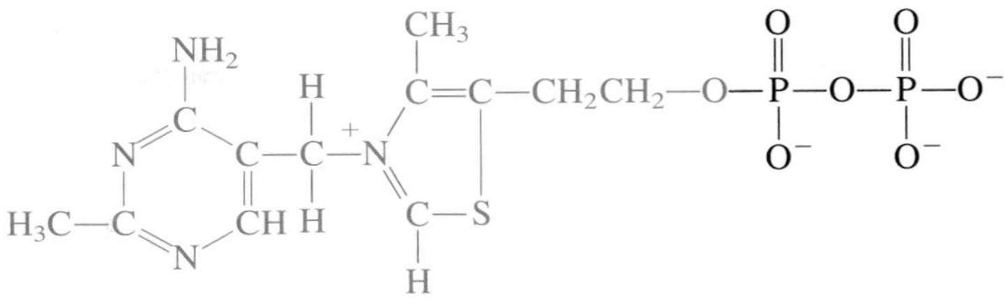

Thiamine pyrophosphate (TPP)

Thiamine is abundant in whole-grain and enriched cereals, meats, legumes, and green leafy vegetables. The RDA for thiamine is about 1.5 mg. Thiamine is lost from whole grains during the refining process. However, thiamine deficiency is largely prevented because many foods, including bread and cereal products, contain thiamine as an additive.

Dietary deficiency leads to *beriberi,* a disease characterized by muscle weakness and mental instability. Sudden recovery from the effects of beriberi is observed within hours of administration of thiamine.

F.5 Pyridoxine

Pyridoxine, also known as *vitamin B_6,* is required for the synthesis and breakdown of amino acids.

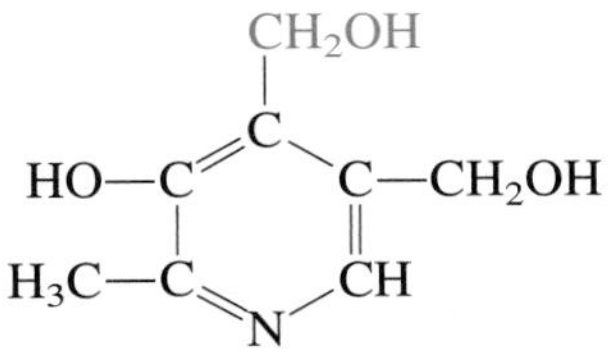

Pyridoxine
Vitamin B_6

It is found in many foods, such as fish, meat, poultry, and leafy green vegetables, which are excellent sources of vitamin B_6. Because this vitamin is so readily available in a variety of foods, its deficiency is relatively rare. When it does occur, the symptoms include nervousness and muscular weakness. The RDA for vitamin B_6 is 2.0 mg.

Because vitamin B_6 is a water-soluble vitamin, we would expect that the ingestion of excessive amounts would result simply in the excretion of the excess. However, excess vitamin B_6 (50–100 times the RDA) taken to reduce the symptoms of premenstrual syndrome has resulted in peripheral neuropathy in several young women. This is characterized by a numbness in the limbs and a clumsy, stumbling walk.

F.6 Folic Acid

Folic acid is a complicated molecule whose structure consists of three components: a heterocyclic ring system known as *pterin, p-aminobenzoic acid,* and the amino acid *glutamic acid.*

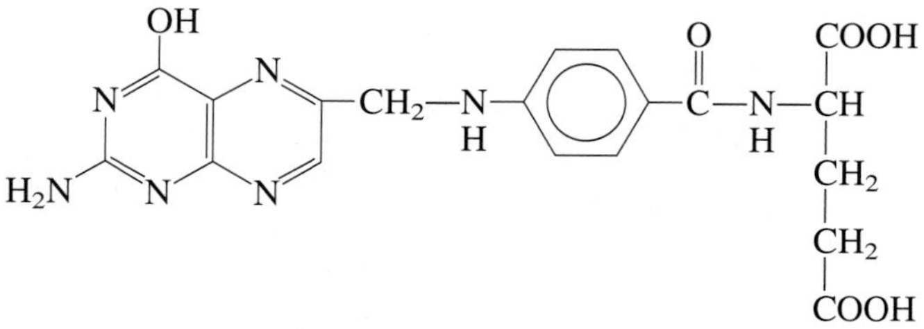

Folic acid

It is required for the synthesis of the amino acid methionine and the nucleic acid precursors: the purines and pyrimidines. The RDA for folic acid is only about 0.4 mg. Because such a small amount is required daily, it might be thought that folic acid deficiency would be rare. The opposite is true: Folic acid deficiency is a very common vitamin deficiency. Green vegetables, whole-grain cereals, and meat contain abundant folic acid, but it is destroyed by cooking. Because a deficiency of folic acid results in anemia and growth failure, folic acid is especially necessary for children and pregnant women.

F.7 Biotin

Biotin, sometimes called vitamin H, is involved in carboxylation and decarboxylation reactions in the metabolism of fats, carbohydrates, and proteins. Liver, egg yolks, cheese, and peanuts are excellent sources of biotin. In addition, it is produced by bacteria in the intestine.

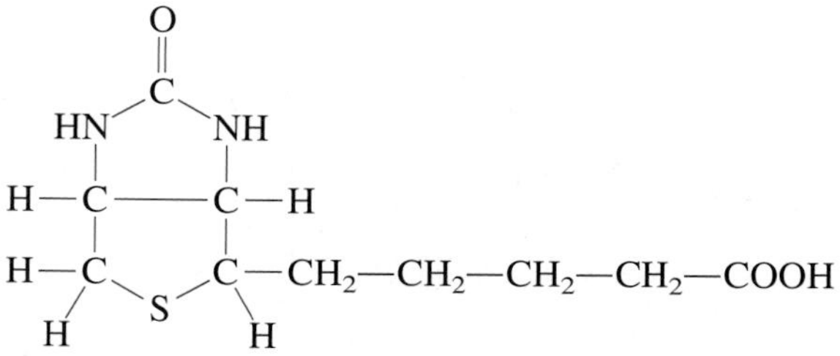

Biotin

The estimated safe and adequate daily dietary intake (ESADDI) for biotin is 0.30 mg, and in a normal diet, biotin deficiency is almost unknown. However, when it does occur, the symptoms include dermatitis (scaling and hardening of the skin), loss of appetite and nausea, muscle pain, and elevated levels of blood cholesterol.

F.8 Vitamin B_{12}

Vitamin B_{12} has an extraordinary chemical structure.

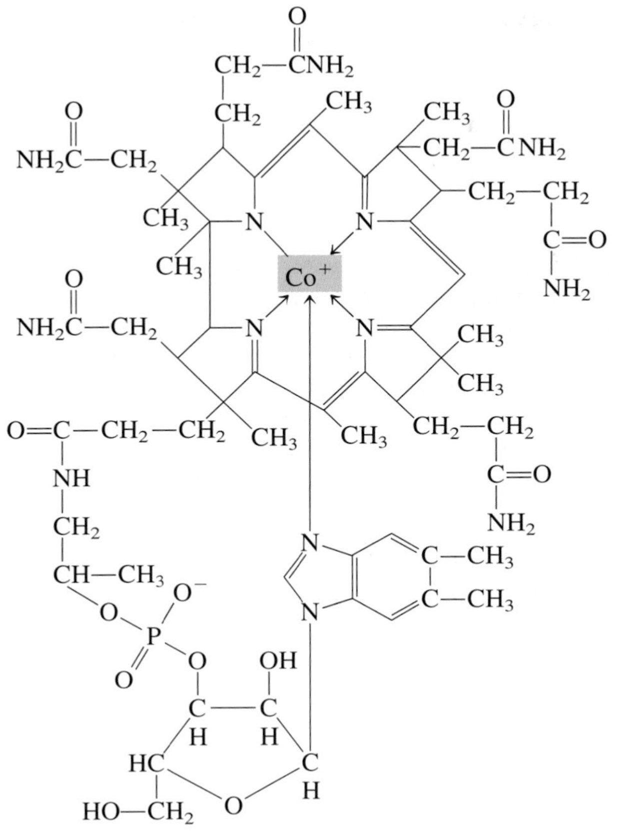

Vitamin B_{12} (cobalamin)

Vitamin B_{12} is a very important vitamin that is needed for the production of red and white blood cells and the normal growth and maintenance of nerve tissue.

A defective mechanism for the uptake of vitamin B_{12} results in *pernicious anemia,* a disease that is characterized by the presence of large, immature red blood cells in the blood. Symptoms include a sore tongue, weight loss, and mental and nervous disorders. The damage to the central nervous system can even cause demyelination of the peripheral nerves in the arms and legs. Eventually, this can progress to the spinal cord. The requirement for vitamin B_{12} is only about 6 μg per day. Because many foods, including meats, eggs, and dairy products, contain this vitamin, nearly all diets, except those that are completely devoid of animal products, provide a sufficient amount of vitamin B_{12}. In fact, bacteria in the human intestine produce enough vitamin B_{12} to satisfy the normal daily requirement.

F.9 Vitamin C

Vitamin C is important in the growth and repair of connective tissue, teeth, bones, and cartilage. In addition, it promotes wound healing, enhances absorption of iron, and functions in the biosynthesis of several hormones. Vitamin C also serves as an antioxidant in many biological processes. It is almost a part of folk medicine that large doses of vitamin C, or *ascorbic acid,* can prevent, or cure, the common cold and a host of other ailments.

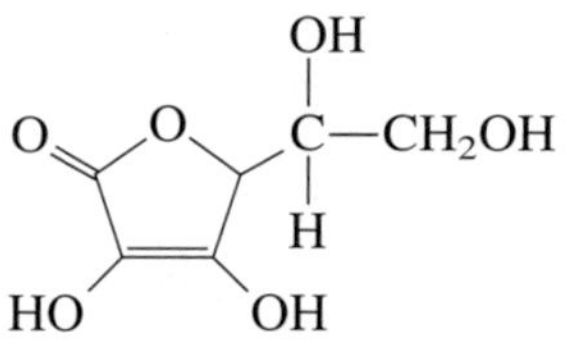

Vitamin C
(ascorbic acid)

The many claims made for the powers of vitamin C have not, however, been substantiated in extensive clinical testing. Although many individuals recommend megadoses of vitamin C, the RDA is only 60 mg. In fact, the ingestion of large doses, more than 1–2 g daily, has been reported to cause intestinal cramps, nausea, diarrhea, and kidney stones. Fresh fruits, especially citrus fruits, and vegetables, among them potatoes, are rich dietary sources of vitamin C.

A deficiency of vitamin C leads to *scurvy,* a disorder that is characterized by bleeding gums, loss of teeth, sore joints, and slow wound healing.

Appendix G

Energy Yields from Aerobic Respiration: Some Alternatives

In Chapter 21 we described catabolic processes as occurring in three stages. In stage I, dietary protein, carbohydrate, and lipid are hydrolyzed into small subunits that can cross the membranes of the cells of the intestine and are transported to the cells of the body. In stage II, these monomers enter cells of the body and are converted into a form that can be completely oxidized. For carbohydrates, glucose is used as a substrate for the glycolysis pathway, the first stage of carbohydrate metabolism. In this pathway, glucose is converted into two pyruvate molecules. In the process, two ATP, net, are produced by substrate level phosphorylation and two NADH are formed by oxidation of glyceraldehyde. Under aerobic conditions, that is, when oxygen is present, the pyruvate is transported into organelles called *mitochondria.* Here the pyruvate dehydrogenase complex is involved in the reaction

$$\text{pyruvate} + \text{coenzyme A} + \text{NAD}^+ \longrightarrow \text{acetyl CoA} + \text{CO}_2 + \text{NADH}$$

In stage III, the two-carbon acetyl group is completely oxidized in the reactions of the citric acid cycle.

When glycolysis occurs under anaerobic conditions, it is followed by fermentation reactions, such as the lactate and alcohol fermentations. These reactions reduce pyruvate—or a molecule produced from pyruvate—and simultaneously oxidize the NADH produced in glycolysis. As a result, the net energy yield from glycolysis under anaerobic conditions is only two ATP. No further ATP energy is harvested from the oxidation of the NADH. It is simply reoxidized in the fermentation reactions.

Under aerobic conditions, the energy yield of glycolysis is much greater because the high-energy electrons carried by NADH are shuttled into mitochondria and used in *oxidative phosphorylation* to produce more ATP.

There are two shuttle systems: the *glycerol-3-phosphate shuttle* found in skeletal muscle and nerve cells and the *oxaloacetate-malate shuttle* found in heart and liver cells. Because skeletal muscle produces the majority of the ATP for the body, it is the glycerol-3-phosphate shuttle that is used most commonly when discussing metabolic energy yields. In Example 22.1, calculation of the ATP harvest of glycolysis is based on this shuttle.

Let's consider the reactions involved in the glycerol-3-phosphate shuttle. In this shuttle, the NADH produced in the cytoplasm is oxidized in a reaction that reduces dihydroxyacetone phosphate to glycerol-3-phosphate.

$$\begin{array}{c} CH_2OH \\ | \\ C{=}O \\ | \\ CH_2OPO_3^{2-} \end{array} + \text{NADH} \longrightarrow \begin{array}{c} CH_2OH \\ | \\ CHOH \\ | \\ CH_2OPO_3^{2-} \end{array} + \text{NAD}^+$$

Dihydroxyacetone phosphate → Glycerol-3-phosphate

The glycerol-3-phosphate then passes through the outer mitochondrial membrane and is oxidized to dihydroxyacetone phosphate by the enzyme glycerol-3-phosphate dehydrogenase. This enzyme, which is located in the inner mitochondrial membrane, simultaneously reduces FAD to $FADH_2$. The electrons from each $FADH_2$ are then used during oxidative phosphorylation to produce two ATP. The dihydroxyacetone phosphate returns to the cytoplasm to continue the shuttle process.

The main drawback of the glycerol-3-phosphate shuttle is that only two ATP are produced for each cytoplasmic NADH. The reason is that the electrons are shuttled to $FADH_2$, which yields only two ATP by oxidative phosphorylation. (The energy yield of the oxidation of mitochondrial NADH is three ATP.) Thus the total energy yield from glycolysis under aerobic conditions in muscle and nerve cells is two ATP, produced by substrate level phosphorylation, plus four ATP (two ATP per NADH), produced by oxidative phosphorylation. This provides an energy yield of six ATP per glucose.

The oxaloacetate-malate shuttle system is more efficient. In this system, cytoplasmic NADH reduces oxaloacetate to malate.

$$\begin{array}{c} COO^- \\ | \\ C{=}O \\ | \\ CH_2 \\ | \\ COO^- \end{array} + NADH \longrightarrow \begin{array}{c} COO^- \\ | \\ CHOH \\ | \\ CH_2 \\ | \\ COO^- \end{array} + NAD^+$$

Oxaloacetate Malate

Malate is then transported into the mitochondrion where it is reoxidized to oxaloacetate. Mitochondrial NAD^+ is reduced in the process. These electrons are then used in oxidative phosphorylation to produce three ATP per NADH. Thus the energy yield of glycolysis in heart and liver cells is two ATP, produced by substrate level phosphorylation, plus six ATP (three ATP per NADH), produced by oxidative phosphorylation. This gives an energy yield of eight ATP per glucose.

The oxaloacetate cannot cross the mitochondrial membrane to return to the cytoplasm to continue the cycle. It is able to return only after a series of reactions involving the amino acids glutamate and aspartate.

In summary, the energy yield of glycolysis depends on the conditions present (aerobic *versus* anaerobic) and the type of cell. The following table summarizes the energy gains from glycolysis under various conditions.

Condition	ATP by Substrate Level Phosphorylation	ATP by Oxidative Phosphorylation
Anaerobic	2	0
Aerobic, muscle (glycerol-3-phosphate shuttle)	2	4
Aerobic, heart (oxaloacetate-malate shuttle)	2	6

Appendix H Minerals and Cellular Function

Many minerals are required in the human diet, and they may be divided into two nutritional classes. The *major minerals* must be consumed in amounts greater than 100 mg/day. The *trace minerals* are required in much smaller amounts (less than 100 mg/day). In some cases the required levels are so small that they cannot be accurately measured.

H.1 The Major Minerals

Calcium and *phosphorus* are major minerals that are needed for the development of healthy bones and teeth. These two minerals are found in a crystalline calcium phosphate mineral known as *hydroxyapatite*, $[Ca_{10}(PO_4)_6(OH)_2]$, that makes up the mineral matrix of bone and teeth. In addition, calcium is required for normal blood clotting and muscle function. The RDA for calcium is 1200 mg/day for adults between nineteen and twenty-four years of age and 800 mg/day for adults over age twenty-five. Milk, cheese, canned salmon, and dark green leafy vegetables are all rich sources of dietary calcium.

Phosphorus is required not only as a component of hydroxyapatite in bone, but also as a component of nucleic acids and many other biologically important molecules. Without phosphorus we would have no energy-storage molecules, such as ATP and creatine phosphate, for the energy derived from glycolysis and the citric acid cycle. The RDA for phosphorus is the same as that for calcium. Because it is abundant in most foods, a deficiency of phosphorus in the presence of an otherwise adequate diet is virtually impossible.

Sodium, potassium, and *chloride* ions are all required in the human diet. When dissolved in water, sodium and potassium are positively charged ions (cations), and chloride is a negatively charged ion (anion). These three minerals are called *blood electrolytes* because the ions can conduct electrical currents. Sodium is found primarily in the extracellular fluids, and potassium is found predominantly within the cell. Both of these elements are needed to maintain a proper fluid balance inside and outside of the cell. Because these three minerals are found in most foods, deficiency is rare.

The Food and Nutrition Board has removed the three electrolytes from its table of estimated safe and adequate daily dietary intake because sufficient information is not available to establish a recommended amount. The major dietary source of sodium and chloride is table salt (40% sodium and 60% chloride). Physicians still recommend that the intake of sodium be restricted to 1–2 g daily. The recommended intake of chloride is approximately 1.7–5.1 g daily. However, getting enough sodium and chloride is not a problem. In fact, sodium intake in the United States is about 5–7 g/day, far in excess of the 1–2 g/day required by a normal adult.

Potassium is the major intracellular cation. It is found in citrus fruits, bananas, and tomatoes. Dietary intake of potassium is about 1.9–5.6 g/day in the United States. Potassium deficiency is rare, but loss of potassium in severe diarrhea, such as can occur in cholera, and the excretion of potassium by a person suffering from diabetes mellitus can lead to a debilitating deficiency. However, potassium deficiency is seen most commonly in individuals who are taking diuretics.

A high intake of table salt, sodium chloride, the major source of sodium in the diet, is one factor that may cause high blood pressure, *hypertension*, in susceptible individuals. There has been considerable emphasis on "low-salt" diets as a means of avoiding hypertension. However, it appears that sodium is not the only culprit. It is the sodium ion-to-potassium ion ratio that appears to be important in controlling blood pressure. Ideally, the Na^+/K^+ ratio should be about 0.6, but the Na^+/K^+ ratio consumed by the average American is greater than 1.0. To avoid hypertension in later life, it is important to reduce the amount of sodium in the diet *and* increase the amount of potassium.

Magnesium ions are vital to cellular metabolism. They are required for the reactions in the liver that convert glycogen to glucose. They are also important in normal muscle function, nerve conductance, and bone development. Mg^{2+} binds to AMP, ADP, and ATP and to nucleic acids. Many enzymes that are involved in the catabolic breakdown of glucose require magnesium ions as cofactors. A typical adult contains about 25 g of magnesium, and the recommended daily intake is about 300 mg/day. Magnesium is plentiful in leafy green vegetables, legumes, cereal grains, and lean meats.

H.2 Trace Minerals

Iron is a required mineral for heme-containing proteins and is an element that is absolutely essential for normal physiological functioning. It is found in the oxygen transport and storage proteins, hemoglobin and myoglobin, and is also a component of the cytochromes that participate in the respiratory electron transport chain. The requirement for iron is so well known that it might be thought that no one would suffer the effects of iron deficiency. In fact, however, iron deficiency is rather common in the United States, especially among women. Iron can be absorbed by the body only in its ferrous, Fe^{2+}, oxidation state. The iron in meat is absorbed more efficiently than that from most other foods. Vegetarians, whose protein intake is mostly in the form of cereal grains, run a risk of iron deficiency because iron in grains is absorbed poorly by the body.

Hemoglobin, myoglobin, and the cytochromes of the respiratory electron transport chain all contain heme. Heme, of course, contains iron, and it is this need that must be satisfied by the diet. Deficiency of iron leads to *iron-deficiency anemia*, a condition in which the amount of hemoglobin in red blood cells is abnormally low.

Copper is a mineral that is required for many essential enzymes. The respiratory electron transport chain contains an enzyme, *cytochrome oxidase*, that contains both heme groups and copper ions. Copper is therefore required in the diet for the function of this essential enzyme. Copper is also required by some of the enzymes that are responsible for the synthesis of connective tissue proteins. Seafood, vegetables, nuts, and meats such as liver are excellent sources of copper ions. The ESADDI for adults is 1.5–3.0 mg. Copper ions in high concentrations are toxic. In fact, mental retardation and death in early adolescence result from an inability to remove excess copper ions from the body. As everywhere in life, the balance of the system is critical to its function.

Iodine is a mineral that is required for the proper function of the thyroid gland. The thyroid gland extracts iodine from nutrients and incorporates it into various hormones. The once-common condition of goiter, an enlargement of the thyroid gland, is an abnormality that results from an effort to compensate for low iodine intake. Goiter can be prevented if iodine is included in the diet. Seafood is one of the best sources of iodine. In areas where seafood is not available, dietary iodine is easily obtained in the form of iodized salt, found in most grocery stores.

Fluoride aids in the prevention of dental caries (cavities). The presence of fluoride ions in the water supplies of many cities has dramatically reduced the incidence of cavities, the most widespread "disease" in the United States. Unfortunately, resistance to water fluoridation, largely as a result of insufficient information, has caused many municipalities to abandon this practice. An excess of fluoride is, in fact, toxic, but at the level found in fluoridated water supplies, approximately 1 part per million, no toxic effects are observed. Fluoride works by displacing hydroxide in calcium hydroxyapatite to give a crystalline mineral in teeth known as *fluorapatite*, $[Ca_3(PO_4)_2 \cdot CaF_2]$, that is far more resistant to the acid produced by oral bacteria than is hydroxyapatite itself.

Many other *trace minerals* are required in the diet. Among them are zinc, nickel, vanadium, tin, silicon, molybdenum, chromium, selenium, and cobalt. Zinc and molybdenum are required by various enzymes, and cobalt is a component of vitamin B_{12}.

Deficiencies of trace minerals are virtually nonexistent, because they are needed in such small quantities that requirements for them are likely to be met in nearly every diet. As in the case of copper, however, most trace minerals are extremely toxic if ingested in large quantities, and "heavy metal poisoning" has been a scourge of industrial cities throughout the world.

Glossary

A

absolute specificity (20.5) the property of an enzyme that allows it to bind and catalyze the reaction of only one substrate

acetal (14.4) the family of organic compounds formed via the reaction of two molecules of alcohol with an aldehyde in the presence of an acid catalyst; acetals have the following general structure:

$$R^1\text{—}\overset{\displaystyle OR^2}{\underset{\displaystyle H}{C}}\text{—}OR^3$$

acetyl coenzyme A (acetyl CoA) (15.4, 22.2) a molecule composed of coenzyme A and an acetyl group; the intermediate that provides acetyl groups for complete oxidation by aerobic respiration

acid (9.1) a substance that behaves as a proton donor

acid anhydride (15.3) the product formed by the combination of an acid chloride and a carboxylate ion; structurally they are two carboxylic acids with a water molecule removed:

$$\text{(Ar) }R\text{—}\overset{\displaystyle O}{\overset{\|}{C}}\text{—}O\text{—}\overset{\displaystyle O}{\overset{\|}{C}}\text{—}R\text{ (Ar)}$$

acid-base reaction (7.2, 9.1) reaction that involves the transfer of a hydrogen ion (H^+) from one reactant to another

acid chloride (15.3) member of the family of organic compounds with the general formula

$$\text{(Ar) }R\text{—}\overset{\displaystyle O}{\overset{\|}{C}}\text{—}Cl$$

actinide series (3.1) the fourteen elements from thorium (Th) through lawrencium (Lr)

activated complex (8.3) the arrangement of atoms at the top of the potential energy barrier as a reaction proceeds

activation energy (8.3) the threshold energy that must be overcome to produce a chemical reaction

active site (20.4) the cleft in the surface of an enzyme that is the site of substrate binding

active transport (18.6) the movement of molecules across a membrane against a concentration gradient

acyl carrier protein (ACP) (23.4) the protein that forms a thioester linkage with fatty acids during fatty acid synthesis

acyl group (15: Intro, 16.3) the functional group found in carboxylic acid derivatives that contains the carbonyl group attached to one alkyl or aryl group:

$$\text{(Ar) }R\text{—}\overset{\displaystyle O}{\overset{\|}{C}}\text{—}$$

addition polymer (12.5) polymers prepared by the sequential addition of a monomer

addition reaction (12.5, 14.4) a reaction in which two molecules add together to form a new molecule; often involves the addition of one molecule to a double or triple bond in an unsaturated molecule; e.g., the addition of alcohol to an aldehyde or ketone to form a hemiacetal or hemiketal

adenosine triphosphate (ATP) (15.4, 21.1) a nucleotide composed of the purine adenine, the sugar ribose, and three phosphoryl groups; the primary energy storage and transport molecule used by the cells in cellular metabolism

adipocyte (23.1) a fat cell

adipose tissue (23.1) fatty tissue that stores most of the body lipids

aerobic respiration (22.3) the oxygen-requiring degradation of food molecules and production of ATP

alcohol (13.1) an organic compound that contains a hydroxyl group (—OH) attached to an alkyl group

aldehyde (14.1) a class of organic molecules characterized by a carbonyl group; the carbonyl carbon is bonded to a hydrogen atom and to another hydrogen or an alkyl or aryl group. Aldehydes have the following general structure:

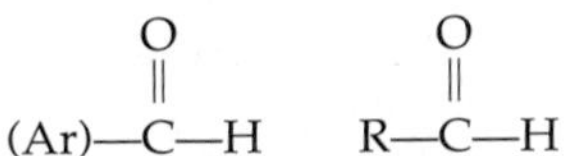

aldol condensation (14.4) a reaction in which aldehydes or ketones react to form a larger molecule

aldose (17.2) a sugar that contains an aldehyde (carbonyl) group

aliphatic hydrocarbon (11.1) any member of the alkanes, alkenes, and alkynes or the substituted alkanes, alkenes, and alkynes

alkali metal (3.1) an element within Group IA (1) of the periodic table

alkaline earth metal (3.1) an element within Group IIA (2) of the periodic table

alkaloid (16.2) a class of naturally occurring compounds that contain one or more nitrogen heterocyclic rings; many of the alkaloids have medicinal and other physiological effects

alkane (11.2) a hydrocarbon that contains only carbon and hydrogen and is bonded together through carbon-hydrogen and carbon-carbon single bonds; a saturated hydrocarbon with the general molecular formula C_nH_{2n+2}

alkene (12.1) a hydrocarbon that contains one or more carbon-carbon double bonds; an unsaturated hydrocarbon with the general formula C_nH_{2n}

alkyl group (11.2) a hydrocarbon group that results from the removal of one hydrogen from the original hydrocarbon (e.g., methyl, CH_3—; ethyl, CH_3CH_2—)

alkyl halide (11.5) a substituted hydrocarbon with the general structure R—X, in which R— represents any alkyl group and X = a halogen (F—, Cl—, Br—, or I—)

alkylammonium ion (16.1) the ion formed when the lone pair of electrons of the nitrogen atom of an amine is shared with a proton (H^+) from a water molecule

alkyne (12.1) a hydrocarbon that contains one or more carbon-carbon triple bonds; an unsaturated hydrocarbon with the general formula C_nH_{2n-2}

allosteric enzyme (20.9) an enzyme that has an effector binding site and an active site; effector binding changes the shape of the active site, rendering it either active or inactive

alpha particle (10.1) a particle consisting of two protons and two neutrons; the alpha particle is identical to a helium nucleus

amide bond (16.3) the bond between the carbonyl carbon of a carboxylic acid and the amino nitrogen of an amine

amides (16.3) the family of organic compounds formed by the reaction between a carboxylic acid derivative and an amine and characterized by the amide group

amines (16.1) the family of organic molecules with the general formula RNH_2, R_2NH, or R_3N (R— can represent either an alkyl or aryl group); they may be viewed as substituted ammonia molecules in which one or more of the ammonia hydrogens has been substituted by a more complex organic group

α-amino acid (19.2) the subunits of proteins composed of an α-carbon bonded to a carboxylate group, a protonated amino group, a hydrogen atom, and a variable R group

aminoacyl group (16.4) the functional group that is characteristic of an amino acid; the aminoacyl group has the following general structure:

```
      H  O
      |  ||
 +
H3N—C—C—
      |
      R
```

aminoacyl tRNA (24.6) the transfer RNA covalently linked to the correct amino acid

aminoacyl tRNA binding site of ribosome (A-site) (24.6) a pocket on the surface of a ribosome that holds the aminoacyl tRNA during translation

aminoacyl tRNA synthetase (24.6) an enzyme that recognizes one tRNA and covalently links the appropriate amino acid to it

amorphous solid (6.3) a solid with no organized, regular structure

amphibolic pathway (22.9) a metabolic pathway that functions in both anabolism and catabolism

amphiprotic substance (9.1) a substance that can behave either as a Brønsted acid or a Brønsted base

amylopectin (17.6) a highly branched form of amylose; the branches are attached to the C-6 hydroxyl by α(1 → 6) glycosidic linkage; a component of starch

amylose (17.6) a linear polymer of α-D-glucose molecules bonded in α(1 → 4) glycosidic linkage that is a major component of starch; a polysaccharide storage form

anabolism (21.1, 22.9) all of the cellular energy-requiring biosynthetic pathways

anaerobic threshold (21.4) the point at which the level of lactate in the exercising muscle inhibits glycolysis and the muscle, deprived of energy, ceases to function

analgesic (16.2) any drug that acts as a painkiller, e.g., aspirin, acetaminophen

anaplerotic reaction (22.9) a reaction that replenishes a substrate needed for a biochemical pathway

anesthetic (16.2) a drug that causes a lack of sensation in part of the body (local anesthetic) or causes unconsciousness (general anesthetic)

angular structure (4.4) a planar molecule with bond angles other than 180°

anion (2.2, 3.3) a negatively charged atom or group of atoms

anode (2.3, 9.5) the positively charged electrode in an electrical cell

anomers (17.4) isomers of cyclic monosaccharides that differ from one another in the arrangement of bonds around the hemiacetal carbon

antibodies (19.1) immunoglobulins; specific glycoproteins produced by cells of the immune system in response to invasion by infectious agents

anticodon (24.4) a sequence of three ribonucleotides on a tRNA that are complementary to a codon on the mRNA; codon-anticodon binding results in delivery of the correct amino acid to the site of protein synthesis

antigen (19.1) any substance that is able to stimulate the immune system; generally a protein or large carbohydrate

antiparallel (24.2) a term describing the polarities of the two strands of the DNA double helix; on one strand the sugar-phosphate backbone advances in the 5′ → 3′ direction; on the opposite, complementary strand the sugar-phosphate backbone advances in the 3′ → 5′ direction

apoenzyme (20.7) the protein portion of an enzyme that requires a cofactor to function in catalysis

aqueous solution (7.3) any solution in which the solvent is water

arachidonic acid (18.2) a fatty acid derived from linoleic acid; the precursor of the prostaglandins

aromatic hydrocarbon (11.1, 12.6) an organic compound that contains the benzene ring or a derivative of the benzene ring

Arrhenius theory (9.1) a theory that describes an acid as a substance that dissociates to produce H^+ and a base as a substance that dissociates to produce OH^-

artificial radioactivity (10.6) radiation that results from the conversion of a stable nucleus to another, unstable nucleus

asymmetric carbon (17.3) a chiral carbon; a carbon bonded to four different groups

atherosclerosis (18.4) deposition of excess plasma cholesterol and other lipids and proteins on the walls of arteries, resulting in decreased artery diameter and increased blood pressure

atom (2.2) the smallest unit of an element that retains the properties of that element

atomic mass (2.2) the mass of an atom expressed in atomic mass units

atomic mass unit (5.1) 1/12 of the mass of a ^{12}C atom, equivalent to 1.661×10^{-24} g

atomic number (2.2) the number of protons in the nucleus of an atom; it is a characteristic identifier of an element

atomic orbital (2.6) a specific region of space where an electron may be found

ATP synthase (22.6) a multiprotein complex within the inner mitochondrial membrane that uses the energy of the proton (H^+) gradient to produce ATP

autoionization (9.1) also known as *self-ionization,* the reaction of a substance, such as water, with itself to produce a positive and a negative ion

Avogadro's law (6.1) a law that states that the volume is directly proportional to the number of moles of gas particles, assuming that the pressure and temperature are constant

Avogadro's number (5.1) 6.022×10^{23} particles of matter contained in 1 mol of a substance

axial atom (11.4) an atom that lies above or below a cycloalkane ring

B

background radiation (10.7) the radiation that emanates from natural sources

barometer (6.1) a device for measuring pressure

base (9.1) a substance that behaves as a proton acceptor

base pair (24.2) a hydrogen-bonded pair of bases within the DNA double helix; the standard base pairs always involve a purine and a pyrimidine; in particular, adenine always base pairs with thymine and cytosine with guanine

Benedict's reagent (17.4) a buffered solution of Cu^{2+} ions that can be used to test for reducing sugars or to distinguish between aldehydes and ketones

Benedict's test (14.4) a test used to determine the presence of reducing sugars or to distinguish between aldehydes and ketones; it requires a buffered solution of Cu^{2+} ions that are reduced to Cu^+, which precipitates as brick-red Cu_2O

beta particle (10.1) an electron formed in the nucleus by the conversion of a neutron into a proton

bile (23.1) micelles of lecithin, cholesterol, bile salts, protein, inorganic ions, and bile pigments that aid in lipid digestion by emulsifying fat droplets

binding energy (10.3) the energy required to break down the nucleus into its component parts

bioinformatics (24.10) an interdisciplinary field that uses computer information sciences and DNA technology to devise methods for understanding, analyzing, and applying DNA sequence information

boat conformation (11.4) a form of a six-member cycloalkane that resembles a rowboat. It is less stable than the chair conformation because the hydrogen atoms are not perfectly staggered

boiling point (4.3) the temperature at which the vapor pressure of a liquid is equal to the atmospheric pressure

bond energy (4.4) the amount of energy necessary to break a chemical bond

Boyle's law (6.1) a law stating that the volume of a gas varies inversely with the pressure exerted if the temperature and number of moles of gas are constant

breeder reactor (10.4) a nuclear reactor that produces its own fuel in the process of providing electrical energy

Brønsted-Lowry theory (9.1) a theory that describes an acid as a proton donor and a base as a proton acceptor

buffer capacity (9.4) a measure of the ability of a solution to resist large changes in pH when a strong acid or strong base is added

buffer solution (9.4) a solution containing a weak acid or base and its salt (the conjugate base or acid) that is resistant to large changes in pH upon addition of strong acids or bases

buret (9.3) a device calibrated to deliver accurately known volumes of liquid, as in a titration

C

C-terminal amino acid (19.3) the amino acid in a peptide that has a free α-CO_2^- group; the last amino acid in a peptide

calorimetry (8.2) the measurement of heat energy changes during a chemical reaction

cap structure (24.4) a 7-methylguanosine unit covalently bonded to the 5′ end of a mRNA by a 5′–5′ triphosphate bridge

carbinol carbon (13.4) that carbon in an alcohol to which the hydroxyl group is attached

carbohydrate (17.1) generally sugars and polymers of sugars; the primary source of energy for the cell

carbonyl group (14: Intro) the functional group that contains a carbon-oxygen double bond: —C=O; the functional group found in aldehydes and ketones

carboxyl group (15.1) the —COOH functional group; the functional group found in carboxylic acids

carboxylic acid (15.1) a member of the family of organic compounds that contain the —COOH functional group

carboxylic acid derivative (15.2) any of several families of organic compounds, including the esters and amides, that are derived from carboxylic acids and have the general formula

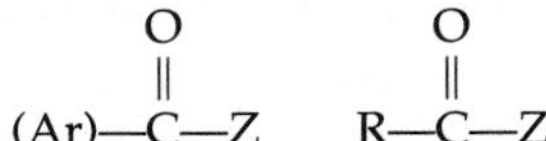

Z = —OR or OAr for the esters, and Z = —NH_2 for the amides

carcinogen (24.7) any chemical or physical agent that causes mutations in the DNA that lead to uncontrolled cell growth or cancer

catabolism (21.1, 22.9) the degradation of fuel molecules and production of ATP for cellular functions

catalyst (8.3) any substance that increases the rate of a chemical reaction (by lowering the activation energy of the reaction) and that is not destroyed in the course of the reaction

cathode (2.3, 9.5) the negatively charged electrode in an electrical cell

cathode rays (2.3) a stream of electrons that is given off by the cathode (negative electrode) in a cathode ray tube

cation (2.2, 3.3) a positively charged atom or group of atoms

cellulose (17.6) a polymer of β-D-glucose linked by β(1 → 4) glycosidic bonds

central dogma (24.4) a statement of the directional transfer of the genetic information in cells: DNA → RNA → Protein

chain reaction (10.4) the reaction in a fission reactor that involves neutron production and causes subsequent reactions accompanied by the production of more neutrons in a continuing process

chair conformation (11.4) the most energetically favorable conformation for a six-member cycloalkane; so-called for its resemblance to a lawn chair

Charles's law (6.1) a law stating that the volume of a gas is directly proportional to the temperature of the gas, assuming that the pressure and number of moles of the gas are constant

chemical bond (4.1) the attractive force holding two atomic nuclei together in a chemical compound

chemical equation (5.4) a record of chemical change, showing the conversion of reactants to products

chemical formula (5.2) the representation of a compound or ion in which elemental symbols represent types of atoms and subscripts show the relative numbers of atoms

chemical property (1.2) characteristics of a substance that relate to the substance's participation in a chemical reaction

chemical reaction (1.2) a process in which atoms are rearranged to produce new combinations

chemistry (1.1) the study of matter and the changes that matter undergoes

chiral molecule (17.3) molecule capable of existing in mirror-image forms

cholesterol (18.4) a twenty-seven-carbon steroid ring structure that serves as the precursor of the steroid hormones

chromosome (24.2) a piece of DNA that carries all the genetic instructions, or genes, of an organism

chylomicron (18.5, 23.1) a plasma lipoprotein (aggregate of protein and triglycerides) that carries triglycerides from the intestine to all body tissues via the bloodstream

***cis-trans* isomers** (11.3) isomers that differ from one another in the placement of substituents on a double bond or ring

citric acid cycle (22.4) a cyclic biochemical pathway that is the final stage of degradation of carbohydrates, fats, and amino acids. It results in the complete oxidation of acetyl groups derived from these dietary fuels

cloning vector (24.8) a DNA molecule that can carry a cloned DNA fragment into a cell and that has a replication origin that allows the DNA to be replicated abundantly within the host cell

coagulation (19.10) the process by which proteins in solution are denatured and aggregate with one another to produce a solid

codon (24.4) a group of three ribonucleotides on the mRNA that specifies the addition of a specific amino acid onto the growing peptide chain

coenzyme (20.7) an organic group required by some enzymes; it generally serves as a donor or acceptor of electrons or a functional group in a reaction

coenzyme A (22.2) a molecule derived from ATP and the vitamin pantothenic acid; coenzyme A functions in the transfer of acetyl groups in lipid and carbohydrate metabolism

cofactor (20.7) an inorganic group, usually a metal ion, that must be bound to an apoenzyme to maintain the correct configuration of the active site

colipase (23.1) a protein that aids in lipid digestion by binding to the surface of lipid droplets and facilitating binding of pancreatic lipase

colligative property (7.6) property of a solution that is dependent only on the concentration of solute particles

colloidal suspension (7.3) a heterogeneous mixture of solute particles in a solvent; distribution of solute particles is not uniform because of the size of the particles

combination reaction (7.1) a reaction in which two substances join to form another substance

combined gas law (6.1) an equation that describes the behavior of a gas when volume, pressure, and temperature may change simultaneously

combustion (11.5) the oxidation of hydrocarbons by burning in the presence of air to produce carbon dioxide and water

competitive inhibitor (20.10) a structural analog; a molecule that has a structure very similar to the natural substrate of an enzyme, competes with the natural substrate for binding to the enzyme active site, and inhibits the reaction

complementary strands (24.2) the opposite strands of the double helix are hydrogen-bonded to one another such that adenine and thymine or guanine and cytosine are always paired

complete protein (19.11) a protein source that contains all the essential and nonessential amino acids

complex lipid (18.5) a lipid bonded to other types of molecules

compound (1.2) a substance that is characterized by constant composition and that can be chemically broken down into elements

concentration (1.5, 7.4) a measure of the quantity of a substance contained in a specified volume of solution

condensation (6.2) the conversion of a gas to a liquid

condensation polymer (15.2) a polymer, which is a large molecule formed by combination of many small molecules (monomers) in a repeating pattern, that resulted from joining of monomers in a reaction that forms a small molecule, such as water or an alcohol

condensed formula (11.2) a structural formula showing all of the atoms in a molecule and placing them in a sequential arrangement that details which atoms are bonded to each other; the bonds themselves are not shown

conformations, conformers (11.4) discrete, distinct isomeric structures that may be converted, one to the other, by rotation about the bonds in the molecule

conjugate acid (9.1) substance that has one more proton than the base from which it is derived

conjugate acid-base pair (9.1) two species related to each other through the gain or loss of a proton

conjugate base (9.1) substance that has one less proton than the acid from which it is derived

constitutional isomers (11.2) two molecules having the same molecular formulas, but different chemical structures

conversion factor (1.3) an equivalence statement or multiplier consisting of a ratio of two equivalent quantities in different units, used to convert a quantity from one unit to another

Cori Cycle (21.6) a metabolic pathway in which the lactate produced by working muscle is taken up by cells in the liver and converted back to glucose by gluconeogenesis

corrosion (9.5) the unwanted oxidation of a metal

covalent bond (4.1) a pair of electrons shared between two atoms

covalent solid (6.3) a collection of atoms held together by covalent bonds

crenation (7.6) the shrinkage of red blood cells caused by water loss to the surrounding medium

cristae (22.1) the folds of the inner membrane of the mitochondria

crystal lattice (4.2) a unit of a solid characterized by a regular arrangement of components

crystalline solid (6.3) a solid having a regular repeating atomic structure

curie (10.8) the quantity of radioactive material that produces 3.7×10^{10} nuclear disintegrations per second

cycloalkane (11.3) a cyclic alkane; a saturated hydrocarbon that has the general formula C_nH_{2n}

D

Dalton's law (6.1) also called the law of partial pressures; states that the total pressure exerted by a gas mixture is the sum of the partial pressures of the component gases

data (1.3) a group of facts resulting from an experiment

decomposition reaction (7.1) the breakdown of a substance into two or more substances

degenerate code (24.5) a term used to describe the fact that several triplet codons may be used to specify a single amino acid in the genetic code

dehydration (of alcohols) (13.5) a reaction that involves the loss of a water molecule, in this case the loss of water from an alcohol and the simultaneous formation of an alkene

deletion mutation (24.7) a mutation that results in the loss of one or more nucleotides from a DNA sequence

denaturation (19.10) the process by which the organized structure of a protein is disrupted, resulting in a completely disorganized, nonfunctional form of the protein

density (1.5) mass per unit volume of a substance

deoxyribonucleic acid (DNA) (24.1) the nucleic acid molecule that carries all of the genetic information of an organism; the DNA molecule is a double helix composed of two strands, each of which is composed of phosphate groups, deoxyribose, and the nitrogenous bases thymine, cytosine, adenine, and guanine

deoxyribonucleotide (24.1) a nucleoside phosphate or nucleotide composed of a nitrogenous base in β-*N*-glycosidic linkage to the 1′ carbon of the sugar 2′-deoxyribose and with one, two, or three phosphoryl groups esterified at the hydroxyl of the 5′ carbon

diabetes mellitus (23.3) a disease caused by the production of insufficient levels of insulin and characterized by the appearance of very high levels of glucose in the blood and urine

dialysis (7.8) the removal of waste material via transport across a membrane

diglyceride (18.3) the product of esterification of glycerol at two positions

dipole-dipole interactions (6.2) attractive forces between polar molecules

disaccharide (17.1) a sugar composed of two monosaccharides joined through an oxygen atom bridge

dissociation (4.3) production of positive and negative ions when an ionic compound dissolves in water

disulfide (13.9) an organic compound that contains a disulfide group (—S—S—)

DNA polymerase (24.3) the enzyme that catalyzes the polymerization of daughter DNA strands using the parental strand as a template

double bond (4.4) a bond in which two pairs of electrons are shared by two atoms

double helix (24.2) the spiral staircase-like structure of the DNA molecule characterized by two sugar-phosphate backbones wound around the outside and nitrogenous bases extending into the center

double-replacement reaction (7.1) a chemical change in which cations and anions "exchange partners"

dynamic equilibrium (8.4) the state that exists when the rate of change in the concentration of products and reactants is equal, resulting in no net concentration change

E

eicosanoid (18.2) any of the derivatives of twenty-carbon fatty acids, including the prostaglandins, leukotrienes, and thromboxanes

electrolysis (9.5) an electrochemical process that uses electrical energy to cause nonspontaneous oxidation-reduction reactions to occur

electrolyte (4.3, 7.3) a material that dissolves in water to produce a solution that conducts an electrical current

electrolytic solution (4.3) a solution composed of an electrolytic solute dissolved in water

electromagnetic radiation (2.4) energy that is propagated as waves at the speed of light

electromagnetic spectrum (2.4) the complete range of electromagnetic waves

electron (2.2) a negatively charged particle outside of the nucleus of an atom

electron affinity (3.4) the energy released when an electron is added to an isolated atom

electron configuration (3.2) the arrangement of electrons around a nucleus of an atom, ion, or a collection of nuclei of a molecule

electron density (2.6) the probability of finding the electron in a particular location

electron transport system (22.6) the series of electron transport proteins embedded in the inner mitochondrial membrane that accept high-energy electrons from NADH and $FADH_2$ and transfer them in stepwise fashion to molecular oxygen (O_2)

electronegativity (4.1) a measure of the tendency of an atom in a molecule to attract shared electrons

electronic transitions (2.5) involves the movement of an electron from one energy level to another within an atom

element (1.2) a substance that cannot be decomposed into simpler substances by chemical or physical means

elimination reaction (13.5) a reaction in which a molecule loses atoms or ions from its structure

elongation factor (24.6) proteins that facilitate the elongation phase of translation

emulsifying agent (18.3) a bipolar molecule that aids in the suspension of fats in water

enantiomers (17.3) stereoisomers that are nonsuperimposable mirror images of one another

endothermic reaction (8.1) a chemical or physical change in which energy is absorbed

energy (1.1) the capacity to do work

energy level (2.5) one of numerous atomic regions where electrons may be found

English System (1.3) a collection of units developed haphazardly in response to the growth of commerce and construction over time

enthalpy (8.1) a term that represents heat energy

entropy (8.1) a measure of randomness or disorder

enzyme (19.1, 20: Intro) a protein that serves as a biological catalyst

enzyme specificity (20.5) the ability of an enzyme to bind to only one, or a very few, substrates and thus catalyze only a single reaction

enzyme-substrate complex (20.4) a molecular aggregate formed when the substrate binds to the active site of the enzyme

equatorial atom (11.4) an atom that lies in the plane of a cycloalkane ring

equilibrium constant (8.4) a numerical quantity that summarizes the relationship between the concentration of reactants and products in an equilibrium reaction

equilibrium reaction (8.4) a reaction that is reversible and the rates of the forward and reverse reactions are equal

equivalence point (9.3) the situation in which reactants have been mixed in the molar ratio corresponding to the balanced equation

equivalent (7.5) the number of grams of an ion corresponding to Avogadro's number of electrical charges

essential amino acid (19.11) an amino acid that cannot be synthesized by the body and must therefore be supplied by the diet

essential fatty acids (18.2) the fatty acids linolenic and linoleic acids that must be supplied in the diet because they cannot be synthesized by the body

ester (15.2) a carboxylic acid derivative formed by the reaction of a carboxylic acid and an alcohol. Esters have the following general formula:

$$\mathrm{R{-}\overset{\displaystyle O}{\overset{\|}{C}}{-}OR} \quad \mathrm{R{-}\overset{\displaystyle O}{\overset{\|}{C}}{-}O(Ar)} \quad \mathrm{(Ar){-}\overset{\displaystyle O}{\overset{\|}{C}}{-}O(Ar)}$$

esterification (18.2) the formation of an ester in the reaction of a carboxylic acid and an alcohol

ether (13.8) an organic compound that contains two alkyl and/or aryl groups attached to an oxygen atom; R—O—R, Ar—O—R, and Ar—O—Ar

eukaryote (24.2) an organism having cells containing a true nucleus enclosed by a nuclear membrane and having a variety of membrane-bound organelles that segregate different cellular functions into different compartments

evaporation (6.2) the conversion of a liquid to a gas below the boiling point of the liquid

excited state (2.5) a condition in which an atom has one or more of its electrons in an energy state higher than the ground state

exon (24.4) protein-coding sequences of a gene found on the final mature mRNA

exothermic reaction (8.1) a chemical or physical change that releases energy

extensive property (1.2) a property of a substance that depends on the quantity of the substance

F

F_0F_1 complex (22.6) an alternative term for the ATP synthase, the multiprotein complex in the inner mitochondrial membrane that uses the energy of the proton gradient to produce ATP

facilitated diffusion (18.6) movement of a solute across a membrane from an area of high concentration to an area of low concentration through a transmembrane protein, or permease

family (3.1) any of the eighteen vertical columns of elements in the periodic table; also called a group

fatty acid (15.1, 18.2) any member of the family of continuous-chain carboxylic acids that generally contain four to twenty carbon atoms; the most concentrated source of energy used by the cell

feedback inhibition (20.9) the process whereby excess product of a biosynthetic pathway turns off the entire pathway for its own synthesis

fermentation (13.3, 21.4) anaerobic (in the absence of oxygen) catabolic reactions that occur with no net oxidation. Pyruvate or an organic compound produced from pyruvate is reduced as NADH is oxidized

fibrous protein (19.5) a protein composed of peptides arranged in long sheets or fibers

Fischer Projection (17.3) a two-dimensional drawing of a molecule, which shows a chiral carbon at the intersection of two lines and horizontal lines representing bonds projecting out of the page and vertical lines representing bonds that project into the page

fission (10.4) the splitting of heavy nuclei into lighter nuclei accompanied by the release of large quantities of energy

fluid mosaic model (18.6) the model of membrane structure that describes the fluid nature of the lipid bilayer and the presence of numerous proteins embedded within the membrane

formula (4.2) the representation of the fundamental compound unit using chemical symbols and numerical subscripts

formula unit (5.2) the smallest collection of atoms from which the formula of a compound can be established

formula weight (5.3) the mass of a formula unit of a compound relative to a standard (carbon-12)

free energy (8.1) the combined contribution of entropy and enthalpy for a chemical reaction

fructose (17.4) a ketohexose that is also called levulose and fruit sugar; the sweetest of all sugars, abundant in honey and fruits

fuel value (8.2) the amount of energy derived from a given mass of material

functional group (11.1) an atom (or group of atoms and their bonds) that imparts specific chemical and physical properties to a molecule

fusion (10.4) the joining of light nuclei to form heavier nuclei, accompanied by the release of large amounts of energy

G

galactose (17.4) an aldohexose that is a component of lactose (milk sugar)

galactosemia (17.5) a human genetic disease caused by the inability to convert galactose to a phosphorylated form of glucose (glucose-1-phosphate) that can be used in cellular metabolic reactions

gamma ray (10.1) a high-energy emission from nuclear processes, traveling at the speed of light; the high-energy region of the electromagnetic spectrum

gaseous state (1.2) a physical state of matter characterized by a lack of fixed shape or volume and ease of compressibility

genome (24.2) the complete set of genetic information in all the chromosomes of an organism

geometric isomer (11.3, 12.3) an isomer that differs from another isomer in the placement of substituents on a double bond or a ring

globular protein (19.6) a protein composed of polypeptide chains that are tightly folded into a compact spherical shape

glucagon (21.7, 23.6) a peptide hormone synthesized by the α-cells of the islets of Langerhans in the pancreas and secreted in response to low blood glucose levels; glucagon promotes glycogenolysis and gluconeogenesis and thereby increases the concentration of blood glucose

gluconeogenesis (21.6) the synthesis of glucose from noncarbohydrate precursors

glucose (17.4) an aldohexose, the most abundant monosaccharide; it is a component of many disaccharides, such as lactose and sucrose, and of polysaccharides, such as cellulose, starch, and glycogen

glyceraldehyde (17.3) an aldotriose that is the simplest carbohydrate; phosphorylated forms of glyceraldehyde are important intermediates in cellular metabolic reactions

glyceride (18.3) a lipid that contains glycerol

glycogen (17.6, 21.7) a long, branched polymer of glucose stored in liver and muscles of animals; it consists of a linear backbone of α-D-glucose in $\alpha(1 \rightarrow 4)$ linkage, with numerous short branches attached to the C-6 hydroxyl group by $\alpha(1 \rightarrow 6)$ linkage

glycogenesis (21.7) the metabolic pathway that results in the addition of glucose to growing glycogen polymers when blood glucose levels are high

glycogen granule (21.7) a core of glycogen surrounded by enzymes responsible for glycogen synthesis and degradation

glycogenolysis (21.7) the biochemical pathway that results in the removal of glucose molecules from glycogen polymers when blood glucose levels are low

glycolysis (21.3) the enzymatic pathway that converts a glucose molecule into two molecules of pyruvate; this anaerobic process generates a net energy yield of two molecules of ATP and two molecules of NADH

glycoprotein (19.7) a protein bonded to sugar groups

glycosidic bond (17.1) the bond between the hydroxyl group of the C-1 carbon of one sugar and a hydroxyl group of another sugar

ground state (2.5) a condition in which an atom is in its lowest energy state

group (3.1) any one of eighteen vertical columns of elements; often referred to as a family

group specificity (20.5) an enzyme that catalyzes reactions involving similar substrate molecules having the same functional groups

guanosine triphosphate (GTP) (21.6) a nucleotide composed of the purine guanosine, the sugar ribose, and three phosphoryl groups

H

half-life ($t_{1/2}$) (10.3) the length of time required for one-half of the initial mass of an isotope to decay to products

halogen (3.1) an element found in Group VIIA (17) of the periodic table

halogenation (11.5, 12.5) a reaction in which one of the C—H bonds of a hydrocarbon is replaced with a C—X bond (X = Br or Cl generally)

Haworth projection (17.4) a means of representing the orientation of substituent groups around a cyclic sugar molecule

α-helix (19.5) a right-handed coiled secondary structure maintained by hydrogen bonds between the amide hydrogen of one amino acid and the carbonyl oxygen of an amino acid four residues away

heme group (19.9) the chemical group found in hemoglobin and myoglobin that is responsible for the ability to carry oxygen

hemiacetal (14.4, 17.4) the family of organic compounds formed via the reaction of one molecule of alcohol with an aldehyde in the presence of an acid catalyst; hemiacetals have the following general structure:

$$R^1—\overset{\displaystyle OH}{\underset{\displaystyle H}{C}}—OR^2$$

hemiketal (14.4, 17.4) the family of organic compounds formed via the reaction of one molecule of alcohol with a ketone in the presence of an acid catalyst; hemiketals have the following general structure:

$$R^1—\overset{\displaystyle OH}{\underset{\displaystyle R^2}{C}}—OR^3$$

hemoglobin (19.9) the major protein component of red blood cells; the function of this red, iron-containing protein is transport of oxygen

hemolysis (7.6) the rupture of red blood cells resulting from movement of water from the surrounding medium into the cell

Henderson-Hasselbalch equation (9.4) an equation for calculating the pH of a buffer system:

$$pH = pKa + \log \frac{[\text{conjugate base}]}{[\text{weak acid}]}$$

Henry's law (7.3) a law stating that the number of moles of a gas dissolved in a liquid at a given temperature is proportional to the partial pressure of the gas

heterocyclic amine (16.2) a heterocyclic compound that contains nitrogen in at least one position in the ring skeleton

heterocyclic aromatic compound (12.7) cyclic aromatic compound having at least one atom other than carbon in the structure of the aromatic ring

heterogeneous mixture (1.2) a mixture of two or more substances characterized by nonuniform composition

hexose (17.2) a six-carbon monosaccharide

high-density lipoprotein (HDL) (18.5) a plasma lipoprotein that transports cholesterol from peripheral tissue to the liver

holoenzyme (20.7) an active enzyme consisting of an apoenzyme bound to a cofactor

homogeneous mixture (1.2) a mixture of two or more substances characterized by uniform composition

hybridization (24.8) a technique for identifying DNA or RNA sequences that is based on specific hydrogen bonding between a radioactive probe and complementary DNA or RNA sequences

hydrate (5.2) any substance that has water molecules incorporated in its structure

hydration (12.5, 13.5) a reaction in which water is added to a molecule, e.g., the addition of water to an alkene to form an alcohol

hydrocarbon (11.1) a compound composed solely of the elements carbon and hydrogen

hydrogen bonding (6.2) the attractive force between a hydrogen atom covalently bonded to a small, highly electronegative atom and another atom containing an unshared pair of electrons

hydrogenation (12.5, 14.4, 18.2) a reaction in which hydrogen (H_2) is added to a double or a triple bond

hydrohalogenation (12.5) the addition of a hydrohalogen (HCl, HBr, or HI) to an unsaturated bond

hydrolase (20.1) an enzyme that catalyzes hydrolysis reactions

hydrolysis (15.2) a chemical change that involves the reaction of a molecule with water; the process by which molecules are broken into their constituents by addition of water

hydronium ion (9.1) a protonated water molecule, H_3O^+

hydrophilic amino acid (19.2) "water loving"; a polar or ionic amino acid that has a high affinity for water

hydrophobic amino acid (19.2) "water fearing"; a nonpolar amino acid that prefers contact with other nonpolar amino acids over contact with water

hydroxyl group (13.1) the —OH functional group that is characteristic of alcohols

hyperammonemia (22.8) a genetic defect in one of the enzymes of the urea cycle that results in toxic or even fatal elevation of the concentration of ammonium ions in the body

hyperglycemia (21.7) blood glucose levels that are higher than normal

hypertonic solution (7.6, 18.6) the more concentrated solution of two separated by a semipermeable membrane

hypoglycemia (21.7) blood glucose levels that are lower than normal

hypothesis (1.1) an attempt to explain observations in a commonsense way

hypotonic solution (7.6, 18.6) the more dilute solution of two separated by a semipermeable membrane

I

ideal gas (6.1) a gas in which the particles do not interact and the volume of the individual gas particles is assumed to be negligible

ideal gas law (6.1) a law stating that for an ideal gas the product of pressure and volume is proportional to the product of the number of moles of the gas and its temperature; the proportionality constant for an ideal gas is symbolized *R*

incomplete protein (19.11) a protein source that does not contain all the essential and nonessential amino acids

indicator (9.2) a solute that shows some condition of a solution (such as acidity or basicity) by its color

induced fit model (20.4) the theory of enzyme-substrate binding that assumes that the enzyme is a flexible molecule and that both the substrate and the enzyme change their shapes to accommodate one another as the enzyme-substrate complex forms

initiation factors (24.6) proteins that are required for formation of the translation initiation complex, which is composed of the large and small ribosomal subunits, the mRNA, and the initiator tRNA, methionyl tRNA

inner mitochondrial membrane (22.1) the highly folded, impermeable membrane within the mitochondrion that is the location of the electron transport system and ATP synthase

insertion mutation (24.7) a mutation that results in the addition of one or more nucleotides to a DNA sequence

insulin (21.7, 23.6) a hormone released from the pancreas in response to high blood glucose levels; insulin stimulates glycogenesis, fat storage, and cellular uptake and storage of glucose from the blood

intensive property (1.2) a property of a substance that is independent of the quantity of the substance

intermembrane space (22.1) the region between the outer and inner mitochondrial membranes, which is the location of the proton (H^+) reservoir that drives ATP synthesis

intermolecular force (4.5) any attractive force that occurs between molecules

intramolecular force (4.5) any attractive force that occurs within molecules

intron (24.4) a noncoding sequence within a eukaryotic gene that must be removed from the primary transcript to produce a functional mRNA

ion (2.2) an electrically charged particle formed by the gain or loss of electrons

ionic bonding (4.1) an electrostatic attractive force between ions resulting from electron transfer

ionic solid (6.3) a solid composed of positive and negative ions in a regular three-dimensional crystalline arrangement

ionization energy (3.4) the energy needed to remove an electron from an atom in the gas phase

ionizing radiation (10.1) radiation that is sufficiently high in energy to cause ion formation upon impact with an atom

ion pair (4.1) the simplest formula unit for an ionic compound

ion product for water (9.1) the product of the hydronium and hydroxide ion concentrations in pure water at a specified temperature; at 25°C, it has a value of 1.0×10^{-14}

irreversible enzyme inhibitor (20.10) a chemical that binds strongly to the R groups of an amino acid in the active site and eliminates enzyme activity

isoelectric point (19.10) a situation in which a protein has an equal number of positive and negative charges and therefore has an overall net charge of zero

isoelectronic (3.3) atoms, ions, and molecules containing the same number of electrons

isomerase (20.1) an enzyme that catalyzes the conversion of one isomer to another

isotonic solution (7.6, 18.6) a solution that has the same solute concentration as another solution with which it is being compared; a solution that has the same osmotic pressure as a solution existing within a cell

isotope (2.2) atom of the same element that differs in mass because it contains different numbers of neutrons

I.U.P.A.C. Nomenclature System (11.2) the International Union of Pure and Applied Chemistry (I.U.P.A.C.) standard, universal system for the nomenclature of organic compounds

K

α-keratin (19.5) a member of the family of fibrous proteins that form the covering of most land animals; major components of fur, skin, beaks, and nails

ketal (14.4) the family of organic compounds formed via the reaction of two molecules of alcohol with a ketone in the presence of an acid catalyst; ketals have the following general structure:

```
     OR³
     |
R¹—C—OR⁴
     |
     R²
```

ketoacidosis (23.3) a drop in the pH of the blood caused by elevated ketone levels

ketone (14.1) a family of organic molecules characterized by a carbonyl group; the carbonyl carbon is bonded to two alkyl groups, two aryl groups, or one alkyl and one aryl group; ketones have the following general structures:

```
  O          O             O
  ‖          ‖             ‖
R—C—R    R—C—(Ar)    (Ar)—C—(Ar)
```

ketone bodies (23.3) acetone, acetoacetone, and β-hydroxybutyrate produced from fatty acids in the liver via acetyl CoA

ketose (17.2) a sugar that contains a ketone (carbonyl) group

ketosis (23.3) an abnormal rise in the level of ketone bodies in the blood

kinetic energy (1.5) the energy resulting from motion of an object [kinetic energy = 1/2 (mass)(velocity)^2]

kinetic-molecular theory (6.1) the fundamental model of particle behavior in the gas phase

kinetics (8.3) the study of rates of chemical reactions

L

lactose (17.5) a disaccharide composed of β-D-galactose and either α- or β-D-glucose in β(1 → 4) glycosidic linkage; milk sugar

lactose intolerance (17.5) the inability to produce the digestive enzyme lactase, which degrades lactose to galactose and glucose

lanthanide series (3.1) the rare earth elements, the fourteen elements from cerium (Ce) to lutetium (Lu)

law (1.1) a summary of a large quantity of information

law of conservation of mass (5.4) a law stating that, in chemical change, matter cannot be created or destroyed

LeChatelier's principle (8.4) a law stating that when a system at equilibrium is disturbed, the equilibrium shifts in the direction that minimizes the disturbance

lethal dose (LD_{50}) (10.8) the quantity of toxic material (such as radiation) that causes the death of 50% of a population of an organism

Lewis symbol (4.1) representation of an atom or ion using the atomic symbol (for the nucleus and core electrons) and dots to represent valence electrons

ligase (20.1) an enzyme that catalyzes the joining of two molecules

linear structure (4.4) the structure of a molecule in which the bond angles about the central atom(s) is (are) 180°

line formula (11.2) the simplest representation of a molecule in which it is assumed that there is a carbon atom at any location where two or more lines intersect, there is a carbon at the end of any line, and each carbon is bonded to the correct number of hydrogen atoms

linkage specificity (20.5) the property of an enzyme that allows it to catalyze reactions involving only one kind of bond in the substrate molecule

lipase (23.1) an enzyme that hydrolyzes the ester linkage between glycerol and the fatty acids of triglycerides

lipid (18.1) a member of the group of biological molecules of varying composition that are classified together on the basis of their solubility in nonpolar solvents

liquid state (1.2) a physical state of matter characterized by a fixed volume and the absence of a fixed shape

lock-and-key model (20.4) the theory of enzyme-substrate binding that depicts enzymes as inflexible molecules; the substrate fits into the rigid active site in the same way a key fits into a lock

London forces (6.2) weak attractive forces between molecules that result from short-lived dipoles that occur because of the continuous movement of electrons in the molecules

lone pair (4.4) an electron pair that is not involved in bonding

low-density lipoprotein (LDL) (18.5) a plasma lipoprotein that carries cholesterol to peripheral tissues and helps to regulate cholesterol levels in those tissues

lyase (20.1) an enzyme that catalyzes a reaction involving double bonds

M

maltose (17.5) a disaccharide composed of α-D-glucose and a second glucose molecule in α(1 → 4) glycosidic linkage

Markovnikov's rule (12.5) the rule stating that a hydrogen atom, adding to a carbon-carbon double bond, will add to the carbon having the larger number of hydrogens attached to it

mass (1.5) a quantity of matter

mass number (2.2) the sum of the number of protons and neutrons in an atom

matrix space (22.1) the region of the mitochondrion within the inner membrane; the location of the enzymes that carry out the reactions of the citric acid cycle and β-oxidation of fatty acids

matter (1.1) the material component of the universe

melting point (4.3, 6.2) the temperature at which a solid converts to a liquid

messenger RNA (24.4) an RNA species produced by transcription and that specifies the amino acid sequence for a protein

metal (3.1) an element located on the left side of the periodic table (left of the "staircase" boundary)

metallic bond (6.3) a bond that results from the orbital overlap of metal atoms

metallic solid (6.3) a solid composed of metal atoms held together by metallic bonds

metalloid (3.1) an element along the "staircase" boundary between metals and nonmetals; metalloids exhibit both metallic and nonmetallic properties

metastable isotope (10.2) an isotope that will give up some energy to produce a more stable form of the same isotope

metric system (1.3) a standards-based, decimal-based system of units developed by the National Assembly of France in 1790

micelle (23.1) an aggregation of molecules having nonpolar and polar regions; the nonpolar regions of the molecules aggregate, leaving the polar regions facing the surrounding water

mitochondria (22.1) the cellular "power plants" in which the reactions of the citric acid cycle, the electron transport system, and ATP synthase function to produce ATP

mixture (1.2) a material composed of two or more substances

molality (7.6) the number of moles of solute per kilogram of solvent

molar mass (5.1) the mass in grams of 1 mol of a substance

molar volume (6.1) the volume occupied by 1 mol of a substance

molarity (7.5) the number of moles of solute per liter of solution

mole (5.1) the amount of substance containing Avogadro's number of particles

molecular formula (11.2) a formula that provides the atoms and number of each type of atom in a molecule but gives no information regarding the bonding pattern involved in the structure of the molecule

molecular solid (6.3) a solid in which the molecules are held together by dipole-dipole and London forces (van der Waals forces)

molecule (4.2) a unit in which the atoms of two or more elements are held together by chemical bonds

monatomic ion (4.2) an ion formed by electron gain or loss from a single atom

monoglyceride (18.3) the product of the esterification of glycerol at one position

monomer (12.5) the individual molecules from which a polymer is formed

monosaccharide (17.1) the simplest type of carbohydrate consisting of a single saccharide unit

movement protein (19.1) a protein involved in any aspect of movement in an organism, for instance actin and myosin in muscle tissue and flagellin that composes bacterial flagella

mutagen (24.7) any chemical or physical agent that causes changes in the nucleotide sequence of a gene

mutation (24.7) any change in the nucleotide sequence of a gene

myoglobin (19.4) the oxygen storage protein found in muscle

myosin (19.5) one of the major proteins of muscle tissue; it has a rodlike structure of two α-helices coiled around one another

N

N-terminal amino acid (19.3) the amino acid in a peptide that has a free $\alpha\text{-N}^+H_3$ group; the first amino acid of a peptide

natural radioactivity (2.3, 10.6) the spontaneous decay of a nucleus to produce high-energy particles or rays

negative allosterism (20.9) effector binding inactivates the active site of an allosteric enzyme

negative ion (3.3) or anion, a negatively charged atom or group of atoms formed by the gain of one or more electrons by the neutral species

neurotransmitter (16.5) a chemical that carries a message, or signal, from a nerve cell to a target cell

neutral glyceride (18.3) the product of the esterification of glycerol at one, two, or three positions

neutralization (9.3) the reaction between an acid and a base

neutron (2.2) an uncharged particle, with the same mass as the proton, in the nucleus of an atom

nicotinamide adenine dinucleotide (NAD^+) (21.3) a molecule synthesized from the vitamin niacin and the nucleotide ATP and that serves as a carrier of hydride anions; a coenzyme that is an oxidizing agent used in a variety of metabolic processes

noble gas (3.1) elements in Group VIIIA (18) of the periodic table

nomenclature (4.2) a system for naming chemical compounds

nonelectrolyte (4.3, 7.3) a substance that, when dissolved in water, produces a solution that does not conduct an electrical current

nonessential amino acid (19.11) any amino acid that can be synthesized by the body

nonmetal (3.1) an element located on the right side of the periodic table (right of the "staircase" boundary)

nonreducing sugar (17.5) a sugar that cannot be oxidized by Benedict's or Tollens' reagent

normal boiling point (6.2) the temperature at which a substance will boil at 1 atm of pressure

nuclear equation (10.2) a balanced equation accounting for the products and reactants in a nuclear reaction

nuclear imaging (10.6) the generation of images of components of the body (organs, tissues) using techniques based on the measurement of radiation

nuclear medicine (10.6) a field of medicine that uses radioisotopes for diagnostic and therapeutic purposes

nuclear reactor (10.6) a device for conversion of nuclear energy into electrical energy

nucleosome (24.2) the first level of chromosome structure consisting of a strand of DNA wrapped around a small disk of histone proteins

nucleotide (21.1, 24.1) a molecule composed of a nitrogenous base, a five-carbon sugar, and one, two, or three phosphoryl groups

nucleus (2.2) the small, dense center of positive charge in the atom

nutrient protein (19.1) a protein that serves as a source of amino acids for embryos or infants

nutritional Calorie (8.2) equivalent to one kilocalorie (1000 calories); also known as a large Calorie

O

octet rule (3.3) a rule predicting that atoms form the most stable molecules or ions when they are surrounded by eight electrons in their highest occupied energy level

oligosaccharide (17.1) an intermediate-sized carbohydrate composed of from three to ten monosaccharides

orbit (2.3) according to the Bohr theory, a region of the atom where electrons may be found

orbital (2.3, 3.2) a specific region of space where an electron may be found; a region of an atom containing a maximum of two electrons with opposite spins

order of the reaction (8.3) the exponent of each concentration term in the rate equation

osmolarity (7.6) molarity of particles in solution; this value is used for osmotic pressure calculations

osmosis (7.6, 18.6) net flow of a solvent across a semipermeable membrane in response to a concentration gradient

osmotic pressure (7.6, 18.6) the net force with which water enters a solution through a semipermeable membrane; alternatively, the pressure required to stop net transfer of solvent across a semipermeable membrane

outer mitochondrial membrane (22.1) the membrane that surrounds the mitochondrion and separates it from the contents of the cytoplasm; it is highly permeable to small "food" molecules

β-oxidation (23.2) the biochemical pathway that results in the oxidation of fatty acids and the production of acetyl CoA

oxidation (9.5, 13.6, 14.4, 15.1) a loss of electrons; in organic compounds it may be recognized as a loss of hydrogen atoms or the gain of oxygen

oxidation-reduction reaction (7.2, 9.5) also called redox reaction, a reaction involving the transfer of one or more electrons from one reactant to another

oxidative deamination (22.7) an oxidation-reduction reaction in which NAD^+ is reduced and the amino acid is deaminated

oxidative phosphorylation (21.3, 22.6) production of ATP using the energy of electrons harvested during biological oxidation-reduction reactions

oxidizing agent (9.5) a substance that oxidizes, or removes electrons from, another substance; the oxidizing agent is reduced in the process

oxidoreductase (20.1) an enzyme that catalyzes an oxidation-reduction reaction

P

pancreatic serine proteases (20.11) a family of proteolytic enzymes, including trypsin, chymotrypsin, and elastase, that arose by divergent evolution

parent compound or parent chain (11.2) in the I.U.P.A.C. Nomenclature System the parent compound is the longest carbon-carbon chain containing the principal functional group in the molecule that is being named

partial pressure (6.1) the pressure exerted by one component of a gas mixture

particle accelerator (10.6) a device for production of high-energy nuclear particles based on the interaction of charged particles with magnetic and electrical fields

passive transport (18.6) the net movement of a solute from an area of high concentration to an area of low concentration

pentose (17.2) a five-carbon monosaccharide

pentose phosphate pathway (21.5) an alternative pathway for glucose degradation that provides the cell with reducing power in the form of NADPH

peptide bond (16.4, 19.3) the amide bond between two amino acids in a peptide chain

peptidyl tRNA binding site of ribosome (P-site) (24.6) a pocket on the surface of the ribosome that holds the tRNA bound to the growing peptide chain

percent yield (5.6) the ratio of the actual and theoretical yields of a chemical reaction multiplied by 100%

period (3.1) any one of seven horizontal rows of elements in the periodic table

periodic law (3.1) a law stating that properties of elements are periodic functions of their atomic numbers (Note that Mendeleev's original statement was based on atomic masses.)

peripheral membrane protein (18.6) a protein bound to either the inner or the outer surface of a membrane

phenol (13.7) an organic compound that contains a hydroxyl group (—OH) attached to a benzene ring

phenyl group (12.6) a benzene ring that has had a hydrogen atom removed, C_6H_5—

pH optimum (20.8) the pH at which an enzyme catalyzes the reaction at maximum efficiency

phosphatidate (18.3) a molecule of glycerol with fatty acids esterified to C-1 and C-2 of glycerol and a free phosphoryl group esterified at C-3

phosphoester (15.4) the product of the reaction between phosphoric acid and an alcohol

phosphoglyceride (18.3) a molecule with fatty acids esterified at the C-1 and C-2 positions of glycerol and a phosphoryl group esterified at the C-3 position

phospholipid (18.3) a lipid containing a phosphoryl group

phosphopantetheine (23.4) the portion of coenzyme A and the acyl carrier protein that is derived from the vitamin pantothenic acid

phosphoric anhydride (15.4) the bond formed when two phosphate groups react with one another and a water molecule is lost

pH scale (9.2) a numerical representation of acidity or basicity of a solution; $pH = -\log [H_3O^+]$

physical change (1.2) a change in the form of a substance but not in its chemical composition; no chemical bonds are broken in a physical change

physical property (1.2) a characteristic of a substance that can be observed without the substance undergoing change (examples include color, density, melting and boiling points)

plasma lipoprotein (18.5) a complex composed of lipid and protein that is responsible for the transport of lipids throughout the body

β-pleated sheet (19.5) a common secondary structure of a peptide chain that resembles the pleats of an Oriental fan

point mutation (24.7) the substitution of one nucleotide pair for another within a gene

polar covalent bonding (4.4) a covalent bond in which the electrons are not equally shared

polar covalent molecule (4.4) a molecule that has a permanent electric dipole moment resulting from an unsymmetrical electron distribution; a dipolar molecule

poly(A) tail (24.4) a tract of 100–200 adenosine monophosphate units covalently attached to the 3′ end of eukaryotic messenger RNA molecules

polyatomic ion (4.2) an ion containing a number of atoms

polymer (12.5) a very large molecule formed by the combination of many small molecules (called monomers) (e.g., polyamides, nylons)

polyprotic substance (9.3) a substance that can accept or donate more than one proton per molecule

polysaccharide (17.1) a large, complex carbohydrate composed of long chains of monosaccharides

polysome (24.6) complexes of many ribosomes all simultaneously translating a single mRNA

positive allosterism (20.9) effector binding activates the active site of an allosteric enzyme

post-transcriptional modification (24.4) alterations of the primary transcripts produced in eukaryotic cells; these include addition of a poly(A) tail to the 3′ end of the mRNA, addition of the cap structure to the 5′ end of the mRNA, and RNA splicing

potential energy (1.5) stored energy or energy caused by position or composition

precipitate (7.3) an insoluble substance formed and separated from a solution

pressure (6.1) a force per unit area

primary (1°) alcohol (13.4) an alcohol with the general formula RCH_2OH

primary (1°) amine (16.1) an amine with the general formula RNH_2

primary (1°) carbon (11.2) a carbon atom that is bonded to only one other carbon atom

primary structure (of a protein) (19.4) the linear sequence of amino acids in a protein chain determined by the genetic information of the gene for each protein

primary transcript (24.4) the RNA product of transcription in eukaryotic cells, before post-transcriptional modifications are carried out

product (1.2, 5.4, 20.2) the chemical species that results from a chemical reaction and that appears on the right side of a chemical equation

prokaryote (24.2) an organism with simple cellular structure in which there is no true nucleus enclosed by a nuclear membrane and there are no true membrane-bound organelles in the cytoplasm

promoter (24.4) the sequence of nucleotides immediately before a gene that is recognized by the RNA polymerase and signals the start point and direction of transcription

promotion (2.5) an electron transition from a lower to a higher energy level, resulting from absorption of energy

properties (1.2) characteristics of matter

prostaglandins (18.2) a family of hormonelike substances derived from the twenty-carbon fatty acid, arachidonic acid; produced by many cells of the body, they regulate many body functions

prosthetic group (19.7) the nonprotein portion of a protein that is essential to the biological activity of the protein; often a complex organic compound

protein (19: Intro) a macromolecule whose primary structure is a linear sequence of α-amino acids and whose final structure results from folding of the chain into a specific three-dimensional structure; proteins serve as catalysts, structural components, and nutritional elements for the cell

protein modification (20.9) a means of enzyme regulation in which a chemical group is covalently added to or removed from a protein. The chemical modification either turns the enzyme on or turns it off

proteolytic enzyme (20.11) an enzyme that hydrolyzes the peptide bonds between amino acids in a protein chain

proton (2.2) a positively charged particle in the nucleus of an atom

pure substance (1.2) a substance with constant composition

purine (24.1) a family of nitrogenous bases (heterocyclic amines) that are components of DNA and RNA and consist of a six-sided ring fused to a five-sided ring; the common purines in nucleic acids are adenine and guanine

pyridoxal phosphate (22.7) a coenzyme derived from vitamin B_6 that is required for all transamination reactions

pyrimidine (24.1) a family of nitrogenous bases (heterocyclic amines) that are components of nucleic acids and consist of a single six-sided ring; the common pyrimidines of DNA are cytosine and thymine; the common pyrimidines of RNA are cytosine and uracil

pyrimidine dimer (24.7) UV-light induced covalent bonding of two adjacent pyrimidine bases in a strand of DNA

pyruvate dehydrogenase complex (22.2) a complex of all the enzymes and coenzymes required for the synthesis of CO_2 and acetyl CoA from pyruvate

Q

quantization (2.5) a characteristic that energy can occur only in discrete units called quanta

quantum level (2.5) a specific energy state or energy level in an atom

quantum number (2.5) an integer used to describe the orbitals of an atom

quaternary ammonium salt (16.1) an amine salt with the general formula $R_4N^+A^-$ (in which R— can be an alkyl or aryl group or a hydrogen atom and A^- can be any anion)

quaternary (4°) carbon (11.2) a carbon atom that is bonded to four other carbon atoms

quaternary structure (of a protein) (19.7) aggregation of more than one folded peptide chain to yield a functional protein

R

rad (10.8) abbreviation for *radiation absorbed dose*, the absorption of 2.4×10^{-3} calories of energy per kilogram of absorbing tissue

radioactivity (10.1) the process by which atoms emit high-energy particles or rays; the spontaneous decomposition of a nucleus to produce a different nucleus

radiocarbon dating (10.5) the estimation of the age of objects through measurement of isotopic ratios of carbon

Raoult's law (7.6) a law stating that the vapor pressure of a component is equal to its mole fraction times the vapor pressure of the pure component

rate constant (8.3) the proportionality constant that relates the rate of a reaction and the concentration of reactants

rate equation (8.3) expresses the rate of a reaction in terms of reactant concentration and a rate constant

rate of chemical reaction (8.3) the change in concentration of a reactant or product per unit time

reactant (1.2, 5.4) starting material for a chemical reaction, appearing on the left side of a chemical equation

reducing agent (9.5) a substance that reduces, or donates electrons to, another substance; the reducing agent is itself oxidized in the process

reducing sugar (17.4) a sugar that can be oxidized by Benedict's or Tollens' reagents; includes all monosaccharides and most disaccharides

reduction (9.5, 13.6) the gain of electrons; in organic compounds it may be recognized by a gain of hydrogen, or loss of oxygen

regulatory proteins (19.1) proteins that control cell functions such as metabolism and reproduction

relaxation (2.5) an electron transition from a higher to a lower energy level, accompanied by a release of energy

release factor (24.6) a protein that binds to the termination codon in the empty A-site of the ribosome and causes the peptidyl transferase to hydrolyze the bond between the peptide and the peptidyl tRNA

rem (10.8) abbreviation for *roentgen equivalent for man,* the product of rad and RBE

replication fork (24.3) the point at which new nucleotides are added to the growing daughter DNA strand

replication origin (24.3) the region of a DNA molecule where DNA replication always begins

representative element (3.1) member of the groups of the periodic table designated as A

resonance (4.4) a condition that occurs when more than one valid Lewis structure can be written for a particular molecule

resonance form (4.4) one of a number of valid Lewis structures for a particular molecule

resonance hybrid (4.4) a description of the bonding in a molecule resulting from a superimposition of all valid Lewis structures (resonance forms)

restriction enzyme (24.8) a bacterial enzyme that recognizes specific nucleotide sequences on a DNA molecule and cuts the sugar-phosphate backbone of the DNA at or near that site

result (1.3) the outcome of a designed experiment, often determined from individual bits of data

reversible, competitive enzyme inhibitor (20.10) a chemical that resembles the structure and charge distribution of the natural substrate and competes with it for the active site of an enzyme

reversible, noncompetitive enzyme inhibitor (20.10) a chemical that binds weakly to an amino acid R group of an enzyme and inhibits activity; when the inhibitor dissociates, the enzyme is restored to its active form

reversible reaction (8.4) a reaction that will proceed in either direction, reactants to products or products to reactants

ribonucleic acid (RNA) (24.1) single-stranded nucleic acid molecules that are composed of phosphoryl groups, ribose, and the nitrogenous bases uracil, cytosine, adenine, and guanine

ribonucleotide (24.1) a ribonucleoside phosphate or nucleotide composed of a nitrogenous base in β-*N*-glycosidic linkage to the 1′ carbon of the sugar ribose and with one, two, or three phosphoryl groups esterified at the hydroxyl of the 5′ carbon of the ribose

ribose (17.4) a five-carbon monosaccharide that is a component of RNA and many coenzymes

ribosomal RNA (rRNA) (24.4) the RNA species that are structural and functional components of the small and large ribosomal subunits

ribosome (24.6) an organelle composed of a large and a small subunit, each of which is made up of ribosomal RNA and proteins; the platform on which translation occurs and that carries the enzymatic activity that forms peptide bonds

RNA polymerase (24.4) the enzyme that catalyzes the synthesis of RNA molecules using DNA as the template

RNA splicing (24.4) removal of portions of the primary transcript that do not encode protein sequences

röentgen (10.8) the dose of radiation producing 2.1×10^9 ions in 1 cm^3 of air at 0°C and 1 atm of pressure

S

saccharide (17.1) a sugar molecule

saponification (15.2, 18.2) a reaction in which a soap is produced; more generally, the hydrolysis of an ester by an aqueous base

saturated fatty acid (18.2) a long-chain monocarboxylic acid in which each carbon of the chain is bonded to the maximum number of hydrogen atoms

saturated hydrocarbon (11.1) an alkane; a hydrocarbon that contains only carbon and hydrogen bonded together through carbon-hydrogen and carbon-carbon single bonds

saturated solution (7.3) one in which undissolved solute is in equilibrium with the solution

scientific method (1.1) the process of studying our surroundings that is based on experimentation

scientific notation (1.4) a system used to represent numbers as powers of ten

secondary (2°) alcohol (13.4) an alcohol with the general formula R_2CHOH

secondary (2°) amine (16.1) an amine with the general formula R_2NH

secondary (2°) carbon (11.2) a carbon atom that is bonded to two other carbon atoms

secondary structure (of a protein) (19.5) folding of the primary structure of a protein into an α-helix or a β-pleated sheet; folding is maintained by hydrogen bonds between the amide hydrogen and the carbonyl oxygen of the peptide bond

semiconservative DNA replication (24.3) DNA polymerase "reads" each parental strand of DNA and produces a complementary daughter strand; thus, all newly synthesized DNA molecules consist of one parental and one daughter strand

semipermeable membrane (7.6, 18.6) a membrane permeable to the solvent but not the solute; a material that allows the transport of certain substances from one side of the membrane to the other

shielding (10.7) material used to provide protection from radiation

sickle cell anemia (19.9) a human genetic disease resulting from inheriting mutant hemoglobin genes from both parents

significant figures (1.4) all digits in a number known with certainty and the first uncertain digit

silent mutation (24.7) a mutation that changes the sequence of the DNA but does not alter the amino acid sequence of the protein encoded by the DNA

single bond (4.4) a bond in which one pair of electrons is shared by two atoms

single-replacement reaction (7.1) also called substitution reaction, one in which one atom in a molecule is displaced by another

soap (15.2) any of a variety of the alkali metal salts of fatty acids

solid state (1.2) a physical state of matter characterized by its rigidity and fixed volume and shape

solubility (4.5, 7.3) the amount of a substance that will dissolve in a given volume of solvent at a specified temperature

solute (7.3) a component of a solution that is present in lesser quantity than the solvent

solution (7.3) a homogeneous (uniform) mixture of two or more substances

solvent (7.3) the solution component that is present in the largest quantity

specific gravity (1.5) the ratio of the density of a substance to the density of water at 4°C or any specified temperature

specific heat (8.2) the quantity of heat (calories) required to raise the temperature of 1 g of a substance one degree Celsius

spectroscopy (2.4) the measurement of intensity and energy of electromagnetic radiation

speed of light (2.4) 2.99×10^8 m/s in a vacuum

sphingolipid (18.4) a phospholipid that is derived from the amino alcohol sphingosine rather than from glycerol

sphingomyelin (18.4) a sphingolipid found in abundance in the myelin sheath that surrounds and insulates cells of the central nervous system

standard solution (9.3) a solution whose concentration is accurately known

standard temperature and pressure (STP) (6.1) defined as 273 K and 1 atm

states of matter (1.2) the three different forms in which matter can exist (solid, liquid, and gas)

stereochemical specificity (20.5) the property of an enzyme that allows it to catalyze reactions involving only one enantiomer of the substrate

stereochemistry (17.3) the study of the spatial arrangement of atoms in a molecule

stereoisomers (17.3) a pair of molecules having the same structural formula and bonding pattern but differing in the arrangement of the atoms in space

steroid (18.4) a lipid derived from cholesterol and composed of one five-sided ring and three six-sided rings; the steroids include sex hormones and anti-inflammatory compounds

structural analog (20.10) a chemical having a structure and charge distribution very similar to those of a natural enzyme substrate

structural formula (11.2) a formula showing all of the atoms in a molecule and exhibiting all bonds as lines

structural isomers (11.2) molecules having the same molecular formula but different chemical structures

structural protein (19.1) a protein that provides mechanical support for large plants and animals

substituted hydrocarbon (11.1) a hydrocarbon in which one or more hydrogen atoms is replaced by another atom or group of atoms

substitution reaction (11.5, 12.6) a reaction that results in the replacement of one group for another

substrate (20.1) the reactant in a chemical reaction that binds to an enzyme active site and is converted to product

substrate-level phosphorylation (21.3) the production of ATP by the transfer of a phosphoryl group from the substrate of a reaction to ADP

sucrose (17.5) a disaccharide composed of α-D-glucose and β-D-fructose in (α1 → β2) glycosidic linkage; table sugar

supersaturated solution (7.3) a solution that is more concentrated than a saturated solution (Note that such a solution is not at equilibrium.)

surface tension (6.2) a measure of the strength of the attractive forces at the surface of a liquid

surfactant (6.2) a substance that decreases the surface tension of a liquid

surroundings (8.1) the universe outside of the system

suspension (7.3) a heterogeneous mixture of particles; the suspended particles are larger than those found in a colloidal suspension

system (8.1) the process under study

T

temperature (1.5) a measure of the relative "hotness" or "coldness" of an object

temperature optimum (20.8) the temperature at which an enzyme functions optimally and the rate of reaction is maximal

terminal electron acceptor (22.6) the final electron acceptor in an electron transport system that removes the low-energy electrons from the system; in aerobic organisms the terminal electron acceptor is molecular oxygen

termination codon (24.6) a triplet of ribonucleotides with no corresponding anticodon on a tRNA; as a result, translation will end, because there is no amino acid to transfer to the peptide chain

terpene (18.4) the general term for lipids that are synthesized from isoprene units; the terpenes include steroids, bile salts, lipid-soluble vitamins, and chlorophyll

tertiary (3°) alcohol (13.4) an alcohol with the general formula R_3COH

tertiary (3°) amine (16.1) an amine with the general formula R_3N

tertiary (3°) carbon (11.2) a carbon atom that is bonded to three other carbon atoms

tertiary structure (of a protein) (19.6) the globular, three-dimensional structure of a protein that results from folding the regions of secondary structure; this folding occurs spontaneously as a result of interactions of the side chains or R groups of the amino acids

tetrahedral structure (4.4) a molecule consisting of four groups attached to a central atom that occupy the four corners of an imagined regular tetrahedron

tetrose (17.2) a four-carbon monosaccharide

theoretical yield (5.6) the maximum amount of product that can be produced from a given amount of reactant

theory (1.1) a hypothesis supported by extensive testing that explains and predicts facts

thermodynamics (8.1) the branch of science that deals with the relationship between energies of systems, work, and heat

thioester (15.4) the product of a reaction between a thiol and a carboxylic acid

thiol (13.9) an organic compound that contains a thiol group (—SH)

titration (9.3) the process of adding a solution from a buret to a sample until a reaction is complete, at which time the volume is accurately measured and the concentration of the sample is calculated

Tollens' test (14.4) a test reagent (silver nitrate in ammonium hydroxide) used to distinguish aldehydes and ketones; also called the Tollens' silver mirror test

tracer (10.6) a radioisotope that is rapidly and selectively transmitted to the part of the body for which diagnosis is desired

transaminase (22.7) an enzyme that catalyzes the transfer of an amino group from one molecule to another

transamination (22.7) a reaction in which an amino group is transferred from one molecule to another

transcription (24.4) the synthesis of RNA from a DNA template

transferase (20.1) an enzyme that catalyzes the transfer of a functional group from one molecule to another

transfer RNA (tRNA) (16.4, 24.4) small RNAs that bind to a specific amino acid at the 3′ end and mediate its addition at the appropriate site in a growing peptide chain; accomplished by recognition of the correct codon on the mRNA by the complementary anticodon on the tRNA

transition element (3.1) any element located between Groups IIA (2) and IIIA (13) in the long periods of the periodic table

transition state (20.6) the unstable intermediate in catalysis in which the enzyme has altered the form of the substrate so that it now shares properties of both the substrate and the product

translation (24.4) the synthesis of a protein from the genetic code carried on the mRNA

translocation (24.6) movement of the ribosome along the mRNA during translation

transmembrane protein (18.6) a protein that is embedded within a membrane and crosses the lipid bilayer, protruding from the membrane both inside and outside the cell

transport protein (19.1) a protein that transports materials across the cell membrane or throughout the body

triglyceride (18.3, 23.1) triacylglycerol; a molecule composed of glycerol esterified to three fatty acids

trigonal pyramidal molecule (4.4) a nonplanar structure involving three groups bonded to a central atom in which each group is equidistant from the central atom

triose (17.2) a three-carbon monosaccharide

triple bond (4.4) a bond in which three pairs of electrons are shared by two atoms

U

unit (1.3) a determinate quantity (of length, time, etc.) that has been adopted as a standard of measurement

unsaturated fatty acid (18.2) a long-chain monocarboxylic acid having at least one carbon-to-carbon double bond

unsaturated hydrocarbon (11.1, 12: Intro) a hydrocarbon containing at least one multiple (double or triple) bond

urea cycle (22.8) a cyclic series of reactions that detoxifies ammonium ions by incorporating them into urea, which is excreted from the body

uridine triphosphate (UTP) (21.7) a nucleotide composed of the pyrimidine uracil, the sugar ribose, and three phosphoryl groups and that serves as a carrier of glucose-1-phosphate in glycogenesis

V

valence electron (3.2) electron in the outermost shell (principal quantum level) of an atom

valence shell electron pair repulsion theory (VSEPR) (4.4) a model that predicts molecular geometry using the premise that electron pairs will arrange themselves as far apart as possible, to minimize electron repulsion

van der Waals forces (6.2) a general term for intermolecular forces that include dipole-dipole and London forces

vapor pressure of a liquid (6.2) the pressure exerted by the vapor at the surface of a liquid at equilibrium

very low density lipoprotein (VLDL) (18.5) a plasma lipoprotein that binds triglycerides synthesized by the liver and carries them to adipose tissue for storage

viscosity (6.2) a measure of the resistance to flow of a substance at constant temperature

vitamin (20.7) an organic substance that is required in the diet in small amounts; water-soluble vitamins are used in the synthesis of coenzymes required for the function of cellular enzymes; lipid-soluble vitamins are involved in calcium metabolism, vision, and blood clotting

voltaic cell (9.5) an electrochemical cell that converts chemical energy into electrical energy

W

wax (18.4) a collection of lipids that are generally considered to be esters of long-chain alcohols

weight (1.5) the force exerted on an object by gravity

weight/volume percent [% (W/V)] (7.4) the concentration of a solution expressed as a ratio of grams of solute to milliliters of solution multiplied by 100%

weight/weight percent [% (W/W)] (7.4) the concentration of a solution expressed as a ratio of mass of solute to mass of solution multiplied by 100%

Z

Zaitsev's rule (13.5) states that in an elimination reaction, the alkene with the greatest number of alkyl groups on the double-bonded carbon (the more highly substituted alkene) is the major product of the reaction

zymogen (20.9) proenzyme; the inactive form of a proteolytic enzyme

Answers to Odd-Numbered Problems

Chapter 1

1.1 **a.** Physical property
b. Chemical property
c. Physical property
d. Physical property
e. Physical property

1.3 **a.** Pure substance
b. Heterogeneous mixture
c. Homogeneous mixture
d. Pure substance

1.5 **a.** 1.0×10^3 mL
b. 1.0×10^6 μL
c. 1.0×10^{-3} kL
d. 1.0×10^2 cL
e. 1.0×10^{-1} daL

1.7 **a.** 1.3×10^{-2} m
b. 0.71 L
c. 2.00 oz
d. 1.5×10^{-4} m^2

1.9 **a.** Three
b. Three
c. Four
d. Two
e. Three

1.11 **a.** 2.4×10^{-3}
b. 1.80×10^{-2}
c. 2.24×10^2

1.13 **a.** 8.09
b. 5.9
c. 20.19

1.15 **a.** 51
b. 8.0×10^1
c. 1.6×10^2

1.17 **a.** 61.4
b. 6.17
c. 6.65×10^{-2}

1.19 **a.** 0°C
b. 273 K

1.21 23.7 g alcohol

1.23 **a.** Potential energy is stored energy or energy caused by position or composition.
b. Kinetic energy is the energy resulting from motion of an object.
c. Energy is the ability to do work.

1.25 **a.** Precision is the degree of agreement among replicate measurements of the same quantity.
b. Accuracy is the nearness of an experimental value to the true value.
c. Data are a group of observations resulting from an experiment.

1.27 **a.** Gram (or kilogram)
b. Liter
c. Meter

1.29 Weight is the force exerted on a body by gravity; mass is a quantity of matter. Mass is an independent quantity whereas weight is dependent on gravity, which may differ from location to location.

1.31 Density is mass per volume. Specific gravity is the ratio of the density of a substance to the density of water at 4°C or any specified temperature.

1.33 The scientific method is an organized way of doing science. It uses carefully planned experimentation to study our surroundings.

1.35 A physical property is a characteristic of a substance that can be observed without the substance undergoing a change in chemical composition.

1.37 **a.** Chemical reaction
b. Physical change
c. Physical change

1.39 **a.** Physical
b. Chemical

1.41 Flammability and toxicity

1.43 A pure substance has constant composition with only a single substance whereas a mixture is composed of two or more substances.

1.45 A homogeneous mixture has uniform composition whereas a heterogeneous mixture has nonuniform composition.

1.47 A gas is made up of particles that are widely separated. A gas will expand to fill any container and has no definite shape or volume.

1.49 An intensive property is a characteristic of a substance that is independent of the quantity of the substance. An extensive property depends on the quantity of the substance.

1.51 An element is a pure substance that cannot be changed into a simpler form of matter by any chemical reaction. An atom is the smallest unit of an element that retains the properties of that element.

1.53 **a.** 32 oz
b. 1.0×10^{-3} t
c. 9.1×10^2 g
d. 9.1×10^5 mg
e. 9.1×10^1 da

1.55 **a.** 6.6×10^{-3} lb
b. 1.1×10^{-1} oz
c. 3.0×10^{-3} kg
d. 3.0×10^2 cg
e. 3.0×10^3 mg

1.57 **a.** 10.0°C
b. 283.2 K

1.59 **a.** 293.2 K
b. 68.0°F
1.61 4 L
1.63 101°F
1.65 **a.** Three
b. Three
c. Three
d. Four
e. Four
f. Three
1.67 **a.** 3.87×10^{-3}
b. 5.20×10^{-2}
c. 2.62×10^{-3}
d. 2.43×10^{1}
e. 2.40×10^{2}
f. 2.41×10^{0}
1.69 **a.** 1.5×10^{4}
b. 2.41×10^{-1}
c. 5.99
d. 1139.42
e. 7.21×10^{3}
1.71 **a.** 1.23×10^{1}
b. 5.69×10^{-2}
c. -1.527×10^{3}
d. 7.89×10^{-7}
e. 9.2×10^{7}
f. 5.280×10^{-3}
g. 1.279×10^{0}
h. -5.3177×10^{2}
1.73 **a.** 3240
b. 0.000150
c. 0.4579
d. −683,000
e. −0.0821
f. 299,790,000
g. 1.50
h. 602,200,000,000,000,000,000,000
1.75 6.00 g/mL
1.77 1.08×10^{3} g
1.79 9.8×10^{-1} g/cm^3, teak
1.81 0.789
1.83 $d_{lead} = 7.9$ g/cm^3
$d_{uranium} = 19$ g/cm^3
$d_{platinum} = 21.4$ g/cm^3
Lead has the lowest density and platinum has the greatest density.

Chapter 2

2.1 **a.** Sixteen protons, sixteen electrons, sixteen neutrons
b. Eleven protons, eleven electrons, twelve neutrons
2.3 20.18 amu
2.5 DeBroglie considered electrons to have both wave and particle properties.
2.7 Polymers and plastics; medicines and medical diagnostic tools; computers, television, and thousands of other items
2.9 Similar mass; both are contained in the nucleus.
2.11 **a.** Eight protons, eight electrons, eight neutrons
b. Fifteen protons, fifteen electrons, sixteen neutrons
2.13 Isotopes are atoms of the same element that differ in mass because they contain different numbers of neutrons.
2.15

Particle	Mass	Charge
a. electron	5.4×10^{-4} amu	−1
b. proton	1.00 amu	+1
c. neutron	1.00 amu	0

2.17 **a.** An ion is a charged atom or group of atoms formed by the loss or gain of electrons.
b. A loss of electrons by a neutral species results in a cation.
c. A gain of electrons by a neutral species results in an anion.
2.19 86
2.21 **a.** 34
b. 46
2.23 **a.** $^{1}_{1}H$
b. $^{14}_{6}C$
2.25 Different masses, due to differing number of neutrons
2.27

Atomic Symbol	# Protons	# Neutrons	# Electrons	Charge
a. $^{23}_{11}Na$	11	12	11	0
b. $^{32}_{16}S^{2-}$	16	16	18	2−
c. $^{16}_{8}O$	8	8	8	0
d. $^{24}_{12}Mg^{2+}$	12	12	10	2+
e. $^{39}_{19}K^{+}$	19	20	18	1+

2.29 **a.** Neutrons
b. Protons
c. Protons, neutrons
d. Ion
e. Nucleus, negative
2.31 All matter consists of tiny particles called atoms. Atoms cannot be created, divided, destroyed, or converted to any other type of atom. All atoms of a particular element have identical properties. Atoms of different elements have different properties. Atoms combine in simple whole-number ratios. Chemical change involves joining, separating, or rearranging atoms.
2.33 **a.** Chadwick demonstrated the existence of the neutron.
b. DeBroglie theorized that electrons had wavelike and particlelike properties.
c. Geiger provided the basic experimental evidence for the existence of a nucleus.
d. The Bohr theory describes electron arrangement in atoms.
2.35 **a.** First atomic theory
b. Characterized electron properties
2.37 Geiger bombarded gold foil with alpha particles, and observed that some alpha particles passed straight through the foil, others were deflected, and some simply bounced back. This led Rutherford to propose that the atom consisted of a small, dense nucleus (alpha particles bounced back), surrounded by a cloud of electrons (some alpha particles were deflected). The size of the nucleus is small when compared to the volume of the atom (alpha particles were able to pass through the foil).
2.39 Crookes used the cathode ray tube. He observed particles emitted by the cathode traveling toward the anode. These rays were deflected by an electric field. Thomson measured the curvature of the rays influenced by the electric field. This measurement provided the mass-to-charge ratio of the negative particle. Thomson also gave the particle the name, electron.
2.41 A cathode ray is the negatively charged particle formed in a cathode ray tube. It was characterized as an electron, with a very small mass and a charge of −1.
2.43 Radiowave
Microwave
Infrared
Visible
Ultraviolet
X ray
Gamma ray
(↑ Increasing wavelength)
2.45 Infrared
2.47 The measurement of intensity and energy of electromagnetic radiation.

2.49 According to Bohr, Planck, and others, electrons exist only in certain allowed regions, quantum levels, outside of the nucleus.

2.51 Electrons are found in orbits at discrete distances from the nucleus. The orbits are quantitized—they are of discrete energies. Electrons can only be found in these orbits, never in between (they are able to jump instantaneously from orbit to orbit). Electrons can undergo transitions—if an electron absorbs energy, it will jump to a higher orbit; when the electron falls back down to a lower orbit, it will release energy.

2.53 Bohr's atomic model was the first to successfully account for electronic properties of atoms, specifically, the interaction of atoms and light (spectroscopy).

2.55 The deBroglie hypothesis stated that the electron has both particlelike and wavelike properties.

Chapter 3

3.1 **a.** Zr (zirconium)
b. 22.99
c. Cr (chromium)
d. Bi (bismuth)

3.3 **a.** Helium, atomic number = 2, mass = 4.00 amu
b. Fluorine, atomic number = 9, mass = 19.00 amu
c. Manganese, atomic number = 25, mass = 54.94 amu

3.5 **a.** Total electrons = 11, valence electrons = 1
b. Total electrons = 12, valence electrons = 2
c. Total electrons = 16, valence electrons = 6
d. Total electrons = 17, valence electrons = 7
e. Total electrons = 18, valence electrons = 8

3.7 **a.** Sulfur: $1s^2, 2s^2, 2p^6, 3s^2, 3p^4$
b. Calcium: $1s^2, 2s^2, 2p^6, 3s^2, 3p^6, 4s^2$

3.9 **a.** [Ne] $3s^2, 3p^4$
b. [Ar] $4s^2$

3.11 **a.** Ca^{2+} and Ar are isoelectronic.
b. Sr^{2+} and Kr are isoelectronic.
c. S^{2-} and Ar are isoelectronic.
d. Mg^{2+} and Ne are isoelectronic.
e. P^{3-} and Ar are isoelectronic.

3.13 **a.** (Smallest) F, N, Be (largest)
b. (Lowest) Be, N, F (highest)
c. (Lowest) Be, N, F (highest)

3.15 **a.** Elemental properties are periodic as a function of their atomic number.
b. A horizontal row across the periodic table
c. A vertical column on the periodic table
d. A charged unit resulting from the gain or loss of electrons from a neutral atom or group of atoms

3.17 **a.** True
b. True

3.19 **a.** Na, Ni, Al
b. Na, Al
c. Na, Ni, Al
d. Ar

3.21 **a.** Sodium
b. Potassium
c. Magnesium

3.23 Group 1A (or 1): lithium, sodium, potassium, rubidium, cesium, and francium

3.25 Group VIIA (or 17): fluorine, chlorine, bromine, iodine, and astatine

3.27 The early periodic table contained many fewer elements and was arranged by atomic weight.

3.29 An element along the "staircase" boundary between metals and nonmetals; metalloids exhibit both metallic and nonmetallic properties.

3.31 **a.** One
b. One
c. Three
d. Seven
e. Zero (or eight)
f. Zero (or two)

3.33 One valence electron located in an *s* orbital and an outermost electron configuration of ns^1.

3.35 $2n^2$

3.37 A principal energy level is designated $n = 1, 2, 3$, and so forth. It is similar to Bohr's orbit in concept. A sublevel is a part of a principal energy level and is designated *s, p, d,* and *f*.

3.39 

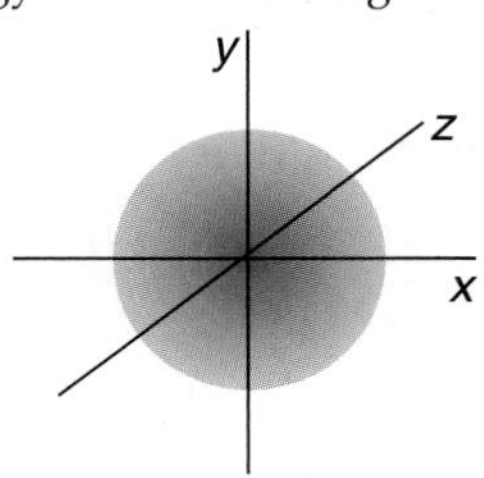

***s* orbital**

The *s* orbital represents the probability of finding an electron in a region of space surrounding the nucleus.

3.41 Three

3.43 A 3*p* orbital is a higher energy orbital than a 2*p* orbital because it is a part of a higher energy principal energy level.

3.45 $2e^-$ for $n = 1$
$8e^-$ for $n = 2$
$18e^-$ for $n = 3$

3.47 **a.** 3*p*
b. 3*s*
c. 3*d*
d. 4*s*
e. 3*d*
f. 3*p*

3.49 **a.** Not possible; $n = 1$ level can have only *s*-level orbitals.
b. Possible; it is the electron configuration of C.
c. Not possible; $n = 2$ level can contain only *s* and *p* orbitals.
d. Not possible; an *s* orbital cannot contain $3e^-$.

3.51 **a.** Li^+
b. O^{2-}
c. Ca^{2+}
d. Br^-
e. S^{2-}
f. Al^{3+}

3.53 **a.** O^{2-}, $10e^-$; Ne, $10e^-$ Isoelectronic
b. S^{2-}, $18e^-$; Cl^-, $18e^-$ Isoelectronic

3.55 Group IA metals form only a 1+ ion because the loss of one electron produces an electron configuration similar to their nearest noble gas. Group IIA metals form only a 2+ ion because the loss of two electrons produces an electron configuration similar to their nearest noble gas.

3.57 **a.** Na^+
b. S^{2-}
c. Cl^-

3.59 **a.** $1s^2, 2s^2, 2p^6, 3s^2, 3p^6$
b. $1s^2, 2s^2, 2p^6$

3.61 **a.** (Smallest) F, O, N (largest)
b. (Smallest) Li, K, Cs (largest)
c. (Smallest) Cl, Br, I (largest)

3.63 Cl

3.65 **a.** (Smallest) O, N, F (largest)
b. (Smallest) Cs, K, Li (largest)
c. (Smallest) I, Br, Cl (largest)

3.67 A positive ion is always smaller than its parent atom because the positive charge of the nucleus is shared among fewer electrons in the ion. As a result, each electron is pulled closer to the nucleus and the volume of the ion decreases.

3.69 The fluoride ion has a completed octet of electrons and an electron configuration resembling its nearest noble gas.

Chapter 4

4.1 **a.** LiBr
b. $CaBr_2$
c. Ca_3N_2

4.3 **a.** Potassium cyanide
b. Magnesium sulfide
c. Magnesium acetate

4.5 **a.** $CaCO_3$
b. $NaHCO_3$
c. Cu_2SO_4

4.7 **a.** Diboron trioxide
b. Nitrogen oxide
c. Iodine chloride
d. Phosphorus trichloride

4.9 **a.** P_2O_5
b. SiO_2

4.11 **a.**
H
H:O:

b.
H
H:C:H
H

4.13 **a.** [H:O:H with H above]$^+$

b. [:O:H]$^-$

4.15 **a.** The bonded nuclei are closer together when a double bond exists, in comparison to a single bond.
b. The bond strength increases as the bond order increases. Therefore, a double bond is stronger than a single bond. The strength of the bond is inversely related to the distance of separation of the bonded nuclei.

4.17 :O:Se::O ⇔ O::Se:O:

4.19 **a.** [:O: / C::O / :O: / H]$^-$ ⇔ [:O: / C:O: / :O: / H]$^-$

b. [:O: / :O:P:O: / :O:]$^{3-}$

4.21 **a.** H:P:H with H below; P with three H (trigonal pyramidal)

b. H:Si:H with H above and below; Si with four H (tetrahedral)

4.23 **a.** Oxygen is more electronegative than sulfur; the bond is polar. The electrons are pulled toward the oxygen atom.
b. Nitrogen is more electronegative than carbon; the bond is polar. The electrons are pulled toward the nitrogen atom.
c. There is no electronegativity difference between two identical atoms; the bond is nonpolar.
d. Chlorine is more electronegative than iodine; the bond is polar. The electrons are pulled toward the chlorine atom.

4.25 **a.** Three groups around the central atom form 120° bond angles. Because of the symmetrical arrangement of the three B—Cl bonds, their polarities cancel and the molecule is nonpolar.
b. Three groups and a lone pair of electrons surround the central atom. Because of the effect of the lone pair, the molecule is polar.
c. The H—Cl bond is polar because of the electronegativity difference between hydrogen and chlorine. Because H—Cl is the only bond in the molecule, the molecule is polar.
d. Four groups, all equivalent, surround the central atom. The structure is tetrahedral and the molecule is nonpolar.

4.27 **a.** H_2O is polar: higher melting and boiling points. C_2H_4 is nonpolar.
b. CO is polar: higher melting and boiling points. CH_4 is nonpolar.
c. NH_3 is polar: higher melting and boiling points. N_2 is nonpolar.
d. Cl_2 is nonpolar. ICl is polar: higher melting and boiling points.

4.29 **a.** Ionic
b. Covalent
c. Covalent
d. Covalent

4.31 **a.** Li· + ·Br: ⟶ Li^+ + :Br:$^-$
b. Mg: + 2·Cl: ⟶ Mg^{2+} + 2:Cl:$^-$

4.33 **a.** :Cl:N:Cl: with :Cl: below
b.
H
H:C:O:H
H
c. S::C::S

4.35 **a.** Sodium ion
b. Copper (I) ion (or cuprous ion)
c. Magnesium ion
d. Iron (II) ion (or ferrous ion)
e. Iron (III) ion (or ferric ion)

4.37 **a.** K^+
b. Br^-
c. Ca^{2+}
d. Cr^{6+}

4.39 **a.** NaCl
b. $MgBr_2$
c. CuO
d. Fe_2O_3
e. $AlCl_3$

4.41 **a.** Magnesium chloride
b. Aluminum chloride
c. Calcium sulfide
d. Sodium oxide
e. Iron (III) hydroxide

4.43 **a.** Al_2O_3
b. Li_2S
c. BH_3
d. Mg_3P_2

4.45 **a.** $NaNO_3$
b. $Mg(NO_3)_2$

c. $Al(NO_3)_3$
d. NH_4NO_3

4.47 **a.** Copper (II) sulfide
b. Copper (II) sulfate
c. Copper (II) hydroxide
d. Copper (II) oxide

4.49 Ionic solid state compounds exist in regular, repeating, three-dimensional structures; the crystal lattice. The crystal lattice is made up of positive and negative ions. Solid state covalent compounds are made up of molecules, which may be arranged in a regular crystalline pattern or in an irregular (amorphous) structure.

4.51 KCl would be a solid; it is an ionic compound, and ionic compounds are characterized by high melting points.

4.53 **a.** H·
b. He:
c. ·Ċ·
d. ·N̈·

4.55 **a.** Li^+
b. Mg^{2+}
c. $:\ddot{Cl}:^-$
d. $:\dot{P}:^{3-}$

4.57 Resonance can occur when more than one valid Lewis structure can be written for a molecule. Each individual structure that can be drawn is a resonance form. The true nature of the structure for the molecule is the resonance hybrid, which consists of the "average" of the resonance forms.

4.59

```
    H  H
    ..  ..  ..
H : C : C : O : H
    ..  ..  ..
    H  H
```

4.61

```
    H : O : H
    ..  ::  ..
H : C : C : C : H
    ..      ..
    H       H
```

4.63 **a.** Polar covalent
b. Polar covalent
c. Ionic
d. Ionic
e. Ionic

4.65 **a.** C and N
[:C≡N:]⁻
b. Si and P
[:Si≡P:]⁻
c., **d.**, and **e.** are ionic compounds.

4.67 A molecule containing no polar bonds *must* be nonpolar. A molecule containing polar bonds may or may not itself be polar. It depends upon the number and arrangement of the bonds.

4.69 Polar compounds have strong intermolecular attractive forces. Higher temperatures are needed to overcome these forces and convert the solid to a liquid; hence, we predict higher melting points for polar compounds when compared to nonpolar compounds.

Chapter 5

5.1 26.98 g

5.3 **a.** 1.51×10^{24} oxygen atoms
b. 3.01×10^{24} oxygen atoms

5.5 14.0 g He

5.7 **a.** The mass of a single unit of NH_3 is 17.04 amu/formula unit. Therefore, the mass of 1 mol of formula units is 17.04 g or 17.04 g/mol.
b. The mass of a single unit of $C_6H_{12}O_6$ is 180.18 amu/formula unit. Therefore, the mass of 1 mol of formula units is 180.18 g or 180.18 g/mol.
c. The mass of a single unit of $CoCl_2 \cdot 6H_2O$ is 237.95 amu/formula unit. Therefore, the mass of 1 mol of formula units is 237.95 g or 237.95 g/mol.

5.9 **a.** $4Fe(s) + 3O_2(g) \rightarrow 2Fe_2O_3(s)$
b. $2C_6H_6(l) + 15O_2(g) \rightarrow 12CO_2(g) + 6H_2O(g)$

5.11 **a.** 90.1 g H_2O
b. 0.590 mol LiCl

5.13 **a.** 3 mol O_2
b. 96.00 g O_2

5.15 **a.** $4Fe(s) + 3O_2(g) \rightarrow 2Fe_2O_3(s)$
b. 3.50 g Fe

5.17 **a.** 132.0 g SnF_2
b. 3.79% yield

5.19 4.00 g He/mol He

5.21 **a.** 5.00 mol He
b. 1.7 mol Na
c. 4.2×10^{-2} mol Cl

5.23 1.62×10^3 g of silver

5.25 A molecule is a single unit composed of atoms joined by covalent bonds. An ion pair is composed of positive and negatively charged ions joined by electrostatic attraction, the ionic bond. The ion pairs, unlike the molecule, do not form single units; the electrostatic charge is directed to other ions in a crystal lattice, as well.

5.27 **a.** 58.44 g/mol
b. 142.04 g/mol
c. 357.49 g/mol

5.29 32.00 g/mol

5.31 **a.** 0.257 mol NaCl
b. 0.106 mol Na_2SO_4

5.33 **a.** 18.02 g H_2O
b. 116.9 g NaCl

5.35 **a.** 40.0 g He
b. 2.02×10^2 g H

5.37 **a.** 2.43 g Mg
b. 10.0 g $CaCO_3$
c. 18.0 g $C_6H_{12}O_6$
d. 5.84 g NaCl

5.39 **a.** 0.420 mol KBr
b. 0.415 mol $MgSO_4$
c. 0.313 mol Br_2
d. 0.935 mol NH_4Cl

5.41 The ultimate basis for a correct chemical equation is the law of conservation of mass.

5.43 The subscript provides the number of atoms or ions in one unit of a compound.

5.45 **a.** $2C_4H_{10}(g) + 13O_2(g) \rightarrow 10H_2O(g) + 8CO_2(g)$
b. $Au_2S_3(s) + 3H_2(g) \rightarrow 2Au(s) + 3H_2S(g)$
c. $Al(OH)_3(s) + 3HCl(aq) \rightarrow AlCl_3(aq) + 3H_2O(l)$
d. $(NH_4)_2Cr_2O_7(s) \rightarrow Cr_2O_3(s) + N_2(g) + 4H_2O(g)$
e. $C_2H_5OH(l) + 3O_2(g) \rightarrow 2CO_2(g) + 3H_2O(g)$

5.47 **a.** $N_2(g) + 3H_2(g) \rightarrow 2NH_3(g)$
b. $HCl(aq) + NaOH(aq) \rightarrow NaCl(aq) + H_2O(l)$

5.49 **a.** $C_6H_{12}O_6(s) + 6O_2(g) \rightarrow 6H_2O(l) + 6CO_2(g)$
b. $Na_2CO_3(s) \xrightarrow{\Delta} Na_2O(s) + CO_2(g)$

5.51 50.3 g B_2O_3

5.53 104 g $CrCl_3$

5.55 **a.** $N_2(g) + 3H_2(g) \rightarrow 2NH_3(g)$
b. Three moles of H_2 will react with one mole of N_2.
c. One mole of N_2 will produce two moles of the product NH_3.

d. 1.50 mol H_2
e. 17.0 g NH_3

5.57 **a.** The mass of a single unit of $C_5H_{11}NO_2S$ is 149.21 amu/formula unit. Therefore the mass of a mole of $C_5H_{11}NO_2S$ formula units is 149.21 g/mol.
b. 1.20×10^{24} O atoms
c. 32.00 g O
d. 10.7 g O

5.59 7.39 g O_2
5.61 6.14×10^4 g O_2
5.63 70.6 g $C_{10}H_{22}$
5.65 9.13×10^2 g N_2
5.67 92.6%
5.69 6.85×10^2 g N_2

Chapter 6

6.1 **a.** 0.954 atm
b. 0.382 atm
c. 0.730 atm
6.3 **a.** 38 atm
b. 25 atm
6.5 **a.** 3.76 L
b. 3.41 L
c. 2.75 L
6.7 0.200 atm
6.9 4.46 mol H_2
6.11 9.00 L
6.13 0.223 mol N_2
6.15 1 atm
6.17 5 L-atm
6.19 5.23 atm
6.21 Charles's law states that the volume of a gas varies directly with the absolute temperature if pressure and number of moles of gas are constant.
6.23 The volume increases from 2.00 L to 2.96 L.
6.25 1.51 L
6.27 Increase
6.29 $V_f = \dfrac{P_iV_iT_f}{P_fT_i}$
6.31 1.82×10^{-2} L
6.33 6.00 L
6.35 Avogadro's law states that equal volumes of a gas contain the same number of moles if measured under the same conditions of temperature and pressure.
6.37 5.94×10^{-2} L
6.39 22.4 L
6.41 9.08×10^3 L
6.43 172°C
6.45 Gases exhibit more ideal behavior at low pressures. At low pressures, gas particles are more widely separated and therefore the attractive forces between particles are less. The ideal gas model assumes negligible attractive forces between gas particles.
6.47 The kinetic molecular theory states that the average kinetic energy of the gas particles increases as the temperature increases. Kinetic energy is proportional to $(\text{velocity})^2$. Therefore, as the temperature increases the gas particle velocity increases and the rate of mixing increases as well.
6.49 Dalton's law states that the total pressure of a mixture of gases is the sum of the partial pressures of the component gases.
6.51 0.74 atm
6.53 Intermolecular forces in liquids are considerably stronger than intermolecular forces in gases. Particles are, on average, much closer together in liquids and the strength of attraction is inversely proportional to the distance of separation.
6.55 The vapor pressure of a liquid increases as the temperature of the liquid increases.
6.57 Evaporation is the conversion of a liquid to a gas at a temperature lower than the boiling point of the liquid. Condensation is the conversion of a gas to a liquid at a temperature lower than the boiling point of the liquid.
6.59 Viscosity is the resistance to flow caused by intermolecular attractive forces. Complex molecules may become entangled and not slide smoothly across one another.
6.61 Solids are essentially incompressible because the average distance of separation among particles in the solid state is small. There is literally no space for the particles to crowd closer together.
6.63 **a.** High melting temperature, brittle
b. High melting temperature, hard
6.65 Beryllium; metallic solids are good electrical conductors. Carbon forms covalent solids, which are poor electrical conductors.

Chapter 7

7.1 **a.** DR
b. SR
c. DR
d. D
7.3 **a.** $KCl(aq) + AgNO_3(aq) \rightarrow KNO_3(aq) + AgCl(s)$
A precipitation reaction occurs.
b. $CH_3COOK(aq) + AgNO_3(aq) \rightarrow$ no reaction
No precipitation reaction occurs.
7.5 16.7% NaCl
7.7 7.50% KCl
7.9 2.56×10^{-2}%
7.11 20.0%
7.13 0.125 mol HCl
7.15 Dilute 1.7×10^{-2} L of 12 *M* HCl with sufficient water to produce 1.0×10^2 mL of total solution.
7.17 1.0×10^{-2} osmol
7.19 0.24 atm
7.21 CO is more soluble than CO_2. CO is a polar molecule as is water and "like dissolves like."
7.23 0.0154 mol/L
7.25 **a.** Heating an alkaline earth metal carbonate,
$MgCO_3(s) \xrightarrow{\Delta} MgO(s) + CO_2(g)$
b. The replacement of copper by zinc in copper sulfate,
$Zn(s) + CuSO_4(aq) \rightarrow ZnSO_4(aq) + Cu(s)$
7.27 Reaction of two soluble substances to form an insoluble product
$2NaOH(aq) + FeCl_2(aq) \rightarrow Fe(OH)_2(s) + 2NaCl(aq)$
7.29 **a.** $2C_2H_6(g) + 7O_2(g) \rightarrow 4CO_2(g) + 6H_2O(g)$
b. $6K_2O(s) + P_4O_{10}(s) \rightarrow 4K_3PO_4(s)$
c. $MgBr_2(aq) + H_2SO_4(aq) \rightarrow 2HBr(g) + MgSO_4(aq)$
7.31 **a.** $Ca(s) + F_2(g) \rightarrow CaF_2(s)$
b. $2Mg(s) + O_2(g) \rightarrow 2MgO(s)$
c. $3H_2(g) + N_2(g) \rightarrow 2NH_3(g)$
7.33 **a.** 2.00% NaCl
b. 6.60% $C_6H_{12}O_6$
7.35 **a.** 5.00% ethanol
b. 10.0% ethanol
7.37 **a.** 21.0% NaCl
b. 3.75% NaCl
7.39 **a.** 2.25 g NaCl
b. 3.13 g CH_3COONa

7.41 a. 0.342 M NaCl
b. 0.367 M $C_6H_{12}O_6$
7.43 a. 1.46 g NaCl
b. 9.00 g $C_6H_{12}O_6$
7.45 0.146 M $C_{12}H_{22}O_{11}$
7.47 5.00×10^{-2} L
7.49 20.0 M
7.51 A colligative property is a solution property that depends on the concentration of solute particles rather than the identity of the particles.
7.53 Salt is an ionic substance that dissociates in water to produce positive and negative ions. These ions (or particles) lower the freezing point of water. If the concentration of salt particles is large, the freezing point may be depressed below the surrounding temperature, and the ice would melt.
7.55 0.5 M sucrose
7.57 0.5 M sucrose
7.59 24 atm at 25°C
7.61 Polar, high boiling point, low vapor pressure, abundant, and easily purified
7.63 The ammonia converts to the extremely soluble and stable ammonium ion.
7.65

```
        H  H
         :O:

         Na+
  ..           ..
H:O:          :O:H
  ..           ..
  H            H
```

Several water molecules "hydrate" each sodium ion.
7.67 A low sodium ion concentration in the dialysis solution favors transport of sodium ions from the blood.
7.69 Confusion, stupor, or coma
7.71 Diabetes, diarrhea, and certain high-protein diets

Chapter 8

8.1 a. Exothermic
b. Exothermic
c. Exothermic
8.3 13°C
8.5 2.7×10^3 J
8.7 $\dfrac{2.1 \times 10^2 \text{ nutritional Cal}}{\text{candy bar}}$
8.9 Heat energy produced by the friction of striking the match provides the activation energy for this combustion process.
8.11 If the enzyme catalyzed a process needed to sustain life, the substance interfering with that enzyme would be classified as a poison.
8.13 a. rate $= k[N_2]^n[O_2]^{n'}$ (n and n' are experimentally determined)
b. rate $= k[C_4H_6]^n$ (n must be experimentally determined)
8.15 At rush hour, approximately the same number of passengers enter and exit the train at each stop. At any time the number of passengers may be essentially unchanged but the identity of the individual passengers is continually changing.
8.17 Measure the concentration of products and reactants at different times until no further concentration change is observed.
8.19 a. $K_{eq} = \dfrac{[N_2][O_2]^2}{[NO_2]^2}$
b. $K_{eq} = [H_2]^2[O_2]$
8.21 A large K_{eq} favors products.
8.23 8.2×10^{-2}
8.25 a. Decrease
b. Increase
c. Decrease
d. Remain the same
8.27 a. An exothermic reaction is one in which energy is released during chemical change.
b. An endothermic reaction is one in which energy is absorbed during chemical change.
c. A calorimeter is a device for measuring heat absorbed or released during chemical change.
8.29 Enthalpy is a measure of heat energy.
8.31 1.20×10^3 cal
8.33 5.02×10^3 J
8.35 a. Entropy increases. Conversion of a solid to a liquid results in an increase in disorder of the substance. Solids retain their shape whereas liquids will flow and their shape is determined by their container.
b. Entropy increases. Conversion of a liquid to a gas results in an increase in disorder of the substance. Gas particles move randomly with very weak interactions between particles, much weaker than those interactions in the liquid state.
8.37 An increase in stability is equated with a decrease in energy (reaching a lower energy state). The energy of products is less than that of the reactants in an exothermic reaction; energy is given off in an exothermic reaction.
8.39 Isopropyl alcohol quickly evaporates after being applied to the skin. Conversion of a liquid to a gas requires heat energy. The heat energy is supplied by the skin. When this heat is lost, the skin temperature drops.
8.41 The activated complex is the arrangement of reactants in an unstable transition state as a chemical reaction proceeds. The activated complex must form to convert reactants to products.
8.43

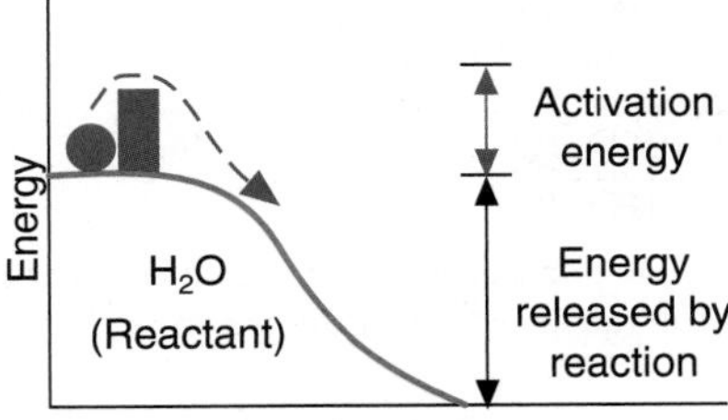

(a) Noncatalyzed reaction

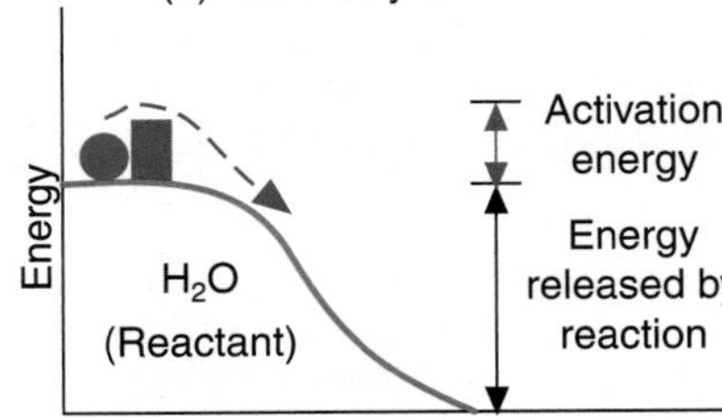

(b) Catalyzed reaction

8.45 Enzymes are biological catalysts. The enzyme lysozyme catalyzes a process that results in the destruction of the cell walls of many harmful bacteria. This helps to prevent disease in organisms.

The breakdown of foods to provide material for construction and repair of body tissue, as well as energy, is catalyzed by a variety of enzymes. For example, amylase begins the hydrolysis of starch in the mouth.
8.47 An increase in concentration of reactants means that there are more molecules in a certain volume. The probability of

collision is enhanced because molecules travel a shorter distance before meeting another molecule. The rate is proportional to the number of collisions per unit time.

8.49 Rate = $k[N_2O_4]^n$

8.51 A catalyst speeds up a chemical reaction by facilitating the formation of the activated complex, thus lowering the activation energy, the energy barrier for the reaction.

8.53 A dynamic equilibrium has fixed concentrations of all reactants and products—these concentrations do not change with time. However, the process is dynamic because products and reactants are continuously being formed and consumed. The concentrations do not change because the *rates* of production and consumption are equal.

8.55 $K_{eq} = \frac{[NO_2]^2}{[N_2O_4]}$

8.57 A physical equilibrium describes physical change; examples include the equilibrium between ice and water, or the equilibrium vapor pressure of a liquid.
A chemical equilibrium describes chemical change; examples include the reactions shown in Questions 8.49 and 8.50.

8.59 **a.** Equilibrium shifts to the left.
b. No change
c. No change

8.61 **a.** False; A slow reaction may go to completion, but take a longer time.
b. False; The rates of forward and reverse reactions are equal in a dynamic equilibrium situation.

8.63 **a.** PCl_3 increases.
b. PCl_3 decreases.
c. PCl_3 decreases.
d. PCl_3 decreases.
e. PCl_3 remains the same.

8.65 Decrease

8.67 $K_{eq} = \frac{[CO][H_2]}{[H_2O]}$

8.69 False. The position of equilibrium is not affected by a catalyst, only the rate at which equilibrium is attained.

Chapter 9

9.1 **a.** $HF(aq) + H_2O(l) \rightleftarrows H_3O^+(aq) + F^-(aq)$
b. $NH_3(aq) + H_2O(l) \rightleftarrows NH_4^+(aq) + OH^-(aq)$

9.3 **a.** HF and F^-; H_2O and H_3O^+
b. NH_3 and NH_4^+; H_2O and OH^-

9.5 **a.** NH_4^+
b. H_2SO_4

9.7 A 1.0×10^{-3} *M* HCl solution corresponds to a pH = 3.00. A solution of hydrochloric acid with a pH = 4.00 corresponds to an HCl concentration of 1.0×10^{-4} *M*.

9.9 12.00.

9.11 $[H_3O^+] = 3.2 \times 10^{-9}$ *M*

9.13 0.1000 *M* NaOH

9.15 $CO_2 + H_2O \rightleftarrows H_2CO_3 \rightleftarrows H_3O^+ + HCO_3^-$
An increase in the partial pressure of CO_2 is a stress on the left side of the equilibrium. The equilibrium will shift to the right in an effort to decrease the concentration of CO_2. This will cause the molar concentration of H_2CO_3 to increase.

9.17 $CO_2 + H_2O \rightleftarrows H_2CO_3 \rightleftarrows H_3O^+ + HCO_3^-$
In Question 9.15, the equilibrium shifts to the right. Therefore the molar concentration of H_3O^+ should increase. In Question 9.16, the equilibrium shifts to the left. Therefore the molar concentration of H_3O^+ should decrease.

9.19 4.87

9.21 4.76

9.23 4.87

9.25 **a.** An Arrhenius acid is a substance that dissociates, producing hydrogen ions.
b. A Brønsted-Lowry acid is a substance that behaves as a proton donor.

9.27 The Brønsted-Lowry theory provides a broader view of acid-base theory than does the Arrhenius theory. Brønsted-Lowry emphasizes the role of the solvent in the dissociation process.

9.29 **a.** $HNO_2(aq) + H_2O(l) \rightleftarrows H_3O^+ (aq) + NO_2^-(aq)$
b. $HCN(aq) + H_2O(l) \rightleftarrows H_3O^+ (aq) + CN^-(aq)$

9.31 **a.** HNO_2 and NO_2^-; H_2O and H_3O^+
b. HCN and CN^{-1}; H_2O and H_3O^+

9.33 **a.** Weak
b. Weak
c. Weak

9.35 **a.** CN^- and HCN; NH_3 and NH_4^+
b. CO_3^{2-} and HCO_3^-; Cl^- and HCl

9.37 Concentration refers to the quantity of acid or base contained in a specified volume of solvent. Strength refers to the degree of dissociation of the acid or base.

9.39 **a.** 1.0×10^{-7} *M*
b. 1.0×10^{-11} *M*

9.41 **a.** Neutral
b. Basic

9.43 **a.** pH = 7.00
b. pH = 5.00

9.45 **a.** $[H_3O^+] = 1.0 \times 10^{-1}$ *M*
$[OH^-] = 1.0 \times 10^{-13}$ *M*
b. $[H_3O^+] = 1.0 \times 10^{-9}$ *M*
$[OH^-] = 1.0 \times 10^{-5}$ *M*

9.47 **a.** $[H_3O^+] = 5.0 \times 10^{-2}$ *M*
$[OH^-] = 2.0 \times 10^{-13}$ *M*
b. $[H_3O^+] = 2.0 \times 10^{-10}$ *M*
$[OH^-] = 5.0 \times 10^{-5}$ *M*

9.49 A neutralization reaction is one in which an acid and a base react to produce water and a salt.

9.51 **a.** $[H_3O^+] = 1.0 \times 10^{-6}$ *M*
$[OH^-] = 1.0 \times 10^{-8}$ *M*
b. $[H_3O^+] = 6.3 \times 10^{-6}$ *M*
$[OH^-] = 1.6 \times 10^{-9}$ *M*
c. $[H_3O^+] = 1.6 \times 10^{-8}$ *M*
$[OH^-] = 6.3 \times 10^{-7}$ *M*

9.53 **a.** 1×10^2
b. 1×10^4
c. 1×10^{10}

9.55 **a.** 1×10^{-5}
b. 1×10^{-12}
c. 3.2×10^{-6}

9.57 **a.** 6.00
b. 8.00
c. 3.25

9.59 **a.** NH_3 and NH_4Cl can form a buffer solution.
b. HNO_3 and KNO_3 cannot form a buffer solution.

9.61 **a.** A buffer solution contains components (a weak acid and its salt or a weak base and its salt) that enable the solution to resist large changes in pH when acids or bases are added.
b. Acidosis is a medical condition characterized by higher than normal levels of CO_2 in the blood and lower than normal blood pH.

9.63 **a.** Addition of strong acid is equivalent to adding H_3O^+. This is a stress on the right side of the equilibrium and the equilibrium will shift to the left. Consequently the $[CH_3COOH]$ increases.

b. Water, in this case, is a solvent and does not appear in the equilibrium expression. Hence, it does not alter the position of the equilibrium.

9.65 $[H_3O^+] = 2.32 \times 10^{-7} M$

9.67 a. Oxidation is the loss of electrons, loss of hydrogen atoms, or gain of oxygen atoms.

b. An oxidizing agent removes electrons from another substance. In doing so the oxidizing agent becomes reduced.

9.69 Loses

9.71 Oxidized

9.73 $Cl_2 + 2KI \longrightarrow 2KCl + I_2$

Cl_2: substance reduced, oxidizing agent

KI: substance oxidized, reducing agent

9.75 An oxidation-reduction reaction must take place to produce electron flow in a voltaic cell.

9.77 Storage battery

Chapter 10

10.1 X-ray, ultraviolet, visible, infrared, microwave, and radiowave

10.3 a. $^{85}_{37}Rb$

b. $^{226}_{88}Ra$

10.5 6.3 ng

10.7 1/4

10.9 Isotopes with short half-lives release their radiation rapidly. There is much more radiation per unit time observed with short half-life substances; hence, the signal is stronger and the sensitivity of the procedure is enhanced.

10.11 The rem takes into account the relative biological effect of the radiation in addition to the quantity of radiation. This provides a more meaningful estimate of potential radiation damage to human tissue.

10.13 a. Natural radioactivity is the spontaneous decay of a nucleus to produce high-energy particles or rays.

b. Background radiation is radiation from natural sources.

10.15 a. A beta particle is an electron formed in the nucleus by the conversion of a neutron into a proton.

b. Gamma radiation is high-energy emission from nuclear processes.

10.17 a. $^{4}_{2}He$

b. $^{-1}_{0}e$

10.19 a. $^{2}_{1}H$

b. $^{3}_{1}H$

10.21 Alpha and beta particles are matter; gamma radiation is pure energy. Alpha particles are large and relatively slow moving. They are the least energetic and least penetrating. Gamma radiation moves at the speed of light, is highly energetic, and is most penetrating.

10.23 A helium atom has two electrons; an α particle has no electrons.

10.25 $^{60}_{27}Co \rightarrow ^{60}_{28}Ni + ^{0}_{-1}\beta + \gamma$

10.27 $^{24}_{11}Na$

10.29 $^{24}_{11}Na$

10.31 $^{140}_{55}Cs$

10.33 Natural radioactivity is a spontaneous process; artificial radioactivity is nonspontaneous and results from a nuclear reaction that produces an unstable nucleus.

10.35 Nuclei for light atoms tend to be most stable if their neutron/proton ratio is close to 1. Nuclei with more than 84 protons tend to be unstable. Isotopes with a "magic number" of protons and neutrons (2, 8, 20, 50, 82, or 126 protons or neutrons) tend to be stable. Isotopes with even numbers of protons or neutrons tend to be more stable.

10.37 0.40 mg of iodine-131 remains after 24 days.

10.39 13 mg of iron-59 remains after 135 days.

10.41 Fission splits nuclei to produce energy.

10.43 a. The fission process involves the breaking down of large, unstable nuclei into smaller, more stable nuclei. This process releases energy in the form of heat and/or light.

b. The heat generated during the fission process could be used to generate steam, which is then used to drive a turbine to generate electricity.

10.45 $^{3}_{1}H + ^{1}_{1}H \rightarrow ^{4}_{2}He + \text{energy}$

10.47 A "breeder" reactor creates the fuel that can be used by a conventional fission reactor during its fission process.

10.49 The reaction in a fission reactor that involves neutron production and causes subsequent reactions accompanied by the production of more neutrons in a continuing process.

10.51 High operating temperatures

10.53 Radiocarbon dating is a process used to determine the age of objects. The ratio of the masses of the stable isotope, carbon-12, and unstable isotope, carbon-14, is measured. Using this value and the half-life of carbon-14, the age of the coffin may be calculated.

10.55 $^{108}_{47}Ag + ^{4}_{2}\alpha \rightarrow ^{112}_{49}In$

$^{112}_{49}In$ is the intermediate isotope of indium.

10.57 a. Technetium-99m is used to study the heart (cardiac output, size, and shape), kidney (follow-up procedure for kidney transplant), and liver and spleen (size, shape, presence of tumors).

b. Xenon-133 is used to locate regions of reduced ventilation and presence of tumors in the lung.

10.59 Radiation therapy provides sufficient energy to destroy molecules critical to the reproduction of cancer cells.

10.61 a. The level of radiation exposure decreases as the distance from the radioactive source increases.

b. Wearing gloves provides a level of shielding that is very efficient for α and β radiation, but totally ineffective for γ radiation.

10.63 Background radiation, radiation from natural sources, is emitted by the sun as cosmic radiation, and from naturally radioactive isotopes found throughout our environment.

10.65 A film badge detects gamma radiation by darkening photographic film in proportion to the amount of radiation exposure over time. Badges are periodically collected and evaluated for their level of exposure. This mirrors the level of exposure of the personnel wearing the badges.

10.67 Relative biological effect is a measure of the damage to biological tissue caused by different forms of radiation.

10.69 a. The curie is the amount of radioactive material needed to produce 3.7×10^{10} atomic disintegrations per second.

b. The roentgen is the amount of radioactive material needed to produce 2×10^9 ion-pairs when passing through 1 cc of air at 0°C.

Chapter 11

11.1 The student could test the solubility of the substance in water and in an organic solvent, such as hexane. Solubility in hexane would suggest an organic substance, whereas solubility in water would suggest an inorganic compound. The student could also determine the melting and boiling points of the substance. If the melting and boiling points are very high, an inorganic substance would be suspected.

11.3 **a.** 2,3-Dimethylbutane
b. 2,2-Dimethylpentane
c. 2,2-Dimethylpropane
d. 1,2,3-Tribromopropane

11.5 **a.** The straight chain isomers of molecular formula C_4H_9Br:

```
   H  H  H  H            H  H  Br H
   |  |  |  |            |  |  |  |
H—C—C—C—C—Br         H—C—C—C—C—H
   |  |  |  |            |  |  |  |
   H  H  H  H            H  H  H  H
```

b. The straight chain isomers of molecular formula $C_4H_8Br_2$:

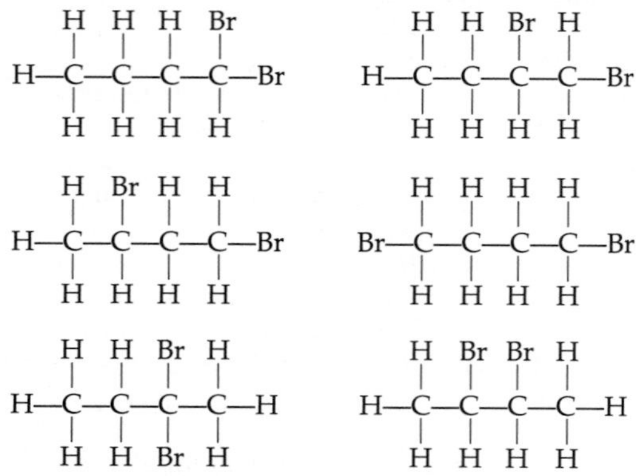

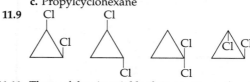

11.7 **a.** 1-Bromo-2-ethylcyclobutane
b. *trans*-1, 2-Dimethylcyclopropane
c. Propylcyclohexane

11.9 Cl Cl

Cl Cl Cl Cl Cl

Cl Cl

11.11 Three of the six axial hydrogen atoms of cyclohexane lie above the ring. The remaining three hydrogen atoms lie below the ring.

11.13 **a.** The combustion of cyclobutane:

$$\text{(cyclobutane)} + 6O_2 \longrightarrow 4CO_2 + 4H_2O + \text{heat energy}$$

b. The monobromination of propane will produce two products as shown in the following two equations:

```
   H  H  H                          H  Br H
   |  |  |       Light or heat      |  |  |
H—C—C—C—H + Br2  ───────────►   H—C—C—C—H + HBr
   |  |  |                          |  |  |
   H  H  H                          H  H  H

   H  H  H                          H  H  Br
   |  |  |       Light or heat      |  |  |
H—C—C—C—H + Br2  ───────────►   H—C—C—C—H + HBr
   |  |  |                          |  |  |
   H  H  H                          H  H  H
```

c. $2CH_3CH_3 + 7O_2 \rightarrow 4CO_2 + 6H_2O + \text{energy}$

d.

```
   H  H  H  H                         H  H  H  H
   |  |  |  |      Heat or light      |  |  |  |
H—C—C—C—C—H + Cl2  ───────────►   H—C—C—C—C—H + HCl
   |  |  |  |                         |  |  |  |
   H  H  H  H                         H  Cl H  H

   H  H  H  H                         H  H  H  H
   |  |  |  |      Heat or light      |  |  |  |
H—C—C—C—C—H + Cl2  ───────────►   H—C—C—C—C—H + HCl
   |  |  |  |                         |  |  |  |
   H  H  H  H                         Cl H  H  H
```

11.15 The products in the reactions in Problem 11.13b are 1-bromopropane and 2-bromopropane. The products of the reactions in Problem 11.13d are 1-chlorobutane and 2-chlorobutane.

11.17 **a.** Water-soluble inorganic compounds
b. Inorganic compounds
c. Organic compounds
d. Inorganic compounds
e. Organic compounds

11.19 **a.**

```
      H           H
      |           |
   H—C—H     H—C—H
   H  |     H     |  H
   |  |     |     |  |
H—C—C———C———C—C—H
   |  |     |     |  |
   H  H     H     H  H
```

b.

```
   H  Br H  H
   |  |  |  |
H—C—C—C—C—H
   |  |  |  |
   H  H  Br H
```

c.

```
               H
               |
            H—C—H
   H  H  H  H  |  H  H
   |  |  |  |  |  |  |
H—C—C—C—C—C—C—C—H
   |  |  |  |  |  |  |
   H  H  |  H  H  H  H
         H—C—H
            |
            H
```

d.

```
   H  H  H  H  H  Br H
   |  |  |  |  |  |  |
H—C—C—C—C—C—C—C—H
   |  |  |  |  |  |  |
   H  H  H  H  H  H  H
```

11.21 Structure b is not possible because there are five bonds to carbon-2. Structure d is not possible because there are five bonds to carbon-3. Structure e is not possible because there are five bonds to carbon-3. Structure f is not possible because there are five bonds to carbon-3.

11.23 An alcohol

```
   H  H
   |  |
H—C—C—OH
   |  |
   H  H
```

An aldehyde

```
   H  O
   |  ||
H—C—C—H
   |
   H
```

A ketone

```
   H  O  H
   |  ||  |
H—C—C—C—H
   |     |
   H     H
```

A carboxylic acid

```
   H  O
   |  ||
H—C—C—OH
   |
   H
```

An amine

```
   H  H    H
   |  |   /
H—C—C—N
   |  |   \
   H  H    H
```

11.25 **a.** C_nH_{2n+2}
b. C_nH_{2n-2}

c. C_nH_{2n}
d. C_nH_{2n}
e. C_nH_{2n-2}

11.27 Alkanes have only carbon-to-carbon and carbon-to-hydrogen single bonds, as in the molecule ethane:

$H_3C{-}CH_3$ (H—C—C—H, each carbon bearing two additional H)

Alkenes have at least one carbon-to-carbon double bond, as in the molecule ethene:

$H_2C{=}CH_2$

Alkynes have at least one carbon-to-carbon triple bond, as in the molecule ethyne:

H—C≡C—H

11.29 a. A carboxylic acid:

$CH_3CH_2{-}C({=}O){-}OH$

b. An amine: $CH_3CH_2CH_2{-}NH_2$
c. An alcohol: $CH_3CH_2CH_2{-}OH$
d. An ether: $CH_3CH_2{-}O{-}CH_2CH_3$

11.31 a. $CH_3CH(Br)CH_2CH_3$

b. $CH_3{-}C(Cl)(CH_3){-}CH_3$

c. $CH_3{-}C(CH_3)_2{-}CH_2CH_2CH_2CH_3$

d. $I{-}C(Cl)(I){-}Cl$

e. CH_3CH_2—(cyclohexane ring)—CH_2CH_3

f. $CH_3{-}C(CH_3)(I){-}CH_2{-}C(CH_3)_2{-}CH_3$

11.33 a. 3-Methylpentane
b. 1-Bromoheptane
c. 3-Ethyl-5-methylheptane
d. 2-Bromo-2-methylpropane
e. 2, 5-Dimethylhexane
f. 1-Chloro-3-methylbutane
g. 1, 4-Dichloropentane

11.35 a. 2,3-Dichloropentane
b. 2,4,4-Trimethylhexane
c. 2,3,4-Trimethylhexane
d. 1,1-Dibromobutane

11.37 a. Identical
b. Identical
c. Identical
d. Isomers

11.39 Structures a and c are incorrect.

11.41 First, determine the name of the parent compound, the longest continuous carbon chain in the compound. Number the parent chain to give the lowest number to the carbon bonded to the first substituent encountered. Place the names and numbers of the substituents before the name of the parent compound. Substituents are listed in alphabetical order.

11.43 a. Chlorocyclopropane
b. *cis*-1, 2-Dichlorocyclopropane
c. *trans*-1, 2-Dichlorocyclopropane
d. Bromocyclobutane
e. *cis*-1, 3-Dibromocyclobutane
f. *cis*-1-Bromo-3-chlorocyclobutane
g. *cis*-1-Chloro-4-methylcyclohexane
h. *trans*-1, 3-Dimethylcyclohexane

11.45 C_nH_{2n}

11.47 a. Incorrect—1, 2-Dibromocyclobutane
b. Incorrect—1, 2-Diethylcyclobutane
c. Correct
d. Incorrect—1, 2, 3-Trichlorocyclohexane

11.49 a. Br, Br (cyclopentane ring with Br on two adjacent carbons)

b. CH_3, CH_3 (cyclobutane ring with CH_3 on two adjacent carbons)

c. Cl, Cl (cyclopropane ring with Cl on two adjacent carbons)

d. CH_2CH_3, CH_2CH_3 (cyclohexane ring with CH_2CH_3 on carbons 1 and 4)

11.51 a. *cis*-1, 2-Dibromocyclopentane
b. *trans*-1, 3-Dibromocyclopentane
c. *cis*-1, 2-Dimethylclohexane
d. *cis*-1, 2-Dimethylcyclopropane

11.53 In the chair conformation the hydrogen atoms, and thus the electron pairs of the C—H bonds, are farther from one another. As a result, there is less electron repulsion and the structure is more stable (more energetically favored). In the boat conformation, the electron pairs are more crowded. This causes greater electron repulsion, producing a less stable, less energetically favored conformation.

11.55 Because conformations are freely and rapidly interconverted, they cannot be separated from one another.

11.57 One conformation is more stable than the other because the electron pairs of the carbon-hydrogen bonds are farther from one another.

11.59 a. $8CO_2 + 10H_2O$

b. $Br{-}C(CH_3)_2{-}CH_3 + CH_3CH(CH_3)CH_2Br + 2\ HBr$

c. Cl_2 + light

11.61 The following molecules are all isomers of C_6H_{14}.

$CH_3CH_2CH_2CH_2CH_2CH_3$
Hexane

$CH_3CH(CH_3)CH_2CH_2CH_3$
2-Methylpentane

$CH_3CH_2CH(CH_3)CH_2CH_3$
3-Methylpentane

$CH_3CH(CH_3)CH(CH_3)CH_3$
2,3-Dimethylbutane

$$\begin{array}{c} CH_3 \\ | \\ CH_3CCH_2CH_3 \\ | \\ CH_3 \end{array}$$

2,2-Dimethylbutane

a. 2, 3-Dimethylbutane produces only two monobrominated derivatives: 1-bromo-2, 3-dimethylbutane and 2-bromo-2, 3-dimethylbutane.
b. Hexane produces three monobrominated products: 1-bromohexane, 2-bromohexane, and 3-bromohexane. 2, 2-Dimethylbutane also produces three monobrominated products: 1-bromo-2, 2-dimethylbutane, 2-bromo-3, 3-dimethylbutane, and 1-bromo-3, 3-dimethylbutane.
c. 3-Methylpentane produces four monobrominated products: 1-bromo-3-methylpentane, 2-bromo-3-methylpentane, 3-bromo-3-methylpentane, and 1-bromo-2-ethylbutane.

11.63 The hydrocarbon is cyclooctane, having a molecular formula of C_8H_{16}.

$$\text{(cyclooctane)} + 12O_2 \longrightarrow 8CO_2 + 8H_2O$$

Chapter 12

12.1 a. $Br{-}CH_2{-}CH_2{-}C{\equiv}C{-}CH_2{-}CH_3$ (Br—C—C—C≡C—C—C—H with H's on each sp³ carbon)

b. $H_3C{-}C{\equiv}C{-}CH_3$ (H—C—C≡C—C—H with H's on each sp³ carbon)

c. Cl—C≡C—Cl

d. $H{-}C{\equiv}C{-}CH_2{-}CH_2{-}CH_2{-}CH_2{-}CH_2{-}CH_2{-}CH_2{-}I$ (H—C≡C—C—C—C—C—C—C—C—I with H's on each sp³ carbon)

12.3 a. cis-3-Hexene: $CH_3CH_2(H)C{=}C(H)CH_2CH_3$ (H, H on the same side); trans-3-Hexene: $CH_3CH_2(H)C{=}C(CH_2CH_3)H$ (H on opposite sides)

cis-3-Hexene *trans*-3-Hexene

b. Br, CH_3 / C=C / CH_3, Br

trans-2,3-Dibromo-2-butene

Br, Br / C=C / CH_3, CH_3

cis-2,3-Dibromo-2-butene

12.5 a. CH_3CH_2, $CH_2CH_2CH_2CH_3$ / C=C / H, H

b. CH_3, H / C=C / H, $CH_2CHClCH_3$

c. CH_3, Cl / C=C / Cl, CH_3

12.7 The hydrogenation of the *cis* and *trans* isomers of 2-pentene would produce the same product, pentane.

12.9 a. $CH_3CH{=}CHCH_3 + H_2O \xrightarrow{H^+} CH_3CH(OH)CH_2CH_3$ (only product)

b. $CH_2{=}CHCH_2CH_2CH(CH_3)CH_3 + H_2O \xrightarrow{H^+} CH_3CH(OH)CH_2CH_2CH(CH_3)CH_3$ (major product)

$CH_2{=}CHCH_2CH_2CH(CH_3)CH_3 + H_2O \xrightarrow{H^+} CH_2(OH)CH_2CH_2CH_2CH(CH_3)CH_3$ (minor product)

c. $CH_3CH_2CH_2CH{=}CHCH_2CH_3 + H_2O \xrightarrow{H^+} CH_3CH_2CH_2CH(OH)CH_2CH_2CH_3$

$CH_3CH_2CH_2CH{=}CHCH_2CH_3 + H_2O \xrightarrow{H^+} CH_3CH_2CH_2CH_2CH(OH)CH_2CH_3$

These products will be formed in approximately equal amounts.

d. $CH_3CHClCH{=}CHCHClCH_3 + H_2O \xrightarrow{H^+} CH_3CHClCH(OH)CH_2CHClCH_3$ (only product)

12.11 a. Reactant—*cis*-2-butene; Only product—butane
b. Reactant—1-butene; Major product—2-butanol
c. Reactant—2-butene; Only product—2, 3-dichlorobutane
d. Reactant—1-pentene; Major product—2-bromopentane

12.13 a. benzene ring with Cl, Cl, Cl (1,3,5-trichlorobenzene)
b. benzene ring with OH and CH_3 (ortho)
c. benzene ring with OH, Br (ortho) and Br (para to Br... 2,5-dibromophenol)
d. benzene ring with NO_2 and NO_2 (para)
e. benzene ring with NH_2 and NO_2 (ortho)
f. benzene ring with CH_3 and NO_2 (meta)

12.15 Alkane: C_nH_{2n+2}. alkene: C_nH_{2n} alkyne: C_nH_{2n-2}.

12.17 Planar

12.19 Linear

12.21 **a.** $(CH_3)_2C{=}CH(CH_2CH_2CH_3)$ — CH_3, CH_3 on one carbon; $CH_2CH_2CH_3$, H on the other
b. CH_3CH_2 and H / H and $CH_2CH_2CH_3$ on C=C (*trans*)
c. $ClCH_2$ and CH_2CH_3 on one side, H and H on the other (C=C)
d. $CH_3CCl(CH_3)$ and $CH_2CH_2CH_3$ on one side, H and H on the other (C=C)
e. $CH_3CH(CH_3)$ and H / H and $CH(Br){-}CH(CH_3)CH_2CH_3$ on C=C

12.23 **a.** 3-Methyl-1-pentene
b. 7-Bromo-1-heptene
c. 5-Bromo-3-heptene
d. 1-*t*-Butyl-4-methylcyclohexene
e. 2, 5-Dimethyl-2-hexene
f. 4-Chloro-3-methyl-1-butyne
g. 6-Chloro-1-heptyne
h. 4-Bromo-3-chlorocyclopentene

12.25 **a.** 2, 3-Dibromobutane could not exist as *cis* and *trans* isomers.
b. CH_3, $CH_2CH_2CH_2CH_3$ / H, H on C=C — *cis*-2-Heptene; CH_3, H / H, $CH_2CH_2CH_2CH_3$ on C=C — *trans*-2-Heptene
c.

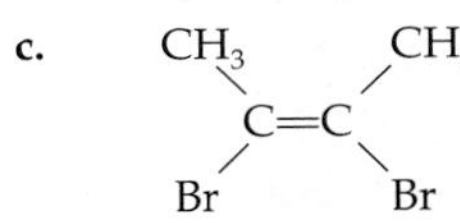

cis-2,3-Dibromo-2-butene

CH_3, Br / Br, CH_3 on C=C — *trans*-2,3-Dibromo-2-butene
d. Propene cannot exist as *cis* and *trans* isomers.
e. 1-Bromo-1-chloro-2-methylpropene cannot exist as *cis* and *trans* isomers.
f. 1, 1-Dichloroethene cannot exist as *cis* and *trans* isomers.
g. H, H / Br, Br on C=C — *cis*-1,2-Dibromoethene; H, Br / Br, H on C=C — *trans*-1,2-Dibromoethene
h. 3-Ethyl-2-methyl-2-hexene cannot exist as *cis* and *trans* isomers.

12.27 **a.** Incorrect. The correct name is 5-methyl-2-hexyne.
b. Correct
c. Incorrect. The correct name is 2, 7-dimethyl-4-nonyne.
d. Incorrect. The correct name is *cis*-6-chloro-3-heptene.
e. Incorrect. The correct name is *trans*-1-chloro-4-methyl-2-hexene.

12.29 **a.** 1,5-Nonadiene
b. 1,4,7-Nonatriene
c. 2,5-Octadiene
d. 4-Methyl-2,5-heptadiene

12.31 Addition of bromine (Br_2) to an alkene results in a color change from red to colorless. If equimolar quantities of Br_2 are added to hexene, the reaction mixture will change from red to colorless. This color change will not occur if cyclohexane is used.

12.33 **a.** H_2
b. H_2O
c. HBr
d. $19O_2 \rightarrow 12CO_2 + 14H_2O$
e. Cl_2
f. cyclopentene

12.35 $CH_2{=}CHCH_2CH_2CH_3$, $CH_3CH{=}CHCH_2CH_3$, $CH_3C(CH_3){=}CHCH_3$, $CH_2{=}C(CH_3)CH_2CH_3$, $CH_2{=}CHCH(CH_3)CH_3$

12.37 **a.** $CH_3CH(Br)CH_2CH_3$
b. $CH_3CH_2{-}C(I)(CH_3){-}CH_2CH_2CH_3$ (major product) + $CH_3CH(I)CH(CH_3)CH_2CH_2CH_3$ (minor product)
c. chlorocyclopentane (Cl on cyclopentane ring)

12.39 A polymer is a macromolecule composed of repeating structural units called *monomers*.

12.41 $n\,F_2C{=}CF_2 \longrightarrow {-}[CF_2{-}CF_2]_n{-}$
Tetrafluoroethene → Teflon

12.43 **a.** $CH_3CH{=}CHCH_2CH_3 + H_2O \xrightarrow{H^+} CH_3CH(OH)CH_2CH_2CH_3$
2-Pentene
$CH_3CH{=}CHCH_2CH_3 + H_2O \xrightarrow{H^+} CH_3CH_2CH(OH)CH_2CH_3$
These products will be formed in approximately equal amounts.
b. $CH_2(Br)CH{=}CH_2 + H_2O \xrightarrow{H^+} CH_2(Br)CH(OH)CH_3$ (major product)
3-Bromo-1-propene
$CH_2(Br)CH{=}CH_2 + H_2O \xrightarrow{H^+} CH_2(Br)CH_2CH_2OH$ (minor product)
c. 3,4-Dimethylcyclohexene $+ H_2O \xrightarrow{H^+}$ 2,3-dimethylcyclohexanol (OH, CH_3, CH_3 on ring)
3,4-Dimethylcyclohexene
3,4-Dimethylcyclohexene $+ H_2O \xrightarrow{H^+}$ 3,4-dimethylcyclohexanol (HO, CH_3, CH_3 on ring)
These products will be formed in approximately equal amounts.

12.45 a. 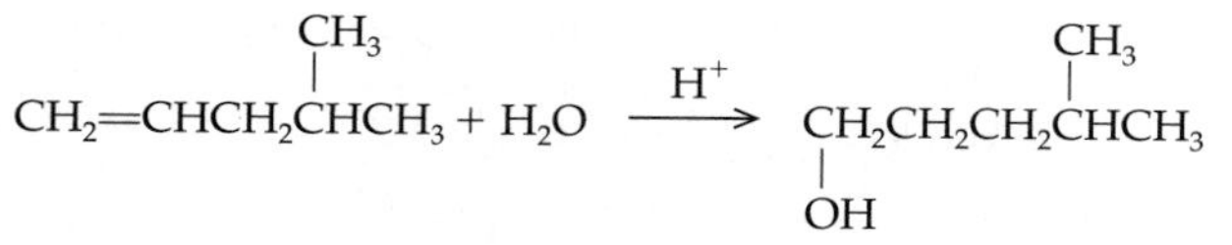 $CH_2{=}CHCH_2CHCH_3 + H_2O \xrightarrow{H^+} CH_2CH_2CH_2CHCH_3$ (CH_3 on C-4 of each; OH on C-1 of the product)

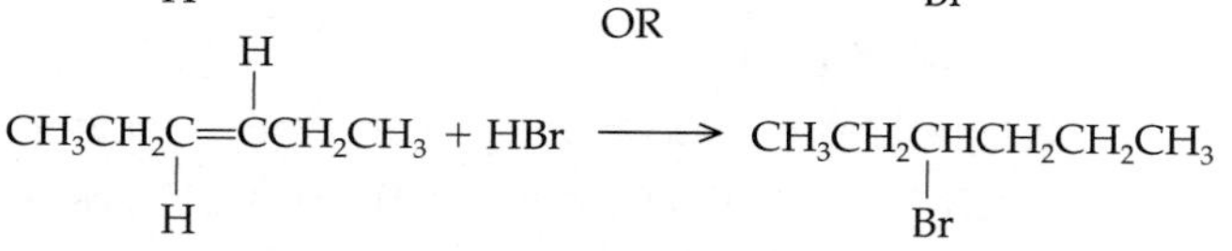

(This is the minor product of this reaction.)

b. 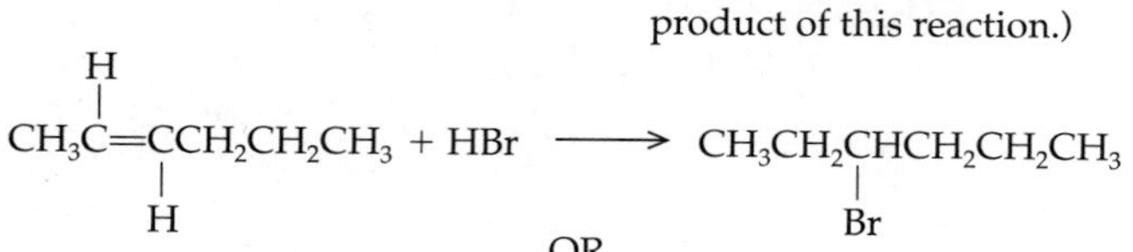 $CH_3C{=}CCH_2CH_2CH_3 + HBr \longrightarrow CH_3CH_2CHCH_2CH_2CH_3$ (H on each double-bond carbon; Br on C-3 of the product)

OR

$CH_3CH_2C{=}CCH_2CH_3 + HBr \longrightarrow CH_3CH_2CHCH_2CH_2CH_3$ (H on each double-bond carbon; Br on C-3 of the product)

c. (methylcyclohexene ring, CH_3) + HBr ⟶ (cyclohexane ring with Br and CH_3)

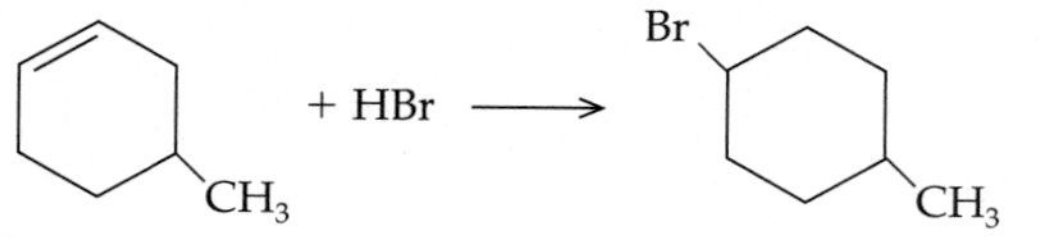

d. 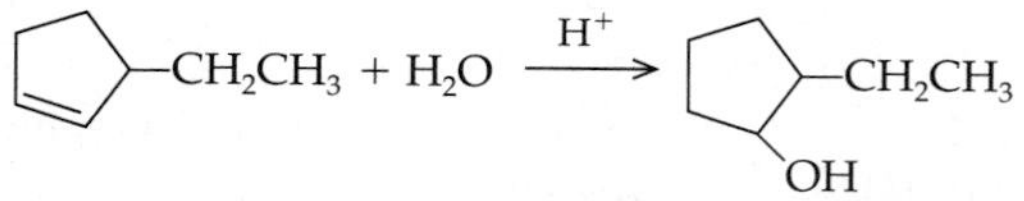 (cyclopentene ring)$-CH_2CH_3 + H_2O \xrightarrow{H^+}$ (cyclopentane ring with OH)$-CH_2CH_3$

12.47 a. 1,4-Hexadiene $CH_2{=}CHCH_2CH{=}CHCH_3$
b. 2,4,6-Octatriene $CH_3CH{=}CHCH{=}CHCH{=}CHCH_3$
c. 1,3-Cyclohexadiene 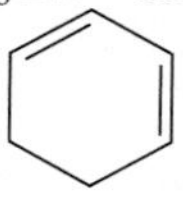
d. 1,3,5-Cyclooctatriene

12.49 a. 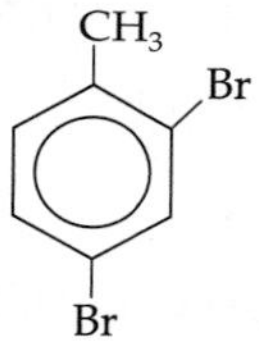 (benzene ring with CH_3, Br, Br)
b. 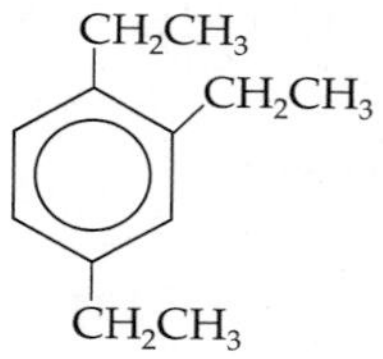 (benzene ring with CH_2CH_3, CH_2CH_3, CH_2CH_3)
c. 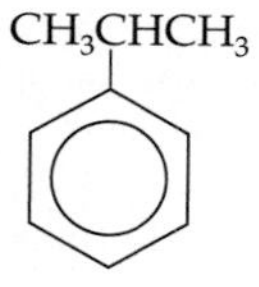 CH_3CHCH_3 (on benzene ring)
d. 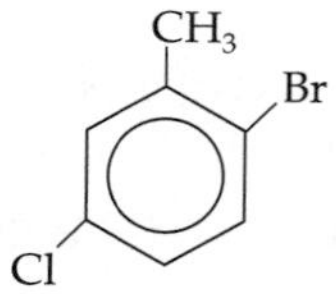 (benzene ring with CH_3, Br, Cl)

12.51 a. 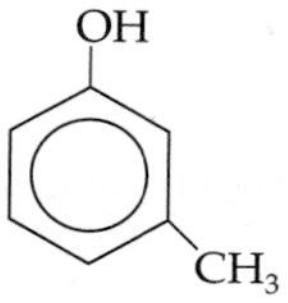 (benzene ring with OH and CH_3)
b. 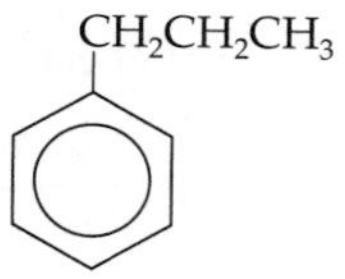 $CH_2CH_2CH_3$ (on benzene ring)
c. 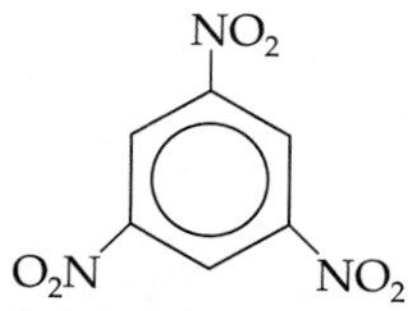 (benzene ring with NO_2, O_2N, NO_2)
d. 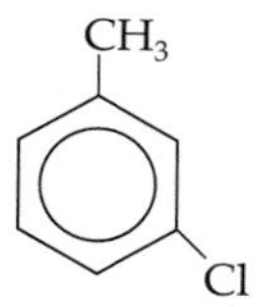 (benzene ring with CH_3 and Cl)

12.53 Kekulé proposed that single and double carbon-carbon bonds alternate around the benzene ring. To explain why benzene does not react like other unsaturated compounds, he proposed that the double and single bonds shift positions rapidly.

12.55 An addition reaction involves addition of a molecule to a double or triple bond in an unsaturated molecule. In a substitution reaction, one chemical group replaces another.

12.57 (benzene) $+ Cl_2 \xrightarrow{FeCl_3}$ (chlorobenzene, Cl) $+ HCl$

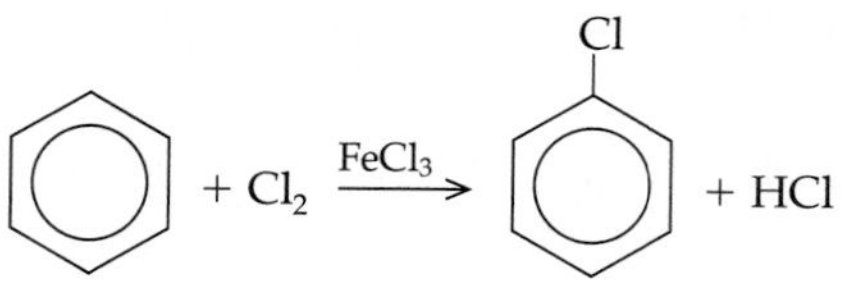

12.59 (ring with N, N)

Pyrimidine

12.61 (fused rings with N, N, N, N–H)

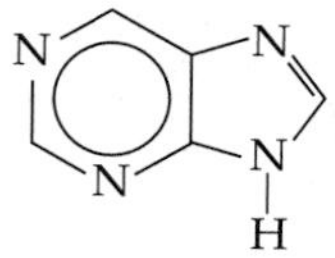

Purine

Chapter 13

13.1 a. 4-Methyl-1-pentanol
b. 4-Methyl-2-hexanol
c. 1, 2, 3-Propanetriol
d. 4-Chloro-3-methyl-1-hexanol

13.3 a. Primary
b. Secondary
c. Tertiary
d. Aromatic (phenol)
e. Secondary

13.5 (cyclopentene) $+ H_2O \xrightarrow{H^+}$ (cyclopentane with HO and H)

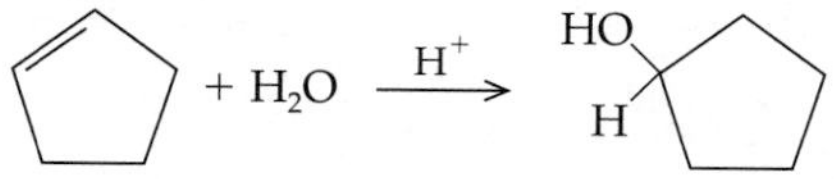

Cyclopentene Cyclopentanol

13.7 a. Ethanol
b. 2-Propanol (major product), 1-Propanol (minor product)
c. 2-Butanol (major product), 1-Butanol (minor product)
d. 2-Butanol
e. 2-Methyl-2-propanol (major product), 2-Methyl-1-propanol (minor product)

13.9

$$CH_3CH_2OH + CH_3CH_2OH \xrightarrow{H^+} CH_3CH_2{-}O{-}CH_2CH_3 + H_2O$$

Ethanol Diethyl ether Water

13.11 $a < d < c < b$

13.13 a. CH_3CH_2OH
b. $CH_3CH_2CH_2CH_2OH$
c. 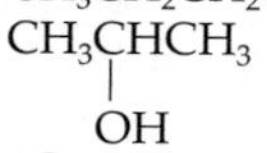 CH_3CHCH_3 (OH on C-2)

13.15 a. 1-Heptanol
b. 2-Propanol
c. 2, 2-Dimethylpropanol
d. 4-Bromo-1-hexanol
e. 3, 3-Dimethyl-2-hexanol
f. 3-Ethyl-3-heptanol

13.17 a. Cyclopentanol
b. Cyclooctanol
c. 3-Methylcyclohexanol

13.19 a. Methyl alcohol
b. Ethyl alcohol
c. Ethylene glycol
d. Propyl alcohol

13.21 **a.** 4-Methyl-2-hexanol

$$CH_3CHCH_2\underset{}{\overset{CH_3}{|}}CHCH_2CH_3 \text{ with OH on C-2 and } CH_3 \text{ on C-4}$$

b. Isobutyl alcohol

$$CH_3\overset{CH_3}{\underset{OH}{C}}CH_3$$

c. 1,5-Pentanediol $\underset{OH}{CH_2}CH_2CH_2CH_2\underset{OH}{CH_2}$

d. 2-Nonanol $CH_3\underset{OH}{CH}CH_2CH_2CH_2CH_2CH_2CH_2CH_3$

e. 1,3,5-Cyclohexanetriol (cyclohexane ring with OH, HO, OH)

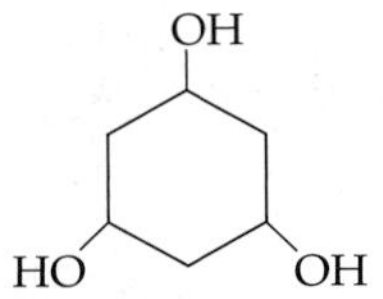

13.23 Denatured alcohol is 100% ethanol to which benzene or methanol is added. The additive makes the ethanol unfit to drink and prevents illegal use of pure ethanol.

13.25 Fermentation is the anaerobic degradation of sugar that involves no net oxidation. The alcohol fermentation, carried out by yeast, produces ethanol and carbon dioxide.

13.27 When the ethanol concentration in a fermentation reaches 12–13%, the yeast producing the ethanol are killed by it. To produce a liquor of higher alcohol concentration, the product of the original fermentation must be distilled.

13.29 **a.** Primary **b.** Secondary
c. Tertiary **d.** Tertiary
e. Tertiary

13.31 **a.** Tertiary
b. Secondary
c. Primary
d. Tertiary

13.33 **a.** 2-Pentanol (major product), 1-pentanol (minor product)
b. 2-Pentanol and 3-pentanol
c. 3-Methyl-2-butanol (major product), 3-methyl-1-butanol (minor product)
d. 3, 3-Dimethyl-2-butanol (major product), 3, 3-dimethyl-1-butanol (minor product)

13.35 **a.**

$$CH_3\overset{H}{C}=\underset{H}{C}CH_2CH_2CH_3 + H_2O \xrightarrow{H^+}$$

2-Hexene

$$CH_3\underset{OH}{CH}CH_2CH_2CH_2CH_3$$

2-Hexanol

or

$$CH_3CH_2\underset{OH}{CH}CH_2CH_2CH_3$$

3-Hexanol

These products will be formed in approximately equal amounts.

b.

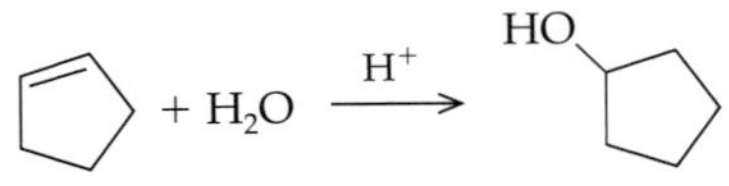

Cyclopentene $+ H_2O \xrightarrow{H^+}$ Cyclopentanol (HO on ring)

c. $CH_2{=}CHCH_2CH_2CH_2CH_2CH_2CH_3 + H_2O \xrightarrow{H^+}$

1-Octene

$$CH_3\underset{OH}{CH}CH_2CH_2CH_2CH_2CH_2CH_3$$

2-Octanol
(major product)

or

$$\underset{OH}{CH_2}CH_2CH_2CH_2CH_2CH_2CH_2CH_3$$

1-Octanol
(minor product)

d.

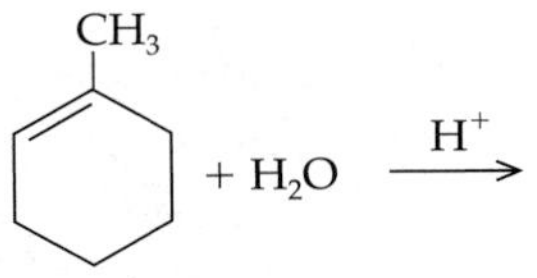

1-Methylcyclohexene (CH_3 on ring) $+ H_2O \xrightarrow{H^+}$

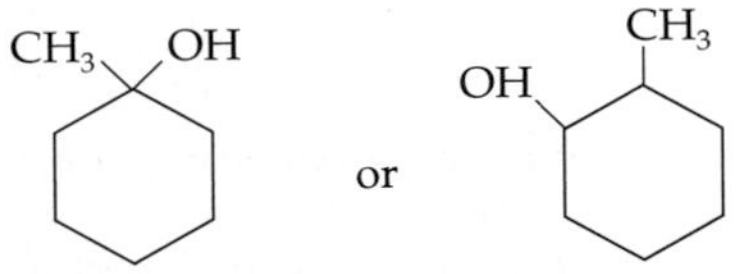

1-Methylcyclohexanol (major product) (CH_3, OH on same carbon) or 2-Methylcyclohexanol (minor product) (OH, CH_3 on adjacent carbons)

13.37 **a.** 2-Butanone
b. N.R.
c. Cyclohexanone
d. N.R.

13.39 **a.** 3-Pentanone
b. Propanal (Upon further oxidation, propanoic acid would be formed.)
c. 4-Methyl-2-pentanone
d. N.R.
e. 3-Phenylpropanal (Upon further oxidation, 3-phenylpropanoic acid will be formed.)

13.41

$$CH_3CH_2OH \xrightarrow{\text{liver enzymes}} CH_3{-}\overset{O}{\overset{\|}{C}}{-}H$$

Ethanol Ethanal

The product, ethanal, is responsible for the symptoms of a hangover.

13.43 The reaction in which a water molecule is added to 1-butene is a hydration reaction.

$$CH_3CH_2CH{=}CH_2 + H_2O \xrightarrow{H^+} CH_3CH_2\overset{OH}{\overset{|}{C}}HCH_3$$

1-Butene 2-Butanol

13.45

$$CH_3CH{=}CH_2 \xrightarrow{H_2O,\ H^+} CH_3{-}\overset{OH}{\overset{|}{C}}H{-}CH_3 \xrightarrow{[O]} CH_3{-}\overset{O}{\overset{\|}{C}}{-}CH_3$$

Propene (propylene) 2-Propanol (isopropanol) Propanone (acetone)

13.47 H_3C CH_3 CH_3 CH_3 CH_3 H O

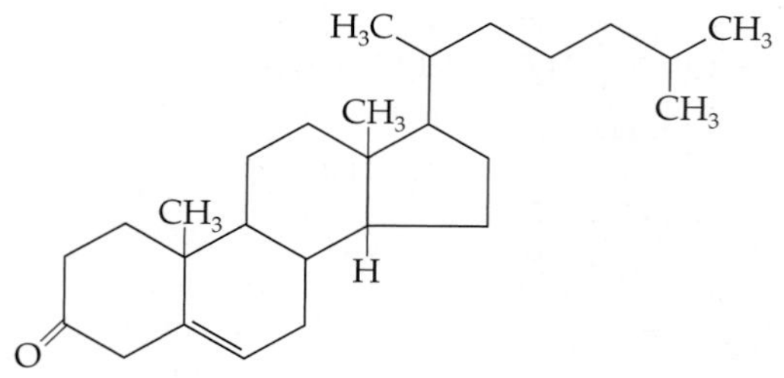

13.49 Oxidation is a loss of electrons, whereas reduction is a gain of electrons.

13.51

$CH_3CH_2CH_3 < CH_3CH_2CH_2OH < CH_3CH_2\overset{O}{\overset{\|}{C}}—H < CH_3CH_2\overset{O}{\overset{\|}{C}}—OH$

13.53 Picric acid: 2,4,6,-Trinitrotoluene:

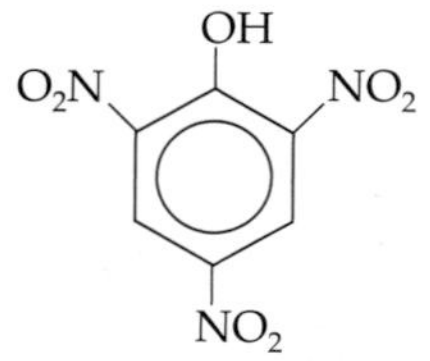

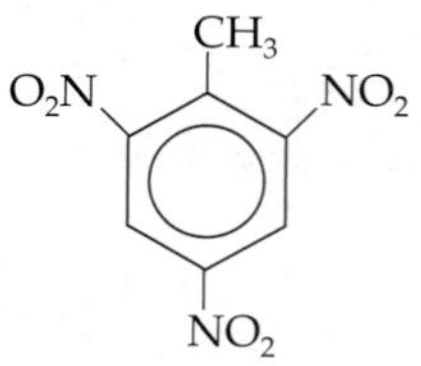

OH, O_2N, NO_2, NO_2 — CH_3, O_2N, NO_2, NO_2

Picric acid is water-soluble because of the polar hydroxyl group that can form hydrogen bonds with water.

13.55 Hexachlorophene, hexylresorcinol, and *o*-phenylphenol are phenol compounds used as antiseptics or disinfectants.

13.57 Alcohols of molecular formula $C_4H_{10}O$

$CH_3CH_2CH_2CH_2OH$, $CH_3\overset{OH}{\overset{|}{C}}HCH_2CH_3$,

$CH_3\underset{CH_3}{\underset{|}{C}}HCH_2OH$, $CH_3—\overset{OH}{\overset{|}{\underset{CH_3}{\underset{|}{C}}}}—CH_3$

Ethers of molecular formula $C_4H_{10}O$

$CH_3—O—CH_2CH_2CH_3$ $CH_3CH_2—O—CH_2CH_3$

$CH_3—O—\underset{CH_3}{\underset{|}{C}}HCH_3$

13.59 Penthrane: 2, 2-Dichloro-1, 1-difluoro-1-methoxyethane
Enthrane: 2-Chloro-1-(difluoromethoxy)-1, 1, 2-trifluoroethane

13.61 a. $CH_3CH_2—O—CH_2CH_3 + H_2O$
b. $CH_3CH_2—O—CH_2CH_3 + CH_3—O—CH_3 + CH_3—O—CH_2CH_3 + H_2O$
c. $CH_3—O—CH_3 + CH_3—O—\underset{CH_3}{\underset{|}{C}}HCH_3 +$

$CH_3\underset{CH_3}{\underset{|}{C}}H—O—\underset{CH_3}{\underset{|}{C}}HCH_3 + H_2O$

d. (cyclopentyl)$—CH_2—O—CH_2—$(cyclopentyl)

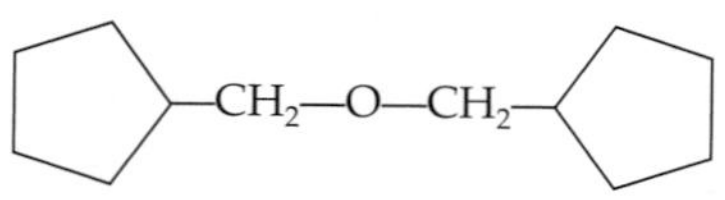

13.63 a. 2-Ethoxypentane
b. 2-Methoxybutane
c. 1-Ethoxybutane
d. Methoxycyclopentane

13.65 Cystine:

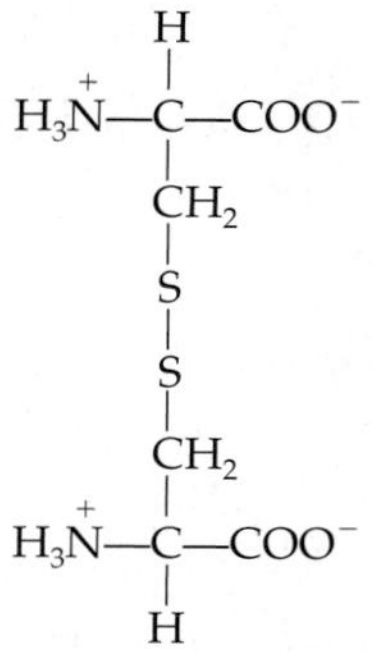

$H_3\overset{+}{N}—\overset{H}{\overset{|}{C}}—COO^-$, CH_2, S, S, CH_2, $H_3\overset{+}{N}—\underset{H}{\underset{|}{C}}—COO^-$

13.67 a. 1-Propanethiol
b. 2-Butanethiol
c. 2-Methyl-2-butanethiol
d. 1, 4-Cyclohexanedithiol

Chapter 14

14.1 a. $CH_3—\overset{O}{\overset{\|}{C}}—CH_3$ b. $CH_3\overset{OH}{\overset{|}{C}}HCH_2CH_2CH_3$

c. (cyclopentyl)$—\overset{O}{\overset{\|}{C}}—H$ d. $\underset{OH}{\underset{|}{C}H_2}—\underset{OH}{\underset{|}{C}H_2}$

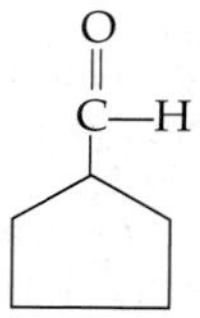

14.3 a. I.U.P.A.C.: 3, 4-Dimethylpentanal
Common: β, γ-Dimethylvaleraldehyde
b. I.U.P.A.C.: 2-Chloropropanal
Common: α-Chloropropionaldehyde
c. I.U.P.A.C.: 2-Ethylpentanal
Common: α-Ethylvaleraldehyde
d. I.U.P.A.C.: 3-Hydroxybutanal
Common: β-Hydroxybutyraldehyde

14.5 a. 3-Iodobutanone
b. 3-Methylbutanone
c. 2-Fluoro-3-pentanone
d. 4-Methyl-2-octanone
e. 2-Methyl-3-pentanone

14.7 $CH_3—\overset{O}{\overset{\|}{C}}—H$

14.9 $\underset{\text{1-Propanol}}{CH_3CH_2CH_2OH} \xrightarrow{H_2Cr_2O_7} \underset{\text{Propanal}}{CH_3CH_2\overset{O}{\overset{\|}{C}}H}$

14.11 $\underset{\text{Ethanal}}{CH_3\overset{O}{\overset{\|}{C}}—H} + \underset{\text{Silver ammonia complex}}{Ag(NH_3)_2^+} \longrightarrow \underset{\text{Ethanoate anion}}{CH_3\overset{O}{\overset{\|}{C}}—O^-} + \underset{\text{Silver metal}}{Ag^0}$

14.13 $\underset{\text{Propanone}}{CH_3—\overset{O}{\overset{\|}{C}}—CH_3} + H_2 \xrightarrow{Ni} \underset{\text{2-Propanol}}{CH_3\overset{OH}{\overset{|}{C}}HCH_3}$

14.15 a. Reduction
b. Reduction
c. Reduction
d. Oxidation
e. Reduction

14.17 $2CH_3CH_2\overset{O}{\overset{\|}{C}}{-}H \xrightarrow{OH^-} CH_3CH_2\overset{OH}{\overset{|}{C}H}\underset{CH_3}{\underset{|}{C}H}\overset{O}{\overset{\|}{C}}{-}H$

Propanal 3-Hydroxy-2-methylpentanal

14.19 A good solvent should dissolve a wide range of compounds. Simple ketones are considered to be universal solvents because they have both a polar carbonyl group and nonpolar side chains. As a result, they dissolve organic compounds and are also miscible in water.

14.21 Water (H–O–H) with each H atom hydrogen-bonded (···) to the carbonyl O of acetaldehyde: $CH_3{-}\overset{O}{\overset{\|}{C}}{-}H \quad H{-}\overset{O}{\overset{\|}{C}}{-}CH_3$

14.23 Alcohols have higher boiling points than aldehydes or ketones of comparable molecular weights because alcohol molecules can form intermolecular hydrogen bonds with one another. Aldehydes and ketones cannot form intermolecular hydrogen bonds.

14.25 a. $H{-}\overset{O}{\overset{\|}{C}}{-}H$

b. $H{-}\overset{H}{\underset{Br}{C}}{-}\overset{H}{\underset{Br}{C}}{-}\overset{H}{\underset{H}{C}}{-}\overset{H}{\underset{H}{C}}{-}\overset{H}{\underset{H}{C}}{-}\overset{H}{\underset{H}{C}}{-}\overset{H}{\underset{H}{C}}{-}\overset{O}{\overset{\|}{C}}{-}H$

c. $H{-}\overset{H}{\underset{H}{C}}{-}\overset{O}{\overset{\|}{C}}{-}\overset{H}{\underset{H}{C}}{-}H$

d. $HO{-}\overset{H}{\underset{H}{C}}{-}\overset{O}{\overset{\|}{C}}{-}H$

14.27 a. Butanone
b. 2-Ethylhexanal
c. 1,1,1-Trichloro-2-propanone
d. 3-Chlorocyclopentanone

14.29 Determine the longest continuous chain containing the carbonyl group. Drop the *-e* ending from the parent alkane and add the suffix *-one*. Number the carbon chain in the direction that gives the carbonyl group the lowest possible number. Name and number the substituents on the chain. Add them as prefixes before the name of the ketone, listing them in alphabetical order.

14.31 a. 3-Bromobutanal
b. 2-Chloro-2-methyl-4-heptanone
c. 4, 4-Diethyl-2-hexanone
d. 4-Chloro-2-hexanone

14.33 a. Acetone
b. Ethyl methyl ketone
c. Acetaldehyde
d. Propionaldehyde
e. Methyl isopropyl ketone

14.35 a. $CH_3\underset{OH}{\underset{|}{C}H}CH_2\overset{O}{\overset{\|}{C}}{-}H$

b. $CH_3CH_2CH_2\underset{CH_3}{\underset{|}{C}H}\overset{O}{\overset{\|}{C}}{-}H$

c. $CH_3CH_2\underset{Br}{\underset{|}{C}H}CH_2CH_2\overset{O}{\overset{\|}{C}}{-}H$

d. $CH_3CH_2\underset{I}{\underset{|}{C}H}CH_2\overset{O}{\overset{\|}{C}}{-}H$

e. $CH_3CH_2CH_2CH_2\underset{OH}{\underset{|}{C}H}\overset{CH_3}{\overset{|}{C}H}\overset{O}{\overset{\|}{C}}{-}H$

14.37 Acetone is a good solvent because it can dissolve a wide range of compounds. It has both a polar carbonyl group and nonpolar side chains. As a result, it dissolves organic compounds and is also miscible in water.

14.39 The liver

14.41 a. $CH_3CH_2\overset{OH}{\overset{|}{C}H}CH_3 \xrightarrow{[O]} CH_3CH_2\overset{O}{\overset{\|}{C}}CH_3$

2-Butanol Butanone

b. $CH_3\overset{CH_3}{\overset{|}{C}H}CH_2OH \xrightarrow{[O]} CH_3\overset{CH_3}{\overset{|}{C}H}{-}\overset{O}{\overset{\|}{C}}{-}H$

2-Methyl-1-propanol Methylpropanal

Note that methylpropanal can be further oxidized to methylpropanoic acid.

c. Cyclopentanol $\xrightarrow{[O]}$ Cyclopentanone

Cyclopentanol Cyclopentanone

14.43 $R{-}CH_2OH \xrightarrow{[O]} R{-}\overset{O}{\overset{\|}{C}}{-}H \xrightarrow{[O]} R{-}\overset{O}{\overset{\|}{C}}{-}OH$

Primary alcohol Aldehyde Carboxylic acid

14.45 a. Reduction reaction

$CH_3{-}\overset{O}{\overset{\|}{C}}{-}H \longrightarrow CH_3CH_2OH$

Ethanal Ethanol

b. Reduction reaction

Cyclohexanone (C=O) $\longrightarrow$ Cyclohexanol (–OH)

Cyclohexanone Cyclohexanol

c. Oxidation reaction

$CH_3\overset{OH}{\overset{|}{C}H}CH_3 \longrightarrow CH_3{-}\overset{O}{\overset{\|}{C}}{-}CH_3$

2-Propanol Propanone

14.47 Only (c) 3-methylbutanal and (f) acetaldehyde would give a positive Tollens' test.

14.49 a. $CH_3{-}\overset{O}{\overset{\|}{C}}{-}CH_3 + CH_3CH_2OH \xrightarrow{H^+} CH_3{-}\overset{OH}{\underset{OCH_2CH_3}{C}}{-}CH_3$

b. $CH_3{-}\overset{O}{\overset{\|}{C}}{-}H + CH_3CH_2OH \xrightarrow{H^+} CH_3{-}\overset{OH}{\underset{OCH_2CH_3}{C}}{-}H$

14.51 Hemiacetal

14.53 Acetal

14.55 a. $CH_3{-}\overset{O}{\overset{\|}{C}}{-}CH_3 + 2\,CH_3OH \xrightarrow{H^+} CH_3{-}\overset{OCH_3}{\underset{OCH_3}{C}}{-}CH_3 + H_2O$

b. $CH_3{-}\overset{O}{\overset{\|}{C}}{-}H + 2\,CH_3OH \xrightarrow{H^+} CH_3{-}\overset{OCH_3}{\underset{OCH_3}{C}}{-}H + H_2O$

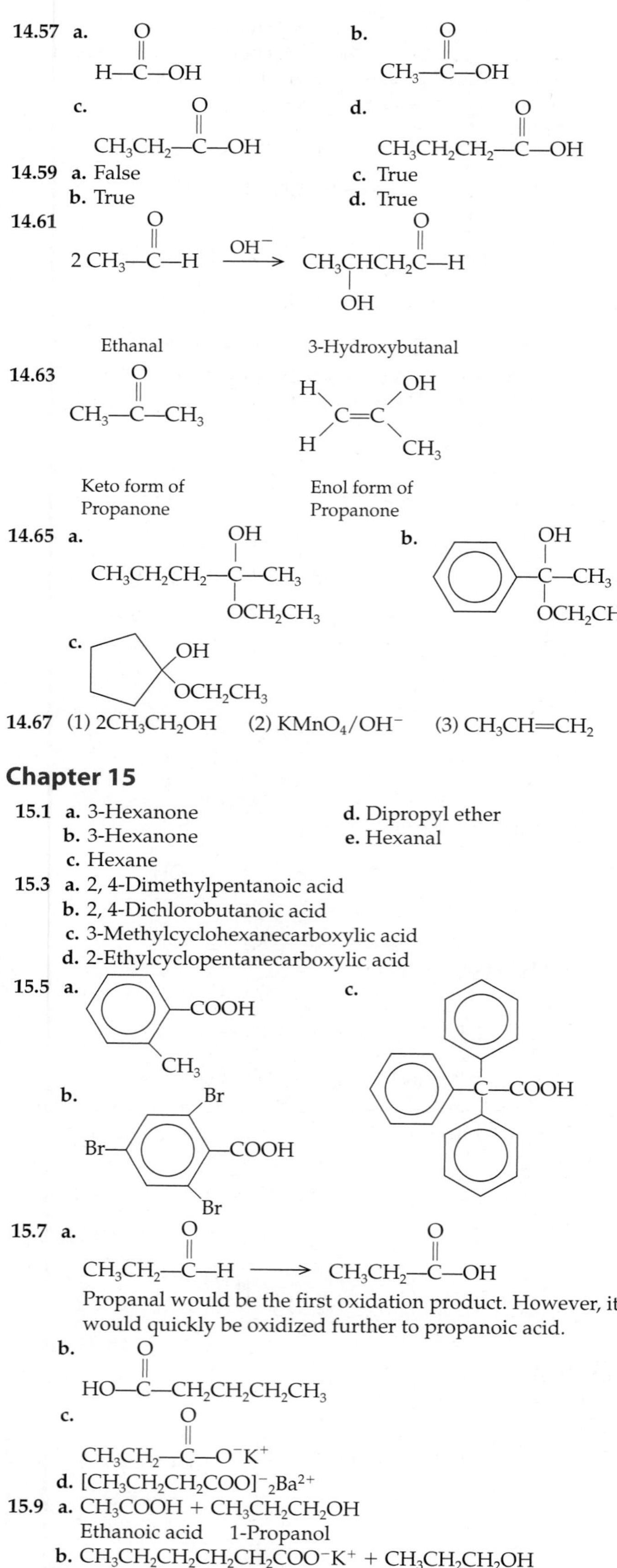

14.57 a. $H{-}\overset{O}{\overset{\|}{C}}{-}OH$ b. $CH_3{-}\overset{O}{\overset{\|}{C}}{-}OH$

c. $CH_3CH_2{-}\overset{O}{\overset{\|}{C}}{-}OH$ d. $CH_3CH_2CH_2{-}\overset{O}{\overset{\|}{C}}{-}OH$

14.59 a. False
b. True
c. True
d. True

14.61

$$2\,CH_3{-}\overset{O}{\overset{\|}{C}}{-}H \xrightarrow{OH^-} CH_3\underset{OH}{\underset{|}{C}}HCH_2\overset{O}{\overset{\|}{C}}{-}H$$

Ethanal 3-Hydroxybutanal

14.63 $CH_3{-}\overset{O}{\overset{\|}{C}}{-}CH_3$ (H)(H)C=C(OH)(CH_3)

Keto form of Propanone Enol form of Propanone

14.65 a. $CH_3CH_2CH_2{-}C(OH)(OCH_2CH_3){-}CH_3$ b. $C_6H_5{-}C(OH)(OCH_2CH_3){-}CH_3$

c. cyclopentane-C(OH)(OCH_2CH_3)

14.67 (1) $2CH_3CH_2OH$ (2) $KMnO_4/OH^-$ (3) $CH_3CH{=}CH_2$

Chapter 15

15.1 a. 3-Hexanone
b. 3-Hexanone
c. Hexane
d. Dipropyl ether
e. Hexanal

15.3 a. 2, 4-Dimethylpentanoic acid
b. 2, 4-Dichlorobutanoic acid
c. 3-Methylcyclohexanecarboxylic acid
d. 2-Ethylcyclopentanecarboxylic acid

15.5 a. benzene ring with —COOH and ortho —CH_3

b. benzene ring with —COOH, Br, Br—, Br (2,4,6-tribromo)

c. $(C_6H_5)_3C{-}COOH$

15.7 a.

$$CH_3CH_2{-}\overset{O}{\overset{\|}{C}}{-}H \longrightarrow CH_3CH_2{-}\overset{O}{\overset{\|}{C}}{-}OH$$

Propanal would be the first oxidation product. However, it would quickly be oxidized further to propanoic acid.

b. $HO{-}\overset{O}{\overset{\|}{C}}{-}CH_2CH_2CH_2CH_3$

c. $CH_3CH_2{-}\overset{O}{\overset{\|}{C}}{-}O^-K^+$

d. $[CH_3CH_2CH_2COO]^-{}_2Ba^{2+}$

15.9 a. $CH_3COOH + CH_3CH_2CH_2OH$
Ethanoic acid 1-Propanol

b. $CH_3CH_2CH_2CH_2CH_2COO^-K^+ + CH_3CH_2CH_2OH$
Potassium hexanoate 1-Propanol

c. $CH_3CH_2CH_2CH_2COO^-Na^+ + CH_3OH$
Sodium pentanoate Methanol

d. $CH_3CH_2CH_2CH_2CH_2COOH + CH_3\underset{OH}{\underset{|}{C}}HCH_2CH_2CH_3$
Hexanoic acid 2-Pentanol

15.11 a.

$$CH_3\underset{CH_3}{\underset{|}{C}}H\overset{O}{\overset{\|}{C}}{-}OH \xrightarrow{PCl_3 \text{ or } SOCl_2} CH_3\underset{CH_3}{\underset{|}{C}}H\overset{O}{\overset{\|}{C}}{-}Cl$$

2-Methylpropanoic acid 2-Methylpropanoyl chloride

b.

$$CH_3\overset{O}{\overset{\|}{C}}{-}OH \xrightarrow{PCl_3 \text{ or } SOCl_2} CH_3\overset{O}{\overset{\|}{C}}{-}Cl$$

Ethanoic acid Ethanoyl chloride

c.

$$CH_3CH_2CH_2CH_2CH_2\overset{O}{\overset{\|}{C}}{-}OH \xrightarrow{PCl_3 \text{ or } SOCl_2} CH_3CH_2CH_2CH_2CH_2\overset{O}{\overset{\|}{C}}{-}Cl$$

Hexanoic acid Hexanoyl chloride

d.

$$CH_3CH_2\underset{Br}{\underset{|}{C}}HCH_2CH_2\overset{O}{\overset{\|}{C}}{-}OH \xrightarrow{PCl_3 \text{ or } SOCl_2} CH_3CH_2\underset{Br}{\underset{|}{C}}HCH_2CH_2\overset{O}{\overset{\|}{C}}{-}Cl$$

4-Bromohexanoic acid 4-Bromohexanoyl chloride

15.13 a.

$$CH_3\underset{CH_3}{\underset{|}{C}}HCH_2\overset{O}{\overset{\|}{C}}{-}Cl \xrightarrow{CH_3\underset{CH_3}{\underset{|}{C}}HCH_2\overset{O}{\overset{\|}{C}}{-}O^-}$$

3-Methylbutanoyl chloride 3-Methylbutanoate ion

$$CH_3\underset{CH_3}{\underset{|}{C}}HCH_2\overset{O}{\overset{\|}{C}}{-}O{-}\overset{O}{\overset{\|}{C}}{-}CH_2\underset{CH_3}{\underset{|}{C}}HCH_3 + Cl^-$$

3-Methylbutanoic anhydride

b.

$$H{-}\overset{O}{\overset{\|}{C}}{-}Cl \xrightarrow{CH_3\overset{O}{\overset{\|}{C}}{-}O^-} H{-}\overset{O}{\overset{\|}{C}}{-}O{-}\overset{O}{\overset{\|}{C}}{-}CH_3 + Cl^-$$

Methanoyl chloride Ethanoate ion Ethanoic methanoic anhydride

15.15 a.

H H H H O
H—C—C—C—C—C—OH
H H H Br

b. $H-CH_2-CH(CH_3)-CHBr-COOH$ (H H—C—H H O; H—C—C—C—C—OH; H H Br)

c. (cyclohexane ring)—COOH, Br

15.17 a. I.U.P.A.C. name: Methanoic acid
Common name: Formic acid
b. I.U.P.A.C. name: 3-Methylbutanoic acid
Common name: β-methylbutyric acid
c. I.U.P.A.C. name: Cyclopentanecarboxylic acid
Common name: cyclovalericcarboxylic acid

15.19 $CH_3CH_2CH_2COOH$ Butanoic acid; $CH_3CH(CH_3)COOH$ Methylpropanoic acid

15.21 a. $CH_3CH_2-C(CH_3)_2-CH_2CH_2COOH$
b. $CH_3CH(CH_3)CH(Br)CH_2COOH$ (CH_3, $CH_3CHCHCH_2COOH$, Br)
c. (benzene ring)—COOH, O_2N, NO_2
d. (cyclohexane ring)—COOH, H_3C

15.23 a. I.U.P.A.C. name: 2-Hydroxypropanoic acid
Common name: α-Hydroxypropionic acid
b. I.U.P.A.C. name: 3-Hydroxybutanoic acid
Common name: β-Hydroxybutyric acid
c. I.U.P.A.C. name: 4, 4-Dimethylpentanoic acid
Common name: γ, γ-Dimethylvaleric acid
d. I.U.P.A.C. name: 3, 3-Dichloropentanoic acid
Common name: β, β-Dichlorovaleric acid

15.25 a. Heptanoic acid c. Pentanoic acid
b. 1-Propanol d. Butanoic acid

15.27 The smaller carboxylic acids are water soluble. They have sharp, sour tastes and unpleasant aromas.

15.29 Citric acid is added to foods to give them a tart flavor or to act as a food preservative and antioxidant. Adipic acid imparts a tart flavor to soft drinks and is a preservative.

15.31 Carboxylic acids are produced commercially by the oxidation of the corresponding alcohol or aldehyde.

15.33 Soaps are made from water, a strong base, and natural fats or oils.

15.35 a. CH_3COOH
b. $CH_3CH_2CH_2-C(=O)-O-CH_3 + H_2O$
c. CH_3OH

15.37 a. The oxidation of 1-pentanol yields pentanal.
b. Continued oxidation of pentanal yields pentanoic acid.

15.39 a. (benzene ring)—$C(=O)-OCH_3$
b. $CH_3CH_2CH_2CH_2CH_2CH_2CH_2CH_2CH_2-C(=O)-O-CH_2CH_2CH_2CH_3$
c. $CH_3CH_2-C(=O)-O-CH_3$
d. $CH_3CH_2-C(=O)-O-CH_2CH_3$

15.41 a. Ethyl ethanoate
b. Methyl propanoate
c. Methyl-3-methylbutanoate
d. Cyclopentyl benzoate

15.43 a. $CH_3CH_2CH_2-C(=O)-O-CH_2CH_3$
b. $CH_3CH_2-C(=O)-OH + CH_3CH_2OH$
c. $CH_3CH_2CH_2OH$
d. $CH_3CH_2CH(Br)CH_2-C(=O)-O^- + CH_3CH_2OH$

15.45 Saponification is a reaction in which a soap is produced. More generally, it is the hydrolysis of an ester in the presence of a base. The following reaction shows the base-catalyzed hydrolysis of an ester:

$$CH_3(CH_2)_{14}-C(=O)-O-CH_3 + NaOH \longrightarrow CH_3(CH_2)_{14}-C(=O)-O^-\,Na^+ + CH_3OH$$

15.47 (benzene ring with OH)—$C(=O)-OH + CH_3OH \xrightarrow{H^+}$ (benzene ring with OH)—$C(=O)-OCH_3 + H_2O$

Salicylic acid — Methyl salicylate

15.49 Compound A is

$CH_3CH_2CH_2CH_2-C(=O)-O-CH_3$

Compound B is

$CH_3CH_2CH_2CH_2-C(=O)-OH$

Compound C is CH_3OH

15.51 a. PCl_3, PCl_5, or $SOCl_2$
b. $CH_3-C(=O)-O^-$
c. (cyclohexane ring)—$C(=O)-O^-$

15.53 **a.** 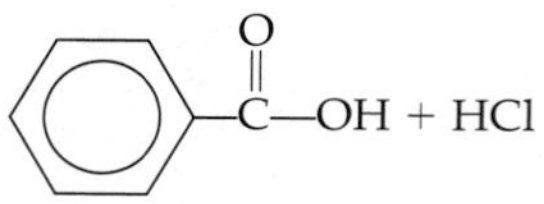 $C_6H_5\text{—}\overset{O}{\overset{\|}{C}}\text{—OH} + HCl$ **b.** 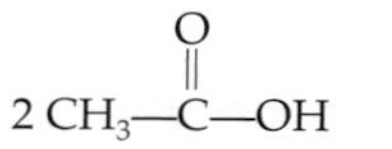 $2\ CH_3\text{—}\overset{O}{\overset{\|}{C}}\text{—OH}$

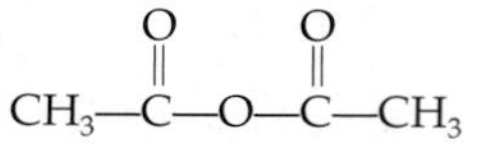

15.55 **a.** 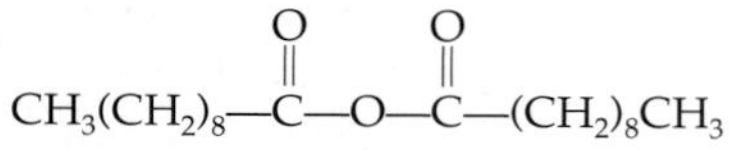 $CH_3(CH_2)_8\text{—}\overset{O}{\overset{\|}{C}}\text{—O—}\overset{O}{\overset{\|}{C}}\text{—}(CH_2)_8CH_3$

b. $CH_3\text{—}\overset{O}{\overset{\|}{C}}\text{—O—}\overset{O}{\overset{\|}{C}}\text{—}CH_3$

c. $CH_3(CH_2)_3\text{—}\overset{O}{\overset{\|}{C}}\text{—O—}\overset{O}{\overset{\|}{C}}\text{—}(CH_2)_3CH_3$

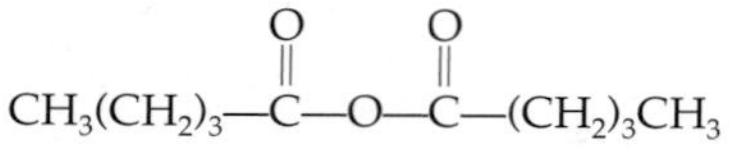

d. 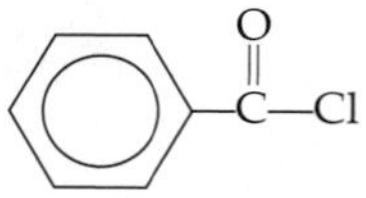 $C_6H_5\text{—}\overset{O}{\overset{\|}{C}}\text{—Cl}$

15.57 Acid chlorides are noxious, irritating chemicals. They are slightly polar and have boiling points similar to comparable aldehydes or ketones. They cannot be dissolved in water because they react violently with it.

15.59 **a.**

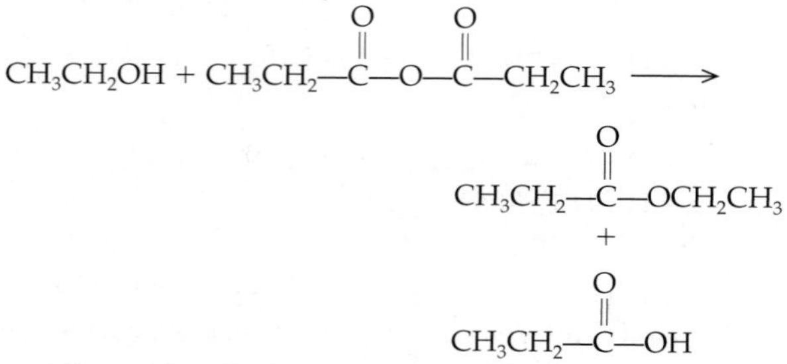

$$CH_3CH_2OH + CH_3CH_2\text{—}\overset{O}{\overset{\|}{C}}\text{—O—}\overset{O}{\overset{\|}{C}}\text{—}CH_2CH_3 \longrightarrow CH_3CH_2\text{—}\overset{O}{\overset{\|}{C}}\text{—}OCH_2CH_3 + CH_3CH_2\text{—}\overset{O}{\overset{\|}{C}}\text{—OH}$$

b.

$$CH_3CH_2OH + CH_3\text{—}\overset{O}{\overset{\|}{C}}\text{—O—}\overset{O}{\overset{\|}{C}}\text{—}CH_3 \longrightarrow CH_3\text{—}\overset{O}{\overset{\|}{C}}\text{—}OCH_2CH_3 + CH_3\text{—}\overset{O}{\overset{\|}{C}}\text{—OH}$$

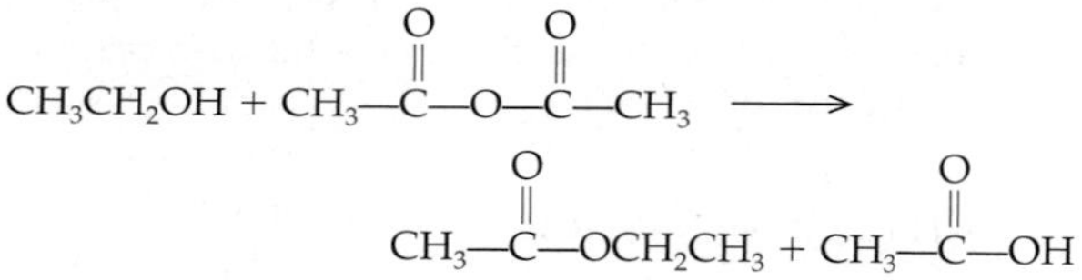

c.

$$CH_3CH_2OH + H\text{—}\overset{O}{\overset{\|}{C}}\text{—O—}\overset{O}{\overset{\|}{C}}\text{—H} \longrightarrow H\text{—}\overset{O}{\overset{\|}{C}}\text{—}OCH_2CH_3 + H\text{—}\overset{O}{\overset{\|}{C}}\text{—OH}$$

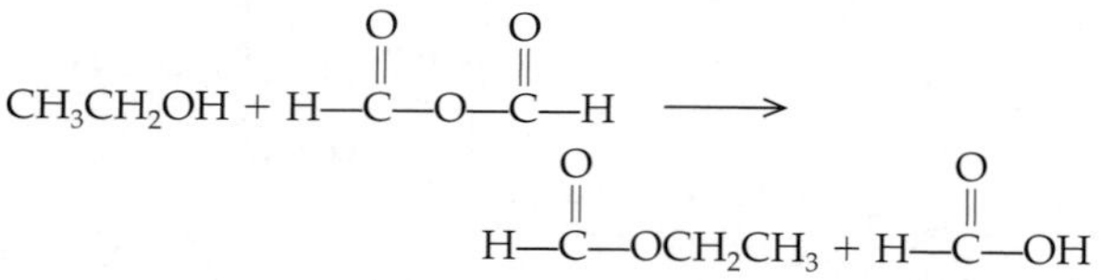

15.61 **a.** Monoester:

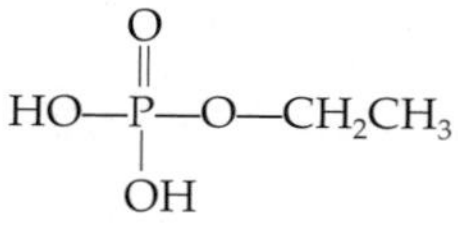

$$HO\text{—}\overset{O}{\overset{\|}{P}}(OH)\text{—O—}CH_2CH_3$$

b. Diester:

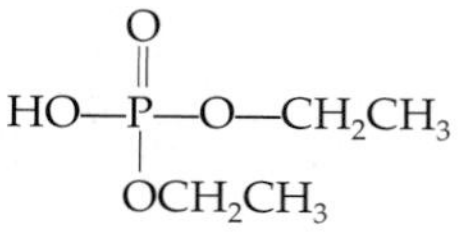

$$HO\text{—}\overset{O}{\overset{\|}{P}}(OCH_2CH_3)\text{—O—}CH_2CH_3$$

c. Triester:

$$CH_3CH_2\text{—O—}\overset{O}{\overset{\|}{P}}(OCH_2CH_3)\text{—O—}CH_2CH_3$$

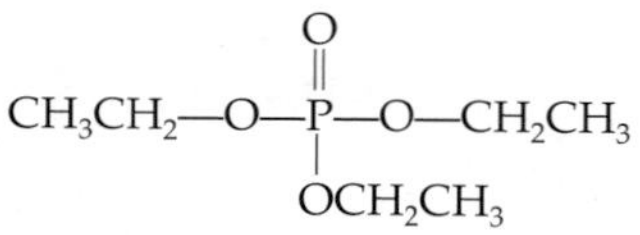

15.63 ATP is the molecule used to store the energy released in metabolic reactions. The energy is stored in the phosphoanhydride bonds between two phosphoryl groups. The energy is released when the bond is hydrolyzed. A portion of the energy can be transferred to another molecule if the phosphoryl group is transferred from ATP to the other molecule.

15.65

$$CH_3\text{—}\overset{O}{\overset{\|}{C}}\sim S\text{—COENZYME A}$$

The squiggle denotes a high energy bond.

15.67

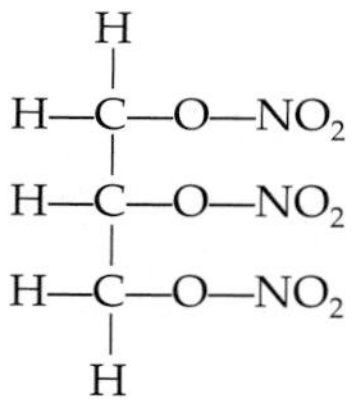

$$H_2C(ONO_2)\text{—}CH(ONO_2)\text{—}CH_2(ONO_2)$$

(H—C—O—NO$_2$ / H—C—O—NO$_2$ / H—C—O—NO$_2$, with H above and below)

Chapter 16

16.1 **a.** Tertiary
b. Primary
c. Secondary

16.3 $CH_3\text{—}NH_2$ hydrogen bonded (N—H···O) to water, water hydrogen bonded to $(CH_3CH_2)(CH_3)N\text{—H}$, whose N—H hydrogen bonds to another water molecule (H—O—H).

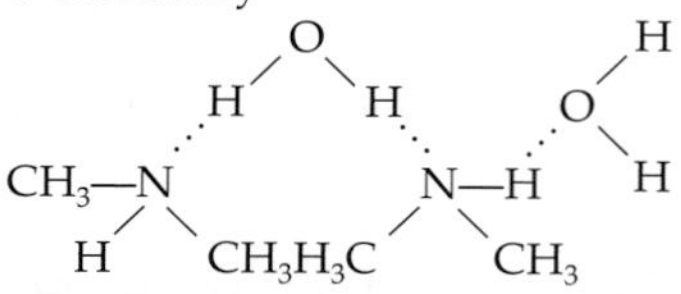

16.5 **a.** Methanol because the intermolecular hydrogen bonds between alcohol molecules will be stronger.
b. Water because the intermolecular hydrogen bonds between water molecules will be stronger.
c. Ethylamine because it has a higher molecular weight.
d. Propylamine because propylamine molecules can form intermolecular hydrogen bonds while the nonpolar butane cannot do so.

16.7 **a.** 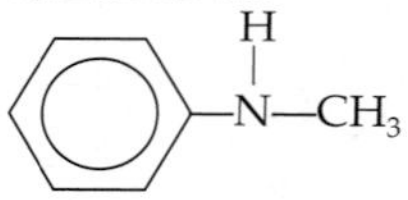 $C_6H_5\text{—}\overset{H}{\overset{|}{N}}\text{—}CH_3$ **b.** 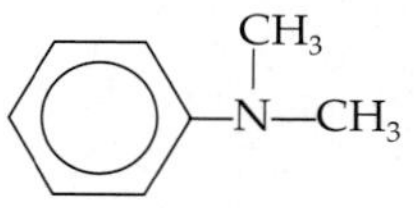 $C_6H_5\text{—}\overset{CH_3}{\overset{|}{N}}\text{—}CH_3$ **c.** $C_6H_5\text{—}\overset{H}{\overset{|}{N}}\text{—}CH_2CH_3$ **d.** 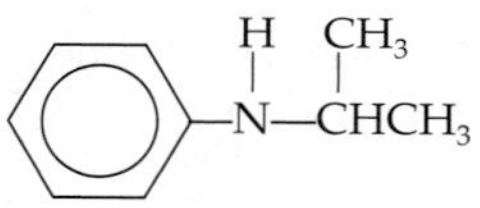 $C_6H_5\text{—}\overset{H}{\overset{|}{N}}\text{—}\overset{CH_3}{\overset{|}{C}}HCH_3$

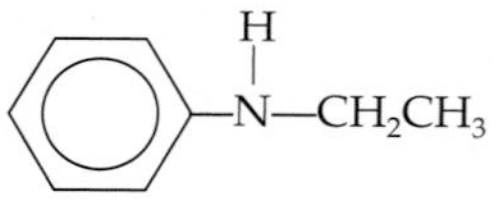

16.9 **a.** 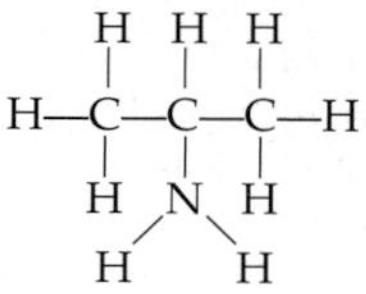 $CH_3\text{—}CH(NH_2)\text{—}CH_3$ (H—C—C—C—H with all H shown; NH_2 on the central carbon)

b. $CH_3\text{—}CH_2\text{—}CH(NH_2)\text{—}CH_2\text{—}CH_2\text{—}CH_2\text{—}CH_2\text{—}CH_3$ (eight-carbon chain with all H shown; NH_2 on the third carbon)

c. $CH_3\text{—}CH(NHCH_2CH_3)\text{—}CH_2\text{—}CH_2\text{—}CH_2\text{—}CH_2\text{—}CH_3$ (seven-carbon chain with all H shown; N—H attached to the second carbon, bearing an H—C—H / H—C—H / H ethyl group)

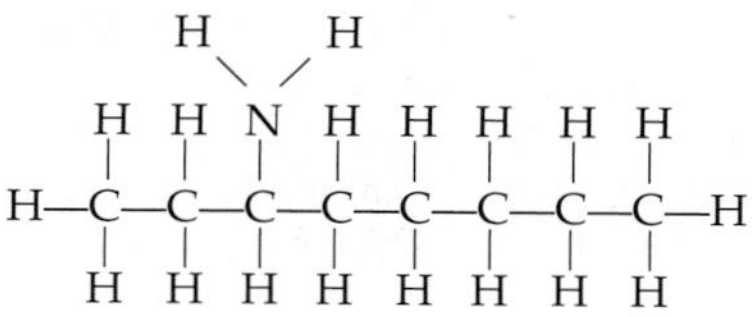

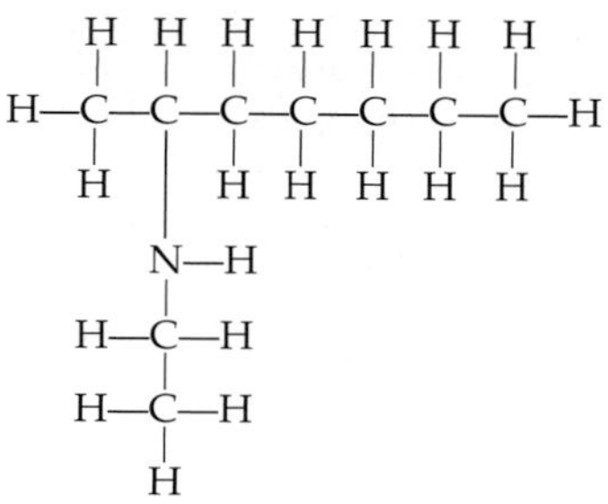

d. Five-carbon chain: C-2 bears NH_2 and a CH_3 branch, and every other position carries H (all bonds drawn out).

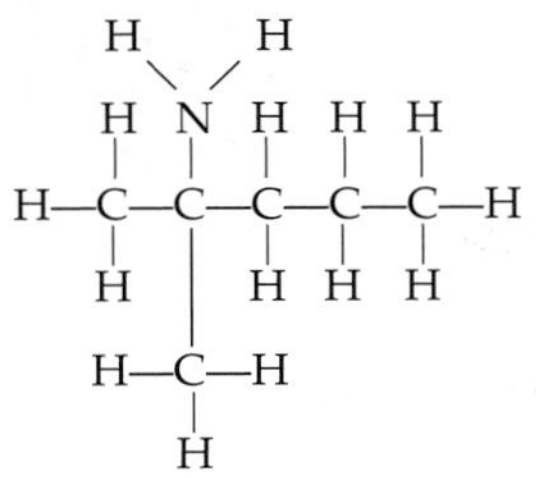

e. Nine-carbon chain: C-1 bears NH_2, C-4 bears Cl and C-5 bears I, and every other position carries H (all bonds drawn out).

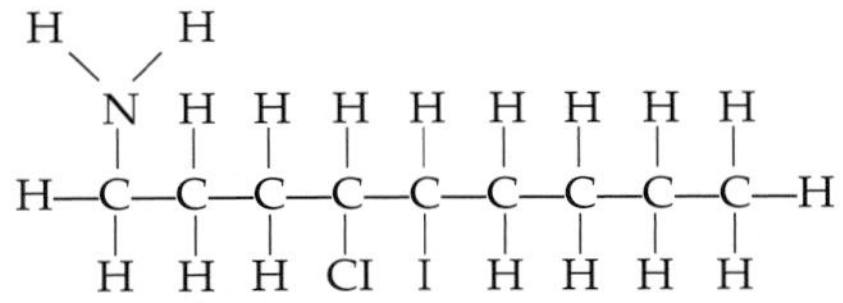

f. N bonded to two ethyl groups and one pentyl group, all bonds drawn out.

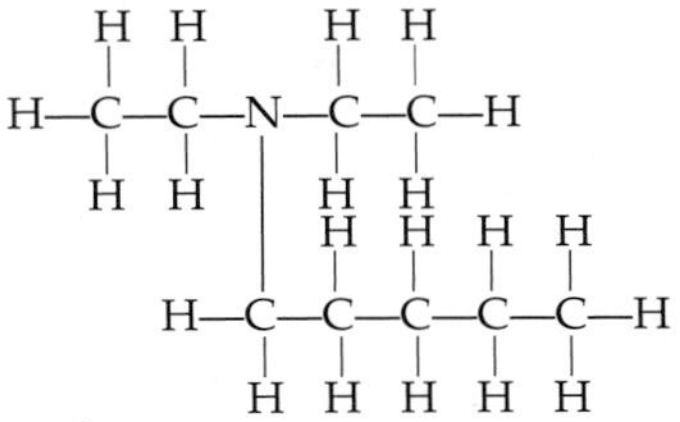

16.11 a. Cyclopentane ring bonded to $-NH_3^+ \, Br^-$

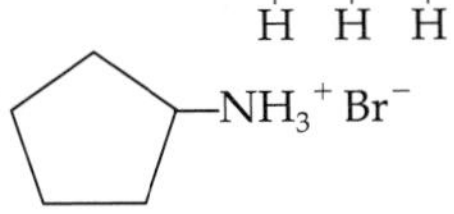

b. $CH_3CH_2—N^+H_2—CH_3 + OH^-$ (the N carries two H atoms, drawn above and below)

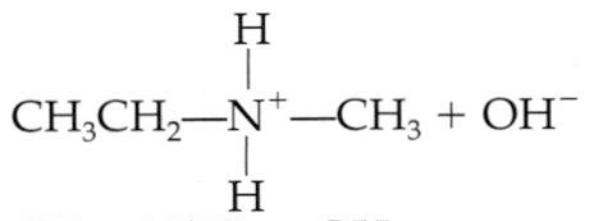

c. $CH_3—N^+H_3 + OH^-$

16.13 a. $CH_3—NH_2$

b. $CH_3—NH—CH_3$ (the CH_3 is drawn above the N)

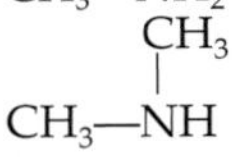

16.15 a. 1-Butanamine would be more soluble in water because it has a polar amine group that can form hydrogen bonds with water molecules.

b. 2-Pentanamine would be more soluble in water because it has a polar amine group that can form hydrogen bonds with water molecules.

16.17 Triethylamine molecules cannot form hydrogen bonds with one another, but 1-hexanamine molecules are able to do so.

16.19 a. 2-Butanamine
b. 3-Hexanamine
c. Cyclopentanamine
d. 2-Methyl-2-propanamine

16.21 a. $CH_3CH_2—NH—CH_2CH_3$

b. $CH_3CH_2CH_2CH_2NH_2$

c. $CH_3CH_2CHCH_2CH_2CH_2CH_2CH_2CH_2CH_3$ with NH_2 on C-3

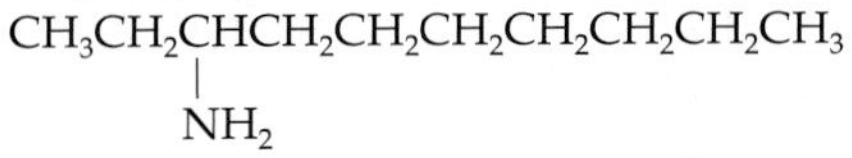

d. $CH_3CHCHCH_2CH_3$ with NH_2 on C-2 and Br on C-3

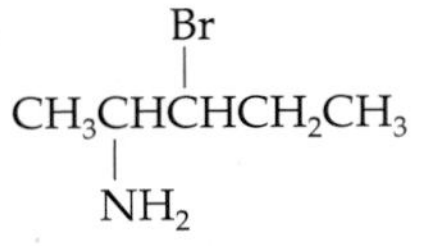

e. N bonded to three benzene rings (triphenylamine)

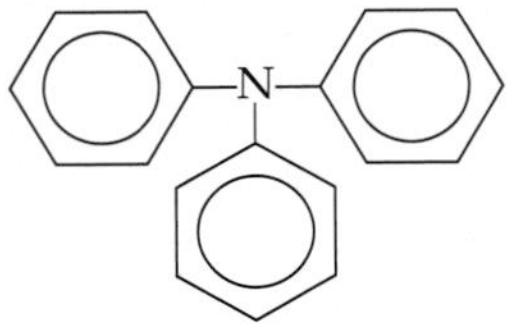

16.23 a. $CH_3CHCH_2CH_2CH_3$ with NH_2 on C-2

b. $CH_3CH_2CHCH_2NH_2$ with Br on C-2 (the $CHBr$ carbon)

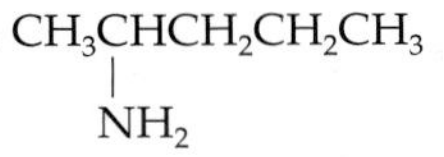

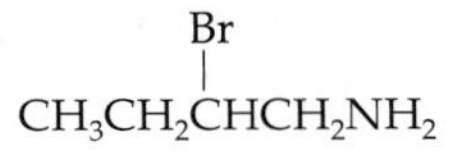

c. $CH_3CH_2—NH—CHCH_3$ with CH_3 on the CH carbon

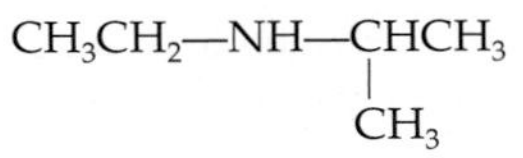

d. Cyclopentane ring bonded to NH_2

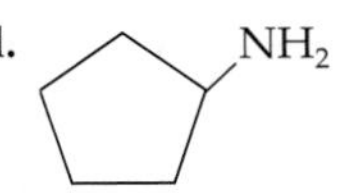

16.25 $CH_3CH_2CH_2CH_2NH_2$
1-Butanamine (Primary amine)

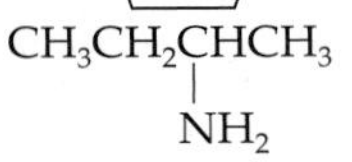

$CH_3CH_2CHCH_3$ with NH_2 on C-2
2-Butanamine (Primary amine)

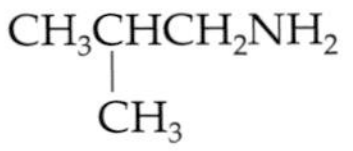

$CH_3CHCH_2NH_2$ with CH_3 on C-2
2-Methyl-1-propanamine (Primary amine)

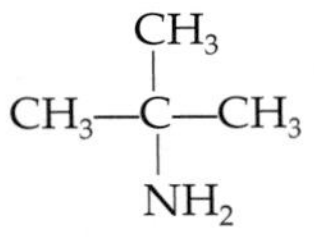

$CH_3—C—CH_3$ with CH_3 and NH_2 on the central carbon
2-Methyl-2-propanamine (Primary amine)

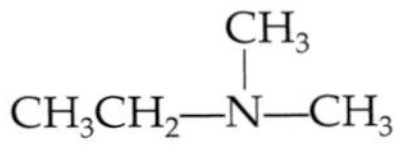

$CH_3CH_2—N—CH_3$ with CH_3 on N
N,N-Dimethylethanamine (Tertiary amine)

$CH_3CH_2—NH—CH_2CH_3$
N-Ethylethanamine (Secondary amine)

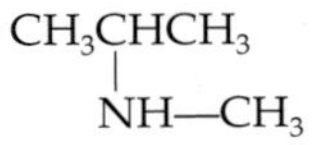

CH_3CHCH_3 with $NH—CH_3$ on C-2
N-Methyl-2-propanamine (Secondary amine)

$CH_3CH_2CH_2—NH—CH_3$
N-Methyl-1-propanamine (Secondary amine)

16.27 a. Primary
b. Secondary
c. Primary
d. Tertiary

16.29 a. Benzene ring with NO_2 and CH_3 (para) $\xrightarrow{[H]}$ benzene ring with NH_2 and CH_3 (para)

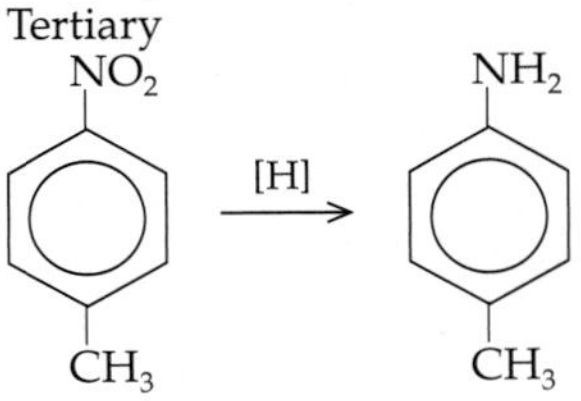

b. Benzene ring with NO_2 and OH (ortho) $\xrightarrow{[H]}$ benzene ring with NH_2 and OH (ortho)

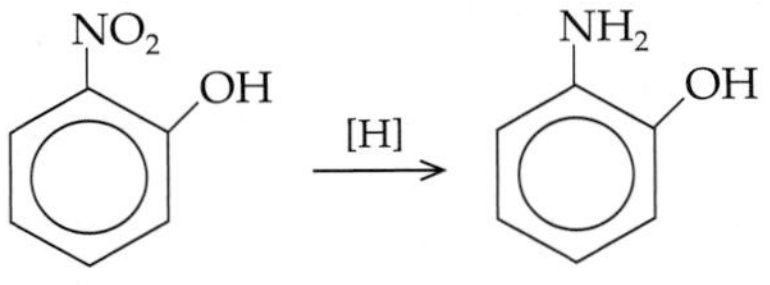

c. Benzene ring with NO_2 $\xrightarrow{[H]}$ benzene ring with NH_2

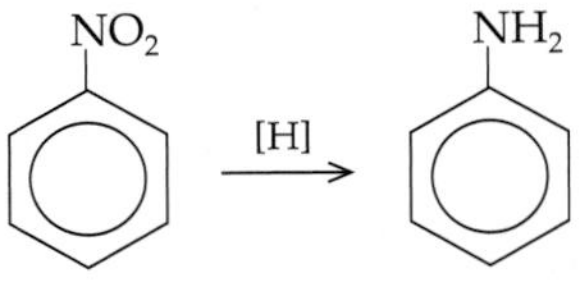

d. Benzene ring with CH_2NO_2 $\xrightarrow{[H]}$ benzene ring with CH_2NH_2

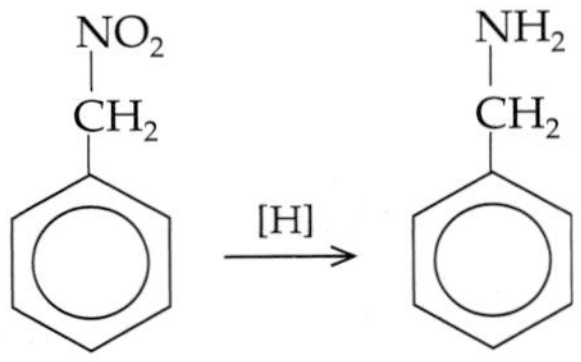

16.31 a. H_2O
b. HBr

c. $CH_3CH_2CH_2—N^+H_3$

d. $CH_3CH_2—N^+H_2Cl^-$ (with CH_2CH_3 on N)

16.33 Lower molecular weight amines are soluble in water because the N—H bond is polar and can form hydrogen bonds with water molecules.

16.35 Drugs containing amine groups are generally administered as ammonium salts because the salt is more soluble in water and, hence, in body fluids.

16.37 Putrescine (1,4-Diaminobutane):

$CH_2(NH_2)CH_2CH_2CH_2(NH_2)$

Cadaverine (1, 5-Diaminopentane):

$CH_2(NH_2)CH_2CH_2CH_2CH_2(NH_2)$

16.39 a.

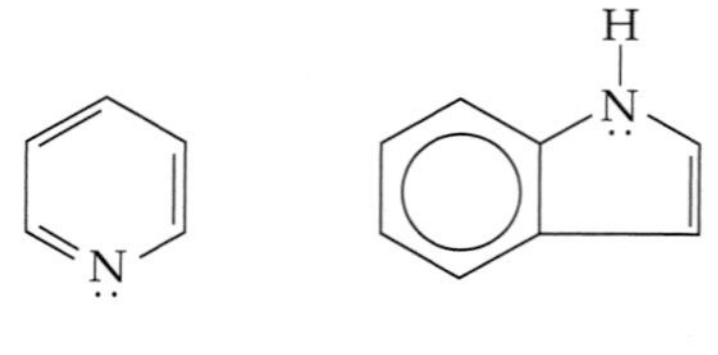

Pyridine Indole

b. The indole ring is found in lysergic acid diethylamide, which is a hallucinogenic drug. The pyridine ring is found in vitamin B_6, an essential water-soluble vitamin.

16.41 Morphine, codeine, quinine, and vitamin B_6

16.43 a. I.U.P.A.C. name: Propanamide
Common name: Propionamide

b. I.U.P.A.C. name: Pentanamide
Common name: Valeramide

c. I.U.P.A.C. name: *N,N*-Dimethylethanamide
Common name: *N,N*-Dimethylacetamide

16.45 a. $CH_3—C(=O)—NH_2$

b. $CH_3CH_2—C(=O)—NH—CH_3$

c. 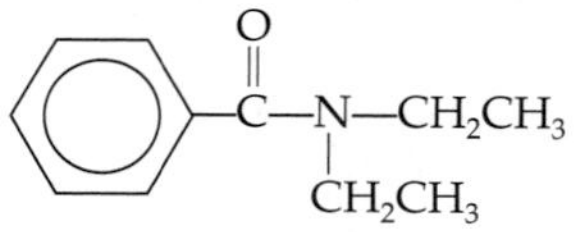 $C_6H_5—C(=O)—N(CH_2CH_3)—CH_2CH_3$

d. $CH_3CH_2CH(CH_3)CH(Br)CH_2—C(=O)—NH_2$

e. $CH_3—C(=O)—N(CH_3)—CH_3$

16.47 *N, N*-Diethyl-*m*-toluamide:

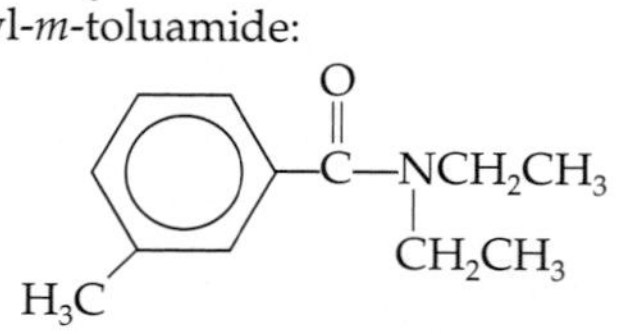

(H_3C on benzene ring, meta) $—C(=O)—NCH_2CH_3$ (with CH_2CH_3 on N)

Hydrolysis of this compound would release the carboxylic acid *m*-toluic acid and the amine *N*-ethylethanamine (diethylamine).

16.49 Amides are not proton acceptors (bases) because the highly electronegative carbonyl oxygen has a strong attraction for the nitrogen lone pair of electrons. As a result they cannot "hold" a proton.

16.51

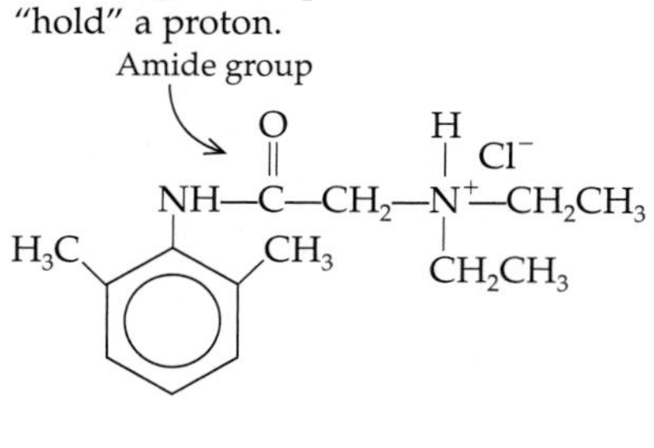

Amide group

$NH—C(=O)—CH_2—N^+H(Cl^-)—CH_2CH_3$ (with CH_2CH_3 on N; NH on benzene ring bearing H_3C and CH_3)

Lidocaine hydrochloride

16.53

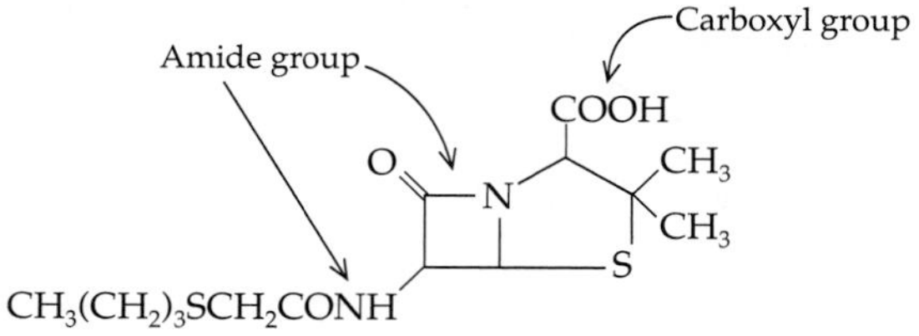

Amide group

Carboxyl group

$CH_3(CH_2)_3SCH_2CONH$ — (β-lactam ring, N, S, COOH, CH_3, CH_3)

Penicillin BT

16.55 a.

$CH_3—C(=O)—NHCH_3 + H_3O^+ \longrightarrow$

N-Methylethanamide

$CH_3COOH + CH_3NH_3^+$

Ethanoic acid Methanamine

b.

$CH_3CH_2CH_2—C(=O)—NH—CH_3 + H_3O^+ \longrightarrow$

N-Methylbutanamide

$CH_3CH_2CH_2COOH + CH_3NH_3^+$

Butanoic acid Methanamine

c.

$CH_3CH(CH_3)CH_2—C(=O)—NH—CH_2CH_3 + H_3O^+ \longrightarrow$

N-Ethyl-3-methylbutanamide Hydronium ion (Strong acid)

$CH_3CH(CH_3)CH_2COOH + CH_3CH_2NH_3^+$

3-Methylbutanoic acid Ethanamine

16.57 a. $CH_3CH_2—C(=O)—O—C(=O)—CH_2CH_3$

b. $CH_3CH_2—C(=O)—NH_2 + NH_4^+Cl^-$

c. $CH_3CH_2CH_2—C(=O)—Cl + 2CH_3CH_2NH_2$

16.59 $H_2N—CH(R)—C(=O)—OH$

16.61

```
                    Amide bond
  H    H  O     ↙ H  O
   \   |  ||      |  ||
    N—C—C—N—C—C—OH
   /   |       |  |
  H    H       H  CH3
```

Glycyl alanine

16.63

```
  H     H  O
   \    |  ||
    N—*C—C—OH
   /    |
  H     CH3
```

16.65 In an acyl group transfer reaction, the acyl group of an acid chloride is transferred from the Cl of the acid chloride to the N of an amine or ammonia. The product is an amide.

16.67 A chemical that carries messages or signals, from a nerve to a target cell

16.69 **a.** Tremors, monotonous speech, loss of memory and problem-solving ability, and loss of motor function
b. Parkinson's disease
c. Schizophrenia, intense satiety sensations

16.71 In proper amounts, dopamine causes a pleasant, satisfied feeling. This feeling becomes intense as the amount of dopamine increases. Several drugs, including cocaine, heroin, amphetamines, alcohol, and nicotine increase the levels of dopamine. It is thought that the intense satiety response this brings about may contribute to addiction to these substances.

16.73 Epinephrine is a component of the flight or fight response. It stimulates glycogen breakdown to provide the body with glucose to supply the needed energy for this stress response.

16.75 The amino acid tryptophan

16.77 Perception of pain, thermoregulation, and sleep

16.79 Promotes the itchy skin rash associated with poison ivy and insect bites; the respiratory symptoms characteristic of hay fever; secretion of stomach acid

16.81 Inhibitory neurotransmitters

16.83 When acetylcholine is released from a nerve cell, it binds to receptors on the surface of muscle cells. This binding stimulates the muscle cell to contract. To stop the contraction, the acetylcholine is then broken down to choline and acetate ion. This is catalyzed by the enzyme acetylcholinesterase.

16.85 Organophosphates inactivate acetylcholinesterase by binding covalently to it. Since acetylcholine is not broken down, nerve transmission continues, resulting in muscle spasm. Pyridine aldoxime methiodide (PAM) is an antidote to organophosphate poisoning because it displaces the organophosphate, thereby allowing acetycholinesterase to function.

Chapter 17

17.1 It is currently recommended that 58% of the calories in the diet should be carbohydrates. Of that amount, no more than 10% should be simple sugars.

17.3 An aldose is a sugar with an aldehyde functional group. A ketose is a sugar with a ketone functional group.

17.5 **a.** Ketose **b.** Aldose **c.** Ketose **d.** Aldose **e.** Ketose **f.** Aldose

17.7

```
        CH2OH                      CH2OH
        |———O                      |———O
  HO  / H     \ OH          HO   / H     \ H
     \ OH   H  /               \ OH   H   /
   H  |       | H            H  |       | OH
      H      OH                 H      OH
```

β-D-Galactose α-D-Galactose

17.9 α-Amylase and β-amylase are digestive enzymes that break down the starch amylose. α-Amylase cleaves glycosidic bonds of the amylose chain at random, producing shorter polysaccharide chains. β-Amylase sequentially cleaves maltose (a disaccharide of glucose) from the reducing end of the polysaccharide chain.

17.11 A monosaccharide is the simplest sugar and consists of a single saccharide unit. A disaccharide is made up of two monosaccharides joined covalently by a glycosidic bond.

17.13 Mashed potato flakes, rice, and corn starch contain amylose and amylopectin, both of which are polysaccharides. A candy bar contains sucrose, a disaccharide. Orange juice contains fructose, a monosaccharide. It may also contain sucrose if the label indicates that sugar has been added.

17.15 Four

17.17

```
      O                     CH2OH
      ||                      |
      C—H                     C=O
      |                       |
   H—C—OH                HO—C—H
      |                       |
  HO—C—H                  H—C—OH
      |                       |
  HO—C—H                  H—C—OH
      |                       |
   H—C—OH                   CH2OH
      |
    CH2OH
```

D-Galactose (An aldohexose) D-Fructose (A ketohexose)

17.19 **a.** β-D-Glucose is a hemiacetal.
b. β-D-Fructose is a hemiketal.
c. α-D-Galactose is a hemiacetal.

17.21

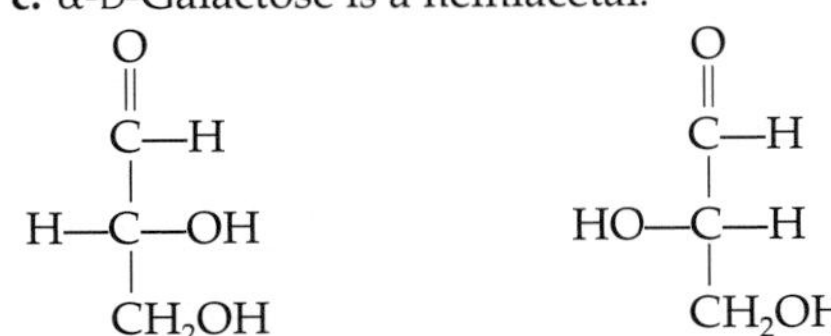

D-Glyceraldehyde L-Glyceraldehyde

17.23 Dextrose is a common name used for D-glucose.

17.25 D- and L-Glyceraldehyde are a pair of enantiomers, that is, they are nonsuperimposable mirror images of one another.

17.27 Stereoisomers are a pair of molecules that have the same structural formula and bonding pattern but that differ in the arrangement of the atoms in space.

17.29 A chiral carbon is one that is bonded to four different chemical groups.

17.31 A polarimeter converts monochromatic light into monochromatic plane-polarized light. This plane-polarized light is passed through a sample and into an analyzer. If the sample is optically active, it will rotate the plane of the light. The degree and angle of rotation are measured by the analyzer.

17.33 a.

O
‖
C—H
HO—*—H
H—*—OH
HO—*—H
HO—*—H
CH_2OH

b.

O
‖
C—H
H—*—OH
H—*—OH
CH_2OH

c.

O
‖
C—H
HO—*—H
H—*—OH
HO—*—H
H—*—OH
HO—*—H
CH_2OH

17.35 When the carbonyl group at C-1 of D-glucose reacts with the C-5 hydroxyl group, a new chiral carbon is created (C-1). In the α-isomer of the cyclic sugar, the C-1 hydroxyl group is below the ring; and in the β-isomer, the C-1 hydroxyl group is above the ring.

17.37 β-Maltose and α-lactose would give positive Benedict's tests. Glycogen would give only a weak reaction because there are fewer reducing ends for a given mass of the carbohydrate.

17.39 Enantiomers are stereoisomers that are nonsuperimposable mirror images of one another. For instance:

O
‖
C—H
H—C—OH
CH_2OH

D-Glyceraldehyde

O
‖
C—H
HO—C—H
CH_2OH

L-Glyceraldehyde

17.41 An aldehyde sugar forms an intramolecular hemiacetal when the carbonyl group of the monosaccharide reacts with a hydroxyl group on one of the other carbon atoms.

17.43

CH_2OH CH_2OH
O O
H H H H H H OH
OH H —O— OH H
HO H
H OH H OH

β-Maltose

17.45 Milk

17.47 Eliminating milk and milk products from the diet

17.49 Lactose intolerance is the inability to produce the enzyme lactase that hydrolyzes the milk sugar lactose into its component monosaccharides, glucose and galactose.

17.51 The glucose units of amylose are joined by α (1 → 4) glycosidic bonds and those of cellulose are bonded together by β (1 → 4) glycosidic bonds.

17.53 Glycogen serves as a storage molecule for glucose.

17.55 The salivary glands and the pancreas

Chapter 18

18.1 a. $CH_3(CH_2)_7CH{=}CH(CH_2)_7COOH$
b. $CH_3(CH_2)_{10}COOH$
c. $CH_3(CH_2)_4CH{=}CH{-}CH_2{-}CH{=}CH(CH_2)_7COOH$
d. $CH_3(CH_2)_{16}COOH$

18.3 a. Esterification of lauric acid and ethanol

$CH_3(CH_2)_{10}{-}C({=}O){-}OH + CH_3CH_2OH$

↓

$CH_3(CH_2)_{10}{-}C({=}O){-}OCH_2CH_3 + H_2O$

b. Reaction of oleic acid with NaOH

$CH_3(CH_2)_7CH{=}CH(CH_2)_7{-}C({=}O){-}OH + NaOH$

↓

$CH_3(CH_2)_7CH{=}CH(CH_2)_7{-}C({=}O){-}O^- Na^+ + H_2O$

c. Hydrogenation of arachidonic acid

$CH_3(CH_2)_4CH{=}CHCH_2CH{=}CHCH_2CH{=}CHCH_2CH{=}CH(CH_2)_3{-}C({=}O){-}OH + 4H_2$

↓ Ni

$CH_3(CH_2)_{18}{-}C({=}O){-}OH$

18.5 a.

$CH_3(CH_2)_7CH{=}CH(CH_2)_7{-}C({=}O){-}O{-}CH_2$
CH—OH
CH_2—OH

$CH_3(CH_2)_7CH{=}CH(CH_2)_7{-}C({=}O){-}O{-}CH_2$
$CH_3(CH_2)_7CH{=}CH(CH_2)_7{-}C({=}O){-}O{-}CH$
CH_2—OH

$CH_3(CH_2)_7CH{=}CH(CH_2)_7{-}C({=}O){-}O{-}CH_2$
$CH_3(CH_2)_7CH{=}CH(CH_2)_7{-}C({=}O){-}O{-}CH$
$CH_3(CH_2)_7CH{=}CH(CH_2)_7{-}C({=}O){-}O{-}CH_2$

b.

$CH_3(CH_2)_8{-}C({=}O){-}O{-}CH_2$
CH—OH
CH_2—OH

$CH_3(CH_2)_8{-}C({=}O){-}O{-}CH_2$
$CH_3(CH_2)_8{-}C({=}O){-}O{-}CH$
CH_2—OH

$$\begin{array}{l} CH_3(CH_2)_8-C(=O)-O-CH_2 \\ CH_3(CH_2)_8-C(=O)-O-CH \\ CH_3(CH_2)_8-C(=O)-O-CH_2 \end{array}$$

c.

$$\begin{array}{l} CH_3(CH_2)_{14}-C(=O)-O-CH_2 \\ CH-OH \\ CH_2-OH \end{array}$$

$$\begin{array}{l} CH_3(CH_2)_{14}-C(=O)-O-CH_2 \\ CH_3(CH_2)_{14}-C(=O)-O-CH \\ CH_2-OH \end{array}$$

$$\begin{array}{l} CH_3(CH_2)_{14}-C(=O)-O-CH_2 \\ CH_3(CH_2)_{14}-C(=O)-O-CH \\ CH_3(CH_2)_{14}-C(=O)-O-CH_2 \end{array}$$

d.

$$\begin{array}{l} CH_3(CH_2)_{10}-C(=O)-O-CH_2 \\ CH-OH \\ CH_2-OH \end{array}$$

$$\begin{array}{l} CH_3(CH_2)_{10}-C(=O)-O-CH_2 \\ CH_3(CH_2)_{10}-C(=O)-O-CH \\ CH_2-OH \end{array}$$

$$\begin{array}{l} CH_3(CH_2)_{10}-C(=O)-O-CH_2 \\ CH_3(CH_2)_{10}-C(=O)-O-CH \\ CH_3(CH_2)_{10}-C(=O)-O-CH_2 \end{array}$$

18.7

HO (rings A, B, C, D; carbons numbered 1–19)

Steroid nucleus

18.9 Receptor-mediated endocytosis

18.11 Membrane transport resembles enzyme catalysis because both processes exhibit a high degree of specificity.

18.13 Fatty acids, glycerides, nonglyceride lipids, and complex lipids

18.15 A saturated fatty acid is one in which the hydrocarbon tail has only carbon-to-carbon single bonds. An unsaturated fatty acid has at least one carbon-to-carbon double bond.

18.17 The melting points increase.

18.19 **a.** Decanoic acid

$CH_3(CH_2)_8COOH$

b. Stearic acid

$CH_3(CH_2)_{16}COOH$

c. *trans*-5-Decenoic acid

$$\begin{array}{l} H \quad \quad CH_2CH_2CH_2-C(=O)-OH \\ \quad C{=}C \\ CH_3CH_2CH_2CH_2 \quad \quad H \end{array}$$

d. *cis*-5-Decenoic acid

$$\begin{array}{l} CH_3CH_2CH_2CH_2 \quad \quad CH_2CH_2CH_2C(=O)-OH \\ \quad C{=}C \\ H \quad \quad H \end{array}$$

18.21 **a.**

$$\begin{array}{l} CH_2OH \\ CHOH + 3CH_3(CH_2)_{12}C(=O)-OH \\ CH_2OH \end{array}$$

$$\downarrow$$

$$\begin{array}{l} CH_3(CH_2)_{12}-C(=O)-O-CH_2 \\ CH_3(CH_2)_{12}-C(=O)-O-CH + 3H_2O \\ CH_3(CH_2)_{12}-C(=O)-O-CH_2 \end{array}$$

b.

$$\begin{array}{l} CH_3(CH_2)_{16}-C(=O)-O-CH_2 \\ CH_3(CH_2)_{16}-C(=O)-O-CH + 3H_2O \\ CH_3(CH_2)_{16}-C(=O)-O-CH_2 \end{array}$$

$$\downarrow$$

$$\begin{array}{l} \quad\quad\quad\quad\quad\quad\quad\quad CH_2OH \\ 3CH_3(CH_2)_{16}-C(=O)-OH + CHOH \\ \quad\quad\quad\quad\quad\quad\quad\quad CH_2OH \end{array}$$

c. $CH_3CH_2CH_2CH_2CH_2CH_2CH_2CH_2CH_2-C(=O)-OH$

$$\downarrow \text{KOH}$$

$CH_3CH_2CH_2CH_2CH_2CH_2CH_2CH_2CH_2-C(=O)-O^- K^+ + H_2O$

d. $CH_3(CH_2)_4CH{=}CHCH_2CH{=}CH(CH_2)_7{-}C({=}O){-}OH + 2H_2 \xrightarrow{Ni} CH_3(CH_2)_{16}{-}C({=}O){-}OH$

18.23 The essential fatty acid linoleic acid is required for the synthesis of arachidonic acid, a precursor for the synthesis of the prostaglandins, a group of hormonelike molecules.

18.25 Aspirin effectively decreases the inflammatory response by inhibiting the synthesis of all prostaglandins. Aspirin works by inhibiting cyclooxygenase, the first enzyme in prostaglandin biosynthesis. This inhibition results from the transfer of an acetyl group from aspirin to the enzyme. Because cyclooxygenase is found in all cells, synthesis of all prostaglandins is inhibited.

18.27 Smooth muscle contraction, enhancement of fever and swelling associated with the inflammatory response, bronchial dilation, inhibition of secretion of acid into the stomach

18.29

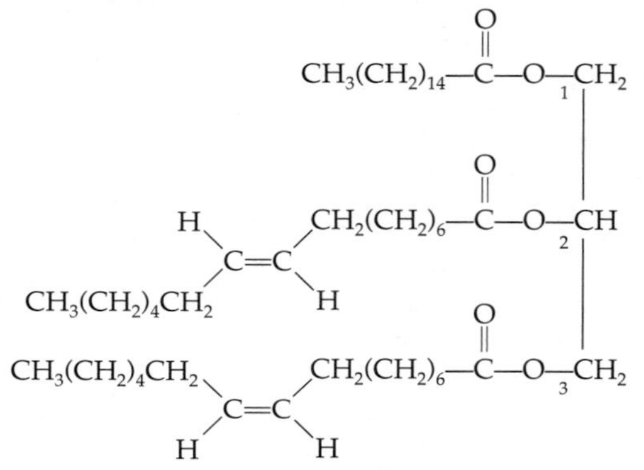

$CH_3(CH_2)_{14}{-}C({=}O){-}O{-}\overset{1}{C}H_2$

$CH_3(CH_2)_4CH_2(H)C{=}C(H)CH_2(CH_2)_6{-}C({=}O){-}O{-}\overset{2}{C}H$

$CH_3(CH_2)_4CH_2(H)C{=}C(H)CH_2(CH_2)_6{-}C({=}O){-}O{-}\overset{3}{C}H_2$

18.31

$CH_3CH_2CH_2CH_2CH_2CH_2CH_2CH_2CH_2{-}C({=}O){-}O{-}\overset{1}{C}H_2$

$CH_3CH_2CH_2CH_2CH_2CH_2CH_2CH_2CH_2CH_2CH_2{-}C({=}O){-}O{-}\overset{2}{C}H$

$\overset{3}{C}H_2{-}O{-}P({=}O)(O^-){-}O^-$

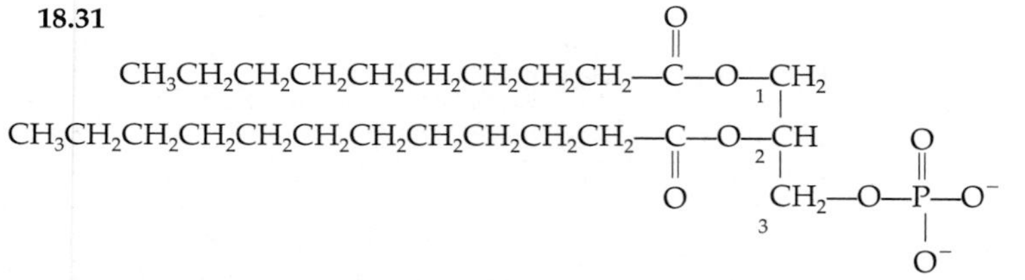

18.33 Triglycerides consist of three fatty acids esterified to the three hydroxyl groups of glycerol. In phospholipids there are only two fatty acids esterified to glycerol. A phosphoryl group is esterified (phosphoester linkage) to the third hydroxyl group.

18.35 Sphingolipids are phospholipids that are derived from sphingosine rather than glycerol. Sphingosine is a nitrogen-containing (amino) alcohol.

18.37 Cholesterol is readily soluble in the hydrophobic region of biological membranes. It is involved in regulating the fluidity of the membrane.

18.39 Progesterone is the most important hormone associated with pregnancy. Testosterone is needed for development of male secondary sexual characteristics. Estrone is required for proper development of female secondary sexual characteristics.

18.41 Cortisone is used to treat rheumatoid arthritis, asthma, gastrointestinal disorders, and many skin conditions.

18.43 Myricyl palmitate (beeswax) is made up of the fatty acid palmitic acid and the alcohol myricyl alcohol—$CH_3(CH_2)_{28}CH_2OH$.

18.45 Isoprenoids are a large, diverse collection of lipids that are synthesized from the isoprene unit:

$CH_2{=}C(CH_3){-}CH{=}CH_2$

18.47 Steroids and bile salts, lipid-soluble vitamins, certain plant hormones, and chlorophyll

18.49 Chylomicrons, high-density lipoproteins, low-density lipoproteins, and very low density lipoproteins

18.51 Atherosclerosis results when cholesterol and other substances coat the arteries causing a narrowing of the passageways. As the passageways become narrower, greater pressure is required to provide adequate blood flow. This results in higher blood pressure (hypertension).

18.53 If the LDL receptor is defective, it cannot function to remove cholesterol-bearing LDL particles from the blood. The excess cholesterol, along with other substances, will accumulate along the walls of the arteries, causing atherosclerosis.

18.55 If the fatty acyl tails of membrane phospholipids are converted from saturated to unsaturated, the fluidity of the membrane will increase.

18.57 The basic structure of a biological membrane is a bilayer of phospholipid molecules arranged so that the hydrophobic hydrocarbon tails are packed in the center and the hydrophilic head groups are exposed on the inner and outer surfaces.

18.59 A peripheral membrane protein is bound to only one surface of the membrane, either inside or outside the cell.

18.61 Cholesterol is freely soluble in the hydrophobic layer of a biological membrane. It moderates the fluidity of the membrane by disrupting the stacking of the fatty acid tails of membrane phospholipids.

18.63 L. Frye and M. Edidin carried out studies in which specific membrane proteins on human and mouse cells were labeled with red and green fluorescent dyes, respectively. The human and mouse cells were fused into single-celled hybrids and were observed using a microscope with an ultraviolet light source. The ultraviolet light caused the dyes to fluoresce. Initially the dyes were localized in regions of the membrane representing the original human or mouse cell. Within an hour, the proteins were evenly distributed throughout the membrane of the fused cell.

18.65 In simple diffusion the molecule moves directly across the membrane, whereas in facilitated diffusion a protein channel through the membrane is required.

18.67 Active transport requires an energy input to transport molecules or ions against the gradient (from an area of lower concentration to an area of higher concentration). Facilitated diffusion is a means of passive transport in which molecules or ions pass from regions of higher concentration to regions of lower concentration through a permease protein. No energy is expended by the cell in facilitated diffusion.

18.69 An antiport transport mechanism is one in which one molecule or ion is transported into the cell while a different molecule or ion is transported out of the cell.

18.71 Each permease or channel protein has a binding site that has a shape and charge distribution that is complementary to the molecule or ion that it can bind and transport across the cell membrane.

18.73 One ATP molecule is hydrolyzed to transport 3 Na^+ out of the cell and 2 K^+ into the cell.

18.75 Active transport is the movement of molecules or ions across a membrane against a concentration gradient (from a region of lower concentration to a region of higher concentration).

Chapter 19

19.1 **a.** Glycine (gly):

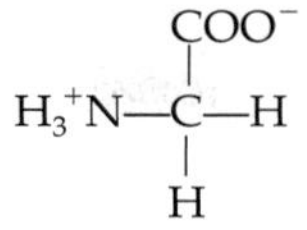

COO^-
$H_3^+N—C—H$
H

b. Proline (pro):

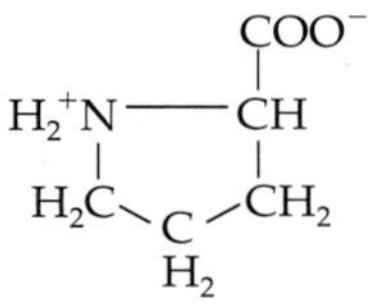

COO^-
$H_2^+N—CH$
H_2C CH_2
C
H_2

c. Threonine (thr):

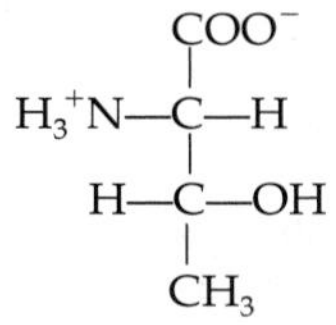

COO^-
$H_3^+N—C—H$
$H—C—OH$
CH_3

d. Aspartate (asp):

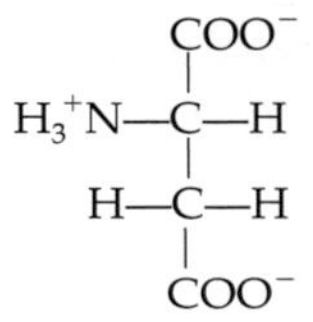

COO^-
$H_3^+N—C—H$
$H—C—H$
COO^-

e. Lysine (lys):

COO^-
$H_3^+N—C—H$
$H—C—H$
$H—C—H$
$H—C—H$
$H—C—H$
N^+H_3

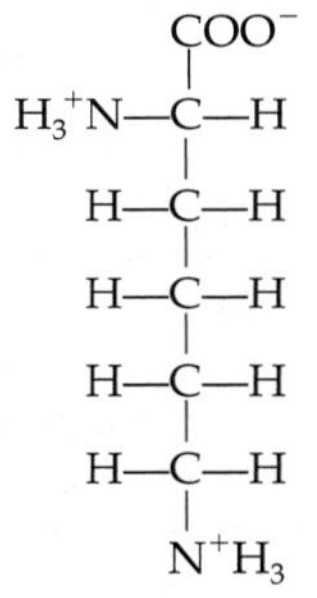

19.3 **a.** Alanyl-phenylalanine:

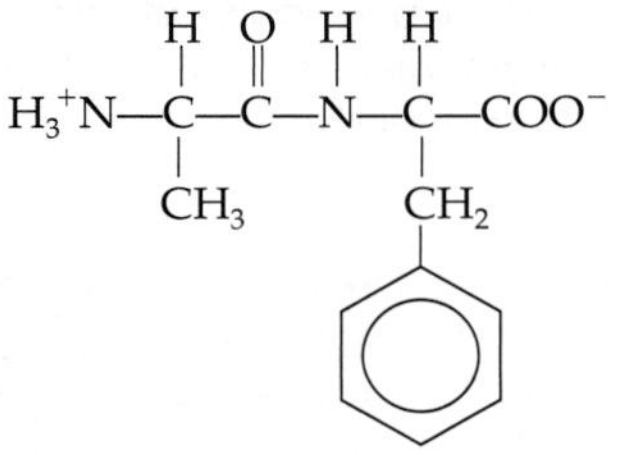

H O H H
$H_3^+N—C—C—N—C—COO^-$
CH_3 CH_2
(phenyl ring)

b. Lysyl-alanine:

H O H H
$H_3^+N—C—C—N—C—COO^-$
CH_2 CH_3
CH_2
CH_2
CH_2
N^+H_3

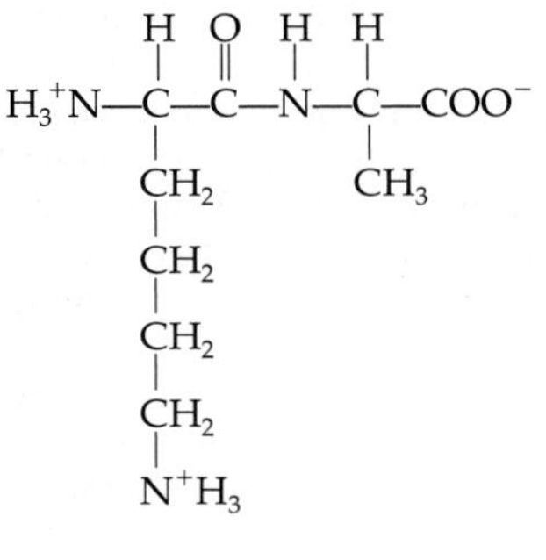

c. Phenylalanyl-tyrosyl-leucine:

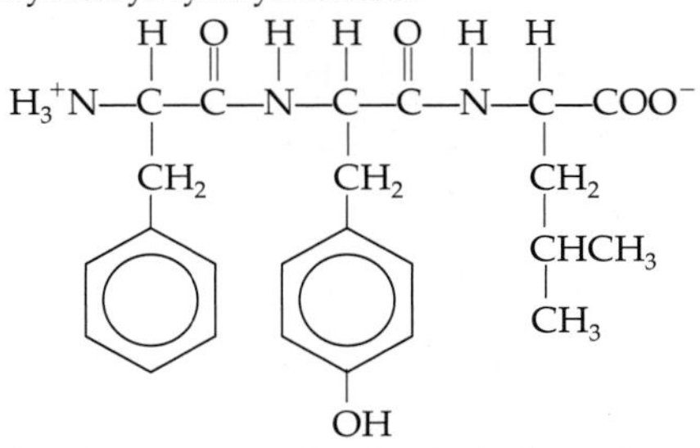

H O H H O H H
$H_3^+N—C—C—N—C—C—N—C—COO^-$
CH_2 CH_2 CH_2
(phenyl ring) (phenol ring) $CHCH_3$
CH_3
OH

19.5 The primary structure of a protein is the amino acid sequence of the protein chain. Regular, repeating folding of the peptide chain caused by hydrogen bonding between the amide nitrogens and carbonyl oxygens of the peptide bond is the secondary structure of a protein. The two most common types of secondary structure are the α-helix and the β-pleated sheet. Tertiary structure is the further folding of the regions of α-helix and β-pleated sheet into a compact, spherical structure. Formation and maintenance of the tertiary structure results from weak attractions between amino acid R groups. The binding of two or more peptides to produce a functional protein defines the quaternary structure.

19.7 Oxygen is efficiently transferred from hemoglobin to myoglobin in the muscle because myoglobin has a greater affinity for oxygen.

19.9 High temperature disrupts the hydrogen bonds and other weak interactions that maintain protein structure.

19.11 Vegetables vary in amino acid composition. No single vegetable can provide all of the amino acid requirements of the body. By eating a variety of different vegetables, all the amino acid requirements of the human body can be met.

19.13 Five of the biological functions carried out by proteins include serving as enzymes to speed up biochemical reactions, acting as antibodies to protect the body against disease, transporting materials throughout the body and into and out of cells, regulating cellular function, and serving as structural support for animals.

19.15

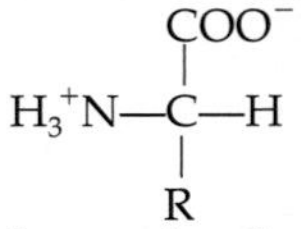

COO^-
$H_3^+N—C—H$
R

19.17 Interactions between the R groups of the amino acids in a polypeptide chain are important for the formation and maintenance of the tertiary and quaternary structures of proteins.

19.19

Glycine: H / $H_3^+N—C—COO^-$ / H

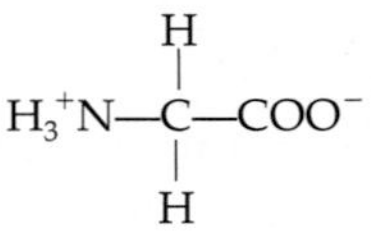

Alanine: H / $H_3^+N—C—COO^-$ / CH_3

Valine: H / $H_3^+N—C—COO^-$ / CH / H_3C CH_3

Leucine: H / $H_3^+N—C—COO^-$ / CH_2 / CH / H_3C CH_3

Isoleucine: H / $H_3^+N—C—COO^-$ / $H—C—CH_3$ / CH_2 / CH_3

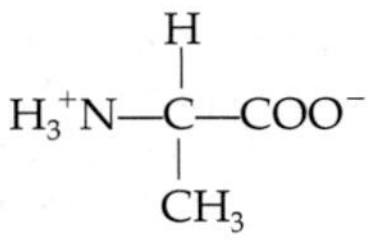

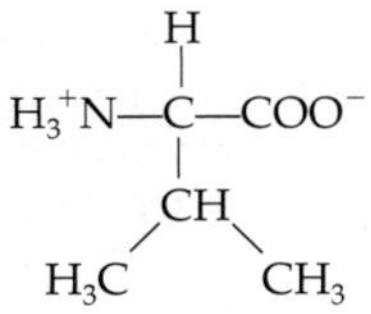

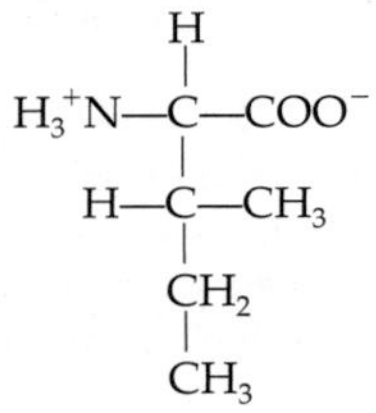

Phenylalanine: H / $H_3^+N—C—COO^-$ / CH_2 / (phenyl ring)

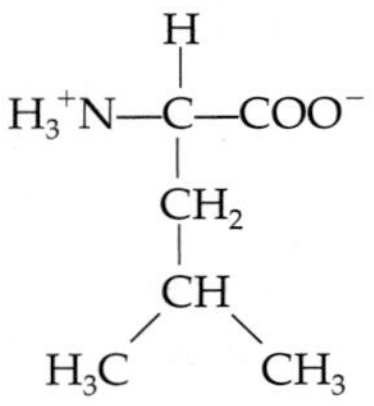

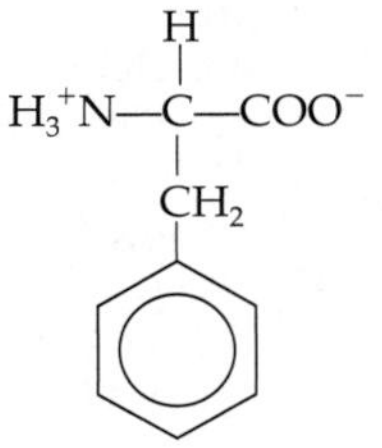

Proline Tryptophan Methionine

```
Proline:
          H
          |
H2N+——————C—COO−
 |        |
CH2      CH2
   \    /
    CH2

Tryptophan:
          H
          |
     H3+N—C—COO−
          |
          CH2
          |
          C
          ||
 (benzene)  CH
        N
        |
        H

Methionine:
          H
          |
     H3+N—C—COO−
          |
          CH2
          |
          CH2
          |
          S
          |
          CH3
```

19.21 **a.** His-trp-cys:

```
        H  O  H  H  O  H  H
        |  ||  |  |  ||  |  |
H3+N—C—C—N—C—C—N—C—COO−
        |           |           |
       CH2         CH2         CH2
  H+N                           |
                                SH
     N            NH
     H
```

b. Gly-leu-ser:

```
        H  O  H  H  O  H  H
        |  ||  |  |  ||  |  |
H3+N—C—C—N—C—C—N—C—COO−
        |           |           |
        H          CH2      H—C—OH
                    |           |
                H—C—CH3         H
                    |
                   CH3
```

c. Arg-ile-val:

```
        H  O  H  H  O  H  H
        |  ||  |  |  ||  |  |
H3+N—C—C—N—C—C—N—C—COO−
        |           |           |
       CH2        CHCH3       CHCH3
        |           |           |
       CH2         CH2         CH3
        |           |
       CH2         CH3
        |
       NH
        |
       C=N+H2
        |
       NH2
```

19.23 The peptide bond consists of an amide group. There is no free rotation around the peptide bond because the lone pair of electrons of the nitrogen atom interacts with the carbon and oxygen of the carbonyl group. This results in a resonance structure with a partially double bonded character.

19.25 The genetic information in the DNA dictates the order in which amino acids will be added to the protein chain. The order of the amino acids is the primary structure of the protein.

19.27 The primary structure of a protein is the linear arrangement of amino acids joined to one another by peptide bonds.

19.29 The secondary structure of a protein is the folding of the primary structure into an α-helix or β-pleated sheet.

19.31 **a.** α-Helix
b. β-Pleated sheet

19.33 A fibrous protein is one that is composed of peptides arranged in long sheets or fibers.

19.35 A parallel β-pleated sheet is one in which the hydrogen bonded peptide chains have their amino-termini aligned head-to-head.

19.37 The tertiary structure of a protein is the globular, three-dimensional structure of a protein that results from folding the regions of secondary structure.

19.39

```
          H
          |
     H3+N—C—COO−
          |
          CH2
          |
          S
          |
          S
          |
          CH2
          |
     H3+N—C—COO−
          |
          H
```

19.41 The tertiary structure is a level of folding of a protein chain that has already undergone secondary folding. The regions of α-helix and β-pleated sheet are folded into a globular structure.

19.43 Quaternary protein structure is the aggregation of two or more folded peptide chains to produce a functional protein.

19.45 A glycoprotein is a protein with covalently attached sugars.

19.47 Hydrogen bonding maintains the secondary structure of a protein and contributes to the stability of the tertiary and quaternary levels of structure.

19.49 The peptide bond exhibits resonance, which results in a partially double bonded character. This causes the rigidity of the peptide bond.

```
 —C       H            —C        H
    \   ..  /              \      /
     C—N     <——>      C=N+
    //    \              /     \
   O       C—          −O       C—
```

19.51 The code for the primary structure of a protein is carried in the genetic information (DNA).

19.53 The function of hemoglobin is to carry oxygen from the lungs to oxygen-demanding tissues throughout the body. Hemoglobin is found in red blood cells.

19.55 Hemoglobin is a protein composed of four subunits—two α-globin and two β-globin subunits. Each subunit holds a heme group, which in turn carries an Fe^{2+} ion.

19.57 The function of the heme group in hemoglobin and myoglobin is to bind to molecular oxygen.

19.59 Because carbon monoxide binds tightly to the heme groups of hemoglobin, it is not easily removed or replaced by oxygen. As a result, the effects of oxygen deprivation (suffocation) occur.

19.61 When sickle cell hemoglobin (HbS) is deoxygenated, the amino acid valine fits into a hydrophobic pocket on the surface of another HbS molecule. Many such sickle cell hemoglobin molecules polymerize into long rods that cause the red blood cell to sickle. In normal hemoglobin, glutamic acid is found in the place of the valine. This negatively charged amino acid will not "fit" into the hydrophobic pocket.

19.63 When individuals have one copy of the sickle cell gene and one copy of the normal gene, they are said to carry the *sickle cell trait.* These individuals will not suffer serious side effects, but may pass the trait to their offspring. Individuals with two copies of the sickle cell globin gene exhibit all the symptoms of the disease and are said to have *sickle cell anemia.*

19.65 *Denaturation* is the process by which the organized structure of a protein is disrupted, resulting in a completely disorganized, nonfunctional form of the protein.

19.67 Heat is an effective means of sterilization because it destroys the proteins of microbial life-forms, including fungi, bacteria, and viruses.

19.69 Even relatively small fluctuations in blood pH can be life threatening. It is likely that these small changes would alter the normal charges on the proteins and modify their interactions. These changes can render a protein incapable of carrying out its functions.

19.71 Proteins become polycations at low pH because the additional protons will protonate the carboxylate groups. As these negative charges are neutralized, the charge on the proteins will be contributed only by the protonated amino groups ($-N^+H_3$).

19.73 The low pH of the yogurt denatures the proteins of microbial contaminants, inhibiting their growth.

19.75 In a vegetarian diet, vegetables are the only source of dietary protein. Because individual vegetable sources do not provide all the needed amino acids, vegetables must be mixed to provide all the essential and nonessential amino acids in the amounts required for biosynthesis.

19.77 Nonessential amino acids can be synthesized by the body and are, therefore, not required in the diet. Essential amino acids cannot be synthesized by the body and must be provided by the diet.

19.79 Synthesis of digestive enzymes must be carefully controlled because the active enzyme would digest, and thus destroy, the cell that produces it.

Chapter 20

20.1 **a.** Transferase
b. Ligase
c. Isomerase
d. Oxidoreductase
e. Isomerase

20.3 **a.** Sucrose
b. Pyruvate
c. Succinate

20.5 The induced fit model assumes that the enzyme is flexible. Both the enzyme and the substrate are able to change shape to form the enzyme-substrate complex. The lock-and-key model assumes that the enzyme is inflexible (the lock) and the substrate (the key) fits into a specific rigid site (the active site) on the enzyme to form the enzyme-substrate complex.

20.7 An enzyme might put pressure on a bond, thereby catalyzing bond breakage. An enzyme could bring two reactants into close proximity and in the proper orientation for the reaction to occur. Finally, an enzyme could alter the pH of the microenvironment of the active site, thereby serving as a transient donor or acceptor of H^+.

20.9 Water-soluble vitamins are required by the body for the synthesis of coenzymes that are required for the function of a variety of enzymes.

20.11 A decrease in pH will change the degree of ionization of the R groups within a peptide chain. This disturbs the weak interactions that maintain the structure of an enzyme, which may denature the enzyme. Less drastic alterations in the charge of R groups in the active site of the enzyme can inhibit enzyme-substrate binding or destroy the catalytic ability of the active site.

20.13 Irreversible inhibitors bind very tightly, sometimes even covalently, to an R group in enzyme active sites. They generally inhibit many different enzymes. The loss of enzyme activity impairs normal cellular metabolism, resulting in death of the cell or the individual.

20.15 A structural analog is a molecule that has a structure and charge distribution very similar to that of the natural substrate of an enzyme. Generally they are able to bind to the enzyme active site. This inhibits enzyme activity because the normal substrate must compete with the structural analog to form an enzyme-substrate complex.

20.17 **a.** Bond cleaved by chymotrypsin

$H_3N^+—C—C—N—C—C—N—C—COO^-$

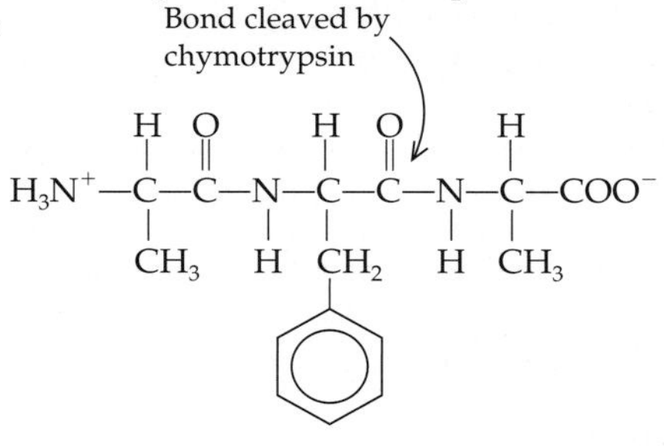

ala-phe-ala

b. Bond cleaved by chymotrypsin

$H_3N^+—C—C—N—C—C—N—C—COO^-$

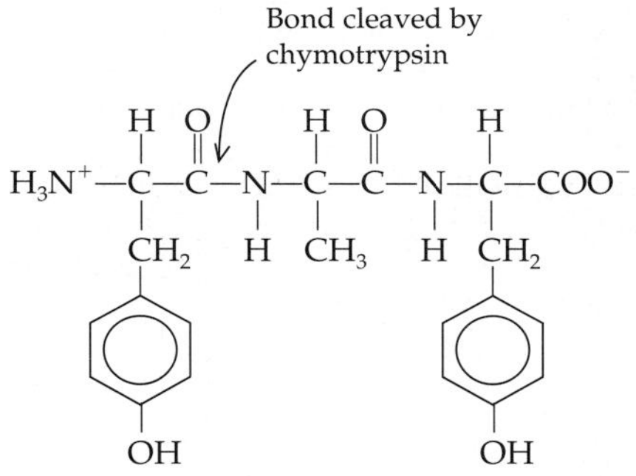

tyr-ala-tyr

c. Bond cleaved by chymotrypsin

$H_3N^+—C—C—N—C—C—N—C—COO^-$

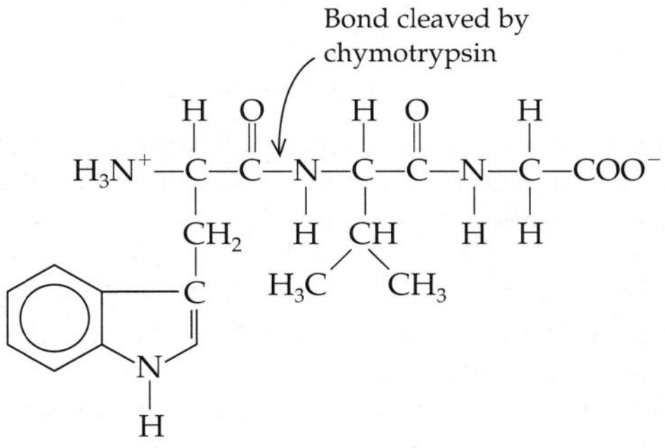

trp-val-gly

d. Bond cleaved by chymotrypsin

$H_3N^+—C—C—N—C—C—N—C—COO^-$

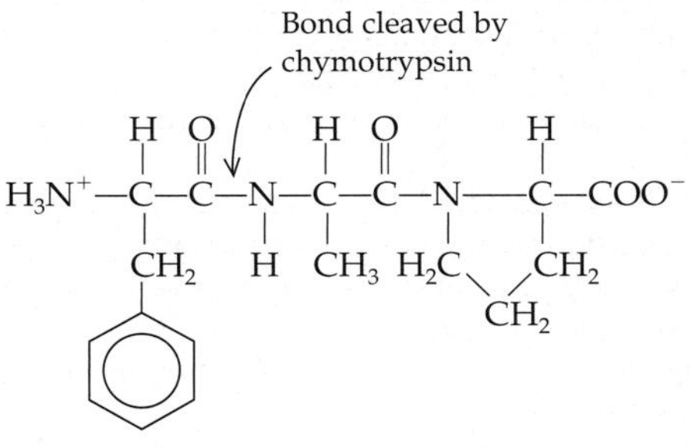

phe-ala-pro

20.19 **1.** Urease
2. Peroxidase
3. Lipase

4. Aspartase
5. Glucose-6-phosphatase
6. Sucrase

20.21 **a.** Citrate decarboxylase catalyzes the cleavage of a carboxyl group from citrate.
b. Adenosine diphosphate phosphorylase catalyzes the addition of a phosphate group to ADP.
c. Oxalate reductase catalyzes the reduction of oxalate.
d. Nitrite oxidase catalyzes the oxidation of nitrite.
e. *cis-trans* Isomerase catalyzes interconversion of *cis* and *trans* isomers.

20.23 The activation energy of a reaction is the energy required for the reaction to occur.

20.25 The equilibrium constant for a chemical reaction is a reflection of the difference in energy of the reactants and products. Consider the following reaction:

$$aA + bB \rightarrow cC + dD$$

The equilibrium constant for this reaction is:

$$K_{eq} = [D]^d[C]^c/[A]^a[B]^b = [products]/[reactants]$$

Because the difference in energy between reactants and products is the same regardless of what path the reaction takes, an enzyme does not alter the equilibrium constant of a reaction.

20.27 The rate of an uncatalyzed chemical reaction typically doubles every time the substrate concentration is doubled.

20.29

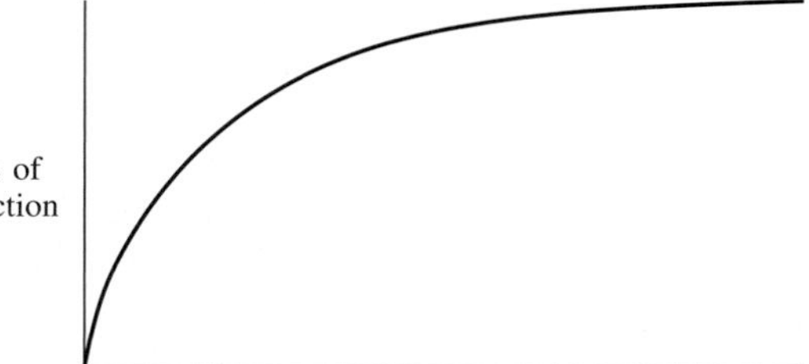

20.31 Enzyme active sites are pockets in the surface of an enzyme that include R groups involved in binding and R groups involved in catalysis. The shape of the active site is complementary to the shape of the substrate. Thus, the conformation of the active site determines the specificity of the enzyme. Enzyme-substrate binding involves weak, noncovalent interactions.

20.33 The lock-and-key model of enzyme-substrate binding was proposed by Emil Fischer in 1894. He thought that the active site was a rigid region of the enzyme into which the substrate fit perfectly. Thus, the model purports that the substrate simply snaps into place within the active site, like two pieces of a jigsaw puzzle fitting together.

20.35 *Absolute specificity*—an enzyme catalyzes the reaction of only one substrate.
Group specificity—an enzyme catalyzes reactions involving similar molecules having the same functional group.
Linkage specificity—an enzyme catalyzes the formation or breakage of only one type of bond.
Stereochemical specificity—an enzyme distinguishes one enantiomer from another.

20.37 The first step of an enzyme-catalyzed reaction is the formation of the enzyme-substrate complex. In the second step, the transition state is formed. This is the state in which the substrate assumes a form intermediate between the original substrate and the product. In step 3 the substrate is converted to product and the enzyme-product complex is formed. Step 4 involves the release of the product and regeneration of the enzyme in its original form.

20.39 In a reaction involving bond breaking, the enzyme might put pressure on a bond, producing a transition state in which the bond is stressed. An enzyme could bring two reactants into close proximity and in the proper orientation for the reaction to occur, producing a transition state in which the proximity of the reactants facilitates bond formation. Finally, an enzyme could alter the pH of the microenvironment of the active site, thereby serving as a transient donor or acceptor of H^+.

20.41 A cofactor helps maintain the shape of the active site of an enzyme.

20.43 NAD^+ serves as a donor or acceptor of hydride anions in biochemical reactions. NAD^+ serves as a coenzyme for oxidoreductases.

20.45 Changes in pH or temperature affect the activity of enzymes, as can changes in the concentration of substrate and the concentrations of certain ions.

20.47 A drastic change in pH above or below the pH optimum for an enzyme will denature the protein. Because a change in the conformation of the protein will drastically alter its active site, it will no longer be able to bind the substrate and catalyze the reaction.

20.49 High temperature denatures bacterial enzymes and structural proteins. Because the life of the cell is dependent on the function of these proteins, the cell dies.

20.51 A lysosome is a membrane-bound vesicle in the cytoplasm of cells that contains approximately fifty types of hydrolytic enzymes.

20.53 Enzymes used for clinical assays in hospitals are typically stored at refrigerator temperatures to ensure that they are not denatured by heat. In this way they retain their activity for long periods.

20.55 **a.** Cells regulate the level of enzyme activity to conserve energy. It is a waste of cellular energy to produce an enzyme if its substrate is not present or if its product is in excess.
b. Production of proteolytic digestive enzymes must be carefully controlled because the active enzyme could destroy the cell that produces it. Thus, they are produced in an inactive form in the cell and are only activated at the site where they carry out digestion.

20.57 In positive allosterism, binding of the effector molecule turns the enzyme on. In negative allosterism, binding of the effector molecule turns the enzyme off.

20.59 A zymogen, or proenzyme, is the inactive form of an enzyme that is converted to the active form of the enzyme at the site of its activity.

20.61 *Competitive enzyme inhibition* occurs when a structural analog of the normal substrate occupies the enzyme active site so that the reaction cannot occur. The structural analog and the normal substrate compete for the active site. Thus, the rate of the reaction will depend on the relative concentrations of the two molecules.

20.63 A structural analog has a shape and charge distribution that are very similar to those of the normal substrate for an enzyme.

20.65 Irreversible inhibitors bind tightly to and block the active site of an enzyme and eliminate catalysis at the site.

20.67 The compound would be a competitive inhibitor of the enzyme.

20.69 The structural similarities among chymotrypsin, trypsin, and elastase suggest that these enzymes evolved from a single ancestral gene that was duplicated. Each copy then evolved independently.

20.71

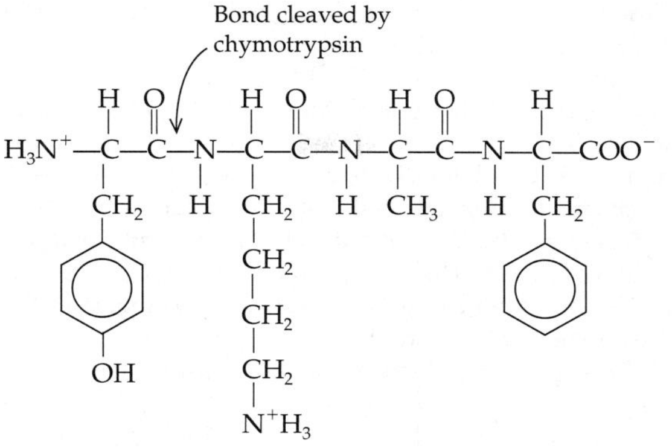

Bond cleaved by chymotrypsin

$$H_3N^+-\underset{\underset{\text{(p-hydroxyphenyl)}}{CH_2}}{\overset{H}{C}}-\overset{O}{\overset{\|}{C}}-\underset{H}{N}-\underset{CH_2CH_2CH_2CH_2N^+H_3}{\overset{H}{C}}-\overset{O}{\overset{\|}{C}}-\underset{H}{N}-\underset{CH_3}{\overset{H}{C}}-\overset{O}{\overset{\|}{C}}-\underset{H}{N}-\underset{\underset{\text{(phenyl)}}{CH_2}}{\overset{H}{C}}-COO^-$$

OH

tyr-lys-ala-phe

20.73 Elastase will cleave the peptide bonds on the carbonyl side of alanine and glycine. Trypsin will cleave the peptide bonds on the carbonyl side of lysine and arginine. Chymotrypsin will cleave the peptide bonds on the carbonyl side of tryptophan and phenylalanine.

20.75 Creatine phosphokinase (CPK), lactate dehydrogenase (LDH), and aspartate aminotransferase (AST/SGOT)

Chapter 21

21.1 ATP is called the universal energy currency because it is the major molecule used by all organisms to store energy.

21.3 The first stage of catabolism is the digestion (hydrolysis) of dietary macromolecules in the stomach and intestine.

In the second stage of catabolism, monosaccharides, amino acids, fatty acids, and glycerol are converted by metabolic reactions into molecules that can be completely oxidized.

In the third stage of catabolism, the two-carbon acetyl group of acetyl CoA is completely oxidized by the reactions of the citric acid cycle. The energy of the electrons harvested in these oxidation reactions is used to make ATP.

21.5 Substrate level phosphorylation is one way the cell can make ATP. In this reaction, a high-energy phosphoryl group of a substrate in the reaction is transferred to ADP to produce ATP.

21.7 Glycolysis is a pathway involving ten reactions. In reactions 1–3, energy is invested in the beginning substrate, glucose. This is done by transferring high-energy phosphoryl groups from ATP to the intermediates in the pathway. The product is fructose-1,6-bisphosphate. In the energy-harvesting reactions of glycolysis, fructose-1,6-bisphosphate is split into two three-carbon molecules that begin a series of rearrangement, oxidation-reduction, and substrate level phosphorylation reactions that produce four ATP, two NADH, and two pyruvate molecules. Because of the investment of two ATP in the early steps of glycolysis, the net yield of ATP is two.

21.9 Both the alcohol and lactate fermentations are anaerobic reactions that use the pyruvate and re-oxidize the NADH produced in glycolysis.

21.11 Gluconeogenesis (synthesis of glucose from noncarbohydrate sources) appears to be the reverse of glycolysis (the first stage of carbohydrate degradation) because the intermediates in the two pathways are the same. However, reactions 1, 3, and 10 of glycolysis are not reversible reactions. Thus, the reverse reactions must be carried out by different enzymes.

21.13 The enzyme glycogen phosphorylase catalyzes the phosphorolysis of a glucose unit at one end of a glycogen molecule. The reaction involves the displacement of the glucose by a phosphate group. The products are glucose-1-phosphate and a glycogen molecule that is one glucose unit shorter.

21.15 Glucokinase traps glucose within the liver cell by phosphorylating it. Because the product, glucose-6-phosphate, is charged, it cannot be exported from the cell.

21.17 Glucagon indirectly stimulates glycogen phosphorylase, the first enzyme of glycogenolysis. This speeds up glycogen degradation. Glucagon also inhibits glycogen synthase, the first enzyme in glycogenesis. This inhibits glycogen synthesis.

21.19 ATP

21.21

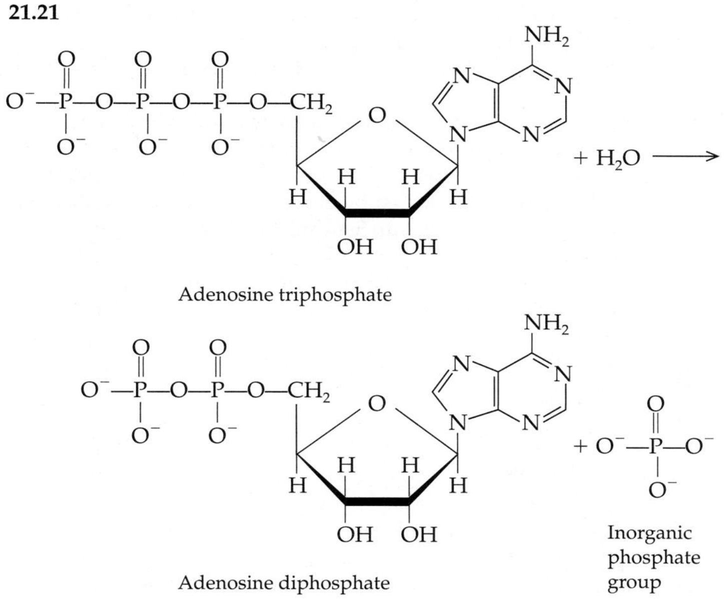

Adenosine triphosphate $+ H_2O \longrightarrow$

Adenosine diphosphate $+ O^-—\overset{O}{\overset{\|}{P}}—O^-$ (with O^- below P)

Inorganic phosphate group

21.23 Glycolysis requires NAD^+ for reaction 6 in which glyceraldehyde-3-phosphate dehydrogenase catalyzes the oxidation of glyceraldehyde-3-phosphate. NAD^+ is reduced.

21.25 Two ATP per glucose

21.27 Although muscle cells have enough ATP stored for only a few seconds of activity, glycolysis speeds up dramatically when there is a demand for more energy. If the cells have a sufficient supply of oxygen, aerobic respiration (the citric acid cycle and oxidative phosphorylation) will contribute large amounts of ATP. If oxygen is limited, the lactate fermentation will speed up. This will use up the pyruvate and re-oxidize the NADH produced by glycolysis and allow continued synthesis of ATP for muscle contraction.

21.29 $\underset{\text{Glucose}}{C_6H_{12}O_6} + 2ADP + 2P_i + 2NAD^+ \rightarrow$

$\underset{\text{Pyruvate}}{2C_3H_3O_3} + 2ATP + 2NADH + 2H_2O$

21.31
a. Hexokinase catalyzes the phosphorylation of glucose.
b. Pyruvate kinase catalyzes the transfer of a phosphoryl group from phosphoenolpyruvate to ADP.
c. Phosphoglycerate mutase catalyzes the isomerization reaction that converts 3-phosphoglycerate to 2-phosphoglycerate.
d. Glyceraldehyde-3-phosphate dehydrogenase catalyzes the oxidation and phosphorylation of glyceraldehyde-3-phosphate and the reduction of NAD^+ to NADH.

21.33 Isomerase

21.35 Enediol

21.37 A kinase transfers a phosphoryl group from one molecule to another

21.39 NAD^+ is reduced, accepting a hydride anion.

21.41 Enolase catalyzes a reaction that produces an enol; in this particular reaction, it is phosphoenolpyruvate.

21.43 To optimize efficiency and minimize waste, it is important that energy-harvesting pathways, such as glycolysis, respond to the energy demands of the cell. If energy in the form of ATP is abundant, there is no need for the pathway to continue at a rapid rate. When this is the case, allosteric enzymes that catalyze the reactions of the pathway are inhibited by binding to their negative effectors. Similarly, when there is a great demand for ATP, the pathway speeds up as a result of the action of allosteric enzymes binding to positive effectors.

21.45 ATP and citrate are allosteric inhibitors of phosphofructokinase, whereas AMP and ADP are allosteric activators.

21.47 Citrate, which is the first intermediate in the citric acid cycle, is an allosteric inhibitor of phosphofructokinase. The citric acid cycle is a pathway that results in the complete oxidation of the pyruvate produced by glycolysis. A high concentration of citrate signals that sufficient substrate is entering the citric acid cycle. The inhibition of phosphofructokinase by citrate is an example of *feedback inhibition:* the product, citrate, allosterically inhibits the activity of an enzyme early in the pathway.

21.49

$$\underset{\text{Acetaldehyde}}{CH_3-\overset{\overset{\displaystyle O}{\|}}{C}-H} \xrightarrow{NADH \quad NAD^+} \underset{\text{Ethanol}}{CH_3CH_2OH}$$

21.51 The lactate fermentation

21.53 Yogurt and some cheeses

21.55 Lactate dehydrogenase

21.57 This child must have the enzymes to carry out the alcohol fermentation. When the child exercised hard, there was not enough oxygen in the cells to maintain aerobic respiration. As a result, glycolysis and the alcohol fermentation were responsible for the majority of the ATP production by the child. The accumulation of alcohol (ethanol) in the child caused the symptoms of drunkenness.

21.59 The first stage of the pentose phosphate pathway is an oxidative stage in which glucose-6-phosphate is converted to ribulose-5-phosphate. Two NADPH molecules and one CO_2 molecule are also produced in these reactions. The second stage of the pentose phosphate pathway involves isomerization reactions that convert ribulose-5-phosphate into other five-carbon sugars, ribose-5-phosphate and xylulose-5-phosphate. The third stage of the pathway involves a complex series of rearrangement reactions that results in the production of two fructose-6-phosphate and one glyceraldehyde-3-phosphate molecules from three molecules of pentose phosphate.

21.61 The ribose-5-phosphate is used for the biosynthesis of nucleotides. The erythrose-4-phosphate is used for the biosynthesis of aromatic amino acids.

21.63 The liver

21.65 Lactate is first converted to pyruvate.

21.67 Because steps 1, 3, and 10 of glycolysis are irreversible, gluconeogenesis is not simply the reverse of glycolysis. The reverse reactions must be carried out by different enzymes.

21.69 Steps 1, 3, and 10 of glycolysis are irreversible. Step 1 is the transfer of a phosphoryl group from ATP to carbon-6 of glucose and is catalyzed by hexokinase. Step 3 is the transfer of a phosphoryl group from ATP to carbon-1 of fructose-6-phosphate and is catalyzed by phosphofructokinase. Step 10 is the substrate level phosphorylation in which a phosphoryl group is transferred from phosphoenolpyruvate to ADP and is catalyzed by pyruvate kinase.

21.71 The liver and pancreas

21.73 *Hypoglycemia* is the condition in which blood glucose levels are too low.

21.75 **a.** Insulin stimulates glycogen synthase, the first enzyme in glycogen synthesis. It also stimulates uptake of glucose from the bloodstream into cells and phosphorylation of glucose by the enzyme glucokinase.
b. This traps glucose within liver cells and increases the storage of glucose in the form of glycogen.
c. These processes decrease blood glucose levels.

21.77 Any defect in the enzymes required to degrade glycogen or export glucose from liver cells will result in a reduced ability of the liver to provide glucose at times when blood glucose levels are low. This will cause hypoglycemia.

Chapter 22

22.1 Mitochondria are the organelles responsible for aerobic respiration.

22.3

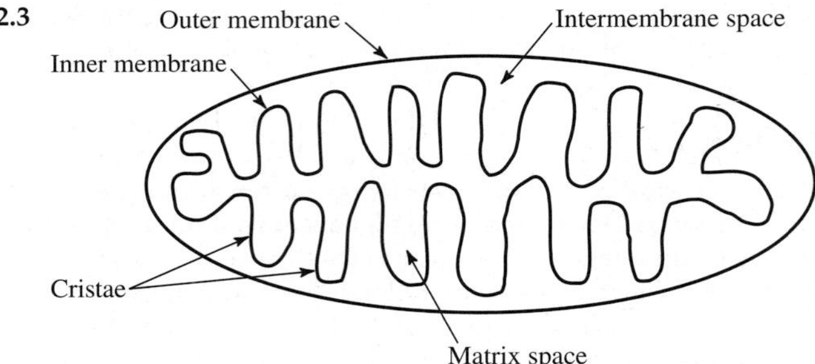

22.5 Pyruvate is converted to acetyl CoA by the pyruvate dehydrogenase complex. This huge enzyme complex requires four coenzymes, each of which is made from a different vitamin. The four coenzymes are thiamine pyrophosphate (made from thiamine), FAD (made from riboflavin), NAD^+(made from niacin), and coenzyme A (made from the vitamin pantothenic acid). The coenzyme lipoamide is also involved in this reaction.

22.7 *Oxidative phosphorylation* is the process by which the energy of electrons harvested from oxidation of a fuel molecule is used to phosphorylate ADP to produce ATP.

22.9 $NAD^+ + H{:}^- \rightarrow NADH$

22.11 During transamination reactions, the α-amino group is transferred to the coenzyme pyridoxal phosphate. In the last part of the reaction, the α-amino group is transferred from pyridoxal phosphate to an α-keto acid.

22.13 The purpose of the urea cycle is to convert toxic ammonium ions to urea, which is excreted in the urine of land animals.

22.15 An amphibolic pathway is a metabolic pathway that functions both in anabolism and catabolism. The citric acid cycle is amphibolic because it has a catabolic function—it completely oxidizes the acetyl group carried by acetyl CoA to provide electrons for ATP synthesis. Because citric acid cycle intermediates are precursors for the biosynthesis of many other molecules, it also serves a function in anabolism.

22.17 The intermembrane compartment is the location of the high-energy proton (H^+) reservoir produced by the electron transport system. The energy of this H^+ reservoir is used to make ATP.

22.19 The outer mitochondrial membrane is freely permeable to substances of molecular weight less than 10,000 g/mol. The inner mitochondrial membrane is highly impermeable.

Embedded within the inner mitochondrial membrane are the electron carriers of the electron transport system, and ATP synthase, the multisubunit enzyme that makes ATP.

22.21 Under aerobic conditions pyruvate is converted to acetyl CoA.

22.23 The coenzymes NAD^+, FAD, thiamine pyrophosphate, and coenzyme A are required by the pyruvate dehydrogenase complex for the conversion of pyruvate to acetyl CoA. These coenzymes are synthesized from the vitamins niacin, riboflavin, thiamine, and pantothenic acid, respectively. If the vitamins are not available, the coenzymes will not be available and pyruvate cannot be converted to acetyl CoA. Because the complete oxidation of the acetyl group of acetyl CoA produces the vast majority of the ATP for the body, ATP production would be severely inhibited by a deficiency of any of these vitamins.

22.25 **a.** False
b. False
c. True
d. True

22.27 **a.** The acetyl group of acetyl CoA is transferred to oxaloacetate.
b. The product is citrate.

22.29 Three

22.31 Two ATP per glucose

22.33 The function of acetyl CoA in the citric acid cycle is to bring the two-carbon remnant (acetyl group) of pyruvate from glycolysis and transfer it to oxaloacetate. In this way the acetyl group enters the citric acid cycle for the final stages of oxidation.

22.35 The high-energy phosphoryl group of the GTP is transferred to ADP to produce ATP. This reaction is catalyzed by the enzyme dinucleotide diphosphokinase.

22.37

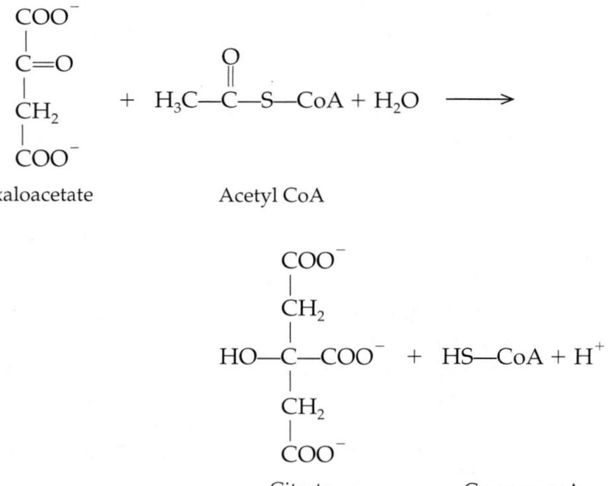

$$\begin{array}{c} COO^- \\ | \\ C{=}O \\ | \\ CH_2 \\ | \\ COO^- \end{array} + H_3C{-}\overset{\overset{\large O}{\|}}{C}{-}S{-}CoA + H_2O \longrightarrow$$

Oxaloacetate Acetyl CoA

$$\begin{array}{c} COO^- \\ | \\ CH_2 \\ | \\ HO{-}C{-}COO^- \\ | \\ CH_2 \\ | \\ COO^- \end{array} + HS{-}CoA + H^+$$

Citrate Coenzyme A

The importance of this reaction is that it brings the acetyl group, the two-carbon remnants of the glucose molecule, into the citric acid cycle to be completely oxidized. Through these reactions, and subsequent oxidative phosphorylation, the majority of the cellular ATP energy is provided.

22.39 The conversion of citrate to *cis*-aconitate is an example of the dehydration of an alcohol to produce an alkene (double bond). The conversion of *cis*-aconitate to isocitrate is an example of the hydration of an alkene, that is, the addition of water to the double bond, to produce an alcohol ($-OH$).

22.41 This reaction is an example of the oxidation of a secondary alcohol to a ketone. The two functional groups are the hydroxyl group of the alcohol and the carbonyl group of the ketone.

22.43 It is a kinase because it transfers a phosphoryl group from one molecule to another. Kinases are a specific type of transferase.

22.45 Energy-harvesting pathways, such as the citric acid cycle, must be responsive to the energy needs of the cell. If the energy requirements are high, as during exercise, the reactions must speed up. If energy demands are low and ATP is in excess, the reactions of the pathway slow down.

22.47 ADP

22.49 Three ATP

22.51 The oxidation of a variety of fuel molecules, including carbohydrates, the carbon skeletons of amino acids, and fatty acids provides the electrons. The energy of these electrons is used to produce an H^+ reservoir. The energy of this proton reservoir is used for ATP synthesis.

22.53 The electron transport system passes electrons harvested during oxidation of fuel molecules to molecular oxygen. At three sites protons are pumped from the mitochondrial matrix into the intermembrane compartment. Thus, the electron transport system builds the high-energy H^+ reservoir that provides energy for ATP synthesis.

22.55 **a.** Two ATP per glucose (net yield) are produced in glycolysis, whereas the complete oxidation of glucose in aerobic respiration (glycolysis, the citric acid cycle, and oxidative phosphorylation) results in the production of thirty-six ATP per glucose.
b. Thus, aerobic respiration harvests nearly 40% of the potential energy of glucose, and anaerobic glycolysis harvests only about 2% of the potential energy of glucose.

22.57 Transaminases transfer amino groups from amino acids to ketoacids.

22.59 The glutamate family of transaminases is very important because the ketoacid corresponding to glutamate is α-ketoglutarate, one of the citric acid cycle intermediates. This provides a link between the citric acid cycle and amino acid metabolism. These transaminases provide amino groups for amino acid synthesis and collect amino groups during catabolism of amino acids.

22.61 **a.** Pyruvate
b. α-Ketoglutarate
c. Oxaloacetate
d. Acetyl CoA
e. Succinate
f. α-Ketoglutarate

22.63

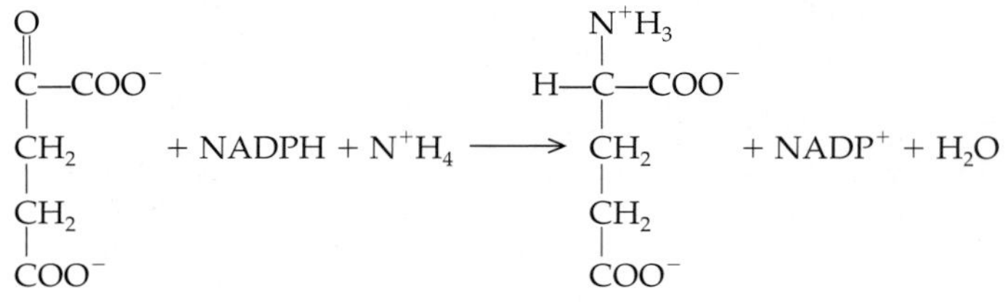

$$\begin{array}{c} O \\ \| \\ C{-}COO^- \\ | \\ CH_2 \\ | \\ CH_2 \\ | \\ COO^- \end{array} + NADPH + N^+H_4 \longrightarrow \begin{array}{c} N^+H_3 \\ | \\ H{-}C{-}COO^- \\ | \\ CH_2 \\ | \\ CH_2 \\ | \\ COO^- \end{array} + NADP^+ + H_2O$$

α-Ketoglutarate Ammonia Glutamate

22.65 Hyperammonemia

22.67 **a.** The source of one amino group of urea is the ammonium ion and the source of the other is the α-amino group of the amino acid aspartate.
b. The carbonyl group of urea is derived from CO_2.

22.69 α-Ketoglutarate

22.71 Citric acid cycle intermediates are the starting materials for the biosynthesis of many biological molecules.

22.73 An essential amino acid is one that cannot be synthesized by the body and must be provided in the diet.

22.75

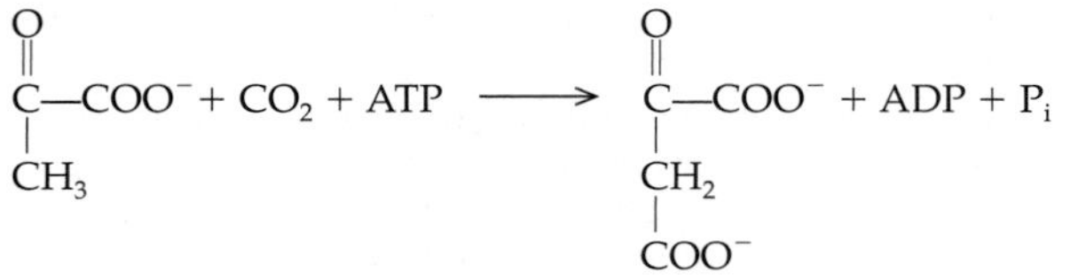

$$\underset{\text{Pyruvate}}{CH_3-\overset{O}{\overset{\|}{C}}-COO^-} + CO_2 + ATP \longrightarrow \underset{\text{Oxaloacetate}}{{}^-OOC-CH_2-\overset{O}{\overset{\|}{C}}-COO^-} + ADP + P_i$$

Chapter 23

23.1 Because dietary lipids are hydrophobic, they arrive in the small intestine as large fat globules. The bile salts emulsify these fat globules into tiny fat droplets. This greatly increases the surface area of the lipids, allowing them to be more accessible to pancreatic lipases and thus more easily digested.

23.3 **a.** Four acetyl CoA, one benzoate, four NADH, and four $FADH_2$

b. Three acetyl CoA, one phenyl acetate, three NADH, and three $FADH_2$

c. Three acetyl CoA, one benzoate, three NADH, and three $FADH_2$

d. Five acetyl CoA, one phenyl acetate, five NADH, and five $FADH_2$

23.5

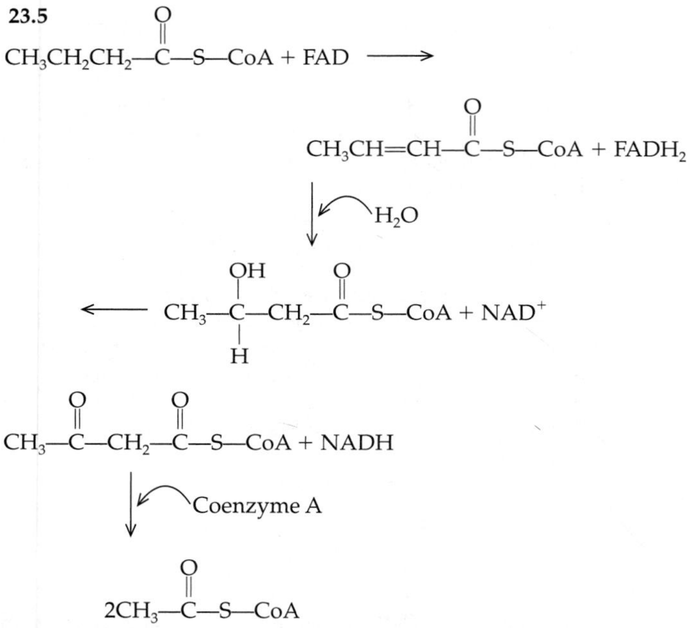

$$CH_3CH_2CH_2-\overset{O}{\overset{\|}{C}}-S-CoA + FAD \longrightarrow CH_3CH{=}CH-\overset{O}{\overset{\|}{C}}-S-CoA + FADH_2$$

$$\xrightarrow{H_2O} CH_3-\underset{H}{\overset{OH}{C}}-CH_2-\overset{O}{\overset{\|}{C}}-S-CoA + NAD^+ \longrightarrow CH_3-\overset{O}{\overset{\|}{C}}-CH_2-\overset{O}{\overset{\|}{C}}-S-CoA + NADH$$

$$\xrightarrow{\text{Coenzyme A}} 2CH_3-\overset{O}{\overset{\|}{C}}-S-CoA$$

23.7 Starvation, a diet low in carbohydrates, and diabetes mellitus are conditions that lead to the production of ketone bodies.

23.9 **(1)** Fatty acid biosynthesis occurs in the cytoplasm whereas β-oxidation occurs in the mitochondria.

(2) The acyl group carrier in fatty acid biosynthesis is acyl carrier protein while the acyl group carrier in β-oxidation is coenzyme A.

(3) The seven enzymes of fatty acid biosynthesis are associated as a multienzyme complex called *fatty acid synthase.* The enzymes involved in β-oxidation are not physically associated with one another.

(4) NADPH is the reducing agent used in fatty acid biosynthesis. NADH and $FADH_2$ are produced by β-oxidation.

23.11 The liver regulates blood glucose levels under the control of the hormones insulin and glucagon. When blood glucose levels are too high, insulin stimulates the uptake of glucose by liver cells and the storage of the glucose in glycogen polymers. When blood glucose levels are too low, the hormone glucagon stimulates the breakdown of glycogen and release of glucose into the bloodstream. Glucagon also stimulates the liver to produce glucose for export into the bloodstream by the process of gluconeogenesis.

23.13 Insulin stimulates uptake of glucose and amino acids by cells, glycogen and protein synthesis, and storage of lipids. It inhibits glycogenolysis, gluconeogenesis, breakdown of stored triglycerides, and ketogenesis.

23.15 Triglycerides

23.17 The large fat globule that takes up nearly the entire cytoplasm

23.19 Lipases catalyze the hydrolysis of the ester bonds of triglycerides.

23.21 Acetyl CoA is the precursor for fatty acids, several amino acids, cholesterol, and other steroids.

23.23 Chylomicrons are plasma lipoproteins (aggregates of protein and triglycerides) that carry dietary triglycerides from the intestine to all tissues via the bloodstream.

23.25 Bile salts serve as detergents. Fat globules stimulate their release from the gallbladder. The bile salts then emulsify the lipids, increasing their surface area and making them more accessible to digestive enzymes (pancreatic lipases).

23.27 When dietary lipids in the form of fat globules reach the duodenum, they are emulsified by bile salts. The triglycerides in the resulting tiny fat droplets are hydrolyzed into monoglycerides and fatty acids by the action of pancreatic lipases, assisted by colipase. The monoglycerides and fatty acids are absorbed by cells lining the intestine.

23.29 Six acetyl CoA, one phenyl acetate, six NADH, and six $FADH_2$

23.31 112 ATP

23.33 Two ATP

23.35 The acetyl CoA produced by β-oxidation will enter the citric acid cycle.

23.37

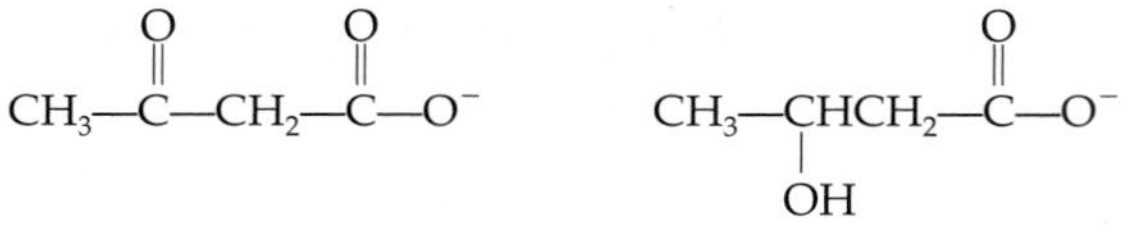

$$\underset{\text{Acetoacetate}}{CH_3-\overset{O}{\overset{\|}{C}}-CH_2-\overset{O}{\overset{\|}{C}}-O^-} \qquad \underset{\beta\text{-Hydroxybutyrate}}{CH_3-\underset{OH}{\underset{|}{C}}HCH_2-\overset{O}{\overset{\|}{C}}-O^-}$$

23.39 In those suffering from uncontrolled diabetes, the glucose in the blood cannot get into the cells of the body. The excess glucose is excreted in the urine. Body cells degrade fatty acids because glucose is not available. β-Oxidation of fatty acids yields enormous quantities of acetyl CoA, so much acetyl CoA, in fact, that it cannot all enter the citric acid cycle because there is not enough oxaloacetate available. Excess acetyl CoA is used for ketogenesis.

23.41 Ketone bodies are the preferred energy source of the heart.

23.43 The phosphopantetheine group allows formation of a high-energy thioester bond with a fatty acid. It is derived from the vitamin pantothenic acid.

23.45 Fatty acid synthase is a huge multienzyme complex consisting of the seven enzymes involved in fatty acid synthesis. It is found in the cell cytoplasm. The enzymes involved in β-oxidation are not physically associated with one another. They are free in the mitochondrial matrix space.

23.47 The major metabolic function of the liver is to regulate blood glucose levels.

23.49 Ketone bodies are the major fuel for the heart. Glucose is the major energy source of the brain, and the liver obtains most of its energy from the oxidation of amino acid carbon skeletons.

23.51 Fatty acids are absorbed from the bloodstream by adipocytes. Using glycerol-3-phosphate, produced as a by-product of glycolysis, triglycerides are synthesized. Triglycerides are constantly being hydrolyzed and resynthesized in adipocytes. The rates of hydrolysis and synthesis are determined by lipases that are under hormonal control.

23.53 Insulin is produced in the β-cells of the islets of Langerhans in the pancreas.

23.55 Insulin stimulates the uptake of glucose from the blood into cells. It enhances glucose storage by stimulating glycogenesis and inhibiting glycogen degradation and gluconeogenesis.

23.57 Insulin stimulates synthesis and storage of triglycerides.

23.59 Untreated diabetes mellitus is starvation in the midst of plenty because blood glucose levels are very high. However, in the absence of insulin, blood glucose can't be taken up into cells. The excess glucose is excreted into the urine while the cells of the body are starved for energy.

Chapter 24

24.1 **a.** Adenosine diphosphate:

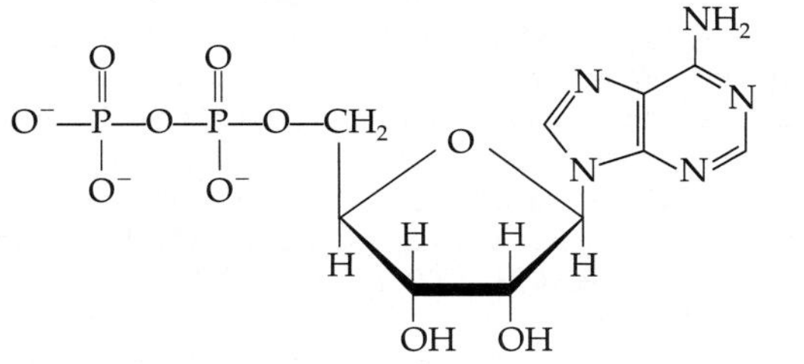

b. Deoxyguanosine triphosphate:

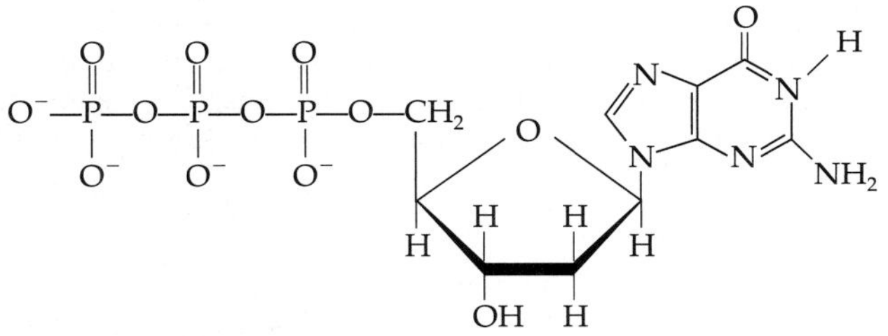

24.3 The deoxyribonucleotides of guanine are:
Deoxyguanosine monophosphate (dGMP)
Deoxyguanosine diphosphate (dGDP)
Deoxyguanosine triphosphate (dGTP)

The ribonucleotides of guanine are:
Guanosine monophosphate (GMP)
Guanosine diphosphate (GDP)
Guanosine triphosphate (GTP)

24.5 The RNA polymerase recognizes the promoter site for a gene, separates the strands of DNA, and catalyzes the polymerization of an RNA strand complementary to the DNA strand that carries the genetic code for a protein. It recognizes a termination site at the end of the gene and releases the RNA molecule.

24.7 The genetic code is said to be degenerate because several different triplet codons may serve as code words for a single amino acid.

24.9 The nitrogenous bases of the codons are complementary to those of the anticodons. As a result they are able to hydrogen bond to one another according to the base pairing rules.

24.11 The ribosomal P-site holds the peptidyl tRNA during protein synthesis. The peptidyl tRNA is the tRNA carrying the growing peptide chain. The only exception to this is during initiation of translation when the P-site holds the initiator tRNA.

24.13 The normal mRNA sequence, AUG-CCC-GAC-UUU, would encode the peptide sequence methionine-proline-aspartate-phenylalanine. The mutant mRNA sequence, AUG-CGC-GAC-UUU, would encode the mutant peptide sequence methionine-arginine-aspartate-phenylalanine. This would not be a silent mutation because a hydrophobic amino acid (proline) has been replaced by a positively charged amino acid (arginine).

24.15 It is the N-9 of the purine that forms the N-glycosidic bond with C-1 of the five-carbon sugar. The general structure of the purine ring is shown below:

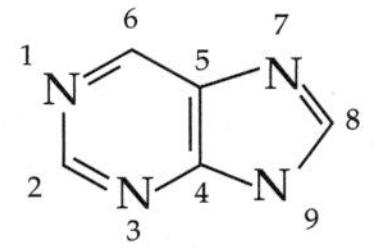

24.17 The ATP nucleotide is composed of the five-carbon sugar ribose, the purine adenine, and a triphosphate group.

24.19 The two strands of DNA in the double helix are said to be *antiparallel* because they run in opposite directions. One strand progresses in the $5' \rightarrow 3'$ direction, and the opposite strand progresses in the $3' \rightarrow 5'$ direction.

24.21 Two

24.23

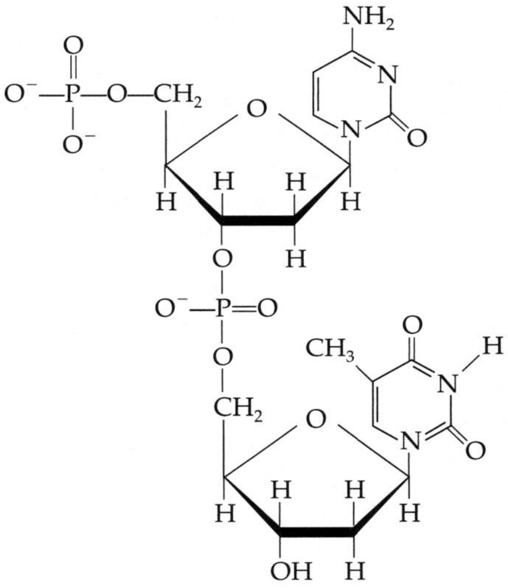

24.25 The term *semiconservative DNA replication* refers to the fact that each parental DNA strand serves as the template for the synthesis of a daughter strand. As a result, each of the daughter DNA molecules is made up of one strand of the original parental DNA and one strand of newly synthesized DNA.

24.27 The two primary functions of DNA polymerase are to read a template DNA strand and catalyze the polymerization of a new daughter strand, and to proofread the newly synthesized strand and correct any errors by removing the incorrectly inserted nucleotide and adding the proper one.

24.29 3′-TACGCCGATCTTATAAGGT-5′

24.31 The *replication origin* of a DNA molecule is the unique sequence on the DNA molecule where DNA replication begins.

24.33 DNA → RNA → Protein

24.35 Anticodons are found on transfer RNA molecules.

24.37 3′-AUGGAUCGAGACCAGUAAUUCCGUCAU-5′.

24.39 *RNA splicing* is the process by which the noncoding sequences (introns) of the primary transcript of a eukaryotic mRNA are removed and the protein coding sequences (exons) are spliced together.

24.41 Messenger RNA, transfer RNA, and ribosomal RNA

24.43 Spliceosomes are small ribonucleoprotein complexes that carry out RNA splicing.

24.45 The *poly(A) tail* is a stretch of 100–200 adenosine nucleotides polymerized onto the 3′ end of a mRNA by the enzyme poly(A) polymerase.

24.47 The *cap structure* is made up of the nucleotide 7-methylguanosine attached to the 5′ end of a mRNA by a 5′-5′ triphosphate bridge. Generally the first two nucleotides of the mRNA are also methylated.

24.49 Sixty-four

24.51 The reading frame of a gene is the sequential set of triplet codons that carries the genetic code for the primary structure of a protein.

24.53 Methionine and tryptophan

24.55 The codon 5′-UUU-3′ encodes the amino acid phenylalanine. The mutant codon 5′-UUA-3′ encodes the amino acid leucine. Both leucine and phenylalanine are hydrophobic amino acids, however, leucine has a smaller R group. It is possible that the smaller R group would disrupt the structure of the protein.

24.57 The ribosomes serve as a platform on which protein synthesis can occur. They also carry the enzymatic activity that forms peptide bonds.

24.59 In the initiation of translation, initiation factors, methionyl tRNA (the initiator tRNA), the mRNA, and the small and large ribosomal subunits form the initiation complex. During the elongation stage of translation, an aminoacyl tRNA binds to the A-site of the ribosome. Peptidyl transferase catalyzes the formation of a peptide bond and the peptide chain is transferred to the tRNA in the A-site. Translocation shifts the peptidyl tRNA from the A-site into the P-site, leaving the A-site available for the next aminoacyl tRNA. In the termination stage of translation, a termination codon is encountered. A release factor binds to the empty A-site and peptidyl transferase catalyzes the hydrolysis of the bond between the peptidyl tRNA and the completed peptide chain.

24.61 An ester bond

24.63 UV light causes the formation of pyrimidine dimers, the covalent bonding of two adjacent pyrimidine bases. Mutations occur when the UV damage repair system makes an error during the repair process. This causes a change in the nucleotide sequence of the DNA.

24.65 **a.** A *carcinogen* is a compound that causes cancer. Cancers are caused by mutations in the genes responsible for controlling cell division.
b. Carcinogens cause DNA damage that results in changes in the nucleotide sequence of the gene. Thus, carcinogens are also mutagens.

24.67 A *restriction enzyme* is a bacterial enzyme that "cuts" the sugar–phosphate backbone of DNA molecules at a specific nucleotide sequence.

24.69 Nucleic acid hybridization is based on the fact that complementary DNA or RNA sequences will hydrogen bond to one another according to the base pairing rules.

24.71 Human insulin, interferon, human growth hormone, and human blood clotting factor VIII

24.73 1024 copies

24.75 The goals of the Human Genome Project are to identify and map all of the genes of the human genome and to determine the DNA sequences of the complete three billion nucleotide pairs.

24.77 A genome library is a set of clones that represents all of the DNA sequences in the genome of an organism.

24.79 A dideoxynucleotide is one that has hydrogen atoms rather than hydroxyl groups bonded to both the 2′ and 3′ carbons of the five-carbon sugar.

24.81 Sequences that these DNA sequences have in common are highlighted in bold.
a. 5′ **AGCTCCT**GATTTCATACAGTTTCTACT**ACCTACTA** 3′
b. 5′ AGACATTCTATCTACCTAGACTATG**TTCAGAA** 3′
c. 5′ **TTCAGAA**CTCATTCAGACCTACTACTATACCTTGGG **AGCTCCT** 3′
d. 5′ **ACCTACTA**GACTATACTACTACTAAGGGGACTATTC CAGACTT 3′

The 5′ end of sequence (a) is identical to the 3′ end of sequence (c). The 3′ end of sequence (a) is identical to the 5′ end of sequence (d). The 3′ end of sequence (b) is identical to the 5′ end of sequence (c). From 5′ to 3′, the sequences would form the following map:

b
c
a
d

Credits

Photographs

Chapter 1

Opener: © Tony Freeman/PhotoEdit; **1.2A:** © The McGraw-Hill Companies, Inc./Louis Rosenstock, photographer; **1.2B:** © The McGraw-Hill Companies, Inc. /Jeff Topping, photographer; **1.2C:** © The McGraw-Hill Companies, Inc./Louis Rosenstock, photographer; **1.3 (both):** © The McGraw-Hill Companies, Inc./Ken Karp, photographer; **1.6A-C, 1.8A-D, 1.10:** © The McGraw-Hill Companies, Inc./Louis Rosenstock, photographer.

Chapter 2

Opener: © Yoav Levy/Phototake; **2.1A:** © Geoff Tompkinson/SPL/Photo Researchers, Inc.; **2.1B:** © T.J. Florian/Rainbow; **2.1C:** © David Parker/Segate Microelectronics, Ltd./Photo Researchers, Inc.; **2.1D:** APHIS, PPQ, Otis Methods Development Center, Otis, MA/USDA; **2.2:** © P. Plaily/SPL/Photo Researchers, Inc.; **2.4A-B:** © The McGraw-Hill Companies, Inc./Louis Rosenstock, photographer; **2.9:** © Yoav Levy/Phototake; **p. 47:** © PhotoDisc/Volume 2; **p. 48:** © Earth Satellite Corp./SPL/Photo Researchers, Inc.; **p. 49 (bottom):** © The McGraw-Hill Companies, Inc./Louis Rosenstock, photographer; **p. 49 (top):** © Dan McCoy/Rainbow.

Chapter 3

Opener: © The McGraw-Hill Companies, Inc./Louis Rosenstock, photographer.

Chapter 4

Opener: © Photri/Stock Market/Corbis; **4.13:** © Charles D. Winters/Photo Researchers, Inc.

Chapter 5

Opener: © The McGraw-Hill Companies, Inc./ Stephen Frisch, photographer; **5.3:** © The McGraw-Hill Companies, Inc./Louis Rosenstock, photographer; **p. 138:** © PhotoDisc/Volume 72.

Chapter 6

Opener: Courtesy of Robert Shoemaker; **p. 144:** © Hulton-Deutsch/Corbis; **6.4:** © Peter Stef Lamberti/Stone/Getty; **p. 151:** © SIU/Visuals Unlimited; **6.5A-B:** © The McGraw-Hill Companies, Inc./Louis Rosenstock, photographer.

Chapter 7

Opener: © Richard Megna/Fundamental Photographs; **7.1:** © Kip and Pat Peticolas/Fundamental Photographs; **p. 179:** © J.W. Mowbray/Photo Researchers, Inc.; **7.2, 7.5A-B:** © The McGraw-Hill Companies, Inc./Louis Rosenstock, photographer; **p. 191 (left):** © RMF/Visuals Unlimited; **p. 191 (right):** Courtesy of Rita Colwell, University of Maryland; **p. 194:** © The McGraw-Hill Companies, Inc./Louis Rosenstock, photographer; **p. 195:** © David Joel.

Chapter 8

Opener: © Richard Megna/Fundamental Photographs; **8.9:** © The McGraw-Hill Companies, Inc./Ken Karp, photographer; **p. 214 (left, right), 8.12:** © The McGraw-Hill Companies, Inc./Louis Rosenstock, photographer; **8.16, 8.17:** © The McGraw-Hill Companies, Inc./Ken Karp, photographer.

Chapter 9

Opener: © Richard Megna/Fundamental Photographs; **9.1:** © Dr. E.R. Degginger/Color Pic Inc.; **9.3A:** © The McGraw-Hill Companies, Inc./Louis Rosenstock, photographer; **9.3B:** © 1992 Richard Megna/Fundamental Photographs, NYC; **9.4A-B:** © The McGraw-Hill Companies, Inc./Louis Rosenstock, photographer; **9.6A-B:** © The McGraw-Hill Companies, Inc./Stephen Frisch, photographer; **p. 249, 9.7, 9.8 (left):** © The McGraw-Hill Companies, Inc./Louis Rosenstock, photographer; **9.8 (right):** © Bonnie Kamin/PhotoEdit; **9.8 (middle):** © Tony Freeman/PhotoEdit; **p. 258:** © AAA Photo/Phototake; **9.9:** © The McGraw-Hill Companies, Inc./Stephen Frisch, photographer.

Chapter 10

Opener: © Bettmann/Corbis; **10.4:** © U.S. Dept. of Energy/Photo Researchers, Inc.; **10.5:** © Gianni Tortoli/Photo Researchers, Inc.; **p. 279:** NASA; **10.6:** © Blair Seitz/Photo Researchers, Inc.; **10.7B:** Bristol-Myers Squibb Medical Imaging; **p. 284 (right):** © SIU/Biomed/ Custom Medical Stock Photo; **p. 284 (left):** © The McGraw-Hill Companies, Inc./Louis Rosenstock, photographer; **10.8:** © U.S. Dept. of Energy/Mark Marten/Photo Researchers, Inc.; **10.9:** © The McGraw-Hill Companies, Inc./Louis Rosenstock, photographer; **10.10:** © Scott Cazmine/Photo Researchers, Inc.

Chapter 11

Opener: © Tom McHugh/Photo Researchers, Inc.; **p. 314:** © Bob Thomason/Stone/Getty; **p. 317:** © Brian Smith/Stock Boston.

Chapter 12

Opener: © Anthony Bannister, Gallo Images/Corbis; **p. 331:** © Buddy Mays/Corbis; **p. 338 (geranium):** © Larry Lefever/Grant Heilman Photography, Inc.; **p. 338 (oranges):** © Michelle Garrett/Corbis; **p. 338 (bayberry):** © Walter H. Hodge/Peter Arnold, Inc.; **p. 338 (lily):** © Hal Horwitz/Corbis; **p. 343 (tomatoes):** © Ed Young/Corbis; **p. 348 (recycling):** © Alan Detrick/Photo Researchers, Inc.; **12 5:** © McGraw-Hill Education/Ken Karp, photographer.

Chapter 13

Opener: © David Young-Wolff/PhotoEdit.

Chapter 14

Opener: © David Addison/Visuals Unlimited; **14.3 (almonds):** © B. Borrell Cassals, Frank Lane Picture Agency/Corbis; **14.3 (cinnamon):** © Rita Maas/Getty Images; **14.3 (berries):** © Charles Krebs/Corbis; **14.3 (vanilla):** © Eisenhut & Mayer/Getty Images; **14.3 (mushroom):** © James Noble/Corbis; **14.3 (lemon grass):** © Corbis/R-F Website; **14.4:** © Fundamental Photographs; **14.5:** © Rob and Ann Simpson/Visuals Unlimited; **p. 409:** © Hanson Carroll/Peter Arnold, Inc.

Chapter 15

Opener: © Christel Rosenfeld/Tony Stone/ Corbis; **p. 430:** © The McGraw-Hill Companies, Inc./Louis Rosenstock, photographer.

Chapter 16

Opener: © Gail Shumway/FPG International.

Chapter 17

Opener: © The McGraw-Hill Companies, Inc./Louis Rosenstock, photographer; **p. 515:** © R. Feldman/Dan McCoy/Rainbow.

Chapter 18

Opener: © Lester Lefkowitz/Corbis; **p. 528:** © Roadsideamerica.com, Kirby, Smith & Wilkins; **p. 541:** © Hans Pfletschinger/Peter Arnold, Inc.; **18.10A:** © James Dennis/PhotoTake; **18.20A-C:** © David M. Phillips/Visuals Unlimited.

Chapter 19

19.16A: © David Scharf/Peter Arnold, Inc.; **19.16B:** © SIU/Peter Arnold, Inc.; **19.16C:** © Jackie Lewin/SPL/Photo Researchers, Inc.

Chapter 20

Opener: Courtesy of David Eisenberg; **p. 604:** © Phil Degginger/Color-Pic, Inc.

Chapter 21

Opener: © The McGraw-Hill Companies, Inc./Louis Rosenstock, photographer; **p. 648:** © The McGraw-Hill Companies, Inc./Louis Rosenstock, photographer.

Chapter 22

Opener: © Zoom Agency/Getty Images; **22.1A:** © CNRI/Phototake.

Chapter 23

Opener: Courtesy of Katherine Denniston.

Chapter 24

Opener: © Douglas Struthers/Stone/Getty; **p. 751:** Courtesy of Orchid Cellmark, Germantown, Maryland.

Text and Line Art

Chapter 1

1.7: From Raymond Chang, *Chemistry,* 6th ed. Copyright © 1998. The McGraw-Hill Companies, Inc., Dubuque, Iowa. All Rights Reserved. Reprinted by permission, p. 21; **Caloric expenditure table:** This table has been reproduced from *The Book of Health,* E.L. Wynder, Editor. American Health Foundation. New York, Franklin Watts, 1981, with permission of the American Health Foundation, p. 25.

Chapter 2

2.3: From Raymond Chang, *Chemistry,* 6th ed. Copyright © 1998. The McGraw-Hill Companies, Inc., Dubuque, Iowa. All Rights Reserved. Reprinted by permission, p. 45; **2.8:** From Raymond Chang, Chemistry, 6th ed. Copyright © 1998. The McGraw-Hill Companies, Inc., Dubuque, Iowa. All Rights Reserved. Reprinted by permission, p. 47; **2.10:** From Martin Silberberg, *Chemistry: The Molecular Nature of Matter and Change,* 2nd ed. Copyright © 2000. The McGraw-Hill Companies, Inc., Dubuque, Iowa. All Rights Reserved. Reprinted by permission, p. 48; **2.11 b,c:** From Martin Silberberg, *Chemistry: The Molecular Nature of Matter and Change,* 2nd ed. Copyright © 2000. The McGraw-Hill Companies, Inc., Dubuque, Iowa. All Rights Reserved. Reprinted by permission, p. 49.

Chapter 4

4.1: From Raymond Chang, *Chemistry,* 6th ed. Copyright © 1998. The McGraw-Hill Companies, Inc., Dubuque, Iowa. All Rights Reserved. Reprinted by permission, p.83; **4.4:** From Raymond Chang, *Chemistry,* 6th ed. Copyright © 1998. The McGraw-Hill Companies, Inc., Dubuque, Iowa. All Rights Reserved. Reprinted by permission, p. 107; **4.5:** From Raymond Chang, *Chemistry,* 6th ed. Copyright © 1998. The McGraw-Hill Companies, Inc., Dubuque, Iowa. All Rights Reserved. Reprinted by permission, p. 107.

Chapter 6

6.9c: From Martin Silberberg, *Chemistry: The Molecular Nature of Matter and Change,* 2nd ed. Copyright © 2000. The McGraw-Hill Companies, Inc., Dubuque, Iowa. All Rights Reserved. Reprinted by permission, p.170; **6.9d:** From Raymond Chang, *Chemistry,* 6th ed. Copyright © 1998. The McGraw-Hill Companies, Inc., Dubuque, Iowa. All Rights Reserved. Reprinted by permission, p. 170.

Chapter 8

8.1: From Raymond Chang, *Chemistry,* 6th ed. Copyright © 1998. The McGraw-Hill Companies, Inc., Dubuque, Iowa. All Rights Reserved. Reprinted by permission, p. 207; **8.3:** From Raymond Chang, *Chemistry,* 6th ed. Copyright © 1998. The McGraw-Hill Companies, Inc., Dubuque, Iowa. All Rights Reserved. Reprinted by permission, p. 210; **8.4:** From Raymond Chang, *Chemistry,* 6th ed. Copyright © 1998. The McGraw-Hill Companies, Inc., Dubuque, Iowa. All Rights Reserved. Reprinted by permission, p. 210; **8.5:** From Raymond Chang, *Chemistry,* 6th ed. Copyright © 1998. The McGraw-Hill Companies, Inc., Dubuque, Iowa. All Rights Reserved. Reprinted by permission, p. 212; **8.7:** From Raymond Chang, *Chemistry,* 6th ed. Copyright © 1998. The McGraw-Hill Companies, Inc., Dubuque, Iowa. All Rights Reserved. Reprinted by permission, p. 215; **8.10:** From Raymond Chang, *Chemistry,* 6th ed. Copyright ©1998. The McGraw-Hill Companies, Inc., Dubuque, Iowa. All Rights Reserved. Reprinted by permission, p. 217; **8.11:** From Raymond Chang, *Chemistry,* 6th ed. Copyright © 1998. The McGraw-Hill Companies, Inc., Dubuque, Iowa. All Rights Reserved. Reprinted by permission, p. 219; **8.13:** From Raymond Chang, *Chemistry,* 6th ed. Copyright © 1998. The McGraw-Hill Companies, Inc., Dubuque, Iowa. All Rights Reserved. Reprinted by permission, p. 220; **8.15:** From Raymond Chang, *Chemistry,* 6th ed. Copyright © 1998. The McGraw-Hill Companies, Inc., Dubuque, Iowa. All Rights Reserved. Reprinted by permission, p. 229.

Chapter 9

9.2: From Martin Silberberg, *Chemistry: The Molecular Nature of Matter and Change,* 2nd ed. Copyright © 2000. The McGraw-Hill Companies, Inc., Dubuque, Iowa. All Rights Reserved. Reprinted by permission, p. 243; **9.9:** From Martin Silberberg, *Chemistry: The Molecular Nature of Matter and Change,* 2nd ed. Copyright © 2000. The McGraw-Hill Companies, Inc., Dubuque, Iowa. All Rights Reserved. Reprinted by permission, p. 263; **9.10:** From Raymond Chang, *Chemistry,* 6th ed. Copyright © 1998. The McGraw-Hill Companies, Inc., Dubuque, Iowa. All Rights Reserved. Reprinted by permission, p. 263; **9.11:** From Martin Silberberg, *Chemistry: The Molecular Nature of Matter and Change,* 2nd ed. Copyright © 2000. The McGraw-Hill Companies, Inc., Dubuque, Iowa. All Rights Reserved. Reprinted by permission, p. 265; **9.12:** From Martin Silberberg, *Chemistry: The Molecular Nature of Matter and Change,* 2nd ed. Copyright © 2000. The McGraw-Hill Companies, Inc., Dubuque, Iowa. All Rights Reserved. Reprinted by permission, p. 265.

Chapter 17

17.2: U.S. Department of Agriculture, p. 469; **17.11b:** From Trudy McKee and James R. McKee, *Biochemistry: An Introduction,* 2nd ed. Copyright © 1999. The McGraw-Hill Companies, Inc., Dubuque, Iowa. All Rights Reserved. Reprinted by permission, p. 486.

Chapter 18

18.8b: From Trudy McKee and James R. McKee, *Biochemistry: The Molecular Basis of Life,* 3rd ed. Copyright © 2003. The McGraw-Hill Companies, Inc., Dubuque, Iowa. All Rights Reserved. Reprinted by permission, p. 351; **18.9:** From Trudy McKee and James R. McKee, *Biochemistry: An Introduction,* 2nd ed. Copyright © 1999. The McGraw-Hill Companies, Inc., Dubuque, Iowa. All Rights Reserved. Reprinted by permission, p. 517; **18.11d:** From Trudy McKee and James R. McKee, *Biochemistry: An Introduction,* 2nd ed. Copyright © 1999. The McGraw-Hill Companies, Inc., Dubuque, Iowa. All Rights Reserved. Reprinted by permission, p. 520; **18.12:** From Trudy McKee and James R. McKee, *Biochemistry: An Introduction,* 2nd ed. Copyright © 1999. The McGraw-Hill Companies, Inc., Dubuque, Iowa. All Rights Reserved. Reprinted by permission, p. 521; **18.13:** From Stuart Ira Fox, *Human Physiology,* 6th ed. Copyright © 1999. The McGraw-Hill Companies, Inc., Dubuque, Iowa. All Rights Reserved. Reprinted by permission, p. 522.

Chapter 19

Opener: From Trudy McKee and James R. McKee, *Biochemistry: The Molecular Basis of Life,* 3rd ed. Copyright © 2003. The McGraw-Hill Companies, Inc., Dubuque, Iowa. All Rights Reserved. Reprinted by permission, p. 148. **19.5:** From Trudy McKee and James R. McKee, *Biochemistry: The Molecular Basis of Life,* 3rd ed. Copyright © 2003. The McGraw-Hill Companies, Inc., Dubuque, Iowa. All Rights Reserved. Reprinted by permission, p. 131. **19.8:** From Trudy McKee and James R. McKee, *Biochemistry: The Molecular Basis of Life,* 3rd ed. Copyright © 2003. The McGraw-Hill Companies, Inc., Dubuque, Iowa. All Rights Reserved. Reprinted by permission, p. 132. **19.11:** From Trudy McKee and James R. McKee, *Biochemistry: An Introduction,* 2nd ed. Copyright © 1999. The McGraw-Hill Companies, Inc., Dubuque, Iowa. All Rights Reserved. Reprinted by permission, p. 550; **19.13:** From Trudy McKee and James R. McKee, *Biochemistry: An Introduction,* 2nd ed. Copyright © 1999. The McGraw-Hill Companies, Inc., Dubuque, Iowa. All Rights Reserved. Reprinted by permission, p. 553; **19.15:** From Trudy McKee and James R. McKee, Biochemistry: An Introduction, 2nd ed. Copyright © 1999. The McGraw-Hill Companies, Inc., Dubuque, Iowa. All Rights Reserved. Reprinted by permission, p. 553.

Chapter 21

21.5: From David Shier, Jackie Butler, and Ricki Lewis, *Hole's Human Anatomy & Physiology,* 8th ed. Copyright © 1999. The McGraw-Hill Companies, Inc., Dubuque, Iowa. All Rights Reserved. Reprinted by permission, p. 606.

Chapter 22

22.1b: From Trudy McKee and James R. McKee, *Biochemistry: An Introduction,* 2nd ed. Copyright © 1999. The McGraw-Hill Companies, Inc., Dubuque, Iowa. All Rights Reserved. Reprinted by permission, p. 637.

Chapter 24

24.6: From Trudy McKee and James R. McKee, *Biochemistry: The Molecular Basis of Life,* 3rd ed. Copyright © 2003. The McGraw-Hill Companies, Inc., Dubuque, Iowa. All Rights Reserved. Reprinted by permission, p. 579. **24.7:** From Trudy McKee and James R. McKee, *Biochemistry: The Molecular Basis of Life,* 3rd ed. Copyright © 2003. The McGraw-Hill Companies, Inc., Dubuque, Iowa. All Rights Reserved. Reprinted by permission, p. 582. **24.8:** From Trudy McKee and James R. McKee, *Biochemistry: The Molecular Basis of Life,* 3rd ed. Copyright © 2003. The McGraw-Hill Companies, Inc., Dubuque, Iowa. All Rights Reserved. Reprinted by permission, p. 611.

Index

Note: Page numbers followed by B indicate boxed material; those followed by F indicate figures; those followed by T indicate tables.

A

B

C

D

E

F

G

H

I

J

K

L

M

N

O

P

Q

R

S

T

U

PRINCIPAL FUNCTIONAL GROUPS IN ORGANIC COMPOUNDS

Type of compound	Structural formula	Condensed formula	Chapter reference	Example: Structural formula	Example: IUPAC name	Example: Common name
Alcohol	R—O—H	ROH	13	CH_3CH_2—O—H	Ethanol	Ethyl alcohol
Aldehyde	R—C(=O)—H	RCHO	14	CH_3C(=O)—H	Ethanal	Acetaldehyde
Amide	R—C(=O)—N(H)—H	$RCONH_2$	16	CH_3C(=O)—N(H)—H	Ethanamide	Acetamide
Amine	R—N(H)—H	RNH_2	16	CH_3CH_2N(H)—H	Aminoethane	Ethyl amine
Carboxylic acid	R—C(=O)—O—H	RCOOH	15	CH_3C(=O)—O—H	Ethanoic acid	Acetic acid
Ester	R—C(=O)—O—R′	RCOOR′	15	CH_3C(=O)—OCH_3	Methyl ethanoate	Methyl acetate
Ether	R—O—R′	ROR′	13	CH_3OCH_3	Methoxymethane	Dimethyl ether
Halide	—Cl (or —Br, —F, —I)	RCl	11	CH_3CH_2Cl	Chloroethane	Ethyl chloride
Ketone	R—C(=O)—R′	RCOR′	14	CH_3CCH_3 (C=O)	Propanone	Acetone